Simply supported beam with load at the middle

$$k_{eq} = \frac{48EI}{l^3}$$

Springs in series

$$\frac{1}{k_{eq}} = \frac{1}{k_1} + \frac{1}{k_2} + \cdots + \frac{1}{k_n}$$

Springs in parallel

$$k_{eq} = k_1 + k_2 + \cdots + k_n$$

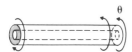

Hollow shaft under torsion (l = length, D = outer diameter, d = inner diameter)

$$k_{eq} = \frac{\pi G}{32l}(D^4 - d^4)$$

Equivalent viscous dampers

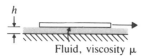

Fluid, viscosity μ

Relative motion between parallel surfaces (A = area of smaller plate)

$$c_{eq} = \frac{\mu A}{h}$$

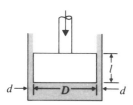

Dashpot (axial motion of a piston in a cylinder)

$$c_{eq} = \mu \frac{3\pi D^3 l}{4d^3}\left(1 + \frac{2d}{D}\right)$$

Torsional damper

$$c_{eq} = \frac{\pi\mu D^2(l - h)}{2d} + \frac{\pi\mu D^3}{32h}$$

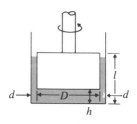

Dry friction (Coulomb damping) (fN = friction force, ω = frequency, X = amplitude of vibration)

$$c_{eq} = \frac{4fN}{\pi\omega X}$$

Mechanical Vibrations

Fourth Edition

Singiresu S. Rao
University of Miami

PEARSON
Prentice
Hall

Pearson Education, Inc.
Upper Saddle River, New Jersey 07458

Library of Congress Cataloging-in-Publication Data

Rao, S. S.
  Mechanical vibrations/Singiresu S. Rao.—4th ed.
    p. cm.
  Includes bibliographical references and index.
  ISBN 0-13-048987-5
  1. Vibration. I. Title.

TA355.R37 2003
620.3—dc21   2002044569

Vice President and Editorial Director, ECS: *Marcia J. Horton*
Acquisitions Editor: *Laura Fischer*
Editorial Assistant: *Erin Katchmar*
Vice President and Director of Production and Manufacturing, ESM: *David W. Riccardi*
Executive Managing Editor: *Vince O'Brien*
Managing Editor: *David A. George*
Production Editor: *Kevin Bradley*
Director of Creative Services: *Paul Belfanti*
Creative Director: *Carole Anson*
Art Director: *Jayne Conte*
Cover Designer: *Bruce Kenselaar*
Art Editor: *Greg Dulles*
Manufacturing Manager: *Trudy Pisciotti*
Manufacturing Buyer: *Lynda Castillo*
Marketing Manager: *Holly Stark*

MATLAB is a registered trademark of The MathWorks, Inc., 3 Apple Hill Drive, Natick, MA 01760–2098.

The author and publisher of this book have used their best efforts in preparing this book. These efforts include the development, research, and testing of the theories and programs to determine their effectiveness. The author and publisher make no warranty of any kind, expressed or implied, with regard to these programs or the documentation contained in this book. The author and publisher shall not be liable in any event for incidental or consequential damages in connection with, or arising out of, the furnishing, performance, or use of these programs.

Pearson Education Ltd., *London*
Pearson Education Australia Pty. Limited, *Sydney*
Pearson Education Singapore, Pte. Ltd.
Pearson Education North Asia Ltd., *Hong Kong*
Pearson Education Canada, Inc., *Toronto*
Pearson Educación de Mexico, S.A. de C.V.
Pearson Education—Japan, *Tokyo*
Pearson Education Malaysia, Pte. Ltd.
Pearson Education, Inc., *Upper Saddle River, New Jersey*

To Lord Sri Venkateswara

Contents

CHAPTER 2

Free Vibration of Single Degree of Freedom Systems

106

CHAPTER 3

Harmonically Excited Vibration   219

CHAPTER 6

Multidegree of Freedom Systems   448

CHAPTER 10

Vibration Measurement and Applications   741

CHAPTER 11

Numerical Integration Methods in Vibration Analysis

808

Preface

This book serves as an introduction to the subject of vibration engineering at the undergraduate level. The style of the prior editions has been retained, with the theory, computational aspects, and applications of vibrations presented in as simple a manner as possible. As in the previous editions, computer techniques of analysis are emphasized. Expanded explanations of the fundamentals are given, emphasizing physical significance and interpretation that build upon previous experiences in undergraduate mechanics. Numerous examples and problems are used to illustrate principles and concepts. Favorable reactions and encouragement from professors and students have provided me with the impetus to write the fourth edition of this book. Several new features have been added and many topics modified and rewritten. Most of the additions were suggested by those who have used the text and by numerous reviewers. Some important changes should be noted:

- More than 900 new review questions have been added to help students in reviewing and testing their understanding of the text material. The review questions include multiple choice questions, questions with brief answers, true-false questions, questions involving matching of related descriptions, and fill-in-the-blank type questions.
- A new appendix has been added to introduce the basic ideas of MATLAB programming.
- Several MATLAB-based examples are included in every chapter.
- General-purpose computer programs using MATLAB, C++, and Fortran programming are given, along with applications, in all chapters for the solution of vibration problems.
- Several new problems—including problems that are based on the use of MATLAB, C++, and Fortran problems—are given at the end of each chapter to expose students to many important computational and programming details.
- Answers to the review questions and the source codes of all MATLAB, C++, and Fortran programs are posted at the website of the book.
- More than 50 new illustrative examples appear throughout the book.
- More than 100 new problems have been added at the ends of various chapters.

Features

Each topic in *Mechanical Vibrations* is self-contained, with all concepts explained fully and the derivations presented with complete details. The computational aspects are emphasized throughout the book. MATLAB-based examples are given in all chapters. Several MATLAB, interactive C++, and Fortran computer programs, most of them in the form of general purpose subroutines, are included in all the chapters. These programs are intended for use by the students. Although the programs have been tested, no warranty is implied as to their accuracy. Examples as well as problems that are based on the use of the various computer programs are given in each chapter to expose students to many important computational and programming details.

Certain subjects are presented in a somewhat unconventional manner. The topics of Chapters 9, 10, and 11 fall in this category. Most textbooks discuss isolators, absorbers, and balancing in different places. Since one of the main purposes of the study of vibrations is to control vibration response, all topics directly related to vibration control are given in Chapter 9. The vibration-measuring instruments, along with vibration exciters, experimental modal analysis procedures, and machine condition monitoring, are presented in Chapter 10. Similarly, all the numerical integration methods applicable to single and multidegree of freedom systems, as well as continuous systems, are unified in Chapter 11.

Specific features include the following:

- More than 200 illustrative examples accompanying most topics.
- More than 900 review questions to help students in reviewing and testing their understanding of the text material.
- More than 1000 problems, with solutions in the instructor's manual.
- More than 30 design project type problems at the ends of various chapters.
- More than 70 MATLAB, C++, and Fortran computer programs to aid students in the numerical implementation of the methods discussed in the text.
- Biographical information about scientists and engineers who contributed to the development of the theory of vibrations are presented on the opening pages of chapters and appendixes.
- The MATLAB, C++, and FORTRAN programs given in the book, answers to problems, and solutions to review questions can be found at the Web site for this book, www.prenhall.com/rao.

Notation and Units

Both the SI and the English system of units have been used in the examples and problems. A list of symbols, along with the associated units in SI and English systems, appears after the Acknowledgments. A brief discussion of SI units as they apply to the field of vibrations is given in Appendix E. Arrows are used over symbols to denote column vectors and square brackets are used to indicate matrices.

Contents

Mechanical Vibrations is organized into 14 chapters and 6 appendixes. The material of the book provides flexible options for different types of vibration courses. For a one-semester senior or dual-level course, Chapters 1 through 5, portions of Chapters 6, 7, 8, and 10, and Chapter 9 may be used. The course can be given a computer orientation by including Chapter 11 in place of Chapter 8. Alternatively, with Chapters 12, 13, and 14, the text has sufficient material for a one-year sequence at the senior level. For shorter courses, the instructor can select the topics, depending on the level and orientation of the course. The relative simplicity with which topics are presented also makes the book useful to practicing engineers for purposes of self-study and as a source of references and computer programs.

Chapter 1 starts with a brief discussion of the history and importance of vibrations. The basic concepts and terminology used in vibration analysis are introduced. The free vibration analysis of single degree of freedom undamped translational and torsional systems is given in Chapter 2. The effects of viscous, Coulomb, and hysteretic damping are also discussed. The harmonic response of single degree of freedom systems is considered in Chapter 3. Chapter 4 is concerned with the response of a single degree of freedom system under general forcing functions. The roles of convolution integral, Laplace transformation, and numerical methods are discussed. The concept of response spectrum is also introduced in this chapter. The free and forced vibration of two degree of freedom systems is considered in Chapter 5. The self-excited vibration and stability of the system are discussed. Chapter 6 presents the vibration analysis of multidegree of freedom systems. Matrix methods of analysis are used for the presentation of the theory. The modal analysis procedure is described for the solution of forced vibration problems. Several methods of determining the natural frequencies of discrete systems are outlined in Chapter 7. The methods of Dunkerley, Rayleigh, Holzer, and Jacobi and matrix iteration are also discussed. The vibration analysis of continuous systems, including strings, bars, shafts, beams, and membranes is given in Chapter 8. The Rayleigh and Rayleigh-Ritz methods of finding the approximate natural frequencies are also described. Chapter 9 discusses the various aspects of vibration control, including the problems of elimination, isolation, and absorption. The balancing of rotating and reciprocating machines and the whirling of shafts are also considered. The vibration-measuring instruments, vibration exciters, and signal analysis are the topics of Chapter 10. Chapter 11 presents several numerical integration techniques for finding the dynamic response of discrete and continuous systems. The central difference, Runge-Kutta, Houbolt, Wilson, and Newmark methods are summarized and illustrated. Finite element analysis, with applications involving one-dimensional elements, is discussed in Chapter 12. An introductory treatment of nonlinear vibration, including a discussion of subharmonic and superharmonic oscillations, limit cycles, systems with time-dependent coefficients and chaos, is given in Chapter 13. The random vibration of linear vibration systems is considered in Chapter 14. Appendixes A and B focus on mathematical relationships and deflection of beams and plates, respectively. The basic relations of matrices, Laplace transforms, and SI units are outlined, respectively, in Appendixes C, D, and E. Finally, Appendix F provides an introduction to MATLAB programming.

S.S. RAO

Acknowledgments

I would like to express my appreciation to the many students and faculty whose comments have helped me improve the book. I am most grateful to the following people for offering their comments, suggestions, and ideas: Richard Alexander, Texas A&M University; C. W. Bert, University of Oklahoma; Raymond M. Brach, University of Notre Dame; Alfonso Diaz-Jimenez, Universidad Distrital "Francisco Jose de Caldas," Colombia; George Doyle, University of Dayton; Hamid Hamidzadeh, South Dakota State University; H. N. Hashemi, Northeastern University; Zhikun Hou, Worchester Polytechnic Institute; J. Richard Houghton, Tennessee Technological University; Faryar Jabbari, University of California, Irvine; Robert Jeffers, University of Connecticut; Richard Keltie, North Carolina State University; J. S. Lamancusa, Pennsylvania State University; Harry Law, Clemson University; Robert Leonard, Virginia Polytechnic Institute and State University; James Li, Columbia University; Sameer Madanshetty, Boston University; M. G. Prasad, Stevens Institute of Technology; F. P. J. Rimrott, University of Toronto; Subhash Sinha, Auburn University; Daniel Stutts, University of Missouri-Rolla; Massoud Tavakoli, Georgia Institute of Technology; Theodore Terry, Lehigh University; Chung Tsui, University of Maryland, College Park; Alexander Vakakis, University of Illinois-Urbana Champaign; Chuck Van Karsen, Michigan Technological University; Aleksandra Vinogradov, Montana State University; K. W. Wang, Pennsylvania State University; William Webster, GMI Engineering and Management Institute. I would like to thank Purdue University for granting me permission to use the Boilermaker Special in Problem 2.91. My sincere thanks to Dr. Qing Liu, a former graduate student at the University of Miami, for helping me in writing some of the MATLAB and C++ programs. Finally, I wish to thank my wife, Kamala, daughters Sridevi and Shobha, and granddaughter, Siriveena Rosa, without whose patience, encouragement, and support this edition might never have been completed.

S.S. RAO

srao@miami.edu

List of Symbols

Symbol	Meaning	English Units	SI Units
$a, a_0, a_1, a_2, \ldots$	constants, lengths		
a_{ij}	flexibility coefficient	in./lb	m/N
$[a]$	flexibility matrix	in./lb	m/N
A	area	in^2	m^2
$A, A_0, A_1, \ldots$	constants		
$b, b_1, b_2, \ldots$	constants, lengths		
$B, B_1, B_2, \ldots$	constants		
$\vec{B}$	balancing weight	lb	N
$c, \underset{\sim}{c}$	viscous damping coefficient	lb-sec/in.	N · s/m
$c, c_0, c_1, c_2, \ldots$	constants		
c	wave velocity	in./sec	m/s
c_c	critical viscous damping constant	lb-sec/in.	N · s/m
c_i	damping constant of ith damper	lb-sec/in.	N · s/m
c_{ij}	damping coefficient	lb-sec/in.	N · s/m
$[c]$	damping matrix	lb-sec/in.	N · s/m
C, C_1, C_2, C_1', C_2'	constants		
d	diameter, dimension	in.	m
D	diameter	in.	m
$[D]$	dynamical matrix	sec^2	s^2
e	base of natural logarithms		
e	eccentricity	in.	m
$\vec{e}_x, \vec{e}_y$	unit vectors parallel to x and y directions		
E	Young's modulus	lb/in^2	Pa
$E[x]$	expected value of x		
f	linear frequency	Hz	Hz
f	force per unit length	lb/in.	N/m

Symbol	Meaning	English Units	SI Units
$\underset{\sim}{f}$	unit impulse	lb-sec	$N \cdot s$
F, F_d	force	lb	N
F_0	amplitude of force $F(t)$	lb	N
F_t, F_T	force transmitted	lb	N
F_t	force acting on ith mass	lb	N
$\vec{F}$	force vector	lb	N
$\underset{\sim}{F}$	impulse	lb-sec	$N \cdot s$
g	acceleration due to gravity	in./sec^2	m/s^2
$g(t)$	impulse response function		
G	shear modulus	lb/in^2	N/m^2
h	hysteresis damping constant	lb/in	N/m
$H(i\omega)$	frequency response function		
i	$\sqrt{-1}$		
I	area moment of inertia	in^4	m^4
$[I]$	identity matrix		
Im()	imaginary part of ()		
j	integer		
J	polar moment of inertia	in^4	m^4
$J, J_0, J_1, J_2, \ldots$	mass moment of inertia	lb-in./sec^2	$kg \cdot m^2$
$k, \underset{\sim}{k}$	spring constant	lb/in.	N/m
k_i	spring constant of ith spring	lb/in.	N/m
k_t	torsional spring constant	lb-in/rad	N-m/rad
k_{ij}	stiffness coefficient	lb/in.	N/m
$[k]$	stiffness matrix	lb/in.	N/m
l, l_i	length	in.	m
$m, \underset{\sim}{m}$	mass	lb-sec^2/in.	kg
m_i	ith mass	lb-sec^2/in.	kg
m_{ij}	mass coefficient	lb-sec^2/in.	kg
$[m]$	mass matrix	lb-sec^2/in.	kg
M	mass	lb-sec^2/in.	kg
M	bending moment	lb-in.	$N \cdot m$
$M_t, M_{t1}, M_{t2}, \ldots$	torque	lb-in.	$N \cdot m$
M_{t0}	amplitude of $M_t(t)$	lb-in.	$N \cdot m$
n	an integer		
n	number of degrees of freedom		
N	normal force	lb	N

Symbol	Meaning	English Units	SI Units
N	total number of time steps		
p	pressure	lb/in^2	N/m^2
$p(x)$	probability density function of x		
$P(x)$	probability distribution function of x		
P	force, tension	lb	N
q_j	jth generalized coordinate		
$\vec{q}$	vector of generalized displacements		
$\dot{\vec{q}}$	vector of generalized velocities		
Q_j	jth generalized force		
r	frequency ratio $= \omega/\omega_n$		
$\vec{r}$	radius vector	in.	m
Re( )	real part of ( )		
$R(\tau)$	autocorrelation function		
R	electrical resistance	ohm	ohm
R	Rayleigh's dissipation function	lb-in/sec	N · m/s
R	Rayleigh's quotient	1/sec^2	1/s^2
s	exponential coefficient, root of equation		
S_a, S_d, S_v	acceleration, displacement, velocity spectrum		
$S_x(\omega)$	spectrum of x		
t	time	sec	s
t_i	ith time station	sec	s
T	torque	lb-in	N-m
T	kinetic energy	in.-lb	J
T_i	kinetic energy of ith mass	in.-lb	J
T_r	transmissibility ratio		
u_{ij}	an element of matrix $[U]$		
U, U_i	axial displacement	in.	m
U	potential energy	in.-lb	J
$\vec{U}$	unbalanced weight	lb	N
$[U]$	upper triangular matrix		
v, v_0	linear velocity	in./sec	m/s
V	shear force	lb	N
V	potential energy	in.-lb	J
V_i	potential energy of ith spring	in.-lb	J
w, w_1, w_2, ω_i	transverse deflections	in.	m
w_0	value of w at $t = 0$	in.	m

Symbol	Meaning	English Units	SI Units
$\dot{w}_0$	value of $\dot{w}$ at $t = 0$	in./sec	m/s
w_n	nth mode of vibration		
W	weight of a mass	lb	N
W	total energy	in.-lb	J
W	transverse deflection	in.	m
W_i	value of W at $t = t_i$	in.	m
$W(x)$	a function of x		
x, y, z	cartesian coordinates, displacements	in.	m
$x_0, x(0)$	value of x at $t = 0$	in.	m
$\dot{x}_0, \dot{x}(0)$	value of $\dot{x}$ at $t = 0$	in./sec	m/s
x_j	displacement of jth mass	in.	m
x_j	value of x at $t = t_j$	in.	m
$\dot{x}_j$	value of $\dot{x}$ at $t = t_j$	in./sec	m/s
x_h	homogeneous part of $x(t)$	in.	m
x_p	particular part of $x(t)$	in.	m
$\vec{x}$	vector of displacements	in.	m
$\vec{x}_i$	value of $\vec{x}$ at $t = t_i$	in.	m
$\dot{\vec{x}}_i$	value of $\dot{\vec{x}}$ at $t = t_i$	in./sec	m/s
$\ddot{\vec{x}}_i$	value of $\ddot{\vec{x}}$ at $t = t_i$	in./sec^2	m/s^2
$\vec{x}^{(i)}(t)$	ith mode		
X	amplitude of $x(t)$	in.	m
X_j	amplitude of $x_j(t)$	in.	m
$\vec{X}^{(i)}$	ith modal vector	in.	m
$X_i^{(j)}$	ith component of jth mode	in.	m
$[X]$	modal matrix	in.	m
$\vec{X}_r$	rth approximation to a mode shape		
y	base displacement	in.	m
Y	amplitude of $y(t)$	in.	m
z	relative displacement, $x - y$	in.	m
Z	amplitude of $z(t)$	in.	m
$Z(i\omega)$	mechanical impedance	lb/in.	N/m
α	angle, constant		
β	angle, constant		
β	hysteresis damping constant		
γ	specific weight	lb/in^3	N/m^3
δ	logarithmic decrement		
$\delta_1, \delta_2, \ldots$	deflections	in.	m

Symbol	Meaning	English Units	SI Units
δ_{st}	static deflection	in.	m
δ_{ij}	Kronecker delta		
Δ	determinant		
ΔF	increment in F	lb	N
Δx	increment in x	in.	m
Δt	increment in time t	sec	s
ΔW	energy dissipated in a cycle	in.-lb	J
ε	a small quantity		
ε	strain		
ζ	damping ratio		
θ	constant, angular displacement		
θ_i	ith angular displacement	rad	rad
θ_0	value of θ at $t = 0$	rad	rad
$\dot{\theta}_0$	value of $\dot{\theta}$ at $t = 0$	rad/sec	rad/s
Θ	amplitude of $\theta(t)$	rad	rad
Θ_i	amplitude of $\theta_i(t)$	rad	rad
λ	eigenvalue $= 1/\omega^2$	sec^2	s^2
$[\lambda]$	transformation matrix		
μ	viscosity of a fluid	lb-sec/in^2	kg/m $\cdot$ s
μ	coefficient of friction		
μ_x	expected value of x		
ρ	mass density	lb-sec^2/in^4	kg/m^3
η	loss factor		
σ_x	standard deviation of x		
σ	stress	lb/in^2	N/m^2
τ	period of oscillation, time	sec	s
τ	shear stress	lb/in^2	N/m^2
ϕ	angle, phase angle	rad	rad
ϕ_i	phase angle in ith mode	rad	rad
ω	frequency of oscillation	rad/sec	rad/s
ω_i	ith natural frequency	rad/sec	rad/s
ω_n	natural frequency	rad/sec	rad/s
ω_d	frequency of damped vibration	rad/sec	rad/s

Subscripts

Symbol	Meaning
cri	critical value
eq	equivalent value
i	ith value
L	left plane
max	maximum value
n	corresponding to natural frequency
R	right plane
0	specific or reference value
t	torsional

Operations

Symbol	Meaning
$\dot{(\ )}$	$\dfrac{d(\ )}{dt}$
$\ddot{(\ )}$	$\dfrac{d^2(\ )}{dt^2}$
$\vec{(\ )}$	column vector ( )
[ ]	matrix
$[\ ]^{-1}$	inverse of [ ]
$[\ ]^T$	transpose of [ ]
$\Delta(\ )$	increment in ( )
$\mathscr{L}(\ )$	Laplace transform of ( )
$\mathscr{L}^{-1}(\ )$	inverse Laplace transform of ( )

Galileo Galilei (1564–1642), an Italian astronomer, philosopher, and professor of mathematics at the Universities of Pisa and Padua, in 1609 became the first man to point a telescope to the sky. He wrote the first treatise on modern dynamics in 1590. His works on the oscillations of a simple pendulum and the vibration of strings are of fundamental significance in the theory of vibrations. (Photo courtesy of Dirk J. Struik, *A Concise History of Mathematics* (2nd rev. ed.), Dover Publications, Inc., New York, 1948.)

C H A P T E R   1

Fundamentals of Vibration

1.1 Preliminary Remarks

The subject of vibrations is introduced here in a relatively simple manner. The chapter begins with a brief history of the subject and continues with an examination of its importance. The various steps involved in vibration analysis of an engineering system are outlined, and essential definitions and concepts of vibration are introduced. There follows a presentation of the concept of harmonic analysis, which can be used for the analysis of general periodic motions. No attempt at exhaustive treatment is made in Chapter 1; subsequent chapters will develop many of the ideas in more detail.

1.2 Brief History of Vibration

1.2.1 Origins of Vibration

People became interested in vibration when the first musical instruments, probably whistles or drums, were discovered. Since then, people have applied ingenuity and critical investigation to study the phenomenon of vibration. Although certain very definite rules were observed in connection with the art of music even in ancient times, they can hardly be called a science. Music was highly developed and was much appreciated by the Chinese, Hindus, Japanese, and, perhaps, the Egyptians as long ago as 4000 B.C. [1.1].

1

From about 3000 B.C., stringed instruments such as harps appeared on the walls of Egyptian tombs [1.2]. In fact, the British Museum exhibits a harp with bull-headed sound-box found on an inlaid panel of a royal tomb at Ur from 2600 B.C. Stringed musical instruments probably originated with the hunter's bow, a weapon favored by the armies of ancient Egypt. One of the most primitive stringed instruments, called the *nanga*, dating back to 1500 B.C., can be seen in the British Museum. Our present system of music is based in ancient Greek civilization. Since ancient times, both musicians and philosophers sought out the rules and laws of sound production, used them in improving musical instruments, and passed them on from generation to generation.

The Greek philosopher and mathematician Pythagoras (582–507 B.C.) is considered to be the first person to investigate musical sounds on a scientific basis (Fig. 1.1). Among other things, Pythagoras conducted experiments on a vibrating string by using a simple apparatus called a monochord. In the monochord shown in Fig. 1.2 the wooden bridges labeled 1 and 3 are fixed. Bridge 2 is made movable while the tension in the string is held constant by the hanging weight. Pythagoras observed that if two like strings of different lengths are subject to the same tension, the shorter one emits a higher note; in addition, if the length of the shorter string is half the longer one, the shorter one will emit a note an octave above the other. However, Pythagoras left no written account of his work (Fig. 1.3). Although the concept of pitch was developed by the time of Pythagoras, its relation with the frequency of vibration of the sounding body was not understood. In fact the relation between the pitch and the frequency was not understood until the time of Galileo in the sixteenth century.

Around 350 B.C., Aristotle wrote treatises on music and sound, making observations such as "the voice is sweeter than the sound of instruments," and "the sound of the flute is sweeter than that of the lyre." In 320 B.C., Aristoxenus, a pupil of Aristotle and a musician, wrote a three-volume work entitled *Elements of Harmony*. These books are perhaps the

FIGURE 1.1   Pythagoras. (Reprinted with permission from L. E. Navia, *Pythagoras: An Annotated Bibliography*, Garland Publishing, Inc., New York, 1990).

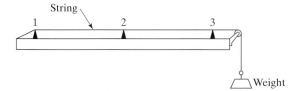

FIGURE 1.2   Monochord.

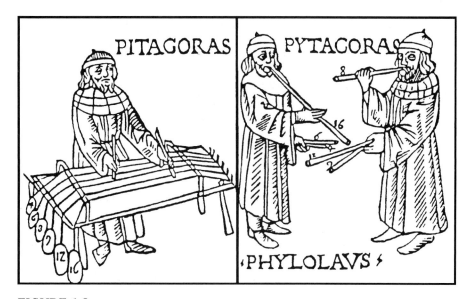

FIGURE 1.3   Pythagoras as a musician. (Reprinted with permission from D. E. Smith, *History of Mathematics*, Vol. I, Dover Publications, Inc., New York, 1958).

oldest ones available on the subject of music written directly by the author. Euclid, in about 300 B.C., wrote briefly about music without any reference to the physical nature of sound in a treatise called *Introduction to Harmonics*. No further scientific additions to sound were made by the Greeks. It appears that the Romans derived their knowledge of music completely from the Greeks, except that Vitruvius, a famous Roman architect, wrote in about 20 B.C. on the acoustic properties of theaters. His treatise, entitled *De Architectura Libri Decem*, was lost for several centuries, to be rediscovered only in the fifteenth century. There appears to have been no development in the theories of sound and vibration for nearly 16 centuries following the works of Vitruvius.

China experienced many earthquakes in ancient times. Zhang Heng, who served as a historian and astronomer in the second century, perceived a need to develop an instrument to measure earthquakes precisely. In A.D. 132 he invented the world's first seismograph to measure earthquakes [1.3, 1.4]. This seismograph was made of fine cast bronze, had a diameter of eight chi (a chi is equal to 0.237 meter), and was shaped like a wine jar (Fig. 1.4). Inside the jar was a mechanism consisting of pendulums surrounded by a group of eight lever mechanisms pointing in eight directions. Eight dragon figures, with a bronze ball in

FIGURE 1.4   The world's first seismo-graph, invented in China in A.D. 132. (Reprinted with permission from R. Taton (Ed.), *History of Science*, Basic Books, Inc., New York, 1957).

the mouth of each, were arranged on the outside of the seismograph. There were toads with mouths open upward underneath each dragon. A strong earthquake in any direction would tilt the pendulum in that direction, triggering the lever in the dragon head. This opened the mouth of the dragon, thereby releasing its bronze ball, which fell in the mouth of the toad with a clanging sound. Thus the seismograph enabled the monitoring personnel to know both the time and direction of occurrence of the earthquake.

**1.2.2
From Galileo to
Rayleigh**

Galileo Galilei (1564–1642) is considered to be the founder of modern experimental science. In fact, the seventeenth century is often considered the "century of genius" since the foundations of modern philosophy and science were laid during that period. Galileo was inspired to study the behavior of a simple pendulum by observing the pendulum movements of a lamp in a church in Pisa. One day, while feeling bored during a sermon, Galileo was staring at the ceiling of the church. A swinging lamp caught his attention. He started measuring the period of the pendulum movements of the lamp with his pulse and found to his amazement that the time period was independent of the amplitude of swings. This led him to conduct more experiments on the simple pendulum. In *Discourses Concerning Two New Sciences*, published in 1638, Galileo discussed vibrating bodies. He described the dependence of the frequency of vibration on the length of a simple pendulum, along with the phenomenon of sympathetic vibrations (resonance). Galileo's writings also indicate that he had a clear understanding of the relationship between the frequency, length, tension, and density of a vibrating stretched string [1.5]. However, the first correct published account of the vibration of strings was given by the French mathematician and theologian, Marin Mersenne (1588–1648) in his book *Harmonicorum Liber*, published in 1636. Mersenne also measured, for the first time, the frequency of vibration of a long string and from that predicted the frequency of a shorter string having the same density and tension. Mersenne is considered by many the father of acoustics. He is often credited with the discovery of the laws of vibrating strings because he published the results in 1636, two years before Galileo. However, the credit belongs to Galileo since the laws were written many years earlier but their publication was prohibited by the orders of the Inquisitor of Rome until 1638.

Inspired by the work of Galileo, the Academia del Cimento was founded in Florence in 1657; this was followed by the formations of the Royal Society of London in 1662 and the Paris Academie des Sciences in 1666. Later, Robert Hooke (1635–1703) also conducted experiments to find a relation between the pitch and frequency of vibration of a string. However, it was Joseph Sauveur (1653–1716) who investigated these experiments thoroughly and coined the word "acoustics" for the science of sound [1.6]. Sauveur in France and John Wallis (1616–1703) in England observed, independently, the phenomenon of mode shapes, and they found that a vibrating stretched string can have no motion at certain points and violent motion at intermediate points. Sauveur called the former points *nodes* and the latter ones *loops*. It was found that such vibrations had higher frequencies compared to the one associated with the simple vibration of the string with no nodes. In fact, the higher frequencies were found to be integral multiples of the frequency of simple vibration, and Sauveur called the higher frequencies harmonics and the frequency of simple vibration the fundamental frequency. Sauveur also found that a string can vibrate with several of its harmonics present at the same time. In addition, he observed the phenomenon of beats when two organ pipes of slightly different pitches are sounded together. In 1700 Sauveur calculated, by a somewhat dubious method, the frequency of a stretched string from the measured sag of its middle point.

Sir Isaac Newton (1642–1727) published his monumental work, *Philosophiae Naturalis Principia Mathematica*, in 1686, describing the law of universal gravitation as well as the three laws of motion and other discoveries. Newton's second law of motion is routinely used in modern books on vibrations to derive the equations of motion of a vibrating body. The theoretical (dynamical) solution of the problem of the vibrating string was found by the English mathematician Brook Taylor (1685–1731) in 1713, who also presented the famous Taylor's theorem on infinite series. The natural frequency of vibration obtained from the equation of motion derived by Taylor agreed with the experimental values observed by Galileo and Mersenne. The procedure adopted by Taylor was perfected through the introduction of partial derivatives in the equations of motion by Daniel Bernoulli (1700–1782), Jean D'Alembert (1717–1783), and Leonard Euler (1707–1783).

The possibility of a string vibrating with several of its harmonics present at the same time (with displacement of any point at any instant being equal to the algebraic sum of displacements for each harmonic) was proved through the dynamic equations of Daniel Bernoulli in his memoir, published by the Berlin Academy in 1755 [1.7]. This characteristic was referred to as the principle of the coexistence of small oscillations, which, in present-day terminology, is the principle of superposition. This principle was proved to be most valuable in the development of the theory of vibrations and led to the possibility of expressing any arbitrary function (i.e., any initial shape of the string) using an infinite series of sines and cosines. Because of this implication, D'Alembert and Euler doubted the validity of this principle. However, the validity of this type of expansion was proved by J. B. J. Fourier (1768–1830) in his *Analytical Theory of Heat* in 1822.

The analytical solution of the vibrating string was presented by Joseph Lagrange (1736–1813) in his memoir published by the Turin Academy in 1759. In his study, Lagrange assumed that the string was made up of a finite number of equally spaced identical mass particles, and he established the existence of a number of independent frequencies equal to the number of mass particles. When the number of particles was allowed to be infinite, the resulting frequencies were found to be the same as the harmonic frequencies of the stretched string. The method of setting up the differential equation of the motion of a string (called the wave equation), presented in most modern books on vibration theory, was first developed by D'Alembert in his memoir published by the Berlin Academy in 1750. The vibration of thin beams supported and clamped in different ways was first studied by Euler in 1744 and Daniel Bernoulli in 1751. Their approach has become known as the Euler-Bernoulli or thin beam theory.

Charles Coulomb did both theoretical and experimental studies in 1784 on the torsional oscillations of a metal cylinder suspended by a wire (Fig. 1.5). By assuming that the resisting torque of the twisted wire is proportional to the angle of twist, he derived the equation of motion for the torsional vibration of the suspended cylinder. By integrating the equation of motion, he found that the period of oscillation is independent of the angle of twist.

There is an interesting story related to the development of the theory of vibration of plates [1.8]. In 1802 the German scientist, E. F. F. Chladni (1756–1824) developed the method of placing sand on a vibrating plate to find its mode shapes and observed the

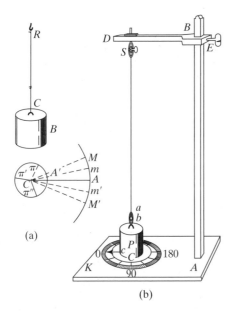

FIGURE 1.5   Coulomb's device for torsional vibration tests. (Reprinted with permission from S. P. Timoshenko, *History of Strength of Materials*, McGraw-Hill Book Company, Inc., New York, 1953).

beauty and intricacy of the modal patterns of the vibrating plates. In 1809 the French Academy invited Chladni to give a demonstration of his experiments. Napoléon Bonaparte, who attended the meeting, was very impressed and presented a sum of 3,000 francs to the academy, to be awarded to the first person to give a satisfactory mathematical theory of the vibration of plates. By the closing date of the competition in October 1811, only one candidate, Sophie Germain, had entered the contest. But Lagrange, who was one of the judges, noticed an error in the derivation of her differential equation of motion. The academy opened the competition again, with a new closing date of October 1813. Sophie Germain again entered the contest, presenting the correct form of the differential equation. However, the academy did not award the prize to her because the judges wanted physical justification of the assumptions made in her derivation. The competition was opened once more. In her third attempt, Sophie Germain was finally awarded the prize in 1815, although the judges were not completely satisfied with her theory. In fact, it was later found that her differential equation was correct but the boundary conditions were erroneous. The correct boundary conditions for the vibration of plates were given in 1850 by G. R. Kirchhoff (1824–1887).

In the meantime, the problem of vibration of a rectangular flexible membrane, which is important for the understanding of the sound emitted by drums, was solved for the first time by Simeon Poisson (1781–1840). The vibration of a circular membrane was studied by R. F. A. Clebsch (1833–1872) in 1862. After this, vibration studies were done on a number of practical mechanical and structural systems. In 1877 Lord Baron Rayleigh published his book on the theory of sound [1.9]; it is considered a classic on the subject of sound and vibration even today. Notable among the many contributions of Rayleigh is the method of finding the fundamental frequency of vibration of a conservative system by making use of the principle of conservation of energy—now known as Rayleigh's method. This method proved to be a helpful technique for the solution of difficult vibration problems. An extension of the method, which can be used to find multiple natural frequencies, is known as the Rayleigh-Ritz method.

1.2.3 Recent Contributions

In 1902 Frahm investigated the importance of torsional vibration study in the design of the propeller shafts of steamships. The dynamic vibration absorber, which involves the addition of a secondary spring-mass system to eliminate the vibrations of a main system, was also proposed by Frahm in 1909. Among the modern contributers to the theory of vibrations, the names of Stodola, De Laval, Timoshenko, and Mindlin are notable. Aurel Stodola (1859–1943) contributed to the study of vibration of beams, plates, and membranes. He developed a method for analyzing vibrating beams that is also applicable to turbine blades. Noting that every major type of prime mover gives rise to vibration problems, C. G. P. De Laval (1845–1913) presented a practical solution to the problem of vibration of an unbalanced rotating disk. After noticing failures of steel shafts in high-speed turbines, he used a bamboo fishing rod as a shaft to mount the rotor. He observed that this system not only eliminated the vibration of the unbalanced rotor but also survived up to speeds as high as 100,000 rpm [1.10].

Stephen Timoshenko (1878–1972) presented an improved theory of vibration of beams, which has become known as the Timoshenko or thick beam theory, by considering the effects of rotary inertia and shear deformation. A similar theory was presented by R. D. Mindlin for the vibration analysis of thick plates by including the effects of rotary inertia and shear deformation.

It has long been recognized that many basic problems of mechanics, including those of vibrations, are nonlinear. Although the linear treatments commonly adopted are quite satisfactory for most purposes, they are not adequate in all cases. In nonlinear systems, phenonmena may occur that are theoretically impossible in linear systems. The mathematical theory of nonlinear vibrations began to develop in the works of Poincaré and Lyapunov at the end of the nineteenth century. Poincaré developed the perturbation method in 1892 in connection with the approximate solution of nonlinear celestial mechanics problems. Lyapunov laid the foundations of modern stability theory in 1892, which is applicable to all types of dynamical systems. After 1920, the studies undertaken by Duffing and van der Pol brought the first definite solutions into the theory of nonlinear vibrations and drew attention to its importance in engineering. In the last 30 years, authors like Minorsky and Stoker have endeavored to collect in monographs the main results concerning nonlinear vibrations. Most practical applications of nonlinear vibration involved the use of some type of a perturbation theory approach. The modern methods of perturbation theory were surveyed by Nayfeh [1.11].

Random characteristics are present in diverse phenomena such as earthquakes, winds, transportation of goods on wheeled vehicles, and rocket and jet engine noise. It became necessary to devise concepts and methods of vibration analysis for these random effects. Although Einstein considered Brownian movement, a particular type of random vibration, as long ago as 1905, no applications were investigated until 1930. The introduction of the correlation function by Taylor in 1920 and of the spectral density by Wiener and Khinchin in the early 1930s opened new prospects for progress in the theory of random vibrations. Papers by Lin and Rice, published between 1943 and 1945, paved the way for the application of random vibrations to practical engineering problems. The monographs of Crandall and Mark, and Robson systematized the existing knowledge in the theory of random vibrations [1.12, 1.13].

Until about 30 years ago, vibration studies, even those dealing with complex engineering systems, were done by using gross models, with only a few degrees of freedom. However, the advent of high-speed digital computers in the 1950s made it possible to treat moderately complex systems and to generate approximate solutions in semidefinite form, relying on classical solution methods but using numerical evaluation of certain terms that cannot be expressed in closed form. The simultaneous development of the finite element method enabled engineers to use digital computers to conduct numerically detailed vibration analysis of complex mechanical, vehicular, and structural systems displaying thousands of degrees of freedom [1.14]. Although the finite element method was not so named until recently, the concept was used several centuries back. For example, ancient mathematicians found the circumference of a circle by approximating it as a polygon, where each side of the polygon, in present-day notation, can be called a finite element. The finite element method as known today was presented by Turner, Clough, Martin, and Topp in connection with the analysis of aircraft structures [1.15]. Figure 1.6 shows the finite element idealization of the body of a bus [1.16].

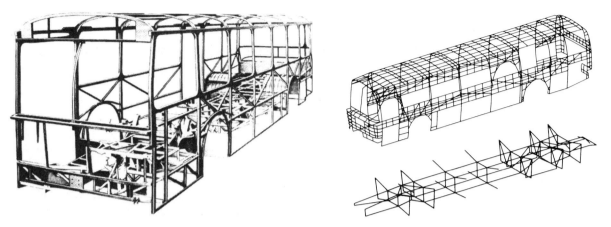

FIGURE 1.6   Finite element idealization of the body of a bus [1.16]. (Reprinted with permission © 1974 Society of Automotive Engineers, Inc.)

1.3   Importance of the Study of Vibration

Most human activities involve vibration in one form or other. For example, we hear because our eardrums vibrate and see because light waves undergo vibration. Breathing is associated with the vibration of lungs and walking involves (periodic) oscillatory motion of legs and hands. We speak due to the oscillatory motion of larynges (and tongues) [1.17]. Early scholars in the field of vibration concentrated their efforts on understanding the natural phenomena and developing mathematical theories to describe the vibration of physical systems. In recent times, many investigations have been motivated by the engineering applications of vibration, such as the design of machines, foundations, structures, engines, turbines, and control systems.

Most prime movers have vibrational problems due to the inherent unbalance in the engines. The unbalance may be due to faulty design or poor manufacture. Imbalance in diesel engines, for example, can cause ground waves sufficiently powerful to create a nuisance in urban areas. The wheels of some locomotives can rise more than a centimeter off the track at high speeds due to imbalance. In turbines, vibrations cause spectacular mechanical failures. Engineers have not yet been able to prevent the failures that result from blade and disk vibrations in turbines. Naturally, the structures designed to support heavy centrifugal machines, like motors and turbines, or reciprocating machines, like steam and gas engines and reciprocating pumps, are also subjected to vibration. In all these situations, the structure or machine component subjected to vibration can fail because of material fatigue resulting from the cyclic variation of the induced stress. Furthermore, the vibration causes more rapid wear of machine parts such as bearings and gears and also creates excessive noise. In machines, vibration causes fasteners such as nuts to become loose. In metal cutting processes, vibration can cause chatter, which leads to a poor surface finish.

Whenever the natural frequency of vibration of a machine or structure coincides with the frequency of the external excitation, there occurs a phenomenon known as *resonance*, which leads to excessive deflections and failure. The literature is full of accounts of system failures brought about by resonance and excessive vibration of components and systems (see Fig. 1.7). Because of the devastating effects that vibrations can have on machines and structures, vibration testing [1.18] has become a standard procedure in the design and development of most engineering systems (see Fig. 1.8).

FIGURE 1.7   Tacoma Narrows bridge during wind-induced vibration. The bridge opened on July 1, 1940, and collapsed on November 7, 1940. (Farquharson photo, Historical Photography Collection, University of Washington Libraries.)

FIGURE 1.8   Vibration testing of the space shuttle *Enterprise*. (Courtesy of NASA.)

In many engineering systems, a human being acts as an integral part of the system. The transmission of vibration to human beings results in discomfort and loss of efficiency. The vibration and noise generated by engines causes annoyance to people and, sometimes, damage to property (see Fig. 1.9). Vibration of instrument panels can cause their malfunction or difficulty in reading the meters [1.19]. Thus one of the important purposes of vibration study is to reduce vibration through proper design of machines and their mountings. In this connection, the mechanical engineer tries to design the engine or machine so as to minimize imbalance, while the structural engineer tries to design the supporting structure so as to ensure that the effect of the imbalance will not be harmful [1.20].

In spite of its detrimental effects, vibration can be utilized profitably in several consumer and industrial applications. In fact, the applications of vibratory equipment have increased considerably in recent years [1.21]. For example, vibration is put to work in vibratory conveyors, hoppers, sieves, compactors, washing machines, electric toothbrushes, dentist's drills, clocks, and electric massaging units. Vibration is also used in pile driving, vibratory testing of materials, vibratory finishing processes, and electronic circuits to filter out the unwanted frequencies (see Fig. 1.10). Vibration has been found to improve the efficiency of certain machining, casting, forging, and welding processes. It is employed to simulate earthquakes for geological research and also to conduct studies in the design of nuclear reactors.

1.4   Basic Concepts of Vibration

1.4.1
Vibration

Any motion that repeats itself after an interval of time is called *vibration* or *oscillation*. The swinging of a pendulum and the motion of a plucked string are typical examples of vibration. The theory of vibration deals with the study of oscillatory motions of bodies and the forces associated with them.

1.4.2
Elementary Parts of Vibrating Systems

A vibratory system, in general, includes a means for storing potential energy (spring or elasticity), a means for storing kinetic energy (mass or inertia), and a means by which energy is gradually lost (damper).

The vibration of a system involves the transfer of its potential energy to kinetic energy and kinetic energy to potential energy, alternately. If the system is damped, some energy is dissipated in each cycle of vibration and must be replaced by an external source if a state of steady vibration is to be maintained.

As an example, consider the vibration of the simple pendulum shown in Fig. 1.11. Let the bob of mass m be released after giving it an angular displacement θ. At position 1 the velocity of the bob and hence its kinetic energy is zero. But it has a potential energy of magnitude $mgl(1 - \cos \theta)$ with respect to the datum position 2. Since the gravitational force mg induces a torque $mgl \sin \theta$ about the point O, the bob starts swinging to the left from position 1. This gives the bob certain angular acceleration in the clockwise direction, and by the time it reaches position 2, all of its potential energy will be converted into kinetic energy. Hence the bob will not stop in position 2 but will continue to swing to position 3. However, as it passes the mean position 2, a counterclockwise torque starts acting on the bob due to gravity and causes the bob to decelerate. The velocity of the bob reduces to zero at the left extreme position. By this time, all the kinetic energy of the bob will be converted to potential energy. Again due to the

FIGURE 1.9   Annoyance caused by vibration and noise. (Reprinted with permission from *Sound and Vibration*, February 1977, Acoustical Publications, Inc.)

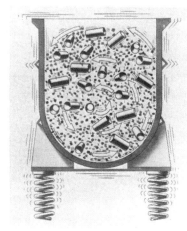

FIGURE 1.10 Vibratory finishing process. (Reprinted courtesy of the Society of Manufacturing Engineers, © 1964 The Tool and Manufacturing Engineer.)

gravity torque, the bob continues to attain a counterclockwise velocity. Hence the bob starts swinging back with progressively increasing velocity and passes the mean position again. This process keeps repeating, and the pendulum will have oscillatory motion. However, in practice, the magnitude of oscillation (θ) gradually decreases and the pendulum ultimately stops due to the resistance (damping) offered by the surrounding medium (air). This means that some energy is dissipated in each cycle of vibration due to damping by the air.

1.4.3 Degree of Freedom

The minimum number of independent coordinates required to determine completely the positions of all parts of a system at any instant of time defines the degree of freedom of the system. The simple pendulum shown in Fig. 1.11, as well as each of the systems

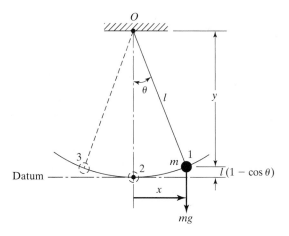

FIGURE 1.11 A simple pendulum.

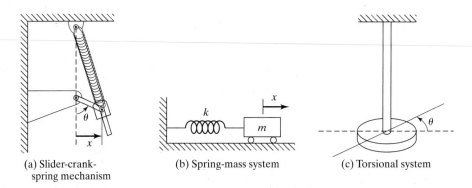

(a) Slider-crank-
      spring mechanism

(b) Spring-mass system

(c) Torsional system

FIGURE 1.12   Single degree of freedom systems.

shown in Fig. 1.12, represents a single degree of freedom system. For example, the motion of the simple pendulum (Fig. 1.11) can be stated either in terms of the angle θ or in terms of the Cartesian coordinates x and y. If the coordinates x and y are used to describe the motion, it must be recognized that these coordinates are not independent. They are related to each other through the relation $x^2 + y^2 = l^2$, where l is the constant length of the pendulum. Thus any one coordinate can describe the motion of the pendulum. In this example, we find that the choice of θ as the independent coordinate will be more convenient than the choice of x or y. For the slider shown in Fig. 1.12(a), either the angular coordinate θ or the coordinate x can be used to describe the motion. In Fig. 1.12(b), the linear coordinate x can be used to specify the motion. For the torsional system (long bar with a heavy disk at the end) shown in Fig. 1.12(c), the angular coordinate θ can be used to describe the motion.

   Some examples of two– and three–degree of freedom systems are shown in Figs. 1.13 and 1.14, respectively. Figure 1.13(a) shows a two mass–two spring system that is described by the two linear coordinates x_1 and x_2. Figure 1.13(b) denotes a two-rotor system whose motion can be specified in terms of θ_1 and θ_2. The motion of the system shown in Fig. 1.13(c) can be described completely either by X and θ or by x, y, and X. In the latter case, x and y are constrained as $x^2 + y^2 = l^2$ where l is a constant.

   For the systems shown in Figs. 1.14(a) and 1.14(c), the coordinates x_i $(i = 1, 2, 3)$ and θ_i $(i = 1, 2, 3)$ can be used, respectively, to describe the motion. In the case of the system shown in Fig. 1.14(b), θ_i $(i = 1, 2, 3)$ specifies the positions of the masses m_i $(i = 1, 2, 3)$.

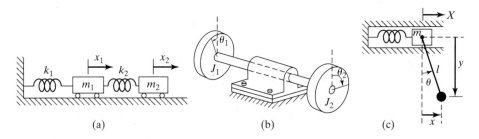

(a)

(b)

(c)

FIGURE 1.13   Two degree of freedom systems.

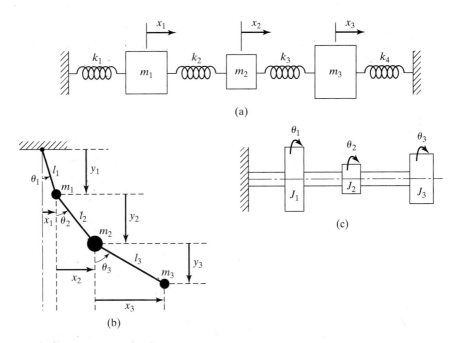

(a)

(c)

(b)

FIGURE 1.14   Three–degree of freedom systems.

An alternate method of describing this system is in terms of x_i and y_i ($i = 1, 2, 3$); but in this case the constraints $x_i^2 + y_i^2 = l_i^2$ ($i = 1, 2, 3$) have to be considered.

The coordinates necessary to describe the motion of a system constitute a set of *generalized coordinates*. The generalized coordinates are usually denoted as $q_1, q_2, \ldots$ and may represent Cartesian and/or non-Cartesian coordinates.

1.4.4 Discrete and Continuous Systems

A large number of practical systems can be described using a finite number of degrees of freedom, such as the simple systems shown in Figs. 1.11 to 1.14. Some systems, especially those involving continuous elastic members, have an infinite number of degrees of freedom. As a simple example, consider the cantilever beam shown in Fig. 1.15. Since the beam has an infinite number of mass points, we need an infinite number of coordinates to specify its deflected configuration. The infinite number of coordinates defines its elastic deflection curve. Thus the cantilever beam has an infinite number of degrees of freedom.

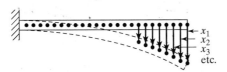

FIGURE 1.15   A cantilever beam (an infinite-number-of-degrees-of-freedom system).

Most structural and machine systems have deformable (elastic) members and therefore have an infinite number of degrees of freedom.

Systems with a finite number of degrees of freedom are called *discrete* or *lumped parameter* systems, and those with an infinite number of degrees of freedom are called *continuous* or *distributed* systems.

Most of the time, continuous systems are approximated as discrete systems, and solutions are obtained in a simpler manner. Although treatment of a system as continuous gives exact results, the analytical methods available for dealing with continuous systems are limited to a narrow selection of problems, such as uniform beams, slender rods, and thin plates. Hence most of the practical systems are studied by treating them as finite lumped masses, springs, and dampers. In general, more accurate results are obtained by increasing the number of masses, springs, and dampers—that is, by increasing the number of degrees of freedom.

1.5   Classification of Vibration

Vibration can be classified in several ways. Some of the important classifications are as follows.

1.5.1
Free and Forced Vibration

Free Vibration.   If a system, after an initial disturbance, is left to vibrate on its own, the ensuing vibration is known as *free vibration*. No external force acts on the system. The oscillation of a simple pendulum is an example of free vibration.

Forced Vibration.   If a system is subjected to an external force (often, a repeating type of force), the resulting vibration is known as *forced vibration*. The oscillation that arises in machines such as diesel engines is an example of forced vibration.

If the frequency of the external force coincides with one of the natural frequencies of the system, a condition known as *resonance* occurs, and the system undergoes dangerously large oscillations. Failures of such structures as buildings, bridges, turbines, and airplane wings have been associated with the occurrence of resonance.

1.5.2
Undamped and Damped Vibration

If no energy is lost or dissipated in friction or other resistance during oscillation, the vibration is known as *undamped vibration*. If any energy is lost in this way, however, it is called *damped vibration*. In many physical systems, the amount of damping is so small that it can be disregarded for most engineering purposes. However, consideration of damping becomes extremely important in analyzing vibratory systems near resonance.

1.5.3
Linear and Nonlinear Vibration

If all the basic components of a vibratory system—the spring, the mass, and the damper—behave linearly, the resulting vibration is known as *linear vibration*. If, however, any of the basic components behave nonlinearly, the vibration is called *nonlinear vibration*. The differential equations that govern the behavior of linear and nonlinear vibratory systems are linear and nonlinear, respectively. If the vibration is linear, the principle of superposition holds, and the mathematical techniques of analysis are well developed. For nonlinear vibration, the superposition principle is not valid, and techniques of analysis are less well-known. Since all vibratory systems tend to behave nonlinearly with increasing amplitude of oscillation, a knowledge of nonlinear vibration is desirable in dealing with practical vibratory systems.

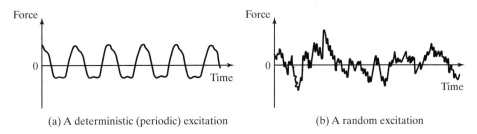

FIGURE 1.16   Deterministic and random excitations.

1.5.4 Deterministic and Random Vibration

If the value or magnitude of the excitation (force or motion) acting on a vibratory system is known at any given time, the excitation is called *deterministic*. The resulting vibration is known as *deterministic vibration.*

In some cases, the excitation is *nondeterministic* or *random;* the value of the excitation at a given time cannot be predicted. In these cases, a large collection of records of the excitation may exhibit some statistical regularity. It is possible to estimate averages such as the mean and mean square values of the excitation. Examples of random excitations are wind velocity, road roughness, and ground motion during earthquakes. If the excitation is random, the resulting vibration is called *random vibration.* In the case of random vibration, the vibratory response of the system is also random; it can be described only in terms of statistical quantities. Figure 1.16 shows examples of deterministic and random excitations.

1.6   Vibration Analysis Procedure

A vibratory system is a dynamic system for which the variables such as the excitations (inputs) and responses (outputs) are time-dependent. The response of a vibrating system generally depends on the initial conditions as well as the external excitations. Most practical vibrating systems are very complex, and it is impossible to consider all the details for a mathematical analysis. Only the most important features are considered in the analysis to predict the behavior of the system under specified input conditions. Often the overall behavior of the system can be determined by considering even a simple model of the complex physical system. Thus the analysis of a vibrating system usually involves mathematical modeling, derivation of the governing equations, solution of the equations, and interpretation of the results.

Step 1: Mathematical Modeling.   The purpose of mathematical modeling is to represent all the important features of the system for the purpose of deriving the mathematical (or analytical) equations governing the system's behavior. The mathematical model should include enough details to be able to describe the system in terms of equations without making it too complex. The mathematical model may be linear or nonlinear, depending on the behavior of the system's components. Linear models permit quick solutions and are simple to handle; however, nonlinear models sometimes reveal certain characteristics of the system that cannot be predicted using linear models. Thus a great deal of engineering judgment is needed to come up with a suitable mathematical model of a vibrating system.

Sometimes the mathematical model is gradually improved to obtain more accurate results. In this approach, first a very crude or elementary model is used to get a quick insight into the overall behavior of the system. Subsequently, the model is refined by including more components and/or details so that the behavior of the system can be observed more closely. To illustrate the procedure of refinement used in mathematical modeling, consider the forging hammer shown in Fig. 1.17(a). The forging hammer consists of a frame, a falling weight known as the tup, an anvil, and a foundation block. The anvil is a massive steel block on which material is forged into desired shape by the repeated blows of the tup. The anvil is usually mounted on an elastic pad to reduce the transmission of vibration to the foundation block and the frame [1.22]. For a first approximation, the frame, anvil, elastic pad, foundation block, and soil are modeled as a single–degree of freedom system as shown in Fig. 1.17(b). For a refined approximation, the weights of the frame and anvil and the foundation block are represented separately with a two–degree of freedom model as shown in Fig. 1.17(c). Further refinement of the model can be made by considering eccentric impacts of the tup, which cause each of the masses shown in Fig. 1.17(c) to have both vertical and rocking (rotation) motions in the plane of the paper.

Step 2: Derivation of Governing Equations.   Once the mathematical model is available, we use the principles of dynamics and derive the equations that describe the vibration of the system. The equations of motion can be derived conveniently by drawing the free-body diagrams of all the masses involved. The free-body diagram of a mass can be obtained by isolating the mass and indicating all externally applied forces, the reactive forces, and the inertia forces. The equations of motion of a vibrating system are usually in the form of a set of ordinary differential equations for a discrete system and partial differential equations for a continuous system. The equations may be linear or nonlinear, depending on the behavior of the components of the system. Several approaches are commonly used to derive the governing equations. Among them are Newton's second law of motion, D'Alembert's principle, and the principle of conservation of energy.

Step 3: Solution of the Governing Equations.   The equations of motion must be solved to find the response of the vibrating system. Depending on the nature of the problem, we can use one of the following techniques for finding the solution: standard methods of solving differential equations, Laplace transform methods, matrix methods,[1] and numerical methods. If the governing equations are nonlinear, they can seldom be solved in closed form. Furthermore, the solution of partial differential equations is far more involved than that of ordinary differential equations. Numerical methods involving computers can be used to solve the equations. However, it will be difficult to draw general conclusions about the behavior of the system using computer results.

Step 4: Interpretation of the Results.   The solution of the governing equations gives the displacements, velocities, and accelerations of the various masses of the system. These results must be interpreted with a clear view of the purpose of the analysis and the possible design implications of the results.

---

[1]The basic definitions and operations of matrix theory are given in Appendix A.

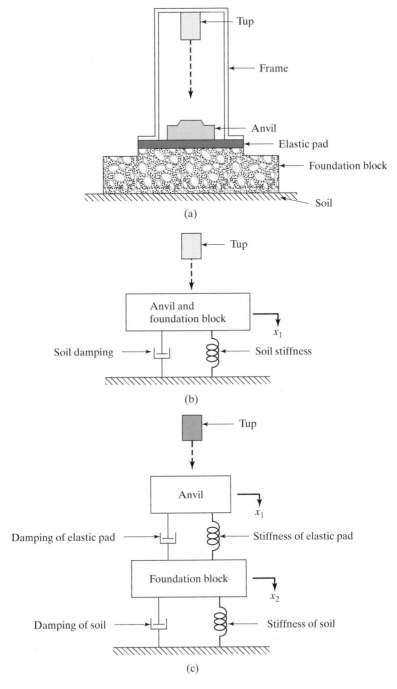

FIGURE 1.17   Modeling of a forging hammer.

EXAMPLE 1.1

Mathematical Model of a Motorcycle

Figure 1.18(a) shows a motorcycle with a rider. Develop a sequence of three mathematical models of the system for investigating vibration in the vertical direction. Consider the elasticity of the tires, elasticity and damping of the struts (in the vertical direction), masses of the wheels, and elasticity, damping, and mass of the rider.

Solution: We start with the simplest model and refine it gradually. When the equivalent values of the mass, stiffness, and damping of the system are used, we obtain a single–degree of freedom model of the motorcycle with a rider as indicated in Fig. 1.18(b). In this model, the equivalent stiffness (k_{eq}) includes the stiffnesses of the tires, struts, and rider. The equivalent damping constant (c_{eq}) includes the damping of the struts and the rider. The equivalent mass includes the masses of the wheels, vehicle body, and the rider. This model can be refined by representing the masses of wheels, elasticity of the tires, and elasticity and damping of the struts separately, as shown in Fig. 1.18(c). In this model, the mass of the vehicle body (m_v) and the mass of the rider (m_r) are shown as a single mass, $m_v + m_r$. When the elasticity (as spring constant k_r) and damping (as damping constant c_r) of the rider are considered, the refined model shown in Fig. 1.18(d) can be obtained.

Note that the models shown in Figs. 1.18(b) to (d) are not unique. For example, by combining the spring constants of both tires, the masses of both wheels, and the spring and damping constants of both struts as single quantities, the model shown in Fig. 1.18(e) can be obtained instead of Fig. 1.18(c).

∎

1.7   Spring Elements

A linear spring is a type of mechanical link that is generally assumed to have negligible mass and damping. A force is developed in the spring whenever there is relative motion between the two ends of the spring. The spring force is proportional to the amount of deformation and is given by

$$F = kx \qquad (1.1)$$

where F is the spring force, x is the deformation (displacement of one end with respect to the other), and k is the *spring stiffness* or *spring constant*. If we plot a graph between F and x, the result is a straight line according to Eq. (1.1). The work done (U) in deforming a spring is stored as strain or potential energy in the spring, and it is given by

$$U = \frac{1}{2}kx^2 \qquad (1.2)$$

Actual springs are nonlinear and follow Eq. (1.1) only up to a certain deformation. Beyond a certain value of deformation (after point A in Fig. 1.19), the stress exceeds the yield point of the material and the force-deformation relation becomes nonlinear [1.23, 1.24]. In many practical applications we assume that the deflections are small and make use of the linear relation in Eq. (1.1). Even if the force-deflection relation of

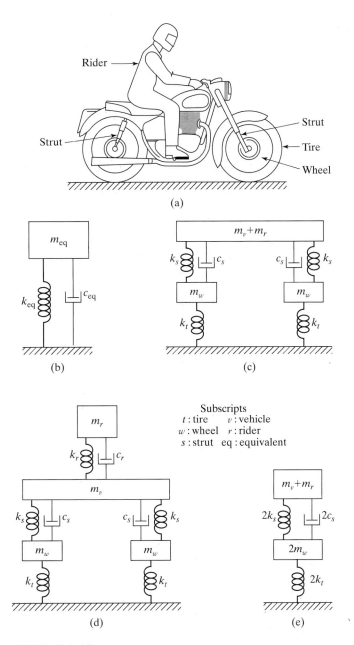

FIGURE 1.18   Motorcycle with a rider—a physical system and
mathematical model.

a spring is nonlinear, as shown in Fig. 1.20, we often approximate it as a linear one by using a linearization process [1.24, 1.25]. To illustrate the linearization process, let the static equilibrium load F acting on the spring cause a deflection of x^*. If an incremental force ΔF is added to F, the spring deflects by an additional quantity Δx. The new spring force $F + \Delta F$ can be expressed using Taylor's series expansion about the static equilibrium position x^* as

$$F + \Delta F = F(x^* + \Delta x)$$

$$= F(x^*) + \left.\frac{dF}{dx}\right|_{x^*} (\Delta x) + \frac{1}{2!} \left.\frac{d^2F}{dx^2}\right|_{x^*} (\Delta x)^2 + \ldots \tag{1.3}$$

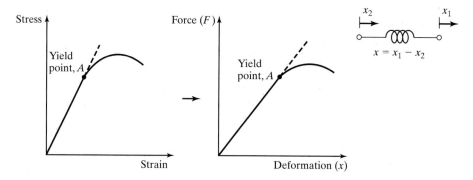

FIGURE 1.19   Nonlinearity beyond proportionality limit.

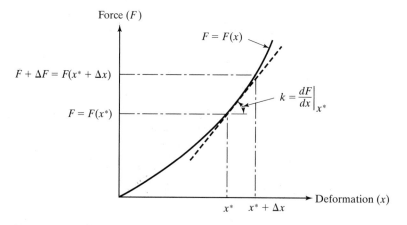

FIGURE 1.20   Linearization process.

For small values of Δx, the higher-order derivative terms can be neglected to obtain

$$F + \Delta F = F(x^*) + \frac{dF}{dx}\bigg|_{x^*}(\Delta x) \tag{1.4}$$

Since $F = F(x^*)$, we can express ΔF as

$$\Delta F = k\Delta x \tag{1.5}$$

where k is the linearized spring constant at x^* given by

$$k = \frac{dF}{dx}\bigg|_{x^*}$$

We may use Eq. (1.5) for simplicity, but sometimes the error involved in the approximation may be very large.

Elastic elements like beams also behave as springs. For example, consider a cantilever beam with an end mass m, as shown in Fig. 1.21. We assume, for simplicity, that the mass of the beam is negligible in comparison with the mass m. From strength of materials [1.26], we know that the static deflection of the beam at the free end is given by

$$\delta_{st} = \frac{Wl^3}{3EI} \tag{1.6}$$

where $W = mg$ is the weight of the mass m, E is Young's modulus, and I is the moment of inertia of the cross section of the beam. Hence the spring constant is

$$k = \frac{W}{\delta_{st}} = \frac{3EI}{l^3} \tag{1.7}$$

Similar results can be obtained for beams with different end conditions.

The formulas given in Appendix B can be used to find the spring constants of beams and plates.

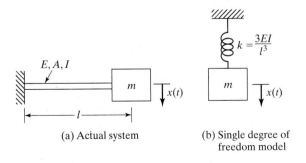

(a) Actual system   (b) Single degree of
                               freedom model

FIGURE 1.21   Cantilever with end mass.

**1.7.1
Combination of
Springs**

In many practical applications, several linear springs are used in combination. These springs can be combined into a single equivalent spring as indicated below.

Case 1: Springs in Parallel.   To derive an expression for the equivalent spring constant of springs connected in parallel, consider the two springs shown in Fig. 1.22(a). When a load W is applied, the system undergoes a static deflection δ_{st} as shown in Fig. 1.22(b). Then the free body diagram, shown in Fig. 1.22(c), gives the equilibrium equation

$$W = k_1 \delta_{st} + k_2 \delta_{st} \tag{1.8}$$

If k_{eq} denotes the equivalent spring constant of the combination of the two springs, then for the same static deflection δ_{st}, we have

$$W = k_{eq} \delta_{st} \tag{1.9}$$

Equations (1.8) and (1.9) give

$$k_{eq} = k_1 + k_2 \tag{1.10}$$

In general, if we have n springs with spring constants $k_1, k_2, \ldots, k_n$ in parallel, then the equivalent spring constant k_{eq} can be obtained:

$$k_{eq} = k_1 + k_2 + \cdots + k_n \tag{1.11}$$

Case 2: Springs in Series.   Next we derive an expression for the equivalent spring constant of springs connected in series by considering the two springs shown in Fig. 1.23(a). Under the action of a load W, springs 1 and 2 undergo elongations δ_1 and δ_2, respectively, as shown in Fig. 1.23(b). The total elongation (or static deflection) of the system, δ_{st}, is given by

$$\delta_{st} = \delta_1 + \delta_2 \tag{1.12}$$

Since both springs are subjected to the same force W, we have the equilibrium shown in Fig. 1.23(c):

$$W = k_1 \delta_1$$
$$W = k_2 \delta_2 \tag{1.13}$$

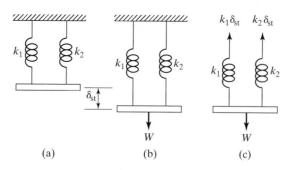

(a)   (b)   (c)

FIGURE 1.22   Springs in parallel.

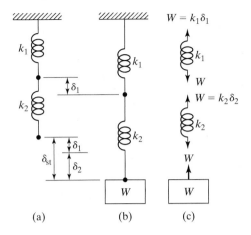

FIGURE 1.23   Springs in series.

If k_{eq} denotes the equivalent spring constant, then for the same static deflection,

$$W = k_{eq} \delta_{st} \tag{1.14}$$

Equations (1.13) and (1.14) give

$$k_1 \delta_1 = k_2 \delta_2 = k_{eq} \delta_{st}$$

or

$$\delta_1 = \frac{k_{eq} \delta_{st}}{k_1} \quad \text{and} \quad \delta_2 = \frac{k_{eq} \delta_{st}}{k_2} \tag{1.15}$$

Substituting these values of δ_1 and δ_2 into Equation (1.12), we obtain

$$\frac{k_{eq} \delta_{st}}{k_1} + \frac{k_{eq} \delta_{st}}{k_2} = \delta_{st}$$

that is,

$$\frac{1}{k_{eq}} = \frac{1}{k_1} + \frac{1}{k_2} \tag{1.16}$$

Equation (1.16) can be generalized to the case of n springs in series:

$$\frac{1}{k_{eq}} = \frac{1}{k_1} + \frac{1}{k_2} + \cdots + \frac{1}{k_n} \tag{1.17}$$

In certain applications, springs are connected to rigid components such as pulleys, levers, and gears. In such cases, an equivalent spring constant can be found using energy equivalence, as illustrated in Example 1.5.

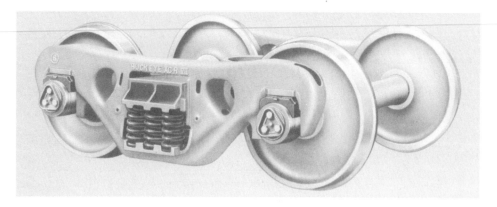

FIGURE 1.24   Parallel arrangement of springs in a freight truck. (Courtesy of Buckeye Steel Castings Company).

EXAMPLE 1.2

Equivalent k of a Suspension System

Figure 1.24 shows the suspension system of a freight truck with a parallel-spring arrangement. Find the equivalent spring constant of the suspension if each of the three helical springs is made of steel with a shear modulus $G = 80 \times 10^9$ N/m^2 and has five effective turns, mean coil diameter $D = 20$ cm, and wire diameter $d = 2$ cm.

Solution: The stiffness of each helical spring is given by

$$k = \frac{d^4 G}{8 D^3 n} = \frac{(0.02)^4 (80 \times 10^9)}{8(0.2)^3 (5)} = 40{,}000.0 \text{ N/m}$$

(See inside front cover for the formula.)
Since the three springs are identical and parallel, the equivalent spring constant of the suspension system is given by

$$k_{eq} = 3k = 3(40{,}000.0) = 120{,}000.0 \text{ N/m}$$

∎

EXAMPLE 1.3

Torsional Spring Constant of a Propeller Shaft

Determine the torsional spring constant of the steel propeller shaft shown in Fig. 1.25.

Solution: We need to consider the segments 12 and 23 of the shaft as springs in combination. From Fig. 1.25, the torque induced at any cross section of the shaft (such as AA or BB) can be seen to be equal to the torque applied at the propeller, T. Hence the elasticities (springs) corresponding to the two segments 12 and 23 are to be considered as series springs. The spring constants of segments 12 and 23 of the shaft ($k_{t_{12}}$ and $k_{t_{23}}$) are given by

$$k_{t_{12}} = \frac{GJ_{12}}{l_{12}} = \frac{G\pi(D_{12}^4 - d_{12}^4)}{32 l_{12}} = \frac{(80 \times 10^9)\pi(0.3^4 - 0.2^4)}{32(2)}$$
$$= 25.5255 \times 10^6 \text{ N-m/rad}$$

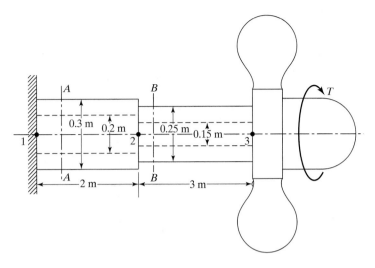

FIGURE 1.25   Propeller shaft.

$$k_{t_{23}} = \frac{GJ_{23}}{l_{23}} = \frac{G\pi(D_{23}^4 - d_{23}^4)}{32l_{23}} = \frac{(80 \times 10^9)\pi(0.25^4 - 0.15^4)}{32(3)}$$
$$= 8.9012 \times 10^6 \text{ N-m/rad}$$

Since the springs are in series, Eq. (1.16) gives

$$k_{t_{eq}} = \frac{k_{t_{12}} k_{t_{23}}}{k_{t_{12}} + k_{t_{23}}} = \frac{(25.5255\ 10^6)(8.9012 \times 10^6)}{(25.5255 \times 10^6 + 8.9012 \times 10^6)} = 6.5997 \times 10^6 \text{ N-m/rad}$$

■

Equivalent k of Hoisting Drum

EXAMPLE 1.4

A hoisting drum, carrying a steel wire rope, is mounted at the end of a cantilever beam as shown in Fig. 1.26(a). Determine the equivalent spring constant of the system when the suspended length of the wire rope is l. Assume that the net cross-sectional diameter of the wire rope is d and the Young's modulus of the beam and the wire rope is E.

Solution: The spring constant of the cantilever beam is given by

$$k_b = \frac{3EI}{b^3} = \frac{3E}{b^3}\left(\frac{1}{12}at^3\right) = \frac{Eat^3}{4b^3} \tag{E.1}$$

The stiffness of the wire rope subjected to axial loading is

$$k_r = \frac{AE}{l} = \frac{\pi d^2 E}{4l} \tag{E.2}$$

Since both the wire rope and the cantilever beam experience the same load W, as shown in Fig. 1.26(b), they can be modeled as springs in series, as shown in Fig. 1.26(c). The equivalent spring constant k_{eq}

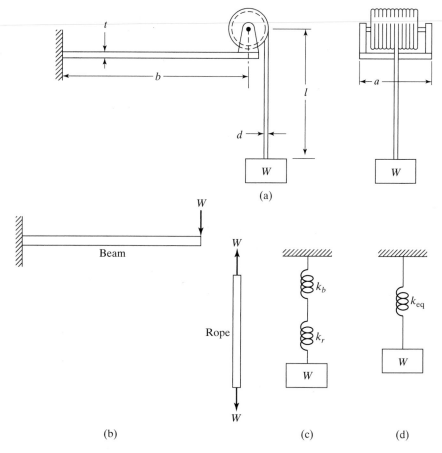

FIGURE 1.26   Hoisting drum.

is given by

$$\frac{1}{k_{eq}} = \frac{1}{k_b} + \frac{1}{k_r} = \frac{4b^3}{Eat^3} + \frac{4l}{\pi d^2 E}$$

or

$$k_{eq} = \frac{E}{4}\left(\frac{\pi at^3 d^2}{\pi d^2 b^3 + lat^3}\right) \qquad\qquad (E.3)$$

∎

EXAMPLE   1.5

Equivalent k of a Crane

The boom AB of the crane shown in Fig. 1.27(a) is a uniform steel bar of length 10 m and area of cross section 2,500 mm^2. A weight W is suspended while the crane is stationary. The cable $CDEBF$ is made of steel and has a cross-sectional area of 100 mm^2. Neglecting the effect of the cable $CDEB$, find the equivalent spring constant of the system in the vertical direction.

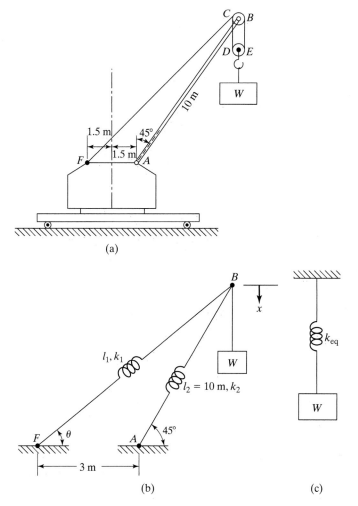

FIGURE 1.27   Crane lifting a load.

Solution: The equivalent spring constant can be found using the equivalence of potential energies of the two systems. Since the base of the crane is rigid, the cable and the boom can be considered to be fixed at points F and A, respectively. Also, the effect of the cable $CDEB$ is negligible; hence the weight W can be assumed to act through point B as shown in Fig. 1.27(b).

A vertical displacement x of point B will cause the spring k_2 (boom) to deform by an amount $x_2 = x \cos 45°$ and the spring k_1 (cable) to deform by an amount $x_1 = x \cos (90° - \theta)$. The length of the cable FB, l_1, is given by Fig. 1.27(b):

$$l_1^2 = 3^2 + 10^2 - 2(3)(10) \cos 135° = 151.426, \quad l_1 = 12.3055 \text{ m}$$

The angle θ satisfies the relation

$$l_1^2 + 3^2 - 2(l_1)(3) \cos \theta = 10^2, \quad \cos \theta = 0.8184, \quad \theta = 35.0736°$$

The total potential energy (U) stored in the springs k_1 and k_2 can be expressed, using Equation (1.2), as

$$U = \tfrac{1}{2}k_1(x \cos 45°)^2 + \tfrac{1}{2}k_2[x \cos(90° - \theta)]^2 \tag{E.1}$$

where

$$k_1 = \frac{A_1 E_1}{l_1} = \frac{(100 \times 10^{-6})(207 \times 10^9)}{12.3055} = 1.6822 \times 10^6 \text{ N/m}$$

and

$$k_2 = \frac{A_2 E_2}{l_2} = \frac{(2500 \times 10^{-6})(207 \times 10^9)}{10} = 5.1750 \times 10^7 \text{ N/m}$$

Since the equivalent spring in the vertical direction undergoes a deformation x, the potential energy of the equivalent spring (U_{eq}) is given by

$$U_{eq} = \tfrac{1}{2}k_{eq}x^2 \tag{E.2}$$

By setting $U = U_{eq}$, we obtain the equivalent spring constant of the system as

$$k_{eq} = 26.4304 \times 10^6 \text{ N/m}$$

∎

1.8   Mass or Inertia Elements

The mass or inertia element is assumed to be a rigid body; it can gain or lose kinetic energy whenever the velocity of the body changes. From Newton's second law of motion, the product of the mass and its acceleration is equal to the force applied to the mass. Work is equal to the force multiplied by the displacement in the direction of the force, and the work done on a mass is stored in the form of the mass's kinetic energy.

In most cases, we must use a mathematical model to represent the actual vibrating system, and there are often several possible models. The purpose of the analysis often determines which mathematical model is appropriate. Once the model is chosen, the mass or inertia elements of the system can be easily identified. For example, consider again the cantilever beam with an end mass shown in Fig. 1.21(a). For a quick and reasonably accurate analysis, the mass and damping of the beam can be disregarded; the system can be modeled as a spring-mass system, as shown in Fig. 1.21(b). The tip mass m represents the mass element, and the elasticity of the beam denotes the stiffness of the spring. Next, consider a multistory building subjected to an earthquake. Assuming that the mass of the frame is negligible compared to the masses of the floors, the building can be modeled as a multi-degree of freedom system, as shown in Fig. 1.28. The masses at the various floor levels represent the mass elements, and the elasticities of the vertical members denote the spring elements.

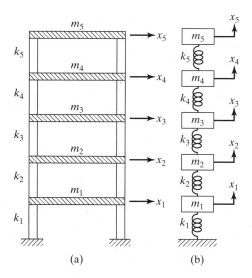

FIGURE 1.28 Idealization of a multistory building as a multi-degree of freedom system.

1.8.1 Combination of Masses

In many practical applications, several masses appear in combination. For a simple analysis, we can replace these masses by a single equivalent mass, as indicated below [1.27].

Case 1: Translational Masses Connected by a Rigid Bar. Let the masses be attached to a rigid bar that is pivoted at one end, as shown in Fig. 1.29(a). The equivalent mass can be assumed to be located at any point along the bar. To be specific, we assume the location of the equivalent mass to be that of mass m_1. The velocities of masses m_2 ($\dot{x}_2$) and m_3 ($\dot{x}_3$) can be expressed in terms of the velocity of mass m_1 ($\dot{x}_1$), by assuming small angular displacements for the bar, as

$$\dot{x}_2 = \frac{l_2}{l_1}\dot{x}_1, \quad \dot{x}_3 = \frac{l_3}{l_1}\dot{x}_1 \tag{1.18}$$

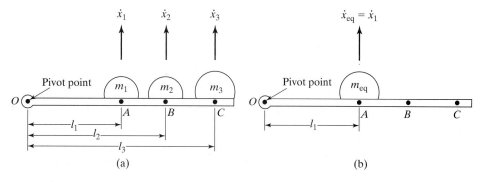

FIGURE 1.29 Translational masses connected by a rigid bar.

and

$$\dot{x}_{eq} = \dot{x}_1 \tag{1.19}$$

By equating the kinetic energy of the three-mass system to that of the equivalent mass system, we obtain

$$\frac{1}{2}m_1\dot{x}_1^2 + \frac{1}{2}m_2\dot{x}_2^2 + \frac{1}{2}m_3\dot{x}_3^2 = \frac{1}{2}m_{eq}\dot{x}_{eq}^2 \tag{1.20}$$

This equation gives, in view of Equations (1.18) and (1.19):

$$m_{eq} = m_1 + \left(\frac{l_2}{l_1}\right)^2 m_2 + \left(\frac{l_3}{l_1}\right)^2 m_3 \tag{1.21}$$

Case 2: Translational and Rotational Masses Coupled Together.   Let a mass m having a translational velocity $\dot{x}$ be coupled to another mass (of mass moment of inertia J_0) having a rotational velocity $\dot{\theta}$, as in the rack and pinion arrangement shown in Fig. 1.30. These two masses can be combined to obtain either (1) a single equivalent translational mass m_{eq} or (2) a single equivalent rotational mass J_{eq}, as shown.

1. *Equivalent translational mass.* The kinetic energy of the two masses is given by

$$T = \frac{1}{2}m\dot{x}^2 + \frac{1}{2}J_0\dot{\theta}^2 \tag{1.22}$$

and the kinetic energy of the equivalent mass can be expressed as

$$T_{eq} = \frac{1}{2}m_{eq}\dot{x}_{eq}^2 \tag{1.23}$$

Since $\dot{x}_{eq} = \dot{x}$ and $\dot{\theta} = \dot{x}/R$, the equivalence of T and T_{eq} gives

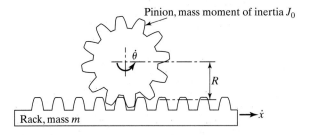

FIGURE 1.30   Translational and rotational masses in a rack and pinion arrangement.

$$\frac{1}{2}m_{eq}\dot{x}^2 = \frac{1}{2}m\dot{x}^2 + \frac{1}{2}J_0\left(\frac{\dot{x}}{R}\right)^2$$

that is,

$$m_{eq} = m + \frac{J_0}{R^2} \qquad (1.24)$$

2. *Equivalent rotational mass.* Here $\dot{\theta}_{eq} = \dot{\theta}$ and $\dot{x} = \dot{\theta}R$, and the equivalence of T and T_{eq} leads to

$$\frac{1}{2}J_{eq}\dot{\theta}^2 = \frac{1}{2}m(\dot{\theta}R)^2 + \frac{1}{2}J_0\dot{\theta}^2$$

or

$$J_{eq} = J_0 + mR^2 \qquad (1.25)$$

Equivalent Mass of a System

EXAMPLE 1.6

Find the equivalent mass of the system shown in Fig. 1.31, where the rigid link 1 is attached to the pulley and rotates with it.

Solution: Assuming small displacements, the equivalent mass (m_{eq}) can be determined using the equivalence of the kinetic energies of the two systems. When the mass m is displaced by a distance

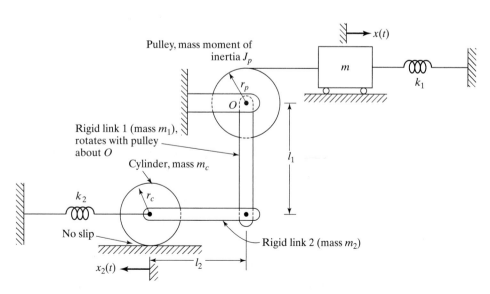

FIGURE 1.31 System considered for finding equivalent mass.

x, the pulley and the rigid link 1 rotate by an angle $\theta_p = \theta_1 = x/r_p$. This causes the rigid link 2 and the cylinder to be displaced by a distance $x_2 = \theta_p l_1 = x l_1/r_p$. Since the cylinder rolls without slippage, it rotates by an angle $\theta_c = x_2/r_c = x l_1/r_p r_c$. The kinetic energy of the system (T) can be expressed (for small displacements) as:

$$T = \frac{1}{2} m \dot{x}^2 + \frac{1}{2} J_p \dot{\theta}_p^2 + \frac{1}{2} J_1 \dot{\theta}_1^2 + \frac{1}{2} m_2 \dot{x}_2^2 + \frac{1}{2} J_c \dot{\theta}_c^2 + \frac{1}{2} m_c \dot{x}_2^2 \tag{E.1}$$

where J_p, J_1, and J_c denote the mass moments of inertia of the pulley, link 1 (about O), and cylinder, respectively, $\dot{\theta}_p$, $\dot{\theta}_1$, and $\dot{\theta}_c$ indicate the angular velocities of the pulley, link 1 (about O), and cylinder, respectively, and $\dot{x}$ and $\dot{x}_2$ represent the linear velocities of the mass m and link 2, respectively. Noting that $J_c = m_c r_c^2/2$ and $J_1 = m_1 l_1^2/3$, Equation (E.1) can be rewritten as

$$T = \frac{1}{2} m \dot{x}^2 + \frac{1}{2} J_p \left(\frac{\dot{x}}{r_p}\right)^2 + \frac{1}{2}\left(\frac{m_1 l_1^2}{3}\right)\left(\frac{\dot{x}}{r_p}\right)^2 + \frac{1}{2} m_2 \left(\frac{\dot{x} l_1}{r_p}\right)^2$$
$$+ \frac{1}{2}\left(\frac{m_c r_c^2}{2}\right)\left(\frac{\dot{x} l_1}{r_p r_c}\right)^2 + \frac{1}{2} m_c \left(\frac{\dot{x} l_1}{r_p}\right)^2 \tag{E.2}$$

By equating Equation (E.2) to the kinetic energy of the equivalent system

$$T = \frac{1}{2} m_{eq} \dot{x}^2 \tag{E.3}$$

we obtain the equivalent mass of the system as

$$m_{eq} = m + \frac{J_p}{r_p^2} + \frac{1}{3}\frac{m_1 l_1^2}{r_p^2} + \frac{m_2 l_1^2}{r_p^2} + \frac{1}{2}\frac{m_c l_1^2}{r_p^2} + m_c \frac{l_1^2}{r_p^2} \tag{E.4}$$

■

Cam-Follower Mechanism

EXAMPLE 1.7

A cam-follower mechanism (Fig. 1.32) is used to convert the rotary motion of a shaft into the oscillating or reciprocating motion of a valve. The follower system consists of a pushrod of mass m_p, a rocker arm of mass m_r, and mass moment of inertia J_r about its C.G., a valve of mass m_v, and a valve spring of negligible mass [1.28–1.30]. Find the equivalent mass (m_{eq}) of this cam-follower system by assuming the location of m_{eq} as (i) point A and (ii) point C.

Solution: The equivalent mass of the cam-follower system can be determined using the equivalence of the kinetic energies of the two systems. Due to a vertical displacement x of the pushrod, the rocker arm rotates by an angle $\theta_r = x/l_1$ about the pivot point, the valve moves downward by

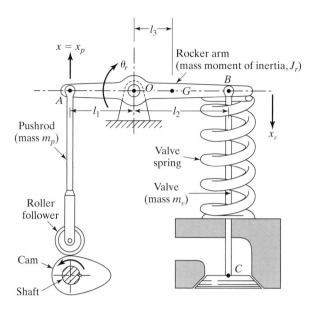

FIGURE 1.32   Cam-follower system.

$x_v = \theta_r l_2 = xl_2/l_1$, and the C.G. of the rocker arm moves downward by $x_r = \theta_r l_3 = xl_3/l_1$. The kinetic energy of the system (T) can be expressed as[2]

$$T = \frac{1}{2} m_p \dot{x}_p^2 + \frac{1}{2} m_v \dot{x}_v^2 + \frac{1}{2} J_r \dot{\theta}_r^2 + \frac{1}{2} m_r \dot{x}_r^2 \tag{E.1}$$

where $\dot{x}_p$, $\dot{x}_r$, and $\dot{x}_v$ are the linear velocities of the pushrod, C.G. of the rocker arm, and the valve, respectively, and $\dot{\theta}_r$ is the angular velocity of the rocker arm.

   (i) If m_{eq} denotes the equivalent mass placed at point A, with $\dot{x}_{eq} = \dot{x}$, the kinetic energy of the equivalent mass system T_{eq} is given by

$$T_{eq} = \frac{1}{2} m_{eq} \dot{x}_{eq}^2 \tag{E.2}$$

By equating T and T_{eq}, and noting that

$$\dot{x}_p = \dot{x}, \quad \dot{x}_v = \frac{\dot{x}l_2}{l_1}, \quad \dot{x}_r = \frac{\dot{x}l_3}{l_1}, \quad \text{and} \quad \dot{\theta}_r = \frac{\dot{x}}{l_1}$$

_____

[2] If the valve spring has a mass m_s, then its equivalent mass will be $\frac{1}{3} m_s$ (see Example 2.8). Thus the kinetic energy of the valve spring will be $\frac{1}{2}(\frac{1}{3}m_s)\dot{x}_v^2$.

we obtain

$$m_{eq} = m_p + \frac{J_r}{l_1^2} + m_v \frac{l_2^2}{l_1^2} + m_r \frac{l_3^2}{l_1^2} \qquad (E.3)$$

(ii) Similarly, if the equivalent mass is located at point C, $\dot{x}_{eq} = \dot{x}_v$ and

$$T_{eq} = \frac{1}{2} m_{eq} \dot{x}_{eq}^2 = \frac{1}{2} m_{eq} \dot{x}_v^2 \qquad (E.4)$$

Equating (E.4) and (E.1) gives

$$m_{eq} = m_v + \frac{J_r}{l_2^2} + m_p \left(\frac{l_1}{l_2}\right)^2 + m_r \left(\frac{l_3}{l_2}\right)^2 \qquad (E.5)$$

■

1.9   Damping Elements

In many practical systems, the vibrational energy is gradually converted to heat or sound. Due to the reduction in the energy, the response, such as the displacement of the system, gradually decreases. The mechanism by which the vibrational energy is gradually converted into heat or sound is known as *damping*. Although the amount of energy converted into heat or sound is relatively small, the consideration of damping becomes important for an accurate prediction of the vibration response of a system. A damper is assumed to have neither mass nor elasticity, and damping force exists only if there is relative velocity between the two ends of the damper. It is difficult to determine the causes of damping in practical systems. Hence damping is modeled as one or more of the following types.

Viscous Damping.   Viscous damping is the most commonly used damping mechanism in vibration analysis. When mechanical systems vibrate in a fluid medium such as air, gas, water, and oil, the resistance offered by the fluid to the moving body causes energy to be dissipated. In this case, the amount of dissipated energy depends on many factors, such as the size and shape of the vibrating body, the viscosity of the fluid, the frequency of vibration, and the velocity of the vibrating body. In viscous damping, the damping force is proportional to the velocity of the vibrating body. Typical examples of viscous damping include (1) fluid film between sliding surfaces, (2) fluid flow around a piston in a cylinder, (3) fluid flow through an orifice, and (4) fluid film around a journal in a bearing.

Coulomb or Dry Friction Damping.   Here the damping force is constant in magnitude but opposite in direction to that of the motion of the vibrating body. It is caused by friction between rubbing surfaces that are either dry or have insufficient lubrication.

Material or Solid or Hysteretic Damping.   When materials are deformed, energy is absorbed and dissipated by the material [1.31]. The effect is due to friction between the internal planes, which slip or slide as the deformations take place. When a body having material damping is subjected to vibration, the stress-strain diagram shows a hysteresis

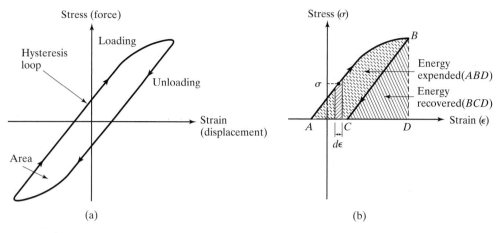

FIGURE 1.33   Hysteresis loop for elastic materials.

loop as indicated in Fig. 1.33(a). The area of this loop denotes the energy lost per unit volume of the body per cycle due to damping.[3]

1.9.1 Construction of Viscous Dampers

A viscous damper can be constructed using two parallel plates separated by a distance h, with a fluid of viscosity μ between the plates (see Fig. 1.34). Let one plate be fixed and let the other plate be moved with a velocity v in its own plane. The fluid layers in contact with the moving plate move with a velocity v, while those in contact with the fixed plate do not move. The velocities of intermediate fluid layers are assumed to vary linearly between 0 and v, as shown in Fig. 1.34. According to Newton's law of viscous flow, the shear stress (τ) developed in the fluid layer at a distance y from the fixed plate is given by

$$\tau = \mu \frac{du}{dy} \tag{1.26}$$

where $du/dy = v/h$ is the velocity gradient. The shear or resisting force (F) developed at the bottom surface of the moving plate is

$$F = \tau A = \frac{\mu A v}{h} = cv \tag{1.27}$$

where A is the surface area of the moving plate and

$$c = \frac{\mu A}{h} \tag{1.28}$$

is called the damping constant.

---

[3]When the load applied to an elastic body is increased, the stress (σ) and the strain (ε) in the body also increase. The area under the $\sigma - \varepsilon$ curve, given by

$$u = \int \sigma \, d\varepsilon$$

denotes the energy expended (work done) per unit volume of the body. When the load on the body is decreased, energy will be recovered. When the unloading path is different from the loading path, the area ABC in Fig. 1.33(b)—the area of the hysteresis loop in Fig. 1.33(a)—denotes the energy lost per unit volume of the body.

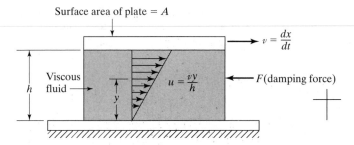

Surface area of plate = A

$v = \dfrac{dx}{dt}$

h Viscous
fluid

$u = \dfrac{vy}{h}$

F(damping force)

FIGURE 1.34   Parallel plates with a viscous fluid in between.

If a damper is nonlinear, a linearization procedure is generally used about the operating velocity (v^*), as in the case of a nonlinear spring. The linearization process gives the equivalent damping constant as

$$c = \dfrac{dF}{dv}\bigg|_{v^*} \tag{1.29}$$

**1.9.2
Combination of
Dampers**

When dampers appear in combination, they can be replaced by an equivalent damper by adopting a procedure similar to the one described in Sections 1.7 and 1.8 (see Problem 1.35).

Clearance in a Bearing

EXAMPLE 1.8

A bearing, which can be approximated as two flat plates separated by a thin film of lubricant (Fig. 1.35), is found to offer a resistance of 400 N when SAE30 oil is used as the lubricant and the relative velocity between the plates is 10 m/s. If the area of the plates (A) is 0.1 m^2, determine the clearance between the plates. Assume the absolute viscosity of SAE30 oil as 50μ reyn or 0.3445 Pa-s.

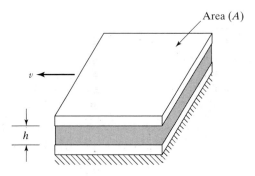

Area (A)

v

h

FIGURE 1.35   Flat plates separated by thin film of lubricant.

Solution: Since the resisting force (F) can be expressed as $F = cv$, where c is the damping constant and v is the velocity, we have

$$c = \frac{F}{v} = \frac{400}{10} = 40 \text{ N} \cdot \text{s/m} \qquad \text{(E.1)}$$

By modeling the bearing as a flat plate-type damper, the damping constant is given by Eq. (1.28):

$$c = \frac{\mu A}{h} \qquad \text{(E.2)}$$

Using the known data, Eq. (E.2) gives

$$c = 40 = \frac{(0.3445)(0.1)}{h} \quad \text{or} \quad h = 0.86125 \text{ mm} \qquad \text{(E.3)}$$

■

Piston-Cylinder Dashpot

EXAMPLE 1.9

Develop an expression for the damping constant of the dashpot shown in Fig. 1.36(a).

Solution: The damping constant of the dashpot can be determined using the shear stress equation for viscous fluid flow and the rate of fluid flow equation. As shown in Fig. 1.36(a), the dashpot consists of a piston of diameter D and length l, moving with velocity v_0 in a cylinder filled with a liquid of viscosity μ [1.24, 1.32]. Let the clearance between the piston and the cylinder wall be d. At a distance y from the moving surface, let the velocity and shear stress be v and τ, and at a distance

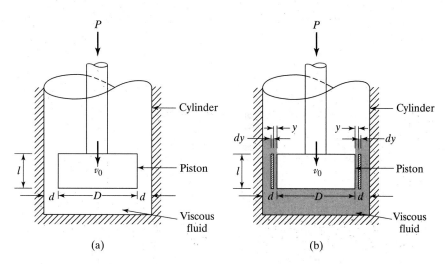

(a)   (b)

FIGURE 1.36   A dashpot.

$(y + dy)$ let the velocity and shear stress be $(v - dv)$ and $(\tau + d\tau)$, respectively (see Fig. 1.36b). The negative sign for dv shows that the velocity decreases as we move toward the cylinder wall. The viscous force on this annular ring is equal to

$$F = \pi D l \, d\tau = \pi D l \frac{d\tau}{dy} \, dy \qquad \text{(E.1)}$$

But the shear stress is given by

$$\tau = -\mu \frac{dv}{dy} \qquad \text{(E.2)}$$

where the negative sign is consistent with a decreasing velocity gradient [1.33]. Using Eq. (E.2) in Eq. (E.1), we obtain

$$F = -\pi D l \, dy \mu \frac{d^2v}{dy^2} \qquad \text{(E.3)}$$

The force on the piston will cause a pressure difference on the ends of the element, given by

$$p = \frac{P}{\left(\dfrac{\pi D^2}{4}\right)} = \frac{4P}{\pi D^2} \qquad \text{(E.4)}$$

Thus the pressure force on the end of the element is

$$p(\pi D \, dy) = \frac{4P}{D} \, dy \qquad \text{(E.5)}$$

where $(\pi D \, dy)$ denotes the annular area between y and $(y + dy)$. If we assume uniform mean velocity in the direction of motion of the fluid, the forces given in Eqs. (E.3) and (E.5) must be equal. Thus we get

$$\frac{4P}{D} \, dy = -\pi D l \, dy \mu \frac{d^2v}{dy^2}$$

or

$$\frac{d^2v}{dy^2} = -\frac{4P}{\pi D^2 l \mu} \qquad \text{(E.6)}$$

Integrating this equation twice and using the boundary conditions $v = -v_0$ at $y = 0$ and $v = 0$ at $y = d$, we obtain

$$v = \frac{2P}{\pi D^2 l \mu}(yd - y^2) - v_0\left(1 - \frac{y}{d}\right) \qquad \text{(E.7)}$$

The rate of flow through the clearance space can be obtained by integrating the rate of flow through an element between the limits $y = 0$ and $y = d$:

$$Q = \int_0^d v\pi D \, dy = \pi D \left[\frac{2Pd^3}{6\pi D^2 l\mu} - \frac{1}{2} v_0 d \right] \tag{E.8}$$

The volume of the liquid flowing through the clearance space per second must be equal to the volume per second displaced by the piston. Hence the velocity of the piston will be equal to this rate of flow divided by the piston area. This gives

$$v_0 = \frac{Q}{\left(\frac{\pi}{4} D^2 \right)} \tag{E.9}$$

Equations (E.9) and (E.8) lead to

$$P = \left[\frac{3\pi D^3 l \left(1 + \frac{2d}{D} \right)}{4d^3} \right] \mu v_0 \tag{E.10}$$

By writing the force as $P = cv_0$, the damping constant c can be found as

$$c = \mu \left[\frac{3\pi D^3 l}{4d^3} \left(1 + \frac{2d}{D} \right) \right] \tag{E.11}$$

∎

Equivalent Spring and Damping Constants of a Machine Tool Support

EXAMPLE 1.10

A precision milling machine is supported on four shock mounts, as shown in Fig. 1.37(a). The elasticity and damping of each shock mount can be modeled as a spring and a viscous damper, as shown in Fig. 1.37(b). Find the equivalent spring constant, k_{eq}, and the equivalent damping constant, c_{eq}, of the machine tool support in terms of the spring constants (k_i) and damping constants (c_i) of the mounts.

Solution: The free-body diagrams of the four springs and four dampers are shown in Fig. 1.37(c). Assuming that the center of mass, G, is located symmetrically with respect to the four springs and dampers, we notice that all the springs will be subjected to the same displacement, x, and all the dampers will be subject to the same relative velocity $\dot{x}$, where x and $\dot{x}$ denote the displacement and velocity, respectively, of the center of mass, G. Hence the forces acting on the springs (F_{si}) and the dampers (F_{di}) can be expressed as

$$F_{si} = k_i x; \quad i = 1, 2, 3, 4$$
$$F_{di} = c_i \dot{x}; \quad i = 1, 2, 3, 4 \tag{E.1}$$

Let the total forces acting on all the springs and all the dampers be F_s and F_d, respectively (see Fig. 1.37d). The force equilibrium equations can thus be expressed as

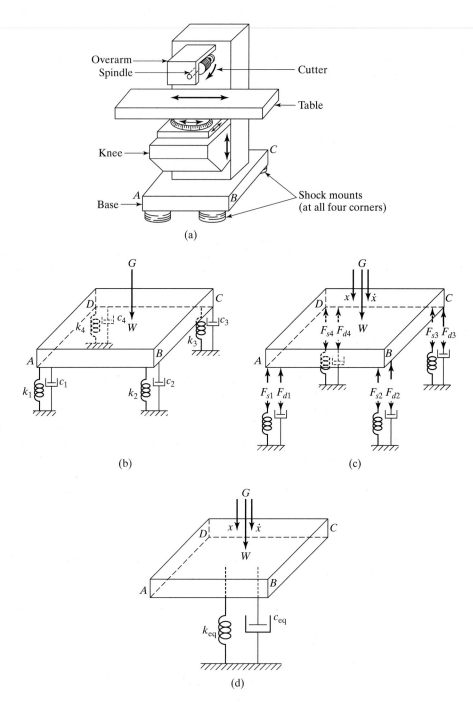

FIGURE 1.37   Horizontal milling machine.

$$F_s = F_{s1} + F_{s2} + F_{s3} + F_{s4}$$
$$F_d = F_{d1} + F_{d2} + F_{d3} + F_{d4} \tag{E.2}$$

where $F_s + F_d = W$, with W denoting the total vertical force (including the inertia force) acting on the milling machine. From Fig. 1.37(d), we have

$$F_s = k_{eq} x$$
$$F_d = c_{eq} \dot{x} \tag{E.3}$$

Equation (E.2) along with Eqs. (E.1) and (E.3), yield

$$k_{eq} = k_1 + k_2 + k_3 + k_4 = 4k$$
$$c_{eq} = c_1 + c_2 + c_3 + c_4 = 4c \tag{E.4}$$

when $k_i = k$ and $c_i = c$ for $i = 1, 2, 3, 4$.

Note: If the center of mass, G, is not located symmetrically with respect to the four springs and dampers, the ith spring experiences a displacement of x_i and the ith damper experiences a velocity of $\dot{x}_i$ where x_i and $\dot{x}_i$ can be related to the displacement x and velocity $\dot{x}$ of the center of mass of the milling machine, G. In such a case, Eqs. (E.1) and (E.4) need to be modified suitably.

■

1.10   Harmonic Motion

Oscillatory motion may repeat itself regularly, as in the case of a simple pendulum, or it may display considerable irregularity, as in the case of ground motion during an earthquake. If the motion is repeated after equal intervals of time, it is called *periodic motion*. The simplest type of periodic motion is *harmonic motion*. The motion imparted to the mass m due to the Scotch yoke mechanism shown in Fig. 1.38 is an example of simple harmonic motion [1.24, 1.34, 1.35]. In this system, a crank of radius A rotates about the point O. The other end of the crank, P, slides in a slotted rod, which reciprocates in the vertical guide R. When the crank rotates at an angular velocity ω, the end point S of the slotted link and hence the mass m of the spring-mass system are displaced from their middle positions by an amount x (in time t) given by

$$x = A \sin \theta = A \sin \omega t \tag{1.30}$$

This motion is shown by the sinusoidal curve in Fig. 1.38. The velocity of the mass m at time t is given by

$$\frac{dx}{dt} = \omega A \cos \omega t \tag{1.31}$$

and the acceleration by

$$\frac{d^2x}{dt^2} = -\omega^2 A \sin \omega t = -\omega^2 x \tag{1.32}$$

It can be seen that the acceleration is directly proportional to the displacement. Such a vibration, with the acceleration proportional to the displacement and directed toward the

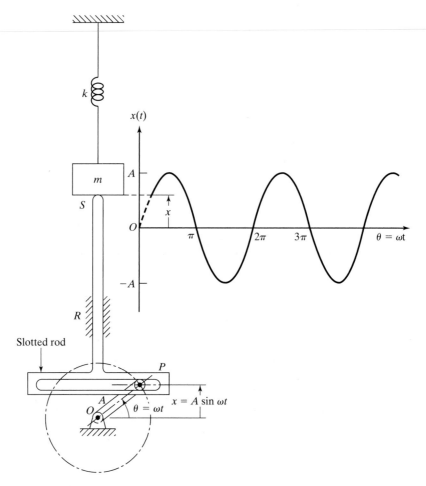

FIGURE 1.38   Scotch yoke mechanism.

mean position, is known as *simple harmonic motion*. The motion given by $x = A \cos \omega t$ is another example of a simple harmonic motion. Figure 1.38 clearly shows the similarity between cyclic (harmonic) motion and sinusoidal motion.

1.10.1
Vectorial
Representation
of Harmonic
Motion

Harmonic motion can be represented conveniently by means of a vector $\overrightarrow{OP}$ of magnitude A rotating at a constant angular velocity ω. In Fig. 1.39, the projection of the tip of the vector $\vec{X} = \overrightarrow{OP}$ on the vertical axis is given by

$$y = A \sin \omega t \tag{1.33}$$

and its projection on the horizontal axis by

$$x = A \cos \omega t \tag{1.34}$$

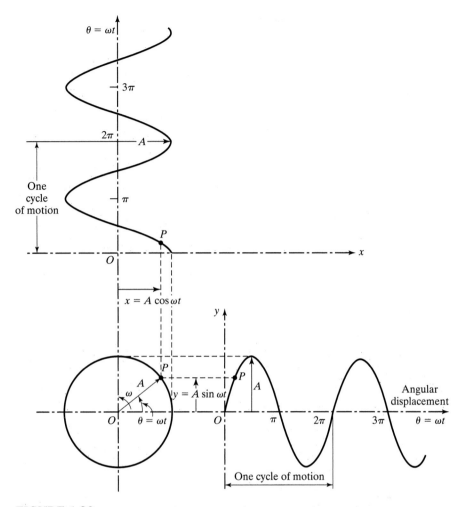

FIGURE 1.39   Harmonic motion as the projection of the end of a rotating vector.

**1.10.2
Complex
Number
Representation
of Harmonic
Motion**

As seen above, the vectorial method of representing harmonic motion requires the description of both the horizontal and vertical components. It is more convenient to represent harmonic motion using a complex number representation. Any vector $\vec{X}$ in the xy plane can be represented as a complex number:

$$\vec{X} = a + ib \tag{1.35}$$

where $i = \sqrt{-1}$ and a and b denote the x and y components of $\vec{X}$, respectively (see Fig. 1.40). Components a and b are also called the *real* and *imaginary* parts of the vector $\vec{X}$. If

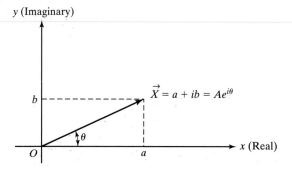

FIGURE 1.40   Representation of a complex number.

A denotes the modulus or absolute value of the vector $\vec{X}$, and θ represents the argument or the angle between the vector and the x-axis, then $\vec{X}$ can also be expressed as

$$\vec{X} = A \cos \theta + iA \sin \theta \tag{1.36}$$

with

$$A = (a^2 + b^2)^{1/2} \tag{1.37}$$

and

$$\theta = \tan^{-1}\frac{b}{a} \tag{1.38}$$

Noting that $i^2 = -1, i^3 = -i, i^4 = 1, \ldots$, $\cos \theta$ and $i \sin \theta$ can be expanded in a series as

$$\cos \theta = 1 - \frac{\theta^2}{2!} + \frac{\theta^4}{4!} - \cdots = 1 + \frac{(i\theta)^2}{2!} + \frac{(i\theta)^4}{4!} + \cdots \tag{1.39}$$

$$i \sin \theta = i\left[\theta - \frac{\theta^3}{3!} + \frac{\theta^5}{5!} - \cdots \right] = i\theta + \frac{(i\theta)^3}{3!} + \frac{(i\theta)^5}{5!} + \cdots \tag{1.40}$$

Equations (1.39) and (1.40) yield

$$(\cos \theta + i \sin \theta) = 1 + i\theta + \frac{(i\theta)^2}{2!} + \frac{(i\theta)^3}{3!} + \cdots = e^{i\theta} \tag{1.41}$$

and

$$(\cos \theta - i \sin \theta) = 1 - i\theta + \frac{(i\theta)^2}{2!} - \frac{(i\theta)^3}{3!} + \cdots = e^{-i\theta} \tag{1.42}$$

Thus Eq. (1.36) can be expressed as

$$\vec{X} = A(\cos\theta + i\sin\theta) = Ae^{i\theta} \tag{1.43}$$

1.10.3 Complex Algebra

Complex numbers are often represented without using a vector notation as

$$z = a + ib \tag{1.44}$$

where a and b denote the real and imaginary parts of z. The addition, subtraction, multiplication, and division of complex numbers can be achieved by using the usual rules of algebra. Let

$$z_1 = a_1 + ib_1 = A_1e^{i\theta_1} \tag{1.45}$$
$$z_2 = a_2 + ib_2 = A_2e^{i\theta_2} \tag{1.46}$$

where

$$A_j = \sqrt{a_j^2 + b_j^2}; j = 1, 2 \tag{1.47}$$

and

$$\theta_j = \tan^{-1}\left(\frac{b_j}{a_j}\right); j = 1, 2 \tag{1.48}$$

The sum and difference of z_1 and z_2 can be found as

$$z_1 + z_2 = A_1e^{i\theta_1} + A_2e^{i\theta_2} = (a_1 + ib_1) + (a_2 + ib_2)$$
$$= (a_1 + a_2) + i(b_1 + b_2) \tag{1.49}$$
$$z_1 - z_2 = A_1e^{i\theta_1} - A_2e^{i\theta_2} = (a_1 + ib_1) - (a_2 + ib_2)$$
$$= (a_1 - a_2) + i(b_1 - b_2) \tag{1.50}$$

1.10.4 Operations on Harmonic Functions

Using complex number representation, the rotating vector $\vec{X}$ of Fig. 1.39 can be written as

$$\vec{X} = Ae^{i\omega t} \tag{1.51}$$

where ω denotes the circular frequency (rad/sec) of rotation of the vector $\vec{X}$ in counterclockwise direction. The differentiation of the harmonic motion given by Eq. (1.51) with respect to time gives

$$\frac{d\vec{X}}{dt} = \frac{d}{dt}(Ae^{i\omega t}) = i\omega Ae^{i\omega t} = i\omega\vec{X} \tag{1.52}$$

$$\frac{d^2\vec{X}}{dt^2} = \frac{d}{dt}(i\omega Ae^{i\omega t}) = -\omega^2 Ae^{i\omega t} = -\omega^2\vec{X} \tag{1.53}$$

Thus the displacement, velocity, and acceleration can be expressed as[4]

$$\text{displacement} = \text{Re}[Ae^{i\omega t}] \qquad = A \cos \omega t \tag{1.54}$$

$$\begin{aligned} \text{velocity} = \text{Re}[i\omega Ae^{i\omega t}] \qquad &= -\omega A \sin \omega t \\ &= \omega A \cos (\omega t + 90°) \end{aligned} \tag{1.55}$$

$$\begin{aligned} \text{acceleration} = \text{Re}[-\omega^2 Ae^{i\omega t}] \qquad &= -\omega^2 A \cos \omega t \\ &= \omega^2 A \cos (\omega t + 180°) \end{aligned} \tag{1.56}$$

where Re denotes the real part. These quantities are shown as rotating vectors in Fig. 1.41. It can be seen that the acceleration vector leads the velocity vector by 90°, and the latter leads the displacement vector by 90°.

Harmonic functions can be added vectorially, as shown in Fig. 1.42. If $\text{Re}(\vec{X}_1) = A_1 \cos \omega t$ and $\text{Re}(\vec{X}_2) = A_2 \cos(\omega t + \theta)$, then the magnitude of the resultant vector $\vec{X}$ is given by

$$A = \sqrt{(A_1 + A_2 \cos \theta)^2 + (A_2 \sin \theta)^2} \tag{1.57}$$

and the angle α by

$$\alpha = \tan^{-1}\left(\frac{A_2 \sin \theta}{A_1 + A_2 \cos \theta}\right) \tag{1.58}$$

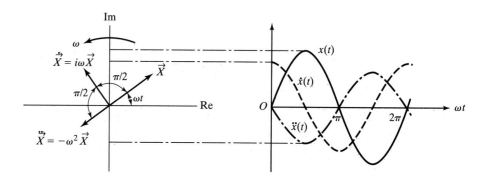

FIGURE 1.41   Displacement, velocity, and accelerations as rotating vectors.

---

[4]If the harmonic displacement is originally given as $x(t) = A \sin \omega t$, then we have

$$\begin{aligned} \text{displacement} &= \text{Im}[Ae^{i\omega t}] = A \sin \omega t \\ \text{velocity} &= \text{Im}[i\omega Ae^{i\omega t}] = \omega A \sin(\omega t + 90°) \\ \text{acceleration} &= \text{Im}[-\omega^2 Ae^{i\omega t}] = \omega^2 A \sin(\omega t + 180°) \end{aligned}$$

where Im denotes the imaginary part.

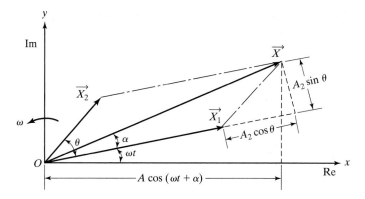

FIGURE 1.42   Vectorial addition of harmonic functions.

Since the original functions are given as real components, the sum $\vec{X}_1 + \vec{X}_2$ is given by $\text{Re}(\vec{X}) = A\cos(\omega t + \alpha)$.

Addition of Harmonic Motions

EXAMPLE 1.11

Find the sum of the two harmonic motions $x_1(t) = 10\cos\omega t$ and $x_2(t) = 15\cos(\omega t + 2)$.

Solution:

Method 1: By using trigonometric relations: Since the circular frequency is the same for both $x_1(t)$ and $x_2(t)$, we express the sum as

$$x(t) = A\cos(\omega t + \alpha) = x_1(t) + x_2(t) \tag{E.1}$$

That is,

$$A(\cos\omega t\cos\alpha - \sin\omega t\sin\alpha) = 10\cos\omega t + 15\cos(\omega t + 2)$$
$$= 10\cos\omega t + 15(\cos\omega t\cos 2 - \sin\omega t\sin 2) \tag{E.2}$$

That is,

$$\cos\omega t(A\cos\alpha) - \sin\omega t(A\sin\alpha) = \cos\omega t(10 + 15\cos 2)$$
$$- \sin\omega t(15\sin 2) \tag{E.3}$$

By equating the corresponding coefficients of $\cos\omega t$ and $\sin\omega t$ on both sides, we obtain

$$A\cos\alpha = 10 + 15\cos 2$$
$$A\sin\alpha = 15\sin 2$$
$$A = \sqrt{(10 + 15\cos 2)^2 + (15\sin 2)^2}$$
$$= 14.1477 \tag{E.4}$$

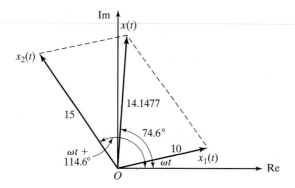

FIGURE 1.43   Addition of harmonic motions.

and

$$\alpha = \tan^{-1}\left(\frac{15 \sin 2}{10 + 15 \cos 2}\right) = 74.5963° \tag{E.5}$$

Method 2: By using vectors: For an arbitrary value of ωt, the harmonic motions $x_1(t)$ and $x_2(t)$ can be denoted graphically as shown in Fig. 1.43. By adding them vectorially, the resultant vector $x(t)$ can be found to be

$$x(t) = 14.1477 \cos(\omega t + 74.5963°) \tag{E.6}$$

Method 3: By using complex number representation: The two harmonic motions can be denoted in terms of complex numbers:

$$\begin{aligned} x_1(t) &= \text{Re}[A_1 e^{i\omega t}] \equiv \text{Re}[10 e^{i\omega t}] \\ x_2(t) &= \text{Re}[A_2 e^{i(\omega t + 2)}] \equiv \text{Re}[15 e^{i(\omega t + 2)}] \end{aligned} \tag{E.7}$$

The sum of $x_1(t)$ and $x_2(t)$ can be expressed as

$$x(t) = \text{Re}[A e^{i(\omega t + \alpha)}] \tag{E.8}$$

where A and α can be determined using Eqs. (1.47) and (1.48) as $A = 14.1477$ and $\alpha = 74.5963°$.

■

1.10.5
Definitions and
Terminology

The following definitions and terminology are useful in dealing with harmonic motion and other periodic functions.

Cycle. The movement of a vibrating body from its undisturbed or equilibrium position to its extreme position in one direction, then to the equilibrium position, then to its extreme position in the other direction, and back to equilibrium position is called a *cycle* of vibration.

One revolution (i.e., angular displacement of 2π radians) of the pin P in Fig. 1.38 or one revolution of the vector $\overrightarrow{OP}$ in Fig. 1.39 constitutes a cycle.

Amplitude. The maximum displacement of a vibrating body from its equilibrium position is called the *amplitude* of vibration. In Figs. 1.38 and 1.39 the amplitude of vibration is equal to A.

Period of oscillation. The time taken to complete one cycle of motion is known as the *period of oscillation* or *time period* and is denoted by τ. It is equal to the time required for the vector $\overrightarrow{OP}$ in Fig. 1.39 to rotate through an angle of 2π and hence

$$\tau = \frac{2\pi}{\omega} \tag{1.59}$$

where ω is called the circular frequency.

Frequency of oscillation. The number of cycles per unit time is called the *frequency of oscillation* or simply the *frequency* and is denoted by f. Thus

$$f = \frac{1}{\tau} = \frac{\omega}{2\pi} \tag{1.60}$$

Here ω is called the circular frequency to distinguish it from the linear frequency $f = \omega/2\pi$. The variable ω denotes the angular velocity of the cyclic motion; f is measured in cycles per second (Hertz) while ω is measured in radians per second.

Phase angle. Consider two vibratory motions denoted by

$$x_1 = A_1 \sin \omega t \tag{1.61}$$

$$x_2 = A_2 \sin(\omega t + \phi) \tag{1.62}$$

The two harmonic motions given by Eqs. (1.61) and (1.62) are called *synchronous* because they have the same frequency or angular velocity, ω. Two synchronous oscillations need not have the same amplitude, and they need not attain their maximum values at the same time. The motions given by Eqs. (1.61) and (1.62) can be represented graphically as shown in Fig. 1.44. In this figure, the second vector $\overrightarrow{OP_2}$ leads the first one $\overrightarrow{OP_1}$ by an angle ϕ, known as the *phase angle*. This means that the maximum of the second vector would occur ϕ radians earlier than that of the first vector. Note that instead of maxima, any other corresponding points can be taken for finding the phase angle. In Eqs. (1.61) and (1.62) or in Fig. 1.44, the two vectors are said to have a *phase difference* of ϕ.

Natural frequency. If a system, after an initial disturbance, is left to vibrate on its own, the frequency with which it oscillates without external forces is known as its *natural frequency*. As will be seen later, a vibratory system having n degrees of freedom will have, in general, n distinct natural frequencies of vibration.

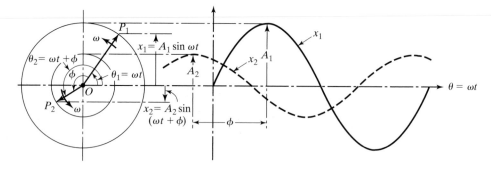

FIGURE 1.44   Phase difference between two vectors.

Beats. When two harmonic motions, with frequencies close to one another, are added, the resulting motion exhibits a phenomenon known as beats. For example, if

$$x_1(t) = X \cos \omega t \tag{1.63}$$
$$x_2(t) = X \cos (\omega + \delta)t \tag{1.64}$$

where δ is a small quantity, the addition of these motions yields

$$x(t) = x_1(t) + x_2(t) = X[\cos \omega t + \cos (\omega + \delta)t] \tag{1.65}$$

Using the relation

$$\cos A + \cos B = 2 \cos \left(\frac{A + B}{2} \right) \cos \left(\frac{A - B}{2} \right) \tag{1.66}$$

Eq. (1.65) can be rewritten as

$$x(t) = 2X \cos \frac{\delta t}{2} \cos \left(\omega + \frac{\delta}{2} \right) t \tag{1.67}$$

This equation is shown graphically in Fig. 1.45. It can be seen that the resulting motion, $x(t)$, represents a cosine wave with frequency $\omega + \delta/2$, which is approximately equal to ω, and with a varying amplitude of $2X \cos \delta t/2$. Whenever the amplitude reaches a maximum, it is called a beat. The frequency (δ) at which the amplitude builds up and dies down between 0 and $2X$ is known as beat frequency. The phenomenon of beats is often observed in machines, structures, and electric power houses. For example, in machines and structures, the beating phenomenon occurs when the forcing frequency is close to the natural frequency of the system (see Section 3.3.2).

Octave. When the maximum value of a range of frequency is twice its minimum value, it is known as an octave band. For example, each of the ranges 75–150 Hz, 150–300 Hz, and 300–600 Hz can be called an octave band. In each case, the maximum and minimum values of frequency, which have a ratio of 2:1, are said to differ by an *octave*.

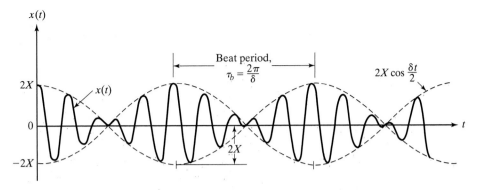

FIGURE 1.45   Phenomenon of beats.

Decibel. The various quantities encountered in the field of vibration and sound (such as displacement, velocity, acceleration, pressure, and power) are often represented using the notation of *decibel*. A decibel (dB) is originally defined as a ratio of electric powers, P/P_0, as

$$dB = 10 \log\left(\frac{P}{P_0}\right) \tag{1.68}$$

where P_0 is some reference value of power. Since electric power is proportional to the square of the voltage (X), the decibel can also be expressed as

$$dB = 10 \log\left(\frac{X}{X_0}\right)^2 = 20 \log\left(\frac{X}{X_0}\right) \tag{1.69}$$

where X_0 is a specified reference voltage. In practice, Eq. (1.69) is also used for expressing the ratios of other quantities such as displacements, velocities, accelerations, and pressures. The reference values of X_0 in Eq. (1.69) are usually taken as 2×10^{-5} N/m^2 for pressure and $1 \mu g = 9.81 \times 10^{-6}$ m/s^2 for acceleration.

1.11   Harmonic Analysis[5]

Although harmonic motion is simplest to handle, the motion of many vibratory systems is not harmonic. However, in many cases the vibrations are periodic—for example, the type shown in Fig. 1.46(a). Fortunately, any periodic function of time can be represented by Fourier series as an infinite sum of sine and cosine terms [1.36].

---

[5]The harmonic analysis forms a basis for Section 4.2.

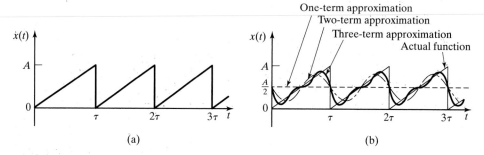

FIGURE 1.46   A periodic function.

1.11.1
Fourier Series
Expansion

If $x(t)$ is a periodic function with period τ, its Fourier series representation is given by

$$x(t) = \frac{a_0}{2} + a_1\cos \omega t + a_2\cos 2\,\omega t + \cdots$$

$$+ b_1\sin \omega t + b_2\sin 2\,\omega t + \cdots$$

$$= \frac{a_0}{2} + \sum_{n=1}^{\infty}(a_n\cos n\omega t + b_n\sin n\omega t) \tag{1.70}$$

where $\omega = 2\pi/\tau$ is the fundamental frequency and $a_0, a_1, a_2, \ldots, b_1, b_2, \ldots$ are constant coefficients. To determine the coefficients a_n and b_n, we multiply Eq. (1.70) by $\cos n\omega t$ and $\sin n\omega t$, respectively, and integrate over one period $\tau = 2\pi/\omega$; for example, from 0 to $2\pi/\omega$. Then we notice that all terms except one on the right-hand side of the equation will be zero, and we obtain

$$a_0 = \frac{\omega}{\pi}\int_0^{2\pi/\omega} x(t)\, dt = \frac{2}{\tau}\int_0^{\tau} x(t)\, dt \tag{1.71}$$

$$a_n = \frac{\omega}{\pi}\int_0^{2\pi/\omega} x(t)\cos n\omega t\, dt = \frac{2}{\tau}\int_0^{\tau} x(t)\cos n\omega t\, dt \tag{1.72}$$

$$b_n = \frac{\omega}{\pi}\int_0^{2\pi/\omega} x(t)\sin n\omega t\, dt = \frac{2}{\tau}\int_0^{\tau} x(t)\sin n\omega t\, dt \tag{1.73}$$

The physical interpretation of Eq. (1.70) is that any periodic function can be represented as a sum of harmonic functions. Although the series in Eq. (1.70) is an infinite sum, we can approximate most periodic functions with the help of only a few harmonic functions. For example, the triangular wave of Fig. 1.46(a) can be represented closely by adding only three harmonic functions, as shown in Fig. 1.46(b).

Fourier series can also be represented by the sum of sine terms only or cosine terms only. For example, the series using cosine terms only can be expressed as

$$x(t) = d_0 + d_1 \cos(\omega t - \phi_1) + d_2 \cos(2\omega t - \phi_2) + \cdots \tag{1.74}$$

where

$$d_0 = a_0/2. \tag{1.75}$$

$$d_n = (a_n^2 + b_n^2)^{1/2} \tag{1.76}$$

and

$$\phi_n = tan^{-1}\left(\frac{b_n}{a_n}\right) \tag{1.77}$$

Gibbs Phenomenon.   When a periodic function is represented by a Fourier series, an anomalous behavior can be observed. For example, Fig. 1.47 shows a triangular wave and its Fourier series representation using a different number of terms. As the number of terms (n) increases, the approximation can be seen to improve everywhere except in the vicinity of the discontinuity (point P in Fig. 1.47). Here the deviation from the true wave form becomes narrower but not any smaller in amplitude. It has been observed that the error in amplitude remains at approximately 9 percent, even when $k \to \infty$. This behavior is known as Gibbs phenomenon, after its discoverer.

**1.11.2
Complex Fourier
Series**

The Fourier series can also be represented in terms of complex numbers. By noting, from Eqs. (1.41) and (1.42), that

$$e^{i\omega t} = \cos \omega t + i \sin \omega t \tag{1.78}$$

and

$$e^{-i\omega t} = \cos \omega t - i \sin \omega t \tag{1.79}$$

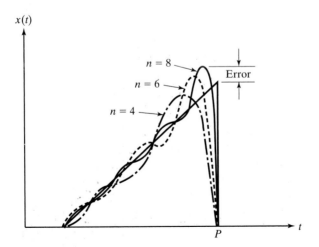

FIGURE 1.47   Gibbs phenomenon.

$\cos \omega t$ and $\sin \omega t$ can be expressed as

$$\cos \omega t = \frac{e^{i\omega t} + e^{-i\omega t}}{2} \tag{1.80}$$

and

$$\sin \omega t = \frac{e^{i\omega t} - e^{-i\omega t}}{2i} \tag{1.81}$$

Thus Eq. (1.70) can be written as

$$x(t) = \frac{a_0}{2} + \sum_{n=1}^{\infty} \left\{ a_n \left(\frac{e^{in\omega t} + e^{-in\omega t}}{2} \right) + b_n \left(\frac{e^{in\omega t} - e^{-in\omega t}}{2i} \right) \right\}$$

$$= e^{i(0)\omega t} \left(\frac{a_0}{2} - \frac{ib_0}{2} \right)$$

$$+ \sum_{n=1}^{\infty} \left\{ e^{in\omega t} \left(\frac{a_n}{2} - \frac{ib_n}{2} \right) + e^{-in\omega t} \left(\frac{a_n}{2} + \frac{ib_n}{2} \right) \right\} \tag{1.82}$$

where $b_0 = 0$. By defining the complex Fourier coefficients c_n and c_{-n} as

$$c_n = \frac{a_n - ib_n}{2} \tag{1.83}$$

and

$$c_{-n} = \frac{a_n + ib_n}{2} \tag{1.84}$$

Eq. (1.82) can be expressed as

$$x(t) = \sum_{n=-\infty}^{\infty} c_n e^{in\omega t} \tag{1.85}$$

The Fourier coefficients c_n can be determined, using Eqs. (1.71) to (1.73), as

$$c_n = \frac{a_n - ib_n}{2} = \frac{1}{\tau} \int_0^{\tau} x(t) [\cos n\omega t - i \sin n\omega t] \, dt$$

$$= \frac{1}{\tau} \int_0^{\tau} x(t) e^{-in\omega t} dt \tag{1.86}$$

**1.11.3
Frequency
Spectrum**

The harmonic functions $a_n \cos n\omega t$ or $b_n \sin n\omega t$ in Eq. (1.70) are called the *harmonics* of order n of the periodic function $x(t)$. The harmonic of order n has a period τ/n. These harmonics can be plotted as vertical lines on a diagram of amplitude (a_n and b_n or d_n and ϕ_n) versus frequency ($n\omega$), called the *frequency spectrum* or *spectral diagram*. Figure 1.48 shows a typical frequency spectrum.

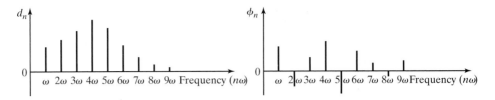

FIGURE 1.48   Frequency spectrum of a typical periodic function of time.

1.11.4
Time and
Frequency
Domain
Representations

The Fourier series expansion permits the description of any periodic function using either a time domain or a frequency domain representation. For example, a harmonic function given by $x(t) = A \sin \omega t$ in time domain (see Fig. 1.49a) can be represented by the amplitude and the frequency ω in the frequency domain (see Fig. 1.49b). Similarly, a periodic function, such as a triangular wave, can be represented in time domain, as shown in Fig. 1.49(c), or in frequency domain, as indicated in Fig. 1.49(d). Note that the amplitudes d_n and the phase angles ϕ_n corresponding to the frequencies ω_n can be used in place of the amplitudes a_n and b_n for representation in the frequency domain. Using a Fourier integral (considered in Section 14.9) permits the representation of even nonperiodic functions in either a time domain or a frequency domain. Figure 1.49 shows that the frequency domain representation does not provide the initial conditions. However, the initial conditions are often considered unnecessary in many practical applications and only the steady state conditions are of main interest.

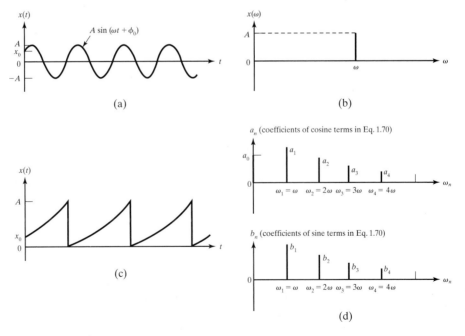

FIGURE 1.49   Representation of a function in time and frequency domains.

1.11.5
Even and Odd
Functions

An even function satisfies the relation

$$x(-t) = x(t) \tag{1.87}$$

In this case, the Fourier series expansion of $x(t)$ contains only cosine terms:

$$x(t) = \frac{a_0}{2} + \sum_{n=1}^{\infty} a_n \cos n\omega t \tag{1.88}$$

where a_0 and a_n are given by Eqs. (1.71) and (1.72), respectively. An odd function satisfies the relation

$$x(-t) = -x(t) \tag{1.89}$$

In this case, the Fourier series expansion of $x(t)$ contains only sine terms:

$$x(t) = \sum_{n=1}^{\infty} b_n \sin n\omega t \tag{1.90}$$

where b_n is given by Eq. (1.73). In some cases, a given function may be considered as even or odd depending on the location of the coordinate axes. For example, the shifting of the vertical axis from (a) to (b) or (c) in Fig. 1.50(i) will make it an odd or even function. This

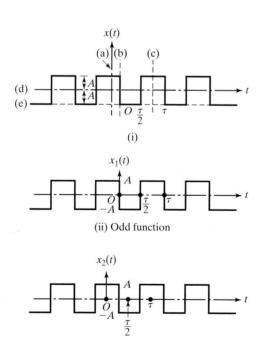

(i)

(ii) Odd function

(iii) Even function

FIGURE 1.50   Even and odd functions.

means that we need to compute only the coefficients b_n or a_n. Similarly, a shift in the time axis from (d) to (e) amounts to adding a constant equal to the amount of shift. In the case of Fig. 1.50(ii), when the function is considered as an odd function, the Fourier series expansion becomes (see Problem 1.65):

$$x_1(t) = \frac{4A}{\pi} \sum_{n=1}^{\infty} \frac{1}{(2n-1)} \sin \frac{2\pi(2n-1)t}{\tau} \tag{1.91}$$

On the other hand, if the function is considered an even function, as shown in Fig. 1.50(iii), its Fourier series expansion becomes (see Problem 1.65):

$$x_2(t) = \frac{4A}{\pi} \sum_{n=1}^{\infty} \frac{(-1)^{n+1}}{(2n-1)} \cos \frac{2\pi(2n-1)t}{\tau} \tag{1.92}$$

Since the functions $x_1(t)$ and $x_2(t)$ represent the same wave, except for the location of the origin, there exists a relationship between their Fourier series expansions also. Noting that

$$x_1\left(t + \frac{\tau}{4}\right) = x_2(t) \tag{1.93}$$

we find from Eq. (1.91),

$$x_1\left(t + \frac{\tau}{4}\right) = \frac{4A}{\pi} \sum_{n=1}^{\infty} \frac{1}{(2n-1)} \sin \frac{2\pi(2n-1)}{\tau}\left(t + \frac{\tau}{4}\right)$$

$$= \frac{4A}{\pi} \sum_{n=1}^{\infty} \frac{1}{(2n-1)} \sin \left\{ \frac{2\pi(2n-1)t}{\tau} + \frac{2\pi(2n-1)}{4} \right\} \tag{1.94}$$

Using the relation $\sin(A + B) = \sin A \cos B + \cos A \sin B$, Eq. (1.94) can be expressed as

$$x_1\left(t + \frac{\tau}{4}\right) = \frac{4A}{\pi} \sum_{n=1}^{\infty} \left\{ \frac{1}{(2n-1)} \sin \frac{2\pi(2n-1)t}{\tau} \cos \frac{2\pi(2n-1)}{4} \right.$$

$$\left. + \cos \frac{2\pi(2n-1)t}{\tau} \sin \frac{2\pi(2n-1)}{4} \right\} \tag{1.95}$$

Since $\cos [2\pi(2n - 1)/4] = 0$ for $n = 1, 2, 3, \ldots$, and $\sin [2\pi(2n - 1)/4] = (-1)^{n+1}$ for $n = 1, 2, 3, \ldots$, Eq. (1.95) reduces to

$$x_1\left(t + \frac{\tau}{4}\right) = \frac{4A}{\pi} \sum_{n=1}^{\infty} \frac{(-1)^{n+1}}{(2n - 1)} \cos \frac{2\pi(2n - 1)t}{\tau} \qquad (1.96)$$

which can be identified to be the same as Eq. (1.92).

**1.11.6
Half-Range
Expansions**

In some practical applications, the function $x(t)$ is defined only in the interval 0 to τ as shown in Fig. 1.51(a). In such a case, there is no condition of periodicity of the function since the function itself is not defined outside the interval 0 to τ. However, we can extend the function arbitrarily to include the interval $-\tau$ to 0 as shown in either Fig. 1.51(b) or Fig. 1.51(c). The extension of the function indicated in Fig. 1.51(b) results in an odd function, $x_1(t)$, while the extension of the function shown in Fig. 1.51(c) results in an even

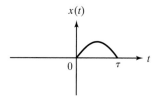

(a) Original function

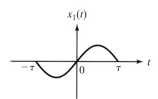

(b) Extension as an odd function

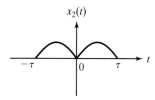

(c) Extension as an even function

FIGURE 1.51   Extension of a function for half-range expansions.

function, $x_2(t)$. Thus the Fourier series expansion of $x_1(t)$ yields only sine terms and that of $x_2(t)$ involves only cosine terms. These Fourier series expansions of $x_1(t)$ and $x_2(t)$ are known as half-range expansions [1.37]. Any of these half-range expansions can be used to find $x(t)$ in the interval 0 to τ.

1.11.7
Numerical
Computation of
Coefficients

For very simple forms of the function $x(t)$, the integrals of Eqs. (1.71) to (1.73) can be evaluated easily. However, the integration becomes involved if $x(t)$ does not have a simple form. In some practical applications, as in the case of experimental determination of the amplitude of vibration using a vibration transducer, the function $x(t)$ is not available in the form of a mathematical expression; only the values of $x(t)$ at a number of points $t_1, t_2, \ldots, t_N$ are available, as shown in Fig. 1.52. In these cases, the coefficients a_n and b_n of Eqs. (1.71) to (1.73) can be evaluated by using a numerical integration procedure like the trapezoidal or Simpson's rule [1.38].

Let's assume that $t_1, t_2, \ldots, t_N$ are an even number of equidistant points over the period τ ($N = $ even) with the corresponding values of $x(t)$ given by $x_1 = x(t_1)$, $x_2 = x(t_2), \ldots, x_N = x(t_N)$, respectively, the application of the trapezoidal rule gives the coefficients a_n and b_n (by setting $\tau = N\Delta t$) as[6]:

$$a_0 = \frac{2}{N} \sum_{i=1}^{N} x_i \tag{1.97}$$

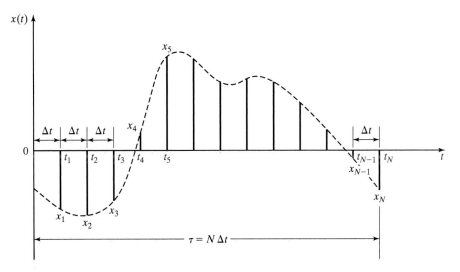

FIGURE 1.52   Values of the periodic function $x(t)$ at discrete points $t_1, t_2, \ldots, t_N$.

---

[6]N needs to be an even number for Simpson's rule but not for the trapezoidal rule. Equations (1.97) to (1.99) assume that the periodicity condition, $x_0 = x_N$, holds true.

$$a_n = \frac{2}{N} \sum_{i=1}^{N} x_i \cos \frac{2n\pi t_i}{\tau} \qquad (1.98)$$

$$b_n = \frac{2}{N} \sum_{i=1}^{N} x_i \sin \frac{2n\pi t_i}{\tau} \qquad (1.99)$$

██████████ Fourier Series Expansion

EXAMPLE 1.12 ───────────────

Determine the Fourier series expansion of the motion of the valve in the cam-follower system shown in Fig. 1.53.

Solution: If $y(t)$ denotes the vertical motion of the pushrod, the motion of the valve, $x(t)$, can be determined from the relation:

$$\tan \theta = \frac{y(t)}{l_1} = \frac{x(t)}{l_2}$$

or

$$x(t) = \frac{l_2}{l_1} y(t) \qquad (E.1)$$

where

$$y(t) = Y\frac{t}{\tau}; \ 0 \le t \le \tau \qquad (E.2)$$

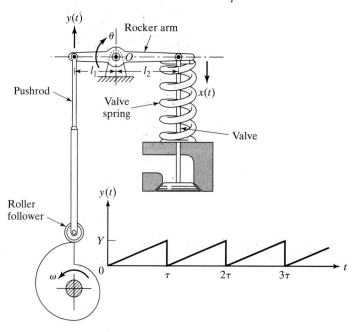

FIGURE 1.53   Cam-follower system.

and the period is given by $\tau = \dfrac{2\pi}{\omega}$. By defining

$$A = \frac{Yl_2}{l_1}$$

$x(t)$ can be expressed as

$$x(t) = A\frac{t}{\tau}; 0 \le t \le \tau \tag{E.3}$$

Equation (E.3) is shown in Fig. 1.46(a). To compute the Fourier coefficients a_n and b_n, we use Eqs. (1.71) to (1.73):

$$a_0 = \frac{\omega}{\pi}\int_0^{2\pi/\omega} x(t)\, dt = \frac{\omega}{\pi}\int_0^{2\pi/\omega} A\frac{t}{\tau}\, dt = \frac{\omega}{\pi}\frac{A}{\tau}\left(\frac{t^2}{2}\right)_0^{2\pi/\omega} = A \tag{E.4}$$

$$a_n = \frac{\omega}{\pi}\int_0^{2\pi/\omega} x(t)\cos n\omega t \cdot dt = \frac{\omega}{\pi}\int_0^{2\pi/\omega} A\frac{t}{\tau}\cos n\omega t \cdot dt$$

$$= \frac{A\omega}{\pi\tau}\int_0^{2\pi/\omega} t\cos n\omega t \cdot dt = \frac{A}{2\pi^2}\left[\frac{\cos n\omega t}{n^2} + \frac{\omega t \sin n\omega t}{n}\right]_0^{2\pi/\omega}$$

$$= 0, n = 1, 2, \ldots \tag{E.5}$$

$$b_n = \frac{\omega}{\pi}\int_0^{2\pi/\omega} x(t)\sin n\omega t \cdot dt = \frac{\omega}{\pi}\int_0^{2\pi/\omega} A\frac{t}{\tau}\sin n\omega t \cdot dt$$

$$= \frac{A\omega}{\pi\tau}\int_0^{2\pi/\omega} t\sin n\omega t \cdot dt = \frac{A}{2\pi^2}\left[\frac{\sin n\omega t}{n^2} - \frac{\omega t \cos n\omega t}{n}\right]_0^{2\pi/\omega}$$

$$= -\frac{A}{n\pi}, n = 1, 2, \ldots \tag{E.6}$$

Therefore the Fourier series expansion of $x(t)$ is

$$x(t) = \frac{A}{2} - \frac{A}{\pi}\sin \omega t - \frac{A}{2\pi}\sin 2\omega t - \ldots$$

$$= \frac{A}{\pi}\left[\frac{\pi}{2} - \left\{\sin \omega t + \frac{1}{2}\sin 2\omega t + \frac{1}{3}\sin 3\omega t + \ldots\right\}\right] \tag{E.7}$$

The first three terms of the series are shown plotted in Fig. 1.46(b). It can be seen that the approximation reaches the sawtooth shape even with a small number of terms.

■

■■■■■■■■■■■■ Numerical Fourier Analysis

EXAMPLE 1.13 ——

The pressure fluctuations of water in a pipe, measured at 0.01 second intervals, are given in Table 1.1. These fluctuations are repetitive in nature. Make a harmonic analysis of the pressure fluctuations and determine the first three harmonics of the Fourier series expansion.

Solution: Since the given pressure fluctuations repeat every 0.12 sec, the period is $\tau = 0.12$ sec and the circular frequency of the first harmonic is 2π radians per 0.12 sec or $\omega = 2\pi/0.12 = 52.36$ rad/sec. As the number of observed values in each wave (N) is 12, we obtain from Eq. (1.97)

$$a_0 = \frac{2}{N} \sum_{i=1}^{N} p_i = \frac{1}{6} \sum_{i=1}^{12} p_i = 68166.7 \tag{E.1}$$

The coefficients a_n and b_n can be determined from Eqs. (1.98) and (1.99):

$$a_n = \frac{2}{N} \sum_{i=1}^{N} p_i \cos \frac{2n\pi t_i}{\tau} = \frac{1}{6} \sum_{i=1}^{12} p_i \cos \frac{2n\pi t_i}{0.12} \tag{E.2}$$

$$b_n = \frac{2}{N} \sum_{i=1}^{N} p_i \sin \frac{2n\pi t_i}{\tau} = \frac{1}{6} \sum_{i=1}^{12} p_i \sin \frac{2n\pi t_i}{0.12} \tag{E.3}$$

The computations involved in Eqs. (E.2) and (E.3) are shown in Table 1.2. From these calculations, the Fourier series expansion of the pressure fluctuations $p(t)$ can be obtained (see Eq. 1.70):

$$p(t) = 34083.3 - 26996.0 \cos 52.36t + 8307.7 \sin 52.36t$$
$$+ 1416.7 \cos 104.72t + 3608.3 \sin 104.72t - 5833.3 \cos 157.08t$$
$$- 2333.3 \sin 157.08t + \cdots \text{N/m}^2 \tag{E.4}$$

TABLE 1.1

Time Station, i	Time (sec), t_i	Pressure (kN/m^2), p_i
0	0	0
1	0.01	20
2	0.02	34
3	0.03	42
4	0.04	49
5	0.05	53
6	0.06	70
7	0.07	60
8	0.08	36
9	0.09	22
10	0.10	16
11	0.11	7
12	0.12	0

TABLE 1.2

			n = 1		n = 2		n = 3	
i	t_i	p_i	$p_i \cos \dfrac{2\pi t_i}{0.12}$	$p_i \sin \dfrac{2\pi t_i}{0.12}$	$p_i \cos \dfrac{4\pi t_i}{0.12}$	$p_i \sin \dfrac{4\pi t_i}{0.12}$	$p_i \cos \dfrac{6\pi t_i}{0.12}$	$p_i \sin \dfrac{6\pi t_i}{0.12}$
1	0.01	20000	17320	10000	10000	17320	0	20000
2	0.02	34000	17000	29444	−17000	29444	−34000	0
3	0.03	42000	0	42000	−42000	0	0	−42000
4	0.04	49000	−24500	42434	−24500	−42434	49000	0
5	0.05	53000	−45898	26500	26500	−45898	0	53000
6	0.06	70000	−70000	0	70000	0	−70000	0
7	0.07	60000	−51960	−30000	30000	51960	0	−60000
8	0.08	36000	−18000	−31176	−18000	31176	36000	0
9	0.09	22000	0	−22000	−22000	0	0	22000
10	0.10	16000	8000	−13856	−8000	−13856	−16000	0
11	0.11	7000	6062	−3500	3500	−6062	0	−7000
12	0.12	0	0	0	0	0	0	0
$\sum_{i=1}^{12}()$		409000	−161976	49846	8500	21650	−35000	−14000
$\dfrac{1}{6}\sum_{i=1}^{12}()$		68166.7	−26996.0	8307.7	1416.7	3608.3	−5833.3	−2333.3

■

1.12   Examples Using MATLAB[7]

Graphical Representation of Fourier Series Using MATLAB

EXAMPLE 1.14

Plot the function

$$x(t) = A\frac{t}{\tau}, \; 0 \le t \le \tau \tag{E.1}$$

and its Fourier series representation with four terms

$$\bar{x}(t) = \frac{A}{\pi}\left\{\frac{\pi}{2} - \left(\sin \omega t + \frac{1}{2}\sin 2\omega t + \frac{1}{3}\sin 3\omega t\right)\right\} \tag{E.2}$$

for $0 \le t \le \tau$ with $A = 1$, $\omega = \pi$, and $\tau = \dfrac{2\pi}{\omega} = 2$.

---

[7]The source codes of all MATLAB programs are given at the web site of the book.

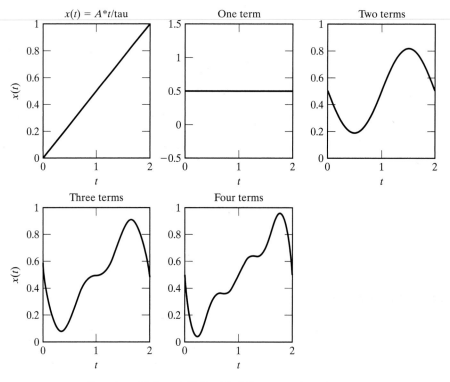

Equations (E.1) and (E.2) with different number of terms.

Solution A MATLAB program is written to plot Eqs. (E.1) and (E.2) with different numbers of terms as shown below.

```
%ex1_14.m
%plot the function x(t) = A * t / tau
A = 1;
w = pi;
tao = 2;
for i = 1: 101
    t(i) = tau * (i-1)/100;
    x(i) = A * t(i) / tau;
end
subplot(231);
plot(t,x);
ylabel('x(t)');
xlabel('t');
title('x(t) = A*t/tau');
for i = 1: 101
    x1(i) = A / 2;
end
subplot(232);
plot(t,x1);
xlabel('t');
title('One term');
for i = 1: 101
    x2(i) = A/2 - A * sin(w*t(i)) / pi;
end
subplot(233);
```

```
plot(t,x2);
xlabel('t');
title('Two terms');
for i = 1: 101
    x3(i) = A/2 - A * sin(w*t(i)) / pi - A * sin(2*w*t(i)) / (2*pi);
end
subplot(234);
plot(t,x3);
ylabel('x(t)');
xlabel('t');
title('Three terms');
for i = 1: 101
    t(i) = tau * (i-1)/100;
    x4(i) = A/2 - A * sin(w*t(i)) / pi - A * sin(2*w*t(i)) / (2*pi)
    - A * sin(3*w*t(i)) / (3*pi);
end
subplot(235);
plot(t,x4);
xlabel('t');
title('Four terms');
```

∎

EXAMPLE 1.15

Graphical Representation of Beats

A mass is subjected to two harmonic motions given by $x_1(t) = X \cos \omega t$ and $x_2(t) = X \cos (\omega + \delta) t$ with $X = 1$ cm, $\omega = 20$ rad/sec, and $\delta = 1$ rad/sec. Plot the resulting motion of the mass using MATLAB and identify the beat frequency.

Solution: The resultant motion of the mass, $x(t)$, is given by

$$x(t) = x_1(t) + x_2(t)$$
$$= X \cos \omega t + X \cos (\omega + \delta)t$$
$$= 2X \cos \frac{\delta t}{2} \cos \left(\omega + \frac{\delta}{2} \right) t \tag{E.1}$$

The motion can be seen to exhibit the phenomenon of beats with a beat frequency $\omega_b = (\omega + \delta) - (\omega) = \delta = 1$ rad/sec. Equation (E.1) is plotted using MATLAB as shown below.

```
% ex1_15.m
% Plot the Phenomenon of beats
A = 1;
w = 20;
delta = 1;
for i = 1: 1001
    t(i) = 15 * (i-1)/1000;
    x(i) = 2 * A * cos (delta*t(i)/2) * cos ((w + delta/2) *t(i));
end
plot (t,x);
xlabel ('t');
ylabel ('x(t)');
title ('Phenomenon of beats');
```

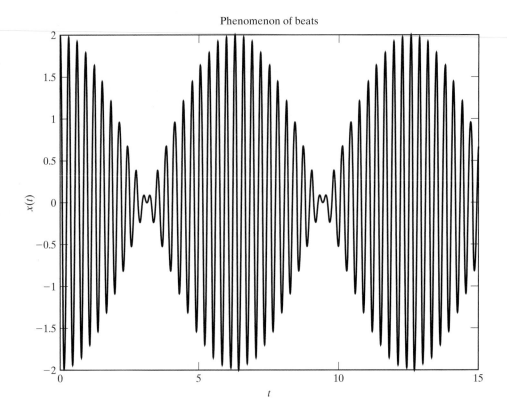

Phenomenon of beats

■

Numerical Fourier Analysis Using MATLAB

EXAMPLE 1.16

Conduct a harmonic analysis of the pressure fluctuations given in Table 1.1 on page 64 and determine the first five harmonics of the Fourier series expansion.

Solution: To find the first five harmonics of the pressure fluctuations (i.e., $a_0, a_1, \ldots, a_5, b_1, \ldots, b_5$), a general-purpose MATLAB program is developed for the harmonic analysis of a function $x(t)$ using Eqs. (1.97) to (1.99). The program, named Program1.m, requires the following input data:

n = number of equidistant points at which the values of $x(t)$ are known

m = number of Fourier coefficients to be computed

time = time period of the function $x(t)$

x = array of dimension n, containing the known values of $x(t)$. $x(i) = x(t_i)$

t = array of dimension n, containing the known values of t. $t(i) = t_i$

The following output is generated by the program:

azero = a_0 of Eq. (1.97)

$i, a(i), b(i); i = 1, 2, \ldots, m$

where $a(i)$ and $b(i)$ denote the computed values of a_i and b_i given by Eqs. (1.98) and (1.99), respectively.

```
>> program1
Fourier series expansion of the function x(t)

Data:

Number of data points in one cycle = 12

Number of Fourier Coefficients required = 5

Time period = 1.200000e-001

Station i        Time at station i: t(i)         x(i) at t(i)
        1            1.000000e-002              2.000000e+004
        2            2.000000e-002              3.400000e+004
        3            3.000000e-002              4.200000e+004
        4            4.000000e-002              4.900000e+004
        5            5.000000e-002              5.300000e+004
        6            6.000000e-002              7.000000e+004
        7            7.000000e-002              6.000000e+004
        8            8.000000e-002              3.600000e+004
        9            9.000000e-002              2.200000e+004
       10            1.000000e-001              1.600000e+004
       11            1.100000e-001              7.000000e+003
       12            1.200000e-001              0.000000e+000

Results of Fourier analysis:

azero=6.816667e+004

values of i            a(i)                    b(i)
        1          -2.699630e+004            8.307582e+003
        2           1.416632e+003            3.608493e+003
        3          -5.833248e+003           -2.333434e+003
        4          -5.834026e+002            2.165061e+003
        5          -2.170284e+003           -6.411708e+002
```

■

1.13   C++ Program

An interactive C++ program, called **Program1.cpp**[8], is given for the harmonic analysis of a function $x(t)$. The input and output parameters of the program are similar to those of the MAT-LAB program given in Example 1.16 and are also described in the comment lines of the program.

EXAMPLE 1.17

▬▬▬▬▬▬ Numerical Fourier Analysis Using C++

Solve Example 1.13 with $M = 5$ using the program **Program1.cpp**.

Solution: The input data are to be entered interactively. The input and output of the program are given below.

```
Results of Program1. cpp:

Please input the data:
Please input n:
12
Please input m:
5
```

---

[8]The source codes of all C++ programs are given at this book's Web site.

```
Please input time:
0.12
Please input the value of x[i], i = 0, ... n-1:
20000.0
34000.0
42000.0
49000.0
53000.0
70000.0
60000.0
36000.0
22000.0
16000.0
7000.0
0.0
Please input the value of t[i], i = 0, ... n-1:
0.01
0.02
0.03
0.04
0.05
0.06
0.07
0.08
0.09
0.10
0.11
0.12

FOURIER SERIES EXPANSION OF THE FUNCTION X(T)

DATA:

NUMBER OF DATA POINTS IN ONE CYCLE = 12
NUMBER OF FOURIER COEFFICIENTS REQUIRED = 5
TIME PERIOD = 1.200000e-001

TIME AT VARIOUS STATIONS, T(I) =
1.000000e-002     2.000000e-002     3.000000e-002     4.000000e-002
5.000000e-002     6.000000e-002     7.000000e-002     8.000000e-002
9.000000e-002     1.000000e-001     1.100000e-001     1.200000e-001

KNOWN VALUES OF X(I) AT T(I) =
2.000000e+004     3.400000e+004     4.200000e+004     4.900000e+004
5.300000e+004     7.000000e+004     6.000000e+004     3.600000e+004
2.200000e+004     1.600000e+004     7.000000e+003     0.000000e+000

RESULTS OF FOURIER ANALYSIS:

AZERO = 6.816667e+004

VALUES OF I, A(I) AND B(I) ARE

1     -2.699630e+004     8.307582e+003
2     1.416632e+003     3.608493e+003
3     -5.833248e+003     -2.333434e+003
4     -5.834026e+002     2.165061e+003
5     -2.170284e+003     -6.411708e+002
```

■

1.14   Fortran Program

A Fortran computer program, in the form of subroutine **FORIER.F**, is given for the harmonic analysis of a function $x(t)$. The arguments of the subroutine are described in the comment lines of the main program that calls **FORIER.F** and are similar to those of the MATLAB program given in Example 1.16.

EXAMPLE 1.18

Numerical Fourier Analysis Using Fortran

Solve Example 1.13 with $M = 5$ using the subroutine **FORIER.F**.

Solution: The main program that calls the subroutine **FORIER.F** and the subroutine **FORIER.F** is given as **PROGRAM1.F**.[9] The output of the program is given below.

```
FOURIER SERIES EXPANSION OF THE FUNCTION X(T)

DATA:

NUMBER OF DATA POINTS IN ONE CYCLE = 12
NUMBER OF FOURIER COEFFICIENTS REQUIRED = 5
TIME PERIOD = 0.12000000E+00

TIME AT VARIOUS STATIONS, T(I) =
0.99999998E-02    0.20000000E-01    0.29999999E-01    0.39999999E-01
0.50000001E-01    0.59999999E-01    0.70000000E-01    0.79999998E-01
0.90000004E-01    0.10000000E+00    0.11000000E+00    0.12000000E+00

KNOWN VALUES OF X(I) AT T(I) =
0.20000000E+05    0.34000000E+05    0.42000000E+05    0.49000000E+05
0.53000000E+05    0.70000000E+05    0.60000000E+05    0.36000000E+05
0.22000000E+05    0.16000000E+05    0.70000000E+04    0.00000000E+00

RESULTS OF FOURIER ANALYSIS:

AZERO = 0.68166664E+05

VALUES OF I, A(I) AND B(I) ARE

     1    -0.26996299E+05      0.83075869E+04
     2     0.14166348E+04      0.36084932E+04
     3    -0.58332480E+04     -0.23334373E+04
     4    -0.58340521E+03      0.21650562E+04
     5    -0.21702822E+04     -0.64117188E+03
```

■

1.15 Vibration Literature

The literature on vibrations is large and diverse. Several textbooks are available [1.39], and dozens of technical periodicals regularly publish papers relating to vibrations. This is primarily because vibration affects so many disciplines, from science of materials to machinery analysis to spacecraft structures. Researchers in many fields must be attentive to vibration research.

The most widely circulated journals that publish papers relating to vibrations are *ASME Journal of Vibration and Acoustics; ASME Journal of Applied Mechanics; Journal of Sound and Vibration; AIAA Journal; ASCE Journal of Engineering Mechanics; Earthquake Engineering and Structural Dynamics; Bulletin of the Japan Society of Mechanical Engineers; International Journal of Solids and Structures; International Journal for Numerical Methods in Engineering; Journal of the Acoustical Society of*

---

[9]The source codes of all Fortran programs are given at this book's Web site.

America; Sound and Vibration; Vibrations, Mechanical Systems and Signal Processing; International Journal of Analytical and Experimental Modal Analysis; JSME International Journal Series III—Vibration Control Engineering; and *Vehicle System Dynamics*. Many of these journals are cited in the chapter references.

In addition, *Shock and Vibration Digest, Applied Mechanics Reviews*, and *Noise and Vibration Worldwide* are monthly abstract journals containing brief discussions of nearly every published vibration paper. Formulas and solutions in vibration engineering can be readily found in references [1.40–1.42].

REFERENCES

1.1 D. C. Miller, *Anecdotal History of the Science of Sound*, Macmillan, New York, 1935.

1.2 N. F. Rieger, "The quest for $\sqrt{k/m}$: Notes on the development of vibration analysis, Part I. genius awakening," *Vibrations*, Vol. 3, No. 3/4, December 1987, pp. 3–10.

1.3 Chinese Academy of Sciences (compiler), *Ancient China's Technology and Science*, Foreign Languages Press, Beijing, 1983.

1.4 R. Taton (ed.), *Ancient and Medieval Science: From the Beginnings to 1450*, A. J. Pomerans (trans.), Basic Books, New York, 1957.

1.5 S. P. Timoshenko, *History of Strength of Materials*, McGraw-Hill, New York, 1953.

1.6 R. B. Lindsay, "The story of acoustics," *Journal of the Acoustical Society of America*, Vol. 39, No. 4, 1966, pp. 629–644.

1.7 J. T. Cannon and S. Dostrovsky, *The Evolution of Dynamics: Vibration Theory from 1687 to 1742*, Springer-Verlag, New York, 1981.

1.8 L. L. Bucciarelli and N. Dworsky, *Sophie Germain: An Essay in the History of the Theory of Elasticity*, D. Reidel Publishing, Dordrecht, Holland, 1980.

1.9 J. W. Strutt (Baron Rayleigh), *The Theory of Sound*, Dover, New York, 1945.

1.10 R. Burton, *Vibration and Impact*, Addison-Wesley, Reading, Mass., 1958.

1.11 A. H. Nayfeh, *Perturbation Methods*, Wiley, New York, 1973.

1.12 S. H. Crandall and W. D. Mark, *Random Vibration in Mechanical Systems*, Academic Press, New York, 1963.

1.13 J. D. Robson, *Random Vibration*, Edinburgh University Press, Edinburgh, 1964.

1.14 S. S. Rao, *The Finite Element Method in Engineering* (2nd ed.), Pergamon Press, Oxford, 1989.

1.15 M. J. Turner, R. W. Clough, H. C. Martin, and L. J. Topp, "Stiffness and deflection analysis of complex structures," *Journal of Aeronautical Sciences*, Vol. 23, 1956, pp. 805–824.

1.16 D. Radaj et al., "Finite element analysis, an automobile engineer's tool," *International Conference on Vehicle Structural Mechanics: Finite Element Application to Design*, Society of Automotive Engineers, Detroit, 1974.

1.17 R. E. D. Bishop, *Vibration* (2nd ed.), Cambridge University Press, Cambridge, 1979.

1.18 M. H. Richardson and K. A. Ramsey, "Integration of dynamic testing into the product design cycle," *Sound and Vibration*, Vol. 15, No. 11, November 1981, pp. 14–27.

1.19 M. J. Griffin and E. M. Whitham, "The discomfort produced by impulsive whole-body vibration," *Journal of the Acoustical Society of America*, Vol. 65, No. 5, 1980, pp. 1277–1284.

1.20 J. E. Ruzicka, "Fundamental concepts of vibration control," *Sound and Vibration*, Vol. 5, No. 7, July 1971, pp. 16–23.

1.21 T. W. Black, "Vibratory finishing goes automatic" (Part 1: Types of machines; Part 2: Steps to automation), *Tool and Manufacturing Engineer*, July 1964, pp. 53–56; and August 1964, pp. 72–76.

1.22 S. Prakash and V. K. Puri, *Foundations for Machines; Analysis and Design*, Wiley, New York, 1988.

1.23 L. Meirovitch, *Fundamentals of Vibrations*, McGraw-Hill, New York, 2001.

1.24 A. Dimarogonas, *Vibration for Engineers*, (2nd ed.), Prentice-Hall, Upper Saddle River, NJ, 1996.

1.25 E. O. Doebelin, *System Modeling and Response*, Wiley, New York, 1980.

1.26 R. W. Fitzgerald, *Mechanics of Materials* (2nd ed.), Addison-Wesley, Reading, Mass., 1982.

1.27 I. Cochin, *Analysis and Design of Dynamic Systems*, Harper & Row, New York, 1980.

1.28 F. Y. Chen, *Mechanics and Design of Cam Mechanisms*, Pergamon Press, New York, 1982.

1.29 W. T. Thomson, *Theory of Vibration with Applications* (4th ed.), Prentice-Hall, Englewood Cliffs, N.J., 1993.

1.30 N. O. Myklestad, *Fundamentals of Vibration Analysis*, McGraw-Hill, New York, 1956.

1.31 C. W. Bert, "Material damping: An introductory review of mathematical models, measures, and experimental techniques," *Journal of Sound and Vibration*, Vol. 29, No. 2, 1973, pp. 129–153.

1.32 J. M. Gasiorek and W. G. Carter, *Mechanics of Fluids for Mechanical Engineers*, Hart Publishing, New York, 1968.

1.33 A. Mironer, *Engineering Fluid Mechanics*, McGraw-Hill, New York, 1979.

1.34 F. P. Beer and E. R. Johnston, *Vector Mechanics for Engineers* (3d ed.), McGraw-Hill, New York, 1962.

1.35 A. Higdon and W. B. Stiles, *Engineering Mechanics* (2nd ed.), Prentice-Hall, New York, 1955.

1.36 E. Kreyszig, *Advanced Engineering Mathematics* (4th ed.), Wiley, New York, 1979.

1.37 M. C. Potter and J. Goldberg, *Mathematical Methods* (2nd ed.), Prentice-Hall, Englewood Cliffs, N.J., 1987.

1.38 S. S. Rao, *Applied Numerical Methods for Engineers and Scientists*, Prentice Hall, Upper Saddle River, NJ, 2002.

1.39 N. F. Rieger, "The literature of vibration engineering," *Shock and Vibration Digest*, Vol. 14, No. 1, January 1982, pp. 5–13.

1.40 R. D. Blevins, *Formulas for Natural Frequency and Mode Shape*, Van Nostrand Reinhold, New York, 1979.

1.41 W. D. Pilkey and P. Y. Chang, *Modern Formulas for Statics and Dynamics*, McGraw-Hill, New York, 1978.

1.42 C. M. Harris (ed.), *Shock and Vibration Handbook* (3rd ed.), McGraw-Hill, New York, 1988.

1.43 J. E. Shigley and C. R. Mischke, *Mechanical Engineering Design* (5th ed.), McGraw-Hill, New York, 1989.

1.44 N. P. Chironis (ed.), *Machine Devices and Instrumentation*, McGraw-Hill, New York, 1966.

1.45 D. Morrey and J. E. Mottershead, "Vibratory bowl feeder design using numerical modelling techniques," in *Modern Practice in Stress and Vibration Analysis*, J. E. Mottershead (ed.), Pergamon Press, Oxford, 1989, pp. 211–217.

1.46 K. McNaughton (ed.), *Solids Handling*, McGraw-Hill, New York, 1981.

1.47 M. M. Kamal and J. A. Wolf, Jr. (eds.), *Modern Automotive Structural Analysis*, Van Nostrand Reinhold, New York, 1982.

1.48 D. J. Inman, *Engineering Vibration*, (2nd ed.), Prentice-Hall, Upper Saddle River, NJ, 2001.

1.49 J. H. Ginsberg, *Mechanical and Structural Vibrations: Theory and Applications*, John Wiley, New York, 2001.

REVIEW QUESTIONS

1.1 Give brief answers to the following:

1. Give two examples each of the bad and the good effects of vibration.
2. What are the three elementary parts of a vibrating system?
3. Define the degree of freedom of a vibrating system.
4. What is the difference between a discrete and a continuous system? Is it possible to solve any vibration problem as a discrete one?
5. In vibration analysis, can damping always be disregarded?
6. Can a nonlinear vibration problem be identified by looking at its governing differential equation?
7. What is the difference between deterministic and random vibration? Give two practical examples of each.
8. What methods are available for solving the governing equations of a vibration problem?
9. How do you connect several springs to increase the overall stiffness?
10. Define spring stiffness and damping constant.
11. What are the common types of damping?
12. State three different ways of expressing a periodic function in terms of its harmonics.
13. Define these terms: cycle, amplitude, phase angle, linear frequency, period, and natural frequency.
14. How are τ, ω, and f related to each other?
15. How can we obtain the frequency, phase, and amplitude of a harmonic motion from the corresponding rotating vector?
16. How do you add two harmonic motions having different frequencies?
17. What are beats?
18. Define the terms *decibel* and *octave*.
19. Explain Gibbs phenomenon.
20. What are half-range expansions?

1.2 Indicate whether each of the following statements is true or false:

1. If energy is lost in any way during vibration, the system can be considered to be damped.
2. The superposition principle is valid for both linear and nonlinear systems.

3. The frequency with which an initially disturbed system vibrates on its own is known as natural frequency.

4. Any periodic function can be expanded into a Fourier series.

5. A harmonic motion is a periodic motion.

6. The equivalent mass of several masses at different locations can be found using the equivalence of kinetic energy.

7. The generalized coordinates are not necessarily Cartesian coordinates.

8. Discrete systems are same as lumped parameter systems.

9. Consider the sum of harmonic motions, $x(t) = x_1(t) + x_2(t) = A \cos(\omega t + \alpha)$, with $x_1(t) = 15 \cos \omega t$ and $x_2(t) = 20 \cos(\omega t + 1)$. The amplitude A is given by 30.8088.

10. Consider the sum of harmonic motions, $x(t) = x_1(t) + x_2(t) = A \cos(\omega t + \alpha)$, with $x_1(t) = 15 \cos \omega t$ and $x_2(t) = 20 \cos(\omega t + 1)$. The phase angle α is given by 1.57 rad.

1.3 Fill in the blank with the proper word:

1. Systems undergo dangerously large oscillations at _____.

2. Undamped vibration is characterized by no loss of _____.

3. A vibratory system consists of a spring, damper, and _____.

4. If a motion repeats after equal intervals of time, it is called a _____ motion.

5. When acceleration is proportional to the displacement and directed towards the mean position, the motion is called _____ harmonic.

6. The time taken to complete one cycle of motion is called the _____ of vibration.

7. The number of cycles per unit time is called the _____ of vibration.

8. Two harmonic motions having the same frequency are said to be _____.

9. The angular difference between the occurrence of similar points of two harmonic motions is called _____.

10. Continuous or distributed systems can be considered to have _____ number of degrees of freedom.

11. Systems with a finite number of degrees of freedom are called _____ systems.

12. The degree of freedom of a system denotes the minimum number of independent _____ necessary to describe the positions of all parts of the system at any insant of time.

13. If a system vibrates due to initial disturbance only, it is called _____ vibration.

14. If a system vibrates due to an external excitation, it is called _____ vibration.

15. Resonance denotes the coincidence of the frequency of external excitation with a _____ frequency of the system.

16. A function $f(t)$ is called an odd function if _____.

17. The _____ range expansions can be used to represent functions defined only in the interval 0 to τ.

18. _____ analysis deals with the Fourier series representation of periodic functions.

1.4 Select the most appropriate answer from the multiple choices given:

1. The world's first seismograph was invented in
   (a) Japan   (b) China   (c) Egypt

2. The first experiments on simple pendulum were conducted by
   (a) Galileo   (b) Pythagoras   (c) Aristotle

3. The *Philosophiae Naturalis Principia Mathematica* was published by
   (a) Galileo   (b) Pythagoras   (c) Newton

4. Mode shapes of plates, by placing sand on vibrating plates, were first observed by
   (a) Chladni   (b) D'Alembert   (c) Galileo

5. The thick beam theory was first presented by
   (a) Mindlin   (b) Einstein   (c) Timoshenko

6. The degree of freedom of a simple pendulum is:
   (a) 0   (b) 1   (c) 2

7. Vibration can be classified in
   (a) one way   (b) two ways   (c) several ways

8. Gibbs phenomenon denotes an anomalous behavior in the Fourier series representation of a
   (a) harmonic function   (b) periodic function   (c) random function

9. The graphical representation of the amplitudes and phase angles of the various frequency components of a periodic function is known as a
   (a) spectral diagram   (b) frequency diagram   (c) harmonic diagram

10. When a system vibrates in a fluid medium, the damping is
    (a) viscous   (b) Coulomb   (c) solid

11. When parts of a vibrating system slide on a dry surface, the damping is
    (a) viscous   (b) Coulomb   (c) solid

12. When the stress-strain curve of the material of a vibrating system exhibits a hysteresis loop, the damping is
    (a) viscous   (b) Coulomb   (c) solid

13. The equivalent spring constant of two parallel springs with stiffnesses k_1 and k_2 is
    (a) $k_1 + k_2$ (b) $\dfrac{1}{\dfrac{1}{k_1} + \dfrac{1}{k_2}}$ (c) $\dfrac{1}{k_1} + \dfrac{1}{k_2}$

14. The equivalent spring constant of two series springs with stiffnesses k_1 and k_2 is
    (a) $k_1 + k_2$ (b) $\dfrac{1}{\dfrac{1}{k_1} + \dfrac{1}{k_2}}$ (c) $\dfrac{1}{k_1} + \dfrac{1}{k_2}$

15. The spring constant of a cantilever beam with an end mass m is
    (a) $\dfrac{3EI}{l^3}$ (b) $\dfrac{l^3}{3EI}$ (c) $\dfrac{Wl^3}{3EI}$

16. If $f(-t) = f(t)$, function $f(t)$ is said to be
    (a) even   (b) odd   (c) continuous

1.5   Match the following:

(1) Pythagoras (582–507 B.C.)   (a) published a book on the theory of sound
(2) Euclid (300 B.C.)   (b) first person to investigate musical sounds on a scientific basis

(3) Zhang Heng (132 A.D.)   (c) wrote a treatise called *Introduction to Harmonics*
(4) Galileo (1564–1642)   (d) founder of modern experimental science
(5) Rayleigh (1877)   (e) invented the world's first seismograph

1.6 Match the following:

(1) Imbalance in diesel engines	(a) can cause failure of turbines and aircraft engines
(2) Vibration in machine tools during metal cutting	(b) cause discomfort in human activity
(3) Blade and disk vibration	(c) can cause wheels of locomotives to rise off the track
(4) Wind-induced vibration	(d) can cause failure of bridges
(5) Transmission of vibration	(e) can give rise to chatter

1.7 Consider four springs with the spring constants:
$k_1 = 20$ lb/in., $k_2 = 50$ lb/in., $k_3 = 100$ lb/in., $k_4 = 200$ lb/in. Match the equivalent spring constants:

(1) k_1, k_2, k_3 and k_4 are in parallel	(a) 18.9189 lb/in.
(2) k_1, k_2, k_3 and k_4 are in series	(b) 370.0 lb/in.
(3) k_1 and k_2 are in parallel ($k_{eq} = k_{12}$)	(c) 11.7647 lb/in.
(4) k_3 and k_4 are in parallel ($k_{eq} = k_{34}$)	(d) 300.0 lb/in.
(5) k_1, k_2 and k_3 are in parallel ($k_{eq} = k_{123}$)	(e) 70.0 lb/in.
(6) k_{123} is in series with k_4	(f) 170.0 lb/in.
(7) $k_2, k_3,$ and k_4 are in parallel ($k_{eq} = k_{234}$)	(g) 350.0 lb/in.
(8) k_1 and k_{234} are in series	(h) 91.8919 lb/in.

PROBLEMS

The problem assignments are organized as follows:

Problems	Section Covered	Topic Covered
1.1–1.6	1.6	Vibration analysis procedure
1.7–1.29	1.7	Spring elements
1.13, 1.29–1.34	1.8	Mass elements
1.35–1.42	1.9	Damping elements
1.43–1.63	1.10	Harmonic motion
1.64–1.75	1.11	Harmonic analysis
1.76–1.80	1.12	MATLAB programs
1.81–1.84	1.13	C++ programs
1.85–1.88	1.14	Fortran programs
1.89–1.94	—	Design projects

1.1* A study of the response of a human body subjected to vibration/shock is important in many applications. In a standing posture, the masses of head, upper torso, hips, and legs and the elasticity/damping of neck, spinal column, abdomen, and legs influence the response characteristics. Develop a sequence of three improved approximations for modeling the human body.

---

*The asterisk denotes a design type problem or a problem with no unique answer.

1.2* Figure 1.54 shows a human body and a restraint system at the time of an automobile collision [1.47]. Suggest a simple mathematical model by considering the elasticity, mass, and damping of the seat, human body, and the restraints for a vibration analysis of the system.

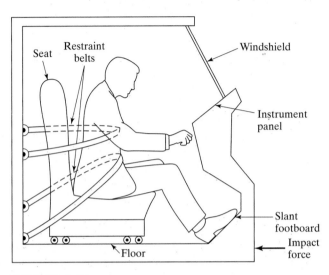

FIGURE 1.54   A human body and a restraint system.

1.3* A reciprocating engine is mounted on a foundation as shown in Fig. 1.55. The unbalanced forces and moments developed in the engine are transmitted to the frame and the foundation. An elastic pad is placed between the engine and the foundation block to reduce the transmission of vibration. Develop two mathematical models of the system using a gradual refinement of the modeling process.

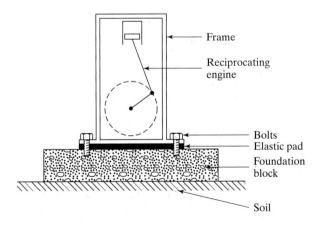

Figure 1.55   A reciprocating engine on a foundation.

1.4* An automobile moving over a rough road (Fig. 1.56) can be modeled considering (a) weight of the car body, passengers, seats, front wheels, and rear wheels; (b) elasticity of tires (suspension), main springs, and seats; and (c) damping of the seats, shock absorbers, and tires. Develop three mathematical models of the system using a gradual refinement in the modeling process.

FIGURE 1.56   An automobile moving on a rough road.

1.5* The consequences of a head-on collision of two automobiles can be studied by considering the impact of the automobile on a barrier, as shown in Fig. 1.57. Construct a mathematical model by considering the masses of the automobile body, engine, transmission, and suspension and the elasticity of the bumpers, radiator, sheet metal body, driveline, and engine mounts.

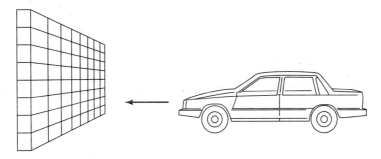

FIGURE 1.57   An automobile colliding on a barrier.

1.6* Develop a mathematical model for the tractor and plow shown in Fig. 1.58 by considering the mass, elasticity, and damping of the tires, shock absorbers, and the plows (blades).

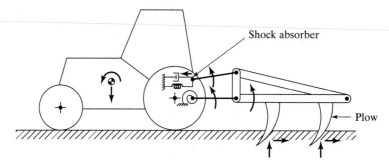

FIGURE 1.58   A tractor and plow.

1.7   Determine the equivalent spring constant of the system shown in Fig. 1.59.

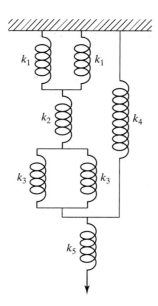

FIGURE 1.59   Springs in
series—parallel.

1.8   In Fig. 1.60, find the equivalent spring constant of the system in the direction of θ.

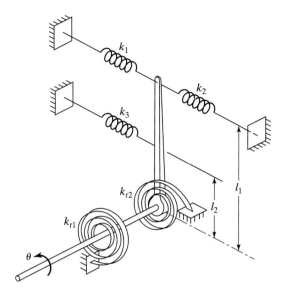

FIGURE 1.60

1.9 Find the equivalent torsional spring constant of the system shown in Fig. 1.61. Assume that $k_1, k_2, k_3,$ and k_4 are torsional and k_5 and k_6 are linear spring constants.

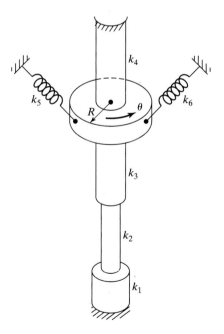

FIGURE 1.61

1.10 A machine of mass $m = 500$ kg is mounted on a simply supported steel beam of length $l = 2$ m having a rectangular cross section (depth $= 0.1$ m, width $= 1.2$ m) and Young's modulus $E = 2.06 \times 10^{11}$ N/m². To reduce the vertical deflection of the beam, a spring of stiffness k is attached at the mid-span, as shown in Fig. 1.62. Determine the value of k needed to reduce the deflection of the beam by

a. 25 percent of its original value
b. 50 percent of its original value
c. 75 percent of its original value.

Assume that the mass of the beam is negligible.

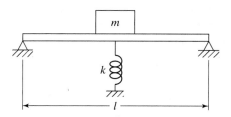

FIGURE 1.62

1.11 Four identical rigid bars—each of length a—are connected to a spring of stiffness k to form a structure for carrying a vertical load P, as shown in Figs. 1.63(a) and (b). Find the equivalent spring constant of the system k_{eq}, for each case, disregarding the masses of the bars and the friction in the joints.

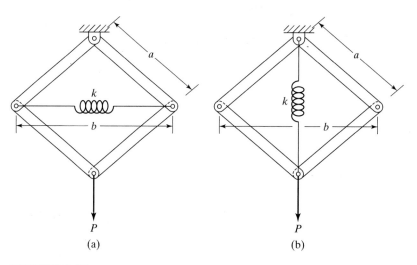

(a)   (b)

FIGURE 1.63

1.12 The tripod shown in Fig. 1.64 is used for mounting an electronic instrument that finds the distance between two points in space. The legs of the tripod are located symmetrically about the mid-vertical axis, each leg making an angle α with the vertical. If each leg has a length l and axial stiffness k, find the equivalent spring stiffness of the tripod in the vertical direction.

FIGURE 1.64

1.13 Find the equivalent spring constant and equivalent mass of the system shown in Fig. 1.65 with references to θ. Assume that the bars AOB and CD are rigid with negligible mass.

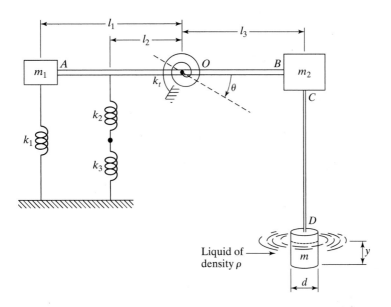

FIGURE 1.65

1.14 Find the length of the equivalent uniform hollow shaft of inner diameter d and thickness t that has the same axial spring constant as that of the solid conical shaft shown in Fig. 1.66.

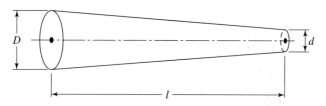

FIGURE 1.66

1.15 The force-deflection characteristic of a spring is described by $F = 500x + 2x^3$, where the force (F) is in Newtons and the deflection (x) is in millimeters. Find (a) the linearized spring constant at $x = 10$ mm and (b) the spring forces at $x = 9$ mm and $x = 11$ mm using the linearized spring constant. Also find the error in the spring forces found in (b).

1.16 Figure 1.67 shows an air spring. This type of spring is generally used for obtaining very low natural frequencies while maintaining zero deflection under static loads. Find the spring constant of this air spring by assuming that the pressure p and volume v change adiabatically when the mass m moves.

Hint:

pv^γ = constant for an adiabatic process, where γ is the ratio of specific heats. For air, $\gamma = 1.4$.

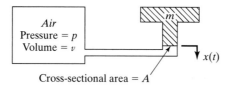

FIGURE 1.67

1.17 Find the equivalent spring constant of the system shown in Fig. 1.68 in the direction of the load P.

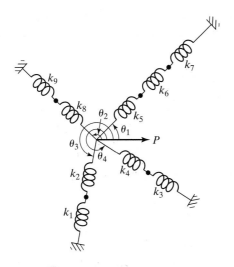

FIGURE 1.68

1.18* Design an air spring using a cylindrical container and a piston to achieve a spring constant of 75 lb/in. Assume that the maximum air pressure available is 200 psi.

1.19 The force (F)–deflection (x) relationship of a nonlinear spring is given by

$$F = ax + bx^3$$

where a and b are constants. Find the equivalent linear spring constant when the deflection is 0.01 m with $a = 20,000$ N/m and $b = 40 \times 10^6$ N/m^3.

1.20 Two nonlinear springs, S_1 and S_2, are connected in two different ways as indicated in Fig. 1.69. The force, F_i, in spring S_i is related to its deflection (x_i) as

$$F_i = a_i x_i + b_i x_i^{3i}, i = 1, 2$$

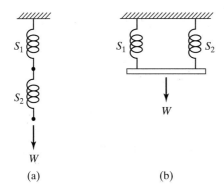

FIGURE 1.69

where a_i and b_i are constants. If an equivalent linear spring constant, k_{eq}, is defined by $W = k_{eq}x$ where x is the total deflection of the system, find an expression for k_{eq} in each case.

1.21* Design a steel helical compression spring to satisfy the following requirements:

Spring stiffness $(k) \geq 8000$ N/mm

Fundamental natural frequency of vibration $(f_1) \geq 0.4$ Hz

Spring index $(D/d) \geq 6$

Number of active turns $(N) \geq 10$.

The stiffness and fundamental natural frequency of the spring are given by [1.43]:

$$ k = \frac{Gd^4}{8D^3N} \quad \text{and} \quad f_1 = \frac{1}{2}\sqrt{\frac{kg}{W}} $$

where G = shear modulus, d = wire diameter, D = coil diameter, W = weight of the spring, and g = acceleration due to gravity.

1.22 Find the spring constant of the bimetallic bar shown in Fig. 1.70 in axial motion.

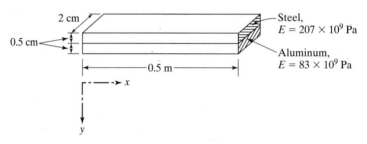

FIGURE 1.70

1.23 A tapered solid steel propeller shaft is shown in Fig. 1.71. Determine the torsional spring constant of the shaft.

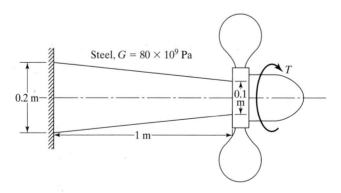

FIGURE 1.71

1.24 A composite propeller shaft, made of steel and aluminum, is shown in Fig. 1.72.

    a. Determine the torsional spring constant of the shaft.
    b. Determine the torsional spring constant of the composite shaft when the inner diameter of the aluminum tube is 5 cm instead of 10 cm.

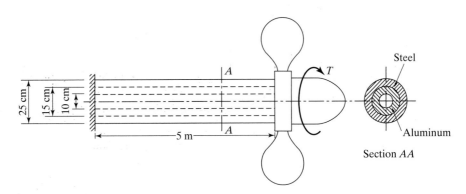

FIGURE 1.72

1.25 Consider two helical springs with the following characteristics:

    Spring 1: material—steel; number of turns—10; mean coil diameter—12 in; wire diameter—2 in; free length—15 in.; shear modulus—12×10^6 psi.

    Spring 2: material—aluminum; number of turns—10; mean coil diameter—10 in.; wire diameter—1 in.; free length—15 in.; shear modulus—4×10^6 psi.

Determine the equivalent spring constant when (a) spring 2 is placed inside spring 1, and (b) spring 2 is placed on top of spring 1.

1.26 Solve Problem 1.25 by assuming the wire diameters of springs 1 and 2 to be 1.0 in. and 0.5 in. instead of 2.0 in. and 1.0 in., respectively.

1.27 The arm *AD* of the excavator shown in Fig. 1.73 can be approximated as a steel tube of outer diameter 10 in., inner diameter 9.5 in., and length 100 in. with a viscous damping coefficient of 0.4. The arm *DE* can be approximated as a steel tube of outer diameter 7 in., inner diameter 6.5 in., and length 75 in. with a viscous damping coefficient of 0.3. Estimate the equivalent spring constant and equivalent damping coefficient of the excavator assuming that the base *AC* is fixed.

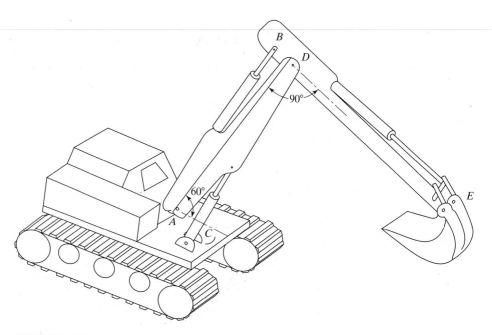

FIGURE 1.73   An excavator.

1.28 A heat exchanger consists of six identical stainless steel tubes connected in parallel as shown in Fig. 1.74. If each tube has an outer diameter 0.30 in., inner diameter 0.29 in., and length 50 in., determine the axial stiffness and the torsional stiffness about the longitudinal axis of the heat exchanger.

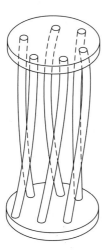

FIGURE 1.74
A heat exchanger.

1.29 Two sector gears, located at the ends of links 1 and 2, are engaged together and rotate about O_1 and O_2, as shown in Fig. 1.75. If links 1 and 2 are connected to springs k_1 to k_4 and k_{t1} and k_{t2} as shown, find the equivalent torsional spring stiffness and equivalent mass moment of inertia of the system with reference to θ_1. Assume (a) the mass moment of inertia of link 1 (including the sector gear) about O_1 is J_1 and that of link 2 (including the sector gear) about O_2 is J_2, and (b) the angles θ_1 and θ_2 are small.

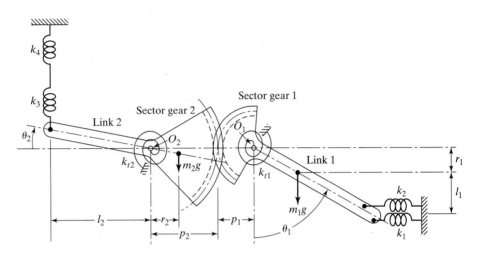

FIGURE 1.75 Two sector gears.

1.30 In Fig. 1.76 find the equivalent mass of the rocker arm assembly with respect to the x coordinate.

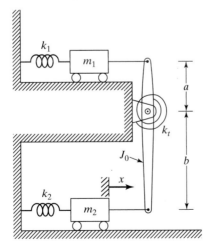

FIGURE 1.76 Rocker arm assembly.

1.31 Find the equivalent mass moment of inertia of the gear train shown in Fig. 1.77 with reference to the driving shaft. In Fig. 1.77, J_i and n_i denote the mass moment of inertia and the number of teeth, respectively, of gear i, $i = 1, 2, \ldots, 2N$.

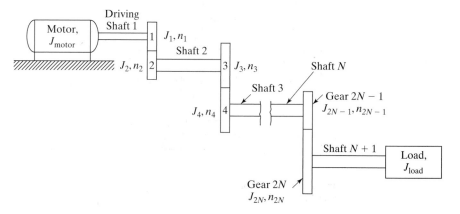

FIGURE 1.77

1.32 Two masses, having mass moments of inertia J_1 and J_2, are placed on rotating rigid shafts that are connected by gears, as shown in Fig. 1.78. If the number of teeth on gears 1 and 2 are n_1 and n_2, respectively, find the equivalent mass moment of inertia corresponding to θ_1.

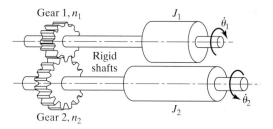

FIGURE 1.78   Rotational masses on geared shafts.

1.33 A simplified model of a petroleum pump is shown in Fig. 1.79, where the rotary motion of the crank is converted to the reciprocating motion of the piston. Find the equivalent mass, m_{eq}, of the system at location A.

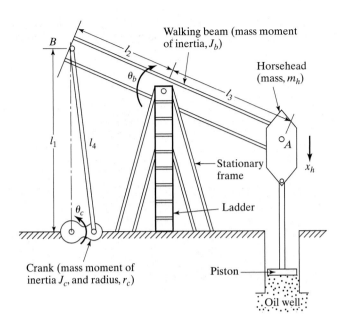

FIGURE 1.79

1.34 Find the equivalent mass of the system shown in Fig. 1.80.

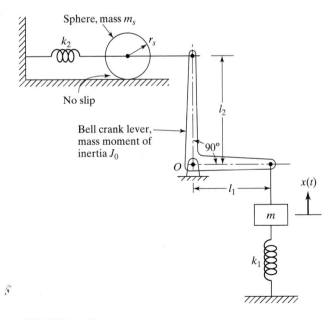

FIGURE 1.80

1.35 Find a single equivalent damping constant for the following cases:

    a. When three dampers are parallel.

    b. When three dampers are in series.

    c. When three dampers are connected to a rigid bar (Fig. 1.81) and the equivalent damper is at site c_1.

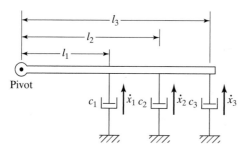

FIGURE 1.81   Dampers connected to a rigid bar.

    d. When three torsional dampers are located on geared shafts (Fig. 1.82) and the equivalent damper is at location c_{t_1}.

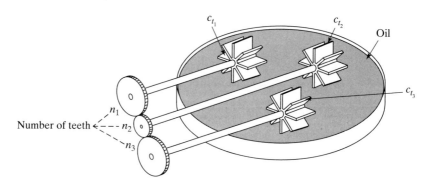

FIGURE 1.82   Dampers located on geared shafts.

Hint: The energy dissipated by a viscous damper in a cycle during harmonic motion is given by $\pi c \omega X^2$, where c is the damping constant, ω is the frequency, and X is the amplitude of oscillation.

1.36* Design a piston-cylinder-type viscous damper to achieve a damping constant of 1 lb-sec/in. using a fluid of viscosity 4μ reyn (1 reyn = 1 lb-sec/in^2).

1.37* Design a shock absorber (piston-cylinder-type dashpot) to obtain a damping constant of 10^5 lb-sec/in. using SAE30 oil at 70°F. The diameter of the piston has to be less than 2.5 in.

1.38 Develop an expression for the damping constant of the rotational damper shown in Fig. 1.83 in terms of D, d, l, h, ω, and μ, where ω denotes the constant angular velocity of the inner cylinder, and d and h represent the radial and axial clearances between the inner and outer cylinders.

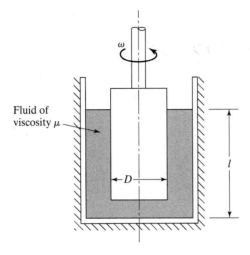

Fluid of viscosity μ

FIGURE 1.83

1.39 The force (F)–velocity $(\dot{x})$ relationship of a nonlinear damper is given by

$$F = a\dot{x} + b\dot{x}^2$$

where a and b are constants. Find the equivalent linear damping constant when the relative velocity is 5 m/s with $a = 5$ N-s/m and $b = 0.2$ N-s^2/m^2.

1.40 The damping constant (c) due to skin friction drag of a rectangular plate moving in a fluid of viscosity μ is given by (see Fig. 1.84):

$$c = 100 \, \mu l^2 d$$

Design a plate-type damper (shown in Fig. 1.35) that provides an identical damping constant for the same fluid.

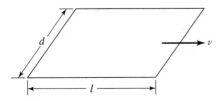

FIGURE 1.84

1.41 The damping constant (c) of the dashpot shown in Fig. 1.85 is given by [1.27]:

$$c = \frac{6\pi\mu l}{h^3}\left[\left(a - \frac{h}{2}\right)^2 - r^2\right]\left[\frac{a^2 - r^2}{a - \frac{h}{2}} - h\right]$$

Determine the damping constant of the dashpot for the following data: $\mu = 0.3445$ Pa-s, $l = 10$ cm, $h = 0.1$ cm, $a = 2$ cm, $r = 0.5$ cm.

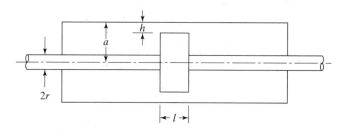

FIGURE 1.85

1.42 In Problem 1.41, using the given data as reference, find the variation of the damping constant c when

   a. r is varied from 0.5 cm to 1.0 cm
   b. h is varied from 0.05 cm to 0.10 cm
   c. a is varied from 2 cm to 4 cm

1.43 Express the complex number $5 + 2i$ in the exponential form $Ae^{i\theta}$.

1.44 Add the two complex numbers $(1 + 2i)$ and $(3 - 4i)$ and express the result in the form $Ae^{i\theta}$.

1.45 Subtract the complex number $(1 + 2i)$ from $(3 - 4i)$ and express the result in the form $Ae^{i\theta}$.

1.46 Find the product of the complex numbers $z_1 = (1 + 2i)$ and $z_2 = (3 - 4i)$ and express the result in the form $Ae^{i\theta}$.

1.47 Find the quotient, z_1/z_2, of the complex numbers $z_1 = (1 + 2i)$ and $z_2 = (3 - 4i)$ and express the result in the form $Ae^{i\theta}$.

1.48 The foundation of a reciprocating engine is subjected to harmonic motions in x and y directions:

$$x(t) = X \cos \omega t$$
$$y(t) = Y \cos(\omega t + \phi)$$

where X and Y are the amplitudes, ω is the angular velocity, and ϕ is the phase difference.

   a. Verify that the resultant of the two motions satisfies the equation of the ellipse given by (see Fig. 1.86):

$$\frac{x^2}{X^2} + \frac{y^2}{Y^2} - 2\frac{xy}{XY}\cos\phi = \sin^2\phi \qquad \text{(E.1)}$$

   b. Discuss the nature of the resultant motion given by Eq. (E.1) for the special cases of $\phi = 0$, $\phi = \dfrac{\pi}{2}$, and $\phi = \pi$.

Note: The elliptic figure represented by Eq. (E.1) is known as a Lissajous figure and is useful in interpreting certain types of experimental results (motions) displayed by oscilloscopes.

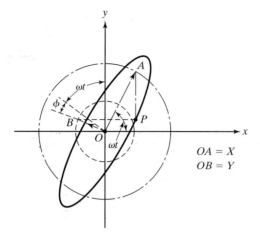

$$OA = X$$
$$OB = Y$$

FIGURE 1.86   Lissajous figure.

1.49 The foundation of an air compressor is subjected to harmonic motions (with the same frequency) in two perpendicular directions. The resultant motion, displayed on an oscilloscope, appears as shown in Fig. 1.87. Find the amplitudes of vibration in the two directions and the phase difference between them.

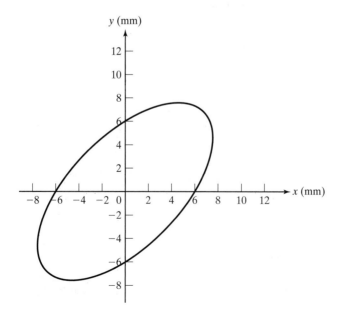

FIGURE 1.87

1.50 A machine is subjected to the motion $x(t) = A \cos(50t + \alpha)$ mm. The initial conditions are given by $x(0) = 3$ mm and $\dot{x}(0) = 1.0$ m/s.

    a. Find the constants A and α.

    b. Express the motion in the form $x(t) = A_1 \cos \omega t + A_2 \sin \omega t$, and identify the constants A_1 and A_2.

1.51 Show that any linear combination of $\sin \omega t$ and $\cos \omega t$ such that $x(t) = A_1 \cos \omega t + A_2 \sin \omega t$ (A_1, A_2 = constants) represents a simple harmonic motion.

1.52 Find the sum of the two harmonic motions $x_1(t) = 5 \cos(3t + 1)$ and $x_2(t) = 10 \cos(3t + 2)$. Use:

    a. Trigonometric relations

    b. Vector addition

    c. Complex number representation

1.53 If one of the components of the harmonic motion $x(t) = 10 \sin(\omega t + 60°)$ is $x_1(t) = 5 \sin(\omega t + 30°)$, find the other component.

1.54 Consider the two harmonic motions $x_1(t) = \frac{1}{2} \cos \frac{\pi}{2} t$ and $x_2(t) = \sin \pi t$. Is the sum $x_1(t) + x_2(t)$ a periodic motion? If so, what is its period?

1.55 Consider two harmonic motions of different frequencies: $x_1(t) = 2 \cos 2t$ and $x_2(t) = \cos 3t$. Is the sum $x_1(t) + x_2(t)$ a harmonic motion? If so, what is its period?

1.56 Consider the two harmonic motions $x_1(t) = \frac{1}{2} \cos \frac{\pi}{2} t$ and $x_2(t) = \cos \pi t$. Is the difference $x(t) = x_1(t) - x_2(t)$ a harmonic motion? If so, what is its period?

1.57 Find the maximum and minimum amplitudes of the combined motion $x(t) = x_1(t) + x_2(t)$ when $x_1(t) = 3 \sin 30t$ and $x_2(t) = 3 \sin 29t$? Also find the frequency of beats corresponding to $x(t)$.

1.58 A machine is subjected to two harmonic motions and the resultant motion, as displayed by an oscilloscope, is shown in Fig. 1.88. Find the amplitudes and frequencies of the two motions.

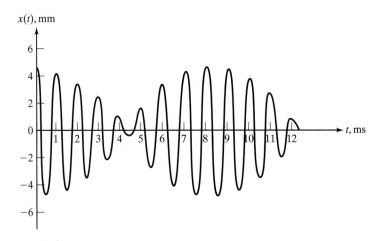

FIGURE 1.88

1.59 A harmonic motion has an amplitude of 0.05 m and a frequency of 10 Hz. Find its period, maximum velocity, and maximum acceleration.

1.60 An accelerometer mounted on a building frame indicates that the frame is vibrating harmonically at 15 cps, with a maximum acceleration of 0.5 g. Determine the amplitude and the maximum velocity of the building frame.

1.61 The maximum amplitude and the maximum acceleration of the foundation of a centrifugal pump were found to be $x_{max} = 0.25$ mm and $\ddot{x}_{max} = 0.4$ g. Find the operating speed of the pump.

1.62 The root mean square (rms) value of a function, $x(t)$, is defined as the square root of the average of the squared value of $x(t)$ over a time period τ:

$$x_{rms} = \sqrt{\frac{1}{\tau} \int_0^\tau [x(t)]^2 \, dt}$$

Using this definition, find the rms value of the function

$$x(t) = X \sin \omega t = X \sin \frac{2\pi t}{\tau}$$

1.63 Using the definition given in Problem 1.58, find the rms value of the function shown in Fig. 1.46(a).

1.64 Prove that the sine Fourier components (b_n) are zero for even functions, that is, when $x(-t) = x(t)$. Also prove that the cosine Fourier components (a_0 and a_n) are zero for odd functions, that is, when $x(-t) = -x(t)$.

1.65 Find the Fourier series expansions of the functions shown in Figs. 1.50(ii) and (iii). Also, find their Fourier series expansions when the time axis is shifted down by a distance A.

1.66 The impact force created by a forging hammer can be modeled as shown in Fig. 1.89. Determine the Fourier series expansion of the impact force.

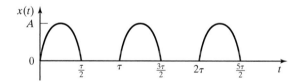

FIGURE 1.89

1.67 Find the Fourier series expansion of the periodic function shown in Fig. 1.90. Also plot the corresponding frequency spectrum.

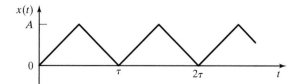

FIGURE 1.90

1.68 Find the Fourier series expansion of the periodic function shown in Fig. 1.91. Also plot the corresponding frequency spectrum.

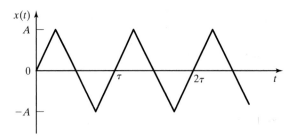

FIGURE 1.91

1.69 Find the Fourier series expansion of the periodic function shown in Fig. 1.92. Also plot the corresponding frequency spectrum.

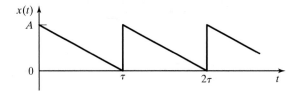

FIGURE 1.92

1.70 The Fourier series of a periodic function, $x(t)$, is an infinite series given by

$$x(t) = \frac{a_0}{2} + \sum_{n=1}^{\infty} (a_n \cos n\omega t + b_n \sin n\omega t) \qquad \text{(E.1)}$$

where

$$a_0 = \frac{\omega}{\pi} \int_0^{\frac{2\pi}{\omega}} x(t)\, dt \qquad \text{(E.2)}$$

$$a_n = \frac{\omega}{\pi} \int_0^{\frac{2\pi}{\omega}} x(t) \cos n\omega t\, dt \qquad \text{(E.3)}$$

$$b_n = \frac{\omega}{\pi} \int_0^{\frac{2\pi}{\omega}} x(t) \sin n\omega t\, dt \qquad \text{(E.4)}$$

ω is the circular frequency and $2\pi/\omega$ is the time period. Instead of including the infinite number of terms in Eq. (E.1), it is often truncated by retaining only k terms as

$$x(t) \approx \tilde{x}(t) = \frac{\tilde{a}_0}{2} + \sum_{n=1}^{k} (\tilde{a}_n \cos n\omega t + \tilde{b}_n \sin n\omega t) \tag{E.5}$$

so that the error, $e(t)$, becomes

$$e(t) = x(t) - \tilde{x}(t) \tag{E.6}$$

Find the coefficients $\tilde{a}_0$, $\tilde{a}_n$, and $\tilde{b}_n$ which minimize the square of the error over a time period:

$$\int_{-\frac{\pi}{\omega}}^{\frac{\pi}{\omega}} e^2(t)\, dt \tag{E.7}$$

Compare the expressions of $\tilde{a}_0$, $\tilde{a}_n$, and $\tilde{b}_n$ with Eqs. (E.2) to (E.4) and state your observation(s).

1.71 Conduct a harmonic analysis, including the first three harmonics, of the function given below:

t_i	0.02	0.04	0.06	0.08	0.10	0.12	0.14	0.16	0.18
x_i	9	13	17	29	43	59	63	57	49

t_i	0.20	0.22	0.24	0.26	0.28	0.30	0.32
x_i	35	35	41	47	41	13	7

1.72 In a centrifugal fan (Fig. 1.93a), the air at any point is subjected to an impulse each time a blade passes the point, as shown in Fig. 1.93(b). The frequency of these impulses is determined by the speed of rotation of the impeller n and the number of blades, N, in the impeller. For $n = 100$ rpm and $N = 4$, determine the first three harmonics of the pressure fluctuation shown in Fig. 1.93(b).

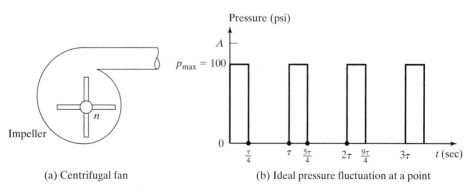

(a) Centrifugal fan

(b) Ideal pressure fluctuation at a point

FIGURE 1.93

1.73 Solve Problem 1.72 by using the values of n and N as 200 rpm and 6 instead of 100 rpm and 4, respectively.

1.74 The torque (M_t) variation with time, of an internal combustion engine, is given in Table 1.3. Make a harmonic analysis of the torque. Find the amplitudes of the first three harmonics.

TABLE 1.3

t(s)	$M_t(N \cdot m)$	t(s)	$M_t(N \cdot m)$	t(s)	$M_t(N \cdot m)$
0.00050	770	0.00450	1890	0.00850	1050
0.00100	810	0.00500	1750	0.00900	990
0.00150	850	0.00550	1630	0.00950	930
0.00200	910	0.00600	1510	0.01000	890
0.00250	1010	0.00650	1390	0.01050	850
0.00300	1170	0.00700	1290	0.01100	810
0.00350	1370	0.00750	1190	0.01150	770
0.00400	1610	0.00800	1110	0.01200	750

1.75 Make a harmonic analysis of the function shown in Fig. 1.94 including the first three harmonics.

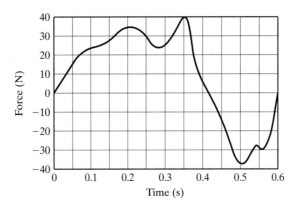

FIGURE 1.94

1.76 Plot the Fourier series expansion of the function $x(t)$ determined in Problem 1.71 using MATLAB.

1.77 Use MATLAB to plot the variation of the force with time using the Fourier series expansion determined in Problem 1.75.

1.78 Use MATLAB to plot the variations of the damping constant c with respect to r, h and a as determined in Problem 1.42.

1.79 Use MATLAB to plot the variation of spring stiffness (k) with deformation (x) given by the relations:

   a. $k = 1000x - 100x^2; 0 \leq x \leq 4$
   b. $k = 500 + 500x^2; 0 \leq x \leq 4$

1.80 A mass is subjected to two harmonic motions given by $x_1(t) = 3 \sin 30t$ and $x_2(t) = 3 \sin 29t$. Plot the resultant motion of the mass using MATLAB and identify the beat frequency and the beat period.

1.81 Solve Problem 1.74 using the C++ program of Section 1.13.

1.82 Solve Problem 1.71 using the C++ program of Section 1.13.

1.83 Find the first six harmonics of the function shown in Fig. 1.94 using the C++ program of Section 1.13.

1.84 Use the C++ program of Section 1.13 to conduct a harmonic analysis of the function shown in Fig. 1.95, including the first 10 harmonics.

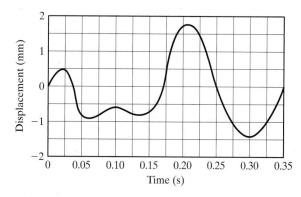

FIGURE 1.95

1.85 Solve Problem 1.74 using the subroutine FORIER of Section 1.14.

1.86 Solve Problem 1.71 using the subroutine FORIER of Section 1.14.

1.87 Find the first six harmonics of the function shown in Fig. 1.94, using the subroutine FORIER of Section 1.14.

1.88 Use the subroutine FORIER of Section 1.14 to conduct a harmonic analysis of the function shown in Fig. 1.95, including the first 10 harmonics.

DESIGN PROJECTS

1.89* A slider crank mechanism is shown in Fig. 1.96. Derive an expression for the motion of the piston P in terms of the crank length r, connecting rod length l, and the constant angular velocity of the crank ω.

    a. Discuss the feasibility of using the mechanism for the generation of harmonic motion.

    b. Find the value of l/r for which the amplitude of every higher harmonic is smaller than that of the first harmonic by a factor of at least 25.

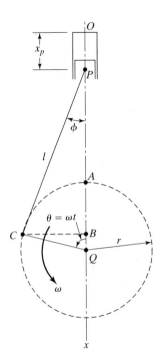

FIGURE 1.96

1.90* The vibration table shown in Fig. 1.97 is used to test certain electronic components for vibration. It consists of two identical mating gears G_1 and G_2 that rotate about the axes O_1 and O_2 attached to the frame F. Two equal masses, m each, are placed symmetrically about the middle vertical axis as shown in Fig. 1.97. During rotation, an unbalanced vertical force of magnitude $P = 2m\omega^2 r \cos\theta$, where $\theta = \omega t$ and ω = angular velocity of gears, will be developed causing the table to vibrate. Design a vibration table that can develop a force in the range $0-100$ N over a frequency range $25-50$ Hz.

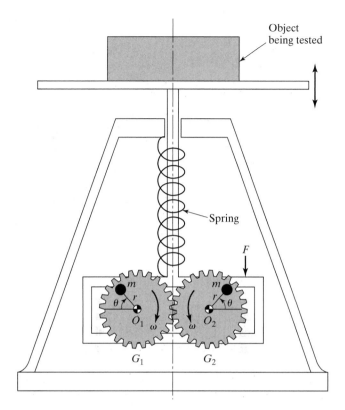

FIGURE 1.97   A vibration table.

1.91* The arrangement shown in Fig. 1.98 is used to regulate the weight of material fed from a hopper to a conveyor [1.44]. The crank imparts a reciprocating motion to the actuating rod through the wedge. The amplitude of motion imparted to the actuating rod can be varied by moving the wedge up or down. Since the conveyor is pivoted about point O, any overload on the conveyor makes the lever OA tilt downward, thereby raising the wedge. This causes a reduction in the amplitude of the actuating rod and hence the feed rate. Design such a weight-regulating system to maintain the weight at 10 ± 0.1 lb per minute.

1.92* Figure 1.99 shows a vibratory compactor. It consists of a plate cam with three profiled lobes and an oscillating roller follower. As the cam rotates, the roller drops after each rise. Correspondingly, the weight attached at the end of the follower also rises and drops. The contact between the roller and the cam is maintained by the spring. Design a vibration compactor that can apply a force of 200 lb at a frequency of 50 Hz.

1.93* Vibratory bowl feeders are widely used in automated processes where a high volume of identical parts are to be oriented and delivered at a steady rate to a workstation for further tooling [1.45, 1.46]. Basically, a vibratory bowl feeder is separated from the base by a set of inclined elastic members (springs), as shown in Fig. 1.100. An electromagnetic coil mounted between the bowl and the base provides the driving force to the bowl. The vibratory motion of the bowl

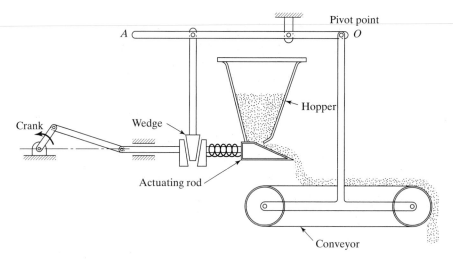

FIGURE 1.98   A vibratory weight-regulating system.

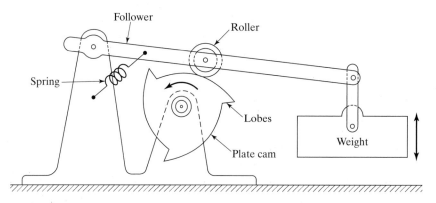

FIGURE 1.99   A vibratory compactor.

causes the components to move along the spiral delivery track located inside the bowl with a hopping motion. Special tooling is fixed at suitable positions along the spiral track in order to reject the parts that are defective or out of tolerance or have incorrect orientation. What factors must be considered in the design of such vibratory bowl feeders?

1.94* The shell and tube exchanger shown in Fig. 1.101(a) can be modeled as shown in Fig. 1.101(b) for a simplified vibration analysis. Find the cross-sectional area of the tubes so that the total stiffness of the heat exchanger exceeds a value of 200×10^6 N/m in the axial direction and 20×10^6 N-m/rad in the tangential direction. Assume that the tubes have the same length and cross section and are spaced uniformly.

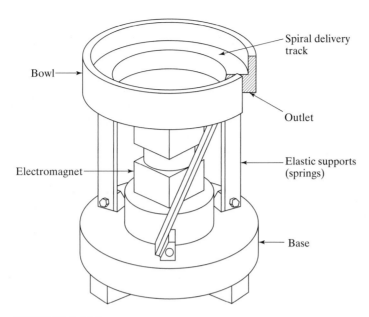

FIGURE 1.100   A vibratory bowl feeder.

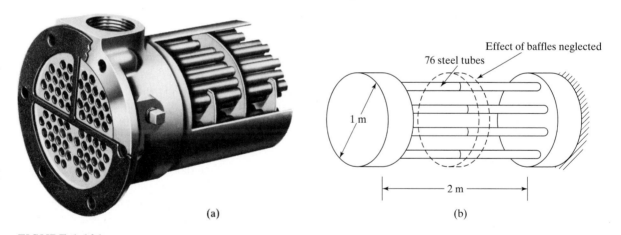

(a)   (b)

FIGURE 1.101   (Part (a) courtesy of Young Radiator Company)

CHAPTER 2

Free Vibration of Single Degree of Freedom Systems

2.1 Introduction

A system is said to undergo free vibration when it oscillates only under an initial disturbance with no external forces acting after the initial disturbance. The oscillations of the pendulum of a grandfather clock, the vertical oscillatory motion felt by a bicyclist after hitting a road bump, and the motion of a child on a swing under an initial push represent a few examples of free vibration.

Figure 2.1(a) shows a spring-mass system that represents the simplest possible vibratory system. It is called a single degree of freedom system since one coordinate (x) is sufficient to specify the position of the mass at any time. There is no external force applied to the mass; hence the motion resulting from an initial disturbance will be free vibration. Since there is no element that causes dissipation of energy during the motion of the mass, the amplitude of motion remains constant with time; it is an *undamped* system. In actual practice, except in a vacuum, the amplitude of free vibration diminishes gradually over time, due to the resistance offered by the surrounding medium (such as air). Such vibrations are said to be *damped*. The study of the free vibration of undamped and damped single degree of freedom systems is fundamental to the understanding of more advanced topics in vibrations.

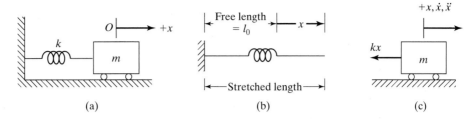

FIGURE 2.1 A spring-mass system in horizontal position.

Several mechanical and structural systems can be idealized as single degree of freedom systems. In many practical systems, the mass is distributed, but for a simple analysis, it can be approximated by a single point mass. Similarly, the elasticity of the system, which may be distributed throughout the system, can also be idealized by a single spring. For the cam-follower system shown in Fig. 1.32, for example, the various masses were replaced by an equivalent mass (m_{eq}) in Example 1.7. The elements of the follower system (pushrod, rocker arm, valve, and valve spring) are all elastic but can be reduced to a single equivalent spring of stiffness k_{eq}. For a simple analysis, the cam-follower system can thus be idealized as a single degree of freedom spring-mass system, as shown in Fig. 2.2.

Similarly, the structure shown in Fig. 2.3 can be considered a cantilever beam that is fixed at the ground. For the study of transverse vibration, the top mass can be considered a point mass and the supporting structure (beam) can be approximated as a spring to obtain the single degree of freedom model shown in Fig. 2.4. The building frame shown in Fig. 2.5(a)

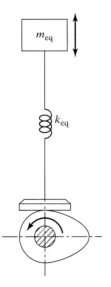

FIGURE 2.2 Equivalent spring-mass system for the cam-follower system of Fig. 1.32.

FIGURE 2.3   The space needle (structure).

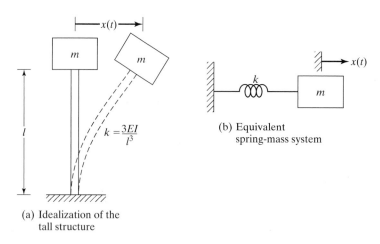

(a) Idealization of the
    tall structure

(b) Equivalent
    spring-mass system

FIGURE 2.4   Modeling of tall structure as spring-mass system.

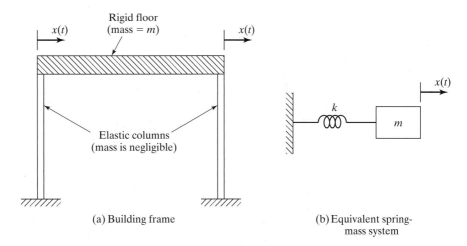

FIGURE 2.5   Idealization of a building frame.

can also be idealized as a spring-mass system, as shown in Fig. 2.5(b). In this case, since the spring constant k is merely the ratio of force to deflection, it can be determined from the geometric and material properties of the columns. The mass of the idealized system is the same as that of the floor if we assume the mass of the columns to be negligible.

2.2   Free Vibration of an Undamped Translational System

**2.2.1
Equation of
Motion Using
Newton's
Second Law of
Motion**

Using Newton's second law of motion, in this section, we will consider the derivation of the equation of motion. The procedure we will use can be summarized as follows:

1. Select a suitable coordinate to describe the position of the mass or rigid body in the system. Use a linear coordinate to describe the linear motion of a point mass or the centroid of a rigid body, and an angular coordinate to describe the angular motion of a rigid body.
2. Determine the static equilibrium configuration of the system and measure the displacement of the mass or rigid body from its static equilibrium position.
3. Draw the free-body diagram of the mass or rigid body when a positive displacement and velocity are given to it. Indicate all the active and reactive forces acting on the mass or rigid body.
4. Apply Newton's second law of motion to the mass or rigid body shown by the free-body diagram. Newton's second law of motion can be stated as follows:

   The rate of change of momentum of a mass is equal to the force acting on it.

Thus, if mass m is displaced a distance $\vec{x}(t)$ when acted upon by a resultant force $\vec{F}(t)$ in the same direction, Newton's second law of motion gives

$$\vec{F}(t) = \frac{d}{dt}\left(m\frac{d\vec{x}(t)}{dt}\right)$$

If mass m is constant, this equation reduces to

$$\vec{F}(t) = m\frac{d^2\vec{x}(t)}{dt^2} = m\ddot{\vec{x}} \qquad (2.1)$$

where

$$\ddot{\vec{x}} = \frac{d^2\vec{x}(t)}{dt^2}$$

is the acceleration of the mass. Equation (2.1) can be stated in words as

Resultant force on the mass = mass × acceleration

For a rigid body undergoing rotational motion, Newton's law gives

$$\vec{M}(t) = J\ddot{\vec{\theta}} \qquad (2.2)$$

where $\vec{M}$ is the resultant moment acting on the body and $\vec{\theta}$ and $\ddot{\vec{\theta}} = d^2\theta(t)/dt^2$ are the resulting angular displacement and angular acceleration, respectively. Equation (2.1) or (2.2) represents the equation of motion of the vibrating system.

The procedure is now applied to the undamped single degree of freedom system shown in Fig. 2.1(a). Here the mass is supported on frictionless rollers and can have translatory motion in the horizontal direction. When the mass is displaced a distance $+x$ from its static equilibrium position, the force in the spring is kx, and the free-body diagram of the mass can be represented as shown in Fig. 2.1(c). The application of Eq. (2.1) to mass m yields the equation of motion

$$F(t) = -kx = m\ddot{x}$$

or

$$m\ddot{x} + kx = 0 \qquad (2.3)$$

2.2.2
Equation of Motion Using Other Methods

As stated in Section 1.6, the equations of motion of a vibrating system can be derived using several methods. The applications of D'Alembert's principle, the principle of virtual displacements, and the principle of conservation of energy are considered in this section.

D'Alembert's Principle.   The equations of motion, Eqs. (2.1) and (2.2), can be rewritten as

$$\vec{F}(t) - m\ddot{\vec{x}} = 0 \qquad (2.4a)$$

$$\vec{M}(t) - J\ddot{\vec{\theta}} = 0 \qquad (2.4b)$$

These equations can be considered equilibrium equations provided that $-m\ddot{\vec{x}}$ and $-J\ddot{\vec{\theta}}$ are treated as a force and a moment. This fictitious force (or moment) is known as the inertia force (or inertia moment) and the artificial state of equilibrium implied by Eq. (2.4a) or (2.4b) is known as dynamic equilibrium. This principle, implied in Eq. (2.4a) or (2.4b), is called D'Alembert's principle. The application of D'Alembert's principle to the system shown in Fig. 2.1(c) yields the equation of motion:

$$-kx - m\ddot{x} = 0 \quad \text{or} \quad m\ddot{x} + kx = 0 \tag{2.3}$$

Principle of Virtual Displacements. The principle of virtual displacements states that "if a system that is in equilibrium under the action of a set of forces is subjected to a virtual displacement, then the total virtual work done by the forces will be zero." Here the virtual displacement is defined as an imaginary infinitesimal displacement given instantaneously. It must be a physically possible displacement that is compatible with the constraints of the system. The virtual work is defined as the work done by all the forces, including the inertia forces for a dynamic problem, due to a virtual displacement.

Consider a spring-mass system in a displaced position as shown in Fig. 2.6(a), where x denotes the displacement of the mass. Figure 2.6(b) shows the free-body diagram of the mass with the reactive and inertia forces indicated. When the mass is given a virtual displacement δx, as shown in Fig. 2.6(b), the virtual work done by each force can be computed as follows:

$$\text{Virtual work done by the spring force} = \delta W_s = -(kx)\delta x$$
$$\text{Virtual work done by the inertia force} = \delta W_i = -(m\ddot{x})\delta x$$

When the total virtual work done by all the forces is set equal to zero, we obtain

$$-m\ddot{x}\delta x - kx\delta x = 0 \tag{2.5}$$

Since the virtual displacement can have an arbitrary value, $\delta x \neq 0$, Eq. (2.5) gives the equation of motion of the spring-mass system as

$$m\ddot{x} + kx = 0 \tag{2.3}$$

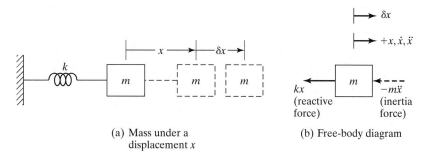

(a) Mass under a
displacement x

(b) Free-body diagram

FIGURE 2.6 Mass under virtual displacement.

Principle of Conservation of Energy.   A system is said to be conservative if no energy is lost due to friction or energy-dissipating nonelastic members. If no work is done on a conservative system by external forces (other than gravity or other potential forces), then the total energy of the system remains constant. Since the energy of a vibrating system is partly potential and partly kinetic, the sum of these two energies remains constant. The kinetic energy T is stored in the mass by virtue of its velocity, and the potential energy U is stored in the spring by virtue of its elastic deformation. Thus the principle of conservation of energy can be expressed as:

$$T + U = \text{constant}$$

or

$$\frac{d}{dt}(T + U) = 0 \qquad (2.6)$$

The kinetic and potential energies are given by

$$T = \tfrac{1}{2}m\dot{x}^2 \qquad (2.7)$$

and

$$U = \tfrac{1}{2}kx^2 \qquad (2.8)$$

Substitution of Eqs. (2.7) and (2.8) into Eq. (2.6) yields the desired equation

$$m\ddot{x} + kx = 0 \qquad (2.3)$$

2.2.3
Equation of Motion of a Spring-Mass System in Vertical Position

Consider the configuration of the spring-mass system shown in Fig. 2.7(a). The mass hangs at the lower end of a spring, which in turn is attached to a rigid support at its upper end. At rest, the mass will hang in a position called the *static equilibrium position*, in which the upward spring force exactly balances the downward gravitational force on the mass. In this position the length of the spring is $l_0 + \delta_{st}$ where δ_{st} is the static deflection—the elongation due to the weight W of the mass m. From Fig. 2.7(a), we find that, for static equilibrium,

$$W = mg = k\delta_{st} \qquad (2.9)$$

where g is the acceleration due to gravity. Let the mass be deflected a distance $+x$ from its static equilibrium position; then the spring force is $-k(x + \delta_{st})$, as shown in Fig. 2.7(c). The application of Newton's second law of motion to mass m gives

$$m\ddot{x} = -k(x + \delta_{st}) + W$$

and since $k\delta_{st} = W$, we obtain

$$m\ddot{x} + kx = 0 \qquad (2.10)$$

Notice that Eqs. (2.3) and (2.10) are identical. This indicates that when a mass moves in a vertical direction, we can ignore its weight, provided we measure x from its static equilibrium position.

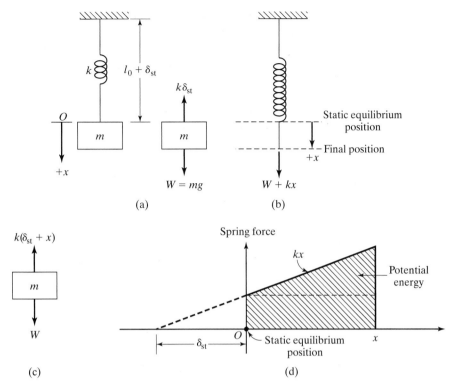

FIGURE 2.7   A spring-mass system in vertical position.

Note: Equation (2.10), the equation of motion of the system shown in Fig. 2.7, can also be derived using D'Alembert's principle, the principle of virtual displacements, or the principle of conservation of energy. For example, if the principle of conservation of energy is to be used, we note that the expression for the kinetic energy, T, remains the same as Eq. (2.7). However, the expression for the potential energy, U, is to be derived by considering the weight of the mass. For this we note that the spring force at static equilibrium position ($x = 0$) is mg. When the spring deflects by an amount x, its potential energy is given by (see Fig. 2.7d):

$$mgx + \frac{1}{2}kx^2$$

Furthermore, the potential energy of the system due to the change in elevation of the mass (note that $+x$ is downward) is $-mgx$. Thus the net potential energy of the system about the static equilibrium position is given by

$$U = \text{potential energy of the spring}$$
$$+ \text{ change in potential energy due to change in elevation of the mass } m$$

$$= mgx + \frac{1}{2}kx^2 - mgx = \frac{1}{2}kx^2$$

Since the expressions of T and U remain unchanged, the application of the principle of conservation of energy gives the same equation of motion, Eq. (2.3).

2.2.4
Solution

The solution of Eq. (2.3) can be found by assuming

$$x(t) = Ce^{st} \tag{2.11}$$

where C and s are constants to be determined. Substitution of Eq. (2.11) into Eq. (2.3) gives

$$C(ms^2 + k) = 0$$

Since C cannot be zero, we have

$$ms^2 + k = 0 \tag{2.12}$$

and hence

$$s = \pm\left(-\frac{k}{m}\right)^{1/2} = \pm i\omega_n \tag{2.13}$$

where $i = (-1)^{1/2}$ and

$$\omega_n = \left(\frac{k}{m}\right)^{1/2} \tag{2.14}$$

Equation (2.12) is called the *auxiliary* or the *characteristic* equation corresponding to the differential Eq. (2.3). The two values of s given by Eq. (2.13) are the roots of the characteristic equation, also known as the *eigenvalues* or the *characteristic values* of the problem. Since both values of s satisfy Eq. (2.12), the general solution of Eq. (2.3) can be expressed as

$$x(t) = C_1 e^{i\omega_n t} + C_2 e^{-i\omega_n t} \tag{2.15}$$

where C_1 and C_2 are constants. By using the identities

$$e^{\pm iat} = \cos \alpha t \pm i \sin \alpha t$$

Eq. (2.15) can be rewritten as

$$x(t) = A_1 \cos \omega_n t + A_2 \sin \omega_n t \tag{2.16}$$

where A_1 and A_2 are new constants. The constants C_1 and C_2 or A_1 and A_2 can be determined from the initial conditions of the system. Two conditions are to be specified to evaluate these constants uniquely. Note that the number of conditions to be specified is the same as the order of the governing differential equation. In the present case, if the values of displacement $x(t)$ and velocity $\dot{x}(t) = (dx/dt)(t)$ are specified as x_0 and $\dot{x}_0$ at $t = 0$, we have, from Eq. (2.16),

$$x(t = 0) = A_1 = x_0$$
$$\dot{x}(t = 0) = \omega_n A_2 = \dot{x}_0 \tag{2.17}$$

Hence $A_1 = x_0$ and $A_2 = \dot{x}_0/\omega_n$. Thus the solution of Eq. (2.3) subject to the initial conditions of Eq. (2.17) is given by

$$x(t) = x_0 \cos \omega_n t + \frac{\dot{x}_0}{\omega_n} \sin \omega_n t \qquad (2.18)$$

**2.2.5
Harmonic
Motion**

Equations (2.15), (2.16), and (2.18) are harmonic functions of time. The motion is symmetric about the equilibrium position of the mass m. The velocity is a maximum and the acceleration is zero each time the mass passes through this position. At the extreme displacements, the velocity is zero and the acceleration is a maximum. Since this represents simple harmonic motion (see Section 1.10), the spring-mass system itself is called a *harmonic oscillator*. The quantity ω_n given by Eq. (2.14), represents the system's natural frequency of vibration.

Equation (2.16) can be expressed in a different form by introducing the notation

$$A_1 = A \cos \phi$$

$$A_2 = A \sin \phi \qquad (2.19)$$

where A and ϕ are the new constants, which can be expressed in terms of A_1 and A_2 as

$$A = (A_1^2 + A_2^2)^{1/2} = \left[x_0^2 + \left(\frac{\dot{x}_0}{\omega_n} \right)^2 \right]^{1/2} = \text{amplitude}$$

$$\phi = \tan^{-1} \left(\frac{A_2}{A_1} \right) = \tan^{-1} \left(\frac{\dot{x}_0}{x_0 \omega_n} \right) = \text{phase angle} \qquad (2.20)$$

Introducing Eq. (2.19) into Eq. (2.16), the solution can be written as

$$x(t) = A \cos (\omega_n t - \phi) \qquad (2.21)$$

By using the relations

$$A_1 = A_0 \sin \phi_0$$
$$A_2 = A_0 \cos \phi_0 \qquad (2.22)$$

Eq. (2.16) can also be expressed as

$$x(t) = A_0 \sin (\omega_n t + \phi_0) \qquad (2.23)$$

where

$$A_0 = A = \left[x_0^2 + \left(\frac{\dot{x}_0}{\omega_n} \right)^2 \right]^{1/2} \qquad (2.24)$$

and

$$\phi_0 = \tan^{-1}\left(\frac{x_0 \omega_n}{\dot{x}_0}\right) \tag{2.25}$$

The nature of harmonic oscillation can be represented graphically as in Fig. 2.8(a). If $\vec{A}$ denotes a vector of magnitude A, which makes an angle $\omega_n t - \phi$ with respect to the vertical (x) axis, then the solution, Eq. (2.21), can be seen to be the projection of the vector $\vec{A}$ on the x-axis. The constants A_1 and A_2 of Eq. (2.16), given by Eq. (2.19), are merely the rectangular components of $\vec{A}$ along two orthogonal axes making angles ϕ and $-(\frac{\pi}{2} - \phi)$ with respect to the vector $\vec{A}$. Since the angle $\omega_n t - \phi$ is a linear function of

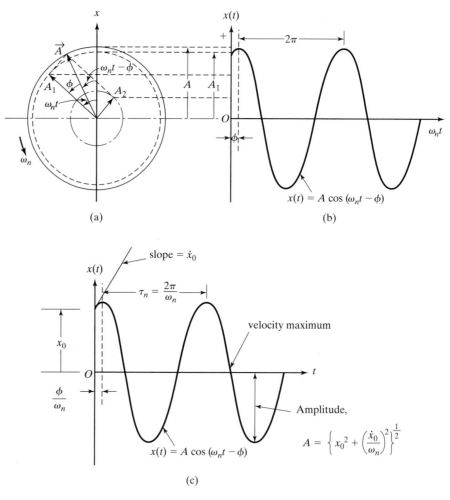

FIGURE 2.8   Graphical representation of the motion of a harmonic oscillator.

time, it increases linearly with time; the entire diagram thus rotates counterclockwise at an angular velocity ω_n. As the diagram (Fig. 2.8a) rotates, the projection of $\vec{A}$ onto the x-axis varies harmonically so that the motion repeats itself every time the vector $\vec{A}$ sweeps an angle of 2π. The projection of $\vec{A}$, namely $x(t)$, is shown plotted in Fig. 2.8(b) as a function of $\omega_n t$, and as a function of t in Fig. 2.8(c). The phase angle ϕ can also be interpreted as the angle between the origin and the first peak.

Note the following aspects of the spring-mass system:

1. If the spring-mass system is in a vertical position, as shown in Fig. 2.7(a), the circular natural frequency can be expressed as

$$\omega_n = \left(\frac{k}{m}\right)^{1/2} \tag{2.26}$$

The spring constant k can be expressed in terms of the mass m from Eq. (2.9) as

$$k = \frac{W}{\delta_{st}} = \frac{mg}{\delta_{st}} \tag{2.27}$$

Substitution of Eq. (2.27) into Eq. (2.14) yields

$$\omega_n = \left(\frac{g}{\delta_{st}}\right)^{1/2} \tag{2.28}$$

Hence the natural frequency in cycles per second and the natural period are given by

$$f_n = \frac{1}{2\pi}\left(\frac{g}{\delta_{st}}\right)^{1/2} \tag{2.29}$$

$$\tau_n = \frac{1}{f_n} = 2\pi\left(\frac{\delta_{st}}{g}\right)^{1/2} \tag{2.30}$$

Thus, when the mass vibrates in a vertical direction, we can compute the natural frequency and the period of vibration by simply measuring the static deflection δ_{st}. It is not necessary that we know the spring stiffness k and the mass m.

2. From Eq. (2.21), the velocity $\dot{x}(t)$ and the acceleration $\ddot{x}(t)$ of the mass m at time t can be obtained as

$$\dot{x}(t) = \frac{dx}{dt}(t) = -\omega_n A \sin(\omega_n t - \phi) = \omega_n A \cos\left(\omega_n t - \phi + \frac{\pi}{2}\right)$$

$$\ddot{x}(t) = \frac{d^2x}{dt^2}(t) = -\omega_n^2 A \cos(\omega_n t - \phi) = \omega_n^2 A \cos(\omega_n t - \phi + \pi) \tag{2.31}$$

Equation (2.31) shows that the velocity leads the displacement by $\pi/2$ and the acceleration leads the displacement by π.

3. If the initial displacement (x_0) is zero, Eq. (2.21) becomes

$$x(t) = \frac{\dot{x}_0}{\omega_n} \cos\left(\omega_n t - \frac{\pi}{2}\right) = \frac{\dot{x}_0}{\omega_n} \sin \omega_n t \qquad (2.32)$$

If the initial velocity ($\dot{x}_0$) is zero, however, the solution becomes

$$x(t) = x_0 \cos \omega_n t \qquad (2.33)$$

4. The response of a single degree of freedom system can be represented in the displacement (x)-velocity ($\dot{x}$) plane, known as the state space or phase plane. For this we consider the displacement given by Eq. (2.21) and the corresponding velocity:

$$x(t) = A \cos(\omega_n t - \phi)$$

or

$$\cos(\omega_n t - \phi) = \frac{x}{A}$$

$$\dot{x}(t) = -A\omega_n \sin(\omega_n t - \phi) \qquad (2.34)$$

or

$$\sin(\omega_n t - \phi) = -\frac{\dot{x}}{A\omega_n} = -\frac{y}{A} \qquad (2.35)$$

where $y = \dot{x}/\omega_n$. By squaring and adding Eqs. (2.34) and (2.35), we obtain

$$\cos^2(\omega_n t - \phi) + \sin^2(\omega_n t - \phi) = 1$$

or

$$\frac{x^2}{A^2} + \frac{y^2}{A^2} = 1 \qquad (2.36)$$

The graph of Eq. (2.36) in the (x, y)-plane is a circle, as shown in Fig. 2.9(a), and it constitutes the phase plane or state space representation of the undamped system. The radius of the circle, A, is determined by the initial conditions of motion. Note that the graph of Eq. (2.36) in the (x, $\dot{x}$) plane will be an ellipse, as shown in Fig. 2.9(b).

Harmonic Response of a Water Tank

EXAMPLE 2.1

The column of the water tank shown in Fig. 2.10(a) is 300 ft. high and is made of reinforced concrete with a tubular cross section of inner diameter 8 ft. and outer diameter 10 ft. The tank weighs 6×10^5 lb with water. By neglecting the mass of the column and assuming the Young's modulus of reinforced concrete as 4×10^6 psi, determine the following:

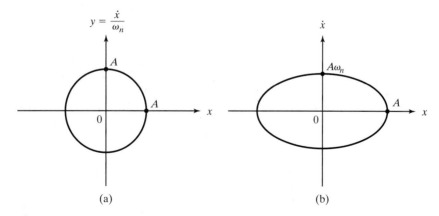

(a)   (b)

FIGURE 2.9   Phase plane representation of an undamped system.

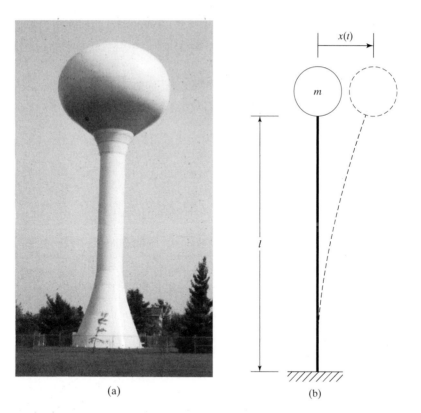

(a)   (b)

FIGURE 2.10   Elevated tank. (Photo courtesy of West Lafayette Water Company.)

a.   the natural frequency and the natural time period of transverse vibration of the water tank
b.   the vibration response of the water tank due to an initial transverse displacement of 10 in.
c.   the maximum values of the velocity and acceleration experienced by the water tank.

Solution: Assuming that the water tank is a point mass, the column has a uniform cross section, and the mass of the column is negligible, the system can be modeled as a cantilever beam with a concentrated load (weight) at the free end as shown in Fig. 2.10(b).

a.   The transverse deflection of the beam, δ, due to a load P is given by $\frac{Pl^3}{3EI}$, where l is the length, E is the Young's modulus, and I is the area moment of inertia of the beam's cross section. The stiffness of the beam (column of the tank) is given by

$$k = \frac{P}{\delta} = \frac{3EI}{l^3}$$

In the present case, $l = 3600$ in., $E = 4 \times 10^6$ psi,

$$I = \frac{\pi}{64}(d_0^4 - d_i^4) = \frac{\pi}{64}(120^4 - 96^4) = 600.9554 \times 10^4 \text{ in.}^4$$

and hence

$$k = \frac{3(4 \times 10^6)(600.9554 \times 10^4)}{3600^3} = 1545.6672 \text{ lb/in.}$$

The natural frequency of the water tank in transverse direction is given by

$$\omega_n = \sqrt{\frac{k}{m}} = \sqrt{\frac{1545.6672 \times 386.4}{6 \times 10^5}} = 0.9977 \text{ rad/sec}$$

The natural time period of transverse vibration of the tank is given by

$$\tau_n = \frac{2\pi}{\omega_n} = \frac{2\pi}{0.9977} = 6.2977 \text{ sec}$$

b.   Using the initial displacement of $x_0 = 10$ in. and the initial velocity of the water tank ($\dot{x}_0$) as zero, the harmonic response of the water tank can be expressed, using Eq. (2.23), as

$$x(t) = A_0 \sin(\omega_n t + \phi_0)$$

where the amplitude of transverse displacement (A_0) is given by

$$A_0 = \left[x_0^2 + \left(\frac{\dot{x}_0}{\omega_n}\right)^2 \right]^{1/2} = x_0 = 10 \text{ in.}$$

and the phase angle (ϕ_0) by

$$\phi_0 = \tan^{-1}\left(\frac{x_0\omega_n}{0}\right) = \frac{\pi}{2}$$

Thus

$$x(t) = 10 \sin\left(0.9977t + \frac{\pi}{2}\right) = 10 \cos 0.9977t \text{ in} \qquad \text{(E.1)}$$

c. The velocity of the water tank can be found by differentiating Eq. (E.1) as

$$\dot{x}(t) = 10(0.9977)\cos\left(0.9977t + \frac{\pi}{2}\right) \qquad \text{(E.2)}$$

and hence

$$\dot{x}_{max} = A_0\omega_n = 10(0.9977) = 9.977 \text{ in./sec}$$

The acceleration of the water tank can be determined by differentiating Eq. (E.2) as

$$\ddot{x}(t) = -10(0.9977)^2 \sin\left(0.9977t + \frac{\pi}{2}\right) \qquad \text{(E.3)}$$

and hence the maximum value of acceleration is given by

$$\ddot{x}_{max} = A_0(\omega_n)^2 = 10(0.9977)^2 = 9.9540 \text{ in./sec}^2$$

∎

Free Vibration Response Due to Impact

EXAMPLE 2.2

A cantilever beam carries a mass M at the free end as shown in Fig. 2.11(a). A mass m falls from a height h on to the mass M and adheres to it without rebounding. Determine the resulting transverse vibration of the beam.

Solution: When the mass m falls through a height h, it will strike the mass M with a velocity of $v_m = \sqrt{2gh}$, where g is the acceleration due to gravity. Since the mass m adheres to M without rebounding, the velocity of the combined mass $(M + m)$ immediately after the impact ($\dot{x}_0$) can be found using the principle of conservation of momentum:

$$mv_m = (M + m)\dot{x}_0$$

or

$$\dot{x}_0 = \left(\frac{m}{M + m}\right)v_m = \left(\frac{m}{M + m}\right)\sqrt{2gh} \qquad \text{(E.1)}$$

The static equilibrium position of the beam with the new mass $(M + m)$ is located at a distance of $\frac{mg}{k}$ below the static equilibrium position of the original mass (M) as shown in Fig. 2.11(c). Here k denotes the stiffness of the cantilever beam, given by

$$k = \frac{3EI}{l^3}$$

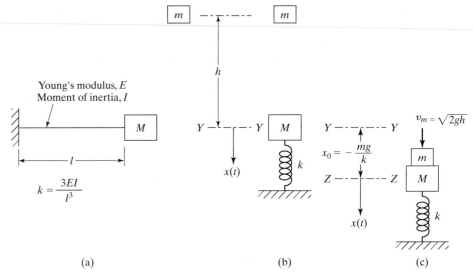

YY = static equilibrium position of M
ZZ = static equilibrium position of $M + m$

FIGURE 2.11   Response due to impact.

Since free vibration of the beam with the new mass $(M + m)$ occurs about its own static equilibrium position, the initial conditions of the problem can be stated as

$$x_0 = -\frac{mg}{k}, \quad \dot{x}_0 = \left(\frac{m}{M + m}\right)\sqrt{2gh} \qquad \text{(E.2)}$$

Thus the resulting free transverse vibration of the beam can be expressed as (see Eq. (2.21)):

$$x(t) = A \cos(\omega_n t - \phi)$$

where

$$A = \left[x_0^2 + \left(\frac{\dot{x}_0}{\omega_n}\right)^2\right]^{1/2}$$

$$\phi = \tan^{-1}\left(\frac{\dot{x}_0}{x_0 \omega_n}\right)$$

$$\omega_n = \sqrt{\frac{k}{M + m}} = \sqrt{\frac{3EI}{l^3(M + m)}}$$

with x_0 and $\dot{x}_0$ given by Eq. (E.2).

∎

EXAMPLE 2.3

Young's Modulus from Natural Frequency Measurement

A simply supported beam of square cross section 5 mm $\times$ 5 mm and length 1 m, carrying a mass of 2.3 kg at the middle, is found to have a natural frequency of transverse vibration of 30 rad/s. Determine the Young's modulus of elasticity of the beam.

Solution: By neglecting the self weight of the beam, the natural frequency of transverse vibration of the beam can be expressed as

$$\omega_n = \sqrt{\frac{k}{m}} \tag{E.1}$$

where

$$k = \frac{192EI}{l^3} \tag{E.2}$$

where E is the Young's modulus, l is the length, and I is the area moment of inertia of the beam:

$$I = \frac{1}{12}(5 \times 10^{-3})(5 \times 10^{-3})^3 = 0.5208 \times 10^{-10}\ \text{m}^4$$

Since $m = 2.3$ kg, $l = 1.0$ m, and $\omega_n = 30.0$ rad/s, Eqs. (E.1) and (E.2) yield

$$k = \frac{192EI}{l^3} = m\omega_n^2$$

or

$$E = \frac{m\omega_n^2 l^3}{192I} = \frac{2.3(30.0)^2(1.0)^3}{192(0.5208 \times 10^{-10})} = 207.0132 \times 10^9\ \text{N/m}^2$$

This indicates that the material of the beam is probably carbon steel.

∎

EXAMPLE 2.4

Natural Frequency of Cockpit of a Firetruck

The cockpit of a firetruck is located at the end of a telescoping boom, as shown in Fig. 2.12(a). The cockpit, along with the fireman, weighs 2000 N. Find the cockpit's natural frequency of vibration in the vertical direction.

Data: Young's modulus of the material: $E = 2.1 \times 10^{11}$ N/m^2; Lengths: $l_1 = l_2 = l_3 = 3$ m; cross-sectional areas: $A_1 = 20$ cm^2, $A_2 = 10$ cm^2, $A_3 = 5$ cm^2.
Solution: To determine the system's natural frequency of vibration, we find the equivalent stiffness of the boom in the vertical direction and use a single degree of freedom idealization. For this we assume that the mass of the telescoping boom is negligible and the telescoping boom can deform only in the axial direction (with no bending). Since the force induced at any cross section O_1O_2 is

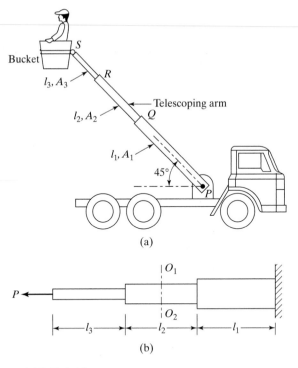

FIGURE 2.12   Telescoping boom of a fire truck.

equal to the axial load applied at the end of the boom, as shown in Fig. 2.12(b), the axial stiffness of the boom (k_b) is given by

$$\frac{1}{k_b} = \frac{1}{k_{b_1}} + \frac{1}{k_{b_2}} + \frac{1}{k_{b_3}} \tag{E.1}$$

where k_{b_i} denotes the axial stiffness of the ith segment of the boom:

$$k_{b_i} = \frac{A_i E_i}{l_i}; i = 1, 2, 3 \tag{E.2}$$

From the known data ($l_1 = l_2 = l_3 = 3$m, $A_1 = 20$ cm², $A_2 = 10$ cm²; $A_3 = 5$ cm², $E_1 = E_2 = E_3 = 2.1 \times 10^{11}$ N/m²),

$$k_{b_1} = \frac{(20 \times 10^{-4})(2.1 \times 10^{11})}{3} = 14 \times 10^7 \text{ N/m}$$

$$k_{b_2} = \frac{(10 \times 10^{-4})(2.1 \times 10^{11})}{3} = 7 \times 10^7 \text{ N/m}$$

$$k_{b_3} = \frac{(5 \times 10^{-4})(2.1 \times 10^{11})}{3} = 3.5 \times 10^7 \text{ N/m}$$

Thus Eq. (E.1) gives

$$\frac{1}{k_b} = \frac{1}{14 \times 10^7} + \frac{1}{7 \times 10^7} + \frac{1}{3.5 \times 10^7} = \frac{1}{2 \times 10^7}$$

or

$$k_b = 2 \times 10^7 \text{ N/m}$$

The stiffness of the telescoping boom in the vertical direction, k, can be determined as

$$k = k_b \cos^2 45° = 10^7 \text{ N/m}$$

The natural frequency of vibration of the cockpit in the vertical direction is given by

$$\omega_n = \sqrt{\frac{k}{m}} = \sqrt{\frac{(10^7)(9.81)}{2000}} = 221.4723 \text{ rad/s}$$

■

Natural Frequency of Pulley System

EXAMPLE 2.5

Determine the natural frequency of the system shown in Fig. 2.13(a). Assume the pulleys to be frictionless and of negligible mass.

Solution: To determine the natural frequency, we find the equivalent stiffness of the system and solve it as a single degree of freedom problem. Since the pulleys are frictionless and massless, the tension in the rope is constant and is equal to the weight W of the mass m. From the static equilibrium of the pulleys and the mass (see Fig. 2.13b), it can be seen that the upward force acting on pulley 1 is $2W$ and the downward force acting on pulley 2 is $2W$. The center of pulley 1 (point A) moves up by a distance $2W/k_1$, and the center of pulley 2 (point B) moves down by $2W/k_2$. Thus the total movement of the mass m (point O) is

$$2\left(\frac{2W}{k_1} + \frac{2W}{k_2}\right)$$

as the rope on either side of the pulley is free to move the mass downward. If k_{eq} denotes the equivalent spring constant of the system,

$$\frac{\text{Weight of the mass}}{\text{Equivalent spring constant}} = \text{Net displacement of the mass}$$

$$\frac{W}{k_{eq}} = 4W\left(\frac{1}{k_1} + \frac{1}{k_2}\right) = \frac{4W(k_1 + k_2)}{k_1 k_2}$$

$$k_{eq} = \frac{k_1 k_2}{4(k_1 + k_2)} \tag{E.1}$$

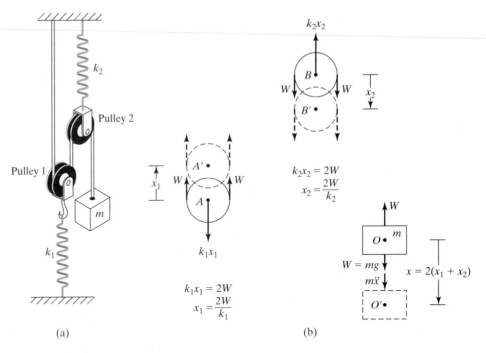

FIGURE 2.13   Pulley system.

By displacing mass m from the static equilibrium position by x, the equation of motion of the mass can be written as

$$m\ddot{x} + k_{eq}x = 0 \tag{E.2}$$

and hence the natural frequency is given by

$$\omega_n = \left(\frac{k_{eq}}{m}\right)^{1/2} = \left[\frac{k_1 k_2}{4m(k_1 + k_2)}\right]^{1/2} \text{rad/sec} \tag{E.3}$$

or

$$f_n = \frac{\omega_n}{2\pi} = \frac{1}{4\pi}\left[\frac{k_1 k_2}{m(k_1 + k_2)}\right]^{1/2} \text{cycles/sec} \tag{E.4}$$

∎

2.3   Free Vibration of an Undamped Torsional System

If a rigid body oscillates about a specific reference axis, the resulting motion is called *torsional vibration*. In this case, the displacement of the body is measured in terms of an angular coordinate. In a torsional vibration problem, the restoring moment may be due to the torsion of an elastic member or to the unbalanced moment of a force or couple.

Figure 2.14 shows a disc, which has a polar mass moment of inertia J_0, mounted at one end of a solid circular shaft, the other end of which is fixed. Let the angular rotation of the disc about the axis of the shaft be θ; θ also represents the shaft's angle of twist. From the theory of torsion of circular shafts [2.1], we have the relation

$$M_t = \frac{GI_o}{l} \tag{2.37}$$

where M_t is the torque that produces the twist θ, G is the shear modulus, l is the length of the shaft, I_o is the polar moment of inertia of the cross section of the shaft, given by

$$I_o = \frac{\pi d^4}{32} \tag{2.38}$$

and d is the diameter of the shaft. If the disc is displaced by θ from its equilibrium position, the shaft provides a restoring torque of magnitude M_t. Thus the shaft acts as a torsional spring with a torsional spring constant

$$k_t = \frac{M_t}{\theta} = \frac{GI_o}{l} = \frac{\pi G d^4}{32l} \tag{2.39}$$

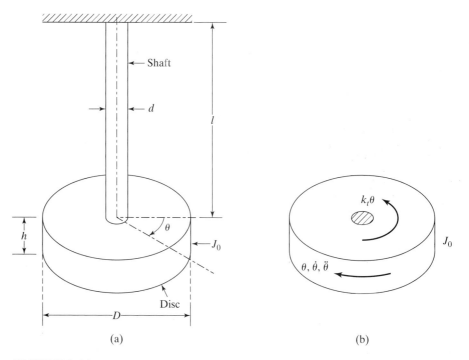

(a)

(b)

FIGURE 2.14   Torsional vibration of a disc.

2.3.1
Equation of
Motion

The equation of the angular motion of the disc about its axis can be derived by using Newton's second law or any of the methods discussed in Section 2.2.2. By considering the free-body diagram of the disc (Fig. 2.14b), we can derive the equation of motion by applying Newton's second law of motion:

$$J_0 \ddot{\theta} + k_t \theta = 0 \tag{2.40}$$

which can be seen to be identical to Eq. (2.3) if the polar mass moment of inertia J_0, the angular displacement θ, and the torsional spring constant k_t are replaced by the mass m, the displacement x, and the linear spring constant k, respectively. Thus the natural circular frequency of the torsional system is

$$\omega_n = \left(\frac{k_t}{J_0} \right)^{1/2} \tag{2.41}$$

and the period and frequency of vibration in cycles per second are

$$\tau_n = 2\pi \left(\frac{J_0}{k_t} \right)^{1/2} \tag{2.42}$$

$$f_n = \frac{1}{2\pi} \left(\frac{k_t}{J_0} \right)^{1/2} \tag{2.43}$$

Note the following aspects of this system:

1. If the cross section of the shaft supporting the disc is not circular, an appropriate torsional spring constant is to be used [2.4, 2.5].
2. The polar mass moment of inertia of a disc is given by

$$J_0 = \frac{\rho h \pi D^4}{32} = \frac{W D^2}{8g}$$

where ρ is the mass density, h is the thickness, D is the diameter, and W is the weight of the disc.
3. The torsional spring-inertia system shown in Fig. 2.14 is referred to as a *torsional pendulum*. One of the most important applications of a torsional pendulum is in a mechanical clock, where a ratchet and pawl convert the regular oscillation of a small torsional pendulum into the movements of the hands.

2.3.2
Solution

The general solution of Eq. (2.40) can be obtained, as in the case of Eq. (2.3):

$$\theta(t) = A_1 \cos \omega_n t + A_2 \sin \omega_n t \tag{2.44}$$

where ω_n is given by Eq. (2.41) and A_1 and A_2 can be determined from the initial conditions. If

$$\theta(t = 0) = \theta_0 \quad \text{and} \quad \dot{\theta}(t = 0) = \frac{d\theta}{dt}(t = 0) = \dot{\theta}_0 \qquad (2.45)$$

the constants A_1 and A_2 can be found:

$$A_1 = \theta_0$$

$$A_2 = \dot{\theta}_0/\omega_n \qquad (2.46)$$

Equation (2.44) can also be seen to represent a simple harmonic motion.

Natural Frequency of Compound Pendulum

EXAMPLE 2.6

Any rigid body pivoted at a point other than its center of mass will oscillate about the pivot point under its own gravitational force. Such a system is known as a compound pendulum (Fig. 2.15). Find the natural frequency of such a system.

Solution: Let O be the point of suspension and G be the center of mass of the compound pendulum, as shown in Fig. 2.15. Let the rigid body oscillate in the xy plane so that the coordinate θ can be used to describe its motion. Let d denote the distance between O and G, and J_0 the mass moment of inertia of the body about the z-axis (perpendicular to both x and y). For a displacement θ, the restoring torque (due to the weight of the body W) is $(Wd \sin \theta)$ and the equation of motion is

$$J_0 \ddot{\theta} + Wd \sin \theta = 0 \qquad (E.1)$$

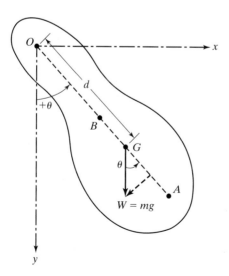

FIGURE 2.15   Compound pendulum.

Note that Eq. (E.1) is a second-order nonlinear ordinary differential equation. Although it is possible to find an exact solution of this equation (see Section 13.3), exact solutions cannot be found for most nonlinear differential equations. An approximate solution of Eq. (E.1) can be found by one of two methods. A numerical procedure can be used to integrate Eq. (E.1). Alternatively, Eq. (E.1) can be approximated by a linear equation whose exact solution can be determined readily. To use the latter approach, we assume small angular displacements so that θ is small and $\sin\theta \approx \theta$. Hence Eq. (E.1) can be approximated by the linear equation:

$$J_0\ddot{\theta} + Wd\theta = 0 \tag{E.2}$$

This gives the natural frequency of the compound pendulum:

$$\omega_n = \left(\frac{Wd}{J_0}\right)^{1/2} = \left(\frac{mgd}{J_0}\right)^{1/2} \tag{E.3}$$

Comparing Eq. (E.3) with the natural frequency of a simple pendulum, $\omega_n = (g/l)^{1/2}$ (see Problem 2.61), we can find the length of the equivalent simple pendulum:

$$l = \frac{J_0}{md} \tag{E.4}$$

If J_0 is replaced by mk_0^2, where k_0 is the radius of gyration of the body about O, Eqs. (E.3) and (E.4) become

$$\omega_n = \left(\frac{gd}{k_0^2}\right)^{1/2} \tag{E.5}$$

$$l = \left(\frac{k_0^2}{d}\right) \tag{E.6}$$

If k_G denotes the radius of gyration of the body about G, we have

$$k_0^2 = k_G^2 + d^2 \tag{E.7}$$

and Eq. (E.6) becomes

$$l = \left(\frac{k_G^2}{d} + d\right) \tag{E.8}$$

If the line OG is extended to point A such that

$$GA = \frac{k_G^2}{d} \tag{E.9}$$

Eq. (E.8) becomes

$$l = GA + d = OA \tag{E.10}$$

Hence, from Eq. (E.5), ω_n is given by

$$\omega_n = \left\{\frac{g}{(k_0^2/d)}\right\}^{1/2} = \left(\frac{g}{l}\right)^{1/2} = \left(\frac{g}{OA}\right)^{1/2} \tag{E.11}$$

This equation shows that, no matter whether the body is pivoted from O or A, its natural frequency is the same. The point A is called the *center of percussion*.

∎

Center of Percussion.   The concepts of compound pendulum and center of percussion can be used in many practical applications:

1. A hammer can be shaped to have the center of percussion at the hammer head while the center of rotation is at the handle. In this case, the impact force at the hammer head will not cause any normal reaction at the handle (Fig. 2.16a).
2. In a baseball bat, if on one hand the ball is made to strike at the center of percussion while the center of rotation is at the hands, no reaction perpendicular to the bat will be experienced by the batter (Fig. 2.16b). On the other hand, if the ball strikes the bat near the free end or near the hands, the batter will experience pain in the hands as a result of the reaction perpendicular to the bat.
3. In Izod (impact) testing of materials, the specimen is suitably notched and held in a vise fixed to the base of the machine (see Fig. 2.16c). A pendulum is released from a standard height, and the free end of the specimen is struck by the pendulum as it passes through its lowest position. The deformation and bending of the pendulum can be reduced if the center of percussion is located near the striking edge. In this case, the pivot will be free of any impulsive reaction.

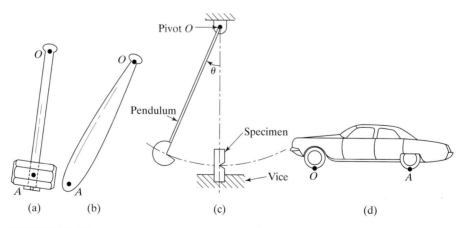

(a)   (b)   (c)   (d)

FIGURE 2.16   Applications of center of percussion.

4. In an automobile (shown in Fig. 2.16d), if the front wheels strike a bump, the passengers will not feel any reaction if the center of percussion of the vehicle is located near the rear axle. Similarly, if the rear wheels strike a bump at point A, no reaction will be felt at the front axle (point O) if the center of percussion is located near the front axle. It is desirable, therefore, to have the center of oscillation of the vehicle at one axle and the center of percussion at the other axle [2.2].

2.4   Stability Conditions

Consider a uniform rigid bar that is pivoted at one end and connected symmetrically by two springs at the other end, as shown in Fig. 2.17(a). Assume that the mass of the bar is m and that the springs are unstretched when the bar is vertical. When the bar is displaced by an angle θ, the spring force in each spring is $kl \sin \theta$; the total spring force is $2kl \sin \theta$. The gravity force $W = mg$ acts vertically downward through the center of gravity, G. The moment about the point of rotation O due to the angular acceleration $\ddot{\theta}$ is $J_0\ddot{\theta} = (ml^2/3)\ddot{\theta}$. Thus the equation of motion of the bar, for rotation about the point O, can be written as

$$\frac{ml^2}{3}\ddot{\theta} + (2kl \sin \theta)\, l \cos \theta - W\frac{l}{2}\sin \theta = 0 \qquad (2.47)$$

For small oscillations, Eq. (2.47) reduces to

$$\frac{ml^2}{3}\ddot{\theta} + 2kl^2\theta - \frac{Wl}{2}\theta = 0$$

or

$$\ddot{\theta} + \left(\frac{12kl^2 - 3Wl}{2ml^2}\right)\theta = 0 \qquad (2.48)$$

The solution of Eq. (2.48) depends on the sign of $(12kl^2 - 3Wl)/2ml^2$, as discussed below.

Case 1.   When $(12kl^2 - 3Wl)/2ml^2 > 0$, the solution of Eq. (2.48) represents stable oscillations and can be expressed as

$$\theta(t) = A_1 \cos \omega_n t + A_2 \sin \omega_n t \qquad (2.49)$$

where A_1 and A_2 are constants and

$$\omega_n = \left(\frac{12kl^2 - 3Wl}{2ml^2}\right)^{1/2} \qquad (2.50)$$

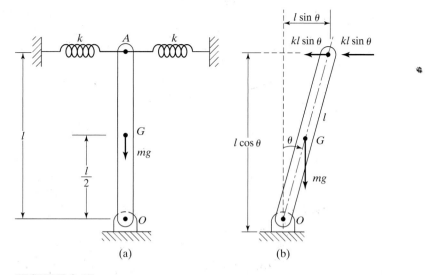

FIGURE 2.17   Stability of a rigid bar.

Case 2.   When $(12kl^2 - 3Wl)/2ml^2 = 0$, Eq. (2.48) reduces to $\ddot{\theta} = 0$ and the solution can be obtained directly by integrating twice as

$$\theta(t) = C_1 t + C_2 \tag{2.51}$$

For the initial conditions $\theta(t = 0) = \theta_0$ and $\dot{\theta}(t = 0) = \dot{\theta}_0$, the solution becomes

$$\theta(t) = \dot{\theta}_0 t + \theta_0 \tag{2.52}$$

Equation (2.52) shows that the angular displacement increases linearly at a constant velocity $\dot{\theta}_0$. However, if $\dot{\theta}_0 = 0$, Eq. (2.52) denotes a static equilibrium position with $\theta = \theta_0$—that is, the pendulum remains in its original position, defined by $\theta = \theta_0$.

Case 3.   When $(12kl^2 - 3Wl)/2ml^2 < 0$, we define

$$\alpha = \left(\frac{3Wl - 12kl^2}{2ml^2} \right)^{1/2}$$

and express the solution of Eq. (2.48) as

$$\theta(t) = B_1 e^{\alpha t} + B_2 e^{-\alpha t} \tag{2.53}$$

where B_1 and B_2 are constants. For the initial conditions $\theta(t = 0) = \theta_0$ and $\dot{\theta}(t = 0) = \dot{\theta}_0$, Eq. (2.53) becomes

$$\theta(t) = \frac{1}{2\alpha}[(\alpha\theta_0 + \dot{\theta}_0)e^{\alpha t} + (\alpha\theta_0 - \dot{\theta}_0)e^{-\alpha t}] \tag{2.54}$$

Equation (2.54) shows that $\theta(t)$ increases exponentially with time; hence the motion is unstable. The physical reason for this is that the restoring moment due to the spring $(2kl^2\theta)$, which tries to bring the system to equilibrium position, is less than the non-restoring moment due to gravity $[-W(l/2)\theta]$, which tries to move the mass away from the equilibrium position. Although the stability conditions are illustrated with reference to Fig. 2.17 in this section, similar conditions need to be examined in the vibration analysis of many engineering systems.

2.5   Rayleigh's Energy Method

For a single degree of freedom system, the equation of motion was derived using the energy method in Section 2.2.2. In this section, we shall use the energy method to find the natural frequencies of single degree of freedom systems. The principle of conservation of energy, in the context of an undamped vibrating system, can be restated as

$$T_1 + U_1 = T_2 + U_2 \tag{2.55}$$

where the subscripts 1 and 2 denote two different instants of time. Specifically, we use the subscript 1 to denote the time when the mass is passing through its static equilibrium position and choose $U_1 = 0$ as reference for the potential energy. If we let the subscript 2 indicate the time corresponding to the maximum displacement of the mass, we have $T_2 = 0$. Thus Eq. (2.55) becomes

$$T_1 + 0 = 0 + U_2 \tag{2.56}$$

If the system is undergoing harmonic motion, then T_1 and U_2 denote the maximum values of T and U, respectively, and Eq. (2.56) becomes

$$T_{max} = U_{max} \tag{2.57}$$

The application of Eq. (2.57), which is also known as *Rayleigh's energy method*, gives the natural frequency of the system directly, as illustrated in the following examples.

Manometer for Diesel Engine

EXAMPLE 2.7

The exhaust from a single-cylinder four-stroke diesel engine is to be connected to a silencer, and the pressure therein is to be measured with a simple U-tube manometer (see Fig. 2.18). Calculate the minimum length of the manometer tube so that the natural frequency of oscillation of the mercury column

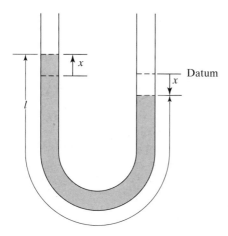

FIGURE 2.18   U-tube manometer.

will be 3.5 times slower than the frequency of the pressure fluctuations in the silencer at an engine speed of 600 rpm. The frequency of pressure fluctuation in the silencer is equal to

$$\frac{\text{Number of cylinders} \times \text{Speed of the engine}}{2}$$

Solution

1. *Natural frequency of oscillation of the liquid column:* Let the datum in Fig. 2.18 be taken as the equilibrium position of the liquid. If the displacement of the liquid column from the equilibrium position is denoted by x, the change in potential energy is given by

U = potential energy of raised liquid column + potential energy of depressed liquid column

= (weight of mercury raised × displacement of the C.G. of the segment) + (weight of mercury depressed × displacement of the C.G. of the segment)

$$= (Ax\gamma)\frac{x}{2} + (Ax\gamma)\frac{x}{2} = A\gamma x^2 \tag{E.1}$$

where A is the cross-sectional area of the mercury column and γ is the specific weight of mercury. The change in kinetic energy is given by

$$T = \frac{1}{2}(\text{mass of mercury})(\text{velocity})^2$$

$$= \frac{1}{2}\frac{Al\gamma}{g}\dot{x}^2 \tag{E.2}$$

where l is the length of the mercury column. By assuming harmonic motion, we can write

$$x(t) = X \cos \omega_n t \tag{E.3}$$

where X is the maximum displacement and ω_n is the natural frequency. By substituting Eq. (E.3) into Eqs. (E.1) and (E.2), we obtain

$$U = U_{\max} \cos^2 \omega_n t \tag{E.4}$$

$$T = T_{\max} \sin^2 \omega_n t \tag{E.5}$$

where

$$U_{\max} = A\gamma X^2 \tag{E.6}$$

and

$$T_{\max} = \frac{1}{2} \frac{A\gamma l \omega_n^2}{g} X^2 \tag{E.7}$$

By equating $U_{\max}$ to $T_{\max}$, we obtain the natural frequency:

$$\omega_n = \left(\frac{2g}{l}\right)^{1/2} \tag{E.8}$$

2. *Length of the mercury column:* The frequency of pressure fluctuations in the silencer

$$= \frac{1 \times 600}{2}$$

$$= 300 \text{ rpm}$$

$$= \frac{300 \times 2\pi}{60} = 10\pi \text{ rad/sec} \tag{E.9}$$

Thus the frequency of oscillations of the liquid column in the manometer is $10\pi/3.5 = 9.0$ rad/sec. By using Eq. (E.8), we obtain

$$\left(\frac{2g}{l}\right)^{1/2} = 9.0 \tag{E.10}$$

or

$$l = \frac{2.0 \times 9.81}{(9.0)^2} = 0.243 \text{ m} \tag{E.11}$$

∎

Effect of Mass on ω_n of a Spring

EXAMPLE 2.8

Determine the effect of the mass of the spring on the natural frequency of the spring-mass system shown in Fig. 2.19.

Solution: To find the effect of the mass of the spring on the natural frequency of the spring-mass system, we add the kinetic energy of the system to that of the attached mass and use the energy

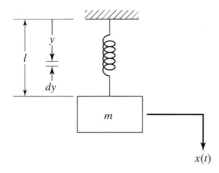

FIGURE 2.19   Equivalent mass of a spring.

method to determine the natural frequency. Let l be the total length of the spring. If x denotes the displacement of the lower end of the spring (or mass m), the displacement at distance y from the support is given by $y(x/l)$. Similarly, if $\dot{x}$ denotes the velocity of the mass m, the velocity of a spring element located at distance y from the support is given by $y(\dot{x}/l)$. The kinetic energy of the spring element of length dy is

$$dT_s = \frac{1}{2}\left(\frac{m_s}{l}dy\right)\left(\frac{y\dot{x}}{l}\right)^2 \tag{E.1}$$

where m_s is the mass of the spring. The total kinetic energy of the system can be expressed as

$$T = \text{kinetic energy of mass } (T_m) + \text{kinetic energy of spring } (T_s)$$

$$= \frac{1}{2}m\dot{x}^2 + \int_{y=0}^{l} \frac{1}{2}\left(\frac{m_s}{l}dy\right)\left(\frac{y^2\dot{x}^2}{l^2}\right)$$

$$= \frac{1}{2}m\dot{x}^2 + \frac{1}{2}\frac{m_s}{3}\dot{x}^2 \tag{E.2}$$

The total potential energy of the system is given by

$$U = \tfrac{1}{2}kx^2 \tag{E.3}$$

By assuming a harmonic motion

$$x(t) = X \cos \omega_n t \tag{E.4}$$

where X is the maximum displacement of the mass and ω_n is the natural frequency, the maximum kinetic and potential energies can be expressed as

$$T_{\max} = \frac{1}{2}\left(m + \frac{m_s}{3}\right)X^2\omega_n^2 \tag{E.5}$$

$$U_{max} = \frac{1}{2}kX^2$$ (E.6)

By equating T_{max} and U_{max}, we obtain the expression for the natural frequency:

$$\omega_n = \left(\frac{k}{m + \dfrac{m_s}{3}}\right)^{1/2}$$ (E.7)

Thus the effect of the mass of the spring can be accounted for by adding one-third of its mass to the main mass [2.3].

■

Effect of Mass of Column on Natural Frequency of Water Tank

EXAMPLE 2.9

Find the natural frequency of transverse vibration of the water tank considered in Example 2.1 and Fig. 2.10 by including the mass of the column.

Solution: To include the mass of the column, we find the equivalent mass of the column at the free end using the equivalence of kinetic energy and use a single degree of freedom model to find the natural frequency of vibration. The column of the tank is considered as a cantilever beam fixed at one end (ground) and carrying a mass M (water tank) at the other end. The static deflection of a cantilever beam under a concentrated end load is given by (see Fig. 2.20):

$$y(x) = \frac{Px^2}{6EI}(3l - x) = \frac{y_{max}\,x^2}{2l^3}(3l - x)$$

$$= \frac{y_{max}}{2l^3}(3x^2l - x^3)$$ (E.1)

The maximum kinetic energy of the beam itself (T_{max}) is given by

$$T_{max} = \frac{1}{2}\int_0^l \frac{m}{l}\{\dot{y}(x)\}^2\,dx$$ (E.2)

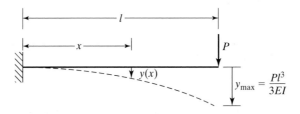

FIGURE 2.20   Equivalent mass of the column.

where m is the total mass and (m/l) is the mass per unit length of the beam. Equation (E.1) can be used to express the velocity variation, $\dot{y}(x)$, as

$$\dot{y}(x) = \frac{\dot{y}_{max}}{2l^3}(3x^2l - x^3) \tag{E.3}$$

and hence Eq. (E.2) becomes

$$T_{max} = \frac{m}{2l}\left(\frac{\dot{y}_{max}}{2l^3}\right)^2 \int_0^l (3x^2l - x^3)^2 dx$$

$$= \frac{1}{2}\frac{m}{l}\frac{\dot{y}_{max}^2}{4l^6}\left(\frac{33}{35}l^7\right) = \frac{1}{2}\left(\frac{33}{140}m\right)\dot{y}_{max}^2 \tag{E.4}$$

If m_{eq} denotes the equivalent mass of the cantilever (water tank) at the free end, its maximum kinetic energy can be expressed as

$$T_{max} = \frac{1}{2}m_{eq}\dot{y}_{max}^2 \tag{E.5}$$

By equating Eqs. (E.4) and (E.5), we obtain

$$m_{eq} = \frac{33}{140}m \tag{E.6}$$

Thus the total effective mass acting at the end of the cantilever beam is given by

$$M_{eff} = M + m_{eq} \tag{E.7}$$

where M is the mass of the water tank. The natural frequency of transverse vibration of the water tank is given by

$$\omega_n = \sqrt{\frac{k}{M_{eff}}} = \sqrt{\frac{k}{M + \dfrac{33}{140}m}} \tag{E.8}$$

∎

2.6   Free Vibration with Viscous Damping

**2.6.1
Equation of
Motion**

As stated in Section 1.9, the viscous damping force F is proportional to the velocity $\dot{x}$ or v and can be expressed as

$$F = -c\dot{x} \tag{2.58}$$

where c is the damping constant or coefficient of viscous damping and the negative sign indicates that the damping force is opposite to the direction of velocity. A single degree of freedom system with a viscous damper is shown in Fig. 2.21. If x is measured from the equilibrium position of the mass m, the application of Newton's law yields the equation of motion:

$$m\ddot{x} = -c\dot{x} - kx$$

or

$$m\ddot{x} + c\dot{x} + kx = 0 \qquad (2.59)$$

2.6.2
Solution

To solve Eq. (2.59), we assume a solution in the form

$$x(t) = Ce^{st} \qquad (2.60)$$

where C and s are undetermined constants. Inserting this function into Eq. (2.59) leads to the characteristic equation

$$ms^2 + cs + k = 0 \qquad (2.61)$$

the roots of which are

$$s_{1,2} = \frac{-c \pm \sqrt{c^2 - 4mk}}{2m} = -\frac{c}{2m} \pm \sqrt{\left(\frac{c}{2m}\right)^2 - \frac{k}{m}} \qquad (2.62)$$

These roots give two solutions to Eq. (2.59):

$$x_1(t) = C_1 e^{s_1 t} \quad \text{and} \quad x_2(t) = C_2 e^{s_2 t} \qquad (2.63)$$

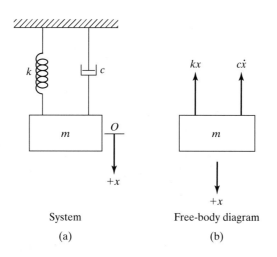

System

(a)

Free-body diagram

(b)

FIGURE 2.21   Single degree of freedom system with viscous damper.

Thus the general solution of Eq. (2.59) is given by a combination of the two solutions $x_1(t)$ and $x_2(t)$:

$$x(t) = C_1 e^{s_1 t} + C_2 e^{s_2 t}$$

$$= C_1 e^{\left\{-\frac{c}{2m} + \sqrt{\left(\frac{c}{2m}\right)^2 - \frac{k}{m}}\right\}t} + C_2 e^{\left\{-\frac{c}{2m} - \sqrt{\left(\frac{c}{2m}\right)^2 - \frac{k}{m}}\right\}t} \tag{2.64}$$

where C_1 and C_2 are arbitrary constants to be determined from the initial conditions of the system.

Critical Damping Constant and the Damping Ratio. The critical damping c_c is defined as the value of the damping constant c for which the radical in Eq. (2.62) becomes zero:

$$\left(\frac{c_c}{2m}\right)^2 - \frac{k}{m} = 0$$

or

$$c_c = 2m\sqrt{\frac{k}{m}} = 2\sqrt{km} = 2m\omega_n \tag{2.65}$$

For any damped system, the damping ratio ζ is defined as the ratio of the damping constant to the critical damping constant:

$$\zeta = c/c_c \tag{2.66}$$

Using Eqs. (2.66) and (2.65), we can write

$$\frac{c}{2m} = \frac{c}{c_c} \cdot \frac{c_c}{2m} = \zeta \omega_n \tag{2.67}$$

and hence

$$s_{1,2} = (-\zeta \pm \sqrt{\zeta^2 - 1})\omega_n \tag{2.68}$$

Thus the solution, Eq. (2.64), can be written as

$$x(t) = C_1 e^{(-\zeta + \sqrt{\zeta^2 - 1})\omega_n t} + C_2 e^{(-\zeta - \sqrt{\zeta^2 - 1})\omega_n t} \tag{2.69}$$

The nature of the roots s_1 and s_2 and hence the behavior of the solution, Eq. (2.69), depends upon the magnitude of damping. It can be seen that the case $\zeta = 0$ leads to the undamped vibrations discussed in Section 2.2. Hence we assume that $\zeta \neq 0$ and consider the following three cases.

Case 1.   *Underdamped system* ($\zeta < 1$ or $c < c_c$ or $c/2m < \sqrt{k/m}$). For this condition, ($\zeta^2 - 1$) is negative and the roots s_1 and s_2 can be expressed as

$$s_1 = \left(-\zeta + i\sqrt{1 - \zeta^2}\right)\omega_n$$

$$s_2 = \left(-\zeta - i\sqrt{1 - \zeta^2}\right)\omega_n$$

and the solution, Eq. (2.69), can be written in different forms:

$$
\begin{aligned}
x(t) &= C_1 e^{\left(-\zeta + i\sqrt{1-\zeta^2}\right)\omega_n t} + C_2 e^{\left(-\zeta - i\sqrt{1-\zeta^2}\right)\omega_n t} \\
&= e^{-\zeta\omega_n t}\left\{ C_1 e^{i\sqrt{1-\zeta^2}\,\omega_n t} + C_2 e^{-i\sqrt{1-\zeta^2}\,\omega_n t} \right\} \\
&= e^{-\zeta\omega_n t}\left\{ (C_1 + C_2)\cos\sqrt{1 - \zeta^2}\,\omega_n t + i(C_1 - C_2)\sin\sqrt{1 - \zeta^2}\,\omega_n t \right\} \\
&= e^{-\zeta\omega_n t}\left\{ C_1' \cos\sqrt{1 - \zeta^2}\,\omega_n t + C_2' \sin\sqrt{1 - \zeta^2}\,\omega_n t \right\} \\
&= X e^{-\zeta\omega_n t} \sin\left(\sqrt{1 - \zeta^2}\,\omega_n t + \phi \right) \\
&= X_0 e^{-\zeta\omega_n t} \cos\left(\sqrt{1 - \zeta^2}\,\omega_n t - \phi_0 \right) \quad\quad\quad (2.70)
\end{aligned}
$$

where (C_1', C_2'), (X, ϕ), and (X_0, ϕ_0) are arbitrary constants to be determined from the initial conditions.

For the initial conditions $x(t = 0) = x_0$ and $\dot{x}(t = 0) = \dot{x}_0$, C_1' and C_2' can be found:

$$C_1' = x_0 \quad \text{and} \quad C_2' = \frac{\dot{x}_0 + \zeta\omega_n x_0}{\sqrt{1 - \zeta^2}\,\omega_n} \quad\quad\quad (2.71)$$

and hence the solution becomes

$$
\begin{aligned}
x(t) = e^{-\zeta\omega_n t}\Bigg\{ &x_0 \cos\sqrt{1 - \zeta^2}\,\omega_n t \\
&+ \frac{\dot{x}_0 + \zeta\omega_n x_0}{\sqrt{1 - \zeta^2}\,\omega_n} \sin\sqrt{1 - \zeta^2}\,\omega_n t \Bigg\} \quad\quad\quad (2.72)
\end{aligned}
$$

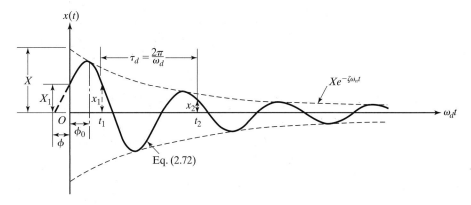

FIGURE 2.22   Underdamped solution.

The constants (X, ϕ) and (X_0, ϕ_0) can be expressed as

$$X = X_0 = \sqrt{(C_1')^2 + (C_2')^2} \tag{2.73}$$

$$\phi = \tan^{-1}(C_1'/C_2') \tag{2.74}$$

$$\phi_0 = \tan^{-1}(-C_2'/C_1') \tag{2.75}$$

The motion described by Eq. (2.72) is a damped harmonic motion of angular frequency $\sqrt{1 - \zeta^2}\, \omega_n$, but because of the factor $e^{-\zeta \omega_n t}$, the amplitude decreases exponentially with time, as shown in Fig. 2.22. The quantity

$$\omega_d = \sqrt{1 - \zeta^2}\, \omega_n \tag{2.76}$$

is called the *frequency of damped vibration*. It can be seen that the frequency of damped vibration ω_d is always less than the undamped natural frequency ω_n. The decrease in the frequency of damped vibration with increasing amount of damping, given by Eq. (2.76), is shown graphically in Fig. 2.23. The underdamped case is very important in the study of mechanical vibrations, as it is the only case that leads to an oscillatory motion [2.10].

Case 2.   *Critically damped system* ($\zeta = 1$ or $c = c_c$ or $c/2m = \sqrt{k/m}$). In this case the two roots s_1 and s_2 in Eq. (2.68) are equal:

$$s_1 = s_2 = -\frac{c_c}{2m} = -\omega_n \tag{2.77}$$

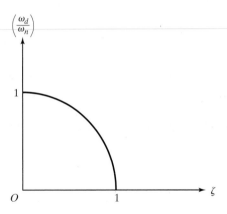

FIGURE 2.23   Variation of ω_d with damping.

Because of the repeated roots, the solution of Eq. (2.59) is given by [2.6][1]

$$x(t) = (C_1 + C_2 t)e^{-\omega_n t} \tag{2.78}$$

The application of the initial conditions $x(t = 0) = x_0$ and $\dot{x}(t = 0) = \dot{x}_0$ for this case gives

$$\begin{aligned} C_1 &= x_0 \\ C_2 &= \dot{x}_0 + \omega_n x_0 \end{aligned} \tag{2.79}$$

and the solution becomes

$$x(t) = [x_0 + (\dot{x}_0 + \omega_n x_0)t]e^{-\omega_n t} \tag{2.80}$$

It can be seen that the motion represented by Eq. (2.80) is *aperiodic* (i.e., nonperiodic). Since $e^{-\omega_n t} \rightarrow 0$ as $t \rightarrow \infty$, the motion will eventually diminish to zero, as indicated in Fig. 2.24.

Case 3.   *Overdamped system* ($\zeta > 1$ or $c > c_c$ or $c/2m > \sqrt{k/m}$). As $\sqrt{\zeta^2 - 1} > 0$, Eq. (2.68) shows that the roots s_1 and s_2 are real and distinct and are given by

$$\begin{aligned} s_1 &= (-\zeta + \sqrt{\zeta^2 - 1})\omega_n < 0 \\ s_2 &= (-\zeta - \sqrt{\zeta^2 - 1})\omega_n < 0 \end{aligned}$$

---

[1]Equation (2.78) can also be obtained by making ζ approach unity in the limit in Eq. (2.72). As $\zeta \rightarrow 1$, $\omega_n \rightarrow 0$; hence $\cos \omega_d t \rightarrow 1$ and $\sin \omega_d t \rightarrow \omega_d t$. Thus Eq. (2.72) yields

$$x(t) = e^{-\omega_n t}(C'_1 + C'_2\omega_d t) = (C_1 + C_2 t)e^{-\omega_n t}$$

where $C_1 = C'_1$ and $C_2 = C'_2\omega_d$ are new constants.

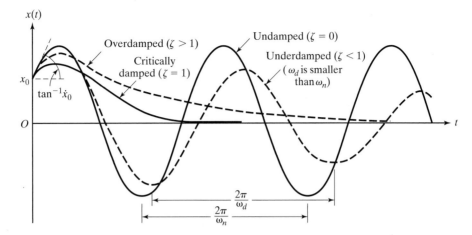

FIGURE 2.24   Comparison of motions with different types of damping.

with $s_2 \ll s_1$. In this case, the solution, Eq. (2.69), can be expressed as

$$x(t) = C_1 e^{(-\zeta + \sqrt{\zeta^2 - 1})\omega_n t} + C_2 e^{(-\zeta - \sqrt{\zeta^2 - 1})\omega_n t} \tag{2.81}$$

For the initial conditions $x(t = 0) = x_0$ and $\dot{x}(t = 0) = \dot{x}_0$, the constants C_1 and C_2 can be obtained:

$$C_1 = \frac{x_0 \omega_n (\zeta + \sqrt{\zeta^2 - 1}) + \dot{x}_0}{2\omega_n \sqrt{\zeta^2 - 1}}$$

$$C_2 = \frac{-x_0 \omega_n (\zeta - \sqrt{\zeta^2 - 1}) - \dot{x}_0}{2\omega_n \sqrt{\zeta^2 - 1}} \tag{2.82}$$

Equation (2.81) shows that the motion is aperiodic regardless of the initial conditions imposed on the system. Since roots s_1 and s_2 are both negative, the motion diminishes exponentially with time, as shown in Fig. 2.24.

Note the following aspects of these systems:

1. The nature of the roots s_1 and s_2 with varying values of damping c or ζ can be shown in a complex plane. In Fig. 2.25, the horizontal and vertical axes are chosen as the real and imaginary axes. The semicircle represents the locus of the roots s_1 and s_2 for different values of ζ in the range $0 < \zeta < 1$. This figure permits us to see instantaneously the effect of the parameter ζ on the behavior of the system. We find that for $\zeta = 0$, we obtain the imaginary roots $s_1 = i\omega_n$ and $s_2 = -i\omega_n$, leading to the solution given in Eq. (2.15). For $0 < \zeta < 1$, the roots s_1 and s_2 are complex conjugate and are located symmetrically about the real axis. As the value of ζ approaches 1, both roots approach the point $-\omega_n$ on the real axis. If $\zeta > 1$,

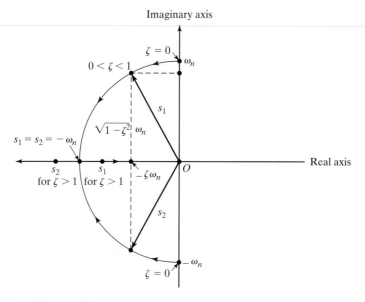

FIGURE 2.25   Locus of s_1 and s_2.

both roots lie on the real axis, one increasing and the other decreasing. In the limit when $\zeta \rightarrow \infty$, $s_1 \rightarrow 0$ and $s_2 \rightarrow -\infty$. The value $\zeta = 1$ can be seen to represent a transition stage, below which both roots are complex and above which both roots are real.

2. A critically damped system will have the smallest damping required for aperiodic motion; hence the mass returns to the position of rest in the shortest possible time without overshooting. The property of critical damping is used in many practical applications. For example, large guns have dashpots with critical damping value, so that they return to their original position after recoil in the minimum time without vibrating. If the damping provided were more than the critical value, some delay would be caused before the next firing.

3. The free damped response of a single degree of freedom system can be represented in phase plane or state space as indicated in Fig. 2.26.

2.6.3 Logarithmic Decrement

The logarithmic decrement represents the rate at which the amplitude of a free-damped vibration decreases. It is defined as the natural logarithm of the ratio of any two successive amplitudes. Let t_1 and t_2 denote the times corresponding to two consecutive amplitudes (displacements), measured one cycle apart for an underdamped system, as in Fig. 2.22. Using Eq. (2.70), we can form the ratio

$$\frac{x_1}{x_2} = \frac{X_0 e^{-\zeta \omega_n t_1} \cos(\omega_d t_1 - \phi_0)}{X_0 e^{-\zeta \omega_n t_2} \cos(\omega_d t_2 - \phi_0)} \qquad (2.83)$$

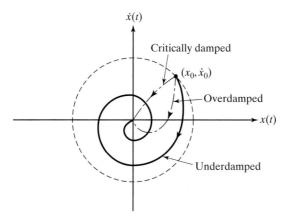

FIGURE 2.26 Phase plane of a damped system.

But $t_2 = t_1 + \tau_d$ where $\tau_d = 2\pi/\omega_d$ is the period of damped vibration. Hence $\cos(\omega_d t_2 - \phi_0) = \cos(2\pi + \omega_d t_1 - \phi_0) = \cos(\omega_d t_1 - \phi_0)$, and Eq. (2.83) can be written as

$$\frac{x_1}{x_2} = \frac{e^{-\zeta\omega_n t_1}}{e^{-\zeta\omega_n(t_1 + \tau_d)}} = e^{\zeta\omega_n \tau_d} \tag{2.84}$$

The logarithmic decrement δ can be obtained from Eq. (2.84):

$$\delta = \ln\frac{x_1}{x_2} = \zeta\omega_n\tau_d = \zeta\omega_n\frac{2\pi}{\sqrt{1 - \zeta^2}\omega_n} = \frac{2\pi\zeta}{\sqrt{1 - \zeta^2}} = \frac{2\pi}{\omega_d}\cdot\frac{c}{2m} \tag{2.85}$$

For small damping, Eq. (2.85) can be approximated:

$$\delta \simeq 2\pi\zeta \quad \text{if} \quad \zeta \ll 1 \tag{2.86}$$

Figure 2.27 shows the variation of the logarithmic decrement δ with ζ as given by Eqs. (2.85) and (2.86). It can be noticed that for values up to $\zeta = 0.3$, the two curves are difficult to distinguish.

The logarithmic decrement is dimensionless and is actually another form of the dimensionless damping ratio ζ. Once δ is known, ζ can be found by solving Eq. (2.85):

$$\zeta = \frac{\delta}{\sqrt{(2\pi)^2 + \delta^2}} \tag{2.87}$$

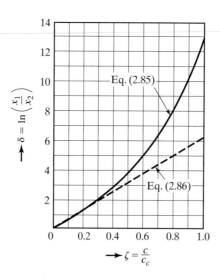

FIGURE 2.27   Variation of logarithmic decrement with damping.

If we use Eq. (2.86) instead of Eq. (2.85), we have

$$\zeta \simeq \frac{\delta}{2\pi} \tag{2.88}$$

If the damping in the given system is not known, we can determine it experimentally by measuring any two consecutive displacements x_1 and x_2. By taking the natural logarithm of the ratio of x_1 and x_2, we obtain δ. By using Eq. (2.87), we can compute the damping ratio ζ. In fact, the damping ratio ζ can also be found by measuring two displacements separated by any number of complete cycles. If x_1 and x_{m+1} denote the amplitudes corresponding to times t_1 and $t_{m+1} = t_1 + m\tau_d$ where m is an integer, we obtain

$$\frac{x_1}{x_{m+1}} = \frac{x_1}{x_2}\frac{x_2}{x_3}\frac{x_3}{x_4}\cdots\frac{x_m}{x_{m+1}} \tag{2.89}$$

Since any two successive displacements separated by one cycle satisfy the equation

$$\frac{x_j}{x_{j+1}} = e^{\zeta\omega_n\tau_d} \tag{2.90}$$

Eq. (2.89) becomes

$$\frac{x_1}{x_{m+1}} = (e^{\zeta\omega_n\tau_d})^m = e^{m\zeta\omega_n\tau_d} \tag{2.91}$$

Equations (2.91) and (2.85) yield

$$\delta = \frac{1}{m} \ln\left(\frac{x_1}{x_{m+1}}\right) \tag{2.92}$$

which can be substituted into Eq. (2.87) or Eq. (2.88) to obtain the viscous damping ratio ζ.

2.6.4
Energy
Dissipated in
Viscous
Damping

In a viscously damped system, the rate of change of energy with time (dW/dt) is given by

$$\frac{dW}{dt} = \text{force} \times \text{velocity} = Fv = -cv^2 = -c\left(\frac{dx}{dt}\right)^2 \tag{2.93}$$

using Eq. (2.58). The negative sign in Eq. (2.93) denotes that energy dissipates with time. Assume a simple harmonic motion as $x(t) = X \sin \omega_d t$, where X is the amplitude of motion and the energy dissipated in a complete cycle is given by[2]

$$\Delta W = \int_{t=0}^{(2\pi/\omega_d)} c\left(\frac{dx}{dt}\right)^2 dt = \int_0^{2\pi} cX^2\omega_d\cos^2\omega_d t \cdot d(\omega_d t)$$

$$= \pi c \omega_d X^2 \tag{2.94}$$

This shows that the energy dissipated is proportional to the square of the amplitude of motion. Note that it is not a constant for given values of damping and amplitude, since ΔW is also a function of the frequency ω_d.

Equation (2.94) is valid even when there is a spring of stiffness k parallel to the viscous damper. To see this, consider the system shown in Fig. 2.28. The total force resisting motion can be expressed as

$$F = -kx - cv = -kx - c\dot{x} \tag{2.95}$$

If we assume simple harmonic motion

$$x(t) = X \sin \omega_d t \tag{2.96}$$

as before, Eq. (2.95) becomes

$$F = -kX \sin \omega_d t - c\omega_d X \cos \omega_d t \tag{2.97}$$

---

[2]In the case of a damped system, simple harmonic motion $x(t) = X \cos \omega_d t$ is possible only when the steady-state response is considered under a harmonic force of frequency ω_d (see Section 3.4). The loss of energy due to the damper is supplied by the excitation under steady-state forced vibration [2.7].

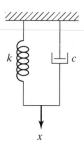

FIGURE 2.28   Spring and damper in parallel.

The energy dissipated in a complete cycle will be

$$\Delta W = \int_{t=0}^{2\pi/\omega_d} Fv \, dt$$

$$= \int_0^{2\pi/\omega_d} kX^2\omega_d \sin \omega_d t \cdot \cos \omega_d t \cdot d(\omega_d t)$$

$$+ \int_0^{2\pi/\omega_d} c\omega_d X^2 \cos^2\omega_d t \cdot d(\omega_d t) = \pi c\omega_d X^2 \qquad (2.98)$$

which can be seen to be identical with Eq. (2.94). This result is to be expected, since the spring force will not do any net work over a complete cycle or any integral number of cycles.

We can also compute the fraction of the total energy of the vibrating system that is dissipated in each cycle of motion ($\Delta W/W$), as follows. The total energy of the system W can be expressed either as the maximum potential energy ($\frac{1}{2}kX^2$) or as the maximum kinetic energy ($\frac{1}{2}mv_{max}^2 = \frac{1}{2}mX^2\omega_d^2$), the two being approximately equal for small values of damping. Thus

$$\frac{\Delta W}{W} = \frac{\pi c\omega_d X^2}{\frac{1}{2}m\omega_d^2 X^2} = 2\left(\frac{2\pi}{\omega_d}\right)\left(\frac{c}{2m}\right) = 2\delta \simeq 4\pi\zeta = \text{constant} \qquad (2.99)$$

using Eqs. (2.85) and (2.88). The quantity $\Delta W/W$ is called the *specific damping capacity* and is useful in comparing the damping capacity of engineering materials. Another quantity known as the *loss coefficient* is also used for comparing the damping capacity of engineering materials. The loss coefficient is defined as the ratio of the energy dissipated per radian and the total strain energy:

$$\text{loss coefficient} = \frac{(\Delta W/2\pi)}{W} = \frac{\Delta W}{2\pi W} \qquad (2.100)$$

**2.6.5
Torsional
Systems with
Viscous
Damping**

The methods presented in Sections 2.6.1 through 2.6.4 for linear vibrations with viscous damping can be extended directly to viscously damped torsional (angular) vibrations. For this, consider a single degree of freedom torsional system with a viscous damper, as shown in Fig. 2.29(a). The viscous damping torque is given by (Fig. 2.29b):

$$T = -c_t \dot{\theta} \tag{2.101}$$

where c_t is the torsional viscous damping constant, $\dot{\theta} = d\theta/dt$ is the angular velocity of the disc, and the negative sign denotes that the damping torque is opposite the direction of angular velocity. The equation of motion can be derived as

$$J_0 \ddot{\theta} + c_t \dot{\theta} + k_t \theta = 0 \tag{2.102}$$

where J_0 = mass moment of inertia of the disc, k_t = spring constant of the system (restoring torque per unit angular displacement), and θ = angular displacement of the disc. The solution of Eq. (2.102) can be found exactly as in the case of linear vibrations. For example, in the underdamped case, the frequency of damped vibration is given by

$$\omega_d = \sqrt{1 - \zeta^2} \, \omega_n \tag{2.103}$$

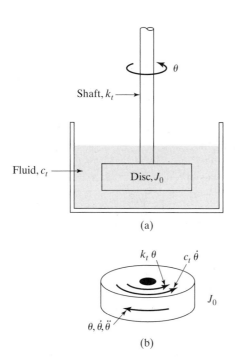

(a)

(b)

FIGURE 2.29   Torsional viscous damper.

where

$$\omega_n = \sqrt{\frac{k_t}{J_0}}$$ (2.104)

and

$$\zeta = \frac{c_t}{c_{tc}} = \frac{c_t}{2J_0\omega_n} = \frac{c_t}{2\sqrt{k_t J_0}}$$ (2.105)

where c_{tc} is the critical torsional damping constant.

Response of Anvil of a Forging Hammer

EXAMPLE 2.10

The anvil of a forging hammer weighs 5,000 N and is mounted on a foundation that has a stiffness of 5×10^6 N/m and a viscous damping constant of 10,000 N-s/m. During a particular forging operation, the tup (i.e., the falling weight or the hammer) weighing 1,000 N, is made to fall from a height of 2 m on to the anvil (Fig. 2.30a). If the anvil is at rest before impact by the tup, determine the response of the anvil after the impact. Assume that the coefficient of restitution between the anvil and the tup is 0.4.

Solution: First we use the principle of conservation of momentum and the definition of the coefficient of restitution to find the initial velocity of the anvil. Let the velocities of the tup just before and just after impact with the anvil be v_{t1} and v_{t2}, respectively. Similarly, let v_{a1} and v_{a2} be the velocities of the anvil just before and just after the impact, respectively (Fig. 2.30b). Note that the displacement of the anvil is measured from its static equilibrium position and all velocities are assumed to be positive when acting downward. The principle of conservation of momentum gives

$$M(v_{a2} - v_{a1}) = m(v_{t1} - v_{t2})$$ (E.1)

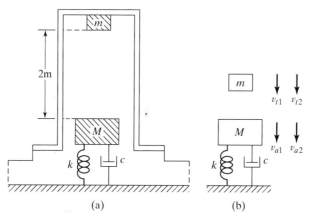

(a)   (b)

FIGURE 2.30   Forging hammer.

where $v_{a1} = 0$ (anvil is at rest before the impact) and v_{t1} can be determined by equating its kinetic energy just before impact to its potential energy before dropping from a height of $h = 2$ m:

$$\frac{1}{2}mv_{t1}^2 = mgh \tag{E.2}$$

or

$$v_{t1} = \sqrt{2gh} = \sqrt{2 \times 9.81 \times 2} = 6.26099 \text{ m/s}$$

Thus Eq. (E.1) becomes

$$\frac{5000}{9.81}(v_{a2} - 0) = \frac{1000}{9.81}(6.26099 - v_{t2})$$

that is,

$$510.204082\, v_{a2} = 638.87653 - 102.040813\, v_{t2} \tag{E.3}$$

The definition of the coefficient of restitution (r) yields:

$$r = -\left(\frac{v_{a2} - v_{t2}}{v_{a1} - v_{t1}}\right) \tag{E.4}$$

that is,

$$0.4 = -\left(\frac{v_{a2} - v_{t2}}{0 - 6.26099}\right)$$

that is,

$$v_{a2} = v_{t2} + 2.504396 \tag{E.5}$$

The solution of Eqs. (E.3) and (E.5) gives

$$v_{a2} = 1.460898 \text{ m/s}; v_{t2} = -1.043498 \text{ m/s}$$

Thus the initial conditions of the anvil are given by

$$x_0 = 0; \dot{x}_0 = 1.460898 \text{ m/s}$$

The damping coefficient is equal to

$$\zeta = \frac{c}{2\sqrt{kM}} = \frac{1000}{2\sqrt{(5 \times 10^6)\left(\dfrac{5000}{9.81}\right)}} = 0.0989949$$

The undamped and damped natural frequencies of the anvil are given by

$$\omega_n = \sqrt{\frac{k}{M}} = \sqrt{\frac{5 \times 10^6}{\left(\dfrac{5000}{9.81}\right)}} = 98.994949 \text{ rad/s}$$

$$\omega_d = \omega_n \sqrt{1 - \zeta^2} = 98.994949 \sqrt{1 - 0.0989949^2} = 98.024799 \text{ rad/s}$$

The displacement response of the anvil is given by Eq. (2.72):

$$x(t) = e^{-\zeta \omega_n t} \left\{ \cos \omega_d t + \frac{\dot{x}_0 + \zeta \omega_n x_0}{\omega_d} \sin \omega_d t \right\}$$

$$= e^{-9.799995t} \{ \cos 98.024799t + 0.01490335 \sin 98.024799t \} \, \text{m}$$

∎

Shock Absorber for a Motorcycle

EXAMPLE 2.11

An underdamped shock absorber is to be designed for a motorcycle of mass 200 kg (Fig. 2.31a). When the shock absorber is subjected to an initial vertical velocity due to a road bump, the resulting displacement-time curve is to be as indicated in Fig. 2.31(b). Find the necessary stiffness and damping constants of the shock absorber if the damped period of vibration is to be 2 s and the amplitude x_1 is to be reduced to one-fourth in one half cycle (i.e., $x_{1.5} = x_1/4$). Also find the minimum initial velocity that leads to a maximum displacement of 250 mm.

Approach: We use the equation for the logarithmic decrement in terms of the damping ratio, equation for the damped period of vibration, time corresponding to maximum displacement for an underdamped system, and envelope passing through the maximum points of an underdamped system.

Solution: Since $x_{1.5} = x_1/4$, $x_2 = x_{1.5}/4 = x_1/16$. Hence the logarithmic decrement becomes

$$\delta = \ln \left(\frac{x_1}{x_2} \right) = \ln(16) = 2.7726 = \frac{2\pi\zeta}{\sqrt{1 - \zeta^2}} \tag{E.1}$$

from which the value of ζ can be found as $\zeta = 0.4037$. The damped period of vibration is given to be 2 s. Hence

$$2 = \tau_d = \frac{2\pi}{\omega_d} = \frac{2\pi}{\omega_n \sqrt{1 - \zeta^2}}$$

$$\omega_n = \frac{2\pi}{2\sqrt{1 - (0.4037)^2}} = 3.4338 \text{ rad/s}$$

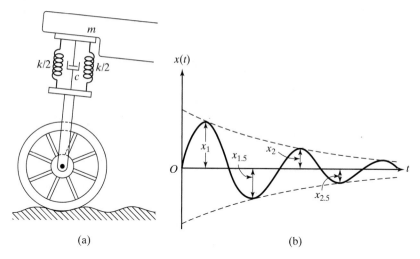

FIGURE 2.31 Shock absorber of a motorcycle.

The critical damping constant can be obtained:

$$c_c = 2m\omega_n = 2(200)(3.4338) = 1373.54 \text{ N-s/m}$$

Thus the damping constant is given by

$$c = \zeta c_c = (0.4037)(1373.54) = 554.4981 \text{ N-s/m}$$

and the stiffness by

$$k = m\omega_n^2 = (200)(3.4338)^2 = 2358.2652 \text{ N/m}$$

The displacement of the mass will attain its maximum value at time t_1, given by

$$\sin \omega_d t_1 = \sqrt{1 - \zeta^2}$$

(See Problem 2.86.) This gives

$$\sin \omega_d t_1 = \sin \pi t_1 = \sqrt{1 - (0.4037)^2} = 0.9149$$

or

$$t_1 = \frac{\sin^{-1}(0.9149)}{\pi} = 0.3678 \text{ sec}$$

The envelope passing through the maximum points (see Problem 2.86) is given by

$$x = \sqrt{1 - \zeta^2} X e^{-\zeta \omega_n t} \qquad \text{(E.2)}$$

Since $x = 250$ mm, Eq. (E.2) gives at t_1

$$0.25 = \sqrt{1 - (0.4037)^2} \; X e^{-(0.4037)(3.4338)(0.3678)}$$

or

$$X = 0.4550 \text{ m.}$$

The velocity of the mass can be obtained by differentiating the displacement

$$x(t) = X e^{-\zeta \omega_n t} \sin \omega_d t$$

as

$$\dot{x}(t) = X e^{-\zeta \omega_n t} (-\zeta \omega_n \sin \omega_d t + \omega_d \cos \omega_d t) \tag{E.3}$$

When $t = 0$, Eq. (E.3) gives

$$\dot{x}(t = 0) = \dot{x}_0 = X \omega_d = X \omega_n \sqrt{1 - \zeta^2} = (0.4550)(3.4338) \sqrt{1 - (0.4037)^2}$$
$$= 1.4294 \text{ m/s}$$

∎

Analysis of Cannon

EXAMPLE 2.12 ─────

The schematic diagram of a large cannon is shown in Fig. 2.32 [2.8]. When the gun is fired, high-pressure gases accelerate the projectile inside the barrel to a very high velocity. The reaction force pushes the gun barrel in the opposite direction of the projectile. Since it is desirable to bring the gun barrel to rest in the shortest time without oscillation, it is made to translate backward against a critically damped spring-damper system called the *recoil mechanism*. In a particular case, the gun barrel and the recoil mechanism have a mass of 500 kg with a recoil spring of stiffness 10,000 N/m. The gun recoils 0.4 m upon firing. Find (1) the critical damping coefficient of the damper, (2) the initial recoil velocity of the gun, and (3) the time taken by the gun to return to a position 0.1 m from its initial position.

Solution
1. The undamped natural frequency of the system is

$$\omega_n = \sqrt{\frac{k}{m}} = \sqrt{\frac{10,000}{500}} = 4.4721 \text{ rad/s}$$

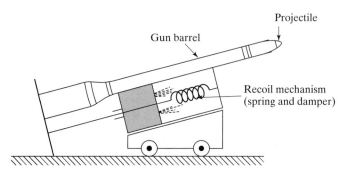

FIGURE 2.32   Recoil of cannon.

and the critical damping coefficient (Eq. 2.65) of the damper is

$$c_c = 2m\omega_n = 2(500)(4.4721) = 4472.1 \text{ N-s/m}$$

2. The response of a critically damped system is given by Eq. (2.78):

$$x(t) = (C_1 + C_2 t)e^{-\omega_n t} \tag{E.1}$$

where $C_1 = x_0$ and $C_2 = \dot{x}_0 + \omega_n x_0$. The time t_1 at which $x(t)$ reaches a maximum value can be obtained by setting $\dot{x}(t) = 0$. The differentiation of Eq. (E.1) gives

$$\dot{x}(t) = C_2 e^{-\omega_n t} - \omega_n (C_1 + C_2 t)e^{-\omega_n t}$$

Hence $\dot{x}(t) = 0$ yields

$$t_1 = \left(\frac{1}{\omega_n} - \frac{C_1}{C_2} \right) \tag{E.2}$$

In this case, $x_0 = C_1 = 0$; hence Eq. (E.2) leads to $t_1 = 1/\omega_n$. Since the maximum value of $x(t)$ or the recoil distance is given to be $x_{max} = 0.4$ m, we have

$$x_{max} = x(t = t_1) = C_2 t_1 e^{-\omega_n t_1} = \frac{\dot{x}_0}{\omega_n} e^{-1} = \frac{\dot{x}_0}{e\omega_n}$$

or

$$\dot{x}_0 = x_{max} \omega_n e = (0.4)(4.4721)(2.7183) = 4.8626 \text{ m/s}$$

3. If t_2 denotes the time taken by the gun to return to a position 0.1 m from its initial position, we have

$$0.1 = C_2 t_2 e^{-\omega_n t_2} = 4.8626 t_2 e^{-4.4721 t_2} \tag{E.3}$$

The solution of Eq. (E.3) gives $t_2 = 0.8258$ s.

■

2.7 Free Vibration with Coulomb Damping

In many mechanical systems, *Coulomb* or *dry friction* dampers are used because of their mechanical simplicity and convenience [2.9]. Also in vibrating structures, whenever the components slide relative to each other, dry friction damping appears internally. As stated in Section 1.9, Coulomb damping arises when bodies slide on dry surfaces. Coulomb's law of dry friction states that, when two bodies are in contact, the force required to produce sliding is proportional to the normal force acting in the plane of contact. Thus the friction force F is given by

$$F = \mu N = \mu W = \mu mg \tag{2.106}$$

where N is the normal force, equal to the weight of the mass ($W = mg$) and μ is the coefficient of sliding or kinetic friction. The value of the coefficient of friction (μ) depends on the materials in contact and the condition of the surfaces in contact. For example, $\mu \simeq 0.1$ for metal on metal (lubricated), 0.3 for metal on metal (unlubricated), and nearly 1.0 for rubber on metal. The friction force acts in a direction opposite to the direction of velocity. Coulomb damping is sometimes called *constant damping*, since the damping force is independent of the displacement and velocity; it depends only on the normal force N between the sliding surfaces.

2.7.1
Equation of Motion

Consider a single degree of freedom system with dry friction as shown in Fig. 2.33(a). Since the friction force varies with the direction of velocity, we need to consider two cases, as indicated in Figs. 2.33(b) and (c).

Case 1.   When x is positive and dx/dt is positive or when x is negative and dx/dt is positive (i.e., for the half cycle during which the mass moves from left to right) the equation of motion can be obtained using Newton's second law (see Fig. 2.33b)

$$m\ddot{x} = -kx - \mu N \quad \text{or} \quad m\ddot{x} + kx = -\mu N \tag{2.107}$$

This is a second-order nonhomogeneous differential equation. The solution can be verified by substituting Eq. (2.108) into Eq. (2.107)

$$x(t) = A_1 \cos \omega_n t + A_2 \sin \omega_n t - \frac{\mu N}{k} \tag{2.108}$$

where $\omega_n = \sqrt{k/m}$ is the frequency of vibration and A_1 and A_2 are constants whose values depend on the initial conditions of this half cycle.

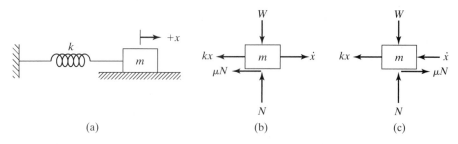

(a)   (b)   (c)

FIGURE 2.33   Spring-mass system with Coulomb damping.

Case 2.   When x is positive and dx/dt is negative or when x is negative and dx/dt is negative (i.e., for the half cycle during which the mass moves from right to left), the equation of motion can be derived from Fig. 2.33(c) as

$$-kx + \mu N = m\ddot{x} \quad \text{or} \quad m\ddot{x} + kx = \mu N \tag{2.109}$$

The solution of Eq. (2.109) is given by

$$x(t) = A_3 \cos \omega_n t + A_4 \sin \omega_n t + \frac{\mu N}{k} \tag{2.110}$$

where A_3 and A_4 are constants to be found from the initial conditions of this half cycle. The term $\mu N/k$ appearing in Eqs. (2.108) and (2.110) is a constant representing the virtual displacement of the spring under the force μN, if it were applied as a static force. Equations (2.108) and (2.110) indicate that in each half cycle the motion is harmonic, with the equilibrium position changing from $\mu N/k$ to $-(\mu N/k)$ every half cycle, as shown in Fig. 2.34.

**2.7.2
Solution**

Equations (2.107) and (2.109) can be expressed as a single equation (using $N = mg$):

$$m\ddot{x} + \mu m g \, \text{sgn}(\dot{x}) + kx = 0 \tag{2.111}$$

where $\text{sgn}(y)$ is called the signum function, whose value is defined as 1 for $y > 0$, -1 for $y < 0$, and 0 for $y = 0$. Equation (2.111) can be seen to be a nonlinear differential equation for which a simple analytical solution does not exist. Numerical methods can be used to solve Eq. (2.111) conveniently (see Example 2.19). Equation (2.111), however, can be

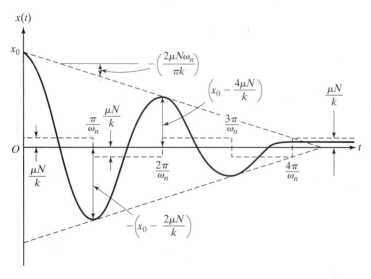

FIGURE 2.34   Motion of the mass with Coulomb damping.

solved analytically if we break the time axis into segments separated by $\dot{x} = 0$ (i.e., time intervals with different directions of motion). To find the solution using this procedure, let us assume the initial conditions as

$$x(t = 0) = x_0$$
$$\dot{x}(t = 0) = 0 \tag{2.112}$$

That is, the system starts with zero velocity and displacement x_0 at $t = 0$. Since $x = x_0$ at $t = 0$, the motion starts from right to left. Let $x_0, x_1, x_2, \ldots$ denote the amplitudes of motion at successive half cycles. Using Eqs. (2.110) and (2.112), we can evaluate the constants A_3 and A_4:

$$A_3 = x_0 - \frac{\mu N}{k}, \quad A_4 = 0$$

Thus Eq. (2.110) becomes

$$x(t) = \left(x_0 - \frac{\mu N}{k} \right) \cos \omega_n t + \frac{\mu N}{k} \tag{2.113}$$

This solution is valid for half the cycle only, that is, for $0 \leq t \leq \pi/\omega_n$. When $t = \pi/\omega_n$, the mass will be at its extreme left position and its displacement from equilibrium position can be found from Eq. (2.113):

$$-x_1 = x \left(t = \frac{\pi}{\omega_n} \right) = \left(x_0 - \frac{\mu N}{k} \right) \cos \pi + \frac{\mu N}{k} = -\left(x_0 - \frac{2\mu N}{k} \right)$$

Since the motion started with a displacement of $x = x_0$ and, in a half cycle, the value of x became $-[x_0 - (2\mu N/k)]$, the reduction in magnitude of x in time π/ω_n is $2\mu N/k$.

In the second half cycle, the mass moves from left to right, so Eq. (2.108) is to be used. The initial conditions for this half cycle are

$$x(t = 0) = \text{value of } x \text{ at } t = \frac{\pi}{\omega_n} \text{ in Eq. (2.113)} = -\left(x_0 - \frac{2\mu N}{k} \right)$$

and

$$\dot{x}(t = 0) = \text{value of } \dot{x} \text{ at } t = \frac{\pi}{\omega_n} \text{ in Eq. (2.113)}$$

$$= \left\{ \text{value of } -\omega_n \left(x_0 - \frac{\mu N}{k} \right) \sin \omega_n t \text{ at } t = \frac{\pi}{\omega_n} \right\} = 0$$

Thus the constants in Eq. (2.108) become

$$-A_1 = -x_0 + \frac{3\mu N}{k}, \qquad A_2 = 0$$

so that Eq. (2.108) can be written as

$$x(t) = \left(x_0 - \frac{3\mu N}{k} \right) \cos \omega_n t - \frac{\mu N}{k} \qquad (2.114)$$

This equation is valid only for the second half cycle—that is, for $\pi/\omega_n \leq t \leq 2\pi/\omega_n$. At the end of this half cycle the value of $x(t)$ is

$$x_2 = x\left(t = \frac{\pi}{\omega_n} \right) \text{ in Eq. (2.114)} = x_0 - \frac{4\mu N}{k}$$

and

$$\dot{x}\left(t = \frac{\pi}{\omega_n} \right) \text{ in Eq. (2.114)} = 0$$

These become the initial conditions for the third half cycle, and the procedure can be continued until the motion stops. The motion stops when $x_n \leq \mu N/k$, since the restoring force exerted by the spring (kx) will then be less than the friction force μN. Thus the number of half cycles (r) that elapse before the motion ceases is given by

$$x_0 - r\frac{2\mu N}{k} \leq \frac{\mu N}{k}$$

that is,

$$r \geq \left\{ \frac{x_0 - \dfrac{\mu N}{k}}{\dfrac{2\mu N}{k}} \right\} \qquad (2.115)$$

Note the following characteristics of a system with Coulomb damping:

1. The equation of motion is nonlinear with Coulomb damping, while it is linear with viscous damping.
2. The natural frequency of the system is unaltered with the addition of Coulomb damping, while it is reduced with the addition of viscous damping.
3. The motion is periodic with Coulomb damping, while it can be nonperiodic in a viscously damped (overdamped) system.

4. The system comes to rest after some time with Coulomb damping, whereas the motion theoretically continues forever (perhaps with an infinitesimally small amplitude) with viscous and hysteresis damping.

5. The amplitude reduces linearly with Coulomb damping, whereas it reduces exponentially with viscous damping.

6. In each successive cycle, the amplitude of motion is reduced by the amount $4\mu N/k$, so the amplitudes at the end of any two consecutive cycles are related:

$$X_m = X_{m-1} - \frac{4\mu N}{k} \tag{2.116}$$

As the amplitude is reduced by an amount $4\mu N/k$ in one cycle (i.e., in time $2\pi/\omega_n$), the slope of the enveloping straight lines (shown dotted) in Fig. 2.34 is

$$-\left(\frac{4\mu N}{k}\right) \bigg/ \left(\frac{2\pi}{\omega_n}\right) = -\left(\frac{2\mu N\omega_n}{\pi k}\right)$$

The final position of the mass is usually displaced from equilibrium ($x = 0$) position and represents a permanent displacement in which the friction force is locked. Slight tapping will usually make the mass come to its equilibrium position.

2.7.3
Torsional
Systems with
Coulomb
Damping

If a constant frictional torque acts on a torsional system, the equation governing the angular oscillations of the system can be derived, similar to Eqs. (2.107) and (2.109), as

$$J_0 \ddot{\theta} + k_t \theta = -T \tag{2.117}$$

and

$$J_0 \ddot{\theta} + k_t \theta = T \tag{2.118}$$

where T denotes the constant damping torque (similar to μN for linear vibrations). The solutions of Eqs. (2.117) and (2.118) are similar to those for linear vibrations. In particular, the frequency of vibration is given by

$$\omega_n = \sqrt{\frac{k_t}{J_0}} \tag{2.119}$$

and the amplitude of motion at the end of the rth half cycle (θ_r) is given by

$$\theta_r = \theta_0 - r\frac{2T}{k_t} \tag{2.120}$$

where θ_0 is the initial angular displacement at $t = 0$ (with $\dot{\theta} = 0$ at $t = 0$). The motion ceases when

$$r \geq \left\{ \frac{\theta_0 - \dfrac{T}{k_t}}{\dfrac{2T}{k_t}} \right\} \tag{2.121}$$

Coefficient of Friction from Measured Positions of Mass

EXAMPLE 2.13

A metal block, placed on a rough surface, is attached to a spring and is given an initial displacement of 10 cm from its equilibrium position. After five cycles of oscillation in 2 s, the final position of the metal block is found to be 1 cm from its equilibrium position. Find the coefficient of friction between the surface and the metal block.

Solution: Since five cycles of oscillation were observed to take place in 2 s, the period (τ_n) is $2/5 = 0.4$ s, and hence the frequency of oscillation is $\omega_n = \sqrt{\frac{k}{m}} = \frac{2\pi}{\tau_n} = \frac{2\pi}{0.4} = 15.708$ rad/s. Since the amplitude of oscillation reduces by

$$\frac{4\mu N}{k} = \frac{4\mu mg}{k}$$

in each cycle, the reduction in amplitude in five cycles is

$$5\left(\frac{4\mu mg}{k}\right) = 0.10 - 0.01 = 0.09 \text{ m}$$

or

$$\mu = \frac{0.09k}{20mg} = \frac{0.09\omega_n^2}{20g} = \frac{0.09(15.708)^2}{20(9.81)} = 0.1132$$

∎

Pulley Subjected to Coulomb Damping

EXAMPLE 2.14

A steel shaft of length 1 m and diameter 50 mm is fixed at one end and carries a pulley of mass moment of inertia 25 kg-m² at the other end. A band brake exerts a constant frictional torque of 400 N-m around the circumference of the pulley. If the pulley is displaced by 6° and released, determine (1) the number of cycles before the pulley comes to rest and (2) the final settling position of the pulley.

Solution: (1) The number of half cycles that elapse before the angular motion of the pulley ceases is given by Eq. (2.121)

$$r \geq \left\{ \frac{\theta_0 - \dfrac{T}{k_t}}{\dfrac{2T}{k_t}} \right\} \tag{E.1}$$

where θ_0 = initial angular displacement = $6°$ = 0.10472 rad, k_t = torsional spring constant of the shaft given by

$$k_t = \frac{GJ}{l} = \frac{(8 \times 10^{10})\left\{\frac{\pi}{32}(0.05)^4\right\}}{1} = 49,087.5 \text{ N-m/rad}$$

and T = constant friction torque applied to the pulley = 400 N-m. Equation (E.1) gives

$$r \geq \frac{0.10472 - \left(\frac{400}{49,087.5}\right)}{\left(\frac{800}{49,087.5}\right)} = 5.926$$

Thus the motion ceases after six half cycles.
   (2) The angular displacement after six half cycles is given by Eq. (2.120).

$$\theta = 0.10472 - 6 \times 2\left(\frac{400}{49,087.5}\right) = 0.006935 \text{ rad} = 0.39734°$$

Thus the pulley stops at $0.39734°$ from the equilibrium position on the same side of the initial displacement.

■

2.8   Free Vibration with Hysteretic Damping

Consider the spring-viscous damper arrangement shown in Fig. 2.35(a). For this system, the force F needed to cause a displacement $x(t)$ is given by

$$F = kx + c\dot{x} \tag{2.122}$$

For a harmonic motion of frequency ω and amplitude X,

$$x(t) = X \sin \omega t \tag{2.123}$$

Equations (2.122) and (2.123) yield

$$F(t) = kX \sin \omega t + cX\omega \cos \omega t$$

$$= kx \pm c\omega\sqrt{X^2 - (X \sin \omega t)^2}$$

$$= kx \pm c\omega\sqrt{X^2 - x^2} \tag{2.124}$$

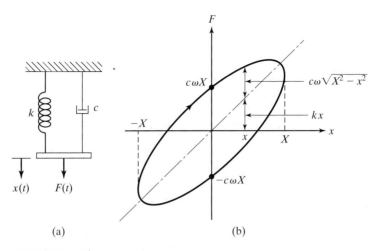

FIGURE 2.35 Spring-viscous damper system.

When F versus x is plotted, Eq. (2.124) represents a closed loop, as shown in Fig. 2.35(b). The area of the loop denotes the energy dissipated by the damper in a cycle of motion and is given by

$$\Delta W = \oint F \, dx = \int_0^{2\pi/\omega} (kX \sin \omega t + cX\omega \cos \omega t)(\omega X \cos \omega t)dt$$
$$= \pi\omega c X^2 \tag{2.125}$$

Equation (2.125) has been derived in Section 2.6.4 also (see Eq. 2.98).

As stated in Section 1.9, the damping caused by the friction between the internal planes that slip or slide as the material deforms is called hysteresis (or solid or structural) damping. This causes a hysteresis loop to be formed in the stress-strain or force-displacement curve (see Fig. 2.36a). The energy loss in one loading and unloading cycle is equal to the area enclosed by the hysteresis loop [2.11–2.13]. The similarity between Figs. 2.35(b) and 2.36(a) can be used to define a hysteresis damping constant. It was found experimentally that the energy loss per cycle due to internal friction is independent of the frequency but approximately proportional to the square of the amplitude. In order to achieve this observed behavior from Eq. (2.125), the damping coefficient c is assumed to be inversely proportional to the frequency as

$$c = \frac{h}{\omega} \tag{2.126}$$

where h is called the hysteresis damping constant. Equations (2.125) and (2.126) give

$$\Delta W = \pi h X^2 \tag{2.127}$$

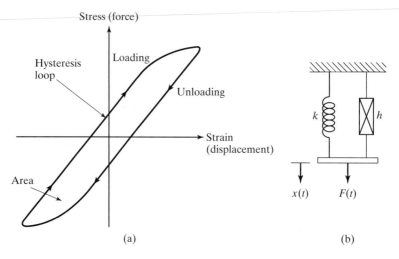

FIGURE 2.36   Hysteresis loop.

Complex Stiffness.   In Fig. 2.35(a), the spring and the damper are connected in parallel, and for a general harmonic motion, $x = Xe^{i\omega t}$, the force is given by

$$F = kXe^{i\omega t} + c\omega i Xe^{i\omega t} = (k + i\omega c)x \qquad (2.128)$$

Similarly, if a spring and a hysteresis damper are connected in parallel, as shown in Fig. 2.36(b), the force-displacement relation can be expressed as

$$F = (k + ih)x \qquad (2.129)$$

where

$$k + ih = k\left(1 + i\frac{h}{k}\right) = k(1 + i\beta) \qquad (2.130)$$

is called the complex stiffness of the system and $\beta = h/k$ is a constant indicating a dimensionless measure of damping.

Response of the System.   In terms of β, the energy loss per cycle can be expressed as

$$\Delta W = \pi k \beta X^2 \qquad (2.131)$$

Under hysteresis damping, the motion can be considered to be nearly harmonic (since ΔW is small), and the decrease in amplitude per cycle can be determined using energy balance. For example, the energies at points P and Q (separated by half a cycle) in Fig. 2.37 are related as

$$\frac{kX_j^2}{2} - \frac{\pi k \beta X_j^2}{4} - \frac{\pi k \beta X_{j+0.5}^2}{4} = \frac{kX_{j+0.5}^2}{2}$$

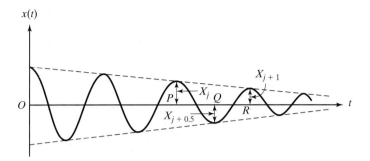

FIGURE 2.37   Response of a hysteretically damped system.

or

$$\frac{X_j}{X_{j+0.5}} = \sqrt{\frac{2 + \pi\beta}{2 - \pi\beta}} \tag{2.132}$$

Similarly, the energies at points Q and R give

$$\frac{X_{j+0.5}}{X_{j+1}} = \sqrt{\frac{2 + \pi\beta}{2 - \pi\beta}} \tag{2.133}$$

Multiplication of Eqs. (2.132) and (2.133) gives

$$\frac{X_j}{X_{j+1}} = \frac{2 + \pi\beta}{2 - \pi\beta} = \frac{2 - \pi\beta + 2\pi\beta}{2 - \pi\beta} \simeq 1 + \pi\beta = \text{constant} \tag{2.134}$$

The hysteresis logarithmic decrement can be defined as

$$\delta = \ln\left(\frac{X_j}{X_{j+1}}\right) \simeq \ln(1 + \pi\beta) \simeq \pi\beta \tag{2.135}$$

Since the motion is assumed to be approximately harmonic, the corresponding frequency is defined by [2.10]

$$\omega = \sqrt{\frac{k}{m}} \tag{2.136}$$

The equivalent viscous damping ratio ζ_{eq} can be found by equating the relation for the logarithmic decrement δ.

$$\delta \simeq 2\pi\zeta_{eq} \simeq \pi\beta = \frac{\pi h}{k}$$

$$\zeta_{eq} = \frac{\beta}{2} = \frac{h}{2k} \tag{2.137}$$

Thus the equivalent damping constant c_{eq} is given by

$$c_{eq} = c_c \cdot \zeta_{eq} = 2\sqrt{mk} \cdot \frac{\beta}{2} = \beta\sqrt{mk} = \frac{\beta k}{\omega} = \frac{h}{\omega} \qquad (2.138)$$

Note that the method of finding an equivalent viscous damping coefficient for a structurally damped system is valid only for harmonic excitation. The above analysis assumes that the system responds approximately harmonically at the frequency ω.

Estimation of Hysteretic Damping Constant

EXAMPLE 2.15

The experimental measurements on a structure gave the force-deflection data shown in Fig. 2.38. From this data, estimate the hysteretic damping constant β and the logarithmic decrement δ.

Solution
Approach: We equate the energy dissipated in a cycle (area enclosed by the hysteresis loop) to ΔW of Eq. (2.127).

The energy dissipated in each full load cycle is given by the area enclosed by the hysteresis curve. Each square in Fig. 2.38 denotes $100 \times 2 = 200$ N-mm. The area enclosed by the loop can be found as area ACB + area $ABDE$ + area $DFE \simeq \frac{1}{2}(AB)(CG) + (AB)(AE) + \frac{1}{2}(DE)(FH) = \frac{1}{2}(1.25)(1.8) + (1.25)(8) + \frac{1}{2}(1.25)(1.8) = 12.25$ square units. This area represents an energy of $12.25 \times 200/1{,}000 = 2.5$ N-m. From Eq. (2.127), we have

$$\Delta W = \pi h X^2 = 2.5 \text{ N-m} \qquad (E.1)$$

Since the maximum deflection X is 0.008 m and the slope of the force-deflection curve (given approximately by the slope of the line OF) is $k = 400/8 = 50$ N/mm $= 50{,}000$ N/m, the hysteretic damping constant h is given by

$$h = \frac{\Delta W}{\pi X^2} = \frac{2.5}{\pi(0.008)^2} = 12433.95 \qquad (E.2)$$

and hence

$$\beta = \frac{h}{k} = \frac{12{,}433.95}{50{,}000} = 0.248679$$

The logarithmic decrement can be found

$$\delta \simeq \pi\beta = \pi(0.248679) = 0.78125 \qquad (E.3)$$

■

Response of a Hysteretically Damped Bridge Structure

EXAMPLE 2.16

A bridge structure is modeled as a single degree of freedom system with an equivalent mass of 5×10^5 kg and an equivalent stiffness of 25×10^6 N/m. During a free vibration test, the ratio of successive amplitudes was found to be 1.04. Estimate the structural damping constant (β) and the approximate free vibration response of the bridge.

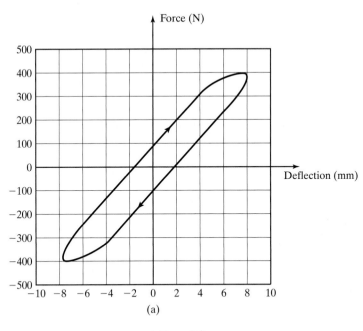

(a)

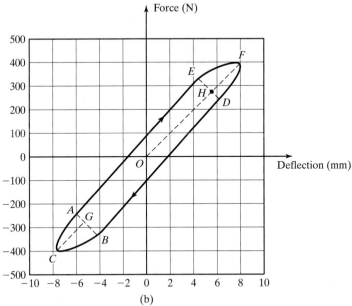

(b)

FIGURE 2.38   Load-deflection curve.

Solution: Using the ratio of successive amplitudes, Eq. (2.135) yields the hysteresis logarithmic decrement (δ) as

$$\delta = \ln\left(\frac{X_j}{X_{j+1}}\right) = \ln(1.04) = \ln(1 + \pi\beta)$$

or

$$1 + \pi\beta = 1.04 \quad \text{or} \quad \beta = \frac{0.04}{\pi} = 0.0127$$

The equivalent viscous damping coefficient (c_{eq}) can be determined from Eq. (2.138) as

$$c_{eq} = \frac{\beta k}{\omega} = \frac{\beta k}{\sqrt{\frac{k}{m}}} = \beta\sqrt{km} \tag{E.1}$$

Using the known values of the equivalent stiffness (k) and the equivalent mass (m) of the bridge, Eq. (E.1) yields

$$c_{eq} = (0.0127)\sqrt{(25 \times 10^6)(5 \times 10^5)} = 44.9013 \times 10^3 \text{ N-s/m}$$

The equivalent critical damping constant of the bridge can be computed using Eq. (2.65) as

$$c_c = 2\sqrt{km} = 2\sqrt{(25 \times 10^6)(5 \times 10^5)} = 7071.0678 \times 10^3 \text{ N-s/m}$$

Since $c_{eq} < c_c$, the bridge is underdamped, and hence its free vibration response is given by Eq. (2.72) as

$$x(t) = e^{-\zeta\omega_n t}\left\{x_0 \cos\sqrt{1 - \zeta^2}\,\omega_n t + \frac{\dot{x}_0 + \zeta\omega_n x_0}{\sqrt{1 - \zeta^2}\,\omega_n} \sin\sqrt{1 - \zeta^2}\,\omega_n t\right\}$$

where

$$\zeta = \frac{c_{eq}}{c_c} = \frac{40.9013 \times 10^3}{7071.0678 \times 10^3} = 0.0063$$

and x_0 and $\dot{x}_0$ denote the initial displacement and initial velocity given to the bridge at the start of free vibration.

∎

2.9 Examples Using MATLAB

Variations of Natural Frequency and Period with Static Deflection

EXAMPLE 2.17

Plot the variations of the natural frequency and the time period with static deflection of an undamped system using MATLAB.

Solution: The natural frequency (ω_n) and the time period (τ_n) are given by Eqs. (2.28) and (2.30):

$$\omega_n = \left(\frac{g}{\delta_{st}}\right)^{1/2}, \tau_n = 2\pi\left(\frac{\delta_{st}}{g}\right)^{1/2}$$

Using $g = 9.81$ m/s^2, ω_n and τ_n are plotted over the range of $\delta_{st} = 0$ to 0.5 using a MATLAB program.

```
% Ex2_17.m
g = 9.81;
for i = 1: 101
    t(i) = 0.01 + (0.5-0.01) * (i-1)/100;
    w(i) = (g/t(i))^0.5;
    tao(i) = 2 * pi * (t(i)/g)^0.5;
end
plot(t,w);
gtext('w_n');
hold on;
plot(t, tao);
gtext('T_n');
xlabel('Delta_s_t');
title('Example 2.17');
```

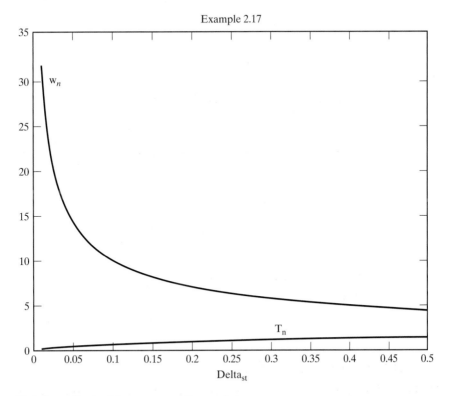

Variations of natural frequency and time period.

▆▆▆▆▆▆▆▆▆▆ Free Vibration Response of a Spring-Mass System

EXAMPLE 2.18 ——————————

A spring-mass system with a mass of $20 \, \frac{\text{lb-sec}^2}{\text{in.}}$ and stiffness 500 lb/in. is subject to an initial displacement of $x_0 = 3.0$ in. and an initial velocity of $\dot{x}_0 = 4.0$ in/sec. Plot the time variations of the mass's displacement, velocity, and acceleration using MATLAB.

Solution: The displacement of an undamped system can be expressed as (see Eq. (2.23)):

$$x(t) = A_0 \sin(\omega_n t + \phi_0) \tag{E.1}$$

where

$$\omega_n = \sqrt{\frac{k}{m}} = \sqrt{\frac{500}{20}} = 5 \text{ rad/sec}$$

$$A_0 = \left[x_0^2 + \left(\frac{\dot{x}_0}{\omega_n} \right)^2 \right]^{1/2} = \left[(3.0)^2 + \left(\frac{4.0}{5.0} \right)^2 \right]^{1/2} = 3.1048 \text{ in.}$$

$$\phi_0 = \tan^{-1}\left(\frac{x_0 \omega_n}{\dot{x}_0} \right) = \tan^{-1}\left(\frac{(3.0)(5.0)}{4.0} \right) = 75.0686° = 1.3102 \text{ rad}$$

Example 2.18

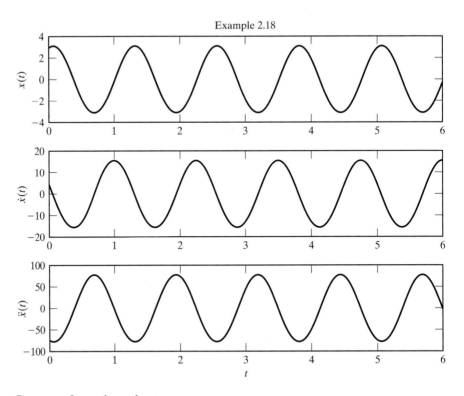

Response of an undamped system.

Thus Eq. (E.1) yields

$$x(t) = 3.1048 \sin(5t + 1.3102) \text{ in.} \tag{E.2}$$

$$\dot{x}(t) = 15.524 \cos(5t + 1.3102) \text{ in./sec} \tag{E.3}$$

$$\ddot{x}(t) = -77.62 \sin(5t + 1.3102) \text{ in./sec}^2 \tag{E.4}$$

Equations (E.2)–(E.4) are plotted using MATLAB in the range $t = 0$ to 6 sec.

```
% EX2_18.m
for i = 1: 101
    t(i) = 6 * (i-1)/100;
    x(i) = 3.1048 * sin(5 * t(i) + 1.3102);
    x1(i) = 15.524 * cos(5 * t(i) + 1.3102);
    x2(i) = -77.62 * sin(5 * t(i) + 1.3102);
end
subplot (311);
plot (t,x);
ylabel ('x(t)');
title ('Example 2.18');
subplot (312);
plot (t,x1);
ylabel ('x^.(t)');
subplot (313);
plot (t,x2);
xlabel ('t');
ylabel ('x^.^.(t)');
```

■

Free Vibration Response of a System with Coulomb Damping

EXAMPLE 2.19

Find the free vibration response of a spring-mass system subject to Coulomb damping for the following initial conditions: $x(0) = 0.5$ m, $\dot{x}(0) = 0$

Data: $m = 10$ kg, $k = 200$ N/m, $\mu = 0.5$
Solution: The equation of motion can be expressed as

$$m\ddot{x} + \mu mg \, \text{sgn}(\dot{x}) + kx = 0 \tag{E.1}$$

In order to solve the second-order differential equation, Eq. (E.1), using the Runge-Kutta method (see Appendix F), we rewrite Eq. (E.1) as a set of two first-order differential equations as follows:

$$x_1 = x, x_2 = \dot{x}_1 = \dot{x}$$

$$\dot{x}_1 = x_2 \equiv f_1(x_1, x_2) \tag{E.2}$$

$$\dot{x}_2 = -\mu g \, \text{sgn}(x_2) - \frac{k}{m}x_1 \equiv f_2(x_1, x_2) \tag{E.3}$$

Equations (E.2) and (E.3) can be expressed in matrix notation as

$$\vec{\dot{X}} = \vec{f}(\vec{X}) \tag{E.4}$$

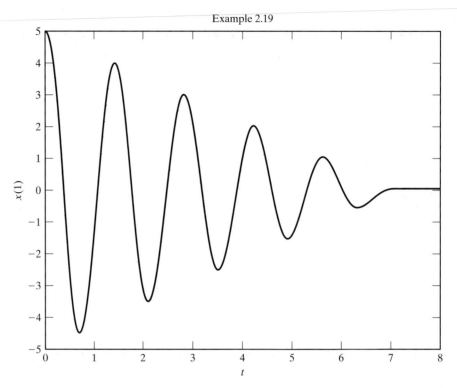

Solution of eq. (E4).

where

$$\vec{X} = \begin{Bmatrix} x_1(t) \\ x_2(t) \end{Bmatrix}, \ \vec{f} = \begin{Bmatrix} f_1(x_1, x_2) \\ f_2(x_1, x_2) \end{Bmatrix}, \ \vec{X}(t = 0) = \begin{Bmatrix} x_1(0) \\ x_2(0) \end{Bmatrix}$$

The MATLAB program **ode23** is used to find the solution of Eq. (E.4) as shown below.

```
% Ex2_19.m
% This program will use dfunc1.m
tspan = [0: 0.05: 8];
x0 = [5.0; 0.0];
[t, x] = ode23 ('dfunc1', tspan, x0);
plot (t, x(:, 1));
xlabel ('t');
ylabel ('x(1)');
title ('Example 2.19');

% dfunc1.m
function f = dfunc1 (t, x)
f = zeros (2, 1);
f(1) = x(2);
f(2) = -0.5 * 9.81 * sign(x(2)) - 200 * x(1) / 10;
```

■

EXAMPLE 2.20

Free Vibration Response of a Viscously Damped System Using MATLAB

Develop a general-purpose MATLAB program, called **Program2.m**, to find the free vibration response of a viscously damped system. Use the program to find the response of a system with the following data:

$$m = 450.0, k = 26519.2, c = 1000.0, x_0 = 0.539657, \dot{x}_0 = 1.0$$

Solution: Program2.m is developed to accept the following input data:

m = mass

k = spring stiffness

c = damping constant

x0 = initial displacement

xd0 = initial velocity

n = number of time steps at which values of $x(t)$ are to be found

delt = time interval between consecutive time steps (Δt)

The program gives the following output:

step number i, time (i), $x(i)$, $\dot{x}(i)$, $\ddot{x}(i)$

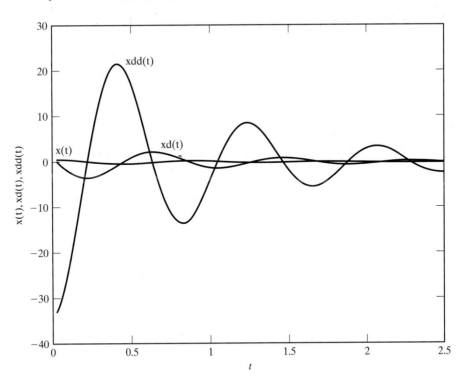

Variations of x, $\dot{x}$, and $\ddot{x}$.

The program also plots the variations of x, $\dot{x}$, and $\ddot{x}$ with time.

```
>> program2
Free vibration analysis of a single degree of freedom analysis

Data:

m=        4.50000000e+002
k=        2.65192000e+004
c=        1.00000000e+003
x0=       5.39657000e-001
xd0=      1.00000000e+000
n=        100
delt=     2.50000000e-002

system is under damped

Results:

i       time(i)            x(i)             xd(i)            xdd(i)

1     2.500000e-002    5.540992e-001     1.596159e-001    -3.300863e+001
2     5.000000e-002    5.479696e-001    -6.410545e-001    -3.086813e+001
3     7.500000e-002    5.225989e-001    -1.375559e+000    -2.774077e+001
4     1.000000e-001    4.799331e-001    -2.021239e+000    -2.379156e+001
5     1.250000e-001    4.224307e-001    -2.559831e+000    -1.920599e+001
6     1.500000e-001    3.529474e-001    -2.977885e+000    -1.418222e+001
.
.
.
96    2.400000e+000    2.203271e-002     2.313895e-001    -1.812621e+000
97    2.425000e+000    2.722809e-002     1.834092e-001    -2.012170e+000
98    2.450000e+000    3.117018e-002     1.314707e-001    -2.129064e+000
99    2.475000e+000    3.378590e-002     7.764312e-002    -2.163596e+000
100   2.500000e+000    3.505350e-002     2.395118e-002    -2.118982e+000
```

■

2.10   C++ Program

An interactive C++ program, called **Program2.cpp**, is given for finding the free vibration analysis of a viscously damped single degree of freedom system. The input and output parameters of the program are similar to those of the MATLAB program given in Example 2.20.

Free Vibration Response of a Viscously Damped System Using C++

EXAMPLE 2.21

Find the free vibration response of a viscously damped system with parameters as indicated in Example 2.20.

Solution: The input data are to be entered interactively. The input and output of the program are given next.

```
Please input M, K, C, X0, XD0, DELT:
450. 26519.2 1000. 0.539567 1.0 0.25

Please input n:
10

FREE VIBRATION ANALYSIS OF A SINGLE DEGREE OF FREEDOM SYSTEM
```

```
DATA

M    = 450
K    = 26519.2
C    = 1000
X0   = 0.539567
XD0  = 1
N    = 10
DELT = 0.25


SYSTEM IS UNDER DAMPED


RESULTS:

I     TIME(I)          X(I)          XD(I)          XDD(I)

0     0.250000      0.019265      -3.350687       6.310661
1     0.500000     -0.318985       1.062303      16.437608
2     0.750000      0.144699       1.403773     -11.646832
3     1.000000      0.112366      -1.294927      -3.744280
4     1.250000     -0.137887      -0.173140       8.510647
5     1.500000      0.002856       0.827508      -2.007241
6     1.750000      0.077718      -0.304712      -3.902928
7     2.000000     -0.039587      -0.326003       3.057362
8     2.250000     -0.025262       0.334008       0.746486
9     2.500000      0.035048       0.023957      -2.118662
```

■

2.11   Fortran Program

A Fortran computer program, in the form of subroutine **FREVIB.F**, is given for the free vibration analysis of a viscously damped single degree of freedom system. The system may be underdamped, critically damped, or overdamped. The arguments of the subroutine are similar to those described for the MATLAB program given in Example 2.20.

EXAMPLE 2.22

Free Vibration of a Viscously Damped System Using Fortran

Solve Example 2.20 using the subroutine **FREVIB.F**.

Solution: The main program that calls the subroutine **FREVIB.F** and the subroutine **FREVIB.F** are given as **PROGRAM2.F**. The output of the program is given below.

```
FREE VIBRATION ANALYSIS OF A SINGLE DEGREE OF FREEDOM SYSTEM

DATA

M    = 0.45000000E+03
K    = 0.26519199E+05
C    = 0.10000000E+04
X0   = 0.53956699E+00
XD0  = 0.10000000E+01
N    =     10
DELT = 0.25000000E+00


SYSTEM IS UNDER DAMPED
```

RESULTS:

I	TIME(I)	X(I)	XD(I)	XDD(I)
1	0.250000E+00	0.192649E-01	-0.335069E+01	0.631066E+01
2	0.500000E+00	-0.318985E+00	0.106230E+01	0.164376E+02
3	0.750000E+00	0.144699E+00	0.140377E+01	-0.116468E+02
4	0.100000E+01	0.112366E+00	-0.129493E+01	-0.374428E+01
5	0.125000E+01	-0.137887E+00	-0.173140E+00	0.851065E+01
6	0.150000E+01	0.285635E-02	0.827508E+00	-0.200724E+01
7	0.175000E+01	0.777184E-01	-0.304712E+00	-0.390293E+01
8	0.200000E+01	-0.395868E-01	-0.326002E+00	0.305736E+01
9	0.225000E+01	-0.252620E-01	0.334008E+00	0.746487E+00
10	0.250000E+01	0.350478E-01	0.239573E-01	-0.211866E+01

REFERENCES

2.1 R. W. Fitzgerald, *Mechanics of Materials* (2nd ed.), Addison-Wesley, Reading, Mass., 1982.

2.2 R. F. Steidel, Jr., *An Introduction to Mechanical Vibrations* (4th ed.), Wiley, New York, 1989.

2.3 W. Zambrano, "A brief note on the determination of the natural frequencies of a spring-mass system," *International Journal of Mechanical Engineering Education*, Vol. 9, October 1981, pp. 331–334; Vol. 10, July 1982, p. 216.

2.4 R. D. Blevins, *Formulas for Natural Frequency and Mode Shape*, Van Nostrand Reinhold, New York, 1979.

2.5 A. D. Dimarogonas, *Vibration Engineering*, West Publishing, Saint Paul, 1976.

2.6 E. Kreyszig, *Advanced Engineering Mathematics* (7th ed.), Wiley, New York, 1993.

2.7 S. H. Crandall, "The role of damping in vibration theory," *Journal of Sound and Vibration*, Vol. 11, 1970, pp. 3–18.

2.8 I. Cochin, *Analysis and Design of Dynamic Systems*, Harper & Row, New York, 1980.

2.9 D. Sinclair, "Frictional vibrations," *Journal of Applied Mechanics*, Vol. 22, 1955, pp. 207–214.

2.10 T. K. Caughey and M. E. J. O'Kelly, "Effect of damping on the natural frequencies of linear dynamic systems," *Journal of the Acoustical Society of America*, Vol. 33, 1961, pp. 1458–1461.

2.11 E. E. Ungar, "The status of engineering knowledge concerning the damping of built-up structures," *Journal of Sound and Vibration*, Vol. 26, 1973, pp. 141–154.

2.12 W. Pinsker, "Structural damping," *Journal of the Aeronautical Sciences*, Vol. 16, 1949, p. 699.

2.13 R. H. Scanlan and A. Mendelson, "Structural damping," *AIAA Journal*, Vol. 1, 1963, pp. 938–939.

REVIEW QUESTIONS

2.1 Give brief answers to the following:

1. Suggest a method for determining the damping constant of a highly damped vibrating system that uses viscous damping.

2. Can you apply the results of Section 2.2 to systems where the restoring force is not proportional to the displacement—that is, where k is not a constant?
3. State the parameters corresponding to m, c, k, and x for a torsional system.
4. What effect does a decrease in mass have on the frequency of a system?
5. What effect does a decrease in the stiffness of the system have on the natural period?
6. Why does the amplitude of free vibration gradually diminish in practical systems?
7. Why is it important to find the natural frequency of a vibrating system?
8. How many arbitrary constants must a general solution to a second-order differential equation have? How are these constants determined?
9. Can the energy method be used to find the differential equation of motion of all single degree of freedom systems?
10. What assumptions are made in finding the natural frequency of a single degree of freedom system using the energy method?
11. Is the frequency of a damped free vibration smaller or greater than the natural frequency of the system?
12. What is the use of the logarithmic decrement?
13. Is hysteresis damping a function of the maximum stress?
14. What is critical damping, and what is its importance?
15. What happens to the energy dissipated by damping?
16. What is equivalent viscous damping? Is the equivalent viscous damping factor a constant?
17. What is the reason for studying the vibration of a single degree of freedom system?
18. How can you find the natural frequency of a system by measuring its static deflection?
19. Give two practical applications of a torsional pendulum.
20. Define these terms: damping ratio, logarithmic decrement, loss coefficient, and specific damping capacity.
21. In what ways is the response of a system with Coulomb damping different from that of systems with other types of damping?
22. What is complex stiffness?
23. Define the hysteresis damping constant.
24. Give three practical applications of the concept of center of percussion.

2.2 Indicate whether each of the following statements is true or false:

1. The amplitude of an undamped system will not change with time.
2. A system vibrating in air can be considered a damped system.
3. The equation of motion of a single degree of freedom system will be the same whether the mass moves in a horizontal plane or an inclined plane.
4. When a mass vibrates in a vertical direction, its weight can always be ignored in deriving the equation of motion.
5. The principle of conservation of energy can be used to derive the equation of motion of both damped and undamped systems.
6. The damped frequency can in some cases be larger than the undamped natural frequency of the system.
7. The damped frequency can be zero in some cases.
8. The natural frequency of vibration of a torsional system is given by $\sqrt{\dfrac{k}{m}}$, where k and m denote the torsional spring constant and the polar mass moment of inertia, respectively.
9. Rayleigh's method is based on the principle of conservation of energy.

10. The final position of the mass is always the equilibrium position in the case of Coulomb damping.

11. The undamped natural frequency of a system is given by $\sqrt{g/\delta_{st}}$, where δ_{st} is the static deflection of the mass.

12. For an undamped system, the velocity leads the displacement by $\pi/2$.

13. For an undamped system, the velocity leads the acceleration by $\pi/2$.

14. Coulomb damping can be called constant damping.

15. The loss coefficient denotes the energy dissipated per radian per unit strain energy.

16. The motion diminishes to zero in both underdamped and overdamped cases.

17. The logarithmic decrement can be used to find the damping ratio.

18. The hysteresis loop of the stress-strain curve of a material causes damping.

19. The complex stiffness can be used to find the damping force in a system with hysteresis damping.

20. Motion in the case of hysteresis damping can be considered harmonic.

2.3 Fill in the blanks with proper words:

1. The free vibration of an undamped system represents interchange of _____ and _____ energies.

2. A system undergoing simple harmonic motion is called a _____ oscillator.

3. The mechanical clock represents a _____ pendulum.

4. The center of _____ can be used advantageously in a baseball bat.

5. With viscous and hysteresis damping, the motion _____ forever, theoretically.

6. The damping force in Coulomb damping is given by _____.

7. The _____ coefficient can be used to compare the damping capacity of different engineering materials.

8. Torsional vibration occurs when a _____ body oscillates about an axis.

9. The property of _____ damping is used in many practical applications, such as large guns.

10. The logarithmic decrement denotes the rate at which the _____ of a free damped vibration decreases.

11. Rayleigh's method can be used to find the _____ frequency of a system directly.

12. Any two successive displacements of the system, separated by a cycle, can be used to find the _____ decrement.

13. The damped natural frequency (ω_d) can be expressed in terms of the undamped natural frequency (ω_n) as _____.

2.4 Select the most appropriate answer out of the multiple choices given:

1. The natural frequency of a system with mass m and stiffness k is given by:

(a) $\dfrac{k}{m}$ (b) $\sqrt{\dfrac{k}{m}}$ (c) $\sqrt{\dfrac{m}{k}}$

2. In Coulomb damping, the amplitude of motion is reduced in each cycle by:

(a) $\dfrac{\mu N}{k}$ (b) $\dfrac{2\mu N}{k}$ (c) $\dfrac{4\mu N}{k}$

3. The amplitude of an undamped system subject to an initial displacement 0 and initial velocity $\dot{x}_0$ is given by:

(a) $\dot{x}_0$ (b) $\dot{x}_0\omega_n$ (c) $\dfrac{\dot{x}_0}{\omega_n}$

4. The effect of the mass of the spring can be accounted for by adding the following fraction of its mass to the vibrating mass:

(a) $\dfrac{1}{2}$ (b) $\dfrac{1}{3}$ (c) $\dfrac{4}{3}$

5. For a viscous damper with damping constant c, the damping force is:

(a) $c\dot{x}$ (b) cx (c) $c\ddot{x}$

6. The relative sliding of components in a mechanical system causes:

(a) dry friction damping   (b) viscous damping   (c) hysteresis damping

7. In torsional vibration, the displacement is measured in terms of a:

(a) linear coordinate   (b) angular coordinate   (c) force coordinate

8. The damping ratio, in terms of the damping constant c and critical damping constant (c_c), is given by:

(a) $\dfrac{c_c}{c}$ (b) $\dfrac{c}{c_c}$ (c) $\sqrt{\dfrac{c}{c_c}}$

9. The amplitude of an underdamped system subject to an initial displacement x_0 and initial velocity 0 is given by:

(a) x_0 (b) $2x_0$ (c) $x_0\omega_n$

10. The phase angle of an undamped system subject to an initial displacement x_0 and initial velocity 0 is given by:

(a) x_0 (b) $2x_0$ (c) 0

11. The energy dissipated due to viscous damping is proportional to the following power of the amplitude of motion:

(a) 1 (b) 2 (c) 3

12. For a critically damping system, the motion will be:

(a) periodic (b) aperiodic (c) harmonic

13. The energy dissipated per cycle in viscous damping with damping constant c during the simple harmonic motion $x(t) = X \sin \omega_d t$, is given by:

(a) $\pi c\omega_d X^2$ (b) $\pi \omega_d X^2$ (c) $\pi c\omega_d X$

14. For a vibrating system with a total energy W and a dissipated energy ΔW per cycle, the specific damping capacity is given by:

(a) $\dfrac{W}{\Delta W}$ (b) $\dfrac{\Delta W}{W}$ (c) ΔW

2.5 Match the following for a single degree of freedom system with $m = 1$, $k = 2$, and $c = 0.5$:

(1) Natural frequency, ω_n	(a) 1.3919
(2) Linear frequency, f_n	(b) 2.8284
(3) Natural time period, τ_n	(c) 2.2571
(4) Damped frequency, ω_d	(d) 0.2251
(5) Critical damping constant, c_c	(e) 0.1768
(6) Damping ratio, ζ	(f) 4.4429
(7) Logarithmic decrement, δ	(g) 1.4142

2.6 Match the following for a mass $m = 5$ kg moving with velocity $v = 10$ m/s:

Damping force	Type of damper
(1) 20 N	(a) Coulomb damping with a coefficient of friction of 0.3
(2) 1.5 N	(b) Viscous damping with a damping coefficient 1 N-s/m
(3) 30 N	(c) Viscous damping with a damping coefficient 2 N-s/m
(4) 25 N	(d) Hysteretic damping with a hysteretic damping coefficient of 12 N/m at a frequency of 4 rad/s
(5) 10 N	(e) Quadratic damping (force $= av^2$) with damping constant $a = 0.25$ N-s^2/m^2

PROBLEMS

The problem assignments are organized as follows:

Problems	Section Covered	Topic Covered
2.1–2.58	2.2	Undamped translational systems
2.59–2.73	2.3	Undamped torsional systems
2.74–2.83	2.5	Energy method
2.84–2.110	2.6	Systems with viscous damping
2.111–2.122	2.7	Systems with Coulomb damping
2.123–2.126	2.8	Systems with hysteretic damping
2.127–2.134	2.9	MATLAB programs
2.135–2.138	2.10	C++ programs
2.139–2.142	2.11	Fortran programs
2.143–2.147	—	Design projects

2.1 An industrial press is mounted on a rubber pad to isolate it from its foundation. If the rubber pad is compressed 5 mm by the self-weight of the press, find the natural frequency of the system.

2.2 A spring-mass system has a natural period of 0.21 sec. What will be the new period if the spring constant is (a) increased by 50 percent and (b) decreased by 50 percent?

2.3 A spring-mass system has a natural frequency of 10 Hz. When the spring constant is reduced by 800 N/m, the frequency is altered by 45 percent. Find the mass and spring constant of the original system.

2.4 A helical spring, when fixed at one end and loaded at the other, requires a force of 100 N to produce an elongation of 10 mm. The ends of the spring are now rigidly fixed, one end vertically above the other, and a mass of 10 kg is attached at the middle point of its length. Determine the time taken to complete one vibration cycle when the mass is set vibrating in the vertical direction.

2.5   An air-conditioning chiller unit weighing 2,000 lb is to be supported by four air springs (Fig. 2.39). Design the air springs such that the natural frequency of vibration of the unit lies between 5 rad/s and 10 rad/s.

FIGURE 2.39   (Courtesy of *Sound and Vibration*)

2.6   The maximum velocity attained by the mass of a simple harmonic oscillator is 10 cm/s, and the period of oscillation is 2 s. If the mass is released with an initial displacement of 2 cm, find (a) the amplitude, (b) the initial velocity, (c) the maximum acceleration, and (d) the phase angle.

2.7   Three springs and a mass are attached to a rigid, weightless bar PQ as shown in Fig. 2.40. Find the natural frequency of vibration of the system.

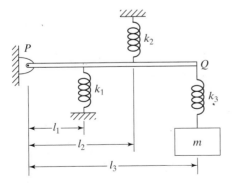

FIGURE 2.40

2.8 An automobile having a mass of 2,000 kg deflects its suspension springs 0.02 m under static conditions. Determine the natural frequency of the automobile in the vertical direction by assuming damping to be negligible.

2.9 Find the natural frequency of vibration of a spring-mass system arranged on an inclined plane, as shown in Fig. 2.41.

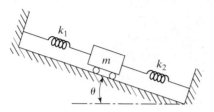

FIGURE 2.41

2.10 A loaded mine cart, weighing 5,000 lb, is being lifted by a frictionless pulley and a wire rope, as shown in Fig. 2.42. Find the natural frequency of vibration of the cart in the given position.

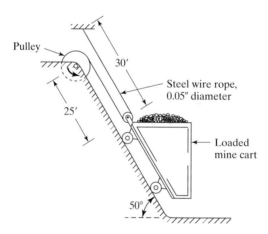

FIGURE 2.42

2.11 An electronic chassis weighing 500 N is isolated by supporting it on four helical springs, as shown in Fig. 2.43. Design the springs so that the unit can be used in an environment in which the vibratory frequency ranges from 0 to 5 Hz.

FIGURE 2.43 An electronic chassis mounted on
vibration isolators. (Courtesy of Titan SESCO.)

2.12 Find the natural frequency of the system shown in Fig. 2.44 with and without the springs k_1 and k_2 in the middle of the elastic beam.

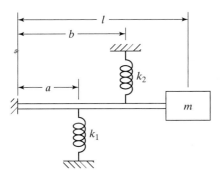

FIGURE 2.44

2.13 Find the natural frequency of the pulley system shown in Fig. 2.45 by neglecting the friction and the masses of the pulleys.

2.14 A weight W is supported by three frictionless and massless pulleys and a spring of stiffness k, as shown in Fig. 2.46. Find the natural frequency of vibration of weight W for small oscillations.

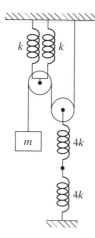

FIGURE 2.45

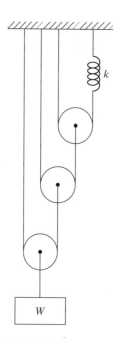

FIGURE 2.46

2.15 A rigid block of mass M is mounted on four elastic supports, as shown in Fig. 2.47. A mass m drops from a height l and adheres to the rigid block without rebounding. If the spring constant of each elastic support is k, find the natural frequency of vibration of the system (a) without the mass m, and (b) with the mass m. Also find the resulting motion of the system in case (b).

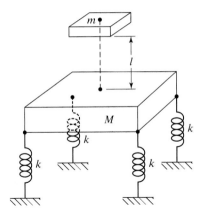

FIGURE 2.47

2.16 A sledgehammer strikes an anvil with a velocity of 50 ft./sec (Fig. 2.48). The hammer and the anvil weigh 12 lb and 100 lb, respectively. The anvil is supported on four springs, each of stiffness $k = 100$ lb/in. Find the resulting motion of the anvil (a) if the hammer remains in contact with the anvil and (b) if the hammer does not remain in contact with the anvil after the initial impact.

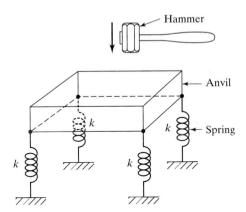

FIGURE 2.48

2.17 Derive the expression for the natural frequency of the system shown in Fig. 2.49. Note that the load W is applied at the tip of beam 1 and midpoint of beam 2.

2.18 A heavy machine weighing 9,810 N is being lowered vertically down by a winch at a uniform velocity of 2 m/s. The steel cable supporting the machine has a diameter of 0.01 m. The winch is suddenly stopped when the steel cable's length is 20 m. Find the period and amplitude of the ensuing vibration of the machine.

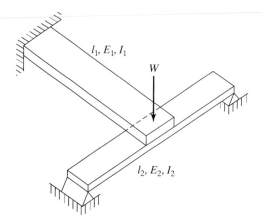

FIGURE 2.49

2.19 The natural frequency of a spring-mass system is found to be 2 Hz. When an additional mass of 1 kg is added to the original mass m, the natural frequency is reduced to 1 Hz. Find the spring constant k and the mass m.

2.20 An electrical switch gear is supported by a crane through a steel cable of length 4 m and diameter 0.01 m (Fig. 2.50). If the natural time period of axial vibration of the switch gear is found to be 0.1 s, find the mass of the switch gear.

FIGURE 2.50   (Photo courtesy of the Institution of Electrical Engineers.)

2.21 Four weightless rigid links and a spring are arranged to support a weight W in two different ways, as shown in Fig. 2.51. Determine the natural frequencies of vibration of the two arrangements.

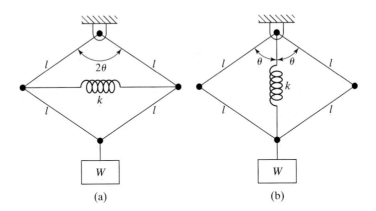

(a)   (b)

FIGURE 2.51

2.22 A scissors jack is used to lift a load W. The links of the jack are rigid and the collars can slide freely on the shaft against the springs of stiffnesses k_1 and k_2 (see Fig. 2.52). Find the natural frequency of vibration of the weight in the vertical direction.

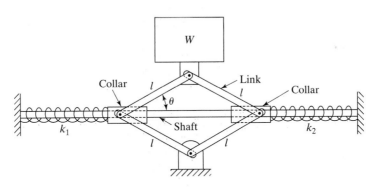

FIGURE 2.52

2.23 A weight is suspended using six rigid links and two springs in two different ways, as shown in Fig. 2.53. Find the natural frequencies of vibration of the two arrangements.

2.24 Figure 2.54 shows a small mass m restrained by four linearly elastic springs, each of which has an unstretched length l, and an angle of orientation of 45° with respect to the x-axis. Determine the equation of motion for small displacements of the mass in the x direction.

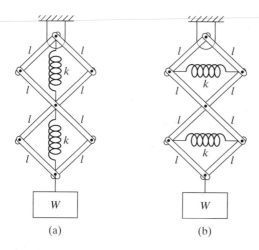

(a)   (b)

FIGURE 2.53

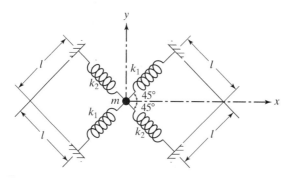

FIGURE 2.54

2.25 A mass m is supported by two sets of springs oriented at 30° and 120° with respect to the X axis, as shown in Fig. 2.55. A third pair of springs, each with a stiffness of k_3 is to be designed so as to make the system have a constant natural frequency while vibrating in any direction x. Determine the necessary spring stiffness k_3 and the orientation of the springs with respect to the X axis.

2.26 A mass m is attached to a cord that is under a tension T, as shown in Fig. 2.56. Assuming that T remains unchanged when the mass is displaced normal to the cord, (a) write the differential equation of motion for small transverse vibrations and (b) find the natural frequency of vibration.

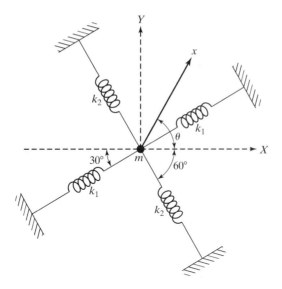

FIGURE 2.55

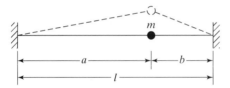

FIGURE 2.56

2.27 A bungee jumper weighing 160 lb ties one end of an elastic rope of length 200 ft. and stiffness 10 lb/in. to bridge and the other end to himself and jumps from the bridge (Fig. 2.57). Assuming the bridge to be rigid, determine the vibratory motion of the jumper about his static equilibrium position.

2.28 An acrobat weighing 120 lb walks on a tightrope, as shown in Fig. 2.58. If the natural frequency of vibration in the given position, in vertical direction, is 10 rad/sec, find the tension in the rope.

2.29 The schematic diagram of a centrifugal governor is shown in Fig. 2.59. The length of each rod is l, the mass of each ball is m and the free length of the spring is h. If the shaft speed is ω, determine the equilibrium position and the frequency for small oscillations about this position.

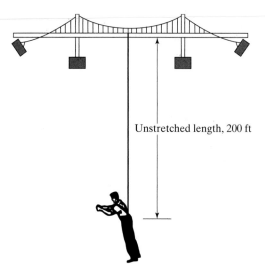

Unstretched length, 200 ft

FIGURE 2.57

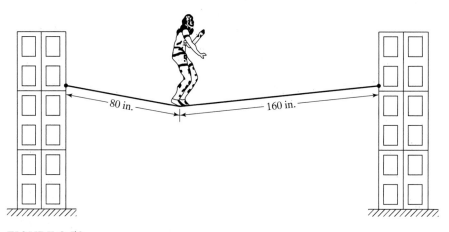

80 in.

160 in.

FIGURE 2.58

2.30 In the Hartnell governor shown in Fig. 2.60, the stiffness of the spring is 10^4 N/m and the weight of each ball is 25 N. The length of the ball arm is 20 cm, and that of the sleeve arm is 12 cm. The distance between the axis of rotation and the pivot of the bell crank lever is 16 cm. The spring is compressed by 1 cm when the ball arm is vertical. Find (a) the speed of the governor at which the ball arm remains vertical and (b) the natural frequency of vibration for small displacements about the vertical position of the ball arms.

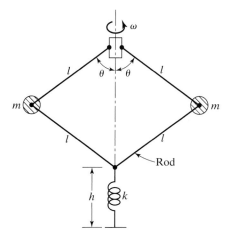

FIGURE 2.59

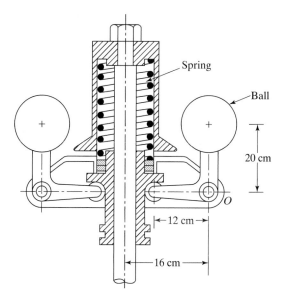

FIGURE 2.60   Hartnell governor.

2.31 A square platform *PQRS* and a car that it is supporting have a combined mass of *M*. The platform is suspended by four elastic wires from a fixed point *O*, as indicated in Fig. 2.61. The vertical distance between the point of suspension *O* and the horizontal equilibrium position of the platform is *h*. If the side of the platform is *a* and the stiffness of each wire is *k*, determine the period of vertical vibration of the platform.

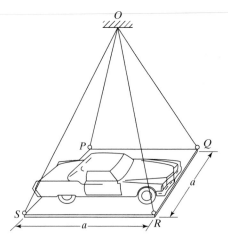

FIGURE 2.61

2.32 The inclined manometer, shown in Fig. 2.62, is used to measure pressure. If the total length of mercury in the tube is L, find an expression for the natural frequency of oscillation of the mercury.

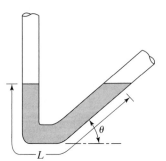

FIGURE 2.62

2.33 The crate, of mass 250 kg, hanging from a helicopter (shown in Fig. 2.63a) can be modeled as shown in Fig. 2.63(b). The rotor blades of the helicopter rotate at 300 rpm. Find the diameter of the steel cables so that the natural frequency of vibration of the crate is at least twice the frequency of the rotor blades.

2.34 A pressure vessel head is supported by a set of steel cables of length 2 m as shown in Fig. 2.64. The time period of axial vibration (in vertical direction) is found to vary from 5 s to 4.0825 s when an additional mass of 5,000 kg is added to the pressure vessel head. Determine the equivalent cross-sectional area of the cables and the mass of the pressure vessel head.

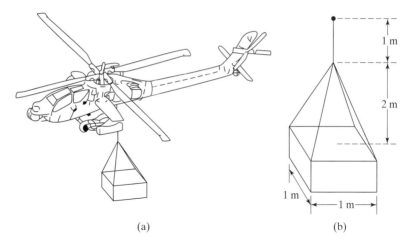

(a)   (b)

FIGURE 2.63

FIGURE 2.64   (Photo courtesy of CBI Industries Inc.)

2.35 A flywheel is mounted on a vertical shaft, as shown in Fig. 2.65. The shaft has a diameter d and length l and is fixed at both ends. The flywheel has a weight of W and a radius of gyration of r. Find the natural frequency of the longitudinal, the transverse, and the torsional vibration of the system.

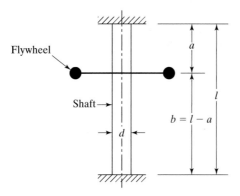

FIGURE 2.65

2.36 A TV antenna tower is braced by four cables, as shown in Fig. 2.66. Each cable is under tension and is made of steel with a cross-sectional area of 0.5 in.^{2}. The antenna tower can be modeled as a steel beam of square section of side 1 in. for estimating its mass and stiffness. Find the tower's natural frequency of bending vibration about the y-axis.

2.37 Figure 2.67(a) shows a steel traffic sign, of thickness $\frac{1}{8}$ in. fixed to a steel post. The post is 72 in. high with a cross section 2 in. × 1/4 in., and it can undergo torsional vibration (about the z-axis) or bending vibration (either in the zx-plane or the yz-plane). Determine the mode of vibration of the post in a storm during which the wind velocity has a frequency component of 1.25 Hz.

Hints:

1. Neglect the weight of the post in finding the natural frequencies of vibration.

2. Torsional stiffness of a shaft with a rectangular section (see Fig. 2.67b) is given by

$$k_t = 5.33 \frac{ab^3 G}{l} \left[1 - 0.63 \frac{b}{a} \left(1 - \frac{b^4}{12a^4} \right) \right]$$

where G is the shear modulus.

3. Mass moment of inertia of a rectangular block about axis OO (see Fig. 2.67c) is given by

$$I_{OO} = \frac{\rho l}{3} (b^3 h + h^3 b)$$

where ρ is the density of the block.

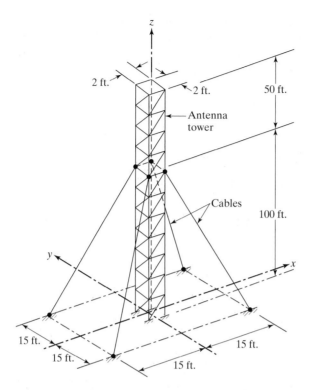

FIGURE 2.66

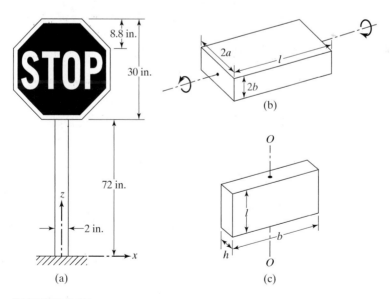

FIGURE 2.67

2.38 A building frame is modeled by four identical steel columns, each of weight w, and a rigid floor of weight W, as shown in Fig. 2.68. The columns are fixed at the ground and have a bending rigidity of EI each. Determine the natural frequency of horizontal vibration of the building frame by assuming the connection between the floor and the columns to be (a) pivoted as shown in Fig. 2.68(a) and (b) fixed against rotation as shown in Fig. 2.68(b). Include the effect of self-weights of the columns.

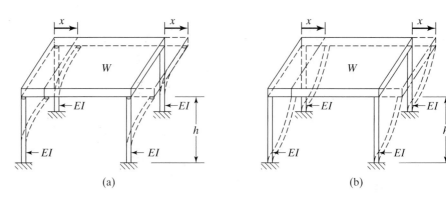

FIGURE 2.68

2.39 A pick-and-place robot arm, shown in Fig. 2.69, carries an object weighing 10 lb. Find the natural frequency of the robot arm in the axial direction for the following data: $l_1 = 12$ in., $l_2 = 10$ in., $l_3 = 8$ in.; $E_1 = E_2 = E_3 = 10^7$ psi; $D_1 = 2$ in., $D_2 = 1.5$ in., $D_3 = 1$ in.; $d_1 = 1.75$ in., $d_2 = 1.25$ in., $d_3 = 0.75$ in.

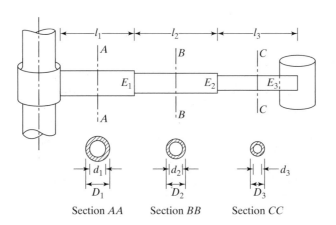

FIGURE 2.69

2.40 A helical spring of stiffness k is cut into two halves and a mass m is connected to the two halves as shown in Fig. 2.70(a). The natural time period of this system is found to be 0.5 s. If an identical spring is cut so that one part is one-fourth and the other part three-fourths of the original length, and the mass m is connected to the two parts as shown in Fig. 2.70(b), what would be the natural period of the system?

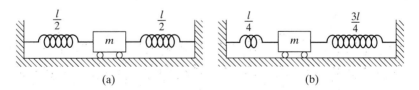

(a)   (b)

FIGURE 2.70

2.41* Figure 2.71 shows a metal block supported on two identical cylindrical rollers rotating in opposite directions at the same angular speed. When the center of gravity of the block is initially displaced by a distance x, the block will be set into simple harmonic motion. If the frequency of motion of the block is found to be ω, determine the coefficient of friction between the block and the rollers.

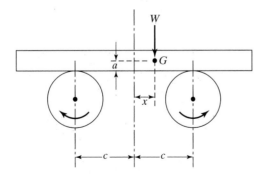

FIGURE 2.71

2.42* If two identical springs of stiffness k each are attached to the metal block of Problem 2.41 as shown in Fig. 2.72, determine the coefficient of friction between the block and the rollers.

2.43 An electromagnet weighing 3,000 lb is at rest while holding an automobile of weight 2,000 lb in a junkyard. The electric current is turned off, and the automobile is dropped. Assuming that the crane and the supporting cable have an equivalent spring constant of 10,000 lb/in., find the following: (a) the natural frequency of vibration of the electromagnet, (b) the resulting motion of the electromagnet, and (c) the maximum tension developed in the cable during the motion.

---

*The asterisk denotes a design problem or a problem with no unique answer.

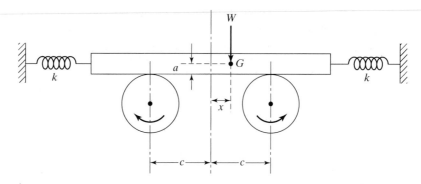

FIGURE 2.72

2.44 Derive the equation of motion of the system shown in Fig. 2.73, using the following methods: (a) Newton's second law of motion, (b) D'Alembert's principle, (c) principle of virtual work, and (d) principle of conservation of energy.

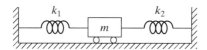

FIGURE 2.73

2.45– Draw the free-body diagram and derive the equation of motion using Newton's second law of
2.46   motion for each of the systems shown in Figs. 2.74 and 2.75.

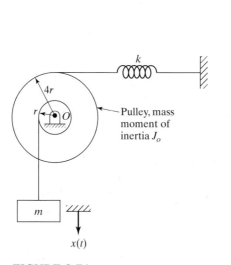

FIGURE 2.74

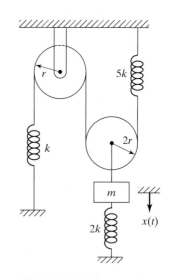

FIGURE 2.75

2.47– Derive the equation of motion using the principle of conservation of energy for each of the
2.48   systems shown in Figs. 2.74 and 2.75.

2.49 A steel beam of length 1 m carries a mass of 50 kg at its free end, as shown in Fig. 2.76. Find
the natural frequency of transverse vibration of the mass by modeling it as a single degree of
freedom system.

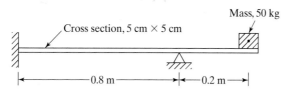

Mass, 50 kg

Cross section, 5 cm × 5 cm

0.8 m   0.2 m

FIGURE 2.76

2.50 A steel beam of length 1 m carries a mass of 50 kg at its free end, as shown in Fig. 2.77. Find
the natural frequency of transverse vibration of the system by modeling it as a single degree
of freedom system.

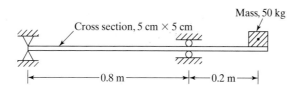

Mass, 50 kg

Cross section, 5 cm × 5 cm

0.8 m   0.2 m

FIGURE 2.77

2.51 Determine the displacement, velocity, and acceleration of the mass of a spring-mass system
with $k = 500$ N/m, $m = 2$ kg, $x_0 = 0.1$ m, and $\dot{x}_0 = 5$ m/s.

2.52 Determine the displacement (x), velocity ($\dot{x}$), and acceleration ($\ddot{x}$) of a spring-mass system
with $\omega_n = 10$ rad/s for the initial conditions $x_0 = 0.05$ m and $\dot{x}_0 = 1$ m/s. Plot $x(t)$, $\dot{x}(t)$,
and $\ddot{x}(t)$ from $t = 0$ to 5 s.

2.53 The free vibration response of a spring-mass system is observed to have a frequency of 2
rad/s, an amplitude of 10 mm, and a phase shift of 1 rad from $t = 0$. Determine the initial
conditions that caused the free vibration. Assume the damping ratio of the system as 0.1.

2.54 An automobile is found to have a natural frequency of 20 rad/s without passengers and 17.32
rad/s with passengers of mass 500 kg. Find the mass and stiffness of the automobile by treating it as a single degree of freedom system.

2.55 A spring-mass system with mass 2 kg and stiffness 3,200 N/m has an initial displacement of
$x_0 = 0$. What is the maximum initial velocity that can be given to the mass without the amplitude of free vibration exceeding a value of 0.1 m?

2.56 A helical spring, made of music wire of diameter d, has a mean coil diameter (D) of 0.5625 in. and N active coils (turns). It is found to have a frequency of vibration (f) of 193 Hz and a spring rate k of 26.4 lb/in. Determine the wire diameter d and the number of coils N, assuming the shear modulus G is 11.5×10^6 psi and weight density ρ is 0.282 lb/in.^{3}. The spring rate (k) and frequency (f) are given by

$$k = \frac{d^4 G}{8 D^3 N}, f = \frac{1}{2}\sqrt{\frac{kg}{W}}$$

where W is the weight of the helical spring and g is the acceleration due to gravity.

2.57 Solve Problem 2.56 if the material of the helical spring is changed from music wire to aluminum with $G = 4 \times 10^6$ psi and $\rho = 0.1$ lb/in.^{3}.

2.58 A steel cantilever beam is used to carry a machine at its free end. To save weight, it is proposed to replace the steel beam by an aluminum beam of identical dimensions. Find the expected change in the natural frequency of the beam-machine system.

2.59 A simple pendulum is set into oscillation from its rest position by giving it an angular velocity of 1 rad/s. It is found to oscillate with an amplitude of 0.5 rad. Find the natural frequency and length of the pendulum.

2.60 A pulley 250 mm in diameter drives a second pulley 1,000 mm in diameter by means of a belt (see Fig. 2.78). The moment of inertia of the driven pulley is 0.2 kg-m^2. The belt connecting these pulleys is represented by two springs, each of stiffness k. For what value of k will the natural frequency be 6 Hz?

2.61 Derive an expression for the natural frequency of the simple pendulum shown in Fig. 1.11. Determine the period of oscillation of a simple pendulum having a mass $m = 5$ kg and a length $l = 0.5$ m.

2.62 A mass m is attached at the end of a bar of negligible mass and is made to vibrate in three different configurations, as indicated in Fig. 2.79. Find the configuration corresponding to the highest natural frequency.

2.63 Figure 2.80 shows a spacecraft with four solar panels. Each panel has the dimensions 5 ft. $\times$ 3 ft. $\times$ 1 ft. with a weight density of 0.1 lb/in.^{3}, and is connected to the body of the spacecraft by aluminum rods of length 12 in. and diameter 1 in. Assuming that the body of the spacecraft is very large (rigid), determine the natural frequency of vibration of each panel about the axis of the connecting aluminum rod.

2.64 One of the blades of an electric fan is removed (as shown by dotted lines in Fig. 2.81). The steel shaft AB, on which the blades are mounted, is equivalent to a uniform shaft of diameter 1 in. and length 6 in. Each blade can be modeled as a uniform slender rod of weight 2 lb and length 12 in. Determine the natural frequency of vibration of the remaining three blades about the y-axis.

2.65 A heavy ring of mass moment of inertia 1.0 kg-m^2 is attached at the end of a two-layered hollow shaft of length 2 m (Fig. 2.82). If the two layers of the shaft are made of steel and brass, determine the natural time period of torsional vibration of the heavy ring.

2.66 Find the natural frequency of the pendulum shown in Fig. 2.83 when the mass of the connecting bar is not negligible compared to the mass of the pendulum bob.

2.67 A steel shaft of 0.05 m diameter and 2 m length is fixed at one end and carries at the other end a steel disc of 1 m diameter and 0.1 m thickness, as shown in Fig. 2.14. Find the system's natural frequency of torsional vibration.

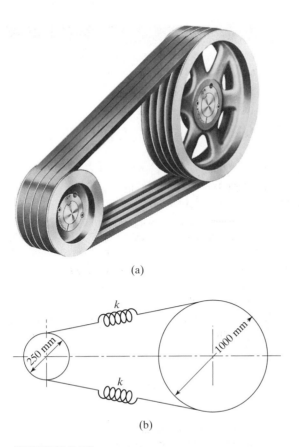

(a)

(b)

FIGURE 2.78 (Photo courtesy of Reliance Electric Company.)

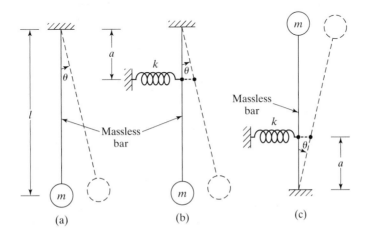

(a)   (b)   (c)

FIGURE 2.79

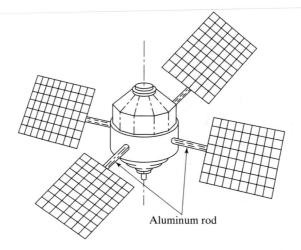

Aluminum rod

FIGURE 2.80

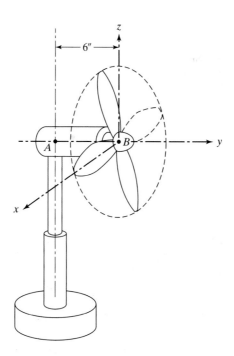

FIGURE 2.81

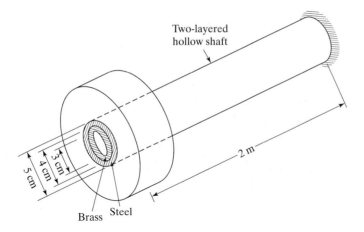

Two-layered
hollow shaft

2 m

3 cm

4 cm

5 cm

Brass Steel

FIGURE 2.82

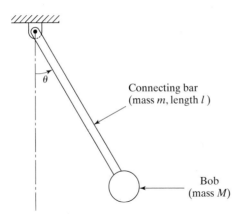

θ

Connecting bar
(mass m, length l)

Bob
(mass M)

FIGURE 2.83

2.68 A uniform slender rod of mass m and length l is hinged at point A and is attached to four linear springs and one torsional spring, as shown in Fig. 2.84. Find the natural frequency of the system if $k = 2000$ N/m, $k_t = 1000$ N-m/rad, $m = 10$ kg, and $l = 5$ m.

2.69 A cylinder of mass m and mass moment of inertia J_0 is free to roll without slipping but is restrained by two springs of stiffnesses k_1 and k_2, as shown in Fig. 2.85. Find its natural frequency of vibration. Also find the value of a that maximizes the natural frequency of vibration.

2.70 If the pendulum of Problem 2.61 is placed in a rocket moving vertically with an acceleration of 5 m/s^2, what will be its period of oscillation?

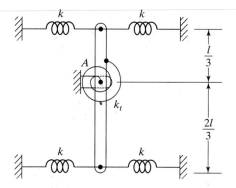

FIGURE 2.84

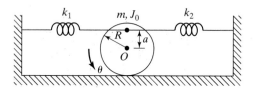

FIGURE 2.85

2.71 Find the equation of motion of the uniform rigid bar OA of length l and mass m shown in Fig. 2.86. Also find its natural frequency.

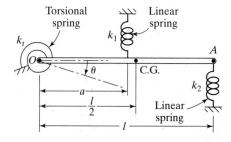

FIGURE 2.86

2.72 A uniform circular disc is pivoted at point O, as shown in Fig. 2.87. Find the natural frequency of the system. Also find the maximum frequency of the system by varying the value of b.

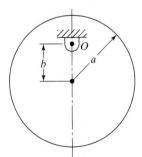

FIGURE 2.87

2.73 Derive the equation of motion of the system shown in Fig. 2.88, using the following methods: (a) Newton's second law of motion, (b) D'Alembert's principle, and (c) principle of virtual work.

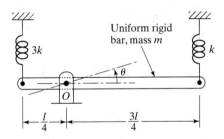

FIGURE 2.88

2.74 Determine the effect of self-weight on the natural frequency of vibration of the pinned-pinned beam shown in Fig. 2.89.

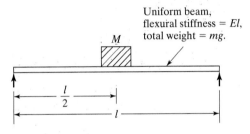

FIGURE 2.89

2.75 Use Rayleigh's method to solve Problem 2.7.

2.76 Use Rayleigh's method to solve Problem 2.13.

2.77 Find the natural frequency of the system shown in Fig. 2.54.

2.78 Use Rayleigh's method to solve Problem 2.26.

2.79 Use Rayleigh's method to solve Problem 2.68.

2.80 Use Rayleigh's method to solve Problem 2.71.

2.81 A wooden rectangular prism of density ρ_w, height h, and cross-section $a \times b$ is initially depressed in an oil tub and made to vibrate freely in the vertical direction (see Fig. 2.90). Use Rayleigh's method to find the natural frequency of vibration of the prism. Assume the density of oil is ρ_0. If the rectangular prism is replaced by a uniform circular cylinder of radius r, height h, and density ρ_w, will there be any change in the natural frequency?

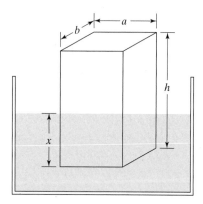

FIGURE 2.90

2.82 Use the energy method to find the natural frequency of the system shown in Fig. 2.85.

2.83 Use the energy method to find the natural frequency of vibration of the system shown in Fig. 2.74.

2.84 A simple pendulum is found to vibrate at a frequency of 0.5 Hz in a vacuum and 0.45 Hz in a viscous fluid medium. Find the damping constant, assuming the mass of the bob of the pendulum is 1 kg.

2.85 The ratio of successive amplitudes of a viscously damped single degree of freedom system is found to be 18:1. Determine the ratio of successive amplitudes if the amount of damping is (a) doubled, and (b) halved.

2.86 Assuming that the phase angle is zero, show that the response $x(t)$ of an underdamped single degree of freedom system reaches a maximum value when

$$\sin \omega_d t = \sqrt{1 - \zeta^2}$$

and a minimum value when

$$\sin \omega_d t = -\sqrt{1 - \zeta^2}$$

Also show that the equations of the curves passing through the maximum and minimum values of $x(t)$ are given, respectively, by

$$x = \sqrt{1 - \zeta^2} X e^{-\zeta \omega_n t}$$

and

$$x = -\sqrt{1 - \zeta^2} X e^{-\zeta \omega_n t}$$

2.87 Derive an expression for the time at which the response of a critically damped system will attain its maximum value. Also find the expression for the maximum response.

2.88 A shock absorber is to be designed to limit its overshoot to 15 percent of its initial displacement when released. Find the damping ratio ζ_0 required. What will be the overshoot if ζ is made equal to (a) $\frac{3}{4}\zeta_0$, and (b) $\frac{5}{4}\zeta_0$?

2.89 The free vibration responses of an electric motor of weight 500 N mounted on different types of foundations are shown in Figs. 2.91 (a) and (b). Identify the following in each case: (i) the nature of damping provided by the foundation, (ii) the spring constant and damping coefficient of the foundation, and (iii) the undamped and damped natural frequencies of the electric motor.

2.90 For a spring-mass-damper system, $m = 50$ kg and $k = 5,000$ N/m. Find the following: (a) critical damping constant c_c, (b) damped natural frequency when $c = c_c/2$, and (c) logarithmic decrement.

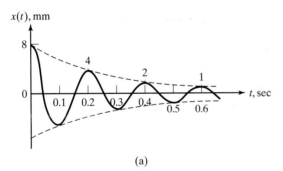

(a)

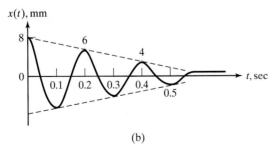

(b)

FIGURE 2.91

2.91   A railroad car of mass 2,000 kg traveling at a velocity $v = 10$ m/s is stopped at the end of the tracks by a spring-damper system, as shown in Fig. 2.92. If the stiffness of the spring is $k = 40$ N/mm and the damping constant is $c = 20$ N-s/mm, determine (a) the maximum displacement of the car after engaging the springs and damper and (b) the time taken to reach the maximum displacement.

FIGURE 2.92

2.92   A torsional pendulum has a natural frequency of 200 cycles/min when vibrating in a vacuum. The mass moment of inertia of the disc is 0.2 kg-m^2. It is then immersed in oil and its natural frequency is found to be 180 cycles/min. Determine the damping constant. If the disc, when placed in oil, is given an initial displacement of 2°, find its displacement at the end of the first cycle.

2.93   A boy riding a bicycle can be modeled as a spring-mass-damper system with an equivalent weight, stiffness, and damping constant of 800 N, 50,000 N/m, and 1,000 N-s/m, respectively. The differential setting of the concrete blocks on the road caused the level surface to decrease suddenly as indicated in Fig. 2.93. If the speed of the bicycle is 5 m/s (18 km/hr), determine the displacement of the boy in the vertical direction. Assume that the bicycle is free of vertical vibration before encountering the step change in the vertical displacement.

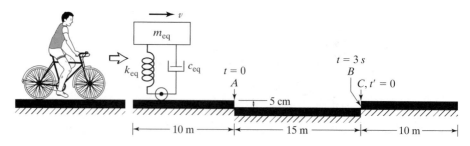

FIGURE 2.93

2.94 A wooden rectangular prism of weight 20 lb, height 3 ft., and cross section 1 ft. $\times$ 2 ft. floats and remains vertical in a tub of oil. The frictional resistance of the oil can be assumed to be equivalent to a viscous damping coefficient ζ. When the prism is depressed by a distance of 6 in. from its equilibrium and released, it is found to reach a depth of 5.5 in. at the end of its first cycle of oscillation. Determine the value of the damping coefficient of the oil.

2.95 A body vibrating with viscous damping makes five complete oscillations per second, and in 50 cycles its amplitude diminishes to 10 percent. Determine the logarithmic decrement and the damping ratio. In what proportion will the period of vibration be decreased if damping is removed?

2.96 The maximum permissible recoil distance of a gun is specified as 0.5 m. If the initial recoil velocity is to be between 8 m/s and 10 m/s, find the mass of the gun and the spring stiffness of the recoil mechanism. Assume that a critically damped dashpot is used in the recoil mechanism and the mass of the gun has to be at least 500 kg.

2.97 A viscously damped system has a stiffness of 5,000 N/m, critical damping constant of 0.2 N-s/mm, and a logarithmic decrement of 2.0. If the system is given an initial velocity of 1 m/s, determine the maximum displacement of the system.

2.98 Explain why an overdamped system never passes through the static equilibrium position when it is given (a) an initial displacement only and (b) an initial velocity only.

2.99– Derive the equation of motion and find the natural frequency of vibration of each of the sys
2.101 tems shown in Figs. 2.94 to 2.96.

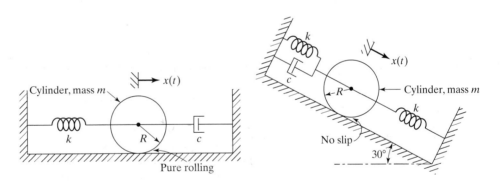

FIGURE 2.94   FIGURE 2.95

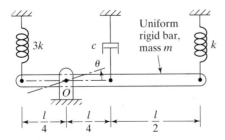

FIGURE 2.96

2.102–
2.104 Using the principle of virtual work, derive the equation of motion for each of the systems shown in Figs. 2.94 to 2.96.

2.105 A wooden rectangular prism of cross section 40 cm × 60 cm, height 120 cm, and mass 40 kg floats in a fluid as shown in Fig. 2.90. When disturbed, it is observed to vibrate freely with a natural period of 0.5 s. Determine the density of the fluid.

2.106 The system shown in Fig. 2.97 has a natural frequency of 5 Hz for the following data: $m = 10$ kg, $J_0 = 5$ kg-m^2, $r_1 = 10$ cm, $r_2 = 25$ cm. When the system is disturbed by giving it an initial displacement, the amplitude of free vibration is reduced by 80 percent in 10 cycles. Determine the values of k and c.

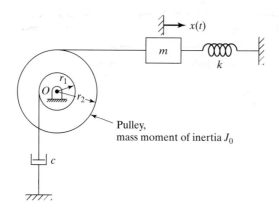

FIGURE 2.97

2.107 The rotor of a dial indicator is connected to a torsional spring and a torsional viscous damper to form a single degree of freedom torsional system. The scale is graduated in equal divisions, and the equilibrium position of the rotor corresponds to zero on the scale. When a torque of 2×10^{-3} N-m is applied, the angular displacement of the rotor is found to be 50° with the pointer showing 80 divisions on the scale. When the rotor is released from this position, the pointer swings first to −20 divisions in one second and then to 5 divisions in another second. Find (a) the mass moment of inertia of the rotor, (b) the undamped natural time period of the rotor, (c) the torsional damping constant, and (d) the torsional spring stiffness.

2.108 Determine the values of ζ and ω_d for the following viscously damped systems:

    a. $m = 10$ kg, $c = 150$ N-s/m, $k = 1000$ N/m
    b. $m = 10$ kg, $c = 200$ N-s/m, $k = 1000$ N/m
    c. $m = 10$ kg, $c = 250$ N-s/m, $k = 1000$ N/m

2.109 Determine the free vibration response of the viscously damped systems described in Problem 2.108 when $x_0 = 0.1$ m and $\dot{x}_0 = 10$ m/s.

2.110 Find the energy dissipated during a cycle of simple harmonic motion given by $x(t) = 0.2 \sin \omega_d t$ m by a viscously damped single degree of freedom system with the following parameters:

    a. $m = 10$ kg, $c = 50$ N-s/m, $k = 1000$ N/m
    b. $m = 10$ kg, $c = 150$ N-s/m, $k = 1000$ N/m

2.111 A single degree of freedom system consists of a mass of 20 kg and a spring of stiffness 4,000 N/m. The amplitudes of successive cycles are found to be 50, 45, 40, 35, ... mm. Determine the nature and magnitude of the damping force and the frequency of the damped vibration.

2.112 A mass of 20 kg slides back and forth on a dry surface due to the action of a spring having a stiffness of 10 N/mm. After four complete cycles, the amplitude has been found to be 100 mm. What is the average coefficient of friction between the two surfaces if the original amplitude was 150 mm? How much time has elapsed during the four cycles?

2.113 A 10-kg mass is connected to a spring of stiffness 3,000 N/m and is released after giving an initial displacement of 100 mm. Assuming that the mass moves on a horizontal surface, as shown in Fig. 2.33(a), determine the position at which the mass comes to rest. Assume the coefficient of friction between the mass and the surface to be 0.12.

2.114 A weight of 25 N is suspended from a spring that has a stiffness of 1,000 N/m. The weight vibrates in the vertical direction under a constant damping force. When the weight is initially pulled downward a distance of 10 cm from its static equilibrium position and released, it comes to rest after exactly two complete cycles. Find the magnitude of the damping force.

2.115 A mass of 20 kg is suspended from a spring of stiffness 10,000 N/m. The vertical motion of the mass is subject to Coulomb friction of magnitude 50 N. If the spring is initially displaced downward by 5 cm from its static equilibrium position, determine (a) the number of half cycles elapsed before the mass comes to rest, (b) the time elapsed before the mass comes to rest, and (c) the final extension of the spring.

2.116 The Charpy impact test is a dynamic test in which a specimen is struck and broken by a pendulum (or hammer) and the energy absorbed in breaking the specimen is measured. The energy values serve as a useful guide for comparing the impact strengths of different materials. As shown in Fig. 2.98, the pendulum is suspended from a shaft, is released from a particular position, and is allowed to fall and break the specimen. If the pendulum is made to oscillate freely (with no specimen), find (a) an expression for the decrease in the angle of swing for each cycle caused by friction, (b) the solution for $\theta(t)$ if the pendulum is released from an angle θ_0, and (c) the number of cycles after which the motion ceases. Assume the mass of the pendulum is m and the coefficient of friction between the shaft and the bearing of the pendulum is μ.

2.117 Find the equivalent viscous damping constant for Coulomb damping for sinusoidal vibration.

2.118 A single degree of freedom system consists of a mass, a spring, and a damper in which both dry friction and viscous damping act simultaneously. The free vibration amplitude is found to decrease by 1 percent per cycle when the amplitude is 20 mm and by 2 percent per cycle when the amplitude is 10 mm. Find the value of $(\mu N/k)$ for the dry friction component of the damping.

2.119 A metal block, placed on a rough surface, is attached to a spring and is given an initial displacement of 10 cm from its equilibrium position. It is found that the natural time period of motion is 1.0 s and that the amplitude reduces by 0.5 cm in each cycle. Find (a) the kinetic coefficient of friction between the metal block and the surface and (b) the number of cycles of motion executed by the block before it stops.

2.120 The mass of a spring-mass system with $k = 10,000$ N/m and $m = 5$ kg is made to vibrate on a rough surface. If the friction force is $F = 20$ N and the amplitude of the mass is observed to decrease by 50 mm in 10 cycles, determine the time taken to complete the 10 cycles.

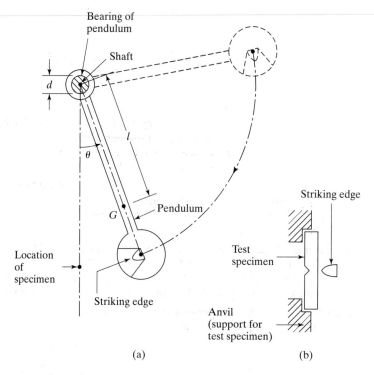

FIGURE 2.98

2.121 The mass of a spring-mass system vibrates on a dry surface inclined at 30° to the horizontal as shown in Fig. 2.99.

    a. Derive the equation of motion.

    b. Find the response of the system for the following data:

$$m = 20 \text{ kg}, \; k = 1000 \text{ N/m}, \; \mu = 0.1, \; x_0 = 0.1 \text{ m}, \; \dot{x}_0 = 5 \text{ m/s}.$$

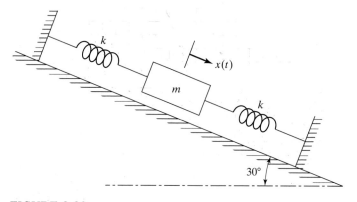

FIGURE 2.99

2.122 The mass of a spring-mass system is initially displaced by 10 cm from its unstressed position by applying a force of 25 N, which is equal to five times the weight of the mass. If the mass is released from this position, how long will the mass vibrate and at what distance will it stop from the unstressed position? Assume a coefficient of friction of 0.2.

2.123 The experimentally observed force-deflection curve for a composite structure is shown in Fig. 2.100. Find the hysteresis damping constant, the logarithmic decrement, and the equivalent viscous damping ratio corresponding to this curve.

2.124 A panel made of fiber-reinforced composite material is observed to behave as a single degree of freedom system of mass 1 kg and stiffness 2 N/m. The ratio of successive amplitudes is found to be 1.1. Determine the value of the hysteresis damping constant β, the equivalent viscous damping constant c_{eq}, and the energy loss per cycle for an amplitude of 10 mm.

2.125 A built-up cantilever beam having a bending stiffness of 200 N/m supports a mass of 2 kg at its free end. The mass is displaced initially by 30 mm and released. If the amplitude is found to be 20 mm after 100 cycles of motion, estimate the hysteresis damping constant β of the beam.

2.126 A mass of 5 kg is attached to the top of a helical spring, and the system is made to vibrate by giving to the mass an initial deflection of 25 mm. The amplitude of the mass is found to reduce to 10 mm after 100 cycles of vibration. Assuming a spring rate of 200 N/m for the helical spring, find the value of the hysteretic damping coefficient (h) of the spring.

2.127 Find the free vibration response of a spring-mass system subject to Coulomb damping using MATLAB for the following data:

$$m = 5 \text{ kg}, k = 100 \text{ N/m}, \mu = 0.5, x_0 = 0.4 \text{ m}, \dot{x}_0 = 0.$$

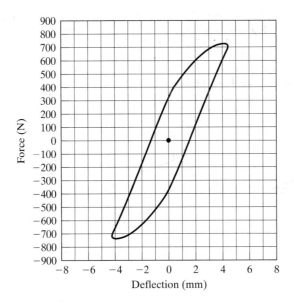

FIGURE 2.100

2.128   Plot the response of a critically damped system (Eq. 2.80) for the following data using MATLAB:

    a. $x_0 = 10$ mm, 50 mm, 100 mm; $\dot{x}_0 = 0$, $\omega_n = 10$ rad/s.
    b. $x_0 = 0$, $\dot{x}_0 = 10$ mm/s, 50 mm/s, 100 mm/s; $\omega_n = 10$ rad/s.

2.129   Plot Eq. (2.81) as well as each of the two terms of Eq. (2.81) as functions of t using MATLAB for the following data:

    $\omega_n = 10$ rad/s, $\zeta = 2.0$, $x_0 = 20$ mm, $\dot{x}_0 = 50$ mm/s.

2.130–   Using the MATLAB Program2.m, plot the free vibration response of a viscously damped system
2.133   with $m = 4$ kg, $k = 2{,}500$ N/m, $x_0 = 100$ mm, $\dot{x}_0 = -10$ m/s, $\Delta t = 0.01$ s, $n = 50$ for the following values of the damping constant:

    a. $c = 0$
    b. $c = 100$ N-s/m
    c. $c = 200$ N-s/m
    d. $c = 400$ N-s/m

2.134   Find the response of the system described in Problem 2.121 using MATLAB.

2.135–
2.138   Solve Problems 2.130–2.133 using the C++ program given in Section 2.10.

2.139–
2.142   Solve Problems 2.130–2.133 using the Fortran program given in Section 2.11.

DESIGN PROJECTS

2.143*   A water turbine of mass 1,000 kg and mass moment of inertia 500 kg-m^2 is mounted on a steel shaft, as shown in Fig. 2.101. The operational speed of the turbine is 2,400 rpm. Assuming the ends of the shaft to be fixed, find the values of l, a, and d, such that the natural frequency of vibration of the turbine in each of the axial, transverse, and circumferential directions is greater than the operational speed of the turbine.

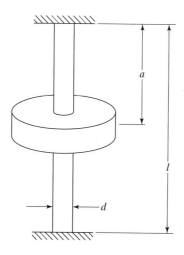

FIGURE 2.101

2.144* Design the columns for each of the building frames shown in Figs. 2.68(a) and (b) for minimum weight such that the natural frequency of vibration is greater than 50 Hz. The weight of the floor (W) is 4,000 lb and the length of the columns (l) is 96 in. Assume that the columns are made of steel and have a tubular cross section with outer diameter d and wall thickness t.

2.145* One end of a uniform rigid bar of mass m is connected to a wall by a hinge joint O, and the other end carries a concentrated mass M, as shown in Fig. 2.102. The bar rotates about the hinge point O against a torsional spring and a torsional damper. It is proposed to use this mechanism, in conjunction with a mechanical counter, to control entrance to an amusement park. Find the masses m and M, the stiffness of the torsional spring (k_t), and the damping force (F_d) necessary to satisfy the following specifications: (1) A viscous damper or a Coulomb damper can be used. (2) The bar has to return to within 5° of closing in less than 2 sec when released from an initial position of $\theta = 75°$.

Amusement park

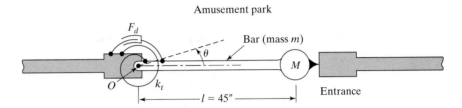

FIGURE 2.102

2.146* The lunar excursion module has been modeled as a mass supported by four symmetrically located legs, each of which can be approximated as a spring-damper system with negligible mass (see Fig. 2.103). Design the springs and dampers of the system in order to have the damped period of vibration between 1 s and 2 s.

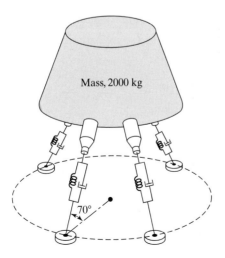

FIGURE 2.103

2.147* Consider the telescoping boom and cockpit of the firetruck shown in Fig. 2.12(a). Assume that the telescoping boom $PQRS$ is supported by a strut QT, as shown in Fig. 2.104. Determine the cross section of the strut QT so that the natural time period of vibration of the cockpit with the fireperson is equal to 1 s for the following data. Assume that each segment of the telescoping boom and the strut is hollow circular in cross section. In addition, assume that the strut acts as a spring that deforms only in the axial direction.

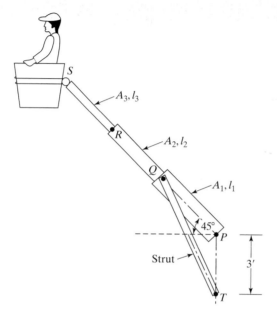

FIGURE 2.104

Data:
Lengths of segments: $PQ = 12$ ft., $QR = 10$ ft., $RS = 8$ ft., $TP = 3$ ft.
Young's modulus of the telescoping arm and strut $= 30 \times 10^6$ psi
Outer diameters of sections: $PQ = 2.0$ in., $QR = 1.5$ in., $RS = 1.0$ in.
Inner diameters of sections: $PQ = 1.75$ in., $QR = 1.25$ in., $RS = 0.75$ in.
Weight of the cockpit $= 100$ lb
Weight of fireperson $= 200$ lb

Charles Augustin de Coulomb (1736–1806) was a French military engineer and physicist. His early work on statics and mechanics was presented in 1779 in his great memoir *The Theory of Simple Machines*, which describes the effect of resistance and the so-called "Coulomb's law of proportionality" between friction and normal pressure. In 1784, he obtained the correct solution to the problem of the small oscillations of a body subjected to torsion. He is well known for his laws of force for electrostatic and magnetic charges. His name is remembered through the unit of electric charge. (Courtesy of *Applied Mechanics Reviews*.)

C H A P T E R 3

Harmonically Excited Vibration

3.1 Introduction

A mechanical or structural system is said to undergo forced vibration whenever external energy is supplied to the system during vibration. External energy can be supplied to the system through either an applied force or an imposed displacement excitation. The applied force or displacement excitation may be harmonic, nonharmonic but periodic, nonperiodic, or random in nature. The response of a system to a harmonic excitation is called *harmonic response*. The nonperiodic excitation may have a long or short duration. The response of a dynamic system to suddenly applied nonperiodic excitations is called *transient response*.

In this chapter, we shall consider the dynamic response of a single degree of freedom system under harmonic excitations of the form $F(t) = F_0 e^{i(\omega t + \phi)}$ or $F(t) = F_0 \cos(\omega t + \phi)$ or $F(t) = F_0 \sin(\omega t + \phi)$, where F_0 is the amplitude, ω is the frequency, and ϕ is the phase angle of the harmonic excitation. The value of ϕ depends on the value of $F(t)$ at $t = 0$ and is usually taken to be zero. Under a harmonic excitation, the response of the system will also be harmonic. If the frequency of excitation coincides with the natural frequency of the system, the response of the system will be very large. This condition, known as resonance, is to be avoided to prevent failure of the system. The vibration produced by an unbalanced rotating machine, the oscillations of a tall chimney due to vortex shedding in a steady wind, and the vertical motion of an automobile on a sinusoidal road surface are examples of harmonically excited vibration.

3.2 Equation of Motion

If a force $F(t)$ acts on a viscously damped spring-mass system as shown in Fig. 3.1, the equation of motion can be obtained using Newton's second law:

$$m\ddot{x} + c\dot{x} + kx = F(t) \tag{3.1}$$

Since this equation is nonhomogeneous, its general solution $x(t)$ is given by the sum of the homogeneous solution, $x_h(t)$, and the particular solution, $x_p(t)$. The homogeneous solution, which is the solution of the homogeneous equation

$$m\ddot{x} + c\dot{x} + kx = 0 \tag{3.2}$$

represents the free vibration of the system and was discussed in Chapter 2. As seen in Section 2.6.2, this free vibration dies out with time under each of the three possible conditions of damping (underdamping, critical damping, and overdamping) and under all possible initial conditions. Thus the general solution of Eq. (3.1) eventually reduces to the particular solution $x_p(t)$, which represents the steady-state vibration. The steady-state motion is present as long as the forcing function is present. The variations of homogeneous, particular, and general solutions with time for a typical case are shown in Fig. 3.2. It can be seen that $x_h(t)$ dies out and $x(t)$ becomes $x_p(t)$ after some time (τ in Fig. 3.2). The part of the motion that dies out due to damping (the free vibration part) is called *transient*. The rate at which the transient motion decays depends on the values of the system parameters k, c, and m. In this chapter, except in Section 3.3, we ignore the transient motion and derive only the particular solution of Eq. (3.1), which represents the steady-state response, under harmonic forcing functions.

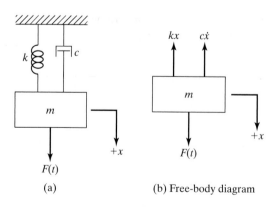

(a)   (b) Free-body diagram

FIGURE 3.1   A spring-mass-damper system.

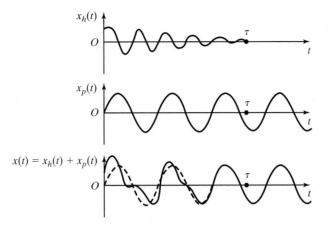

FIGURE 3.2   Homogenous, particular, and general solutions of Eq. (3.1) for an underdamped case.

3.3   Response of an Undamped System Under Harmonic Force

Before studying the response of a damped system, we consider an undamped system subjected to a harmonic force, for the sake of simplicity. If a force $F(t) = F_0 \cos \omega t$ acts on the mass m of an undamped system, the equation of motion, Eq. (3.1), reduces to

$$m\ddot{x} + kx = F_0 \cos \omega t \qquad (3.3)$$

The homogeneous solution of this equation is given by

$$x_h(t) = C_1 \cos \omega_n t + C_2 \sin \omega_n t \qquad (3.4)$$

where $\omega_n = (k/m)^{1/2}$ is the natural frequency of the system. Because the exciting force $F(t)$ is harmonic, the particular solution $x_p(t)$ is also harmonic and has the same frequency ω. Thus we assume a solution in the form

$$x_p(t) = X \cos \omega t \qquad (3.5)$$

where X is a constant that denotes the maximum amplitude of $x_p(t)$. By substituting Eq. (3.5) into Eq. (3.3) and solving for X, we obtain

$$X = \frac{F_0}{k - m\omega^2} = \frac{\delta_{st}}{1 - \left(\dfrac{\omega}{\omega_n}\right)^2} \qquad (3.6)$$

where $\delta_{st} = F_0/k$ denotes the deflection of the mass under a force F_0 and is sometimes called *static deflection* because F_0 is a constant (static) force. Thus the total solution of Eq. (3.3) becomes

$$x(t) = C_1 \cos \omega_n t + C_2 \sin \omega_n t + \frac{F_0}{k - m\omega^2} \cos \omega t \qquad (3.7)$$

Using the initial conditions $x(t = 0) = x_0$ and $\dot{x}(t = 0) = \dot{x}_0$, we find that

$$C_1 = x_0 - \frac{F_0}{k - m\omega^2}, \qquad C_2 = \frac{\dot{x}_0}{\omega_n} \tag{3.8}$$

and hence

$$x(t) = \left(x_0 - \frac{F_0}{k - m\omega^2} \right) \cos \omega_n t + \left(\frac{\dot{x}_0}{\omega_n} \right) \sin \omega_n t$$

$$+ \left(\frac{F_0}{k - m\omega^2} \right) \cos \omega t \tag{3.9}$$

The maximum amplitude X in Eq. (3.6) can be expressed as

$$\frac{X}{\delta_{st}} = \frac{1}{1 - \left(\dfrac{\omega}{\omega_n} \right)^2} \tag{3.10}$$

The quantity X/δ_{st} represents the ratio of the dynamic to the static amplitude of motion and is called the *magnification factor, amplification factor*, or *amplitude ratio*. The variation of the amplitude ratio, X/δ_{st}, with the frequency ratio $r = \omega/\omega_n$ (Eq. 3.10) is shown in Fig. 3.3. From this figure, the response of the system can be identified to be of three types.

Case 1.   When $0 < \omega/\omega_n < 1$, the denominator in Eq. (3.10) is positive and the response is given by Eq. (3.5) without change. The harmonic response of the system $x_p(t)$ is said to be in phase with the external force as shown in Fig. 3.4.

Case 2.   When $\omega/\omega_n > 1$, the denominator in Eq. (3.10) is negative, and the steady-state solution can be expressed as

$$x_p(t) = -X \cos \omega t \tag{3.11}$$

where the amplitude of motion X is redefined to be a positive quantity as

$$X = \frac{\delta_{st}}{\left(\dfrac{\omega}{\omega_n} \right)^2 - 1} \tag{3.12}$$

The variations of $F(t)$ and $x_p(t)$ with time are shown in Fig. 3.5. Since $x_p(t)$ and $F(t)$ have opposite signs, the response is said to be 180° out of phase with the external force. Further, as $\omega/\omega_n \to \infty$, $X \to 0$. Thus the response of the system to a harmonic force of very high frequency is close to zero.

Case 3.   When $\omega/\omega_n = 1$, the amplitude X given by Eq. (3.10) or (3.12) becomes infinite. This condition, for which the forcing frequency ω is equal to the natural frequency of the

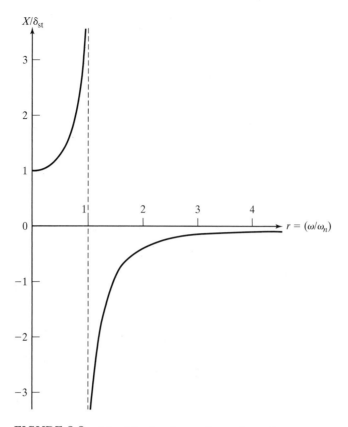

FIGURE 3.3   Magnification factor of an undamped system.

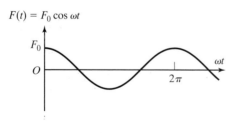

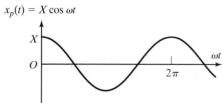

FIGURE 3.4   Harmonic response when $0 < \omega/\omega_n < 1$.

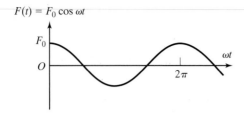

$F(t) = F_0 \cos \omega t$

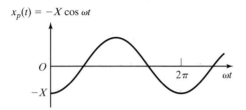

$x_p(t) = -X \cos \omega t$

FIGURE 3.5   Harmonic response when $\omega/\omega_n > 1$.

system ω_n, is called *resonance*. To find the response for this condition, we rewrite Eq. (3.9) as

$$x(t) = x_0 \cos \omega_n t + \frac{\dot{x}_0}{\omega_n} \sin \omega_n t + \delta_{st} \left[\frac{\cos \omega t - \cos \omega_n t}{1 - \left(\dfrac{\omega}{\omega_n}\right)^2} \right] \tag{3.13}$$

Since the last term of this equation takes an indefinite form for $\omega = \omega_n$, we apply L'Hospital's rule [3.1] to evaluate the limit of this term:

$$\lim_{\omega \to \omega_n} \left[\frac{\cos \omega t - \cos \omega_n t}{1 - \left(\dfrac{\omega}{\omega_n}\right)^2} \right] = \lim_{\omega \to \omega_n} \left[\frac{\dfrac{d}{d\omega}(\cos \omega t - \cos \omega_n t)}{\dfrac{d}{d\omega}\left(1 - \dfrac{\omega^2}{\omega_n^2}\right)} \right]$$

$$= \lim_{\omega \to \omega_n} \left[\frac{t \sin \omega t}{2 \dfrac{\omega}{\omega_n^2}} \right] = \frac{\omega_n t}{2} \sin \omega_n t \tag{3.14}$$

Thus the response of the system at resonance becomes

$$x(t) = x_0 \cos \omega_n t + \frac{\dot{x}_0}{\omega_n} \sin \omega_n t + \frac{\delta_{st} \omega_n t}{2} \sin \omega_n t \tag{3.15}$$

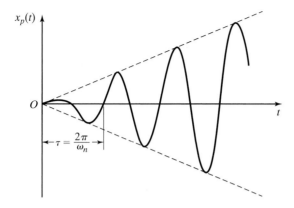

FIGURE 3.6   Response when $\omega/\omega_n = 1$.

It can be seen from Eq. (3.15) that at resonance, $x(t)$ increases indefinitely. The last term of Eq. (3.15) is shown in Fig. 3.6, from which the amplitude of the response can be seen to increase linearly with time.

3.3.1
Total Response

The total response of the system, Eq. (3.7) or Eq. (3.9), can also be expressed as

$$x(t) = A \cos (\omega_n t - \phi) + \frac{\delta_{st}}{1 - \left(\dfrac{\omega}{\omega_n}\right)^2} \cos \omega t; \quad \text{for } \frac{\omega}{\omega_n} < 1 \qquad (3.16)$$

$$x(t) = A \cos (\omega_n t - \phi) - \frac{\delta_{st}}{1 - \left(\dfrac{\omega}{\omega_n}\right)^2} \cos \omega t; \quad \text{for } \frac{\omega}{\omega_n} > 1 \qquad (3.17)$$

where A and ϕ can be determined as in the case of Eq. (2.21). Thus the complete motion can be expressed as the sum of two cosine curves of different frequencies. In Eq. (3.16), the forcing frequency ω is smaller than the natural frequency, and the total response is shown in Fig. 3.7(a). In Eq. (3.17), the forcing frequency is greater than the natural frequency, and the total response appears as shown in Fig. 3.7(b).

3.3.2
Beating
Phenomenon

If the forcing frequency is close to, but not exactly equal to, the natural frequency of the system, a phenomenon known as *beating* may occur. In this kind of vibration, the amplitude builds up and then diminishes in a regular pattern (see Section 1.10.5). The phenomenon of

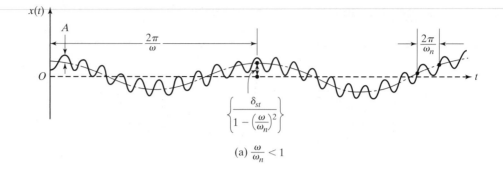

(a) $\dfrac{\omega}{\omega_n} < 1$

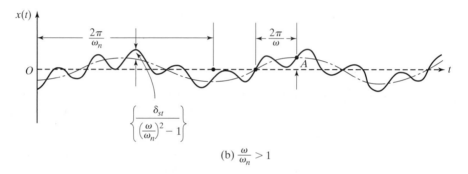

(b) $\dfrac{\omega}{\omega_n} > 1$

FIGURE 3.7   Total response.

beating can be explained by considering the solution given by Eq. (3.9). If the initial conditions are taken as $x_0 = \dot{x}_0 = 0$, Eq. (3.9) reduces to

$$x(t) = \frac{(F_0/m)}{\omega_n^2 - \omega^2}(\cos \omega t - \cos \omega_n t)$$

$$= \frac{(F_0/m)}{\omega_n^2 - \omega^2}\left[2 \sin \frac{\omega + \omega_n}{2} t \cdot \sin \frac{\omega_n - \omega}{2} t\right] \qquad (3.18)$$

Let the forcing frequency ω be slightly less than the natural frequency:

$$\omega_n - \omega = 2\varepsilon \qquad (3.19)$$

where ε is a small positive quantity. Then $\omega_n \approx \omega$ and

$$\omega + \omega_n \simeq 2\omega \qquad (3.20)$$

Multiplication of Eqs. (3.19) and (3.20) gives

$$\omega_n^2 - \omega^2 = 4\varepsilon\omega \qquad (3.21)$$

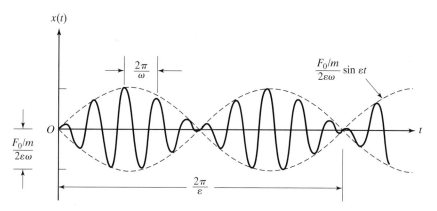

FIGURE 3.8   Phenomenon of beats.

The use of Eqs. (3.19) to (3.21) in Eq. (3.18) gives

$$x(t) = \left(\frac{F_0/m}{2\varepsilon\omega} \sin \varepsilon t \right) \sin \omega t \qquad (3.22)$$

Since ε is small, the function $\sin \varepsilon t$ varies slowly; its period, equal to $2\pi/\varepsilon$, is large. Thus Eq. (3.22) may be seen as representing vibration with period $2\pi/\omega$ and of variable amplitude equal to

$$\left(\frac{F_0/m}{2\varepsilon\omega} \right) \sin \varepsilon t$$

It can also be observed that the $\sin \omega t$ curve will go through several cycles, while the $\sin \varepsilon t$ wave goes through a single cycle, as shown in Fig. 3.8. Thus the amplitude builds up and dies down continuously. The time between the points of zero amplitude or the points of maximum amplitude is called the *period of beating* (τ_b) and is given by

$$\tau_b = \frac{2\pi}{2\varepsilon} = \frac{2\pi}{\omega_n - \omega} \qquad (3.23)$$

with the frequency of beating defined as

$$\omega_b = 2\varepsilon = \omega_n - \omega$$

EXAMPLE   3.1

Plate Supporting a Pump

A reciprocating pump, weighing 150 lb, is mounted at the middle of a steel plate of thickness 0.5 in., width 20 in., and length 100 in., clamped along two edges as shown in Fig. 3.9. During operation of the pump, the plate is subjected to a harmonic force, $F(t) = 50 \cos 62.832\, t$ lb. Find the amplitude of vibration of the plate.

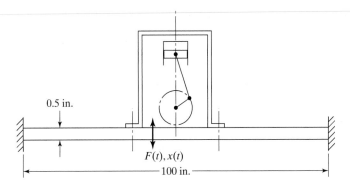

0.5 in.

$F(t), x(t)$

100 in.

FIGURE 3.9   Plate supporting an unbalanced pump.

Solution: The plate can be modeled as a fixed-fixed beam having Young's modulus (E) = 30 × 10⁶ psi, length (l) = 100 in., and area moment of inertia (I) = $\frac{1}{12}(20)(0.5)^3$ = 0.2083 in⁴. The bending stiffness of the beam is given by

$$k = \frac{192EI}{l^3} = \frac{192(30 \times 10^6)(0.2083)}{(100)^3} = 1200.0 \text{ lb/in.} \tag{E.1}$$

The amplitude of harmonic response is given by Eq. (3.6) with F_0 = 50 lb, m = 150/386.4 lb-sec²/in. (neglecting the weight of the steel plate), k = 1200.0 lb/in., and ω = 62.832 rad/s. Thus Eq. (3.6) gives

$$X = \frac{F_0}{k - m\omega^2} = \frac{50}{1200.0 - (150/386.4)(62.832)^2} = -0.1504 \text{ in.} \tag{E.2}$$

The negative sign indicates that the response $x(t)$ of the plate is out of phase with the excitation $F(t)$.

■

3.4   Response of a Damped System Under Harmonic Force

If the forcing function is given by $F(t) = F_0 \cos \omega t$, the equation of motion becomes

$$m\ddot{x} + c\dot{x} + kx = F_0 \cos \omega t \tag{3.24}$$

The particular solution of Eq. (3.24) is also expected to be harmonic; we assume it in the form[1]

$$x_p(t) = X \cos (\omega t - \phi) \tag{3.25}$$

where X and ϕ are constants to be determined. X and ϕ denote the amplitude and phase angle of the response, respectively. By substituting Eq. (3.25) into Eq. (3.24), we arrive at

$$X[(k - m\omega^2) \cos (\omega t - \phi) - c\omega \sin (\omega t - \phi)] = F_0 \cos \omega t \tag{3.26}$$

---

[1]Alternatively, we can assume $x_p(t)$ to be of the form $x_p(t) = C_1 \cos \omega t + C_2 \sin \omega t$, which also involves two constants C_1 and C_2. But the final result will be the same in both the cases.

Using the trigonometric relations

$$\cos(\omega t - \phi) = \cos \omega t \cos \phi + \sin \omega t \sin \phi$$
$$\sin(\omega t - \phi) = \sin \omega t \cos \phi - \cos \omega t \sin \phi$$

in Eq. (3.26) and equating the coefficients of $\cos \omega t$ and $\sin \omega t$ on both sides of the resulting equation, we obtain

$$X[(k - m\omega^2) \cos \phi + c\omega \sin \phi] = F_0$$
$$X[(k - m\omega^2) \sin \phi - c\omega \cos \phi] = 0 \qquad (3.27)$$

Solution of Eq. (3.27) gives

$$X = \frac{F_0}{[(k - m\omega^2)^2 + c^2\omega^2]^{1/2}} \qquad (3.28)$$

and

$$\phi = \tan^{-1}\left(\frac{c\omega}{k - m\omega^2}\right) \qquad (3.29)$$

By inserting the expressions of X and ϕ from Eqs. (3.28) and (3.29) into Eq. (3.25), we obtain the particular solution of Eq. (3.24). Figure 3.10(a) shows typical plots of the forcing function and (steady-state) response. The various terms of Eq. (3.26) are shown vectorially in Fig. 3.10(b). Dividing both the numerator and denominator of Eq. (3.28) by k and making the following substitutions

$$\omega_n = \sqrt{\frac{k}{m}} = \text{undamped natural frequency,}$$

$$\zeta = \frac{c}{c_c} = \frac{c}{2m\omega_n} = \frac{c}{2\sqrt{mk}}; \frac{c}{m} = 2\zeta\omega_n,$$

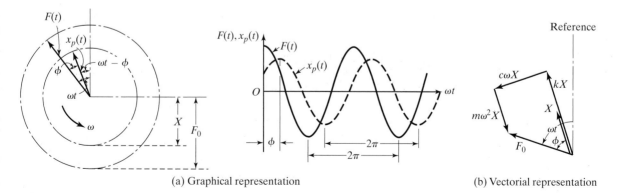

(a) Graphical representation   (b) Vectorial representation

FIGURE 3.10   Representation of forcing function and response.

$$\delta_{st} = \frac{F_0}{k} = \text{deflection under the static force } F_0, \text{ and}$$

$$r = \frac{\omega}{\omega_n} = \text{frequency ratio}$$

we obtain

$$\frac{X}{\delta_{st}} = \frac{1}{\left\{\left[1 - \left(\dfrac{\omega}{\omega_n}\right)^2\right]^2 + \left[2\zeta\dfrac{\omega}{\omega_n}\right]^2\right\}^{1/2}} = \frac{1}{\sqrt{(1 - r^2)^2 + (2\zeta r)^2}} \qquad (3.30)$$

and

$$\phi = \tan^{-1}\left\{\frac{2\zeta\dfrac{\omega}{\omega_n}}{1 - \left(\dfrac{\omega}{\omega_n}\right)^2}\right\} = \tan^{-1}\left(\frac{2\zeta r}{1 - r^2}\right) \qquad (3.31)$$

As stated in Section 3.3, the quantity $M = X/\delta_{st}$ is called the *magnification factor, amplification factor*, or *amplitude ratio*. The variations of X/δ_{st} and ϕ with the frequency ratio r and the damping ratio ζ are shown in Fig. 3.11.

The following characteristics of the magnification factor (M) can be noted from Eq. (3.30) and Fig. 3.11(a):

1. For an undamped system $(\zeta = 0)$, Eq. (3.30) reduces to Eq. (3.10), and $M \rightarrow \infty$ as $r \rightarrow 1$.

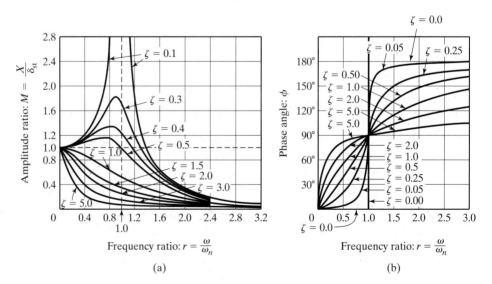

FIGURE 3.11   Variation of X and ϕ with frequency ratio r.

2. Any amount of damping ($\zeta > 0$) reduces the magnification factor (M) for all values of the forcing frequency.
3. For any specified value of r, a higher value of damping reduces the value of M.
4. In the degenerate case of a constant force (when $r = 0$), the value of $M = 1$.
5. The reduction in M in the presence of damping is very significant at or near resonance.
6. The amplitude of forced vibration becomes smaller with increasing values of the forcing frequency (that is, $M \to 0$ as $r \to \infty$).
7. For $0 < \zeta < \frac{1}{\sqrt{2}}$, the maximum value of M occurs when (see Problem 3.27)

$$r = \sqrt{1 - 2\zeta^2} \quad \text{or} \quad \omega = \omega_n\sqrt{1 - 2\zeta^2} \tag{3.32}$$

which can be seen to be lower than the undamped natural frequency ω_n and the damped natural frequency $\omega_d = \omega_n\sqrt{1 - \zeta^2}$.

8. The maximum value of X (when $r = \sqrt{1 - 2\zeta^2}$) is given by

$$\left(\frac{X}{\delta_{st}}\right)_{max} = \frac{1}{2\zeta\sqrt{1 - \zeta^2}} \tag{3.33}$$

and the value of X at $\omega = \omega_n$ by

$$\left(\frac{X}{\delta_{st}}\right)_{\omega = \omega_n} = \frac{1}{2\zeta} \tag{3.34}$$

Equation (3.33) can be used for the experimental determination of the measure of damping present in the system. In a vibration test, if the maximum amplitude of the response $(X)_{max}$ is measured, the damping ratio of the system can be found using Eq. (3.33). Conversely, if the amount of damping is known, one can make an estimate of the maximum amplitude of vibration.

9. For $\zeta = \frac{1}{\sqrt{2}}, \frac{dM}{dr} = 0$ when $r = 0$. For $\zeta > \frac{1}{\sqrt{2}}$, the graph of M monotonically decreases with increasing values of r.

The following characteristics of the phase angle can be observed from Eq. (3.31) and Fig. 3.11(b):

1. For an undamped system ($\zeta = 0$), Eq. (3.31) shows that the phase angle is 0 for $0 < r < 1$ and $180°$ for $r > 1$. This implies that the excitation and response are in phase for $0 < r < 1$ and out of phase for $r > 1$ when $\zeta = 0$.
2. For $\zeta > 0$ and $0 < r < 1$, the phase angle is given by $0 < \phi < 90°$, implying that the response lags the excitation.
3. For $\zeta > 0$ and $r > 1$, the phase angle is given by $90° < \phi < 180°$, implying that the response leads the excitation.

4. For $\zeta > 0$ and $r = 1$, the phase angle is given by $\phi = 90°$, implying that the phase difference between the excitation and the response is 90°.

5. For $\zeta > 0$ and large values of r, the phase angle approaches 180°, implying that the response and the excitation are out of phase.

3.4.1
Total Response

The complete solution is given by $x(t) = x_h(t) + x_p(t)$ where $x_h(t)$ is given by Eq. (2.70). Thus, for an underdamped system, we have

$$x(t) = X_0 e^{-\zeta \omega_n t} \cos (\omega_d t - \phi_0) + X \cos (\omega t - \phi) \qquad (3.35)$$

where

$$\omega_d = \sqrt{1 - \zeta^2} \, \omega_n \qquad (3.36)$$

X and ϕ are given by Eqs. (3.30) and (3.31), respectively, and X_0 and ϕ_0 can be determined from the initial conditions. For the initial conditions, $x(t = 0) = x_0$ and $\dot{x}(t = 0) = \dot{x}_0$, Eq. (3.35) yields

$$x_0 = X_0 \cos \phi_0 + X \cos \phi$$
$$\dot{x}_0 = -\zeta \omega_n X_0 \cos \phi_0 + \omega_d X_0 \sin \phi_0 + \omega X \sin \phi \qquad (3.37)$$

The solution of Eq. (3.37) gives X_0 and ϕ_0, as illustrated in the following example.

Total Response of a System

EXAMPLE 3.2

Find the total response of a single degree of freedom system with $m = 10$ kg, $c = 20$ N-s/m, $k = 4000$ N/m, $x_0 = 0.01$ m, and $\dot{x}_0 = 0$ under the following conditions:

a. An external force $F(t) = F_0 \cos \omega t$ acts on the system with $F_0 = 100$ N and $\omega = 10$ rad/s.

b. Free vibration with $F(t) = 0$.

Solution:

a. From the given data, we obtain

$$\omega_n = \sqrt{\frac{k}{m}} = \sqrt{\frac{4000}{10}} = 20 \text{ rad/s}$$

$$\delta_{st} = \frac{F_0}{k} = \frac{100}{4000} = 0.025 \text{ m}$$

$$\zeta = \frac{c}{c_c} = \frac{c}{2\sqrt{km}} = \frac{20}{2\sqrt{(4000)(10)}} = 0.05$$

$$\omega_d = \sqrt{1 - \zeta^2} \omega_n = \sqrt{1 - (0.05)^2}(20) = 19.974984 \text{ rad/s}$$

$$r = \frac{\omega}{\omega_n} = \frac{10}{20} = 0.5$$

$$X = \frac{\delta_{st}}{\sqrt{(1 - r^2)^2 + (2\,\zeta\,r)^2}} = \frac{0.025}{\left[(1 - 0.05^2)^2 + (2\cdot 0.5 \cdot 0.5)^2\right]^{1/2}} = 0.03326 \text{ m} \qquad (E.1)$$

$$\phi = \tan^{-1}\left(\frac{2\zeta r}{1 - r^2}\right) = \tan^{-1}\left(\frac{2\cdot 0.05 \cdot 0.5}{1 - 0.5^2}\right) = 3.814075° \qquad (E.2)$$

Using the initial conditions, $x_0 = 0.01$ and $\dot{x}_0 = 0$, Eq. (3.37) yields:

$$0.01 = X_0 \cos\phi_0 + (0.03326)(0.997785)$$

or

$$X_0 \cos\phi_0 = -0.023186 \qquad (E.3)$$

$$0 = -(0.05)(20)\,X_0 \cos\phi_0 + X_0\,(19.974984)\sin\phi_0 + (0.03326)(10)\sin(3.814075°) \qquad (E.4)$$

Substituting the value of $X_0 \cos\phi_0$ from Eq. (E.3) into (E.4), we obtain

$$X_0 \sin\phi_0 = -0.002268 \qquad (E.5)$$

Solution of Eqs. (E.3) and (E.5) yields

$$X_0 = \left[(X_0 \cos\phi_0)^2 + (X_0 \sin\phi_0)^2\right]^{1/2} = 0.023297 \qquad (E.6)$$

and

$$\tan\phi_0 = \frac{X_0 \sin\phi_0}{X_0 \cos\phi_0} = 0.0978176$$

or

$$\phi_0 = 5.586765° \qquad (E.7)$$

b.   For free vibration, the total response is given by

$$x(t) = X_0 e^{-\zeta \omega_n t} \cos(\omega_d t - \phi_0) \qquad (E.8)$$

Using the initial conditions $x(0) = x_0 = 0.01$ and $\dot{x}(0) = \dot{x}_0 = 0$, X_0 and ϕ_0 of Eq. (E.8) can be determined as (see Eqs. 2.73 and 2.75):

$$X_0 = \left[x_0^2 + \left(\frac{\zeta \omega_n x_0}{\omega_d}\right)^2\right]^{1/2} = \left[0.01^2 + \left(\frac{0.05 \cdot 20 \cdot 0.01}{19.974984}\right)^2\right]^{1/2} = 0.010012 \qquad (E.9)$$

$$\phi_0 = \tan^{-1}\left(-\frac{\dot{x}_0 + \zeta\,\omega_n\,x_0}{\omega_d\,x_0}\right) = \tan^{-1}\left(-\frac{0.05 \cdot 20}{19.974984}\right) = -2.865984° \qquad (E.10)$$

Note that the constants X_0 and ϕ_0 in cases (a) and (b) are very different.

■

3.4.2
Quality Factor
and Bandwidth

For small values of damping ($\zeta < 0.05$), we can take

$$\left(\frac{X}{\delta_{st}}\right)_{max} \simeq \left(\frac{X}{\delta_{st}}\right)_{\omega=\omega_n} = \frac{1}{2\zeta} = Q \tag{3.38}$$

The value of the amplitude ratio at resonance is also called Q *factor* or *quality factor* of the system, in analogy with some electrical-engineering applications, such as the tuning circuit of a radio, where the interest lies in an amplitude at resonance that is as large as possible [3.2]. The points R_1 and R_2, where the amplification factor falls to $Q/\sqrt{2}$, are called *half power points* because the power absorbed (ΔW) by the damper (or by the resistor in an electrical circuit), responding harmonically at a given frequency, is proportional to the square of the amplitude (see Eq. 2.94):

$$\Delta W = \pi c \omega X^2 \tag{3.39}$$

The difference between the frequencies associated with the half power points R_1 and R_2 is called the *bandwidth* of the system (see Fig. 3.12). To find the values of R_1 and R_2, we set $X/\delta_{st} = Q/\sqrt{2}$ in Eq. (3.30) so that

$$\frac{1}{\sqrt{(1-r^2)^2 + (2\zeta r)^2}} = \frac{Q}{\sqrt{2}} = \frac{1}{2\sqrt{2}\zeta}$$

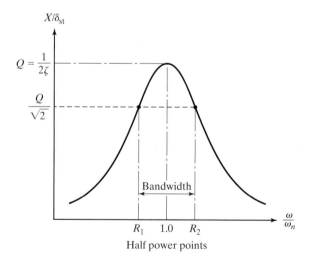

FIGURE 3.12   Harmonic response curve showing half power points and bandwidth.

or

$$r^4 - r^2(2 - 4\zeta^2) + (1 - 8\zeta^2) = 0 \tag{3.40}$$

The solution of Eq. (3.40) gives

$$r_1^2 = 1 - 2\zeta^2 - 2\zeta\sqrt{1 + \zeta^2}, \qquad r_2^2 = 1 - 2\zeta^2 + 2\zeta\sqrt{1 + \zeta^2} \tag{3.41}$$

For small values of ζ, Eq. (3.41) can be approximated as

$$r_1^2 = R_1^2 = \left(\frac{\omega_1}{\omega_n}\right)^2 \simeq 1 - 2\zeta, \qquad r_2^2 = R_2^2 = \left(\frac{\omega_2}{\omega_n}\right)^2 \simeq 1 + 2\zeta \tag{3.42}$$

where $\omega_1 = \omega|_{R_1}$ and $\omega_2 = \omega|_{R_2}$. From Eq. (3.42),

$$\omega_2^2 - \omega_1^2 = (\omega_2 + \omega_1)(\omega_2 - \omega_1) = (R_2^2 - R_1^2)\omega_n^2 \simeq 4\zeta\omega_n^2 \tag{3.43}$$

Using the relation

$$\omega_2 + \omega_1 = 2\omega_n \tag{3.44}$$

in Eq. (3.43), we find that the bandwidth $\Delta\omega$ is given by

$$\Delta\omega = \omega_2 - \omega_1 \simeq 2\zeta\omega_n \tag{3.45}$$

Combining Eqs. (3.38) and (3.45), we obtain

$$Q \simeq \frac{1}{2\zeta} \simeq \frac{\omega_n}{\omega_2 - \omega_1} \tag{3.46}$$

It can be seen that the quality factor Q can be used for estimating the equivalent viscous damping in a mechanical system.[2]

3.5   Response of a Damped System Under $F(t) = F_0 e^{i\omega t}$

Let the harmonic forcing function be represented in complex form as $F(t) = F_0 e^{i\omega t}$ so that the equation of motion becomes

$$m\ddot{x} + c\dot{x} + kx = F_0 e^{i\omega t} \tag{3.47}$$

Since the actual excitation is given only by the real part of $F(t)$, the response will also be given only by the real part of $x(t)$ where $x(t)$ is a complex quantity satisfying the differential

---

[2]The determination of the system parameters (m, c, and k) based on half power points and other response characteristics of the system is considered in Section 10.8.

equation (3.47). F_0 in Eq. (3.47) is, in general, a complex number. By assuming the particular solution $x_p(t)$

$$x_p(t) = Xe^{i\omega t} \tag{3.48}$$

we obtain, by substituting Eq. (3.48) into Eq. (3.47),[3]

$$X = \frac{F_0}{(k - m\omega^2) + ic\omega} \tag{3.49}$$

Multiplying the numerator and denominator on the right side of Eq. (3.49) by $[(k - m\omega^2) - ic\omega]$ and separating the real and imaginary parts, we obtain

$$X = F_0 \left[\frac{k - m\omega^2}{(k - m\omega^2)^2 + c^2\omega^2} - i \frac{c\omega}{(k - m\omega^2)^2 + c^2\omega^2} \right] \tag{3.50}$$

Using the relation, $x + iy = Ae^{i\phi}$ where $A = \sqrt{x^2 + y^2}$ and $\tan\phi = y/x$, Eq. (3.50) can be expressed as

$$X = \frac{F_0}{[(k - m\omega^2)^2 + c^2\omega^2]^{1/2}} e^{-i\phi} \tag{3.51}$$

where

$$\phi = \tan^{-1}\left(\frac{c\omega}{k - m\omega^2}\right) \tag{3.52}$$

Thus the steady-state solution, Eq. (3.48), becomes

$$x_p(t) = \frac{F_0}{[(k - m\omega^2)^2 + (c\omega)^2]^{1/2}} e^{i(\omega t - \phi)} \tag{3.53}$$

Frequency Response.   Equation (3.49) can be rewritten in the form

$$\frac{kX}{F_0} = \frac{1}{1 - r^2 + i2\zeta r} \equiv H(i\omega) \tag{3.54}$$

where $H(i\omega)$ is known as the *complex frequency response* of the system. The absolute value of $H(i\omega)$ given by

$$|H(i\omega)| = \left|\frac{kX}{F_0}\right| = \frac{1}{[(1 - r^2)^2 + (2\zeta r)^2]^{1/2}} \tag{3.55}$$

---

[3]Equation (3.49) can be written as $Z(i\omega)X = F_0$, where $Z(i\omega) = -m\omega^2 + i\omega c + k$ is called the *mechanical impedance* of the system [3.8].

denotes the magnification factor defined in Eq. (3.30). Recalling that $e^{i\phi} = \cos\phi + i\sin\phi$, we can show that Eqs. (3.54) and (3.55) are related:

$$H(i\omega) = |H(i\omega)|e^{-i\phi} \tag{3.56}$$

where ϕ is given by Eq. (3.52), which can also be expressed as

$$\phi = \tan^{-1}\left(\frac{2\zeta r}{1 - r^2}\right) \tag{3.57}$$

Thus Eq. (3.53) can be expressed as

$$x_p(t) = \frac{F_0}{k}|H(i\omega)|e^{i(\omega t - \phi)} \tag{3.58}$$

It can be seen that the complex frequency response function, $H(i\omega)$, contains both the magnitude and phase of the steady-state response. The use of this function in the experimental determination of the system parameters (m, c, and k) is discussed in Section 10.8. If $F(t) = F_0 \cos\omega t$, the corresponding steady-state solution is given by the real part of Eq. (3.53):

$$x_p(t) = \frac{F_0}{\left[(k - m\omega^2)^2 + (c\omega)^2\right]^{1/2}} \cos(\omega t - \phi)$$

$$= \text{Re}\left[\frac{F_0}{k}H(i\omega)e^{i\omega t}\right] = \text{Re}\left[\frac{F_0}{k}|H(i\omega)|e^{i(\omega t - \phi)}\right] \tag{3.59}$$

which can be seen to be the same as Eq. (3.25). Similarly, if $F(t) = F_0 \sin\omega t$, the corresponding steady-state solution is given by the imaginary part of Eq. (3.53):

$$x_p(t) = \frac{F_0}{\left[(k - m\omega^2)^2 + (c\omega)^2\right]^{1/2}} \sin(\omega t - \phi)$$

$$= \text{Im}\left[\frac{F_0}{k}|H(i\omega)|e^{i(\omega t - \phi)}\right] \tag{3.60}$$

Complex Vector Representation of Harmonic Motion.   The harmonic excitation and the response of the damped system to that excitation can be represented graphically in the complex plane, and interesting interpretation can be given to the resulting diagram. We first differentiate Eq. (3.58) with respect to time and obtain

$$\text{Velocity} = \dot{x}_p(t) = i\omega\frac{F_0}{k}|H(i\omega)|e^{i(\omega t - \phi)} = i\omega x_p(t)$$

$$\text{Acceleration} = \ddot{x}_p(t) = (i\omega)^2\frac{F_0}{k}|H(i\omega)|e^{i(\omega t - \phi)} = -\omega^2 x_p(t) \tag{3.61}$$

Because i can be expressed as

$$i = \cos \frac{\pi}{2} + i \sin \frac{\pi}{2} = e^{i\frac{\pi}{2}} \tag{3.62}$$

we can conclude that the velocity leads the displacement by the phase angle $\pi/2$ and that it is multiplied by ω. Similarly, -1 can be written as

$$-1 = \cos \pi + i \sin \pi = e^{i\pi} \tag{3.63}$$

Hence the acceleration leads the displacement by the phase angle π, and it is multiplied by ω^2.

Thus the various terms of the equation of motion (3.47) can be represented in the complex plane, as shown in Fig. 3.13. The interpretation of this figure is that the sum of the complex vectors $m\ddot{x}(t)$, $c\dot{x}(t)$, and $kx(t)$ balances $F(t)$, which is precisely what is required to satisfy Eq. (3.47). It is to also be noted that the entire diagram rotates with angular velocity ω in the complex plane. If only the real part of the response is to be considered, then the entire diagram must be projected onto the real axis. Similarly, if only the imaginary part of the response is to be considered, then the diagram must be projected onto the imaginary axis. In Fig. 3.13, notice that the force $F(t) = F_0 e^{i\omega t}$ is represented as a vector located at an angle ωt to the real axis. This implies that F_0 is real. If F_0 is also complex, then the force vector $F(t)$ will be located at an angle of $(\omega + \psi)$, where ψ is some phase angle introduced by F_0. In such a case, all the other vectors, namely, $m\ddot{x}$, $c\dot{x}$, and kx will be shifted by the same angle ψ. This is equivalent to multiplying both sides of Eq. (3.47) by $e^{i\psi}$.

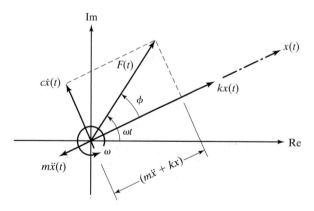

FIGURE 3.13   Representation of Eq. (3.47) in a complex plane.

3.6   Response of a Damped System Under the Harmonic Motion of the Base

Sometimes the base or support of a spring-mass-damper system undergoes harmonic motion, as shown in Fig. 3.14(a). Let $y(t)$ denote the displacement of the base and $x(t)$ the displacement of the mass from its static equilibrium position at time t. Then the net elongation of the spring is $x - y$ and the relative velocity between the two ends of the damper is $\dot{x} - \dot{y}$. From the free-body diagram shown in Fig. 3.14(b), we obtain the equation of motion:

$$m\ddot{x} + c(\dot{x} - \dot{y}) + k(x - y) = 0 \tag{3.64}$$

If $y(t) = Y \sin \omega t$, Eq. (3.64) becomes

$$m\ddot{x} + c\dot{x} + kx = ky + c\dot{y} = kY \sin \omega t + c\omega Y \cos \omega t$$
$$= A \sin (\omega t - \alpha) \tag{3.65}$$

where $A = Y\sqrt{k^2 + (c\omega)^2}$ and $\alpha = \tan^{-1}\left[-\frac{c\omega}{k}\right]$. This shows that giving excitation to the base is equivalent to applying a harmonic force of magnitude A to the mass. By using the solution indicated by Eq. (3.60), the steady-state response of the mass, $x_p(t)$, can be expressed as

$$x_p(t) = \frac{Y\sqrt{k^2 + (c\omega)^2}}{\left[(k - m\omega^2)^2 + (c\omega)^2\right]^{1/2}} \sin (\omega t - \phi_1 - \alpha) \tag{3.66}$$

where

$$\phi_1 = \tan^{-1}\left(\frac{c\omega}{k - m\omega^2}\right)$$

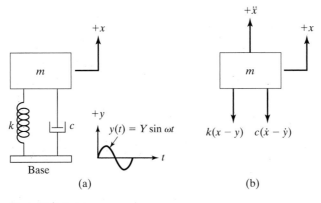

FIGURE 3.14   Base excitation.

Using trigonometric identities, Eq. (3.66) can be rewritten in a more convenient form as

$$x_p(t) = X \sin(\omega t - \phi) \tag{3.67}$$

where X and ϕ are given by

$$\frac{X}{Y} = \left[\frac{k^2 + (c\omega)^2}{(k - m\omega^2)^2 + (c\omega)^2}\right]^{1/2} = \left[\frac{1 + (2\zeta r)^2}{(1 - r^2)^2 + (2\zeta r)^2}\right]^{1/2} \tag{3.68}$$

and

$$\phi = \tan^{-1}\left[\frac{mc\omega^3}{k(k - m\omega^2) + (\omega c)^2}\right] = \tan^{-1}\left[\frac{2\zeta r^3}{1 + (4\zeta^2 - 1)r^2}\right] \tag{3.69}$$

The ratio of the amplitude of the response $x_p(t)$ to that of the base motion $y(t)$, $\frac{X}{Y}$, is called the *displacement transmissibility*. The variations of $\frac{X}{Y} \equiv T_d$ and ϕ, given by Eqs. (3.68) and (3.69), are shown in Figs. 3.15(a) and (b), respectively, for different values of r and ζ.

Note that if the harmonic excitation of the base is expressed in complex form as $y(t) = \text{Re}(Ye^{i\omega t})$, the response of the system can be expressed using the analysis of Section 3.5, as

$$x_p(t) = \text{Re}\left\{\left(\frac{1 + i2\zeta r}{1 - r^2 + i2\zeta r}\right)Ye^{i\omega t}\right\} \tag{3.70}$$

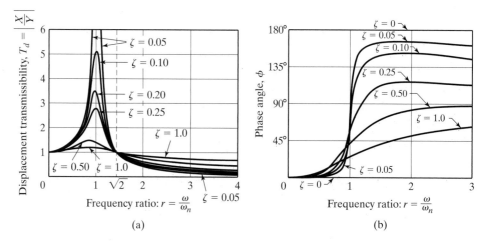

(a)   (b)

FIGURE 3.15   Variations of T_d and ϕ with r.

and the displacement transmissibility as

$$\frac{X}{Y} = T_d = \left[1 + (2\zeta r)^2\right]^{1/2} |H(i\omega)| \tag{3.71}$$

where $|H(i\omega)|$ is given by Eq. (3.55).

The following aspects of *displacement transmissibility*, $T_d = \frac{X}{Y}$, can be noted from Fig. 3.15(a):

1. The value of T_d is unity at $r = 0$ and close to unity for small values of r.
2. For an undamped system ($\zeta = 0$), $T_d \rightarrow \infty$ at resonance ($r = 1$).
3. The value of T_d is less than unity ($T_d < 1$) for values of $r > \sqrt{2}$ (for any amount of damping ζ).
4. The value of $T_d = 1$ for all values of ζ at $r = \sqrt{2}$.
5. For $r < \sqrt{2}$, smaller damping ratios lead to larger values of T_d. On the other hand, for $r > \sqrt{2}$, smaller values of damping ratio lead to smaller values of T_d.
6. The displacement transmissibility, T_d, attains a maximum for $0 < \zeta < 1$ at the frequency ratio $r = r_m < 1$ given by (see Problem 3.49):

$$r_m = \frac{1}{2\zeta} \left[\sqrt{1 + 8\zeta^2} - 1\right]^{1/2}$$

3.6.1
Force
Transmitted

In Fig. 3.14, a force, F, is transmitted to the base or support due to the reactions from the spring and the dashpot. This force can be determined as

$$F = k(x - y) + c(\dot{x} - \dot{y}) = -m\ddot{x} \tag{3.72}$$

From Eq. (3.67), Eq. (3.72) can be written as

$$F = m\omega^2 X \sin(\omega t - \phi) = F_T \sin(\omega t - \phi) \tag{3.73}$$

where F_T is the amplitude or maximum value of the force transmitted to the base given by

$$\frac{F_T}{kY} = r^2 \left[\frac{1 + (2\zeta r)^2}{(1 - r^2)^2 + (2\zeta r)^2}\right]^{1/2} \tag{3.74}$$

The ratio (F_T/kY) is known as the *force transmissibility*.[4] Note that the transmitted force is in phase with the motion of the mass $x(t)$. The variation of the force transmitted to the base with the frequency ratio r is shown in Fig. 3.16 for different values of ζ.

---

[4]The use of the concept of transmissibility in the design of vibration isolation systems is discussed in Chapter 9.

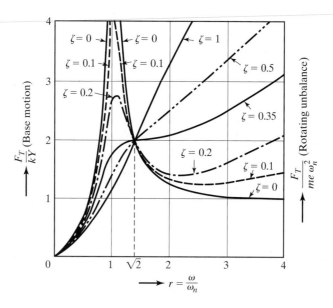

FIGURE 3.16   Force transmissibility.

**3.6.2
Relative Motion**

If $z = x - y$ denotes the motion of the mass relative to the base, the equation of motion, Eq. (3.64), can be rewritten as

$$m\ddot{z} + c\dot{z} + kz = -m\ddot{y} = m\omega^2 Y \sin \omega t \qquad (3.75)$$

The steady-state solution of Eq. (3.75) is given by

$$z(t) = \frac{m\omega^2 Y \sin(\omega t - \phi_1)}{[(k - m\omega^2)^2 + (c\omega)^2]^{1/2}} = Z \sin(\omega t - \phi_1) \qquad (3.76)$$

where Z, the amplitude of $z(t)$, can be expressed as

$$Z = \frac{m\omega^2 Y}{\sqrt{(k - m\omega^2)^2 + (c\omega)^2}} = Y \frac{r^2}{\sqrt{(1 - r^2)^2 + (2\zeta r)^2}} \qquad (3.77)$$

and ϕ_1 by

$$\phi_1 = \tan^{-1}\left(\frac{c\omega}{k - m\omega^2}\right) = \tan^{-1}\left(\frac{2\zeta r}{1 - r^2}\right)$$

The ratio Z/X is shown graphically in Fig. 3.17. The variation of ϕ_1 is same as that of ϕ shown in Fig. 3.11(b).

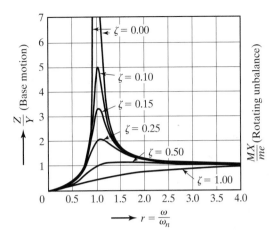

FIGURE 3.17 Variation of (Z/Y) or (MX/me) with frequency ratio $r = (\omega/\omega_n)$.

EXAMPLE 3.3

Vehicle Moving on a Rough Road

Figure 3.18 shows a simple model of a motor vehicle that can vibrate in the vertical direction while traveling over a rough road. The vehicle has a mass of 1200 kg. The suspension system has a spring constant of 400 kN/m and a damping ratio of $\zeta = 0.5$. If the vehicle speed is 20 km/hr, determine the displacement amplitude of the vehicle. The road surface varies sinusoidally with an amplitude of $Y = 0.05$ m and a wavelength of 6 m.

Solution: The frequency ω of the base excitation can be found by dividing the vehicle speed v km/hr by the length of one cycle of road roughness:

$$\omega = 2\pi f = 2\pi \left(\frac{v \times 1000}{3600} \right) \frac{1}{6} = 0.290889v \text{ rad/s}$$

For $v = 20$ km/hr, $\omega = 5.81778$ rad/s. The natural frequency of the vehicle is given by

$$\omega_n = \sqrt{\frac{k}{m}} = \left(\frac{400 \times 10^3}{1200} \right)^{1/2} = 18.2574 \text{ rad/s}$$

and hence the frequency ratio r is

$$r = \frac{\omega}{\omega_n} = \frac{5.81778}{18.2574} = 0.318653$$

The amplitude ratio can be found from Eq. (3.68):

$$\frac{X}{Y} = \left\{ \frac{1 + (2\zeta r)^2}{(1 - r^2)^2 + (2\zeta r)^2} \right\}^{1/2} = \left\{ \frac{1 + (2 \times 0.5 \times 0.318653)^2}{(1 - 0.318653)^2 + (2 \times 0.5 \times 0.318653)^2} \right\}^{1/2}$$

$$= 1.469237$$

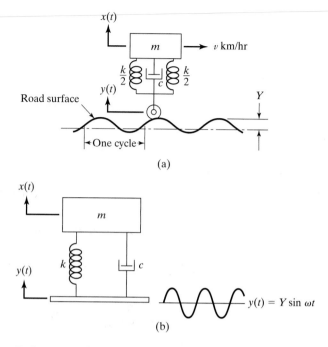

FIGURE 3.18   Vehicle moving over a rough road.

Thus the displacement amplitude of the vehicle is given by

$$X = 1.469237 \, Y = 1.469237 \, (0.05) = 0.073462 \text{ m}$$

This indicates that a 5 cm bump in the road is transmitted as a 7.3 cm bump to the chassis and the passengers of the car. Thus the passengers feel an amplified motion in the present case (see Problem 3.79 for other situations).

■

Machine on Resilient Foundation

EXAMPLE 3.4

A heavy machine, weighing 3000 N, is supported on a resilient foundation. The static deflection of the foundation due to the weight of the machine is found to be 7.5 cm. It is observed that the machine vibrates with an amplitude of 1 cm when the base of the foundation is subjected to harmonic oscillation at the undamped natural frequency of the system with an amplitude of 0.25 cm. Find (a) the damping constant of the foundation, (b) the dynamic force amplitude on the base, and (c) the amplitude of the displacement of the machine relative to the base.

Solution
a.   The stiffness of the foundation can be found from its static deflection: k = weight of machine/δ_{st} = 3000/0.075 = 40,000 N/m

At resonance ($\omega = \omega_n$ or $r = 1$), Eq. (3.68) gives

$$\frac{X}{Y} = \frac{0.010}{0.0025} = 4 = \left[\frac{1 + (2\zeta)^2}{(2\zeta)^2}\right]^{1/2} \tag{E.1}$$

The solution of Eq. (E.1) gives $\zeta = 0.1291$. The damping constant is given by

$$c = \zeta \cdot c_c = \zeta 2\sqrt{km} = 0.1291 \times 2 \times \sqrt{40,000 \times (3000/9.81)}$$
$$= 903.0512 \text{ N-s/m} \tag{E.2}$$

b.   The dynamic force amplitude on the base at $r = 1$ can be found from Eq. (3.74):

$$F_T = Yk\left[\frac{1 + 4\zeta^2}{4\zeta^2}\right]^{1/2} = kX = 40,000 \times 0.01 = 400 \text{ N} \tag{E.3}$$

c.   The amplitude of the relative displacement of the machine at $r = 1$ can be obtained from Eq. (3.77):

$$Z = \frac{Y}{2\zeta} = \frac{0.0025}{2 \times 0.1291} = 0.00968 \text{ m} \tag{E.4}$$

It can be noticed that $X = 0.01$ m, $Y = 0.0025$ m, and $Z = 0.00968$ m; therefore, $Z \neq X - Y$. This is due to the phase differences between x, y, and z.

■

3.7   Response of a Damped System Under Rotating Unbalance

Unbalance in rotating machinery is one of the main causes of vibration. A simplified model of such a machine is shown in Fig. 3.19. The total mass of the machine is M, and there are two eccentric masses $m/2$ rotating in opposite directions with a constant angular velocity ω. The centrifugal force $(me\omega^2)/2$ due to each mass will cause excitation of the mass M. We consider two equal masses $m/2$ rotating in opposite directions in order to have the horizontal components of excitation of the two masses cancel each other. However, the vertical components of excitation add together and act along the axis of symmetry $A - A$ in Fig. 3.19. If the angular position of the masses is measured from a horizontal position, the total vertical component of the excitation is always given by $F(t) = me\omega^2 \sin \omega t$. The equation of motion can be derived by the usual procedure:

$$M\ddot{x} + c\dot{x} + kx = me\omega^2 \sin \omega t \tag{3.78}$$

The solution of this equation will be identical to Eq. (3.60) if we replace m and F_0 by M and $me\omega^2$ respectively. This solution can also be expressed as

$$x_p(t) = X \sin(\omega t - \phi) = \text{Im}\left[\frac{me}{M}\left(\frac{\omega}{\omega_n}\right)^2 |H(i\omega)|e^{i(\omega t - \phi)}\right] \tag{3.79}$$

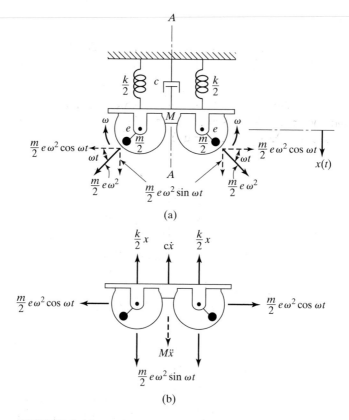

(a)

(b)

FIGURE 3.19   Rotating unbalanced masses.

where $\omega_n = \sqrt{k/M}$ and X and ϕ denote the amplitude and the phase angle of vibration given by

$$X = \frac{me\omega^2}{\left[(k - M\omega^2)^2 + (c\omega)^2\right]^{1/2}} = \frac{me}{M}\left(\frac{\omega}{\omega_n}\right)^2 |H(i\omega)| \qquad (3.80)$$

and

$$\phi = \tan^{-1}\left(\frac{c\omega}{k - M\omega^2}\right) \qquad (3.81)$$

By defining $\zeta = c/c_c$ and $c_c = 2M\omega_n$, Eqs. (3.80) and (3.81) can be rewritten as

$$\frac{MX}{me} = \frac{r^2}{\left[(1 - r^2)^2 + (2\zeta r)^2\right]^{1/2}} = r^2 |H(i\omega)| \qquad (3.82)$$

and

$$\phi = \tan^{-1}\left(\frac{2\zeta r}{1 - r^2}\right) \tag{3.83}$$

The variation of MX/me with r for different values of ζ is shown in Fig. 3.17. On the other hand, the graph of ϕ versus r remains as in Fig. 3.11(b). The following observations can be made from Eq. (3.82) and Fig. 3.17:

1. All the curves begin at zero amplitude. The amplitude near resonance ($\omega = \omega_n$) is markedly affected by damping. Thus if the machine is to be run near resonance, damping should be introduced purposefully to avoid dangerous amplitudes.
2. At very high speeds (ω large), MX/me is almost unity, and the effect of damping is negligible.
3. For $0 < \zeta < \dfrac{1}{\sqrt{2}}$, the maximum of $\dfrac{MX}{me}$ occurs when

$$\frac{d}{dr}\left(\frac{MX}{me}\right) = 0 \tag{3.84}$$

The solution of Eq. (3.84) gives

$$r = \frac{1}{\sqrt{1 - 2\zeta^2}} > 1$$

with the corresponding maximum value of $\dfrac{MX}{me}$ given by

$$\left(\frac{MX}{me}\right)_{\text{max}} = \frac{1}{2\zeta\sqrt{1 - \zeta^2}}$$

Thus the peaks occur to the right of the resonance value of $r = 1$.

4. For $\zeta > \dfrac{1}{\sqrt{2}}$, $\left[\dfrac{MX}{me}\right]$ does not attain a maximum. Its value grows from 0 at $r = 0$ to 1 at $r \to \infty$.

Francis Water Turbine

EXAMPLE 3.5

The schematic diagram of a Francis water turbine is shown in Fig. 3.20 in which water flows from A into the blades B and down into the tail race C. The rotor has a mass of 250 kg and an unbalance (me) of 5 kg-mm. The radial clearance between the rotor and the stator is 5 mm. The turbine operates in the speed range 600 to 6000 rpm. The steel shaft carrying the rotor can be assumed to be

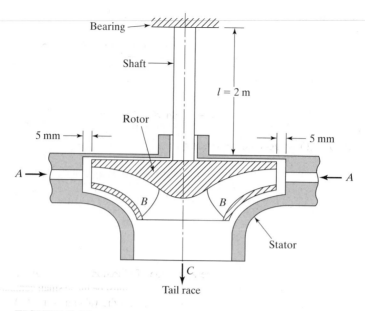

FIGURE 3.20   Francis water turbine.

clamped at the bearings. Determine the diameter of the shaft so that the rotor is always clear of the stator at all the operating speeds of the turbine. Assume damping to be negligible.

Solution: The maximum amplitude of the shaft (rotor) due to rotating unbalance can be obtained from Eq. (3.80) by setting $c = 0$ as

$$X = \frac{me\omega^2}{(k - M\omega^2)} = \frac{me\omega^2}{k(1 - r^2)} \tag{E.1}$$

where $me = 5$ kg-mm, $M = 250$ kg, and the limiting value of $X = 5$ mm. The value of ω ranges from

$$600 \text{ rpm} = 600 \times \frac{2\pi}{60} = 20\pi \text{ rad/s}$$

to

$$6000 \text{ rpm} = 6000 \times \frac{2\pi}{60} = 200\pi \text{ rad/s}$$

while the natural frequency of the system is given by

$$\omega_n = \sqrt{\frac{k}{M}} = \sqrt{\frac{k}{250}} = 0.625\sqrt{k} \text{ rad/s} \tag{E.2}$$

if k is in N/m. For $\omega = 20\pi$ rad/s, Eq. (E.1) gives

$$0.005 = \frac{(5.0 \times 10^{-3}) \times (20\pi)^2}{k\left[1 - \dfrac{(20\pi)^2}{0.004\,k}\right]} = \frac{2\pi^2}{k - 10^5\pi^2}$$

$$k = 10.04 \times 10^4\pi^2 \text{ N/m} \tag{E.3}$$

For $\omega = 200\pi$ rad/s, Eq. (E.1) gives

$$0.005 = \frac{(5.0 \times 10^{-3}) \times (200\pi)^2}{k\left[1 - \dfrac{(200\pi)^2}{0.004k}\right]} = \frac{200\pi^2}{k - 10^7\pi^2}$$

$$k = 10.04 \times 10^6\pi^2 \text{ N/m} \tag{E.4}$$

From Fig. 3.17, we find that the amplitude of vibration of the rotating shaft can be minimized by making $r = \omega/\omega_n$ very large. This means that ω_n must be made small compared to ω—that is, k must be made small. This can be achieved by selecting the value of k as $10.04 \times 10^4\pi^2$ N/m. Since the stiffness of a cantilever beam (shaft) supporting a load (rotor) at the end is given by

$$k = \frac{3EI}{l^3} = \frac{3E}{l^3}\left(\frac{\pi d^4}{64}\right) \tag{E.5}$$

the diameter of the beam (shaft) can be found:

$$d^4 = \frac{64kl^3}{3\pi E} = \frac{(64)(10.04 \times 10^4\pi^2)(2^3)}{3\pi(2.07 \times 10^{11})} = 2.6005 \times 10^{-4} \text{ m}^4$$

or

$$d = 0.1270 \text{ m} = 127 \text{ mm} \tag{E.6}$$

■

3.8   Forced Vibration with Coulomb Damping

For a single degree of freedom system with Coulomb or dry friction damping, subjected to a harmonic force $F(t) = F_0 \sin \omega t$ as in Fig. 3.21, the equation of motion is given by

$$m\ddot{x} + kx \pm \mu N = F(t) = F_0 \sin \omega t \tag{3.85}$$

where the sign of the friction force ($\mu N = \mu mg$) is positive (negative) when the mass moves from left to right (right to left). The exact solution of Eq. (3.85) is quite involved. However, we can expect that if the dry friction damping force is large, the motion of the

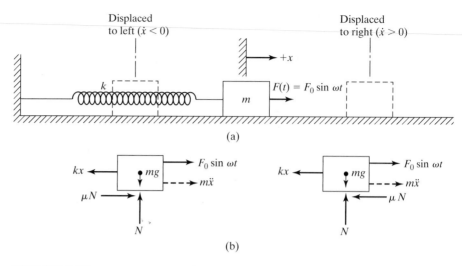

FIGURE 3.21 Single degree of freedom system with Coulomb damping.

mass will be discontinuous. On the other hand, if the dry friction force is small compared to the amplitude of the applied force F_0, the steady-state solution is expected to be nearly harmonic. In this case, we can find an approximate solution of Eq. (3.85) by finding an equivalent viscous damping ratio. To find such a ratio, we equate the energy dissipated due to dry friction to the energy dissipated by an equivalent viscous damper during a full cycle of motion. If the amplitude of motion is denoted as X, the energy dissipated by the friction force μN in a quarter cycle is $\mu N X$. Hence in a full cycle, the energy dissipated by dry friction damping is given by

$$\Delta W = 4\mu N X \tag{3.86}$$

If the equivalent viscous damping constant is denoted as c_{eq}, the energy dissipated during a full cycle (see Eq. 2.94) will be

$$\Delta W = \pi c_{eq}\omega X^2 \tag{3.87}$$

By equating Eqs. (3.86) and (3.87), we obtain

$$c_{eq} = \frac{4\mu N}{\pi \omega X} \tag{3.88}$$

Thus the steady-state response is given by

$$x_p(t) = X \sin(\omega t - \phi) \tag{3.89}$$

where the amplitude X can be found from Eq. (3.60):

$$X = \frac{F_0}{\left[(k - m\omega^2)^2 + (c_{eq}\omega)^2 \right]^{1/2}} = \frac{(F_0/k)}{\left[\left(1 - \frac{\omega^2}{\omega_n^2} \right)^2 + \left(2\zeta_{eq}\frac{\omega}{\omega_n} \right)^2 \right]^{1/2}} \qquad (3.90)$$

with

$$\zeta_{eq} = \frac{c_{eq}}{c_c} = \frac{c_{eq}}{2m\omega_n} = \frac{4\mu N}{2m\omega_n\pi\omega X} = \frac{2\mu N}{\pi m\omega\omega_n X} \qquad (3.91)$$

Substitution of Eq. (3.91) into Eq. (3.90) gives

$$X = \frac{(F_0/k)}{\left[\left(1 - \frac{\omega^2}{\omega_n^2} \right)^2 + \left(\frac{4\mu N}{\pi k X} \right)^2 \right]^{1/2}} \qquad (3.92)$$

The solution of this equation gives the amplitude X as

$$X = \frac{F_0}{k} \left[\frac{1 - \left(\frac{4\mu N}{\pi F_0} \right)^2}{\left(1 - \frac{\omega^2}{\omega_n^2} \right)^2} \right]^{1/2} \qquad (3.93)$$

As stated earlier, Eq. (3.93) can be used only if the friction force is small compared to F_0. In fact, the limiting value of the friction force μN can be found from Eq. (3.93). To avoid imaginary values of X, we need to have

$$1 - \left(\frac{4\mu N}{\pi F_0} \right)^2 > 0 \qquad \text{or} \qquad \frac{F_0}{\mu N} > \frac{4}{\pi}.$$

If this condition is not satisfied, the exact analysis, given in Ref. [3.3], is to be used. The phase angle ϕ appearing in Eq. (3.89) can be found using Eq. (3.52):

$$\phi = \tan^{-1}\left(\frac{c_{eq}\omega}{k - m\omega^2} \right) = \tan^{-1}\left[\frac{2\zeta_{eq}\frac{\omega}{\omega_n}}{1 - \frac{\omega^2}{\omega_n^2}} \right] = \tan^{-1}\left\{ \frac{\frac{4\mu N}{\pi k X}}{1 - \frac{\omega^2}{\omega_n^2}} \right\} \qquad (3.94)$$

Substituting Eq. (3.93) into Eq. (3.94) for X, we obtain

$$\phi = \tan^{-1}\left[\frac{\dfrac{4\mu N}{\pi F_0}}{\left\{1 - \left(\dfrac{4\mu N}{\pi F_0}\right)^2\right\}^{1/2}}\right] \tag{3.95}$$

Equation (3.94) shows that $\tan\phi$ is a constant for a given value of $F_0/\mu N$. ϕ is discontinuous at $\omega/\omega_n = 1$ (resonance) since it takes a positive value for $\omega/\omega_n < 1$ and a negative value for $\omega/\omega_n > 1$. Thus Eq. (3.95) can also be expressed as

$$\phi = \tan^{-1}\left[\frac{\pm\dfrac{4\mu N}{\pi F_0}}{\left\{1 - \left(\dfrac{4\mu N}{\pi F_0}\right)^2\right\}^{1/2}}\right] \tag{3.96}$$

Equation (3.93) shows that friction serves to limit the amplitude of forced vibration for $\omega/\omega_n \neq 1$. However, at resonance ($\omega/\omega_n = 1$), the amplitude becomes infinite. This can be explained as follows. The energy directed into the system over one cycle when it is excited harmonically at resonance is

$$\Delta W' = \int_{\text{cycle}} F \cdot dx = \int_0^\tau F\frac{dx}{dt}dt$$

$$= \int_0^{\tau = 2\pi/\omega} F_0 \sin \omega t \cdot [\omega X \cos(\omega t - \phi)]\, dt \tag{3.97}$$

Since Eq. (3.94) gives $\phi = 90°$ at resonance, Eq. (3.97) becomes

$$\Delta W' = F_0 X \omega \int_0^{2\pi/\omega} \sin^2 \omega t\, dt = \pi F_0 X \tag{3.98}$$

The energy dissipated from the system is given by Eq. (3.86). Since $\pi F_0 X > 4\mu N X$ for X to be real-valued, $\Delta W' > \Delta W$ at resonance (see Fig. 3.22). Thus more energy is directed into the system per cycle than is dissipated per cycle. This extra energy is used to

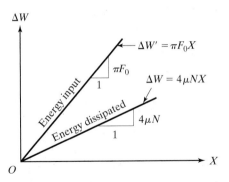

FIGURE 3.22   Energy input and energy dissipated with Coulomb damping.

build up the amplitude of vibration. For the nonresonant condition ($\omega/\omega_n \neq 1$), the energy input can be found from Eq. (3.97):

$$\Delta W' = \omega F_0 X \int_0^{2\pi/\omega} \sin \omega t \cos(\omega t - \phi) \, dt = \pi F_0 X \sin \phi \qquad (3.99)$$

Due to the presence of $\sin \phi$ in Eq. (3.99), the input energy curve in Fig. 3.22 is made to coincide with the dissipated energy curve, so the amplitude is limited. Thus the phase of the motion ϕ can be seen to limit the amplitude of the motion.

   The periodic response of a spring-mass system with Coulomb damping subjected to base excitation is given in Refs. [3.10, 3.11].

Spring-Mass System with Coulomb Damping

EXAMPLE 3.6

A spring-mass system, having a mass of 10 kg and a spring of stiffness of 4000 N/m, vibrates on a horizontal surface. The coefficient of friction is 0.12. When subjected to a harmonic force of frequency 2 Hz, the mass is found to vibrate with an amplitude of 40 mm. Find the amplitude of the harmonic force applied to the mass.

Solution: The vertical force (weight) of the mass is $N = mg = 10 \times 9.81 = 98.1$ N. The natural frequency is

$$\omega_n = \sqrt{\frac{k}{m}} = \sqrt{\frac{4000}{10}} = 20 \text{ rad/s}$$

and the frequency ratio is

$$\frac{\omega}{\omega_n} = \frac{2 \times 2\pi}{20} = 0.6283$$

The amplitude of vibration X is given by Eq. (3.93):

$$X = \frac{F_0}{k} \left[\frac{1 - \left(\frac{4\mu N}{\pi F_0}\right)^2}{\left\{1 - \left(\frac{\omega}{\omega_n}\right)^2\right\}^2} \right]^{1/2}$$

$$0.04 = \frac{F_0}{4000} \left[\frac{1 - \left\{\frac{4(0.12)(98.1)}{\pi F_0}\right\}^2}{(1 - 0.6283^2)^2} \right]^{1/2}$$

The solution of this equation gives $F_0 = 97.9874$ N.

■

3.9   Forced Vibration with Hysteresis Damping

Consider a single degree of freedom system with hysteresis damping and subjected to a harmonic force $F(t) = F_0 \sin \omega t$, as indicated in Fig. 3.23. The equation of motion of the mass can be derived, using Eq. (2.138) as

$$m\ddot{x} + \frac{\beta k}{\omega}\dot{x} + kx = F_0 \sin \omega t \tag{3.100}$$

where $(\beta k/\omega)\dot{x} = (h/\omega)\dot{x}$ denotes the damping force.[5] Although the solution of Eq. (3.100) is quite involved for a general forcing function $F(t)$, our interest is to find the response under a harmonic force.

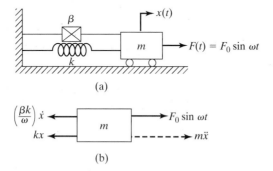

(a)

(b)

FIGURE 3.23   System with hysteresis damping.

---

[5]In contrast to viscous damping, the damping force here can be seen to be a function of the forcing frequency ω (see Section 2.8).

The steady-state solution of Eq. (3.100) can be assumed:

$$x_p(t) = X \sin (\omega t - \phi).$$ (3.101)

By substituting Eq. (3.101) into Eq. (3.100), we obtain

$$X = \frac{F_0}{k \left[\left(1 - \dfrac{\omega^2}{\omega_n^2} \right)^2 + \beta^2 \right]^{1/2}}$$ (3.102)

and

$$\phi = \tan^{-1} \left[\frac{\beta}{\left(1 - \dfrac{\omega^2}{\omega_n^2} \right)} \right]$$ (3.103)

Equations (3.102) and (3.103) are shown plotted in Fig. 3.24 for several values of β. A comparison of Fig. 3.24 with Fig. 3.11 for viscous damping reveals the following:

1. The amplitude ratio

$$\frac{X}{(F_0/k)}$$

   attains its maximum value of $F_0/k\beta$ at the resonant frequency ($\omega = \omega_n$) in the case of hysteresis damping, while it occurs at a frequency below resonance ($\omega < \omega_n$) in the case of viscous damping.
2. The phase angle ϕ has a value of $\tan^{-1}(\beta)$ at $\omega = 0$ in the case of hysteresis damping, while it has a value of zero at $\omega = 0$ in the case of viscous damping. This indicates that the response can never be in phase with the forcing function in the case of hysteresis damping.

Note that if the harmonic excitation is assumed to be $F(t) = F_0 e^{i\omega t}$ in Fig. 3.23, the equation of motion becomes

$$m\ddot{x} + \frac{\beta k}{\omega}\dot{x} + kx = F_0 e^{i\omega t}$$ (3.104)

In this case, the response $x(t)$ is also a harmonic function involving the factor $e^{i\omega t}$. Hence $\dot{x}(t)$ is given by $i\omega x(t)$, and Eq. (3.104) becomes

$$m\ddot{x} + k(1 + i\beta)\, x = F_0 e^{i\omega t}$$ (3.105)

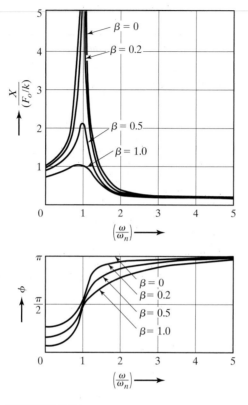

FIGURE 3.24   Steady state response.

where the quantity $k(1 + i\beta)$ is called the *complex stiffness* or *complex damping* [3.7]. The steady-state solution of Eq. (3.105) is given by the real part of

$$x(t) = \frac{F_0 e^{i\omega t}}{k\left[1 - \left(\dfrac{\omega}{\omega_n}\right)^2 + i\beta\right]} \tag{3.106}$$

3.10   Forced Motion with Other Types of Damping

Viscous damping is the simplest form of damping to use in practice, since it leads to linear equations of motion. In the cases of Coulomb and hysteretic damping, we defined equivalent viscous damping coefficients to simplify the analysis. Even for a more complex form of damping, we define an equivalent viscous damping coefficient, as illustrated in the following examples. The practical use of equivalent damping is discussed in Ref. [3.12].

Quadratic Damping

EXAMPLE 3.7

Find the equivalent viscous damping coefficient corresponding to *quadratic* or *velocity squared damping* that is present when a body moves in a turbulent fluid flow.

Solution: The damping force is assumed to be

$$F_d = \pm a(\dot{x})^2 \tag{E.1}$$

where a is a constant, $\dot{x}$ is the relative velocity across the damper, and the negative (positive) sign must be used in Eq. (E.1) when $\dot{x}$ is positive (negative). The energy dissipated per cycle during harmonic motion $x(t) = X \sin \omega t$ is given by

$$\Delta W = 2 \int_{-x}^{x} a(\dot{x})^2 \, dx = 2X^3 \int_{-\pi/2}^{\pi/2} a\omega^2 \cos^3 \omega t \, d(\omega t) = \frac{8}{3} \omega^2 a X^3 \tag{E.2}$$

By equating this energy to the energy dissipated in an equivalent viscous damper (see Eq. 2.94)

$$\Delta W = \pi c_{eq} \omega X^2 \tag{E.3}$$

we obtain the equivalent viscous damping coefficient (c_{eq})

$$c_{eq} = \frac{8}{3\pi} a\omega X \tag{E.4}$$

It can be noted that c_{eq} is not a constant but varies with ω and X. The amplitude of the steady-state response can be found from Eq. (3.30):

$$\frac{X}{\delta_{st}} = \frac{1}{\sqrt{(1 - r^2)^2 + (2\zeta_{eq}r)^2}} \tag{E.5}$$

where $r = \omega/\omega_n$ and

$$\zeta_{eq} = \frac{c_{eq}}{c_c} = \frac{c_{eq}}{2m\omega_n} \tag{E.6}$$

Using Eqs. (E.4) and (E.6), Eq. (E.5) can be solved to obtain

$$X = \frac{3\pi m}{8ar^2} \left[-\frac{(1 - r^2)^2}{2} + \sqrt{\frac{(1 - r^2)^4}{4} + \left(\frac{8ar^2\delta_{st}}{3\pi m}\right)^2} \right]^{1/2} \tag{E.7}$$

■

3.11   Self-Excitation and Stability Analysis

The force acting on a vibrating system is usually external to the system and independent of the motion. However, there are systems for which the exciting force is a function of the motion parameters of the system, such as displacement, velocity, or acceleration. Such systems are called self-excited vibrating systems since the motion itself produces the exciting force (see Problem 3.74). The instability of rotating shafts, the flutter of turbine blades, the flow-induced vibration of pipes, and the automobile wheel shimmy and aerodynamically induced motion of bridges are typical examples of self-excited vibrations.

**3.11.1
Dynamic
Stability
Analysis**

A system is dynamically stable if the motion (or displacement) converges or remains steady with time. On the other hand, if the amplitude of displacement increases continuously (diverges) with time, it is said to be dynamically unstable. The motion diverges and the system becomes unstable if energy is fed into the system through self-excitation. To see the circumstances that lead to instability, we consider the equation of motion of a single degree of freedom system:

$$m\ddot{x} + c\dot{x} + kx = 0 \tag{3.107}$$

If a solution of the form $x(t) = Ce^{st}$, where C is a constant, is assumed, Eq. (3.107) leads to the characteristic equation

$$s^2 + \frac{c}{m}s + \frac{k}{m} = 0 \tag{3.108}$$

The roots of this equation are

$$s_{1,2} = -\frac{c}{2m} \pm \frac{1}{2}\left[\left(\frac{c}{m}\right)^2 - 4\left(\frac{k}{m}\right)\right]^{1/2} \tag{3.109}$$

Since the solution is assumed to be $x(t) = Ce^{st}$, the motion will be diverging and aperiodic if the roots s_1 and s_2 are real and positive. This situation can be avoided if c/m and k/m are positive. The motion will also diverge if the roots s_1 and s_2 are complex conjugates with positive real parts. To analyze the situation, let the roots s_1 and s_2 of Eq. (3.108) be expressed as

$$s_1 = p + iq, \qquad s_2 = p - iq \tag{3.110}$$

where p and q are real numbers so that

$$(s - s_1)(s - s_2) = s^2 - (s_1 + s_2)s + s_1 s_2 = s^2 + \frac{c}{m}s + \frac{k}{m} = 0 \tag{3.111}$$

Equations (3.111) and (3.110) give

$$\frac{c}{m} = -(s_1 + s_2) = -2p, \qquad \frac{k}{m} = s_1 s_2 = p^2 + q^2 \tag{3.112}$$

Equations (3.112) show that for negative p, c/m must be positive and for positive $p^2 + q^2$, k/m must be positive. Thus the system will be dynamically stable if c and k are positive (assuming that m is positive).

Instability of Spring-Supported Mass on Moving Belt

EXAMPLE 3.8

Consider a spring-supported mass on a moving belt, as shown in Fig. 3.25(a). The kinetic coefficient of friction between the mass and the belt varies with the relative (rubbing) velocity, as shown in Fig. 3.25(b). As rubbing velocity increases, the coefficient of friction first decreases from its static value linearly and then starts to increase. Assuming that the rubbing velocity, v, is less than the transition value, v_Q, the coefficient of friction can be expressed as

$$\mu = \mu_0 - \frac{a}{W} v$$

where a is a constant and $W = mg$ is the weight of the mass. Determine the nature of free vibration about the equilibrium position of the mass.

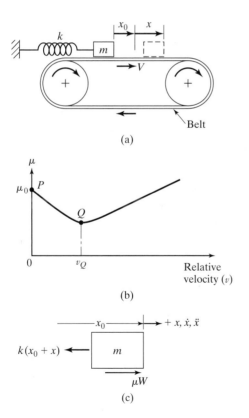

(a)

(b)

(c)

FIGURE 3.25 Motion of a spring-supported mass due to belt friction.

Solution: Let the equilibrium position of mass m correspond to an extension of x_0 of the spring. Then

$$\mu W = k x_0$$

or

$$x_0 = \frac{\mu W}{k} = \frac{\mu_0 W}{k} - \frac{aV}{k}$$

where V is the velocity of the belt. If the mass is displaced by a distance x from its equilibrium position (x_0), the rubbing velocity v is given by

$$v = V - \dot{x}$$

The equation of motion for free vibration can be written, using Newton's second law of motion, as (see Fig. 3.25c):

$$m\ddot{x} = -k(x_0 + x) + \mu W = -k(x_0 + x) + W\left(\mu_0 - \frac{a}{W}(V - \dot{x})\right)$$

i.e.,

$$m\ddot{x} - a\dot{x} + kx = 0 \tag{E.1}$$

Since the coefficient of $\dot{x}$ is negative, the motion given by Eq. (E.1) will be unstable. The solution of Eq. (E.1) is given by

$$x(t) = e^{(a/2m)t}\{C_1 e^{r_1 t} + C_2 e^{r_2 t}\} \tag{E.2}$$

where C_1 and C_2 are constants and

$$r_1 = \frac{1}{2}\left[\left(\frac{a}{m}\right)^2 - 4\left(\frac{k}{m}\right)\right]^{1/2}$$

$$r_2 = -\frac{1}{2}\left[\left(\frac{a}{m}\right)^2 - 4\left(\frac{k}{m}\right)\right]^{1/2}$$

As can be seen from Eq. (E.2), the value of x increases with time. The value of x increases until either $V - \dot{x} = 0$ or $V + \dot{x} = v_Q$. After this, the μ will have a positive slope and hence the nature of motion will be different [3.13].

Note: A similar motion can be observed in belt and pulley-type absorption brakes and in machine tool slides [3.14]. In machine tools, for example, a work table is mounted on suitable guideways and a feed screw is used to impart motion to the work table, as shown in Fig. 3.26. In some cases, the work table may slide in a jerky fashion even when the feed screw has a uniform and smooth motion. Such a motion is known as stick-slip motion. A simplified analysis of the stick-slip motion can be conducted by modeling the work table as a mass (m) and the connection between the work table and the feed screw (which is never perfectly rigid) as a spring (k) and viscous damper (c). The coefficient of friction between the mass and the sliding surface varies as a function of the sliding speed, as indicated

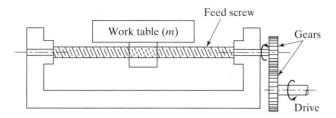

FIGURE 3.26 Motion of work table on feed screw in a machine tool.

in Fig. 3.25(b). The equation of motion of the mass (work table) can be derived as in the case of Eq. (E.1) of Example 3.8 as

$$m\ddot{x} + c\dot{x} + kx = \mu W = W \left[\mu_0 - \frac{a}{W}(V - \dot{x}) \right]$$

i.e.,

$$m\ddot{x} + (c - a)\dot{x} + kx = 0$$

It can be seen that dynamic instability occurs if $c < a$.

■

3.11.2 Dynamic Instability Caused by Fluid Flow

The vibration caused by a fluid flowing around a body is known as flow-induced vibration [3.4]. For example, tall chimneys, submarine periscopes, electric transmission lines, and nuclear fuel rods are found to vibrate violently under certain conditions of fluid flow around them. Similarly, water and oil pipelines and tubes in air compressors undergo severe vibration under certain conditions of fluid flow through them. In all these examples, the vibration of the system continuously extracts energy from the source, leading to larger and larger amplitudes of vibration.

The flow-induced vibration may be caused by various phenomena. For example, in ice-covered electric transmission lines, low-frequency vibration (1 to 2 Hz) known as *galloping*, occurs as a result of the lift and drag forces developed by air flowing around the ice-covered transmission lines. The unstable vibration, known as *flutter*, of airfoil sections is also due to the lift and drag forces developed by the air flowing around the airfoil. In addition, a high-frequency vibration known as *singing of transmission lines*, occurs as a result of the phenomenon of vortex shedding.

To see the phenomenon of galloping of wires, consider a cylindrical section with wind blowing against it at a velocity U, as shown in Fig. 3.27(a) [3.3]. Due to symmetry of the section, the direction of force due to wind will be same as that of the wind. If a small downward velocity u is given to the cylinder, the wind will have an upward component of velocity u (relative to the cylinder) along with the horizontal component U. Thus the direction

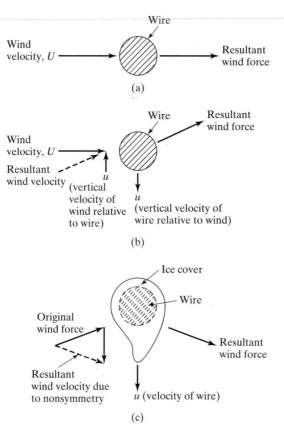

FIGURE 3.27   Galloping of a wire.

of the resultant force due to wind on the cylinder will be upward, as shown in Fig. 3.27(b). Since this force (upward) is opposite to the direction of motion of the cylinder (downward), the motion of the cylinder will be damped. In contrast, if a noncircular section such as an ice-covered cylindrical wire is considered, the resultant wind force may not always oppose the motion of the wire, as shown in Fig. 3.27(c). In such a case, the motion of the wire is aided by the wind forces implying a negative damping in the system.

To visualize the phenomenon of singing of wires, consider a fluid flowing past a smooth cylinder. Under certain conditions, a regular pattern of alternating vortices are formed downstream, as shown in Fig. 3.28. These vortices are called Karman vortices, in honor of the prominent fluid mechanician, Theodor von Karman, who was first to predict the stable spacing of the vortices on theoretical grounds in 1911. The Karman vortices are alternately clockwise and counterclockwise and thus cause harmonically varying lift forces on the cylinder perpendicular to the velocity of the fluid. Experimental data show

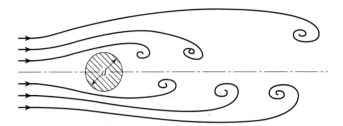

FIGURE 3.28   Fluid flow past a cylinder.

that regular vortex shedding occurs strongly in the range of Reynolds number (Re) from about 60 to 5000. In this case

$$\text{Re} = \frac{\rho V d}{\mu} \tag{3.113}$$

where d is the diameter of the cylinder, ρ is the density, V is the velocity, and μ is the absolute viscosity of the fluid. For Re $>$ 1000, the dimensionless frequency of vortex shedding, expressed as a Strouhal number (St), is approximately equal to 0.21 [3.15]

$$\text{St} \equiv \frac{f d}{V} = 0.21 \tag{3.114}$$

where f is the frequency of vortex shedding. The harmonically varying lift force (F) is given by

$$F(t) = \frac{1}{2} c \rho V^2 A \sin \omega t \tag{3.115}$$

where c is a constant ($c \approx 1$ for a cylinder), A is the projected area of the cylinder perpendicular to the direction of V, ω is the circular frequency ($\omega = 2\pi f$) and t is time. The mechanism of vortex shedding from a cylinder can be called a self-excited one since the fluid flow (V) has no alternating component. From a design point of view, we have to ensure the following:

1. The magnitude of the force exerted on the cylinder, given by Eq. (3.115), is less than the static failure load.
2. Even if the magnitude of force F is small, the frequency of oscillation (f) should not cause fatigue failure during the expected lifetime of the structure (or cylinder).
3. The frequency of vortex shedding (f) does not coincide with the natural frequency of the structure or cylinder to avoid resonance.

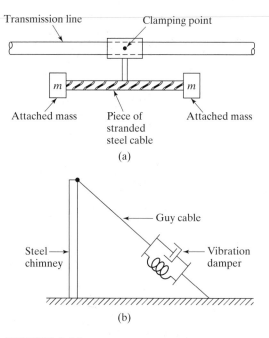

FIGURE 3.29   Stockbridge damper.

Reduction of Flow-Induced Vibration.   Several methods can be used to reduce failures caused by flow-induced vibration.

1. To reduce the singing vibration of transmission lines due to vortex shedding, a damped vibration absorber, known as Stockbridge damper, can be used. A typical Stockbridge damper consists of a short steel cable with two masses attached at the ends. This damper is clamped to the transmission line, as shown in Fig. 3.29(a). The device thus acts as a spring-mass system and can be tuned to the frequency of flow-induced vibration by adjusting its length (the length of the cable) or value of the masses. The Stockbridge damper is clamped to the transmission line at a point where the amplitude of vibration is expected to be large.

2. For tall steel chimneys, the effect of flow-induced vibration can be minimized by attaching vibration dampers through guy cables between the top of the chimney and the ground, as shown in Fig. 3.29(b).

3. For tall chimneys, helical spoilers or strakes can be provided around the chimney, as shown in Fig. 3.30. The helical spoilers break the vortex pattern so that no well-defined excitation is applied to the chimney wall.

4. In high-speed (racing) cars, the flow-induced lift forces can unload the tires, thereby causing problems with steering control and stability of the vehicle. Although lift forces can be countered partly by adding spoilers, the drag force will increase. In recent years, movable inverted airfoils are being used to develop a downward aero-dynamic force with improved stability characteristics (see Fig. 3.31).

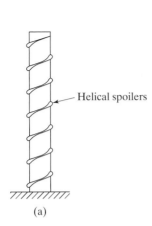

(a)   (b)

FIGURE 3.30   Helical spoilers. (Photo courtesy of Bethlehem Steel Corporation).

FIGURE 3.31   Contemporary sports-racing car with aerodynamic features for low drag and high stability. (Photo courtesy of Goodyear Tire & Rubber Co. Inc.)

Dynamic Instability of an Airfoil

EXAMPLE 3.9

Find the value of free stream velocity u at which the airfoil section (single degree of freedom system) shown in Fig. 3.32 becomes unstable.

Solution

Approach: Find the vertical force acting on the airfoil (or mass m) and obtain the condition that leads to zero damping.

The vertical force acting on the airfoil (or mass m) due to fluid flow can be expressed as [3.4]

$$F = \frac{1}{2} \rho u^2 \, D C_x \qquad \text{(E.1)}$$

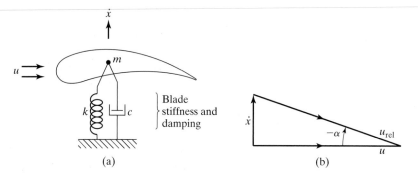

FIGURE 3.32   Modeling of airfoil as a single degree of freedom system.

where ρ = density of the fluid, u = free stream velocity, D = width of the cross section normal to the fluid flow direction, and C_x = vertical force coefficient, which can be expressed as

$$C_x = \frac{u_{rel}^2}{u^2} (C_L \cos \alpha + C_D \sin \alpha) \tag{E.2}$$

where u_{rel} is the relative velocity of the fluid, C_L is the lift coefficient, C_D is the drag coefficient, and α is the angle of attack (see Fig. 3.32)

$$\alpha = -\tan^{-1}\left(\frac{\dot{x}}{u}\right) \tag{E.3}$$

For small angles of attack,

$$\alpha = -\frac{\dot{x}}{u} \tag{E.4}$$

and C_x can be approximated, using Taylor's series expansion, about $\alpha = 0$, as

$$C_x \simeq C_x\bigg|_{\alpha=0} + \frac{\partial C_x}{\partial \alpha}\bigg|_{\alpha=0} \cdot \alpha \tag{E.5}$$

where, for small values of α, $u_{rel} \simeq u$ and Eq. (E.2) becomes

$$C_x = C_L \cos \alpha + C_D \sin \alpha \tag{E.6}$$

Equation (E.5) can be rewritten, using Eqs. (E.6) and (E.4), as

$$C_x = (C_L \cos \alpha + C_D \sin \alpha)\bigg|_{\alpha=0}$$

$$+ \alpha \left[\frac{\partial C_L}{\partial \alpha} \cos \alpha - C_L \sin \alpha + \frac{\partial C_D}{\partial \alpha} \sin \alpha + C_D \cos \alpha \right]\bigg|_{\alpha=0}$$

$$= C_L \bigg|_{\alpha=0} + \alpha \frac{\partial C_x}{\partial \alpha} \bigg|_{\alpha=0}$$

$$= C_L \bigg|_{\alpha=0} - \frac{\dot{x}}{u} \left\{ \frac{\partial C_L}{\partial \alpha} \bigg|_{\alpha=0} + C_D \bigg|_{\alpha=0} \right\} \tag{E.7}$$

Substitution of Eq. (E.7) into Eq. (E.1) gives

$$F = \frac{1}{2} \rho u^2 D C_L \bigg|_{\alpha=0} - \frac{1}{2} \rho u D \frac{\partial C_x}{\partial \alpha} \bigg|_{\alpha=0} \dot{x} \tag{E.8}$$

The equation of motion of the airfoil (or mass m) is

$$m\ddot{x} + c\dot{x} + kx = F = \frac{1}{2} \rho u^2 D C_L \bigg|_{\alpha=0} - \frac{1}{2} \rho u D \frac{\partial C_x}{\partial \alpha} \bigg|_{\alpha=0} \dot{x} \tag{E.9}$$

The first term on the right-hand side of Eq. (E.9) produces a static displacement and hence only the second term can cause instability of the system. The equation of motion, considering only the second term on the right-hand side, is

$$m\ddot{x} + c\dot{x} + kx \equiv m\ddot{x} + \left[c + \frac{1}{2} \rho u D \frac{\partial C_x}{\partial \alpha} \bigg|_{\alpha=0} \right] \dot{x} + kx = 0 \tag{E.10}$$

Note that m includes the mass of the entrained fluid. We can see from Eq. (E.10) that the displacement of the airfoil (or mass m) will grow without bound (i.e., the system becomes unstable) if c is negative. Hence the minimum velocity of the fluid for the onset of unstable oscillations is given by $c = 0$, or,

$$u = - \left\{ \frac{2c}{\rho D \dfrac{\partial C_x}{\partial \alpha} \bigg|_{\alpha=0}} \right\} \tag{E.11}$$

The value of $\dfrac{\partial C_x}{\partial \alpha} \bigg|_{\alpha=0} = -2.7$ for a square section in a steady flow [3.4].

■

Note: An analysis similar to that of Example 3.9 is applicable to other vibrating structures such as water tanks (Fig. 3.33a) and galloping of ice-coated power lines (Fig. 3.33b) under wind loading.

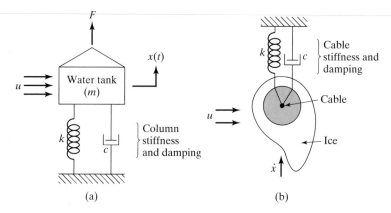

FIGURE 3.33   Instability of typical vibrating structures.

<hr>

EXAMPLE 3.10

Flow-Induced Vibration of a Chimney

A steel chimney has a height 2 m, an inner diameter 0.75 m, and an outer diameter 0.80 m. Find the velocity of the wind flowing around the chimney which will induce transverse vibration of the chimney in the direction of airflow.

Solution

Approach: Model the chimney as a cantilever beam and equate the natural frequency of the transverse vibration of the chimney to the frequency of vortex shedding.

To find the natural frequency of the transverse vibration of the chimney, the Rayleigh's energy method can be used by assuming a suitable transverse deflection of the cantilever beam (see Section 8.7). However, in this case, we use the natural frequencies of the cantilever beam given in Fig. 8.15. Figure 8.15 gives the fundamental natural frequency of transverse vibration (ω_1) of a cantilever (fixed-free) beam as

$$\omega_1 = (\beta_1 l)^2 \sqrt{\frac{EI}{\rho A l^4}} \tag{E.1}$$

where

$$\beta_1 l = 1.875104 \tag{E.2}$$

For the chimney, $E = 207 \times 10^9$ Pa, ρg = unit weight = 76.5×10^3 N/m³, $l = 20$ m, $d = 0.75$ m, $D = 0.80$ m,

$$A = \frac{\pi}{4}(D^2 - d^2) = \frac{\pi}{4}(0.80^2 - 0.75^2) = 0.0608685 \text{ m}^2$$

and

$$I = \frac{\pi}{64}(D^4 - d^4) = \frac{\pi}{64}(0.80^4 - 0.75^4) = 0.004574648 \text{ m}^4$$

Thus

$$\omega_1 = (1.875104)^2 \left\{ \frac{(207 \times 10^9)(0.004574648)}{\left(\dfrac{76.5 \times 10^3}{9.81}\right)(0.0608685)\,(20)^4} \right\}^{1/2}$$

$$= 12.415417 \text{ rad/s} = 1.975970 \text{ Hz}$$

The frequency of vortex shedding (f) is given by the Strouhal number:

$$\text{St} = \frac{fd}{V} = 0.21$$

Using $d = 0.80$ m and $f = f_1 = 1.975970$ Hz, the velocity of wind (V) which causes resonance can be determined as

$$V = \frac{f_1 d}{0.21} = \frac{1.975970\,(0.80)}{0.21} = 7.527505 \text{ m/s}$$

3.12 Examples Using MATLAB

Total Response of an Undamped System

EXAMPLE 3.11

Using MATLAB, plot the response of a spring-mass system under a harmonic force for the following data:

$$m = 5 \text{ kg}, k = 2000 \text{ N/m}, F(t) = 100 \cos 30\, t \text{ N}, x_0 = 0.1 \text{ m}, \dot{x}_0 = 0.1 \text{ m/s}.$$

Solution: The response of the system is given by Eq. (3.9), which can be rewritten as

$$x(t) = \frac{\dot{x}_0}{\omega_n} \sin \omega_n t + \left(x_0 - \frac{f_0}{\omega_n^2 - \omega^2} \right) \cos \omega_n t + \frac{f_0}{\omega_n^2 - \omega^2} \cos \omega t \qquad \text{(E.1)}$$

where $f_0 = \dfrac{F_0}{m} = \dfrac{100}{5} = 20$, $\omega_n = \sqrt{\dfrac{k}{m}} = 20$ rad/s, and $\omega = 30$ rad/s.

Equation (E.1) is plotted using the following MATLAB program:

```
% Ex3_11.m
F0  =  100;
wn  =  20;
m   =  5;
w   =  30;
```

```
x0   =    0.1;
x0_dot = 0.1;
f_0 = F0/m;
for i = 1: 101
    t(i) = 2 * (i−1)/100;
    x(i) = x0_dot*sin(wn*t(i))/wn + (x0 − f_0/(wn^2−w^2))*cos
        (wn*t(i)).. + f_0/ (wn^2−w^2)*cos(w*t(i));
end
plot (t, x);
xlabel ('t');
ylabel ('x(t)');
title ('Ex3.11')
```

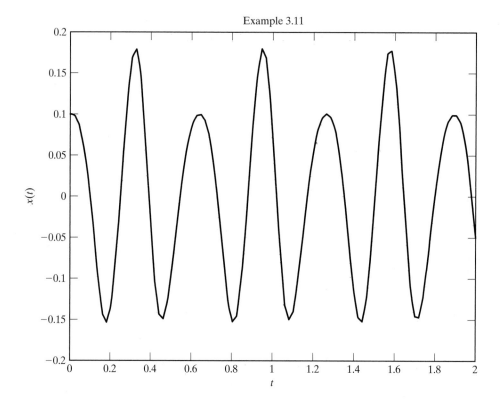

Example 3.11

■

Forced Response of a System with Coulomb Damping

EXAMPLE 3.12

Using MATLAB, plot the forced response of a spring-mass system with Coulomb damping for the following data: $m = 5$ kg, $k = 2000$ N/m, $\mu = 0.5$, $F(t) = 100 \sin 30\, t$ N, $x_0 = 0.1$ m, $\dot{x}_0 = 0.1$ m/s.

Solution: The equation of motion of the system can be expressed as

$$m\ddot{x} + kx + \mu mg \, \text{sgn}(\dot{x}) = F_0 \sin \omega t \qquad (E.1)$$

which can be rewritten as a system of two first-order differential equations (using $x_1 = x$ and $x_2 = \dot{x}$) as

$$\dot{x}_1 = x_2$$

$$\dot{x}_2 = \frac{F_0}{m} \sin \omega t - \frac{k}{m} x_1 - \mu g \, \text{sgn}(x_2) \qquad (E.2)$$

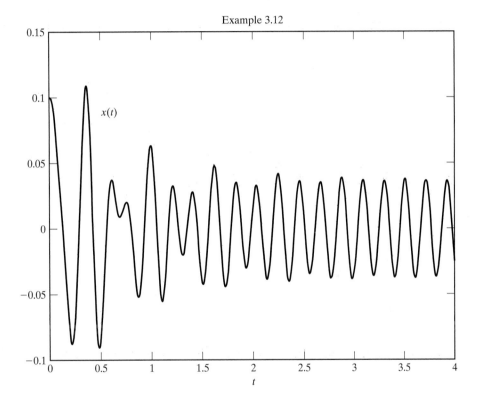

Example 3.12

with the initial conditions $x_1(0) = 0.1$ and $x_2(0) = 0.1$. The MATLAB solution of Eq. (E.2), using **ode23**, is given next.

```
% Ex3_12.m
% This program will use the function dfunc3_12.m, they should
% be in the same folder
tspan   =  [0: 0.01: 4];
```

```
x0    =   [0.1; 0.1];
[t, x]    =   ode23 ('dfunc3_12', tspan, x0);
disp ('         t          x(t)       xd(t)');
disp ([t x]);
plot (t, x(:, 1));
xlabel ('t');
gtext ('x(t)');
title ('Ex3.12');


% dfunc3_12.m
function f = dfunc3_12 (t, x)
f = zeros (2, 1);
f(1) = x(2);
f(2) = 100*sin(30*t)/5 - 9.81*0.5*sign(x(2)) - (2000/5)*x(1);
```

```
>> Ex3_12
      t              x(t)           xd(t)
         0           0.1000         0.1000
    0.0100           0.0991        -0.2427
    0.0200           0.0954        -0.4968
    0.0300           0.0894        -0.6818
    0.0400           0.0819        -0.8028
    0.0500           0.0735        -0.8704
      .
      .
      .
    3.9500           0.0196        -0.9302
    3.9600           0.0095        -1.0726
    3.9700          -0.0016        -1.1226
    3.9800          -0.0126        -1.0709
    3.9900          -0.0226        -0.9171
    4.0000          -0.0307        -0.6704
```

∎

Response of a System Under Base Excitation

EXAMPLE 3.13

Using MATLAB, find and plot the response of a viscously damped spring-mass system under the base excitation $y(t) = Y \sin \omega t$ for the following data: $m = 1200$ kg, $k = 4 \times 10^5$ N/m, $\zeta = 0.5$, $Y = 0.05$ m, $\omega = 29.0887$ rad/s, $x_0 = 0$, $\dot{x}_0 = 0.1$ m/s.

Solution: The equation of motion, Eq. (3.64),

$$m \ddot{x} + c \dot{x} + k x = k y + c \dot{y} \tag{E.1}$$

can be expressed as a system of two first-order ordinary differential equations (using $x_1 = x$ and $x_2 = \dot{x}$) as

$$\dot{x}_1 = x_2$$

$$\dot{x}_2 = -\frac{c}{m} x_2 - \frac{k}{m} x_1 + \frac{k}{m} y + \frac{c}{m} \dot{y} \tag{E.2}$$

Example 3.13

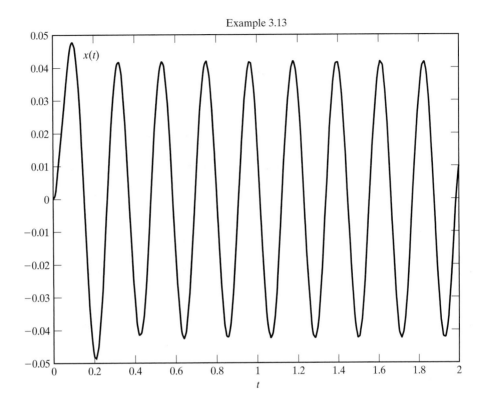

with $c = \zeta\, c_c = 2\, \zeta\, \sqrt{k\, m} = 2\,(0.5)\,\sqrt{(4 \times 10^5)(1200)}$, $y = 0.5 \sin 29.0887\, t$, and $\dot{y} = (29.0887)$ $(0.05) \cos 29.0887\, t$. The MATLAB solution of Eq. (E.2), using **ode23**, is given below.

```
% Ex3_13.m
% This program will use the function dfunc3_13.m, they should
% be in the same folder
tspan = [0: 0.01: 2];
x0 = [0; 0.1];
[t, x] = ode23 ('dfunc3_13', tspan, x0);
disp ('     t          x(t)    xd(t)');
disp ([t x]);
plot (t, x (:, 1));
xlabel ('t');
gtext ('x(t)');
title ('Ex3.13');

% dfunc3_13.m
function f = dfunc3_13 (t, x)
f = zeros (2, 1);
f(1) = x(2);
f(2) = 400000*0.05*sin(29.0887*t)/1200 + ...
    sqrt (400000*1200)*29.0887*0.05*cos(29.0887*t)/1200 ...
    - sqrt(400000*1200)*x(2)/1200 - (400000/1200)*x(1);
```

```
>> Ex3_13
        t              x(t)            xd(t)
        0               0             0.1000
   0.0100           0.0022           0.3422
   0.0200           0.0067           0.5553
   0.0300           0.0131           0.7138
   0.0400           0.0208           0.7984
   0.0500           0.0288           0.7976
        .
        .
        .
   1.9500          -0.0388           0.4997
   1.9600          -0.0322           0.8026
   1.9700          -0.0230           1.0380
   1.9800          -0.0118           1.1862
   1.9900           0.0004           1.2348
   2.0000           0.0126           1.1796
```

■

Steady-State Response of a Viscously Damped System

EXAMPLE 3.14

Develop a general-purpose MATLAB program, called **Program3.m**, to find the steady-state response of a viscously damped single degree of freedom system under the harmonic force $F_0 \cos \omega t$ or $F_0 \sin \omega t$. Use the program to find and plot the response of a system with the following data:

$$m = 5 \text{ kg}, c = 20 \text{ N-s/m}, k = 500 \text{ N/m}, F_0 = 250 \text{ N}, \omega = 40 \text{ rad/s}, n = 40 \text{ and } ic = 0.$$

Solution: **Program3.m** is developed to accept the following input data:

xm = mass

xc = damping constant

xk = spring constant

f0 = amplitude of the forcing function

om = forcing frequency

n = number of time steps in a cycle at which the response is to be computed

ic = 1 for cosine-type forcing function; 0 for sine-type forcing function

The program gives the following output:

$$\text{step number i, } x(i), \dot{x}(i), \ddot{x}(i)$$

The program also plots the variations of x, $\dot{x}$, and $\ddot{x}$ with time.

```
>> program3
Steady state response of an undamped
Single degree of freedom system under harmonic force

Given data
xm  =  5.00000000e+000
xc  =  2.00000000e+001
xk  =  5.00000000e+002
f0  =  2.50000000e+002
om  =  4.00000000e+001
ic  =  0
n   =  20
```

Response:

i	x(i)	xd(i)	xdd(i)
1	1.35282024e−002	1.21035472e+000	−2.16451238e+001
2	2.22166075e−002	9.83897315e−001	−3.55465721e+001
3	2.87302863e−002	6.61128738e−001	−4.59684581e+001
4	3.24316314e−002	2.73643972e−001	−5.18906102e+001
5	3.29583277e−002	−1.40627096e−001	−5.27333244e+001
6	3.02588184e−002	−5.41132540e−001	−4.84141094e+001
7	2.45973513e−002	−8.88667916e−001	−3.93557620e+001
8	1.65281129e−002	−1.14921388e+000	−2.64449806e+001
9	6.84098018e−003	−1.29726626e+000	−1.09455683e+001
10	−3.51579846e−003	−1.31833259e+000	5.62527754e+000
11	−1.35284247e−002	−1.21035075e+000	2.16454794e+001
12	−2.22167882e−002	−9.83890787e−001	3.55468612e+001
13	−2.87304077e−002	−6.61120295e−001	4.59686523e+001
14	−3.24316817e−002	−2.73634442e−001	5.18906907e+001
15	−3.29583019e−002	1.40636781e−001	5.27332831e+001
16	−3.02587190e−002	5.41141432e−001	4.84139504e+001
17	−2.45971881e−002	8.88675144e−001	3.93555009e+001
18	−1.65279018e−002	1.14921874e+000	2.64446429e+001
19	−6.84074192e−003	1.29726827e+000	1.09451871e+001
20	3.51604059e−003	1.31833156e+000	−5.62566494e+000

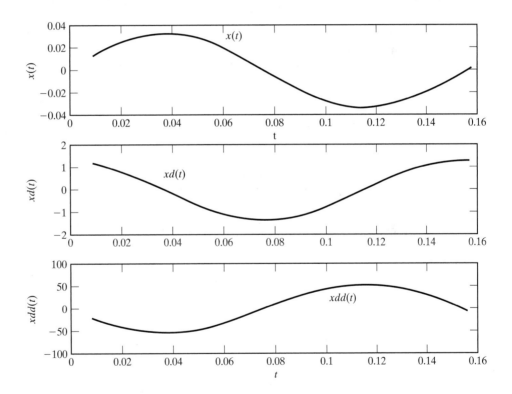

3.13 C++ Program

An interactive C++ program, called **Program3.cpp**, is used to find the steady-state response of a viscously damped single degree of freedom system. The input and output parameters of the program are similar to those of the MATLAB program, **Program3.m**, given in Example 3.14.

EXAMPLE 3.15

Steady-State Response of a Viscously Damped System Using C++

Find the steady-state response of the viscously damped system described in Example 3.14.

Solution: The input data are to be entered interactively. The input and output of the program are given below.

```
Please input n and ic:
20 0

Please input xm, xc, xk, f0, om:
5.0 20.0 500.0 250.0 40.0

STEADY STATE RESPONSE OF AN UNDERDAMPED
SINGLE DEGREE OF FREEDOM SYSTEM UNDER HARMONIC FORCE

GIVEN DATA:
XM = 5
XC = 20
F0 = 250
OM = 40
IC = 0
N = 20

RESONSE:
    I          X(I)            XD(I)           XDD(I)
    0       0.01352820      1.21035472      -21.64512378
    1       0.02221661      0.98389732      -35.54657206
    2       0.02873029      0.66112874      -45.96845807
    3       0.03243163      0.27364397      -51.89061025
    4       0.03295833     -0.14062710      -52.73332439
    5       0.03025882     -0.54113254      -48.41410938
    6       0.02459735     -0.88866792      -39.35576205
    7       0.01652811     -1.14921388      -26.44498065
    8       0.00684098     -1.29726626      -10.94556828
    9      -0.00351580     -1.31833259        5.62527754
   10      -0.01352842     -1.21035075       21.64547945
   11      -0.02221679     -0.98389079       35.54686119
   12      -0.02873041     -0.66112029       45.96865235
   13      -0.03243168     -0.27363444       51.89069066
   14      -0.03295830      0.14063678       52.73328306
   15      -0.03025872      0.54114143       48.41395037
   16      -0.02459719      0.88867514       39.35550091
   17      -0.01652790      1.14921874       26.44464294
   18      -0.00684074      1.29726827       10.94518707
   19       0.00351604      1.31833156       -5.62566494
```

■

3.14   Fortran Program

A Fortran computer program, in the form of the subroutine **HARESP.F**, is given for finding the steady-state response of a viscously damped system. The arguments of the subroutine are similar to those described for the MATLAB program, **Program3.m**, given in Example 3.14.

EXAMPLE 3.16

Steady-State Response of a Viscously Damped System Using Fortran

Find the steady-state response of the viscously damped system described in Example 3.14 using the subroutine **HARESP.F**.

Solution: The main program that calls the subroutine HARESP.F and the subroutine HARESP.F are given as **PROGRAM3.F**. The output of the program is given below.

```
STEADY STATE RESPONSE OF AN UNDERDAMPED
SINGLE DEGREE OF FREEDOM SYSTEM UNDER HARMONIC FORCE

GIVEN DATA:
XM = 0.50000000E+01
XC = 0.20000000E+02
XK = 0.50000000E+03
FO = 0.25000000E+03
OM = 0.40000000E+02
IC = 0
N  =20

RESPONSE:
   I         X(I)            XD(I)            XDD(I)
   1    0.13528203E-01   0.12103548E+01   -0.21645124E+02
   2    0.22216609E-01   0.98389733E+00   -0.35546574E+02
   3    0.28730286E-01   0.66112888E+00   -0.45968460E+02
   4    0.32431632E-01   0.27364409E+00   -0.51890614E+02
   5    0.32958329E-01  -0.14062698E+00   -0.52733330E+02
   6    0.30258821E-01  -0.54113245E+00   -0.48414116E+02
   7    0.24597352E-01  -0.88866800E+00   -0.39355762E+02
   8    0.16528117E-01  -0.11492138E+01   -0.26444986E+02
   9    0.68409811E-02  -0.12972662E+01   -0.10945570E+02
  10   -0.35157942E-02  -0.13183327E+01    0.56252708E+01
  11   -0.13528424E-01  -0.12103508E+01    0.21645479E+02
  12   -0.22216786E-01  -0.98389095E+00    0.35546856E+02
  13   -0.28730409E-01  -0.66112041E+00    0.45968655E+02
  14   -0.32431684E-01  -0.27363408E+00    0.51890697E+02
  15   -0.32958303E-01   0.14063700E+00    0.52733284E+02
  16   -0.30258721E-01   0.54114151E+00    0.48413952E+02
  17   -0.24597190E-01   0.88867509E+00    0.39355507E+02
  18   -0.16527895E-01   0.11492189E+01    0.26444633E+02
  19   -0.68407385E-02   0.12972684E+01    0.10945182E+02
  20    0.35160405E-02   0.13183316E+01   -0.56256652E+01
```

REFERENCES

3.1 G. B. Thomas and R. L. Finney, *Calculus and Analytic Geometry* (6th ed.), Addison-Wesley, Reading, Mass., 1984.

3.2 J. W. Nilsson, *Electric Circuits*, Addison-Wesley, Reading, Mass., 1983.

3.3 J. P. Den Hartog, "Forced vibrations with combined Coulomb and viscous friction," *Journal of Applied Mechanics* (Transactions of ASME), Vol. 53, 1931, pp.APM 107–115.

3.4 R. D. Blevins, *Flow-Induced Vibration* (2nd ed.), Van Nostrand Reinhold, New York, 1990.

3.5 J. C. R. Hunt and D. J. W. Richards, "Overhead line oscillations and the effect of aerodynamic dampers," *Proceedings of the Institute of Electrical Engineers*, London, Vol. 116, 1969, pp. 1869–1874.

3.6 K. P. Singh and A. I. Soler, *Mechanical Design of Heat Exchangers and Pressure Vessel Components*, Arcturus Publishers, Cherry Hill, N. J., 1984.

3.7 N. O. Myklestad, "The concept of complex damping," *Journal of Applied Mechanics*, Vol. 19, 1952, pp. 284–286.

3.8 R. Plunkett (ed.), *Mechanical Impedance Methods for Mechanical Vibrations*, American Society of Mechanical Engineers, New York, 1958.

3.9 A. D. Dimarogonas, *Vibration Engineering*, West Publishing, St. Paul, 1976.

3.10 B. Westermo and F. Udwadia, "Periodic response of a sliding oscillator system to harmonic excitation," *Earthquake Engineering and Structural Dynamics*, Vol. 11, No. 1, 1983, pp. 135–146.

3.11 M. S. Hundal, "Response of a base excited system with Coulomb viscous friction," *Journal of Sound and Vibration*, Vol. 64, 1979, pp. 371–378.

3.12 J. P. Bandstra, "Comparison of equivalent viscous damping and nonlinear damping in discrete and continuous vibrating systems," *Journal of Vibration, Acoustics, Stress, and Reliability in Design*, Vol. 105, 1983, pp. 382–392.

3.13 W. G. Green, *Theory of Machines* (2nd ed.), Blackie & Son, London, 1962.

3.14 S. A. Tobias, *Machine-Tool Vibration*, Wiley, New York, 1965.

3.15 R. W. Fox and A. T. McDonald, *Introduction to Fluid Mechanics* (4th ed.), Wiley, New York, 1992.

REVIEW QUESTIONS

3.1 Give brief answers to the following:

1. How are the amplitude, frequency, and phase of a steady-state vibration related to those of the applied harmonic force for an undamped system?
2. Explain why a constant force on the vibrating mass has no effect on the steady-state vibration.
3. Define the term *magnification factor*. How is the magnification factor related to the frequency ratio?
4. What will be the frequency of the applied force with respect to the natural frequency of the system if the magnification factor is less than unity?
5. What are the amplitude and the phase angle of the response of a viscously damped system in the neighborhood of resonance?
6. Is the phase angle corresponding to the peak amplitude of a viscously damped system ever larger than 90°?
7. Why is damping considered only in the neighborhood of resonance in most cases?
8. Show the various terms in the forced equation of motion of a viscously damped system in a vector diagram.

9. What happens to the response of an undamped system at resonance?
10. Define the following terms: *beating, quality factor, transmissibility, complex stiffness, quadratic damping*.
11. Give a physical explanation of why the magnification factor is nearly equal to 1 for small values of *r* and is small for large values of *r*.
12. Will the force transmitted to the base of a spring-mounted machine decrease with the addition of damping?
13. How does the force transmitted to the base change as the speed of the machine increases?
14. If a vehicle vibrates badly while moving on a uniformly bumpy road, will a change in the speed improve the condition?
15. Is it possible to find the maximum amplitude of a damped forced vibration for any value of *r* by equating the energy dissipated by damping to the work done by the external force?
16. What assumptions are made about the motion of a forced vibration with nonviscous damping in finding the amplitude?
17. Is it possible to find the approximate value of the amplitude of a damped forced vibration without considering damping at all? If so, under what circumstances?
18. Is dry friction effective in limiting the reasonant amplitude?
19. How do you find the response of a viscously damped system under rotating unbalance?
20. What is the frequency of the response of a viscously damped system when the external force is $F_0 \sin \omega t$? Is this response harmonic?
21. What is the difference between the peak amplitude and the resonant amplitude?
22. Why is viscous damping used in most cases rather than other types of damping?
23. What is self-excited vibration?

3.2 Indicate whether each of the following statements is true or false:

1. The magnification factor is the ratio of maximum amplitude and static deflection.
2. The response will be harmonic if excitation is harmonic.
3. The phase angle of the response depends on the system parameter *m, c, k,* and *ω*.
4. The phase angle of the response depends on the amplitude of the forcing function.
5. During beating, the amplitude of the response builds up and then diminishes in a regular pattern.
6. The *Q*-factor can be used to estimate the damping in a system.
7. The half-power points denote the values of frequency ratio where the amplification factor falls to $Q/\sqrt{2}$ where Q is the Q-factor.
8. The amplitude ratio attains its maximum value at resonance in the case of viscous damping.
9. The response is always in phase with the harmonic forcing function in the case of hysteresis damping.
10. Damping reduces the amplitude ratio for all values of the forcing frequency.
11. The unbalance in a rotating machine causes vibration.
12. The steady-state solution can be assumed to be harmonic for small values of dry friction force.
13. In a system with rotational unbalance, the effect of damping becomes negligibly small at higher speeds.

3.3 Fill in each of the following blanks with the appropriate word:

1. The excitation can be _____, periodic, nonperiodic, or random in nature.

2. The response of a system to a harmonic excitation is called _____ response.
3. The response of a system to suddenly applied nonperiodic excitation is called _____ response.
4. When the frequency of excitation coincides with the natural frequency of the system, the condition is known as _____.
5. The magnification factor is also known as _____ factor.
6. The phenomenon of _____ can occur when the forcing frequency is close to the natural frequency of the system.
7. When the base of system is subject to harmonic motion with amplitude Y resulting in a response amplitude X, the ratio $\frac{X}{Y}$ is called the displacement _____.
8. $Z(i\omega) = -m\omega^2 + i\omega c + k$ is called the mechanical _____ of the system.
9. The difference between the frequencies associated with half-power points is called the _____ of the system.
10. The value of the amplitude ratio at resonance is called _____ factor.
11. The dry friction damping is also known as _____ damping.
12. For _____ values of dry friction damping, the motion of the mass will be discontinuous.
13. The quantity $k(1 + i\beta)$ in hysteresis damping is called _____ stiffness.
14. Quadratic or velocity squared damping is present whenever a body moves in a _____ fluid flow.
15. In self-excited systems, the _____ itself produces the exciting force.
16. The flutter of turbine blades is an example of _____ vibration.
17. The motion _____ and the system becomes unstable during self-excitation.

3.4 Select the most appropriate answer out of the choices given:

1. The response of an undamped system under resonance will be
   (a) very large   (b) infinity   (c) zero
2. The reduction of the amplitude ratio in the presence of damping is very significant
   (a) near $\omega = \omega_n$ (b) near $\omega = 0$ (c) near $\omega = \infty$
3. The frequency of beating is
   (a) $\omega_n - \omega$ (b) ω_n (c) ω
4. The energy dissipated in a cycle by dry friction damping is given by
   (a) $4\mu N X$ (b) $4\mu N$ (c) $4\mu N X^2$
5. The complex frequency response, $H(i\omega)$, is defined by
   (a) $\dfrac{kX}{F_0}$ (b) $\dfrac{X}{F_0}$ (c) $\left|\dfrac{kX}{F_0}\right|$
6. The energy dissipated over the following duration is considered in finding the equivalent viscous damping constant of a system with Coulomb damping:
   (a) half cycle   (b) full cycle   (c) one second
7. The damping force depends on the frequency of the applied force in the case of
   (a) viscous damping   (b) Coulomb damping   (c) hysteresis damping
8. The system governed by the equation $m\ddot{x} + c\dot{x} + kx = 0$ is dynamically stable if
   (a) k is positive   (b) c and k are positive   (c) c is positive
9. Complex stiffness or complex damping is defined in the case of
   (a) hysteresis damping   (b) Coulomb damping   (c) viscous damping

10. The equation of motion of a machine (rotating at frequency ω) of mass M, with an unbalanced mass m, at radius e, is given by

(a) $m\ddot{x} + c\dot{x} + kx = me\omega^2 \sin \omega t$

(b) $M\ddot{x} + c\dot{x} + kx = me\omega^2 \sin \omega t$

(c) $M\ddot{x} + c\dot{x} + kx = Me\omega^2 \sin \omega t$

11. The force transmissibility of a system, subjected to base excitation (with amplitude Y) resulting in a transmitted force F_T, is defined as

(a) $\dfrac{F_T}{kY}$ (b) $\dfrac{X}{kY}$ (c) $\dfrac{F_T}{k}$

3.5 Using the notation:

$r = $ frequency ratio $= \dfrac{\omega}{\omega_n}$

$\omega = $ forcing frequency

$\omega_n = $ natural frequency

$\zeta = $ damping ratio

$\omega_1, \omega_2 = $ frequencies corresponding to half-power points

match the items in the two columns below:

1. Magnification factor of an undamped system   (a) $\dfrac{2\pi}{\omega_n - \omega}$

2. Period of beating   (b) $\left[\dfrac{1 + (2\zeta r)^2}{(1 - r^2)^2 + (2\zeta r)^2} \right]^{1/2}$

3. Magnification factor of a damped system   (c) $\dfrac{\omega_n}{\omega_2 - \omega_1}$

4. Damped frequency   (d) $\dfrac{1}{1 - r^2}$

5. Quality factor   (e) $\omega_n \sqrt{1 - \zeta^2}$

6. Displacement transmissibility   (f) $\left[\dfrac{1}{(1 - r^2)^2 + (2\zeta r)^2} \right]^{1/2}$

3.6 Match the following equations of motion:

1. $m\ddot{z} + c\dot{z} + kz = -m\ddot{y}$ (a) System with Coulomb damping

2. $M\ddot{x} + c\dot{x} + kx = me\omega^2 \sin \omega t$ (b) System with viscous damping

3. $m\ddot{x} + kx \pm \mu N = F(t)$ (c) System subject to base excitation

4. $m\ddot{x} + k(1 + i\beta)x = F_0 \sin \omega t$ (d) System with hysteresis damping

5. $m\ddot{x} + c\dot{x} + kx = F_0 \sin \omega t$ (e) System with rotating unbalance

PROBLEMS

The problem assignments are organized as follows:

Problems	Section Covered	Topic Covered
3.1–3.20	3.3	Undamped systems
3.21–3.40	3.4	Damped systems
3.41–3.50	3.6	Base excitation
3.51–3.61	3.7	Rotating unbalance
3.62–3.64	3.8	Response under Coulomb damping
3.65–3.66	3.9	Response under hysteresis damping
3.67–3.70	3.10	Response under other types of damping
3.71–3.74	3.11	Self-excitation and stability
3.75–3.79	3.12	Examples using MATLAB
3.80–3.81	3.13	C++ Program
3.82–3.85	3.14	Fortran Program
3.86–3.87	—	Design projects

3.1 A weight of 50 N is suspended from a spring of stiffness 4000 N/m and is subjected to a harmonic force of amplitude 60 N and frequency 6 Hz. Find (a) the extension of the spring due to the suspended weight, (b) the static displacement of the spring due to the maximum applied force, and (c) the amplitude of forced motion of the weight.

3.2 A spring-mass system is subjected to a harmonic force whose frequency is close to the natural frequency of the system. If the forcing frequency is 39.8 Hz and the natural frequency is 40.0 Hz, determine the period of beating.

3.3 Consider a spring-mass system, with $k = 4000$ N/m and $m = 10$ kg, subject to a harmonic force $F(t) = 400 \cos 10\,t$ N. Find and plot the total response of the system under the following initial conditions:

    a. $x_0 = 0.1$ m, $\dot{x}_0 = 0$
    b. $x_0 = 0$, $\dot{x}_0 = 10$ m/s
    c. $x_0 = 0.1$ m, $\dot{x}_0 = 10$ m/s

3.4 Consider a spring-mass system, with $k = 4000$ N/m and $m = 10$ kg, subject to a harmonic force $F(t) = 400 \cos 20\,t$ N. Find and plot the total response of the system under the following initial conditions:

    a. $x_0 = 0.1$ m, $\dot{x}_0 = 0$
    b. $x_0 = 0$, $\dot{x}_0 = 10$ m/s
    c. $x_0 = 0.1$ m, $\dot{x}_0 = 10$ m/s

3.5 Consider a spring-mass system, with $k = 4000$ N/m and $m = 10$ kg, subject to a harmonic force $F(t) = 400 \cos 20.1\,t$ N. Find and plot the total response of the system under the following initial conditions:

    a. $x_0 = 0.1$ m, $\dot{x}_0 = 0$

    b. $x_0 = 0$, $\dot{x}_0 = 10$ m/s

    c. $x_0 = 0.1$ m, $\dot{x}_0 = 10$ m/s

3.6 Consider a spring-mass system, with $k = 4000$ N/m and $m = 10$ kg, subject to a harmonic force $F(t) = 400 \cos 30\,t$ N. Find and plot the total response of the system under the following initial conditions:

    a. $x_0 = 0.1$ m, $\dot{x}_0 = 0$

    b. $x_0 = 0$, $\dot{x}_0 = 10$ m/s

    c. $x_0 = 0.1$ m, $\dot{x}_0 = 10$ m/s

3.7 A spring-mass system consists of a mass weighing 100 N and a spring with a stiffness of 2000 N/m. The mass is subjected to resonance by a harmonic force $F(t) = 25 \cos \omega t$ N. Find the amplitude of the forced motion at the end of (a) $\frac{1}{4}$ cycle, (b) $2\frac{1}{2}$ cycles, and (c) $5\frac{3}{4}$ cycles.

3.8 A mass m is suspended from a spring of stiffness 4000 N/m and is subjected to a harmonic force having an amplitude of 100 N and a frequency of 5 Hz. The amplitude of the forced motion of the mass is observed to be 20 mm. Find the value of m.

3.9 A spring-mass system with $m = 10$ kg and $k = 5000$ N/m is subjected to a harmonic force of amplitude 250 N and frequency ω. If the maximum amplitude of the mass is observed to be 100 mm, find the value of ω.

3.10 In Fig. 3.1(a), a periodic force $F(t) = F_0 \cos \omega t$ is applied at a point on the spring that is located at a distance of 25 percent of its length from the fixed support. Assuming that $c = 0$, find the steady-state response of the mass m.

3.11 An aircraft engine has a rotating unbalanced mass m at radius r. If the wing can be modeled as a cantilever beam of uniform cross section $a \times b$, as shown in Fig. 3.34(b), determine the maximum deflection of the engine at a speed of N rpm. Assume damping and effect of the wing between the engine and the free end to be negligible.

3.12 A three-bladed wind turbine (Fig. 3.35a) has a small unbalanced mass m located at a radius r in the plane of the blades. The blades are located from the central vertical (y) axis at a distance R and rotate at an angular velocity of ω. If the supporting truss can be modeled as a hollow steel shaft of outer diameter 0.1 m and inner diameter 0.08 m, determine the maximum stresses developed at the base of the support (point A). The mass moment of inertia of the turbine system about the vertical (y) axis is J_0. Assume $R = 0.5$ m, $m = 0.1$ kg, $r = 0.1$ m, $J_0 = 100$ kg-m^2, $h = 8$ m, and $\omega = 31.416$ rad/s.

3.13 An electromagnetic fatigue-testing machine is shown in Fig. 3.36 in which an alternating force is applied to the specimen by passing an alternating current of frequency f through the armature. If the weight of the armature is 40 lb, the stiffness of the spring (k_1) is 10,217.0296 lb/in. and the stiffness of the steel specimen is 75×10^4 lb/in., determine the frequency of the a.c. current that induces a stress in the specimen that is twice the amount generated by the magnets.

3.14 The spring actuator shown in Fig. 3.37 operates by using the air pressure from a pneumatic controller (p) as input and providing an output displacement to a valve (x) proportional to the input air pressure. The diaphragm, made of a fabric-base rubber, has an area A and deflects under the input air pressure against a spring of stiffness k. Find the response of the valve under a harmonically fluctuating input air pressure $p(t) = p_0 \sin \omega t$ for the following data: $p_0 = 10$ psi, $\omega = 8$ rad/s, $A = 100$ in^2, $k = 400$ lb/in, weight of spring $= 15$ lb, and weight of valve and valve rod $= 20$ lb.

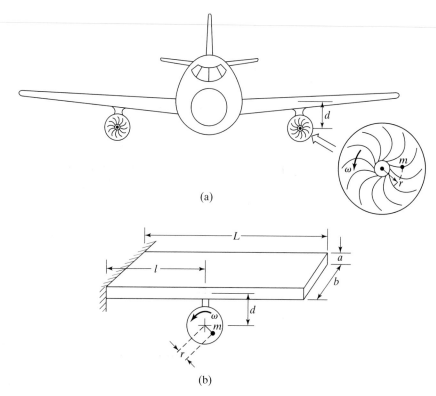

(a)

(b)

FIGURE 3.34

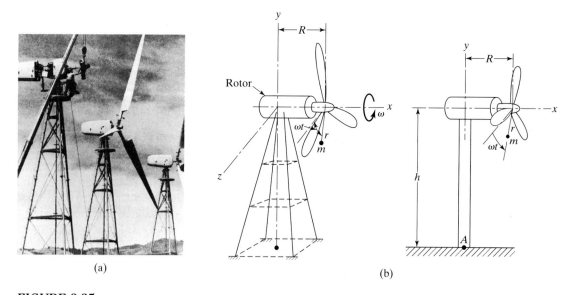

(a)

(b)

FIGURE 3.35 Three-bladed wind turbine. (Photo courtesy of *Power Transmission Design*.)

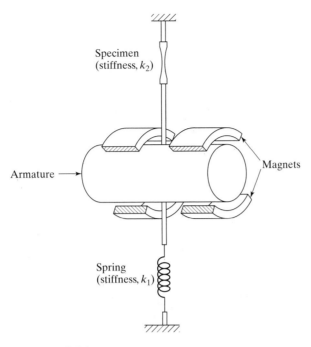

FIGURE 3.36   Electromagnetic fatigue testing machine.

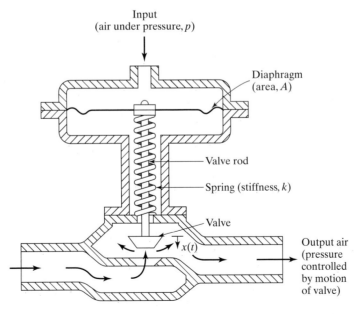

FIGURE 3.37   A spring actuator.

3.15 In the cam-follower system shown in Fig. 3.38, the rotation of the cam imparts a vertical motion to the follower. The pushrod, which acts as a spring, has been compressed by an amount x_0 before assembly. Determine the following: (a) equation of motion of the follower, including the gravitational force; (b) force exerted on the follower by the cam; and (c) conditions under which the follower loses contact with the cam.

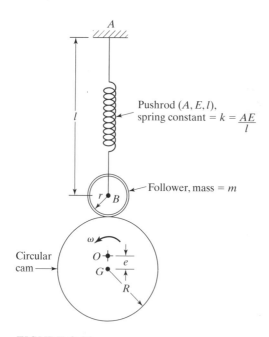

FIGURE 3.38

3.16* Design a solid steel shaft supported in bearings which carries the rotor of a turbine at the middle. The rotor weighs 500 lb and delivers a power of 200 hp at 3000 rpm. In order to keep the stress due to the unbalance in the rotor small, the critical speed of the shaft is to be made one-fifth of the operating speed of the rotor. The length of the shaft is to be made equal to at least 30 times its diameter.

3.17 A hollow steel shaft, of length 100 in., outer diameter 4 in. and inner diameter 3.5 in., carries the rotor of a turbine, weighing 500 lb, at the middle and is supported at the ends in bearings. The clearance between the rotor and the stator is 0.5 in. The rotor has an eccentricity equivalent to a weight of 0.5 lb at a radius of 2 in. A limit switch is installed to stop the rotor whenever the rotor touches the stator. If the rotor operates at resonance, how long will it take to activate the limit switch? Assume the initial displacement and velocity of the rotor perpendicular to the shaft to be zero.

3.18 A steel cantilever beam, carrying a weight of 0.1 lb at the free end, is used as a frequency meter.[6] The beam has a length of 10 in., width of 0.2 in., and thickness of 0.05 in. The internal friction is equivalent to a damping ratio of 0.01. When the fixed end of the beam is subjected to a harmonic displacement $y(t) = 0.05 \cos \omega t$, the maximum tip displacement has been observed to be 2.5 in. Find the forcing frequency.

---

*The asterisk denotes a design type problem or a problem with no unique answer.
[6]The use of cantilever beams as frequency meters is discussed in detail in Section 10.4.

3.19 Derive the equation of motion and find the steady-state response of the system shown in Fig. 3.39 for rotational motion about the hinge O for the following data: $k_1 = k_2 = 5000$ N/m, $a = 0.25$ m, $b = 0.5$ m, $l = 1$ m, $M = 50$ kg, $m = 10$ kg, $F_0 = 500$ N, $\omega = 1000$ rpm.

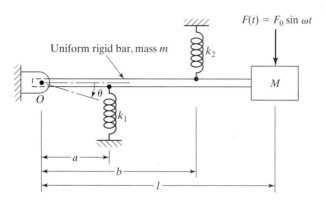

FIGURE 3.39

3.20 Derive the equation of motion and find the steady-state solution of the system shown in Fig. 3.40 for rotational motion about the hinge O for the following data: $k = 5000$ N/m, $l = 1$ m, $m = 10$ kg, $M_0 = 100$ N-m, $\omega = 1000$ rpm.

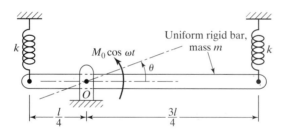

FIGURE 3.40

3.21 Consider a spring-mass-damper system with $k = 4000$ N/m, $m = 10$ kg, and $c = 40$ N-s/m. Find the steady-state and total responses of the system under the harmonic force $F(t) = 200 \cos 10\, t$ N and the initial conditions $x_0 = 0.1$ m and $\dot{x}_0 = 0$.

3.22 Consider a spring-mass-damper system with $k = 4000$ N/m, $m = 10$ kg, and $c = 40$ N-s/m. Find the steady-state and total responses of the system under the harmonic force $F(t) = 200 \cos 10\, t$ N and the initial conditions $x_0 = 0$ and $\dot{x}_0 = 10$ m/s.

3.23 Consider a spring-mass-damper system with $k = 4000$ N/m, $m = 10$ kg, and $c = 40$ N-s/m. Find the steady-state and total responses of the system under the harmonic force $F(t) = 200 \cos 20\, t$ N and the initial conditions $x_0 = 0.1$ m and $\dot{x}_0 = 0$.

3.24 Consider a spring-mass-damper system with $k = 4000$ N/m, $m = 10$ kg, and $c = 40$ N-s/m. Find the steady-state and total responses of the system under the harmonic force $F(t) = 200 \cos 20\, t$ N and the initial conditions $x_0 = 0$ and $\dot{x}_0 = 10$ m/s.

3.25 A four-cylinder automobile engine is to be supported on three shock mounts, as indicated in Fig. 3.41. The engine block assembly weighs 500 lb. If the unbalanced force generated by the engine is given by $200 \sin 100 \pi t$ lb, design the three shock mounts (each of stiffness k and viscous damping constant c) such that the amplitude of vibration is less than 0.1 in.

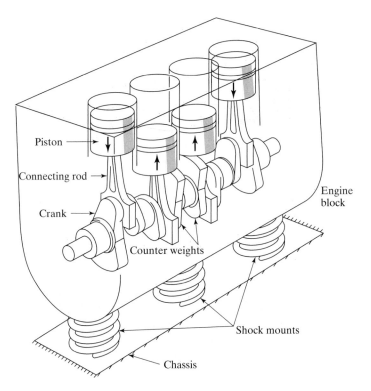

Piston

Connecting rod

Crank

Counter weights

Engine block

Shock mounts

Chassis

FIGURE 3.41   Four-cylinder automobile engine.

3.26 The propeller of a ship, of weight 10^5 N and polar mass moment of inertia 10,000 kg-m^2, is connected to the engine through a hollow stepped steel propeller shaft, as shown in Fig. 3.42. Assuming that water provides a viscous damping ratio of 0.1, determine the torsional vibratory response of the propeller when the engine induces a harmonic angular displacement of $0.05 \sin 314.16 \, t$ rad at the base (point A) of the propeller shaft.

3.27 Find the frequency ratio $r = \omega/\omega_n$ at which the amplitude of a single degree of freedom damped system attains the maximum value. Also find the value of the maximum amplitude.

3.28 Figure 3.43 shows a permanent-magnet moving coil ammeter. When current (I) flows through the coil wound on the core, the core rotates by an angle proportional to the magnitude of the current that is indicated by the pointer on a scale. The core, with the coil, has a mass moment of inertia J_0, the torsional spring constant of k_t, and the torsional damper has a damping constant of c_t. The scale of the ammeter is calibrated such that when a d.c. current of magnitude 1 ampere is passed through the coil, the pointer indicates a current of 1 ampere. The meter has to be recalibrated for measuring the magnitude of a.c. current. Determine the steady-state

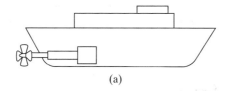

(a)

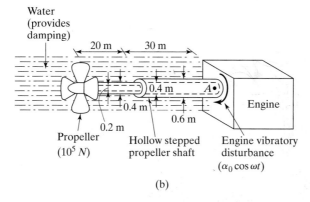

(b)

FIGURE 3.42 Propeller of a ship.

value of the current indicated by the pointer when an a.c. current of magnitude 5 amperes and frequency 50 Hz is passed through the coil. Assume $J_0 = 0.001$ N-m^2, $k_t = 62.5$ N-m/rad and $c_t = 0.5$ N-m-s/rad.

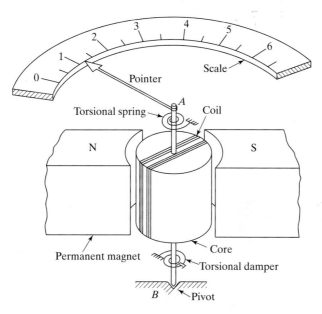

FIGURE 3.43 Permanent-magnet moving coil ammeter.

3.29 A spring-mass-damper system is subjected to a harmonic force. The amplitude is found to be 20 mm at resonance and 10 mm at a frequency 0.75 times the resonant frequency. Find the damping ratio of the system.

3.30 For the system shown in Fig. 3.44, x and y denote, respectively, the absolute displacements of the mass m and the end Q of the dashpot c_1. (a) Derive the equation of motion of the mass m, (b) find the steady state displacement of the mass m, and (c) find the force transmitted to the support at P, when the end Q is subjected to the harmonic motion $y(t) = Y \cos \omega t$.

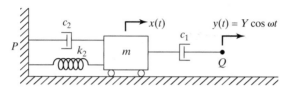

FIGURE 3.44

3.31 Show that, for small values of damping, the damping ratio ζ can be expressed as

$$\zeta = \frac{\omega_2 - \omega_1}{\omega_2 + \omega_1}$$

where ω_1 and ω_2 are the frequencies corresponding to the half-power points.

3.32 A torsional system consists of a disc of mass moment of inertia $J_0 = 10$ kg-m^2, a torsional damper of damping constant $c_t = 300$ N-m-s/rad, and a steel shaft of diameter 4 cm and length 1 m (fixed at one end and attached to the disc at the other end). A steady angular oscillation of amplitude 2° is observed when a harmonic torque of magnitude 1000 N-m is applied to the disc. (a) Find the frequency of the applied torque, and (b) find the maximum torque transmitted to the support.

3.33 For a vibrating system, $m = 10$ kg, $k = 2500$ N/m, and $c = 45$ N-s/m. A harmonic force of amplitude 180 N and frequency 3.5 Hz acts on the mass. If the initial displacement and velocity of the mass are 15 mm and 5 m/s, find the complete solution representing the motion of the mass.

3.34 The peak amplitude of a single degree of freedom system, under a harmonic excitation, is observed to be 0.2 in. If the undamped natural frequency of the system is 5 Hz, and the static deflection of the mass under the maximum force is 0.1 in., (a) estimate the damping ratio of the system, and (b) find the frequencies corresponding to the amplitudes at half power.

3.35 The landing gear of an airplane can be idealized as the spring-mass-damper system shown in Fig. 3.45. If the runway surface is described $y(t) = y_0 \cos \omega t$, determine the values of k and c that limit the amplitude of vibration of the airplane (x) to 0.1 m. Assume $m = 2000$ kg, $y_0 = 0.2$ m, and $\omega = 157.08$ rad/s.

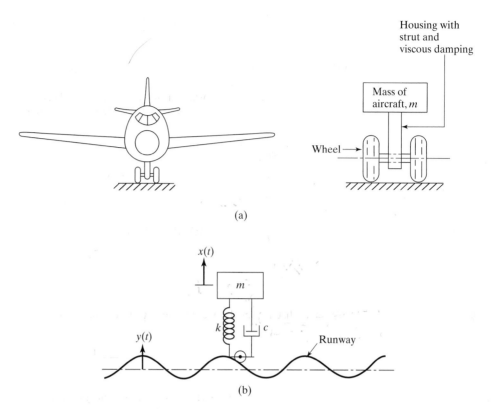

FIGURE 3.45   Modeling of landing gear.

3.36 A precision grinding machine (Fig. 3.46) is supported on an isolator that has a stiffness of 1 MN/m and a viscous damping constant of 1 kN-s/m. The floor on which the machine is mounted is subjected to a harmonic disturbance due to the operation of an unbalanced engine in the vicinity of the grinding machine. Find the maximum acceptable displacement amplitude of the floor if the resulting amplitude of vibration of the grinding wheel is to be restricted to 10^{-6} m. Assume that the grinding machine and the wheel are a rigid body of weight 5000 N.

3.37 Derive the equation of motion and find the steady-state response of the system shown in Fig. 3.47 for rotational motion about the hinge O for the following data: $k = 5000$ N/m, $l = 1$ m, $c = 1000$ N-s/m, $m = 10$ kg, $M_0 = 100$ N-m, $\omega = 1000$ rpm.

3.38 An air compressor of mass 100 kg is mounted on an elastic foundation. It has been observed that, when a harmonic force of amplitude 100 N is applied to the compressor, the maximum steady-state displacement of 5 mm occurred at a frequency of 300 rpm. Determine the equivalent stiffness and damping constant of the foundation.

3.39 Find the steady-state response of the system shown in Fig. 3.48 for the following data: $k_1 = 1000$ N/m, $k_2 = 500$ N/m, $c = 500$ N-s/m, $m = 10$ kg, $r = 5$ cm, $J_0 = 1$ kg-m^2, $F_0 = 50$ N, $\omega = 20$ rad/s.

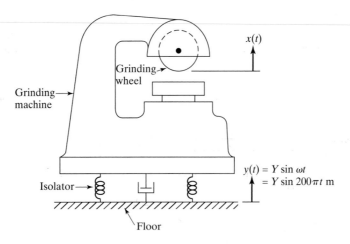

FIGURE 3.46

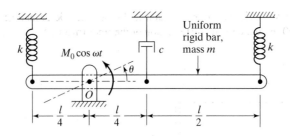

FIGURE 3.47

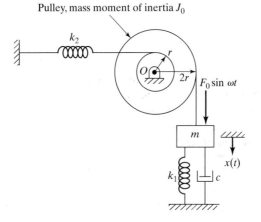

FIGURE 3.48

3.40 A uniform slender bar of mass m may be supported in one of two ways indicated in Fig. 3.49. Determine the arrangement that results in a reduced steady-state response of the bar under a harmonic force, $F_0 \sin \omega t$, applied at the middle of the bar, as shown in the figure.

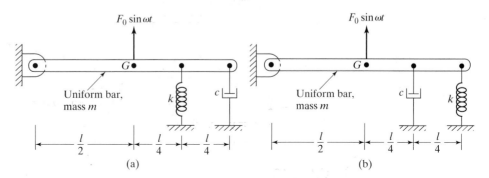

FIGURE 3.49

3.41 A single-story building frame is subjected to a harmonic ground acceleration, as shown in Fig. 3.50. Find the steady-state motion of the floor (mass m).

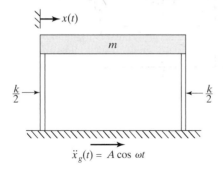

FIGURE 3.50

3.42 Find the horizontal displacement of the floor (mass m) of the building frame show in Fig. 3.50 when the ground acceleration is given by $\ddot{x}_g = 100 \sin \omega t$ mm/sec: Assume $m = 2000$ kg, $k = 0.1$ MN/m, $\omega = 25$ rad/s, and $x_g(t = 0) = \dot{x}_g(t = 0) = x(t = 0) = \dot{x}(t = 0) = 0$.

3.43 If the ground is subjected to a horizontal harmonic displacement with frequency $\omega = 200$ rad/s and amplitude $X_g = 15$ mm in Fig. 3.50, find the amplitude of vibration of the floor (mass m). Assume the mass of the floor as 2000 kg and the stiffness of the columns as 0.5 MN/m.

3.44 An automobile is modeled as a single degree of freedom system vibrating in the vertical direction. It is driven along a road whose elevation varies sinusoidally. The distance from peak to trough is 0.2 m and the distance along the road between the peaks is 35 m. If the natural frequency of the automobile is 2 Hz and the damping ratio of the shock absorbers is 0.15, determine the amplitude of vibration of the automobile at a speed of 60 km/hour. If the speed of the automobile is varied, find the most unfavorable speed for the passengers.

3.45 Derive Eq. (3.74).

3.46 A single-story building frame is modeled by a rigid floor of mass m and columns of stiffness k, as shown in Fig. 3.51. It is proposed that a damper shown in the figure is attached to absorb vibrations due to a horizontal ground motion $y(t) = Y \cos \omega t$. Derive an expression for the damping constant of the damper that absorbs maximum power.

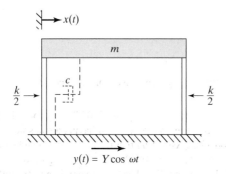

FIGURE 3.51

3.47 A uniform bar of mass m is pivoted at point O and supported at the ends by two springs, as shown in Fig. 3.52. End P of spring PQ is subjected to a sinusoidal displacement, $x(t) = x_0 \sin \omega t$. Find the steady-state angular displacement of the bar when $l = 1$ m, $k = 1000$ N/m, $m = 10$ kg, $x_0 = 1$ cm, and $\omega = 10$ rad/s.

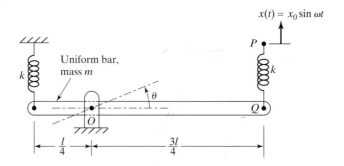

FIGURE 3.52

3.48 A uniform bar of mass m is pivoted at point O and supported at the ends by two springs, as shown in Fig. 3.53. End P of spring PQ is subjected to a sinusoidal displacement, $x(t) = x_0 \sin \omega t$. Find the steady-state angular displacement of the bar when $l = 1$ m, $k = 1000$ N/m, $c = 500$ N-s/m, $m = 10$ kg, $x_0 = 1$ cm, and $\omega = 10$ rad/s.

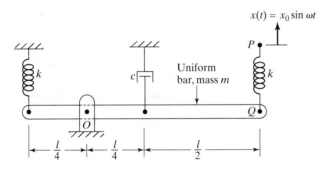

FIGURE 3.53

3.49 Find the frequency ratio, $r = r_m$, at which the displacement transmissibility given by Eq. (3.68) attains a maximum value.

3.50 An automobile, weighing 1000 lb empty and 3000 lb fully loaded, vibrates in a vertical direction while traveling at 55 mph on a rough road having a sinusoidal wave form with an amplitude Y ft and a period 12 ft. Assuming that the automobile can be modeled as a single degree of freedom system with stiffness 30,000 lb/ft and damping ratio $\zeta = 0.2$, determine the amplitude of vibration of the automobile when it is (a) empty and (b) fully loaded.

3.51 A single-cylinder air compressor of mass 100 kg is mounted on rubber mounts, as shown in Fig. 3.54. The stiffness and damping constants of the rubber mounts are given by 10^6 N/m and 2000 N-s/m, respectively. If the unbalance of the compressor is equivalent to a mass 0.1 kg located at the end of the crank (point A), determine the response of the compressor at a crank speed of 3000 rpm. Assume $r = 10$ cm and $l = 40$ cm.

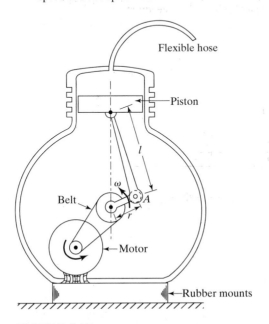

FIGURE 3.54

3.52   One of the tail rotor blades of a helicopter has an unbalanced mass of $m = 0.5$ kg at a distance of $e = 0.15$ m from the axis of rotation, as shown in Fig. 3.55. The tail section has a length of 4 m, a mass of 240 kg, a flexural stiffness (EI) of 2.5 MN $-$ m^2, and a damping ratio of 0.15. The mass of the tail rotor blades, including their drive system, is 20 kg. Determine the forced response of the tail section when the blades rotate at 1500 rpm.

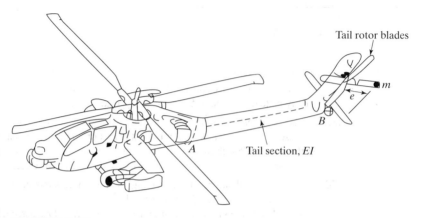

FIGURE 3.55

3.53   When an exhaust fan of mass 380 kg is supported on springs with negligible damping, the resulting static deflection is found to be 45 mm. If the fan has a rotating unbalance of 0.15 kg-m, find (a) the amplitude of vibration at 1750 rpm, and (b) the force transmitted to the ground at this speed.

3.54   A fixed-fixed steel beam, of length 5 m, width 0.5 m, and thickness 0.1 m, carries an electric motor of mass 75 kg and speed 1200 rpm at its mid-span, as shown in Fig. 3.56. A rotating force of magnitude $F_0 = 5000$ N is developed due to the unbalance in the rotor of the motor. Find the amplitude of steady-state vibrations by disregarding the mass of the beam. What will be the amplitude if the mass of the beam is considered?

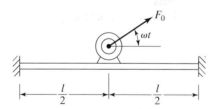

FIGURE 3.56

3.55*   If the electric motor of Problem 3.54 is to be mounted at the free end of a steel cantilever beam of length 5 m (Fig. 3.57), and the amplitude of vibration is to be limited to 0.5 cm, find the necessary cross-sectional dimensions of the beam. Include the weight of the beam in the computations.

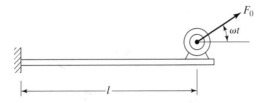

FIGURE 3.57

3.56 A centrifugal pump, weighing 600 N and operating at 1000 rpm, is mounted on six springs of stiffness 6000 N/m each. Find the maximum permissible unbalance in order to limit the steady-state deflection to 5 mm peak-to-peak.

3.57* An air compressor, weighing 1000 lb and operating at 1500 rpm, is to be mounted on a suitable isolator. A helical spring with a stiffness of 45,000 lb/in., another helical spring with a stiffness of 15,000 lb/in., and a shock absorber with a damping ratio of 0.15 are available for use. Select the best possible isolation system for the compressor.

3.58 A variable-speed electric motor, having an unbalance, is mounted on an isolator. As the speed of the motor is increased from zero, the amplitudes of vibration of the motor have been observed to be 0.55 in. at resonance and 0.15 in. beyond resonance. Find the damping ratio of the isolator.

3.59 An electric motor weighing 750 lb and running at 1800 rpm is supported on four steel helical springs, each of which has eight active coils with a wire diameter of 0.25 in. and a coil diameter of 3 in. The rotor has a weight of 100 lb with its center of mass located at a distance of 0.01 in. from the axis of rotation. Find the amplitude of vibration of the motor and the force transmitted through the springs to the base.

3.60 A small exhaust fan, rotating at 1500 rpm, is mounted on a 0.2 in. steel shaft. The rotor of the fan weighs 30 lb and has an eccentricity of 0.01 in. from the axis of rotation. Find (a) the maximum force transmitted to the bearings, and (b) the horsepower needed to drive the shaft.

3.61 A rigid plate, weighing 100 lb, is hinged along an edge (*P*) and is supported on a dashpot with $c = 1$ lb-sec/in. at the opposite edge (*Q*), as shown in Fig. 3.58. A small fan weighing 50 lb and rotating at 750 rpm is mounted on the plate through a spring with $k = 200$ lb/in. If the center of gravity of the fan is located at 0.1 in. from its axis of rotation, find the steady-state motion of the edge *Q* and the force transmitted to the point *S*.

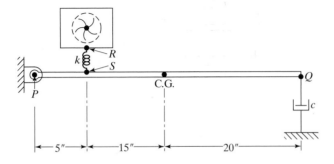

FIGURE 3.58

3.62 Derive Eq. (3.99).

3.63 Derive the equation of motion of the mass m shown in Fig. 3.59 when the pressure in the cylinder fluctuates sinusoidally. The two springs with stiffnesses k_1 are initially under a tension of T_0, and the coefficient of friction between the mass and the contacting surfaces is μ.

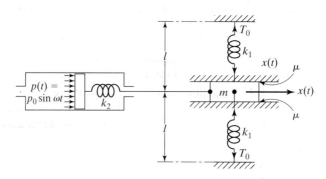

FIGURE 3.59

3.64 A spring-mass system is subjected to Coulomb damping. When a harmonic force of amplitude 120 N and frequency 2.5173268 Hz is applied, the system is found to oscillate with an amplitude of 75 mm. Determine the coefficient of dry friction if $m = 2$ kg and $k = 2100$ N/m.

3.65 A load of 5000 N resulted in a static displacement of 0.05 m in a composite structure. A harmonic force of amplitude 1000 N is found to cause a resonant amplitude of 0.1 m. Find (a) the hysteresis damping constant of the structure, (b) the energy dissipated per cycle at resonance, (c) the steady-state amplitude at one-quarter of the resonant frequency, and (d) the steady-state amplitude at thrice the resonant frequency.

3.66 The energy dissipated in hysteresis damping per cycle under harmonic excitation can be expressed in the general form

$$\Delta W = \pi \beta k X^\gamma \tag{E.1}$$

where γ is an exponent ($\gamma = 2$ was considered in Eq. 2.131), and β is a coefficient of dimension $(\text{meter})^{2-\gamma}$. A spring-mass system having $k = 60$ kN/m vibrates under hysteresis damping. When excited harmonically at resonance, the steady-state amplitude is found to be 40 mm for an energy input of 3.8 N-m. When the resonant energy input is increased to 9.5 N-m, the amplitude is found to be 60 mm. Determine the values of β and γ in Eq. (E.1).

3.67 When a spring-mass-damper system is subjected to a harmonic force $F(t) = 5 \cos 3\pi t$ lb, the resulting displacement is given by $x(t) = 0.5 \cos (3\pi t - \pi/3)$ in. Find the work done (a) during the first second, and (b) during the first 4 seconds.

3.68 Find the equivalent viscous damping coefficient of a damper that offers a damping force of $F_d = c(\dot{x})^n$, where c and n are constants and $\dot{x}$ is the relative velocity across the damper. Also, find the amplitude of vibration.

3.69 Show that for a system with both viscous and Coulomb damping the approximate value of the steady-state amplitude is given by

$$X^2[k^2(1 - r^2)^2 + c^2\omega^2] + X\frac{8\mu Nc\omega}{\pi} + \left(\frac{16\mu^2 N^2}{\pi^2} - F_0^2\right) = 0$$

3.70 The equation of motion of a spring-mass-damper system is given by

$$m\ddot{x} \pm \mu N + c\dot{x}^3 + kx = F_0 \cos \omega t$$

Derive expressions for (a) the equivalent viscous damping constant, (b) the steady-state amplitude, and (c) the amplitude ratio at resonance.

3.71 A fluid, with density ρ, flows through a cantilevered steel pipe of length l and cross-sectional area A (Fig. 3.60). Determine the velocity (v) of the fluid at which instability occurs. Assume that the total mass and the bending stiffness of the pipe are m and EI, respectively.

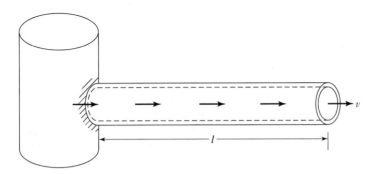

FIGURE 3.60

3.72 The first two natural frequencies of the telescoping car antenna shown in Fig. 3.61 are given by 3.0 Hz and 7.0 Hz. Determine whether the vortex shedding around the antenna causes instability over the speed range 50–75 mph of the automobile.

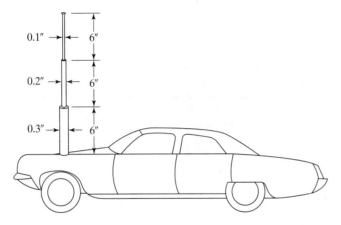

FIGURE 3.61

3.73 The signpost of a fast food restaurant consists of a hollow steel cylinder of height h, inside diameter d, and outside diameter D, fixed to the ground and carries a concentrated mass M at the top. It can be modeled as a single degree of freedom spring-mass-damper system with an equivalent viscous damping ratio of 0.1 for analyzing its transverse vibration characteristics under wind excitation. Determine the following: (a) the natural frequency of transverse vibration of the signpost; (b) the wind velocity at which the signpost undergoes maximum steady-state displacement; and (c) the maximum wind induced steady-state displacement of the signpost. Data: $h = 10$ m, $D = 25$ cm, $d = 20$ cm, $M = 200$ kg.

3.74 Consider the equation of motion of a single degree of freedom system:

$$m\ddot{x} + c\dot{x} + kx = F$$

Derive the condition that leads to divergent oscillations in each of the following cases: (a) when the forcing function is proportional to the displacement, $F(t) = F_0 x(t)$; (b) when the forcing function is proportional to the velocity, $F(t) = F_0 \dot{x}(t)$; and (c) when the forcing function is proportional to the acceleration, $F(t) = F_0 \ddot{x}(t)$.

3.75 Plot the forced response of an undamped spring-mass system under the following conditions using MATLAB: $m = 10$ kg, $k = 4000$ N/m, $F(t) = 200 \cos 10\,t$ N, $x_0 = 0.1$ m, $\dot{x}_0 = 10$ m/s.

3.76 Plot the forced response of a spring-mass system subject to Coulomb damping using MATLAB. Assume the following data: $m = 10$ kg, $k = 4000$ N/m, $F(t) = 200 \sin 10\,t$ N, $\mu = 0.3$, $x_0 = 0.1$ m, $\dot{x}_0 = 10$ m/s.

3.77 Plot the response of a viscously damped system under harmonic base excitation, $y(t) = Y \sin \omega\,t$ m, using MATLAB for the following data: $m = 100$ kg, $k = 4 \times 10^4$ N/m, $\zeta = 0.25$, $Y = 0.05$ m, $\omega = 10$ rad/s, $x_0 = 1$ m, $\dot{x}_0 = 0$.

3.78 Plot the steady-state response of a viscously damped system under the harmonic force $F(t) = F_0 \cos \omega\,t$ using MATLAB. Assume the following data: $m = 10$ kg, $k = 1000$ N/m, $\zeta = 0.1$, $F_0 = 100$ N, $\omega = 20$ rad/s.

3.79 Consider an automobile traveling over a rough road at a speed of v km/hr. The suspension system has a spring constant of 40 kN/m and a damping ratio of $\zeta = 0.1$. The road surface varies sinusoidally with an amplitude of $Y = 0.05$ m and a wavelength of 6 m. Write a MATLAB program to find the displacement amplitude of the automobile for the following conditions: (a) mass of the automobile = 600 kg (empty), 1000 kg (loaded) (b) velocity of the automobile (v): 10 km/hr, 50 km/hr, 100 km/hr

3.80 Using Program3.cpp, find the steady-state response of a torsional system with $J_0 = 6$ kg-m^2, $c_t = 210$ N-m-s/rad, $k_t = 14{,}000$ N-m/rad, and $F(t) = 450 \sin 10\,t$ N-m.

3.81 Using Program3.cpp, find the steady-state solution of a single degree of freedom system with $m = 10$ kg, $c = 45$ N-s/m, $k = 2500$ N/m, $F(t) = 180 \cos 20\,t$ N, $x_0 = 0$, and $\dot{x}_0 = 10$ m/s.

3.82 Use subroutine HARESP to find the steady-state response of a torsional system with $J_0 = 6$ kg-m^2, $c_t = 210$ N-m-s/rad, $k_t = 14{,}000$ N-m/rad, and $F(t) = 450 \sin 10t$ N-m.

3.83 Write a subroutine called TOTALR for finding the complete solution (homogeneous part plus particular integral) of a single degree of freedom system. Use this program to find the solution of Problem 3.33.

3.84 Find the steady-state solution of a single degree of freedom system with $m = 10$ kg, $c = 45$ N-s/m, $k = 2500$ N/m, $F(t) = 180 \cos 20t$ N, $x_0 = 0$, and $\dot{x}_0 = 10$ m/s, using subroutine HARESP.

3.85 Write a computer program for finding the total response of a spring-mass-viscous damper system subjected to base excitation. Use this program to find the solution of a problem with $m = 2$ kg, $c = 10$ N-s/m, $k = 100$ N/m, $y(t) = 0.1 \sin 25t$ m, $x_0 = 10$ mm, and $\dot{x}_0 = 5$ m/s.

DESIGN PROJECTS

3.86 The arrangement shown in Fig. 3.62 consists of two eccentric masses rotating in opposite directions at the same speed ω. It is to be used as a mechanical shaker over the frequency range 20 to 30 Hz. Find the values of ω, e, M, m, k, and c to satisfy the following requirements: (a) The mean power output of the shaker should be at least 1 hp over the specified frequency range. (b) The amplitude of vibration of the masses should be between 0.1 and 0.2 in. (c) The mass of the shaker (M) should be at least 50 times that of the eccentric mass (m).

3.87 Design a minimum weight, hollow circular steel column for the water tank shown in Fig. 3.63. The weight of the tank (W) is 100,000 lb and the height is 50 ft. The stress induced in the column should not exceed the yield strength of the material, which is 30,000 psi, when subjected to a harmonic ground acceleration (due to an earthquake) of amplitude 0.5 g and frequency 15 Hz. In addition, the natural frequency of the water tank should be greater than 15 Hz. Assume a damping ratio of 0.15 for the column.

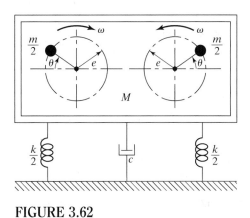

FIGURE 3.62

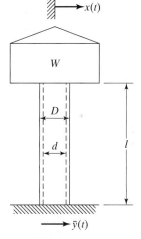

FIGURE 3.63

C H A P T E R   4

Vibration Under General Forcing Conditions

4.1 Introduction

The response of a single degree of freedom system under general, nonharmonic, forcing functions is considered in this chapter. A general forcing function may be periodic (non-harmonic) or nonperiodic. A nonperiodic forcing function may be acting for a short, long, or infinite duration. If the duration of the forcing function or excitation is small compared to the natural time period of the system, the forcing function or excitation is called a shock. The motion imparted by a cam to the follower, the vibration felt by an instrument when its package is dropped from a height, the force applied to the foundation of a forging press, the motion of an automobile when it hits a pothole, and the ground vibration of a building frame during an earthquake are examples of general forcing functions.

If the forcing function is periodic but not harmonic, it can be replaced by a sum of harmonic functions using the harmonic analysis procedure discussed in Section 1.11. Using the principle of superposition, the response of the system can then be determined by superposing the responses due to the individual harmonic forcing functions. However, if the system is subjected to a suddenly applied nonperiodic force, the response will involve transient vibration. The transient response of a system can be found by using what is known as the *convolution integral*.

4.2   Response Under a General Periodic Force

When the external force $F(t)$ is periodic with period $\tau = 2\pi/\omega$, it can be expanded in a Fourier series (see Section 1.11):

$$F(t) = \frac{a_0}{2} + \sum_{j=1}^{\infty} a_j \cos j\omega t + \sum_{j=1}^{\infty} b_j \sin j\omega t \tag{4.1}$$

where

$$a_j = \frac{2}{\tau} \int_0^{\tau} F(t) \cos j\omega t \, dt, \qquad j = 0, 1, 2, \dots \tag{4.2}$$

and

$$b_j = \frac{2}{\tau} \int_0^{\tau} F(t) \sin j\omega t \, dt, \qquad j = 1, 2, \dots \tag{4.3}$$

The equation of motion of the system can be expressed as

$$m\ddot{x} + c\dot{x} + kx = F(t) = \frac{a_0}{2} + \sum_{j=1}^{\infty} a_j \cos j\omega t + \sum_{j=1}^{\infty} b_j \sin j\omega t \tag{4.4}$$

The right-hand side of this equation is a constant plus a sum of harmonic functions. Using the principle of superposition, the steady-state solution of Eq. (4.4) is the sum of the steady-state solutions of the following equations:

$$m\ddot{x} + c\dot{x} + kx = \frac{a_0}{2} \tag{4.5}$$

$$m\ddot{x} + c\dot{x} + kx = a_j \cos j\omega t \tag{4.6}$$

$$m\ddot{x} + c\dot{x} + kx = b_j \sin j\omega t \tag{4.7}$$

Noting that the solution of Eq. (4.5) is given by

$$x_p(t) = \frac{a_0}{2k} \tag{4.8}$$

and using the results of Section 3.4, we can express the solutions of Eqs. (4.6) and (4.7), respectively, as

$$x_p(t) = \frac{(a_j/k)}{\sqrt{(1 - j^2 r^2)^2 + (2\zeta j r)^2}} \cos(j\omega t - \phi_j) \tag{4.9}$$

$$x_p(t) = \frac{(b_j/k)}{\sqrt{(1 - j^2 r^2)^2 + (2\zeta j r)^2}} \sin(j\omega t - \phi_j) \tag{4.10}$$

where

$$\phi_j = \tan^{-1}\left(\frac{2\zeta jr}{1 - j^2 r^2}\right) \tag{4.11}$$

and

$$r = \frac{\omega}{\omega_n} \tag{4.12}$$

Thus the complete steady-state solution of Eq. (4.4) is given by

$$x_p(t) = \frac{a_0}{2k} + \sum_{j=1}^{\infty} \frac{(a_j/k)}{\sqrt{(1 - j^2 r^2)^2 + (2\zeta jr)^2}} \cos(j\omega t - \phi_j)$$

$$+ \sum_{j=1}^{\infty} \frac{(b_j/k)}{\sqrt{(1 - j^2 r^2)^2 + (2\zeta jr)^2}} \sin(j\omega t - \phi_j) \tag{4.13}$$

It can be seen from the solution, Eq. (4.13), that the amplitude and phase shift corresponding to the jth term depend on j. If $j\omega = \omega_n$, for any j, the amplitude of the corresponding harmonic will be comparatively large. This will be particularly true for small values of j and ζ. Further, as j becomes larger, the amplitude becomes smaller and the corresponding terms tend to zero. Thus the first few terms are usually sufficient to obtain the response with reasonable accuracy.

The solution given by Eq. (4.13) denotes the steady-state response of the system. The transient part of the solution arising from the initial conditions can also be included to find the complete solution. To find the complete solution, we need to evaluate the arbitrary constants by setting the value of the complete solution and its derivative to the specified values of initial displacement $x(0)$ and the initial velocity $\dot{x}(0)$. This results in a complicated expression for the transient part of the total solution.

EXAMPLE 4.1

Periodic Vibration of a Hydraulic Valve

In the study of vibrations of valves used in hydraulic control systems, the valve and its elastic stem are modeled as a damped spring-mass system, as shown in Fig. 4.1(a). In addition to the spring force and damping force, there is a fluid pressure force on the valve that changes with the amount of opening or closing of the valve. Find the steady-state response of the valve when the pressure in the chamber varies as indicated in Fig. 4.1(b). Assume $k = 2500$ N/m, $c = 10$ N-s/m, and $m = 0.25$ kg.

Solution: The valve can be considered as a mass connected to a spring and a damper on one side and subjected to a forcing function $F(t)$ on the other side. The forcing function can be expressed as

$$F(t) = Ap(t) \tag{E.1}$$

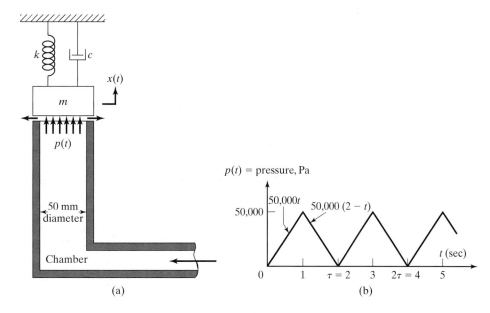

FIGURE 4.1   Periodic vibration of a hydraulic valve.

where A is the cross-sectional area of the chamber, given by

$$A = \frac{\pi(50)^2}{4} = 625\,\pi \text{ mm}^2 = 0.000625\,\pi \text{ m}^2 \tag{E.2}$$

and $p(t)$ is the pressure acting on the valve at any instant t. Since $p(t)$ is periodic with period $\tau = 2$ seconds and A is a constant, $F(t)$ is also a periodic function of period $\tau = 2$ seconds. The frequency of the forcing function is $\omega = (2\pi/\tau) = \pi$ rad/s. $F(t)$ can be expressed in a Fourier series as

$$F(t) = \frac{a_0}{2} + a_1 \cos \omega t + a_2 \cos 2\omega t + \cdots$$
$$+ b_1 \sin \omega t + b_2 \sin 2\omega t + \cdots \tag{E.3}$$

where a_j and b_j are given by Eqs. (4.2) and (4.3). Since the function $F(t)$ is given by

$$F(t) = \begin{cases} 50\,000 At & \text{for } 0 \leqslant t \leqslant \dfrac{\tau}{2} \\[2mm] 50\,000 A(2 - t) & \text{for } \dfrac{\tau}{2} \leqslant t \leqslant \tau \end{cases} \tag{E.4}$$

the Fourier coefficients a_j and b_j can be computed with the help of Eqs. (4.2) and (4.3):

$$a_0 = \frac{2}{2}\left[\int_0^1 50\,000 At \, dt + \int_1^2 50\,000 A(2 - t) \, dt \right] = 50\,000 A \tag{E.5}$$

$$a_1 = \frac{2}{2}\left[\int_0^1 50\,000At\cos \pi t\,dt + \int_1^2 50\,000A(2 - t)\cos \pi t\,dt\right]$$

$$= -\frac{2 \times 10^5 A}{\pi^2} \tag{E.6}$$

$$b_1 = \frac{2}{2}\left[\int_0^1 50\,000At\sin \pi t\,dt + \int_1^2 50\,000A(2 - t)\sin \pi t\,dt\right] = 0 \tag{E.7}$$

$$a_2 = \frac{2}{2}\left[\int_0^1 50\,000At\cos 2\pi t\,dt + \int_1^2 50\,000A(2 - t)\cos 2\pi t\,dt\right] = 0 \tag{E.8}$$

$$b_2 = \frac{2}{2}\left[\int_0^1 50\,000At\sin 2\pi t\,dt + \int_1^2 50\,000A(2 - t)\sin 2\pi t\,dt\right] = 0 \tag{E.9}$$

$$a_3 = \frac{2}{2}\left[\int_0^1 50\,000At\cos 3\pi t\,dt + \int_1^2 50\,000A(2 - t)\cos 3\pi t\,dt\right]$$

$$= -\frac{2 \times 10^5 A}{9\pi^2} \tag{E.10}$$

$$b_3 = \frac{2}{2}\left[\int_0^1 50\,000At\sin 3\pi t\,dt + \int_1^2 50\,000A(2 - t)\sin 3\pi t\,dt\right] = 0 \tag{E.11}$$

Likewise, we can obtain $a_4 = a_6 = + \cdots = b_4 = b_5 = b_6 = \cdots = 0$. By considering only the first three harmonics, the forcing function can be approximated:

$$F(t) \simeq 25\,000A - \frac{2 \times 10^5 A}{\pi^2}\cos \omega t - \frac{2 \times 10^5 A}{9\pi^2}\cos 3\omega t \tag{E.12}$$

The steady-state response of the valve to the forcing function of Eq. (E.12) can be expressed as

$$x_p(t) = \frac{25\,000A}{k} - \frac{(2 \times 10^5 A/(k\pi^2))}{\sqrt{(1 - r^2)^2 + (2\zeta r)^2}}\cos (\omega t - \phi_1)$$

$$- \frac{(2 \times 10^5 A/(9k\pi^2))}{\sqrt{(1 - 9r^2)^2 + (6\zeta r)^2}}\cos (3\omega t - \phi_3) \tag{E.13}$$

The natural frequency of the valve is given by

$$\omega_n = \sqrt{\frac{k}{m}} = \sqrt{\frac{2500}{0.25}} = 100 \text{ rad/s} \tag{E.14}$$

and the forcing frequency ω by

$$\omega = \frac{2\pi}{\tau} = \frac{2\pi}{2} = \pi \text{ rad/s} \qquad (E.15)$$

Thus the frequency ratio can be obtained

$$r = \frac{\omega}{\omega_n} = \frac{\pi}{100} = 0.031416 \qquad (E.16)$$

and the damping ratio:

$$\zeta = \frac{c}{c_c} = \frac{c}{2m\omega_n} = \frac{10.0}{2(0.25)(100)} = 0.2 \qquad (E.17)$$

The phase angles ϕ_1 and ϕ_3 can be computed as follows:

$$\phi_1 = \tan^{-1}\left(\frac{2\zeta r}{1 - r^2}\right)$$

$$= \tan^{-1}\left(\frac{2 \times 0.2 \times 0.031416}{1 - 0.031416^2}\right) = 0.0125664 \text{ rad} \qquad (E.18)$$

and

$$\phi_3 = \tan^{-1}\left(\frac{6\zeta r}{1 - 9r^2}\right)$$

$$= \tan^{-1}\left(\frac{6 \times 0.2 \times 0.031416}{1 - 9(0.031416)^2}\right) = 0.0380483 \text{ rad} \qquad (E.19)$$

In view of Eqs. (E.2) and (E.14) to (E.19), the solution can be written as

$$x_p(t) = 0.019635 - 0.015930 \cos(\pi t - 0.0125664)$$

$$- 0.0017828 \cos(3\pi t - 0.0380483) \text{ m} \qquad (E.20)$$

∎

Total Response Under Harmonic Base Excitation

EXAMPLE 4.2

Find the total response of a viscously damped single degree of freedom system subjected to a harmonic base excitation for the following data: $m = 10$ kg, $c = 20$ N-m/s, $k = 4000$ N/m, $y(t) = 0.05 \sin 5\, t$ m, $x_0 = 0.02$ m, $\dot{x}_0 = 10$ m/s.

Solution: The equation of motion of the system is given by (see Eq. 3.65):

$$m\ddot{x} + c\dot{x} + kx = ky + c\dot{y} = kY \sin \omega t + c\omega Y \cos \omega t \qquad (E.1)$$

Noting that Eq. (E.1) is similar to Eq. (4.4) with $a_0 = 0$, $a_1 = c \omega Y$, $b_1 = k Y$ and $a_i = b_i = 0$; $i = 2, 3, \ldots$, the steady-state response of the system can be expressed, using Eq. (4.13), as

$$x_p(t) = \frac{1}{\sqrt{(1 - r^2)^2 + (2\zeta r)^2}} \left[\frac{a_1}{k} \cos (\omega t - \phi_1) + \frac{b_1}{k} \sin (\omega t - \phi_1) \right] \tag{E.2}$$

For the given data, we find

$$Y = 0.05 \text{ m}, \ \omega = 5 \text{ rad/s}, \ \omega_n = \sqrt{\frac{k}{m}} = \sqrt{\frac{4000}{10}} = 20 \text{ rad/s},$$

$$r = \frac{\omega}{\omega_n} = \frac{5}{20} = 0.25, \ \zeta = \frac{c}{c_c} = \frac{c}{2\sqrt{km}} = \frac{20}{2\sqrt{(4000)(10)}} = 0.05,$$

$$\omega_d = \sqrt{1 - \zeta^2} \, \omega_n = 19.975 \text{ rad/s},$$

$$a_1 = c \omega Y = (20)(5)(0.05) = 5, \ b_1 = k Y = (4000)(0.05) = 200,$$

$$\phi_1 = \tan^{-1} \left(\frac{2\zeta r}{1 - r^2} \right) = \tan^{-1} \left(\frac{2(0.05)(0.25)}{1 - (0.25)^2} \right) = 0.02666 \text{ rad}$$

$$\sqrt{(1 - r^2)^2 + (2\zeta r)^2} = \sqrt{(1 - 0.25^2)^2 + (2(0.05)(0.25))^2} = 0.937833.$$

The solution of the homogeneous equation is given by (see Eq. 2.70):

$$x_h(t) = X_0 e^{-\zeta \omega_n t} \cos (\omega_d t - \phi_0) = X_0 e^{-t} \cos (19.975 t - \phi_0) \tag{E.3}$$

where X_0 and ϕ_0 are unknown constants. The total solution can be expressed as the superposition of $x_h(t)$ and $x_p(t)$ as

$$x(t) = X_0 e^{-t} \cos (19.975 t - \phi_0) + \frac{1}{0.937833} \left[\frac{5}{4000} \cos (5t - \phi_1) + \frac{200}{4000} \sin (5t - \phi_1) \right]$$

$$= X_0 e^{-t} \cos (19.975 t - \phi_0) + 0.001333 \cos (5t - 0.02666)$$

$$+ 0.053314 \sin (5t - 0,02666) \tag{E.4}$$

where the unknowns X_0 and ϕ_0 are to be found from the initial conditions. The velocity of the mass can be expressed from Eq. (E.4) as

$$\dot{x}(t) = \frac{dx}{dt}(t) = -X_0 e^{-t} \cos (19.975 t - \phi_0) - 19.975 X_0 e^{-t} \sin (19.975 t - \phi_0)$$

$$- 0.006665 \sin (5t - 0.02666) + 0.266572 \cos (5t - 0.02666) \tag{E.5}$$

Using Eqs. (E.4) and (E.5), we find

$$x_0 = x(t = 0) = 0.02 = X_0 \cos \phi_0 + 0.001333 \cos(0.02666) - 0.053314 \sin (0.02666)$$

or

$$X_0 \cos \phi_0 = 0.020088 \tag{E.6}$$

and

$$\dot{x}_0 = \dot{x}(t = 0) = 10 = -X_0 \cos \phi_0 + 19.975 \, X_0 \sin \phi_0$$
$$+ \, 0.006665 \sin (0.02666) + 0.266572 \cos (0.02666)$$

or

$$- \, X_0 \cos \phi_0 + 19.975 \sin \phi_0 = 9.733345 \tag{E.7}$$

The solution of Eqs. (E.6) and (E.7) yields $X_0 = 0.488695$ and $\phi_0 = 1.529683$ rad. Thus the total response of the mass under base excitation, in meters, is given by

$$x(t) = 0.488695 \, e^{-t} \cos (19.975 \, t - 1.529683)$$
$$+ \, 0.001333 \cos (5t - 0.02666) + 0.053314 \sin (5t - 0.02666) \tag{E.8}$$

Note: Equation (E.8) is plotted in Example 4.18.

■

4.3   Response Under a Periodic Force of Irregular Form

In some cases, the force acting on a system may be quite irregular and may be determined only experimentally. Examples of such forces include wind and earth quake-induced forces. In such cases, the forces will be available in graphical form and no analytical expression can be found to describe $F(t)$. Sometimes, the value of $F(t)$ may be available only at a number of discrete points $t_1, t_2, \ldots, t_N$. In all these cases, it is possible to find the Fourier coefficients by using a numerical integration procedure, as described in Section 1.11. If $F_1, F_2, \ldots, F_N$ denote the values of $F(t)$ at $t_1, t_2, \ldots, t_N$, respectively, where N denotes an even number of equidistant points in one time period $\tau(\tau = N\Delta t)$, as shown in Fig. 4.2, the application of trapezoidal rule [4.1] gives

$$a_0 = \frac{2}{N} \sum_{i=1}^{N} F_i \tag{4.14}$$

$$a_j = \frac{2}{N} \sum_{i=1}^{N} F_i \cos \frac{2j\pi t_i}{\tau}, \qquad j = 1, 2, \ldots \tag{4.15}$$

$$b_j = \frac{2}{N} \sum_{i=1}^{N} F_i \sin \frac{2j\pi t_i}{\tau}, \qquad j = 1, 2, \ldots \tag{4.16}$$

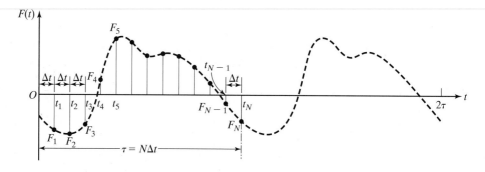

FIGURE 4.2   An irregular forcing function.

Once the Fourier coefficients a_0, a_j, and b_j are known, the steady-state response of the system can be found using Eq. (4.13) with

$$r = \left(\frac{2\pi}{\tau \omega_n} \right)$$

EXAMPLE 4.3

Steady-State Vibration of a Hydraulic Valve

Find the steady-state response of the valve in Example 4.1 if the pressure fluctuations in the chamber are found to be periodic. The values of pressure measured at 0.01 second intervals in one cycle are given below.

Time, t_i (seconds)	0	0.01	0.02	0.03	0.04	0.05	0.06	0.07	0.08	0.09	0.10	0.11	0.12
$p_i = p(t_i)$ (kN/m²)	0	20	34	42	49	53	70	60	36	22	16	7	0

Solution: Since the pressure fluctuations on the valve are periodic, the Fourier analysis of the given data of pressures in a cycle gives

$$p(t) = 34083.3 - 26996.0 \cos 52.36t + 8307.7 \sin 52.36t$$
$$+ 1416.7 \cos 104.72t + 3608.3 \sin 104.72t$$
$$- 5833.3 \cos 157.08t + 2333.3 \sin 157.08t + \ldots \text{ N/m}^2 \qquad \text{(E.1)}$$

(See Example 1.13.) Other quantities needed for the computation are

$$\omega = \frac{2\pi}{\tau} = \frac{2\pi}{0.12} = 52.36 \text{ rad/s}$$

$$\omega_n = 100 \text{ rad/s}$$

$$r = \frac{\omega}{\omega_n} = 0.5236$$

$$\zeta = 0.2$$

$$A = 0.000625\pi \text{ m}^2$$

$$\phi_1 = \tan^{-1}\left(\frac{2\zeta r}{1 - r^2}\right) = \tan^{-1}\left(\frac{2 \times 0.2 \times 0.5236}{1 - 0.5236^2}\right) = 16.1°$$

$$\phi_2 = \tan^{-1}\left(\frac{4\zeta r}{1 - 4r^2}\right) = \tan^{-1}\left(\frac{4 \times 0.2 \times 0.5236}{1 - 4 \times 0.5236^2}\right) = -77.01°$$

$$\phi_3 = \tan^{-1}\left(\frac{6\zeta r}{1 - 9r^2}\right) = \tan^{-1}\left(\frac{6 \times 0.2 \times 0.5236}{1 - 9 \times 0.5236^2}\right) = -23.18°$$

The steady-state response of the valve can be expressed, using Eq. (4.13), as

$$x_p(t) = \frac{34083.3\,A}{k} - \frac{(26996.0\,A/k)}{\sqrt{(1 - r^2)^2 + (2\zeta r)^2}}\cos(52.36t - \phi_1)$$

$$+ \frac{(8309.7\,A/k)}{\sqrt{(1 - r^2)^2 + (2\zeta r)^2}}\sin(52.36t - \phi_1)$$

$$+ \frac{(1416.7\,A/k)}{\sqrt{(1 - 4r^2)^2 + (4\zeta r)^2}}\cos(104.72t - \phi_2)$$

$$+ \frac{(3608.3\,A/k)}{\sqrt{(1 - 4r^2)^2 + (4\zeta r)^2}}\sin(104.72t - \phi_2)$$

$$- \frac{(5833.3\,A/k)}{\sqrt{(1 - 9r^2)^2 + (6\zeta r)^2}}\cos(157.08t - \phi_3)$$

$$+ \frac{(2333.3\,A/k)}{\sqrt{(1 - 9r^2)^2 + (6\zeta r)^2}}\sin(157.08t - \phi_3)$$

■

4.4   Response Under a Nonperiodic Force

We have seen that periodic forces of any general wave form can be represented by Fourier series as a superposition of harmonic components of various frequencies. The response of a linear system is then found by superposing the harmonic response to each of the exciting forces. When the exciting force $F(t)$ is nonperiodic, such as that due to the blast from an explosion, a different method of calculating the response is required. Various methods can be used to find the response of the system to an arbitrary excitation. Some of these methods are as follows:

1. Representing the excitation by a Fourier integral
2. Using the method of convolution integral
3. Using the method of Laplace transforms

4. First approximating $F(t)$ by a suitable interpolation model and then using a numerical procedure
5. Numerically integrating the equations of motion

We shall discuss Methods 2, 3, and 4 in the following sections and Method 5 in Chapter 11.

4.5   Convolution Integral

A nonperiodic exciting force usually has a magnitude that varies with time; it acts for a specified period of time and then stops. The simplest form is the impulsive force—a force that has a large magnitude F and acts for a very short period of time Δt. From dynamics we know that impulse can be measured by finding the change in momentum of the system caused by it [4.2]. If $\dot{x}_1$ and $\dot{x}_2$ denote the velocities of the mass m before and after the application of the impulse, we have

$$\text{Impulse} = F\Delta t = m\dot{x}_2 - m\dot{x}_1 \tag{4.17}$$

By designating the magnitude of the impulse $F\Delta t$ by $\underset{\sim}{F}$, we can write, in general,

$$\underset{\sim}{F} = \int_t^{t+\Delta t} F\, dt \tag{4.18}$$

A unit impulse $(\underset{\sim}{f})$ is defined as

$$\underset{\sim}{f} = \lim_{\Delta t \to 0} \int_t^{t+\Delta t} F\, dt = F\, dt = 1 \tag{4.19}$$

It can be seen that in order for $F\, dt$ to have a finite value, F tends to infinity (since dt tends to zero). Although the unit impulse function has no physical meaning, it is a convenient tool in our present analysis.[1]

4.5.1
Response to an Impulse

We first consider the response of a single degree of freedom system to an impulse excitation; this case is important in studying the response under more general excitations. Consider a viscously damped spring-mass system subjected to a unit impulse at $t = 0$, as shown in Figs. 4.3(a) and (b). For an underdamped system, the solution of the equation of motion

$$m\ddot{x} + c\dot{x} + kx = 0 \tag{4.20}$$

---

[1]The unit impulse, $\underset{\sim}{f}$, acting at $t = 0$ is also denoted by the Dirac delta function, $\delta(t)$. The Dirac delta function at time $t = \tau$, denoted as $\delta(t - \tau)$, has the properties

$$\delta(t - \tau) = 0 \quad \text{for } t \neq \tau;$$

$$\int_0^\infty \delta(t - \tau)\, dt = 1; \qquad \int_0^\infty \delta(t - \tau)\, F(t)\, dt = F(\tau)$$

where $0 < \tau < \infty$. Thus an impulsive force acting at $t = \tau$ can be denoted as $F(t) = \underset{\sim}{F}\,\delta(t - \tau)$.

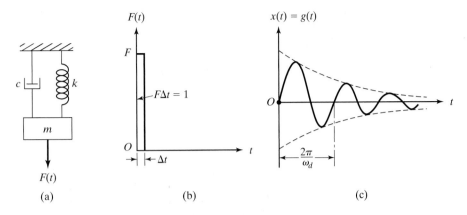

FIGURE 4.3 A single degree of freedom system subjected to an impulse.

is given by Eq. (2.72) as

$$x(t) = e^{-\zeta\omega_n t}\left\{x_0\cos\omega_d t + \frac{\dot{x}_0 + \zeta\omega_n x_0}{\omega_d}\sin\omega_d t\right\} \tag{4.21}$$

where

$$\zeta = \frac{c}{2m\omega_n} \tag{4.22}$$

$$\omega_d = \omega_n\sqrt{1 - \zeta^2} = \sqrt{\frac{k}{m} - \left(\frac{c}{2m}\right)^2} \tag{4.23}$$

$$\omega_n = \sqrt{\frac{k}{m}} \tag{4.24}$$

If the mass is at rest before the unit impulse is applied ($x = \dot{x} = 0$ for $t < 0$ or at $t = 0^-$), we obtain, from the impulse-momentum relation,

$$\text{Impulse} = \underset{\sim}{f} = 1 = m\dot{x}(t = 0) - m\dot{x}(t = 0^-) = m\dot{x}_0 \tag{4.25}$$

Thus the initial conditions are given by

$$x(t = 0) = x_0 = 0$$

$$\dot{x}(t = 0) = \dot{x}_0 = \frac{1}{m} \tag{4.26}$$

In view of Eq. (4.26), Eq. (4.21) reduces to

$$x(t) = g(t) = \frac{e^{-\zeta\omega_n t}}{m\omega_d}\sin\omega_d t \tag{4.27}$$

Equation (4.27) gives the response of a single degree of freedom system to a unit impulse, which is also known as the *impulse response function*, denoted by $g(t)$. The function $g(t)$, Eq. (4.27), is shown in Fig. 4.3(c).

If the magnitude of the impulse is $\underset{\sim}{F}$ instead of unity, the initial velocity $\dot{x}_0$ is $\underset{\sim}{F}/m$ and the response of the system becomes

$$x(t) = \frac{\underset{\sim}{F}e^{-\zeta\omega_n t}}{m\omega_d}\sin\omega_d t = \underset{\sim}{F}g(t) \tag{4.28}$$

If the impulse $\underset{\sim}{F}$ is applied at an arbitrary time $t = \tau$, as shown in Fig. 4.4(a), it will change the velocity at $t = \tau$ by an amount $\underset{\sim}{F}/m$. Assuming that $x = 0$ until the impulse is applied, the displacement x at any subsequent time t, caused by a change in the velocity at time τ, is given by Eq. (4.28) with t replaced by the time elapsed after the application of the impulse, that is, $t - \tau$. Thus we obtain

$$x(t) = \underset{\sim}{F}g(t - \tau) \tag{4.29}$$

This is shown in Fig. 4.4(b).

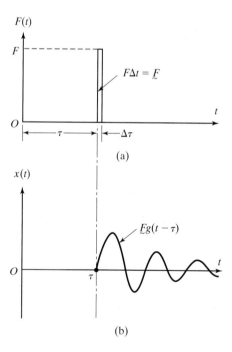

(a)

(b)

FIGURE 4.4 Impulse response.

Response of a Structure Under Impact

EXAMPLE 4.4

In the vibration testing of a structure, an impact hammer with a load cell to measure the impact force is used to cause excitation, as shown in Fig. 4.5(a). Assuming $m = 5$ kg, $k = 2000$ N/m, $c = 10$ N-s/m and $\underset{\sim}{F} = 20$ N-s, find the response of the system.

Solution: From the known data, we can compute

$$\omega_n = \sqrt{\frac{k}{m}} = \sqrt{\frac{2000}{5}} = 20 \text{ rad/s}, \quad \zeta = \frac{c}{c_c} = \frac{c}{2\sqrt{km}} = \frac{10}{2\sqrt{2000}\,(5)} = 0.05,$$

$$\omega_d = \sqrt{1 - \zeta^2}\,\omega_n = 19.975 \text{ rad/s}$$

Assuming that the impact is given at $t = 0$, we find (from Eq. 4.28) the response of the system as

$$x_1(t) = \underset{\sim}{F}\,\frac{e^{-\zeta \omega_n t}}{m \omega_d} \sin \omega_d t$$

$$= \frac{20}{(5)(19.975)} e^{-0.05\,(20)\,t} \sin 19.975\,t = 0.20025\,e^{-t} \sin 19.975\,t\text{m} \qquad \text{(E.1)}$$

Note: The graph of Eq. (E.1) is shown in Example 4.19.

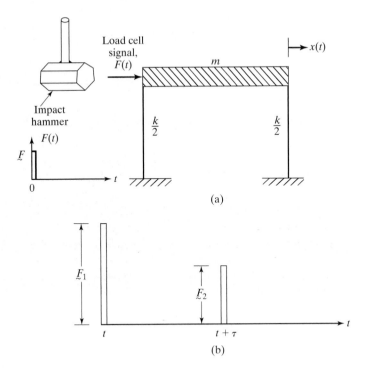

(a)

(b)

FIGURE 4.5   Structural testing using an impact hammer.

Response of a Structure Under Double Impact

EXAMPLE 4.5

In many cases, providing only one impact to the structure using an impact hammer is difficult. Sometimes a second impact takes place after the first, as shown in Fig. 4.5(b) and the applied force, $F(t)$, can be expressed as

$$F(t) = F_1 \, \delta(t) + F_2 \, \delta(t - \tau)$$

where $\delta(t)$ is the Dirac delta function and τ indicates the time between the two impacts of magnitudes F_1 and F_2. For a structure with $m = 5$ kg, $k = 2000$ N/m, $c = 10$ N-s/m and $F(t) = 20 \, \delta(t) + 10 \, \delta(t - 0.2)$ N, find the response of the structure.

Solution: From the known data, we find $\omega_n = 20$ rad/s (see the solution for Example 4.4), $\zeta = 0.05$, and $\omega_d = 19.975$ rad/s. The response due to the impulse $F_1 \, \delta(t)$ is given by Eq. (E.1) of Example 4.4 while the response due to the impulse $F_2 \, \delta(t - 0.2)$ can be determined from Eqs. (4.29) and (4.28) as

$$x_2(t) = F_2 \, \frac{e^{-\zeta \omega_n (t - \tau)}}{m \, \omega_d} \sin \omega_d \, (t - \tau) \tag{E.1}$$

For $\tau = 0.2$, Eq. (E.1) becomes

$$
\begin{aligned}
x_2(t) &= \frac{10}{(5)(19.975)} \, e^{-0.05 \, (20)(t - 0.2)} \sin 19.975 \, (t - 0.2) \\
&= 0.100125 \, e^{-(t - 0.2)} \sin 19.975 \, (t - 0.2); \, t > 0.2 \tag{E.2}
\end{aligned}
$$

Using the superposition of the two responses $x_1(t)$ and $x_2(t)$, the response due to two impacts, in meters, can be expressed as

$$x(t) = \begin{cases} 0.20025 \, e^{-t} \sin 19.975 \, t; \, 0 \le t \le 0.2 \\ 0.20025 \, e^{-t} \sin 19.975 \, t + 0.100125 \, e^{-(t - 0.2)} \sin 19.975 \, (t - 0.2); \, t > 0.2 \end{cases} \tag{E.3}$$

Note: The graph of Eq. (E.3) is shown in Example 4.19.

■

4.5.2 Response to a General Forcing Condition

Now we consider the response of the system under an arbitrary external force $F(t)$, shown in Fig. 4.6. This force may be assumed to be made up of a series of impulses of varying magnitude. Assuming that at time τ, the force $F(\tau)$ acts on the system for a short period of time $\Delta\tau$, the impulse acting at $t = \tau$ is given by $F(\tau) \, \Delta\tau$. At any time t, the elapsed time since the impulse is $t - \tau$, so the response of the system at t due to this impulse alone is given by Eq. (4.29) with $F = F(\tau) \, \Delta\tau$:

$$\Delta x(t) = F(\tau) \, \Delta\tau g \, (t - \tau) \tag{4.30}$$

The total response at time t can be found by summing all the responses due to the elementary impulses acting at all times τ:

$$x(t) \simeq \Sigma F(\tau) g(t - \tau) \, \Delta\tau \tag{4.31}$$

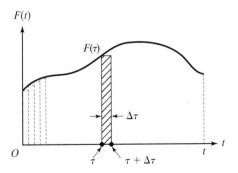

FIGURE 4.6 An arbitrary (nonperiodic) forcing function.

Letting $\Delta\tau \to 0$ and replacing the summation by integration, we obtain

$$x(t) = \int_0^t F(\tau)g(t - \tau)\, d\tau \tag{4.32}$$

By substituting Eq. (4.27) into Eq. (4.32), we obtain

$$x(t) = \frac{1}{m\omega_d}\int_0^t F(\tau)e^{-\zeta\omega_n(t-\tau)}\sin\omega_d(t - \tau)\, d\tau \tag{4.33}$$

which represents the response of an underdamped single degree of freedom system to the arbitrary excitation $F(t)$. Note that Eq. (4.33) does not consider the effect of initial conditions of the system. The integral in Eq. (4.32) or Eq. (4.33) is called the *convolution* or *Duhamel integral*. In many cases the function $F(t)$ has a form that permits an explicit integration of Eq. (4.33). In case such integration is not possible, it can be evaluated numerically without much difficulty, as illustrated in Section 4.8 and in Chapter 11. An elementary discussion of the Duhamel integral in vibration analysis is given in Ref. [4.6].

4.5.3 Response to Base Excitation

If a spring-mass-damper system is subjected to an arbitrary base excitation described by its displacement, velocity, or acceleration, the equation of motion can be expressed in terms of the relative displacement of the mass $z = x - y$ as follows (see Section 3.6.2)

$$m\ddot{z} + c\dot{z} + kz = -m\ddot{y} \tag{4.34}$$

This equation is similar to the equation

$$m\ddot{x} + c\dot{x} + kx = F \tag{4.35}$$

with the variable z replacing x and the term $-m\ddot{y}$ replacing the forcing function F. Hence all of the results derived for the force-excited system are applicable to the base-excited system also for z when the term F is replaced by $-m\ddot{y}$. For an underdamped system subjected to base excitation, the relative displacement can be found from Eq. (4.33):

$$z(t) = -\frac{1}{\omega_d}\int_0^t \ddot{y}(\tau)\, e^{-\zeta\omega_n(t-\tau)} \sin \omega_d\,(t - \tau)\,d\tau \qquad (4.36)$$

EXAMPLE 4.6

Step Force on a Compacting Machine

A compacting machine, modeled as a single degree of freedom system, is shown in Fig. 4.7(a). The force acting on the mass m (m includes the masses of the piston, the platform, and the material being compacted) due to a sudden application of the pressure can be idealized as a step force, as shown in Fig. 4.7(b). Determine the response of the system.

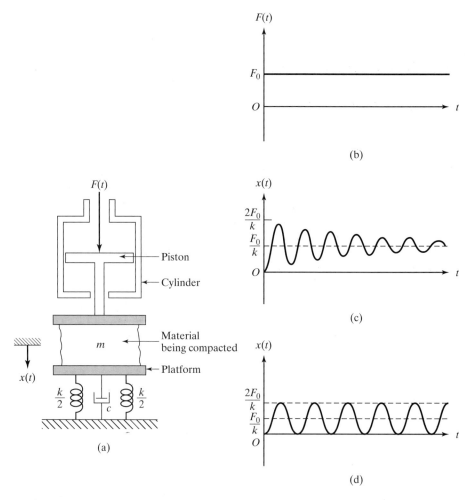

FIGURE 4.7   Step force applied to a compacting machine.

Solution: Since the compacting machine is modeled as a mass-spring-damper system, the problem is to find the response of a damped single degree of freedom system subjected to a step force. By noting that $F(t) = F_0$, we can write Eq. (4.33) as

$$x(t) = \frac{F_0}{m\omega_d}\int_0^t e^{-\zeta\omega_n(t-\tau)}\sin\omega_d(t-\tau)\,d\tau$$

$$= \frac{F_0}{m\omega_d}\left[e^{-\zeta\omega_n(t-\tau)}\left\{\frac{\zeta\omega_n\sin\omega_d(t-\tau)+\omega_d\cos\omega_d(t-\tau)}{(\zeta\omega_n)^2+(\omega_d)^2}\right\}\right]_{\tau=0}^t$$

$$= \frac{F_0}{k}\left[1 - \frac{1}{\sqrt{1-\zeta^2}}\cdot e^{-\zeta\omega_n t}\cos(\omega_d t - \phi)\right] \tag{E.1}$$

where

$$\phi = \tan^{-1}\left(\frac{\zeta}{\sqrt{1-\zeta^2}}\right) \tag{E.2}$$

This response is shown in Fig. 4.7(c). If the system is undamped ($\zeta = 0$ and $\omega_d = \omega_n$), Eq. (E.1) reduces to

$$x(t) = \frac{F_0}{k}[1 - \cos\omega_n t] \tag{E.3}$$

Equation (E.3) is shown graphically in Fig. 4.7(d). It can be seen that if the load is instantaneously applied to an undamped system, a maximum displacement of twice the static displacement will be attained, that is, $x_{max} = 2F_0/k$.

■

EXAMPLE 4.7

Time-Delayed Step Force

Find the response of the compacting machine shown in Fig. 4.7(a) when it is subjected to the force shown in Fig. 4.8.

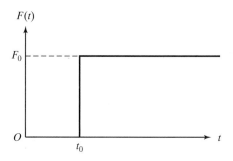

FIGURE 4.8 Step force applied with a time-delay.

Solution: Since the forcing function starts at $t = t_0$ instead of at $t = 0$, the response can be obtained from Eq. (E.1) of Example 4.6 by replacing t by $t - t_0$. This gives

$$x(t) = \frac{F_0}{k\sqrt{1 - \zeta^2}}\left[\sqrt{1 - \zeta^2} - e^{-\zeta\omega_n(t-t_0)}\cos\{\omega_d(t - t_0) - \phi\}\right] \qquad (E.1)$$

If the system is undamped, Eq. (E.1) reduces to

$$x(t) = \frac{F_0}{k}[1 - \cos\omega_n(t - t_0)] \qquad (E.2)$$

∎

Rectangular Pulse Load

EXAMPLE 4.8

If the compacting machine shown in Fig. 4.7(a) is subjected to a constant force only during the time $0 \le t \le t_0$ (Fig. 4.9a), determine the response of the machine.

Solution: The given forcing function, $F(t)$, can be considered as the sum of a step function $F_1(t)$ of magnitude $+ F_0$ beginning at $t = 0$ and a second step function $F_2(t)$ of magnitude $- F_0$ starting at time $t = t_0$, as shown in Fig. 4.9(b).

Thus the response of the system can be obtained by subtracting Eq. (E.1) of Example 4.7 from Eq. (E.1) of Example 4.6. This gives

$$x(t) = \frac{F_0 e^{-\zeta\omega_n t}}{k\sqrt{1 - \zeta^2}}\left[-\cos(\omega_d t - \phi) + e^{\zeta\omega_n t_0}\cos\{\omega_d(t - t_0) - \phi\}\right] \qquad (E.1)$$

with

$$\phi = \tan^{-1}\left(\frac{\zeta}{\sqrt{1 - \zeta^2}}\right) \qquad (E.2)$$

To see the vibration response graphically, we consider the system as undamped, so that Eq. (E.1) reduces to

$$x(t) = \frac{F_0}{k}\left[\cos\omega_n(t - t_0) - \cos\omega_n t\right] \qquad (E.3)$$

The response is shown in Fig. 4.9(c) for two different pulse widths of t_0 for the following data (Problem 4.58): $m = 100$ kg, $c = 50$ N-s/m, $k = 1200$ N/m, and $F_0 = 100$ N. The responses will be different for the two cases $t_0 > \tau_n/2$ and $t_0 > \tau_n/2$, where τ_n is the undamped natural time period of the system. If $t_0 > \tau_n/2$, the peak will be larger and occurs during the forced vibration era (that is, during 0 to t_0) while the peak will be smaller and occurs in the residual vibration era (that is, after t_0) if $t_0 > \tau_n/2$. In Fig. 4.9(c), $\tau_n = 1.8138$ s and the peak corresponding to $t_0 = 1.5$ s is about six times larger than the one with $t_0 = 0.1$ s.

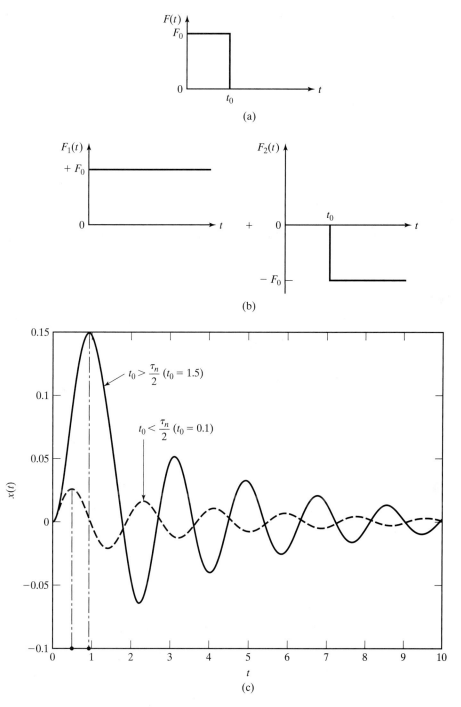

FIGURE 4.9   Response due to a pulse load.

EXAMPLE 4.9

◼◼◼◼◼◼◼◼ **Compacting Machine Under Linear Force**

Determine the response of the compacting machine shown in Fig. 4.10(a) when a linearly varying force (shown in Fig. 4.10(b)) is applied due to the motion of the cam.

Solution: The linearly varying force shown in Fig. 4.10(b) is known as the ramp function. This forcing function can be represented as $F(\tau) = \delta F \cdot \tau$, where δF denotes the rate of increase of the force F per unit time. By substituting this into Eq. (4.33), we obtain

$$x(t) = \frac{\delta F}{m\omega_d} \int_0^t \tau e^{-\zeta\omega_n(t-\tau)} \sin \omega_d (t - \tau) \, d\tau$$

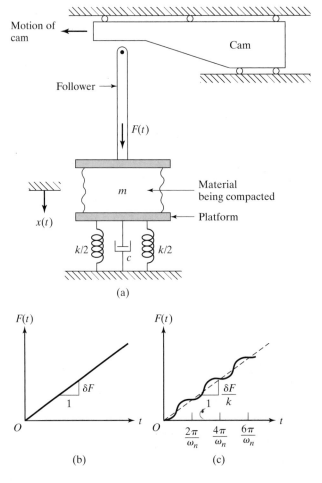

(a)

(b)   (c)

FIGURE 4.10 Compacting machine subjected to a linear force.

$$= \frac{\delta F}{m\omega_d} \int_0^t (t - \tau) e^{-\zeta\omega_n(t-\tau)} \sin \omega_d (t - \tau)(-d\tau)$$

$$- \frac{\delta F \cdot t}{m\omega_d} \int_0^t e^{-\zeta\omega_n(t-\tau)} \sin \omega_d (t - \tau)(-d\tau)$$

These integrals can be evaluated and the response expressed as follows:

$$x(t) = \frac{\delta F}{k} \left[t - \frac{2\zeta}{\omega_n} + e^{-\zeta\omega_n t} \left(\frac{2\zeta}{\omega_n} \cos \omega_d t - \left\{ \frac{\omega_d^2 - \zeta^2\omega_n^2}{\omega_n^2\omega_d} \right\} \sin \omega_d t \right) \right] \qquad \text{(E.1)}$$

(See Problem 4.28)
For an undamped system, Eq. (E.1) reduces to

$$x(t) = \frac{\delta F}{\omega_n k}[\omega_n t - \sin \omega_n t] \qquad \text{(E.2)}$$

Figure 4.10(c) shows the response given by Eq. (E.2).

■

Blast Load on a Building Frame

EXAMPLE 4.10

A building frame is modeled as an undamped single degree of freedom system (Fig. 4.11a). Find the response of the frame if it is subjected to a blast loading represented by the triangular pulse shown in Fig. 4.11(b).

Solution: The forcing function is given by

$$F(\tau) = F_0 \left(1 - \frac{\tau}{t_0} \right) \text{ for } 0 \leq \tau \leq t_0 \qquad \text{(E.1)}$$

$$F(\tau) = 0 \qquad \text{for } \tau > t_0 \qquad \text{(E.2)}$$

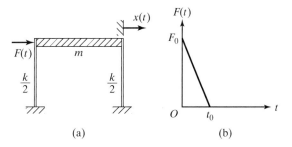

(a)   (b)

FIGURE 4.11 Building frame subjected to a blast load.

Equation (4.33) gives, for an undamped system,

$$x(t) = \frac{1}{m\omega_n} \int_0^t F(\tau) \sin \omega_n(t - \tau) d\tau \tag{E.3}$$

Response during $0 \leq t \leq t_0$:   Using Eq. (E.1) for $F(\tau)$ in Eq. (E.3) gives

$$x(t) = \frac{F_0}{m\omega_n^2} \int_0^t \left(1 - \frac{\tau}{t_0}\right) [\sin \omega_n t \cos \omega_n \tau - \cos \omega_n t \sin \omega_n \tau] \, d(\omega_n \tau)$$

$$= \frac{F_0}{k} \sin \omega_n t \int_0^t \left(1 - \frac{\tau}{t_0}\right) \cos \omega_n \tau \cdot d(\omega_n \tau)$$

$$- \frac{F_0}{k} \cos \omega_n t \int_0^t \left(1 - \frac{\tau}{t_0}\right) \sin \omega_n \tau \cdot d(\omega_n \tau) \tag{E.4}$$

By noting that integration by parts gives

$$\int \tau \cos \omega_n \tau \cdot d(\omega_n \tau) = \tau \sin \omega_n \tau + \frac{1}{\omega_n} \cos \omega_n \tau \tag{E.5}$$

and

$$\int \tau \sin \omega_n \tau \cdot d(\omega_n \tau) = -\tau \cos \omega_n \tau + \frac{1}{\omega_n} \sin \omega_n \tau \tag{E.6}$$

Eq. (E.4) can be written as

$$x(t) = \frac{F_0}{k} \left\{ \sin \omega_n t \left[\sin \omega_n t - \frac{t}{t_0} \sin \omega_n t - \frac{1}{\omega_n t_0} \cos \omega_n t + \frac{1}{\omega_n t_0} \right] \right.$$

$$\left. - \cos \omega_n t \left[-\cos \omega_n t + 1 + \frac{t}{t_0} \cos \omega_n t - \frac{1}{\omega_n t_0} \sin \omega_n t \right] \right\} \tag{E.7}$$

Simplifying this expression, we obtain

$$x(t) = \frac{F_0}{k} \left[1 - \frac{t}{t_0} - \cos \omega_n t + \frac{1}{\omega_n t_0} \sin \omega_n t \right] \tag{E.8}$$

Response during $t > t_0$:   Here also we use Eq. (E.1) for $F(\tau)$, but the upper limit of integration in Eq. (E.3) will be t_0, since $F(\tau) = 0$ for $\tau > t_0$. Thus the response can be found from Eq. (E.7) by setting $t = t_0$ within the square brackets. This results in

$$x(t) = \frac{F_0}{k\omega_n t_0} \left[(1 - \cos \omega_n t_0) \sin \omega_n t - (\omega_n t_0 - \sin \omega_n t_0) \cos \omega_n t \right] \tag{E.9}$$

■

4.6   Response Spectrum

The graph showing the variation of the maximum response (maximum displacement, velocity, acceleration, or any other quantity) with the natural frequency (or natural period) of a single degree of freedom system to a specified forcing function is known as the *response spectrum*. Since the maximum response is plotted against the natural frequency (or natural period), the response spectrum gives the maximum response of all possible single degree of freedom systems. The response spectrum is widely used in earthquake engineering design [4.2, 4.5]. A review of recent literature on shock and seismic response spectra in engineering design is given in Ref. [4.7].

Once the response spectrum corresponding to a specified forcing function is available, we need to know just the natural frequency of the system to find its maximum response. Example 4.11 illustrates the construction of a response spectrum.

EXAMPLE 4.11

Response Spectrum of Sinusoidal Pulse

Find the undamped response spectrum for the sinusoidal pulse force shown in Fig. 4.12(a) using the initial conditions $x(0) = \dot{x}(0) = 0$.

Solution

Approach: Find the response and express its maximum value in terms of its natural time period.

The equation of motion of an undamped system can be expressed as

$$m\ddot{x} + kx = F(t) = \begin{cases} F_0 \sin \omega t, & 0 \le t \le t_0 \\ 0, & t > t_0 \end{cases} \tag{E.1}$$

where

$$\omega = \frac{\pi}{t_0} \tag{E.2}$$

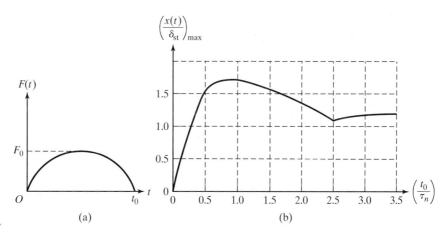

(a)

(b)

FIGURE 4.12   Response spectrum due to a sinusoidal pulse.

The solution of Eq. (E.1) can be obtained by superposing the homogeneous solution $x_c(t)$ and the particular solution $x_p(t)$ as

$$x(t) = x_c(t) + x_p(t) \tag{E.3}$$

That is,

$$x(t) = A \cos \omega_n t + B \sin \omega_n t + \left(\frac{F_0}{k - m\omega^2} \right) \sin \omega t \tag{E.4}$$

where A and B are constants and ω_n is the natural frequency of the system:

$$\omega_n = \frac{2\pi}{\tau_n} = \sqrt{\frac{k}{m}} \tag{E.5}$$

Using the initial conditions $x(0) = \dot{x}(0) = 0$ in Eq. (E.4), we can find the constants A and B as

$$A = 0, \qquad B = -\frac{F_0 \omega}{\omega_n(k - m\omega^2)} \tag{E.6}$$

Thus the solution becomes

$$x(t) = \frac{F_0/k}{1 - (\omega/\omega_n)^2} \left\{ \sin \omega t - \frac{\omega}{\omega_n} \sin \omega_n t \right\}, \qquad 0 \le t \le t_0 \tag{E.7}$$

which can be rewritten as

$$\frac{x(t)}{\delta_{st}} = \frac{1}{1 - \left(\dfrac{\tau_n}{2t_0} \right)^2} \left\{ \sin \frac{\pi t}{t_0} - \frac{\tau_n}{2t_0} \sin \frac{2\pi t}{\tau_n} \right\}, \qquad 0 \le t \le t_0 \tag{E.8}$$

where

$$\delta_{st} = \frac{F_0}{k} \tag{E.9}$$

The solution given by Eq. (E.8) is valid only during the period of force application, $0 \le t \le t_0$. Since there is no force applied for $t > t_0$, the solution can be expressed as a free vibration solution

$$x(t) = A' \cos \omega_n t + B' \sin \omega_n t, \qquad t > t_0 \tag{E.10}$$

where the constants A' and B' can be found by using the values of $x(t = t_0)$ and $\dot{x}(t = t_0)$, given by Eq. (E.8), as initial conditions for the duration $t > t_0$. This gives

$$x(t = t_0) = \alpha \left[-\frac{\tau_n}{2t_0} \sin \frac{2\pi t_0}{\tau_n} \right] = A' \cos \omega_n t_0 + B' \sin \omega_n t_0 \tag{E.11}$$

$$\dot{x}(t = t_0) = \alpha \left\{ \frac{\pi}{t_0} - \frac{\pi}{t_0} \cos \frac{2\pi t_0}{\tau_n} \right\}$$

$$= -\omega_n A' \sin \omega_n t + \omega_n B' \cos \omega_n t \qquad (E.12)$$

where

$$\alpha = \frac{\delta_{st}}{1 - \left(\dfrac{\tau_n}{2t_0}\right)^2} \qquad (E.13)$$

Equations (E.11) and (E.12) can be solved to find A' and B' as

$$A' = \frac{\alpha \pi}{\omega_n t_0} \sin \omega_n t_0, \qquad B' = -\frac{\alpha \pi}{\omega_n t_0}[1 + \cos \omega_n t_0] \qquad (E.14)$$

Equations (E.14) can be substituted into Eq. (E.10) to obtain

$$\frac{x(t)}{\delta_{st}} = \frac{(\tau_n/t_0)}{2\left\{1 - (\tau_n/2t_0)^2\right\}} \left[\sin 2\pi \left(\frac{t_0}{\tau_n} - \frac{t}{\tau_n} \right) - \sin 2\pi \frac{t}{\tau_n} \right], \qquad t \geq t_0 \qquad (E.15)$$

Equations (E.8) and (E.15) give the response of the system in nondimensional form—that is, x/δ_{st} is expressed in terms of t/τ_n. Thus for any specified value of t_0/τ_n, the maximum value of x/δ_{st} can be found. This maximum value of x/δ_{st}, when plotted against t_0/τ_n, gives the response spectrum shown in Fig. 4.12(b). It can be observed that the maximum value of $(x/\delta_{st})_{max} \simeq 1.75$ occurs at a value of $t_0/\tau_n \simeq 0.75$.

∎

In Example 4.11, the input force is simple and hence a closed form solution has been obtained for the response spectrum. However, if the input force is arbitrary, we can find the response spectrum only numerically. In such a case, Eq. (4.33) can be used to express the peak response of an undamped single degree of freedom system due to an arbitrary input force $F(t)$ as

$$x(t)\Big|_{max} = \frac{1}{m\omega_n} \int_0^t F(\tau) \sin \omega_n (t - \tau) \, d\tau \Big|_{max} \qquad (4.37)$$

4.6.1
Response Spectrum for Base Excitation

In the design of machinery and structures subjected to a ground shock, such as that caused by an earthquake, the response spectrum corresponding to the base excitation is useful. If the base of a damped single degree of freedom system is subjected to an acceleration $\ddot{y}(t)$, the equation of motion, in terms of the relative displacement $z = x - y$, is given by Eq. (4.34) and the response $z(t)$ by Eq. (4.36). In the case of a ground shock, the velocity response spectrum is generally used. The displacement and acceleration spectra are then expressed in terms of the velocity spectrum. For a harmonic oscillator (an undamped system under free vibration), we notice that

$$\ddot{x}|_{max} = -\omega_n^2 x|_{max} \quad \text{and} \quad \dot{x}|_{max} = \omega_n x|_{max}$$

Thus the acceleration and displacement spectra S_a and S_d can be obtained in terms of the velocity spectrum (S_v):

$$S_d = \frac{S_v}{\omega_n}, \quad S_a = \omega_n S_v \tag{4.38}$$

To consider damping in the system, if we assume that the maximum relative displacement occurs after the shock pulse has passed, the subsequent motion must be harmonic. In such a case we can use Eq. (4.38). The fictitious velocity associated with this apparent harmonic motion is called the *pseudo velocity* and its response spectrum, S_v, is called the *pseudo spectrum*. The velocity spectra of damped systems are used extensively in earthquake analysis.

To find the relative velocity spectrum, we differentiate Eq. (4.36) and obtain[2]

$$\dot{z}(t) = -\frac{1}{\omega_d} \int_0^t \ddot{y}(\tau) e^{-\zeta\omega_n(t-\tau)} [-\zeta\omega_n \sin \omega_d (t - \tau)$$
$$+ \omega_d \cos \omega_d (t - \tau)] \, d\tau \tag{4.39}$$

Equation (4.39) can be rewritten as

$$\dot{z}(t) = \frac{e^{-\zeta\omega_n t}}{\sqrt{1 - \zeta^2}} \sqrt{P^2 + Q^2} \sin(\omega_d t - \phi) \tag{4.40}$$

where

$$P = \int_0^t \ddot{y}(\tau) e^{\zeta\omega_n t} \cos \omega_d \tau \, d\tau \tag{4.41}$$

$$Q = \int_0^t \ddot{y}(\tau) e^{\zeta\omega_n t} \sin \omega_d \tau \, d\tau \tag{4.42}$$

and

$$\phi = \tan^{-1} \left\{ \frac{-(P\sqrt{1 - \zeta^2} + Q\zeta)}{(P\zeta - Q\sqrt{1 - \zeta^2})} \right\} \tag{4.43}$$

---

[2]The following relation is used in deriving Eq. (4.39) from Eq. (4.36):
$$\frac{d}{dt} \int_0^t f(t, \tau) \, d\tau = \int_0^t \frac{\partial f}{\partial t}(t, \tau) d\tau + f(t, \tau)|_{\tau=t}$$

The velocity response spectrum, S_v, can be obtained from Eq. (4.40):

$$S_v = |\dot{z}(t)|_{max} = \left| \frac{e^{-\zeta \omega_n t}}{\sqrt{1 - \zeta^2}} \sqrt{P^2 + Q^2} \right|_{max} \tag{4.44}$$

Thus the pseudo response spectra are given by

$$S_d = |z|_{max} = \frac{S_v}{\omega_n}; \qquad S_v = |\dot{z}|_{max}; \qquad S_a = |\ddot{z}|_{max} = \omega_n S_v \tag{4.45}$$

EXAMPLE 4.12

Water Tank Subjected to Base Acceleration

The water tank, shown in Fig. 4.13(a), is subjected to a linearly varying ground acceleration as shown in Fig. 4.13(b) due to an earthquake. The mass of the tank is m, the stiffness of the column is k, and damping is negligible. Find the response spectrum for the relative displacement, $z = x - y$, of the water tank.

Solution

Approach: Model the water tank as an undamped single degree of freedom system. Find the maximum relative displacement of the tank and express it as a function of ω_n.

The base acceleration can be expressed as

$$\ddot{y}(t) = \ddot{y}_{max} \left(1 - \frac{t}{t_0} \right) \qquad \text{for} \qquad 0 \le t \le 2t_0 \tag{E.1}$$

$$\ddot{y}(t) = 0 \qquad \text{for} \qquad t > 2t_0 \tag{E.2}$$

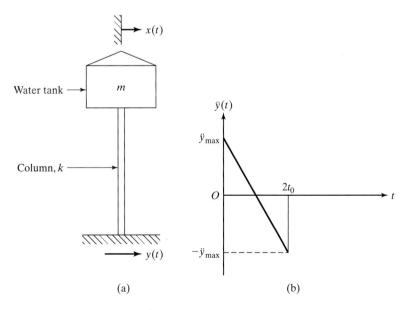

(a)   (b)

FIGURE 4.13   Water tank subjected to base motion.

Response during $0 \leq t \leq 2t_0$:   By substituting Eq. (E.1) into Eq. (4.36), the response can be expressed, for an undamped system, as

$$z(t) = -\frac{1}{\omega_n}\ddot{y}_{max}\left[\int_0^t \left(1 - \frac{\tau}{t_0}\right)\right.$$

$$\left.(\sin \omega_n t \cos \omega_n \tau - \cos \omega_n t \sin \omega_n \tau) \, d\tau\right] \tag{E.3}$$

This equation is the same as Eq. (E.4) of Example 4.10 except that $(-\ddot{y}_{max})$ appears in place of F_0/m. Hence $z(t)$ can be written, using Eq. (E.8) of Example 4.10, as

$$z(t) = -\frac{\ddot{y}_{max}}{\omega_n^2}\left[1 - \frac{t}{t_0} - \cos \omega_n t + \frac{1}{\omega_n t_0}\sin \omega_n t\right] \tag{E.4}$$

To find the maximum response z_{max}, we set

$$\dot{z}(t) = -\frac{\ddot{y}_{max}}{t_0\omega_n^2}\left[-1 + \omega_n t_0 \sin \omega_n t + \cos \omega_n t\right] = 0 \tag{E.5}$$

This equation gives the time t_m at which z_{max} occurs:

$$t_m = \frac{2}{\omega_n}\tan^{-1}(\omega_n t_0) \tag{E.6}$$

By substituting Eq. (E.6) into Eq. (E.4), the maximum response of the tank can be found:

$$z_{max} = -\frac{\ddot{y}_{max}}{\omega_n^2}\left[1 - \frac{t_m}{t_0} - \cos \omega_n t_m + \frac{1}{\omega_n t_0}\sin \omega_n t_m\right] \tag{E.7}$$

Response during $t > 2t_0$:   Since there is no excitation during this time, we can use the solution of the free vibration problem (Eq. 2.18)

$$z(t) = z_0 \cos \omega_n t + \left(\frac{\dot{z}_0}{\omega_n}\right)\sin \omega_n t \tag{E.8}$$

provided that we take the initial displacement and initial velocity as

$$z_0 = z(t = 2t_0) \quad \text{and} \quad \dot{z}_0 = \dot{z}(t = 2t_0) \tag{E.9}$$

using Eq. (E.7). The maximum of $z(t)$ given by Eq. (E.8) can be identified as

$$z_{max} = \left[z_0^2 + \left(\frac{\dot{z}_0}{\omega_n}\right)^2\right]^{1/2} \tag{E.10}$$

where z_0 and $\dot{z}_0$ are computed as indicated in Eq. (E.9).

■

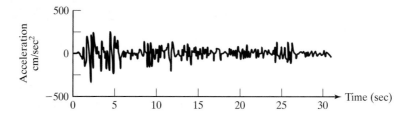

FIGURE 4.14
A typical accelerogram.

**4.6.2
Earthquake
Response
Spectra**

The most direct description of an earthquake motion in time domain is provided by accelerograms that are recorded by instruments called *strong motion accelerographs*. The accelerograph records three orthogonal components of ground acceleration at a certain location. A typical accelerogram is shown in Fig. 4.14. The accelerograms are generally recorded on photographic paper or film and are digitized for engineering applications. The peak ground acceleration, duration, and frequency content of the earthquake can be obtained from an accelerogram. An accelerogram can be integrated to obtain the time variations of the ground velocity and ground displacement.

A response spectrum is used to provide the most descriptive representation of the influence of a given earthquake on a structure or machine. It is possible to plot the maximum response of a single degree of freedom system in terms of the acceleration, relative pseudo velocity, and relative displacement using logarithmic scales. A typical response spectrum, plotted on a four-way logarithmic paper, is shown in Fig. 4.15. In this figure, the vertical axis denotes the spectral velocity, the horizontal axis represents the natural time period, the 45° inclined axis indicates the spectral acceleration, and the 135° inclined axis shows the spectral displacement.

As can be seen from Fig. 4.15, the response spectrum of a particular accelerogram (earthquake) exhibits considerable irregularities in the frequency domain. However, spectra corresponding to an ensemble of accelerograms produced by ground shakings of sites with similar geological and seismological features are smooth functions of time and provide statistical trends that characterize them collectively. This idea has led to the development of the concept of a design spectrum, a typical one shown in Fig. 4.16, for use in earthquake-resistent design of structures and machines. The following examples illustrate the use of the response and design spectra of earthquakes.

Response of a Building Frame to an Earthquake

EXAMPLE 4.13

A building frame has a weight of 15,000 lb and two columns of total stiffness k, as indicated in Fig. 4.17. It has a damping ratio of 0.05 and a natural time period of 1.0 sec. For the earthquake characterized in Fig. 4.15, determine the following:

a.   Maximum relative displacement of the mass, x_{max}
b.   Maximum shear force in the columns
c.   Maximum bending stress in the columns

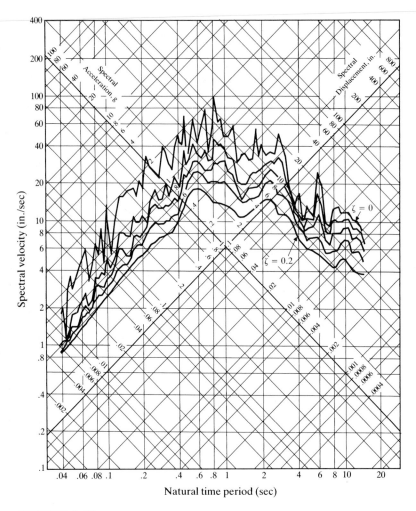

FIGURE 4.15   Response spectrum of a typical earthquake [4.12]. (Imperial Valley Earthquake, May 18, 1940; ζ = 0, .02, 0.05, 0.10, and 0.20.) (Reprinted with permission from *The Shock and Vibration Digest.*)

Solution

Approach: Find the spectral displacement, spectral velocity and spectral acceleration corresponding to the given natural time period.

For $\tau_n = 1.0$ sec and $\zeta = 0.05$, Fig. 4.15 gives $S_v = 25$ in./sec, $S_d = 4.2$ in, and $S_a = 0.42$ $g = 162.288$ in./sec^2.

a.   Maximum relative displacement of the mass, $x_{max} = S_d = 4.2$ in.

b.   Maximum shear force in both columns:

$$|kx_{max}| = m\ddot{x}_{max} = \frac{W}{g} S_a = \left(\frac{15,000}{386.4}\right)(162.288) = 6300 \text{ lb}$$

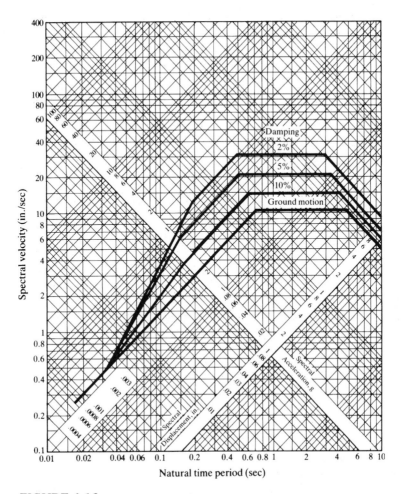

FIGURE 4.16   Design spectrum [4.12]. (Reprinted with permission from *The Shock and Vibration Digest.*)

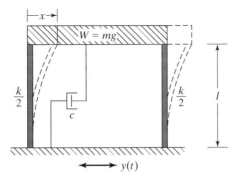

FIGURE 4.17   Building frame subjected to base motion.

Thus the maximum shear force in each column is given by

$$F_{max} = 6,300/2 = 3,150 \text{ lb}$$

c.   Maximum bending moment in each column $= M_{max} = F_{max} \, l$. Thus the maximum bending stress is given by the beam formula

$$\sigma_{max} = \frac{M_{max} \, c}{I}$$

where I is the area moment of inertia and c is the distance of the outer fiber from the neutral axis of the column section.

∎

EXAMPLE 4.14

Derailment of Trolley of a Crane During Earthquake

The trolley of an electric overhead traveling (EOT) crane travels horizontally on the girder as indicated in Fig. 4.18. Assuming the trolley as a point mass, the crane can be modeled as a single degree of freedom system with a period 2 s and a damping ratio 2%. Determine whether the trolley derails under a vertical earthquake excitation whose design spectrum is given by Fig. 4.16.

Solution

Approach: Determine whether the spectral acceleration of the trolley (mass) exceeds a value of 1 g.

For $\tau_n = 2$ s and $\zeta = 0.02$, Fig. 4.16 gives the spectral acceleration as $S_a = 0.25$ g and hence the trolley will not derail.

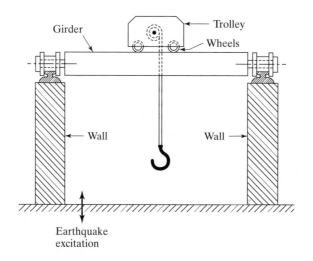

FIGURE 4.18   A crane subjected to an earthquake excitation.

∎

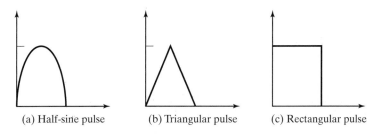

(a) Half-sine pulse   (b) Triangular pulse   (c) Rectangular pulse

FIGURE 4.19   Typical shock pulses.

**4.6.3
Design Under
a Shock
Environment**

When a force is applied for short duration, usually for a period of less than one natural time period, it is called a *shock load*. A shock causes a significant increase in the displacement, velocity, acceleration, or stress in a mechanical system. Although fatigue is a major cause of failure under harmonic forces, usually it is not very important under shock loads. A shock may be described by a pulse shock, velocity shock, or a shock response spectrum. The pulse shocks are introduced by suddenly applied forces or displacements in the form of a square, half sine, triangular, or similar shape (see Fig. 4.19). A velocity shock is caused by sudden changes in the velocity such as those caused when packages are dropped from a height. The shock response spectrum describes the way in which a machine or structure responds to a specific shock instead of describing the shock itself. Different types of shock pulses are used in qualifying most commercial, industrial, and military products. Many military specifications such as MIL-E-5400 and MIL-STD-810 define different types of shock pulses and detailed methods for testing with these pulses. The following example illustrates the method of limiting dynamic stresses in mechanical systems under a shock environment.

▬▬▬▬▬▬▬▬▬▬   Design of a Bracket for Shock Loads

EXAMPLE 4.15 ───

A printed circuit board (PCB) is mounted on a cantilevered aluminum bracket, as shown in Fig. 4.20(a). The bracket is placed in a container that is expected to be dropped from a low-flying helicopter. The resulting shock can be approximated as a half-sine wave pulse, as shown in Fig. 4.20(b). Design the bracket to withstand an acceleration level of 100 g under the half-sine wave pulse shown in Fig. 4.20(b). Assume a specific weight of 0.1 lb/in.^{3}, a Young's modulus of 10^7 psi, and a permissible stress of 26,000 psi for aluminum.

Solution: The self-weight of the beam (w) is given by

$$w = (10)\left(\frac{1}{2} \times d\right)(0.1) = 0.5\,d$$

and the total weight, W, assumed to be a concentrated load at the free end of the beam, is given by

$$W = \text{Weight of beam} + \text{Weight of PCB} = 0.5\,d + 0.4$$

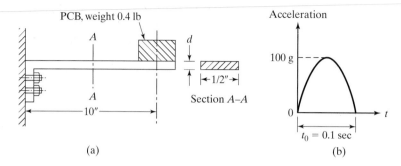

FIGURE 4.20   A cantilever subjected to an acceleration pulse.

The area moment of inertia (I) of the cross section of the beam is

$$I = \frac{1}{12} \times \frac{1}{2} \times d^3 = 0.04167 \, d^3$$

The static deflection of the beam under the end load W, δ_{st}, can be computed as

$$\delta_{st} = \frac{Wl^3}{3 \, EI} = \frac{(0.5 \, d + 0.4)(10^3)}{3 \times 10^7 \, (0.04167 \, d^3)} = \frac{(0.5 \, d + 0.4)}{d^3} 7.9994 \times 10^{-4}$$

Since the shock amplification factor (the ordinate in Fig. 4.12b) cannot be determined unless the value of t_0/τ_n is known, we adopt a trial-and-error procedure to determine the value of τ_n and hence that of t_0/τ_n. If d is assumed as $\frac{1}{2}$ in,

$$\delta_{st} = \left(\frac{0.5 \times 0.5 + 0.4}{0.5^3} \right) 7.9997 \times 10^{-4} = 41.5969 \times 10^{-4} \text{ in.}$$

Eq. (2.30) gives

$$\tau_n = 2\pi \sqrt{\frac{\delta_{st}}{g}} = 2\pi \sqrt{\left(\frac{41.5969 \times 10^{-4}}{386.4} \right)} = 0.020615 \text{ sec}$$

Hence

$$\frac{t_0}{\tau_n} = \frac{0.1}{0.020615} = 4.8508$$

The shock amplification factor (A_a) can be found from Fig. 4.12(b) as 1.1. The dynamic load (P_d) acting on the cantilever is given by

$$P_d = A_a \, M a_s = (1.1) \left(\frac{0.65}{g} \right)(100 \, g) = 71.5 \text{ lb}$$

where a_s is the acceleration corresponding to the shock, M is the mass at the end of the beam, and Ma_s is the inertia force on the beam. Noting that $I = 0.04167 \, d^3 = 0.005209$ in.^{4}, the maximum bending stress at the root of the cantilever bracket can be computed as

$$\sigma_{max} = \frac{M_b c}{I} = \frac{(71.5 \times 10)\dfrac{0.5}{2}}{0.005209} = 34315.6076 \text{ lb/in.}^2$$

Since this stress exceeds the permissible value, we assume the next trial value of d as 0.6 in. This yields

$$\delta_{st} = \left(\frac{0.5 \times 0.6 + 0.4}{0.6^3} \right) 7.9994 \times 10^{-4} = 25.9240 \times 10^{-4} \text{ in.}$$

$$\tau_n = 2\pi \sqrt{\frac{\delta_{st}}{g}} = 2\pi \sqrt{\frac{25.9240 \times 10^{-4}}{386.4}} = 0.01627 \text{ sec}$$

$$\frac{t_0}{\tau_n} = \frac{0.1}{0.01627} = 6.1445$$

From Fig. 4.12 (b), the shock amplification factor is found as $A_a \approx 1.1$ and hence the dynamic load acting on the beam can be determined as

$$P_d = (1.1)\left(\frac{0.7}{g} \right)(100 \text{ g}) = 77.0 \text{ lb}$$

Since $d = 0.6$ in. gives $I = 0.04167 \, d^3 = 0.009001$ in.^{4}, the maximum bending stress at the root of the bracket will be

$$\sigma_{max} = \frac{M_b c}{I} = \frac{(77.0 \times 10)\left(\dfrac{0.6}{2} \right)}{0.009001} = 25663.8151 \text{ lb/in.}^2$$

Since this stress is within the permissible limit, the thickness of the bracket can be taken as $d = 0.6$ in.

∎

4.7   Laplace Transforms

The Laplace transform method can be used to find the response of a system under any type of excitation, including the harmonic and periodic types. This method can be used for the efficient solution of linear differential equations, particularly those with constant coefficients [4.3]. It permits the conversion of differential equations into algebraic ones, which are easier to manipulate. The major advantages of the method are that it can treat discontinuous functions without any particular difficulty and that it automatically takes into account the initial conditions.

The Laplace transform of a function $x(t)$, denoted symbolically as $\overline{x}(s) = \mathcal{L}x(t)$, is defined as

$$\overline{x}(s) = \mathcal{L}x(t) = \int_0^\infty e^{-st} x(t) \, dt \tag{4.46}$$

where s is, in general, a complex quantity and is called the *subsidiary variable*. The function e^{-st} is called the *kernel* of the transformation. Since the integration is with respect to t, the transformation gives a function of s. In order to solve a vibration problem using the Laplace transform method, the following steps are necessary:

1. Write the equation of motion of the system.
2. Transform each term of the equation, using known initial conditions.
3. Solve for the transformed response of the system.
4. Obtain the desired solution (response) by using inverse Laplace transformation.

To solve the forced vibration equation

$$m\ddot{x} + c\dot{x} + kx = F(t) \tag{4.47}$$

by the Laplace transform method, it is necessary to find the transforms of the derivatives

$$\dot{x}(t) = \frac{dx}{dt}(t) \quad \text{and} \quad \ddot{x}(t) = \frac{d^2x}{dt^2}(t)$$

These can be found as follows:

$$\mathcal{L}\frac{dx}{dt}(t) = \int_0^\infty e^{-st} \frac{dx}{dt}(t) \, dt \tag{4.48}$$

This can be integrated by parts to obtain

$$\mathcal{L}\frac{dx}{dt}(t) = e^{-st} x(t) \Big|_0^\infty + s \int_0^\infty e^{-st} x(t) \, dt = s\overline{x}(s) - x(0) \tag{4.49}$$

where $x(0) = x_0$ is the initial displacement of the mass m. Similarly, the Laplace transform of the second derivative of $x(t)$ can be obtained

$$\mathcal{L}\frac{d^2x}{dt^2}(t) = \int_0^\infty e^{-st} \frac{d^2x}{dt^2}(t) \, dt = s^2\overline{x}(s) - sx(0) - \dot{x}(0) \tag{4.50}$$

where $\dot{x}(0) = \dot{x}_0$ is the initial velocity of the mass m. Since the Laplace transform of the force $F(t)$ is given by

$$\overline{F}(s) = \mathcal{L}F(t) = \int_0^\infty e^{-st} F(t) \, dt \tag{4.51}$$

we can transform both sides of Eq. (4.47) and obtain, using Eqs. (4.46) and (4.48) in (4.51),

$$m\mathscr{L}\ddot{x}(t) + c\mathscr{L}\dot{x}(t) + k\mathscr{L}x(t) = \mathscr{L}F(t)$$

or

$$(ms^2 + cs + k)\,\overline{x}(s) = \overline{F}(s) + m\dot{x}(0) + (ms + c)x(0) \qquad (4.52)$$

where the right-hand side of Eq. (4.52) can be regarded as generalized transformed excitation.

For the present, we take $\dot{x}(0)$ and $x(0)$ as zero, which is equivalent to ignoring the homogeneous solution of the differential equation (4.47). Then the ratio of the transformed excitation to the transformed response $\overline{Z}(s)$ can be expressed as

$$\overline{Z}(s) = \frac{\overline{F}(s)}{\overline{x}(s)} = ms^2 + cs + k \qquad (4.53)$$

The function $\overline{Z}(s)$ is known as the *generalized impedance* of the system. The reciprocal of the function $\overline{Z}(s)$ is called the *admittance* or *transfer function* of the system and is denoted as $\overline{Y}(s)$:

$$\overline{Y}(s) = \frac{1}{\overline{Z}(s)} = \frac{\overline{x}(s)}{\overline{F}(s)} = \frac{1}{ms^2 + cs + k} = \frac{1}{m(s^2 + 2\zeta\omega_n s + \omega_n^2)} \qquad (4.54)$$

It can be seen that by letting $s = i\omega$ in $\overline{Y}(s)$ and multiplying by k, we obtain the complex frequency response $H(i\omega)$ defined in Eq. (3.54). Equation (4.54) can also be expressed as

$$\overline{x}(s) = \overline{Y}(s)\,\overline{F}(s) \qquad (4.55)$$

which indicates that the transfer function can be regarded as an algebraic operator that operates on the transformed force to yield the transformed response.

To find the desired response $x(t)$ from $\overline{x}(s)$, we have to take the inverse Laplace transform of $\overline{x}(s)$, which can be defined symbolically as

$$x(t) = \mathscr{L}^{-1}\overline{x}(s) = \mathscr{L}^{-1}\overline{Y}(s)\,\overline{F}(s) \qquad (4.56)$$

In general, the operator $\mathscr{L}^{-1}$ involves a line integral in the complex domain, [4.9, 4.10]. Fortunately, we need not evaluate these integrals separately for each problem; such integrations have been carried out for various common forms of the function $F(t)$ and tabulated [4.4]. One such table is given in Appendix D. In order to find the solution using Eq. (4.56), we usually look for ways of decomposing $\overline{x}(s)$ into a combination of simple functions whose inverse transformations are available in Laplace transform tables. We can decompose $\overline{x}(s)$ conveniently by the method of partial fractions.

In the above discussion, we ignored the homogeneous solution by assuming $x(0)$ and $\dot{x}(0)$ as zero. We now consider the general solution by taking the initial conditions as

$x(0) = x_0$ and $\dot{x}(0) = \dot{x}_0$. From Eq. (4.52), the transformed response $\bar{x}(s)$ can be obtained:

$$\bar{x}(s) = \frac{\overline{F}(s)}{m(s^2 + 2\zeta\omega_n s + \omega_n^2)} + \frac{s + 2\zeta\omega_n}{s^2 + 2\zeta\omega_n s + \omega_n^2} x_0$$
$$+ \frac{1}{s^2 + 2\zeta\omega_n s + \omega_n^2} \dot{x}_0 \tag{4.57}$$

We can obtain the inverse transform of $\bar{x}(s)$ by considering each term on the right side of Eq. (4.57) separately. We also make use of the following relation [4.4]:

$$\mathcal{L}^{-1}\overline{f}_1(s)\,\overline{f}_2(s) = \int_0^t f_1(\tau)f_2(t - \tau)\,d\tau \tag{4.58}$$

By considering the first term on the right side of Eq. (4.57) as $\overline{f}_1(s)\overline{f}_2(s)$, where

$$\overline{f}_1(s) = \overline{F}(s) \quad \text{and} \quad \overline{f}_2(s) = \frac{1}{m(s^2 + 2\zeta\omega_n s + \omega_n^2)}$$

and by noting that $f_1(t) = \mathcal{L}^{-1}\overline{f}_1(s) = F(t)$, we obtain[3]

$$\mathcal{L}^{-1}\overline{f}_1(s)\overline{f}_2(s) = \frac{1}{m\omega_d}\int_0^t F(\tau)\,e^{-\zeta\omega_n(t-\tau)}\,\sin\omega_d(t - \tau)\,d\tau \tag{4.59}$$

Considering the second term on the right side of Eq. (4.57), we find the inverse transform of the coefficient of x_0 from the table in Appendix D:

$$\mathcal{L}^{-1}\left(\frac{s + 2\zeta\omega_n}{s^2 + 2\zeta\omega_n s + \omega_n^2}\right) = \frac{1}{\sqrt{1 - \zeta^2}}\,e^{-\zeta\omega_n t}\sin(\omega_d t + \phi_1) \tag{4.60}$$

where

$$\phi_1 = \cos^{-1}(\zeta) \tag{4.61}$$

Finally, the inverse transform of the coefficient of $\dot{x}_0$ in the third term on the right side of Eq. (4.57) can be obtained from the table in Appendix D:

$$\mathcal{L}^{-1}\left[\frac{1}{(s^2 + 2\zeta\omega_n s + \omega_n^2)}\right] = \frac{1}{\omega_d}\,e^{-\zeta\omega_n t}\,\sin\omega_d t \tag{4.62}$$

---

[3]The inverse transform of $\overline{f}_2(s)$ is obtained from the Laplace transform table in Appendix D.

Using Eqs. (4.57), (4.59), (4.60), and (4.62), the general solution of Eq. (4.47) can be expressed as

$$x(t) = \frac{x_0}{(1 - \zeta^2)^{1/2}} e^{-\zeta \omega_n t} \sin(\omega_d t + \phi_1) + \frac{\dot{x}_0}{\omega_d} e^{-\zeta \omega_n t} \sin \omega_d t$$

$$+ \frac{1}{m\omega_d} \int_0^t F(\tau) e^{-\zeta \omega_n (t - \tau)} \sin \omega_d (t - \tau) \, d\tau \qquad (4.63)$$

Response of a Compacting Machine

EXAMPLE 4.16

Find the response of the compacting machine of Example 4.6 assuming the system to be under-damped (i.e., $\zeta < 1$).

Approach: Use a spring-mass-damper model of the compacting machine and use Laplace transform technique.

Solution: The forcing function is given by

$$F(t) = \begin{cases} F_0 & \text{for } 0 \le t \le t_0 \\ 0 & \text{for } t > t_0 \end{cases} \qquad (E.1)$$

By taking the Laplace transform of the governing differential equation, Eq. (4.47), we obtain Eq. (4.57), using Appendix D, with

$$\overline{F}(s) = \mathscr{L}F(t) = \frac{F_0(1 - e^{-t_0 s})}{s} \qquad (E.2)$$

Thus Eq. (4.57) can be written as

$$\overline{x}(s) = \frac{F_0(1 - e^{-t_0 s})}{ms(s^2 + 2\zeta \omega_n s + \omega_n^2)} + \frac{s + 2\zeta \omega_n}{s^2 + 2\zeta \omega_n s + \omega_n^2} x_0$$

$$+ \frac{1}{s^2 + 2\zeta \omega_n + \omega_n^2} \dot{x}_0$$

$$= \frac{F_0}{m\omega_n^2} \frac{1}{s \left(\dfrac{s^2}{\omega_n^2} + \dfrac{2\zeta s}{\omega_n} + 1 \right)} - \frac{F_0}{m\omega_n^2} \frac{e^{-t_0 s}}{s \left(\dfrac{s^2}{\omega_n^2} + \dfrac{2\zeta s}{\omega_n} + 1 \right)}$$

$$+ \frac{x_0}{\omega_n^2} \frac{s}{\left(\dfrac{s^2}{\omega_n^2} + \dfrac{2\zeta s}{\omega_n} + 1 \right)} + \left(\frac{2\zeta x_0}{\omega_n} + \frac{\dot{x}_0}{\omega_n^2} \right) \frac{1}{\left(\dfrac{s^2}{\omega_n^2} + \dfrac{2\zeta s}{\omega_n} + 1 \right)} \qquad (E.3)$$

The inverse transform of Eq. (E.3) can be expressed by using the results in Appendix D as

$$
x(t) = \frac{F_0}{m\omega_n^2}\left[1 - \frac{e^{-\zeta\omega_n t}}{\sqrt{1 - \zeta^2}}\sin\{\omega_n\sqrt{1 - \zeta^2}t + \phi_1\}\right]
$$

$$
- \frac{F_0}{m\omega_n^2}\left[1 - \frac{e^{-\zeta\omega_n(t-t_0)}}{\sqrt{1 - \zeta^2}}\sin\{\omega_n\sqrt{1 - \zeta^2}(t - t_0) + \phi_1\}\right]
$$

$$
- \frac{x_0}{\omega_n^2}\left[\frac{\omega_n^2 e^{-\zeta\omega_n t}}{\sqrt{1 - \zeta^2}}\sin\{\omega_n\sqrt{1 - \zeta^2}t - \phi_1\}\right]
$$

$$
+ \left(\frac{2\zeta x_0}{\omega_n} + \frac{\dot{x}_0}{\omega_n^2}\right)\left[\frac{\omega_n}{\sqrt{1 - \zeta^2}}e^{-\zeta\omega_n t}\sin(\omega_n\sqrt{1 - \zeta^2}t)\right] \tag{E.4}
$$

where ϕ_1 is given by Eq. (4.61). Thus the response of the compacting machine can be expressed as

$$
x(t) = \frac{F_0}{m\omega_n^2\sqrt{1 - \zeta^2}}[- e^{-\zeta\omega_n t}\sin(\omega_n\sqrt{1 - \zeta^2}t + \phi_1)
$$

$$
+ e^{-\zeta\omega_n(t-t_0)}\sin\{\omega_n\sqrt{1 - \zeta^2}(t - t_0) + \phi_1\}]
$$

$$
- \frac{x_0}{\sqrt{1 - \zeta^2}}e^{-\zeta\omega_n t}\sin(\omega_n\sqrt{1 - \zeta^2}t - \phi_1)
$$

$$
+ \frac{(2\zeta\omega_n x_0 + \dot{x}_0)}{\omega_n\sqrt{1 - \zeta^2}}e^{-\zeta\omega_n t}\sin(\omega_n\sqrt{1 - \zeta^2}t) \tag{E.5}
$$

Although the first part of Eq. (E.5) is expected to be the same as Eq. (E.1) of Example 4.8, it is difficult to see the equivalence in the present form of Eq. (E.5). However, for the undamped system, Eq. (E.5) reduces to

$$
x(t) = \frac{F_0}{m\omega_n^2}\left[- \sin\left(\omega_n t + \frac{\pi}{2}\right) + \sin\left\{\omega_n(t - t_0) + \frac{\pi}{2}\right\}\right]
$$

$$
- x_0\sin\left(\omega_n t - \frac{\pi}{2}\right) + \frac{\dot{x}_0}{\omega_n}\sin\omega_n t
$$

$$
= \frac{F_0}{k}[\cos\omega_n(t - t_0) - \cos\omega_n t] + x_0\cos\omega_n t + \frac{\dot{x}_0}{\omega_n}\sin\omega_n t \tag{E.6}
$$

The first or steady-state part of Eq. (E.6) can be seen to be identical to Eq. (E.3) of Example 4.8.

∎

4.8 Response to Irregular Forcing Conditions Using Numerical Methods

In the previous sections, it was assumed that the forcing functions $F(t)$ are available as functions of time in an explicit manner. In many practical problems, however, the forcing functions $F(t)$ are not available in the form of analytical expressions. When a forcing function is determined experimentally, $F(t)$ may be known as an irregular curve. Sometimes only the values of $F(t) = F_i$ at a series of points $t = t_i$ may be available, in the form of a diagram or a table. In such cases, we can fit polynomials or some such curves to the data and use them in the Duhamel integral, Eq. (4.33), to find the response of the system. Another more common method of finding the response involves dividing the time axis into a number of discrete points and using a simple variation of $F(t)$ during each time step. We shall present this numerical approach in this section, using several types of interpolation functions for $F(t)$ [4.8]. The direct numerical integration of the equations of motion is discussed in Chapter 11.

Method 1.   Let the function $F(t)$ vary with time in an arbitrary manner, as indicated in Fig. 4.21. This forcing function can be approximated by a series of step functions having different magnitudes starting at different instants, as shown in Fig. 4.22. In this figure, the first step function starts at time $t = t_1 = 0$ and has a magnitude of ΔF_1, the second step function starts at time $t = t_2$ and has a magnitude of ΔF_2, and so forth. The response of the system in any time interval $t_{j-1} \leq t \leq t_j$ due to the step functions ΔF_i $(i = 1, 2, \ldots, j - 1)$ can be found, using the results of Example 4.6:

$$x(t) = \frac{1}{k} \sum_{i=1}^{j-1} \Delta F_i \left[1 - e^{-\zeta \omega_n (t - t_i)} \right.$$

$$\left. \times \left\{ \cos \omega_d (t - t_i) + \frac{\zeta \omega_n}{\omega_d} \sin \omega_d (t - t_i) \right\} \right] \qquad (4.64)$$

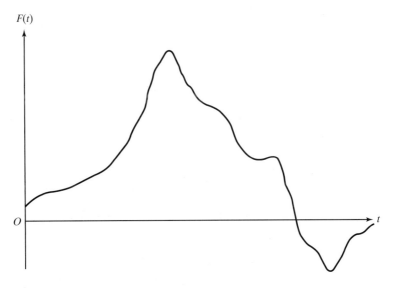

$F(t)$

O

t

FIGURE 4.21   Arbitrary forcing function.

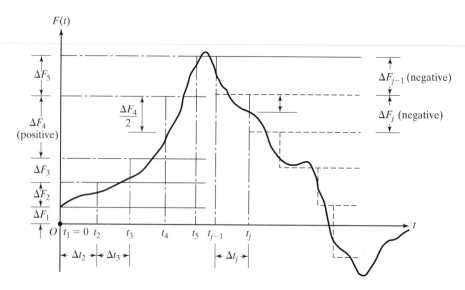

FIGURE 4.22   Approximation of forcing function as a series of step functions.

Thus the response of the system at $t = t_j$ becomes

$$x_j = \frac{1}{k}\sum_{i=1}^{j-1}\Delta F_i\left[1 - e^{-\zeta\omega_n(t_j-t_i)}\right.$$

$$\left.\times\left\{\cos\omega_d(t_j - t_i) + \frac{\zeta\omega_n}{\omega_d}\sin\omega_d(t_j - t_i)\right\}\right] \tag{4.65}$$

Notice that the step function ΔF_i of step i is positive if the slope of the F-versus-t curve is positive, and it is negative if the slope of the F-versus-t curve is negative, as indicated in Fig. 4.22. For higher accuracy, the time steps taken should be small. In addition, it is desirable to make the force steps start, after the first one, at instants when the ordinates of the $F(t)$ curve are at the midheights of the steps, as shown in Fig. 4.22. In this case, the errors involved in approximating the $F(t)$ curve will be self-compensatory; that is, the areas lying above the $F(t)$ curve will be approximately equal to the areas lying below the $F(t)$ curve.

Method 2.   Instead of approximating the $F(t)$ curve by a succession of step functions, we can approximate it by a series of rectangular impulses F_i, as shown in Fig. 4.23. These impulses F_i are positive or negative, depending on whether the curve $F(t)$ lies above or below the time (t) axis. As in the previous case, the magnitudes of F_i should be selected as the values of $F(t)$ at the midpoints of the time intervals Δt_i, as shown in Fig. 4.23, to make the errors self-compensating. The response of the system in any time interval

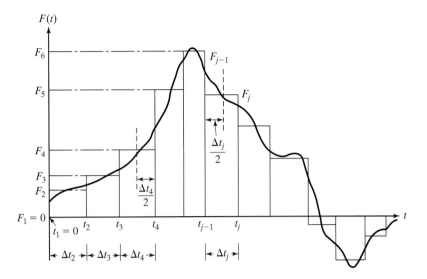

FIGURE 4.23   Approximation of forcing function as a series of impulses.

$t_{j-1} \leq t \leq t_j$ can be found by adding the response due to F_j (applied in the interval Δt_j) to the response existing at $t = t_{j-1}$ (initial condition). This gives

$$
x(t) = \frac{F_j}{k}\left[1 - e^{-\zeta\omega_n(t-t_{j-1})}\left\{\cos\omega_d(t - t_{j-1}) + \frac{\zeta\omega_n}{\omega_d}\sin\omega_d(t - t_{j-1})\right\}\right]
$$

$$
+ e^{-\zeta\omega_n(t-t_{j-1})}\left\{x_{j-1}\cos\omega_d(t - t_{j-1})\right.
$$

$$
\left. + \frac{\dot{x}_{j-1} + \zeta\omega_n x_{j-1}}{\omega_d}\sin\omega_d(t - t_{j-1})\right\} \tag{4.66}
$$

By substituting $t = t_j$ in Eq. (4.66), the response of the system at the end of the interval Δt_j can be obtained:

$$
x_j = \frac{F_j}{k}\left[1 - e^{-\zeta\omega_n\Delta t_j}\left\{\cos\omega_d\cdot\Delta t_j + \frac{\zeta\omega_n}{\omega_d}\sin\omega_d\cdot\Delta t_j\right\}\right]
$$

$$
+ e^{-\zeta\omega_n\Delta t_j}\left\{x_{j-1}\cos\omega_d\cdot\Delta t_j + \frac{\dot{x}_{j-1} + \zeta\omega_n x_{j-1}}{\omega_d}\sin\omega_d\cdot\Delta t_j\right\} \tag{4.67}
$$

By differentiating Eq. (4.66) with respect to t and substituting $t = t_j$, we obtain the velocity $\dot{x}_j$ at the end of the interval Δt_j:

$$\dot{x}_j = \frac{F_j \omega_d}{k} e^{-\zeta\omega_n \Delta t_j}\left(1 + \frac{\zeta^2\omega_n^2}{\omega_d^2}\right)\sin\omega_d \cdot \Delta t_j + \omega_d\, e^{-\zeta\omega_n \cdot \Delta t_j}$$

$$\times \left\{ -x_{j-1}\sin\omega_d \cdot \Delta t_j + \frac{\dot{x}_{j-1} + \zeta\omega_n x_{j-1}}{\omega_d}\cos\omega_d \cdot \Delta t_j \right.$$

$$\left. -\frac{\zeta\omega_n}{\omega_d}\left[x_{j-1}\cos\omega_d \cdot \Delta t_j + \frac{\dot{x}_{j-1} + \zeta\omega_n x_{j-1}}{\omega_d}\sin\omega_d \cdot \Delta t_j\right]\right\} \quad (4.68)$$

Equations (4.67) and (4.68) represent recurrence relations for computing the response at the end of jth time step. They also provide the initial conditions of x_j and $\dot{x}_j$ at the beginning of step $j + 1$. These equations may be sequentially applied to find the variations of displacement and velocity of the system with time.

Method 3. In the piecewise-constant types of approximations used in Methods 1 and 2, it is not always possible to make the areas above and below the $F(t)$ curve equal and make the errors self-compensating. Hence it is desirable to use a higher order interpolation, such as a piecewise linear or a piecewise quadratic approximation, for $F(t)$. In the piecewise linear interpolation, the variation of $F(t)$ in any time interval is assumed to be linear, as shown in Fig. 4.24. In this case, the response of the system in the time interval $t_{j-1} \le t \le t_j$ can

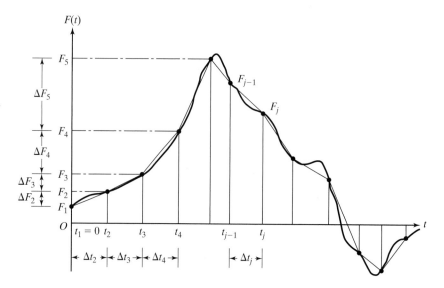

FIGURE 4.24 Approximation of forcing function as a piecewise linear function.

be found by adding the response due to the linear (ramp) function applied during the current interval to the response existing at $t = t_{j-1}$ (initial condition). This gives

$$
\begin{aligned}
x(t) = \frac{\Delta F_j}{k \Delta t_j} & \left[t - t_{j-1} - \frac{2\zeta}{\omega_n} + e^{-\zeta \omega_n (t - t_{j-1})} \right. \\
& \times \left. \left\{ \frac{2\zeta}{\omega_n} \cos \omega_d (t - t_{j-1}) - \frac{\omega_d^2 - \zeta^2 \omega_n^2}{\omega_n^2 \omega_d} \sin \omega_d (t - t_{j-1}) \right\} \right] \\
+ \frac{F_{j-1}}{k} & \left[1 - e^{-\zeta \omega_n (t - t_{j-1})} \left\{ \cos \omega_d (t - t_{j-1}) \right. \right. \\
& \left. \left. + \frac{\zeta \omega_n}{\omega_d} \sin \omega_d (t - t_{j-1}) \right\} \right] \\
+ e^{-\zeta \omega_n (t - t_{j-1})} & \left[x_{j-1} \cos \omega_d (t - t_{j-1}) \right. \\
& \left. + \frac{\dot{x}_{j-1} + \zeta \omega_n x_{j-1}}{\omega_d} \sin \omega_d (t - t_{j-1}) \right] \quad (4.69)
\end{aligned}
$$

where $\Delta F_j = F_j - F_{j-1}$. By setting $t = t_j$ in Eq. (4.69), we obtain the response at the end of the interval Δt_j:

$$
\begin{aligned}
x_j = \frac{\Delta F_j}{k \Delta t_j} & \left[\Delta t_j - \frac{2\zeta}{\omega_n} + e^{-\zeta \omega_n \Delta t_j} \left\{ \frac{2\zeta}{\omega_n} \cos \omega_d \Delta t_j - \frac{\omega_d^2 - \zeta^2 \omega_n^2}{\omega_n^2 \omega_d} \sin \omega_d \Delta t_j \right\} \right] \\
+ \frac{F_{j-1}}{k} & \left[1 - e^{-\zeta \omega_n \cdot \Delta t_j} \left\{ \cos \omega_d \Delta t_j + \frac{\zeta \omega_n}{\omega_d} \sin \omega_d \Delta t_j \right\} \right] \\
+ e^{-\zeta \omega_n \Delta t_j} & \left[x_{j-1} \cos \omega_d \Delta t_j + \frac{\dot{x}_{j-1} + \zeta \omega_n x_{j-1}}{\omega_d} \sin \omega_d \Delta t_j \right] \quad (4.70)
\end{aligned}
$$

By differentiating Eq. (4.69) with respect to t and substituting $t = t_j$, we obtain the velocity at the end of the interval:

$$
\begin{aligned}
\dot{x}_j = \frac{\Delta F_j}{k \Delta t_j} & \left[1 - e^{-\zeta \omega_n \Delta t_j} \left\{ \cos \omega_d \Delta t_j + \frac{\zeta \omega_n}{\omega_d} \sin \omega_d \Delta t_j \right\} \right] \\
+ \frac{F_{j-1}}{k} & e^{-\zeta \omega_n \Delta t_j} \frac{\omega_n^2}{\omega_d} \cdot \sin \omega_d \Delta t_j + e^{-\zeta \omega_n \cdot \Delta t_j} \\
& \times \left[\dot{x}_{j-1} \cos \omega_d \Delta t_j - \frac{\zeta \omega_n}{\omega_d} \left(\dot{x}_{j-1} + \frac{\omega_n}{\zeta} x_{j-1} \right) \sin \omega_d \Delta t_j \right] \quad (4.71)
\end{aligned}
$$

Equations (4.70) and (4.71) are the recurrence relations for finding the response of the system at the end of jth time step.

Damped Response Using Numerical Methods

EXAMPLE 4.17

Find the response of a spring-mass-damper system subjected to the forcing function

$$F(t) = F_0\left(1 - \sin\frac{\pi t}{2t_0}\right) \tag{E.1}$$

in the interval $0 \le t \le t_0$, using a numerical procedure. Assume $F_0 = 1, k = 1, m = 1, \zeta = 0.1$, and $t_0 = \tau_n/2$, where τ_n denotes the natural period of vibration given by

$$\tau_n = \frac{2\pi}{\omega_n} = \frac{2\pi}{(k/m)^{1/2}} = 2\pi \tag{E.2}$$

The values of x and $\dot{x}$ at $t = 0$ are zero.

Approach: Use numerical methods.
Solution: Figure 4.25 shows the forcing function of Eq. (E.1). For the numerical computations, the time interval 0 to t_0 is divided into 10 equal steps with

$$\Delta t_i = \frac{t_0}{10} = \frac{\pi}{10}; \quad i = 2,3, \dots, 11 \tag{E.3}$$

Four different methods are used to approximate the forcing function $F(t)$. In Fig. 4.26, $F(t)$ is approximated by a series of rectangular impulses, each starting at the beginning of the corresponding time

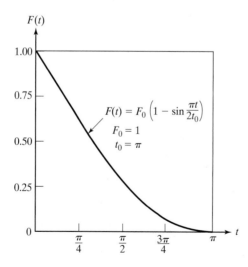

FIGURE 4.25   Forcing function.

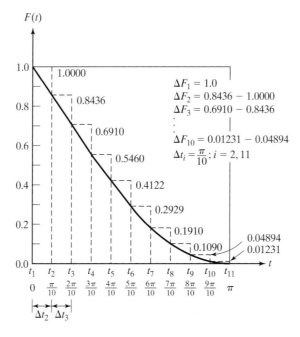

FIGURE 4.26   Approximation as a series of rectangular impulses.

step. A similar approximation, with the magnitude of the impulse at the end of the time step, is used in Fig. 4.27. The value of $F(t)$ at the middle of the time step is used as an impulse in Fig. 4.28. In Fig. 4.29, piecewise linear (trapezoidal) impulses are used to approximate the forcing function $F(t)$. The numerical results are given in Table 4.1. As can be expected from the idealizations, the results obtained by idealizations 1 and 2 (Figs. 4.26 and 4.27) overestimate and underestimate the true response, respectively. The results given by idealizations 3 and 4 are expected to lie between those given by idealizations 1 and 2. Furthermore, the results obtained from idealization 4 will be the most accurate ones.

■

4.9   Examples Using MATLAB

Total Response of a System Under Base Excitation

EXAMPLE 4.18

Using MATLAB, plot the total response of the viscously damped system subject to harmonic base excitation considered in Example 4.2.

Solution: The total response of the system is given by Eq. (E.8) of Example 4.2:

$$x(t) = 0.488695\, e^{-t} \cos(19.975\, t - 1.529683)$$
$$+ 0.001333 \cos(5\, t - 0.02666) + 0.053314 \sin(5\, t - 0.02666)$$

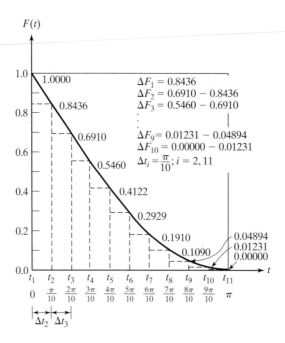

FIGURE 4.27   Approximation as a series of rectangular impulses (magnitudes different).

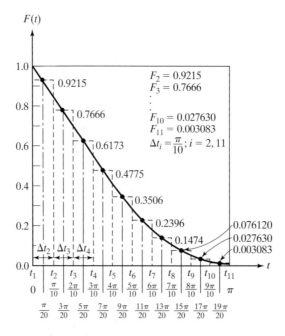

FIGURE 4.28   Rectangular impulses at middle of time steps.

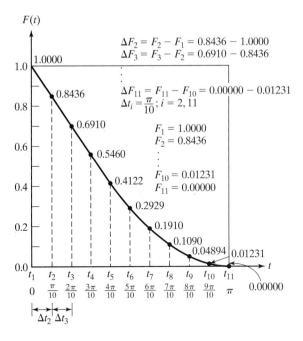

FIGURE 4.29 Piecewise linear approximation.

TABLE 4.1 Response of the system with different approximations.

		$x(t_i)$ Obtained According to			
i	t_i	Fig. 4.26 (Idealization 1)	Fig. 4.27 (Idealization 2)	Fig. 4.28 (Idealization 3)	Fig. 4.29 (Idealization 4)
1	0	0.00000	0.00000	0.00000	0.00000
2	0.1π	0.04794	0.04044	0.04417	0.04541
3	0.2π	0.17578	0.14729	0.16147	0.16377
4	0.3π	0.35188	0.29228	0.32190	0.32499
5	0.4π	0.54248	0.44609	0.49392	0.49746
6	0.5π	0.71540	0.58160	0.64790	0.65151
7	0.6π	0.84330	0.67659	0.75906	0.76238
8	0.7π	0.90630	0.71578	0.80986	0.81255
9	0.8π	0.89367	0.69214	0.79142	0.79323
10	0.9π	0.80449	0.60717	0.70403	0.70482
11	π	0.64730	0.47152	0.55672	0.55647

The MATLAB program to plot this equation is given below.

```
% Ex4_18.m
for i = 1: 1001
    t(i) = (i - 1)*10/1000;
    x(i) = 0.488695 * exp(-t(i)) * cos(19.975*t(i)-1.529683) + ...
           0.001333*cos(5*t(i)-0.02666) + 0.053314 * sin(5*t(i)
           - 0.02666);
end
plot(t,x);
xlabel('t');
ylabel('x(t)');
```

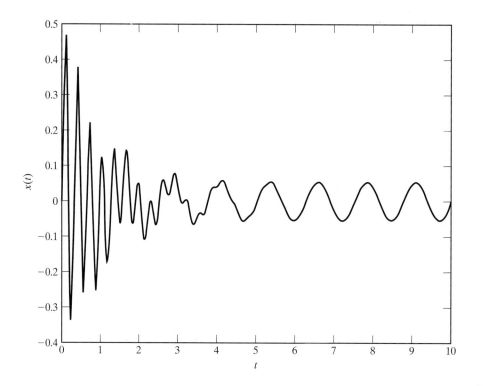

■

Impulse Response of a Structure

EXAMPLE 4.19

Using MATLAB, plot the impulse response of the single degree of freedom structure due to (a) a single impact and (b) a double impact considered in Examples 4.4 and 4.5.

Solution: The impulse responses of the structure due to single and double impacts are given by Eqs. (E.1) and (E.3) of Examples 4.4 and 4.5, respectively:

$$x(t) = 0.20025 \, e^{-t} \sin 19.975 \, t \tag{E.1}$$

$$x(t) = \begin{cases} 0.20025\, e^{-t} \sin 19.975\, t;\ 0 \le t \le 0.2 \\ 0.20025\, e^{-t} \sin 19.975\, t + 0.100125\, e^{-(t-0.2)} \sin 19.975\, (t - 0.2);\ t \ge 0.2 \end{cases} \qquad (\text{E.2})$$

The MATLAB program to plot Eqs. (E.1) and (E.2) is given below.

```
% Ex4_19.m
for i = 1: 1001
    t (i) = (i-1)*5/1000;
    x1(i) = 0.20025 * exp(-t(i)) * sin(19.975*t(i));
    if t(i) > 0.2
        a = 0.100125;
    else
        a = 0.0;
    end
    x2(i) = 0.20025 * exp(-t(i)) * sin(19.975*t(i)) + ...
        a * exp(-(t(i)-0.2)) * sin(19.975*(t(i)-0.2));
end
plot(t,x1);
gtext('Eq. (E.1): solid line');
hold on;
plot(t,x2,'-');
gtext('Eq. (E.2): dash line');
xlabel('t');
```

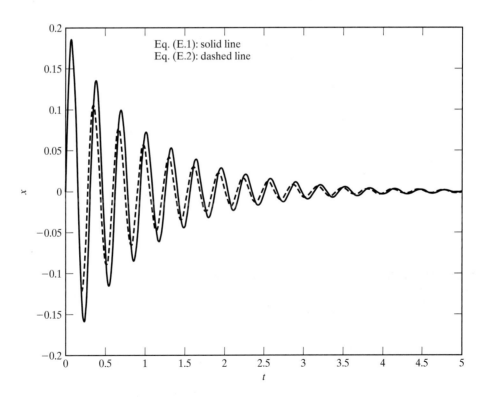

Response Under a Periodic Force

EXAMPLE 4.20

Develop a general purpose MATLAB program, called **Program4.m**, to find the steady-state response of a viscously damped single degree of freedom system under a periodic force. Use the program to find the response of a system that is subject to the force shown in the following figure with the following data: $m = 100$ kg, $k = 10^5$ N/m, $\zeta = 0.1$.

Solution: Program4.m is developed to accept the values of the periodic force at n discrete values of time. The input data of the program are as follows:

xm = mass of the system

xk = stiffness of the system

xai = damping ratio (ζ)

n = number of equidistant points at which the values of the force $F(t)$ are known

m = number of Fourier coefficients to be considered in the solution

time = time period of the function F(t)

f = array of dimension n that contains the known values of $F(t)$; $f(i) = F(t_i), i = 1, 2, \ldots, n$

t = array of dimension n that contains the known discrete values of time t; $t(i) = t_i$, $i = 1, 2, \ldots, n$

The program gives the following output:

$$\text{step number } i, t(i), f(i), x(i)$$

where $x(i) = x(t = t_i)$ is the response at time step i. The program also plots the variation of x with time.

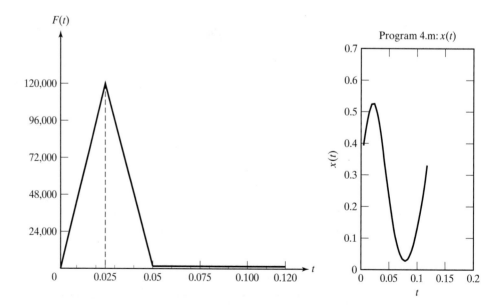

■

Response Under Arbitrary Forcing Function

EXAMPLE 4.21

Develop a general-purpose MATLAB program, called **Program5.m**, to find the response of a viscously damped spring-mass system under an arbitrary forcing function using the methods of Section 4.8. Use the program to find the solution of Example 4.17.

Solution: **Program5.m** is developed to accept the values of the applied force at n discrete values of time. The program requires the following input data:

n = number of time stations at which the values of the forcing function are known

t = array of size n containing the values of time at which the forcing function is known

f = array of size n containing the values of the forcing function at various time stations according to idealization of Fig. 4.22 (Figs. 4.26 or 4.27 for Example 4.17)

ff = array of size n containing the values of the forcing function at various time stations according to the idealization of Fig. 4.24 (Figs. 4.28 or 4.29 for Example 4.17)

xai = damping factor (ζ)

omn = undamped natural frequency of the system

$delt$ = incremental time between consecutive time stations

xk = spring stiffness

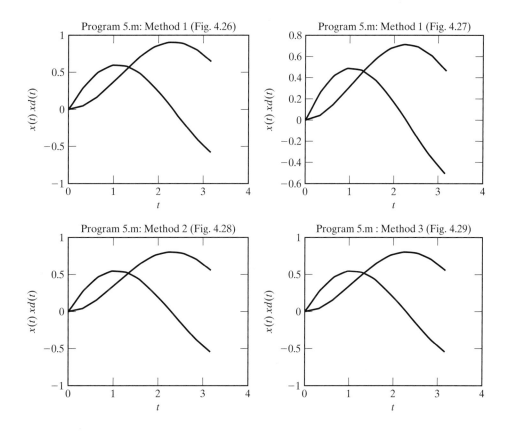

The program gives the following output: time station i, $x(i)$ given by Method 1 (Figs. 4.26 and 4.27), $x(i)$ given by Method 2 (Fig. 4.28), $x(i)$ given by Method 3 (Fig. 4.29), where $x(i)$ is the response of the system at time station i. The program also plots the variation of x with time according to different methods.

■

4.10   C++ Programs

4.10.1
Response Under an Arbitrary Periodic Force

An interactive C++ program called **Program4.cpp** is given for finding the dynamic response of a damped oscillator excited by any periodic external force applied to the mass. The input and output of the program are similar to those of the MATLAB program, **Program4.m**, given in Example 4.20.

■■■■■■■■ Response Under a Periodic Force Using C++

EXAMPLE 4.22 ──

Using **Program4.cpp**, find the steady-state response of the viscously damped single degree of freedom system described in Example 4.20.

Solution: The input data are to be entered interactively. The input and output of the program are given below.

```
Please input n and m:
24
20

Please input xm, xk, xai, time:
100.0 100000.0 0.1 0.12

Please input f[i]: i = 0, 1, ..., n-1
24000.0
48000.0
72000.0
96000.0
120000.0
96000.0
72000.0
48000.0
24000.0
0.0
0.0
0.0
0.0
0.0
0.0
0.0
0.0
0.0
0.0
0.0
0.0
0.0
0.0
0.0

RESPONSE OF A SINGLE D.O.F SYSTEM UNDER PERIODIC FORCE

XM = 100
XK = 100000
XAI = 0.1
N = 24
```

```
M = 20
TIME = 0.12

APPLIED FORCE RESPONSE
        I             T(I)                    F(I)                X(I)
        0         0.00500000        24000.00000000          0.39311979
        1         0.01000000        48000.00000000          0.45115592
        2         0.01500000        72000.00000000          0.49675477
        3         0.02000000        96000.00000000          0.52336489
        4         0.02500000       120000.00000000          0.52511284
        5         0.03000000        96000.00000000          0.49745120
        6         0.03500000        72000.00000000          0.44728028
        7         0.04000000        48000.00000000          0.38235024
        8         0.04500000        24000.00000000          0.31053464
        9         0.05000000            0.00000000          0.23964565
       10         0.05500000            0.00000000          0.17698370
       11         0.06000000            0.00000000          0.12413921
       12         0.06500000            0.00000000          0.08215259
       13         0.07000000            0.00000000          0.05174979
       14         0.07500000            0.00000000          0.03332517
       15         0.08000000            0.00000000          0.02694470
       16         0.08500000            0.00000000          0.03236967
       17         0.09000000            0.00000000          0.04908952
       18         0.09500000            0.00000000          0.07635072
       19         0.10000000            0.00000000          0.11317605
       20         0.10500000            0.00000000          0.15837841
       21         0.11000000            0.00000000          0.21058025
       22         0.11500000            0.00000000          0.26824905
       23         0.12000000            0.00000000          0.32974721
```

∎

4.10.2
Response Under an Arbitrary Forcing Function

An interactive C++ program called **Program5.cpp** is used to find the response of a viscously damped single degree of freedom system under arbitrary forcing function using the methods described in Section 4.8. The input and output parameters of the program are similar to those of the MATLAB program, **Program5.m**, given in Example 4.21.

EXAMPLE 4.23

Response Under an Arbitrary Forcing Function Using C++

Using **Program5.cpp**, find the steady-state response of the viscously damped spring-mass system described in Example 4.17.

Solution: The input data are to be entered interactively. The output of the program is given below.

VALUE OF	METHOD #1 (FIG. 4.26)	METHOD #1 (FIG. 4.27)	METHOD #2 (FIG. 4.28)	METHOD #3 (FIG. 4.29)
I	X(I)	X(I)	X(I)	X(I)
1	0.047936	0.040437	0.044175	0.045415
2	0.175781	0.147295	0.161471	0.163773
3	0.351883	0.292278	0.321877	0.324989
4	0.542483	0.446091	0.493842	0.497464
5	0.715396	0.581603	0.647699	0.651514
6	0.843296	0.676586	0.758676	0.762379
7	0.906301	0.715783	0.809225	0.812552
8	0.893674	0.692145	0.790486	0.793232
9	0.804490	0.607167	0.702788	0.704821
10	0.647299	0.469170	0.555199	0.556465

I	XD(I)	XD(I)	XD(I)	XD(I)
1	0.298008	0.251389	0.276010	0.275640
2	0.502976	0.418148	0.462605	0.461687
3	0.602270	0.491876	0.549249	0.547683
4	0.595174	0.474576	0.536630	0.534405
5	0.492171	0.377744	0.435845	0.433036
6	0.313187	0.220601	0.266613	0.263378
7	0.085019	0.027665	0.054767	0.051327
8	−0.161754	−0.174042	−0.170711	−0.174093
9	−0.396047	−0.357870	−0.380784	−0.383830
10	−0.589414	−0.507466	−0.549290	−0.551730

■

4.11   Fortran Programs

4.11.1 Response Under an Arbitrary Periodic Force

A Fortran computer program, in the form of a subroutine called **PERIOD.F**, is used to find the dynamic response of a damped single degree of freedom system excited by any periodic external force applied to the mass. The arguments of the subroutine are similar to those described for **Program4.m** given in Example 4.20.

EXAMPLE 4.24

Response Under a Periodic Force Using Fortran

Using the subroutine **PERIOD.F**, find the steady-state response of the viscously damped spring-mass system described in Example 4.20.

Solution: The main program that calls the subroutine **PERIOD.F** and the subroutine **PERIOD.F** are given as **PROGRAM4.F**. The output of the program is given below.

```
RESPONSE OF A SINGLE D.O.F. SYSTEM UNDER PERIODIC FORCE

XM  = 0.100000E+03
XK  = 0.100000E+06
XAI = 0.100000E+00
N   =   24
M   =   20
TIME= 0.120000E+00

APPLIED FORCE AND RESPONSE
  I       T(I)           F(I)            X(I)

  1    0.500000E-02    0.240000E+05    0.393120E+00
  2    0.100000E-01    0.480000E+05    0.451156E+00
  3    0.150000E-01    0.720000E+05    0.496755E+00
  4    0.200000E-01    0.960000E+05    0.523365E+00
  5    0.250000E-01    0.120000E+06    0.525113E+00
  6    0.300000E-01    0.960000E+05    0.497451E+00
  7    0.350000E-01    0.720000E+05    0.447280E+00
  8    0.400000E-01    0.480000E+05    0.382350E+00
  9    0.450000E-01    0.240000E+05    0.310534E+00
 10    0.500000E-01    0.000000E+00    0.239646E+00
 11    0.550000E-01    0.000000E+00    0.176984E+00
 12    0.600000E-01    0.000000E+00    0.124139E+00
 13    0.650000E-01    0.000000E+00    0.821526E-01
 14    0.700000E-01    0.000000E+00    0.517498E-01
 15    0.750000E-01    0.000000E+00    0.333252E-01
 16    0.800000E-01    0.000000E+00    0.269447E-01
 17    0.850000E-01    0.000000E+00    0.323697E-01
 18    0.900000E-01    0.000000E+00    0.490896E-01
```

19	0.950000E−01	0.000000E+00	0.763507E−01
20	0.100000E+00	0.000000E+00	0.113176E+00
21	0.105000E+00	0.000000E+00	0.158378E+00
22	0.110000E+00	0.000000E+00	0.210580E+00
23	0.115000E+00	0.000000E+00	0.268249E+00
24	0.120000E+00	0.000000E+00	0.329747E+00

∎

4.11.2 Response Under an Arbitrary Forcing Function

A Fortran computer program called **PROGRAM5.F** is used to find the response of a damped single degree of freedom system under an arbitrary forcing function using the methods of Section 4.8. The arguments of the subroutine are similar to those described for **Program5.m** given in Example 4.21.

Response Under an Arbitrary Force Using Fortran

EXAMPLE 4.25

Using **PROGRAM5.F**, find the steady-state response of the viscously damped single degree of freedom system described in Example 4.17.

Solution: The output of the program is given below.

VALUE OF	METHOD #1 (FIG. 4.26)	METHOD #1 (FIG. 4.27)	METHOD #2 (FIG. 4.28)	METHOD #3 (FIG. 4.29)
I	X(I)	X(I)	X(I)	X(I)
2	0.479360E−01	0.404372E−01	0.441750E−01	0.454151E−01
3	0.175781E+00	0.147294E+00	0.161471E+00	0.163773E+00
4	0.351883E+00	0.292277E+00	0.321896E+00	0.324989E+00
5	0.542483E+00	0.446091E+00	0.493921E+00	0.497464E+00
6	0.715396E+00	0.581603E+00	0.647897E+00	0.651514E+00
7	0.843296E+00	0.676586E+00	0.759062E+00	0.762379E+00
8	0.906301E+00	0.715783E+00	0.809863E+00	0.812552E+00
9	0.893674E+00	0.692145E+00	0.791421E+00	0.793231E+00
10	0.804490E+00	0.607167E+00	0.704031E+00	0.704820E+00
11	0.647299E+00	0.469170E+00	0.556719E+00	0.556465E+00
I	XD(I)	XD(I)	XD(I)	XD(I)
2	0.298008E+00	0.251389E+00	0.276010E+00	0.275640E+00
3	0.502976E+00	0.418148E+00	0.462668E+00	0.461687E+00
4	0.602270E+00	0.491876E+00	0.549451E+00	0.547683E+00
5	0.595174E+00	0.474576E+00	0.537041E+00	0.534405E+00
6	0.492171E+00	0.377744E+00	0.436504E+00	0.433036E+00
7	0.313187E+00	0.220601E+00	0.267517E+00	0.263378E+00
8	0.850188E−01	0.276649E−01	0.558601E−01	0.513272E−01
9	−0.161754E+00	−0.174042E+00	−0.169532E+00	−0.174093E+00
10	−0.396047E+00	−0.357870E+00	−0.379660E+00	−0.383830E+00
11	−0.589414E+00	−0.507466E+00	−0.548380E+00	−0.551730E+00

∎

REFERENCES

4.1 S. S. Rao, *Applied Numerical Methods for Engineers and Scientists*, Prentice Hall, Upper Saddle River, NJ, 2002.

4.2 M. Paz, *Structural Dynamics: Theory and Computation* (2nd ed.), Van Nostrand Reinhold, New York, 1985.

4.3 E. Kreyszig, *Advanced Engineering Mathematics* (7th ed.), Wiley, New York, 1993.

4.4 F. Oberhettinger and L. Badii, *Tables of Laplace Transforms*, Springer Verlag, New York, 1973.

4.5 G. M. Hieber et al., "Understanding and measuring the shock response spectrum. Part I," *Sound and Vibration*, Vol. 8, March 1974, pp. 42–49.

4.6 R. E. D. Bishop, A. G. Parkinson, and J. W. Pendered, "Linear analysis of transient vibration," *Journal of Sound and Vibration*, Vol. 9, 1969, pp. 313–337.

4.7 Y. Matsuzaki and S. Kibe, "Shock and seismic response spectra in design problems." *Shock and Vibration Digest*, Vol. 15, October 1983, pp. 3–10.

4.8 S. Timoshenko, D. H. Young, and W. Weaver, Jr., *Vibration Problems in Engineering* (4th ed.), Wiley, New York, 1974.

4.9 R. A. Spinelli, "Numerical inversion of a Laplace transform," *SIAM Journal of Numerical Analysis*, Vol. 3, 1966, pp. 636–649.

4.10 R. Bellman, R. E. Kalaba, and J. A. Lockett, *Numerical Inversion of the Laplace Transform*, American Elsevier, New York, 1966.

4.11 S. G. Kelly, *Fundamentals of Mechanical Vibrations*, McGraw-Hill, New York, 1993.

4.12 P-T. D. Spanos, "Digital synthesis of response-design spectrum compatible earthquake records for dynamic analyses," *Shock and Vibration Digest*, Vol. 15, No. 3, March 1983, pp. 21–30.

REVIEW QUESTIONS

4.1 Give brief answers to the following:

1. What is the basis for expressing the response of a system under periodic excitation as a summation of several harmonic responses?
2. Indicate some methods for finding the response of a system under nonperiodic forces.
3. What is the Duhamel integral? What is its use?
4. How are the initial conditions determined for a single degree of freedom system subjected to an impulse at $t = 0$?
5. Derive the equation of motion of a system subjected to base excitation.
6. What is a response spectrum?
7. What are the advantages of the Laplace transform method?
8. What is the use of a pseudo spectrum?
9. How is the Laplace transform of a function $x(t)$ defined?
10. Define the terms *generalized impedance* and *admittance of a system*.
11. State the interpolation models that can be used for approximating an arbitrary forcing function.
12. How many resonant conditions are there when the external force is not harmonic?
13. How do you compute the frequency of the first harmonic of a periodic force?
14. What is the relation between the frequencies of higher harmonics and frequency of the first harmonic for a periodic excitation?

4.2 Indicate whether each of the following statements is true or false:

1. The change in momentum is called impulse.
2. The response of a system under arbitrary force can be found by summing the responses due to several elementary impulses.

3. The response spectrum corresponding to base excitation is useful in the design of machinery subject to earthquakes.
4. Some periodic functions cannot be replaced by a sum of harmonic functions.
5. The amplitudes of higher harmonics will be smaller in the response of a system.
6. The Laplace transform method takes the initial conditions into account automatically.
7. The equation of motion can be integrated numerically even when the exciting force is nonperiodic.
8. The response spectrum gives the maximum response of all possible single degree of freedom systems.
9. For a harmonic oscillator, the acceleration and displacement spectra can be obtained from the velocity spectrum.

4.3 Fill in each of the following blanks with the appropriate word:

1. The response of a linear system under any periodic force can be found by _____ appropriate harmonic responses.
2. Any nonperiodic function can be represented by a _____ integral.
3. An impulse force has a large magnitude and acts for a very _____ period of time.
4. The response of a single degree of freedom system to a unit _____ is known as the impulse response function.
5. The Duhamel integral is also known as the _____ integral.
6. The variation of the maximum response with the natural frequency of a single degree of freedom system is known as _____ spectrum.
7. The transient response of a system can be found using the _____ integral.
8. The complete solution of a vibration problem is composed of the _____ state and transient solutions.
9. The Laplace transform method converts a differential equation into an _____ equation.
10. The transfer function is the _____ of the generalized impedance.
11. An impulse can be measured by finding the change in _____ of the system.
12. The Duhamel integral is based on the _____ response function of the system.
13. The Duhamel integral can be used to find the response of _____ single degree of freedom systems under arbitrary excitations.
14. The velocity response spectrum, determined from the acceleration spectrum, is known as the _____ spectrum.
15. Any periodic forcing function can be expanded in _____ series.

4.4 Select the most appropriate answer out of the choices given:

1. The transient part of the solution arises from
   (a) forcing function   (b) initial conditions   (c) bounding conditions

2. If a system is subjected to a suddenly applied nonperiodic force, the response will be
   (a) periodic   (b) transient   (c) steady

3. The initial conditions are to be applied to a
   (a) steady-state solution   (b) transient solution   (c) total solution

4. The acceleration spectrum (S_a) can be expressed in terms of the displacement spectrum (S_d) as
   (a) $S_a = -\omega_n^2 S_d$ (b) $S_a = \omega_n S_d$ (c) $S_a = \omega_n^2 S_d$

5. The pseudo spectrum is associated with
(a) pseudo acceleration   (b) pseudo velocity   (c) pseudo displacement

6. The Fourier coefficients are to be found numerically when the values of the function $f(t)$ are available
(a) in analytical form
(b) at discrete values of t
(c) in the form of a complex equation

7. The response of a single degree of freedom system under base excitation, $y(t)$, can be determined by using the external force as
(a) $-m\ddot{y}$ (b) $m\ddot{y}$ (c) $m\ddot{y} + c\dot{y} + ky$

8. The response spectrum is widely used in
(a) building design under large live loads
(b) earthquake design
(c) design of machinery under fatigue

9. The equation of motion of a system subjected to base excitation, $y(t)$, is given by
(a) $m\ddot{x} + c\dot{x} + kx = -m\ddot{y}$

(b) $m\ddot{z} + c\dot{z} + kz = -m\ddot{y}; z = x - y$

(c) $m\ddot{x} + c\dot{x} + kx = -m\ddot{z}; z = x - y$

10. The function e^{-st} used in the Laplace transform is known as
(a) kernal   (b) integrand   (c) subsidiary term

11. The Laplace transform of $x(t)$ is defined by

(a) $\bar{x}(s) = \int_0^\infty e^{-st} x(t)\,dt$ (b) $\bar{x}(s) = \int_{-\infty}^\infty e^{-st} x(t)\,dt$ (c) $\bar{x}(s) = \int_0^\infty e^{st} x(t)\,dt$

4.5   Match the items in the two columns below:

1. $x(t) = \dfrac{1}{m\omega_d} e^{-\zeta\omega_n t} \sin \omega_d t$ (a) Inverse Laplace transform of $\bar{x}(s)$

2. $x(t) = \displaystyle\int_0^t F(\tau)g(t - \tau)\,d\tau$ (b) Generalized impedance function

3. $x(t) = \mathcal{L}^{-1}\bar{Y}(s)\bar{F}(s)$ (c) Unit impulse response function

4. $\bar{Y}(s) = \dfrac{1}{ms^2 + cs + k}$ (d) Laplace transform

5. $\bar{z}(s) = ms^2 + cs + k$ (e) Convolution integral

6. $\bar{x}(s) = \displaystyle\int_0^\infty e^{-st} x(t)\,dt$ (f) Admittance function

PROBLEMS

The problem assignments are organized as follows:

Problems	Section Covered	Topic Covered
4.1–4.11	4.2	Response under general periodic force
4.12–4.14	4.3	Periodic force of irregular form
4.15–4.38	4.5	Convolution integral
4.39–4.48	4.6	Response spectrum
4.49–4.51	4.7	Laplace transforms
4.52–4.55	4.8	Irregular forcing conditions using numerical methods
4.56–4.63	4.9	MATLAB examples
4.64–4.65	4.10	C++ programs
4.66–4.70	4.11	Fortran programs
4.71–4.73	—	Design projects

4.1–
4.4 Find the steady-state response of the hydraulic control valve shown in Fig. 4.1(a) to the forcing functions obtained by replacing $x(t)$ with $F(t)$ and A with F_0 in Figs. 1.89–1.92.

4.5 Find the steady-state response of a viscously damped system to the forcing function obtained by replacing $x(t)$ and A with $F(t)$ and F_0, respectively, in Fig. 1.46(a).

4.6 The torsional vibrations of a driven gear mounted on a shaft (see Fig. 4.30) under steady conditions are governed by the equation

$$J_0\ddot{\theta} + k_t\theta = M_t$$

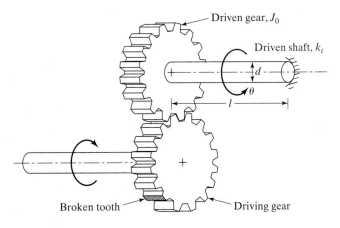

Driven gear, J_0

Driven shaft, k_t

d

θ

l

Broken tooth

Driving gear

FIGURE 4.30

where k_t is the torsional stiffness of the driven shaft, M_t is the torque transmitted, J_0 is the mass moment of inertia, and θ is the angular deflection of the driven gear. If one of the 16 teeth on the driving gear breaks, determine the resulting torsional vibration of the driven gear for the following data.

Driven gear: $J_0 = 0.1$ N-m-s^2, speed $= 1000$ rpm, driven shaft: material - steel, solid circular section with diameter 5 cm and length 1 m, $M_{t0} = 1000$ N-m.

4.7 A slider crank mechanism is used to impart motion to the base of a spring-mass-damper system, as shown in Fig. 4.31. Approximating the base motion $y(t)$ as a series of harmonic functions, find the response of the mass for $m = 1$ kg, $c = 10$ N-s/m, $k = 100$ N/m, $r = 10$ cm, $l = 1$ m, and $\omega = 100$ rad/s.

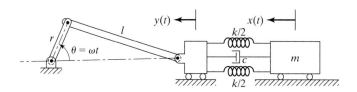

FIGURE 4.31

4.8 The base of a spring-mass-damper system is subjected to the periodic displacement shown in Fig. 4.32. Determine the response of the mass using the principle of superposition.

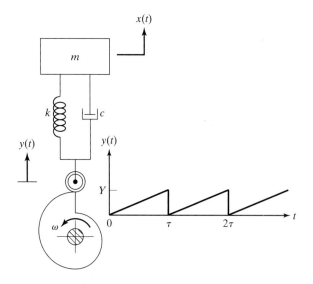

FIGURE 4.32

4.9 The base of a spring-mass system, with Coulomb damping, is connected to the slider crank mechanism shown in Fig. 4.33. Determine the response of the system for a coefficient of friction μ between the mass and the surface by approximating the motion $y(t)$ as a series of harmonic functions for $m = 1$ kg, $k = 100$ N/m, $r = 10$ cm, $l = 1$ m, $\mu = 0.1$, and $\omega = 100$ rad/s. Discuss the limitations of your solution.

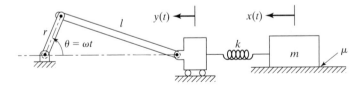

FIGURE 4.33

4.10 A roller cam is used to impart a periodic motion to the base of the spring-mass system shown in Fig. 4.34. If the coefficient of friction between the mass and the surface is μ, find the response of the system using the principle of superposition. Discuss the validity of the result.

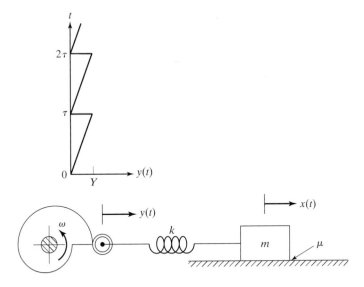

FIGURE 4.34

4.11 Find the total response of a viscously damped single degree of freedom system subjected to a harmonic base excitation for the following data: $m = 10$ kg, $c = 20$ N-s/m, $k = 4000$ N/m, $y(t) = 0.05 \cos 5t$ m, $x_0 = 0.1$ m, $\dot{x}_0 = 1$ m/s.

4.12 Find the response of a damped system with $m = 1$ kg, $k = 15$ kN/m. and $\zeta = 0.1$ under the action of a periodic forcing function, as shown in Fig. 1.94.

4.13 Find the response of a viscously damped system under the periodic force whose values are given in Problem 1.74. Assume that M_i denotes the value of the force in Newtons at time t_i seconds. Use $m = 0.5$ kg, $k = 8000$ N/m, and $\zeta = 0.06$.

4.14 Find the displacement of the water tank shown in Fig. 4.35(a) under the periodic force shown in Fig. 4.35(b) by treating it as an undamped single degree of freedom system. Use the numerical procedure described in Section 4.3.

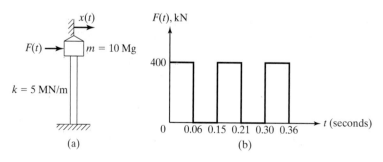

(a)   (b)

FIGURE 4.35

4.15 Sandblasting is a process in which an abrasive material, entrained in a jet, is directed onto the surface of a casting to clean its surface. In a particular setup for sandblasting, the casting of mass m is placed on a flexible support of stiffness k as shown in Fig. 4.36(a). If the force exerted on the casting due to the sandblasting operation varies as shown in Fig. 4.36(b), find the response of the casting.

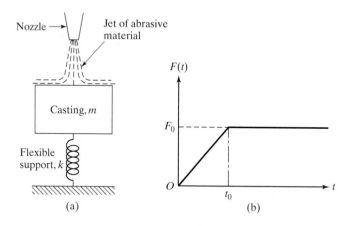

(a)   (b)

FIGURE 4.36

4.16 The frame, anvil, and the base of the forging hammer, shown in Fig. 4.37(a), have a total mass of m. The support elastic pad has a stiffness of k. If the force applied by the hammer is given by Fig. 4.37(b), find the response of the anvil.

4.17 Find the displacement of a damped single degree of freedom system under the forcing function $F(t) = F_0 e^{-\alpha t}$ where α is a constant.

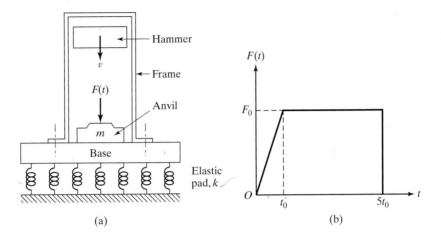

(a)

(b)

FIGURE 4.37

4.18 A compressed air cylinder is connected to the spring-mass system shown in Fig. 4.38(a). Due to a small leak in the valve, the pressure on the piston, $p(t)$, builds up as indicated in Fig. 4.38(b). Find the response of the piston for the following data: $m = 10$ kg, $k = 1000$ N/m, and $d = 0.1$ m.

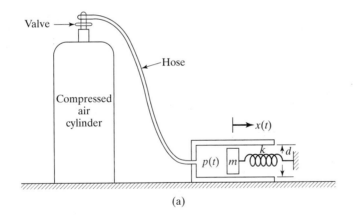

(a)

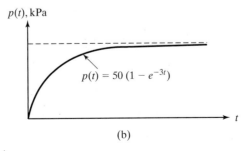

$$p(t) = 50 \, (1 - e^{-3t})$$

(b)

FIGURE 4.38

4.19 Find the transient response of an undamped spring-mass system for $t > \pi/\omega$ when the mass is subjected to a force

$$F(t) = \begin{cases} \dfrac{F_0}{2}(1 - \cos \omega t) & \text{for } 0 \le t \le \dfrac{\pi}{\omega} \\[3mm] F_0 & \text{for } t > \dfrac{\pi}{\omega} \end{cases}$$

Assume that the displacement and velocity of the mass are zero at $t = 0$.

4.20–
4.22 Use the Dahamel integral method to derive expressions for the response of an undamped system subjected to the forcing functions shown in Figs. 4.39(a) to (c).

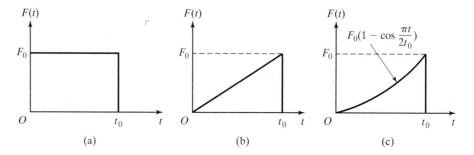

(a)   (b)   (c)

FIGURE 4.39

4.23 Figure 4.40 shows a one degree of freedom model of a motor vehicle traveling in the horizontal direction. Find the relative displacement of the vehicle as it travels over a road bump of the form $y(s) = Y \sin \pi s/\delta$.

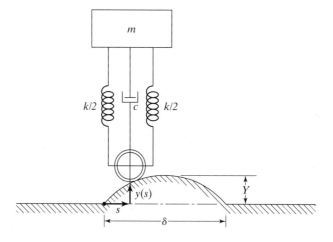

FIGURE 4.40

4.24 A vehicle traveling at a constant speed v in the horizontal direction encounters a triangular road bump, as shown in Fig. 4.41. Treating the vehicle as an undamped spring-mass system, determine the response of the vehicle in the vertical direction.

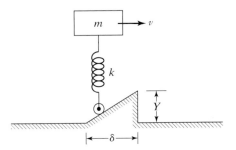

FIGURE 4.41

4.25 An automobile, having a mass of 1000 kg, runs over a road bump of the shape shown in Fig. 4.42. The speed of the automobile is 50 km/hr. If the undamped natural period of vibration in the vertical direction is 1.0 sec, find the response of the car by assuming it as a single degree of freedom undamped system vibrating in the vertical direction.

Height of bump (m)

0.1

0   0.25   0.50   Distance along
                                    road (m)

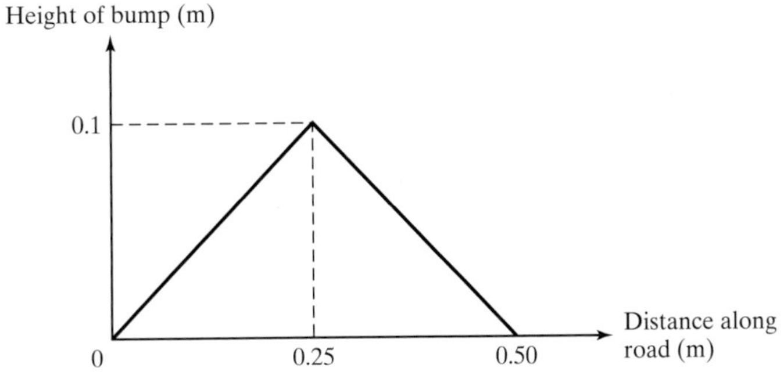

FIGURE 4.42

4.26 A camcorder of mass m is packed in a container using a flexible packing material. The stiffness and damping constant of the packing material are given by k and c, respectively, and the mass of the container is negligible. If the container is dropped accidentally from a height of h onto a rigid floor (see Fig. 4.43), find the motion of the camcorder.

4.27 An airplane, taxiing on a runway, encounters a bump. As a result, the root of the wing is subjected to a displacement that can be expressed as

$$y(t) = \begin{cases} Y(t^2/t_0^2), & 0 \le t \le t_0 \\ 0, & t > t_0 \end{cases}$$

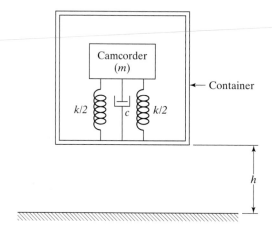

FIGURE 4.43

Find the response of the mass located at the tip of the wing if the stiffness of the wing is k (see Fig. 4.44).

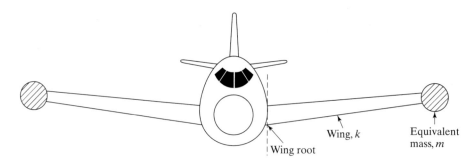

FIGURE 4.44

4.28 Derive Eq. (E.1) of Example 4.9.

4.29 In a static firing test of a rocket the rocket is anchored to a rigid wall by a spring-damper system, as shown in Fig. 4.45(a). The thrust acting on the rocket reaches its maximum value F in a negligibly short time and remains constant until the burnout time t_0, as indicated in Fig. 4.45(b). The thrust acting on the rocket is given by $F = m_0 v$, where m_0 is the constant rate at which fuel is burnt and v is the velocity of the jet stream. The initial mass of the rocket is M, so that its mass at any time t is given by $m = M - m_0 t$, $0 \le t \le t_0$. If the data are $k = 7.5 \times 10^6$ N/m, $c = 0.1 \times 10^6$ N-s/m, $m_0 = 10$ kg/s, $v = 2000$ m/s, $M = 2000$ kg, and $t_0 = 100$ s, (1) derive the equation of motion of the rocket, and (2) find the maximum steady-state displacement of the rocket by assuming an average (constant) mass of $(M - \frac{1}{2} m_0 t_0)$.

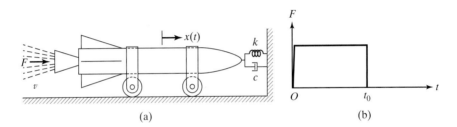

(a)   (b)

FIGURE 4.45

4.30 Show that the response to a unit step function $h(t)$ ($F_0 = 1$ in Fig. 4.7b) is related to the impulse response function $g(t)$, Eq. (4.27), as follows:

$$g(t) = \frac{dh(t)}{dt}$$

4.31 Show that the convolution integral, Eq. (4.33), can also be expressed in terms of the response to a unit step function $h(t)$ as

$$x(t) = F(0)h(t) + \int_0^t \frac{dF(\tau)}{d\tau} h(t - \tau)\, d\tau$$

4.32 Find the response of the rigid bar shown in Fig. 4.46 using convolution integral for the following data: $k_1 = k_2 = 5000$ N/m, $a = 0.25$ m, $b = 0.5$ m, $l = 1.0$ m, $M = 50$ kg, $m = 10$ kg, $F_0 = 500$ N.

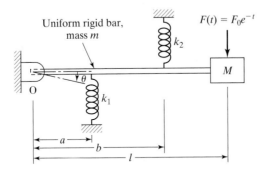

FIGURE 4.46

4.33 Find the response of the rigid bar shown in Fig. 4.47 using convolution integral for the following data: $k = 5000$ N/m, $l = 1$ m, $m = 10$ kg, $M_0 = 100$ N-m.

4.34 Find the response of the rigid bar shown in Fig. 4.48 using convolution integral when the end P of the spring PQ is subjected to the displacement, $x(t) = x_0\, e^{-t}$. Data: $k = 5000$ N/m, $l = 1$ m, $m = 10$ kg, $x_0 = 1$ cm.

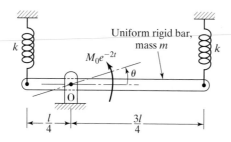

FIGURE 4.47

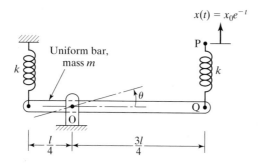

FIGURE 4.48

4.35 Find the response of the mass shown in Fig. 4.49 under the force $F(t) = F_0 e^{-t}$ using convolution integral. Data: $k_1 = 1000$ N/m, $k_2 = 500$ N/m, $r = 5$ cm, $m = 10$ kg, $J_0 = 1$ kg-m^2, $F_0 = 50$ N.

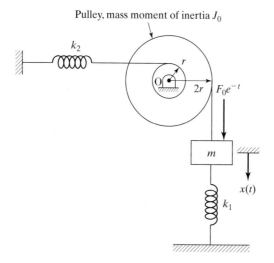

FIGURE 4.49

4.36 Find the impulse response functions of a viscously damped spring-mass system for the following cases:

  a. Undamped ($c = 0$)
  b. Underdamped ($c < c_c$)
  c. Critically damped ($c = c_c$)
  d. Overdamped ($c > c_c$)

4.37 Find the response of a single degree of freedom system under an impulse F for the following data: $m = 2$ kg, $c = 4$ N-s/m, $k = 32$ N/m, $\underset{\sim}{F} = 4\,\delta(t)$, $x_0 = 0.01$ m, $\dot{x}_0 = 1$ m/s.

4.38 The wing of a fighter aircraft, carrying a missile at its tip, as shown in Fig. 4.50, can be approximated as an equivalent cantilever beam with $EI = 15 \times 10^9$ N $-$ m^2 about the vertical axis and length $l = 10$ m. If the equivalent mass of the wing, including the mass of the missile and its carriage system, at the tip of the wing is $m = 2500$ kg, determine the vibration response of the wing (of m) due to the release of the missile. Assume that the force on m due to the release of the missile can be approximated as an impulse function of magnitude $\underset{\sim}{F} = 50$ N-s.

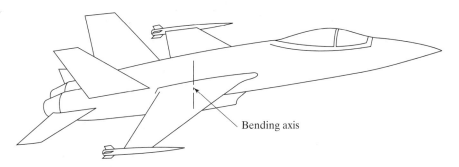

Bending axis

FIGURE 4.50

4.39 Derive the response spectrum of an undamped system for the rectangular pulse shown in Fig. 4.39(a). Plot $(x/\delta_{st})_{max}$ with respect to (t_0/τ_n).

4.40 Find the displacement response spectrum of an undamped system for the pulse shown in Fig. 4.39(c).

4.41 The base of an undamped spring-mass system is subjected to an acceleration excitation given by $a_0[1 - \sin(\pi t/2t_0)]$. Find the relative displacement of the mass z.

4.42 Find the response spectrum of the system considered in Example 4.10. Plot $\left(\dfrac{kx}{F_0}\right)_{max}$ versus $\omega_n t_0$ in the range $0 \le \omega_n t_0 \le 15$.

4.43* A building frame is subjected to a blast load and the idealization of the frame and the load are shown in Fig. 4.11. If $m = 5000$ kg, $F_0 = 4$ MN, and $t_0 = 0.4$ s, find the minimum stiffness required if the displacement is to be limited to 10 mm.

---

*The asterisk denotes a problem with no unique answer.

4.44 Consider the printed circuit board (PCB) mounted on a cantilevered aluminum bracket shown in Fig. 4.20(a). Design the bracket to withstand an acceleration level of 100 g under the rectangular pulse shown in Fig. 4.51. Assume the specific weight, Young's modulus, and permissible stress of aluminum as 0.1 lb/in³, 10^7 lb/in², and 26,000 lb/in², respectively.

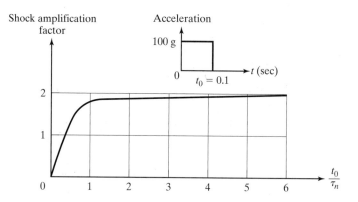

FIGURE 4.51

4.45 Consider the printed circuit board (PCB) mounted on a cantilevered aluminum bracket shown in Fig. 4.20(a). Design the bracket to withstand an acceleration level of 100 g under the triangular pulse shown in Fig. 4.52. Assume the material properties as given in Problem 4.44.

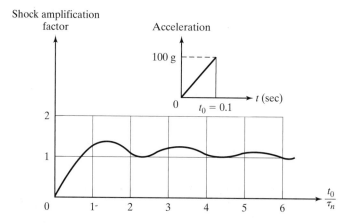

FIGURE 4.52

4.46 An electronic box, weighing 1 lb, is to be shock-tested using a 100 g half-sine pulse with a 0.1 sec time base for a qualification test. The box is mounted at the middle of a fixed-fixed beam as shown in Fig. 4.53. The beam, along with the box, is placed in a container and subjected to the shock test. Design the beam to withstand the stated shock pulse. Assume the material properties as given in Problem 4.44.

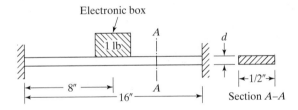

Electronic box

1 lb

A

d

8"

16"

A

←1/2"→

Section *A*–*A*

FIGURE 4.53

4.47* The water tank shown in Fig. 4.54 is subjected to an earthquake whose response spectrum is indicated in Fig. 4.15. The weight of the tank with water is 100,000 lb. Design a uniform steel hollow circular column of height 50 ft so that the maximum bending stress does not exceed the yield stress of the material. Assume a damping ratio of 0.05 and a factor of safety of 2.

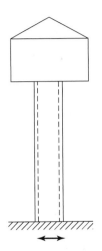

FIGURE 4.54

4.48 Consider the overhead traveling crane shown in Fig. 4.18. Assuming the weight of the trolley as 5000 lb and the overall damping ratio as 2%, determine the overall stiffness of the system necessary in order to avoid derailment of the trolley under a vertical earthquake excitation whose design spectrum is given by Fig. 4.16.

4.49 Find the steady-state response of an undamped single degree of freedom system subjected to the force $F(t) = F_0 e^{i\omega t}$ by using the method of Laplace transformation.

4.50 Find the response of a damped spring-mass system subjected to a step function of magnitude F_0 by using the method of Laplace transformation.

4.51 Find the response of an undamped system subjected to a square pulse $F(t) = F_0$ for $0 \le t \le t_0$ and 0 for $t > t_0$ by using the Laplace transformation method. Assume the initial conditions as zero.

4.52 Determine the expression for the velocity $\dot{x}_j$ for the damped response represented by Eq. (4.64).

4.53 Derive Eqs. (4.68) and (4.71).

4.54 Compare the values of $\dot{x}_j$ given by Eqs. (4.68) and (4.71) in the case of Example 4.17.

4.55 Derive the expressions for x_j and $\dot{x}_j$ according to the three interpolation functions considered in Section 4.8 for the undamped case. Using these expressions, find the solution of Example 4.17 by assuming the damping to be zero.

4.56 A machine is given an impact force by an impact hammer. If the machine can be modeled as a single degree of freedom system with $m = 10$ kg, $k = 4000$ N/m, and $c = 40$ N-s/m, and the magnitude of the impact is $F = 100$ N-s, determine the response of the machine. Also plot the response using MATLAB.

4.57 If the machine described in Problem 4.56 is given a double impact by the impact hammer, find the response of the machine. Assume the impact force, $F(t)$, as $F(t) = 100\ \delta(t) + 50\ \delta(t - 0.5)$ N where $\delta(t)$ is the Dirac delta function. Also plot the response of the machine using MATLAB.

4.58 Using MATLAB, plot the response of a viscously damped spring-mass system subject to the rectangular pulse shown in Fig. 4.9(a) with (a) $t_0 = 0.1$ s and (b) $t_0 = 1.5$ s. Assume the following data: $m = 100$ kg, $k = 1200$ N/m, $c = 50$ N-s/m, $F_0 = 100$ N.

4.59 Using **Program4.m**, find the steady-state response of a viscously damped system with $m = 1$ kg, $k = 400$ N/m, and $c = 5$ N-s/m subject to the periodic force shown in Fig. 4.55.

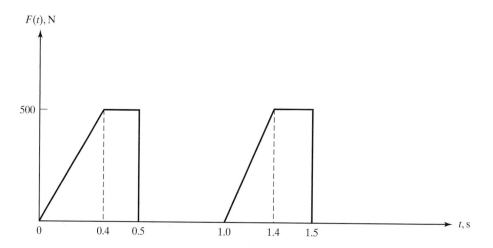

FIGURE 4.55

4.60 Using **Program5.m**, find the response of a viscously damped system with $m = 100$ kg, $k = 10^5$ N, and $\zeta = 0.1$ subject to the force $F(t) = 1000(1 - \cos \pi t)$ N.

4.61 A damped single degree of freedom system has a mass $m = 2$, a spring of stiffness $k = 50$, and a damper with $c = 2$. A forcing function $F(t)$, whose magnitude is indicated in the table below, acts on the mass for 1 sec. Find the response of the system by using the piecewise linear interpolation method described in Section 4.8 using **Program5.m**.

Time (t_i)	$F(t_i)$
0.0	−8.0
0.1	−12.0
0.2	−15.0
0.3	−13.0
0.4	−11.0
0.5	−7.0
0.6	−4.0
0.7	3.0
0.8	10.0
0.9	15.0
1.0	18.0

4.62 The equation of motion of an undamped system is given by $2\ddot{x} + 1500x = F(t)$ where the forcing function is defined by the curve shown in Fig. 4.56. Find the response of the system numerically for $0 \le t \le 0.5$. Assume the initial conditions as $x_0 = \dot{x}_0 = 0$ and the step size as $\Delta t = 0.01$. Use the MATLAB program **ode23**.

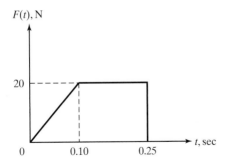

FIGURE 4.56

4.63 Solve Problem 4.62 using MATLAB program **ode23** if the system is viscously damped so that the equation of motion is

$$2\ddot{x} + 10\dot{x} + 1500x = F(t).$$

4.64 Using **Program4.cpp**, find the solution of the problem described in Problem 4.59.

4.65 Using **Program5.cpp**, find the solution of the problem described in Problem 4.60.

4.66 Using **PROGRAM4.F**, find the solution of the problem described in Problem 4.59.

4.67 Using **PROGRAM5.F**, find the solution of the problem described in Problem 4.60.

4.68 Write a computer program for finding the steady-state response of a single degree of freedom system subjected to an arbitrary force, by numerically evaluating the Duhamel integral. Using this program, solve Example 4.17.

4.69 Find the relative displacement of the water tank shown in Fig. 4.35(a) when its base is subjected to the earthquake acceleration record shown in Fig. 1.95 by assuming the ordinate represents acceleration in g's. Use the program of Problem 4.68.

4.70 The differential equation of motion of an undamped system is given by $2\ddot{x} + 150x = F(t)$ with the initial conditions $x_0 = \dot{x}_0 = 0$. If $F(t)$ is as shown in Fig. 4.57, find the response of the problem using the computer program of Problem 4.68.

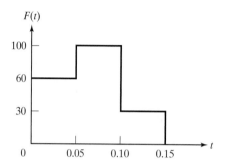

FIGURE 4.57

DESIGN PROJECTS

4.71 Design a seismometer of the type shown in Fig. 4.58(a) (by specifying the values of a, m, and k) to measure earthquakes. The seismometer should have a natural frequency of 10 Hz and the maximum relative displacement of the mass should be at least 2 cm when its base is subjected to the displacement shown in Fig. 4.58(b).

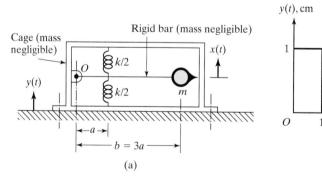

FIGURE 4.58

4.72 The cutting forces developed during two different machining operations are shown in Figs. 4.59(a) and (b). The inaccuracies (in the vertical direction) in the surface finish in the two cases were observed to be 0.1 mm and 0.05 mm, respectively. Find the equivalent mass and stiffness of the cutting head (Fig. 4.60), assuming it to be an undamped single degree of freedom system.

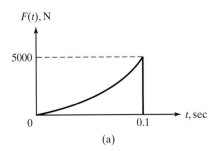 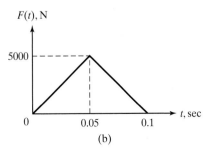

FIGURE 4.59

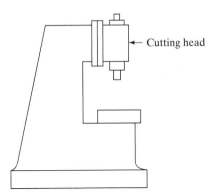

FIGURE 4.60

4.73 A milling cutter, mounted at the middle of an arbor, is used to remove metal from a workpiece (Fig. 4.61). A torque of 500 N-m is developed in the cutter under steady-state cutting conditions. One of the 16 teeth on the cutter breaks during the cutting operation. Determine the cross section of the arbor to limit the amplitude of angular displacement of the cutter to 1°. Assume that the arbor can be modeled as a hollow steel shaft fixed at both ends.

Data: Length of arbor = 0.5 m, mass moment of inertia of the cutter = 0.1 N-m^2, speed of cutter = 1000 rpm.

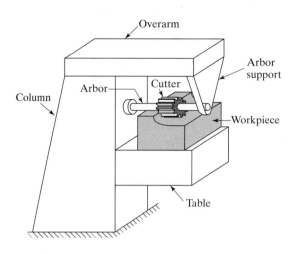

FIGURE 4.61

Daniel Bernoulli (1700–1782) was a Swiss who became a professor of mathematics at Saint Petersburg in 1725 after receiving his doctorate in medicine for his thesis on the action of lungs. He later became professor of anatomy and botany at Basel. He developed the theory of hydrostatics and hydrodynamics and "Bernoulli's theorem" is well known to engineers. He derived the equation of motion for the vibration of beams (the Euler-Bernoulli theory) and studied the problem of vibrating strings. Bernoulli was the first person to propose the principle of superposition of harmonics in free vibration.
(Photo courtesy of David Eugene Smith, *History of Mathematics*, Volume 1—*General Survey of the History of Elementary Mathematics*. Dover Publications, New York, 1958.)

CHAPTER 5

Two Degree of Freedom Systems

5.1 Introduction

Systems that require two independent coordinates to describe their motion are called *two degree of freedom systems*. Some examples of systems having two degrees of freedom were shown in Fig. 1.13. We shall consider only two degree of freedom systems in this chapter, so as to provide a simple introduction to the behavior of systems with an arbitrarily large number of degrees of freedom, which is the subject of Chapter 6.

Consider the motor-pump system shown in Fig. 5.1(a). Assuming that the system vibrates in a vertical plane, it can be idealized as a bar of mass m and mass moment of inertia J_0 supported on two springs of stiffnesses k_1 and k_2, as shown in Fig. 5.1(b). The displacement of the system at any time can be specified by a linear coordinate $x(t)$, indicating the vertical displacement of the center of gravity (C.G.) of the mass, and an angular coordinate $\theta(t)$, denoting the rotation of the mass m about its C.G. Instead of $x(t)$ and $\theta(t)$, we can also use $x_1(t)$ and $x_2(t)$ as independent coordinates to specify the motion of the system. Thus the system has two degrees of freedom. It is important to note that in this case the mass m is not treated as a point mass but as a rigid body having two possible types of motion. (If it is a particle, there is no need to specify the rotation of the mass about its C.G.)

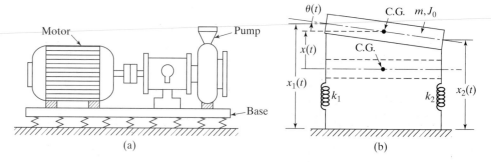

FIGURE 5.1   Motor-pump system on springs.

Next, consider the system shown in Fig. 5.2(a), which illustrates the packaging of an instrument of mass m. Assuming that the motion of the instrument is confined to the xy plane, the system can be modeled as a mass m supported by springs in the x and y directions, as indicated in Fig. 5.2(b). Thus the system has one point mass m and two degrees of freedom, because the mass has two possible types of motion (translations along the x and y directions). The general rule for the computation of the number of degrees of freedom can be stated as follows:

$$\begin{array}{c} \text{Number of} \\ \text{degrees of freedom} = \\ \text{of the system} \end{array} \begin{array}{c} \text{Number of masses in the system} \\ \times \text{ number of possible types} \\ \text{of motion of each mass} \end{array}$$

There are two equations of motion for a two degree of freedom system, one for each mass (more precisely, for each degree of freedom). They are generally in the form of *coupled differential equations*—that is, each equation involves all the coordinates. If a harmonic solution is assumed for each coordinate, the equations of motion lead to a frequency equation that gives two natural frequencies for the system. If we give suitable initial excitation, the system vibrates at one of these natural frequencies. During free vibration at one of the natural frequencies, the amplitudes of the two degrees of freedom (coordinates) are related

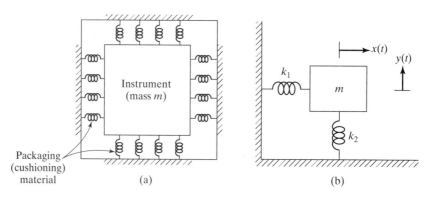

FIGURE 5.2   Packaging of an instrument.

in a specific manner and the configuration is called a *normal mode, principal mode,* or *natural mode* of vibration. Thus a two degree of freedom system has two normal modes of vibration corresponding to the two natural frequencies.

If we give an arbitrary initial excitation to the system, the resulting free vibration will be a superposition of the two normal modes of vibration. However, if the system vibrates under the action of an external harmonic force, the resulting forced harmonic vibration takes place at the frequency of the applied force. Under harmonic excitation, resonance occurs (i.e., the amplitudes of the two coordinates will be maximum) when the forcing frequency is equal to one of the natural frequencies of the system.

As is evident from the systems shown in Figs. 5.1 and 5.2, the configuration of a system can be specified by a set of independent coordinates such as length, angle, or some other physical parameters. Any such set of coordinates is called *generalized coordinates.* Although the equations of motion of a two degree of freedom system are generally coupled so that each equation involves all the coordinates, it is always possible to find a particular set of coordinates such that each equation of motion contains only one coordinate. The equations of motion are then *uncoupled* and can be solved independently of each other. Such a set of coordinates, which leads to an uncoupled system of equations, is called *principal coordinates.*

5.2   Equations of Motion for Forced Vibration

Consider a viscously damped two degree of freedom spring-mass system, shown in Fig. 5.3(a). The motion of the system is completely described by the coordinates $x_1(t)$ and $x_2(t)$, which define the positions of the masses m_1 and m_2 at any time t from the respective equilibrium positions. The external forces $F_1(t)$ and $F_2(t)$ act on the masses m_1 and m_2 respectively. The free-body diagrams of the masses m_1 and m_2 are shown in Fig. 5.3(b). The application of Newton's second law of motion to each of the masses gives the equations of motion:

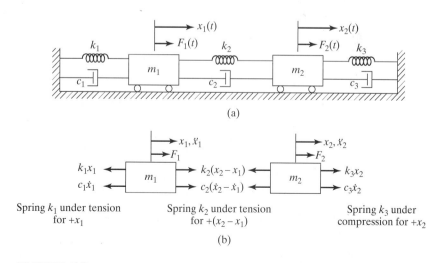

(a)

(b)

Spring k_1 under tension for $+x_1$

Spring k_2 under tension for $+(x_2 - x_1)$

Spring k_3 under compression for $+x_2$

FIGURE 5.3   A two degree of freedom spring-mass-damper system.

$$m_1 \ddot{x}_1 + (c_1 + c_2)\dot{x}_1 - c_2 \dot{x}_2 + (k_1 + k_2)x_1 - k_2 x_2 = F_1 \qquad (5.1)$$

$$m_2 \ddot{x}_2 - c_2 \dot{x}_1 + (c_2 + c_3)\dot{x}_2 - k_2 x_1 + (k_2 + k_3)x_2 = F_2 \qquad (5.2)$$

It can be seen that Eq. (5.1) contains terms involving x_2 (namely, $-c_2 \dot{x}_2$ and $-k_2 x_2$), whereas Eq. (5.2) contains terms involving x_1 (namely, $-c_2 \dot{x}_1$ and $-k_2 x_1$). Hence they represent a system of two coupled second-order differential equations. We can therefore expect that the motion of the mass m_1 will influence the motion of the mass m_2, and vice versa. Equations (5.1) and (5.2) can be written in matrix form as

$$[m]\, \ddot{\vec{x}}(t) + [c]\, \dot{\vec{x}}(t) + [k]\, \vec{x}(t) = \vec{F}(t) \qquad (5.3)$$

where $[m]$, $[c]$, and $[k]$ are called the *mass, damping,* and *stiffness matrices*, respectively, and are given by

$$[m] = \begin{bmatrix} m_1 & 0 \\ 0 & m_2 \end{bmatrix}$$

$$[c] = \begin{bmatrix} c_1 + c_2 & -c_2 \\ -c_2 & c_2 + c_3 \end{bmatrix}$$

$$[k] = \begin{bmatrix} k_1 + k_2 & -k_2 \\ -k_2 & k_2 + k_3 \end{bmatrix}$$

and $\vec{x}(t)$ and $\vec{F}(t)$ are called the *displacement* and *force vectors*, respectively, and are given by

$$\vec{x}(t) = \begin{Bmatrix} x_1(t) \\ x_2(t) \end{Bmatrix}$$

and

$$\vec{F}(t) = \begin{Bmatrix} F_1(t) \\ F_2(t) \end{Bmatrix}$$

It can be seen that the matrices $[m]$, $[c]$, and $[k]$ are all 2×2 matrices whose elements are the known masses, damping coefficients, and stiffnesses of the system, respectively. Further, these matrices can be seen to be symmetric, so that

$$[m]^T = [m], \qquad [c]^T = [c], \qquad [k]^T = [k]$$

where the superscript T denotes the transpose of the matrix.

Notice that the equations of motion (5.1) and (5.2) become uncoupled (independent of one another) only when $c_2 = k_2 = 0$, which implies that the two masses m_1 and m_2 are not physically connected. In such a case, the matrices $[m]$, $[c]$, and $[k]$ become diagonal. The solution of the equations of motion (5.1) and (5.2) for any arbitrary forces $F_1(t)$ and $F_2(t)$ is difficult to obtain, mainly due to the coupling of the variables $x_1(t)$ and $x_2(t)$. The solution of Eqs. (5.1) and (5.2) involves four constants of integration (two for each equation).

Usually the initial displacements and velocities of the two masses are specified as $x_1(t = 0) = x_1(0)$, $\dot{x}_1(t = 0) = \dot{x}_1(0)$, $x_2(t = 0) = x_2(0)$, and $\dot{x}_2(t = 0) = \dot{x}_2(0)$. We shall first consider the free vibration solution of Eqs. (5.1) and (5.2).

5.3 Free Vibration Analysis of an Undamped System

For the free vibration analysis of the system shown in Fig. 5.3(a), we set $F_1(t) = F_2(t) = 0$. Further, if damping is disregarded, $c_1 = c_2 = c_3 = 0$, and the equations of motion (5.1) and (5.2) reduce to

$$m_1\ddot{x}_1(t) + (k_1 + k_2)x_1(t) - k_2 x_2(t) = 0 \tag{5.4}$$

$$m_2\ddot{x}_2(t) - k_2 x_1(t) + (k_2 + k_3)x_2(t) = 0 \tag{5.5}$$

We are interested in knowing whether m_1 and m_2 can oscillate harmonically with the same frequency and phase angle but with different amplitudes. Assuming that it is possible to have harmonic motion of m_1 and m_2 at the same frequency ω and the same phase angle ϕ, we take the solutions of Eqs. (5.4) and (5.5) as

$$x_1(t) = X_1 \cos(\omega t + \phi)$$
$$x_2(t) = X_2 \cos(\omega t + \phi) \tag{5.6}$$

where X_1 and X_2 are constants that denote the maximum amplitudes of $x_1(t)$ and $x_2(t)$, and ϕ is the phase angle. Substituting Eq. (5.6) into Eqs. (5.4) and (5.5), we obtain

$$[\{-m_1\omega^2 + (k_1 + k_2)\} X_1 - k_2 X_2]\cos(\omega t + \phi) = 0$$
$$[-k_2 X_1 + \{-m_2\omega^2 + (k_2 + k_3)\} X_2]\cos(\omega t + \phi) = 0 \tag{5.7}$$

Since Eq. (5.7) must be satisfied for all values of the time t, the terms between brackets must be zero. This yields

$$\{-m_1\omega^2 + (k_1 + k_2)\} X_1 - k_2 X_2 = 0$$
$$-k_2 X_1 + \{-m_2\omega^2 + (k_2 + k_3)\} X_2 = 0 \tag{5.8}$$

which represent two simultaneous homogenous algebraic equations in the unknowns X_1 and X_2. It can be seen that Eq. (5.8) is satisfied by the trivial solution $X_1 = X_2 = 0$, which implies that there is no vibration. For a nontrivial solution of X_1 and X_2, the determinant of the coefficients of X_1 and X_2 must be zero:

$$\det \begin{bmatrix} \{-m_1\omega^2 + (k_1 + k_2)\} & -k_2 \\ -k_2 & \{m_2\omega^2 + (k_2 + k_3)\} \end{bmatrix} = 0$$

or

$$(m_1 m_2)\omega^4 - \{(k_1 + k_2)m_2 + (k_2 + k_3)m_1\}\omega^2$$
$$+ \{(k_1 + k_2)(k_2 + k_3) - k_2^2\} = 0 \tag{5.9}$$

Equation (5.9) is called the *frequency* or *characteristic equation* because solution of this equation yields the frequencies or the characteristic values of the system. The roots of Eq. (5.9) are given by

$$\omega_1^2, \omega_2^2 = \frac{1}{2} \left\{ \frac{(k_1 + k_2)m_2 + (k_2 + k_3)m_1}{m_1 m_2} \right\}$$

$$\mp \frac{1}{2} \left[\left\{ \frac{(k_1 + k_2)m_2 + (k_2 + k_3)m_1}{m_1 m_2} \right\}^2 \right.$$

$$\left. - 4 \left\{ \frac{(k_1 + k_2)(k_2 + k_3) - k_2^2}{m_1 m_2} \right\} \right]^{1/2} \qquad (5.10)$$

This shows that it is possible for the system to have a nontrivial harmonic solution of the form of Eq. (5.6) when ω is equal to ω_1 and ω_2 given by Eq. (5.10). We call ω_1 and ω_2 the *natural frequencies* of the system.

The values of X_1 and X_2 remain to be determined. These values depend on the natural frequencies ω_1 and ω_2. We shall denote the values of X_1 and X_2 corresponding to ω_1 as $X_1^{(1)}$ and $X_2^{(1)}$ and those corresponding to ω_2 as $X_1^{(2)}$ and $X_2^{(2)}$. Further, since Eq. (5.8) is homogenous, only the ratios $r_1 = \{X_2^{(1)}/X_1^{(1)}\}$ and $r_2 = \{X_2^{(2)}/X_1^{(2)}\}$ can be found. For $\omega^2 = \omega_1^2$ and $\omega^2 = \omega_2^2$, Eq. (5.8) gives

$$r_1 = \frac{X_2^{(1)}}{X_1^{(1)}} = \frac{-m_1\omega_1^2 + (k_1 + k_2)}{k_2} = \frac{k_2}{-m_2\omega_1^2 + (k_2 + k_3)}$$

$$r_2 = \frac{X_2^{(2)}}{X_1^{(2)}} = \frac{-m_1\omega_2^2 + (k_1 + k_2)}{k_2} = \frac{k_2}{-m_2\omega_2^2 + (k_2 + k_3)} \qquad (5.11)$$

Notice that the two ratios given for each r_i $(i = 1, 2)$ in Eq. (5.11) are identical. The normal modes of vibration corresponding to ω_1^2 and ω_2^2 can be expressed, respectively, as

$$\vec{X}^{(1)} = \begin{Bmatrix} X_1^{(1)} \\ X_2^{(1)} \end{Bmatrix} = \begin{Bmatrix} X_1^{(1)} \\ r_1 X_1^{(1)} \end{Bmatrix}$$

and

$$\vec{X}^{(2)} = \begin{Bmatrix} X_1^{(2)} \\ X_2^{(2)} \end{Bmatrix} = \begin{Bmatrix} X_1^{(2)} \\ r_2 X_1^{(2)} \end{Bmatrix} \qquad (5.12)$$

The vectors $\vec{X}^{(1)}$ and $\vec{X}^{(2)}$, which denote the normal modes of vibration, are known as the *modal vectors* of the system. The free vibration solution or the motion in time can be expressed, using Eq. (5.6), as

$$\vec{x}^{(1)}(t) = \begin{Bmatrix} x_1^{(1)}(t) \\ x_2^{(1)}(t) \end{Bmatrix} = \begin{Bmatrix} X_1^{(1)} \cos(\omega_1 t + \phi_1) \\ r_1 X_1^{(1)} \cos(\omega_1 t + \phi_1) \end{Bmatrix} = \text{first mode}$$

$$\vec{x}^{(2)}(t) = \begin{Bmatrix} x_1^{(2)}(t) \\ x_2^{(2)}(t) \end{Bmatrix} = \begin{Bmatrix} X_1^{(2)} \cos(\omega_2 t + \phi_2) \\ r_2 X_1^{(2)} \cos(\omega_2 t + \phi_2) \end{Bmatrix} = \text{second mode} \qquad (5.13)$$

where the constants $X_1^{(1)}$, $X_1^{(2)}$, ϕ_1, and ϕ_2 are determined by the initial conditions.

Initial Conditions.   As stated earlier, each of the two equations of motion, Eqs. (5.1) and (5.2), involves second-order time derivatives; hence we need to specify two initial conditions for each mass. As stated in Section 5.1, the system can be made to vibrate in its ith normal mode ($i = 1, 2$) by subjecting it to the specific initial conditions

$$x_1(t = 0) = X_1^{(i)} = \text{some constant}, \qquad \dot{x}_1(t = 0) = 0,$$

$$x_2(t = 0) = r_i X_1^{(i)}, \qquad \dot{x}_2(t = 0) = 0$$

However, for any other general initial conditions, both modes will be excited. The resulting motion, which is given by the general solution of Eqs. (5.4) and (5.5), can be obtained by a linear superposition of the two normal modes, Eq. (5.13)

$$\vec{x}(t) = c_1 \vec{x}_1(t) + c_2 \vec{x}_2(t) \qquad (5.14)$$

where c_1 and c_2 are constants. Since $\vec{x}^{(1)}(t)$ and $\vec{x}^{(2)}(t)$ already involve the unknown constants $X_1^{(1)}$ and $X_1^{(2)}$ (see Eq. 5.13), we can choose $c_1 = c_2 = 1$ with no loss of generality. Thus the components of the vector $\vec{x}(t)$ can be expressed, using Eq. (5.14) with $c_1 = c_2 = 1$ and Eq. (5.13), as

$$x_1(t) = x_1^{(1)}(t) + x_1^{(2)}(t) = X_1^{(1)}\cos(\omega_1 t + \phi_1) + X_1^{(2)}\cos(\omega_2 t + \phi_2)$$

$$x_2(t) = x_2^{(1)}(t) + x_2^{(2)}(t)$$

$$= r_1 X_1^{(1)} \cos(\omega_1 t + \phi_1) + r_2 X_1^{(2)} \cos(\omega_2 t + \phi_2) \qquad (5.15)$$

where the unknown constants $X_1^{(1)}$, $X_1^{(2)}$, ϕ_1, and ϕ_2 can be determined from the initial conditions:

$$x_1(t = 0) = x_1(0), \qquad \dot{x}_1(t = 0) = \dot{x}_1(0),$$

$$x_2(t = 0) = x_2(0), \qquad \dot{x}_2(t = 0) = \dot{x}_2(0) \qquad (5.16)$$

Substitution of Eq. (5.16) into Eq. (5.15) leads to

$$x_1(0) = X_1^{(1)} \cos \phi_1 + X_1^{(2)} \cos \phi_2$$

$$\dot{x}_1(0) = -\omega_1 X_1^{(1)} \sin \phi_1 - \omega_2 X_1^{(2)} \sin \phi_2$$

$$x_2(0) = r_1 X_1^{(1)} \cos \phi_1 + r_2 X_1^{(2)} \cos \phi_2$$

$$\dot{x}_2(0) = -\omega_1 r_1 X_1^{(1)} \sin \phi_1 - \omega_2 r_2 X_1^{(2)} \sin \phi_2 \qquad (5.17)$$

Equation (5.17) can be regarded as four algebraic equations in the unknowns $X_1^{(1)} \cos \phi_1$, $X_1^{(2)} \cos \phi_2$, $X_1^{(1)} \sin \phi_1$, and $X_1^{(2)} \sin \phi_2$. The solution of Eq. (5.17) can be expressed as

$$X_1^{(1)} \cos \phi_1 = \left\{ \frac{r_2 x_1(0) - x_2(0)}{r_2 - r_1} \right\}, \quad X_1^{(2)} \cos \phi_2 = \left\{ \frac{-r_1 x_1(0) + x_2(0)}{r_2 - r_1} \right\}$$

$$X_1^{(1)} \sin \phi_1 = \left\{ \frac{-r_2 \dot{x}_1(0) + \dot{x}_2(0)}{\omega_1(r_2 - r_1)} \right\}, \quad X_1^{(2)} \sin \phi_2 = \left\{ \frac{r_1 \dot{x}_1(0) - \dot{x}_2(0)}{\omega_2(r_2 - r_1)} \right\}$$

from which we obtain the desired solution

$$X_1^{(1)} = [\{X_1^{(1)} \cos \phi_1\}^2 + \{X_1^{(1)} \sin \phi_1\}^2]^{1/2}$$

$$= \frac{1}{(r_2 - r_1)} \left[\{r_2 x_1(0) - x_2(0)\}^2 + \frac{\{-r_2 \dot{x}_1(0) + \dot{x}_2(0)\}^2}{\omega_1^2} \right]^{1/2}$$

$$X_1^{(2)} = [\{X_1^{(2)} \cos \phi_2\}^2 + \{X_1^{(2)} \sin \phi_2\}^2]^{1/2}$$

$$= \frac{1}{(r_2 - r_1)} \left[\{-r_1 x_1(0) + x_2(0)\}^2 + \frac{\{r_1 \dot{x}_1(0) - \dot{x}_2(0)\}^2}{\omega_2^2} \right]^{1/2}$$

$$\phi_1 = \tan^{-1} \left\{ \frac{X_1^{(1)} \sin \phi_1}{X_1^{(1)} \cos \phi_1} \right\} = \tan^{-1} \left\{ \frac{-r_2 \dot{x}_1(0) + \dot{x}_2(0)}{\omega_1[r_2 x_1(0) - x_2(0)]} \right\}$$

$$\phi_2 = \tan^{-1} \left\{ \frac{X_1^{(2)} \sin \phi_2}{X_1^{(2)} \cos \phi_2} \right\} = \tan^{-1} \left\{ \frac{r_1 \dot{x}_1(0) - \dot{x}_2(0)}{\omega_2[-r_1 x_1(0) + x_2(0)]} \right\} \qquad (5.18)$$

Frequencies of Spring-Mass System

EXAMPLE 5.1

Find the natural frequencies and mode shapes of a spring-mass system, shown in Fig. 5.4, which is constrained to move in the vertical direction only. Take $n = 1$.

Solution: If we measure x_1 and x_2 from the static equilibrium positions of the masses m_1 and m_2, respectively, the equations of motion and the solution obtained for the system of Fig. 5.3(a) are also

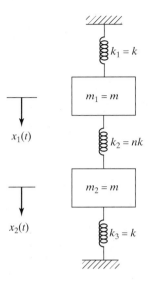

FIGURE 5.4   Two degree of freedom system.

applicable to this case if we substitute $m_1 = m_2 = m$ and $k_1 = k_2 = k_3 = k$. Thus the equations of motion, Eqs. (5.4) and (5.5), are given by

$$m\ddot{x}_1 + 2kx_1 - kx_2 = 0$$
$$m\ddot{x}_2 - kx_1 + 2kx_2 = 0 \tag{E.1}$$

By assuming harmonic solution as

$$x_i(t) = X_i \cos(\omega t + \phi); i = 1, 2 \tag{E.2}$$

the frequency equation can be obtained by substituting Eq. (E.2) into Eq. (E.1):

$$\begin{vmatrix} (-m\omega^2 + 2k) & (-k) \\ (-k) & (-m\omega^2 + 2k) \end{vmatrix} = 0$$

or

$$m^2\omega^4 - 4km\omega^2 + 3k^2 = 0 \tag{E.3}$$

The solution of Eq. (E.3) gives the natural frequencies

$$\omega_1 = \left\{ \frac{4km - [16k^2m^2 - 12m^2k^2]^{1/2}}{2m^2} \right\}^{1/2} = \sqrt{\frac{k}{m}} \tag{E.4}$$

$$\omega_2 = \left\{ \frac{4km + [16k^2m^2 - 12m^2k^2]^{1/2}}{2m^2} \right\}^{1/2} = \sqrt{\frac{3k}{m}} \tag{E.5}$$

From Eq. (5.11), the amplitude ratios are given by

$$r_1 = \frac{X_2^{(1)}}{X_1^{(1)}} = \frac{-m\omega_1^2 + 2k}{k} = \frac{k}{-m\omega_1^2 + 2k} = 1 \tag{E.6}$$

$$r_2 = \frac{X_2^{(2)}}{X_1^{(2)}} = \frac{-m\omega_2^2 + 2k}{k} = \frac{k}{-m\omega_2^2 + 2k} = -1 \tag{E.7}$$

The natural modes are given by Eq. (5.13):

$$\text{First mode} = \vec{x}^{(1)}(t) = \left\{ \begin{array}{c} X_1^{(1)} \cos\left(\sqrt{\dfrac{k}{m}}\, t + \phi_1\right) \\[4mm] X_1^{(1)} \cos\left(\sqrt{\dfrac{k}{m}}\, t + \phi_1\right) \end{array} \right\} \tag{E.8}$$

$$\text{Second mode} = \vec{x}^{(2)}(t) = \left\{ \begin{array}{c} X_1^{(2)} \cos\left(\sqrt{\dfrac{3k}{m}}\, t + \phi_2\right) \\[4mm] -X_1^{(2)} \cos\left(\sqrt{\dfrac{3k}{m}}\, t + \phi_2\right) \end{array} \right\} \tag{E.9}$$

It can be seen from Eq. (E.8) that when the system vibrates in its first mode, the amplitudes of the two masses remain the same. This implies that the length of the middle spring remains constant. Thus the motions of m_1 and m_2 are in phase (see Fig. 5.5a). When the system vibrates in its second mode, Eq. (E.9) shows that the displacements of the two masses have the same magnitude with opposite signs. Thus the motions of m_1 and m_2 are 180° out of phase (see Fig. 5.5b). In this case the midpoint

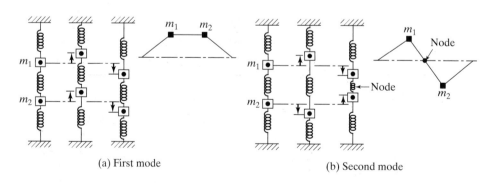

(a) First mode   (b) Second mode

FIGURE 5.5   Modes of vibration.

of the middle spring remains stationary for all time t. Such a point is called a *node*. Using Eq. (5.15), the motion (general solution) of the system can be expressed as

$$x_1(t) = X_1^{(1)} \cos\left(\sqrt{\frac{k}{m}}\, t + \phi_1\right) + X_1^{(2)} \cos\left(\sqrt{\frac{3k}{m}}\, t + \phi_2\right)$$

$$x_2(t) = X_1^{(1)} \cos\left(\sqrt{\frac{k}{m}}\, t + \phi_1\right) - X_1^{(2)} \cos\left(\sqrt{\frac{3k}{m}}\, t + \phi_2\right) \tag{E.10}$$

∎

Note: It can be seen that the computation of the natural frequencies and mode shapes is lengthy and tedious. Computer programs can be used conveniently for the numerical computation of the natural frequencies and mode shapes of multidegree of freedom systems (see Section 5.9).

Initial Conditions to Excite Specific Mode

EXAMPLE 5.2

Find the initial conditions that need to be applied to the system shown in Fig. 5.4 so as to make it vibrate in (a) the first mode, and (b) the second mode.

Solution:
Approach: Specify the solution to be obtained for the first or second mode from the general solution for arbitrary initial conditions and solve the resulting equations.

For arbitrary initial conditions, the motion of the masses is described by Eq. (5.15). In the present case, $r_1 = 1$ and $r_2 = -1$, so Eq. (5.15) reduces to Eq. (E.10) of Example 5.1:

$$x_1(t) = X_1^{(1)} \cos\left(\sqrt{\frac{k}{m}}\, t + \phi_1\right) + X_1^{(2)} \cos\left(\sqrt{\frac{3k}{m}}\, t + \phi_2\right)$$

$$x_2(t) = X_1^{(1)} \cos\left(\sqrt{\frac{k}{m}}\, t + \phi_1\right) - X_1^{(2)} \cos\left(\sqrt{\frac{3k}{m}}\, t + \phi_2\right) \tag{E.1}$$

Assuming the initial conditions as in Eq. (5.16), the constants $X_1^{(1)}$, $X_1^{(2)}$, ϕ_1, and ϕ_2 can be obtained from Eq. (5.18), using $r_1 = 1$ and $r_2 = -1$:

$$X_1^{(1)} = -\frac{1}{2}\left\{[x_1(0) + x_2(0)]^2 + \frac{m}{k}[\dot{x}_1(0) + \dot{x}_2(0)]^2\right\}^{1/2} \tag{E.2}$$

$$X_1^{(2)} = -\frac{1}{2}\left\{[-x_1(0) + x_2(0)]^2 + \frac{m}{3k}[\dot{x}_1(0) - \dot{x}_2(0)]^2\right\}^{1/2} \tag{E.3}$$

$$\phi_1 = \tan^{-1}\left\{\frac{-\sqrt{m}\,[\dot{x}_1(0) + \dot{x}_2(0)]}{\sqrt{k}\,[x_1(0) + x_2(0)]}\right\} \tag{E.4}$$

$$\phi_2 = \tan^{-1}\left\{\frac{\sqrt{m}\,[\dot{x}_1(0) - \dot{x}_2(0)]}{\sqrt{3k}\,[-x_1(0) + x_2(0)]}\right\} \tag{E.5}$$

a.   The first normal mode of the system is given by Eq. (E.8) of Example 5.1:

$$\vec{x}^{(1)}(t) = \left\{ \begin{array}{c} X_1^{(1)}\cos\left(\sqrt{\dfrac{k}{m}}\,t + \phi_1\right) \\[4mm] X_1^{(1)}\cos\left(\sqrt{\dfrac{k}{m}}\,t + \phi_1\right) \end{array} \right\} \tag{E.6}$$

Comparison of Eqs. (E.1) and (E.6) shows that the motion of the system is identical with the first normal mode only if $X_1^{(2)} = 0$. This requires that (from Eq. E.3)

$$x_1(0) = x_2(0) \qquad \text{and} \qquad \dot{x}_1(0) = \dot{x}_2(0) \tag{E.7}$$

b.   The second normal mode of the system is given by Eq. (E.9) of Example 5.1:

$$\vec{x}^{(2)}(t) = \left\{ \begin{array}{c} X_1^{(2)}\cos\left(\sqrt{\dfrac{3k}{m}}\,t + \phi_2\right) \\[4mm] -X_1^{(2)}\cos\left(\sqrt{\dfrac{3k}{m}}\,t + \phi_2\right) \end{array} \right\} \tag{E.8}$$

Comparison of Eqs. (E.1) and (E.8) shows that the motion of the system coincides with the second normal mode only if $X_1^{(1)} = 0$. This implies that (from Eq. E.2)

$$x_1(0) = -x_2(0) \qquad \text{and} \qquad \dot{x}_1(0) = -\dot{x}_2(0) \tag{E.9}$$

∎

Free Vibration Response of a Two Degree of Freedom System

EXAMPLE 5.3

Find the free vibration response of the system shown in Fig. 5.3(a) with $k_1 = 30$, $k_2 = 5$, $k_3 = 0$, $m_1 = 10$, $m_2 = 1$ and $c_1 = c_2 = c_3 = 0$ for the initial conditions $x_1(0) = 1$, $\dot{x}_1(0) = x_2(0) = \dot{x}_2(0) = 0$.

Solution: For the given data, the eigenvalue problem, Eq. (5.8), becomes

$$\begin{bmatrix} -m_1\omega^2 + k_1 + k_2 & -k_2 \\ -k_2 & -m_2\omega^2 + k_2 + k_3 \end{bmatrix} \begin{Bmatrix} X_1 \\ X_2 \end{Bmatrix} = \begin{Bmatrix} 0 \\ 0 \end{Bmatrix}$$

or

$$\begin{bmatrix} -10\omega^2 + 35 & -5 \\ -5 & -\omega^2 + 5 \end{bmatrix} \begin{Bmatrix} X_1 \\ X_2 \end{Bmatrix} = \begin{Bmatrix} 0 \\ 0 \end{Bmatrix} \tag{E.1}$$

By setting the determinant of the coefficient matrix in Eq. (E.1) to zero, we obtain the frequency equation (see Eq. 5.9):

$$10\,\omega^4 - 85\,\omega^2 + 150 = 0 \tag{E.2}$$

from which the natural frequencies can be found as

$$\omega_1^2 = 2.5, \qquad \omega_2^2 = 6.0$$

or

$$\omega_1 = 1.5811, \qquad \omega_2 = 2.4495 \tag{E.3}$$

The substitution of $\omega^2 = \omega_1^2 = 2.5$ in Eq. (E.1) leads to $X_2^{(1)} = 2X_1^{(1)}$, while $\omega^2 = \omega_2^2 = 6.0$ in Eq. (E.1) yields $X_2^{(2)} = -5X_1^{(2)}$. Thus the normal modes (or eigenvectors) are given by

$$\vec{X}^{(1)} = \begin{Bmatrix} X_1^{(1)} \\ X_2^{(1)} \end{Bmatrix} = \begin{Bmatrix} 1 \\ 2 \end{Bmatrix} X_1^{(1)} \tag{E.4}$$

$$\vec{X}^{(2)} = \begin{Bmatrix} X_1^{(2)} \\ X_2^{(2)} \end{Bmatrix} = \begin{Bmatrix} 1 \\ -5 \end{Bmatrix} X_1^{(2)} \tag{E.5}$$

The free vibration responses of the masses m_1 and m_2 are given by (see Eq. 5.15):

$$x_1(t) = X_1^{(1)} \cos(1.5811t + \phi_1) + X_1^{(2)} \cos(2.4495t + \phi_2) \tag{E.6}$$

$$x_2(t) = 2X_1^{(1)} \cos(1.5811t + \phi_1) - 5X_1^{(2)} \cos(2.4495t + \phi_2) \tag{E.7}$$

where $X_1^{(1)}$, $X_1^{(2)}$, ϕ_1, and ϕ_2 are constants to be determined from the initial conditions. By using the given intial conditions in Eqs. (E.6) and (E.7), we obtain

$$x_1(t = 0) = 1 = X_1^{(1)} \cos\phi_1 + X_1^{(2)} \cos\phi_2 \tag{E.8}$$

$$x_2(t = 0) = 0 = 2X_1^{(1)} \cos\phi_1 - 5X_1^{(2)} \cos\phi_2 \tag{E.9}$$

$$\dot{x}_1(t = 0) = 0 = -1.5811X_1^{(1)} \sin\phi_1 - 2.4495X_1^{(2)} \sin\phi_2 \tag{E.10}$$

$$\dot{x}_2(t = 0) = -3.1622X_1^{(1)} + 12.2475X_1^{(2)} \sin\phi_2 \tag{E.11}$$

The solution of Eqs. (E.8) and (E.9) yields

$$X_1^{(1)} \cos\phi_1 = \frac{5}{7}, \quad X_1^{(2)} \cos\phi_2 = \frac{2}{7} \tag{E.12}$$

while the solution of Eqs. (E.10) and (E.11) leads to

$$X_1^{(1)} \sin \phi_1 = 0, \ X_1^{(2)} \sin \phi_2 = 0 \tag{E.13}$$

Equations (E.12) and (E.13) give

$$X_1^{(1)} = \frac{5}{7}, \ X_1^{(2)} = \frac{2}{7}, \ \phi_1 = 0, \ \phi_2 = 0 \tag{E.14}$$

Thus the free vibration responses of m_1 and m_2 are given by

$$x_1(t) = \frac{5}{7} \cos 1.5811t + \frac{2}{7} \cos 2.4495t \tag{E.15}$$

$$x_2(t) = \frac{10}{7} \cos 1.5811t - \frac{10}{7} \cos 2.4495t \tag{E.16}$$

The graphical representation of Eqs. (E.15) and (E.16) is considered in Example 5.12.

■

5.4   Torsional System

Consider a torsional system consisting of two discs mounted on a shaft, as shown in Fig. 5.6. The three segments of the shaft have rotational spring constants k_{t1}, k_{t2}, and k_{t3}, as indicated in the figure. Also shown are the discs of mass moments of inertia J_1 and J_2, the applied torques M_{t1} and M_{t2}, and the rotational degrees of freedom θ_1 and θ_2. The differential equations of rotational motion for the discs J_1 and J_2 can be derived as

$$J_1 \ddot{\theta}_1 = -k_{t1}\theta_1 + k_{t2}(\theta_2 - \theta_1) + M_{t1}$$
$$J_2 \ddot{\theta}_2 = -k_{t2}(\theta_2 - \theta_1) - k_{t3}\theta_2 + M_{t2}$$

which upon rearrangement become

$$J_1 \ddot{\theta}_1 + (k_{t1} + k_{t2})\,\theta_1 - k_{t2}\theta_2 = M_{t1}$$
$$J_2 \ddot{\theta}_2 - k_{t2}\theta_1 + (k_{t2} + k_{t3})\,\theta_2 = M_{t2} \tag{5.19}$$

For the free vibration analysis of the system, Eq. (5.19) reduces to

$$J_1 \ddot{\theta}_1 + (k_{t1} + k_{t2})\,\theta_1 - k_{t2}\theta_2 = 0$$
$$J_2 \ddot{\theta}_2 - k_{t2}\theta_1 + (k_{t2} + k_{t3})\,\theta_2 = 0 \tag{5.20}$$

Note that Eq. (5.20) is similar to Eqs. (5.4) and (5.5). In fact, Eq. (5.20) can be obtained by substituting θ_1, θ_2, J_1, J_2, k_{t1}, k_{t2}, and k_{t3} for x_1, x_2, m_1, m_2, k_1, k_2, and k_3, respectively.

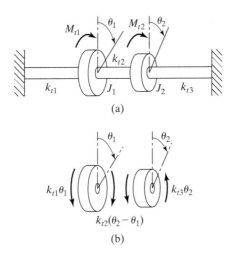

FIGURE 5.6   Torsional system with discs mounted on a shaft.

Thus the analysis presented in Section 5.3 is also applicable to torsional systems with proper substitutions. The following two examples illustrate the procedure.

Natural Frequencies of a Torsional System

EXAMPLE 5.4

Find the natural frequencies and mode shapes for the torsional system shown in Fig. 5.7 for $J_1 = J_0$, $J_2 = 2J_0$, and $k_{t1} = k_{t2} = k_t$.

Solution: The differential equations of motion, Eq. (5.20), reduce to (with $k_{t3} = 0$, $k_{t1} = k_{t2} = k_t$, $J_1 = J_0$, and $J_2 = 2J_0$):

$$J_0\ddot{\theta}_1 + 2k_t\theta_1 - k_t\theta_2 = 0$$
$$2J_0\ddot{\theta}_2 - k_t\theta_1 + k_t\theta_2 = 0 \tag{E.1}$$

Rearranging and substituting the harmonic solution

$$\theta_i(t) = \Theta_i \cos(\omega t + \phi); \qquad i = 1, 2 \tag{E.2}$$

gives the frequency equation:

$$2\omega^4 J_0^2 - 5\omega^2 J_0 k_t + k_t^2 = 0 \tag{E.3}$$

The solution of Eq. (E.3) gives the natural frequencies

$$\omega_1 = \sqrt{\frac{k_t}{4J_0}(5 - \sqrt{17})} \qquad \text{and} \qquad \omega_2 = \sqrt{\frac{k_t}{4J_0}(5 + \sqrt{17})} \tag{E.4}$$

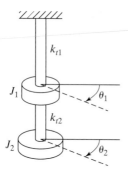

FIGURE 5.7
Torsional system.

The amplitude ratios are given by

$$r_1 = \frac{\Theta_2^{(1)}}{\Theta_1^{(1)}} = 2 - \frac{(5 - \sqrt{17})}{4}$$

$$r_2 = \frac{\Theta_2^{(2)}}{\Theta_1^{(2)}} = 2 - \frac{(5 + \sqrt{17})}{4} \tag{E.5}$$

Equations (E.4) and (E.5) can also be obtained by substituting $k_1 = k_{t1} = k_t$, $k_2 = k_{t2} = k_t$, $m_1 = J_1 = J_0$, $m_2 = J_2 = 2 J_0$, and $k_3 = 0$ in Eqs. (5.10) and (5.11).

∎

Note: For a two degree of freedom system, the two natural frequencies ω_1 and ω_2 are not equal to either of the natural frequencies of the two single degree of freedom systems constructed from the same components. In Example 5.4, the single degree of freedom systems k_{t1} and J_1

$$\left(\text{with } \overline{\omega}_1 = \sqrt{\frac{k_{t1}}{J_1}} = \sqrt{\frac{k_t}{J_0}} \right)$$

and k_{t2} and J_2

$$\left(\text{with } \overline{\omega}_2 = \sqrt{\frac{k_{t2}}{J_2}} = \frac{1}{\sqrt{2}} \sqrt{\frac{k_t}{J_0}} \right)$$

are combined to obtain the system shown in Fig. 5.7. It can be seen that ω_1 and ω_2 are different from $\overline{\omega}_1$ and $\overline{\omega}_2$.

EXAMPLE 5.5

Natural Frequencies of a Marine Engine Propeller

The schematic diagram of a marine engine connected to a propeller through gears is shown in Fig. 5.8(a). The mass moments of inertia of the flywheel, engine, gear 1, gear 2, and the propeller (in kg-m^2) are 9000, 1000, 250, 150, and 2000, respectively. Find the natural frequencies and mode shapes of the system in torsional vibration.

Solution

Approach: Find the equivalent mass moments of inertia of all rotors with respect to one rotor and use a two degree of freedom model.

Assumptions

1. The flywheel can be considered to be stationary (fixed) since its mass moment of inertia is very large compared to that of other rotors.
2. The engine and gears can be replaced by a single equivalent rotor.

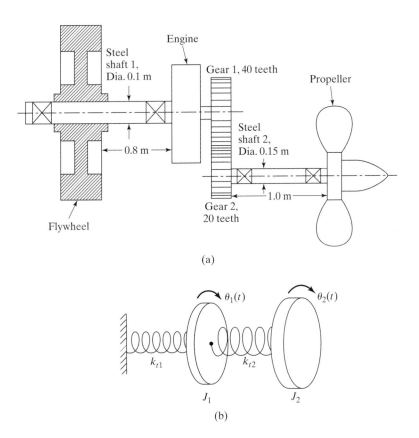

(a)

(b)

FIGURE 5.8   Marine engine propeller system.

Since gears 1 and 2 have 40 and 20 teeth, shaft 2 rotates at twice the speed of shaft 1. Thus the mass moments of inertia of gear 2 and the propeller, referred to the engine, are given by

$$(J_{G2})_{eq} = (2)^2(150) = 600 \text{ kg-m}^2$$

$$(J_P)_{eq} = (2)^2(2000) = 8000 \text{ kg-m}^2$$

Since the distance between the engine and the gear unit is small, the engine and the two gears can be replaced by a single rotor with a mass moment of inertia of

$$J_1 = J_E + J_{G1} + (J_{G2})_{eq} = 1000 + 250 + 600 = 1850 \text{ kg-m}^2$$

Assuming a shear modulus of 80×10^9 N/m^2 for steel, the torsional stiffnesses of shafts 1 and 2 can be determined as

$$k_{t1} = \frac{G I_{01}}{l_1} = \frac{G}{l_1}\left(\frac{\pi d_1^4}{32}\right) = \frac{(80 \times 10^9)(\pi)(0.10)^4}{(0.8)(32)} = 981,750.0 \text{ N-m/rad}$$

$$k_{t2} = \frac{G I_{02}}{l_2} = \frac{G}{l_2}\left(\frac{\pi d_2^4}{32}\right) = \frac{(80 \times 10^9)(\pi)(0.15)^4}{(1.0)(32)} = 3,976,087.5 \text{ N-m/rad}$$

Since the length of shaft 2 is not negligible, the propeller is assumed to be a rotor connected at the end of shaft 2. Thus the system can be represented as a two degree of freedom torsional system, as indicated in Fig. 5.8(b). By setting $k_3 = 0$, $k_1 = k_{t1}$, $k_2 = k_{t2}$, $m_1 = J_1$, and $m_2 = J_2$ in Eq. (5.10), the natural frequencies of the system can be found as

$$\omega_1^2, \omega_2^2 = \frac{1}{2}\left\{\frac{(k_{t1} + k_{t2})J_2 + k_{t2}J_1}{J_1 J_2}\right\}$$

$$\pm \left[\left\{\frac{(k_{t1} + k_{t2})J_2 + k_{t2}J_1}{J_1 J_2}\right\}^2 - 4\left\{\frac{(k_{t1} + k_{t2})k_{t2} - k_{t2}^2}{J_1 J_2}\right\}\right]^{1/2}$$

$$= \left\{\frac{(k_{t1} + k_{t2})}{2 J_1} + \frac{k_{t2}}{2 J_2}\right\}$$

$$\pm \left[\left\{\frac{(k_{t1} + k_{t2})}{2 J_1} + \frac{k_{t2}}{2 J_2}\right\}^2 - \frac{k_{t1}k_{t2}}{J_1 J_2}\right]^{1/2} \tag{E.1}$$

Since

$$\frac{(k_{t1} + k_{t2})}{2 J_1} + \frac{k_{t2}}{2 J_2} = \frac{(98.1750 + 397.6087) \times 10^4}{2 \times 1850} + \frac{397.6087 \times 10^4}{2 \times 8000}$$

$$= 1588.46$$

and

$$\frac{k_{t1} k_{t2}}{J_1 J_2} = \frac{(98.1750 \times 10^4)(397.6087 \times 10^4)}{(1850)(8000)} = 26.3750 \times 10^4$$

Eq. (E.1) gives

$$\omega_1^2, \omega_2^2 = 1588.46 \pm [(1588.46)^2 - 26.3750 \times 10^4]^{1/2}$$

$$= 1588.46 \pm 1503.1483$$

Thus

$$\omega_1^2 = 85.3117 \qquad \text{or} \qquad \omega_1 = 9.2364 \text{ rad/sec}$$

$$\omega_2^2 = 3091.6083 \qquad \text{or} \qquad \omega_2 = 55.6022 \text{ rad/sec}$$

For the mode shapes, we set $k_1 = k_{t1}$, $k_2 = k_{t2}$, $k_3 = 0$, $m_1 = J_1$, and $m_2 = J_2$ in Eq. (5.11) to obtain

$$r_1 = \frac{-J_1\omega_1^2 + (k_{t1} + k_{t2})}{k_{t2}}$$

$$= \frac{-(1850)\,(85.3117) + (495.7837 \times 10^4)}{397.6087 \times 10^4} = 1.2072$$

and

$$r_2 = \frac{-J_1\omega_2^2 + (k_{t1} + k_{t2})}{k_{t2}}$$

$$= \frac{-(1850)\,(3091.6083) + (495.7837 \times 10^4)}{397.6087 \times 10^4} = -0.1916$$

Thus the mode shapes can be determined from an equation similar to Eq. (5.12) as

$$\left\{ \begin{matrix} \Theta_1 \\ \Theta_2 \end{matrix} \right\}^{(1)} = \left\{ \begin{matrix} 1 \\ r_1 \end{matrix} \right\} = \frac{1}{1.2072}$$

and

$$\left\{ \begin{matrix} \Theta_1 \\ \Theta_2 \end{matrix} \right\}^{(2)} = \left\{ \begin{matrix} 1 \\ r_2 \end{matrix} \right\} = \frac{1}{-0.1916}$$

■

5.5   Coordinate Coupling and Principal Coordinates

As stated earlier, an n degree of freedom system requires n independent coordinates to describe its configuration. Usually, these coordinates are independent geometrical quantities measured from the equilibrium position of the vibrating body. However, it is possible to select some other set of n coordinates to describe the configuration of the system. The latter set may be, for example, different from the first set in that the coordinates may have their origin away from the equilibrium position of the body. There could be still other sets

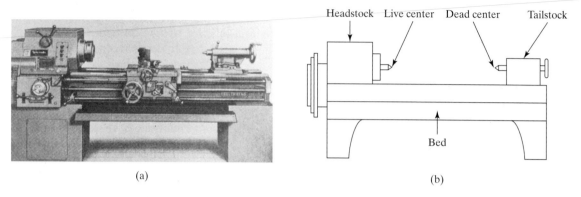

(a)   (b)

FIGURE 5.9   Lathe. (Photo courtersy of South Bend Lathe Corp.)

of coordinates to describe the configuration of the system. Each of these sets of n coordinates is called the *generalized coordinates*.

As an example, consider the lathe shown in Fig. 5.9(a). An accurate model of this machine tool would involve the consideration of the lathe bed as an elastic beam supported on short elastic columns with lumped masses attached to the beam, as indicated in Fig. 5.9(b) [5.1–5.3]. However, for simplified vibration analysis, the lathe bed can be considered as a rigid body having mass and inertia, and the headstock and tailstock can each be replaced by lumped masses. The bed can be assumed to be supported on springs at the ends. Thus the final model will be a rigid body of total mass m and mass moment of inertia J_0 about its C.G., resting on springs of stiffnesses k_1 and k_2, as shown in Fig. 5.10(a). For this two degree of freedom system, any of the following sets of coordinates may be used to describe the motion:

1. Deflections $x_1(t)$ and $x_2(t)$ of the two ends of the lathe AB
2. Deflection $x(t)$ of the C.G. and rotation $\theta(t)$
3. Deflection $x_1(t)$ of the end A and rotation $\theta(t)$
4. Deflection $y(t)$ of point P located at a distance e to the left of the C.G. and rotation $\theta(t)$, as indicated in Fig. 5.10(b).

Thus any set of these coordinates—(x_1, x_2), (x, θ), (x_1, θ), and (y, θ)—represents the generalized coordinates of the system. Now we shall derive the equations of motion of the lathe using two different sets of coordinates to illustrate the concept of coordinate coupling.

Equations of Motion Using x(t) and θ(t).   From the free-body diagram shown in Fig. 5.10(a), with the positive values of the motion variables as indicated, the force equilibrium equation in the vertical direction can be written as

$$m\ddot{x} = -k_1(x - l_1\theta) - k_2(x + l_2\theta) \tag{5.21}$$

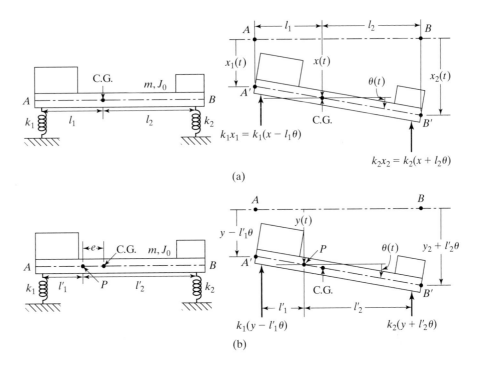

FIGURE 5.10 Modeling of a lathe.

and the moment equation about the C.G. can be expressed as

$$J_0\ddot{\theta} = k_1(x - l_1\theta)l_1 - k_2(x + l_2\theta)l_2 \tag{5.22}$$

Equations (5.21) and (5.22) can be rearranged and written in matrix form as

$$\begin{bmatrix} m & 0 \\ 0 & J_0 \end{bmatrix} \begin{Bmatrix} \ddot{x} \\ \ddot{\theta} \end{Bmatrix} + \begin{bmatrix} (k_1 + k_2) & -(k_1l_1 - k_2l_2) \\ -(k_1l_1 - k_2l_2) & (k_1l_1^2 + k_2l_2^2) \end{bmatrix} \begin{Bmatrix} x \\ \theta \end{Bmatrix}$$

$$= \begin{Bmatrix} 0 \\ 0 \end{Bmatrix} \tag{5.23}$$

It can be seen that each of these equations contain x and θ. They become independent of each other if the coupling term $(k_1l_1 - k_2l_2)$ is equal to zero—that is, if $k_1l_1 = k_2l_2$. If $k_1l_1 \neq k_2l_2$, the resultant motion of the lathe AB is both translational and ratational when either a displacement or torque is applied through the C.G. of the body as an initial condition. In other words, the lathe rotates in the vertical plane and has vertical motion as well, unless $k_1l_1 = k_2l_2$. This is known as *elastic* or *static coupling*.

Equations of Motion Using y(t) and θ(t).   From Fig. 5.10(b), where $y(t)$ and $\theta(t)$ are used as the generalized coordinates of the system, the equations of motion for translation and rotation can be written as

$$m\ddot{y} = -k_1(y - l_1'\theta) - k_2(y + l_2'\theta) - me\ddot{\theta}$$

$$J_p\ddot{\theta} = k_1(y - l_1'\theta)l_1' - k_2(y + l_2'\theta)l_2' - me\ddot{y} \qquad (5.24)$$

These equations can be rearranged and written in matrix form as

$$\begin{bmatrix} m & me \\ me & J_p \end{bmatrix} \begin{Bmatrix} \ddot{y} \\ \ddot{\theta} \end{Bmatrix} + \begin{bmatrix} (k_1 + k_2) & (k_2l_2' - k_1l_1') \\ (-k_1l_1' + k_2l_2') & (k_1l_1'^2 + k_2l_2'^2) \end{bmatrix} \begin{Bmatrix} y \\ \theta \end{Bmatrix}$$

$$= \begin{Bmatrix} 0 \\ 0 \end{Bmatrix} \qquad (5.25)$$

Both the equations of motion represented by Eq. (5.25) contain y and θ, so they are coupled equations. They contain static (or elastic) as well as dynamic (or mass) coupling terms. If $k_1l_1' = k_2l_2'$, the system will have *dynamic* or *inertia coupling* only. In this case, if the lathe moves up and down in the y direction, the inertia force $m\ddot{y}$, which acts through the center of gravity of the body, induces a motion in the θ direction, by virtue of the moment $m\ddot{y}e$. Similarly, a motion in the θ direction induces a motion of the lathe in the y direction due to the force $me\ddot{\theta}$.

Note the following characteristics of these systems:

1. In the most general case, a viscously damped two degree of freedom system has equations of motion in the following form:

$$\begin{bmatrix} m_{11} & m_{12} \\ m_{12} & m_{22} \end{bmatrix} \begin{Bmatrix} \ddot{x}_1 \\ \ddot{x}_2 \end{Bmatrix} + \begin{Bmatrix} c_{11} & c_{12} \\ c_{12} & c_{22} \end{Bmatrix} \begin{Bmatrix} \dot{x}_1 \\ \dot{x}_2 \end{Bmatrix} + \begin{bmatrix} k_{11} & k_{12} \\ k_{12} & k_{22} \end{bmatrix} \begin{Bmatrix} x_1 \\ x_2 \end{Bmatrix} = \begin{Bmatrix} 0 \\ 0 \end{Bmatrix} \qquad (5.26)$$

This equation reveals the type of coupling present. If the stiffness matrix is not diagonal, the system has elastic or static coupling. If the damping matrix is not diagonal, the system has damping or velocity coupling. Finally, if the mass matrix is not diagonal, the system has mass or inertial coupling. Both velocity and mass coupling come under the heading of dynamic coupling.

2. The system vibrates in its own natural way regardless of the coordinates used. The choice of the coordinates is a mere convenience.

3. From Eqs. (5.23) and (5.25), it is clear that the nature of the coupling depends on the coordinates used and is not an inherent property of the system. It is possible to choose a system of coordinates $q_1(t)$ and $q_2(t)$ which give equations of motion that are uncoupled both statically and dynamically. Such coordinates are called *principal* or *natural coordinates*. The main advantage of using principal coordinates is that the resulting uncoupled equations of motion can be solved independently of one another.

The following example illustrates the method of finding the principal coordinates in terms of the geometrical coordinates.

EXAMPLE 5.6

Principal Coordinates of Spring-Mass System

Determine the principal coordinates for the spring-mass system shown in Fig. 5.4.

Solution

Approach: Define two independent solutions as principal coordinates and express them in terms of the solutions $x_1(t)$ and $x_2(t)$.

The general motion of the system shown in Fig. 5.4 is given by Eq. (E.10) of Example 5.1

$$x_1(t) = B_1 \cos\left(\sqrt{\frac{k}{m}}\,t + \phi_1\right) + B_2 \cos\left(\sqrt{\frac{3k}{m}}\,t + \phi_2\right)$$

$$x_2(t) = B_1 \cos\left(\sqrt{\frac{k}{m}}\,t + \phi_1\right) - B_2 \cos\left(\sqrt{\frac{3k}{m}}\,t + \phi_2\right) \tag{E.1}$$

where $B_1 = X_1^{(1)}$, $B_2 = X_1^{(2)}$, ϕ_1, and ϕ_2 are constants. We define a new set of coordinates $q_1(t)$ and $q_2(t)$ such that

$$q_1(t) = B_1 \cos\left(\sqrt{\frac{k}{m}}\,t + \phi_1\right)$$

$$q_2(t) = B_2 \cos\left(\sqrt{\frac{3k}{m}}\,t + \phi_2\right) \tag{E.2}$$

Since $q_1(t)$ and $q_2(t)$ are harmonic functions, their corresponding equations of motion can be written as[1]

$$\ddot{q}_1 + \left(\frac{k}{m}\right)q_1 = 0$$

$$\ddot{q}_2 + \left(\frac{3k}{m}\right)q_2 = 0 \tag{E.3}$$

These equations represent a two degree of freedom system whose natural frequencies are $\omega_1 = \sqrt{k/m}$ and $\omega_2 = \sqrt{3k/m}$. Because there is neither static nor dynamic coupling in the equations of motion (E.3), $q_1(t)$ and $q_2(t)$ are principal coordinates. From Eqs. (E.1) and (E.2), we can write

$$x_1(t) = q_1(t) + q_2(t)$$
$$x_2(t) = q_1(t) - q_2(t) \tag{E.4}$$

---

[1]Note that the equation of motion corresponding to the solution $q = B \cos(\omega t + \phi)$ is given by $\ddot{q} + \omega^2 q = 0$.

The solution of Eqs. (E.4) gives the principal coordinates:

$$q_1(t) = \frac{1}{2}[x_1(t) + x_2(t)]$$

$$q_2(t) = \frac{1}{2}[x_1(t) - x_2(t)] \qquad \text{(E.5)}$$

■

Frequencies and Modes of an Automobile

EXAMPLE 5.7

Determine the pitch (angular motion) and bounce (up and down linear motion) frequencies and the location of oscillation centers (nodes) of an automobile with the following data (see Fig. 5.11):

Mass $(m) = 1000$ kg
Radius of gyration $(r) = 0.9$ m
Distance between front axle and C.G. $(l_1) = 1.0$ m
Distance between rear axle and C.G. $(l_2) = 1.5$ m
Front spring stiffness $(k_f) = 18$ kN/m
Rear spring stiffness $(k_r) = 22$ kN/m

Solution: If x and θ are used as independent coordinates, the equations of motion are given by Eq. (5.23) with $k_1 = k_f$, $k_2 = k_r$, and $J_0 = mr^2$. For free vibration, we assume a harmonic solution:

$$x(t) = X \cos(\omega t + \phi), \qquad \theta(t) = \Theta \cos(\omega t + \phi) \qquad \text{(E.1)}$$

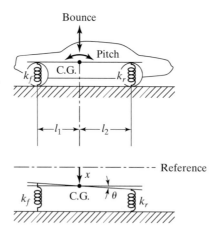

FIGURE 5.11   Pitch and bounce motions of an automobile.

Using Eqs. (E.1) and (5.23), we obtain

$$\begin{bmatrix} (-m\omega^2 + k_1 + k_2) & (-k_1 l_1 + k_2 l_2) \\ (-k_1 l_1 + k_2 l_2) & (-J_0\omega^2 + k_1 l_1^2 + k_2 l_2^2) \end{bmatrix} \begin{Bmatrix} X \\ \Theta \end{Bmatrix} = \begin{Bmatrix} 0 \\ 0 \end{Bmatrix} \tag{E.2}$$

For the known data, Eq. (E.2) becomes

$$\begin{bmatrix} (-1000\omega^2 + 40{,}000) & 15{,}000 \\ 15{,}000 & (-810\omega^2 + 67{,}500) \end{bmatrix} \begin{Bmatrix} X \\ \Theta \end{Bmatrix} = \begin{Bmatrix} 0 \\ 0 \end{Bmatrix} \tag{E.3}$$

from which the frequency equation can be derived:

$$8.1\omega^4 - 999\omega^2 + 24{,}750 = 0 \tag{E.4}$$

The natural frequencies can be found from Eq. (E.4):

$$\omega_1 = 5.8593 \text{ rad/s}, \qquad \omega_2 = 9.4341 \text{ rad/s} \tag{E.5}$$

With these values, the ratio of amplitudes can be found from Eq. (E.3):

$$\frac{X^{(1)}}{\Theta^{(1)}} = -2.6461, \qquad \frac{X^{(2)}}{\Theta^{(2)}} = 0.3061 \tag{E.6}$$

The node locations can be obtained by noting that the tangent of a small angle is approximately equal to the angle itself. Thus, from Fig. 5.12, we find that the distance between the C.G. and the node is -2.6461 m for ω_1 and 0.3061 m for ω_2. The mode shapes are shown by dashed lines in Fig. 5.12.

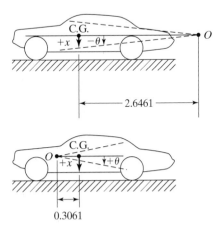

FIGURE 5.12   Mode shapes of an automobile.

5.6   Forced Vibration Analysis

The equations of motion of a general two degree of freedom system under external forces can be written as

$$
\begin{bmatrix} m_{11} & m_{12} \\ m_{12} & m_{22} \end{bmatrix} \begin{Bmatrix} \ddot{x}_1 \\ \ddot{x}_2 \end{Bmatrix} + \begin{bmatrix} c_{11} & c_{12} \\ c_{12} & c_{22} \end{bmatrix} \begin{Bmatrix} \dot{x}_1 \\ \dot{x}_2 \end{Bmatrix}
$$

$$
+ \begin{bmatrix} k_{11} & k_{12} \\ k_{12} & k_{22} \end{bmatrix} \begin{Bmatrix} x_1 \\ x_2 \end{Bmatrix} = \begin{Bmatrix} F_1 \\ F_2 \end{Bmatrix} \tag{5.27}
$$

Equations (5.1) and (5.2) can be seen to be special cases of Eq. (5.27), with $m_{11} = m_1$, $m_{22} = m_2$, and $m_{12} = 0$. We shall consider the external forces to be harmonic:

$$
F_j(t) = F_{j0}e^{i\omega t}, \qquad j = 1, 2 \tag{5.28}
$$

where ω is the forcing frequency. We can write the steady-state solutions as

$$
x_j(t) = X_j e^{i\omega t}, \qquad j = 1, 2 \tag{5.29}
$$

where X_1 and X_2 are, in general, complex quantities that depend on ω and the system parameters. Substitution of Eqs. (5.28) and (5.29) into Eq. (5.27) leads to

$$
\begin{bmatrix} (-\omega^2 m_{11} + i\omega c_{11} + k_{11}) & (-\omega^2 m_{12} + i\omega c_{12} + k_{12}) \\ (-\omega^2 m_{12} + i\omega c_{12} + k_{12}) & (-\omega_2 m_{22} + i\omega c_{22} + k_{22}) \end{bmatrix} \begin{Bmatrix} X_1 \\ X_2 \end{Bmatrix}
$$

$$
= \begin{Bmatrix} F_{10} \\ F_{20} \end{Bmatrix} \tag{5.30}
$$

As in Section 3.5, we define the mechanical impedance $Z_{rs}(i\omega)$ as

$$
Z_{rs}(i\omega) = -\omega^2 m_{rs} + i\omega c_{rs} + k_{rs}, \quad r, s = 1, 2 \tag{5.31}
$$

and write Eq. (5.30) as

$$
[Z(i\omega)]\vec{X} = \vec{F}_0 \tag{5.32}
$$

where

$$
[Z(i\omega)] = \begin{bmatrix} Z_{11}(i\omega) & Z_{12}(i\omega) \\ Z_{12}(i\omega) & Z_{22}(i\omega) \end{bmatrix} = \text{Impedance matrix}
$$

$$
\vec{X} = \begin{Bmatrix} X_1 \\ X_2 \end{Bmatrix}
$$

and

$$\vec{F}_0 = \begin{Bmatrix} F_{10} \\ F_{20} \end{Bmatrix}$$

Equation (5.32) can be solved to obtain

$$\vec{X} = [Z(i\omega)]^{-1}\,\vec{F}_0 \tag{5.33}$$

where the inverse of the impedance matrix is given by

$$[Z(i\omega)]^{-1} = \frac{1}{Z_{11}(i\omega)Z_{22}(i\omega) - Z_{12}^2(i\omega)}\begin{bmatrix} Z_{22}(i\omega) & -Z_{12}(i\omega) \\ -Z_{12}(i\omega) & Z_{11}(i\omega) \end{bmatrix} \tag{5.34}$$

Equations (5.33) and (5.34) lead to the solution

$$X_1(i\omega) = \frac{Z_{22}(i\omega)F_{10} - Z_{12}(i\omega)F_{20}}{Z_{11}(i\omega)Z_{22}(i\omega) - Z_{12}^2(i\omega)}$$

$$X_2(i\omega) = \frac{-Z_{12}(i\omega)F_{10} + Z_{11}(i\omega)F_{20}}{Z_{11}(i\omega)Z_{22}(i\omega) - Z_{12}^2(i\omega)} \tag{5.35}$$

By substituting Eq. (5.35) into Eq. (5.29) we can find the complete solution, $x_1(t)$ and $x_2(t)$.

The analysis of a two degree of freedom system used as a vibration absorber is given in Section 9.11. Reference [5.4] deals with the impact response of a two degree of freedom system, while Ref. [5.5] considers the steady-state response under harmonic excitation.

EXAMPLE 5.8

Steady-State Response of Spring-Mass System

Find the steady-state response of the system shown in Fig. 5.13 when the mass m_1 is excited by the force $F_1(t) = F_{10} \cos \omega t$. Also, plot its frequency response curve.

Solution: The equations of motion of the system can be expressed as

$$\begin{bmatrix} m & 0 \\ 0 & m \end{bmatrix}\begin{Bmatrix} \ddot{x}_1 \\ \ddot{x}_2 \end{Bmatrix} + \begin{bmatrix} 2k & -k \\ -k & 2k \end{bmatrix}\begin{Bmatrix} x_1 \\ x_2 \end{Bmatrix} = \begin{Bmatrix} F_{10} \cos \omega t \\ 0 \end{Bmatrix} \tag{E.1}$$

Comparison of Eq. (E.1) with Eq. (5.27) shows that

$$m_{11} = m_{22} = m, \qquad m_{12} = 0, \qquad c_{11} = c_{12} = c_{22} = 0,$$

$$k_{11} = k_{22} = 2k, \qquad k_{12} = -k, \qquad F_1 = F_{10} \cos \omega t, \qquad F_2 = 0$$

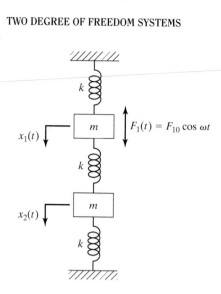

FIGURE 5.13   A two-mass system subjected to harmonic force.

We assume the solution to be as follows:[2]

$$x_j(t) = X_j \cos \omega t; \qquad j = 1, 2 \tag{E.2}$$

Equation (5.31) gives

$$Z_{11}(\omega) = Z_{22}(\omega) = -m\omega^2 + 2k, \qquad Z_{12}(\omega) = -k \tag{E.3}$$

Hence X_1 and X_2 are given by Eq. (5.35):

$$X_1(\omega) = \frac{(-\omega^2 m + 2k)F_{10}}{(-\omega^2 m + 2k)^2 - k^2} = \frac{(-\omega^2 m + 2k)F_{10}}{(-m\omega^2 + 3k)(-m\omega^2 + k)} \tag{E.4}$$

$$X_2(\omega) = \frac{kF_{10}}{(-m\omega^2 + 2k)^2 - k^2} = \frac{kF_{10}}{(-m\omega^2 + 3k)(-m\omega^2 + k)} \tag{E.5}$$

By defining $\omega_1^2 = k/m$ and $\omega_2^2 = 3k/m$, Eqs. (E.4) and (E.5) can be expressed as

$$X_1(\omega) = \frac{\left\{2 - \left(\dfrac{\omega}{\omega_1}\right)^2\right\}F_{10}}{k\left[\left(\dfrac{\omega_2}{\omega_1}\right)^2 - \left(\dfrac{\omega}{\omega_1}\right)^2\right]\left[1 - \left(\dfrac{\omega}{\omega_1}\right)^2\right]} \tag{E.6}$$

---

[2]Since $F_{10} \cos \omega t = \text{Real}(F_{10}e^{i\omega t})$, we shall assume the solution also to be $x_j = \text{Real}(X_j e^{i\omega t}) = X_j \cos \omega t, j = 1, 2$. It can be verified that X_j are real for an undamped system.

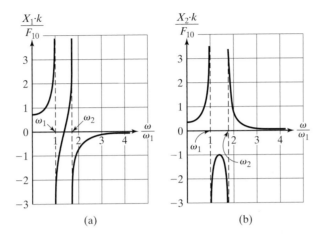

FIGURE 5.14   Frequency response curves of Example 5.8.

$$X_2(\omega) = \frac{F_{10}}{k\left[\left(\dfrac{\omega_2}{\omega_1}\right)^2 - \left(\dfrac{\omega}{\omega_1}\right)^2\right]\left[1 - \left(\dfrac{\omega}{\omega_1}\right)^2\right]} \tag{E.7}$$

The responses X_1 and X_2 are shown in Fig. 5.14 in terms of the dimensionless parameter ω/ω_1. In the dimensionless parameter, ω/ω_1, ω_1 was selected arbitrarily; ω_2 could have been selected just as easily. It can be seen that the amplitudes X_1 and X_2 become infinite when $\omega^2 = \omega_1^2$ or $\omega^2 = \omega_2^2$. Thus there are two resonance conditions for the system: one at ω_1 and another at ω_2. At all other values of ω, the amplitudes of vibration are finite. It can be noted from Fig. 5.14 that there is a particular value of the frequency ω at which the vibration of the first mass m_1, to which the force $F_1(t)$ is applied, is reduced to zero. This characteristic forms the basis of the dynamic vibration absorber discussed in Chapter 9.

∎

5.7   Semidefinite Systems

Semidefinite systems are also known as *unrestrained* or *degenerate systems*. Two examples of such systems are shown in Fig. 5.15. The arrangement in Fig. 5.15(a) may be considered to represent two railway cars of masses m_1 and m_2 with a coupling spring k. The arrangement in Fig. 5.15(b) may be considered to represent two rotors of mass moments of inertia J_1 and J_2 connected by a shaft of torsional stiffness k_t. For the system shown in Fig. 5.15(a), the equations of motion can be written as

$$m_1\ddot{x}_1 + k(x_1 - x_2) = 0$$
$$m_2\ddot{x}_2 + k(x_2 - x_1) = 0 \tag{5.36}$$

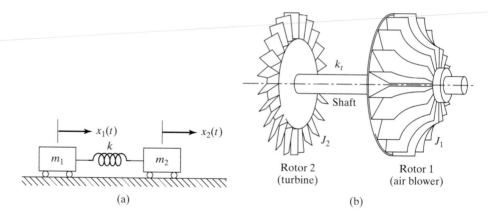

FIGURE 5.15   Semidefinite systems.

For free vibration, we assume the motion to be harmonic:

$$x_j(t) = X_j \cos(\omega t + \phi_j), \qquad j = 1, 2 \tag{5.37}$$

Substitution of Eq. (5.37) into Eq. (5.36) gives

$$(-m_1\omega^2 + k)X_1 - kX_2 = 0$$
$$-kX_1 + (-m_2\omega^2 + k)X_2 = 0 \tag{5.38}$$

By equating the determinant of the coefficients of X_1 and X_2 to zero, we obtain the frequency equation as

$$\omega^2[m_1m_2\omega^2 - k(m_1 + m_2)] = 0 \tag{5.39}$$

from which the natural frequencies can be obtained:

$$\omega_1 = 0 \quad \text{and} \quad \omega_2 = \sqrt{\frac{k(m_1 + m_2)}{m_1 m_2}} \tag{5.40}$$

It can be seen that one of the natural frequencies of the system is zero, which means that the system is not oscillating. In other words, the system moves as a whole without any relative motion between the two masses (rigid body translation). Such systems, which have one of the natural frequencies equal to zero, are called *semidefinite systems*. We can verify, by substituting ω_2 into Eq. (5.38), that $X_1^{(2)}$ and $X_2^{(2)}$ are opposite in phase. There would thus be a node at the middle of the spring.

Response Under Impulse Using Laplace Transform Method

EXAMPLE 5.9

Two railway cars, of masses $m_1 = M$ and $m_2 = m$, are connected by a spring of stiffness k, as shown in Fig. 5.15(a). If the car of mass M is subjected to an impulse $F_0 \delta(t)$, determine the time responses of the cars using the Laplace transform method.

Solution: The responses of the cars can be determined using either of the following approaches:

a. Consider the system to be undergoing free vibration due to the initial velocity caused by the impulse applied to car M.

b. Consider the system to be undergoing forced vibration due to the force $F(t) = F_0 \delta(t)$ applied to car M (with the displacements and velocities of cars M and m considered to be zero initially).

Using the second approach, the equations of motion of the cars can be expressed as

$$M\ddot{x}_1 + k x_1 - k x_2 = F_0 \delta(t) \tag{E.1}$$

$$-k x_1 + m\ddot{x}_2 + k x_2 = 0 \tag{E.2}$$

Using Laplace transforms, Eqs. (E.1) and (E.2) can be written as

$$(Ms^2 + k)\overline{x}_1(s) - k\overline{x}_2(s) = F_0 \tag{E.3}$$

$$-k\overline{x}_1(s) + (ms^2 + k)\overline{x}_2(s) = 0 \tag{E.4}$$

Equations (E.3) and (E.4) can be solved for $\overline{x}_1(s)$ and $\overline{x}_2(s)$ as

$$\overline{x}_1(s) = \frac{F_0(ms^2 + k)}{s^2\{Mms^2 + k(M + m)\}} \tag{E.5}$$

$$\overline{x}_2(s) = \frac{F_0 k}{s^2\{Mms^2 + k(M + m)\}} \tag{E.6}$$

Using partial fractions, Eqs. (E.5) and (E.6) can be rewritten as

$$\overline{x}_1(s) = \frac{F_0}{M + m}\left(\frac{1}{s^2} + \frac{m}{\omega M}\frac{\omega}{s^2 + \omega^2}\right) \tag{E.7}$$

$$\overline{x}_2(s) = \frac{F_0}{M + m}\left(\frac{1}{s^2} - \frac{1}{\omega}\frac{\omega}{s^2 + \omega^2}\right) \tag{E.8}$$

where

$$\omega^2 = k\left(\frac{1}{M} + \frac{1}{m}\right) \tag{E.9}$$

The inverse transforms of Eqs. (E.7) and (E.8), using the results of Appendix D, yield the time responses of the cars as

$$x_1(t) = \frac{F_0}{M + m}\left(t + \frac{m}{\omega M}\sin \omega t\right) \tag{E.10}$$

$$x_2(t) = \frac{F_0}{M + m}\left(t - \frac{1}{\omega}\sin \omega t\right) \tag{E.11}$$

Note: Equations (E.10) and (E.11) are plotted in Example 5.13.

■

5.8 Self-Excitation and Stability Analysis

In Section 3.11, the stability conditions of a single degree of freedom system have been expressed in terms of the physical constants of the system. The procedure is extended to a two degree of freedom system in this section. When the system is subjected to self-exciting forces, the force terms can be combined with the damping/stiffness terms, and the resulting equations of motion can be expressed in matrix notation as

$$\begin{bmatrix} m_{11} & m_{12} \\ m_{21} & m_{22} \end{bmatrix}\begin{Bmatrix} \ddot{x}_1 \\ \ddot{x}_2 \end{Bmatrix} + \begin{bmatrix} c_{11} & c_{12} \\ c_{21} & c_{22} \end{bmatrix}\begin{Bmatrix} \dot{x}_1 \\ \dot{x}_2 \end{Bmatrix}$$

$$+ \begin{bmatrix} k_{11} & k_{12} \\ k_{21} & k_{22} \end{bmatrix}\begin{Bmatrix} x_2 \\ x_2 \end{Bmatrix} = \begin{Bmatrix} 0 \\ 0 \end{Bmatrix} \tag{5.41}$$

By substituting the solution

$$x_j(t) = X_j e^{st}, \quad j = 1, 2 \tag{5.42}$$

in Eq. (5.41) and setting the determinant of the coefficient matrix to zero, we obtain the characteristic equation of the form

$$a_0 s^4 + a_1 s^3 + a_2 s^2 + a_3 s + a_4 = 0 \tag{5.43}$$

The coefficients $a_0, a_1, a_2, a_3,$ and a_4 are real numbers, since they are derived from the physical parameters of the system. If $s_1, s_2, s_3,$ and s_4 denote the roots of Eq. (5.43), we have

$$(s - s_1)(s - s_2)(s - s_3)(s - s_4) = 0$$

or

$$s^4 - (s_1 + s_2 + s_3 + s_4)s^3$$
$$+ (s_1s_2 + s_1s_3 + s_1s_4 + s_2s_3 + s_2s_4 + s_3s_4)s^2$$
$$- (s_1s_2s_3 + s_1s_2s_4 + s_1s_3s_4 + s_2s_3s_4)s + (s_1s_2s_3s_4) = 0 \qquad (5.44)$$

A comparison of Eqs. (5.43) and (5.44) yields

$$a_0 = 1$$
$$a_1 = -(s_1 + s_2 + s_3 + s_4)$$
$$a_2 = s_1s_2 + s_1s_3 + s_1s_4 + s_2s_3 + s_2s_4 + s_3s_4$$
$$a_3 = -(s_1s_2s_3 + s_1s_2s_4 + s_1s_3s_4 + s_2s_3s_4)$$
$$a_4 = s_1s_2s_3s_4 \qquad (5.45)$$

The criterion for stability is that the real parts of $s_i (i = 1, 2, 3, 4)$ must be negative to avoid increasing exponentials in Eq. (5.42). Using the properties of a quartic equation, it can be derived that a necessary and sufficient condition for stability is that all the coefficients of the equation (a_0, a_1, a_2, a_3, and a_4) be positive and that the condition

$$a_1a_2a_3 > a_0a_3^2 + a_4a_1^2 \qquad (5.46)$$

be fulfilled [5.8, 5.9]. A more general technique, which can be used to investigate the stability of an n degree of freedom system, is known as the Routh-Hurwitz criterion [5.10]. For the system under consideration, Eq. (5.43), the Routh-Hurwitz criterion states that the system will be stable if all the coefficients $a_0, a_1, \ldots, a_4$ are positive and the determinants defined below are positive:

$$T_1 = |a_1| > 0 \qquad (5.47)$$

$$T_2 = \begin{vmatrix} a_1 & a_3 \\ a_0 & a_2 \end{vmatrix} = a_1a_2 - a_0a_3 > 0 \qquad (5.48)$$

$$T_3 = \begin{vmatrix} a_1 & a_3 & 0 \\ a_0 & a_2 & a_4 \\ 0 & a_1 & a_3 \end{vmatrix} = a_1a_2a_3 - a_1^2a_4 - a_0a_3^2 > 0 \qquad (5.49)$$

Equation (5.47) simply states that the coefficient a_1 must be positive, while the satisfaction of Eq. (5.49), coupled with the satisfaction of the conditions $a_3 > 0$ and $a_4 > 0$, implies the satisfaction of Eq. (5.48). Thus the necessary and sufficient condition for the stability of the system is that all the coefficients a_0, a_1, a_2, a_3, and a_4 be positive and that the inequality stated in Eq. (5.46) be satisfied.

5.9   Examples Using MATLAB

▆▆▆▆▆▆▆▆▆▆ Solution of the Eigenvalue Problem

EXAMPLE 5.10 ──────────────────────────────────

────────────── Using MATLAB, determine the natural frequencies and mode shapes of the following problem:

$$\left[-\omega^2 m \begin{bmatrix} 1 & 0 \\ 0 & 1 \end{bmatrix} + k \begin{bmatrix} 2 & -1 \\ -1 & 2 \end{bmatrix} \right] \vec{X} = \vec{0} \qquad (E.1)$$

Solution: The eigenvalue problem, Eq. (E.1), can be rewritten as

$$\begin{bmatrix} 2 & -1 \\ -1 & 2 \end{bmatrix} \vec{X} = \lambda \begin{bmatrix} 1 & 0 \\ 0 & 1 \end{bmatrix} \vec{X} \qquad (E.2)$$

where $\lambda = m\,\omega^2/k$ is the eigenvalue, ω is the natural frequency, and $\vec{X}$ is the eigenvector or mode shape. The solution of Eq. (E.2) can be found using MATLAB as follows:

```
>> A=[2 -1; -1 2]

A =

       2      -1
      -1       2

>> [V, D] = eig(A)

V =

     -0.7071    -0.7071
      0.7071    -0.7071


D =

      3.0000          0
           0     1.0000
```

Thus the eigenvalues are $\lambda_1 = 1.0$ and $\lambda_2 = 3.0$ and the corresponding eigenvectors are

$$\vec{X}_1 = \begin{Bmatrix} -0.7071 \\ -0.7071 \end{Bmatrix} \text{ and } \vec{X}_2 = \begin{Bmatrix} -0.7071 \\ 0.7071 \end{Bmatrix}$$

■

▆▆▆▆▆▆▆▆▆▆ Roots of a Quartic Equation

EXAMPLE 5.11 ──────────────────────────────────

────────────── Using MATLAB, find the roots of the quartic equation

$$f(x) = x^4 - 8x + 12 = 0$$

Solution: The MATLAB command **roots** is used to obtain the roots of the fourth-degree polynomial as

$$x_{1,2} = -1.37091 \pm 1.82709i$$
$$x_{3,4} = 1.37091 \pm 0.648457i$$

```
>> roots ([1 0 0 -8 12])

ans =

 -1.3709 + 1.8271i
 -1.3709 - 1.8271i
  1.3709 + 0.6485i
  1.3709 - 0.6485i

>>
```

■

Plotting a Free Vibration Response

EXAMPLE 5.12

Using MATLAB, plot the free vibration response of the masses m_1 and m_2 of Example 5.3.

Solution: The time responses of the masses m_1 and m_2 are given by Eqs. (E.15) and (E.16) of Example 5.3. The MATLAB program to plot the responses is given below.

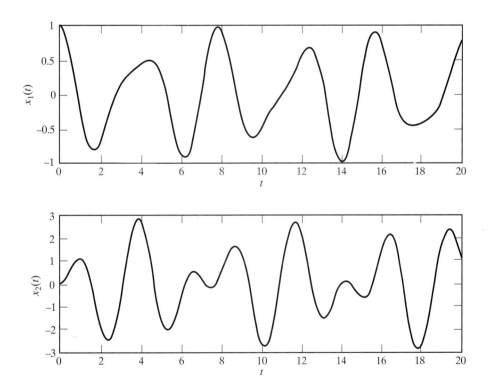

```
% E5_3.m
for i = 1: 501
    t(i) = 20 * (i-1)/500;
    x1(i) = (5/7) * cos(1.5811*t(i)) + (2/7) * cos(2.4495*t(i));
    x2(i) = (10/7) * cos(1.5811*t(i)) - (10/7) * cos(2.4495*t(i));
end
subplot(211);
plot(t, x1);
xlabel('t');
ylabel('x1(t)');
subplot(212);
plot(t, x2);
xlabel('t');
ylabel('x2(t)');
```

■

Time Response of Railway Cars

EXAMPLE 5.13

Using MATLAB plot the time responses of the two railway cars considered in Example 5.9 for the following data: $F_0 = 1500$ N, $M = 5000$ kg, $m = 2500$ kg, $k = 10^4$ N/m.

Solution: For the given data, the time responses of the railway cars can be expressed as (from Eqs. E.10 and E.11 of Example 5.9):

$$x_1(t) = 0.2(t + 0.204124 \sin 2.44949 t) \tag{E.1}$$

$$x_2(t) = 0.2(t - 0.408248 \sin 2.44949 t) \tag{E.2}$$

where

$$\omega^2 = 10^4 \left(\frac{1}{5000} + \frac{1}{2500} \right) \text{ or } \omega = 2.44949 \text{ rad/s} \tag{E.3}$$

The MATLAB program to plot Eqs. (E.1) and (E.2) is given below.

```
% Ex5_13.m
for i=1 : 101
    t(i) = 6* (i - 1) / 100;
    x1(i) = 0.2* (t(i) + 0.204124*sin(2.44949*t(i)));
    x2(i) = 0.2* (t(i) - 0.408248*sin(2.44949*t(i)));
end
plot (t, x1);
xlabel ('t');
ylabel ('x1(t), x2(t)');
hold on;
plot (t, x2, '--');
gtext ('x1: Solid line');
gtext ('x2: Dotted line');
```

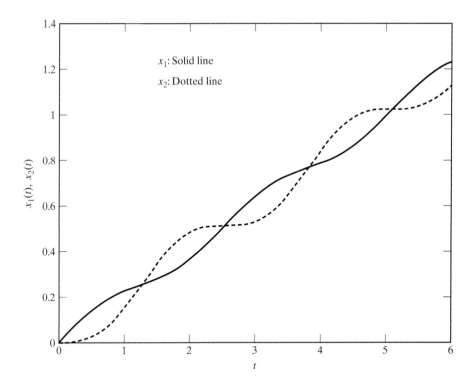

x_1: Solid line

x_2: Dotted line

■

EXAMPLE 5.14

Plotting of Frequency Response of a Two Degree of Freedom System

Using MATLAB, plot the frequency response functions of the system considered in Example 5.8

Solution: The frequency response functions $X_1(\omega)$ and $X_2(\omega)$, given by Eqs. (E.6) and (E.7) of Example 5.8, are

$$\frac{X_1(\omega)k}{F_{10}} = \frac{(2 - \lambda^2)}{(\lambda_2^2 - \lambda^2)(1 - \lambda^2)} \tag{E.1}$$

$$\frac{X_2(\omega)k}{F_{10}} = \frac{1}{(\lambda_2^2 - \lambda^2)(1 - \lambda^2)} \tag{E.2}$$

where $\lambda = \omega/\omega_1$ and $\lambda_2 = \omega_2/\omega_1$. From the results of Example 5.8, we find that $\lambda_2 = \omega_2/\omega_1 = (3k/m)/(k/m) = 3$. The MATLAB program to plot Eqs. (E.1) and (E.2) is given below.

```
% Ex5_14.m
for i = 1: 101
    w_w1 (i) = 5 * (i - 1) / 100; % 0 to 5
```

```
        x1 (i) = (2-w_w1 (i) ^2) / ( (3-w_w1 (i) ^2) * (1-w_w1 (i) ^2) );
        x2 (i) = 1 / ( (3-w_w1 (i) ^2) * (1-w_w1 (i) ^2) );
end
subplot (211);
plot (w_w1, x1);
xlabel ('w/w_1');
ylabel ('X_1*K/F_1_0');
grid on;
subplot (212);
plot (w_w1, x2);
xlabel ('w/w_1');
ylabel ('X_2*K/F_1_0');
grid on
```

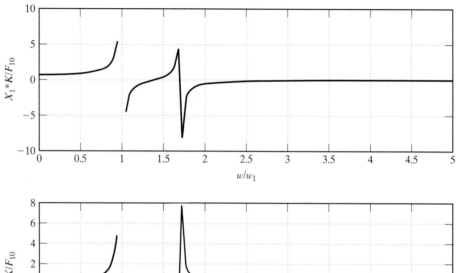

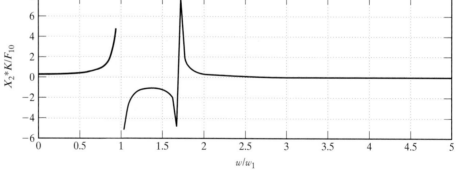

Forced Response of a Two Degree of Freedom System

EXAMPLE 5.15

Determine and plot the time response of a two degree of freedom system with equations of motion

$$\begin{bmatrix} 1 & 0 \\ 0 & 2 \end{bmatrix} \begin{Bmatrix} \ddot{x}_1 \\ \ddot{x}_2 \end{Bmatrix} + \begin{bmatrix} 4 & -1 \\ -1 & 2 \end{bmatrix} \begin{Bmatrix} \dot{x}_1 \\ \dot{x}_2 \end{Bmatrix} + \begin{bmatrix} 5 & -2 \\ -2 & 3 \end{bmatrix} \begin{Bmatrix} x_1 \\ x_2 \end{Bmatrix} = \begin{Bmatrix} 1 \\ 2 \end{Bmatrix} \cos 3t \qquad (E.1)$$

with the initial conditions

$$x_1(0) = 0.2, \; \dot{x}_1(0) = 1.0, \; x_2(0) = 0, \; \dot{x}_2(0) = 0 \tag{E.2}$$

Solution: In order to use the MATLAB program **ode23**, the two coupled second-order differential equations, Eq. (E.1), are to be expressed as a system of coupled first-order differential equations. For this, we introduce new variables y_1, y_2, y_3, and y_4 as

$$y_1 = x_1, \qquad y_2 = \dot{x}_1, \qquad y_3 = x_2, \qquad y_4 = \dot{x}_2$$

and express Eq. (E.1) as

$$\ddot{x}_1 + 4\dot{x}_1 - \dot{x}_2 + 5x_1 - 2x_2 = \cos 3t \tag{E.3}$$

or

$$\dot{y}_2 = \cos 3t - 4y_2 + y_4 - 5y_1 + 2y_3 \tag{E.4}$$

and

$$2\ddot{x}_2 - \dot{x}_1 + 2\dot{x}_2 - 2x_1 + 3x_2 = 2\cos 3t \tag{E.5}$$

or

$$\dot{y}_4 = \cos 3t + \frac{1}{2}y_2 - y_4 + y_1 - \frac{3}{2}y_3 \tag{E.6}$$

Thus Eq. (E.1) can be restated as

$$\begin{Bmatrix} \dot{y}_1 \\ \dot{y}_2 \\ \dot{y}_3 \\ \dot{y}_4 \end{Bmatrix} = \begin{Bmatrix} y_2 \\ \cos 3t - 4y_2 + y_4 - 5y_1 + 2y_3 \\ y_4 \\ \cos 3t + \frac{1}{2}y_2 - y_4 + y_1 - \frac{3}{2}y_3 \end{Bmatrix} \tag{E.7}$$

with the initial conditions

$$\vec{y}(0) = \begin{Bmatrix} y_1(0) \\ y_2(0) \\ y_3(0) \\ y_4(0) \end{Bmatrix} = \begin{Bmatrix} 0.2 \\ 1.0 \\ 0.0 \\ 0.0 \end{Bmatrix} \tag{E.8}$$

The MATLAB program to solve Eqs. (E.7) with the initial conditions of Eq. (E.8) is given below.

```
% Ex5_15.m
tspan = [0: 0.01: 20];
y0 = [0.2; 1.0; 0.0; 0.0];
[t,y] = ode23('dfunc5_15', tspan, y0);
subplot (211)
plot (t,y (:, 1));
xlabel ('t');
ylabel ('x1 (t)');
subplot (212)
plot (t,y (:, 3));
xlabel ('t');
ylabel ('x2 (t)');

%dfunc5_15.m
function f = dfunc5_15(t,y)
f = zeros(4, 1);
f(1) = y(2);
f(2) = cos(3*t) − 4*y(2) + y(4) − 5*y(1) + 2*y(3);
f(3) = y(4);
f(4) = cos(3*t) + 0.5*y(2) − y(4) + y(1) − 1.5*y(3);
```

Program to Find the Roots of a Quartic Equation

EXAMPLE 5.16

Develop a general program, called **Program6.m**, to find the roots of a quartic equation. Use the program to find the roots of the equation

$$f(x) = x^4 - 8x + 12 = 0$$

Solution: **Program6.m** is developed to solve the equation a1*(x^4) + a2*(x^3) + a3*(x^2) + a4*x + a5 = 0 with a1, a2, a3, a4, and a5 as input data. The program gives the polynomial coefficients as well as the roots of the equation as output.

```
>> program6

Solution of a quartic equation

Data:
a(1)  =    1.000000e+000
a(2)  =    0.000000e+000
a(3)  =    0.000000e+000
a(4)  =   -8.000000e+000
a(5)  =    1.200000e+001

  Roots:

  Root No.      Real part        Imaginary part

     1        -1.370907e+000      1.827094e+000
     2        -1.370907e+000     -1.827094e+000
     3         1.370907e+000      6.484572e-001
     4         1.370907e+000     -6.484572e-001
```

■

5.10   C++ Program

An interactive C++ program, called **Program6.cpp**, is given for finding the roots of a quartic equation. The program requires the coefficients of the polynomial equation as input and gives the roots of the equation as output.

Roots of a Quartic Equation

EXAMPLE 5.17

Using **Program6.cpp**, find the roots of the equation $f(x) = x^4 - 8x + 12 = 0$

Solution: The input data are to be entered interactively. The input and output of the program are given below.

```
Please input the coefficient array a (a[0] is the coefficient of x^4
and nonzero): 1.0 0.0 0.0 -8.0 12.0

SOLUTION OF QUARTIC EQUATION

DATA:
A[0]  =   1.000000
A[1]  =   0.000000
A[2]  =   0.000000
```

```
A[3]  =  -8.000000
A[4]  =  12.000000
```

```
ROOTS:
```

ROOT NO.	REAL PART	IMAGINARY PART
1	-1.37090672	1.82709433
2	-1.37090672	-1.82709433
3	1.37090672	0.64845723
4	1.37090672	-0.64845723

■

5.11   Fortran Program

A Fortran subroutine, called **QUART.F**, is given for the solution of a quartic equation. This subroutine requires the coefficients of the polynomial as input and gives the roots of the equation as output.

■ Roots of a Quartic Equation

EXAMPLE 5.18

Using **QUART.F**, find the roots of the equation $x^4 - 8x + 12 = 0$

Solution: The main program that calls the subroutine **QUART.F** and the subroutine **QUART.F** are given as **PROGRAM6.F**. The output of the program is given below.

```
SOLUTION OF A QUARTIC EQUATION

DATA:
A(1)  =   0.100000E+01
A(2)  =   0.000000E+00
A(3)  =   0.000000E+00
A(4)  =  -0.800000E+01
A(5)  =   0.120000E+02

ROOTS:
```

ROOT NO.	REAL PART	IMAGINARY PART
1	-0.137091E+01	0.182709E+01
2	-0.137091E+01	-0.182709E+01
3	0.137091E+01	0.648457E+00
4	0.137091E+01	-0.648457E+00

■

REFERENCES

5.1   H. Sato, Y. Kuroda, and M. Sagara, "Development of the finite element method for vibration analysis of machine tool structure and its application," *Proceedings of the Fourteenth International Machine Tool Design and Research Conference*, Macmillan, London, 1974, pp. 545–552.

5.2 F. Koenigsberger and J. Tlusty, *Machine Tool Structures*, Pergamon Press, Oxford, 1970.

5.3 C. P. Reddy and S. S. Rao, "Automated optimum design of machine tool structures for static rigidity, natural frequencies and regenerative chatter stability," *Journal of Engineering for Industry*, Vol. 100, 1978, pp. 137–146.

5.4 M. S. Hundal, "Effect of damping on impact response of a two degree of freedom system," *Journal of Sound and Vibration*, Vol. 68, 1980, pp. 407–412.

5.5 J. A. Linnett, "The effect of rotation on the steady-state response of a spring-mass system under harmonic excitation," *Journal of Sound and Vibration*, Vol. 35, 1974, pp. 1–11.

5.6 A. Hurwitz, "On the conditions under which an equation has only roots with negative real parts," in *Selected Papers on Mathematical Trends in Control Theory*, Dover Publications, New York, 1964, pp. 70–82.

5.7 R. C. Dorf, *Modern Control Systems* (6th ed.), Addison-Wesley, Reading, Mass., 1992.

5.8 J. P. Den Hartog, *Mechanical Vibrations* (4th ed.), McGraw-Hill, New York, 1956.

5.9 R. H. Scanlan and R. Rosenbaum, *Introduction to the Study of Aircraft Vibration and Flutter*, Macmillan, New York, 1951.

5.10 L. A. Pipes and L. R. Harvill, *Applied Mathematics for Engineers and Physicists* (3rd ed.), McGraw-Hill, New York, 1970.

REVIEW QUESTIONS

5.1 Give brief answers to the following:

1. How do you determine the number of degrees of freedom of a lumped-mass system?
2. Define these terms: *mass coupling, velocity coupling, elastic coupling.*
3. Is the nature of the coupling dependent on the coordinates used?
4. How many degrees of freedom does an airplane in flight have if it is treated as (a) a rigid body, and (b) an elastic body?
5. What are principal coordinates? What is their use?
6. Why are the mass, damping, and stiffness matrices symmetrical?
7. What is a node?
8. What is meant by static and dynamic coupling? How can you eliminate coupling of the equations of motion?
9. Define the impedance matrix.
10. How can we make a system vibrate in one of its natural modes?
11. What is a degenerate system? Give two examples of physical systems that are degenerate.
12. How many degenerate modes can a vibrating system have?

5.2 Indicate whether each of the following statements is true or false:

1. The normal modes can also be called principal modes.
2. The generalized coordinates are linearly dependent.
3. Principal coordinates can be considered as generalized coordinates.
4. The vibration of a system depends on the coordinate system.
5. The nature of coupling depends in the coordinate system.
6. The principal coordinates avoid both static and dynamic coupling.
7. The use of principal coordinates helps in finding the response of the system.

8. The mass, stiffness, and damping matrices of a two degree of freedom system are symmetric.
9. The characteristics of a two degree of freedom system are used in the design of dynamic vibration absorber.
10. Semidefinite systems are also known as degenerate systems.
11. A semidefinite system cannot have nonzero natural frequencies.
12. The generalized coordinates are always measured form the equilibrium position of the body.
13. During free vibration, different degrees of freedom oscillate with different phase angles.
14. During free vibration, different degrees of freedom oscillate at different frequencies.
15. During free vibration, different degrees of freedom oscillate with different amplitudes.
16. The relative amplitudes of different degrees of freedom in a two degree of freedom system depend on the natural frequency.
17. The modal vectors of a system denote the normal modes of vibration.

5.3. Fill in each of the following blanks with the appropriate word:

1. The free vibration of a two degree of freedom system under arbitrary initial excitation can be found by superposing the two _____ modes of vibration.
2. The motion of a two degree of freedom system is described by two _____ coordinates.
3. When the forcing frequency is equal to one of the natural frequencies of the system, a phenomenon known as _____ occurs.
4. The amplitudes and phase angles are determined from the _____ conditions of the system.
5. For a torsional system, _____ and _____ are analogous to the masses and linear springs, respectively, of a mass-spring system.
6. The use of different generalized coordinates lead to different types of _____.
7. A semidefinite system has at least one _____ body motion.
8. The elastic coupling is also known as _____ coupling.
9. The inertia coupling is also known as _____ coupling.
10. The damping coupling is also known as _____ coupling.
11. The equations of motion of a system will be _____ when principal coordinates are used.
12. The Routh-Hurwitz criterion can be used to investigate the _____ of a system.
13. The equations of motion of a two degree of freedom system are uncoupled only when the two masses are not _____ connected.
14. The vibration of a system under initial conditions only is called _____ vibration.
15. The vibration of a system under external forces is called _____ vibration.

5.4 Select the most appropriate answer out of the choices given:

1. When a two degree of freedom system is subjected to a harmonic force, the system vibrates at the
   (a) frequency of applied force
   (b) smaller natural frequency
   (c) larger natural frequency

2. The number of degrees of freedom of a vibrating system depends on
   (a) number of masses
   (b) number of masses and the degrees of freedom of each of the masses
   (c) number of coordinates used to describe the position of each mass

3. A two degree of freedom system has
   (a) one normal mode
   (b) two normal modes
   (c) many normal modes

4. The equations of motion of a two degree of freedom system are in general
   (a) coupled
   (b) uncoupled
   (c) linear

5. Mechanical impedance $Z_{rs}(i\omega)$ is

   (a) $[m_{rs}]\ddot{\vec{x}} + [c_{rs}]\dot{\vec{x}} + [k_{rs}]\vec{x}$

   (b) $\begin{Bmatrix} X_r(i\omega) \\ X_s(i\omega) \end{Bmatrix}$

   (c) $-\omega^2 m_{rs} + i\omega c_{rs} + k_{rs}$

6. The impedance matrix, $[Z(i\omega)]$, can be used to find the solution as

   (a) $\vec{X} = [Z(i\omega)]^{-1}\vec{F}_0$

   (b) $\vec{X} = [Z(i\omega)]\vec{F}_0$

   (c) $\vec{X} = [Z(i\omega)]\vec{X}_0$

7. The configuration of a system vibrating at one of its natural frequencies is called
   (a) natural mode   (b) natural frequency   (c) solution

8. The equations of motion of a two degree of freedom system are in general in the form of
   (a) coupled algebraic equations
   (b) coupled differential equations
   (c) uncoupled equations

5.5 Match the items in the two columns below:

1. Static coupling   (a) Only the mass matrix is nondiagonal
2. Inertial coupling   (b) The mass and damping matrices are nondiagonal
3. Velocity coupling   (c) Only the stiffness matrix is nondiagonal
4. Dynamic coupling   (d) Only the damping matrix is nondiagonal

5.6 Match the data given in the left column with the frequency equations given in the right column for a two degree of freedom system governed by the equations of motion:

$$J_0 \ddot{\theta}_1 - 2k_t\,\theta_1 - k_t\,\theta_2 = 0$$
$$2J_0\ddot{\theta}_2 - k_t\,\theta_1 + k_t\,\theta_2 = 0$$

1. $J_0 = 1, k_t = 2$ (a) $32\omega^4 - 20\omega^2 + 1 = 0$
2. $J_0 = 2, k_t = 1$ (b) $\omega^4 - 5\omega^2 + 2 = 0$

3. $J_0 = 2, k_t = 2$ (c) $\omega^4 - 10\omega^2 + 8 = 0$

4. $J_0 = 1, k_t = 4$ (d) $8\omega^4 - 10\omega^2 + 1 = 0$

5. $J_0 = 4, k_t = 1$ (e) $2\omega^4 - 5\omega^2 + 1 = 0$

PROBLEMS

The problem assignments are organized as follows:

Problems	Section Covered	Topic Covered
5.1–5.27	5.3	Free vibration of undamped systems
5.28–5.31	5.4	Torsional systems
5.32–5.43	5.5	Coordinate coupling
5.44–5.56	5.6	Forced vibrations
5.57–5.64	5.7	Semidefinite systems
5.65–6.66	5.8	Stability analysis
5.67–5.69	—	Computer programs
5.70–5.77	5.9	Examples using MATLAB
5.78–5.79	5.10	C++ programs
5.80–5.81	5.11	Fortran programs
5.82–5.83	—	Design projects

5.1 Find the natural frequencies of the system shown in Fig. 5.16, with $m_1 = m$, $m_2 = 2m$, $k_1 = k$, and $k_2 = 2k$. Determine the response of the system when $k = 1000$ N/m, $m = 20$ kg, and the initial values of the displacements of the masses m_1 and m_2 are 1 and -1, respectively.

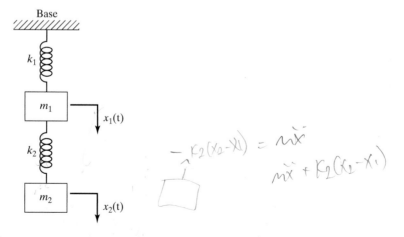

FIGURE 5.16

5.2   Set up the differential equations of motion for the double pendulum shown in Fig. 5.17, using the coordinates x_1 and x_2 and assuming small amplitudes. Find the natural frequencies, the ratios of amplitudes, and the locations of nodes for the two modes of vibration when $m_1 = m_2 = m$ and $l_1 = l_2 = l$.

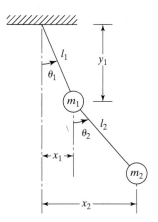

FIGURE 5.17

5.3   Determine the natural modes of the system shown in Fig. 5.18 when $k_1 = k_2 = k_3 = k$.

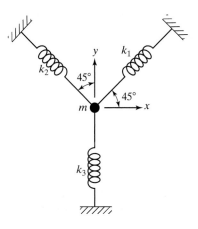

FIGURE 5.18

5.4   A machine tool, having a mass of $m = 1000$ kg and a mass moment of inertia of $J_0 = 300$ kg-m^2, is supported on elastic supports, as shown in Fig. 5.19. If the stiffnesses of the supports are given by $k_1 = 3000$ N/mm, and $k_2 = 2000$ N/mm, and the supports are located at $l_1 = 0.5$ m and $l_2 = 0.8$ m, find the natural frequencies and mode shapes of the machine tool.

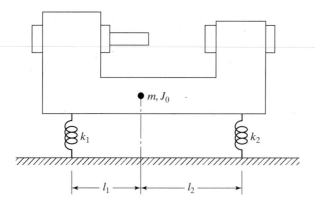

FIGURE 5.19

5.5 An electric overhead traveling crane, consisting of a girder, trolley, and wire rope, is shown in Fig. 5.20. The girder has a flexural rigidity (EI) of 6×10^{12} lb-in.^{2} and a span (L) of 30 ft. The rope is made of steel and has a length (l) of 20 ft. The weights of the trolley and the load lifted are 8000 lb and 2000 lb, respectively. Find the area of cross section of the rope such that the fundamental natural frequency is greater than 20 Hz.

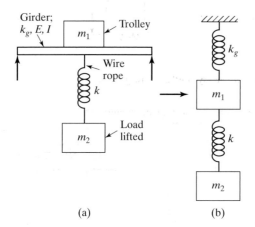

(a)   (b)

FIGURE 5.20

5.6 An overhead traveling crane can be modeled as indicated in Fig. 5.20. Assuming that the girder has a span of 40 m, an area moment of inertia (I) of 0.02 m^4, and a modulus of elasticity (E) of 2.06×10^{11} N/m^2, the trolley has a mass (m_1) of 1000 kg, the load being lifted has a mass (m_2) of 5000 kg, and the cable through which the mass (m_2) is lifted has a stiffness (k) of 3.0×10^5 N/m, determine the natural frequencies and mode shapes of the system.

5.7 The drilling machine shown in Fig. 5.21(a) can be modeled as a two degree of freedom system as indicated in Fig. 5.21(b). Since a transverse force applied to mass m_1 or mass m_2

causes both the masses to deflect, the system exhibits elastic coupling. The bending stiffnesses of the column are given by (see Section 6.4 for the definition of stiffness influence coefficients)

$$k_{11} = \frac{768}{7}\frac{EI}{l^3}, \qquad k_{12} = k_{21} = -\frac{240}{7}\frac{EI}{l^3}, \qquad k_{22} = \frac{96}{7}\frac{EI}{l^3}$$

Determine the natural frequencies of the drilling machine.

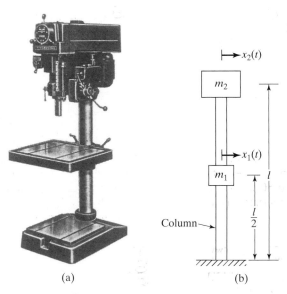

(a)   (b)

FIGURE 5.21   (Photo courtesy of Atlas-Clausing division.)

5.8   One of the wheels and leaf springs of an automobile, traveling over a rough road, is shown in Fig. 5.22. For simplicity, all the wheels can be assumed to be identical and the system can be idealized as shown in Fig. 5.23. The automobile has a mass of $m_1 = 1000$ kg and the leaf springs have a total stiffness of $k_1 = 400$ kN/m. The wheels and axles have a mass of $m_2 = 300$ kg and the tires have a stiffness of $k_2 = 500$ kN/m. If the road surface varies

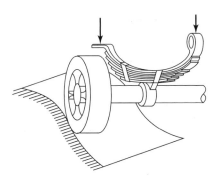

FIGURE 5.22

sinusoidally with an amplitude of $Y = 0.1$ m and a period of $l = 6$ m, find the critical velocities of the automobile.

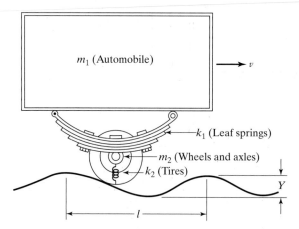

FIGURE 5.23

5.9 Derive the equations of motion of the double pendulum shown in Fig. 5.17, using the coordinates θ_1 and θ_2. Also find the natural frequencies and mode shapes of the system for
$m_1 = m_2 = m$ and $l_1 = l_2 = l$.

5.10 Find the natural frequencies and mode shapes of the system shown in Fig. 5.16 for
$m_1 = m_2 = m$ and $k_1 = k_2 = k$.

5.11 The normal modes of a two degree of freedom system are orthogonal if $X^{(1)^T}[m]\,\overrightarrow{X}^{(2)} = 0$.
Prove that the mode shapes of the system shown in Fig. 5.3(a) are orthogonal.

5.12 Find the natural frequencies of the system shown in Fig. 5.4 for $k_1 = 300$ N/m,
$k_2 = 500$ N/m, $k_3 = 200$ N/m, $m_1 = 2$ kg, and $m_2 = 1$ kg.

5.13 Find the natural frequencies and mode shapes of the system shown in Fig. 5.16 for
$m_1 = m_2 = 1$ kg, $k_1 = 2000$ N/m, and $k_2 = 6000$ N/m.

5.14 Derive expressions for the displacements of the masses in Fig. 5.4 when $m_i = 25$ lb-sec^2/in.,
$i = 1, 2$, and $k_i = 50{,}000$ lb/in., $i = 1, 2, 3$.

5.15 For the system shown in Fig. 5.4, $m_1 = 1$ kg, $m_2 = 2$ kg, $k_1 = 2000$ N/m, $k_2 = 1000$ N/m,
$k_3 = 3000$ N/m, and an initial velocity of 20 m/s is imparted to mass m_1. Find the resulting
motion of the two masses.

5.16 For Problem 5.13, calculate $x_1(t)$ and $x_2(t)$ for the following initial conditions:

    a. $x_1(0) = 0.2$, $\dot{x}_1(0) = x_2(0) = \dot{x}_2(0) = 0$
    b. $x_1(0) = 0.2$, $\dot{x}_1(0) = x_2(0) = 0$, $\dot{x}_2(0) = 5.0$.

5.17 A two-story building frame is modeled as shown in Fig. 5.24. The girders are assumed to
be rigid, and the columns have flexural rigidities EI_1 and EI_2, with negligible masses. The
stiffness of each column can be computed as

$$\frac{24EI_i}{h_i^3}, \qquad i = 1, 2$$

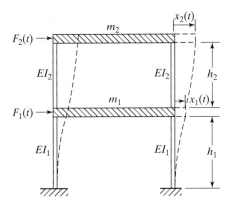

FIGURE 5.24

For $m_1 = 2m$, $m_2 = m$, $h_1 = h_2 = h$, and $EI_1 = EI_2 = EI$, determine the natural frequencies and mode shapes of the frame.

5.18 Figure 5.25 shows a system of two masses attached to a tightly stretched string, fixed at both ends. Determine the natural frequencies and mode shapes of the system for $m_1 = m_2 = m$ and $l_1 = l_2 = l_3 = l$.

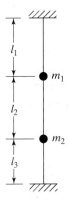

FIGURE 5.25

5.19 Find the normal modes of the two-story building shown in Fig. 5.24 when $m_1 = 3m$, $m_2 = m$, $k_1 = 3k$, and $k_2 = k$, where k_1 and k_2 represent the total equivalent stiffnesses of the lower and upper columns, respectively.

5.20 A hoisting drum, having a weight W_1, is mounted at the end of a steel cantilever beam of thickness t, width a, and length b, as shown in Fig. 5.26. The wire rope is made of steel and has a diameter of d and a suspended length of l. If the load hanging at the end of the rope is W_2, derive expressions for the natural frequencies of the system.

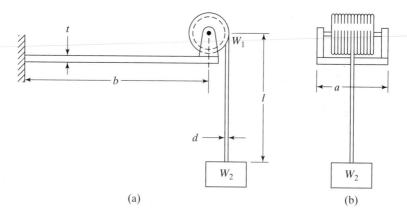

(a)   (b)

FIGURE 5.26

5.21 Determine the initial conditions of the system shown in Fig. 5.16 for which the system vibrates only at its lowest natural frequency for the following data: $k_1 = k$, $k_2 = 2k$, $m_1 = m$, $m_2 = 2m$.

5.22 The system shown in Fig. 5.16 is initially disturbed by holding the mass m_1 stationary and giving the mass m_2 a downward displacement of 0.1 m. Discuss the nature of the resulting motion of the system.

5.23 Design the cantilever beam supporting the hoisting drum and the wire rope carrying the load in Problem 5.20 in order to have the natural frequencies of the system greater than 10 Hz when $W_1 = 1000$ lb and $W_2 = 500$ lb, $b = 30$ in., and $l = 60$ in.

5.24 Find the free vibration response of the two degree of freedom system shown in Fig. 5.4 with $n = 1, k = 8$, and $m = 2$ for the initial conditions $x_1(0) = 1$, $x_2(0) = \dot{x}_1(0) = 0$, and $\dot{x}_2(0) = 1$.

5.25 Find the free vibration response of the two degree of freedom system shown in Fig. 5.4 with $n = 1, k = 8$, and $m = 2$ for the initial conditions $x_1(0) = 1$ and $x_2(0) = \dot{x}_1(0) = \dot{x}_2(0) = 0$.

5.26 Using the results of Example 5.1, verify that the mode shapes satisfy the following relations, known as orthogonality relations:

$$\vec{X}^{(1)^T}\vec{X}^{(2)} = 0, \ \vec{X}^{(1)^T}[m]\vec{X}^{(2)} = 0, \ \vec{X}^{(1)^T}[m]\vec{X}^{(1)} = c_1 = \text{constant},$$

$$\vec{X}^{(2)^T}[m]\vec{X}^{(2)} = c_2 = \text{constant}$$

$$\vec{X}^{(1)^T}[k]\vec{X}^{(1)} = c_1\omega_1^2, \ \vec{X}^{(2)^T}[k]\vec{X}^{(2)} = c_2\omega_2^2$$

5.27 Two identical pendulums, each with mass m and length l, are connected by a spring of stiffness k at a distance d from the fixed end, as shown in Fig. 5.27.

    a. Derive the equations of motion of the two masses.

    b. Find the natural frequencies and mode shapes of the system.

    c. Find the free vibration response of the system for the initial conditions $\theta_1(0) = a$, $\theta_2(0) = 0$, $\dot{\theta}_1(0) = 0$, and $\dot{\theta}_2(0) = 0$.

    d. Determine the condition(s) under which the system exhibits a beating phenomenon.

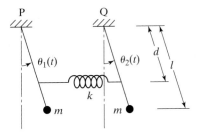

FIGURE 5.27

5.28 Determine the natural frequencies and normal modes of the torsional system shown in Fig. 5.28 for $k_{t2} = 2k_{t1}$ and $J_2 = 2J_1$.

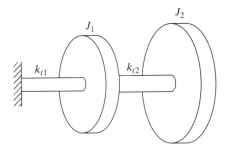

FIGURE 5.28

5.29 Determine the natural frequencies of the system shown in Fig. 5.29 by assuming that the rope passing over the cylinder does not slip.

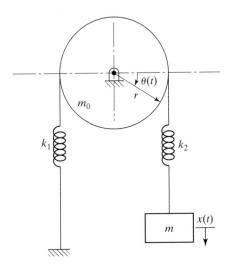

FIGURE 5.29

5.30 Find the natural frequencies and mode shapes of the system shown in Fig. 5.6(a) by assuming that $J_1 = J_0$, $J_2 = 2J_0$, and $k_{t1} = k_{t2} = k_{t3} = k_t$.

5.31 Determine the normal modes of the torsional system shown in Fig. 5.7 when $k_{t1} = k_t$, $k_{t2} = 5k_t$, $J_1 = J_0$, and $J_2 = 5J_0$.

5.32 A simplified ride model of the military vehicle in Fig. 5.30(a) is shown in Fig. 5.30(b). This model can be used to obtain information about the bounce and pitch modes of the vehicle. If the total mass of the vehicle is m and the mass moment of inertia about its C.G. is J_0, derive the equations of motion of the vehicle using two different sets of coordinates, as indicated in Section 5.5.

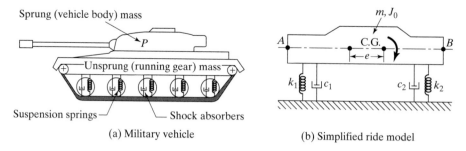

(a) Military vehicle

(b) Simplified ride model

FIGURE 5.30

5.33 Find the natural frequencies and the amplitude ratios of the system shown in Fig. 5.31.

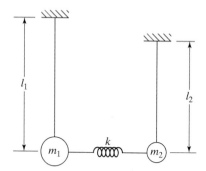

FIGURE 5.31

5.34 A rigid rod of negligible mass and length $2l$ is pivoted at the middle point and is constrained to move in the vertical plane by springs and masses, as shown in Fig. 5.32. Find the natural frequencies and mode shapes of the system.

5.35 An airfoil of mass m is suspended by a linear spring of stiffness k and a torsional spring of stiffness k_t in a wind tunnel, as shown in Fig. 5.33. The C.G. is located at a distance of e from point O. The mass moment of inertia of the airfoil about an axis passing through point O is J_0. Find the natural frequencies of the airfoil.

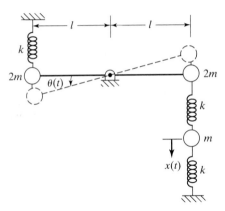

FIGURE 5.32

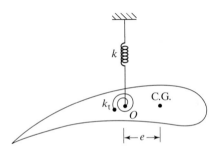

FIGURE 5.33

5.36 The expansion joints of a concrete highway, which are located at 15 m intervals, cause a series of impulses to affect cars running at a constant speed. Determine the speeds at which bounce motion and pitch motion are most likely to arise for the automobile of Example 5.7.

5.37 Consider the overhead traveling crane described in Problem 5.5 (Fig. 5.20). If the rails on both sides of the girder have a sinusoidally varying surface in the z-direction (perpendicular to the page), as shown in Fig. 5.34, set up the equations and the initial conditions for finding the vibration response of the load lifted (m) in the vertical direction. Assume that the velocity of the crane is 30 ft/min in the z-direction.

FIGURE 5.34

5.38 An automobile is modeled with a capability of pitch and bounce motions, as shown in Fig. 5.35. It travels on a rough road whose surface varies sinusoidally with an amplitude of 0.05 m and a wavelength of 10 m. Derive the equations of motion of the automobile for the following data: Mass = 1,000 kg, radius of gyration = 0.9 m, l_1 = 1.0 m, l_2 = 1.5 m, k_f = 18 kN/m, k_r = 22 kN/m, velocity = 50 km/hr.

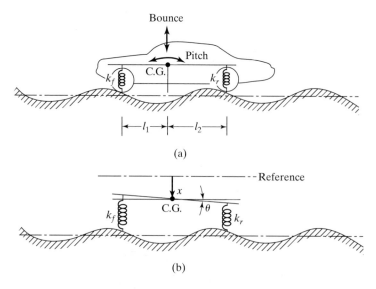

(a)

(b)

FIGURE 5.35

5.39 A steel shaft, of diameter 2 in., is supported on two bearings and carries a pulley and a motor, as shown in Fig. 5.36. The weights of the pulley and the motor are 200 lb and 500 lb, respectively. A transverse load applied at any point along the length of the shaft results in the deflection of all

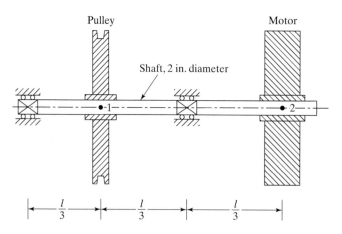

FIGURE 5.36

points on the shaft and hence the system exhibits elastic coupling. The stiffness coefficients are given by (see Section 6.4 for the definition of stiffness influence coefficients)

$$k_{11} = \frac{1296}{5} \frac{EI}{l^3}, \qquad k_{12} = k_{21} = \frac{324}{5} \frac{EI}{l^3}, \qquad k_{22} = \frac{216}{5} \frac{EI}{l^3}$$

Determine the natural frequencies of the system in bending vibration for $l = 90$ inches.

5.40 A simplified model of a mountain bike with a rider is shown in Fig. 5.37. Discuss methods of finding the vibratory response of the bicycle due to the unevenness of the terrain using a two degree of freedom model.

FIGURE 5.37

5.41 A uniform rigid bar of length l and mass m is supported on two springs and is subjected to a force $F(t) = F_0 \sin \omega t$, as shown in Fig. 5.38. (a) Derive the equations of motion of the bar for small displacements. (b) Discuss the nature of coupling in the system.

5.42 A trailer of mass M, connected to a wall through a spring of stiffness k and a damper of damping coefficient c, slides on a frictionless surface, as shown in Fig. 5.39. A uniform rigid bar, pin-connected to the trailer, can oscillate about the hinge point, O. Derive the equations of motion of the system under the applied forces $F(t)$ and $M_t(t)$ indicated in Fig. 5.39.

5.43 A trailer of mass M is connected to a wall through a spring of stiffness k_1 and can move on a frictionless horizontal surface, as shown in Fig. 5.40. A uniform cylinder of mass m, connected to the wall of the trailer by a spring of stiffness k_2, can roll on the floor of the trailer without slipping. Derive the equations of motion of the system and discuss the nature of coupling present in the system.

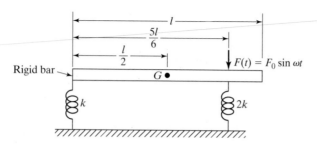

FIGURE 5.38

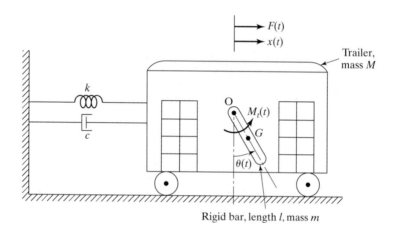

FIGURE 5.39

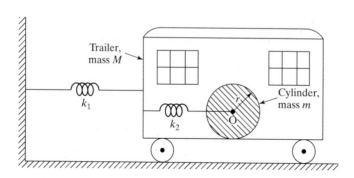

FIGURE 5.40

5.44 The weights of the tup, frame, anvil (along with the workpiece), and the foundation block in a forging hammer (Fig. 5.41) are 5000 lb, 40,000 lb, 60,000 lb and 140,000 lb, respectively. The stiffnesses of the elastic pad placed between the anvil and the foundation block and the isolation placed underneath the foundation (including the elasticity of the soil) are 6×10^6 lb/in. and 3×10^6 lb/in., respectively. If the velocity of the tup before it strikes the anvil is 15 ft/sec, find (a) the natural frequencies of the system, and (b) the magnitudes of displacement of the anvil and the foundation block. Assume the coefficient of restitution as 0.5 and damping to be negligible in the system.

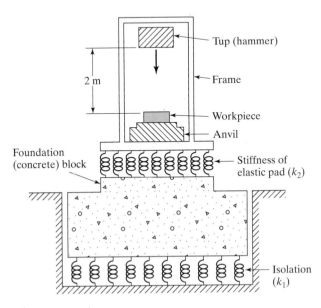

FIGURE 5.41

5.45 Find (a) the natural frequencies of the system, and (b) the responses of the anvil and the foundation block of the forging hammer shown in Fig. 5.41 when the time history of the force applied to the anvil is as shown in Fig. 5.42. Assume the following data:

Mass of anvil and frame = 200 Mg
Mass of foundation block = 250 Mg
Stiffness of the elastic pad = 150 MN/m
Stiffness of the soil = 75 MN/m
$F_0 = 10^5$ N and $T = 0.5$ s

5.46 Derive the equations of motion for the free vibration of the system shown in Fig. 5.43. Assuming the solution as $x_i(t) = C_i e^{st}$, $i = 1, 2$, express the characteristic equation in the form

$$a_0 s^4 + a_1 s^3 + a_2 s^2 + a_3 s + a_4 = 0$$

Discuss the nature of possible solutions, $x_1(t)$ and $x_2(t)$.

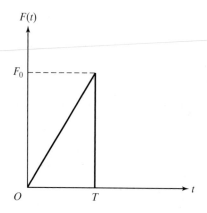

FIGURE 5.42

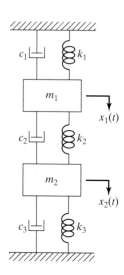

FIGURE 5.43

5.47 Find the displacements $x_1(t)$ and $x_2(t)$ in Fig. 5.43 for $m_1 = 1$ kg, $m_2 = 2$ kg, $k_1 = k_2 = k_3 = 10{,}000$ N/m, and $c_1 = c_2 = c_3 = 2000$ N-s/m using the initial conditions $x_1(0) = 0.2$ m, $x_2(0) = 0.1$ m, and $\dot{x}_1(0) = \dot{x}_2(0) = 0$.

5.48 A centrifugal pump, having an unbalance of me, is supported on a rigid foundation of mass m_2 through isolator springs of stiffness k_1, as shown in Fig. 5.44. If the soil stiffness and damping are k_2 and c_2, find the displacements of the pump and the foundation for the following data: $mg = 0.5$ lb, $e = 6$ in., $m_1 g = 800$ lb, $k_1 = 2000$ lb/in., $m_2 g = 2000$ lb, $k_2 = 1000$ lb/in., $c_2 = 200$ lb-sec/in., and speed of pump $= 1200$ rpm.

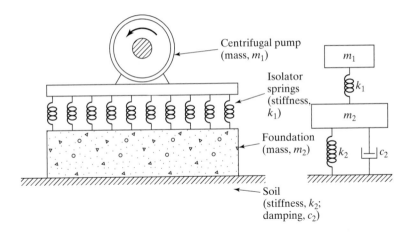

FIGURE 5.44

5.49 A reciprocating engine of mass m_1 is mounted on a fixed-fixed beam of length l, width a, thickness t, and Young's modulus E, as shown in Fig. 5.45. A spring-mass system (k_2, m_2) is suspended from the beam as indicated in the figure. Find the relation between m_2 and k_2 that leads to no steady-state vibration of the beam when a harmonic force, $F_1(t) = F_0 \cos \omega t$, is developed in the engine during its operation.[3]

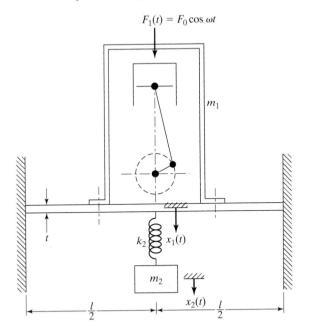

FIGURE 5.45

[3]The spring-mass system (k_2, m_2) added to make the amplitude of the first mass zero is known as a "vibration absorber." A detailed discussion of vibration absorbers is given in Section 9.11.

5.50 Find the steady-state response of the system shown in Fig. 5.16 by using the mechanical impedance method, when the mass m_1 is subjected to the force $F(t) = F_0 \sin \omega t$ in the direction of $x_1(t)$.

5.51 Find the steady-state response of the system shown in Fig. 5.16 when the base is subjected to a displacement $y(t) = Y_0 \cos \omega t$.

5.52 The mass m_1 of the two degree of freedom system shown in Fig. 5.16 is subjected to a force $F_0 \cos \omega t$. Assuming that the surrounding air damping is equivalent to $c = 200 \, \text{N} \cdot \text{s/m}$, find the steady-state response of the two masses. Assume $m_1 = m_2 = 1 \, \text{kg}$, $k_1 = k_2 = 500 \, \text{N/m}$, and $\omega = 1 \, \text{rad/s}$.

5.53 Determine the steady-state vibration of the system shown in Fig. 5.3(a), assuming that $c_1 = c_2 = c_3 = 0$, $F_1(t) = F_{10} \cos \omega t$, and $F_2(t) = F_{20} \cos \omega t$.

5.54 In the system shown in Fig. 5.16, the mass m_1 is excited by a harmonic force having a maximum value of 50 N and a frequency of 2 Hz. Find the forced amplitude of each mass for $m_1 = 10 \, \text{kg}$, $m_2 = 5 \, \text{kg}$, $k_1 = 8000 \, \text{N/m}$, and $k_2 = 2000 \, \text{N/m}$.

5.55 Find the response of the two masses of the two-story building frame shown in Fig. 5.24 under the ground displacement $y(t) = 0.2 \sin \pi t$ m. Assume the equivalent stiffness of the lower and upper columns to be 800 N/m and 600 N/m, respectively, and $m_1 = m_2 = 50 \, \text{kg}$.

5.56 Find the forced vibration response of the system shown in Fig. 5.13 when $F_1(t)$ is a step force of magnitude 5 N using the Laplace transform method. Assume $x_1(0) = \dot{x}_1(0) = x_2(0) = \dot{x}_2(0) = 0$, $m = 1 \, \text{kg}$, and $k = 100 \, \text{N/m}$.

5.57 Determine the equations of motion and the natural frequencies of the system shown in Fig. 5.46.

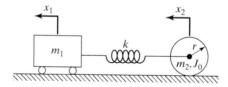

FIGURE 5.46

5.58 Two identical circular cylinders, of radius r and mass m each, are connected by a spring, as shown in Fig. 5.47. Determine the natural frequencies of oscillation of the system.

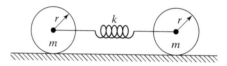

FIGURE 5.47

5.59 The differential equations of motion for a two degree of freedom system are given by

$$a_1\ddot{x}_1 + b_1x_1 + c_1x_2 = 0$$
$$a_2\ddot{x}_2 + b_2x_1 + c_2x_2 = 0$$

Derive the condition to be satisfied for the system to be degenerate.

5.60 Find the angular displacements $\theta_1(t)$ and $\theta_2(t)$ of the system shown in Fig. 5.48 for the initial conditions $\theta_1(t = 0) = \theta_1(0)$, $\theta_2(t = 0) = \theta_2(0)$, and $\dot{\theta}_1(t = 0) = \dot{\theta}_2(t = 0) = 0$.

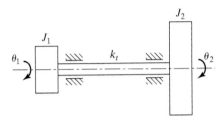

FIGURE 5.48

5.61 Determine the normal modes of the system shown in Fig. 5.7 with $k_{t1} = 0$. Show that the system with $k_{t1} = 0$ can be treated as a single degree of freedom system by using the coordinate $\alpha = \theta_1 - \theta_2$.

5.62 A turbine is connected to an electric generator through gears, as shown in Fig. 5.49. The mass moments of inertia of the turbine, generator, gear 1, and gear 2 are given, respectively, by 3000, 2000, 500, and 1000 kg-m². Shafts 1 and 2 are made of steel and have diameters 30 cm and 10 cm, and lengths 2 cm and 1.0 m, respectively. Find the natural frequencies of the system.

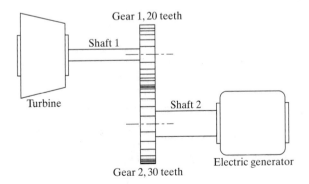

FIGURE 5.49

5.63 A hot air balloon of mass m is used to lift a load, Mg, through 12 equally spaced elastic ropes, each of stiffness k (see Fig. 5.50). Find the natural frequencies of vibration of the balloon in vertical direction. State the assumptions made in your solution and discuss their validity.

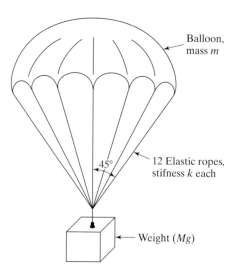

Balloon, mass m

45°

12 Elastic ropes, stifness k each

Weight (Mg)

FIGURE 5.50

5.64 A turbine of mass moment of inertia of 4 lb-in.-sec² is connected to an electric generator of mass moment of inertia of 2 lb-in.-sec² by a hollow steel shaft of inner diameter 1 in., outer diameter 2 in., and length 15 in., (similar to the system in Fig. 5.15b). If the turbine is suddenly stopped while delivering horsepower of 100 at a speed of 6000 rpm, the transmitted torque drops to zero. Find the resulting angular displacements of the turbine and the generator. Assume damping to be negligible in the system.

5.65 The transient vibrations of the drive line developed during the application of a cone (friction) clutch lead to unpleasant noise. To reduce the noise, a flywheel having a mass moment of inertia J_2 is attached to the drive line through a torsional spring k_{t2} and a viscous torsional damper c_{t2}, as shown in Fig. 5.51. If the mass moment of inertia of the cone clutch is J_1 and the stiffness and damping constant of the drive line are given by k_{t1} and c_{t1}, respectively, derive the relations to be satisfied for the stable operation of the system.

5.66 A uniform rigid bar of mass m is connected to the wall of a trailer by a spring of stiffness k (see Fig. 5.52). The trailer has a mass $5m$, is connected to a spring of stiffness $2k$, and moves on a frictionless surface. Derive the conditions necessary for the stability of the system.

5.67 Find the response of the system shown in Fig. 5.3(a) using a numerical procedure when $k_1 = k, k_2 = 2k, k_3 = k, m_1 = 2m, m_2 = m, F_2(t) = 0$, and $F_1(t)$ is a rectangular pulse of magnitude 500 N and duration 0.5 sec. Assume $m = 10$ kg, $c_1 = c_2 = c_3 = 0$, and $k = 2000$ N/m, and zero initial conditions.

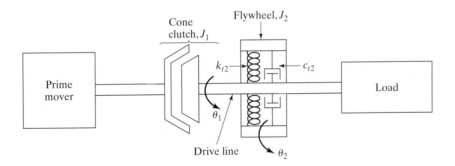

FIGURE 5.51

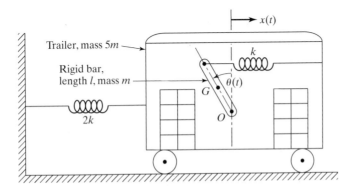

FIGURE 5.52

5.68 (a) Find the roots of the frequency equation of the system shown in Fig. 5.3 using subroutine QUART.F with the following data: $m_1 = m_2 = 0.2$ lb-s^2/in., $k_1 = k_2 = 18$ lb/in., $k_3 = 0$, $c_1 = c_2 = c_3 = 0$. (b) If the initial conditions are $x_1(0) = x_2(0) = 2$ in., $\dot{x}_1(0) = \dot{x}_2(0) = 0$, determine the displacements $x_1(t)$ and $x_2(t)$ of the masses.

5.69 Write a computer program for finding the steady-state response of a two degree of free-dom system under the harmonic excitation $F_j(t) = F_{j0}e^{i\omega t}$ and $j = 1, 2$ using Eqs. (5.29) and (5.35). Use this program to find the response of a system with $m_{11} = m_{22} = 0.1$ lb-s^2/in., $m_{12} = 0$, $c_{11} = 1.0$ lb-s/in., $c_{12} = c_{22} = 0$, $k_{11} = 40$ lb/in., $k_{22} = 20$ lb/in., $k_{12} = -20$ lb/in., $F_{10} = 1$ lb, $F_{20} = 2$ lb, and $\omega = 5$ rad/s.

5.70 Find and plot the free vibration response of the system shown in Fig. 5.16 for the following data: $k_1 = 1000$ N/m, $k_2 = 500$ N/m, $m_1 = 2$ kg, $m_2 = 1$ kg, $x_1(0) = 1$, $x_2(0) = 0$, $\dot{x}_2(0) = 0$.

5.71 Find and plot the free vibration response of the system shown in Fig. 5.16 for the following data: $k_1 = 1000$ N/m, $k_2 = 500$ N/m, $m_1 = 2$ kg, $m_2 = 1$ kg, $x_1(0) = 1$, $x_2(0) = 2$, $\dot{x}_1(0) = 1$, $\dot{x}_2(0) = -2$.

5.72 Solve the following eigenvalue problem using MATLAB:

$$\begin{bmatrix} 25 \times 10^6 & -5 \times 10^6 \\ -5 \times 10^6 & 5 \times 10^6 \end{bmatrix} \begin{Bmatrix} x_1 \\ x_2 \end{Bmatrix} = \omega^2 \begin{bmatrix} 10000 & 0 \\ 0 & 5000 \end{bmatrix} \begin{Bmatrix} x_1 \\ x_2 \end{Bmatrix}$$

5.73 Find and plot the response of the following two degree of freedom system using MATLAB:

$$\begin{bmatrix} 2 & 0 \\ 0 & 10 \end{bmatrix} \begin{Bmatrix} \ddot{x}_1 \\ \ddot{x}_2 \end{Bmatrix} + \begin{bmatrix} 20 & -5 \\ -5 & 5 \end{bmatrix} \begin{Bmatrix} \dot{x}_1 \\ \dot{x}_2 \end{Bmatrix} + \begin{bmatrix} 50 & -10 \\ -10 & 10 \end{bmatrix} \begin{Bmatrix} x_1 \\ x_2 \end{Bmatrix} = \begin{Bmatrix} 2 \sin 3t \\ 5 \cos 5t \end{Bmatrix}$$

The initial conditions are $x_1(0) = 1$, $\dot{x}_1(0) = 0$, $x_2(0) = -1$, and $\dot{x}_2(0) = 0$.

5.74 Using MATLAB, solve Problem 5.67. Use the MATLAB function stepfun for the rectangular pulse.

5.75 Using MATLAB, solve Problem 5.68(a).

5.76 Using MATLAB, solve Problem 5.69. Plot the steady-state responses of masses m_{11} and m_{22}.

5.77 Using MATLAB, find the roots of the equation $x^4 - 32x^3 + 244x^2 - 20x - 1200 = 0$.

5.78 Using **Program6.cpp**, solve Problem 5.68(a).

5.79 Using **Program6.cpp**, find the roots of the quartic equation given in Problem 5.77.

5.80 Using **PROGRAM6.F**, solve Problem 5.68(a).

5.81 Using **PROGRAM6.F**, find the roots of the quartic equation given in Problem 5.77.

DESIGN PROJECTS

5.82 A step-cone pulley with a belt drive (Fig. 5.53) is used to change the cutting speeds in a lathe. The speed of the driving shaft is 350 rpm and the speeds of the output shaft are 150, 250, 450, and 750 rpm. The diameters of the driving and the driven pulleys, corresponding to 150 rpm output speed, are 250 mm and 1000 mm, respectively. The center distance between the shafts is 5 m. The mass moments of inertia of the driving and driven step cones are 0.1 and 0.2 kg-m², respectively. Find the cross-sectional area of the belt to avoid resonance with any of the input/output speeds of the system. Assume the Young's modulus of the belt material as 10^{10} N/m².

5.83 The masses of the tup, frame (along with the anvil and the workpiece), and the concrete block in the forging hammer shown in Fig. 5.41 are 1000 kg, 5000 kg, and 25000 kg, respectively. The tup drops on to the workpiece from a height of 2 m. Design suitable springs k_1 and k_2 for the following conditions: (a) The impact is inelastic—that is, the tup will not rebound after

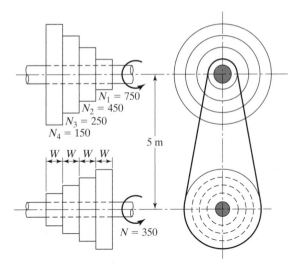

$N_1 = 750$
$N_2 = 450$
$N_3 = 250$
$N_4 = 150$

5 m

W W W W

$N = 350$

FIGURE 5.53

striking the workpiece. (b) The natural frequencies of vibration of the forging hammer should be greater than 5 Hz. (c) The stresses in the springs should be smaller than the yield stress of the material with a factor of safety of at least 1.5. Assume that the elasticity of the soil is negligible.

C H A P T E R 6

Multidegree of Freedom Systems

6.1 Introduction

As stated in Chapter 1, most engineering systems are continuous and have an infinite number of degrees of freedom. The vibration analysis of continuous systems requires the solution of partial differential equations, which is quite difficult. In fact, analytical solutions do not exist for many partial differential equations. The analysis of a multidegree of freedom, on the other hand, requires the solution of a set of ordinary differential equations, which is relatively simple. Hence, for simplicity of analysis, continuous systems are often approximated as multidegree of freedom systems.

All the concepts introduced in the preceding chapter can be directly extended to the case of multidegree of freedom systems. For example, there is one equation of motion for each degree of freedom; if generalized coordinates are used, there is one generalized coordinate for each degree of freedom. The equations of motion can be obtained from Newton's second law of motion or by using the influence coefficients defined in Section 6.4. However, it is often more convenient to derive the equations of motion of a multidegree of freedom system by using Lagrange's equations.

There are *n* natural frequencies, each associated with its own mode shape, for a system having *n* degrees of freedom. The method of determining the natural frequencies from the characteristic equation obtained by equating the determinant to zero also applies to these systems. However, as the number of degrees of freedom increases, the solution of the characteristic equation becomes more complex. The mode shapes exhibit a property known as *orthogonality*, which often enables us to simplify the analysis of multidegree of freedom systems.

6.2 Modeling of Continuous Systems as Multidegree of Freedom Systems

Different methods can be used to approximate a continuous system as a multidegree of freedom system. A simple method involves replacing the distributed mass or inertia of the system by a finite number of lumped masses or rigid bodies. The lumped masses are assumed to be connected by massless elastic and damping members. Linear (or angular) coordinates are used to describe the motion of the lumped masses (or rigid bodies). Such models are called lumped-parameter or lumped-mass or discrete-mass systems. The minimum number of coordinates necessary to describe the motion of the lumped masses and rigid bodies defines the number of degrees of freedom of the system. Naturally, the larger the number of lumped masses used in the model, the higher the accuracy of the resulting analysis.

Some problems automatically indicate the type of lumped-parameter model to be used. For example, the three-story building shown in Fig. 6.1(a) automatically suggests using a three lumped-mass model, as indicated in Fig. 6.1(b). In this model, the inertia of the system is assumed to be concentrated as three point masses located at the floor levels, and the elasticities of the columns are replaced by the springs. Similarly, the radial drilling machine shown in Fig. 6.2(a) can be modeled using four lumped masses and four spring elements (elastic beams), as shown in Fig. 6.2(b).

Another popular method of approximating a continuous system as a multidegree of freedom system involves replacing the geometry of the system by a large number of small

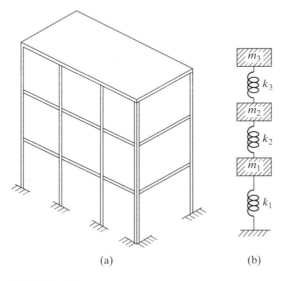

(a)   (b)

FIGURE 6.1   Three-story building.

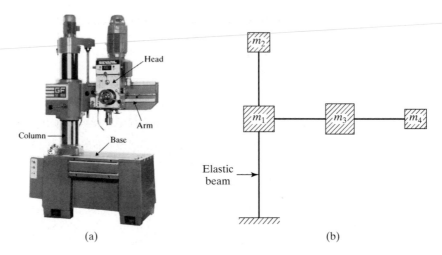

(a)   (b)

FIGURE 6.2   Radial drilling machine. (Photo courtesy of South Bend Lathe Corp.)

elements. By assuming a simple solution within each element, the principles of compatibility and equilibrium are used to find an approximate solution to the original system. This method, known as the finite element method, is considered in detail in Chapter 12.

6.3   Using Newton's Second Law to Derive Equations of Motion

The following procedure can be adopted to derive the equations of motion of a multidegree of freedom system using Newton's second law of motion:

1. Set up suitable coordinates to describe the positions of the various point masses and rigid bodies in the system. Assume suitable positive directions for the displacements, velocities, and accelerations of the masses and rigid bodies.
2. Determine the static equilibrium configuration of the system and measure the displacements of the masses and rigid bodies from their respective static equilibrium positions.
3. Draw the free-body diagram of each mass or rigid body in the system. Indicate the spring, damping, and external forces acting on each mass or rigid body when positive displacement and velocity are given to that mass or rigid body.
4. Apply Newton's second law of motion to each mass or rigid body shown by the free-body diagram as

$$m_i \ddot{x}_i = \sum_j F_{ij} \text{ (for mass } m_i) \qquad (6.1)$$

or

$$J_i \ddot{\theta}_i = \sum_j M_{ij} \text{ (for rigid body of inertia } J_i) \qquad (6.2)$$

where $\sum_j F_{ij}$ denotes the sum of all forces acting on mass m_i and $\sum_j M_{ij}$ indicates the sum of moments of all forces (about a suitable axis) acting on the rigid body of mass moment of inertia J_i.

The procedure is illustrated in the following examples.

EXAMPLE 6.1

Equations of Motion of a Spring-Mass-Damper System

Derive the equations of motion of the spring-mass-damper system shown in Fig. 6.3(a).

Solution

Approach: Draw free-body diagrams of masses and apply Newton's second law of motion. The coordinates describing the positions of the masses, $x_i(t)$, are measured from their respective static equilibrium positions, as indicated in Fig. 6.3(a). The free-body diagram of a typical interior mass m_i is shown in Fig. 6.3(b) along with the assumed positive directions for its displacement, velocity, and acceleration. The application of Newton's second law of motion to mass m_i gives

$$m_i \ddot{x}_i = -k_i(x_i - x_{i-1}) + k_{i+1}(x_{i+1} - x_i) - c_i(\dot{x}_i - \dot{x}_{i-1})$$
$$+ c_{i+1}(\dot{x}_{i+1} - \dot{x}_i) + F_i; i = 2, 3, \ldots, n - 1$$

or

$$m_i \ddot{x}_i - c_i \dot{x}_{i-1} + (c_i + c_{i+1}) \dot{x}_i - c_{i+1} \dot{x}_{i+1} - k_i x_{i-1}$$
$$+ (k_i + k_{i+1}) x_i - k_{i+1} x_{i+1} = F_i; i = 2, 3, \ldots, n - 1 \tag{E.1}$$

The equations of motion of the masses m_1 and m_n can be derived from Eq. (E.1) by setting $i = 1$ along with $x_0 = 0$ and $i = n$ along with $x_{n+1} = 0$, respectively:

$$m_1 \ddot{x}_1 + (c_1 + c_2) \dot{x}_1 - c_2 \dot{x}_2 + (k_1 + k_2) x_1 - k_2 x_2 = F_1 \tag{E.2}$$

$$m_n \ddot{x}_n - c_n \dot{x}_{n-1} + (c_n + c_{n+1}) \dot{x}_n - k_n x_{n-1} + (k_n + k_{n+1}) x_n = F_n \tag{E.3}$$

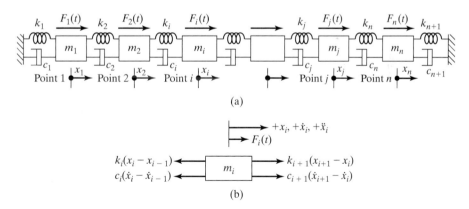

(a)

(b)

FIGURE 6.3   Spring-mass-damper system.

Notes

1. The equations of motion, Eqs. (E.1) to (E.3), of Example 6.1 can be expressed in matrix form as

$$[m] \, \ddot{\vec{x}} + [c] \, \dot{\vec{x}} + [k] \, \vec{x} = \vec{F} \tag{6.3}$$

where $[m]$, $[c]$, and $[k]$ are called the mass, damping, and stiffness matrices, respectively, and are given by

$$[m] = \begin{bmatrix} m_1 & 0 & 0 & \cdots & 0 & 0 \\ 0 & m_2 & 0 & \cdots & 0 & 0 \\ 0 & 0 & m_3 & \cdots & 0 & 0 \\ \vdots & & & & & \\ 0 & 0 & 0 & \cdots & 0 & m_n \end{bmatrix} \tag{6.4}$$

$$[c] = \begin{bmatrix} (c_1 + c_2) & -c_2 & 0 & \cdots & 0 & 0 \\ -c_2 & (c_2 + c_3) & -c_3 & \cdots & 0 & 0 \\ 0 & -c_3 & (c_3 + c_4) & \cdots & 0 & 0 \\ \cdot & \cdot & \cdot & \cdots & \cdot & \cdot \\ \cdot & \cdot & \cdot & \cdots & \cdot & \cdot \\ \cdot & \cdot & \cdot & \cdots & \cdot & \cdot \\ 0 & 0 & 0 & \cdots & -c_n & (c_n + c_{n+1}) \end{bmatrix} \tag{6.5}$$

$$[k] = \begin{bmatrix} (k_1 + k_2) & -k_2 & 0 & \cdots & 0 & 0 \\ -k_2 & (k_2 + k_3) & -k_3 & \cdots & 0 & 0 \\ 0 & -k_3 & (k_3 + k_4) & \cdots & 0 & 0 \\ \cdot & & & & & \\ \cdot & & & & & \\ \cdot & & & & & \\ 0 & 0 & 0 & \cdots & -k_n & (k_n + k_{n+1}) \end{bmatrix} \tag{6.6}$$

and $\vec{x}, \dot{\vec{x}}, \ddot{\vec{x}}$, and $\vec{F}$ are the displacement, velocity, acceleration, and force vectors, given by

$$\vec{x} = \begin{Bmatrix} x_1(t) \\ x_2(t) \\ \cdot \\ \cdot \\ \cdot \\ x_n(t) \end{Bmatrix}, \qquad \dot{\vec{x}} = \begin{Bmatrix} \dot{x}_1(t) \\ \dot{x}_2(t) \\ \cdot \\ \cdot \\ \cdot \\ \dot{x}_n(t) \end{Bmatrix},$$

$$\ddot{\vec{x}} = \begin{Bmatrix} \ddot{x}_1(t) \\ \ddot{x}_2(t) \\ \cdot \\ \cdot \\ \cdot \\ \ddot{x}_n(t) \end{Bmatrix}, \qquad \vec{F} = \begin{Bmatrix} F_1(t) \\ F_2(t) \\ \cdot \\ \cdot \\ \cdot \\ F_n(t) \end{Bmatrix} \tag{6.7}$$

2. For an undamped system (with all $c_i = 0$, $i = 1, 2, \ldots, n + 1$), the equations of motion reduce to

$$[m]\ddot{\vec{x}} + [k]\vec{x} = \vec{F} \tag{6.8}$$

3. The spring-mass-damper system considered above is a particular case of a general n degree of freedom spring-mass-damper system. In their most general form, the mass, damping, and stiffness matrices are given by

$$[m] = \begin{bmatrix} m_{11} & m_{12} & m_{13} & \cdots & m_{1n} \\ m_{12} & m_{22} & m_{23} & \cdots & m_{2n} \\ \cdot & & & & \\ \cdot & & & & \\ \cdot & & & & \\ m_{1n} & m_{2n} & m_{3n} & \cdots & m_{nn} \end{bmatrix} \tag{6.9}$$

$$[c] = \begin{bmatrix} c_{11} & c_{12} & c_{13} & \cdots & c_{1n} \\ c_{12} & c_{22} & c_{23} & \cdots & c_{2n} \\ \cdot & \cdot & \cdot & \cdots & \cdot \\ \cdot & \cdot & \cdot & \cdots & \cdot \\ \cdot & \cdot & \cdot & \cdots & \cdot \\ c_{1n} & c_{2n} & c_{3n} & \cdots & c_{nn} \end{bmatrix} \tag{6.10}$$

and

$$[k] = \begin{bmatrix} k_{11} & k_{12} & k_{13} & \cdots & k_{1n} \\ k_{12} & k_{22} & k_{23} & \cdots & k_{2n} \\ \cdot & & & & \\ \cdot & & & & \\ \cdot & & & & \\ k_{1n} & k_{2n} & k_{3n} & \cdots & k_{nn} \end{bmatrix} \tag{6.11}$$

4. The differential equations of the spring-mass system considered in Example 6.1 (Fig. 6.3a) can be seen to be coupled; each equation involves more than one coordinate. This means that the equations cannot be solved individually one at a time; they can only be solved simultaneously. In addition, the system can be seen to be statically coupled since stiffnesses are coupled—that is, the stiffness matrix has at least one nonzero off-diagonal term. On the other hand, if the mass

matrix has at least one off-diagonal term nonzero, the system is said to be dynamically coupled. Further, if both the stiffness and mass matrices have nonzero off-diagonal terms, the system is said to be coupled both statically and dynamically.

∎

EXAMPLE 6.2

Equations of Motion of a Trailer–Compound Pendulum System

Derive the equations of motion of the trailer–compound pendulum system shown in Fig. 6.4(a).

Solution

Approach: Draw the free-body diagrams and apply Newton's second law of motion.

The coordinates $x(t)$ and $\theta(t)$ are used to describe, respectively, the linear displacement of the trailer and the angular displacement of the compound pendulum from their respective static equilibrium

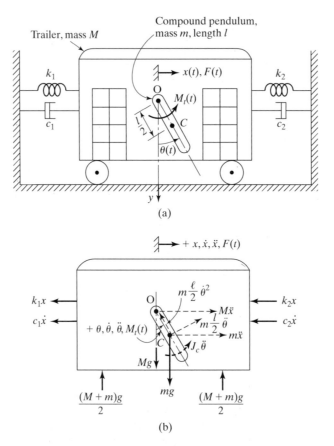

FIGURE 6.4   Compound pendulum and trailer system.

positions. When positive values are assumed for the displacements $x(t)$ and $\theta(t)$, velocities $\dot{x}(t)$ and $\dot{\theta}(t)$, and accelerations $\ddot{x}(t)$ and $\ddot{\theta}(t)$, the external forces on the trailer will be the applied force $F(t)$, the spring forces $k_1 x$ and $k_2 x$, and the damping forces $c_1 \dot{x}$ and $c_2 \dot{x}$, as shown in Fig. 6.4(b). The external forces on the compound pendulum will be the applied torque $M_t(t)$ and the gravitational force mg, as shown in Fig. 6.4(b). The inertia forces that act on the trailer and the compound pendulum are indicated by the dashed lines in Fig. 6.4(b). Note that the rotational motion of the compound pendulum about the hinge O induces a radially inward force (toward O) $m\frac{l}{2}\dot{\theta}^2$ and a normal force (perpendicular to OC) $m\frac{l}{2}\ddot{\theta}$ as shown in Fig. 6.4(b). The application of Newton's second law for translatory motion in the horizontal direction gives

$$M\ddot{x} + m\ddot{x} + m\frac{l}{2}\ddot{\theta}\cos\theta - m\frac{l}{2}\dot{\theta}^2\sin\theta =$$
$$- k_1 x - k_2 x - c_1 \dot{x} - c_2 \dot{x} + F(t) \tag{E.1}$$

Similarly the application of Newton's second law for rotational motion about hinge O yields

$$\left(m\frac{l}{2}\ddot{\theta}\right)\frac{l}{2} + \left(m\frac{l^2}{12}\right)\ddot{\theta} + (m\ddot{x})\frac{l}{2}\cos\theta = -(mg)\frac{l}{2}\sin\theta + M_t(t) \tag{E.2}$$

Notes

1. The equations of motion, Eqs. (E.1) and (E.2), can be seen to be nonlinear due to the presence of the terms involving $\sin\theta$, $\cos\theta$, and $(\dot{\theta})^2\sin\theta$.
2. Equations (E.1) and (E.2) can be linearized if the term involving $(\dot{\theta})^2\sin\theta$ is assumed negligibly small and the displacements are assumed small so that $\cos\theta \approx 1$ and $\sin\theta \approx 0$. The linearized equations can be derived as

$$(M + m)\ddot{x} + \left(m\frac{l}{2}\right)\ddot{\theta} + (k_1 + k_2)x$$
$$+ (c_1 + c_2)\dot{x} = F(t) \tag{E.3}$$

and

$$\left(\frac{ml}{2}\right)\ddot{x} + \left(\frac{ml^2}{3}\right)\ddot{\theta} + \left(\frac{mgl}{2}\right)\theta = M_t(t) \tag{E.4}$$

∎

6.4   Influence Coefficients

The equations of motion of a multidegree of freedom system can also be written in terms of influence coefficients, which are extensively used in structural engineering. Basically, one set of influence coefficients can be associated with each of the matrices involved in the equations of motion. The influence coefficients associated with the stiffness and mass matrices are, respectively, known as the stiffness and inertia influence coefficients. In some cases, it is more convenient to rewrite the equations of motion using the inverse of the stiffness matrix (known as the flexibility matrix) or the inverse of the mass matrix. The influence coefficients corresponding to the inverse stiffness matrix are called the flexibility influence coefficients, and those corresponding to the inverse mass matrix are known as the inverse inertia coefficients.

6.4.1 Stiffness Influence Coefficients

For a simple linear spring, the force necessary to cause a unit elongation is called the stiffness of the spring. In more complex systems, we can express the relation between the displacement at a point and the forces acting at various other points of the system by means of stiffness influence coefficients. The stiffness influence coefficient, denoted as k_{ij}, is defined as the force at point i due to a unit displacement at point j when all the points other than the point j are fixed. Using this definition, for the spring-mass system shown in Fig. 6.5, the total force at point i, F_i, can be found by summing up the forces due to all displacements $x_j(j = 1, 2, \ldots, n)$ as

$$F_i = \sum_{j=1}^{n} k_{ij} x_j, \qquad i = 1, 2, \ldots, n \tag{6.12}$$

Equation (6.12) can be stated in matrix form as

$$\vec{F} = [k]\vec{x} \tag{6.13}$$

where $\vec{x}$ and $\vec{F}$ are the displacement and force vectors defined in Eq. (6.7) and $[k]$ is the stiffness matrix given by

$$[k] = \begin{bmatrix} k_{11} & k_{12} & \cdots & k_{1n} \\ k_{21} & k_{22} & \cdots & k_{2n} \\ \vdots & & & \\ k_{n1} & k_{n2} & \cdots & k_{nn} \end{bmatrix} \tag{6.14}$$

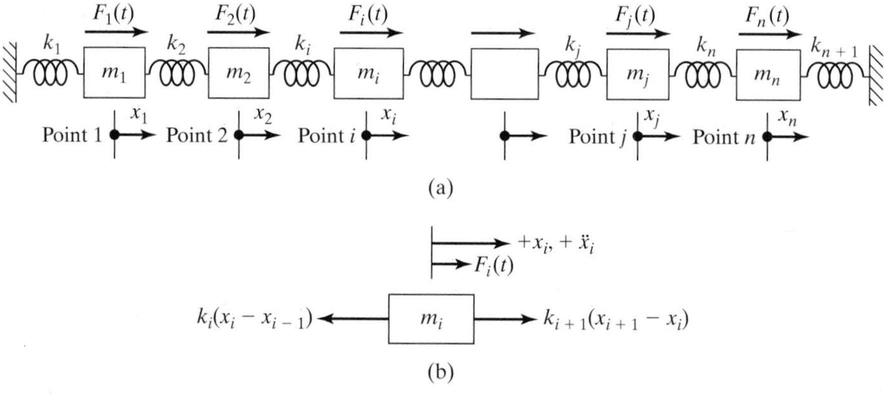

FIGURE 6.5   Multidegree of freedom spring-mass system.

The following aspects of stiffness influence coefficients are to be noted:

1. Since the force required at point i to cause a unit deflection at point j and zero deflection at all other points is the same as the force required at point j to cause a unit deflection at point i and zero deflection at all other points (Maxwell's reciprocity theorem [6.1]), we have $k_{ij} = k_{ji}$.
2. The stiffness influence coefficients can be calculated by applying the principles of statics and solid mechanics.
3. The stiffness influence coefficients for torsional systems can be defined in terms of unit angular displacement and the torque that causes the angular displacement. For example, in a multirotor torsional system, k_{ij} can be defined as the torque at point i (rotor i) due to a unit angular displacement at point j and zero angular displacement at all other points.

The stiffness influence coefficients of a multidegree of freedom system can be determined as follows:

1. Assume a value of one for the displacement x_j ($j = 1$ to start with) and a value of zero for all other displacements $x_1, x_2, \ldots, x_{j-1}, x_{j+1}, \ldots, x_n$. By definition, the set of forces $k_{ij}(i = 1, 2, \ldots, n)$ will maintain the system in the assumed configuration $(x_j = 1, x_1 = x_2 = \cdots = x_{j-1} = x_{j+1} = \cdots = x_n = 0)$. Then the static equilibrium equations are written for each mass and the resulting set of n equations solved to find the n influence coefficients $k_{ij}(i = 1, 2, \ldots, n)$.
2. After completing step 1 for $j = 1$, the procedure is repeated for $j = 2, 3, \ldots, n$.

The following examples illustrate the procedure.

EXAMPLE 6.3

Stiffness Influence Coefficients

Find the stiffness influence coefficients of the system shown in Fig. 6.6(a).

Solution

Approach: Use the definition of k_{ij} and static equilibrium equations.

Let x_1, x_2, and x_3 denote the displacements of the masses m_1, m_2, and m_3, respectively. The stiffness influence coefficients k_{ij} of the system can be determined in terms of the spring stiffnesses k_1, k_2, and k_3 as follows. First, we set the displacement of m_1 equal to one ($x_1 = 1$) and the displacements of m_2 and m_3 equal to zero ($x_2 = x_3 = 0$), as shown in Fig. 6.6(b). The set of forces $k_{i1}(i = 1, 2, 3)$ is assumed to maintain the system in this configuration. The free-body diagrams of the masses corresponding to the configuration of Fig. 6.6(b) are indicated in Fig. 6.6(c). The equilibrium of forces for the masses m_1, m_2, and m_3 in the horizontal direction yields

$$\text{Mass } m_1: k_1 = -k_2 + k_{11} \tag{E.1}$$

$$\text{Mass } m_2: k_{21} = -k_2 \tag{E.2}$$

$$\text{Mass } m_3: k_{31} = 0 \tag{E.3}$$

The solution of Eqs. (E.1) to (E.3) gives

$$k_{11} = k_1 + k_2, k_{21} = -k_2, k_{31} = 0 \tag{E.4}$$

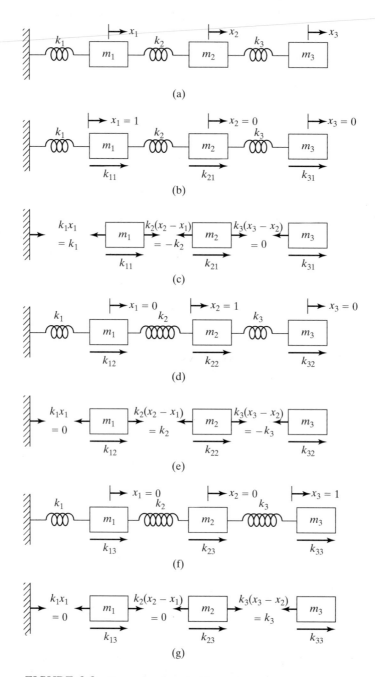

FIGURE 6.6   Determination of stiffness influence coefficients.

Next the displacements of the masses are assumed as $x_1 = 0$, $x_2 = 1$, and $x_3 = 0$, as shown in Fig. 6.6(d). Since the forces $k_{i2}(i = 1, 2, 3)$ are assumed to maintain the system in this configuration, the free-body diagrams of the masses can be developed as indicated in Fig. 6.6(e). The force equilibrium equations of the masses are:

$$\text{Mass } m_1: k_{12} + k_2 = 0 \tag{E.5}$$

$$\text{Mass } m_2: k_{22} - k_3 = k_2 \tag{E.6}$$

$$\text{Mass } m_3: k_{32} = -k_3 \tag{E.7}$$

The solution of Eqs. (E.5) to (E.7) yields

$$k_{12} = -k_2, \, k_{22} = k_2 + k_3, \, k_{32} = -k_3 \tag{E.8}$$

Finally the set of forces $k_{i3}(i = 1, 2, 3)$ is assumed to maintain the system with $x_1 = 0$, $x_2 = 0$, and $x_3 = 1$ (Fig. 6.6f). The free-body diagrams of the various masses in this configuration are shown in Fig. 6.6(g) and the force equilibrium equations lead to

$$\text{Mass } m_1: k_{13} = 0 \tag{E.9}$$

$$\text{Mass } m_2: k_{23} + k_3 = 0 \tag{E.10}$$

$$\text{Mass } m_3: k_{33} = k_3 \tag{E.11}$$

The solution of Eqs. (E.9) to (E.11) yields

$$k_{13} = 0, \, k_{23} = -k_3, \, k_{33} = k_3 \tag{E.12}$$

Thus the stiffness matrix of the system is given by

$$[k] = \begin{bmatrix} (k_1 + k_2) & -k_2 & 0 \\ -k_2 & (k_2 + k_3) & -k_3 \\ 0 & -k_3 & k_3 \end{bmatrix} \tag{E.13}$$

■

EXAMPLE 6.4

Stiffness Matrix of a Frame

Determine the stiffness matrix of the frame shown in Fig. 6.7(a). Neglect the effect of axial stiffness of the members *AB* and *BC*.

Solution: Since the segments *AB* and *BC* of the frame can be considered as beams, the beam force-deflection formulas can be used to generate the stiffness matrix of the frame. The forces necessary to cause a displacement along one coordinate while maintaining zero displacements along other coordinates of a beam are indicated in Fig. 6.7(b) [6.1, 6.8]. In Fig. 6.7(a), the ends *A* and *C* are fixed and hence the joint *B* will have three possible displacements—*x, y,* and *θ*, as indicated. The forces

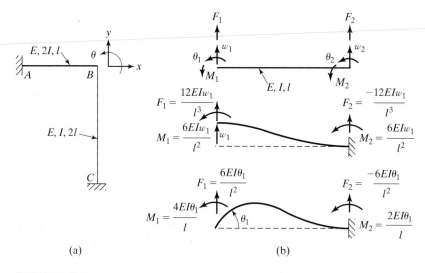

(a)   (b)

FIGURE 6.7   Stiffness matrix of a frame.

necessary to maintain a unit displacement along x direction and zero displacement along y and θ directions at the joint B are given by (from Fig. 6.7b)

$$F_x = \left(\frac{12\,EI}{l^3}\right)_{BC} = \frac{3\,EI}{2\,l^3}, \qquad F_y = 0, \; M_\theta = \left(\frac{6\,EI}{l^2}\right)_{BC} = \frac{3\,EI}{2l^2}$$

Similarly, when a unit displacement is given along y direction at joint B with zero displacements along x and θ directions, the forces required to maintain the configuration can be found from Fig. 6.7(b) as

$$F_x = 0, \; F_y = \left(\frac{12\,EI}{l^3}\right)_{BA} = \frac{24\,EI}{l^3}, \qquad M_\theta = -\left(\frac{6\,EI}{l^2}\right)_{BA} = -\frac{12\,EI}{l^2}$$

Finally, the forces necessary to maintain a unit displacement along θ direction and zero displacements along x and y directions at joint B can be seen, from Fig. 6.7(b), as

$$F_x = \left(\frac{6\,EI}{l^2}\right)_{BC} = \frac{3\,EI}{2\,l^2}, \qquad F_y = -\left(\frac{6\,EI}{l^2}\right)_{BA} = -\frac{12\,EI}{l^3},$$

$$M_\theta = \left(\frac{4\,EI}{l}\right)_{BC} + \left(\frac{4\,EI}{l}\right)_{BA} = \frac{2\,EI}{l} + \frac{8\,EI}{l} = \frac{10\,EI}{l}$$

Thus the stiffness matrix, $[k]$, is given by

$$\vec{F} = [k]\vec{x}$$

where

$$\vec{F} = \begin{Bmatrix} F_x \\ F_y \\ M_\theta \end{Bmatrix}, \vec{x} = \begin{Bmatrix} x \\ y \\ \theta \end{Bmatrix}, [k] = \frac{EI}{l^3} \begin{bmatrix} \dfrac{3}{2} & 0 & \dfrac{3l}{2} \\ 0 & 24 & -12l \\ \dfrac{3l}{2} & -12l & 10l^2 \end{bmatrix}$$

■

6.4.2 Flexibility Influence Coefficients

As seen in Examples 6.3 and 6.4, the computation of stiffness influence coefficients requires the application of the principles of statics and some algebraic manipulation. In fact, the generation of n stiffness influence coefficients $k_{1j}, k_{2j}, \ldots, k_{nj}$ for any specific j requires the solution of n simultaneous linear equations. Thus n sets of linear equations (n equations in each set) are to be solved to generate all the stiffness influence coefficients of an n degree of freedom system. This implies a significant computational effort for large values of n. The generation of the flexibility influence coefficients, on the other hand, proves to be simpler and more convenient. To illustrate the concept, consider again the spring-mass system shown in Fig. 6.5.

Let the system be acted on by just one force F_j, and let the displacement at point i (i.e., mass m_i) due to F_j be x_{ij}. The flexibility influence coefficient, denoted by a_{ij}, is defined as the deflection at point i due to a unit load at point j. Since the deflection increases proportionately with the load for a linear system, we have

$$x_{ij} = a_{ij}F_j \tag{6.15}$$

If several forces F_j ($j = 1, 2, \ldots, n$) act at different points of the system, the total deflection at any point i can be found by summing up the contributions of all forces F_j:

$$x_i = \sum_{j=1}^{n} x_{ij} = \sum_{j=1}^{n} a_{ij}F_j, \quad i = 1, 2, \ldots, n \tag{6.16}$$

Equation (6.16) can be expressed in matrix form as

$$\vec{x} = [a]\,\vec{F} \tag{6.17}$$

where $\vec{x}$ and $\vec{F}$ are the displacement and force vectors defined in Eq. (6.7) and $[a]$ is the flexibility matrix given by

$$[a] = \begin{bmatrix} a_{11} & a_{12} & \cdots & a_{1n} \\ a_{21} & a_{22} & \cdots & a_{2n} \\ \vdots & & & \\ a_{n1} & a_{n2} & \cdots & a_{nn} \end{bmatrix} \tag{6.18}$$

The following characteristics of flexibility influence coefficients can be noted:

1. An examination of Eqs. (6.17) and (6.13) indicates that the flexibility and stiffness matrices are related. If we substitute Eq. (6.13) into Eq. (6.17), we obtain

$$\vec{x} = [a]\,\vec{F} = [a][k]\,\vec{x} \qquad (6.19)$$

from which we can obtain the relation

$$[a][k] = [I] \qquad (6.20)$$

where $[I]$ denotes the unit matrix. Equation (6.20) is equivalent to

$$[k] = [a]^{-1}, \qquad [a] = [k]^{-1} \qquad (6.21)$$

That is, the stiffness and flexibility matrices are the inverse of one another. The use of dynamic stiffness influence coefficients in the vibration of nonuniform beams is discussed in Ref. [6.10].

2. Since the deflection at point i due to a unit load at point j is the same as the deflection at point j due to a unit load at point i for a linear system (Maxwell's reciprocity theorem [6.1]), we have $a_{ij} = a_{ji}$.

3. The flexibility influence coefficients of a torsional system can be defined in terms of unit torque and the angular deflection it causes. For example, in a multirotor torsional system, a_{ij} can be defined as the angular deflection of point i (rotor i) due to a unit torque at point j (rotor j).

The flexibility influence coefficients of a multidegree of freedom system can be determined as follows:

1. Assume a unit load at point j ($j = 1$ to start with). By definition, the displacements of the various points i ($i = 1, 2, \ldots, n$) resulting from this load give the flexibility influence coefficients, a_{ij}, $i = 1, 2, \ldots, n$. Thus a_{ij} can be found by applying the simple principles of statics and solid mechanics.

2. After completing Step 1 for $j = 1$, the procedure is repeated for $j = 2, 3, \ldots, n$.

3. Instead of applying Steps 1 and 2, the flexibility matrix, $[a]$, can be determined by finding the inverse of the stiffness matrix, $[k]$, if the stiffness matrix is available.

The following examples illustrate the procedure.

■■■■■■■ **Flexibility Influence Coefficients**

EXAMPLE 6.5 ────────────────────────────────────

Find the flexibility influence coefficients of the system shown in Fig. 6.8(a).

Solution: Let x_1, x_2, and x_3 denote the displacements of the masses m_1, m_2, and m_3, respectively. The flexibility influence coefficients a_{ij} of the system can be determined in terms of the spring stiffnesses k_1, k_2, and k_3 as follows. Apply a unit force at mass m_1 and no force at other masses

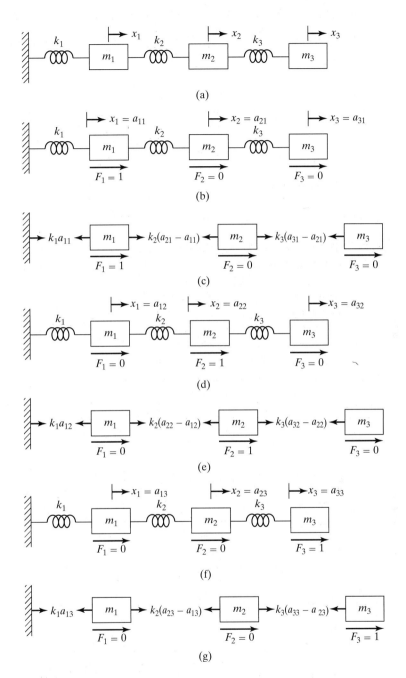

FIGURE 6.8 Determination of flexibility influence coefficients.

($F_1 = 1$, $F_2 = F_3 = 0$), as shown in Fig. 6.8(b). The resulting deflections of the masses m_1, m_2, and m_3 (x_1, x_2, and x_3) are, by definition, a_{11}, a_{21}, and a_{31}, respectively (see Fig. 6.8b). The free-body diagrams of the masses are shown in Fig. 6.8(c). The equilibrium of forces in the horizontal direction for the various masses gives the following:

$$\text{Mass } m_1: k_1 a_{11} = k_2 (a_{21} - a_{11}) + 1 \tag{E.1}$$

$$\text{Mass } m_2: k_2 (a_{21} - a_{11}) = k_3 (a_{31} - a_{21}) \tag{E.2}$$

$$\text{Mass } m_3: k_3 (a_{31} - a_{21}) = 0 \tag{E.3}$$

The solution of Eqs. (E.1) to (E.3) yields

$$a_{11} = \frac{1}{k_1}, \quad a_{21} = \frac{1}{k_1}, \quad a_{31} = \frac{1}{k_1} \tag{E.4}$$

Next, we apply a unit force at mass m_2 and no force at masses m_1 and m_3, as shown in Fig. 6.8(d). These forces cause the masses m_1, m_2, and m_3 to deflect by $x_1 = a_{12}$, $x_2 = a_{22}$, and $x_3 = a_{32}$, respectively (by definition of a_{i2}), as shown in Fig. 6.8(d). The free-body diagrams of the masses, shown in Fig. 6.8(e), yield the following equilibrium equations:

$$\text{Mass } m_1: k_1 (a_{12}) = k_2 (a_{22} - a_{12}) \tag{E.5}$$

$$\text{Mass } m_2: k_2 (a_{22} - a_{12}) = k_3 (a_{32} - a_{22}) + 1 \tag{E.6}$$

$$\text{Mass } m_3: k_3 (a_{32} - a_{22}) = 0 \tag{E.7}$$

The solution of Eqs. (E.5) to (E.7) gives

$$a_{12} = \frac{1}{k_1}, a_{22} = \frac{1}{k_1} + \frac{1}{k_2}, a_{32} = \frac{1}{k_1} + \frac{1}{k_2} \tag{E.8}$$

Finally, when we apply a unit force to mass m_3 and no force to masses m_1 and m_2, the masses deflect by $x_1 = a_{13}$, $x_2 = a_{23}$, and $x_3 = a_{33}$, as shown in Fig. 6.8(f). The resulting free-body diagrams of the various masses (Fig. 6.8g) yield the following equilibrium equations:

$$\text{Mass } m_1: k_1 a_{13} = k_2 (a_{23} - a_{13}) \tag{E.9}$$

$$\text{Mass } m_2: k_2 (a_{23} - a_{13}) = k_3 (a_{33} - a_{23}) \tag{E.10}$$

$$\text{Mass } m_3: k_3 (a_{33} - a_{23}) = 1 \tag{E.11}$$

The solution of Eqs. (E.9) to (E.11) gives the flexibility influence coefficients a_{i3} as

$$a_{13} = \frac{1}{k_1}, \quad a_{23} = \frac{1}{k_1} + \frac{1}{k_2}, \quad a_{33} = \frac{1}{k_1} + \frac{1}{k_2} + \frac{1}{k_3} \tag{E.12}$$

It can be verified that the stiffness matrix of the system, given by Eq. (E.13) of Example 6.3, can also be found from the relation $[k] = [a]^{-1}$.

∎

Flexibility Matrix of a Beam

EXAMPLE 6.6

Derive the flexibility matrix of the weightless beam shown in Fig. 6.9(a). The beam is simply supported at both ends, and the three masses are placed at equal intervals. Assume the beam to be uniform with stiffness EI.

Solution: Let x_1, x_2, and x_3 denote the total transverse deflection of the masses $m_1, m_2,$ and m_3, respectively. From the known formula for the deflection of a pinned-pinned beam [6.2], the influence coefficients $a_{1j}(j = 1, 2, 3)$ can be found by applying a unit load at the location of m_1 and zero load at the locations of m_2 and m_3 (see Fig. 6.9b):

$$a_{11} = \frac{9}{768} \frac{l^3}{EI}, \qquad a_{12} = \frac{11}{768} \frac{l^3}{EI}, \qquad a_{13} = \frac{7}{768} \frac{l^3}{EI} \qquad \text{(E.1)}$$

Similarly, by applying a unit load at the locations of m_2 and m_3 separately (with zero load at other locations), we obtain

$$a_{21} = a_{12} = \frac{11}{768} \frac{l^3}{EI}, \qquad a_{22} = \frac{1}{48} \frac{l^3}{EI}, \qquad a_{23} = \frac{11}{768} \frac{l^3}{EI} \qquad \text{(E.2)}$$

and

$$a_{31} = a_{13} = \frac{7}{768} \frac{l^3}{EI}, \qquad a_{32} = a_{23} = \frac{11}{768} \frac{l^3}{EI}, \qquad a_{33} = \frac{9}{768} \frac{l^3}{EI} \qquad \text{(E.3)}$$

Thus the flexibility matrix of the system is given by

$$[a] = \frac{l^3}{768EI} \begin{bmatrix} 9 & 11 & 7 \\ 11 & 16 & 11 \\ 7 & 11 & 9 \end{bmatrix} \qquad \text{(E.4)}$$

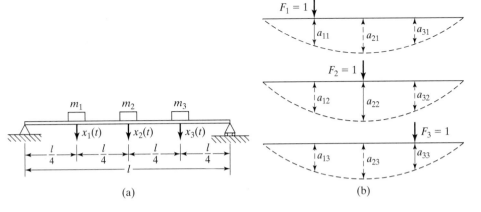

(a)   (b)

FIGURE 6.9 Beam deflections.

6.4.3
Inertia Influence
Coefficients

The elements of the mass matrix, m_{ij}, are known as the inertia influence coefficients. Although it is more convenient to derive the inertia influence coefficients from the expression for kinetic energy of the system (see Section 6.5), the coefficients m_{ij} can be computed using the impulse-momentum relations. The inertia influence coefficients $m_{1j}, m_{2j}, \ldots, m_{nj}$ are defined as the set of impulses applied at points $1, 2, \ldots, n$, respectively, to produce a unit velocity at point j and zero velocity at every other point (that is, $\dot{x}_j = 1, \dot{x}_1 = \dot{x}_2 = \cdots = \dot{x}_{j-1} = \dot{x}_{j+1} = \cdots = \dot{x}_n = 0$). Thus, for a multidegree of freedom system, the total impulse at point i, F_i, can be found by summing up the impulses causing the velocities $\dot{x}_j (j = 1, 2, \ldots, n)$ as

$$\underset{\sim}{F_i} = \sum_{j=1}^{n} m_{ij} \dot{x}_j \tag{6.22}$$

Equation (6.22) can be stated in matrix form as

$$\vec{F} = [m]\vec{x} \tag{6.23}$$

where $\vec{x}$ and $\vec{F}$ are the velocity and impulse vectors given by

$$\vec{x} = \begin{Bmatrix} \dot{x}_1 \\ \dot{x}_2 \\ . \\ . \\ . \\ \dot{x}_n \end{Bmatrix}, \quad \vec{F} = \begin{Bmatrix} F_1 \\ F_2 \\ . \\ . \\ . \\ F_n \end{Bmatrix} \tag{6.24}$$

and $[m]$ is the mass matrix given by

$$[m] = \begin{bmatrix} m_{11} & m_{12} & \cdots & m_{1n} \\ m_{21} & m_{22} & \cdots & m_{2n} \\ . & . & \cdots & . \\ . & . & \cdots & . \\ . & . & \cdots & . \\ m_{n1} & m_{n2} & \cdots & m_{nn} \end{bmatrix} \tag{6.25}$$

It can be verified easily that the inertia influence coefficients are symmetric for a linear system—that is, $m_{ij} = m_{ji}$. The following procedure can be used to derive the inertia influence coefficients of a multidegree of freedom system.

1. Assume that a set of impulses f_{ij} are applied at various points $i (i = 1, 2, \ldots, n)$ so as to produce a unit velocity at point j ($\dot{x}_j = 1$ with $j = 1$ to start with) and a zero velocity at all other points ($\dot{x}_1 = \dot{x}_2 = \cdots \dot{x}_{j-1} = \dot{x}_{j+1} = \cdots = \dot{x}_n = 0$). By definition, the set of impulses $f_{ij} (i = 1, 2, \ldots, n)$ denote the inertia influence coefficients $m_{ij} (i = 1, 2, \ldots, n)$.

2. After completing step 1 for $j = 1$, the procedure is repeated for $j = 2, 3, \ldots, n$.

Note that if x_j denotes an angular coordinate, then $\dot{x}_j$ represents an angular velocity and $\underset{\sim}{F}_j$ indicates an angular impulse. The following example illustrates the procedure of generating m_{ij}.

■■■■■■■■■■■ **Inertia Influence Coefficients**

EXAMPLE 6.7 ───

Find the inertia influence coefficients of the system shown in Fig. 6.4(a).

Solution

Approach: Use the definition of m_{ij} along with impulse-momentum relations.

Let $x(t)$ and $\theta(t)$ denote the coordinates to define the linear and angular positions of the trailer (M) and the compound pendulum (m). To derive the inertia influence coefficients, impulses of magnitudes m_{11} and m_{21} are applied along the directions $x(t)$ and $\theta(t)$ to result in the velocities $\dot{x} = 1$, and $\dot{\theta} = 0$. Then the linear impulse-linear momentum equation gives

$$m_{11} = (M + m)(1) \tag{E.1}$$

and the angular impulse-angular momentum equation (about O) yields

$$m_{21} = m(1)\frac{l}{2} \tag{E.2}$$

Next, impulses of magnitudes m_{12} and m_{22} are applied along the directions $x(t)$ and $\theta(t)$ to obtain the velocities $\dot{x} = 0$ and $\dot{\theta} = 1$. Then the linear impulse-linear momentum relation provides

$$m_{12} = m(1)\left(\frac{l}{2}\right) \tag{E.3}$$

and the angular impulse-angular momentum equation (about O) gives

$$m_{22} = \left(\frac{ml^2}{3}\right)(1) \tag{E.4}$$

Thus the mass or inertia matrix of the system is given by

$$[m] = \begin{bmatrix} M + m & \dfrac{ml}{2} \\ \dfrac{ml}{2} & \dfrac{ml^2}{3} \end{bmatrix} \tag{E.5}$$

■

6.5   Potential and Kinetic Energy Expressions in Matrix Form

Let x_i denote the displacement of mass m_i and F_i the force applied in the direction of x_i at mass m_i in an n degree of freedom system similar to the one shown in Fig. 6.5.

The elastic potential energy (also known as *strain energy* or *energy of deformation*) of the ith spring is given by

$$V_i = \frac{1}{2}F_i x_i \tag{6.26}$$

The total potential energy can be expressed as

$$V = \sum_{i=1}^{n} V_i = \frac{1}{2}\sum_{i=1}^{n} F_i x_i \tag{6.27}$$

Since

$$F_i = \sum_{j=1}^{n} k_{ij} x_j \tag{6.28}$$

Eq. (6.27) becomes

$$V = \frac{1}{2}\sum_{i=1}^{n}\left(\sum_{j=1}^{n} k_{ij} x_j\right)x_i = \frac{1}{2}\sum_{i=1}^{n}\sum_{j=1}^{n} k_{ij} x_i x_j \tag{6.29}$$

Equation (6.29) can also be written in matrix form as[1]

$$V = \frac{1}{2}\vec{x}^{T}[k]\vec{x} \tag{6.30}$$

where the displacement vector is given by Eq. (6.7) and the stiffness matrix is given by

$$[k] = \begin{bmatrix} k_{11} & k_{12} & \cdots & k_{1n} \\ k_{21} & k_{22} & \cdots & k_{2n} \\ \vdots & & & \\ k_{n1} & k_{n2} & \cdots & k_{nn} \end{bmatrix} \tag{6.31}$$

The kinetic energy associated with mass m_i is, by definition, equal to

$$T_i = \frac{1}{2}m_i \dot{x}_i^2 \tag{6.32}$$

The total kinetic energy of the system can be expressed as

$$T = \sum_{i=1}^{n} T_i = \frac{1}{2}\sum_{i=1}^{n} m_i \dot{x}_i^2 \tag{6.33}$$

---

[1]Since the indices i and j can be interchanged in Eq. (6.29), we have the relation $k_{ij} = k_{ji}$.

which can be written in matrix form as

$$T = \frac{1}{2}\vec{\dot{x}}^T [m] \vec{\dot{x}} \tag{6.34}$$

where the velocity vector $\vec{\dot{x}}$ is given by

$$\vec{\dot{x}} = \begin{Bmatrix} \dot{x}_1 \\ \dot{x}_2 \\ \vdots \\ \dot{x}_n \end{Bmatrix}$$

and the mass matrix $[m]$ is a diagonal matrix given by

$$[m] = \begin{bmatrix} m_1 & & & 0 \\ & m_2 & & \\ & & \ddots & \\ 0 & & & m_n \end{bmatrix} \tag{6.35}$$

If generalized coordinates (q_i), discussed in Section 6.6, are used instead of the physical displacements (x_i), the kinetic energy can be expressed as

$$T = \frac{1}{2}\vec{\dot{q}}^T [m] \vec{\dot{q}} \tag{6.36}$$

where $\vec{\dot{q}}$ is the vector of generalized velocities, given by

$$\vec{\dot{q}} = \begin{Bmatrix} \dot{q}_1 \\ \dot{q}_2 \\ \vdots \\ \dot{q}_n \end{Bmatrix} \tag{6.37}$$

and $[m]$ is called the *generalized mass matrix*, given by

$$[m] = \begin{bmatrix} m_{11} & m_{12} & \cdots & m_{1n} \\ m_{21} & m_{22} & \cdots & m_{2n} \\ \vdots & & & \\ m_{n1} & m_{n2} & \cdots & m_{nn} \end{bmatrix} \tag{6.38}$$

with $m_{ij} = m_{ji}$. The generalized mass matrix given by Eq. (6.38) is full, as opposed to the diagonal mass matrix of Eq. (6.35).

It can be seen that the potential energy is a quadratic function of the displacements, and the kinetic energy is a quadratic function of the velocities. Hence they are said to be in quadratic form. Since kinetic energy, by definition, cannot be negative and vanishes only when all the velocities vanish, Eqs. (6.34) and (6.36) are called *positive definite quadratic forms* and the mass matrix $[m]$ is called a *positive definite matrix*. On the other hand, the potential energy expression, Eq. (6.30), is a positive definite quadratic form, but the matrix $[k]$ is positive definite only if the system is a stable one. There are systems for which the potential energy is zero without the displacements or coordinates $x_1, x_2, \ldots, x_n$ being zero. In these cases the potential energy will be a positive quadratic function rather than positive definite; correspondingly, the matrix $[k]$ is said to be positive. A system for which $[k]$ is positive and $[m]$ is positive definite is called a semidefinite system (see Section 6.12).

6.6   Generalized Coordinates and Generalized Forces

The equations of motion of a vibrating system can be formulated in a number of different coordinate systems. As stated earlier, n independent coordinates are necessary to describe the motion of a system having n degrees of freedom. Any set of n independent coordinates is called generalized coordinates, usually designated by $q_1, q_2, \ldots, q_n$. The generalized coordinates may be lengths, angles, or any other set of numbers that define the configuration of the system at any time uniquely. They are also independent of the conditions of constraint.

To illustrate the concept of generalized coordinates, consider the triple pendulum shown in Fig. 6.10. The configuration of the system can be specified by the six coordinates (x_j, y_j), $j = 1, 2, 3$. However, these coordinates are not independent but are constrained by the relations

$$x_1^2 + y_1^2 = l_1^2$$
$$(x_2 - x_1)^2 + (y_2 - y_1)^2 = l_2^2$$
$$(x_3 - x_2)^2 + (y_3 - y_2)^2 = l_3^2 \tag{6.39}$$

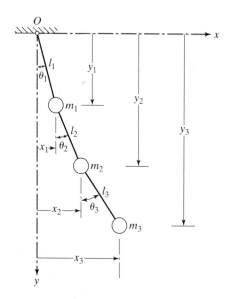

FIGURE 6.10   Triple pendulum.

Since the coordinates (x_j, y_j), $j = 1, 2, 3$ are not independent, they cannot be called generalized coordinates. Without the constraints of Eq. (6.39), each of the masses m_1, m_2, and m_3 will be free to occupy any position in the x, y plane. The constraints eliminate three degrees of freedom from the six coordinates (two for each mass) and the system thus has only three degrees of freedom. If the angular displacements $\theta_j (j = 1, 2, 3)$ are used to specify the locations of the masses $m_j (j = 1, 2, 3)$ at any time, there will be no constraints on θ_j. Thus they form a set of generalized coordinates and are denoted as $q_j = \theta_j$, $j = 1, 2, 3$.

When external forces act on the system, the configuration of the system changes. The new configuration of the system can be obtained by changing the generalized coordinates q_j by δq_j, $j = 1, 2, \ldots, n$, where n denotes the number of generalized coordinates (or degrees of freedom) of the system. If U_j denotes the work done in changing the generalized coordinate q_j by the amount δq_j, the corresponding generalized force Q_j can be defined as

$$Q_j = \frac{U_j}{\delta q_j}, \qquad j = 1, 2, \ldots, n \tag{6.40}$$

where Q_j will be a force (moment) when q_j is a linear (angular) displacement.

6.7   Using Lagrange's Equations to Derive Equations of Motion

The equations of motion of a vibrating system can often be derived in a simple manner in terms of generalized coordinates by the use of Lagrange's equations [6.3]. Lagrange's equations can be stated, for an n degree of freedom system, as

$$\frac{d}{dt}\left(\frac{\partial T}{\partial \dot{q}_j}\right) - \frac{\partial T}{\partial q_j} + \frac{\partial V}{\partial q_j} = Q_j^{(n)}, \qquad j = 1, 2, \ldots, n \tag{6.41}$$

where $\dot{q}_j = \partial q_j / \partial t$ is the generalized velocity and $Q_j^{(n)}$ is the nonconservative generalized force corresponding to the generalized coordinate q_j. The forces represented by $Q_j^{(n)}$ may be dissipative (damping) forces or other external forces that are not derivable from a potential function. For example, if F_{xk}, F_{yk}, and F_{zk} represent the external forces acting on the kth mass of the system in the x, y, and z directions, respectively, then the generalized force $Q_j^{(n)}$ can be computed as follows:

$$Q_j^{(n)} = \sum_k \left(F_{xk}\frac{\partial x_k}{\partial q_j} + F_{yk}\frac{\partial y_k}{\partial q_j} + F_{zk}\frac{\partial z_k}{\partial q_j}\right) \tag{6.42}$$

where x_k, y_k, and z_k are the displacements of the kth mass in the x, y, and z directions, respectively. Note that for a torsional system, the force F_{xk}, for example, is to be replaced by the moment acting about the x axis (M_{xk}), and the displacement x_k by the angular displacement about the x axis (θ_{xk}) in Eq. (6.42). For a conservative system, $Q_j^{(n)} = 0$, so Eq. (6.41) takes the form

$$\frac{d}{dt}\left(\frac{\partial T}{\partial \dot{q}_j}\right) - \frac{\partial T}{\partial q_j} + \frac{\partial V}{\partial q_j} = 0, \qquad j = 1, 2, \dots, n \qquad (6.43)$$

Equations (6.41) or (6.43) represent a system of n differential equations, one corresponding to each of the n generalized coordinates. Thus the equations of motion of the vibrating system can be derived, provided the energy expressions are available.

Equations of Motion of a Torsional System

EXAMPLE 6.8

The arrangement of the compressor, turbine, and generator in a thermal power plant is shown in Fig. 6.11. This arrangement can be considered as a torsional system where J_i denote the mass moments of inertia of the three components (compressor, turbine, and generator), M_{ti} indicate the external moments acting on the components, and k_{ti} represent the torsional spring constants of the shaft between the components, as indicated in Fig. 6.11. Derive the equations of motion of the system using Lagrange's equations by treating the angular displacements of the components θ_i as generalized coordinates.

Solution: Here $q_1 = \theta_1$, $q_2 = \theta_2$, and $q_3 = \theta_3$, and the kinetic energy of the system is given by

$$T = \frac{1}{2}J_1 \dot{\theta}_1^2 + \frac{1}{2}J_2 \dot{\theta}_2^2 + \frac{1}{2}J_3 \dot{\theta}_3^2 \qquad (E.1)$$

For the shaft, the potential energy is equal to the work done by the shaft as it returns from the dynamic configuration to the reference equilibrium position. Thus if θ denotes the angular displacement, for a shaft having a torsional spring constant k_t, the potential energy is equal to the work done in causing an angular displacement θ of the shaft:

$$V = \int_0^\theta (k_t \theta)\, d\theta = \frac{1}{2}k_t \theta^2 \qquad (E.2)$$

Thus the total potential energy of the system can be expressed as

$$V = \frac{1}{2}k_{t1}\theta_1^2 + \frac{1}{2}k_{t2}(\theta_2 - \theta_1)^2 + \frac{1}{2}k_{t3}(\theta_3 - \theta_2)^2 \qquad (E.3)$$

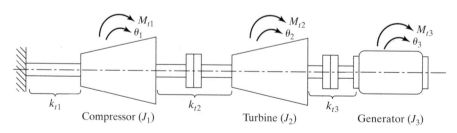

FIGURE 6.11   Torsional system.

There are external moments applied to the components, so Eq. (6.42) gives

$$Q_j^{(n)} = \sum_{k=1}^{3} M_{tk} \frac{\partial \theta_k}{\partial q_j} = \sum_{k=1}^{3} M_{tk} \frac{\partial \theta_k}{\partial \theta_j} \tag{E.4}$$

from which we can obtain

$$Q_1^{(n)} = M_{t1} \frac{\partial \theta_1}{\partial \theta_1} + M_{t2} \frac{\partial \theta_2}{\partial \theta_1} + M_{t3} \frac{\partial \theta_3}{\partial \theta_1} = M_{t1}$$

$$Q_2^{(n)} = M_{t1} \frac{\partial \theta_1}{\partial \theta_2} + M_{t2} \frac{\partial \theta_2}{\partial \theta_2} + M_{t3} \frac{\partial \theta_3}{\partial \theta_2} = M_{t2}$$

$$Q_3^{(n)} = M_{t1} \frac{\partial \theta_1}{\partial \theta_3} + M_{t2} \frac{\partial \theta_2}{\partial \theta_3} + M_{t3} \frac{\partial \theta_3}{\partial \theta_3} = M_{t3} \tag{E.5}$$

Substituting Eqs. (E.1), (E.3), and (E.5) in Lagrange's equations, Eq. (6.41), we obtain for $j = 1, 2, 3$ the equations of motion

$$J_1 \ddot{\theta}_1 + (k_{t1} + k_{t2})\theta_1 - k_{t2}\theta_2 = M_{t1}$$
$$J_2 \ddot{\theta}_2 + (k_{t2} + k_{t3})\theta_2 - k_{t2}\theta_1 - k_{t3}\theta_3 = M_{t2}$$
$$J_3 \ddot{\theta}_3 + k_{t3}\theta_3 - k_{t3}\theta_2 = M_{t3} \tag{E.6}$$

which can be expressed in matrix form as

$$\begin{bmatrix} J_1 & 0 & 0 \\ 0 & J_2 & 0 \\ 0 & 0 & J_3 \end{bmatrix} \begin{Bmatrix} \ddot{\theta}_1 \\ \ddot{\theta}_2 \\ \ddot{\theta}_3 \end{Bmatrix} + \begin{bmatrix} (k_{t1} + k_{t2}) & -k_{t2} & 0 \\ -k_{t2} & (k_{t2} + k_{t3}) & -k_{t3} \\ 0 & -k_{t3} & k_{t3} \end{bmatrix} \begin{Bmatrix} \theta_1 \\ \theta_2 \\ \theta_3 \end{Bmatrix}$$

$$= \begin{Bmatrix} M_{t1} \\ M_{t2} \\ M_{t3} \end{Bmatrix} \tag{E.7}$$

$\blacksquare$

EXAMPLE 6.9

Lagrange's Equations

Derive the equations of motion of the trailer–compound pendulum system shown in Fig. 6.4(a).

Solution: The coordinates $x(t)$ and $\theta(t)$ can be used as generalized coordinates to describe, respectively, the linear displacement of the trailer and the angular displacement of the compound pendulum. If a y-coordinate is introduced, for convenience, as shown in Fig. 6.4(a), the displacement components of point C can be expressed as

$$x_C = x + \frac{l}{2} \sin \theta \tag{E.1}$$

$$y_C = \frac{l}{2} \cos \theta \tag{E.2}$$

Differentiation of Eqs. (E.1) and (E.2) with respect to time gives the velocities of point C as

$$\dot{x}_C = \dot{x} + \frac{l}{2} \dot{\theta} \cos \theta \tag{E.3}$$

$$\dot{y}_C = -\frac{l}{2} \dot{\theta} \sin \theta \tag{E.4}$$

The kinetic energy of the system, T, can be expressed as

$$T = \frac{1}{2} M \dot{x}^2 + \frac{1}{2} m (\dot{x}_C^2 + \dot{y}_C^2) + J_C \dot{\theta}^2 \tag{E.5}$$

where $J_C = \frac{1}{12} m l^2$. Using Eqs. (E.3) and (E.4), Eq. (E.5) can be rewritten as

$$
\begin{aligned}
T &= \frac{1}{2} M \dot{x}^2 + \frac{1}{2} m \left(\dot{x}^2 + \frac{l^2 \dot{\theta}^2}{4} + \dot{x} \dot{\theta} l \cos \theta \right) + \frac{1}{2} \left(\frac{m l^2}{12} \right) \dot{\theta}^2 \\
&= \frac{1}{2} (M + m) \dot{x}^2 + \frac{1}{2} \left(\frac{m l^2}{3} \right) \dot{\theta}^2 + \frac{1}{2} (m l \cos \theta) \dot{x} \dot{\theta} \tag{E.6}
\end{aligned}
$$

The potential energy of the system, V, due to the strain energy of the springs and the gravitational potential, can be expressed as

$$V = \frac{1}{2} k_1 x^2 + \frac{1}{2} k_2 x^2 + mg \frac{l}{2} (1 - \cos \theta) \tag{E.7}$$

where the lowest position of point C is taken as the datum. Since there are nonconservative forces acting on the system, the generalized forces corresponding to $x(t)$ and $\theta(t)$ are to be computed. The force, $X(t)$, acting in the direction of $x(t)$ can be found from Eq. (6.42) as

$$X(t) = Q_1^{(n)} = F(t) - c_1 \dot{x}(t) - c_2 \dot{x}(t) \tag{E.8}$$

where the negative sign for the terms $c_1 \dot{x}$ and $c_2 \dot{x}$ indicates that the damping forces oppose the motion. Similarly, the force $\Theta(t)$ acting in the direction of $\theta(t)$ can be determined as

$$\Theta(t) = Q_2^{(n)} = M_t(t) \tag{E.9}$$

where $q_1 = x$ and $q_2 = \theta$. By differentiating the expressions of T and V as required by Eqs. (6.41) and substituting the resulting expressions, along with Eqs. (E.8) and (E.9), we obtain the equations of motion of the system as

$$(M + m) \ddot{x} + \frac{1}{2} (m l \cos \theta) \ddot{\theta} - \frac{1}{2} m l \sin \theta \dot{\theta}^2 + k_1 x + k_2 x$$

$$= F(t) - c_1 \dot{x} - c_2 \dot{x} \tag{E.10}$$

$$\left(\frac{1}{3}ml^2\right)\ddot{\theta} + \frac{1}{2}(ml\cos\theta)\ddot{x} - \frac{1}{2}ml\sin\theta\,\dot{\theta}\,\dot{x} + \frac{1}{2}ml\sin\theta\,\dot{\theta}\,\dot{x}$$

$$+ \frac{1}{2}mgl\sin\theta = M_t(t) \tag{E.11}$$

Equations (E.10) and (E.11) can be seen to be identical to those obtained using Newton's second law of motion (Eqs. E.1 and E.2 in Example 6.2).

∎

6.8   Equations of Motion of Undamped Systems in Matrix Form

We can derive the equations of motion of a multidegree of freedom system in matrix form from Lagrange's equations.[2]

$$\frac{d}{dt}\left(\frac{\partial T}{\partial \dot{x}_i}\right) - \frac{\partial T}{\partial x_i} + \frac{\partial V}{\partial x_i} = F_i, \quad i = 1, 2, \dots n \tag{6.44}$$

where F_i is the nonconservative generalized force corresponding to the ith generalized coordinate x_i and $\dot{x}_i$ is the time derivative of x_i (generalized velocity). The kinetic and potential energies of a multidegree of freedom system can be expressed in matrix form as indicated in Section 6.5

$$T = \frac{1}{2}\vec{\dot{x}}^T[m]\vec{\dot{x}} \tag{6.45}$$

$$V = \frac{1}{2}\vec{x}^T[k]\vec{x} \tag{6.46}$$

where $\vec{x}$ is the column vector of the generalized coordinates

$$\vec{x} = \begin{Bmatrix} x_1 \\ x_2 \\ \cdot \\ \cdot \\ \cdot \\ x_n \end{Bmatrix} \tag{6.47}$$

From the theory of matrices, we obtain, by taking note of the symmetry of $[m]$,

$$\frac{\partial T}{\partial \dot{x}_i} = \frac{1}{2}\vec{\delta}^T[m]\vec{\dot{x}} + \frac{1}{2}\vec{\dot{x}}^T[m]\vec{\delta} = \vec{\delta}^T[m]\vec{\dot{x}}$$

$$= \vec{m}_i^T\vec{\dot{x}}, \quad i = 1, 2, \dots, n \tag{6.48}$$

---

[2]The generalized coordinates are denoted as x_i instead of q_i and the generalized forces as F_i instead of $Q_i^{(n)}$ in Eq. (6.44).

where δ_{ji} is the Kronecker delta ($\delta_{ji} = 1$ if $j = i$ and $= 0$ if $j \neq i$), $\vec{\delta}$ is the column vector of Kronecker deltas whose elements in the rows for which $j \neq i$ are equal to zero and whose element in the row $i = j$ is equal to 1, and $\vec{m}_i^T$ is a row vector which is identical to the ith row of the matrix $[m]$. All the relations represented by Eq. (6.48) can be expressed as

$$\frac{\partial T}{\partial \dot{x}_i} = \vec{m}_i^T \, \dot{\vec{x}} \tag{6.49}$$

Differentiation of Eq. (6.49) with respect to time gives

$$\frac{d}{dt}\left(\frac{\partial T}{\partial \dot{x}_i}\right) = \vec{m}_i^T \, \ddot{\vec{x}}, \qquad i = 1, 2, \ldots, n \tag{6.50}$$

since the mass matrix is not a function of time. Further, the kinetic energy is a function of only the velocities $\dot{x}_i$, and so

$$\frac{\partial T}{\partial x_i} = 0, \qquad i = 1, 2, \ldots, n \tag{6.51}$$

Similarly, we can differentiate Eq. (6.46), taking note of the symmetry of $[k]$,

$$\frac{\partial V}{\partial x_i} = \frac{1}{2}\vec{\delta}^T[k]\vec{x} + \frac{1}{2}\vec{x}^T[k]\vec{\delta} = \vec{\delta}^T[k]\vec{x}$$

$$= \vec{k}_i^T\vec{x}, \qquad i = 1, 2, \ldots, n \tag{6.52}$$

where $\vec{k}_i^T$ is a row vector identical to the ith row of the matrix $[k]$. By substituting Eqs. (6.50) to (6.52) into Eq. (6.44), we obtain the desired equations of motion in matrix form

$$[m]\ddot{\vec{x}} + [k]\vec{x} = \vec{F} \tag{6.53}$$

where

$$\vec{F} = \begin{Bmatrix} F_1 \\ F_2 \\ \cdot \\ \cdot \\ \cdot \\ F_n \end{Bmatrix} \tag{6.54}$$

Note that if the system is conservative, there are no nonconservative forces F_i, so the equations of motion become

$$[m]\ddot{\vec{x}} + [k]\vec{x} = \vec{0} \tag{6.55}$$

Note also that if the generalized coordinates x_i are same as the actual (physical) displacements, the mass matrix $[m]$ is a diagonal matrix.

6.9   Eigenvalue Problem

The solution of Eq. (6.55) corresponds to the undamped free vibration of the system. In this case, if the system is given some energy in the form of initial displacements or initial velocities or both, it vibrates indefinitely because there is no dissipation of energy. We can find the solution of Eq. (6.55) by assuming a solution of the form

$$x_i(t) = X_i T(t), \qquad i = 1, 2, \ldots, n \tag{6.56}$$

where X_i is a constant and T is a function of time t. Equation (6.56) shows that the amplitude ratio of two coordinates

$$\left\{ \frac{x_i(t)}{x_j(t)} \right\}$$

is independent of time. Physically, this means that all coordinates have synchronous motions. The configuration of the system does not change its shape during motion, but its amplitude does. The configuration of the system, given by the vector

$$\vec{X} = \begin{Bmatrix} X_1 \\ X_2 \\ \cdot \\ \cdot \\ \cdot \\ X_n \end{Bmatrix}$$

is known as the *mode shape* of the system. Substituting Eq. (6.56) into Eq. (6.55), we obtain

$$[m]\vec{X}\ddot{T}(t) + [k]\vec{X}T(t) = \vec{0} \tag{6.57}$$

Equation (6.57) can be written in scalar form as n separate equations

$$\left(\sum_{j=1}^{n} m_{ij}X_j \right) \ddot{T}(t) + \left(\sum_{j=1}^{n} k_{ij}X_j \right) T(t) = 0, \qquad i = 1, 2, \ldots, n \tag{6.58}$$

from which we can obtain the relations

$$-\frac{\ddot{T}(t)}{T(t)} = \frac{\left(\sum\limits_{j=1}^{n} k_{ij}X_j \right)}{\left(\sum\limits_{j=1}^{n} m_{ij}X_j \right)}, \qquad i = 1, 2, \ldots, n \tag{6.59}$$

Since the left side of Eq. (6.59) is independent of the index i, and the right side is independent of t, both sides must be equal to a constant. By assuming this constant[3] as ω^2, we can write Eq. (6.59) as

$$\ddot{T}(t) + \omega^2 T(t) = 0 \tag{6.60}$$

$$\sum_{j=1}^{n} (k_{ij} - \omega^2 m_{ij}) X_j = 0, \qquad i = 1, 2, \ldots, n$$

---

[3]The constant is assumed to be a positive number, ω^2, so as to obtain a harmonic solution to the resulting Eq. (6.60). Otherwise, the solution of $T(t)$ and hence that of $x(t)$ become exponential, which violates the physical limitations of finite total energy.

or

$$\left[[k] - \omega^2[m] \right]\vec{X} = \vec{0} \tag{6.61}$$

The solution of Eq. (6.60) can be expressed as

$$T(t) = C_1 \cos(\omega t + \phi) \tag{6.62}$$

where C_1 and ϕ are constants, known as the *amplitude* and the *phase angle*, respectively. Equation (6.62) shows that all the coordinates can perform a harmonic motion with the same frequency ω and the same phase angle ϕ. However, the frequency ω cannot take any arbitrary value; it has to satisfy Eq. (6.61). Since Eq. (6.61) represents a set of n linear homogeneous equations in the unknowns $X_i (i = 1, 2, \ldots, n)$, the trivial solution is $X_1 = X_2 = \cdots = X_n = 0$. For a nontrivial solution of Eq. (6.61), the determinant Δ of the coefficient matrix must be zero. That is,

$$\Delta = |k_{ij} - \omega^2 m_{ij}| = |[k] - \omega^2[m]| = 0 \tag{6.63}$$

Equation (6.61) represents what is known as the *eigenvalue* or *characteristic value* problem, Eq. (6.63) is called the *characteristic equation*, ω^2 is known as the *eigenvalue* or the *characteristic value*, and ω is called the *natural frequency* of the system.

The expansion of Eq. (6.63) leads to an nth order polynomial equation in ω^2. The solution (roots) of this polynomial or characteristic equation gives n values of ω^2. It can be shown that all the n roots are real and positive when the matrices $[k]$ and $[m]$ are symmetric and positive definite [6.4], as in the present case. If $\omega_1^2, \omega_2^2, \ldots, \omega_n^2$ denote the n roots in ascending order of magnitude, their positive square roots give the n natural frequencies of the system $\omega_1 \le \omega_2 \le \cdots \le \omega_n$. The lowest value ($\omega_1$) is called the *fundamental* or *first natural frequency*. In general, all the natural frequencies ω_i are distinct, although in some cases two natural frequencies might possess the same value.

6.10 Solution of the Eigenvalue Problem

Several methods are available to solve an eigenvalue problem. We shall consider an elementary method in this section.

6.10.1 Solution of the Characteristic (Polynomial) Equation

Equation (6.61) can also be expressed as

$$\left[\lambda[k] - [m] \right]\vec{X} = \vec{0} \tag{6.64}$$

where

$$\lambda = \frac{1}{\omega^2} \tag{6.65}$$

By premultiplying Eq. (6.64) by $[k]^{-1}$, we obtain

$$\left[\lambda[I] - [D] \right]\vec{X} = \vec{0}$$

or

$$\lambda[I]\vec{X} = [D]\vec{X} \tag{6.66}$$

where $[I]$ is the identity matrix and

$$[D] = [k]^{-1}[m] \tag{6.67}$$

is called the *dynamical matrix*. The eigenvalue problem of Eq. (6.66) is known as the *standard eigenvalue problem*. For a nontrivial solution of $\vec{X}$, the characteristic determinant must be zero—that is,

$$\Delta = |\lambda[I] - [D]| = 0 \tag{6.68}$$

On expansion, Eq. (6.68) gives an nth degree polynomial in λ, known as the *characteristic* or *frequency equation*. If the degree of freedom of the system (n) is large, the solution of this polynomial equation becomes quite tedious. We must use some numerical method, several of which are available to find the roots of a polynomial equation [6.5].

Natural Frequencies of a Three Degree of Freedom System

EXAMPLE 6.10

Find the natural frequencies and mode shapes of the system shown in Fig. 6.8(a) for $k_1 = k_2 = k_3 = k$ and $m_1 = m_2 = m_3 = m$.

Solution: The dynamical matrix is given by

$$[D] = [k]^{-1}[m] \equiv [a][m] \tag{E.1}$$

where the flexibility and mass matrices can be obtained from Example 6.5:

$$[a] = \frac{1}{k}\begin{bmatrix} 1 & 1 & 1 \\ 1 & 2 & 2 \\ 1 & 2 & 3 \end{bmatrix} \tag{E.2}$$

and

$$[m] = m\begin{bmatrix} 1 & 0 & 0 \\ 0 & 1 & 0 \\ 0 & 0 & 1 \end{bmatrix} \tag{E.3}$$

Thus

$$[D] = \frac{m}{k}\begin{bmatrix} 1 & 1 & 1 \\ 1 & 2 & 2 \\ 1 & 2 & 3 \end{bmatrix} \tag{E.4}$$

By setting the characteristic determinant equal to zero, we obtain the frequency equation

$$\Delta = |\lambda[I] - [D]| = \left| \begin{bmatrix} \lambda & 0 & 0 \\ 0 & \lambda & 0 \\ 0 & 0 & \lambda \end{bmatrix} - \frac{m}{k}\begin{bmatrix} 1 & 1 & 1 \\ 1 & 2 & 2 \\ 1 & 2 & 3 \end{bmatrix} \right| = 0 \tag{E.5}$$

where

$$\lambda = \frac{1}{\omega^2} \tag{E.6}$$

By dividing throughout by λ, Eq. (E.5) gives

$$\begin{vmatrix} 1 - \alpha & -\alpha & -\alpha \\ -\alpha & 1 - 2\alpha & -2\alpha \\ -\alpha & -2\alpha & 1 - 3\alpha \end{vmatrix} = \alpha^3 - 5\alpha^2 + 6\alpha - 1 = 0 \tag{E.7}$$

where

$$\alpha = \frac{m}{k\lambda} = \frac{m\omega^2}{k} \tag{E.8}$$

The roots of the cubic equation (E.7) are given by

$$\alpha_1 = \frac{m\omega_1^2}{k} = 0.19806, \qquad \omega_1 = 0.44504\sqrt{\frac{k}{m}} \tag{E.9}$$

$$\alpha_2 = \frac{m\omega_2^2}{k} = 1.5553, \qquad \omega_2 = 1.2471\sqrt{\frac{k}{m}} \tag{E.10}$$

$$\alpha_3 = \frac{m\omega_3^2}{k} = 3.2490, \qquad \omega_3 = 1.8025\sqrt{\frac{k}{m}} \tag{E.11}$$

Once the natural frequencies are known, the mode shapes or eigenvectors can be calculated using Eq. (6.66):

$$[\lambda_i[I] - [D]]\, \vec{X}^{(i)} = \vec{0}, \qquad i = 1, 2, 3 \tag{E.12}$$

where

$$\vec{X}^{(i)} = \begin{Bmatrix} X_1^{(i)} \\ X_2^{(i)} \\ X_3^{(i)} \end{Bmatrix}$$

denotes the ith mode shape. The procedure is outlined below.

First Mode: By substituting the value of ω_1 $\left(\text{i.e., } \lambda_1 = 5.0489\frac{m}{k}\right)$ in Eq. (E.12), we obtain

$$\left[5.0489\frac{m}{k}\begin{bmatrix} 1 & 0 & 0 \\ 0 & 1 & 0 \\ 0 & 0 & 1 \end{bmatrix} - \frac{m}{k}\begin{bmatrix} 1 & 1 & 1 \\ 1 & 2 & 2 \\ 1 & 2 & 3 \end{bmatrix} \right] \begin{Bmatrix} X_1^{(1)} \\ X_2^{(1)} \\ X_3^{(1)} \end{Bmatrix} = \begin{Bmatrix} 0 \\ 0 \\ 0 \end{Bmatrix}$$

That is,

$$\begin{bmatrix} 4.0489 & -1.0 & -1.0 \\ -1.0 & 3.0489 & -2.0 \\ -1.0 & -2.0 & 2.0489 \end{bmatrix} \begin{Bmatrix} X_1^{(1)} \\ X_2^{(1)} \\ X_3^{(1)} \end{Bmatrix} = \begin{Bmatrix} 0 \\ 0 \\ 0 \end{Bmatrix} \qquad (E.13)$$

Equation (E.13) denotes a system of three homogeneous linear equations in the three unknowns $X_1^{(1)}$, $X_2^{(1)}$, and $X_3^{(1)}$. Any two of these unknowns can be expressed in terms of the remaining one. If we choose, arbitrarily, to express $X_2^{(1)}$ and $X_3^{(1)}$ in terms of $X_1^{(1)}$, we obtain from the first two rows of Eq. (E.13)

$$X_2^{(1)} + X_3^{(1)} = 4.0489 \, X_1^{(1)}$$

$$3.0489 X_2^{(1)} - 2.0 \, X_3^{(1)} = X_1^{(1)} \qquad (E.14)$$

Once Eqs. (E.14) are satisfied, the third row of Eq. (E.13) is satisfied automatically. The solution of Eqs. (E.14) can be obtained:

$$X_2^{(1)} = 1.8019 \, X_1^{(1)} \qquad \text{and} \qquad X_3^{(1)} = 2.2470 \, X_1^{(1)} \qquad (E.15)$$

Thus the first mode shape is given by

$$\vec{X}^{(1)} = X_1^{(1)} \begin{Bmatrix} 1.0 \\ 1.8019 \\ 2.2470 \end{Bmatrix} \qquad (E.16)$$

where the value of $X_1^{(1)}$ can be chosen arbitrarily.

Second Mode: The substitution of the value of $\omega_2 \left(\text{i.e., } \lambda_2 = 0.6430 \frac{m}{k} \right)$ in Eq. (E.12) leads to

$$\begin{bmatrix} 0.6430 \frac{m}{k} \begin{bmatrix} 1 & 0 & 0 \\ 0 & 1 & 0 \\ 0 & 0 & 1 \end{bmatrix} - \frac{m}{k} \begin{bmatrix} 1 & 1 & 1 \\ 1 & 2 & 2 \\ 1 & 2 & 3 \end{bmatrix} \end{bmatrix} \begin{Bmatrix} X_1^{(2)} \\ X_2^{(2)} \\ X_3^{(2)} \end{Bmatrix} = \begin{Bmatrix} 0 \\ 0 \\ 0 \end{Bmatrix}$$

that is,

$$\begin{bmatrix} -0.3570 & -1.0 & -1.0 \\ -1.0 & -1.3570 & -2.0 \\ -1.0 & -2.0 & -2.3570 \end{bmatrix} \begin{Bmatrix} X_1^{(2)} \\ X_2^{(2)} \\ X_3^{(2)} \end{Bmatrix} = \begin{Bmatrix} 0 \\ 0 \\ 0 \end{Bmatrix} \qquad (E.17)$$

As before, the first two rows of Eq. (E.17) can be used to obtain

$$-X_2^{(2)} - X_3^{(2)} = 0.3570 \, X_1^{(2)}$$

$$- 1.3570 X_2^{(2)} - 2.0 \, X_3^{(2)} = X_1^{(2)} \qquad (E.18)$$

The solution of Eqs. (E.18) leads to

$$X_2^{(2)} = 0.4450 \, X_1^{(2)} \qquad \text{and} \qquad X_3^{(2)} = -0.8020 \, X_1^{(2)} \qquad (E.19)$$

Thus the second mode shape can be expressed as

$$\vec{X}^{(2)} = X_1^{(2)} \begin{Bmatrix} 1.0 \\ 0.4450 \\ -0.8020 \end{Bmatrix} \tag{E.20}$$

where the value of $X_1^{(2)}$ can be chosen arbitrarily.

Third Mode: To find the third mode, we substitute the value of $\omega_3 \left(\text{i.e., } \lambda_3 = 0.3078 \frac{m}{k} \right)$ in Eq. (E.12) and obtain

$$\left[0.3078 \frac{m}{k} \begin{bmatrix} 1 & 0 & 0 \\ 0 & 1 & 0 \\ 0 & 0 & 1 \end{bmatrix} - \frac{m}{k} \begin{bmatrix} 1 & 1 & 1 \\ 1 & 2 & 2 \\ 1 & 2 & 3 \end{bmatrix} \right] \begin{Bmatrix} X_1^{(3)} \\ X_2^{(3)} \\ X_3^{(3)} \end{Bmatrix} = \begin{Bmatrix} 0 \\ 0 \\ 0 \end{Bmatrix}$$

that is,

$$\begin{bmatrix} -0.6922 & -1.0 & -1.0 \\ -1.0 & -1.6922 & -2.0 \\ -1.0 & -2.0 & -2.6922 \end{bmatrix} \begin{Bmatrix} X_1^{(3)} \\ X_2^{(3)} \\ X_3^{(3)} \end{Bmatrix} = \begin{Bmatrix} 0 \\ 0 \\ 0 \end{Bmatrix} \tag{E.21}$$

The first two rows of Eq. (E.21) can be written as

$$-X_2^{(3)} - X_3^{(3)} = 0.6922 \, X_1^{(3)}$$
$$-1.6922 \, X_2^{(3)} - 2.0 \, X_3^{(3)} = X_1^{(3)} \tag{E.22}$$

Equations (E.22) give

$$X_2^{(3)} = -1.2468 \, X_1^{(3)} \quad \text{and} \quad X_3^{(3)} = 0.5544 \, X_1^{(3)} \tag{E.23}$$

Hence the third mode shape can be written as

$$\vec{X}^{(3)} = X_1^{(3)} \begin{Bmatrix} 1.0 \\ -1.2468 \\ 0.5544 \end{Bmatrix} \tag{E.24}$$

where the value of $X_1^{(3)}$ is arbitrary. The values of $X_1^{(1)}$, $X_1^{(2)}$, and $X_1^{(3)}$ are usually taken as 1, and the mode shapes are shown in Fig. 6.12.

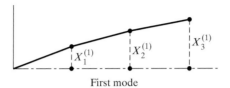

First mode

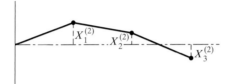

Second mode

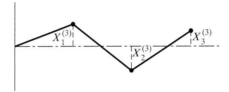

Third mode

FIGURE 6.12   Mode shapes of three
degree of freedom system.

■

**6.10.2
Orthogonality of
Normal Modes**

In the previous section we considered a method of finding the n natural frequencies ω_i and
the corresponding normal modes or modal vectors $\vec{X}^{(i)}$. We shall now see an important
property of the normal modes—*orthogonality*. The natural frequency ω_i and the corresponding modal vector $\vec{X}^{(i)}$ satisfy Eq. (6.61) so that

$$\omega_i^2 [m] \, \vec{X}^{(i)} = [k] \, \vec{X}^{(i)} \tag{6.69}$$

If we consider another natural frequency ω_j and the corresponding modal vector $\vec{X}^{(j)}$, they
also satisfy Eq. (6.61) so that

$$\omega_j^2 [m] \, \vec{X}^{(j)} = [k] \, \vec{X}^{(j)} \tag{6.70}$$

By premultiplying Eqs. (6.69) and (6.70) by $\vec{X}^{(j)^T}$ and $\vec{X}^{(i)^T}$ respectively, we obtain, by
considering the symmetry of the matrices $[k]$ and $[m]$,

$$\omega_i^2 \, \vec{X}^{(j)T} \, [m] \, \vec{X}^{(i)} = \vec{X}^{(j)T} \, [k] \, \vec{X}^{(i)} \equiv \vec{X}^{(i)T} \, [k] \vec{X}^{(j)} \tag{6.71}$$

$$\omega_j^2 \, \vec{X}^{(i)T} [m] \, \vec{X}^{(j)} \equiv \omega_j^2 \vec{X}^{(j)T} [m] \, \vec{X}^{(i)} = \vec{X}^{(i)T} \, [k] \, \vec{X}^{(j)} \tag{6.72}$$

By subtracting Eq. (6.72) from Eq. (6.71), we obtain

$$(\omega_i^2 - \omega_j^2) \, \vec{X}^{(j)T} \, [m] \, \vec{X}^{(i)} = 0 \tag{6.73}$$

In general, $\omega_i^2 \neq \omega_j^2$, so Eq. (6.73) leads to[4]

$$\vec{X}^{(j)T} \, [m] \, \vec{X}^{(i)} = 0, \qquad i \neq j \tag{6.74}$$

From Eqs. (6.71) and (6.72), we obtain, in view of Eq. (6.74),

$$\vec{X}^{(j)T} \, [k] \, \vec{X}^{(i)} = 0, \qquad i \neq j \tag{6.75}$$

Equations (6.74) and (6.75) indicate that the modal vectors $\vec{X}^{(i)}$ and $\vec{X}^{(j)}$ are orthogonal with respect to both mass and stiffness matrices.

When $i = j$, the left sides of Eqs. (6.74) and (6.75) are not equal to zero, but they do yield the generalized mass and stiffness coefficients of the ith mode:

$$M_{ii} = \vec{X}^{(i)T} \, [m] \, \vec{X}^{(i)}, \qquad i = 1, 2, \ldots, n \tag{6.76}$$

$$K_{ii} = \vec{X}^{(i)T} \, [k] \, \vec{X}^{(i)}, \qquad i = 1, 2, \ldots, n \tag{6.77}$$

Equations (6.76) and (6.77) can be written in matrix form as

$$\begin{bmatrix} \ddots M \ddots \end{bmatrix} = \begin{bmatrix} M_{11} & & & 0 \\ & M_{22} & & \\ & & \ddots & \\ 0 & & & M_{nn} \end{bmatrix} = [X]^T [m][X] \tag{6.78}$$

$$\begin{bmatrix} \ddots K \ddots \end{bmatrix} = \begin{bmatrix} K_{11} & & & 0 \\ & K_{22} & & \\ & & \ddots & \\ 0 & & & K_{nn} \end{bmatrix} = [X]^T [k][X] \tag{6.79}$$

where $[X]$ is called the *modal matrix*, in which the ith column corresponds to the ith modal vector:

$$[X] = [\vec{X}^{(1)} \, \vec{X}^{(2)} \cdots \vec{X}^{(n)}] \tag{6.80}$$

---

[4]In the case of repeated eigenvalues, $\omega_i = \omega_j$, the associated modal vectors are orthogonal to all the remaining modal vectors but are not usually orthogonal to each other.

In many cases, we normalize the modal vectors $\vec{X}^{(i)}$ such that $\left[\diagdown M \diagdown\right] = [I]$—that is,

$$\vec{X}^{(i)T} [m] \; \vec{X}^{(i)} = 1, \qquad i = 1, 2, \ldots, n \tag{6.81}$$

In this case the matrix $\left[\diagdown K \diagdown\right]$ reduces to

$$\left[\diagdown K \diagdown\right] = \left[\diagdown \omega_i^2 \diagdown\right] = \begin{bmatrix} \omega_1^2 & & & 0 \\ & \omega_2^2 & & \\ & & \ddots & \\ 0 & & & \omega_n^2 \end{bmatrix} \tag{6.82}$$

Orthonormalization of Eigenvectors

EXAMPLE 6.11

Orthonormalize the eigenvectors of Example 6.10 with respect to the mass matrix.

Solution

Approach: Multiply each eigenvector by a constant and find its value from the relation $\vec{X}^{(i)T}[m]\vec{X}^{(i)} = 1$, $i = 1, 2, 3$.

The eigenvectors of Example 6.10 are given by

$$\vec{X}^{(1)} = X_1^{(1)} \begin{Bmatrix} 1.0 \\ 1.8019 \\ 2.2470 \end{Bmatrix}$$

$$\vec{X}^{(2)} = X_1^{(2)} \begin{Bmatrix} 1.0 \\ 0.4450 \\ -0.8020 \end{Bmatrix}$$

$$\vec{X}^{(3)} = X_1^{(3)} \begin{Bmatrix} 1.0 \\ -1.2468 \\ 0.5544 \end{Bmatrix}$$

The mass matrix is given by

$$[m] = m \begin{bmatrix} 1 & 0 & 0 \\ 0 & 1 & 0 \\ 0 & 0 & 1 \end{bmatrix}$$

The eigenvector $\vec{X}^{(i)}$ is said to be $[m]$-orthonormal if the following condition is satisfied:

$$\vec{X}^{(i)T} [m] \; \vec{X}^{(i)} = 1 \tag{E.1}$$

Thus for $i = 1$, Eq. (E.1) leads to

$$m(X_1^{(1)})^2 (1.0^2 + 1.8019^2 + 2.2470^2) = 1$$

or

$$X_1^{(1)} = \frac{1}{\sqrt{m(9.2959)}} = \frac{0.3280}{\sqrt{m}}$$

Similarly, for $i = 2$ and $i = 3$, Eq. (E.1) gives

$$m(X_1^{(2)})^2 (1.0^2 + 0.4450^2 + \{-0.8020\}^2) = 1 \quad \text{or} \quad X_1^{(2)} = \frac{0.7370}{\sqrt{m}}$$

and

$$m(X_1^{(3)})^2 (1.0^2 + \{-1.2468\}^2 + 0.5544^2) = 1 \quad \text{or} \quad X_1^{(3)} = \frac{0.5911}{\sqrt{m}}$$

∎

6.10.3
Repeated
Eigenvalues

When the characteristic equation possesses repeated roots, the corresponding mode shapes are not unique. To see this, let $\vec{X}^{(1)}$ and $\vec{X}^{(2)}$ be the mode shapes corresponding to the repeated eigenvalue $\lambda_1 = \lambda_2 = \lambda$ and let $\vec{X}^{(3)}$ be the mode shape corresponding to a different eigenvalue λ_3. Equation (6.66) can be written as

$$[D]\,\vec{X}^{(1)} = \lambda\,\vec{X}^{(1)} \tag{6.83}$$

$$[D]\,\vec{X}^{(2)} = \lambda\,\vec{X}^{(2)} \tag{6.84}$$

$$[D]\,\vec{X}^{(3)} = \lambda_3\,\vec{X}^{(3)} \tag{6.85}$$

By multiplying Eq. (6.83) by a constant p and adding to Eq. (6.84), we obtain

$$[D]\,(p\vec{X}^{(1)} + \vec{X}^{(2)}) = \lambda\,(p\vec{X}^{(1)} + \vec{X}^{(2)}) \tag{6.86}$$

This shows that the new mode shape, $(p\vec{X}^{(1)} + \vec{X}^{(2)})$, which is a linear combination of the first two, also satisfies Eq. (6.66), so the mode shape corresponding to λ is not unique. Any $\vec{X}$ corresponding to λ must be orthogonal to $\vec{X}^{(3)}$ if it is to be a normal mode. If all three modes are orthogonal, they will be linearly independent and can be used to describe the free vibration resulting from any initial conditions.

The response of a multidegree of freedom system with repeated natural frequencies to force and displacement excitation was presented by Mahalingam and Bishop [6.16].

Repeated Eigenvalues

EXAMPLE 6.12

Determine the eigenvalues and eigenvectors of a vibrating system for which

$$[m] = \begin{bmatrix} 1 & 0 & 0 \\ 0 & 2 & 0 \\ 0 & 0 & 1 \end{bmatrix} \quad \text{and} \quad [k] = \begin{bmatrix} 1 & -2 & 1 \\ -2 & 4 & -2 \\ 1 & -2 & 1 \end{bmatrix}$$

Solution: The eigenvalue equation $[[k] - \lambda[m]] \vec{X} = \vec{0}$ can be written in the form

$$\begin{bmatrix} (1 - \lambda) & -2 & 1 \\ -2 & 2(2 - \lambda) & -2 \\ 1 & -2 & (1 - \lambda) \end{bmatrix} \begin{Bmatrix} X_1 \\ X_2 \\ X_3 \end{Bmatrix} = \begin{Bmatrix} 0 \\ 0 \\ 0 \end{Bmatrix} \tag{E.1}$$

where $\lambda = \omega^2$. The characteristic equation gives

$$|[k] - \lambda[m]| = \lambda^2(\lambda - 4) = 0$$

so

$$\lambda_1 = 0, \lambda_2 = 0, \lambda_3 = 4 \tag{E.2}$$

Eigenvector for $\lambda_3 = 4$:   Using $\lambda_3 = 4$, Eq. (E.1) gives

$$-3X_1^{(3)} - 2X_2^{(3)} + X_3^{(3)} = 0$$
$$-2X_1^{(3)} - 4X_2^{(3)} - 2X_3^{(3)} = 0$$
$$X_1^{(3)} - 2 X_2^{(3)} - 3 X_3^{(3)} = 0 \tag{E.3}$$

If $X_1^{(3)}$ is set equal to 1, Eqs. (E.3) give the eigenvector $\vec{X}^{(3)}$:

$$\vec{X}^{(3)} = \begin{Bmatrix} 1 \\ -1 \\ 1 \end{Bmatrix} \tag{E.4}$$

Eigenvector for $\lambda_1 = \lambda_2 = 0$:   The value $\lambda_1 = 0$ or $\lambda_2 = 0$ indicates that the system is degenerate (see Section 6.12). Using $\lambda_1 = 0$ in Eq. (E.1), we obtain

$$X_1^{(1)} - 2X_2^{(1)} + X_3^{(1)} = 0$$
$$-2X_1^{(1)} + 4X_2^{(1)} - 2X_3^{(1)} = 0$$
$$X_1^{(1)} - 2X_2^{(1)} + X_3^{(1)} = 0 \tag{E.5}$$

All these equations are of the form

$$X_1^{(1)} = 2 X_2^{(1)} - X_3^{(1)}$$

Thus the eigenvector corresponding to $\lambda_1 = \lambda_2 = 0$ can be written as

$$\vec{X}^{(1)} = \begin{Bmatrix} 2X_2^{(1)} - X_3^{(1)} \\ X_2^{(1)} \\ X_3^{(1)} \end{Bmatrix} \tag{E.6}$$

If we choose $X_2^{(1)} = 1$ and $X_3^{(1)} = 1$, we obtain

$$\vec{X}^{(1)} = \begin{Bmatrix} 1 \\ 1 \\ 1 \end{Bmatrix} \tag{E.7}$$

If we select $X_2^{(1)} = 1$ and $X_3^{(1)} = -1$, Eq. (E.6) gives

$$\vec{X}^{(1)} = \begin{Bmatrix} 3 \\ 1 \\ -1 \end{Bmatrix} \tag{E.8}$$

As shown earlier in Eq. (6.86), $\vec{X}^{(1)}$ and $\vec{X}^{(2)}$ are not unique: Any linear combination of $\vec{X}^{(1)}$ and $\vec{X}^{(2)}$ will also satisfy the original Eq. (E.1). Note that $\vec{X}^{(1)}$ given by Eq. (E.6) is orthogonal to $\vec{X}^{(3)}$ of Eq. (E.4) for all values of $X_2^{(1)}$ and $X_3^{(1)}$, since

$$\vec{X}^{(3)T} [m] \, \vec{X}^{(1)} = (1 \quad -1 \quad 1) \begin{bmatrix} 1 & 0 & 0 \\ 0 & 2 & 0 \\ 0 & 0 & 1 \end{bmatrix} \begin{Bmatrix} 2\,X_2^{(1)} - X_3^{(1)} \\ X_2^{(1)} \\ X_3^{(1)} \end{Bmatrix} = 0$$

∎

6.11 Expansion Theorem

The eigenvectors, due to their property of orthogonality, are linearly independent.[5] Hence they form a basis in the n-dimensional space.[6] This means that any vector in the n-dimensional space can be expressed by a linear combination of the n linearly independent vectors. If $\vec{x}$ is an arbitrary vector in n-dimensional space, it can be expressed as

$$\vec{x} = \sum_{i=1}^{n} c_i \vec{X}^{(i)} \tag{6.87}$$

where c_i are constants. By premultiplying Eq. (6.87) throughout by $\vec{X}^{(i)T}[m]$, the value of the constant c_i can be determined as

$$c_i = \frac{\vec{X}^{(i)T} [m] \, \vec{x}}{\vec{X}^{(i)T} [m] \, \vec{X}^{(i)}} = \frac{\vec{X}^{(i)T} [m] \, \vec{x}}{M_{ii}}, \qquad i = 1, 2, \dots, n \tag{6.88}$$

where M_{ii} is the generalized mass in the ith normal mode. If the modal vectors $\vec{X}^{(i)}$ are normalized according to Eq. (6.81), c_i is given by

$$c_i = \vec{X}^{(i)T}[m] \, \vec{x}, \qquad i = 1, 2, \dots, n \tag{6.89}$$

---

[5]A set of vectors is called linearly independent if no vector in the set can be obtained by a linear combination of the remaining ones.
[6]Any set of n linearly independent vectors in an n-dimensional space is called a *basis* in that space.

Equation (6.89) represents what is known as the *expansion theorem* [6.6]. It is very useful in finding the response of multidegree of freedom systems subjected to arbitrary forcing conditions according to a procedure called *modal analysis*.

6.12   Unrestrained Systems

As stated in Section 5.7, an unrestrained system is one that has no restraints or supports and that can move as a rigid body. It is not uncommon to see, in practice, systems that are not attached to any stationary frame. A common example is the motion of two railway cars with masses m_1 and m_2 and a coupling spring k. Such systems are capable of moving as rigid bodies, which can be considered as modes of oscillation with zero frequency. For a conservative system, the kinetic and potential energies are given by Eqs. (6.34) and (6.30), respectively. By definition, the kinetic energy is always positive, so the mass matrix $[m]$ is a positive definite matrix. However, the stiffness matrix $[k]$ is a semidefinite matrix: V is zero without the displacement vector $\vec{x}$ being zero for unrestrained systems. To see this, consider the equation of motion for free vibration in normal coordinates:

$$\ddot{q}(t) + \omega^2 q(t) = 0 \tag{6.90}$$

For $\omega = 0$, the solution of Eq. (6.90) can be expressed as

$$q(t) = \alpha + \beta t \tag{6.91}$$

where α and β are constants. Equation (6.91) represents a rigid body translation. Let the modal vector of a multidegree of freedom system corresponding to the rigid body mode be denoted by $\vec{X}^{(0)}$. The eigenvalue problem, Eq. (6.64) can be expressed as

$$\omega^2 [m]\, \vec{X}^{(0)} = [k]\, \vec{X}^{(0)} \tag{6.92}$$

With $\omega = 0$, Eq. (6.92) gives

$$[k]\, \vec{X}^{(0)} = \vec{0}$$

That is,

$$k_{11} X_1^{(0)} + k_{12} X_2^{(0)} + \cdots + k_{1n} X_n^{(0)} = 0$$
$$k_{21} X_1^{(0)} + k_{22} X_2^{(0)} + \cdots + k_{2n} X_n^{(0)} = 0$$
$$\cdot$$
$$\cdot$$
$$\cdot$$
$$k_{n1} X_1^{(0)} + k_{n2} X_2^{(0)} + \cdots + k_{nn} X_n^{(0)} = 0 \tag{6.93}$$

If the system undergoes rigid body translation, not all the components $X_i^{(0)}$, $i = 1, 2, \ldots, n$ are zero—that is, the vector $\vec{X}^{(0)}$ is not equal to $\vec{0}$. Hence, in order to satisfy Eq. (6.93), the determinant of $[k]$ must be zero. Thus the stiffness matrix of an unrestrained system (having zero natural frequency) is singular. If $[k]$ is singular, the potential energy is given by

$$V = \frac{1}{2}\vec{X}^{(0)^T} [k]\, \vec{X}^{(0)} \tag{6.94}$$

by virtue of Eq. (6.93). The mode $\vec{X}^{(0)}$ is called a *zero mode* or *rigid body mode*. If we substitute any vector $\vec{X}$ other than $\vec{X}^{(0)}$ and $\vec{0}$ for $\vec{x}$ in Eq. (6.30), the potential energy V becomes a positive quantity. The matrix $[k]$ is then a positive semidefinite matrix. This is why an unrestrained system is also called a *semidefinite system*.

Note that a multidegree of freedom system can have at most six rigid body modes with the corresponding frequencies equal to zero. There can be three modes for rigid body translation, one for translation along each of the three Cartesian coordinates, and three modes for rigid body rotation, one for rotation about each of the three Cartesian coordinates. We can determine the mode shapes and natural frequencies of a semidefinite system by the procedures outlined in Section 6.10.

EXAMPLE 6.13

Natural Frequencies of a Free System

Three freight cars are coupled by two springs, as shown in Fig. 6.13. Find the natural frequencies and mode shapes of the system for $m_1 = m_2 = m_3 = m$ and $k_1 = k_2 = k$.

Solution: The kinetic energy of the system can be written as

$$T = \frac{1}{2}(m_1\dot{x}_1^2 + m_2\dot{x}_2^2 + m_3\dot{x}_3^2) = \frac{1}{2}\dot{\vec{x}}^T [m]\, \dot{\vec{x}} \tag{E.1}$$

where

$$\vec{x} = \begin{Bmatrix} x_1 \\ x_2 \\ x_3 \end{Bmatrix}, \qquad \dot{\vec{x}} = \begin{Bmatrix} \dot{x}_1 \\ \dot{x}_2 \\ \dot{x}_3 \end{Bmatrix}$$

and

$$[m] = \begin{bmatrix} m_1 & 0 & 0 \\ 0 & m_2 & 0 \\ 0 & 0 & m_3 \end{bmatrix} \tag{E.2}$$

The elongations of the springs k_1 and k_2 are $(x_2 - x_1)$ and $(x_3 - x_2)$, respectively, so the potential energy of the system is given by

$$V = \frac{1}{2}\{k_1(x_2 - x_1)^2 + k_2(x_3 - x_2)^2\} = \frac{1}{2}\vec{x}^T [k]\, \vec{x} \tag{E.3}$$

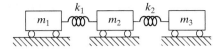

FIGURE 6.13   Semidefinite system.

where

$$[k] = \begin{bmatrix} k_1 & -k_1 & 0 \\ -k_1 & k_1 + k_2 & -k_2 \\ 0 & -k_2 & k_2 \end{bmatrix} \tag{E.4}$$

It can be verified that the stiffness matrix $[k]$ is singular. Furthermore, if we take all the displacement components to be the same as $x_1 = x_2 = x_3 = c$ (rigid body motion), the potential energy V can be seen to be equal to zero.

To find the natural frequencies and the mode shapes of the system, we express the eigenvalue problem as

$$[[k] - \omega^2[m]]\,\vec{X} = \vec{0} \tag{E.5}$$

Since $[k]$ is singular, we cannot find its inverse $[k]^{-1}$ and the dynamical matrix $[D] = [k]^{-1}[m]$. Hence we set the determinant of the coefficient matrix of $\vec{X}$ in Eq. (E.5) equal to zero. For $k_1 = k_2 = k$ and $m_1 = m_2 = m_3 = m$, this yields

$$\begin{vmatrix} (k - \omega^2 m) & -k & 0 \\ -k & (2k - \omega^2 m) & -k \\ 0 & -k & (k - \omega^2 m) \end{vmatrix} = 0 \tag{E.6}$$

The expansion of the determinant in Eq. (E.6) leads to

$$m^3\omega^6 - 4m^2 k\omega^4 + 3mk^2\omega^2 = 0 \tag{E.7}$$

By setting

$$\lambda = \omega^2 \tag{E.8}$$

Eq. (E.7) can be rewritten as

$$m\lambda\left(\lambda - \frac{k}{m}\right)\left(\lambda - \frac{3k}{m}\right) = 0 \tag{E.9}$$

As $m \neq 0$, the roots of Eq. (E.9) are

$$\lambda_1 = \omega_1^2 = 0$$

$$\lambda_2 = \omega_2^2 = \frac{k}{m}$$

$$\lambda_3 = \omega_3^2 = \frac{3k}{m} \tag{E.10}$$

The first natural frequency ω_1 can be observed to be zero in Eq. (E.10). To find the mode shapes, we substitute the values of ω_1, ω_2, and ω_3 into Eq. (E.5) and solve for $\vec{X}^{(1)}$, $\vec{X}^{(2)}$, and $\vec{X}^{(3)}$, respectively. For $\omega_1 = 0$, Eq. (E.5) gives

$$kX_1^{(1)} - kX_2^{(1)} = 0$$

$$-kX_1^{(1)} + 2kX_2^{(1)} - kX_3^{(1)} = 0$$

$$-kX_2^{(1)} + kX_3^{(1)} = 0 \tag{E.11}$$

By fixing the value of one component of $\vec{X}^{(1)}$—say, $X_1^{(1)}$ as 1—Eq. (E.11) can be solved to obtain

$$X_2^{(1)} = X_1^{(1)} = 1 \quad \text{and} \quad X_3^{(1)} = X_2^{(1)} = 1$$

Thus the first (rigid body) mode $\vec{X}^{(1)}$ corresponding to $\omega_1 = 0$ is given by

$$\vec{X}^{(1)} = \begin{Bmatrix} 1 \\ 1 \\ 1 \end{Bmatrix} \tag{E.12}$$

For $\omega_2 = (k/m)^{1/2}$, Eq. (E.5) yields

$$-kX_2^{(2)} = 0$$

$$-kX_1^{(2)} + kX_2^{(2)} - kX_3^{(2)} = 0$$

$$-kX_2^{(2)} = 0 \tag{E.13}$$

By fixing the value of one component of $\vec{X}^{(2)}$—say, $X_1^{(2)}$ as 1—Eq. (E.13) can be solved to obtain

$$X_2^{(2)} = 0 \quad \text{and} \quad X_3^{(2)} = -X_1^{(2)} = -1$$

Thus the second mode $\vec{X}^{(2)}$ corresponding to $\omega_2 = (k/m)^{1/2}$ is given by

$$\vec{X}^{(2)} = \begin{Bmatrix} 1 \\ 0 \\ -1 \end{Bmatrix} \tag{E.14}$$

For $\omega_3 = (3k/m)^{1/2}$, Eq. (E.5) gives

$$-2kX_1^{(3)} - kX_2^{(3)} = 0$$

$$-kX_1^{(3)} - kX_2^{(3)} - kX_3^{(3)} = 0$$

$$-kX_2^{(3)} - 2kX_3^{(3)} = 0 \tag{E.15}$$

By fixing the value of one component of $\vec{X}^{(3)}$—say, $X_1^{(3)}$ as 1—Eq. (E.15) can be solved to obtain

$$X_2^{(3)} = -2X_1^{(3)} = -2 \quad \text{and} \quad X_3^{(3)} = -\frac{1}{2}X_2^{(3)} = 1$$

Thus the third mode $\vec{X}^{(3)}$ corresponding to $\omega_3 = (3k/m)^{1/2}$ is given by

$$\vec{X}^{(3)} = \begin{Bmatrix} 1 \\ -2 \\ 1 \end{Bmatrix} \tag{E.16}$$

■

6.13   Free Vibration of Undamped Systems

The equation of motion for the free vibration of an undamped system can be expressed in matrix form as

$$[m]\ddot{\vec{x}} + [k]\vec{x} = \vec{0} \tag{6.95}$$

The most general solution of Eq. (6.95) can be expressed as a linear combination of all possible solutions given by Eqs. (6.56) and (6.62) as

$$\vec{x}(t) = \sum_{i=1}^{n} \vec{X}^{(i)} A_i \cos(\omega_i t + \phi_i) \tag{6.96}$$

where $\vec{X}^{(i)}$ is the ith modal vector and ω_i is the corresponding natural frequency, and A_i and ϕ_i are constants. The constants A_i and $\phi_i (i = 1, 2, \ldots, n)$ can be evaluated from the specified initial conditions of the system. If

$$\vec{x}(0) = \begin{Bmatrix} x_1(0) \\ x_2(0) \\ \cdot \\ \cdot \\ \cdot \\ x_n(0) \end{Bmatrix} \quad \text{and} \quad \dot{\vec{x}}(0) = \begin{Bmatrix} \dot{x}_1(0) \\ \dot{x}_2(0) \\ \cdot \\ \cdot \\ \cdot \\ \dot{x}_n(0) \end{Bmatrix} \tag{6.97}$$

denote the initial displacements and velocities given to the system, Eqs. (6.96) give

$$\vec{x}(0) = \sum_{i=1}^{n} \vec{X}^{(i)} A_i \cos \phi_i \tag{6.98}$$

$$\dot{\vec{x}}(0) = -\sum_{i=1}^{n} \vec{X}^{(i)} A_i \omega_i \sin \phi_i \tag{6.99}$$

Equations (6.98) and (6.99) represent, in scalar form, $2n$ simultaneous equations which can be solved to find the n values of $A_i (i = 1, 2, \ldots, n)$ and n values of $\phi_i (i = 1, 2, \ldots, n)$.

EXAMPLE 6.14

Free Vibration Analysis of a Spring-Mass System

Find the free vibration response of the spring-mass system shown in Fig. 6.8(a) corresponding to the initial conditions $\dot{x}_i(0) = 0$ ($i = 1, 2, 3$), $x_1(0) = x_{10}$, $x_2(0) = x_3(0) = 0$. Assume that $k_i = k$ and $m_i = m$ for $i = 1, 2, 3$.

Solution

Approach: Assume free vibration response as a sum of natural modes.

The natural frequencies and mode shapes of the system are given by (see Example 6.10):

$$\omega_1 = 0.44504\sqrt{\frac{k}{m}}, \qquad \omega_2 = 1.2471\sqrt{\frac{k}{m}}, \qquad \omega_3 = 1.8025\sqrt{\frac{k}{m}}$$

$$\vec{X}^{(1)} = \begin{Bmatrix} 1.0 \\ 1.8019 \\ 2.2470 \end{Bmatrix}, \qquad \vec{X}^{(2)} = \begin{Bmatrix} 1.0 \\ 0.4450 \\ -0.8020 \end{Bmatrix}, \qquad \vec{X}^{(3)} = \begin{Bmatrix} 1.0 \\ -1.2468 \\ 0.5544 \end{Bmatrix}$$

where the first component of each mode shape is assumed as unity for simplicity. The application of the initial conditions, Eqs. (6.98) and (6.99), leads to

$$A_1 \cos \phi_1 + A_2 \cos \phi_2 + A_3 \cos \phi_3 = x_{10} \tag{E.1}$$

$$1.8019\, A_1 \cos \phi_1 + 0.4450\, A_2 \cos \phi_2 - 1.2468\, A_3 \cos \phi_3 = 0 \tag{E.2}$$

$$2.2470\, A_1 \cos \phi_1 - 0.8020\, A_2 \cos \phi_2 + 0.5544\, A_3 \cos \phi_3 = 0 \tag{E.3}$$

$$-0.44504\sqrt{\frac{k}{m}}\, A_1 \sin \phi_1 - 1.2471\sqrt{\frac{k}{m}} A_2 \sin \phi_2$$
$$- 1.8025\sqrt{\frac{k}{m}} A_3 \sin \phi_3 = 0 \tag{E.4}$$

$$-0.80192\sqrt{\frac{k}{m}}\, A_1 \sin \phi_1 - 0.55496\sqrt{\frac{k}{m}} A_2 \sin \phi_2$$
$$+ 2.2474\sqrt{\frac{k}{m}} A_3 \sin \phi_3 = 0 \tag{E.5}$$

$$-1.0\sqrt{\frac{k}{m}}\, A_1 \sin \phi_1 + 1.0\sqrt{\frac{k}{m}} A_2 \sin \phi_2 - 1.0\sqrt{\frac{k}{m}} A_3 \sin \phi_3 = 0 \tag{E.6}$$

The solution of Eqs. (E.1) to (E.6) is given by[7] $A_1 = 0.1076\, x_{10}$, $A_2 = 0.5431\, x_{10}$, $A_3 = 0.3493\, x_{10}$, $\phi_1 = 0$, $\phi_2 = 0$, and $\phi_3 = 0$. Thus the free vibration solution of the system can be expressed as

---

[7]Note that Eqs. (E.1) to (E.3) can be considered as a system of linear equations in the unknowns $A_1 \cos \phi_1$, $A_2 \cos \phi_2$, and $A_3 \cos \phi_3$, while Eqs. (E.4) to (E.6) can be considered as a set of linear equations in the unknowns

$$\sqrt{\frac{k}{m}} A_1 \sin \phi_1, \qquad \sqrt{\frac{k}{m}} A_2 \sin \phi_2, \qquad \text{and} \qquad \sqrt{\frac{k}{m}} A_3 \sin \phi_3.$$

$$x_1(t) = x_{10} \left[0.1076 \cos \left(0.44504 \sqrt{\frac{k}{m}} \, t \right) \right.$$

$$+ 0.5431 \cos \left(1.2471 \sqrt{\frac{k}{m}} \, t \right)$$

$$\left. + 0.3493 \cos \left(1.8025 \sqrt{\frac{k}{m}} \, t \right) \right] \tag{E.7}$$

$$x_2(t) = x_{10} \left[0.1939 \cos \left(0.44504 \sqrt{\frac{k}{m}} \, t \right) \right.$$

$$+ 0.2417 \cos \left(1.2471 \sqrt{\frac{k}{m}} \, t \right)$$

$$\left. - 0.4355 \cos \left(1.8025 \sqrt{\frac{k}{m}} \, t \right) \right] \tag{E.8}$$

$$x_3(t) = x_{10} \left[0.2418 \cos \left(0.44504 \sqrt{\frac{k}{m}} \, t \right) \right.$$

$$- 0.4356 \cos \left(1.2471 \sqrt{\frac{k}{m}} \, t \right)$$

$$\left. + 0.1937 \cos \left(1.8025 \sqrt{\frac{k}{m}} \, t \right) \right] \tag{E.9}$$

∎

6.14   Forced Vibration of Undamped Systems Using Modal Analysis

When external forces act on a multidegree of freedom system, the system undergoes forced vibration. For a system with n coordinates or degrees of freedom, the governing equations of motion are a set of n coupled ordinary differential equations of second order. The solution of these equations becomes more complex when the degree of freedom of the system (n) is large and/or when the forcing functions are nonperiodic.[8] In such cases, a more convenient method known as *modal analysis* can be used to solve the problem. In this method, the expansion theorem is used, and the displacements of the masses are expressed as a linear combination of the normal modes of the system. This linear transformation uncouples the equations of motion so that we obtain a set of n uncoupled differential equations of second order. The solution of these equations, which is equivalent to the solution of the equations of n single degree of freedom systems, can be readily obtained. We shall now consider the procedure of modal analysis.

---

[8]The dynamic response of multidegree of freedom systems with statistical properties is considered in Ref. [6.15].

Modal Analysis.   The equations of motion of a multidegree of freedom system under external forces are given by

$$[m]\ddot{\vec{x}} + [k]\vec{x} = \vec{F} \tag{6.100}$$

where $\vec{F}$ is the vector of arbitrary external forces. To solve Eq. (6.100) by modal analysis, it is necessary first to solve the eigenvalue problem.

$$\omega^2[m]\vec{X} = [k]\vec{X} \tag{6.101}$$

and find the natural frequencies $\omega_1, \omega_2, \ldots, \omega_n$ and the corresponding normal modes $\vec{X}^{(1)}, \vec{X}^{(2)}, \ldots, \vec{X}^{(n)}$. According to the expansion theorem, the solution vector of Eq. (6.100) can be expressed by a linear combination of the normal modes

$$\vec{x}(t) = q_1(t)\vec{X}^{(1)} + q_2(t)\vec{X}^{(2)} + \cdots + q_n(t)\vec{X}^{(n)} \tag{6.102}$$

where $q_1(t), q_2(t), \ldots, q_n(t)$ are time-dependent generalized coordinates, also known as the *principal coordinates* or *modal participation coefficients*. By defining a modal matrix $[X]$ in which the jth column is the vector $\vec{X}^{(j)}$—that is,

$$[X] = [\vec{X}^{(1)}\ \vec{X}^{(2)}\ \cdots\ \vec{X}^{(n)}] \tag{6.103}$$

Eq. (6.102) can be rewritten as

$$\vec{x}(t) = [X]\vec{q}(t) \tag{6.104}$$

where

$$\vec{q}(t) = \begin{Bmatrix} q_1(t) \\ q_2(t) \\ \vdots \\ q_n(t) \end{Bmatrix} \tag{6.105}$$

Since $[X]$ is not a function of time, we obtain from Eq. (6.104)

$$\ddot{\vec{x}}(t) = [X]\ddot{\vec{q}}(t) \tag{6.106}$$

Using Eqs. (6.104) and (6.106), we can write Eq. (6.100) as

$$[m][X]\ddot{\vec{q}} + [k][X]\vec{q} = \vec{F} \tag{6.107}$$

Premultiplying Eq. (6.107) throughout by $[X]^T$, we obtain

$$[X]^T[m][X]\ddot{\vec{q}} + [X]^T[k][X]\vec{q} = [X]^T\vec{F} \tag{6.108}$$

If the normal modes are normalized according to Eqs. (6.74) and (6.75), we have

$$[X]^T[m][X] = [I] \tag{6.109}$$

$$[X]^T[k][X] = [\diagdown\omega^2\diagdown] \tag{6.110}$$

By defining the vector of generalized forces $\vec{Q}(t)$ associated with the generalized coordinates $\vec{q}(t)$ as

$$\vec{Q}(t) = [X]^T \vec{F}(t) \tag{6.111}$$

Eq. (6.108) can be expressed, using Eqs. (6.109) and (6.110), as

$$\ddot{\vec{q}}(t) + [\diagdown\omega^2\diagdown]\vec{q}(t) = \vec{Q}(t) \tag{6.112}$$

Equation (6.112) denotes a set of n uncoupled differential equations of second order[9]

$$\ddot{q}_i(t) + \omega_i^2 q_i(t) = Q_i(t), \qquad i = 1, 2, \ldots, n \tag{6.113}$$

It can be seen that Eqs. (6.113) have precisely the form of the differential equation describing the motion of an undamped single degree of freedom system. The solution of Eqs. (6.113) can be expressed (see Eq. (4.33)) as

$$q_i(t) = q_i(0) \cos \omega_i t + \left(\frac{\dot{q}(0)}{\omega_i} \right) \sin \omega_i t$$

$$+ \frac{1}{\omega_i} \int_0^t Q_i(\tau) \sin \omega_i (t - \tau) \, d\tau,$$

$$i = 1, 2, \ldots, n \tag{6.114}$$

The initial generalized displacements $q_i(0)$ and the initial generalized velocities $\dot{q}_i(0)$ can be obtained from the initial values of the physical displacements $x_i(0)$ and physical velocities

---

[9]It is possible to approximate the solution vector $\vec{x}(t)$ by only the first $r(r < n)$ modal vectors (instead of n vectors as in Eq. 6.102):

$$\underset{n\times 1}{\vec{x}(t)} = \underset{n\times r}{[X]}\underset{r\times 1}{\vec{q}(t)}$$

where

$$[X] = [\vec{X}^{(1)} \, \vec{X}^{(2)} \, \cdots \, \vec{X}^{(r)}] \qquad \text{and} \qquad \vec{q}(t) = \begin{Bmatrix} q_1(t) \\ q_2(t) \\ \vdots \\ q_r(t) \end{Bmatrix}$$

This leads to only r uncoupled differential equations

$$\ddot{q}_i(t) + \omega_i^2 q_i(t) = Q_i(t), \qquad i = 1, 2, \ldots, r$$

instead of n equations. The resulting solution $\vec{x}(t)$ will be an approximate solution. This procedure is called the mode displacement method. An alternate procedure, known as the mode acceleration method for finding an approximate solution is indicated in Problem 6.77.

$\dot{x}_i(0)$ as (see Problem 6.79):

$$\vec{q}(0) = [X]^T [m] \vec{x}(0) \tag{6.115}$$

$$\dot{\vec{q}}(0) = [X]^T [m] \dot{\vec{x}}(0) \tag{6.116}$$

where

$$\vec{q}(0) = \begin{Bmatrix} q_1(0) \\ q_2(0) \\ \vdots \\ q_n(0) \end{Bmatrix},$$

$$\dot{\vec{q}}(0) = \begin{Bmatrix} \dot{q}_1(0) \\ \dot{q}_2(0) \\ \vdots \\ \dot{q}_n(0) \end{Bmatrix},$$

$$\vec{x}(0) = \begin{Bmatrix} x_1(0) \\ x_2(0) \\ \vdots \\ x_n(0) \end{Bmatrix},$$

$$\dot{\vec{x}}(0) = \begin{Bmatrix} \dot{x}_1(0) \\ \dot{x}_2(0) \\ \vdots \\ \dot{x}_n(0) \end{Bmatrix}$$

Once the generalized displacements $q_i(t)$ are found, using Eqs. (6.114) to (6.116), the physical displacements $x_i(t)$ can be found with the help of Eq. (6.104).

Free Vibration Response Using Modal Analysis

EXAMPLE 6.15

Using modal analysis, find the free vibration response of a two degree of freedom system with equations of motion

$$\begin{bmatrix} m_1 & 0 \\ 0 & m_2 \end{bmatrix} \begin{Bmatrix} \ddot{x}_1 \\ \ddot{x}_2 \end{Bmatrix} + \begin{bmatrix} k_1 + k_2 & -k_2 \\ -k_2 & k_2 + k_3 \end{bmatrix} \begin{Bmatrix} x_1 \\ x_2 \end{Bmatrix} = \vec{F} = \begin{Bmatrix} 0 \\ 0 \end{Bmatrix} \tag{E.1}$$

Assume the following data: $m_1 = 10$, $m_2 = 1$, $k_1 = 30$, $k_2 = 5$, $k_3 = 0$, and

$$\vec{x}(0) = \begin{Bmatrix} x_1(0) \\ x_2(0) \end{Bmatrix} = \begin{Bmatrix} 1 \\ 0 \end{Bmatrix}, \ \dot{\vec{x}}(0) = \begin{Bmatrix} \dot{x}_1(0) \\ \dot{x}_2(0) \end{Bmatrix} = \begin{Bmatrix} 0 \\ 0 \end{Bmatrix} \tag{E.2}$$

Solution: The natural frequencies and normal modes of the system are given by (see Example 5.3)

$$\omega_1 = 1.5811, \qquad \vec{X}^{(1)} = \begin{Bmatrix} 1 \\ 2 \end{Bmatrix} X_1^{(1)}$$

$$\omega_2 = 2.4495, \qquad \vec{X}^{(2)} = \begin{Bmatrix} 1 \\ -5 \end{Bmatrix} X_1^{(2)}$$

where $X_1^{(1)}$ and $X_1^{(2)}$ are arbitrary constants. By orthogonalizing the normal modes with respect to the mass matrix, we can find the values of $X_1^{(1)}$ and $X_1^{(2)}$ as

$$\vec{X}^{(1)^T} [m] \, \vec{X}^{(1)} = 1 \Rightarrow (X_1^{(1)})^2 \, \{1 \quad 2\} \begin{bmatrix} 10 & 0 \\ 0 & 1 \end{bmatrix} \begin{Bmatrix} 1 \\ 2 \end{Bmatrix} = 1$$

or $X_1^{(1)} = 0.2673$

$$\vec{X}^{(2)^T} [m] \, \vec{X}^{(2)} = 1 \Rightarrow (X_1^{(2)})^2 \, \{1 \quad -5\} \begin{bmatrix} 10 & 0 \\ 0 & 1 \end{bmatrix} \begin{Bmatrix} 1 \\ -5 \end{Bmatrix} = 1$$

or $X_1^{(2)} = 0.1690$

Thus the modal matrix becomes

$$[X] = \begin{bmatrix} \vec{X}^{(1)} & \vec{X}^{(2)} \end{bmatrix} = \begin{bmatrix} 0.2673 & 0.1690 \\ 0.5346 & -0.8450 \end{bmatrix} \tag{E.3}$$

Using

$$\vec{x}(t) = [X] \, \vec{q}(t) \tag{E.4}$$

Equation (E.1) can be expressed as (see Eq. 6.112):

$$\ddot{\vec{q}}(t) + [\diagdown \omega^2 \diagdown] \, \vec{q}(t) = \vec{Q}(t) = \vec{0} \tag{E.5}$$

where $\vec{Q}(t) = [X]^T \vec{F} = \vec{0}$. Equation (E.5) can be written in scalar form as

$$\ddot{q}_i(t) + \omega_i^2 \, q_i(t) = 0, \qquad i = 1, 2 \tag{E.6}$$

The solution of Eq. (E.6) is given by (see Eq. 2.18):

$$q_i(t) = q_{i0} \cos \omega_i t + \frac{\dot{q}_{i0}}{\omega_i} \sin \omega_i t \tag{E.7}$$

where q_{i0} and $\dot{q}_{i0}$ denote the initial values of $q_i(t)$ and $\dot{q}_i(t)$, respectively. Using the initial conditions of Eq. (E.2), we can find (see Eqs. 6.115 and 6.116):

$$\vec{q}(0) = \begin{Bmatrix} q_{10}(0) \\ q_{20}(0) \end{Bmatrix} = [X]^T [m] \, \vec{x}(0) = \begin{bmatrix} 0.2673 & 0.5346 \\ 0.1690 & -0.8450 \end{bmatrix} \begin{bmatrix} 10 & 0 \\ 0 & 1 \end{bmatrix} \begin{Bmatrix} 1 \\ 0 \end{Bmatrix} = \begin{Bmatrix} 2.673 \\ 1.690 \end{Bmatrix} \tag{E.8}$$

$$\vec{\dot{q}}(0) = \begin{Bmatrix} \dot{q}_{10}(0) \\ \dot{q}_{20}(0) \end{Bmatrix} = [X]^T [m] \, \vec{\dot{x}}(0) = \begin{Bmatrix} 0 \\ 0 \end{Bmatrix} \qquad (E.9)$$

Equations (E.7) to (E.9) lead to

$$q_1(t) = 2.673 \cos 1.5811\,t \qquad (E.10)$$

$$q_2(t) = 1.690 \cos 2.4495\,t \qquad (E.11)$$

Using Eqs. (E.4), we obtain the displacements of the masses m_1 and m_2 as

$$\vec{x}(t) = \begin{bmatrix} 0.2673 & 0.1690 \\ 0.5346 & -0.8450 \end{bmatrix} \begin{Bmatrix} 2.673 \cos 1.5811\,t \\ 1.690 \cos 2.4495\,t \end{Bmatrix}$$

or

$$\begin{Bmatrix} x_1(t) \\ x_2(t) \end{Bmatrix} = \begin{Bmatrix} 0.7145 \cos 1.5811\,t + 0.2856 \cos 2.4495\,t \\ 1.4280 \cos 1.5811\,t - 1.4280 \cos 2.4495\,t \end{Bmatrix} \qquad (E.12)$$

It can be seen that this solution is identical to the one obtained in Example 5.3 and plotted in Example 5.12.

■

EXAMPLE 6.16

Forced Vibration Response of a Forging Hammer

The force acting on the workpiece of the forging hammer shown in Fig. 5.41 due to impact by the hammer can be approximated as a rectangular pulse, as shown in Fig. 6.14(a). Find the resulting vibration of the system for the following data: mass of the workpiece, anvil and frame $(m_1) = 200$ Mg, mass of the foundation block $(m_2) = 250$ Mg, stiffness of the elastic pad $(k_1) = 150$ MN/m, and stiffness of the soil $(k_2) = 75$ MN/m. Assume the initial displacements and initial velocities of the masses as zero.

Solution: The forging hammer can be modeled as a two degree of freedom system as indicated in Fig. 6.14(b). The equations of motion of the system can be expressed as

$$[m] \, \vec{\ddot{x}} + [k] \, \vec{x} = \vec{F}(t) \qquad (E.1)$$

where

$$[m] = \begin{bmatrix} m_1 & 0 \\ 0 & m_2 \end{bmatrix} = \begin{bmatrix} 200 & 0 \\ 0 & 250 \end{bmatrix} \text{Mg}$$

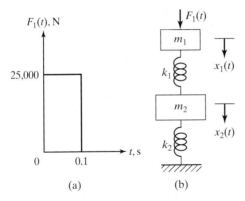

FIGURE 6.14   Impact caused by forging
hammer.

$$[k] = \begin{bmatrix} k_1 & -k_1 \\ -k_1 & k_1 + k_2 \end{bmatrix} = \begin{bmatrix} 150 & -150 \\ -150 & 225 \end{bmatrix} \text{MN/m}$$

$$\vec{F}(t) = \begin{Bmatrix} F_1(t) \\ 0 \end{Bmatrix}$$

Natural Frequencies and Mode Shapes: The natural frequencies of the system can be found by solving the frequency equation

$$|-\omega^2 [m] + [k]| = \left| -\omega^2 \begin{bmatrix} 2 & 0 \\ 0 & 2.5 \end{bmatrix} 10^5 + \begin{bmatrix} 150 & -150 \\ -150 & 225 \end{bmatrix} 10^6 \right| = 0 \qquad \text{(E.2)}$$

as

$$\omega_1 = 12.2474 \text{ rad/s, and } \omega_2 = 38.7298 \text{ rad/s}$$

The mode shapes can be found as

$$\vec{X}^{(1)} = \begin{Bmatrix} 1 \\ 0.8 \end{Bmatrix} \qquad \text{and} \qquad \vec{X}^{(2)} = \begin{Bmatrix} 1 \\ -1 \end{Bmatrix}$$

Orthonormalization of Mode Shapes: The mode shapes are assumed as

$$\vec{X}^{(1)} = a \begin{Bmatrix} 1 \\ 0.8 \end{Bmatrix} \qquad \text{and} \qquad \vec{X}^{(2)} = b \begin{Bmatrix} 1 \\ -1 \end{Bmatrix}$$

where a and b are constants. The constants a and b can be determined by normalizing the vectors $\vec{X}^{(1)}$ and $\vec{X}^{(2)}$ as

$$[X]^T [m] [X] = [I] \qquad \text{(E.3)}$$

where $[X] = [\vec{X}^{(1)}\vec{X}^{(2)}]$ denotes the modal matrix. Equation (E.3) gives $a = 1.6667 \times 10^{-3}$ and $b = 1.4907 \times 10^{-3}$, which means that the new modal matrix (with normalized mode shapes) becomes

$$[X] = [\vec{X}^{(1)}\vec{X}^{(2)}] = \begin{bmatrix} 1.6667 & 1.4907 \\ 1.3334 & -1.4907 \end{bmatrix} \times 10^{-3}$$

Response in Terms of Generalized Coordinates: Since the two masses m_1 and m_2 are at rest at $t = 0$, the initial conditions are $x_1(0) = x_2(0) = \dot{x}_1(0) = \dot{x}_2(0) = 0$ and hence Eqs. (6.115) and (6.116) give $q_1(0) = q_2(0) = \dot{q}_1(0) = \dot{q}_2(0) = 0$. Thus the generalized coordinates are given by the solution of the equations

$$q_i(t) = \frac{1}{\omega_i} \int_0^t Q_i(\tau) \sin \omega_i(t - \tau)d\tau, i = 1, 2 \tag{E.4}$$

where

$$\vec{Q}(t) = [X]^T \vec{F}(t) \tag{E.5}$$

or

$$\begin{Bmatrix} Q_1(t) \\ Q_2(t) \end{Bmatrix} = \begin{bmatrix} 1.6667 & 1.3334 \\ 1.4907 & -1.4907 \end{bmatrix} 10^{-3} \begin{Bmatrix} F_1(t) \\ 0 \end{Bmatrix}$$

$$= \begin{Bmatrix} 1.6667 \times 10^{-3} F_1(t) \\ 1.4907 \times 10^{-3} F_1(t) \end{Bmatrix} \tag{E.6}$$

with $F_1(t) = 25000$ N for $0 \le t \le 0.1$s and 0 for $t > 0.1$ s. Using Eq. (6.104), the displacements of the masses can be found as

$$\begin{Bmatrix} x_1(t) \\ x_2(t) \end{Bmatrix} = [X]\,\vec{q}(t) = \begin{Bmatrix} 1.6667 \, q_1(t) + 1.4907 \, q_2(t) \\ 1.3334 \, q_1(t) - 1.4907 \, q_2(t) \end{Bmatrix} 10^{-3} \text{ m} \tag{E.7}$$

where

$$q_1(t) = 3.4021 \int_0^t \sin 12.2474 \, (t - \tau) \, d\tau = 0.2778 \, (1 - \cos 12.2474 \, t)$$

$$q_2(t) = 0.9622 \int_0^t \sin 38.7298 \, (t - \tau) \, d\tau = 0.02484 \, (1 - \cos 38.7298 \, t) \tag{E.8}$$

Note that the solution given by Eqs. (E.8) is valid for $0 \le 0.1$ s. For $t > 0.1$ s, there is no applied force and hence the response is given by the free vibration solution of an undamped single degree of freedom system (Eq. 2.18) for $q_1(t)$ and $q_2(t)$ with $q_1(0.1)$ and $\dot{q}_1(0.1)$, and $q_2(0.1)$ and $\dot{q}_2(0.1)$ as initial conditions for $q_1(t)$ and $q_2(t)$, respectively.

∎

6.15   Forced Vibration of Viscously Damped Systems

Modal analysis, as presented in Section 6.14, applies only to undamped systems. In many cases, the influence of damping upon the response of a vibratory system is minor and can be disregarded. However, the effect of damping must be considered if the response of the system is required for a relatively long period of time compared to the natural periods of the system. Further, if the frequency of excitation (in the case of a periodic force) is at or near one of the natural frequencies of the system, damping is of primary importance and must be taken into account. In general, since the effects are not known in advance, damping must be considered in the vibration analysis of any system. In this section, we shall consider the equations of motion of a damped multidegree of freedom system and their solution using Lagrange's equations. If the system has viscous damping, its motion will be resisted by a force whose magnitude is proportional to that of the velocity but in the opposite direction. It is convenient to introduce a function R, known as Rayleigh's dissipation function, in deriving the equations of motion by means of Lagrange's equations [6.7]. This function is defined as

$$R = \frac{1}{2}\vec{\dot{x}}^T [c]\, \vec{\dot{x}} \tag{6.117}$$

where the matrix $[c]$ is called the *damping matrix* and is positive definite, like the mass and stiffness matrices. Lagrange's equations, in this case [6.8], can be written as

$$\frac{d}{dt}\left(\frac{\partial T}{\partial \dot{x}_i}\right) - \frac{\partial T}{\partial x_i} + \frac{\partial R}{\partial \dot{x}_i} + \frac{\partial V}{\partial x_i} = F_i, \qquad i = 1, 2, \ldots, n \tag{6.118}$$

where F_i is the force applied to mass m_i. By substituting Eqs. (6.30), (6.34), and (6.117) into Eq. (6.118), we obtain the equations of motion of a damped multidegree of freedom system in matrix form:

$$[m]\, \vec{\ddot{x}} + [c]\, \vec{\dot{x}} + [k]\, \vec{x} = \vec{F} \tag{6.119}$$

For simplicity, we shall consider a special system for which the damping matrix can be expressed as a linear combination of the mass and stiffness matrices

$$[c] = \alpha\,[m] + \beta\,[k] \tag{6.120}$$

where α and β are constants. This type of damping is known as *proportional damping* because $[c]$ is proportional to a linear combination of $[m]$ and $[k]$. By substituting Eq. (6.120) into Eq. (6.119), we obtain

$$[m]\, \vec{\ddot{x}} + [\alpha[m] + \beta[k]]\, \vec{\dot{x}} + [k]\, \vec{x} = \vec{F} \tag{6.121}$$

By expressing the solution vector $\vec{x}$ as a linear combination of the natural modes of the undamped system, as in the case of Eq. (6.104),

$$\vec{x}(t) = [X]\,\vec{q}(t) \tag{6.122}$$

Eq. (6.121) can be rewritten as

$$[m][X]\ddot{\vec{q}}(t) + [\alpha[m] + \beta[k]][X]\dot{\vec{q}}(t)$$

$$+ [k][X]\vec{q}(t) = \vec{F}(t) \tag{6.123}$$

Premultiplication of Eq. (6.123) by $[X]^T$ leads to

$$[X]^T[m][X]\ddot{\vec{q}} + [\alpha[X]^T[m][X] + \beta[X]^T[k][X]]\dot{\vec{q}}$$

$$+ [X]^T[k][X]\vec{q} = [X]^T\vec{F} \tag{6.124}$$

If the eigenvectors $\vec{X}^{(j)}$ are normalized according to Eqs. (6.74) and (6.75), Eq. (6.124) reduces to

$$[I]\ddot{\vec{q}}(t) + \left[\alpha[I] + \beta\left[\diagdown\omega^2\diagdown\right]\right]\dot{\vec{q}}(t) + \left[\diagdown\omega^2\diagdown\right]\vec{q}(t) = \vec{Q}(t)$$

that is,

$$\ddot{q}_i(t) + (\alpha + \omega_i^2\beta)\,\dot{q}_i(t) + \omega_i^2 q_i(t) = Q_i(t),$$

$$i = 1, 2, \ldots, n \tag{6.125}$$

where ω_i is the ith natural frequency of the undamped system and

$$\vec{Q}(t) = [X]^T\vec{F}(t) \tag{6.126}$$

By writing

$$\alpha + \omega_i^2\beta = 2\zeta_i\omega_i \tag{6.127}$$

where ζ_i is called the *modal damping ratio* for the ith normal mode, Eqs. (6.125) can be rewritten as

$$\ddot{q}_i(t) + 2\zeta_i\omega_i\dot{q}_i(t) + \omega_i^2 q_i(t) = Q_i(t), \qquad i = 1, 2, \ldots, n \tag{6.128}$$

It can be seen that each of the n equations represented by this expression is uncoupled from all of the others. Hence we can find the response of the ith mode in the same manner as that of a viscously damped single degree of freedom system. The solution of Eqs. (6.128), when $\zeta_i < 1$, can be expressed as

$$q_i(t) = e^{-\zeta_i\omega_i t}\left\{\cos\omega_{di}t + \frac{\zeta_i}{\sqrt{1 - \zeta_i^2}}\sin\omega_{di}t\right\}q_i(0)$$

$$+ \left\{\frac{1}{\omega_{di}}e^{-\zeta_i\omega_i t}\sin\omega_{di}t\right\}\dot{q}_0(0)$$

$$+ \frac{1}{\omega_{di}}\int_0^t Q_i(\tau)\,e^{-\zeta_i\omega_i(t-\tau)}\sin\omega_{di}(t - \tau)\,d\tau,$$

$$i = 1, 2, \ldots, n \tag{6.129}$$

where

$$\omega_{di} = \omega_i \sqrt{1 - \zeta_i^2} \qquad (6.130)$$

Note the following aspects of these systems:

1. It has been shown by Caughey [6.9] that the condition given by Eq. (6.120) is sufficient but not necessary for the existence of normal modes in damped systems. The necessary condition is that the transformation that diagonalizes the damping matrix also uncouples the coupled equations of motion. This condition is less restrictive than Eq. (6.120) and covers more possibilities.

2. In the general case of damping, the damping matrix cannot be diagonalized simultaneously with the mass and stiffness matrices. In this case, the eigenvalues of the system are either real and negative or complex with negative real parts. The complex eigenvalues exist as conjugate pairs: the associated eigenvectors also consist of complex conjugate pairs. A common procedure for finding the solution of the eigenvalue problem of a damped system involves the transformation of the n coupled second order equations of motion into $2n$ uncoupled first order equations [6.6].

3. The error bounds and numerical methods in the modal analysis of dynamic systems are discussed in Refs. [6.11, 6.12].

Equations of Motion of a Dynamic System

EXAMPLE 6.17

Derive the equations of motion of the system shown in Fig. 6.15.

Solution

Approach: Use Lagrange's equations in conjunction with Rayleigh's dissipation function.
The kinetic energy of the system is

$$T = \frac{1}{2}(m_1 \dot{x}_1^2 + m_2 \dot{x}_2^2 + m_3 \dot{x}_3^2) \qquad (E.1)$$

The potential energy has the form

$$V = \frac{1}{2}[k_1 x_1^2 + k_2 (x_2 - x_1)^2 + k_3 (x_3 - x_2)^2] \qquad (E.2)$$

and Rayleigh's dissipation function is

$$R = \frac{1}{2}[c_1 \dot{x}_1^2 + c_2 (\dot{x}_2 - \dot{x}_1)^2 + c_3 (\dot{x}_3 - \dot{x}_2)^2 + c_4 \dot{x}_2^2 + c_5 (\dot{x}_3 - \dot{x}_1)^2] \qquad (E.3)$$

Lagrange's equations can be written as

$$\frac{d}{dt}\left(\frac{\partial T}{\partial \dot{x}_i}\right) - \frac{\partial T}{\partial x_i} + \frac{\partial R}{\partial \dot{x}_i} + \frac{\partial V}{\partial x_i} = F_i, \qquad i = 1, 2, 3 \qquad (E.4)$$

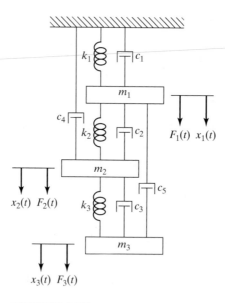

FIGURE 6.15   Three degree of freedom dynamic system.

By substituting Eqs. (E.1) to (E.3) into Eq. (E.4), we obtain the differential equations of motion

$$[m]\,\ddot{\vec{x}} + [c]\,\dot{\vec{x}} + [k]\vec{x} = \vec{F} \tag{E.5}$$

where

$$[m] = \begin{bmatrix} m_1 & 0 & 0 \\ 0 & m_2 & 0 \\ 0 & 0 & m_3 \end{bmatrix} \tag{E.6}$$

$$[c] = \begin{bmatrix} c_1 + c_2 + c_5 & -c_2 & -c_5 \\ -c_2 & c_2 + c_3 + c_4 & -c_3 \\ -c_5 & -c_3 & c_3 + c_5 \end{bmatrix} \tag{E.7}$$

$$[k] = \begin{bmatrix} k_1 + k_2 & -k_2 & 0 \\ -k_2 & k_2 + k_3 & -k_3 \\ 0 & -k_3 & k_3 \end{bmatrix} \tag{E.8}$$

$$\vec{x} = \begin{Bmatrix} x_1(t) \\ x_2(t) \\ x_3(t) \end{Bmatrix} \quad \text{and} \quad \vec{F} = \begin{Bmatrix} F_1(t) \\ F_2(t) \\ F_3(t) \end{Bmatrix} \tag{E.9}$$

∎

Steady-State Response of a Forced System

EXAMPLE 6.18

Find the steady-state response of the system shown in Fig. 6.15 when the masses are subjected to the simple harmonic forces $F_1 = F_2 = F_3 = F_0 \cos \omega t$, where $\omega = 1.75\sqrt{k/m}$. Assume that $m_1 = m_2 = m_3 = m$, $k_1 = k_2 = k_3 = k$, $c_4 = c_5 = 0$, and the damping ratio in each normal mode is given by $\zeta_i = 0.01$, $i = 1, 2, 3$.

Solution: The (undamped) natural frequencies of the system (see Example 6.10) are given by

$$\omega_1 = 0.44504 \sqrt{\frac{k}{m}}$$

$$\omega_2 = 1.2471 \sqrt{\frac{k}{m}}$$

$$\omega_3 = 1.8025 \sqrt{\frac{k}{m}} \tag{E.1}$$

and the corresponding $[m]$-orthonormal mode shapes (see Example 6.11) are given by

$$\vec{X}^{(1)} = \frac{0.3280}{\sqrt{m}} \begin{Bmatrix} 1.0 \\ 1.8019 \\ 2.2470 \end{Bmatrix}, \qquad \vec{X}^{(2)} = \frac{0.7370}{\sqrt{m}} \begin{Bmatrix} 1.0 \\ 1.4450 \\ -0.8020 \end{Bmatrix},$$

$$\vec{X}^{(3)} = \frac{0.5911}{\sqrt{m}} \begin{Bmatrix} 1.0 \\ -1.2468 \\ 0.5544 \end{Bmatrix}, \tag{E.2}$$

Thus the modal vector can be expressed as

$$[X] = [\vec{X}^{(1)} \vec{X}^{(2)} \vec{X}^{(3)}] = \frac{1}{\sqrt{m}} \begin{bmatrix} 0.3280 & 0.7370 & 0.5911 \\ 0.5911 & 0.3280 & -0.7370 \\ 0.7370 & -0.5911 & 0.3280 \end{bmatrix} \tag{E.3}$$

The generalized force vector

$$\vec{Q}(t) = [X]^T \vec{F}(t) = \frac{1}{\sqrt{m}} \begin{bmatrix} 0.3280 & 0.5911 & 0.7370 \\ 0.7370 & 0.3280 & -0.5911 \\ 0.5911 & -0.7370 & 0.3280 \end{bmatrix} \begin{Bmatrix} F_0 \cos \omega t \\ F_0 \cos \omega t \\ F_0 \cos \omega t \end{Bmatrix}$$

$$= \begin{Bmatrix} Q_{10} \\ Q_{20} \\ Q_{30} \end{Bmatrix} \cos \omega t \tag{E.4}$$

can be obtained where

$$Q_{10} = 1.6561 \frac{F_0}{\sqrt{m}}, \qquad Q_{20} = 0.4739 \frac{F_0}{\sqrt{m}}, \qquad Q_{30} = 0.1821 \frac{F_0}{\sqrt{m}} \tag{E.5}$$

If the generalized coordinates or the modal participation factors for the three principal modes are denoted as $q_1(t)$, $q_2(t)$, and $q_3(t)$, the equations of motion can be expressed as

$$\ddot{q}_i(t) + 2\zeta_i \omega_i \dot{q}_i(t) + \omega_i^2 q_i(t) = Q_i(t), \qquad i = 1, 2, 3 \tag{E.6}$$

The steady-state solution of Eqs. (E.6) can be written as

$$q_i(t) = q_{i0} \cos(\omega t - \phi), \qquad i = 1, 2, 3 \tag{E.7}$$

where

$$q_{i0} = \frac{Q_{i0}}{\omega_i^2} \frac{1}{\left[\left\{ 1 - \left(\dfrac{\omega}{\omega_i} \right)^2 \right\}^2 + \left(2\zeta_i \dfrac{\omega}{\omega_i} \right)^2 \right]^{1/2}} \tag{E.8}$$

and

$$\phi_i = \tan^{-1} \left\{ \frac{2\zeta_i \dfrac{\omega}{\omega_i}}{1 - \left(\dfrac{\omega}{\omega_i} \right)^2} \right\} \tag{E.9}$$

By substituting the values given in Eqs. (E.5) and (E.1) into Eqs. (E.8) and (E.9), we obtain

$$q_{10} = 0.57815 \frac{F_0 \sqrt{m}}{k}, \qquad \phi_1 = \tan^{-1}(-0.00544)$$

$$q_{20} = 0.31429 \frac{F_0 \sqrt{m}}{k}, \qquad \phi_2 = \tan^{-1}(-0.02988)$$

$$q_{30} = 0.92493 \frac{F_0 \sqrt{m}}{k}, \qquad \phi_3 = \tan^{-1}(0.33827) \tag{E.10}$$

Finally the steady-state response can be found using Eq. (6.122).

■

6.16 Self-Excitation and Stability Analysis

In a number of damped vibratory systems, friction leads to negative damping instead of positive damping. This leads to the instability (or self-excited vibration) of the system. In general, for an n degree of freedom system shown in Fig. 6.16, the equations of motion will be a set of second-order linear differential equations (as given by Eqs. 6.119 or 6.128):

$$[m]\ddot{\vec{x}} + [c]\dot{\vec{x}} + [k]\vec{x} = \vec{F} \tag{6.131}$$

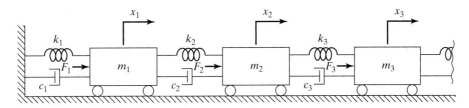

FIGURE 6.16   Multidegree of freedom system.

The method presented in Section 5.8 can be extended to study the stability of the system governed by Eq. (6.131). Accordingly, we assume a solution of the form

$$x_j(t) = C_j e^{st}, \qquad j = 1, 2, \ldots, n$$

or

$$\vec{x}(t) = \vec{C} e^{st} \tag{6.132}$$

where s is a complex number to be determined, C_j is the amplitude of x_j, and

$$\vec{C} = \begin{Bmatrix} C_1 \\ C_2 \\ \vdots \\ C_n \end{Bmatrix}$$

The real part of s determines the damping and its imaginary part gives the natural frequency of the system. The substitution of Eq. (6.132) into the free vibration equations (obtained by setting $\vec{F} = \vec{0}$ in Eq. 6.131) leads to

$$([m] s^2 + [c]s + [k]) \vec{C} e^{st} = \vec{0} \tag{6.133}$$

For a nontrivial solution of C_j, the determinant of the coefficients of C_j is set equal to zero, which leads to the "characteristic equation," similar to Eq. (6.63):

$$D(s) = \|[m] s^2 + [c] s + [k]\| = 0 \tag{6.134}$$

The expansion of Eq. (6.134) leads to a polynomial in s of order $m = 2n$, which can be expressed in the form

$$D(s) = a_0 s^m + a_1 s^{m-1} + a_2 s^{m-2} + \cdots + a_{m-1}s + a_m = 0 \tag{6.135}$$

The stability or instability of the system depends on the roots of the polynomial equation, $D(s) = 0$. Let the roots of Eq. (6.135) be denoted as

$$s_j = b_j + i\omega_j, \qquad j = 1, 2, \ldots, m \tag{6.136}$$

If the real parts of all the roots b_j are negative numbers, there will be decaying time functions, $e^{b_j t}$, in Eq. (6.132), and hence the solution (system) will be stable. On the other hand, if one or more roots s_j have a positive real part, then the solution of Eq. (6.131) will contain one or more exponentially increasing time functions $e^{b_j t}$, and hence the solution (system) will be unstable. If there is a purely imaginary root of the form $s_j = i\omega_j$, it will lead to an oscillatory solution $e^{i\omega_j t}$, which represents a borderline case between stability and instability. If s_j is a multiple root, the above conclusion still holds unless it is a pure imaginary number, such as $s_j = i\omega_j$. In this case, the solution contains functions of the type $e^{i\omega_j t}, te^{i\omega_j t}, t^2 e^{i\omega_j t}, \ldots,$ which increase with time. Thus the multiple roots with purely imaginary values indicate the instability of the system. Thus, in order for a linear system governed by Eq. (6.131) to be stable, it is necessary and sufficient that the roots of Eq. (6.135) should have nonpositive real parts, and that, if any purely imaginary root exists, it should not appear as a multiple root.

Since finding the roots of the polynomial equation (6.135) is a lengthy procedure, a simplified procedure, known as the Routh-Hurwitz stability criterion [6.13, 6.14], can be used to investigate the stability of the system. In order to apply this procedure, the following mth order determinant T_m is defined in terms of the coefficients of the polynomial equation (6.135) as

$$T_m = \begin{vmatrix} a_1 & a_3 & a_5 & a_7 & \cdots & a_{2m-1} \\ a_0 & a_2 & a_4 & a_6 & \cdots & a_{2m-2} \\ 0 & a_1 & a_3 & a_5 & \cdots & a_{2m-3} \\ 0 & a_0 & a_2 & a_4 & \cdots & a_{2m-4} \\ 0 & 0 & a_1 & a_3 & \cdots & a_{2m-5} \\ \vdots & & & & & \\ \cdot & \cdot & \cdot & \cdot & \cdots & a_m \end{vmatrix} \tag{6.137}$$

Then the following subdeterminants, indicated by the dashed lines in Eq. (6.137) are defined:

$$T_1 = a_1 \tag{6.138}$$

$$T_2 = \begin{vmatrix} a_1 & a_3 \\ a_0 & a_2 \end{vmatrix} \tag{6.139}$$

$$T_3 = \begin{vmatrix} a_1 & a_3 & a_5 \\ a_0 & a_2 & a_4 \\ 0 & a_1 & a_3 \end{vmatrix}$$

$$\vdots \tag{6.140}$$

In constructing these subdeterminants, all the coefficients a_i with $i > m$ or $i < 0$ are to be replaced by zeros. According to the Routh-Hurwitz criterion, a necessary and sufficient condition for the stability of the system is that all the coefficients $a_0, a_1, \ldots, a_m$ must be positive and also that all the determinants $T_1, T_2, \ldots, T_m$ must be positive.

6.17 Examples Using MATLAB

■■■■■■■■■■■ Solution of Eigenvalue Problem

EXAMPLE 6.19

Find the eigenvalues and eigenvectors of the matrix (see Example 6.10):

$$[A] = \begin{bmatrix} 1 & 1 & 1 \\ 1 & 2 & 2 \\ 1 & 2 & 3 \end{bmatrix}$$

Solution

```
% Ex 6.19
>> A = [1 1 1; 1 2 2; 1 2 3]

A =

         1        1        1
         1        2        2
         1        2        3

>> [V, D] = eig(A)

V =

    0.5910    0.7370    0.3280
   -0.7370    0.3280    0.5910
    0.3280   -0.5910    0.7370

D =

    0.3080         0         0
         0    0.6431         0
         0         0    5.0489
```

■

■■■■■■■■■■■ Free Vibration Response of a Multidegree of Freedom System

EXAMPLE 6.20

Plot the free vibration response, $x_1(t)$, $x_2(t)$ and $x_3(t)$, of the system considered in Example 6.14 for the following data: $x_{10} = 1.0$, $k = 4000$, and $m = 10$.

Solution: The free vibration response of the masses, $x_1(t)$, $x_2(t)$, and $x_3(t)$, is given by Eqs. (E.7) to (E.9) of Example 6.14.

```
% Ex6_20.m
x10 = 1.0;
k = 4000;
m = 10;
for i = 1: 1001
    t(i) = 5* (i-1) / 1000;
    x1 (i) = x10 * ( 0.1076 * cos (0.44504 * sqrt (k/m) * t(i)) +
        0.5431 * cos (1.2471*sqrt(k/m) *t(i)) + 0.3493 *
        cos (1.8025*sqrt (k/m) *t(i)) );
    x2 (i) = x10 * ( 0.1939 * cos (0.44504 * sqrt (k/m) * t(i)) +
        0.2417 * cos (1.2471*sqrt(k/m) *t(i)) - 0.4355 *
```

```
                        cos (1.8025*sqrt (k/m) *t(i)) );
          x3 (i) = x10 * ( 0.2418 * cos (0.44504 * sqrt(k/m) * t(i)) –
                   0.4356 * cos (1.2471*sqrt (k/m) *t(i)) + 0.1937 *
                        cos (1.8025*sqrt (k/m) *t(i)) );
          end
          subplot (311);
          plot (t, x1);
          ylabel ('x1 (t) ');
          subplot (312);
          plot (t, x2);
          ylabel ('x2 (t) ');
          subplot (313);
          plot (t, x3);
          ylabel ('x3 (t) ');
          xlabel ('t');
```

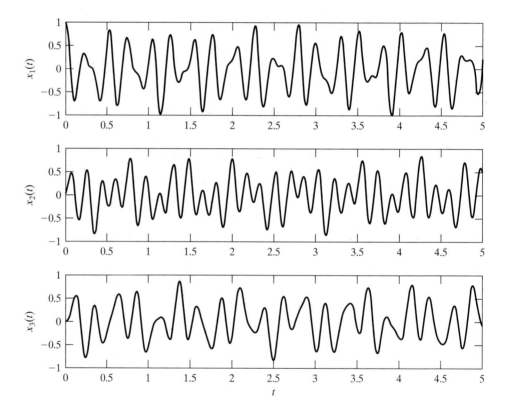

Forced Vibration Response of a Multidegree of Freedom System

EXAMPLE 6.21

Find and plot the forced vibration response of the forging hammer considered in Example 6.16 by solving the governing differential equations. Assume the initial conditions to be zero.

Solution: The governing equations are given by

$$[m]\,\ddot{\vec{x}}(t) + [k]\,\vec{x}(t) = \vec{F}(t) \qquad \text{(E.1)}$$

with

$$[m] = 10^5 \begin{bmatrix} 2 & 0 \\ 0 & 2.5 \end{bmatrix}, [k] = 10^6 \begin{bmatrix} 150 & -150 \\ -150 & 225 \end{bmatrix}, \vec{F}(t) = \begin{Bmatrix} F_1(t) \\ 0 \end{Bmatrix}$$

where $F_1(t)$ is a step function of magnitude 25000 N and duration $0 \le t \le 0.1$ s.

Equations (E.1) can be expressed as a set of four coupled first-order differential equations as

$$\dot{y}_1 = y_2$$
$$\dot{y}_2 = \frac{F_1}{m_1} - \frac{k_1}{m_1} y_1 + \frac{k_1}{m_1} y_3$$
$$\dot{y}_3 = y_4$$
$$\dot{y}_4 = \frac{k_1}{m_2} y_1 - \frac{k_2}{m_2} y_3$$

where $y_1 = x_1$, $y_2 = \dot{x}_1$, $y_3 = x_2$, $y_4 = \dot{x}_2$, $m_1 = 2 \times 10^5$, $m_2 = 2.5 \times 10^5$, $k_1 = 150 \times 10^6$ and $k_2 = 225 \times 10^6$.

Using the initial values of all $y_i = 0$, the following results can be obtained.

```
% Ex6_21.m
% This program will use the function dfunc6_21.m, they should
% be in the same folder
tspan = [0: 0.001: 10];
y0 = [0; 0; 0; 0];
[t, y] = ode23 ('dfunc6_21', tspan, y0);
subplot (211);
plot (t, y(:, 1));
xlabel ('t');
ylabel ('x1 (t) ');
subplot (212);
plot (t, y(:, 3));
xlabel ('t');
ylabel ('x2 (t) ');

% dfunc6_21.m
function f = dfunc6_21 (t, y)
f = zeros (4, 1);
m1 = 2*1e5;
m2 = 2.5*1e5;
k1 = 150 * 1e6;
k2 = 225 * 1e6;
F1 = 25000 * (stepfun (t, 0) - stepfun (t, 0.1));
f(1) = y(2);
f(2) = F1/m1 + k1 * y(3) /m1 - k1 * y(1) /m1;
f(3) = y(4);
f(4) = -k2 * y(3) /m2 + k1 * y(1) /m2;
```

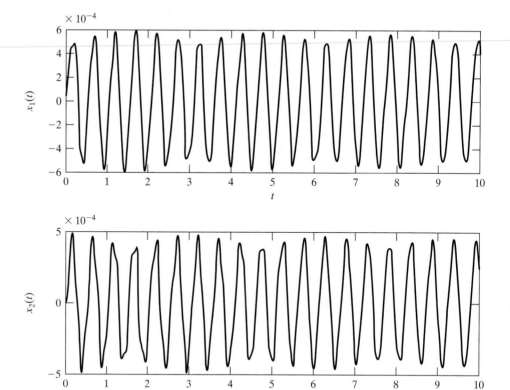

Roots of a Polynomial Equation

EXAMPLE 6.22

Using MATLAB, find the roots of the polynomial

$$f(x) = x^3 - 6x^2 + 11x - 6 = 0$$

Solution

```
>> roots ([1 -6 11 -6])

ans =

        3.0000
        2.0000
        1.0000

>>
```

EXAMPLE 6.23

<hr>

Forced Vibration Response of a Damped System

<hr>

Find the forced vibration response of a damped multidegree of freedom system with equations of motion

$$[m]\ddot{\vec{x}} + [c]\dot{\vec{x}} + [k]\vec{x} = \vec{f} \qquad \text{(E.1)}$$

with

$$[m] = \begin{bmatrix} 100 & 0 & 0 \\ 0 & 10 & 0 \\ 0 & 0 & 10 \end{bmatrix}, \ [c] = 100\begin{bmatrix} 4 & -2 & 0 \\ -2 & 4 & -2 \\ 0 & -2 & 2 \end{bmatrix}, \ [k] = 1000\begin{bmatrix} 8 & -4 & 0 \\ -4 & 8 & -4 \\ 0 & -4 & 4 \end{bmatrix},$$

$$\vec{f} = \begin{Bmatrix} 1 \\ 1 \\ 1 \end{Bmatrix} F_0 \cos \omega t$$

with $F_0 = 50$ and $\omega = 50$. Assume zero initial conditions.

Solution: Equations (E.1) can be rewritten as a set of six first-order differential equations

$$\dot{y}_1 = y_2$$

$$\dot{y}_2 = \frac{F_0}{10}\cos \omega t - \frac{400}{10}y_2 + \frac{200}{10}y_4 - \frac{8000}{10}y_1 + \frac{4000}{10}y_3$$

$$\dot{y}_3 = y_4$$

$$\dot{y}_4 = \frac{F_0}{10}\cos \omega t + \frac{200}{10}y_2 - \frac{400}{10}y_4 + \frac{200}{10}y_6 + \frac{4000}{10}y_1 - \frac{8000}{10}y_3 + \frac{4000}{10}y_5$$

$$\dot{y}_5 = y_6$$

$$\dot{y}_6 = \frac{F_0}{10}\cos \omega t + \frac{200}{10}y_4 - \frac{200}{10}y_6 + \frac{4000}{10}y_3 - \frac{4000}{10}y_5$$

where $y_1 = x_1$, $y_2 = \dot{x}_1$, $y_3 = x_2$, $y_4 = \dot{x}_2$, $y_5 = x_3$, and $y_6 = \dot{x}_3$.

Using zero initial values of all y_i, the solution can be found as follows.

```
% Ex6_23.m
% This program will use the function dfunc6_23.m, they should
% be in the same folder
tspan = [0: 0.01: 10];
y0 = [0; 0; 0; 0; 0; 0];
[t, y] = ode23 ('dfunc6_23', tspan, y0);
subplot (311);
plot (t, y (:, 1));
xlabel ('t');
ylabel ('x1 (t)');
subplot (312);
plot (t, y (:, 3));
xlabel ('t');
ylabel ('x2 (t)');
subplot (313);
plot (t, y (:, 5));
xlabel ('t');
ylabel ('x3 (t)');
```

```
% dfunc6_23.m
function f = dfunc6_23 (t, y)
f = zeros (6, 1);
F0 = 50.0;
w = 50.0;
f(1) = y(2);
f(2) = F0*cos(w*t)/100 - 400*y(2)/100 + 200*y(4)/100 - 8000*y(1)/100
     + 4000*y(3)/100;
f(3) = y(4);
f(4) = F0*cos(w*t)/10 + 200*y(2)/10 - 400*y(4)/10 + 200*y(6)/10
     + 4000*y(1)/10 - 8000*y(3)/10 + 4000*y(5)/10;
f(5) = y(6);
f(6) = F0*cos(w*t)/10 + 200*y(4)/10 - 200*y(6)/10 + 4000*y(3)/10
     - 4000*y(5)/10;
```

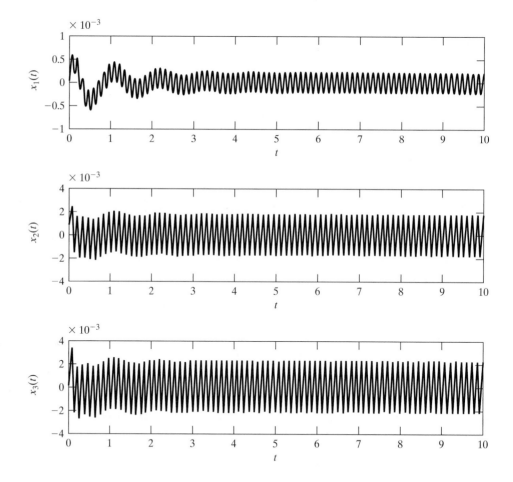

EXAMPLE 6.24

Program to Generate Characteristic Polynomial

Develop a general computer program, called **Program7.m**, to generate the characteristic polynomial corresponding to a given square matrix. Use the program to generate the characteristic polynomial corresponding to the matrix

$$[A] = \begin{bmatrix} 2 & -1 & 0 \\ -1 & 2 & -1 \\ 0 & -1 & 2 \end{bmatrix}$$

Solution: **Program7.m** is developed to accept the following input data:

n = order of the matrix [A]

$[a]$ = given matrix [A]

The following output is generated by the program:

pcf = vector of polynomial coefficients starting from the constant term.

```
>> program7

polynomial expansion of a determinantal equation

data: determinant A:
      2.000000e+000     -1.000000e+000      0.000000e+000
     -1.000000e+000      2.000000e+000     -1.000000e+000
      0.000000e+000     -1.000000e+000      2.000000e+000

result: polynomial coefficients in
pcf(np)*(x^n)+pcf(n)*(x^(n-1))+...+pcf(2)+pcf(1)=0

 -4.000000e+000   1.000000e+001    -6.000000e+000   1.000000e+000
```

∎

EXAMPLE 6.25

Program for Modal Analysis of Multidegree of Freedom Systems

Develop a MATLAB program, called **Program8.m**, to find the response of a multidegree of freedom system using modal analysis. Use the program to find the solution of a system with the following data:

Mass matrix:

$$[m] = \begin{bmatrix} 1 & 0 & 0 \\ 0 & 1 & 0 \\ 0 & 0 & 1 \end{bmatrix}$$

Modal matrix (with modes as columns; modes are not made m-orthogonal):

$$[ev] = \begin{bmatrix} 1.0000 & 1.0000 & 1.0000 \\ 1.8019 & 0.4450 & -1.2468 \\ 2.2470 & -0.8020 & 0.5544 \end{bmatrix}$$

Natural frequencies: $\omega_1 = 0.89008, \omega_2 = 1.4942, \omega_3 = 3.6050$
Modal damping ratios: $\zeta_i = 0.01, i = 1, 2, 3$
Vector of forces applied to different masses:

$$\vec{F}(t) = \begin{Bmatrix} F_0 \\ F_0 \\ F_0 \end{Bmatrix} \cos \omega t; F_0 = 2.0, \omega = 3.5$$

Initial conditions: $\vec{x}(0) = \vec{0}, \dot{\vec{x}}(0) = \vec{0}$

Solution: `Program8.m` is developed to accept the following input data:

n = degree of freedom of the system

nvec = number of modes to be used in the modal analysis

xm = mass matrix of size $n \times n$

ev = modal matrix of size $n \times nvec$

z = vector of size $nvec$ = modal damping ratios vector

om = vector of size $nvec$ = natural frequencies vector

f = vector of forces applied to masses, of size n

x_0 = initial displacements of masses, vector of size n

xd_0 = initial velocities of masses, vector of size n

nstep = number of time stations or integration points $t_1, t_2, \ldots, t_{nstep}$

delt = interval between consecutive time stations

t = array of size nstep containing times $t_1, t_2, \ldots, t_{nstep}$

The program gives the following output:

x = matrix of size $n \times$ nstep = displacements of masses $m_1, m_2, \ldots, m_n$ at various time stations $t_1, t_2, \ldots, t_{nstep}$

```
>> program8

Response of system using modal analysis

    Coordinate 1
    1.21920e-002    4.62431e-002    9.57629e-002    1.52151e-001    2.05732e-001
    2.47032e-001    2.68028e-001    2.63214e-001    2.30339e-001    1.70727e-001
    8.91432e-002   -6.79439e-003   -1.07562e-001   -2.02928e-001   -2.83237e-001
   -3.40630e-001   -3.70023e-001   -3.69745e-001   -3.41725e-001   -2.91231e-001

    Coordinate 2
    1.67985e-002    6.40135e-002    1.33611e-001    2.14742e-001    2.94996e-001
    3.61844e-001    4.04095e-001    4.13212e-001    3.84326e-001    3.16843e-001
    2.14565e-001    8.53051e-002   -5.99475e-002   -2.08242e-001   -3.46109e-001
   -4.61071e-001   -5.43061e-001   -5.85566e-001   -5.86381e-001   -5.47871e-001

    Coordinate 3
    1.99158e-002    7.57273e-002    1.57485e-001    2.51794e-001    3.43491e-001
    4.17552e-001    4.60976e-001    4.64416e-001    4.23358e-001    3.38709e-001
    2.16699e-001    6.81361e-002   -9.29091e-002   -2.50823e-001   -3.90355e-001
   -4.98474e-001   -5.65957e-001   -5.88490e-001   -5.67173e-001   -5.08346e-001
```

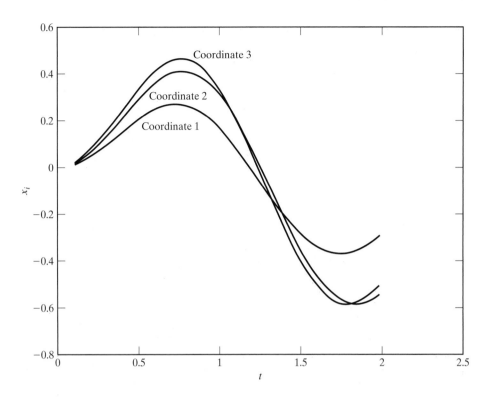

6.18   C++ Programs

Two interactive C++ programs, called **Program7.cpp** and **Program8.cpp**, are given for generating the characteristic polynomial corresponding to a given square matrix and for finding the response of a multidgree of freedom system using modal analysis, respectively. The inputs and outputs of these programs are similar to those of **Program7.m** and **Program8.m**.

Generation of a Characteristic Polynomial

EXAMPLE 6.26

Generate the characteristic polynomial corresponding to the matrix given in Example 6.24 using **Program7.cpp**.

Solution: The input data are to be entered interactively. The input and output of the program are given below.

```
Please input the dimension n of matrix A:
3

Please input matrix A row by row:
```

```
 2   -1    0
-1    2   -1
 0   -1    2
```

POLYNOMIAL EXPANSION OF A DETERMINANTAL EQUATION

DATA: DETERMINANT A:

```
  2.000000    -1.000000     0.000000
 -1.000000     2.000000    -1.000000
  0.000000    -1.000000     2.000000
```

RESULTS: POLYNOMIAL COEFFICIENTS IN
PCF (NP) * (X ** N) + PCF (N) * (X ** (N - 1)) + . . . + PCF(2) * X
+ PCF(1) = 0

```
    -4.000000   10.000000   -6.000000   1.000000
```

■

Response of a System Using Modal Analysis

EXAMPLE 6.27

Using **Program8.cpp**, find the response of the multidegree of freedom system considered in Example 6.25.

Solution: The input data are to be given interactively. The input and output of the program are given below.

```
Please input N, NVEC and NSTEP:
3 3 20

please input delt, omf and f0:
0.1  3.5  2.0

Please input mass matrix xm row by row:
1.0 0.0 0.0
0.0 1.0 0.0
0.0 0.0 1.0

Please input matrix ev row by row:
1.0 1.0 1.0
1.8019 0.4450 -1.2468
2.2470 -0.8020 0.5544

Please input array om:
0.89008 1.4942 3.6050

RESPONSE OF SYSTEM USING MODAL ANALYSIS


COORDINATE    1
    0.009363     0.035438     0.073118     0.115538     0.155010
    0.184085     0.196572     0.188401     0.158198     0.107508
    0.040629    -0.035924    -0.114247    -0.186060    -0.243782
   -0.281509    -0.295776    -0.286010    -0.254588    -0.206525

COORDINATE  2
    0.009395     0.035874     0.075128     0.121321     0.167711
    0.207370     0.233936     0.242311     0.229239     0.193705
    0.137094     0.063112    -0.022556    -0.112803    -0.199814
   -0.275895    -0.334293    -0.369942    -0.380052    -0.364457

COORDINATE  3
    0.009379     0.035648     0.074087     0.118353     0.161294
    0.195874     0.216076     0.217673     0.198772     0.160053
    0.104682     0.037905    -0.033627    -0.102705    -0.162484
   -0.207353    -0.233644    -0.240091    -0.227975    -0.200947
```

■

6.19   Fortran Programs

Two Fortran programs are given. **PROGRAM7.F** finds the characteristic polynomial corresponding to a given square matrix. **PROGRAM8.F** finds the response of a multidegree of freedom system using modal analysis. The inputs and outputs of the programs are similar to those of **Program7.m** and **Program8.m**, respectively.

Characteristic Polynomial Corresponding to a Matrix

EXAMPLE 6.28

Using **PROGRAM7.F**, generate the characteristic polynomial of the matrix given in Example 6.24.

Solution: The output of the program is given below.

```
POLYNOMIAL EXPANSION OF A DETERMINANTAL EQUATION

DATA: DETERMINANT A:
    0.200000E+01    -0.100000E+01     0.000000E+00
   -0.100000E+01     0.200000E+01    -0.100000E+01
    0.000000E+00    -0.100000E+01     0.200000E+01

RESULT: POLYNOMIAL COEFFICIENTS IN
PCF(NP)*(X**N)+PCF(N)*(X**N-1)+...+PCF(2)*X+PCF(1)=0

   -0.400000E+01     0.100000E+02    -0.600000E+01     0.100000E+01
```

■

Forced Response of a System Using Modal analysis

EXAMPLE 6.29

Using **PROGRAM8.F**, find the response of the three degree of freedom system considered in Example 6.25.

Solution: The output of the program is given below.

```
RESPONSE OF SYSTEM USING MODAL ANALYSIS

COORDINATE     1
    0.936322E-02    0.354378E-01    0.731183E-01    0.115538E+00    0.155010E+00
    0.184085E+00    0.196572E+00    0.188401E+00    0.158198E+00    0.107508E+00
    0.406290E-01   -0.359245E-01   -0.114247E+00   -0.186060E+00   -0.243782E+00
   -0.281509E+00   -0.295776E+00   -0.286009E+00   -0.254587E+00   -0.206525E+00

COORDINATE     2
    0.939528E-02    0.358739E-01    0.751277E-01    0.121321E+00    0.167711E+00
    0.207370E+00    0.233936E+00    0.242311E+00    0.229239E+00    0.193704E+00
    0.137094E+00    0.631125E-01   -0.225565E-01   -0.112803E+00   -0.199814E+00
   -0.275895E+00   -0.334293E+00   -0.369942E+00   -0.380052E+00   -0.364457E+00

COORDINATE     3
    0.937892E-02    0.356481E-01    0.740876E-01    0.118353E+00    0.161294E+00
    0.195874E+00    0.216076E+00    0.217673E+00    0.198772E+00    0.160053E+00
    0.104682E+00    0.379048E-01   -0.336272E-01   -0.102705E+00   -0.162484E+00
   -0.207353E+00   -0.233644E+00   -0.240091E+00   -0.227975E+00   -0.200947E+00
```

■

REFERENCES

6.1 F. W. Beaufait, *Basic Concepts of Structural Analysis*, Prentice Hall, Englewood Cliffs, N.J., 1977.

6.2 R. J. Roark and W. C. Young, *Formulas for Stress and Strain* (5th ed.), McGraw-Hill, New York, 1975.

6.3 D. A. Wells, *Theory and Problems of Lagrangian Dynamics*, Schaum's Outline Series, McGraw-Hill, New York, 1967.

6.4 J. H. Wilkinson, *The Algebraic Eigenvalue Problem*, Clarendon Press. Oxford, 1965.

6.5 A. Ralston, *A First Course in Numerical Analysis*, McGraw-Hill, New York, 1965.

6.6 L. Meirovitch, *Analytical Methods in Vibrations*, Macmillan, New York, 1967.

6.7 J. W. Strutt, Lord Rayleigh, *The Theory of Sound*, Macmillan, London, 1877 (reprinted by Dover Publications, New York in 1945).

6.8 W. C. Hurty and M. F. Rubinstein, *Dynamics of Structures*, Prentice Hall, Englewood Cliffs, N.J., 1964.

6.9 T. K. Caughey, "Classical normal modes in damped linear dynamic systems," *Journal of Applied Mechanics*, Vol. 27, 1960, pp. 269–271.

6.10 A. Avakian and D. E. Beskos, "Use of dynamic stiffness influence coefficients in vibrations of non-uniform beams," letter to the editor, *Journal of Sound and Vibration*, Vol. 47, 1976, pp. 292–295.

6.11 I. Gladwell and P. M. Hanson, "Some error bounds and numerical experiments in modal methods for dynamics of systems," *Earthquake Engineering and Structural Dynamics*, Vol. 12, 1984, pp. 9–36.

6.12 R. Bajan, A. R. Kukreti, and C. C. Feng, "Method for improving incomplete modal coupling," *Journal of Engineering Mechanics*, Vol. 109, 1983, pp. 937–949.

6.13 E. J. Routh, *Advanced Rigid Dynamics*, Macmillan, New York, 1905.

6.14 D. W. Nicholson and D. J. Inman, "Stable response of damped linear systems," *Shock and Vibration Digest*, Vol. 15, November 1983, pp. 19–25.

6.15 P. C. Chen and W. W. Soroka, "Multidegree dynamic response of a system with statistical properties," *Journal of Sound and Vibration*, Vol. 37, 1974, pp. 547–556.

6.16 S. Mahalingam and R. E. D. Bishop, "The response of a system with repeated natural frequencies to force and displacement excitation," *Journal of Sound and Vibration*, Vol. 36, 1974, pp. 285–295.

6.17 S. G. Kelly, *Fundamentals of Mechanical Vibrations*, McGraw-Hill, New York, 1993.

REVIEW QUESTIONS

6.1 Give brief answers to the following:

　　1. Define the flexibility and stiffness influence coefficients. What is the relation between them?

　　2. Write the equations of motion of a multidegree of freedom system in matrix form using
　　　(a) the flexibility matrix and
　　　(b) the stiffness matrix.

3. Express the potential and kinetic energies of an n degree of freedom system, using matrix notation.
4. What is a generalized mass matrix?
5. Why is the mass matrix $[m]$ always positive definite?
6. Is the stiffness matrix $[k]$ always positive definite? Why?
7. What is the difference between generalized coordinates and Cartesian coordinates?
8. State Lagrange's equations.
9. What is an eigenvalue problem?
10. What is a mode shape? How is it computed?
11. How many distinct natural frequencies can exist for an n degree of freedom system?
12. What is a dynamical matrix? What is its use?
13. How is the frequency equation derived for a multidegree of freedom system?
14. What is meant by the orthogonality of normal modes? What are orthonormal modal vectors?
15. What is a basis in n-dimensional space?
16. What is the expansion theorem? What is its importance?
17. Explain the modal analysis proceduce.
18. What is a rigid body mode? How is it determined?
19. What is a degenerate system?
20. How can we find the response of a multidegree of freedom system using the first few modes only?
21. Define Rayleigh's dissipation function.
22. Define these terms: *proportional damping, modal damping ratio, modal participation factor.*
23. When do we get complex eigenvalues?
24. What is the use of Routh-Hurwitz criterion?

6.2 Indicate whether each of the following statements is true or false:

1. For a multidegree of freedom system, one equation of motion can be written for each degree of freedom.
2. Lagrange's equation cannot be used to derive the equations of motion of a multidegree of freedom system.
3. The mass, stiffness, and damping matrices of a multidegree of freedom are always symmetric.
4. The product of stiffness and flexibility matrices of a system is always an identity matrix.
5. The modal analysis of a n-degree of freedom system can be conducted using r modes with $r < n$.
6. For a damped multidegree of freedom system, all the eigenvalues can be complex.
7. The modal damping ratio denotes damping in a particular normal mode.
8. A multidegree of freedom system can have six of the natural frequencies equal to zero.
9. The generalized coordinates will always have the unit of length.
10. The generalized coordinates are independent of the conditions of constraint of the system.
11. The generalized mass matrix of a multidegree of freedom system is always diagonal.
12. The potential and kinetic energies of a multidegree of freedom system are always quadratic functions.
13. The mass matrix of a system is always symmetric and positive definite.
14. The stiffness matrix of a system is always symmetric and positive definite.

15. The rigid body mode is also called the zero mode.

16. An unrestrained system is also known as a semidefinite system.

17. Newton's second law of motion can always be used to derive the equations of motion of a vibrating system.

6.3 Fill in each of the following blanks with the appropriate word:

1. The spring constant denotes the_____necessary to cause a unit elongation.

2. The flexibility influence coefficient a_{ij} denotes the deflection at point _____due to a unit load at point_____.

3. The force at point i due to a unit displacement at point j, when all the points other than the point j are fixed, is known as _____ influence coefficient.

4. The mode shapes of a multidegree of freedom system are _____.

5. The equations of motion of a multidegree of freedom system can be expressed in terms of_____coefficients.

6. Lagrange's equations are expressed in terms of_____coordinates.

7. The value of the Kronecker delta (δ_{ij}) is 1 for $i = j$ and _____ for $i \neq j$.

8. The stiffness matrix of a semidefinite system is_____.

9. A multidegree of freedom system can have at most_____rigid body modes.

10. When the solution vector is denoted as a linear combination of the normal modes as $\vec{x}(t) = \sum_{i=1}^{n} q_i(t)\vec{X}^{(i)}$, the generalized coordinates $q_i(t)$ are also known as the_____ participation coefficients.

11. Any set of n linearly independent vectors in an n-dimensional space is called a _____.

12. The representation of an arbitrary n-dimensional vector as a linear combination of n-linearly independent vectors is known as _____theorem.

13. The_____analysis is based on the expansion theorem.

14. The modal analysis basically_____the equations of motion.

15. The eigenvalues of an n degree of freedom system form a_____in the n-dimensional space.

16. The application of Lagrange's equations requires the availability of _____ expressions.

17. The determinantal equation, $|[k] - \omega^2[m]| = 0$, is known as the_____equation.

18. The symmetry of stiffness and flexibility matrices is due to the_____reciprocity theorem.

19. Maxwell's reciprocity theorem states that the influence coefficients are_____.

20. The stiffness matrix is positive definite only if the system is_____.

21. During free vibration of an undamped system, all coordinates will have _____ motion.

22. In proportional damping, the damping matrix is assumed to be a linear combination of the_____and_____matrices.

6.4 Select the most appropriate answer out of the choices given:

1. The number of distinct natural frequencies for an n-degree of freedom system can be
(a) 1   (b) ∞ (c) n

2. The dynamical matrix, $[D]$, is given by
(a) $[k]^{-1}[m]$ (b) $[m]^{-1}[k]$ (c) $[k][m]$

3. The orthognolity of modes implies
(a) $\vec{X}^{(i)^T}[m]\vec{X}^{(j)} = 0$ only

(b) $\vec{X}^{(i)^T}[k]\vec{X}^{(j)} = 0$ only

(c) $\vec{X}^{(i)^T}[m]\vec{X}^{(j)} = 0$ and $\vec{X}^{(i)^T}[k]\vec{X}^{(j)} = 0$

4. The modal matrix, $[X]$, is given by

(a) $[X] = \begin{bmatrix} \vec{X}^{(1)} & \vec{X}^{(2)} & \cdots & \vec{X}^{(n)} \end{bmatrix}$

(b) $[X] = \begin{bmatrix} \vec{X}^{(1)^T} \\ \vec{X}^{(2)^T} \\ \vdots \\ \vec{X}^{(n)^T} \end{bmatrix}$

(c) $[X] = [k]^{-1}[m]$

5. Rayleigh's dissipation function is used to generate a
   (a) stiffness matrix   (b) damping matrix   (c) mass matrix

6. The characteristic equation of an n-degree of freedom system is a
   (a) transcendental equation
   (b) polynomial of degree n
   (c) differential equation of order n

7. The fundamental natural frequency of a system is
   (a) the largest value   (b) the smallest value   (c) any value

8. Negative damping leads to
   (a) instability   (b) fast convergence   (c) oscillations

9. The Routh-Hurwitz criterion can be used to investigate the
   (a) convergence of a system
   (b) oscillations of a system
   (c) stability of a system

10. The stiffness and flexibility matrices are related as
   (a) $[k] = [a]$ (b) $[k] = [a]^{-1}$ (c) $[k] = [a]^T$

11. A system for which $[k]$ is positive and $[m]$ is positive definite is called a
   (a) semidefinite system
   (b) positive-definite system
   (c) indefinite system

12. $[m]$ − orthogonality of modal vectors implies

(a) $\vec{X}^{(i)^T}[m]\vec{X}^{(i)} = 0$

(b) $\vec{X}^{(i)^T}[m]\vec{X}^{(j)} = 0$

(c) $[X]^T[m][X] = [\omega_i^2]$

13. Modal analysis can be used conveniently to find the response of a multidegree of freedom system
   (a) under arbitrary forcing conditions
   (b) under free vibration conditions
   (c) involving several modes

6.5 Match the items in the two columns below:

1. $\dfrac{1}{2}\dot{\vec{X}}^T[m]\dot{\vec{X}}$ (a) equal to zero yields the characteristic values

2. $\dfrac{1}{2}\vec{X}^T[m]\vec{X}$ (b) equal to $[\omega_i^2]$ when modes are normalized

3. $\vec{X}^{(i)^T}[m]\vec{X}^{(j)}$ (c) kinetic energy of the system

4. $\vec{X}^{(i)^T}[m]\vec{X}^{(i)}$ (d) equal to zero when modes are orthogonal

5. $[X]^T[k][X]$ (e) equal to the dynamical matrix $[D]$

6. $[m]\ddot{\vec{x}} + [k]\vec{x}$ (f) strain energy of the system

7. $|[k] - \omega^2[m]|$ (g) equal to the applied force vector $\vec{F}$

8. $[k]^{-1}[m]$ (h) equal to one when modes are orthonormal

PROBLEMS

The problem assignments are organized as follows:

Problems	Section Covered	Topic Covered
6.1–6.6	6.3	Derivation of equations of motion
6.7–6.32	6.4	Influence coefficients
6.33	6.6	Generalized coordinates
6.34–6.43	6.7	Lagrange's equations
6.44–6.45	6.9	Eigenvalue problem
6.46–6.63	6.10	Solution of eigenvalue problem
6.64–6.65	6.12	Unrestrained systems
6.66–6.73	6.13	Free vibration of undamped systems
6.74–6.82	6.14	Forced vibration of undamped systems
6.83–6.84	6.15	Forced vibration of viscously damped systems
6.85–6.92	6.17	MATLAB problems
6.93–6.94	6.18	C++ problems
6.95–6.96	6.19	Fortran problems
6.97–6.100	—	Computer programs
6.101	—	Design project

6.1–
6.5   Derive the equations of motion, using Newton's second law of motion, for each of the systems
         shown in Figs. 6.17 to 6.21.

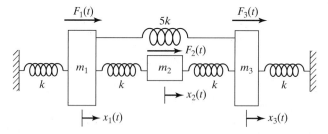

FIGURE 6.17

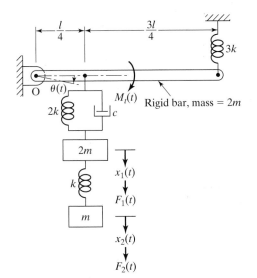

FIGURE 6.18

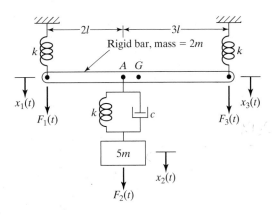

FIGURE 6.19

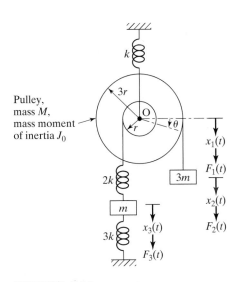

FIGURE 6.20

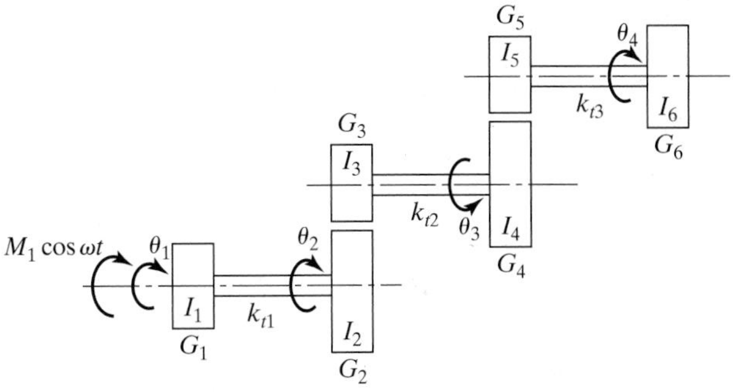

Number of teeth on gear $G_i = n_i$ (i = 1 to 6)
Mass moment of inertia of gear $G_i = I_i$ (i = 1 to 6)

FIGURE 6.21

6.6 An automobile is modeled as shown in Fig. 6.22. Derive the equations of motion using Newton's second law of motion.

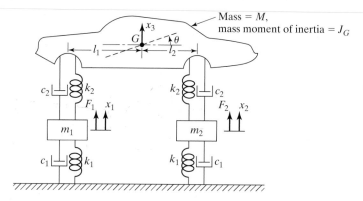

FIGURE 6.22

6.7–

6.12 Derive the stiffness matrix of each of the systems shown in Figs. 6.17 to 6.22 using the indicated coordinates.

6.13 Derive the flexibility matrix of the system shown in Fig. 5.29.

6.14 Derive the stiffness matrix of the system shown in Fig. 5.29.

6.15 Derive the flexibility matrix of the system shown in Fig. 5.32.

6.16 Derive the stiffness matrix of the system shown in Fig. 5.32.

6.17 Derive the mass matrix of the system shown in Fig. 5.32.

6.18 Find the flexibility and stiffness influence coefficients of the torsional system shown in Fig. 6.23. Also write the equations of motion of the system.

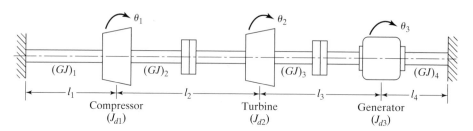

FIGURE 6.23

6.19 Find the flexibility and stiffness influence coefficients of the system shown in Fig. 6.24. Also, derive the equations of motion of the system.

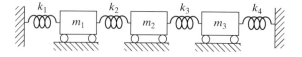

FIGURE 6.24

6.20 An airplane wing, Fig. 6.25(a), is modeled as a three degree of freedom lumped mass system, as shown in Fig. 6.25(b). Derive the flexibility matrix and the equations of motion of the wing by assuming that all $A_i = A$, $(EI)_i = EI$, $l_i = l$ and that the root is fixed.

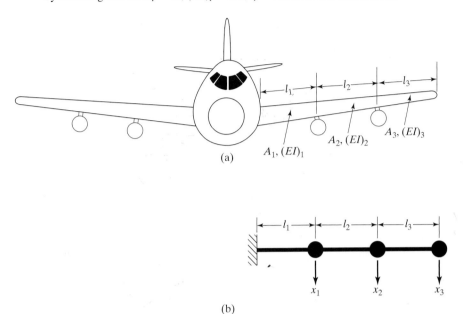

(a)

(b)

FIGURE 6.25

6.21 Determine the flexibility matrix of the uniform beam shown in Fig. 6.26. Disregard the mass of the beam compared to the concentrated masses placed on the beam and assume all $l_i = l$.

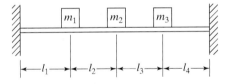

FIGURE 6.26

6.22 Derive the flexibility and stiffness matrices of the spring-mass system shown in Fig. 6.27 assuming that all the contacting surfaces are frictionless.

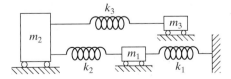

FIGURE 6.27

6.23 Derive the equations of motion for the tightly stretched string carrying three masses, as shown in Fig. 6.28. Assume the ends of the string to be fixed.

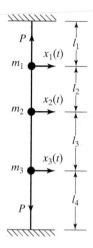

FIGURE 6.28

6.24 Derive the equations of motion of the system shown in Fig. 6.29.

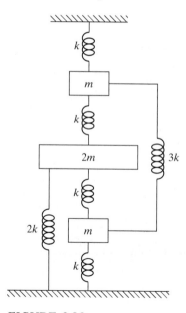

FIGURE 6.29

6.25 Four identical springs, each having a stiffness k, are arranged symmetrically at 90° from each other, as shown in Fig. 2.54. Find the influence coefficient of the junction point in an arbitrary direction.

6.26 Show that the stiffness matrix of the spring-mass system shown in Fig. 6.3(a) is a band matrix along the diagonal.

6.27–
6.31 Derive the mass matrix of each of the systems shown in Figs. 6.17 to 6.21 using the indicated coordinates.

6.32 The inverse influence coefficient b_{ij} is defined as the velocity induced at point i due to a unit impulse applied at point j. Using this definition, derive the inverse mass matrix of the system shown in Fig. 6.4(a).

6.33 For the four-story shear building shown in Fig. 6.30, there is no rotation of the horizontal section at the level of floors. Assuming that the floors are rigid and the total mass is concentrated at the levels of the floors, derive the equations of motion of the building using (a) Newton's second law of motion and (b) Lagrange's equations.

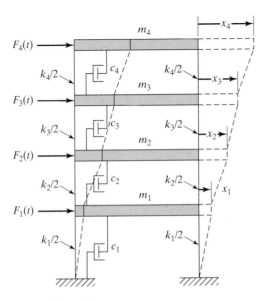

FIGURE 6.30

6.34 Derive the equations of motion of the system shown in Fig. 6.31 by using Lagrange's equations with x and θ as generalized coordinates.

6.35 Derive the equations of motion of the system shown in Fig. 5.10(a), using Lagrange's equations with (a) x_1 and x_2 as generalized coordinates and (b) x and θ as generalized coordinates.

6.36 Derive the equations of motion of the system shown in Fig. 6.24 using Lagrange's equations.

6.37 Derive the equations of motion of the triple pendulum shown in Fig. 6.10 using Lagrange's equations.

6.38 When an airplane (see Fig. 6.32a) undergoes symmetric vibrations, the fuselage can be idealized as a concentrated central mass M_0 and the wings can be modeled as rigid bars carrying end masses M, as shown in Fig. 6.32(b). The flexibility between the wings and the fuselage can be represented by two torsional springs of stiffness k_t each. (a) Derive the equations of motion of the airplane, using Lagrange's equations with x and θ as generalized coordinates.

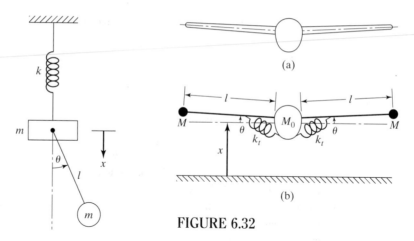

FIGURE 6.31

FIGURE 6.32

(b) Find the natural frequencies and mode shapes of the airplane. (c) Find the torsional spring constant in order to have the natural frequency of vibration, in torsional mode, greater than 2 Hz when $M_0 = 1000$ kg, $M = 500$ kg, and $l = 6$ m.

6.39–

6.43 Use Lagrange's equations to derive the equations of motion of each of the systems shown in Figs. 6.17 to 6.21.

6.44 Set up the eigenvalue problem of Example 6.10 in terms of the coordinates $q_1 = x_1$, $q_2 = x_2 - x_1$, and $q_3 = x_3 - x_2$, and solve the resulting problem. Compare the results obtained with those of Example 6.10 and draw conclusions.

6.45 Derive the frequency equation of the system shown in Fig. 6.24.

6.46 Find the natural frequencies and mode shapes of the system shown in Fig. 6.6(a) when $k_1 = k$, $k_2 = 2k$, $k_3 = 3k$, $m_1 = m$, $m_2 = 2m$, and $m_3 = 3m$. Plot the mode shapes.

6.47 Set up the matrix equation of motion and determine the three principal modes of vibration for the system shown in Fig. 6.6(a) with $k_1 = 3k$, $k_2 = k_3 = k$, $m_1 = 3m$, and $m_2 = m_3 = m$. Check the orthogonality of the modes found.

6.48 Find the natural frequencies of the system shown in Fig. 6.10 with $l_1 = 20$ cm, $l_2 = 30$ cm, $l_3 = 40$ cm, $m_1 = 1$ kg, $m_2 = 2$ kg, and $m_3 = 3$ kg.

6.49* (a) Find the natural frequencies of the system shown in Fig. 6.26 with $m_1 = m_2 = m_3 = m$ and $l_1 = l_2 = l_3 = l_4 = l/4$. (b) Find the natural frequencies of the beam when $m = 10$ kg, $l = 0.5$ m, cross section is solid circular section with diameter 2.5 cm, and the material is steel. (c) Consider using hollow circular, solid rectangular, or hollow rectangular cross section for the beam to achieve the same natural frequencies as in (b). Identify the cross section corresponding to the least weight of the beam.

---

*The asterisk denotes a problem with no unique answer.

6.50 The frequency equation of a three degree of freedom system is given by

$$\begin{vmatrix} \lambda - 5 & -3 & -2 \\ -3 & \lambda - 6 & -4 \\ -1 & -2 & \lambda - 6 \end{vmatrix} = 0$$

Find the roots of this equation.

6.51 Determine the eigenvalues and eigenvectors of the system shown in Fig. 6.24, taking $k_1 = k_2 = k_3 = k_4 = k$ and $m_1 = m_2 = m_3 = m$.

6.52 Find the natural frequencies and mode shapes of the system shown in Fig. 6.24 for $k_1 = k_2 = k_3 = k_4 = k$, $m_1 = 2m$, $m_2 = 3m$, and $m_3 = 2m$.

6.53 Find the natural frequencies and principal modes of the triple pendulum shown in Fig. 6.10, assuming that $l_1 = l_2 = l_3 = l$ and $m_1 = m_2 = m_3 = m$.

6.54 Find the natural frequencies and mode shapes of the system considered in Problem 6.22 with $m_1 = m$, $m_2 = 2m$, $m_3 = m$, $k_1 = k_2 = k$, and $k_3 = 2k$.

6.55 Show that the natural frequencies of the system shown in Fig. 6.6(a), with $k_1 = 3k$, $k_2 = k_3 = k$, $m_1 = 4m$, $m_2 = 2m$, and $m_3 = m$, are given by $\omega_1 = 0.46\sqrt{k/m}$, $\omega_2 = \sqrt{k/m}$, and $\omega_3 = 1.34\sqrt{k/m}$. Find the eigenvectors of the system.

6.56 Find the natural frequencies of the system considered in Problem 6.23 with $m_1 = 2m$, $m_2 = m$, $m_3 = 3m$, and $l_1 = l_2 = l_3 = l_4 = l$.

6.57 Find the natural frequencies and principal modes of the torsional system shown in Fig. 6.23 for $(GJ)_i = GJ$, $i = 1, 2, 3, 4$, $J_{d1} = J_{d2} = J_{d3} = J_0$, and $l_1 = l_2 = l_3 = l_4 = l$.

6.58 The mass matrix $[m]$ and the stiffness matrix $[k]$ of a uniform bar are

$$[m] = \frac{\rho A l}{4}\begin{bmatrix} 1 & 0 & 0 \\ 0 & 2 & 0 \\ 0 & 0 & 1 \end{bmatrix} \quad \text{and} \quad [k] = \frac{2AE}{l}\begin{bmatrix} 1 & -1 & 0 \\ -1 & 2 & -1 \\ 0 & -1 & 1 \end{bmatrix}$$

where ρ is the density, A is the cross-sectional area, E is Young's modulus, and l is the length of the bar. Find the natural frequencies of the system by finding the roots of the characteristic equation. Also find the principal modes.

6.59 The mass matrix of a vibrating system is given by

$$[m] = \begin{bmatrix} 1 & 0 & 0 \\ 0 & 2 & 0 \\ 0 & 0 & 1 \end{bmatrix}$$

and the eigenvectors by

$$\begin{Bmatrix} 1 \\ -1 \\ 1 \end{Bmatrix}, \quad \begin{Bmatrix} 1 \\ 1 \\ 1 \end{Bmatrix} \quad \text{and} \quad \begin{Bmatrix} 0 \\ 1 \\ 2 \end{Bmatrix}$$

Find the $[m]$-orthonormal modal matrix of the system.

6.60 For the system shown in Fig. 6.33, (a) determine the characteristic polynomial $\Delta(\omega^2) = \det|[k] - \omega^2[m]|$, (b) plot $\Delta(\omega^2)$ from $\omega^2 = 0$ to $\omega^2 = 4.0$ (using increments $\Delta\omega^2 = 0.2$), and (c) find ω_1^2, ω_2^2, and ω_3^2.

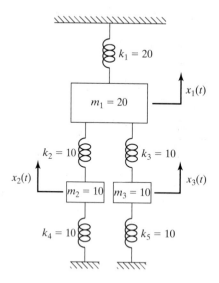

FIGURE 6.33

6.61 (a) Two of the eigenvectors of a vibrating system are known to be

$$\begin{Bmatrix} 0.2754946 \\ 0.3994672 \\ 0.4490562 \end{Bmatrix} \quad \text{and} \quad \begin{Bmatrix} 0.6916979 \\ 0.2974301 \\ -0.3389320 \end{Bmatrix}$$

Prove that these are orthogonal with respect to the mass matrix

$$[m] = \begin{bmatrix} 1 & 0 & 0 \\ 0 & 2 & 0 \\ 0 & 0 & 3 \end{bmatrix}$$

Find the remaining $[m]$-orthogonal eigenvector. (b) If the stiffness matrix of the system is given by

$$\begin{bmatrix} 6 & -4 & 0 \\ -4 & 10 & 0 \\ 0 & 0 & 6 \end{bmatrix}$$

determine all the natural frequencies of the system, using the eigenvectors of part (a).

6.62 Find the natural frequencies of the system shown in Fig. 6.17 for $m_i = m$, $i = 1, 2, 3$.

6.63 Find the natural frequencies of the system shown in Fig. 6.18 with $m = 1$ kg, $l = 1$ m, $k = 1000$ N/m and $c = 100$ N-s/m.

6.64 Find the natural frequencies and mode shapes of the system shown in Fig. 6.13 with $m_1 = m$, $m_2 = 2m$, $m_3 = 3m$, and $k_1 = k_2 = k$.

6.65 Find the modal matrix for the semi-definite system shown in Fig. 6.34 for $J_1 = J_2 = J_3 = J_0$, $k_{t1} = k_t$, and $k_{t2} = 2k_t$.

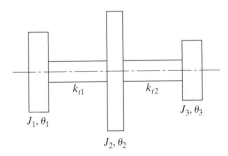

FIGURE 6.34

6.66 Find the free vibration response of the spring-mass system shown in Fig. 6.24 for $k_i = k (i = 1, 2, 3, 4)$, $m_1 = 2m$, $m_2 = 3m$, $m_3 = 2m$ for the initial conditions $x_1(0) = x_{10}$ and $x_2(0) = x_3(0) = \dot{x}_1(0) = \dot{x}_2(0) = \dot{x}_3(0) = 0$.

6.67 Find the free vibration response of the triple pendulum shown in Fig. 6.10 for $l_i = l (i = 1, 2, 3)$ and $m_i = m (i = 1, 2, 3)$ for the initial conditions $\theta_1(0) = \theta_2(0) = 0$, $\theta_3(0) = \theta_{30}$, $\dot{\theta}_i(0) = 0 (i = 1, 2, 3)$.

6.68 Find the free vibration response of the tightly stretched string shown in Fig. 6.28 for $m_1 = 2m$, $m_2 = m$, $m_3 = 3m$ and $l_i = l (i = 1, 2, 3, 4)$. Assume the initial conditions as $x_1(0) = x_3(0) = 0$, $x_2(0) = x_{20}$, $\dot{x}_i(0) = 0 (i = 1, 2, 3)$.

6.69 Find the free vibration response of the spring-mass system shown in Fig. 6.6(a) for $k_1 = k$, $k_2 = 2k$, $k_3 = 3k$, $m_1 = m$, $m_2 = 2m$, and $m_3 = 3m$ corresponding to the initial conditions $\dot{x}_1(0) = \dot{x}_{10}$, $x_i(0) = 0 (i = 1, 2, 3)$, and $\dot{x}_2(0) = \dot{x}_3(0) = 0$.

6.70 Find the free vibration response of the spring-mass system shown in Fig. 6.27 for $m_1 = m$, $m_2 = 2m$, $m_3 = m$, $k_1 = k_2 = k$, and $k_3 = 2k$ corresponding to the initial conditions $\dot{x}_3(0) = \dot{x}_{30}$, and $x_1(0) = x_2(0) = x_3(0) = \dot{x}_1(0) = \dot{x}_2(0) = 0$.

6.71 In the freight car system shown in Fig. 6.13, the first car acquires a velocity of $\dot{x}_0$ due to an impact. Find the resulting free vibration of the system. Assume $m_i = m (i = 1, 2, 3)$ and $k_1 = k_2 = k$.

6.72 Find the free vibration response of a three degree of freedom system governed by the equations

$$10 \begin{bmatrix} 1 & 0 & 0 \\ 0 & 1 & 0 \\ 0 & 0 & 1 \end{bmatrix} \ddot{\vec{x}}(t) + 100 \begin{bmatrix} 2 & -1 & 0 \\ -1 & 2 & -1 \\ 0 & -1 & 1 \end{bmatrix} \vec{x}(t) = \vec{0}$$

Assume the initial conditions as $x_i(0) = 0.1$ and $\dot{x}_i(0) = 0$; $i = 1, 2, 3$.

Note: The natural frequencies and mode shapes of the system are given in Examples 6.10 and 6.11.

6.73 Using modal analysis, determine the free vibration response of a two degree of freedom system with equations of motion

$$2\begin{bmatrix} 1 & 0 \\ 0 & 1 \end{bmatrix}\ddot{\vec{x}}(t) + 8\begin{bmatrix} 2 & -1 \\ -1 & 2 \end{bmatrix}\vec{x}(t) = \vec{0}$$

with initial conditions

$$\vec{x}(0) = \begin{Bmatrix} 1 \\ 0 \end{Bmatrix}, \quad \dot{\vec{x}}(0) = \begin{Bmatrix} 0 \\ 1 \end{Bmatrix}$$

6.74 Determine the amplitudes of motion of the three masses in Fig. 6.35 when a harmonic force $F(t) = F_0 \sin \omega t$ is applied to the lower left mass with $m = 1$ kg, $k = 1000$ N/m, $F_0 = 5$ N and $\omega = 10$ rad/s using the mode superposition method.

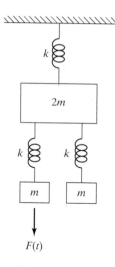

FIGURE 6.35

6.75 (a) Determine the natural frequencies and mode shapes of the torsional system shown in Fig. 6.11 for $k_{t1} = k_{t2} = k_{t3} = k_t$ and $J_1 = J_2 = J_3 = J_0$. (b) If a torque $M_{t3}(t) = M_{t0} \cos \omega t$, with $M_{t0} = 500$ N-m and $\omega = 100$ rad/s acts on the generator (J_3), find the amplitude of each component. Assume $M_{t1} = M_{t2} = 0$, $k_t = 100$ N-m/rad, and $J_0 = 1$ kg-m^2.

6.76 Using the results of Problems 6.19 and 6.51, determine the modal matrix $[X]$ of the system shown in Fig. 6.24 and derive the uncoupled equations of motion.

6.77 An approximate solution of a multidegree of freedom system can be obtained using the mode acceleration method. According to this method, the equations of motion of an undamped system, for example, are expressed as

$$\vec{x} = [k]^{-1}(\vec{F} - [m]\,\ddot{\vec{x}}) \tag{E.1}$$

and $\ddot{\vec{x}}$ is approximated using the first r modes ($r < n$) as

$$\underset{n \times 1}{\ddot{\vec{x}}} = \underset{n \times r}{[X]}\,\underset{r \times 1}{\ddot{\vec{q}}} \tag{E.2}$$

Since $([k] - \omega_i^2[m])\vec{X}^{(i)} = \vec{0}$, Eq. (E.1) can be written as

$$\vec{x}(t) = [k]^{-1}\vec{F}(t) - \sum_{i=1}^{r} \frac{1}{\omega_i^2}\vec{X}^{(i)}\ddot{q}_i(t) \tag{E.3}$$

Find the approximate response of the system described in Example 6.18 (without damping), using the mode acceleration method with $r = 1$.

6.78 Determine the response of the system in Problem 6.46 to the initial conditions $x_1(0) = 1$, $\dot{x}_1(0) = 0$, $x_2(0) = 2$, $\dot{x}_2(0) = 1$, $x_3(0) = 1$, and $\dot{x}_3(0) = -1$. Assume $k/m = 1$.

6.79 Show that the initial conditions of the generalized coordinates $q_i(t)$ can be expressed in terms of those of the physical coordinates $x_i(t)$ in modal analysis as

$$\vec{q}(0) = [X]^T [m]\vec{x}(0), \quad \dot{\vec{q}}(0) = [X]^T [m]\dot{\vec{x}}(0)$$

6.80 A simplified model of a bicycle with its rider is shown in Fig. 6.36. Find the vertical motion of the rider when the bicycle hits an elevated road, as shown in the figure.

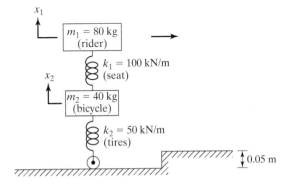

FIGURE 6.36

6.81 Find the response of the triple pendulum shown in Fig. 6.10 for $l_i = 0.5$ m ($i = 1, 2, 3$) and $m_i = 1$ kg ($i = 1, 2, 3$) when a moment, in the form of a rectangular pulse of magnitude 0.1 N-m and duration 0.1 s, is applied along the direction of θ_3. Assume that the pendulum is at rest at $t = 0$.

6.82 Find the response of the spring-mass system shown in Fig. 6.6(a) for $k_1 = k$, $k_2 = 2k$, $k_3 = 3k$, $m_1 = m$, $m_2 = 2m$, and $m_3 = 3m$ with $k = 10^4$ N/m and $m = 2$ kg when a force, in the form of a rectangular pulse of magnitude 1000 N and duration 0.25 s, is applied to mass m_1 in the direction of x_1.

6.83 Find the steady-state response of the system shown in Fig. 6.16 with $k_1 = k_2 = k_3 = k_4 = 100$ N/m, $c_1 = c_2 = c_3 = c_4 = 1$ N-s/m, $m_1 = m_2 = m_3 = 1$ kg, $F_1(t) = F_0 \cos \omega t$, $F_0 = 10$ N and $\omega = 1$ rad/s. Assume that the spring k_4 and the damper c_4 are connected to a rigid wall at the right end. Use the mechanical impedance method described in Section 5.6 for solution.

6.84 An airplane wing, Fig. 6.37(a), is modeled as a twelve degree of freedom lumped mass system as shown in Fig. 6.37(b). The first three natural mode shapes, obtained experimentally, are given below.

Mode Shape	**Degree of Freedom**												
	0	1	2	3	4	5	6	7	8	9	10	11	12
$\vec{X}^{(1)}$	0.0	0.126	0.249	0.369	0.483	0.589	0.686	0.772	0.846	0.907	0.953	0.984	1.000
$\vec{X}^{(2)}$	0.0	−0.375	−0.697	−0.922	−1.017	−0.969	−0.785	−0.491	−0.127	0.254	0.599	0.860	1.000
$\vec{X}^{(3)}$	0.0	0.618	1.000	1.000	0.618	0.000	−0.618	−1.000	−1.000	−0.618	0.000	0.618	1.000

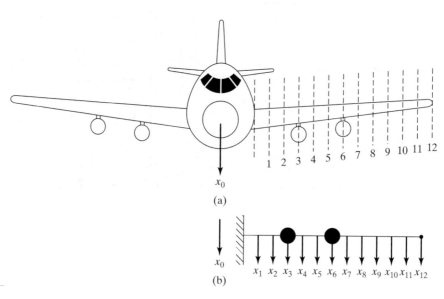

FIGURE 6.37

The natural frequencies corresponding to the three mode shapes are given by $\omega_1 = 225$ rad/s, $\omega_2 = 660$ rad/s, and $\omega_3 = 1100$ rad/s. If the fuselage of the airplane is subjected to a known vertical motion $x_0(t)$, derive the uncoupled equations for determining the dynamic response of the wing by approximating it as a linear combination of the first three normal modes. *Hint:* The equation of motion of the airplane wing can be written, similar to Eq. (3.64), as

$$[m]\,\ddot{\vec{x}} + [c]\,(\dot{\vec{x}} - \dot{x}_0\vec{u}_1) + [k]\,(\vec{x} - x_0\vec{u}_1) = \vec{0}$$

or

$$[m]\,\ddot{\vec{x}} + [c]\,\dot{\vec{x}} + [k]\,\vec{x} = -x_0\,[m]\,\vec{u}_1$$

where $\vec{u}_1 = \{1, 0, 0, \ldots, 0\}^T$ is a unit vector.

6.85 Using MATLAB, find the eigenvalues and eigenvectors of a system with mass and stiffness matrices given in Example 6.12.

6.86 Using MATLAB, find and plot the free vibration response of the system described in Problem 6.68 for the following data: $x_{20} = 0.5$, $P = 100$, $l = 5$, $m = 2$.

6.87 Using the MATLAB function **ode23**, find and plot the forced vibration response of the system described in Problem 6.74.

6.88 Using the MATLAB function **roots**, find the roots of the following equation:

$$f(x) = x^{12} - 2 = 0$$

6.89 Find the forced vibration response of a viscously damped three degree of freedom system with equations of motion:

$$10\begin{bmatrix} 1 & 0 & 0 \\ 0 & 2 & 0 \\ 0 & 0 & 3 \end{bmatrix}\ddot{\vec{x}}(t) + 5\begin{bmatrix} 3 & -1 & 0 \\ -1 & 4 & -3 \\ 0 & -3 & 3 \end{bmatrix}\dot{\vec{x}}(t) + 20\begin{bmatrix} 7 & -3 & 0 \\ -3 & 5 & -2 \\ 0 & -2 & 2 \end{bmatrix}\vec{x}(t) = \begin{Bmatrix} 5\cos 2t \\ 0 \\ 0 \end{Bmatrix}$$

Assume zero initial conditions.

6.90 Using the MATLAB function **ode23**, solve Problem 6.83 and plot $x_1(t)$, $x_2(t)$, and $x_3(t)$.

6.91 Using **Program7.m**, generate the characteristic polynomial corresponding to the matrix

$$[A] = \begin{bmatrix} 5 & 3 & 2 \\ 3 & 6 & 4 \\ 1 & 2 & 6 \end{bmatrix}$$

6.92 Using **Program8.m**, find the steady-state response of a three degree of freedom system with the following data:

$$\omega_1 = 25.076 \text{ rad/s}, \ \omega_2 = 53.578 \text{ rad/s}, \ \omega_3 = 110.907 \text{ rad/s}$$

$$\zeta_i = 0.001, \ i = 1, 2, 3$$

$$[m] = \begin{bmatrix} 41.4 & 0 & 0 \\ 0 & 38.8 & 0 \\ 0 & 0 & 25.88 \end{bmatrix}, \ [ev] = \begin{bmatrix} 1 & 1.0 & 1.0 \\ 1.303 & 0.860 & -1.000 \\ 1.947 & -1.685 & 0.183 \end{bmatrix}$$

$$\vec{F}(t) = \begin{Bmatrix} F_1(t) \\ F_2(t) \\ F_3(t) \end{Bmatrix} = \begin{Bmatrix} 5000 \cos 5\,t \\ 10000 \cos 10\,t \\ 20000 \cos 20\,t \end{Bmatrix}$$

6.93 Using `Program7.cpp`, solve Problem 6.91.

6.94 Using `Program8.cpp`, solve Problem 6.92.

6.95 Using `PROGRAM7.F`, solve Problem 6.91.

6.96 Using `PROGRAM8.F`, solve Problem 6.92.

6.97 Write a computer program for finding the eigenvectors using the known eigenvalues in Eq. (6.61). Find the mode shapes of Problem 6.52 using this program.

6.98 Write a computer program for generating the [m]-orthonormal modal matrix [X]. The program should accept the number of degrees of freedom, the normal modes, and the mass matrix as input. Solve Problem 6.59 using this program.

6.99 The equations of motion of an undamped system in SI units are given by

$$\begin{bmatrix} 2 & 0 & 0 \\ 0 & 2 & 0 \\ 0 & 0 & 2 \end{bmatrix} \ddot{x} + \begin{bmatrix} 16 & -8 & 0 \\ -8 & 16 & -8 \\ 0 & -8 & 16 \end{bmatrix} \vec{x} = \begin{Bmatrix} 10 \sin \omega t \\ 0 \\ 0 \end{Bmatrix}$$

Using subroutine MODAL, find the steady-state response of the system when $\omega = 5$ rad/s.

6.100 Find the response of the system in Problem 6.99 by varying ω between 1 and 10 rad/s in increments of 1 rad/s. Plot the graphs showing the variations of magnitudes of the first peaks of $x_i(t)$, $i = 1, 2, 3$ with respect to ω.

DESIGN PROJECT

6.101 A heavy machine tool mounted on the first floor of a building, Fig. 6.38(a), has been modeled as a three degree of freedom system as indicated in Fig. 6.38(b). (a) For $k_1 = 5000$ lb/in., $k_2 = 500$ lb/in., $k_3 = 2000$ lb/in., $c_1 = c_2 = c_3 = 10$ lb-sec/in., $m_f = 50$ lb-sec^2/in., $m_b = 10$ lb-sec^2/in., $m_h = 2$ lb-sec^2/in., and $F(t) = 1000 \cos 60t$ lb, find the steady-state vibration of the system using the mechanical impedance method described in Section 5.6. (b) If the maximum response of the machine tool head (x_3) has to be reduced by 25 percent, how should the stiffness of the mounting (k_2) be changed? (c) Is there any better way of achieving the goal stated in (b)? Provide details.

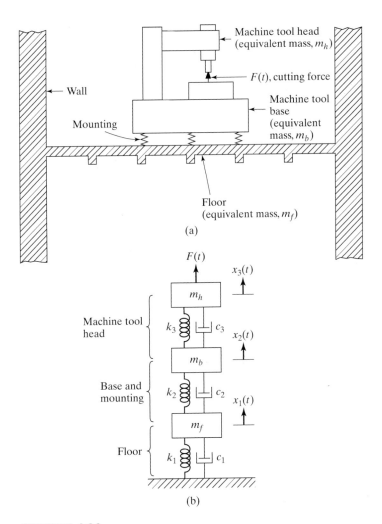

(a)

(b)

FIGURE 6.38

J ohn William Strutt, Lord Rayleigh (1842–1919), was an English physicist who held the positions of professor of experimental physics at Cambridge University, professor of natural philosophy at the Royal Institution in London, president of the Royal Society, and chancellor of Cambridge University. His works in optics and acoustics are well known, with *Theory of Sound* (1877) considered as a standard reference even today. The method of computing approximate natural frequencies of vibrating bodies using an energy approach has become known as "Rayleigh's method." (Courtesy of *Applied Mechanics Reviews*.)

CHAPTER 7

Determination of Natural Frequencies and Mode Shapes

7.1 Introduction

In the preceding chapter, the natural frequencies (eigenvalues) and the natural modes (eigenvectors) of a multidegree of freedom system were found by setting the characteristic determinant equal to zero. Although this is an exact method, the expansion of the characteristic determinant and the solution of the resulting nth degree polynomial equation to obtain the natural frequencies can become quite tedious for large values of n. Several analytical and numerical methods have been developed to compute the natural frequencies and mode shapes of multidegree of freedom systems. In this chapter, we shall consider Dunkerley's formula, Rayleigh's method, Holzer's method, the matrix iteration method, and Jacobi's method. Dunkerley's formula and Rayleigh's method are useful for estimating the fundamental natural frequency only. Holzer's method is essentially a tabular method that can be used to find partial or full solutions to eigenvalue problems. The matrix iteration method finds one natural frequency at a time, usually starting from the lowest value. The method can thus be terminated after finding the required number of natural frequencies and mode shapes. When all the natural frequencies and mode shapes are required, Jacobi's method can be used; it finds all the eigenvalues and eigenvectors simultaneously.

7.2   Dunkerley's Formula

Dunkerley's formula gives the approximate value of the fundamental frequency of a composite system in terms of the natural frequencies of its component parts. It is derived by making use of the fact that the higher natural frequencies of most vibratory systems are large compared to their fundamental frequencies [7.1–7.3]. To derive Dunkerley's formula, consider a general n degree of freedom system whose eigenvalues can be determined by solving the frequency equation, Eq. (6.63):

$$| - [k] + \omega^2 [m]| = 0$$

or

$$\left| - \frac{1}{\omega^2}[I] + [a][m] \right| = 0 \tag{7.1}$$

For a lumped mass system with a diagonal mass matrix, Eq. (7.1) becomes

$$\left| -\frac{1}{\omega^2} \begin{bmatrix} 1 & 0 & \cdots & 0 \\ 0 & 1 & \cdots & 0 \\ \vdots & & & \\ 0 & 0 & \cdots & 1 \end{bmatrix} \right.$$

$$\left. + \begin{bmatrix} a_{11} & a_{12} & \cdots & a_{1n} \\ a_{21} & a_{22} & \cdots & a_{2n} \\ \vdots & & & \\ a_{n1} & a_{n2} & \cdots & a_{nn} \end{bmatrix} \begin{bmatrix} m_1 & 0 & \cdots & 0 \\ 0 & m_2 & \cdots & 0 \\ \vdots & & & \\ 0 & 0 & \cdots & m_n \end{bmatrix} \right| = 0$$

that is,

$$\begin{vmatrix} \left(-\dfrac{1}{\omega^2} + a_{11}m_1 \right) & a_{12}m_2 & \cdots & a_{1n}m_n \\[2mm] a_{21}m_1 & \left(-\dfrac{1}{\omega^2} + a_{22}m_2 \right) & \cdots & a_{2n}m_n \\[2mm] \vdots & \vdots & & \vdots \\[2mm] a_{n1}m_1 & a_{n2}m_2 & \cdots & \left(-\dfrac{1}{\omega^2} + a_{nn}m_n \right) \end{vmatrix} = 0 \tag{7.2}$$

The expansion of Eq. (7.2) leads to

$$\left(\frac{1}{\omega^2}\right)^n - (a_{11}m_1 + a_{22}m_2 + \cdots + a_{nn}m_n)\left(\frac{1}{\omega^2}\right)^{n-1}$$

$$+ (a_{11}a_{22}m_1m_2 + a_{11}a_{33}m_1m_3 + \cdots + a_{n-1,n-1}a_{nn}m_{n-1}m_n$$

$$- a_{12}a_{21}m_1m_2 - \cdots - a_{n-1,n}a_{n,n-1}m_{n-1}m_n)\left(\frac{1}{\omega_2}\right)^{n-2}$$

$$- \cdots = 0 \tag{7.3}$$

This is a polynomial equation of nth degree in $(1/\omega^2)$. Let the roots of Eq. (7.3) be denoted as $1/\omega_1^2, 1/\omega_2^2, \cdots, 1/\omega_n^2$. Thus

$$\left(\frac{1}{\omega^2} - \frac{1}{\omega_1^2}\right)\left(\frac{1}{\omega^2} - \frac{1}{\omega_2^2}\right)\cdots\left(\frac{1}{\omega^2} - \frac{1}{\omega_n^2}\right)$$

$$= \left(\frac{1}{\omega^2}\right)^n - \left(\frac{1}{\omega_1^2} + \frac{1}{\omega_2^2} + \cdots + \frac{1}{\omega_n^2}\right)\left(\frac{1}{\omega^2}\right)^{n-1} - \cdots = 0 \tag{7.4}$$

Equating the coefficient of $(1/\omega^2)^{n-1}$ in Eqs. (7.4) and (7.3) gives

$$\frac{1}{\omega_1^2} + \frac{1}{\omega_2^2} + \cdots + \frac{1}{\omega_n^2} = a_{11}m_1 + a_{22}m_2 + \cdots + a_{nn}m_n \tag{7.5}$$

In most cases, the higher frequencies $\omega_2, \omega_3, \cdots, \omega_n$ are considerably larger than the fundamental frequency ω_1, and so

$$\frac{1}{\omega_i^2} \ll \frac{1}{\omega_1^2}, \qquad i = 2, 3, \cdots, n$$

Thus, Eq. (7.5) can be approximately written as

$$\frac{1}{\omega_1^2} \simeq a_{11}m_1 + a_{22}m_2 + \cdots + a_{nn}m_n \tag{7.6}$$

This equation is known as *Dunkerley's formula*. The fundamental frequency given by Eq. (7.6) will always be smaller than the exact value. In some cases, it will be more convenient to rewrite Eq. (7.6) as

$$\frac{1}{\omega_1^2} \simeq \frac{1}{\omega_{1n}^2} + \frac{1}{\omega_{2n}^2} + \cdots + \frac{1}{\omega_{nn}^2} \tag{7.7}$$

where $\omega_{in} = (1/a_{ii}m_i)^{1/2} = (k_{ii}/m_i)^{1/2}$ denotes the natural frequency of a single degree of freedom system consisting of mass m_i and spring of stiffness k_{ii}, $i = 1, 2, \cdots, n$. The use of Dunkerley's formula for finding the lowest frequency of elastic systems is presented in Refs. [7.4, 7.5].

EXAMPLE   7.1

Fundamental Frequency of a Beam

Estimate the fundamental natural frequency of a simply supported beam carrying three identical equally spaced masses, as shown in Fig. 7.1.

Solution: The flexibility influence coefficients (see Example 6.6) required for the application of Dunkerley's formula are given by

$$a_{11} = a_{33} = \frac{3}{256} \frac{l^3}{EI}, \qquad a_{22} = \frac{1}{48} \frac{l^3}{EI} \qquad (E.1)$$

Using $m_1 = m_2 = m_3 = m$, Eq. (7.6) thus gives

$$\frac{1}{\omega_1^2} \simeq \left(\frac{3}{256} + \frac{1}{48} + \frac{3}{256} \right) \frac{ml^3}{EI} = 0.04427 \frac{ml^3}{EI}$$

$$\omega_1 \simeq 4.75375 \sqrt{\frac{EI}{ml^3}}$$

This value can be compared with the exact value of the fundamental frequency

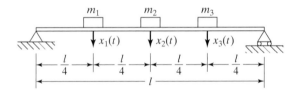

FIGURE 7.1   Beam carrying masses.

■

7.3   Rayleigh's Method

Rayleigh's method was presented in Section 2.5 to find the natural frequencies of single degree of freedom systems. The method can be extended to find the approximate value of the fundamental natural frequency of a discrete system.[1] The method is based on *Rayleigh's principle*, which can be stated as follows [7.6]:

> *The frequency of vibration of a conservative system vibrating about an equilibrium position has a stationary value in the neighborhood of a natural mode. This stationary value, in fact, is a minimum value in the neighborhood of the fundamental natural mode.*

We shall now derive an expression for the approximate value of the first natural frequency of a multidegree of freedom system according to Rayleigh's method.

---

[1] Rayleigh's method for continuous systems is presented in Section 8.7.

The kinetic and potential energies of an n degree of freedom discrete system can be expressed as

$$T = \frac{1}{2} \dot{\vec{x}}^T [m] \dot{\vec{x}} \qquad (7.8)$$

$$V = \frac{1}{2} \vec{x}^T [k] \vec{x} \qquad (7.9)$$

To find the natural frequencies, we assume harmonic motion to be

$$\vec{x} = \vec{X} \cos \omega t \qquad (7.10)$$

where $\vec{X}$ denotes the vector of amplitudes (mode shape) and ω represents the natural frequency of vibration. If the system is conservative, the maximum kinetic energy is equal to the maximum potential energy:

$$T_{max} = V_{max} \qquad (7.11)$$

By substituting Eq. (7.10) into Eqs. (7.8) and (7.9), we find

$$T_{max} = \frac{1}{2} \vec{X}^T [m] \vec{X} \omega^2 \qquad (7.12)$$

$$V_{max} = \frac{1}{2} \vec{X}^T [k] \vec{X} \qquad (7.13)$$

By equating T_{max} and V_{max}, we obtain[2]

$$\omega^2 = \frac{\vec{X}^T [k] \vec{X}}{\vec{X}^T [m] \vec{X}} \qquad (7.14)$$

The right-hand side of Eq. (7.14) is known as *Rayleigh's quotient* and is denoted as $R(\vec{X})$.

**7.3.1
Properties of
Rayleigh's
Quotient**

As stated earlier, $R(\vec{X})$ has a stationary value when the arbitrary vector $\vec{X}$ is in the neighborhood of any eigenvector $\vec{X}^{(r)}$. To prove this, we express the arbitrary vector $\vec{X}$ in terms of the normal modes of the system, $\vec{X}^{(i)}$, as

$$\vec{X} = c_1 \vec{X}^{(1)} + c_2 \vec{X}^{(2)} + c_3 \vec{X}^{(3)} + \cdots \qquad (7.15)$$

Then

$$\vec{X}^T [k] \vec{X} = c_1^2 \vec{X}^{(1)^T} [k] \vec{X}^{(1)} + c_2^2 \vec{X}^{(2)^T} [k] \vec{X}^{(2)}$$
$$+ c_3^2 \vec{X}^{(3)^T} [k] \vec{X}^{(3)} + \cdots \qquad (7.16)$$

---

[2]Equation (7.14) can also be obtained from the relation $[k]\vec{X} = \omega^2 [m]\vec{X}$. Premultiplying this equation by $\vec{X}^T$ and solving the resulting equation gives Eq. (7.14).

and

$$\vec{X}^T [m] \vec{X} = c_1^2 \vec{X}^{(1)^T} [m] \vec{X}^{(1)} + c_2^2 \vec{X}^{(2)^T} [m] \vec{X}^{(2)}$$
$$+ c_3^2 \vec{X}^{(3)^T} [m] \vec{X}^{(3)} + \cdots \qquad (7.17)$$

as the cross terms of the form $c_i c_j \vec{X}^{(i)^T} [k] \vec{X}^{(j)}$ and $c_i c_j \vec{X}^{(i)^T} [m] \vec{X}^{(j)}, i \neq j$, are zero by the orthogonality property. Using Eqs. (7.16) and (7.17) and the relation

$$\vec{X}^{(i)^T} [k] \vec{X}^{(i)} = \omega_i^2 \vec{X}^{(i)^T} [m] \vec{X}^{(i)} \qquad (7.18)$$

the Rayleigh's quotient of Eq. (7.14) can be expressed as

$$\omega^2 = R(\vec{X}) = \frac{c_1^2 \omega_1^2 \vec{X}^{(1)^T} [m] \vec{X}^{(1)} + c_2^2 \omega_2^2 \vec{X}^{(2)^T} [m] \vec{X}^{(2)} + \cdots}{c_1^2 \vec{X}^{(1)^T} [m] \vec{X}^{(1)} + c_2^2 \vec{X}^{(2)^T} [m] \vec{X}^{(2)} + \cdots} \qquad (7.19)$$

If the normal modes are normalized, this equation becomes

$$\omega^2 = R(\vec{X}) = \frac{c_1^2 \omega_1^2 + c_2^2 \omega_2^2 + \cdots}{c_1^2 + c_2^2 + \cdots} \qquad (7.20)$$

If $\vec{X}$ differs little from the eigenvector $\vec{X}^{(r)}$, the coefficient c_r will be much larger than the remaining coefficients c_i ($i \neq r$) and Eq. (7.20) can be written as

$$R(\vec{X}) = \frac{c_r^2 \omega_r^2 + c_r^2 \displaystyle\sum_{\substack{i=1, 2, \cdots \\ i \neq r}} \left(\frac{c_i}{c_r}\right)^2 \omega_i^2}{c_r^2 + c_r^2 \displaystyle\sum_{\substack{i=1, 2, \cdots \\ i \neq r}} \left(\frac{c_i}{c_r}\right)^2} \qquad (7.21)$$

Since $|c_i/c_r| = \varepsilon_i \ll 1$ where ε_i is a small number for all $i \neq r$, Eq. (7.21) gives

$$R(\vec{X}) = \omega_r^2 \{1 + 0(\varepsilon^2)\} \qquad (7.22)$$

where $0(\varepsilon^2)$ represents an expression in ε of the second order or higher. Equation (7.22) indicates that if the arbitrary vector $\vec{X}$ differs from the eigenvector $\vec{X}^{(r)}$ by a small quantity of the first order, $R(\vec{X})$ differs from the eigenvalue ω_r^2 by a small quantity of the second order. This means that Rayleigh's quotient has a stationary value in the neighborhood of an eigenvector.

The stationary value is actually a minimum value in the neighborhood of the fundamental mode, $\overrightarrow{X}^{(1)}$. To see this, let $r = 1$ in Eq. (7.21) and write

$$R(\overrightarrow{X}) = \frac{\omega_1^2 + \displaystyle\sum_{i=2, 3, \ldots} \left(\frac{c_i}{c_1}\right)^2 \omega_i^2}{\left\{1 + \displaystyle\sum_{i=2, 3, \ldots} \left(\frac{c_i}{c_1}\right)^2\right\}}$$

$$\simeq \omega_1^2 + \sum_{i=2, 3, \ldots} \varepsilon_i^2 \omega_i^2 - \omega_1^2 \sum_{i=2, 3, \ldots} \varepsilon_i^2$$

$$\simeq \omega_1^2 + \sum_{i=2, 3, \ldots} (\omega_i^2 - \omega_1^2) \varepsilon_i^2 \tag{7.23}$$

Since, in general, $\omega_i^2 > \omega_1^2$ for $i = 2, 3, \ldots,$ Eq. (7.23) leads to

$$R(\overrightarrow{X}) \geq \omega_1^2 \tag{7.24}$$

which shows that Rayleigh's quotient is never lower than the first eigenvalue. By proceeding in a similar manner, we can show that

$$R(\overrightarrow{X}) \leq \omega_n^2 \tag{7.25}$$

which means that Rayleigh's quotient is never higher than the highest eigenvalue. Thus Rayleigh's quotient provides an upper bound for ω_1^2 and a lower bound for ω_n^2.

7.3.2 Computation of the Fundamental Natural Frequency

Equation (7.14) can be used to find an approximate value of the first natural frequency (ω_1) of the system. For this, we select a trial vector $\overrightarrow{X}$ to represent the first natural mode $\overrightarrow{X}^{(1)}$ and substitute it on the right-hand side of Eq. (7.14). This yields the approximate value of ω_1^2. Because Rayleigh's quotient is stationary, remarkably good estimates of ω_1^2 can be obtained even if the trial vector $\overrightarrow{X}$ deviates greatly from the true natural mode $\overrightarrow{X}^{(1)}$. Obviously, the estimated value of the fundamental frequency ω_1 is more accurate if the trial vector ($\overrightarrow{X}$) chosen resembles the true natural mode $\overrightarrow{X}^{(1)}$ closely. Rayleigh's method is compared with Dunkerley's and other methods in Refs. [7.7–7.9].

Fundamental Frequency of a Three Degree of Freedom System

EXAMPLE 7.2

Estimate the fundamental frequency of vibration of the system shown in Fig. 7.2. Assume that $m_1 = m_2 = m_3 = m$, $k_1 = k_2 = k_3 = k$, and the mode shape is

$$\overrightarrow{X} = \begin{Bmatrix} 1 \\ 2 \\ 3 \end{Bmatrix}$$

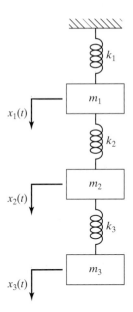

FIGURE 7.2 Three degree of freedom spring-mass system.

Solution: The stiffness and mass matrices of the system are

$$[k] = k \begin{bmatrix} 2 & -1 & 0 \\ -1 & 2 & -1 \\ 0 & -1 & 1 \end{bmatrix} \tag{E.1}$$

$$[m] = m \begin{bmatrix} 1 & 0 & 0 \\ 0 & 1 & 0 \\ 0 & 0 & 1 \end{bmatrix} \tag{E.2}$$

By substituting the assumed mode shape in the expression for Rayleigh's quotient, we obtain

$$R(\vec{X}) = \omega^2 = \frac{\vec{X}^T [k] \vec{X}}{\vec{X}^T [m] \vec{X}} = \frac{(1 \quad 2 \quad 3)k \begin{bmatrix} 2 & -1 & 0 \\ -1 & 2 & -1 \\ 0 & -1 & 1 \end{bmatrix} \begin{Bmatrix} 1 \\ 2 \\ 3 \end{Bmatrix}}{(1 \quad 2 \quad 3)m \begin{bmatrix} 1 & 0 & 0 \\ 0 & 1 & 0 \\ 0 & 0 & 1 \end{bmatrix} \begin{Bmatrix} 1 \\ 2 \\ 3 \end{Bmatrix}}$$

$$= 0.2143 \frac{k}{m} \tag{E.3}$$

$$\omega_1 = 0.4629 \sqrt{\frac{k}{m}} \qquad\qquad\qquad\qquad (E.4)$$

This value is 4.0225 percent larger than the exact value of $0.4450\sqrt{k/m}$. The exact fundamental mode shape (see Example 6.10) in this case is

$$\vec{X}^{(1)} = \begin{Bmatrix} 1.0000 \\ 1.8019 \\ 2.2470 \end{Bmatrix} \qquad\qquad\qquad\qquad (E.5)$$

∎

**7.3.3
Fundamental
Frequency of
Beams and
Shafts**

Although the procedure outlined above is applicable to all discrete systems, a simpler equation can be derived for the fundamental frequency of the lateral vibration of a beam or a shaft carrying several masses such as pulleys, gears, or flywheels. In these cases, the static deflection curve is used as an approximation of the dynamic deflection curve.

Consider a shaft carrying several masses, as shown in Fig. 7.3. The shaft is assumed to have negligible mass. The potential energy of the system is the strain energy of the deflected shaft, which is equal to the work done by the static loads. Thus

$$V_{max} = \frac{1}{2}(m_1 g w_1 + m_2 g w_2 + \cdots) \qquad\qquad\qquad (7.26)$$

where $m_i g$ is the static load due to the mass m_i, and w_i is the total static deflection of mass m_i due to all the masses. For harmonic oscillation (free vibration), the maximum kinetic energy due to the masses is

$$T_{max} = \frac{\omega^2}{2}(m_1 w_1^2 + m_2 w_2^2 + \cdots) \qquad\qquad\qquad (7.27)$$

where ω is the frequency of oscillation. Equating V_{max} and T_{max}, we obtain

$$\omega = \left\{ \frac{g(m_1 w_1 + m_2 w_2 + \cdots)}{(m_1 w_1^2 + m_2 w_2^2 + \cdots)} \right\}^{1/2} \qquad\qquad (7.28)$$

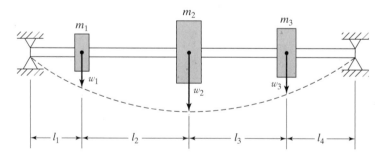

FIGURE 7.3 Shaft carrying masses.

Fundamental Frequency of a Shaft with Rotors

EXAMPLE 7.3

Estimate the fundamental frequency of the lateral vibration of a shaft carrying three rotors (masses), as shown in Fig. 7.3, with $m_1 = 20$ kg, $m_2 = 50$ kg, $m_3 = 40$ kg, $l_1 = 1$ m, $l_2 = 3$ m, $l_3 = 4$ m, and $l_4 = 2$ m. The shaft is made of steel with a solid circular cross section of diameter 10 cm.

Solution: From strength of materials, the deflection of the beam shown in Fig. 7.4 due to a static load P [7.10] is given by

$$w(x) = \begin{cases} \dfrac{Pbx}{6EIl}(l^2 - b^2 - x^2); & 0 \le x \le a \quad\quad (E.1) \\[4mm] -\dfrac{Pa(l - x)}{6EIl}[a^2 + x^2 - 2lx]; & a \le x \le l \quad\quad (E.2) \end{cases}$$

Deflection Due to the Weight of m_1: At the location of mass m_1 (with $x = 1$ m, $b = 9$ m, and $l = 10$ m in Eq. E.1):

$$w_1' = \frac{(20 \times 9.81)(9)(1)}{6EI(10)}(100 - 81 - 1) = \frac{529.74}{EI} \quad\quad (E.3)$$

At the location of m_2 (with $a = 1$ m, $x = 4$ m, and $l = 10$ m in Eq. E.2):

$$w_2' = -\frac{(20 \times 9.81)(1)(6)}{6EI(10)}[1 + 16 - 2(10)(4)] = \frac{1236.06}{EI} \quad\quad (E.4)$$

At the location of m_3 (with $a = 1$ m, $x = 8$ m, and $l = 10$ m in Eq. E.2):

$$w_3' = -\frac{(20 \times 9.81)(1)(2)}{6EI(10)}[1 + 64 - 2(10)(8)] = \frac{621.3}{EI} \quad\quad (E.5)$$

Deflection Due to the Weight of m_2: At the location of m_1 (with $x = 1$ m, $b = 6$ m, and $l = 10$ m in Eq. E.1):

$$w_1'' = \frac{(50 \times 9.81)(6)(1)}{6EI(10)}(100 - 36 - 1) = \frac{3090.15}{EI} \quad\quad (E.6)$$

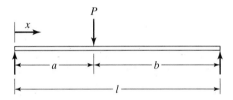

P

FIGURE 7.4   Beam under static load.

At the location of m_2 (with $x = 4$ m, $b = 6$ m, and $l = 10$ m in Eq. E.1):

$$w_2'' = \frac{(50 \times 9.81)(6)(4)}{6EI(10)}(100 - 36 - 16) = \frac{9417.6}{EI} \tag{E.7}$$

At the location of m_3 (with $a = 4$ m, $x = 8$ m, and $l = 10$ m in Eq. E.2):

$$w_3'' = -\frac{(50 \times 9.81)(4)(2)}{6EI(10)}[16 + 64 - 2(10)(8)] = \frac{5232.0}{EI} \tag{E.8}$$

Deflection Due to the Weight of m_3: At the location of m_1 (with $x = 1$ m, $b = 2$ m, and $l = 10$ m in Eq. E.1):

$$w_1''' = \frac{(40 \times 9.81)(2)(1)}{6EI(10)}(100 - 4 - 1) = \frac{1242.6}{EI} \tag{E.9}$$

At the location of m_2 (with $x = 4$ m, $b = 2$ m, and $l = 10$ m in Eq. E.1):

$$w_2''' = \frac{(40 \times 9.81)(2)(4)}{6EI(10)}(100 - 4 - 16) = \frac{4185.6}{EI} \tag{E.10}$$

At the location of m_3 (with $x = 8$ m, $b = 2$ m, and $l = 10$ m in Eq. E.1):

$$w_3''' = \frac{(40 \times 9.81)(2)(8)}{6EI(10)}(100 - 4 - 64) = \frac{3348.48}{EI} \tag{E.11}$$

The total deflections of the masses m_1, m_2, and m_3 are

$$w_1 = w_1' + w_1'' + w_1''' = \frac{4862.49}{EI}$$

$$w_2 = w_2' + w_2'' + w_2''' = \frac{14839.26}{EI}$$

$$w_3 = w_3' + w_3'' + w_3''' = \frac{9201.78}{EI}$$

Substituting into Eq. (7.28), we find the fundamental natural frequency:

$$\omega = \left\{ \frac{9.81(20 \times 4862.49 + 50 \times 14839.26 + 40 \times 9201.78)\, EI}{20 \times (4862.49)^2 + 50 \times (14839.26)^2 + 40 \times (9201.78)^2} \right\}^{1/2}$$

$$= 0.028222 \sqrt{EI} \tag{E.12}$$

For the shaft, $E = 2.07 \times 10^{11}$ N/m^2 and $I = \pi(0.1)^4/64 = 4.90875 \times 10^{-6}$ m^4 and hence Eq. (E.12) gives

$$\omega = 28.4482 \text{ rad/s}$$

■

7.4 Holzer's Method

Holzer's method is essentially a trial-and-error scheme to find the natural frequencies of undamped, damped, semidefinite, fixed, or branched vibrating systems involving linear and angular displacements [7.11, 7.12]. The method can also be programmed for computer applications. A trial frequency of the system is first assumed, and a solution is found when the assumed frequency satisfies the constraints of the system. This generally requires several trials. Depending on the trial frequency used, the fundamental as well as the higher frequencies of the system can be determined. The method also gives the mode shapes.

**7.4.1
Torsional
Systems**

Consider the undamped torsional semidefinite system shown in Fig. 7.5. The equations of motion of the discs can be derived as follows:

$$J_1\ddot{\theta}_1 + k_{t1}(\theta_1 - \theta_2) = 0 \tag{7.29}$$

$$J_2\ddot{\theta}_2 + k_{t1}(\theta_2 - \theta_1) + k_{t2}(\theta_2 - \theta_3) = 0 \tag{7.30}$$

$$J_3\ddot{\theta}_3 + k_{t2}(\theta_3 - \theta_2) = 0 \tag{7.31}$$

Since the motion is harmonic in a natural mode of vibration, we assume that $\theta_i = \Theta_i \cos(\omega t + \phi)$ in Eqs. (7.29) to (7.31) and obtain

$$\omega^2 J_1 \Theta_1 = k_{t1}(\Theta_1 - \Theta_2) \tag{7.32}$$

$$\omega^2 J_2 \Theta_2 = k_{t1}(\Theta_2 - \Theta_1) + k_{t2}(\Theta_2 - \Theta_3) \tag{7.33}$$

$$\omega^2 J_3 \Theta_3 = k_{t2}(\Theta_3 - \Theta_2) \tag{7.34}$$

Summing these equations gives

$$\sum_{i=1}^{3} \omega^2 J_i \Theta_i = 0 \tag{7.35}$$

Equation (7.35) states that the sum of the inertia torques of the semidefinite system must be zero. This equation can be treated as another form of the frequency equation, and the trial frequency must satisfy this requirement.

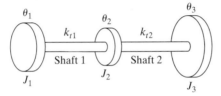

FIGURE 7.5   Torsional semidefinite system.

In Holzer's method, a trial frequency ω is assumed, and Θ_1 is arbitrarily chosen as unity. Next, Θ_2 is computed from Eq. (7.32), and then Θ_3 is found from Eq. (7.33). Thus we obtain

$$\Theta_1 = 1 \tag{7.36}$$

$$\Theta_2 = \Theta_1 - \frac{\omega^2 J_1 \Theta_1}{k_{t1}} \tag{7.37}$$

$$\Theta_3 = \Theta_2 - \frac{\omega^2}{k_{t2}} (J_1 \Theta_1 + J_2 \Theta_2) \tag{7.38}$$

These values are substituted in Eq. (7.35) to verify whether the constraint is satisfied. If Eq. (7.35) is not satisfied, a new trial value of ω is assumed and the process repeated. Equations (7.35), (7.37), and (7.38) can be generalized for an n disc system as follows:

$$\sum_{i=1}^{n} \omega^2 J_i \Theta_i = 0 \tag{7.39}$$

$$\Theta_i = \Theta_{i-1} - \frac{\omega^2}{k_{ti-1}} \left(\sum_{k=1}^{i-1} J_k \Theta_k \right), \qquad i = 2, 3, \cdots, n \tag{7.40}$$

Thus the method uses Eqs. (7.39) and (7.40) repeatedly for different trial frequencies. If the assumed trial frequency is not a natural frequency of the system, Eq. (7.39) is not satisfied. The resultant torque in Eq. (7.39) represents a torque applied at the last disc. This torque M_t is then plotted for the chosen ω. When the calculation is repeated with other values of ω, the resulting graph appears as shown in Fig. 7.6. From this graph, the natural frequencies of the system can be identified as the values of ω at which $M_t = 0$. The amplitudes $\Theta_i (i = 1, 2, \cdots, n)$ corresponding to the natural frequencies are the mode shapes of the system.

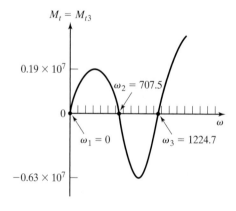

FIGURE 7.6   Resultant torque versus frequency.

Holzer's method can also be applied to systems with fixed ends. At a fixed end, the amplitude of vibration must be zero. In this case, the natural frequencies can be found by plotting the resulting amplitude (instead of the resultant torque) against the assumed frequencies. For a system with one end free and the other end fixed, Eq. (7.40) can be used for checking the amplitude at the fixed end. An improvement of Holzer's method is presented in Refs. [7.13, 7.14].

EXAMPLE 7.4

Natural Frequencies of a Torsional System

The arrangement of the compressor, turbine, and generator in a thermal power plant is shown in Fig. 7.7. Find the natural frequencies and mode shapes of the system.

Solution: This system represents an unrestrained or free-free torsional system. Table 7.1 shows its parameters and the sequence of computations. The calculations for the trial frequencies

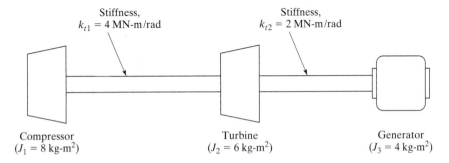

Stiffness,
$k_{t1} = 4$ MN-m/rad

Stiffness,
$k_{t2} = 2$ MN-m/rad

Compressor
($J_1 = 8$ kg-m^2)

Turbine
($J_2 = 6$ kg-m^2)

Generator
($J_3 = 4$ kg-m^2)

FIGURE 7.7   Free-free torsional system.

TABLE 7.1

Parameters of the System	Quantity	Trial 1	2	3	⋯	71	72
	ω	0	10	20		700	710
	ω^2	0	100	400		490000	504100
Station 1:							
$J_1 = 8$	Θ_1	1.0	1.0	1.0		1.0	1.0
$k_{t1} = 4 \times 10^6$	$M_{t1} = \omega^2 J_1 \Theta_1$	0	800	3200		0.392E7	0.403E7
Station 2:							
$J_2 = 6$	$\Theta_2 = 1 - \dfrac{M_{t1}}{k_{t1}}$	1.0	0.9998	0.9992		0.0200	− 0.0082
$k_{t2} = 2 \times 10^6$	$M_{t2} = M_{t1} + \omega^2 J_2 \Theta_2$	0	1400	5598		0.398E7	0.401E7
Station 3:							
$J_3 = 4$	$\Theta_3 = \Theta_2 - \dfrac{M_{t2}}{k_{t2}}$	1.0	0.9991	0.9964		− 1.9690	− 2.0120
$k_{t3} = 0$	$M_{t3} = M_{t2} + \omega^2 J_3 \Theta_3$	0	1800	7192		0.119E6	− 0.494E5

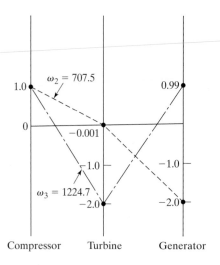

FIGURE 7.8   First two flexible modes.

$\omega = 0$, 10, 20, 700, and 710 are shown in this table. The quantity M_{t3} denotes the torque to the right of Station 3 (generator), which must be zero at the natural frequencies. Figure 7.6 shows the graph of M_{t3} versus ω. Closely spaced trial values of ω are used in the vicinity of $M_{t3} = 0$ to obtain accurate values of the first two flexible mode shapes, shown in Fig. 7.8. Note that the value $\omega = 0$ corresponds to the rigid body rotation.

■

**7.4.2
Spring-Mass
Systems**

Although Holzer's method has been extensively applied to torsional systems, the procedure is equally applicable to the vibration analysis of spring-mass systems. The equations of motion of a spring-mass system (see Fig. 7.9) can be expressed as

$$m_1 \ddot{x}_1 + k_1 (x_1 - x_2) = 0 \tag{7.41}$$

$$m_2 \ddot{x}_2 + k_1 (x_2 - x_1) + k_2 (x_2 - x_3) = 0$$

$$\cdots \tag{7.42}$$

For harmonic motion, $x_i(t) = X_i \cos \omega t$, where X_i is the amplitude of mass m_i, and Eqs. (7.41) and (7.42) can be written as

$$\omega^2 m_1 X_1 = k_1 (X_1 - X_2) \tag{7.43}$$

$$\omega^2 m_2 X_2 = k_1 (X_2 - X_1) + k_2 (X_2 - X_3)$$

$$= -\omega^2 m_1 X_1 + k_2 (X_2 - X_3)$$

$$\cdots \tag{7.44}$$

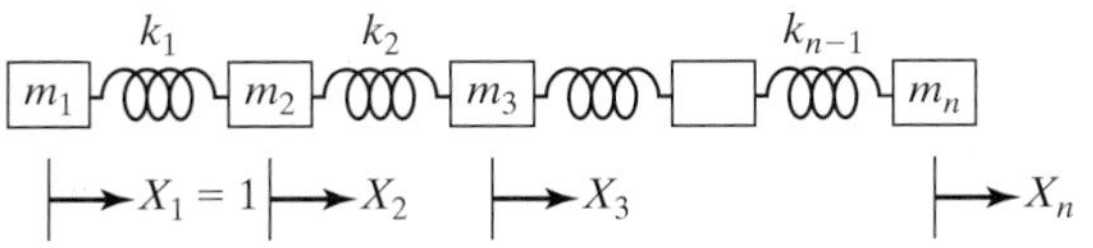

FIGURE 7.9   Free-free spring mass system.

The procedure for Holzer's method starts with a trial frequency ω and the amplitude of mass m_1 as $X_1 = 1$. Equations (7.43) and (7.44) can then be used to obtain the amplitudes of the masses $m_2, m_3, \cdots, m_i$:

$$X_2 = X_1 - \frac{\omega^2 m_1 X_1}{k_1} \tag{7.45}$$

$$X_3 = X_2 - \frac{\omega^2}{k_2}(m_1 X_1 + m_2 X_2) \tag{7.46}$$

$$X_i = X_{i-1} - \frac{\omega^2}{k_{i-1}}\left(\sum_{k=1}^{i-1} m_k X_k\right), \qquad i = 2, 3, \cdots, n \tag{7.47}$$

As in the case of torsional systems, the resultant force applied to the last (nth) mass can be computed as follows:

$$F = \sum_{i=1}^{n} \omega^2 m_i X_i \tag{7.48}$$

The calculations are repeated with several other trial frequencies ω. The natural frequencies are identified as those values of ω that give $F = 0$ for a free-free system. For this, it is convenient to plot a graph between F and ω, using the same procedure for spring-mass systems as for torsional systems.

7.5   Matrix Iteration Method

The matrix iteration method assumes that the natural frequencies are distinct and well separated such that $\omega_1 < \omega_2 < \cdots < \omega_n$. The iteration is started by selecting a trial vector $\vec{X}_1$, which is then premultiplied by the dynamical matrix $[D]$. The resulting column vector is then normalized, usually by making one of its components to unity. The normalized column vector is premultiplied by $[D]$ to obtain a third column vector, which is normalized in the same way as before and becomes still another trial column vector. The process is repeated until the successive normalized column vectors converge to a common vector: the fundamental eigenvector. The normalizing factor gives the largest value of $\lambda = 1/\omega^2$—that is, the smallest or the fundamental natural frequency [7.15]. The convergence of the process can be explained as follows.

According to the expansion theorem, any arbitrary n-dimensional vector $\vec{X}_1$ can be expressed as a linear combination of the n orthogonal eigenvectors of the system $\vec{X}^{(i)}, i = 1, 2, \cdots, n$:

$$\vec{X}_1 = c_1\vec{X}^{(1)} + c_2\vec{X}^{(2)} + \cdots + c_n \vec{X}^{(n)} \tag{7.49}$$

where $c_1, c_2, \cdots, c_n$ are constants. In the iteration method, the trial vector $\vec{X}_1$ is selected arbitrarily and is therefore a known vector. The modal vectors $\vec{X}^{(i)}$, although unknown, are constant vectors because they depend upon the properties of the system. The constants c_i are unknown numbers to be determined. According to the iteration method, we premultiply $\vec{X}_1$ by the matrix $[D]$. In view of Eq. (7.49), this gives

$$[D] \vec{X}_1 = c_1[D] \vec{X}^{(1)} + c_2[D] \vec{X}^{(2)} + \cdots + c_n[D] \vec{X}^{(n)} \tag{7.50}$$

Now, according to Eq. (6.66), we have

$$[D] \vec{X}^{(i)} = \lambda_i[I] \vec{X}^{(i)} = \frac{1}{\omega_i^2}\vec{X}^{(i)}; \qquad i = 1, 2, \cdots, n \tag{7.51}$$

Substitution of Eq. (7.51) into Eq. (7.50) yields

$$[D]\vec{X}_1 = \vec{X}_2$$

$$= \frac{c_1}{\omega_1^2} \vec{X}^{(1)} + \frac{c_2}{\omega_2^2} \vec{X}^{(2)} + \cdots + \frac{c_n}{\omega_n^2} \vec{X}^{(n)} \tag{7.52}$$

where $\vec{X}_2$ is the second trial vector. We now repeat the process and premultiply $\vec{X}_2$ by $[D]$ to obtain, by Eqs. (7.49) and (6.66),

$$[D] \vec{X}_2 = \vec{X}_3$$

$$= \frac{c_1}{\omega_1^4} \vec{X}^{(1)} + \frac{c_2}{\omega_2^4} \vec{X}^{(2)} + \cdots + \frac{c_n}{\omega_n^4} \vec{X}^{(n)} \tag{7.53}$$

By repeating the process we obtain, after the rth iteration,

$$[D] \vec{X}_r = \vec{X}_{r+1}$$

$$= \frac{c_1}{\omega_1^{2r}} \vec{X}^{(1)} + \frac{c_2}{\omega_2^{2r}} \vec{X}^{(2)} + \cdots + \frac{c_n}{\omega_n^{2r}} \vec{X}^{(n)} \tag{7.54}$$

Since the natural frequencies are assumed to be $\omega_1 < \omega_2 < \cdots < \omega_n$, a sufficiently large value of r yields

$$\frac{1}{\omega_1^{2r}} \gg \frac{1}{\omega_2^{2r}} \gg \cdots \gg \frac{1}{\omega_n^{2r}} \tag{7.55}$$

Thus the first term on the right-hand side of Eq. (7.54) becomes the only significant one. Hence we have

$$\vec{X}_{r+1} = \frac{c_1}{\omega_1^{2r}} \vec{X}^{(1)} \tag{7.56}$$

which means that the $(r + 1)$th trial vector becomes identical to the fundamental modal vector to within a multiplicative constant. Since

$$\vec{X}_r = \frac{c_1}{\omega_1^{2(r-1)}} \vec{X}^{(1)} \tag{7.57}$$

the fundamental natural frequency ω_1 can be found by taking the ratio of any two corresponding components in the vectors $\vec{X}_r$ and $\vec{X}_{r+1}$

$$\omega_1^2 \simeq \frac{X_{i,r}}{X_{i,r+1}}, \qquad \text{for any } i = 1, 2, \cdots, n \tag{7.58}$$

where $X_{i,r}$ and $X_{i,r+1}$ are the ith elements of the vectors $\vec{X}_r$ and $\vec{X}_{r+1}$, respectively.

Discussion

1. In the above proof, nothing has been said about the normalization of the successive trial vectors $\vec{X}_i$. Actually, it is not necessary to establish the proof of convergence of the method. The normalization amounts to a readjustment of the constants $c_1, c_2, \cdots, c_n$ in each iteration.

2. Although it is theoretically necessary to have $r \to \infty$ for the convergence of the method, in practice only a finite number of iterations suffices to obtain a reasonably good estimate of ω_1.

3. The actual number of iterations necessary to find the value of ω_1 to within a desired degree of accuracy depends on how closely the arbitrary trial vector $\vec{X}_1$ resembles the fundamental mode $\vec{X}^{(1)}$ and on how well ω_1 and ω_2 are separated. The required number of iterations is less if ω_2 is very large compared to ω_1.

4. The method has a distinct advantage in that any computational errors made do not yield incorrect results. Any error made in premultiplying $\vec{X}_i$ by $[D]$ results in a vector other than the desired one, $\vec{X}_{i+1}$. But this wrong vector can be considered as a new trial vector. This may delay the convergence but does not produce wrong results.

5. One can take any set of n numbers for the first trial vector $\vec{X}_1$ and still achieve convergence to the fundamental modal vector. Only in the unusual case in which the trial vector $\vec{X}_1$ is exactly proportional to one of the modes $\vec{X}^{(i)}$ ($i \neq 1$) does the method fail to converge to the first mode. In such a case, the premultiplication of $\vec{X}^{(i)}$ by $[D]$ results in a vector proportional to $\vec{X}^{(i)}$ itself.

**7.5.1
Convergence to
the Highest
Natural
Frequency**

To obtain the highest natural frequency ω_n and the corresponding mode shape or eigenvector $\overrightarrow{X}^{(n)}$ by the matrix iteration method, we first rewrite Eq. (6.66) as

$$[D]^{-1} \overrightarrow{X} = \omega^2 [I] \overrightarrow{X} = \omega^2 \overrightarrow{X} \tag{7.59}$$

where $[D]^{-1}$ is the inverse of the dynamical matrix $[D]$ given by

$$[D]^{-1} = [m]^{-1}[k] \tag{7.60}$$

Now we select any arbitrary trial vector $\overrightarrow{X}_1$ and premultiply it by $[D]^{-1}$ to obtain an improved trial vector $\overrightarrow{X}_2$. The sequence of trial vectors $\overrightarrow{X}_{i+1}$ ($i = 1, 2, \cdots$) obtained by premultiplying by $[D]^{-1}$ converges to the highest normal mode $\overrightarrow{X}^{(n)}$. It can be seen that the procedure is similar to the one already described. The constant of proportionality in this case is ω^2 instead of $1/\omega^2$.

**7.5.2
Computation of
Intermediate
Natural
Frequencies**

Once the first natural frequency ω_1 (or the largest eigenvalue $\lambda_1 = 1/\omega_1^2$) and the corresponding eigenvector $\overrightarrow{X}^{(1)}$ are determined, we can proceed to find the higher natural frequencies and the corresponding mode shapes by the matrix iteration method. Before we proceed, it should be remembered that any arbitrary trial vector premultiplied by $[D]$ would lead again to the largest eigenvalue. It is thus necessary to remove the largest eigenvalue from the matrix $[D]$. The succeeding eigenvalues and eigenvectors can be obtained by eliminating the root λ_1 from the characteristic or frequency equation

$$|[D] - \lambda[I]| = 0 \tag{7.61}$$

A procedure known as *matrix deflation* can be used for this purpose [7.16]. To find the eigenvector $\overrightarrow{X}^{(i)}$ by this procedure, the previous eigenvector $\overrightarrow{X}^{(i-1)}$ is normalized with respect to the mass matrix such that

$$\overrightarrow{X}^{(i-1)T} [m] \overrightarrow{X}^{(i-1)} = 1 \tag{7.62}$$

The deflated matrix $[D_i]$ is then constructed as

$$[D_i] = [D_{i-1}] - \lambda_{i-1} \overrightarrow{X}^{(i-1)} \overrightarrow{X}^{(i-1)T} [m], \qquad i = 2, 3, \cdots, n \tag{7.63}$$

where $[D_1] = [D]$. Once $[D_i]$ is constructed, the iterative scheme

$$\overrightarrow{X}_{r+1} = [D_i] \overrightarrow{X}_r \tag{7.64}$$

is used, where $\overrightarrow{X}_1$ is an arbitrary trial eigenvector.

EXAMPLE 7.5

Natural Frequencies of a Three Degree of Freedom System

Find the natural frequencies and mode shapes of the system shown in Fig. 7.2 for $k_1 = k_2 = k_3 = k$ and $m_1 = m_2 = m_3 = m$ by the matrix iteration method.

Solution: The mass and stiffness matrices of the system are given in Example 7.2. The flexibility matrix is

$$[a] = [k]^{-1} = \frac{1}{k}\begin{bmatrix} 1 & 1 & 1 \\ 1 & 2 & 2 \\ 1 & 2 & 3 \end{bmatrix} \tag{E.1}$$

and so the dynamical matrix is

$$[k]^{-1}[m] = \frac{m}{k}\begin{bmatrix} 1 & 1 & 1 \\ 1 & 2 & 2 \\ 1 & 2 & 3 \end{bmatrix} \tag{E.2}$$

The eigenvalue problem can be stated as

$$[D]\vec{X} = \lambda\vec{X} \tag{E.3}$$

where

$$[D] = \begin{bmatrix} 1 & 1 & 1 \\ 1 & 2 & 2 \\ 1 & 2 & 3 \end{bmatrix} \tag{E.4}$$

and

$$\lambda = \frac{k}{m} \cdot \frac{1}{\omega^2} \tag{E.5}$$

First Natural Frequency: By assuming the first trial eigenvector or mode shape to be

$$\vec{X}_1 = \begin{Bmatrix} 1 \\ 1 \\ 1 \end{Bmatrix} \tag{E.6}$$

the second trial eigenvector can be obtained:

$$\vec{X}_2 = [D]\vec{X}_1 = \begin{Bmatrix} 3 \\ 5 \\ 6 \end{Bmatrix} \tag{E.7}$$

By making the first element equal to unity, we obtain

$$\vec{X}_2 = 3.0\begin{Bmatrix} 1.0000 \\ 1.6667 \\ 2.0000 \end{Bmatrix} \tag{E.8}$$

and the corresponding eigenvalue is given by

$$\lambda_1 \simeq 3.0 \quad \text{or} \quad \omega_1 \simeq 0.5773\sqrt{\frac{k}{m}} \tag{E.9}$$

The subsequent trial eigenvector can be obtained from the relation

$$\vec{X}_{i+1} = [D]\vec{X}_i \tag{E.10}$$

and the corresponding eigenvalues are given by

$$\lambda_1 \simeq X_{1,i+1} \tag{E.11}$$

where $X_{1,i+1}$ is the first component of the vector $\vec{X}_{i+1}$ before normalization. The various trial eigenvectors and eigenvalues obtained by using Eqs. (E.10) and (E.11) are shown in the table below.

i	$\vec{X}_i$ with $X_{1,i} = 1$	$\vec{X}_{i+1} = [D]\vec{X}_i$	$\lambda_1 \simeq X_{1,i+1}$	ω_1
1	$\begin{Bmatrix} 1 \\ 1 \\ 1 \end{Bmatrix}$	$\begin{Bmatrix} 3 \\ 5 \\ 6 \end{Bmatrix}$	3.0	$0.5773\sqrt{\dfrac{k}{m}}$
2	$\begin{Bmatrix} 1.00000 \\ 1.66667 \\ 2.00000 \end{Bmatrix}$	$\begin{Bmatrix} 4.66667 \\ 8.33333 \\ 10.33333 \end{Bmatrix}$	4.66667	$0.4629\sqrt{\dfrac{k}{m}}$
3	$\begin{Bmatrix} 1.0000 \\ 1.7857 \\ 2.2143 \end{Bmatrix}$	$\begin{Bmatrix} 5.00000 \\ 9.00000 \\ 11.2143 \end{Bmatrix}$	5.00000	$0.4472\sqrt{\dfrac{k}{m}}$
.				
.				
.				
7	$\begin{Bmatrix} 1.00000 \\ 1.80193 \\ 2.24697 \end{Bmatrix}$	$\begin{Bmatrix} 5.04891 \\ 9.09781 \\ 11.34478 \end{Bmatrix}$	5.04891	$0.44504\sqrt{\dfrac{k}{m}}$
8	$\begin{Bmatrix} 1.00000 \\ 1.80194 \\ 2.24698 \end{Bmatrix}$	$\begin{Bmatrix} 5.04892 \\ 9.09783 \\ 11.34481 \end{Bmatrix}$	5.04892	$0.44504\sqrt{\dfrac{k}{m}}$

It can be seen that the mode shape and the natural frequency converged (to the fourth decimal place) in eight iterations. Thus the first eigenvalue and the corresponding natural frequency and mode shape are given by

$$\lambda_1 = 5.04892, \quad \omega_1 = 0.44504\sqrt{\frac{k}{m}}$$

$$\vec{X}^{(1)} = \begin{Bmatrix} 1.00000 \\ 1.80194 \\ 2.24698 \end{Bmatrix} \tag{E.12}$$

Second Natural Frequency: To compute the second eigenvalue and the eigenvector, we must first produce a deflated matrix:

$$[D_2] = [D_1] - \lambda_1 \vec{X}^{(1)} \vec{X}^{(1)T} [m] \tag{E.13}$$

This equation, however, calls for a normalized vector $\vec{X}^{(1)}$ satisfying $\vec{X}^{(1)T}[m]\vec{X}^{(1)} = 1$. Let the normalized vector be denoted as

$$\vec{X}^{(1)} = \alpha \begin{Bmatrix} 1.00000 \\ 1.80194 \\ 2.24698 \end{Bmatrix}$$

where α is a constant whose value must be such that

$$\vec{X}^{(1)T} [m] \vec{X}^{(1)} = \alpha^2 m \begin{Bmatrix} 1.00000 \\ 1.80194 \\ 2.24698 \end{Bmatrix}^T \begin{bmatrix} 1 & 0 & 0 \\ 0 & 1 & 0 \\ 0 & 0 & 1 \end{bmatrix} \begin{Bmatrix} 1.00000 \\ 1.80194 \\ 2.24698 \end{Bmatrix}$$
$$= \alpha^2 m(9.29591) = 1 \tag{E.14}$$

from which we obtain $\alpha = 0.32799 m^{-1/2}$. Hence the first normalized eigenvector is

$$\vec{X}^{(1)} = m^{-1/2} \begin{Bmatrix} 0.32799 \\ 0.59102 \\ 0.73699 \end{Bmatrix} \tag{E.15}$$

Next we use Eq. (E.13) and form the first deflated matrix

$$[D_2] = \begin{bmatrix} 1 & 1 & 1 \\ 1 & 2 & 2 \\ 1 & 2 & 3 \end{bmatrix} - 5.04892 \begin{Bmatrix} 0.32799 \\ 0.59102 \\ 0.73699 \end{Bmatrix} \begin{Bmatrix} 0.32799 \\ 0.59102 \\ 0.73699 \end{Bmatrix}^T \begin{bmatrix} 1 & 0 & 0 \\ 0 & 1 & 0 \\ 0 & 0 & 1 \end{bmatrix}$$
$$= \begin{bmatrix} 0.45684 & 0.02127 & -0.22048 \\ 0.02127 & 0.23641 & -0.19921 \\ -0.22048 & -0.19921 & 0.25768 \end{bmatrix} \tag{E.16}$$

Since the trial vector can be chosen arbitrarily, we again take

$$\vec{X}_1 = \begin{Bmatrix} 1 \\ 1 \\ 1 \end{Bmatrix} \tag{E.17}$$

By using the iterative scheme

$$\vec{X}_{i+1} = [D_2] \, \vec{X}_i \tag{E.18}$$

we obtain $\vec{X}_2$

$$\vec{X}_2 = \left\{ \begin{array}{c} 0.25763 \\ 0.05847 \\ -0.16201 \end{array} \right\} = 0.25763 \left\{ \begin{array}{c} 1.00000 \\ 0.22695 \\ -0.62885 \end{array} \right\} \tag{E.19}$$

Hence λ_2 can be found from the general relation

$$\lambda_2 \simeq X_{1,i+1} \tag{E.20}$$

as 0.25763. Continuation of this procedure gives the results shown in the table below.

i	$\vec{X}_i$ with $X_{1,i} = 1$	$\vec{X}_{i+1} = [D_2]\vec{X}_i$	$\lambda_2 \simeq X_{1,i+1}$	ω_2
1	$\left\{\begin{array}{c} 1 \\ 1 \\ 1 \end{array}\right\}$	$\left\{\begin{array}{c} 0.25763 \\ 0.05847 \\ -0.16201 \end{array}\right\}$	0.25763	$1.97016\sqrt{\dfrac{k}{m}}$
2	$\left\{\begin{array}{c} 1.00000 \\ 0.22695 \\ -0.62885 \end{array}\right\}$	$\left\{\begin{array}{c} 0.60032 \\ 0.20020 \\ -0.42773 \end{array}\right\}$	0.60032	$1.29065\sqrt{\dfrac{k}{m}}$
.				
10	$\left\{\begin{array}{c} 1.00000 \\ 0.44443 \\ -0.80149 \end{array}\right\}$	$\left\{\begin{array}{c} 0.64300 \\ 0.28600 \\ -0.51554 \end{array}\right\}$	0.64300	$1.24708\sqrt{\dfrac{k}{m}}$
11	$\left\{\begin{array}{c} 1.00000 \\ 0.44479 \\ -0.80177 \end{array}\right\}$	$\left\{\begin{array}{c} 0.64307 \\ 0.28614 \\ -0.51569 \end{array}\right\}$	0.64307	$1.24701\sqrt{\dfrac{k}{m}}$

Thus the converged second eigenvalue and the eigenvector are

$$\lambda_2 = 0.64307, \qquad \omega_2 = 1.24701\sqrt{\dfrac{k}{m}}$$

$$\vec{X}^{(2)} = \left\{ \begin{array}{c} 1.00000 \\ 0.44496 \\ -0.80192 \end{array} \right\} \tag{E.21}$$

Third Natural Frequency: For the third eigenvalue and the eigenvector, we use a similar procedure. The detailed calculations are left as an exercise to the reader. Note that before computing the deflated matrix $[D_3]$, we need to normalize $\vec{X}^{(2)}$ by using Eq. (7.62), which gives

$$\vec{X}^{(2)} = m^{-1/2} \left\{ \begin{array}{r} 0.73700 \\ 0.32794 \\ -0.59102 \end{array} \right\} \tag{E.22}$$

■

7.6   Jacobi's Method

The matrix iteration method described in the preceding section produces the eigenvalues and eigenvectors of matrix $[D]$ one at a time. Jacobi's method is also an iterative method but produces all the eigenvalues and eigenvectors of $[D]$ simultaneously, where $[D] = [d_{ij}]$ is a real symmetric matrix of order $n \times n$. The method is based on a theorem in linear algebra stating that a real symmetric matrix $[D]$ has only real eigenvalues and that there exists a real orthogonal matrix $[R]$ such that $[R]^T[D][R]$ is diagonal [7.17]. The diagonal elements are the eigenvalues, and the columns of the matrix $[R]$ are the eigenvectors. According to Jacobi's method, the matrix $[R]$ is generated as a product of several rotation matrices [7.18] of the form

$$\underset{n \times n}{[R_1]} = \begin{array}{cc} & \overset{\textit{i}\text{th column} \quad \textit{j}\text{th column}}{} \\ \begin{bmatrix} 1 & 0 & & & & & \\ 0 & 1 & & & & & \\ & & \ddots & & & & \\ & & & \cos\theta & & -\sin\theta & \\ & & & & \ddots & & \\ & & & \sin\theta & & \cos\theta & \\ & & & & & & \ddots \\ & & & & & & & 1 \end{bmatrix} & \begin{array}{l} \\ \\ \\ \textit{i}\text{th row} \\ \\ \textit{j}\text{th row} \\ \\ \end{array} \end{array} \tag{7.65}$$

where all elements other than those appearing in columns and rows i and j are identical with those of the identity matrix $[I]$. If the sine and cosine entries appear in positions (i, i), (i, j), (j, i), and (j, j), then the corresponding elements of $[R_1]^T[D][R_1]$ can be computed as follows:

$$\underline{d}_{ii} = d_{ii} \cos^2\theta + 2d_{ij}\sin\theta\cos\theta + d_{jj}\sin^2\theta \tag{7.66}$$

$$\underline{d}_{ij} = \underline{d}_{ji} = (d_{jj} - d_{ii})\sin\theta\cos\theta + d_{ij}(\cos^2\theta - \sin^2\theta) \tag{7.67}$$

$$\underline{d}_{ji} = d_{ii}\sin^2\theta - 2d_{ij}\sin\theta\cos\theta + d_{jj}\cos^2\theta \tag{7.68}$$

If θ is chosen to be

$$\tan 2\theta = \left(\frac{2d_{ij}}{d_{ii} - d_{jj}}\right) \tag{7.69}$$

then it makes $\underline{d}_{ij} = \underline{d}_{ji} = 0$. Thus each step of Jacobi's method reduces a pair of off-diagonal elements to zero. Unfortunately, in the next step, while the method reduces a new pair of zeros, it introduces nonzero contributions to formerly zero positions. However, successive matrices of the form

$$[R_2]^T[R_1]^T[D][R_1][R_2], \qquad [R_3]^T[R_2]^T[R_1]^T[D][R_1][R_2][R_3], \ \ldots$$

converge to the required diagonal form; the final matrix $[R]$, whose columns give the eigenvectors, then becomes

$$[R] = [R_1][R_2][R_3]\ldots \tag{7.70}$$

Eigenvalue Solution Using Jacobi Method

EXAMPLE 7.6

Find the eigenvalues and eigenvectors of the matrix

$$[D] = \begin{bmatrix} 1 & 1 & 1 \\ 1 & 2 & 2 \\ 1 & 2 & 3 \end{bmatrix}$$

using Jacobi's method.

Solution: We start with the largest off-diagonal term $d_{23} = 2$ in the matrix $[D]$ and try to reduce it to zero. From Eq. (7.69),

$$\theta_1 = \frac{1}{2}\tan^{-1}\left(\frac{2d_{23}}{d_{22} - d_{33}}\right) = \frac{1}{2}\tan^{-1}\left(\frac{4}{2 - 3}\right) = -37.981878°$$

$$[R_1] = \begin{bmatrix} 1.0 & 0.0 & 0.0 \\ 0.0 & 0.7882054 & 0.6154122 \\ 0.0 & -0.6154122 & 0.7882054 \end{bmatrix}$$

$$[D'] = [R_1]^T[D][R_1] = \begin{bmatrix} 1.0 & 0.1727932 & 1.4036176 \\ 0.1727932 & 0.4384472 & 0.0 \\ 1.4036176 & 0.0 & 4.5615525 \end{bmatrix}$$

Next we try to reduce the largest off-diagonal term of $[D']$ namely, $d'_{13} = 1.4036176$ to zero. Equation (7.69) gives

$$\theta_2 = \frac{1}{2}\tan^{-1}\left(\frac{2d'_{13}}{d'_{11} - d'_{33}}\right) = \frac{1}{2}\tan^{-1}\left(\frac{2.8072352}{1.0 - 4.5615525}\right) = -19.122686°$$

$$[R_2] = \begin{bmatrix} 0.9448193 & 0.0 & 0.3275920 \\ 0.0 & 1.0 & 0.0 \\ -0.3275920 & 0.0 & 0.9448193 \end{bmatrix}$$

$$[D''] = [R_2]^T[D'][R_2] = \begin{bmatrix} 0.5133313 & 0.1632584 & 0.0 \\ 0.1632584 & 0.4384472 & 0.0566057 \\ 0.0 & 0.0566057 & 5.0482211 \end{bmatrix}$$

The largest off-diagonal element is $[D'']$ is $d''_{12} = 0.1632584$. θ_3 can be obtained from Eq. (7.69) as

$$\theta_3 = \frac{1}{2} \tan^{-1}\left(\frac{2d''_{12}}{d''_{11} - d''_{22}}\right) = \frac{1}{2} \tan^{-1}\left(\frac{0.3265167}{0.5133313 - 0.4384472}\right) = 38.541515°$$

$$[R_3] = \begin{bmatrix} 0.7821569 & -0.6230815 & 0.0 \\ 0.6230815 & 0.7821569 & 0.0 \\ 0.0 & 0.0 & 1.0 \end{bmatrix}$$

$$[D'''] = [R_3]^T[D''][R_3] = \begin{bmatrix} 0.6433861 & 0.0 & 0.0352699 \\ 0.0 & 0.3083924 & 0.0442745 \\ 0.0352699 & 0.0442745 & 5.0482211 \end{bmatrix}$$

Assuming that all the off-diagonal terms in $[D''']$ are close to zero, we can stop the process here. The diagonal elements of $[D''']$ give the eigenvalues (values of $1/\omega^2$) as 0.6433861, 0.3083924, and 5.0482211. The corresponding eigenvectors are given by the columns of the matrix $[R]$ where

$$[R] = [R_1][R_2][R_3] = \begin{bmatrix} 0.7389969 & -0.5886994 & 0.3275920 \\ 0.3334301 & 0.7421160 & 0.5814533 \\ -0.5854125 & -0.3204631 & 0.7447116 \end{bmatrix}$$

The iterative process can be continued for obtaining a more accurate solution. The present eigenvalues can be compared with the exact values: 0.6431041, 0.3079786, and 5.0489173.

∎

7.7   Standard Eigenvalue Problem

In the preceding chapter, the eigenvalue problem was stated as

$$[k]\vec{X} = \omega^2[m]\,\vec{X} \tag{7.71}$$

which can be rewritten in the form of a standard eigenvalue problem [7.19] as

$$[D]\,\vec{X} = \lambda\,\vec{X} \tag{7.72}$$

where

$$[D] = [k]^{-1}[m] \tag{7.73}$$

and

$$\lambda = \frac{1}{\omega^2} \tag{7.74}$$

In general, the matrix $[D]$ is nonsymmetric, although the matrices $[k]$ and $[m]$ are both symmetric. Since Jacobi's method (described in Section 7.6) is applicable only to symmetric matrices $[D]$, we can adopt the following procedure [7.18] to derive a standard eigenvalue problem with a symmetric matrix $[D]$.

Assuming that the matrix $[k]$ is symmetric and positive definite, we can use Choleski decomposition (see Section 7.7.1) and express $[k]$ as

$$[k] = [U]^T[U] \tag{7.75}$$

where $[U]$ is an upper triangular matrix. Using this relation, the eigenvalue problem of Eq. (7.71) can be stated as

$$\lambda[U]^T [U]\vec{X} = [m]\vec{X} \tag{7.76}$$

Premultiplying this equation by $([U]^T)^{-1}$, we obtain

$$\lambda[U]\vec{X} = ([U]^T)^{-1} [m]\vec{X} = ([U]^T)^{-1} [m][U]^{-1}[U]\vec{X} \tag{7.77}$$

By defining a new vector $\vec{Y}$ as

$$\vec{Y} = [U]\vec{X} \tag{7.78}$$

Eq. (7.77) can be written as a standard eigenvalue problem

$$[D]\vec{Y} = \lambda\vec{Y} \tag{7.79}$$

where

$$[D] = ([U]^T)^{-1} [m][U]^{-1} \tag{7.80}$$

Thus, to formulate $[D]$ according to Eq. (7.80), we first decompose the symmetric matrix $[k]$ as shown in Eq. (7.75), find $[U]^{-1}$ and $([U]^T)^{-1} = ([U]^{-1})^T$ as outlined in the next section, and then carry out the matrix multiplication as stated in Eq. (7.80). The solution of the eigenvalue problem stated in Eq. (7.79) yields λ_i and $\vec{Y}^{(i)}$. We then apply inverse transformation and find the desired eigenvectors:

$$\vec{X}^{(i)} = [U]^{-1} \vec{Y}^{(i)} \tag{7.81}$$

**7.7.1
Choleski
Decomposition**

Any symmetric and positive definite matrix $[A]$ of order $n \times n$ can be decomposed uniquely [7.20]

$$[A] = [U]^T [U] \tag{7.82}$$

where $[U]$ is an upper triangular matrix given by

$$[U] = \begin{bmatrix} u_{11} & u_{12} & u_{13} & \cdots & u_{1n} \\ 0 & u_{22} & u_{23} & \cdots & u_{2n} \\ 0 & 0 & u_{33} & \cdots & u_{3n} \\ \vdots & & & & \\ 0 & 0 & 0 & \cdots & u_{nn} \end{bmatrix} \tag{7.83}$$

with

$$u_{11} = (a_{11})^{1/2}$$

$$u_{1j} = \frac{a_{1j}}{u_{11}}, \qquad j = 2, 3, \cdots, n$$

$$u_{ij} = \frac{1}{u_{ii}} \left(a_{ij} - \sum_{k=1}^{i-1} u_{ki} u_{kj} \right), \qquad i = 2, 3, \cdots, n \text{ and } j = i+1, i+2, \cdots, n$$

$$u_{ii} = \left(a_{ii} - \sum_{k=1}^{i-1} u_{ki}^2 \right)^{1/2}, \qquad i = 2, 3, \cdots, n$$

$$u_{ij} = 0, \quad i > j \tag{7.84}$$

Inverse of the Matrix [U]. If the inverse of the upper triangular matrix $[U]$ is denoted as $[\alpha_{ij}]$, the elements α_{ij} can be determined from the relation

$$[U][U]^{-1} = [I] \tag{7.85}$$

which gives

$$\alpha_{ii} = \frac{1}{u_{ii}}$$

$$\alpha_{ij} = \frac{-1}{u_{ii}} \left(\sum_{k=i+1}^{j} u_{ik} \alpha_{kj} \right), \quad i < j$$

$$\alpha_{ij} = 0, \quad i > j \tag{7.86}$$

Thus the inverse of $[U]$ is also an upper triangular matrix.

Decomposition of a Symmetric Matrix

EXAMPLE 7.7

Decompose the matrix

$$[A] = \begin{bmatrix} 5 & 1 & 0 \\ 1 & 3 & 2 \\ 0 & 2 & 8 \end{bmatrix}$$

into the form of Eq. (7.82).

Solution: Equation (7.84) gives

$$u_{11} = \sqrt{a_{11}} = \sqrt{5} = 2.2360680$$

$$u_{12} = a_{12}/u_{11} = 1/2.236068 = 0.4472136$$

$$u_{13} = a_{13}/u_{11} = 0$$

$$u_{22} = [a_{22} - u_{12}^2]^{1/2} = (3 - 0.4472136^2)^{1/2} = 1.6733201$$

$$u_{33} = [a_{33} - u_{13}^2 - u_{23}^2]^{1/2}$$

where

$$u_{23} = (a_{23} - u_{12}u_{13})/u_{22} = (2 - 0.4472136 \times 0)/1.6733201 = 1.1952286$$

$$u_{33} = (8 - 0^2 - 1.1952286^2)^{1/2} = 2.5634799$$

Since $u_{ij} = 0$ for $i > j$, we have

$$[U] = \begin{bmatrix} 2.2360680 & 0.4472136 & 0.0 \\ 0.0 & 1.6733201 & 1.1952286 \\ 0.0 & 0.0 & 2.5634799 \end{bmatrix}$$

∎

7.7.2
Other Solution Methods

Several other methods have been developed for finding the numerical solution of an eigenvalue problem [7.18, 7.21]. Bathe and Wilson [7.22] have done a comparative study of some of these methods. Recent emphasis has been on the economical solution of large eigenproblems [7.23, 7.24]. The estimation of natural frequencies by the use of Sturm sequences is presented in Refs. [7.25] and [7.26]. An alternative way to solve a class of lumped mechanical vibration problems using topological methods is presented in Ref. [7.27].

7.8   Examples Using MATLAB

Solution of an Eigenvalue Problem

EXAMPLE 7.8

Using MATLAB, find the eigenvalues and eigenvectors of the matrix

$$[A] = \begin{bmatrix} 3 & -1 & 0 \\ -2 & 4 & -3 \\ 0 & -1 & 1 \end{bmatrix}$$

Solution

```
>> A=[3 -1 0; -2 4 -3; 0 -1 1]

A =
```

$$
\begin{array}{rrr}
3 & -1 & 0 \\
-2 & 4 & -3 \\
0 & -1 & 1
\end{array}
$$

```
>> [V, D] = eig (A)

V =

   -0.3665   -0.8305    0.2262
    0.9080   -0.4584    0.6616
   -0.2028    0.3165    0.7149

D =

    5.4774         0         0
         0    2.4481         0
         0         0    0.0746

>>
```

∎

Using a Program for Jacobi's Method to Solve an Eigenvalue Problem

EXAMPLE 7.9

Develop a general program, called **Program9.m**, to implement Jacobi's method to find the eigenvalues and eigenvectors of a symmetric matrix. Use the program to find the eigenvalues and eigenvectors of the matrix

$$
[A] = \begin{bmatrix} 1 & 1 & 1 \\ 1 & 2 & 2 \\ 1 & 2 & 3 \end{bmatrix}
$$

Solution: Program9.m is developed to accept the following data:

n = order of the matrix

d = given matrix of order n × n

eps = convergence specification, a small quantity on the order of 10^{-5}

itmax = maximum number of iterations permitted

The program gives the eigenvalues and eigenvectors of the matrix d.

```
>> program9
Eigenvalue solution by Jacobi Method

Given matrix
1.00000000e+000    1.00000000e+000    1.00000000e+000
1.00000000e+000    2.00000000e+000    2.00000000e+000
1.00000000e+000    2.00000000e+000    3.00000000e+000

Eigen values are
5.04891734e+000    6.43104132e-001    3.07978528e-001

Eigen vectors are
   First              Second              Third
3.27984948e-001   -7.36976229e-001    5.91009231e-001
5.91009458e-001   -3.27985278e-001   -7.36975900e-001
7.36976047e-001    5.91009048e-001    3.27985688e-001
```

∎

EXAMPLE 7.10

Program for an Eigenvalue Solution Using the Matrix Iteration Method

Develop a general computer program, called **Program10.m**, to implement the matrix iteration method. Use the program to find the eigenvalues and eigenvectors of the matrix [A] given in Example 7.9.

Solution: Program10.m is developed to accept the following input data:

n = order of the matrix d

d = given matrix of order n × n

xs = initial guess vector of order n

nvec = number of eigenvalues and eigenvectors to be determined

xm = mass matrix of order n × n

eps = convergence requirement, a small quantity on the order of 10^{-5}

The program gives the following output:

freq = array of size nvec, containing the computed natural frequencies

eig = array of size n × nvec, containing the computed eigenvectors (columns)

```
>> program10
Solution of eigenvalue problem by
matrix iteration method

Natural frequencies:

     4.450424e-001        1.246983e+000        1.801938e+000

Mode shapes (Columnwise):

     1.000000e+000        1.000000e+000        1.000000e+000
     1.801937e+000        4.450328e-001       -1.247007e+000
     2.246979e+000       -8.019327e-001        5.549798e-001
```

∎

EXAMPLE 7.11

Program for Solving a General Eigenvalue Problem

Develop a general program, called **Program11.m**, to solve a general eigenvalue problem. Use the program to find the solution of the general eigenvalue problem

$$[k] \, \vec{X} = \omega^2 \, [m] \, \vec{X}$$

where

$$[k] = \begin{bmatrix} 2 & -1 & 0 \\ -1 & 2 & -1 \\ 0 & -1 & 1 \end{bmatrix}, \qquad [m] = \begin{bmatrix} 1 & 0 & 0 \\ 0 & 1 & 0 \\ 0 & 0 & 1 \end{bmatrix}$$

Solution: `Program11.m` is developed to solve the problem $[k] \vec{X} = \omega^2 [m] \vec{X}$ by first converting it to the form of a special eigenvalue problem $[D] \vec{Y} = \frac{1}{\omega^2} [I] \vec{Y}$, where $[D]$ is equal to $([U]^T)^{-1}[m][U]^{-1}$ and $[k] = [U]^T [U]$. The program is developed to accept the following input data:

nd = size of the problem (size of mass and stiffness matrices)

bk = stiffness matrix of size nd × nd

bm = mass matrix of size nd × nd

The program gives the upper triangular matrix of $[bk]$, the inverse of the upper triangular matrix $[ui]$, the matrix $[uti]\ [bm]\ [ui]$ where $[uti]$ is the transpose of $[ui]$, and the eigenvalues and eigenvectors of the problem.

```
>> program11
Upper triangular matrix [U]:

   1.414214e+000    -7.071068e-001     0.000000e+000
   0.000000e+000     1.224745e+000    -8.164966e-001
   0.000000e+000     0.000000e+000     5.773503e-001

Inverse of the upper triangular matrix:

   7.071068e-001     4.082483e-001     5.773503e-001
   0.000000e+000     8.164966e-001     1.154701e+000
   0.000000e+000     0.000000e+000     1.732051e+000

Matrix [UMU] = [UTI] [M] [UI]:

   5.000000e-001     2.886751e-001     4.082483e-001
   2.886751e-001     8.333333e-001     1.178511e+000
   4.082483e-001     1.178511e+000     4.666667e+000

Eigenvectors:

   5.048917e+000     6.431041e-001     3.079785e-001

Eigenvectors (Columnwise):

   7.369762e-001    -5.910090e-001     3.279853e-001
   1.327985e+000    -2.630237e-001    -4.089910e-001
   1.655971e+000     4.739525e-001     1.820181e-001
```

■

7.9 C++ Programs

Three interactive C++ programs are given for the solution of an eigenvalue problem. `Program9.cpp` is based on Jacobi's method and `Program10.cpp` implements the matrix iteration method for finding the eigenvalues and eigenvectors of a given matrix. `Program11.m` solves a general eigenvalue problem by first reducing the problem to a special eigenvalue problem. The inputs and outputs of the three programs are similar to those of `Program9.m`, `Program10.m`, and `Program11.m`.

Eigenvalue Solution Using Program9.cpp

EXAMPLE 7.12

Using `Program9.cpp` (Jacobi's method), find the eigenvalues and eigenvectors of the matrix given in Example 7.9.

Solution: The input data are to be entered interactively. The input and output of the program are shown below.

```
Please input N:
3

Please input matrix D row by row:
1.0   1.0   1.0
1.0   2.0   2.0
1.0   2.0   3.0

EIGENVALUE SOLUTION BY JACOBI METHOD

  GIVEN MATRIX

       1.000000    1.000000    1.000000
       1.000000    2.000000    2.000000
       1.000000    2.000000    3.000000

EIGEN VALUES ARE

          5.04891734         0.64310413         0.30797853

EIGEN VECTORS
                FIRST           SECOND            THIRD
            0.32798495      -0.73697623       0.59100923
            0.59100946      -0.32798528      -0.73697590
            0.73697605       0.59100905       0.32798569
```

■

EXAMPLE 7.13

Eigenvalue Solution Using Program10.cpp

Using **Program10.cpp** (matrix iteration method), find the eigenvalues and eigenvectors of the matrix given in Example 7.10.

Solution: The input data are to be given interactively. The input and output of the program are given below.

```
Please input N and NVEC:
3 3

Please input matrix D row by row
1 1 1
1 2 2
1 2 3

Please input matrix XM row by row
1 0 0
0 1 0
0 0 1

SOLUTION OF EIGENVALUE PROBLEM BY
MATRIX ITERATION METHOD


NATURAL FREQUENCIES

      0.445042     1.246981     1.801938

MODE SHAPES (COLUMNWISE):

       1.000000     1.000000     1.000000
       1.801937     0.445037    -1.246993
       2.246979    -0.801936     0.554969
```

■

EXAMPLE 7.14

Solution of General Eigenvalue Problem Using Program11.cpp

Using **Program11.cpp**, find the eigenvalues and eigenvectors of the general eigenvalue problem given in Example 7.11.

Solution: The input data are to be entered interactively. The input and output of the program are shown below.

```
Please input ND:
3
Please input BK matrix row by row:
 2.0  -1.0   0.0
-1.0   2.0  -1.0
 0.0  -1.0   1.0
Please input BM matrix row by row:
 1.0   0.0   0.0
 0.0   1.0   0.0
 0.0   0.0   1.0

UPPER TRIANGULAR MATRIX [U]:

     1.41421356    -0.70710678     0.00000000
     0.00000000     1.22474487    -0.81649658
     0.00000000     0.00000000     0.57735027

INVERSE OF THE UPPER TRIANGULAR MATRIX, [UI],

     0.70710678     0.40824829     0.57735027
     0.00000000     0.81649658     1.15470054
     0.00000000     0.00000000     1.73205081

MATRIX [UMU] = [UTI] [M] [UI]:

     0.50000000     0.28867513     0.40824829
     0.28867513     0.83333333     1.17851130
     0.40824829     1.17851130     4.66666667

EIGENVALUES:

     5.04891734     0.64310413     0.30797853

EIGENVECTORS (COLUMNWISE):

     0.73697624    -0.59100904     0.32798528
     1.32798528    -0.26302375    -0.40899095
     1.65597055     0.47395249     0.18201810
```

■

7.10 Fortran Programs

Three Fortran programs—**PROGRAM9.F**, **PROGRAM10.F** and **PROGRAM11.F**—are given for the solution of an eigenvalue problem. Whereas **PROGRAM9.F** implements Jacobi's method for symmetric matrices, **PROGRAM10.F** is based on the matrix iteration method. **PROGRAM11.F** can be used to solve a general eigenvalue problem. The inputs and outputs of these programs are similar to those of **Program9.m**, **Program10.m**, and **Program11.m**.

Eigenvalue Solution Using Jacobi's Method

EXAMPLE 7.15

Using **PROGRAM9.F** (Jacobi's method), find the eigenvalues and eigenvectors of the matrix given in Example 7.9.

Solution: The output of the program is given below.

```
        EIGENVALUE SOLUTION BY JACOBI METHOD

GIVEN MATRIX
     0.100000E+01       0.100000E+01       0.100000E+01
     0.100000E+01       0.200000E+01       0.200000E+01
     0.100000E+01       0.200000E+01       0.300000E+01

EIGEN VALUES ARE
     0.504892E+01       0.643104E+00       0.307979E+00

EIGEN VECTORS
        FIRST             SECOND              THIRD
     0.327985E+00      -0.736984E+00       0.590999E+00
     0.591007E+00      -0.327977E+00      -0.736981E+00
     0.736978E+00       0.591004E+00       0.327991E+00
```

■

Eigenvalue Solution Using the Matrix Iteration Method

EXAMPLE 7.16

Using **PROGRAM10.F** (the matrix iteration method), find the eigenvalues and eigenvectors of the matrix given in Example 7.10.

Solution: The output of the program is given below.

```
SOLUTION OF EIGENVALUE PROBLEM BY
MATRIX ITERATION METHOD

NATURAL FREQUENCIES

0.44504240E+00      0.12469811E+01      0.18019377E+01

MODE SHAPES (COLUMNWISE):

0.10000000E+01      0.10000000E+01      0.10000000E+01

0.18019372E+01      0.44503731E+00     -0.12469926E+01

0.22469788E+01     -0.80193609E+00      0.55496848E+00
```

■

Solution of a General Eigenvalue Problem

EXAMPLE 7.17

Using **PROGRAM11.F** (based on the decomposition of the stiffness matrix of the system), find the solution of the general eigenvalue problem considered in Example 7.11.

Solution: The output of the program is shown below.

```
UPPER TRIANGULAR MATRIX [U]:

   0.141421E+01     -0.707107E+00      0.000000E+00
   0.000000E+00      0.122474E+01     -0.816497E+00
   0.000000E+00      0.000000E+00      0.577350E+00

INVERSE OF THE UPPER TRIANGULAR MATRIX, [UI],

   0.707107E+00      0.408248E+00      0.577350E+00
   0.000000E+00      0.816497E+00      0.115470E+01
   0.000000E+00      0.000000E+00      0.173205E+01

MATRIX [UMU] = [UTI] [M] [UI]:

   0.500000E+00      0.288675E+00      0.408248E+00
   0.288675E+00      0.833333E+00      0.117851E+01
   0.408248E+00      0.117851E+01      0.466667E+01

EIGENVALUES:

0.504892E+01          0.643104E+00      0.307979E+00

EIGENVECTORS (COLUMNWISE):

   0.736973E+00     -0.590976E+00      0.328051E+00
   0.132799E+01     -0.263064E+00     -0.408952E+00
   0.165597E+01      0.473971E+00      0.181988E+00
```

■

REFERENCES

7.1 S. Dunkerley, "On the whirling and vibration of shafts," *Philosophical Transactions of the Royal Society of London*, 1894, Series A, Vol. 185, Part I, pp. 279–360.

7.2 B. Atzori, "Dunkerley's formula for finding the lowest frequency of vibration of elastic systems," letter to the editor, *Journal of Sound and Vibration*, 1974, Vol. 36, pp. 563–564.

7.3 H. H. Jeffcott, "The periods of lateral vibration of loaded shafts—The rational derivation of Dunkerley's empirical rule for determining whirling speeds," *Proceedings of the Royal Society of London*, 1919, Series A, Vol. 95, No. A666, pp. 106–115.

7.4 M. Endo and O. Taniguchi, "An extension of the Southwell-Dunkerley methods for synthesizing frequencies," *Journal of Sound and Vibration*, 1976, "Part I: Principles," Vol. 49, pp. 501–516, and "Part II: Applications," Vol. 49, pp. 517–533.

7.5 A Rutenberg, "A lower bound for Dunkerley's formula in continuous elastic systems," *Journal of Sound and Vibration*, 1976, Vol. 45, pp. 249–252.

7.6 G. Temple and W. G. Bickley, *Rayleigh's Principle and Its Applications to Engineering*, Dover, New York, 1956.

7.7 N. G. Stephen, "Rayleigh's, Dunkerley's, and Southwell's methods," *International Journal of Mechanical Engineering Education*, January 1983, Vol. 11, pp. 45–51.

7.8 A. Rutenberg, "Dunkerley's formula and alternative approximations," letter to the editor, *Journal of Sound and Vibration*, 1975, Vol. 39, pp. 530–531.

7.9 R. Jones, "Approximate expressions for the fundamental frequency of vibration of several dynamic systems," *Journal of Sound and Vibration*, 1976, Vol. 44, pp. 475–478.

7.10 R. W. Fitzgerald, *Mechanics of Materials* (2nd ed.), Addison-Wesley, Reading, Mass., 1982.

7.11 H. Holzer, *Die Berechnung der Drehschwin gungen*, Julius Springer, Berlin, 1921.

7.12 H. E. Fettis, "A modification of the Holzer method for computing uncoupled torsion and bending modes," *Journal of the Aeronautical Sciences*, October 1949, pp. 625–634; May 1954, pp. 359–360.

7.13 S. H. Crandall and W. G. Strang, "An improvement of the Holzer table based on a suggestion of Rayleigh's," *Journal of Applied Mechanics*, 1957, Vol. 24, p. 228.

7.14 S. Mahalingam, "An improvement of the Holzer method," *Journal of Applied Mechanics*, 1958, Vol. 25, p. 618.

7.15 S. Mahalingam, "Iterative procedures for torsional vibration analysis and their relationships," *Journal of Sound and Vibration*, 1980, Vol. 68, pp. 465–467.

7.16 L. Meirovitch, *Computational Methods in Structural Dynamics*, Sijthoff and Noordhoff, The Netherlands, 1980.

7.17 J. H. Wilkinson and G. Reinsch, *Linear Algebra*, Springer Verlag, New York, 1971.

7.18 J. W. Wilkinson, *The Algebraic Eigenvalue Problem*, Oxford University Press, London, 1965.

7.19 R. S. Martin and J. H. Wilkinson, "Reduction of a symmetric eigenproblem $Ax = \lambda Bx$ and related problems to standard form," *Numerical Mathematics*, 1968, Vol. 11, pp. 99–110.

7.20 G. B. Haggerty, *Elementary Numerical Analysis with Programming*, Allyn and Bacon, Boston, 1972.

7.21 A. Jennings, "Eigenvalue methods for vibration analysis," *Shock and Vibration Digest*, Part I, February 1980, Vol. 12, pp. 3–16; Part II, January 1984, Vol. 16, pp. 25–33.

7.22 K. Bathe and E. L. Wilson, "Solution methods for eigenvalue problems in structural mechanics," *International Journal for Numerical Methods in Engineering*, 1973, Vol. 6, pp. 213–226.

7.23 E. Cohen and H. McCallion, "Economical methods for finding eigenvalues and eigenvectors," *Journal of Sound and Vibration*, 1967, Vol. 5, pp. 397–406.

7.24 A. J. Fricker, "A method for solving high-order real symmetric eigenvalue problems," *International Journal for Numerical Methods in Engineering*, 1983, Vol. 19, pp. 1131–1138.

7.25 G. Longbottom and K. F. Gill, "The estimation of natural frequencies by use of Sturm sequences," *International Journal of Mechanical Engineering Education*, 1976, Vol. 4, pp. 319–329.

7.26 K. K. Gupta, "Solution of eigenvalue problems by Sturm sequence method," *International Journal for Numerical Methods in Engineering*, 1972, Vol. 4, pp. 379–404.

7.27 W. K. Chen and F. Y. Chen, "Topological analysis of a class of lumped vibrational systems," *Journal of Sound and Vibration*, 1969, Vol. 10, pp. 198–207.

REVIEW QUESTIONS

7.1   Give brief answers to the following:

1. Name a few methods for finding the fundamental natural frequency of a multidegree of freedom system.
2. What is the basic assumption made in deriving Dunkerley's formula?

3. What is Rayleigh's principle?

4. State whether we get a lower bound or an upper bound to the fundamental natural frequency if we use (a) Dunkerley's formula and (b) Rayleigh's method.

5. What is Rayleigh's quotient?

6. What is the basic principle used in Holzer's method?

7. What is the matrix iteration method?

8. Can we use any trial vector $\vec{X}_1$ in the matrix iteration method to find the largest natural frequency?

9. Using the matrix iteration method, how do you find the intermediate natural frequencies?

10. What is the difference between the matrix iteration method and Jacobi's method?

11. What is a rotation matrix? What is its purpose in Jacobi's method?

12. What is a standard eigenvalue problem?

13. What is the role of Choleski decomposition in deriving a standard eigenvalue problem?

14. How do you find the inverse of an upper triangular matrix?

7.2 Indicate whether each of the following statements is true or false:

1. The fundamental frequency given by Durkerley's formula will always be larger than the exact value.

2. The fundamental frequency given by Rayleigh's method will always be larger than the exact value.

3. $[A]\, \vec{X} = \lambda[B]\, \vec{X}$ is a standard eigenvalue problem.

4. $[A]\, \vec{X} = \lambda[I][B]\, \vec{X}$ is a standard eigenvalue problem.

5. Jacobi's method can find the eigenvalues of only symmetric matrices.

6. Jacobi's method uses rotation matrices.

7. The matrix iteration method requires the natural frequencies to be distinct and well separated.

8. In the matrix iteration method, any computational error will not yield incorrect results.

9. The matrix iteration method will never fail to converge to higher frequencies.

10. When Rayleigh's method is used for a shaft carrying several rotors, the static deflection curve can be used as the appropriate mode shape.

11. Rayleigh's method can be considered to be same as the conservation of energy for a vibrating system.

7.3 Fill in each of the following blanks with the appropriate word:

1. Any symmetric positive definite matrix $[A]$ can be decomposed as $[A] = [U]^T[U]$ where $[U]$ is _____ triangular matrix.

2. The method of decomposing a symmetric positive definite matrix $[A]$ as $[A] = [U]^T[U]$ is known as _____ method.

3. Each step of Jacobi's method reduces a pair of off-diagonal elements to _____.

4. The _____ theorem permits the representation of any vector as a linear combination of the eigenvectors of the system.

5. If the matrix iteration method converges to the smallest eigenvalue with $[D]\vec{X} = \lambda\vec{X}$, the method converges to the _____ eigenvalue with $[D]^{-1}\vec{X} = \mu\vec{X}$.

6. Rayleigh's quotient provides _____ bound to ω_1^2 and _____ bound to ω_n^2.

7. Rayleigh's quotient has a stationary value in the neighborhood of an _____.

8. For a shaft carrying masses $m_1, m_2, \cdots$, Rayleigh's method gives the natural frequency

as $\omega = \left\{ \dfrac{g(m_1 w_1 + m_2 w_2 + \cdots)}{m_1 w_1^2 + m_2 w_2^2 + \cdots} \right\}^{1/2}$, where $w_1, w_2, \cdots$ denote the _____ deflections

of $m_1, m_2, \cdots$, respectively.

9. Holzer's method is basically a _____ method.

10. _____ method is more extensively applied to torsional systems, although the method is equally applicable to linear systems.

11. The computation of higher natural frequencies, based on the matrix iteration method, involves a process known as matrix _____.

7.4 Select the most appropriate answer out of the choices given:

1. When the trial vector

$$\vec{X}^{(1)} = \begin{Bmatrix} 1 \\ 1 \\ 1 \end{Bmatrix}$$

is used for the solution of the eigenvalue problem,

$$\begin{bmatrix} 1 & 1 & 2 \\ 1 & 2 & 2 \\ 1 & 2 & 3 \end{bmatrix} \vec{X} = \lambda \vec{X}$$

the next trial vector, $\vec{X}^{(2)}$, given by the matrix iteration method is

(a) $\begin{Bmatrix} 3 \\ 5 \\ 6 \end{Bmatrix}$ (b) $\begin{Bmatrix} 1 \\ 1 \\ 1 \end{Bmatrix}$ (c) $\begin{Bmatrix} 3 \\ 3 \\ 3 \end{Bmatrix}$

2. For a semidefinite system, the final equation in Holzer's method denotes the
(a) amplitude at the end as zero
(b) sum of inertia forces as zero
(c) equation of motion

3. Dunkerley's formula is given by

(a) $\omega_1^2 \approx a_{11} m_1 + a_{22} m_2 + \cdots + a_{nn} m_n$

(b) $\dfrac{1}{\omega_1^2} \approx a_{11} m_1 + a_{22} m_2 + \cdots + a_{nn} m_n$

(c) $\dfrac{1}{\omega_1^2} \approx k_{11} m_1 + k_{22} m_2 + \cdots + k_{nn} m_n$

4. Rayleigh's quotient is given by

(a) $\dfrac{\vec{X}^T [k] \vec{X}}{\vec{X}^T [m] \vec{X}}$ (b) $\dfrac{\vec{X}^T [m] \vec{X}}{\vec{X}^T [k] \vec{X}}$ (c) $\dfrac{\vec{X}^T [k] \vec{X}}{\dot{\vec{X}}^T [m] \dot{\vec{X}}}$

5. Rayleigh's quotient satisfies the following relation:

(a) $R(\vec{X}) \le \omega_1^2$ (b) $R(\vec{X}) \ge \omega_n^2$ (c) $R(\vec{X}) \ge \omega_1^2$

6. For a vibrating system with $[k] = \begin{bmatrix} 2 & -1 \\ -1 & 2 \end{bmatrix}$ and $[m] = \begin{bmatrix} 1 & 0 \\ 0 & 1 \end{bmatrix}$, the mode shape closest to the fundamental mode, according to the Rayleigh's quotient, $R(\vec{X}) = \dfrac{\vec{X}^T [k]\vec{X}}{\vec{X}^T [m]\vec{X}}$, is given by

(a) $\vec{X} = \begin{Bmatrix} 1 \\ 1 \end{Bmatrix}$ (b) $\begin{Bmatrix} 1 \\ -1 \end{Bmatrix}$ (c) $\begin{Bmatrix} -1 \\ 1 \end{Bmatrix}$

7.5 Match the items in the two columns below:

1. Dunkerley's formula

2. Rayleigh's method

3. Holzer's method

4. Matrix iteration method

5. Jacobi's method

(a) Finds the natural frequencies and mode shapes of the system, one at a time, using several trial values for each frequency.

(b) Finds all the natural frequencies using trial vectors and matrix deflation procedure.

(c) Finds all the eigenvalues and eigenvectors simultaneously without using trial vectors.

(d) Finds the approximate value of the fundamental frequency of a composite system.

(e) Finds the approximate value of the fundamental frequency of a system that is always larger than the true value.

PROBLEMS

The problem assignments are organized as follows:

Problems	Section Covered	Topic Covered
7.1–7.6, 7.32	7.2	Dunkerley's formula
7.7–7.12	7.3	Rayleigh's method
7.13–7.17	7.4	Holzer's method
7.18–7.23	7.5	Matrix iteration method
7.24–7.25, 7.29	7.6	Jacobi's method
7.26–7.28, 7.30–7.31	7.7	Standard eigenvalue problem
7.33–7.37	7.8	MATLAB programs
7.38–7.40	7.9	C++ programs
7.41–7.48	7.10	Fortran programs
7.49–7.50	—	Design projects

7.1 Estimate the fundamental frequency of the beam shown in Fig. 6.9 using Dunkerley's formula for the following data: (a) $m_1 = m_3 = 5m$, $m_2 = m$ and (b) $m_1 = m_3 = m$, $m_2 = 5m$.

7.2 Find the fundamental frequency of the torsional system shown in Fig. 6.11, using Dunkerley's formula for the following data: (a) $J_1 = J_2 = J_3 = J_0$; $k_{t1} = k_{t2} = k_{t3} = k_t$; and (b) $J_1 = J_0$, $J_2 = 2J_0$, $J_3 = 3J_0$; $k_{t1} = k_t$, $k_{t2} = 2k_t$, $k_{t3} = 3k_t$.

7.3 Estimate the fundamental frequency of the shaft shown in Fig. 7.3, using Dunkerley's formula for the following data: $m_1 = m$, $m_2 = 2m$, $m_3 = 3m$, $l_1 = l_2 = l_3 = l_4 = l/4$.

7.4 The natural frequency of vibration, in bending, of the wing of a military aircraft is found to be 20 Hz. Find the new frequency of bending vibration of the wing when a weapon, weighing 2000 lb, is attached at the tip of the wing, as shown in Fig. 7.10. The stiffness of the wing tip, in bending, is known to be 50,000 lb/ft.

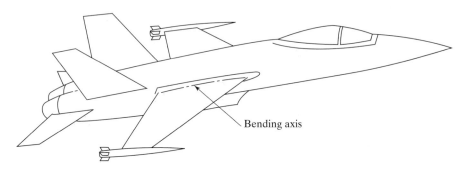

Bending axis

FIGURE 7.10

7.5 In an overhead crane (see Fig. 7.11) the trolley weighs ten times the weight of the girder. Using Dunkerley's formula, estimate the fundamental frequency of the system.

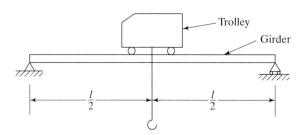

Trolley

Girder

$\frac{l}{2}$ $\frac{l}{2}$

FIGURE 7.11

7.6 Using Dunkerley's formula, determine the fundamental natural frequency of the stretched string system shown in Fig. 5.25 with $m_1 = m_2 = m$ and $l_1 = l_2 = l_3 = l$.

7.7 Using Rayleigh's method, determine the first natural frequency of vibration of the system shown in Fig. 7.2. Assume $k_1 = k$, $k_2 = 2k$, $k_3 = 3k$, and $m_1 = m$, $m_2 = 2m$, $m_3 = 3m$.

7.8 Using Rayleigh's method, find the fundamental natural frequency of the torsional system shown in Fig. 6.11. Assume that $J_1 = J_0$, $J_2 = 2J_0$, $J_3 = 3J_0$, and $k_{t1} = k_{t2} = k_{t3} = k_t$.

7.9 Using Rayleigh's method, solve Problem 7.6.

7.10 Using Rayleigh's method, determine the fundamental natural frequency of the system shown in Fig. 5.25 when $m_1 = m$, $m_2 = 5m$, $l_1 = l_2 = l_3 = l$.

7.11 A two-story shear building is shown in Fig. 7.12 in which the floors are assumed to be rigid. Using Rayleigh's method, compute the first natural frequency of the building for $m_1 = 2m$, $m_2 = m$, $h_1 = h_2 = h$, and $k_1 = k_2 = 3EI/h^3$. Assume the first mode configuration to be the same as the static equilibrium shape due to loads proportional to the floor weights.

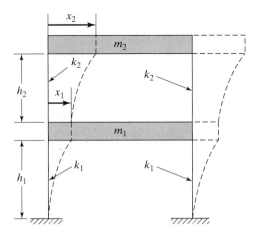

FIGURE 7.12

7.12 Prove that Rayleigh's quotient is never higher than the highest eigenvalue.

7.13 Using Holzer's method, find the natural frequencies and mode shapes of the system shown in Fig. 6.13, with $m_1 = 100$ kg, $m_2 = 20$ kg, $m_3 = 200$ kg, $k_1 = 8000$ N/m, and $k_2 = 4000$ N/m.

7.14 The stiffness and mass matrices of a vibrating system are given by

$$[k] = k \begin{bmatrix} 2 & -1 & 0 \\ -1 & 2 & -1 \\ 0 & -1 & 3 \end{bmatrix}, \qquad [m] = m \begin{bmatrix} 1 & 0 & 0 \\ 0 & 1 & 0 \\ 0 & 0 & 2 \end{bmatrix}$$

Using Holzer's method, determine all the principal modes and the natural frequencies.

7.15 For the torsional system shown in Fig. 6.11, determine a principal mode and the corresponding frequency by Holzer's method. Assume that $k_{t1} = k_{t2} = k_{t3} = k_t$ and $J_1 = J_2 = J_3 = J_0$.

7.16 Using Holzer's method, find the natural frequencies and mode shapes of the shear building shown in Fig. 7.12. Assume that $m_1 = 2m$, $m_2 = m$, $h_1 = h_2 = h$, $k_1 = 2k$, $k_2 = k$, and $k = 3EI/h^3$.

7.17 Using Holzer's method, find the natural frequencies and mode shapes of the system shown in Fig. 6.34. Assume that $J_1 = 10$ kg-m^2, $J_2 = 5$ kg-m^2, $J_3 = 1$ kg-m^2, and $k_{t1} = k_{t2} = 1 \times 10^6$ N-m/rad.

7.18 The largest eigenvalue of the matrix

$$[D] = \begin{bmatrix} 2.5 & -1 & 0 \\ -1 & 5 & -\sqrt{2} \\ 0 & -\sqrt{2} & 10 \end{bmatrix}$$

is given by $\lambda_1 = 10.38068$. Using the matrix iteration method, find the other eigenvalues and all the eigenvectors of the matrix. Assume $[m] = [I]$.

7.19 The mass and stiffness matrices of a spring-mass system are known to be

$$[m] = m \begin{bmatrix} 1 & 0 & 0 \\ 0 & 1 & 0 \\ 0 & 0 & 2 \end{bmatrix} \quad \text{and} \quad [k] = k \begin{bmatrix} 2 & -1 & 0 \\ -1 & 3 & -2 \\ 0 & -2 & 2 \end{bmatrix}$$

Using the matrix iteration method, find the natural frequencies and mode shapes of the system.

7.20 Using the matrix iteration method, find the natural frequencies and mode shapes of the system shown in Fig. 6.6 with $k_1 = k$, $k_2 = 2k$, $k_3 = 3k$, and $m_1 = m_2 = m_3 = m$.

7.21 Using the matrix iteration method, find the natural frequencies of the system shown in Fig. 6.23. Assume that $J_{d1} = J_{d2} = J_{d3} = J_0$, $l_i = l$, and $(GJ)_i = GJ$ for $i = 1$ to 4.

7.22 Using the matrix iteration method, solve problem 7.6.

7.23 The stiffness and mass matrices of a vibrating system are given by

$$[k] = k \begin{bmatrix} 4 & -2 & 0 & 0 \\ -2 & 3 & -1 & 0 \\ 0 & -1 & 2 & -1 \\ 0 & 0 & -1 & 1 \end{bmatrix} \quad \text{and} \quad [m] = m \begin{bmatrix} 3 & 0 & 0 & 0 \\ 0 & 2 & 0 & 0 \\ 0 & 0 & 1 & 0 \\ 0 & 0 & 0 & 1 \end{bmatrix}$$

Using the matrix iteration method, find the fundamental frequency and the mode shape of the system.

7.24 Using Jacobi's method, find the eigenvalues and eigenvectors of the matrix

$$[D] = \begin{bmatrix} 3 & -2 & 0 \\ -2 & 5 & -3 \\ 0 & -3 & 3 \end{bmatrix}$$

7.25 Using Jacobi's method, find the eigenvalues and eigenvectors of the matrix

$$[D] = \begin{bmatrix} 3 & 2 & 1 \\ 2 & 2 & 1 \\ 1 & 1 & 1 \end{bmatrix}$$

7.26 Using the Choleski decomposition technique, decompose the matrix

$$[A] = \begin{bmatrix} 4 & -2 & 6 & 4 \\ -2 & 2 & -1 & 3 \\ 6 & -1 & 22 & 13 \\ 4 & 3 & 13 & 46 \end{bmatrix}$$

7.27 Using the decomposition $[A] = [U]^T[U]$, find the inverse of the following matrix:

$$[A] = \begin{bmatrix} 5 & -1 & 1 \\ -1 & 6 & -4 \\ 1 & -4 & 3 \end{bmatrix}$$

7.28 Using Choleski decomposition, find the inverse of the following matrix:

$$[A] = \begin{bmatrix} 2 & 5 & 8 \\ 5 & 16 & 28 \\ 8 & 28 & 54 \end{bmatrix}$$

7.29 Using Jacobi's method, find the eigenvalues of the matrix $[A]$ given in Problem 7.26.

7.30 Convert Problem 7.23 to a standard eigenvalue problem with a symmetric matrix.

7.31 Using the Choleski decomposition technique, express the following matrix as the product of two triangular matrices:

$$[A] = \begin{bmatrix} 16 & -20 & -24 \\ -20 & 89 & -50 \\ -24 & -50 & 280 \end{bmatrix}$$

7.32* Design a minimum weight tubular section for the shaft shown in Fig. 7.3 to achieve a fundamental frequency of vibration of 0.5 Hz. Assume $m_1 = 20$ kg, $m_2 = 50$ kg, $m_3 = 40$ kg, $l_1 = 1$ m, $l_2 = 3$ m, $l_3 = 4$ m, $l_4 = 2$ m, and $E = 2.07 \times 10^{11}$ N/m^2.

7.33 Using MATLAB, find the eigenvalues and eigenvectors of the following matrix:

$$[A] = \begin{bmatrix} 3 & -2 & 0 \\ -2 & 5 & -3 \\ 0 & -1 & 1 \end{bmatrix}$$

7.34 Using MATLAB, find the eigenvalues and eigenvectors of the following matrix:

$$[A] = \begin{bmatrix} -5 & 2 & 1 \\ 1 & -9 & -1 \\ 2 & -1 & 7 \end{bmatrix}$$

7.35 Using **Program9.m**, find the eigenvalues and eigenvectors of the matrix $[D]$ given in Problem 7.18.

---

*The asterisk denotes a problem with no unique answer.

7.36 Using `Program10.m`, determine the eigenvalues and eigenvectors of the matrix $[D]$ given in Problem 7.25.

7.37 Using `Program11.m`, find the solution of the general eigenvalue problem given in Problem 7.23 with $k = m = 1$.

7.38 Using `Program9.cpp`, find the eigenvalues and eigenvectors of the following matrix:

$$[A] = \begin{bmatrix} 2 & 2 & 2 \\ 2 & 5 & 5 \\ 2 & 5 & 12 \end{bmatrix}$$

7.39 Using `Program10.cpp`, solve Problem 7.38.

7.40 Using `Program11.cpp`, solve the following eigenvalue problem:

$$\omega^2 \begin{bmatrix} 3 & 0 & 0 \\ 0 & 2 & 0 \\ 0 & 0 & 1 \end{bmatrix} \vec{X} = \begin{bmatrix} 10 & -4 & 0 \\ -4 & 6 & -2 \\ 0 & -2 & 2 \end{bmatrix} \vec{X}$$

7.41 Using `PROGRAM9.F`, solve Problem 7.38.

7.42 Using `PROGRAM10.F`, solve Problem 7.38.

7.43 Using `PROGRAM11.F`, solve Problem 7.40.

7.44 Using `PROGRAM9.F`, solve Problem 7.27.

7.45 Using `PROGRAM9.F`, solve Problem 7.31.

7.46 Use `PROGRAM11.F` to solve the general eigenvalue problem stated in Problem 7.14.

7.47 Use `PROGRAM11.F` to solve the general eigenvalue problem stated in Problem 7.19.

7.48 Use `PROGRAM11.F` to solve the general eigenvalue problem stated in Problem 7.23.

DESIGN PROJECTS

7.49 A flywheel of mass $m_1 = 100$ kg and a pulley of mass $m_2 = 50$ kg are to be mounted on a shaft of length $l = 2$ m, as shown in Fig. 7.13. Determine their locations l_1 and l_2 to maximize the fundamental frequency of vibration of the system.

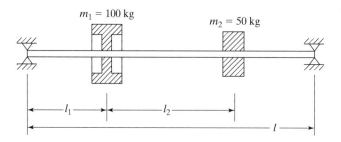

FIGURE 7.13

7.50 A simplified diagram of an overhead traveling crane is shown in Fig. 7.14. The girder, with square cross section, and the wire rope, with circular cross section, are made up of steel. Design the girders and the wire rope such that the natural frequencies of the system are greater than the operating speed, 1500 rpm, of an electric motor located in the trolley.

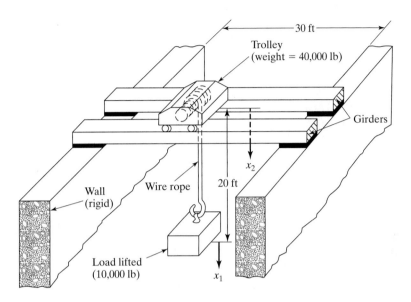

FIGURE 7.14

Stephen Prokf'yevich Timoshenko (1878–1972), a Russian-born engineer who emigrated to the United States, was one of the most widely known authors of books in the field of elasticity, strength of materials, and vibrations. He held the chair of mechanics at the University of Michigan and later at Stanford University, and he is regarded as the father of engineering mechanics in the United States. The improved theory he presented in 1921 for the vibration of beams has become known as the Timoshenko beam theory. (Courtesy of *Applied Mechanics Reviews*.)

CHAPTER 8

Continuous Systems

8.1 Introduction

We have so far dealt with discrete systems where mass, damping, and elasticity were assumed to be present only at certain discrete points in the system. There are many cases, known as *distributed* or *continuous systems*, in which it is not possible to identify discrete masses, dampers, or springs. We must then consider the continuous distribution of the mass, damping, and elasticity and assume that each of the infinite number of points of the system can vibrate. This is why a continuous system is also called a *system of infinite degrees of freedom*.

If a system is modeled as a discrete one, the governing equations are ordinary differential equations, which are relatively easy to solve. On the other hand, if the system is modeled as a continuous one, the governing equations are partial differential equations, which are more difficult. However, the information obtained from a discrete model of a system may not be as accurate as that obtained from a continuous model. The choice between the two models must be made carefully, with due consideration of factors such as the purpose of the analysis, the influence of the analysis on design, and the computational time available.

In this chapter, we shall consider the vibration of simple continuous systems—strings, bars, shafts, beams, and membranes. A more specialized treatment of the vibration of continuous structural elements is given in Refs. [8.1–8.3]. In general, the frequency equation of a continuous system is a transcendental equation that yields an infinite number of natural frequencies and normal modes. This is in contrast to the behavior of discrete systems, which yield a finite number of such frequencies and modes. We need to apply boundary conditions to find the natural frequencies of a continuous system. The question of boundary conditions does not arise in the case of discrete systems except in an indirect way, because the influence coefficients depend on the manner in which the system is supported.

8.2   Transverse Vibration of a String or Cable

8.2.1
Equation of
Motion

Consider a tightly stretched elastic string or cable of length l subjected to a transverse force $f(x, t)$ per unit length, as shown in Fig. 8.1(a). The transverse displacement of the string, $w(x, t)$, is assumed to be small. Equilibrium of the forces in the z direction gives (see Fig. 8.1b):
The net force acting on an element is equal to the inertia force acting on the element or

$$(P + dP)\sin(\theta + d\theta) + f\,dx - P\sin\theta = \rho\,dx\frac{\partial^2 w}{\partial t^2} \tag{8.1}$$

where P is the tension, ρ is the mass per unit length, and θ is the angle the deflected string makes with the x axis. For an elemental length dx,

$$dP = \frac{\partial P}{\partial x}\,dx \tag{8.2}$$

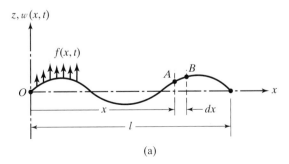

(a)

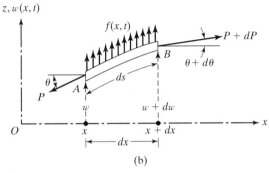

(b)

FIGURE 8.1   A vibrating string.

$$\sin \theta \simeq \tan \theta = \frac{\partial w}{\partial x} \tag{8.3}$$

and

$$\sin (\theta + d\theta) \simeq \tan (\theta + d\theta) = \frac{\partial w}{\partial x} + \frac{\partial^2 w}{\partial x^2} dx \tag{8.4}$$

Hence the forced vibration equation of the nonuniform string, Eq. (8.1), can be simplified to

$$\frac{\partial}{\partial x}\left[P\frac{\partial w(x, t)}{\partial x} \right] + f(x, t) = \rho(x)\frac{\partial^2 w(x, t)}{\partial t^2} \tag{8.5}$$

If the string is uniform and the tension is constant, Eq. (8.5) reduces to

$$P\frac{\partial^2 w(x, t)}{\partial x^2} + f(x, t) = \rho\frac{\partial^2 w(x, t)}{\partial t^2} \tag{8.6}$$

If $f(x,t) = 0$, we obtain the free vibration equation

$$P\frac{\partial^2 w(x, t)}{\partial x^2} = \rho\frac{\partial^2 w(x, t)}{\partial t^2} \tag{8.7}$$

or

$$c^2\frac{\partial^2 w}{\partial x^2} = \frac{\partial^2 w}{\partial t^2} \tag{8.8}$$

where

$$c = \left(\frac{P}{\rho} \right)^{1/2} \tag{8.9}$$

Equation (8.8) is also known as the *wave equation*.

8.2.2 Initial and Boundary Conditions

The equation of motion, Eq. (8.5) or its special forms (8.6) and (8.7), is a partial differential equation of the second order. Since the order of the highest derivative of w with respect to x and t in this equation is two, we need to specify two boundary and two initial conditions in finding the solution $w(x, t)$. If the string has a known deflection $w_0(x)$ and velocity $\dot{w}_0(x)$ at time $t = 0$, the initial conditions are specified as

$$w(x, t = 0) = w_0(x)$$

$$\frac{\partial w}{\partial t}(x, t = 0) = \dot{w}_0(x) \tag{8.10}$$

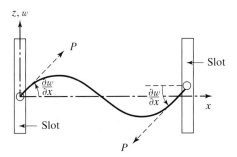

FIGURE 8.2   String connected to pins at the ends.

If the string is fixed at an end, say $x = 0$, the displacement w must always be zero, and so the boundary condition is

$$w(x = 0, t) = 0, \qquad t \geq 0 \tag{8.11}$$

If the string or cable is connected to a pin that can move in a perpendicular direction as shown in Fig. 8.2, the end cannot support a transverse force. Hence the boundary condition becomes

$$P(x) \frac{\partial w(x, t)}{\partial x} = 0 \tag{8.12}$$

If the end $x = 0$ is free and P is a constant, then Eq. (8.12) becomes

$$\frac{\partial w(0, t)}{\partial x} = 0, \qquad t \geq 0 \tag{8.13}$$

If the end $x = l$ is constrained elastically as shown in Fig. 8.3, the boundary condition becomes

$$P(x) \frac{\partial w(x, t)}{\partial x} \bigg|_{x=l} = -\underset{\sim}{k} w(x, t)|_{x=l}, \qquad t \geq 0 \tag{8.14}$$

where $\underset{\sim}{k}$ is the spring constant.

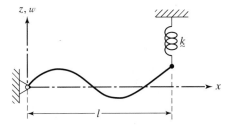

FIGURE 8.3   String with elastic constraint.

**8.2.3
Free Vibration of
a Uniform
String**

The free vibration equation, Eq. (8.8), can be solved by the method of separation of variables. In this method, the solution is written as the product of a function $W(x)$ (which depends only on x) and a function $T(t)$ (which depends only on t) [8.4]:

$$w(x, t) = W(x)T(t) \tag{8.15}$$

Substitution of Eq. (8.15) into Eq. (8.8) leads to

$$\frac{c^2}{W}\frac{d^2W}{dx^2} = \frac{1}{T}\frac{d^2T}{dt^2} \tag{8.16}$$

Since the left-hand side of this equation depends only on x and the right-hand side depends only on t, their common value must be a constant—say, a—so that

$$\frac{c^2}{W}\frac{d^2W}{dx^2} = \frac{1}{T}\frac{d^2T}{dt^2} = a \tag{8.17}$$

The equations implied in Eq. (8.17) can be written as

$$\frac{d^2W}{dx^2} - \frac{a}{c^2}W = 0 \tag{8.18}$$

$$\frac{d^2T}{dt^2} - aT = 0 \tag{8.19}$$

Since the constant a is generally negative (see Problem 8.9), we can set $a = -\omega^2$ and write Eqs. (8.18) and (8.19) as

$$\frac{d^2W}{dx^2} + \frac{\omega^2}{c^2}W = 0 \tag{8.20}$$

$$\frac{d^2T}{dt^2} + \omega^2T = 0 \tag{8.21}$$

The solutions of these equations are given by

$$W(x) = A\cos\frac{\omega x}{c} + B\sin\frac{\omega x}{c} \tag{8.22}$$

$$T(t) = C\cos\omega t + D\sin\omega t \tag{8.23}$$

where ω is the frequency of vibration and the constants A, B, C, and D can be evaluated from the boundary and initial conditions.

**8.2.4
Free Vibration of
a String with
Both Ends Fixed**

If the string is fixed at both ends, the boundary conditions are $w(0, t) = w(l, t) = 0$ for all time $t \geq 0$. Hence, from Eq. (8.15), we obtain

$$W(0) = 0 \tag{8.24}$$

$$W(l) = 0 \tag{8.25}$$

In order to satisfy Eq. (8.24), A must be zero in Eq. (8.22). Equation (8.25) requires that

$$B \sin \frac{\omega l}{c} = 0 \tag{8.26}$$

Since B cannot be zero for a nontrivial solution, we have

$$\sin \frac{\omega l}{c} = 0 \tag{8.27}$$

Equation (8.27) is called the *frequency* or *characteristic equation* and is satisfied by several values of ω. The values of ω are called the *eigenvalues* (or *natural frequencies* or *characteristic values*) of the problem. The nth natural frequency is given by

$$\frac{\omega_n l}{c} = n\pi, \qquad n = 1, 2, \ldots$$

or

$$\omega_n = \frac{nc\pi}{l}, \qquad n = 1, 2, \ldots \tag{8.28}$$

The solution $w_n(x, t)$ corresponding to ω_n can be expressed as

$$w_n(x, t) = W_n(x)T_n(t) = \sin \frac{n\pi x}{l} \left[C_n \cos \frac{nc\pi t}{l} + D_n \sin \frac{nc\pi t}{l} \right] \tag{8.29}$$

where C_n and D_n are arbitrary constants. The solution $w_n(x, t)$ is called the *n*th *mode of vibration* or *n*th *harmonic* or *n*th *normal mode* of the string. In this mode, each point of the string vibrates with an amplitude proportional to the value of W_n at that point, with the circular frequency $\omega_n = (nc\pi)/l$. The function $W_n(x)$ is called the *n*th *normal mode*, or *characteristic function*. The first three modes of vibration are shown in Fig. 8.4. The mode corresponding to $n = 1$ is called the *fundamental mode*, and ω_1 is called the *fundamental frequency*. The fundamental period is

$$\tau_1 = \frac{2\pi}{\omega_1} = \frac{2l}{c}$$

The points at which $w_n = 0$ for all times are called *nodes*. Thus the fundamental mode has two nodes, at $x = 0$ and $x = l$; the second mode has three nodes, at $x = 0$, $x = l/2$, and $x = l$; etc.

The general solution of Eq. (8.8), which satisfies the boundary conditions of Eqs. (8.24) and (8.25), is given by the superposition of all $w_n(x, t)$:

$$w(x, t) = \sum_{n=1}^{\infty} w_n(x, t)$$

$$= \sum_{n=1}^{\infty} \sin \frac{n\pi x}{l} \left[C_n \cos \frac{nc\pi t}{l} + D_n \sin \frac{nc\pi t}{l} \right] \tag{8.30}$$

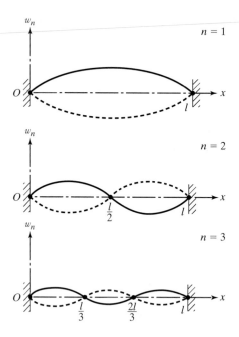

FIGURE 8.4   Mode shapes of a string.

This equation gives all possible vibrations of the string; the particular vibration that occurs is uniquely determined by the specified initial conditions. The initial conditions give unique values of the constants C_n and D_n. If the initial conditions are specified as in Eq. (8.10), we obtain

$$\sum_{n=1}^{\infty} C_n \sin \frac{n\pi x}{l} = w_0(x) \tag{8.31}$$

$$\sum_{n=1}^{\infty} \frac{nc\pi}{l} D_n \sin \frac{n\pi x}{l} = \dot{w}_0(x) \tag{8.32}$$

which can be seen to be Fourier sine series expansions of $w_0(x)$ and $\dot{w}_0(x)$ in the interval $0 \leq x \leq l$. The values of C_n and D_n can be determined by multiplying Eqs. (8.31) and (8.32) by $\sin(n\pi x/l)$ and integrating with respect to x from 0 to l:

$$C_n = \frac{2}{l} \int_0^l w_0(x) \sin \frac{n\pi x}{l} \, dx \tag{8.33}$$

$$D_n = \frac{2}{nc\pi} \int_0^l \dot{w}_0(x) \sin \frac{n\pi x}{l} \, dx \tag{8.34}$$

Note: The solution given by Eq. (8.30) can be identified as the *mode superposition method* since the response is expressed as a superposition of the normal modes. The procedure is applicable in finding not only the free vibration solution but also the forced vibration solution of continuous systems.

EXAMPLE   8.1

Dynamic Response of a Plucked String

If a string of length l, fixed at both ends, is plucked at its midpoint as shown in Fig. 8.5 and then released, determine its subsequent motion.

Solution: The solution is given by Eq. (8.30) with C_n and D_n given by Eqs. (8.33) and (8.34), respectively. Since there is no initial velocity, $\dot{w}_0(x) = 0$, and so $D_n = 0$. Thus the solution of Eq. (8.30) reduces to

$$w(x, t) = \sum_{n=1}^{\infty} C_n \sin \frac{n\pi x}{l} \cos \frac{nc\pi t}{l} \tag{E.1}$$

where

$$C_n = \frac{2}{l} \int_0^l w_0(x) \sin \frac{n\pi x}{l} \, dx \tag{E.2}$$

The initial deflection $w_0(x)$ is given by

$$w_0(x) = \begin{cases} \dfrac{2hx}{l} & \text{for } 0 \leq x \leq \dfrac{l}{2} \\[2mm] \dfrac{2h(l - x)}{l} & \text{for } \dfrac{l}{2} \leq x \leq l \end{cases} \tag{E.3}$$

By substituting Eq. (E.3) into Eq. (E.2), C_n can be evaluated:

$$C_n = \frac{2}{l} \left\{ \int_0^{l/2} \frac{2hx}{l} \sin \frac{n\pi x}{l} \, dx + \int_{l/2}^{l} \frac{2h}{l} (l - x) \sin \frac{n\pi x}{l} \, dx \right\}$$

$$= \begin{cases} \dfrac{8h}{\pi^2 n^2} \sin \dfrac{n\pi}{2} & \text{for } n = 1, 3, 5, \ldots \\[2mm] 0 & \text{for } n = 2, 4, 6, \ldots \end{cases} \tag{E.4}$$

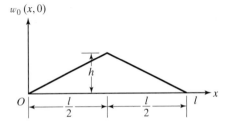

$w_0(x, 0)$

FIGURE 8.5   Initial deflection of the string.

By using the relation

$$\sin \frac{n\pi}{2} = (-1)^{(n-1)/2}, \qquad n = 1, 3, 5, \ldots \tag{E.5}$$

the desired solution can be expressed as

$$w(x, t) = \frac{8h}{\pi^2} \left\{ \sin \frac{\pi x}{l} \cos \frac{\pi ct}{l} - \frac{1}{9} \sin \frac{3\pi x}{l} \cos \frac{3\pi ct}{l} + \cdots \right\} \tag{E.6}$$

In this case, no even harmonics are excited.

■

8.2.5
Traveling-Wave
Solution

The solution of the wave equation, Eq. (8.8), for a string of infinite length can be expressed as [8.5]

$$w(x, t) = w_1(x - ct) + w_2(x + ct) \tag{8.35}$$

where w_1 and w_2 are arbitrary functions of $(x - ct)$ and $(x + ct)$, respectively. To show that Eq. (8.35) is the correct solution of Eq. (8.8), we first differentiate Eq. (8.35):

$$\frac{\partial^2 w(x, t)}{\partial x^2} = w_1'' (x - ct) + w_2'' (x + ct) \tag{8.36}$$

$$\frac{\partial^2 w(x, t)}{\partial t^2} = c^2 w_1'' (x - ct) + c^2 w_2'' (x + ct) \tag{8.37}$$

Substitution of these equations into Eq. (8.8) reveals that the wave equation is satisfied. In Eq. (8.35), $w_1(x - ct)$ and $w_2(x + ct)$ represent waves that propagate in the positive and negative directions of the x axis, respectively, with a velocity c.

For a given problem, the arbitrary functions w_1 and w_2 are determined from the initial conditions, Eq. (8.10). Substitution of Eq. (8.35) into Eq. (8.10) gives, at $t = 0$,

$$w_1(x) + w_2(x) = w_0(x) \tag{8.38}$$

$$- cw_1'(x) + cw_2'(x) = \dot{w}_0(x) \tag{8.39}$$

where the prime indicates differentiation with respect to the respective argument at $t = 0$ (that is, with respect to x). Integration of Eq. (8.39) yields

$$- w_1(x) + w_2(x) = \frac{1}{c} \int_{x_0}^{x} \dot{w}_0(x') \, dx' \tag{8.40}$$

where x_0 is a constant. Solution of Eqs. (8.38) and (8.40) gives w_1 and w_2:

$$w_1(x) = \frac{1}{2}\left[w_0(x) - \frac{1}{c}\int_{x_0}^{x} \dot{w}_0(x')\,dx' \right] \tag{8.41}$$

$$w_2(x) = \frac{1}{2}\left[w_0(x) + \frac{1}{c}\int_{x}^{x} \dot{w}_0(x')\,dx' \right] \tag{8.42}$$

By replacing x by $(x - ct)$ and $(x + ct)$, respectively, in Eqs. (8.41) and (8.42), we obtain the total solution:

$$\begin{aligned} w(x, t) &= w_1(x - ct) + w_2(x + ct) \\ &= \frac{1}{2}[w_0(x - ct) + w_0(x + ct)] + \frac{1}{2c}\int_{x-ct}^{x+ct} {}_0(x')\,dx' \end{aligned} \tag{8.43}$$

The following points should be noted:

1. As can be seen from Eq. (8.43), there is no need to apply boundary conditions to the problem.
2. The solution given by Eq. (8.43) can be expressed as

$$w(x, t) = w_D(x, t) + w_V(x, t) \tag{8.44}$$

where $w_D(x, t)$ denotes the waves propagating due to the known initial displacement $w_0(x)$ with zero initial velocity, and $w_V(x, t)$ represents waves traveling due only to the known initial velocity $\dot{w}_0(x)$ with zero initial displacement.

The transverse vibration of a string fixed at both ends excited by the transverse impact of an elastic load at an intermediate point was considered in [8.6]. A review of the literature on the dynamics of cables and chains was given by Triantafyllou [8.7].

8.3   Longitudinal Vibration of a Bar or Rod

**8.3.1
Equation of
Motion and
Solution**

Consider an elastic bar of length l with varying cross-sectional area $A(x)$, as shown in Fig. 8.6. The forces acting on the cross sections of a small element of the bar are given by P and $P + dP$ with

$$P = \sigma A = EA\frac{\partial u}{\partial x} \tag{8.45}$$

where σ is the axial stress, E is Young's modulus, u is the axial displacement, and $\partial u/\partial x$ is the axial strain. If $f(x, t)$ denotes the external force per unit length, the summation of the

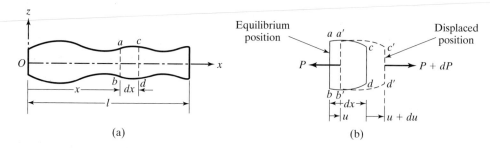

FIGURE 8.6   Longitudinal vibration of a bar.

forces in the x direction gives the equation of motion

$$(P + dP) + f\,dx - P = \rho A\,dx\,\frac{\partial^2 u}{\partial t^2} \tag{8.46}$$

where ρ is the mass density of the bar. By using the relation $dP = (\partial P/\partial x)\,dx$ and Eq. (8.45), the equation of motion for the forced longitudinal vibration of a nonuniform bar, Eq. (8.46), can be expressed as

$$\frac{\partial}{\partial x}\left[EA(x)\frac{\partial u(x,t)}{\partial x}\right] + f(x,t) = \rho(x)\,A(x)\frac{\partial^2 u}{\partial t^2}(x,t) \tag{8.47}$$

For a uniform bar, Eq. (8.47) reduces to

$$EA\frac{\partial^2 u}{\partial x^2}(x,t) + f(x,t) = \rho A\frac{\partial^2 u}{\partial t^2}(x,t) \tag{8.48}$$

The free vibration equation can be obtained from Eq. (8.48), by setting $f = 0$, as

$$c^2\frac{\partial^2 u}{\partial x^2}(x,t) = \frac{\partial^2 u}{\partial t^2}(x,t) \tag{8.49}$$

where

$$c = \sqrt{\frac{E}{\rho}} \tag{8.50}$$

Note that Eqs. (8.47) to (8.50) can be seen to be similar to Eqs. (8.5), (8.6), (8.8), and (8.9), respectively. The solution of Eq. (8.49), which can be obtained as in the case of Eq. (8.8),

End Conditions of Bar	Bounty Conditions	Frequency Equation	Mode Shape (Normal Function)	Natural Frequencies
Fixed-free 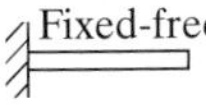	$u(0, t) = 0$ $\dfrac{\partial u}{\partial x}(l, t) = 0$	$\cos \dfrac{\omega l}{c} = 0$	$U_n(x) = C_n \sin \dfrac{(2n + 1)\,\pi x}{2l}$	$\omega_n = \dfrac{(2n + 1)\,\pi c}{2l}$; $n = 0, 1, 2, \ldots$
Free-free 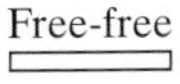	$\dfrac{\partial u}{\partial x}(0, t) = 0$ $\dfrac{\partial u}{\partial x}(l, t) = 0$	$\sin \dfrac{\omega l}{c} = 0$	$U_n(x) = C_n \cos \dfrac{n\pi x}{l}$	$\omega_n = \dfrac{n\pi c}{l}$; $n = 0, 1, 2, \ldots$
Fixed-fixed 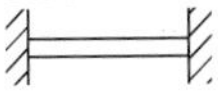	$u(0, t) = 0$ $u(l, t) = 0$	$\sin \dfrac{\omega l}{c} = 0$	$U_n(x) = C_n \cos \dfrac{n\pi x}{l}$	$\omega_n = \dfrac{n\pi c}{l}$; $n = 1, 2, 3, \ldots$

FIGURE 8.7   Common boundary conditions for a bar in longitudinal vibration.

can thus be written as

$$u(x, t) = U(x)T(t) \equiv \left(\underset{\sim}{A} \cos \frac{\omega x}{c} + \underset{\sim}{B} \sin \frac{\omega x}{c} \right) (C \cos \omega t + D \sin \omega t)^{1} \quad (8.51)$$

where the function $U(x)$ represents the normal mode and depends only on x and the function $T(t)$ depends only on t. If the bar has known initial axial displacement $u_0(x)$ and initial velocity $\dot{u}_0(x)$, the initial conditions can be stated as

$$u(x, t = 0) = u_0(x)$$

$$\frac{\partial u}{\partial t}(x, t = 0) = \dot{u}_0(x) \quad (8.52)$$

The common boundary conditions and the corresponding frequency equations for the longitudinal vibration of uniform bars are shown in Fig. 8.7.

Boundary Conditions for a Bar

EXAMPLE 8.2

A uniform bar of cross-sectional area A, length l, and Young's modulus E is connected at both ends by springs, dampers, and masses, as shown in Fig. 8.8(a). State the boundary conditions.

Solution: The free-body diagrams of the masses m_1 and m_2 are shown in Fig. 8.8(b). From this, we find that at the left end ($x = 0$), the force developed in the bar due to positive u and $\partial u/\partial x$ must be equal to the sum of spring, damper, and inertia forces:

$$AE \frac{\partial u}{\partial x}(0, t) = k_1 u(0, t) + c_1 \frac{\partial u}{\partial t}(0, t) + m_1 \frac{\partial^2 u}{\partial t^2}(0, t) \quad (E.1)$$

---

[1]We use $\underset{\sim}{A}$ and $\underset{\sim}{B}$ in this section; A is used to denote the cross-sectional area of the bar.

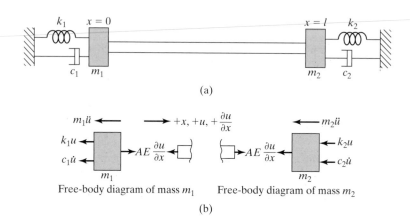

FIGURE 8.8   Bar connected to springs-masses-dampers at ends.

Similarly at the right end ($x = l$), the force developed in the bar due to positive u and $\partial u/\partial x$ must be equal to the negative sum of spring, damper, and inertia forces:

$$AE\frac{\partial u}{\partial x}(l, t) = -k_2 u(l, t) - c_2\frac{\partial u}{\partial t}(l, t) - m_2\frac{\partial^2 u}{\partial t^2}(l, t) \qquad \text{(E.2)}$$

■

8.3.2 Orthogonality of Normal Functions

The normal functions for the longitudinal vibration of bars satisfy the orthogonality relation

$$\int_0^l U_i(x)U_j(x)\,dx = 0 \qquad (8.53)$$

where $U_i(x)$ and $U_j(x)$ denote the normal functions corresponding to the ith and jth natural frequencies ω_i and ω_j, respectively. When $u(x, t) = U_i(x)T(t)$ and $u(x, t) = U_j(x)T(t)$ are assumed as solutions, Eq. (8.49) gives

$$c^2\frac{d^2 U_i(x)}{dx^2} + \omega_i^2 U_i(x) = 0 \qquad \text{or} \qquad c^2 U_i''(x) + \omega_i^2 U_i(x) = 0 \qquad (8.54)$$

and

$$c^2\frac{d^2 U_j(x)}{dx^2} + \omega_j^2 U_j(x) = 0 \qquad \text{or} \qquad c^2 U_j''(x) + \omega_j^2 U_j(x) = 0 \qquad (8.55)$$

where $U_i'' = \dfrac{d^2 U_i}{dx^2}$ and $U_j'' = \dfrac{d^2 U_j}{dx^2}$. Multiplication of Eq. (8.54) by U_j and Eq. (8.55) by U_i gives

$$c^2 U_i'' U_j + \omega_i^2 U_i U_j = 0 \qquad (8.56)$$

$$c^2 U_j'' U_i + \omega_j^2 U_j U_i = 0 \tag{8.57}$$

Subtraction of Eq. (8.57) from Eq. (8.56) and integration from 0 to l results in

$$\int_0^l U_i U_j \, dx = -\frac{c^2}{\omega_i^2 - \omega_j^2} \int_0^l (U_i'' U_j - U_j'' U_i) \, dx$$

$$= -\frac{c^2}{\omega_i^2 - \omega_j^2} \left[U_i' U_j - U_j' U_i \right] \Big|_0^l \tag{8.58}$$

The right-hand side of Eq. (8.58) can be proved to be zero for any combination of boundary conditions. For example, if the bar is fixed at $x = 0$ and free at $x = l$,

$$u(0, t) = 0, \qquad t \geq 0 \qquad \text{or} \qquad U(0) = 0 \tag{8.59}$$

$$\frac{\partial u}{\partial x}(l, t) = 0, \qquad t \geq 0 \qquad \text{or} \qquad U'(l) = 0 \tag{8.60}$$

Thus $(U_i' U_j - U_j' U_i)\big|_{x=l} = 0$ due to U' being zero (Eq. 8.60) and $(U_i' U_j - U_j' U_i)\big|_{x=0} = 0$ due to U being zero (Eq. 8.59). Equation (8.58) thus reduces to Eq. (8.53), which is also known as the *orthogonality principle for the normal functions*.

EXAMPLE 8.3

Free Vibrations of a Fixed-Free Bar

Find the natural frequencies and the free vibration solution of a bar fixed at one end and free at the other.

Solution: Let the bar be fixed at $x = 0$ and free at $x = l$, so that the boundary conditions can be expressed as

$$u(0, t) = 0, \qquad t \geq 0 \tag{E.1}$$

$$\frac{\partial u}{\partial x}(l, t) = 0, \qquad t \geq 0 \tag{E.2}$$

The use of Eq. (E.1) in Eq. (8.51) gives $\underset{\sim}{A} = 0$, while the use of Eq. (E.2) gives the frequency equation

$$\underset{\sim}{B} \frac{\omega}{c} \cos \frac{\omega l}{c} = 0 \qquad \text{or} \qquad \cos \frac{\omega l}{c} = 0 \tag{E.3}$$

The eigenvalues or natural frequencies are given by

$$\frac{\omega_n l}{c} = (2n + 1)\frac{\pi}{2}, \qquad n = 0, 1, 2, \ldots$$

or

$$\omega_n = \frac{(2n + 1)\pi c}{2l}, \qquad n = 0, 1, 2, \ldots \tag{E.4}$$

Thus the total (free vibration) solution of Eq. (8.49) can be written, using the mode superposition method, as

$$u(x, t) = \sum_{n=0}^{\infty} u_n(x, t)$$

$$= \sum_{n=0}^{\infty} \sin\frac{(2n + 1)\,\pi x}{2l} \left[C_n \cos\frac{(2n + 1)\,\pi ct}{2l} + D_n \sin\frac{(2n + 1)\,\pi ct}{2l} \right] \tag{E.5}$$

where the values of the constants C_n and D_n can be determined from the initial conditions, as in Eqs. (8.33) and (8.34):

$$C_n = \frac{2}{l} \int_0^l u_0(x)\sin\frac{(2n + 1)\,\pi x}{2l}\,dx \tag{E.6}$$

$$D_n = \frac{4}{(2n + 1)\,\pi c} \int_0^l \dot{u}_0(x)\sin\frac{(2n + 1)\pi x}{2l}\,dx \tag{E.7}$$

■

EXAMPLE 8.4

Natural Frequencies of a Bar Carrying a Mass

Find the natural frequencies of a bar with one end fixed and a mass attached at the other end, as in Fig. 8.9.

Solution: The equation governing the axial vibration of the bar is given by Eq. (8.49) and the solution by Eq. (8.51). The boundary condition at the fixed end ($x = 0$)

$$u(0, t) = 0 \tag{E.1}$$

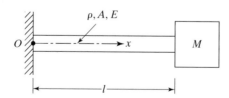

FIGURE 8.9   Bar carrying an end mass.

leads to $A = 0$ in Eq. (8.51). At the end $x = l$, the tensile force in the bar must be equal to the inertia force of the vibrating mass M, and so

$$AE\frac{\partial u}{\partial x}(l, t) = -M\frac{\partial^2 u}{\partial t^2}(l, t) \tag{E.2}$$

With the help of Eq. (8.51), this equation can be expressed as

$$AE\frac{\omega}{c}\cos\frac{\omega l}{c}(C\cos\omega t + D\sin\omega t) = M\omega^2\sin\frac{\omega l}{c}(C\cos\omega t + D\sin\omega t)$$

That is,

$$\frac{AE\omega}{c}\cos\frac{\omega l}{c} = M\omega^2\sin\frac{\omega l}{c}$$

or

$$\alpha\tan\alpha = \beta \tag{E.3}$$

where

$$\alpha = \frac{\omega l}{c} \tag{E.4}$$

and

$$\beta = \frac{AEl}{c^2 M} = \frac{A\rho l}{M} = \frac{m}{M} \tag{E.5}$$

where m is the mass of the bar. Equation (E.3) is the frequency equation (in the form of a transcendental equation) whose solution gives the natural frequencies of the system. The first two natural frequencies are given in Table 8.1 for different values of the parameter β.

Note: If the mass of the bar is negligible compared to the mass attached, $m \simeq 0$,

$$c = \left(\frac{E}{\rho}\right)^{1/2} = \left(\frac{EAl}{m}\right)^{1/2} \to \infty \qquad \text{and} \qquad \alpha = \frac{\omega l}{c} \to 0$$

In this case

$$\tan\frac{\omega l}{c} \simeq \frac{\omega l}{c}$$

and the frequency equation (E.3) can be taken as

$$\left(\frac{\omega l}{c}\right)^2 = \beta$$

TABLE 8.1

	Values of the Mass Ratio β				
	0.01	0.1	1.0	10.0	100.0
Value of $\alpha_1 \left(\omega_1 = \dfrac{\alpha_1 c}{l} \right)$	0.1000	0.3113	0.8602	1.4291	1.5549
Value of $\alpha_2 \left(\omega_2 = \dfrac{\alpha_2 c}{l} \right)$	3.1448	3.1736	3.4267	4.3063	4.6658

This gives the approximate value of the fundamental frequency

$$\omega_1 = \frac{c}{l} \beta^{1/2} = \frac{c}{l} \left(\frac{\rho A l}{M} \right)^{1/2} = \left(\frac{EA}{lM} \right)^{1/2} = \left(\frac{g}{\delta_s} \right)^{1/2}$$

where

$$\delta_s = \frac{Mgl}{EA}$$

represents the static elongation of the bar under the action of the load Mg.

■

EXAMPLE   8.5

Vibrations of a Bar Subjected to Initial Force

A bar of uniform cross-sectional area A, density ρ, modulus of elasticity E, and length l is fixed at one end and free at the other end. It is subjected to an axial force F_0 at its free end, as shown in Fig. 8.10(a). Study the resulting vibrations if the force F_0 is suddenly removed.

Solution: The tensile strain induced in the bar due to F_0 is

$$\varepsilon = \frac{F_0}{EA}$$

Thus the displacement of the bar just before the force F_0 is removed (initial displacement) is given by (see Fig. 8.10b)

$$u_0 = u(x, 0) = \varepsilon x = \frac{F_0 x}{EA}, \qquad 0 \le x \le l \tag{E.1}$$

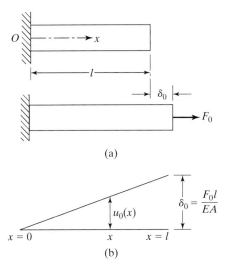

(a)

(b)

FIGURE 8.10 Bar subjected to an axial force at end.

Since the initial velocity is zero, we have

$$\dot{u}_0 = \frac{\partial u}{\partial t}(x, 0) = 0, \qquad 0 \leq x \leq l \tag{E.2}$$

The general solution of a bar fixed at one end and free at the other end is given by Eq. (E.5) of Example 8.3:

$$u(x, t) = \sum_{n=0}^{\infty} u_n(x, t)$$

$$= \sum_{n=0}^{\infty} \sin\frac{(2n + 1)\pi x}{2l} \left[C_n \cos\frac{(2n + 1)\pi ct}{2l} + D_n \sin\frac{(2n + 1)\pi ct}{2l} \right] \tag{E.3}$$

where C_n and D_n are given by Eqs. (E.6) and (E.7) of Example 8.3. Since $\dot{u}_0 = 0$, we obtain $D_n = 0$. By using the initial displacement of Eq. (E.1) in Eq. (E.6) of Example 8.3, we obtain

$$C_n = \frac{2}{l}\int_0^l \frac{F_0 x}{EA} \cdot \sin\frac{(2n + 1)\pi x}{2l} dx = \frac{8F_0 l}{EA\pi^2} \frac{(-1)^n}{(2n + 1)^2} \tag{E.4}$$

Thus the solution becomes

$$u(x, t) = \frac{8F_0 l}{EA\pi^2}\sum_{n=0}^{\infty} \frac{(-1)^n}{(2n + 1)^2} \sin\frac{(2n + 1)\pi x}{2l} \cos\frac{(2n + 1)\pi ct}{2l} \tag{E.5}$$

Equations (E.3) and (E.5) indicate that the motion of a typical point at $x = x_0$ on the bar is composed of the amplitudes

$$C_n \sin \frac{(2n + 1)\pi x_0}{2l}$$

corresponding to the circular frequencies

$$\frac{(2n + 1)\pi c}{2l}$$

∎

8.4 Torsional Vibration of a Shaft or Rod

Figure 8.11 represents a nonuniform shaft subjected to an external torque $f(x, t)$ per unit length. If $\theta(x, t)$ denotes the angle of twist of the cross section, the relation between the torsional deflection and the twisting moment $M_t(x, t)$ is given by [8.8]

$$M_t(x, t) = GJ(x) \frac{\partial \theta}{\partial x}(x, t) \tag{8.61}$$

where G is the shear modulus and $GJ(x)$ is the torsional stiffness, with $J(x)$ denoting the polar moment of inertia of the cross section in the case of a circular section. If the mass

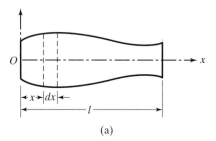

(a)

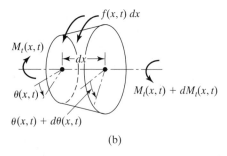

(b)

FIGURE 8.11   Torsional vibration of a shaft.

polar moment of inertia of the shaft per unit length is I_o, the inertia torque acting on an element of length dx becomes

$$I_o dx \frac{\partial^2 \theta}{\partial t^2}$$

If an external torque $f(x, t)$ acts on the shaft per unit length, the application of Newton's second law yields the equation of motion:

$$(M_t + dM_t) + f\, dx - M_t = I_o\, dx\, \frac{\partial^2 \theta}{\partial t^2} \tag{8.62}$$

By expressing dM_t as

$$\frac{\partial M_t}{\partial x}\, dx$$

and using Eq. (8.61), the forced torsional vibration equation for a nonuniform shaft can be obtained:

$$\frac{\partial}{\partial x}\left[GJ(x) \frac{\partial \theta}{\partial x}(x, t) \right] + f(x, t) = I_o(x) \frac{\partial^2 \theta}{\partial t^2}(x, t) \tag{8.63}$$

For a uniform shaft, Eq. (8.63) takes the form

$$GJ \frac{\partial^2 \theta}{\partial x^2}(x, t) + f(x, t) = I_o \frac{\partial^2 \theta}{\partial t^2}(x, t) \tag{8.64}$$

which, in the case of free vibration, reduces to

$$c^2 \frac{\partial^2 \theta}{\partial x^2}(x, t) = \frac{\partial^2 \theta}{\partial t^2}(x, t) \tag{8.65}$$

where

$$c = \sqrt{\frac{GJ}{I_o}} \tag{8.66}$$

Notice that Eqs. (8.63) to (8.66) are similar to the equations derived in the cases of transverse vibration of a string and longitudinal vibration of a bar. If the shaft has a uniform cross section, $I_0 = \rho J$. Hence Eq. (8.66) becomes

$$c = \sqrt{\frac{G}{\rho}} \tag{8.67}$$

If the shaft is given an angular displacement $\theta_0(x)$ and an angular velocity $\dot{\theta}_0(x)$ at $t = 0$, the initial conditions can be stated as

$$\theta(x, t = 0) = \theta_0(x)$$

End Conditions of Shaft	Bounty Conditions	Frequency Equation	Mode Shape (Normal Function)	Natural Frequencies
Fixed-free	$\theta(0, t) = 0$ $\dfrac{\partial \theta}{\partial x}(l, t) = 0$	$\cos \dfrac{\omega l}{c} = 0$	$\Theta(x) = C_n \sin \dfrac{(2n + 1)\,\pi x}{2l}$	$\omega_n = \dfrac{(2n + 1)\,\pi c}{2l};$ $n = 0, 1, 2, \ldots$
Free-free	$\dfrac{\partial \theta}{\partial x}(0, t) = 0$ $\dfrac{\partial \theta}{\partial x}(l, t) = 0$	$\sin \dfrac{\omega l}{c} = 0$	$\Theta(x) = C_n \cos \dfrac{n\pi x}{l}$	$\omega_n = \dfrac{n\pi c}{l};$ $n = 0, 1, 2, \ldots$
Fixed-fixed	$\theta(0, t) = 0$ $\theta(l, t) = 0$	$\sin \dfrac{\omega l}{c} = 0$	$\Theta(x) = C_n \cos \dfrac{n\pi x}{l}$	$\omega_n = \dfrac{n\pi c}{l};$ $n = 1, 2, 3, \ldots$

FIGURE 8.12   Boundary conditions for uniform shafts (rods) subjected to torsional vibration.

$$\frac{\partial \theta}{\partial t}(x, t = 0) = \dot{\theta}_0(x) \tag{8.68}$$

The general solution of Eq. (8.65) can be expressed as

$$\theta(x, t) = \left(A \cos \frac{\omega x}{c} + B \sin \frac{\omega x}{c} \right)(C \cos \omega t + D \sin \omega t) \tag{8.69}$$

The common boundary conditions for the torsional vibration of uniform shafts are indicated in Fig. 8.12 along with the corresponding frequency equations and the normal functions.

Natural Frequencies of a Milling Cutter

EXAMPLE 8.6

Find the natural frequencies of the plane milling cutter shown in Fig. 8.13 when the free end of the shank is fixed. Assume the torsional rigidity of the shank as GJ and the mass moment of inertia of the cutter as I_0.

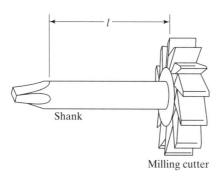

Milling cutter

FIGURE 8.13   Plane milling cutter.

Solution: The general solution is given by Eq. (8.61). By using the fixed boundary condition $\theta(0, t) = 0$, we obtain from Eq. (8.61), $A = 0$. The boundary condition at $x = l$ can be stated as

$$GJ \frac{\partial \theta}{\partial x} (l, t) = -I_0 \frac{\partial^2 \theta}{\partial t^2} (l, t) \tag{E.1}$$

That is,

$$BGJ \frac{\omega}{c} \cos \frac{\omega l}{c} = BI_0 \omega^2 \sin \frac{\omega l}{c}$$

or

$$\frac{\omega l}{c} \tan \frac{\omega l}{c} = \frac{J \rho l}{I_0} = \frac{J_{\text{rod}}}{I_0} \tag{E.2}$$

where $J_{\text{rod}} = J \rho l$. Equation (E.2) can be expressed as

$$\alpha \tan \alpha = \beta \quad \text{where} \quad \alpha = \frac{\omega l}{c} \quad \text{and} \quad \beta = \frac{J_{\text{rod}}}{I_0} \tag{E.3}$$

The solution of Eq. (E.3), and thus the natural frequencies of the system, can be obtained as in the case of Example 8.4.

∎

8.5   Lateral Vibration of Beams

8.5.1
Equation of
Motion

Consider the free-body diagram of an element of a beam shown in Fig. 8.14 where $M(x, t)$ is the bending moment, $V(x, t)$ is the shear force, and $f(x, t)$ is the external force per unit length of the beam. Since the inertia force acting on the element of the beam is

$$\rho A(x) \, dx \frac{\partial^2 w}{\partial t^2} (x, t)$$

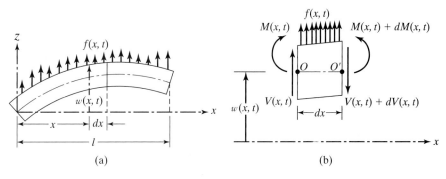

(a)   (b)

FIGURE 8.14   A beam in bending.

the force equation of motion in the z direction gives

$$-(V + dV) + f(x, t) \, dx + V = \rho A(x) \, dx \frac{\partial^2 w}{\partial t^2}(x, t) \qquad (8.70)$$

where ρ is the mass density and $A(x)$ is the cross-sectional area of the beam. The moment equation of motion about the y axis passing through point 0 in Fig. 8.14 leads to

$$(M + dM) - (V + dV) \, dx + f(x, t) \, dx \frac{dx}{2} - M = 0 \qquad (8.71)$$

By writing

$$dV = \frac{\partial V}{\partial x} \, dx \qquad \text{and} \qquad dM = \frac{\partial M}{\partial x} \, dx$$

and disregarding terms involving second powers in dx, Eqs. (8.70) and (8.71) can be written as

$$-\frac{\partial V}{\partial x}(x, t) + f(x, t) = \rho A(x) \frac{\partial^2 w}{\partial t^2}(x, t) \qquad (8.72)$$

$$\frac{\partial M}{\partial x}(x, t) - V(x, t) = 0 \qquad (8.73)$$

By using the relation $V = \partial M / \partial x$ from Eq. (8.73), Eq. (8.72) becomes

$$-\frac{\partial^2 M}{\partial x^2}(x, t) + f(x, t) = \rho A(x) \frac{\partial^2 w}{\partial t^2}(x, t) \qquad (8.74)$$

From the elementary theory of bending of beams (also known as the *Euler-Bernoulli* or *thin beam theory*), the relationship between bending moment and deflection can be expressed as [8.8]

$$M(x, t) = EI(x) \frac{\partial^2 w}{\partial x^2}(x, t) \qquad (8.75)$$

where E is Young's modulus and $I(x)$ is the moment of inertia of the beam cross section about the y axis. Inserting Eq. (8.75) into Eq. (8.74), we obtain the equation of motion for the forced lateral vibration of a nonuniform beam:

$$\frac{\partial^2}{\partial x^2} \left[EI(x) \frac{\partial^2 w}{\partial x^2}(x, t) \right] + \rho A(x) \frac{\partial^2 w}{\partial t^2}(x, t) = f(x, t) \qquad (8.76)$$

For a uniform beam, Eq. (8.76) reduces to

$$EI \frac{\partial^4 w}{\partial x^4}(x, t) + \rho A \frac{\partial^2 w}{\partial t^2}(x, t) = f(x, t) \qquad (8.77)$$

For free vibration, $f(x, t) = 0$, and so the equation of motion becomes

$$c^2 \frac{\partial^4 w}{\partial x^4}(x, t) + \frac{\partial^2 w}{\partial t^2}(x, t) = 0 \tag{8.78}$$

where

$$c = \sqrt{\frac{EI}{\rho A}} \tag{8.79}$$

**8.5.2
Initial
Conditions**

Since the equation of motion involves a second-order derivative with respect to time and a fourth-order derivative with respect to x, two initial conditions and four boundary conditions are needed for finding a unique solution for $w(x, t)$. Usually, the values of lateral displacement and velocity are specified as $w_0(x)$ and $\dot{w}_0(x)$ at $t = 0$, so that the initial conditions become

$$w(x, t = 0) = w_0(x)$$

$$\frac{\partial w}{\partial t}(x, t = 0) = w_0(x) \tag{8.80}$$

**8.5.3
Free Vibration**

The free vibration solution can be found using the method of separation of variables as

$$w(x, t) = W(x)T(t) \tag{8.81}$$

Substituting Eq. (8.81) into Eq. (8.78) and rearranging leads to

$$\frac{c^2}{W(x)} \frac{d^4 W(x)}{dx^4} = -\frac{1}{T(t)} \frac{d^2 T(t)}{dt^2} = a = \omega^2 \tag{8.82}$$

where $a = \omega^2$ is a positive constant (see Problem 8.43). Equation (8.82) can be written as two equations:

$$\frac{d^4 W(x)}{dx^4} - \beta^4 W(x) = 0 \tag{8.83}$$

$$\frac{d^2 T(t)}{dt^2} + \omega^2 T(t) = 0 \tag{8.84}$$

where

$$\beta^4 = \frac{\omega^2}{c^2} = \frac{\rho A \omega^2}{EI} \tag{8.85}$$

The solution of Eq. (8.84) can be expressed as

$$T(t) = A \cos \omega t + B \sin \omega t \tag{8.86}$$

where A and B are constants that can be found from the initial conditions. For the solution of Eq. (8.83), we assume

$$W(x) = Ce^{sx} \tag{8.87}$$

where C and s are constants, and derive the auxiliary equation as

$$s^4 - \beta^4 = 0 \tag{8.88}$$

The roots of this equation are

$$s_{1,2} = \pm\beta, \qquad s_{3,4} = \pm i\beta \tag{8.89}$$

Hence the solution of Eq. (8.83) becomes

$$W(x) = C_1 e^{\beta x} + C_2 e^{-\beta x} + C_3 e^{i\beta x} + C_4 e^{-i\beta x} \tag{8.90}$$

where C_1, C_2, C_3, and C_4 are constants. Equation (8.90) can also be expressed as

$$W(x) = C_1 \cos \beta x + C_2 \sin \beta x + C_3 \cosh \beta x + C_4 \sinh \beta x \tag{8.91}$$

or

$$\begin{aligned} W(x) = {} & C_1(\cos \beta x + \cosh \beta x) + C_2(\cos \beta x - \cosh \beta x) \\ & + C_3(\sin \beta x + \sinh \beta x) + C_4(\sin \beta x - \sinh \beta x) \end{aligned} \tag{8.92}$$

where C_1, C_2, C_3, and C_4, in each case, are different constants. The constants C_1, C_2, C_3, and C_4 can be found from the boundary conditions. The natural frequencies of the beam are computed from Eq. (8.85) as

$$\omega = \beta^2 \sqrt{\frac{EI}{\rho A}} = (\beta l)^2 \sqrt{\frac{EI}{\rho A l^4}} \tag{8.93}$$

The function $W(x)$ is known as the normal mode or characteristic function of the beam and ω is called the natural frequency of vibration. For any beam, there will be an infinite number of normal modes with one natural frequency associated with each normal mode. The unknown constants C_1 to C_4 in Eq. (8.91) or (8.92) and the value of β in Eq. (8.93) can be determined from the boundary conditions of the beam as indicated below.

8.5.4 Boundary Conditions

The common boundary conditions are as follows:

1. *Free end:*

$$\text{Bending moment} = EI\frac{\partial^2 w}{\partial x^2} = 0,$$

$$\text{Shear force} = \frac{\partial}{\partial x}\left(EI\frac{\partial^2 w}{\partial x^2}\right) = 0 \tag{8.94}$$

2. *Simply supported (pinned) end:*

$$\text{Deflection} = w = 0, \qquad \text{Bending moment} = EI\frac{\partial^2 w}{\partial x^2} = 0 \qquad (8.95)$$

3. *Fixed (clamped) end:*

$$\text{Deflection} = 0, \qquad \text{Slope} = \frac{\partial w}{\partial x} = 0 \qquad (8.96)$$

The frequency equations, the mode shapes (normal functions), and the natural frequencies for beams with common boundary conditions are given in Fig. 8.15 [8.13, 8.17]. We shall now consider some other possible boundary conditions for a beam.

End Conditions of Beam	Frequency Equation	Mode Shape (Normal Function)	Value of $\beta_n l$
Pinned-pinned 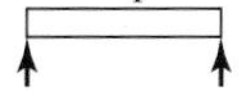	$\sin \beta_n l = 0$	$W_n(x) = C_n[\sin\beta_n x]$	$\beta_1 l = \pi$ $\beta_2 l = 2\pi$ $\beta_3 l = 3\pi$ $\beta_4 l = 4\pi$
Free-free 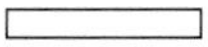	$\cos \beta_n l \cdot \cosh \beta_n l = 1$	$W_n(x) = C_n[\sin\beta_n x + \sinh\beta_n x$ $+ \alpha_n(\cos\beta_n x + \cosh\beta_n x)]$ where $\alpha_n = \left(\dfrac{\sin\beta_n l - \sinh\beta_n l}{\cosh\beta_n l - \cos\beta_n l}\right)$	$\beta_1 l = 4.730041$ $\beta_2 l = 7.853205$ $\beta_3 l = 10.995608$ $\beta_4 l = 14.137165$ ($\beta l = 0$ for rigid body mode)
Fixed-fixed 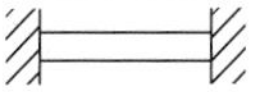	$\cos \beta_n l \cdot \cosh \beta_n l = 1$	$W_n(x) = C_n[\sinh\beta_n x - \sin\beta_n x$ $+ \alpha_n(\cosh\beta_n x - \cos\beta_n x)]$ where $\alpha_n = \left(\dfrac{\sinh\beta_n l - \sin\beta_n l}{\cos\beta_n l - \cosh\beta_n l}\right)$	$\beta_1 l = 4.730041$ $\beta_2 l = 7.853205$ $\beta_3 l = 10.995608$ $\beta_4 l = 14.137165$
Fixed-free 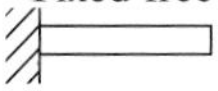	$\cos \beta_n l \cdot \cosh \beta_n l = -1$	$W_n(x) = C_n[\sin\beta_n x - \sinh\beta_n x$ $- \alpha_n(\cos\beta_n x - \cosh\beta_n x)]$ where $\alpha_n = \left(\dfrac{\sin\beta_n l + \sinh\beta_n l}{\cos\beta_n l + \cosh\beta_n l}\right)$	$\beta_1 l = 1.875104$ $\beta_2 l = 4.694091$ $\beta_3 l = 7.854757$ $\beta_4 l = 10.995541$
Fixed-pinned 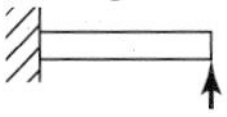	$\tan \beta_n l - \tanh \beta_n l = 0$	$W_n(x) = C_n[\sin\beta_n x - \sinh\beta_n x$ $+ \alpha_n(\cosh\beta_n x - \cos\beta_n x)]$ where $\alpha_n = \left(\dfrac{\sin\beta_n l - \sinh\beta_n l}{\cos\beta_n l - \cosh\beta_n l}\right)$	$\beta_1 l = 3.926602$ $\beta_2 l = 7.068583$ $\beta_3 l = 10.210176$ $\beta_4 l = 13.351768$
Pinned-free 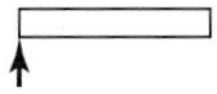	$\tan \beta_n l - \tanh \beta_n l = 0$	$W_n(x) = C_n[\sin\beta_n x + \alpha_n \sinh\beta_n x]$ where $\alpha_n = \left(\dfrac{\sin\beta_n l}{\sinh\beta_n l}\right)$	$\beta_1 l = 3.926602$ $\beta_2 l = 7.068583$ $\beta_3 l = 10.210176$ $\beta_4 l = 13.351768$ ($\beta l = 0$ for rigid body mode)

FIGURE 8.15 Common boundary conditions for the transverse vibration of a beam.

4. *End connected to a linear spring, damper, and mass* (Fig. 8.16a): When the end of a beam undergoes a transverse displacement w and slope $\partial w/\partial x$. with velocity $\partial w/\partial t$ and acceleration $\partial^2 w/\partial t^2$, the resisting forces due to the spring, damper, and mass are proportional to w, $\partial w/\partial t$, and $\partial^2 w/\partial t^2$, respectively. This resisting force is balanced by the shear force at the end. Thus

$$\frac{\partial}{\partial x}\left(EI\frac{\partial^2 w}{\partial x^2}\right) = a\left[kw + c\frac{\partial w}{\partial t} + m\frac{\partial^2 w}{\partial t^2}\right] \tag{8.97}$$

where $a = -1$ for left end and $+1$ for right end of the beam. In addition, the bending moment must be zero; hence

$$EI\frac{\partial^2 w}{\partial x^2} = 0 \tag{8.98}$$

5. *End connected to a torsional spring, torsional damper, and rotational inertia* (Fig. 8.16b): In this case, the boundary conditions are

$$EI\frac{\partial^2 w}{\partial x^2} = a\left[k_t\frac{\partial w}{\partial x} + c_t\frac{\partial^2 w}{\partial x\partial t} + I_0\frac{\partial^3 w}{\partial x\partial t^2}\right] \tag{8.99}$$

where $a = -1$ for left end and $+1$ for the right end of the beam, and

$$\frac{\partial}{\partial x}\left[EI\frac{\partial^2 w}{\partial x^2}\right] = 0 \tag{8.100}$$

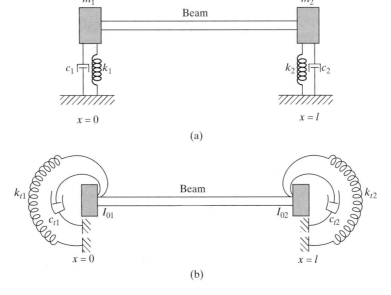

FIGURE 8.16   Beams connected with springs-dampers-masses at ends.

**8.5.5
Orthogonality of
Normal
Functions**

The normal functions $W(x)$ satisfy Eq. (8.83):

$$c^2 \frac{d^4 W}{dx^4}(x) - \omega^2 W(x) = 0 \qquad (8.101)$$

Let $W_i(x)$ and $W_j(x)$ be the normal functions corresponding to the natural frequencies ω_i and $\omega_j (i \neq j)$ so that

$$c^2 \frac{d^4 W_i}{dx^4} - \omega_i^2 W_i = 0 \qquad (8.102)$$

and

$$c^2 \frac{d^4 W_j}{dx^4} - \omega_j^2 W_j = 0 \qquad (8.103)$$

Multiplying Eq. (8.102) by W_j and Eq. (8.103) by W_i, subtracting the resulting equations one from the other, and integrating from 0 to l gives

$$\int_0^l \left[c^2 \frac{d^4 W_i}{dx^4} W_j - \omega_i^2 W_i W_j \right] dx - \int_0^l \left[c^2 \frac{d^4 W_j}{dx^4} W_i - \omega_j^2 W_j W_i \right] dx = 0$$

or

$$\int_0^l W_i W_j \, dx = -\frac{c^2}{\omega_i^2 - \omega_j^2} \int_0^l (W_i''''W_j - W_i W_j'''') \, dx \qquad (8.104)$$

where a prime indicates differentiation with respect to x. The right-hand side of Eq. (8.104) can be evaluated using integration by parts to obtain

$$\int_0^l W_i W_j \, dx = -\frac{c^2}{\omega_i^2 - \omega_j^2} [W_i W_j''' - W_j W_i''' + W_j' W_i'' - W_i' W_j'']\bigg|_0^l \qquad (8.105)$$

The right-hand side of Eq. (8.105) can be shown to be zero for any combination of free, fixed, or simply supported end conditions. At a free end, the bending moment and shear force are equal to zero so that

$$W'' = 0, \qquad W''' = 0 \qquad (8.106)$$

For a fixed end, the deflection and slope are zero:

$$W = 0, \qquad W' = 0 \qquad (8.107)$$

At a simply supported end, the bending moment and deflection are zero:

$$W'' = 0, \qquad W = 0 \qquad (8.108)$$

Since each term on the right-hand side of Eq. (8.105) is zero at $x = 0$ or $x = l$ for any combination of the boundary conditions in Eqs. (8.106) to (8.108), Eq. (8.105) reduces to

$$\int_0^l W_i W_j \, dx = 0 \tag{8.109}$$

which proves the orthogonality of normal functions for the transverse vibration of beams.

Natural Frequencies of a Fixed-Pinned Beam

EXAMPLE 8.7

Determine the natural frequencies of vibration of a uniform beam fixed at $x = 0$ and simply supported at $x = l$.

Solution: The boundary conditions can be stated as

$$W(0) = 0 \tag{E.1}$$

$$\frac{dW}{dx}(0) = 0 \tag{E.2}$$

$$W(l) = 0 \tag{E.3}$$

$$EI\frac{d^2W}{dx^2}(l) = 0 \quad \text{or} \quad \frac{d^2W}{dx^2}(l) = 0 \tag{E.4}$$

Condition (E.1) leads to

$$C_1 + C_3 = 0 \tag{E.5}$$

in Eq. (8.91), while Eqs. (E.2) and (8.91) give

$$\frac{dW}{dx}\bigg|_{x=0} = \beta[- C_1 \sin \beta x + C_2 \cos \beta x + C_3 \sinh \beta x + C_4 \cosh \beta x]_{x=0} = 0$$

or

$$\beta[C_2 + C_4] = 0 \tag{E.6}$$

Thus the solution, Eq. (8.91), becomes

$$W(x) = C_1(\cos \beta x - \cosh \beta x) + C_2(\sin \beta x - \sinh \beta x) \tag{E.7}$$

Applying conditions (E.3) and (E.4) to Eq. (E.7) yields

$$C_1(\cos \beta l - \cosh \beta l) + C_2(\sin \beta l - \sinh \beta l) = 0 \tag{E.8}$$

$$- C_1(\cos \beta l + \cosh \beta l) - C_2(\sin \beta l + \sinh \beta l) = 0 \tag{E.9}$$

For a nontrivial solution of C_1 and C_2, the determinant of their coefficients must be zero—that is,

$$\begin{vmatrix} (\cos \beta l - \cosh \beta l) & (\sin \beta l - \sinh \beta l) \\ -(\cos \beta l + \cosh \beta l) & -(\sin \beta l + \sinh \beta l) \end{vmatrix} = 0 \tag{E.10}$$

Expanding the determinant gives the frequency equation

$$\cos \beta l \sinh \beta l - \sin \beta l \cosh \beta l = 0$$

or

$$\tan \beta l = \tanh \beta l \tag{E.11}$$

The roots of this equation, $\beta_n l$, give the natural frequencies of vibration

$$\omega_n = (\beta_n l)^2 \left(\frac{EI}{\rho A l^4} \right)^{1/2}, \qquad n = 1, 2, \ldots \tag{E.12}$$

where the values of $\beta_n l, n = 1, 2, \ldots$ satisfying Eq. (E.11) are given in Fig. 8.15. If the value of C_2 corresponding to β_n is denoted as C_{2n}, it can be expressed in terms of C_{1n} from Eq. (E.8) as

$$C_{2n} = -C_{1n} \left(\frac{\cos \beta_n l - \cosh \beta_n l}{\sin \beta_n l - \sinh \beta_n l} \right) \tag{E.13}$$

Hence Eq. (E.7) can be written as

$$W_n(x) = C_{1n} \left[(\cos \beta_n x - \cosh \beta_n x) - \left(\frac{\cos \beta_n l - \cosh \beta_n l}{\sin \beta_n l - \sinh \beta_n l} \right) (\sin \beta_n x - \sinh \beta_n x) \right] \tag{E.14}$$

The normal modes of vibration can be obtained by the use of Eq. (8.81)

$$w_n(x, t) = W_n(x)(A_n \cos \omega_n t + B_n \sin \omega_n t) \tag{E.15}$$

with $W_n(x)$ given by Eq. (E.14). The general or total solution of the fixed-simply supported beam can be expressed by the sum of the normal modes:

$$w(x, t) = \sum_{n=1}^{\infty} w_n(x, t) \tag{E.16}$$

∎

8.5.6 Forced Vibration

The forced vibration solution of a beam can be determined using the mode superposition principle. For this, the deflection of the beam is assumed as

$$w(x, t) = \sum_{n=1}^{\infty} W_n(x) q_n(t) \tag{8.110}$$

where $W_n(x)$ is the nth normal mode or characteristic function satisfying the differential equation (Eq. 8.101)

$$EI\frac{d^4W_n(x)}{dx^4} - \omega_n^2\,\rho\,A\,W_n(x) = 0; \ n = 1, 2, \ldots \tag{8.111}$$

and $q_n(t)$ is the generalized coordinate in the nth mode. By substituting Eq. (8.110) into the forced vibration equation, Eq. (8.77), we obtain

$$EI\sum_{n=1}^{\infty}\frac{d^4W_n(x)}{dx^4}q_n(t) + \rho A\sum_{n=1}^{\infty}W_n(x)\frac{d^2q_n(t)}{dt^2} = f(x, t) \tag{8.112}$$

In view of Eq. (8.111), Eq. (8.112) can be written as

$$\sum_{n=1}^{\infty}\omega_n^2\,W_n(x)\,q_n(t) + \sum_{n=1}^{\infty}W_n(x)\frac{d^2q_n(t)}{dt^2} = \frac{1}{\rho A}f(x, t) \tag{8.113}$$

By multiplying Eq. (8.113) throughout by $W_m(x)$, integrating from 0 to l, and using the orthogonality condition, Eq. (8.109), we obtain

$$\frac{d^2q_n(t)}{dt^2} + \omega_n^2\,q_n(t) = \frac{1}{\rho Ab}Q_n(t) \tag{8.114}$$

where $Q_n(t)$ is called the generalized force corresponding to $q_n(t)$

$$Q_n(t) = \int_0^l f(x, t)\,W_n(x)\,dx \tag{8.115}$$

and the constant b is given by

$$b = \int_0^l W_n^2(x)\,dx \tag{8.116}$$

Equation (8.114) can be identified to be, essentially, the same as the equation of motion of an undamped single degree of freedom system. Using the Duhamel integral, the solution of Eq. (8.114) can be expressed as

$$q_n(t) = A_n \cos \omega_n t + B_n \sin \omega_n t$$
$$+ \frac{1}{\rho Ab\omega_n}\int_0^t Q_n(\tau) \sin \omega_n (t - \tau)\,d\tau \tag{8.117}$$

where the first two terms on the right-hand side of Eq. (8.117) represent the transient or free vibration (resulting from the initial conditions) and the third term denotes the steady-state vibration (resulting from the forcing function). Once Eq. (8.117) is solved for $n = 1, 2, \ldots$, the total solution can be determined from Eq. (8.110).

EXAMPLE 8.8

Forced Vibration of a Simply Supported Beam

Find the steady-state response of a pinned-pinned beam subject to a harmonic force $f(x, t) = f_0 \sin \omega t$ applied at $x = a$, as shown in Fig. 8.17.

Solution

Approach: Mode superposition method.

The normal mode functions of a pinned-pinned beam are given by (see Fig. 8.15; also Problem 8.31)

$$W_n(x) = \sin \beta_n x = \sin \frac{n\pi x}{l} \tag{E.1}$$

where

$$\beta_n l = n\pi \tag{E.2}$$

The generalized force $Q_n(t)$, given by Eq. (8.115), becomes

$$Q_n(t) = \int_0^l f(x, t) \sin \beta_n x \, dx = f_0 \sin \frac{n\pi a}{l} \sin \omega t \tag{E.3}$$

The steady-state response of the beam is given by Eq. (8.117)

$$q_n(t) = \frac{1}{\rho A b \omega_n} \int_0^t Q_n(\tau) \sin \omega_n (t - \tau) \, d\tau \tag{E.4}$$

where

$$b = \int_0^l W_n^2(x) \, dx = \int_0^l \sin^2 \beta_n x \, dx = \frac{l}{2} \tag{E.5}$$

The solution of Eq. (E.4) can be expressed as

$$q_n(t) = \frac{2 f_0}{\rho A l} \frac{\sin \frac{n\pi a}{l}}{\omega_n^2 - \omega^2} \sin \omega t \tag{E.6}$$

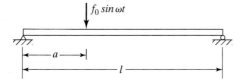

FIGURE 8.17 Pinned-pinned beam under harmonic force.

Thus the response of the beam is given by Eq. (8.110):

$$w(x, t) = \frac{2 f_0}{\rho \, Al} \sum_{n=1}^{\infty} \frac{1}{\omega_n^2 - \omega^2} \sin \frac{n \pi a}{l} \sin \frac{n \pi x}{l} \sin \omega t \qquad \text{(E.7)}$$

∎

8.5.7
Effect of Axial
Force

The problem of vibrations of a beam under the action of axial force finds application in the study of vibrations of cables and guy wires. For example, although the vibrations of a cable can be found by treating it as an equivalent string, many cables have failed due to fatigue caused by alternating flexure. The alternating flexure is produced by the regular shedding of vortices from the cable in a light wind. We must therefore consider the effects of axial force and bending stiffness on lateral vibrations in the study of fatigue failure of cables.

To find the effect of an axial force $P(x, t)$ on the bending vibrations of a beam, consider the equation of motion of an element of the beam, as shown in Fig. 8.18. For the vertical motion, we have

$$-(V + dV) + f \, dx + V + (P + dP) \sin(\theta + d\theta) - P \sin \theta = \rho A \, dx \frac{\partial^2 w}{\partial t^2}$$

$$\text{(8.118)}$$

and for the rotational motion about 0,

$$(M + dM) - (V + dV) \, dx + f \, dx \frac{dx}{2} - M = 0 \qquad \text{(8.119)}$$

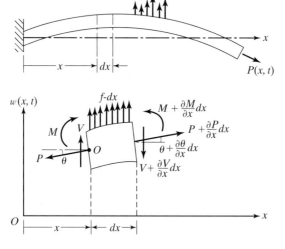

FIGURE 8.18   An element of a beam under axial load.

For small deflections,

$$\sin(\theta + d\theta) \simeq \theta + d\theta = \theta + \frac{\partial\theta}{\partial x}\,dx = \frac{\partial w}{\partial x} + \frac{\partial^2 w}{\partial x^2}\,dx$$

With this, Eqs. (8.118), (8.119), and (8.75) can be combined to obtain a single differential equation of motion:

$$\frac{\partial^2}{\partial x^2}\left[EI\frac{\partial^2 w}{\partial x^2}\right] + \rho A\frac{\partial^2 w}{\partial t^2} - P\frac{\partial^2 w}{\partial x^2} = f \tag{8.120}$$

For the free vibration of a uniform beam, Eq. (8.120) reduces to

$$EI\frac{\partial^4 w}{\partial x^4} + \rho A\frac{\partial^2 w}{\partial t^2} - P\frac{\partial^2 w}{\partial x^2} = 0 \tag{8.121}$$

The solution of Eq. (8.121) can be obtained using the method of separation of variables as

$$w(x, t) = W(x)(A\cos\omega t + B\sin\omega t) \tag{8.122}$$

Substitution of Eq. (8.122) into Eq. (8.121) gives

$$EI\frac{d^4 W}{dx^4} - P\frac{d^2 W}{dx^2} - \rho A\omega^2 W = 0 \tag{8.123}$$

By assuming the solution $W(x)$ to be

$$W(x) = Ce^{sx} \tag{8.124}$$

in Eq. (8.123), the auxiliary equation can be obtained:

$$s^4 - \frac{P}{EI}s^2 - \frac{\rho A\omega^2}{EI} = 0 \tag{8.125}$$

The roots of Eq. (8.125) are

$$s_1^2,\, s_2^2 = \frac{P}{2EI} \pm \left(\frac{P^2}{4E^2 I^2} + \frac{\rho A\omega^2}{EI}\right)^{1/2} \tag{8.126}$$

and so the solution can be expressed as (with absolute value of s_2)

$$W(x) = C_1\cosh s_1 x + C_2\sinh s_1 x + C_3\cos s_2 x + C_4\sin s_2 x \tag{8.127}$$

where the constants C_1 to C_4 are to be determined from the boundary conditions.

EXAMPLE 8.9

Beam Subjected to an Axial Compressive Force

Find the natural frequencies of a simply supported beam subjected to an axial compressive force.

Solution: The boundary conditions are

$$W(0) = 0 \tag{E.1}$$

$$\frac{d^2W}{dx^2}(0) = 0 \tag{E.2}$$

$$W(l) = 0 \tag{E.3}$$

$$\frac{d^2W}{dx^2}(l) = 0 \tag{E.4}$$

Equations (E.1) and (E.2) require that $C_1 = C_3 = 0$ in Eq. (8.127), and so

$$W(x) = C_2 \sinh s_1 x + C_4 \sin s_2 x \tag{E.5}$$

The application of Eqs. (E.3) and (E.4) to Eq. (E.5) leads to

$$\sinh s_1 l \cdot \sin s_2 l = 0 \tag{E.6}$$

Since $\sinh s_1 l > 0$ for all values of $s_1 l \neq 0$, the only roots to this equation are

$$s_2 l = n\pi, \qquad n = 0, 1, 2, \ldots \tag{E.7}$$

Thus Eqs. (E.7) and (8.126) give the natural frequencies of vibration:

$$\omega_n = \frac{\pi^2}{l^2}\sqrt{\frac{EI}{\rho A}}\left(n^4 + \frac{n^2 P l^2}{\pi^2 EI}\right)^{1/2} \tag{E.8}$$

Since the axial force P is compressive, P is negative. Further, from strength of materials, the smallest Euler buckling load for a simply supported beam is given by [8.9]

$$P_{\text{cri}} = \frac{\pi^2 EI}{l^2} \tag{E.9}$$

Thus Eq. (E.8) can be written as

$$\omega_n = \frac{\pi^2}{l^2}\left(\frac{EI}{\rho A}\right)^{1/2}\left(n^4 - n^2\frac{P}{P_{\text{cri}}}\right)^{1/2} \tag{E.10}$$

The following observations can be made from the present example:

1. If $P = 0$, the natural frequency will be same as that of a simply supported beam given in Fig. 8.15.
2. If $EI = 0$, the natural frequency (see Eq. E.8) reduces to that of a taut string.
3. If $P > 0$, the natural frequency increases as the tensile force stiffens the beam.
4. As $P \to P_{\text{cri}}$, the natural frequency approaches zero for $n = 1$.

■

8.5.8
Effects of Rotary Inertia and Shear Deformation

If the cross-sectional dimensions are not small compared to the length of the beam, we need to consider the effects of rotary inertia and shear deformation. The procedure, presented by Timoshenko [8.10], is known as the *thick beam theory* or *Timoshenko beam theory*. Consider the element of the beam shown in Fig. 8.19. If the effect of shear deformation is disregarded, the tangent to the deflected center line $O'T$ coincides with the normal to the face $Q'R'$ (since cross sections normal to the center line remain normal even after deformation). Due to shear deformation, the tangent to the deformed center line $O'T$ will not be perpendicular to the face $Q'R'$. The angle γ between the tangent to the deformed center line ($O'T$) and the normal to the face ($O'N$) denotes the shear deformation of the element. Since positive shear on the right face $Q'R'$ acts downward, we have, from Fig. 8.19,

$$\gamma = \phi - \frac{\partial w}{\partial x} \tag{8.128}$$

where ϕ denotes the slope of the deflection curve due to bending deformation alone. Note that because of shear alone, the element undergoes distortion but no rotation.

The bending moment M and the shear force V are related to ϕ and w by the formulas[2]

$$M = EI\frac{\partial \phi}{\partial x} \tag{8.129}$$

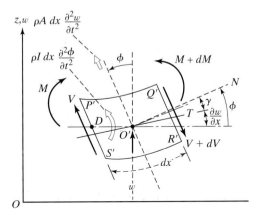

FIGURE 8.19   An element of Timoshenko beam.

---

[2]Equation (8.129) is similar to Eq. (8.75). Eq. (8.130) can be obtained as follows:

$$\text{Shear force} = \text{Shear stress} \times \text{Area} = \text{Shear strain} \times \text{Shear modulus} \times \text{Area}$$

or

$$V = \gamma G A$$

This equation is modified as $V = kAG\gamma$ by introducing a factor k on the right-hand side to take care of the shape of the cross section.

and

$$V = kAG\gamma = kAG\left(\phi - \frac{\partial w}{\partial x}\right) \tag{8.130}$$

where G denotes the modulus of rigidity of the material of the beam and k is a constant, also known as *Timoshenko's shear coefficient*, which depends on the shape of the cross section. For a rectangular section the value of k is 5/6; for a circular section it is 9/10 [8.11].

The equations of motion for the element shown in Fig. 8.19 can be derived as follows:

1. For translation in the z direction:

$$-[V(x, t) + dV(x, t)] + f(x, t)\, dx + V(x, t)$$

$$= \rho A(x)\, dx\, \frac{\partial^2 w}{\partial t^2}(x, t)$$

$$\equiv \text{Translational inertia of the element} \tag{8.131}$$

2. For rotation about a line passing through point D and parallel to the y axis:

$$[M(x, t) + dM(x, t)] + [V(x, t) + dV(x, t)]\, dx$$

$$+ f(x, t)\, dx\, \frac{dx}{2} - M(x, t)$$

$$= \rho I(x)\, dx\, \frac{\partial^2 \phi}{\partial t^2} \equiv \text{Rotary inertia of the element} \tag{8.132}$$

Using the relations

$$dV = \frac{\partial V}{\partial x}\, dx \qquad \text{and} \qquad dM = \frac{\partial M}{\partial x}\, dx$$

along with Eqs. (8.129) and (8.130) and disregarding terms involving second powers in dx, Eqs. (8.131) and (8.132) can be expressed as

$$-kAG\left(\frac{\partial \phi}{\partial x} - \frac{\partial^2 w}{\partial x^2}\right) + f(x, t) = \rho A \frac{\partial^2 w}{\partial t^2} \tag{8.133}$$

$$EI\frac{\partial^2 \phi}{\partial x^2} - kAG\left(\phi - \frac{\partial w}{\partial x}\right) = \rho I \frac{\partial^2 \phi}{\partial t^2} \tag{8.134}$$

By solving Eq. (8.133) for $\partial\phi/\partial x$ and substituting the result in Eq. (8.134), we obtain the desired equation of motion for the forced vibration of a uniform beam:

$$EI\frac{\partial^4 w}{\partial x^4} + \rho A\frac{\partial^2 w}{\partial t^2} - \rho I\left(1 + \frac{E}{kG}\right)\frac{\partial^4 w}{\partial x^2 \partial t^2} + \frac{\rho^2 I}{kG}\frac{\partial^4 w}{\partial t^4}$$

$$+ \frac{EI}{kAG}\frac{\partial^2 f}{\partial x^2} - \frac{\rho I}{kAG}\frac{\partial^2 f}{\partial t^2} - f = 0 \qquad (8.135)$$

For free vibration, $f = 0$, and Eq. (8.135) reduces to

$$EI\frac{\partial^4 w}{\partial x^4} + \rho A\frac{\partial^2 w}{\partial t^2} - \rho I\left(1 + \frac{E}{kG}\right)\frac{\partial^4 w}{\partial x^2 \partial t^2} + \frac{\rho^2 I}{kG}\frac{\partial^4 w}{\partial t^4} = 0 \qquad (8.136)$$

The following boundary conditions are to be applied in the solution of Eq. (8.135) or (8.136):

1. Fixed end:

$$\phi = w = 0$$

2. Simply supported end:

$$EI\frac{\partial\phi}{\partial x} = w = 0$$

3. Free end:

$$kAG\left(\frac{\partial w}{\partial x} - \phi\right) = EI\frac{\partial\phi}{\partial x} = 0$$

Natural Frequencies of a Simply Supported Beam

EXAMPLE 8.10

Determine the effects of rotary inertia and shear deformation on the natural frequencies of a simply supported uniform beam.

Solution: By defining

$$\alpha^2 = \frac{EI}{\rho A} \quad \text{and} \quad r^2 = \frac{I}{A} \qquad (E.1)$$

Eq. (8.136) can be written as

$$\alpha^2\frac{\partial^4 w}{\partial x^4} + \frac{\partial^2 w}{\partial t^2} - r^2\left(1 + \frac{E}{kG}\right)\frac{\partial^4 w}{\partial x^2 \partial t^2} + \frac{\rho r^2}{kG}\frac{\partial^4 w}{\partial t^4} = 0 \qquad (E.2)$$

We can express the solution of Eq. (E.2) as

$$w(x, t) = C \sin \frac{n\pi x}{l} \cos \omega_n t \tag{E.3}$$

which satisfies the necessary boundary conditions at $x = 0$ and $x = l$. Here, C is a constant and ω_n is the nth natural frequency. By substituting Eq. (E.3) into Eq. (E.2), we obtain the frequency equation:

$$\omega_n^4 \left(\frac{\rho r^2}{kG} \right) - \omega_n^2 \left(1 + \frac{n^2 \pi^2 r^2}{l^2} + \frac{n^2 \pi^2 r^2}{l^2} \frac{E}{kG} \right) + \left(\frac{\alpha^2 n^4 \pi^4}{l^4} \right) = 0 \tag{E.4}$$

It can be seen that Eq. (E.4) is a quadratic equation in ω_n^2 and for any given n; there are two values of ω_n that satisfy Eq. (E.4). The smaller value corresponds to the bending deformation mode, while the larger one corresponds to the shear deformation mode.

The values of the ratio of ω_n given by Eq. (E.4) to the natural frequency given by the classical theory (in Fig. 8.15) are plotted for three values of E/kG in Fig. 8.20 [8.22].[3]

Note the following aspects of rotary inertia and shear deformation:

1. If the effect of rotary inertia alone is considered, the resulting equation of motion does not contain any term involving the shear coefficient k. Hence we obtain (from Eq. 8.136):

$$EI \frac{\partial^4 w}{\partial x^4} + \rho A \frac{\partial^2 w}{\partial t^2} - \rho I \frac{\partial^4 w}{\partial x^2 \partial t^2} = 0 \tag{E.5}$$

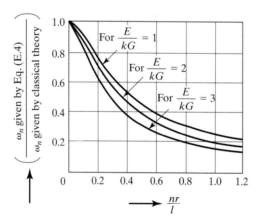

FIGURE 8.20   Variation of frequency.

---

[3]The theory used for the derivation of the equation of motion (8.76), which disregards the effects of rotary inertia and shear deformation, is called the *classical* or *Euler-Bernoulli* or *thin beam theory*.

In this case the frequency equation (E.4) reduces to

$$\omega_n^2 = \frac{\alpha^2 n^4 \pi^4}{l^4 \left(1 + \dfrac{n^2 \pi^2 r^2}{l^2} \right)} \tag{E.6}$$

2. If the effect of shear deformation alone is considered, the resulting equation of motion does not contain the terms originating from $\rho I (\partial^2 \phi / \partial t^2)$ in Eq. (8.134). Thus we obtain the equation of motion

$$EI \frac{\partial^4 w}{\partial x^4} + \rho A \frac{\partial^2 w}{\partial t^2} - \frac{EI\rho}{kG} \frac{\partial^4 w}{\partial x^2 \partial t^2} = 0 \tag{E.7}$$

and the corresponding frequency equation

$$\omega_n^2 = \frac{\alpha^2 n^4 \pi^4}{l^4 \left(1 + \dfrac{n^2 \pi^2 r^2}{l^2} \dfrac{E}{kG} \right)} \tag{E.8}$$

3. If both the effects of rotary inertia and shear deformation are disregarded, Eq. (8.136) reduces to the classical equation of motion, Eq. (8.78)

$$EI \frac{\partial^4 w}{\partial x^4} + \rho A \frac{\partial^2 w}{\partial t^2} = 0 \tag{E.9}$$

and Eq. (E.4) to

$$\omega_n^2 = \frac{\alpha^2 n^4 \pi^4}{l^4} \tag{E.10}$$

∎

8.5.9
Other Effects

The transverse vibration of tapered beams is presented in Refs. [8.12, 8.14]. The natural frequencies of continuous beams are discussed by Wang [8.15]. The dynamic response of beams resting on elastic foundation is considered in Ref. [8.16]. The effect of support flexibility on the natural frequencies of beams is presented in [8.18, 8.19]. A treatment of the problem of natural vibrations of a system of elastically connected Timoshenko beams is given in Ref. [8.20]. A comparison of the exact and approximate solutions of vibrating beams is made by Hutchinson [8.30]. The steady-state vibration of damped beams is considered in Ref. [8.21].

8.6   Vibration of Membranes

A membrane is a plate that is subjected to tension and has negligible bending resistance. Thus a membrane bears the same relationship to a plate as a string bears to a beam. A drumhead is an example of a membrane.

8.6.1
Equation of
Motion

To derive the equation of motion of a membrane, consider the membrane to be bounded by a plane curve S in the xy plane, as shown in Fig. 8.21. Let $f(x, y, t)$ denote the pressure loading acting in the z direction and P the intensity of tension at a point that is equal to the product of the tensile stress and the thickness of the membrane. The magnitude of P is usually constant throughout the membrane, as in a drumhead. If we consider an elemental area $dx\ dy$, forces of magnitude $P\ dx$ and $P\ dy$ act on the sides parallel to the y and x axes, respectively, as shown in Fig. 8.21. The net forces acting along the z direction due to these forces are

$$\left(P\,\frac{\partial^2 w}{\partial y^2}\,dx\,dy\right) \qquad \text{and} \qquad \left(P\,\frac{\partial^2 w}{\partial x^2}\,dx\,dy\right)$$

The pressure force along the z direction is $f(x, y, t)dx\,dy$, and the inertia force is

$$\rho(x, y)\frac{\partial^2 w}{\partial t^2}dx\,dy$$

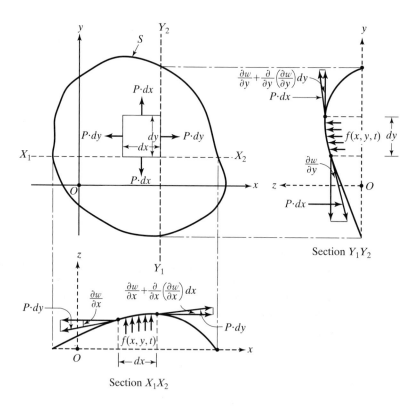

FIGURE 8.21   A membrane under uniform tension.

where $\rho(x, y)$ is the mass per unit area. The equation of motion for the forced transverse vibration of the membrane can be obtained

$$P\left(\frac{\partial^2 w}{\partial x^2} + \frac{\partial^2 w}{\partial y^2}\right) + f = \rho \frac{\partial^2 w}{\partial t^2} \tag{8.137}$$

If the external force $f(x, y, t) = 0$, Eq. (8.137) gives the free vibration equation

$$c^2\left(\frac{\partial^2 w}{\partial x^2} + \frac{\partial^2 w}{\partial y^2}\right) = \frac{\partial^2 w}{\partial t^2} \tag{8.138}$$

where

$$c = \left(\frac{P}{\rho}\right)^{1/2} \tag{8.139}$$

Equations (8.137) and (8.138) can be expressed as

$$P\nabla^2 w + f = \rho \frac{\partial^2 w}{\partial t^2} \tag{8.140}$$

and

$$c^2\nabla^2 w = \frac{\partial^2 w}{\partial t^2} \tag{8.141}$$

where

$$\nabla^2 = \frac{\partial^2}{\partial x^2} + \frac{\partial^2}{\partial y^2} \tag{8.142}$$

is the Laplacian operator.

8.6.2 Initial and Boundary Conditions

Since the equation of motion, Eq. (8.137) or (8.138), involves second-order partial derivatives with respect to each of t, x, and y, we need to specify two initial conditions and four boundary conditions to find a unique solution of the problem. Usually, the displacement and velocity of the membrane at $t = 0$ are specified as $w_0(x, y)$ and $\dot{w}_0(x, y)$. Hence the initial conditions are given by

$$w(x, y, 0) = w_0(x, y)$$

$$\frac{\partial w}{\partial t}(x, y, 0) = \dot{w}_0(x, y) \tag{8.143}$$

The boundary conditions are of the following types:

1. If the membrane is fixed at any point (x_1, y_1) on a segment of the boundary, we have

$$w(x_1, y_1, t) = 0, \qquad t \geq 0 \qquad\qquad (8.144)$$

2. If the membrane is free to deflect transversely (in the z direction) at a different point (x_2, y_2) of the boundary, then the force component in the z direction must be zero. Thus

$$P \frac{\partial w}{\partial n}(x_2, y_2, t) = 0, \qquad t \geq 0 \qquad\qquad (8.145)$$

where $\partial w/\partial n$ represents the derivative of w with respect to a direction n normal to the boundary at the point (x_2, y_2).

The solution of the equation of motion of the vibrating membrane was presented in Refs. [8.23–8.25].

EXAMPLE 8.11

■■■■■■ **Free Vibrations of a Rectangular Membrane**

Find the free vibration solution of a rectangular membrane of sides a and b along the x and y axes, respectively.

Solution: By using the method of separation of variables, $w(x, y, t)$ can be assumed to be

$$w(x, y, t) = W(x, y)\, T(t) = X(x)\, Y(y)\, T(t) \qquad\qquad (E.1)$$

By using Eqs. (E.1) and (8.138), we obtain

$$\frac{d^2 X(x)}{dx^2} + \alpha^2 X(x) = 0 \qquad\qquad (E.2)$$

$$\frac{d^2 Y(y)}{dy^2} + \beta^2 Y(y) = 0 \qquad\qquad (E.3)$$

$$\frac{d^2 T(t)}{dt^2} + \omega^2 T(t) = 0 \qquad\qquad (E.4)$$

where α^2 and β^2 are constants related to ω^2 as follows:

$$\beta^2 = \frac{\omega^2}{c^2} - \alpha^2 \qquad\qquad (E.5)$$

The solutions of Eqs. (E.2) to (E.4) are given by

$$X(x) = C_1 \cos \alpha x + C_2 \sin \alpha x \qquad\qquad (E.6)$$

$$Y(y) = C_3 \cos \beta y + C_4 \sin \beta y \qquad\qquad (E.7)$$

$$T(t) = A \cos \omega t + B \sin \omega t \qquad\qquad (E.8)$$

where the constants C_1 to C_4, A, and B can be determined from the boundary and initial conditions.

■

8.7 Rayleigh's Method

Rayleigh's method can be applied to find the fundamental natural frequency of continuous systems. This method is much simpler than exact analysis for systems with varying distributions of mass and stiffness. Although the method is applicable to all continuous systems, we shall apply it only to beams in this section.[4] Consider the beam shown in Fig. 8.14. In order to apply Rayleigh's method, we need to derive expressions for the maximum kinetic and potential energies and Rayleigh's quotient. The kinetic energy of the beam can be expressed as

$$T = \frac{1}{2} \int_0^l \dot{w}^2 \, dm = \frac{1}{2} \int_0^l \dot{w}^2 \rho A(x) \, dx \qquad\qquad (8.146)$$

The maximum kinetic energy can be found by assuming a harmonic variation $w(x, t) = W(x) \cos \omega t$:

$$T_{\max} = \frac{\omega^2}{2} \int_0^l \rho A(x) \, W^2(x) \, dx \qquad\qquad (8.147)$$

The potential energy of the beam V is the same as the work done in deforming the beam. By disregarding the work done by the shear forces, we have

$$V = \frac{1}{2} \int_0^l M \, d\theta \qquad\qquad (8.148)$$

where M is the bending moment given by Eq. (8.75) and θ is the slope of the deformed beam given by $\theta = \dfrac{\partial w}{\partial x}$. Thus Eq. (8.148) can be rewritten as

$$V = \frac{1}{2} \int_0^l \left(EI \frac{\partial^2 w}{\partial x^2} \right) \frac{\partial^2 w}{\partial x^2} \, dx = \frac{1}{2} \int_0^l EI \left(\frac{\partial^2 w}{\partial x^2} \right)^2 dx \qquad\qquad (8.149)$$

Since the maximum value of $w(x, t)$ is $W(x)$, the maximum value of V is given by

$$V_{\max} = \frac{1}{2} \int_0^l EI(x) \left(\frac{d^2 W(x)}{dx^2} \right)^2 dx \qquad\qquad (8.150)$$

---

[4]An integral equation approach for the determination of the fundamental frequency of vibrating beams is presented by Penny and Reed [8.26].

By equating T_{max} to V_{max}, we obtain Rayleigh's quotient:

$$R(\omega) = \omega^2 = \frac{\int_0^l EI\left(\frac{d^2W(x)}{dx^2}\right)^2 dx}{\int_0^l \rho A(W(x))^2\, dx} \tag{8.151}$$

Thus the natural frequency of the beam can be found once the deflection $W(x)$ is known. In general, $W(x)$ is not known and must therefore be assumed. Generally, the static equilibrium shape is assumed for $W(x)$ to obtain the fundamental frequency. It is to be noted that the assumed shape $W(x)$ unintentionally introduces a constraint on the system (which amounts to adding additional stiffness to the system), and so the frequency given by Eq. (8.151) is higher than the exact value [8.27].

For a stepped beam, Eq. (8.151) can be more conveniently written as

$$R(\omega) = \omega^2$$

$$= \frac{E_1 I_1 \int_0^{l_1}\left(\frac{d^2W}{dx^2}\right)^2 dx + E_2 I_2 \int_{l_1}^{l_2}\left(\frac{d^2W}{dx^2}\right)^2 dx + \cdots}{\rho A_1 \int_0^{l_1} W^2\, dx + \rho A_2 \int_{l_1}^{l_2} W^2\, dx + \cdots} \tag{8.152}$$

where E_i, I_i, A_i, and l_i correspond to the ith step ($i = 1, 2, \ldots$).

Fundamental Frequency of a Tapered Beam

EXAMPLE 8.12

Find the fundamental frequency of transverse vibration of the nonuniform cantilever beam shown in Fig. 8.22, using the deflection shape $W(x) = (1 - x/l)^2$.

Solution: The given deflection shape can be verified to satisfy the boundary conditions of the beam. The cross-sectional area A and the moment of inertia I of the beam can be expressed as

$$A(x) = \frac{hx}{l}, \qquad I(x) = \frac{1}{12}\left(\frac{hx}{l}\right)^3 \tag{E.1}$$

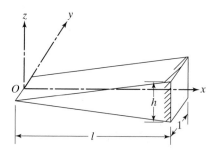

FIGURE 8.22   Tapered cantilever beam.

Thus Rayleigh's quotient gives

$$\omega^2 = \frac{\int_0^l E\left(\dfrac{h^3 x^3}{12 l^3}\right)\left(\dfrac{2}{l^2}\right)^2 dx}{\int_0^l \rho\left(\dfrac{hx}{l}\right)\left(1 - \dfrac{x}{l}\right)^4 dx} = 2.5 \frac{Eh^2}{\rho l^4}$$

or

$$\omega = 1.5811 \left(\frac{Eh^2}{\rho l^4}\right)^{1/2} \tag{E.2}$$

The exact value of the frequency, for this case [8.2], is known to be

$$\omega_1 = 1.5343 \left(\frac{Eh^2}{\rho l^4}\right)^{1/2} \tag{E.3}$$

Thus the value of ω_1 given by Rayleigh's method can be seen to be 3.0503 percent higher than the exact value.

∎

8.8   The Rayleigh-Ritz Method

The Rayleigh-Ritz method can be considered an extension of Rayleigh's method It is based on the premise that a closer approximation to the exact natural mode can be obtained by superposing a number of assumed functions than by using a single assumed function, as in Rayleigh's method. If the assumed functions are suitably chosen, this method provides not only the approximate value of the fundamental frequency but also the approximate values of the higher natural frequencies and the mode shapes. An arbitrary number of functions can be used, and the number of frequencies that can be obtained is equal to the number of functions used. A large number of functions, although it involves more computational work, leads to more accurate results.

In the case of transverse vibration of beams, if n functions are chosen for approximating the deflection $W(x)$, we can write

$$W(x) = c_1 w_1(x) + c_2 w_2(x) + \cdots + c_n w_n(x) \tag{8.153}$$

where $w_1(x), w_2(x), \ldots, w_n(x)$ are known linearly independent functions of the spatial coordinate x, which satisfy all the boundary conditions of the problem, and $c_1, c_2, \ldots, c_n$ are coefficients to be found. The coefficients c_i are to be determined so that the assumed functions $w_i(x)$ provide the best possible approximation to the natural modes. To obtain such approximations, the coefficients c_i are adjusted and the natural frequency is made stationary at the natural modes. For this we substitute Eq. (8.153) in Rayleigh's quotient, Eq. (8.151), and the resulting expression is partially differentiated with respect to each of the coefficients

c_i. To make the natural frequency stationary, we set each of the partial derivatives equal to zero and obtain

$$\frac{\partial(\omega^2)}{\partial c_i} = 0, \qquad i = 1, 2, \ldots, n \tag{8.154}$$

Equation (8.154) denotes a set of n linear algebraic equations in the coefficients $c_1, c_2, \ldots, c_n$ and also contains the undertermined quantity ω^2. This defines an algebraic eigenvalue problem similar to the ones that arise in multidegree of freedom systems. The solution of this eigenvalue problem generally gives n natural frequencies $\omega_i^2, i = 1, 2, \ldots, n$ and n eigenvectors, each containing a set of numbers for $c_1, c_2, \ldots, c_n$. For example, the ith eigenvector corresponding to ω_i may be expressed as

$$\vec{C}^{(i)} = \begin{Bmatrix} c_1^{(i)} \\ c_2^{(i)} \\ \vdots \\ c_n^{(i)} \end{Bmatrix} \tag{8.155}$$

When this eigenvector—the values of $c_1^{(i)}, c_2^{(i)}, \ldots, c_n^{(i)}$—is substituted into Eq. (8.153), we obtain the best possible approximation to the ith mode of the beam. A method of reducing the size of the eigenproblem in the Rayleigh-Ritz method is presented in Ref. [8.28]. A new approach, which combines the advantages of the Rayleigh-Ritz analysis and the finite element method is given in Ref. [8.29]. The basic Rayleigh-Ritz procedure is illustrated with the help of the following example.

▬▬▬▬▬▬▬ **First Two Frequencies of a Tapered Beam**

EXAMPLE 8.13
──────────────
Find the natural frequencies of the tapered cantilever beam of Example 8.12 by using the Rayleigh-Ritz method.

Solution: We assume the deflection functions $w_i(x)$ to be

$$w_1(x) = \left(1 - \frac{x}{l}\right)^2 \tag{E.1}$$

$$w_2(x) = \frac{x}{l}\left(1 - \frac{x}{l}\right)^2 \tag{E.2}$$

$$w_3(x) = \frac{x^2}{l^2}\left(1 - \frac{x}{l}\right)^2 \tag{E.3}$$

If we use the one-term approximation

$$W(x) = c_1 \left(1 - \frac{x}{l} \right)^2 \tag{E.4}$$

the fundamental frequency will be the same as the one found in Example 8.12. Now we use the two-term approximation

$$W(x) = c_1 \left(1 - \frac{x}{l} \right)^2 + c_2 \frac{x}{l} \left(1 - \frac{x}{l} \right)^2 \tag{E.5}$$

Rayleigh's quotient is given by

$$R[W(x)] = \omega^2 = \frac{X}{Y} \tag{E.6}$$

where

$$X = \int_0^l EI(x) \left(\frac{d^2 W(x)}{dx^2} \right)^2 dx \tag{E.7}$$

and

$$Y = \int_0^l \rho A(x)[W(x)]^2 \, dx \tag{E.8}$$

If Eq. (E.5) is substituted, Eq. (E.6) becomes a function of c_1 and c_2. The conditions that make ω^2 or $R[W(x)]$ stationary are

$$\frac{\partial(\omega^2)}{\partial c_1} = \frac{Y \dfrac{\partial X}{\partial c_1} - X \dfrac{\partial Y}{\partial c_1}}{Y^2} = 0 \tag{E.9}$$

$$\frac{\partial(\omega^2)}{\partial c_2} = \frac{Y \dfrac{\partial X}{\partial c_2} - X \dfrac{\partial Y}{\partial c_2}}{Y^2} = 0 \tag{E.10}$$

These equations can be rewritten as

$$\frac{\partial X}{\partial c_1} - \frac{X}{Y} \frac{\partial Y}{\partial c_1} = \frac{\partial X}{\partial c_1} - \omega^2 \frac{\partial Y}{\partial c_1} = 0 \tag{E.11}$$

$$\frac{\partial X}{\partial c_2} - \frac{X}{Y} \frac{\partial Y}{\partial c_2} = \frac{\partial X}{\partial c_2} - \omega^2 \frac{\partial Y}{\partial c_2} = 0 \tag{E.12}$$

By substituting Eq. (E.5) into Eqs. (E.7) and (E.8), we obtain

$$X = \frac{Eh^3}{3l^3}\left(\frac{c_1^2}{4} + \frac{c_2^2}{10} + \frac{c_1 c_2}{5}\right) \tag{E.13}$$

$$Y = \rho h l \left(\frac{c_1^2}{30} + \frac{c_2^2}{280} + \frac{2c_1 c_2}{105}\right) \tag{E.14}$$

With the help of Eqs. (E.13) and (E.14), Eqs. (E.11) and (E.12) can be expressed as

$$\begin{bmatrix} \left(\dfrac{1}{2} - \tilde{\omega}^2 \cdot \dfrac{1}{15}\right) & \left(\dfrac{1}{5} - \tilde{\omega}^2 \cdot \dfrac{2}{105}\right) \\ \left(\dfrac{1}{5} - \tilde{\omega}^2 \cdot \dfrac{2}{105}\right) & \left(\dfrac{1}{5} - \tilde{\omega}^2 \cdot \dfrac{1}{140}\right) \end{bmatrix} \begin{Bmatrix} c_1 \\ c_2 \end{Bmatrix} = \begin{Bmatrix} 0 \\ 0 \end{Bmatrix} \tag{E.15}$$

where

$$\tilde{\omega}^2 = \frac{3\omega^2 \rho l^4}{Eh^2} \tag{E.16}$$

By setting the determinant of the matrix in Eq. (E.15) equal to zero, we obtain the frequency equation

$$\frac{1}{8820}\tilde{\omega}^4 - \frac{13}{1400}\tilde{\omega}^2 + \frac{3}{50} = 0 \tag{E.17}$$

The roots of Eq. (E.17) are given by $\tilde{\omega}_1 = 2.6599$ and $\tilde{\omega}_2 = 8.6492$. Thus the natural frequencies of the tapered beam are

$$\omega_1 \simeq 1.5367 \left(\frac{Eh^2}{\rho l^4}\right)^{1/2} \tag{E.18}$$

and

$$\omega_2 \simeq 4.9936 \left(\frac{Eh^2}{\rho l^4}\right)^{1/2} \tag{E.19}$$

∎

8.9   Examples Using MATLAB

Plotting the Forced Vibration Response of a Simply Supported Beam

EXAMPLE 8.14

Using MATLAB, plot the steady-state response of the pinned-pinned beam considered in Example 8.8, Eq. (E.7), for $n = 1, 2,$ and 5.

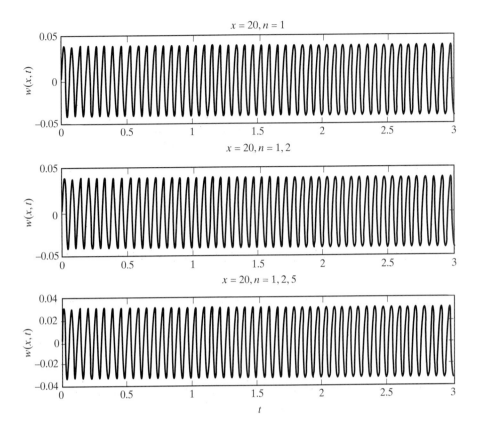

Solution: The MATLAB program to plot Eq. (E.7) of Example 8.8 at $x = 20$ with $n = 1$, 2, and 5 is given below.

```
%Ex8_14.m
x = 20;
f0 = 100;
a = 10;
A = 1;
1 = 40;
ro = 0.283/386.4;
w = 100;
n = 1;
wn = (n^2) * 360.393674;
for i = 1: 1001
    t(i) = 3 * (i-1)/1000;
    w1(i) = ( 2*f0/(ro*A*1) )*sin (n*pi*a/1)*sin (n*pi*x/1)*sin
        (w*t(i))/(wn^2-w^2);
end
n = 2;
for i = 1: 1001
    t(i) = 3 * (i-1)/1000;
    w2(i) = ( 2*f0/(ro*A*1) )*( sin (pi*a/1)*sin (pi*x/1)*sin
        (w*t(i))/(360.393674^2-w^2)+sin (2*pi*a/1)*sin
        (2*pi*x/1)*sin (w*t(i))/((2*360.393674)^2-w^2) );
end
```

```
for i = 1: 1001
    t(i) = 3 * (i-1)/1000;
    w3(i) = ( 2*f0/(ro*A*1) )*( sin (pi*a/1)*sin (pi*x/1)*sin
    (w*t(i))/(360.393674^2-w^2)+sin (2*pi*a/1)*sin
    (2*pi*x/1)*sin (w*t(i))/((2*360.393674)^2-w^2)+sin
    (5*pi*a/1)*sin (5*pi*x/1)*sin (w*t(i))/
    ((5*360.393674)^2-w^2) );
end
subplot ('311');
plot(t,w1);
ylabel('w(x,t)');
title('x = 20, n = 1');
subplot ('312');
plot(t,w2);
ylabel('w(x,t)');
title('x = 20, n = 1, 2');
subplot('313');
plot(t,w3);
xlabel('t');
ylabel('w(x,t)');
title('x = 20, n = 1, 2, 5');
```

∎

Solution of a Frequency Equation

EXAMPLE 8.15

Using MATLAB, find the root(s) of the frequency equation corresponding to the fixed-pinned beam

$$\tan \beta_n l - \tanh \beta_n l = 0$$

with the starting value of $\beta_n l = 3.0$.

Solution

```
>> x = fzero(inline('tan(y)-tanh(y)'), 3.0)

x =

    3.92660231204792

>> tan (x) - tanh (x)

ans =

    -4.440892098500626e-016
```

∎

Program to Find the Roots of Transcendental and Nonlinear Equations

EXAMPLE 8.16

Develop a general MATLAB program, called **Program12.m**, for finding the roots of nonlinear and transcendental equations. Use the program to find the root of the equation

$$\tan \beta l - \tanh \beta l = 0 \qquad (E.1)$$

Solution: **Program12.m** is developed to accept the following input data:

n = number of roots to be determined

xs = initial guess for the first root

xinc = initial increment to be used in searching for the root

nint = maximum number of subintervals to be used (usual value: 50)

iter = maximum number of iterations permitted in finding a root (usual value: 100)

eps = convergence requirement (usual value: 10^{-6})

The given nonlinear equation is to be defined in a subprogram called function.m. The program gives the computed roots as output.

```
>> programs12
Roots of nonlinear equation

Data:
n    =   5
xs   =   2.000000e+000
xinc =   1.000000e-001
nint =   50
iter =   100
eps  =   1.000000e-006

Roots
   3.926602e+000
   7.068583e+000
   1.021018e+001
   1.335177e+001
   1.649336e+001
```

◾

8.10   C++ Program

Program to Solve Nonlinear Equations

EXAMPLE 8.17

Develop a general C++ computer program, called **Program12.cpp**, to find the roots of nonlinear equations. Use the program to find the roots of Eq. (E.1) of Example 8.16.

Solution: **Program12.cpp** is developed to accept the input data interactively. The input and outputs of the program are similar to those of **Program12.m**.

```
Please input N:
5
Please input XS and XINC:
2.0   0.1

ROOTS OF NONLINEAR EQUATION

DATA:
N = 5
XS = 2
XINC = 0.1
NINT = 50
ITER = 100
EPS = 1e-006

ROOTS:
      3.92660231
      7.06858276
```

```
10.21017612
13.35176878
16.49336143
```

■

8.11   Fortran Program

■■■■■■■■■■   **Program to solve nonlinear equations**

EXAMPLE 8.18 ───

Develop a Fortran subroutine called **NONEQN.F** to find the roots of nonlinear and transcendental equations. Use the program to find the roots of Eq. (E.1) of Example 8.16.

Solution: The main program that calls the subroutine **NONEQN.F** is labeled **PROGRAM12.F**. The input and outputs of the program are similar to those of **Program12.m**. The output of **PROGRAM12.F** is given below.

```
ROOTS OF NONLINEAR EQUATION

DATA:
N     =     5
XS    =     0.200000E+01
XINC  =     0.100000E+00
NINT  =    50
ITER  =   100
EPS   =     0.100000E-05

ROOTS:
   0.392660E+01
   0.706858E+01
   0.102102E+02
   0.133518E+02
   0.164934E+02
```

■

REFERENCES

8.1   S. K. Clark, *Dynamics of Continuous Elements*, Prentice Hall, Englewood Cliffs, N.J., 1972.

8.2   S. Timoshenko, D. H. Young, and W. Weaver, Jr., *Vibration Problems in Engineering* (4th ed.), Wiley, New York, 1974.

8.3   A. Leissa, *Vibration of Plates*, NASA SP-160, Washington, D.C., 1969.

8.4   I. S. Habib, *Engineering Analysis Methods*, Lexington Books, Lexington, Mass., 1975.

8.5   J. D. Achenbach, *Wave Propagation in Elastic Solids*, North-Holland Publishing, Amsterdam, 1973.

8.6   K. K. Deb, "Dynamics of a string and an elastic hammer," *Journal of Sound and Vibration*, Vol. 40, 1975, pp. 243–248.

8.7   M. S. Triantafyllou, "Linear dynamics of cables and chains," *Shock and Vibration Digest*, Vol. 16, March 1984, pp. 9–17.

8.8 R. W. Fitzgerald, *Mechanics of Materials* (2nd ed.), Addison-Wesley, Reading, Mass., 1982.

8.9 S. P. Timoshenko and J. Gere, *Theory of Elastic Stability* (2nd ed.), McGraw-Hill, New York, 1961.

8.10 S. P. Timoshenko, "On the correction for shear of the differential equation for transverse vibration of prismatic bars," *Philosophical Magazine*, Series 6, Vol. 41, 1921, pp. 744–746.

8.11 G. R. Cowper, "The shear coefficient in Timoshenko's beam theory," *Journal of Applied Mechanics*, Vol. 33, 1966, pp. 335–340.

8.12 G. W. Housner and W. O. Keightley, "Vibrations of linearly tapered beams," Part I, *Transactions of ASCE*, Vol. 128, 1963, pp. 1020–1048.

8.13 C. M. Harris (ed.), *Shock and Vibration Handbook* (3rd ed.), McGraw-Hill, New York, 1988.

8.14 J. H. Gaines and E. Volterra, "Transverse vibrations of cantilever bars of variable cross section," *Journal of the Acoustical Society of America*, Vol. 39, 1966, pp. 674–679.

8.15 T. M. Wang, "Natural frequencies of continuous Timoshenko beams," *Journal of Sound and Vibration*, Vol. 13, 1970, pp. 409–414.

8.16 S. L. Grassie, R. W. Gregory, D. Harrison, and K. L. Johnson, "The dynamic response of railway track to high frequency vertical excitation," *Journal of Mechanical Engineering Science*, Vol. 24, June 1982, pp. 77–90.

8.17 A. Dimarogonas, *Vibration Engineering*, West Publishing, Saint Paul, 1976.

8.18 T. Justine and A. Krishnan, "Effect of support flexibility on fundamental frequency of beams," *Journal of Sound and Vibration*, Vol. 68, 1980, pp. 310–312.

8.19 K. A. R. Perkins, "The effect of support flexibility on the natural frequencies of a uniform cantilever," *Journal of Sound and Vibration*, Vol. 4, 1966, pp. 1–8.

8.20 S. S. Rao, "Natural frequencies of systems of elastically connected Timoshenko beams," *Journal of the Acoustical Society of America*, Vol. 55, 1974, pp. 1232–1237.

8.21 A. M. Ebner and D. P. Billington, "Steady-state vibration of damped Timoshenko beams," *Journal of Structural Division* (ASCE), Vol. 3, 1968, p. 737.

8.22 M. Levinson and D. W. Cooke, "On the frequency spectra of Timoshenko beams," *Journal of Sound and Vibration*, Vol. 84, 1982, pp. 319–326.

8.23 N. Y. Olcer, "General solution to the equation of the vibrating membrane," *Journal of Sound and Vibration*, Vol. 6, 1967, pp. 365–374.

8.24 G. R. Sharp, "Finite transform solution of the vibrating annular membrane," *Journal of Sound and Vibration*, Vol. 6, 1967, pp. 117–128.

8.25 J. Mazumdar, "A review of approximate methods for determining the vibrational modes of membranes," *Shock and Vibration Digest*, Vol. 14, February 1982, pp. 11–17.

8.26 J. E. Penny and J. R. Reed, "An integral equation approach to the fundamental frequency of vibrating beams," *Journal of Sound and Vibration*, Vol. 19, 1971, pp. 393–400.

8.27 G. Temple and W. G. Bickley, *Rayleigh's Principle and Its Application to Engineering*, Dover, New York, 1956.

8.28 W. L. Craver, Jr. and D. M. Egle, "A method for selection of significant terms in the assumed solution in a Rayleigh-Ritz analysis," *Journal of Sound and Vibration*, Vol. 22, 1972, pp. 133–142.

8.29 L. Klein, "Transverse vibrations of non-uniform beams," *Journal of Sound and Vibration*, Vol. 37, 1974, pp. 491–505.

8.30 J. R. Hutchinson, "Transverse vibrations of beams: Exact versus approximate solutions," *Journal of Applied Mechanics*, Vol. 48, 1981, pp. 923–928.

REVIEW QUESTIONS

8.1 Give brief answers to the following:

1. How does a continuous system differ from a discrete system in the nature of its equation of motion?
2. How many natural frequencies does a continuous system have?
3. Are the boundary conditions important in a discrete system? Why?
4. What is a wave equation? What is a traveling-wave solution?
5. What is the significance of wave velocity?
6. State the boundary conditions to be specified at the simply supported end of a beam if (a) thin-beam theory is used and (b) Timoshenko beam theory is used.
7. State the possible boundary conditions at the ends of a string.
8. What is the main difference in the nature of the frequency equations of a discrete system and a continuous system?
9. What is the effect of a tensile force on the natural frequencies of a beam?
10. Under what circumstances does the frequency of vibration of a beam subjected to an axial load become zero?
11. Why does the natural frequency of a beam become lower if the effects of shear deformation and rotary inertia are considered?
12. Give two practical examples of the vibration of membranes.
13. What is the basic principle used in Rayleigh's method?
14. Why is the natural frequency given by Rayleigh's method always larger than the true value of ω_1?
15. What is the difference between Rayleigh's method and the Rayleigh-Ritz method?
16. What is Rayleigh's quotient?

8.2 Indicate whether each of the following statements is true or false:

1. Continuous systems are the same as distributed systems.
2. Continuous systems can be considered to have an infinite number of degrees of freedom.
3. The governing equation of a continuous system is an ordinary differential equation.
4. The free vibration equations corresponding to the transverse motion of a string, the longitudinal motion of a bar, and the torsional motion of a shaft have the same form.
5. The normal modes of a continuous system are orthogonal.
6. A membrane has zero bending resistance.
7. Rayleigh's method can be considered as a method of conservation of energy.
8. The Rayleigh-Ritz method assumes that the solution is a series of functions that satisfy the boundary conditions of the problem.
9. For a discrete system, the boundary conditions are to be applied explicitly.
10. The Euler-Bernoulli beam theory is more accurate than the Timoshenko theory.

8.3 Fill in each of the following blanks with appropriate words:

1. The free vibration equation of a string is also called a _____ equation.
2. The frequency equation is also known as the _____ equation.

3. The method of separation of variables is used to express the free vibration solution of a string as a _____ of function of x and function of t.

4. Both boundary and _____ conditions are to be specified to find the solution of a vibrating continuous system.

5. In the wave-solution $w(x, t) = w_1(x - ct) + w_2(x + ct)$, the first term represents the wave that propagates in the _____ directions of x.

6. The quantities EI and GJ are called the _____ and _____ stiffnesses, respectively.

7. The thin-beam theory is also known as the _____ theory.

8. The lateral vibration of a thin beam is governed by a _____ order partial differential equation in spatial variable.

9. When a beam is subjected to an axial force (tension), it _____ the natural frequency.

10. The Timoshenko beam theory can be considered as _____ beam theory.

11. A drumhead can be considered as a _____.

12. A string has the same relationship to a beam as a membrane bears to a _____.

13. Rayleigh's method can be used to estimate the _____ natural frequency of a continuous system.

14. $EI \dfrac{\partial^2 w}{\partial x^2}$ denotes the _____ in a beam.

15. For a discrete system, the governing equations are _____ differential equations.

16. An axial tensile load increases the bending _____ of a beam.

17. The _____ energy of a beam is denoted by $\dfrac{1}{2} \displaystyle\int_0^l \rho A \left(\dfrac{\partial w}{\partial t} \right)^2 dx$.

18. The _____ energy of a beam is denoted by $\dfrac{1}{2} \displaystyle\int_0^l EI \left(\dfrac{\partial^2 w}{\partial x^2} \right)^2 dx$.

8.4 Select the most appropriate answer out of the choices given:

1. The frequency equation of a continuous system is a
   (a) polynomial equation
   (b) transcendental equation
   (c) differential equation

2. The number of natural frequencies of a continuous system is
   (a) infinite   (b) one   (c) finite

3. When the axial force approaches the Euler buckling load, the fundamental frequency of the beam reaches
   (a) infinity   (b) the frequency of a taut string   (c) zero

4. The value of the Timoshenko shear coefficient depends on the following:
   (a) shape of the cross section
   (b) size of the cross section
   (c) length of the beam

5. A Laplacian operator is given by

   (a) $\dfrac{\partial^2}{\partial x \partial y}$

(b) $\dfrac{\partial^2}{\partial x^2} + \dfrac{\partial^2}{\partial y^2} + 2\,\dfrac{\partial^2}{\partial x \partial y}$

(c) $\dfrac{\partial^2}{\partial x^2} + \dfrac{\partial^2}{\partial y^2}$

6. The boundary condition corresponding to the free end of a bar in longitudinal vibration is given by

(a) $u(0, t) = 0$

(b) $\dfrac{\partial u}{\partial x}(0, t) = 0$

(c) $AE\,\dfrac{\partial u}{\partial x}(0, t) - u(0, t) = 0$

7. The orthogonality of normal functions of the longitudinal vibration of a bar is given by

(a) $\displaystyle\int_0^l U_i(x)U_j(x)\,dx = 0$

(b) $\displaystyle\int_0^l \left(U_i'U_j - U_j'U_i\right) dx = 0$

(c) $\displaystyle\int_0^l (U_i(x) + U_j(x))\,dx = 0$

8.5 Match the items in the two columns below regarding boundary conditions for a thin beam:

1. Free end	(a) Bending moment $= 0$; shear force equal the spring force
2. Pinned end	(b) Deflection $= 0$; slope $= 0$
3. Fixed end	(c) Deflection $= 0$; bending moment $= 0$
4. Elastically restrained end	(d) Bending moment $= 0$; shear force $= 0$

8.6 Match the items in the two columns below regarding a uniform beam:

1. $W = 0$	(a) Zero bending moment
2. $W' = 0$	(b) Zero transverse displacement
3. $W'' = 0$	(c) Zero shear force
4. $W''' = 0$	(d) Zero slope

8.7 Match the items in the two columns below regarding the wave equation $c^2\,\dfrac{\partial^2 w}{\partial x^2} = \dfrac{\partial^2 w}{\partial t^2}$:

1. $c = \left(\dfrac{P}{\rho}\right)^{1/2}$	(a) Longitudinal vibration of a bar
2. $c = \left(\dfrac{E}{\rho}\right)^{1/2}$	(b) Torsional vibration of a shaft
3. $c = \left(\dfrac{G}{\rho}\right)^{1/2}$	(c) Transverse vibration of a string

PROBLEMS

The problem assignments are organized as follows:

Problems	Section Covered	Topic Covered
8.1–8.13	8.2	Transverse vibration of strings
8.14–8.19	8.3	Longitudinal vibration of bars
8.19–8.27	8.4	Torsional vibration of shafts
8.28–8.48, 8.56	8.5	Beams
8.49–8.55	8.6	Membranes
8.57–8.66	8.7	Rayleigh's method
8.67–8.71	8.8	The Rayleigh-Ritz method
8.72–8.74	8.9	MATLAB programs
8.75–8.76	8.10	C++ program
8.77–8.79	8.11	Fortran programs
8.80	—	Design project

8.1 Determine the velocity of wave propagation in a cable of mass $\rho = 5$ kg/m when stretched by a tension $P = 4000$ N.

8.2 A steel wire of 2 mm diameter is fixed between two points located 2 m apart. The tensile force in the wire is 250 N. Determine (a) the fundamental frequency of vibration and (b) the velocity of wave propagation in the wire.

8.3 A stretched cable of length 2 m has a fundamental frequency of 3000 Hz. Find the frequency of the third mode. How are the fundamental and third mode frequencies changed if the tension is increased by 20 percent?

8.4 Find the time it takes for a transverse wave to travel along a transmission line from one tower to another one 300 m away. Assume the horizontal component of the cable tension as 30,000 N and the mass of the cable as 2 kg/m of length.

8.5 A cable of length l and mass ρ per unit length is stretched under a tension P. One end of the cable is connected to a mass m, which can move in a frictionless slot and the other end is fastened to a spring of stiffness k, as shown in Fig. 8.23. Derive the frequency equation for the transverse vibration of the cable.

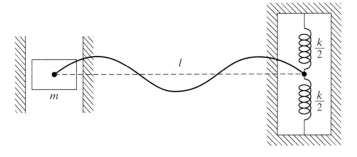

FIGURE 8.23

8.6   The cord of a musical instrument is fixed at both ends and has a length 2 m, diameter 0.5 mm, and density 7800 kg/m^3. Find the tension required in order to have a fundamental frequency of (a) 1 Hz and (b) 5 Hz.

8.7   A cable of length l and mass ρ per unit length is stretched under a tension P. One end of the cable is fixed and the other end is connected to a pin, which can move in a frictionless slot. Find the natural frequencies of vibration of the cable.

8.8   Find the free vibration solution of a cord fixed at both ends when its initial conditions are given by

$$w(x, 0) = 0, \qquad \frac{\partial w}{\partial t}(x, 0) = \frac{2ax}{l} \qquad \text{for } 0 \le x \le \frac{l}{2}$$

and

$$\frac{\partial w}{\partial t}(x, 0) = 2a\left(1 - \frac{x}{l}\right) \qquad \text{for } \frac{l}{2} \le x \le l$$

8.9   Prove that the constant a in Eqs. (8.18) and (8.19) is negative for common boundary conditions. *Hint:* Multiply Eq. (8.18) by $W(x)$ and integrate with respect to x from 0 to l.

8.10* The cable between two electric transmission towers has a length of 2000 m. It is clamped at its ends under a tension P (Fig. 8.24). The density of the cable material is 8890 kg/m^3. If the first four natural frequencies are required to lie between 0 and 20 Hz, determine the necessary cross-sectional area of the cable and the initial tension.

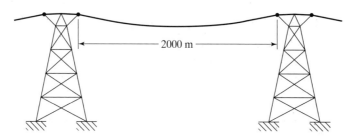

FIGURE 8.24

8.11 If a string of length l, fixed at both ends, is given an initial transverse displacement of h at $x = l/3$ and then released, determine its subsequent motion. Compare the deflection shapes of the string at times $t = 0, l/(4c), l/(3c), l/(2c),$ and l/c by considering the first four terms of the series solution.

8.12 A cord of length l is made to vibrate in a viscous medium. Derive the equation of motion considering the viscous damping force.

8.13 Determine the free vibration solution of a string fixed at both ends under the initial conditions $w(x, 0) = w_0 \sin(\pi x/l)$ and $(\partial w/\partial t)(x, 0) = 0$.

---

*The asterisk denotes a problem with no unique answer.

8.14 Derive an equation for the principal modes of longitudinal vibration of a uniform bar having both ends free.

8.15 Derive the frequency equation for the longitudinal vibration of the systems shown in Fig. 8.25.

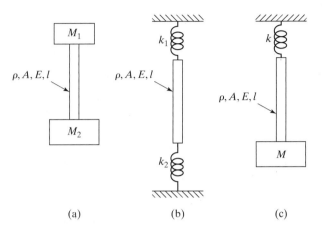

(a)   (b)   (c)

FIGURE 8.25

8.16* A thin bar of length l and mass m is clamped at one end and free at the other. What mass M must be attached to the free end in order to decrease the fundamental frequency of longitudinal vibration by 50 percent from its fixed-free value?

8.17 Show that the normal functions corresponding to the longitudinal vibration of the bar shown in Fig. 8.26 are orthogonal.

$x = 0$ $x = l$

FIGURE 8.26

8.18 Derive the frequency equation for the longitudinal vibration of a stepped bar having two different cross-sectional areas A_1 and A_2 over lengths l_1 and l_2, respectively. Assume fixed-free end conditions.

8.19 A steel shaft of diameter d and length l is fixed at one end and carries a propeller of mass m and mass moment of inertia J_0 at the other end (Fig. 8.27). Determine the fundamental natural frequency of vibration of the shaft in (a) axial vibration, and (b) torsional vibration. Data: $d = 5$ cm, $l = 1$ m, $m = 100$ kg, $J_0 = 10$ kg-m^2.

8.20 A torsional system consists of a shaft with a disc of mass moment of inertia I_0 mounted at its center. If both ends of the shaft are fixed, find the response of the system in free torsional vibration of the shaft. Assume that the disc is given a zero initial angular displacement and an initial velocity of $\dot{\theta}_0$.

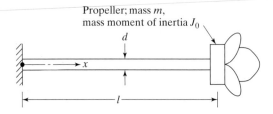

FIGURE 8.27

8.21 Find the natural frequencies for torsional vibration of a fixed-fixed shaft.

8.22 A uniform shaft of length l and torsional stiffness GJ is connected at both ends by torsional springs, torsional dampers, and discs with inertias, as shown in Fig. 8.28. State the boundary conditions.

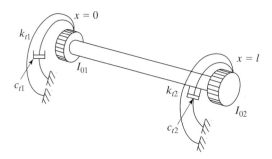

FIGURE 8.28

8.23 Solve Problem 8.21 if one end of the shaft is fixed and the other free.

8.24 Derive the frequency equation for the torsional vibration of a uniform shaft carrying rotors of mass moment of inertia I_{01} and I_{02} one at each end.

8.25 An external torque $M_t(t) = M_{t0} \cos \omega t$ is applied at the free end of a fixed-free uniform shaft. Find the steady-state vibration of the shaft.

8.26 Find the fundamental frequency for torsional vibration of a shaft of length 2 m and diameter 50 mm when both the ends are fixed. The density of the material is 7800 kg/m^3 and the modulus of rigidity is 0.8×10^{11} N/m^2.

8.27 A uniform shaft, supported at $x = 0$ and rotating at an angular velocity ω, is suddenly stopped at the end $x = 0$. If the end $x = l$ is free, determine the subsequent angular displacement response of the shaft.

8.28 Compute the first three natural frequencies and the corresponding mode shapes of the transverse vibrations of a uniform beam of rectangular cross section (100 mm $\times$ 300 mm) with $l = 2$ m, $E = 20.5 \times 10^{10}$ N/m^2, and $\rho = 7.83 \times 10^3$ kg/m^3 for the following cases: (a) when both ends are simply supported; (b) when both ends are built-in (clamped); (c) when one end is fixed and the other end is free; and (d) when both ends are free. Plot the mode shapes.

8.29 Derive an expression for the natural frequencies for the lateral vibration of a uniform fixed-free beam.

8.30 Prove that the normal functions of a uniform beam, whose ends are connected by springs as shown in Fig. 8.29, are orthogonal.

FIGURE 8.29

8.31 Derive an expression for the natural frequencies for the transverse vibration of a uniform beam with both ends simply supported.

8.32 Derive the expression for the natural frequencies for the lateral vibration of a uniform beam suspended as a pendulum, neglecting the effect of dead weight.

8.33 Find the cross-sectional area (A), and the area moment of inertia (I) of a simply supported steel beam of length 1 m for which the first three natural frequencies lie in the range 1500 Hz–5000 Hz.

8.34 A uniform beam, simply supported at both ends, is found to vibrate in its first mode with an amplitude of 10 mm at its center. If $A = 120$ mm^2, $I = 1000$ mm^4, $E = 20.5 \times 10^{10}$ N/m^2, $\rho = 7.83 \times 10^3$ kg/m^3, and $l = 1$ m, determine the maximum bending moment in the beam.

8.35 Derive the frequency equation for the transverse vibration of a uniform beam resting on springs at both ends, as shown in Fig. 8.30. The springs can deflect vertically only and the beam is horizontal in the equilibrium position.

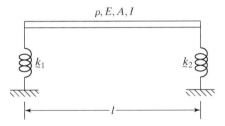

FIGURE 8.30

8.36 A simply supported uniform beam of length l carries a mass M at the center of the beam. Assuming the mass M to be a point mass, obtain the frequency equation of the system.

8.37 A uniform fixed-fixed beam of length $2l$ is simply supported at the middle point. Derive the frequency equation for the transverse vibration of the beam.

8.38 A simply supported beam carries initially a uniformly distributed load of intensity f_0. Find the vibration response of the beam if the load is suddenly removed.

8.39 Estimate the fundamental frequency of a cantilever beam whose cross-sectional area and moment of inertia vary as

$$A(x) = A_0 \frac{x}{l} \qquad \text{and} \qquad I(x) = \bar{I} \frac{x}{l}$$

where x is measured from the free end.

8.40 (a) Derive a general expression for the response of a uniform beam subjected to an arbitrary force. (b) Use the result of part (a) to find the response of a uniform simply supported beam under the harmonic force $F_0 \sin \omega t$ applied at $x = a$. Assume the initial conditions as $w(x, 0) = (\partial w/\partial t)(x, 0) = 0$.

8.41 Derive Eqs. (E.5) and (E.6) of Example 8.10.

8.42 Derive Eqs. (E.7) and (E.8) of Example 8.10.

8.43 Prove that the constant a in Eq. (8.82) is positive for common boundary conditions. *Hint:* Multiply Eq. (8.83) by $W(x)$ and integrate with respect to x from 0 to l.

8.44 Find the response of a simply supported beam subject to a uniformly distributed harmonically varying load.

8.45 A fixed-fixed beam carries an electric motor, of mass 100 kg and operational speed 3000 rpm at its midspan, as shown in Fig. 8.31. If the motor has a rotational unbalance of 0.5 kg-m, determine the steady-state response of the beam. Assume the length of the beam as $l = 2$ m, cross section as 10 cm $\times$ 10 cm, and the material as steel.

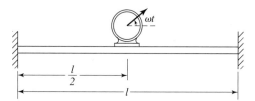

FIGURE 8.31

8.46 A steel cantilever beam of diameter 2 cm and length 1 m is subjected to an exponentially decaying force $100\, e^{-0.1t}$ N at the free end, as shown in Fig. 8.32. Determine the steady-state response of the beam. Assume the density and Young's modulus of steel as 7500 kg/m^3 and 210×10^9 N/m^2, respectively.

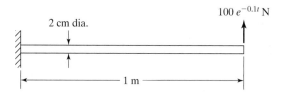

FIGURE 8.32

8.47 Find the steady-state response of a cantilever beam that is subjected to a suddenly applied step bending moment of magnitude M_0 at its free end.

8.48 A cantilever beam of length l, density ρ, Young's modulus E, area of cross section A, and area moment of inertia I carries a concentrated mass M at its free end. Derive the frequency equation for the transverse vibration of the beam.

8.49 Starting from fundamentals, show that the equation for the lateral vibration of a circular membrane is given by

$$\frac{\partial^2 w}{\partial r^2} + \frac{1}{r}\frac{\partial w}{\partial r} + \frac{1}{r^2}\frac{\partial^2 w}{\partial \theta^2} = \frac{\rho}{P}\frac{\partial^2 w}{\partial t^2}$$

8.50 Using the equation of motion given in Problem 8.49, find the natural frequencies of a circular membrane of radius R clamped around the boundary at $r = R$.

8.51 Consider a rectangular membrane of sides a and b supported along all the edges. (a) Derive an expression for the deflection $w(x,y,t)$ under an arbitrary pressure $f(x,y,t)$. (b) Find the response when a uniformly distributed pressure f_0 is applied to a membrane that is initially at rest.

8.52 Find the free vibration solution and the natural frequencies of a rectangular membrane that is clamped along all the sides. The membrane has dimensions a and b along the x and y directions, respectively.

8.53 Find the free vibration response of a rectangular membrane of sides a and b subject to the following initial conditions:

$$w(x, y, 0) = w_0 \sin\frac{\pi x}{a} \sin\frac{\pi y}{b}, \qquad 0 \le x \le a, \qquad 0 \le y \le b$$

$$\frac{\partial w}{\partial t}(x, y, 0) = 0, \qquad 0 \le x \le a, \qquad 0 \le y \le b$$

8.54 Find the free vibration response of a rectangular membrane of sides a and b subjected to the following initial conditions:

$$\left.\begin{array}{c} w(x, y, 0) = 0 \\[2mm] \dfrac{\partial w}{\partial t}(x, y, 0) = \dot{w}_0 \sin\dfrac{\pi x}{a} \sin\dfrac{2\pi y}{b} \end{array}\right\}, \qquad \begin{array}{c} 0 \le x \le a \\[2mm] 0 \le y \le b \end{array}$$

Assume that the edges of the membrane are fixed.

8.55 Compare the fundamental natural frequencies of transverse vibration of membranes of the following shapes: (a) square; (b) circular; and (c) rectangular with sides in the ratio of 2:1. Assume that all the membranes are clamped around their edges and have the same area, material, and tension.

8.56 Consider a railway car moving on a railroad track as shown in Fig. 8.33(a). The track can be modeled as an infinite beam resting on an elastic foundation and the car can be idealized as a moving load $F_0(x, t)$ (see Fig. 8.33b). If the soil stiffness per unit length is k, and the constant velocity of the car is v_0, show that the equation of motion of the beam can be expressed as

$$EI\frac{\partial^4 w(x, t)}{\partial x^4} + \rho A\frac{\partial^2 w(x, t)}{\partial t^2} + kw(x, t) = F_0(x - v_0 t)$$

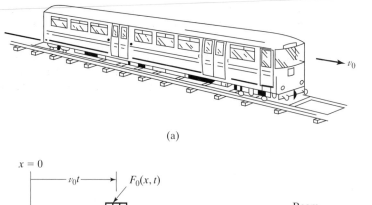

(a)

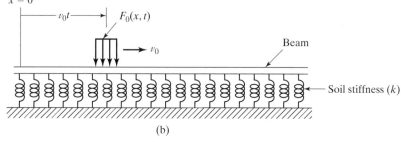

(b)

FIGURE 8.33

Indicate a method of solving the equation of motion if the moving load is assumed to be constant in magnitude.

8.57 Find the fundamental natural frequency of a fixed-fixed beam using the static deflection curve

$$W(x) = \frac{c_0 x^2}{24EI}(l - x)^2$$

where c_0 is a constant.

8.58 Solve Problem 8.57 using the deflection shape $W(x) = c_0\left(1 - \cos\dfrac{2\pi x}{l}\right)$, where c_0 is a constant.

8.59 Find the fundamental natural frequency of vibration of a uniform beam of length l that is fixed at one end and simply supported at the other end. Assume the deflection shape of the beam to be same as the static deflection curve under its self weight. *Hint:* The static deflection of a uniform beam under self weight is governed by

$$EI\frac{d^4W(x)}{dx^4} = \rho g A$$

where ρ is the density, g is the acceleration due to gravity, and A is the area of cross section of the beam. This equation can be integrated for any known boundary conditions of the beam.

8.60 Determine the fundamental frequency of a uniform fixed-fixed beam carrying a mass M at the middle by applying Rayleigh's method. Use the static deflection curve for $W(x)$.

8.61 Applying Rayleigh's method, determine the fundamental frequency of a cantilever beam (fixed at $x = l$) whose cross-sectional area $A(x)$ and moment of inertia $I(x)$ vary as $A(x) = A_0 x/l$ and $I(x) = I_0 x/l$.

8.62 Using Rayleigh's method, find the fundamental frequency for the lateral vibration of the beam shown in Fig. 8.34. The restoring force in the spring k is proportional to the deflection, and the restoring moment in the spring k_t is proportional to the angular deflection.

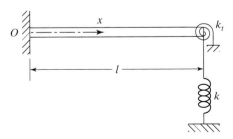

FIGURE 8.34

8.63 Using Rayleigh's method, estimate the fundamental frequency for the lateral vibration of a uniform beam fixed at both the ends. Assume the deflection curve to be

$$W(x) = c_1\left(1 - \cos\frac{2\pi x}{l}\right)$$

8.64 Find the fundamental frequency of longitudinal vibration of the tapered bar shown in Fig. 8.35, using Rayleigh's method with the mode shape

$$U(x) = c_1 \sin\frac{\pi x}{2l}$$

The mass per unit length is given by

$$m(x) = 2m_0\left(1 - \frac{x}{l}\right)$$

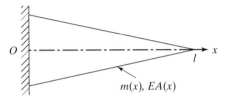

FIGURE 8.35

and the stiffness by

$$EA(x) = 2EA_0\left(1 - \frac{x}{l}\right)$$

8.65 Approximate the fundamental frequency of a rectangular membrane supported along all the edges by using Rayleigh's method with

$$W(x, y) = c_1 xy(x - a)(y - b).$$

Hint:

$$V = \frac{P}{2}\iint\left[\left(\frac{\partial w}{\partial x}\right)^2 + \left(\frac{\partial w}{\partial y}\right)^2\right]dx\,dy \qquad \text{and} \qquad T = \frac{\rho}{2}\iint\left(\frac{\partial w}{\partial t}\right)^2 dx\,dy$$

8.66 Using Rayeigh's method, determine the fundamental natural frequency of the system shown in Fig. 8.36.

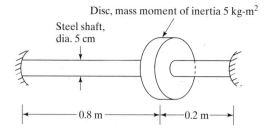

Disc, mass moment of inertia 5 kg-m²

Steel shaft, dia. 5 cm

|←——— 0.8 m ———→|←—0.2 m—→|

FIGURE 8.36

8.67 Estimate the fundamental frequency of a fixed-fixed string, assuming the mode shape (a) $W(x) = c_1 x(l - x)$ and (b) $W(x) = c_1 x(l - x) + c_2 x^2(l - x)^2$.

8.68 Estimate the fundamental frequency for the longitudinal vibration of a uniform bar fixed at $x = 0$ and free at $x = l$ by assuming the mode shapes as (a) $U(x) = c_1(x/l)$ and (b) $U(x) = c_1(x/l) + c_2(x/l)^2$.

8.69 A stepped bar, fixed at $x = 0$ and free at $x = l$, has a cross-sectional area of $2A$ for $0 \le x < l/3$ and A for $l/3 \le x \le l$. Assuming the mode shape

$$U(x) = c_1 \sin\frac{\pi x}{2l} + c_2 \sin\frac{3\pi x}{2l}$$

estimate the first two natural frequencies of longitudinal vibration.

8.70 Solve Problem 8.64 using the Rayleigh-Ritz method with the mode shape

$$U(x) = c_1 \sin\frac{\pi x}{2l} + c_2 \sin\frac{3\pi x}{2l}$$

8.71 Find the first two natural frequencies of a fixed-fixed uniform string of mass density ρ per unit length stretched between $x = 0$ and $x = l$ with an initial tension P. Assume the deflection functions

$$w_1(x) = x(l - x)$$
$$w_2(x) = x^2(l - x)^2$$

8.72 Using `Program12.m`, solve Example 8.4.

8.73 Using `Program12.m`, find the first five natural frequencies of a thin fixed-fixed beam.

8.74 Using MATLAB, plot the dynamic response of the plucked string, Eq. (E.6) of Example 8.1, at $x = l/2$. Data: $h = 0.1$ m, $l = 1.0$ m, $c = 100$ m/s.

8.75 Using `Program12.cpp`, solve Example 8.4.

8.76 Using `Program12.cpp`, find the first five natural frequencies of a thin fixed-fixed beam.

8.77 Using `PROGRAM12.F`, solve Example 8.4.

8.78 Using `PROGRAM12.F`, find the first five natural frequencies of a thin fixed-fixed beam.

8.79 Write a computer program for finding numerically the mode shapes of thin fixed-simply supported beams by using the known values of the natural frequencies.

DESIGN PROJECT

8.80 A vehicle, of weight F_0, moving at a constant speed on a bridge (Fig. 8.37a) can be modeled as a concentrated load travelling on a simply supported beam as shown in Fig. 8.37(b). The

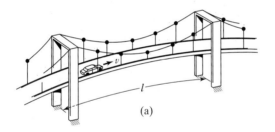

(a)

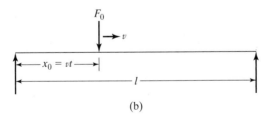

(b)

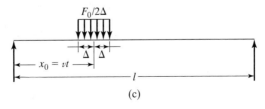

(c)

FIGURE 8.37

concentrated load F_0 can be considered as a uniformly distributed load over an infinitesimal length 2Δ and can be expressed as a sum of sine terms using Fourier sine series expansion (of the distributed load). Find the transverse displacement of the bridge as a sum of the responses due to each of the moving harmonic load components. Assume the initial conditions of the bridge as $w(x, 0) = \partial w/\partial t(x, 0) = 0$.

Leonhard Euler (1707–1783) was a Swiss mathematician who became a court mathematician and later a professor of mathematics in Saint Petersburg, Russia. He produced many works in algebra and geometry and was interested in the geometrical form of deflection curves in strength of materials. Euler's column buckling load is quite familiar to mechanical and civil engineers, and Euler's constant and Euler's coordinate system are well known to mathematicians. He derived the equation of motion for the bending vibrations of a rod (Euler-Bernoulli theory) and presented a series form of solution, as well as studying the dynamics of a vibrating ring. (Photo courtesy of Dirk J. Struik, *A Concise History of Mathematics*, 2nd ed. Dover Publications, New York, 1948.)

CHAPTER 9

Vibration Control

9.1 Introduction

There are numerous sources of vibration in an industrial environment: impact processes such as pile driving and blasting; rotating or reciprocating machinery such as engines, compressors, and motors; transportation vehicles such as trucks, trains, and aircraft; the flow of fluids; and many others. The presence of vibration often leads to excessive wear of bearings, formation of cracks, loosening of fasteners, structural and mechanical failures, frequent and costly maintenance of machines, electronic malfunctions through fracture of solder joints, and abrasion of insulation around electric conductors causing shorts. The occupational exposure of humans to vibration leads to pain, discomfort, and reduced efficiency. Vibration can sometimes be eliminated on the basis of theoretical analysis. However, the manufacturing costs involved in eliminating the vibration may be too high; a designer must compromise between an acceptable amount of vibration and a reasonable manufacturing cost. In some cases the excitation or shaking force is inherent in the machine. As seen earlier, even a relatively small excitation force can cause an undesirably large response near resonance, especially in lightly damped systems. In these cases, the magnitude of the response can be significantly reduced by the use of isolators and auxiliary mass absorbers [9.1]. In this chapter, we shall consider various techniques of vibration control—that is, methods involving the elimination or reduction of vibration.

9.2   Vibration Nomograph and Vibration Criteria

The acceptable levels of vibration are often specified in terms of the response of an undamped single degree of freedom system undergoing harmonic vibration. The bounds are shown in a graph, called the *vibration nomograph*, which displays the variations of displacement, velocity, and acceleration amplitudes with respect to the frequency of vibration. For the harmonic motion

$$x(t) = X \sin \omega t \tag{9.1}$$

the velocity and accelerations are given by

$$v(t) = \dot{x}(t) = \omega X \cos \omega t = 2\pi f X \cos \omega t \tag{9.2}$$

$$a(t) = \ddot{x}(t) = -\omega^2 X \sin \omega t = -4\pi^2 f^2 X \sin \omega t \tag{9.3}$$

where ω is the circular frequency (rad/s), f is the linear frequency (Hz), and X is the amplitude of displacement. The amplitudes of displacement (X), velocity (v_{max}) and acceleration (a_{max}) are related as

$$v_{max} = 2\pi f X \tag{9.4}$$

$$a_{max} = -4\pi^2 f^2 X = -2\pi f v_{max} \tag{9.5}$$

By taking logarithms of Eqs. (9.4) and (9.5), we obtain the following linear relations:

$$\ln v_{max} = \ln(2\pi f) + \ln X \tag{9.6}$$

$$\ln v_{max} = -\ln a_{max} - \ln(2\pi f) \tag{9.7}$$

It can be seen that for a constant value of the displacement amplitude (X), Eq. (9.6) shows that $\ln v_{max}$ varies with $\ln(2\pi f)$ as a straight line with slope $+1$. Similarly, for a constant value of the acceleration amplitude (a_{max}), Eq. (9.7) indicates that $\ln v_{max}$ varies with $\ln(2\pi f)$ as a straight line with slope -1. These variations are shown as a nomograph in Fig. 9.1. Thus every point on the nomograph denotes a specific sinusoidal (harmonic) vibration.

Since the vibration imparted to a human or machine is composed of many frequencies—rarely of just one frequency—the root mean square values of $x(t)$, $v(t)$, and $a(t)$ are used in the specification of vibration levels.

The usual ranges of vibration encountered in different scientific and engineering applications are given below [9.2]:

1. Atomic vibrations: Frequency $= 10^{12}$ Hz, displacement amplitude $= 10^{-8}$ to 10^{-6} mm.
2. Microseisms or minor tremors of earth's crust: Frequency $= 0.1$ to 1 Hz, displacement amplitude $= 10^{-5}$ to 10^{-3} mm. This vibration also denotes the threshold of disturbance of optical, electronic and computer equipment.
3. Machinery and building vibration: Frequency $= 10$ to 100 Hz, displacement amplitude $= 0.01$ to 1 mm. The threshold of human perception falls in the frequency range 1 to 8 Hz.

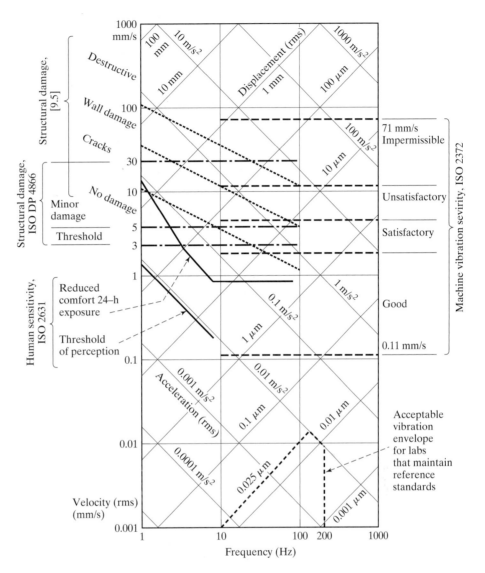

FIGURE 9.1   Vibration nomograph and vibration criteria [9.2].

4. Swaying of tall buildings: Frequency range = 0.1 to 5 Hz, displacement amplitude = 10 to 1000 mm.

Vibration severity of machinery is defined in terms of the rms value of the vibration velocity in ISO 2372 [9.3]. The ISO definition identifies 15 vibration severity ranges in the velocity range 0.11–71 mm/s for four classes of machines: (1) small, (2) medium, (3) large, and (4) turbomachine. The vibration severity of class 3 machines, including large primemovers, is shown in Fig. 9.1. In order to apply these criteria, the vibration is to be measured on machine surfaces such as bearing caps in the frequency range 10–1000 Hz.

ISO DP 4866 [9.4] gives the vibration severity for whole building vibration under blasting and steady-state vibration in the frequency range 1–100 Hz. For the vibration from blasting, the velocity is to be measured at the building foundation nearest the blast and for the steady-state vibration, the peak velocity is to be measured on the top floor. The limits given are 3–5 mm/s for threshold of damage and 5–30 mm/s for minor damage. The vibration results reported by Steffens [9.5] on structural damage are also shown in Fig. 9.1.

The vibration limits recommended in ISO 2631 [9.6] on human sensitivity to vibration are also shown in Fig. 9.1. In the United States an estimated 8 million workers are exposed to either whole-body vibration or segmented vibration to specific body parts. The whole-body vibration may be due to transmission through a supporting structure such as the seat of a helicopter, and the vibration to specific body parts may be due to work processes such as compacting, drilling, and chain-saw operations. Human tolerance of whole-body vibration is found to be lowest in the 4–8 Hz frequency range. The segmental vibration is found to cause localized stress injuries to different body parts at different frequencies, as indicated in Fig. 9.2. In addition, the following effects have been observed at different frequencies [9.7]: motion sickness (0.1–1 Hz), blurring vision (2–20 Hz), speech disturbance (1–20 Hz), interference with tasks (0.5–20 Hz) and after-fatigue (0.2–15 Hz).

The acceptable vibration levels for laboratories that maintain reference standards are also shown in Fig. 9.1.

Helicopter Seat Vibration Reduction

EXAMPLE 9.1

The seat of a helicopter, with the pilot, weighs 1000 N and is found to have a static deflection of 10 mm under self-weight. The vibration of the rotor is transmitted to the base of the seat as harmonic motion with frequency 4 Hz and amplitude 0.2 mm.

(a) What is the level of vibration felt by the pilot?
(b) How can the seat be redesigned to reduce the effect of vibration?

Solution

(a) By modeling the seat as an undamped single degree of freedom system, we can compute the following:

Mass $= m = 1000/9.81 = 101.9368$ kg

Stiffness $= k = \dfrac{W}{\delta_{st}} = \dfrac{1000}{0.01} = 10^5$ N/m

Natural frequency $= \omega_n = \sqrt{\dfrac{k}{m}} = \sqrt{\dfrac{10^5}{101.9368}} = 31.3209$ rad/s $= 4.9849$ Hz

Frequency ratio $= r = \dfrac{\omega}{\omega_n} = \dfrac{4.9849}{4.0} = 1.2462$

Since the seat is subject to harmonic base excitation, the amplitude of vibration felt by the pilot (mass of the seat) is given by Eq. (3.68) with $\zeta = 0$:

$$X = \pm \frac{Y}{1 - r^2} \qquad \text{(E.1)}$$

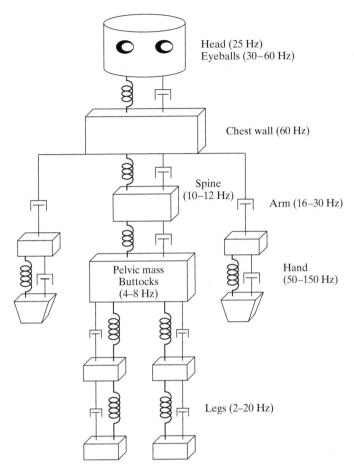

FIGURE 9.2   Vibration frequency sensitivity of different parts of human body.

where Y is the amplitude of base displacement. Equation (E.1) yields

$$X = \frac{0.2}{1 - 1.2462^2} = 0.3616 \text{ mm}$$

The amplitudes of velocity and acceleration felt by the pilot are given by $\omega X = 2\pi f X = 2(\pi)$ (5)(0.3616) = 9.0887 mm/s, and $\omega^2 X = (2\pi f)^2 X$ = 228.4074 mm/s^2 = 0.2284 m/s^2. Corresponding to the frequency 4 Hz, Fig. 9.1 shows that the amplitude of motion of 0.3616 mm may not cause much discomfort. However, the velocity and acceleration levels at the same frequency (4 Hz) are not acceptable for a comfortable ride.

(b) To bring the vibration level to an acceptable level, let us try to bring the acceleration felt by the pilot from the level 0.2284 m/s^2 to 0.01 m/s^2. Using $a_{max} = 10$ mm/s$^2 = -(2\pi f)^2 X = -(8\pi)^2 X$, we obtain $X = 0.01583$ mm. This leads to

$$\frac{X}{Y} = \frac{0.01583}{0.2} = \pm \frac{1}{1 - r^2} \qquad \text{or} \qquad r = 3.6923$$

This gives the new natural frequency of the seat as

$$\omega_n = \frac{\omega}{3.6923} = \frac{8\pi}{3.6923} = 6.8068 \text{ rad/s}$$

Using the relation $\omega_n = \sqrt{k/m}$ with $m = 101.9368$ kg, the new stiffness is given by $k = 4722.9837$ N/m. This implies that the stiffness of the seat is to be reduced from 10^5 N/m to 4722.9837 N/m. This can be accomplished by using a softer material for the seat or by using a different spring design. Alternatively, the desired acceleration level can be achieved by increasing the mass of the seat. However, this solution is not usually acceptable as it increases the weight of the helicopter.

∎

9.3   Reduction of Vibration at the Source

The first thing to be explored to control vibrations is to try to alter the source of vibration so that it produces less vibration. This method may not always be feasible. Some examples of the sources of vibration that cannot be altered are earthquake excitation, atmospheric turbulence, road roughness, and engine combustion instability. On the other hand, certain sources of vibration such as unbalance in rotating or reciprocating machines can be altered to reduce the vibrations. This can be achieved, usually, by using either internal balancing or an increase in the precision of machine elements. The use of close tolerances and better surface finish for machine parts (which have relative motion with respect to one another) make the machine less susceptible to vibration. Of course, there may be economic and manufacturing constraints on the degree of balancing that can be achieved or the precision with which the machine parts can be made. We shall consider the analysis of rotating and reciprocating machines in the presence of unbalance as well as the means of controlling the vibrations that result from unbalanced forces.

9.4   Balancing of Rotating Machines

The presence of an eccentric or unbalanced mass in a rotating disc causes vibration, which may be acceptable up to a certain level. If the vibration caused by an unbalanced mass is not acceptable, it can be eliminated either by removing the eccentric mass or by adding an equal mass in such a position that it cancels the effect of the unbalance. In order to use this procedure, we need to determine the amount and location of the eccentric mass experimentally. The unbalance in practical machines can be attributed to such irregularities as machining errors and variations in sizes of bolts, nuts, rivets, and welds. In this section, we shall consider two types of balancing: *single-plane* or *static balancing* and *two-plane* or *dynamic balancing* [9.9, 9.10].

9.4.1 Single-Plane Balancing

Consider a machine element in the form of a thin circular disc such as a fan, flywheel, gear, and a grinding wheel mounted on a shaft. When the center of mass is displaced from the axis of rotation due to manufacturing errors, the machine element is said to be statically unbalanced. To determine whether a disc is balanced or not, mount the shaft on two low friction bearings, as shown in Fig. 9.3(a). Rotate the disc and permit it to come to rest. Mark the lowest point on the circumference of the disc with chalk. Repeat the process several times, each time marking the lowest point on the disc with chalk. If the disc is balanced, the chalk marks will be scattered randomly all over the circumference. On the other hand, if the disc is unbalanced, all the chalk marks will coincide.

The unbalance detected by this procedure is known as *static unbalance*. The static unbalance can be corrected by removing (drilling) metal at the chalk mark or by adding a weight at 180° from the chalk mark. Since the magnitude of unbalance is not known, the

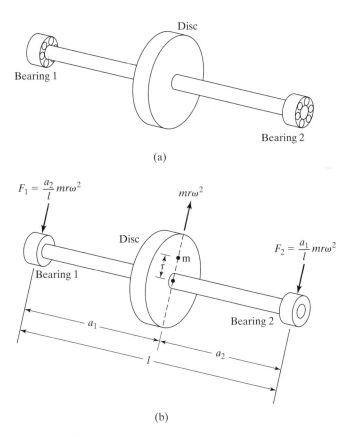

FIGURE 9.3 Single plane balancing of a disc.

amount of material to be removed or added must be determined by trial and error. This procedure is called "single-plane balancing" since all the mass lies practically in a single plane. The amount of unbalance can be found by rotating the disc at a known speed ω and measuring the reactions at the two bearings (see Fig. 9.3b). If an unbalanced mass m is located at a radius r of the disc, the centrifugal force will be $mr\omega^2$. Thus the measured bearing reactions F_1 and F_2 give m and r:

$$F_1 = \frac{a_2}{l}\, mr\omega^2, \qquad F_2 = \frac{a_1}{l}\, mr\omega^2 \tag{9.8}$$

Another procedure for single-plane balancing, using a vibration analyzer, is illustrated in Fig. 9.4. Here, a grinding wheel (disc) is attached to a rotating shaft that has bearing at A and is driven by an electric motor rotating at an angular velocity ω.

Before starting the procedure, *reference marks*, also known as *phase marks*, are placed both on the rotor (wheel) and the stator, as shown in Fig. 9.5(a). A vibration pickup is placed in contact with the bearing, as shown in Fig. 9.4, and the vibration analyzer is set to a frequency corresponding to the angular velocity of the grinding wheel. The vibration signal (the displacement amplitude) produced by the unbalance can be read from the indicating meter of the vibration analyzer. A stroboscopic light is fired by the vibration analyzer at the frequency of the rotating wheel. When the rotor rotates at speed ω, the phase mark on the rotor appears stationary under the stroboscopic light but is positioned at an angle θ from the mark on the stator, as shown in Fig. 9.5(b), due to phase lag in the response. Both the angle θ and the amplitude A_u (read from the vibration analyzer) caused by the original unbalance are noted. The rotor is then stopped, and a known trial weight W is attached to the rotor, as shown in Fig. 9.5(b). When the rotor runs at speed ω, the new angular position of the rotor phase mark ϕ and the vibration amplitude A_{u+w}, caused by the combined unbalance of rotor and trial weight, are noted (see Fig. 9.5c).[1]

Now we construct a vector diagram to find the magnitude and location of the correction mass for balancing the wheel. The original unbalance vector $\overrightarrow{A_u}$ is drawn in an arbitrary

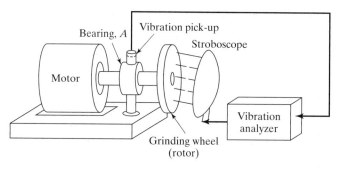

FIGURE 9.4   Single plane balancing using vibration analyzer.

---

[1]Note that if the trial weight is placed in a position that shifts the net unbalance in a clockwise direction, the stationary position of the phase mark will be shifted by exactly the same amount in the counterclockwise direction, and vice versa.

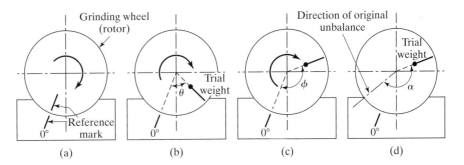

FIGURE 9.5 Use of phase marks.

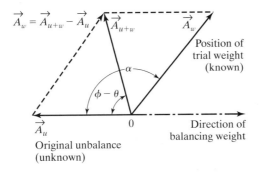

FIGURE 9.6 Unbalance due to trial weight W.

direction, with its length equal to A_u, as shown in Fig. 9.6. Then the combined unbalance vector is drawn as $\vec{A}_{u+w}$ at an angle $\phi - \theta$ from the direction of $\vec{A}_u$ with a length of A_{u+w}. The difference vector $\vec{A}_w = \vec{A}_{u+w} - \vec{A}_u$ in Fig. 9.6 then represents the unbalance vector due to the trial weight W. The magnitude of $\vec{A}_w$ can be computed using the law of cosines:

$$A_w = [A_u^2 + A_{u+w}^2 - 2A_u A_{u+w} \cos(\phi - \theta)]^{1/2} \tag{9.9}$$

Since the magnitude of the trial weight W and its direction relative to the original unbalance (α in Fig. 9.6) are known, the original unbalance itself must be at an angle α away from the position of the trial weight, as shown in Fig. 9.5(d). The angle α can be obtained from the law of cosines:

$$\alpha = \cos^{-1}\left[\frac{A_u^2 + A_w^2 - A_{u+w}^2}{2A_u A_w}\right] \tag{9.10}$$

The magnitude of the original unbalance is $W_o = (A_u/A_w) \cdot W$, located at the same radial distance from the rotation axis of the rotor as the weight W. Once the location and magnitude of the original unbalance are known, correction weight can be added to balance the wheel properly.

**9.4.2
Two-Plane
Balancing**

The single-plane balancing procedure can be used for balancing in one plane—that is, for rotors of the rigid disc type. If the rotor is an elongated rigid body as shown in Fig. 9.7, the unbalance can be anywhere along the length of the rotor. In this case, the rotor can be balanced by adding balancing weights in any two planes [9.10, 9.11]. For convenience, the two planes are usually chosen as the end planes of the rotor (shown by dashed lines in Fig. 9.7).

To see that any unbalanced mass in the rotor can be replaced by two equivalent unbalanced masses (in any two planes), consider a rotor with an unbalanced mass m at a distance $l/3$ from the right end, as shown in Fig. 9.8(a). When the rotor rotates at a speed of ω, the force due to the unbalance will be $F = m\omega^2 R$, where R is the radius of the rotor. The unbalanced mass m can be replaced by two masses m_1 and m_2, located at the ends of the rotor, as shown in Fig. 9.8(b). The forces exerted on the rotor by these masses are

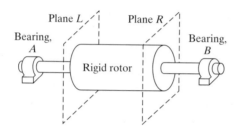

FIGURE 9.7   Two-plane balancing of a rotor.

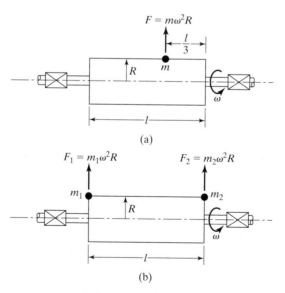

FIGURE 9.8   Representation of an unbalanced mass as two equivalent unbalanced masses.

$F_1 = m_1 \omega^2 R$ and $F_2 = m_2 \omega^2 R$. For the equivalence of force in Figs. 9.8(a) and (b), we have

$$m \omega^2 R = m_1 \omega^2 R + m_2 \omega^2 R \qquad \text{or} \qquad m = m_1 + m_2 \tag{9.11}$$

For the equivalence of moments in the two cases, we consider moments about the right end so that

$$m \omega^2 R \frac{l}{3} = m_1 \omega^2 R l \qquad \text{or} \qquad m = 3m_1 \tag{9.12}$$

Equations (9.11) and (9.12) give $m_1 = m/3$ and $m_2 = 2m/3$. Thus any unbalanced mass can be replaced by two equivalent unbalanced masses in the end planes of the rotor.

We now consider the two-plane balancing procedure using a vibration analyzer. In Fig. 9.9, the total unbalance in the rotor is replaced by two unbalanced weights U_L and U_R in the left and the right planes, respectively. At the rotor's operating speed ω, the vibration amplitude and phase due to the original unbalance are measured at the two bearings A and B, and the results are recorded as vectors $\vec{V}_A$ and $\vec{V}_B$. The magnitude of the vibration vector is taken as the vibration amplitude, while the direction of the vector is taken as the negative of the phase angle observed under stroboscopic light with reference to the stator reference line. The measured vectors $\vec{V}_A$ and $\vec{V}_B$ can be expressed as

$$\vec{V}_A = \vec{A}_{AL} \vec{U}_L + \vec{A}_{AR} \vec{U}_R \tag{9.13}$$

$$\vec{V}_B = \vec{A}_{BL} \vec{U}_L + \vec{A}_{BR} \vec{U}_R \tag{9.14}$$

where $\vec{A}_{ij}$ can be considered as a vector, reflecting the effect of the unbalance in plane j $(j = L, R)$ on the vibration at bearing i $(i = A, B)$. Note that $\vec{U}_L$, $\vec{U}_R$, and all the vectors $\vec{A}_{ij}$ are unknown in Eqs. (9.13) and (9.14).

As in the case of single-plane balancing, we add known trial weights and take measurements to obtain information about the unbalanced masses. We first add a known weight $\vec{W}_L$ in the left plane at a known angular position and measure the displacement and phase

FIGURE 9.9   Two-plane balancing.

of vibration at the two bearings while the rotor is rotating at speed ω. We denote these measured vibrations as vectors as

$$\vec{V_A'} = \vec{A}_{AL}\,(\vec{U_L} + \vec{W_L}) + \vec{A}_{AR}\vec{U_R} \tag{9.15}$$

$$\vec{V_B'} = \vec{A}_{BL}\,(\vec{U_L} + \vec{W_L}) + \vec{A}_{BR}\vec{U_R} \tag{9.16}$$

By subtracting Eqs. (9.13) and (9.14) from Eqs. (9.15) and (9.16), respectively, and solving, we obtain[2]

$$\vec{A}_{AL} = \frac{\vec{V_A'} - \vec{V_A}}{\vec{W_L}} \tag{9.17}$$

$$\vec{A}_{BL} = \frac{\vec{V_B'} - \vec{V_B}}{\vec{W_L}} \tag{9.18}$$

We then remove $\vec{W_L}$ and add a known weight $\vec{W_R}$ in the right plane at a known angular position and measure the resulting vibrations while the rotor is running at speed ω. The measured vibrations can be denoted as vectors:

$$\vec{V_A''} = \vec{A}_{AR}\,(\vec{U_R} + \vec{W_R}) + \vec{A}_{AL}\vec{U_L} \tag{9.19}$$

$$\vec{V_B''} = \vec{A}_{BR}\,(\vec{U_R} + \vec{W_R}) + \vec{A}_{BL}\vec{U_L} \tag{9.20}$$

As before, we subtract Eqs. (9.13) and (9.14) from Eqs. (9.19) and (9.20), respectively, to find

$$\vec{A}_{AR} = \frac{\vec{V_A''} - \vec{V_A}}{\vec{W_R}} \tag{9.21}$$

---

[2]It can be seen that complex subtraction, division, and multiplication are often used in the computation of the balancing weights. If

$$\vec{A} = a\angle\theta_A \qquad \text{and} \qquad \vec{B} = b\angle\theta_B$$

we can rewrite $\vec{A}$ and $\vec{B}$ as $\vec{A} = a_1 + ia_2$ and $\vec{B} = b_1 + ib_2$, where $a_1 = a\cos\theta_A$, $a_2 = a\sin\theta_A$, $b_1 = b\cos\theta_B$, and $b_2 = b\sin\theta_B$. Then the formulas for complex subtraction, division, and multiplication are [9.12]:

$$\vec{A} - \vec{B} = (a_1 - b_1) + i(a_2 - b_2)$$

$$\frac{\vec{A}}{\vec{B}} = \frac{(a_1 b_1 + a_2 b_2) + i(a_2 b_1 - a_1 b_2)}{(b_1^2 + b_2^2)}$$

$$\vec{A} \cdot \vec{B} = (a_1 b_1 - a_2 b_2) + i(a_2 b_1 + a_1 b_2)$$

$$\vec{A}_{BR} = \frac{\vec{V}_B'' - \vec{V}_B}{\vec{W}_R} \tag{9.22}$$

Once the vector operators $\vec{A}_{ij}$ are known, Eqs. (9.13) and (9.14) can be solved to find the unbalance vectors $\vec{U}_L$ and $\vec{U}_R$:

$$\vec{U}_L = \frac{\vec{A}_{BR}\vec{V}_A - \vec{A}_{AR}\vec{V}_B}{\vec{A}_{BR}\vec{A}_{AL} - \vec{A}_{AR}\vec{A}_{BL}} \tag{9.23}$$

$$\vec{U}_R = \frac{\vec{A}_{BL}\vec{V}_A - \vec{A}_{AL}\vec{V}_B}{\vec{A}_{BL}\vec{A}_{AR} - \vec{A}_{AL}\vec{A}_{BR}} \tag{9.24}$$

The rotor can now be balanced by adding equal and opposite balancing weights in each plane. The balancing weights in the left and right planes can be denoted vectorially as $\vec{B}_L = -\vec{U}_L$ and $\vec{B}_R = -\vec{U}_R$. It can be seen that the two-plane balancing procedure is a straightforward extension of the single-plane balancing procedure. Although high-speed rotors are balanced during manufacture, usually it becomes necessary to rebalance them in the field due to slight unbalances introduced due to creep, high temperature operation, and the like. Figure 9.10 shows a practical example of two-plane balancing.

FIGURE 9.10   Two-plane balancing. (Courtesy of Bruel and Kjaer Instruments, Inc., Marlborough. Mass.)

EXAMPLE 9.2

Two-Plane Balancing of Turbine Rotor

In the two-plane balancing of a turbine rotor, the data obtained from measurement of the original unbalance, the right-plane trial weight, and the left-plane trial weight are shown below. The displacement amplitudes are in mils (1/1000 inch.) Determine the size and location of the balance weights required.

	Vibration (Displacement) Amplitude		Phase Angle	
Condition	At Bearing A	At Bearing B	At Bearing A	At Bearing B
Original unbalance	8.5	6.5	60°	205°
W_L = 10.0 oz added at 270° from reference mark	6.0	4.5	125°	230°
W_R = 12.0 oz added at 180° from reference mark	6.0	10.5	35°	160°

Solution: The given data can be expressed in vector notation as

$$\vec{V_A} = 8.5 \,\underline{/60°} = 4.2500 + i\,7.3612$$

$$\vec{V_B} = 6.5 \,\underline{/205°} = -5.8910 - i\,2.7470$$

$$\vec{V_A'} = 6.0 \,\underline{/125°} = -3.4415 + i\,4.9149$$

$$\vec{V_B'} = 4.5 \,\underline{/230°} = -2.8926 - i\,3.4472$$

$$\vec{V_A''} = 6.0 \,\underline{/35°} = 4.9149 + i\,3.4472$$

$$\vec{V_B''} = 10.5 \,\underline{/160°} = -9.8668 + i\,3.5912$$

$$\vec{W_L} = 10.0 \,\underline{/270°} = 0.0000 - i\,10.0000$$

$$\vec{W_R} = 12.0 \,\underline{/180°} = -12.0000 + i\,0.0000$$

Equations (9.17) and (9.18) give

$$\vec{A}_{AL} = \frac{\vec{V_A'} - \vec{V_A}}{\vec{W_L}} = \frac{-7.6915 - i\,2.4463}{0.0000 - i\,10.0000} = 0.2446 - i\,0.7691$$

$$\vec{A}_{BL} = \frac{\vec{V_B'} - \vec{V_B}}{\vec{W_L}} = \frac{2.9985 - i\,0.7002}{0.0000 - i\,10.0000} = 0.0700 + i\,0.2998$$

The use of Eqs. (9.21) and (9.22) leads to

$$\vec{A}_{AR} = \frac{\vec{V_A''} - \vec{V_A}}{\vec{W_R}} = \frac{0.6649 - i\,3.9198}{-12.0000 + i\,0.0000} = -0.0554 + i\,0.3266$$

$$\vec{A}_{BR} = \frac{\vec{V}_B'' - \vec{V}_B}{\vec{W}_R} = \frac{-3.9758 + i\,6.3382}{-12.0000 + i\,0.0000} = 0.3313 - i\,0.5282$$

The unbalance weights can be determined from Eqs. (9.23) and (9.24):

$$\vec{U}_L = \frac{(5.2962 + i\,0.1941) - (1.2237 - i\,1.7721)}{(-0.3252 - i\,0.3840) - (-0.1018 + i\,0.0063)} = \frac{(4.0725 + i\,1.9661)}{(-0.2234 - i\,0.3903)}$$

$$= -8.2930 + i\,5.6879$$

$$\vec{U}_R = \frac{(-1.9096 + i\,1.7898) - (3.5540 + i\,3.8590)}{(-0.1018 + i\,0.0063) - (-0.3252 - i\,0.3840)} = \frac{(1.6443 - i\,2.0693)}{(0.2234 + i\,0.3903)}$$

$$= -2.1773 - i\,5.4592$$

Thus the required balance weights are given by

$$\vec{B}_L = -\vec{U}_L = (8.2930 - i\,5.6879) = 10.0561\,\underline{/145.5548^\circ}$$

$$\vec{B}_R = -\vec{U}_R = (2.1773 + i\,5.4592) = 5.8774\,\underline{/248.2559^\circ}$$

This shows that the addition of a 10.0561 oz weight in the left plane at 145.5548° and a 5.8774 oz weight in the right plane at 248.2559° from the reference position will balance the turbine rotor. It is implied that the balance weights are added at the same radial distance as the trial weights. If a balance weight is to be located at a different radial position, the required balance weight is to be modified in inverse proportion to the radial distance from the axis of rotation.

■

9.5 Whirling of Rotating Shafts

In the previous section, the rotor system—the shaft as well as the rotating body—was assumed to be rigid. However, in many practical applications such as turbines, compressors, electric motors and pumps, a heavy rotor is mounted on a lightweight, flexible shaft that is supported in bearings. There will be unbalance in all rotors due to manufacturing errors. These unbalances as well as other effects such as the stiffness and damping of the shaft, gyroscopic effects, and fluid friction in bearings will cause a shaft to bend in a complicated manner at certain rotational speeds, known as the whirling, whipping, or critical speeds. Whirling is defined as the rotation of the plane made by the line of centers of the bearings and the bent shaft. We consider the aspects of modeling the rotor system, critical speeds, response of the system, and stability in this section [9.13–9.14].

9.5.1 Equations of Motion

Consider a shaft supported by two bearings and carrying a rotor or disc of mass m at the middle, as shown in Fig. 9.11. We shall assume that the rotor is subjected to a steady-state excitation due to mass unbalance. The forces acting on the rotor are the inertia force due

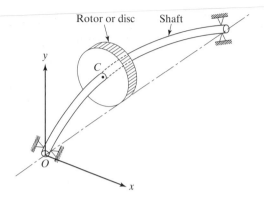

FIGURE 9.11   Shaft carrying a rotor.

to the acceleration of the mass center, the spring force due to the elasticity of the shaft, and the external and internal damping forces.[3]

Let O denote the equilibrium position of the shaft when balanced perfectly, as shown in Fig. 9.12. The shaft (line CG) is assumed to rotate with a constant angular velocity ω. During rotation, the rotor deflects radially by a distance $A = OC$ (in steady state). The rotor (disc) is assumed to have an eccentricity a so that its mass center (center of gravity) G is at a distance a from the geometric center, C. We use a fixed coordinate system (x and y fixed to the earth) with O as the origin for describing the motion of the system. The angular

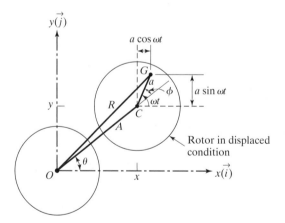

FIGURE 9.12   Rotor with eccentricity.

---

[3]Any rotating system responds in two different ways to damping or friction forces, depending upon whether the forces rotate with the shaft or not. When the positions at which the forces act remain fixed in space, as in the case of damping forces (which cause energy losses) in the bearing support structure, the damping is called *stationary* or *external damping*. On the other hand, if the positions at which they act rotate with the shaft in space, as in the case of internal friction of the shaft material, the damping is called *rotary* or *internal damping*.

velocity of the line OC, $\dot{\theta} = d\theta/dt$, is known as the whirling speed and, in general, is not equal to ω. The equations of motion of the rotor (mass m) can be written as

$$\text{Inertia force } (\vec{F}_i) = \text{Elastic force } (\vec{F}_e)$$

$$+ \text{ Internal damping force } (\vec{F}_{di})$$

$$+ \text{ External damping force } (\vec{F}_{de}) \qquad (9.25)$$

The various forces in Eq. (9.25) can be expressed as follows:

$$\text{Inertia force: } \vec{F}_i = m\ddot{\vec{R}} \qquad (9.26)$$

where $\vec{R}$ denotes the radius vector of the mass center G given by

$$\vec{R} = (x + a \cos \omega t)\,\vec{i} + (y + a \sin \omega t)\,\vec{j} \qquad (9.27)$$

with x and y representing the coordinates of the geometric center C and $\vec{i}$ and $\vec{j}$ denoting the unit vectors along the x and y coordinates, respectively. Equations (9.26) and (9.27) lead to

$$\vec{F}_i = m\,[(\ddot{x} - a\omega^2 \cos \omega t)\,\vec{i} + (\ddot{y} - a\omega^2 \sin \omega t)\,\vec{j}] \qquad (9.28)$$

$$\text{Elastic force: } \vec{F}_e = -k(x\vec{i} + y\vec{j}) \qquad (9.29)$$

where k is the stiffness of the shaft.

$$\text{Internal damping force: } \vec{F}_{di} = -c_i\,[(\dot{x} + \omega y)\,\vec{i} + (\dot{y} + \omega x)\,\vec{j}] \qquad (9.30)$$

where c_i is the internal or rotary damping coefficient.

$$\text{External damping force: } \vec{F}_{de} = -c\,(\dot{x}\vec{i} + \dot{y}\vec{j}) \qquad (9.31)$$

where c is the external damping coefficient. By substituting Eqs. (9.28) to (9.31) into Eq. (9.25), we obtain the equations of motion in scalar form:

$$m\ddot{x} + (c_i + c)\,\dot{x} + kx - c_i\,\omega y = m\omega^2 a \cos \omega t \qquad (9.32)$$

$$m\ddot{y} + (c_i + c)\,\dot{y} + ky - c_i\omega x = m\omega^2 a \sin \omega t \qquad (9.33)$$

These equations of motion, which describe the lateral vibration of the rotor, are coupled and are dependent on the speed of the steady-state rotation of the shaft, ω. By defining a complex quantity w as

$$w = x + iy \qquad (9.34)$$

where $i = (-1)^{1/2}$ and by adding Eq. (9.32) to Eq. (9.33) multiplied by i, we obtain a single equation of motion:

$$m\ddot{w} + (c_i + c)\dot{w} + kw - i\omega c_i w = m\omega^2 a e^{i\omega t} \tag{9.35}$$

9.5.2 Critical Speeds

A critical speed is said to exist when the frequency of the rotation of a shaft equals one of the natural frequencies of the shaft. The undamped natural frequency of the rotor system can be obtained by solving Eqs. (9.32), (9.33), or (9.35), retaining only the homogeneous part with $c_i = c = 0$. This gives the natural frequency of the system (or critical speed of the undamped system):

$$\omega_n = \left(\frac{k}{m}\right)^{1/2} \tag{9.36}$$

When the rotational speed is equal to this critical speed, the rotor undergoes large deflections, and the force transmitted to the bearings can cause bearing failures. A rapid transition of the rotating shaft through a critical speed is expected to limit the whirl amplitudes, while a slow transition through the critical speed aids the development of large amplitudes. Reference [9.15] investigates the behavior of the rotor during acceleration and deceleration through critical speeds. A FORTRAN computer program for calculating the critical speeds of rotating shafts is given in Ref. [9.16].

9.5.3 Response of the System

To determine the response of the rotor, we assume the excitation to be a harmonic force due to the unbalance of the rotor. In addition, we assume the internal damping to be negligible ($c_i = 0$). Then, we can solve Eqs. (9.32) and (9.33) (or equivalently, Eq. 9.35) and find the rotor's dynamic whirl amplitudes resulting from the mass unbalance. With $c_i = 0$, Eq. (9.35) reduces to

$$m\ddot{w} + c\dot{w} + kw = m\omega^2 a e^{i\omega t} \tag{9.37}$$

The solution of Eq. (9.37) can be expressed as

$$w(t) = C e^{-(\alpha t + \beta)} + A e^{i(\omega t - \phi)} \tag{9.38}$$

where C, β, A, and ϕ are constants. Note that the first term on the right-hand side of Eq. (9.38) contains a decaying exponential term representing a transient solution and the second term denotes a steady-state circular motion (whirl). By substituting the steady-state part of Eq. (9.38) into Eq. (9.37), we can find the amplitude of the circular motion (whirl) as

$$A = \frac{m\omega^2 a}{[(k - m\omega^2)^2 + \omega^2 c^2]^{1/2}} = \frac{a r^2}{[(1 - r^2)^2 + (2\zeta r)^2]^{1/2}} \tag{9.39}$$

and the phase angle as

$$\phi = \tan^{-1}\left(\frac{c\omega}{k - m\omega^2}\right) = \tan^{-1}\left(\frac{2\zeta r}{1 - r^2}\right) \tag{9.40}$$

where

$$r = \frac{\omega}{\omega_n}, \ \omega_n = \sqrt{\frac{k}{m}}, \text{ and } \zeta = \frac{c}{2\sqrt{km}}.$$

By differentiating Eq. (9.39) with respect to ω and setting the result equal to zero, we can find the rotational speed ω at which the whirl amplitude becomes a maximum

$$\omega = \frac{\omega_n}{\left\{ 1 - \frac{1}{2}\left(\frac{c}{\omega_n}\right)^2 \right\}^{1/2}} \qquad (9.41)$$

where ω_n is given by Eq. (9.36). It can be seen that the critical speed corresponds exactly to the natural frequency ω_n only when the damping (c) is zero. Furthermore, Eq. (9.41) shows that the presence of damping, in general, increases the value of the critical speed compared to ω_n. A plot of Eqs. (9.39) and (9.40) is shown in Fig. 9.13 [9.14]. Since the forcing function is proportional to ω^2, we normally expect the vibration amplitude to increase with the speed ω. However, the actual amplitude appears as shown in Fig. 9.13. From Eq. (9.39), we note that the amplitude of circular whirl A at low speeds is determined by the spring constant k, since the other two terms, $m\omega^2$ and $c^2\omega^2$, are small. Also, the value of the phase angle ϕ can be seen to be 0° from Eq. (9.40) for small values of ω. As ω increases, the amplitude of the response reaches a peak, since resonance occurs at $k - m\omega^2 = 0$. Around resonance, the response is essentially limited by the damping term. The phase lag is 90° at resonance. As the speed ω increases beyond ω_n, the response is dominated by the mass term $m^2\omega^4$ in Eq. (9.39). Since this term is 180° out of phase

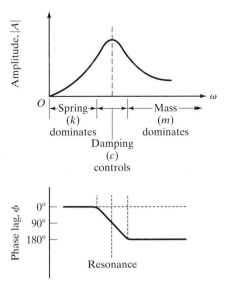

FIGURE 9.13   Plots of Eqs. (9.39) and (9.40).

with the unbalanced force, the shaft rotates in a direction opposite to that of the unbalanced force and hence the response of the shaft will be limited.

Notes

1. Equation (9.38) implicitly assumes a condition of forward synchronous whirl under steady state (that is, $\dot\theta = \omega$). As a general case, if the steady-state solution of Eq. (9.37) is assumed as $w(t) = A\,e^{i(\gamma t - \phi)}$, the solution can be obtained as $\gamma = \pm\,\omega$, with $\gamma = +\omega$ representing the forward synchronous whirl and $\gamma = -\omega$ denoting a backward synchronous whirl. For simple rotors, such as the one shown in Fig. 9.11, only forward synchronous whirl occurs in practice.

2. To determine the bearing reactions, we first find the deflection of the mass center of the disc from the bearing axis, R in Fig. 9.12, as

$$R^2 = A^2 + a^2 + 2\,A\,a\,\cos\phi \tag{9.42}$$

In view of Eqs. (9.39) and (9.40), Eq. (9.42) can be rewritten as

$$R = a\left[\frac{1 + (2\zeta r)^2}{(1 - r^2)^2 + (2\zeta r)^2}\right]^{1/2} \tag{9.43}$$

The bearing reactions can then be determined from the centrifugal force, $m\omega^2 R$. The vibration and balancing of unbalanced flexible rotors are presented in Refs. [9.17, 9.18].

9.5.4
Stability
Analysis

Instability in a flexible rotor system can occur due to several factors like internal friction, eccentricity of the rotor, and the oil whip in the bearings. As seen earlier, the stability of the system can be investigated by considering the equation governing the dynamics of the system. Assuming $w(t) = e^{st}$, the characteristic equation corresponding to the homogeneous part of Eq. (9.35) can be written as

$$ms^2 + (c_i + c)\,s + k - i\omega c_i = 0 \tag{9.44}$$

With $s = i\lambda$, Eq. (9.44) becomes

$$-m\lambda^2 + (c_i + c)i\lambda + k - i\omega c_i = 0 \tag{9.45}$$

This equation is a particular case of the more general equation

$$(p_2 + iq_2)\lambda^2 + (p_1 + iq_1)\lambda + (p_0 + iq_0) = 0 \tag{9.46}$$

A necessary and sufficient condition for the system governed by Eq. (9.46) to be stable, according to Routh-Hurwitz criterion, is that the following inequalities are satisfied:

$$-\begin{vmatrix} p_2 & p_1 \\ q_2 & q_1 \end{vmatrix} > 0 \tag{9.47}$$

and

$$\begin{vmatrix} p_2 & p_1 & p_0 & 0 \\ q_2 & q_1 & q_0 & 0 \\ 0 & p_2 & p_1 & p_0 \\ 0 & q_2 & q_1 & q_0 \end{vmatrix} > 0 \tag{9.48}$$

Noting that $p_2 = -m$, $q_2 = 0$, $p_1 = 0$, $q_1 = c_i + c$, $p_0 = k$, and $q_0 = -\omega c_i$, from Eq. (9.45), the application of Eqs. (9.47) and (9.48) leads to

$$m(c_i + c) > 0 \tag{9.49}$$

and

$$km(c_i + c)^2 - m^2(\omega^2 c_i^2) > 0 \tag{9.50}$$

Equation (9.49) is automatically satisfied, while Eq. (9.50) yields the condition

$$\sqrt{\frac{k}{m}}\left(1 + \frac{c}{c_i}\right) - \omega > 0 \tag{9.51}$$

This equation also shows that internal and external friction can cause instability at rotating speeds above the first critical speed of $\omega = \sqrt{\dfrac{k}{m}}$.

Whirl Amplitude of a Shaft Carrying an Unbalanced Rotor

EXAMPLE 9.3

A shaft, carrying a rotor of weight 100 lb and eccentricity 0.1 in, rotates at 1200 rpm. Determine (a) the steady-state whirl amplitude and (b) the maximum whirl amplitude during start-up conditions of the system. Assume the stiffness of the shaft as 2×10^5 lb/in and the external damping ratio as 0.1.

Solution: The forcing frequency of the rotor (rotational speed of the shaft) is given by

$$\omega = \frac{1200 \times 2\pi}{60} = 40\pi = 125.6640 \text{ rad/s}$$

The natural frequency of the system can be determined as

$$\omega_n = \sqrt{\frac{k}{m}} = \sqrt{\frac{2.0 \times 10^5}{(100/386.4)}} = 87.9090 \text{ rad/s}$$

and the frequency ratio as

$$r = \frac{\omega}{\omega_n} = \frac{125.6640}{87.9090} = 1.4295$$

(a) The steady-state amplitude is given by Eq. (9.39):

$$A = \frac{ar^2}{\sqrt{(1 - r^2)^2 + (2\zeta r)^2}} \tag{E.1}$$

$$= \frac{(0.1)(1.4295)^2}{\sqrt{(1 - 1.4295^2)^2 + (2 \times 0.1 \times 1.4295)^2}} = 0.18887 \text{ in.} \tag{E.2}$$

(b) During start-up conditions, the frequency (speed) of the rotor, ω, passes through the natural frequency of the system. Thus, using $r = 1$ in Eq. (E.1), we obtain the whirl amplitude as

$$A\big|_{r=1} = \frac{a}{2\zeta} = \frac{0.1}{2(0.1)} = 0.5 \text{ in.}$$

∎

9.6 Balancing of Reciprocating Engines

The essential moving elements of a reciprocating engine are the piston, the crank, and the connecting rod. Vibrations in reciprocating engines arise due to (1) periodic variations of the gas pressure in the cylinder and (2) inertia forces associated with the moving parts [9.19]. We shall now analyze a reciprocating engine and find the unbalanced forces caused by these factors.

9.6.1
Unbalanced
Forces Due to
Fluctuations in
Gas Pressure

Figure 9.14(a) is a schematic diagram of a cylinder of a reciprocating engine. The engine is driven by the expanding gas in the cylinder. The expanding gas exerts on the piston a pressure force F, which is transmitted to the crankshaft through the connecting rod. The reaction to the force F can be resolved into two components: one of magnitude $F/\cos \phi$, acting along the connecting rod, and the other of magnitude $F \tan \phi$, acting in a horizontal direction. The force $F/\cos \phi$ induces a torque M_t, which tends to rotate the crankshaft. (In Fig. 9.14b, M_t acts about an axis perpendicular to the plane of the paper and passes through point Q.)

$$M_t = \left(\frac{F}{\cos\phi}\right) r \cos \theta \tag{9.52}$$

For force equilibrium of the overall system, the forces at the bearings of the crankshaft will be F in the vertical direction and $F \tan \phi$ in the horizontal direction.

Thus the forces transmitted to the stationary parts of the engine are as follows:

1. Force F acting upward at the cylinder head
2. Force $F \tan \phi$ acting toward the right at the cylinder head
3. Force F acting downward at the crankshaft bearing Q
4. Force $F \tan \phi$ acting toward the left at the crankshaft bearing

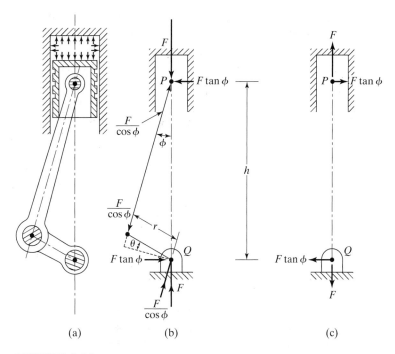

(a)   (b)   (c)

FIGURE 9.14   Forces in a reciprocating engine.

These forces are shown in Fig. 9.14(c). Although the total resultant force is zero, there is a resultant torque $M_Q = Fh \tan \phi$ on the body of the engine, where h can be found from the geometry of the system:

$$h = \frac{r \cos \theta}{\sin \phi} \tag{9.53}$$

Thus the resultant torque is given by

$$M_Q = \frac{Fr \cos \theta}{\cos \phi} \tag{9.54}$$

As expected, M_t and M_Q given by Eqs. (9.52) and (9.54) can be seen to be identical, which indicates that the torque induced on the crankshaft due to the gas pressure on the piston is felt at the support of the engine. Since the magnitude of the gas force F varies with time, the torque M_Q also varies with time. The magnitude of force F changes from a maximum to a minimum at a frequency governed by the number of cylinders in the engine, the type of the operating cycle, and the rotating speed of the engine.

9.6.2
Unbalanced
Forces Due to
Inertia of the
Moving Parts

Acceleration of the Piston.   Figure 9.15 shows the crank (of length r), the connecting rod (of length l), and the piston of a reciprocating engine. The crank is assumed to rotate in an anticlockwise direction at a constant angular speed of ω, as shown in Fig. 9.15. If we consider the origin of the x axis (O) as the uppermost position of the piston, the displacement of the piston P corresponding to an angular displacement of the crank of $\theta = \omega t$ can be expressed as in Fig. 9.15. The displacement of the piston P corresponding to an angular displacement of the crank $\theta = \omega t$ from its topmost position (origin O) can be expressed as

$$x_p = r + l - r \cos \theta - l \cos \phi$$

$$= r + l - r \cos \omega t - l\sqrt{1 - \sin^2 \phi} \tag{9.55}$$

But

$$l \sin \phi = r \sin \theta = r \sin \omega t \tag{9.56}$$

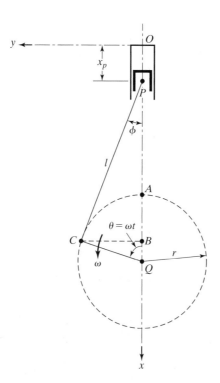

FIGURE 9.15   Motions of crank, connecting rod, and piston.

and hence

$$\cos \phi = \left(1 - \frac{r^2}{l^2} \sin^2 \omega t \right)^{1/2} \tag{9.57}$$

By substituting Eq. (9.57) into Eq. (9.55), we obtain

$$x_p = r + l - r \cos \omega t - l\sqrt{1 - \frac{r^2}{l^2} \sin^2 \omega t} \tag{9.58}$$

Due to the presence of the term involving the square root, Eq. (9.58) is not very convenient in further calculation. Equation (9.58) can be simplified by noting that, in general, $r/l < \frac{1}{4}$ and by using the expansion relation

$$\sqrt{1 - \varepsilon} \simeq 1 - \frac{\varepsilon}{2} \tag{9.59}$$

Hence Eq. (9.58) can be approximated as

$$x_p \simeq r(1 - \cos \omega t) + \frac{r^2}{2l} \sin^2 \omega t \tag{9.60}$$

or, equivalently,

$$x_p = r\left(1 + \frac{r}{2l} \right) - r\left(\cos \omega t + \frac{r}{4l} \cos 2\omega t \right) \tag{9.61}$$

Equation (9.61) can be differentiated with respect to time to obtain expressions for the velocity and the acceleration of the piston:

$$\dot{x}_p = r\omega \left(\sin \omega t + \frac{r}{2l} \sin 2\omega t \right) \tag{9.62}$$

$$\ddot{x}_p = r\omega^2 \left(\cos \omega t + \frac{r}{l} \cos 2\omega t \right) \tag{9.63}$$

Acceleration of the Crankpin.   With respect to the xy coordinate axes shown in Fig. 9.15, the vertical and horizontal displacements of the crankpin C are given by

$$x_c = OA + AB = l + r(1 - \cos \omega t) \tag{9.64}$$

$$y_c = CB = r \sin \omega t \tag{9.65}$$

Differentiation of Eqs. (9.64) and (9.65) with respect to time gives the velocity and acceleration components of the crankpin as

$$\dot{x}_c = r\omega \sin \omega t \qquad (9.66)$$

$$\dot{y}_c = r\omega \cos \omega t \qquad (9.67)$$

$$\ddot{x}_c = r\omega^2 \cos \omega t \qquad (9.68)$$

$$\ddot{y}_c = -r\omega^2 \sin \omega t \qquad (9.69)$$

Inertia Forces.   Although the mass of the connecting rod is distributed throughout its length, it is generally idealized as a massless link with two masses concentrated at its ends—the piston end and the crankpin end. If m_p and m_c denote the total mass of the piston and of the crankpin (including the concentrated mass of the connecting rod) respectively, the vertical component of the inertia force (F_x) for one cylinder is given by

$$F_x = m_p\ddot{x}_p + m_c\ddot{x}_c \qquad (9.70)$$

By substituting Eqs. (9.63) and (9.68) for the accelerations of P and C, Eq. (9.70) becomes

$$F_x = (m_p + m_c)r\omega^2 \cos \omega t + m_p\frac{r^2\omega^2}{l}\cos 2\omega t \qquad (9.71)$$

It can be observed that the vertical component of the inertia force consists of two parts. One part, known as the *primary part*, has a frequency equal to the rotational frequency of the crank ω. The other part, known as the *secondary part*, has a frequency equal to twice the rotational frequency of the crank.

Similarly, the horizontal component of inertia force for a cylinder can be obtained

$$F_y = m_p\ddot{y}_p + m_c\ddot{y}_c \qquad (9.72)$$

where $\ddot{y}_p = 0$ and $\ddot{y}_c$ is given by Eq. (9.69). Thus

$$F_y = -m_c r\omega^2 \sin \omega t \qquad (9.73)$$

The horizontal component of the inertia force can be observed to have only a primary part.

9.6.3
Balancing of
Reciprocating
Engines

The unbalanced or inertia forces on a single cylinder are given by Eqs. (9.71) and (9.73). In these equations, m_p and m_c represent the equivalent reciprocating and rotating masses, respectively. The mass m_p is always positive, but m_c can be made zero by counterbalancing the crank. It is therefore possible to reduce the horizontal inertia force F_y to zero, but the vertical unbalanced force always exists. Thus a single cylinder engine is inherently unbalanced.

In a multicylinder engine, it is possible to balance some or all of the inertia forces and torques by proper arrangement of the cranks. Figure 9.16(a) shows the general arrangement of an N-cylinder engine (only six cylinders, $N = 6$, are shown in the figure). The

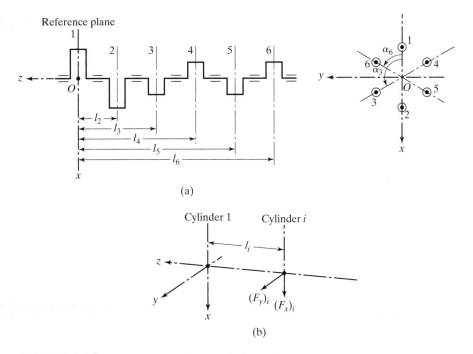

FIGURE 9.16   Arrangement of an N-cylinder engine.

lengths of all the cranks and connecting rods are assumed to be r and l, respectively, and the angular velocity of all the cranks is taken to be a constant, ω. The axial displacement and angular orientation of ith cylinder from those of the first cylinder are assumed to be α_i and l_i, respectively; $i = 2, 3, \ldots, N$. For force balance, the total inertia force in the x and y directions must be zero. Thus

$$(F_x)_{\text{total}} = \sum_{i=1}^{N} (F_x)_i = 0 \tag{9.74}$$

$$(F_y)_{\text{total}} = \sum_{i=1}^{N} (F_y)_i = 0 \tag{9.75}$$

where $(F_x)_i$ and $(F_y)_i$ are the vertical and horizontal components of inertia force of cylinder i given by (see Eqs. 9.71 and 9.73):

$$(F_x)_i = (m_p + m_c)_i\, r\omega^2 \cos(\omega t + \alpha_i) + (m_p)_i\, \frac{r^2\omega^2}{l} \cos(2\omega t + 2\alpha_i) \tag{9.76}$$

$$(F_y)_i = -(m_c)_i\, r\omega^2 \sin(\omega t + \alpha_i) \tag{9.77}$$

For simplicity, we assume the reciprocating and rotating masses for each cylinder to be same—that is, $(m_p)_i = m_p$ and $(m_c)_i = m_c$ for $i = 1, 2, \ldots, N$. Without loss of generality, Eqs. (9.74) and (9.75) can be applied at time $t = 0$. Thus the conditions necessary for the total force balance are given by

$$\sum_{i=1}^{N} \cos \alpha_i = 0 \quad \text{and} \quad \sum_{i=1}^{N} \cos 2\alpha_i = 0 \tag{9.78}$$

$$\sum_{i=1}^{N} \sin \alpha_i = 0 \tag{9.79}$$

The inertia forces $(F_x)_i$ and $(F_y)_i$ of the ith cylinder induce moments about the y and x axes, respectively, as shown in Fig. 9.16(b). The moments about the z and x axes are given by

$$M_z = \sum_{i=2}^{N} (F_x)_i l_i = 0 \tag{9.80}$$

$$M_x = \sum_{i=2}^{N} (F_y)_i l_i = 0 \tag{9.81}$$

By substituting Eqs. (9.76) and (9.77) into Eqs. (9.80) and (9.81) and assuming $t = 0$, we obtain the necessary conditions to be satisfied for the balancing of moments about z and x axes as

$$\sum_{i=2}^{N} l_i \cos \alpha_i = 0 \quad \text{and} \quad \sum_{i=2}^{N} l_i \cos 2\alpha_i = 0 \tag{9.82}$$

$$\sum_{i=2}^{N} l_i \sin \alpha_i = 0 \tag{9.83}$$

Thus we can arrange the cylinders of a multicylinder reciprocating engine so as to satisfy Eqs. (9.78), (9.79), (9.82), and (9.83); it will be completely balanced against the inertia forces and moments.

9.7   Control of Vibration

In many practical situations, it is possible to reduce but not eliminate the dynamic forces that cause vibrations. Several methods can be used to control vibrations. Among them, the following are important:

1. By controlling the natural frequencies of the system and avoiding resonance under external excitations.
2. By preventing excessive response of the system, even at resonance by introducing a damping or energy-dissipating mechanism.

3. By reducing the transmission of the excitation forces from one part of the machine to another, by the use of vibration isolators.

4. By reducing the response of the system, by the addition of an auxiliary mass neutralizer or vibration absorber.

We shall now consider the details of these methods.

9.8   Control of Natural Frequencies

It is well known that whenever the frequency of excitation coincides with one of the natural frequencies of the system, resonance occurs. The most prominent feature of resonance is a large displacement. In most mechanical and structural systems, large displacements indicate undesirably large strains and stresses, which can lead to the failure of the system. Hence resonance conditions must be avoided in any system. In most cases, the excitation frequency cannot be controlled, because it is imposed by the functional requirements of the system or machine. We must concentrate on controlling the natural frequencies of the system to avoid resonance.

As indicated by Eq. (2.14), the natural frequency of a system can be changed either by varying the mass m or the stiffness k.[4] In many practical cases, however, the mass cannot be changed easily, since its value is determined by the functional requirements of the system. For example, the mass of a flywheel on a shaft is determined by the amount of energy it must store in one cycle. Therefore, the stiffness of the system is the factor that is most often changed to alter its natural frequencies. For example, the stiffness of a rotating shaft can be altered by varying one or more of its parameters, such as materials or number and location of support points (bearings).

9.9   Introduction of Damping

Although damping is disregarded so as to simplify the analysis, especially in finding the natural frequencies, most systems possess damping to some extent. The presence of damping is helpful in many cases. In systems such as automobile shock absorbers and many vibration-measuring instruments, damping must be introduced to fulfill the functional requirements [9.20–9.21].

If the system undergoes forced vibration, its response or amplitude of vibration tends to become large near resonance if there is no damping. The presence of damping always limits the amplitude of vibration. If the forcing frequency is known, it may be possible to avoid resonance by changing the natural frequency of the system. However, the system or the machine may be required to operate over a range of speeds, as in the case of a variable-speed electric motor or an internal combustion engine. It may not be possible to avoid resonance under all operating conditions. In such cases, we can introduce damping into the

---

[4]Although this statement is made with reference to a single degree of freedom system, it is generally true even for multidegree of freedom and continuous systems.

system to control its response, by the use of structural materials having high internal damping, such as cast iron or laminated or sandwich materials.

In some structural applications, damping is introduced through joints. For example, bolted and riveted joints, which permit slip between surfaces, dissipate more energy compared to welded joints, which do not permit slip. Hence a bolted or riveted joint is desirable to increase the damping of the structure. However, bolted and riveted joints reduce the stiffness of the structure, produce debris due to joint slip, and cause fretting corrosion. In spite of this, if a highly damped structure is desired, bolted or riveted joints should not be ignored.

Use of Viscoelastic Materials.   The equation of motion of a single degree of freedom system with internal damping, under harmonic excitation $F(t) = F_0 e^{i\omega t}$, can be expressed as

$$m\ddot{x} + k(1 + i\eta)x = F_0 e^{i\omega t} \tag{9.84}$$

where η is called the loss factor (or loss coefficient), which is defined as (see Section 2.6.4)

$$\eta = \frac{(\Delta W/2\pi)}{W} = \left(\frac{\text{Energy dissipated during 1 cycle of harmonic displacement/radian}}{\text{Maximum strain energy in cycle}} \right)$$

The amplitude of the response of the system at resonance ($\omega = \omega_n$) is given by

$$\frac{F_0}{k\eta} = \frac{F_0}{aE\eta} \tag{9.85}$$

since the stiffness is proportional to the Young's modulus ($k = aE$; $a = $ constant).

The viscoelastic materials have larger values of the loss factor and hence are used to provide internal damping. When viscoelastic materials are used for vibration control, they are subjected to shear or direct strains. In the simplest arrangement, a layer of viscoelastic material is attached to an elastic one. In another arrangement, a viscoelastic layer is sandwiched between the elastic layers. This arrangement is known as constrained layer damping.[5] Damping tapes, consisting of thin metal foil covered with a viscoelastic adhesive, are used on existing vibrating structures. A disadvantage with the use of viscoelastic materials is that their properties change with temperature, frequency, and strain. Equation (9.85) shows that a material with the highest value of ($E\eta$) gives the smallest resonance amplitude. Since the strain is proportional to the displacement x and the stress is proportional to

---

[5]It appears that constrained layer damping was used, possibly unknowingly, as far back as the seventeenth century, in the manufacture of violins [9.22]. Antonio Stradivari (1644–1737), the renowned Italian violin manufacturer, bought the wood necessary for the manufacture of violins from Venice. The varnish used for sealing the wood was made from a mixture of resin and ground gem stones. This varnish—stone particles in resin matrix—acted as a form of constrained layer (friction mechanism) that created enough damping to explain why many of his violins had a rich, full tone.

Ex, the material with the largest value of the loss factor will be subjected to the smallest stresses. The values of loss coefficient for some materials are given below:

Material	Loss Factor (η)
Polystyrene	2.0
Hard rubber	1.0
Fiber mats with matrix	0.1
Cork	0.13–0.17
Aluminum	1×10^{-4}
Iron and steel	2–6×10^{-4}

The damping ratios obtainable with different types of construction/arrangement are indicated below:

Type of Construction/Arrangement	Equivalent Viscous Damping Ratio (%)
Welded construction	1–4
Bolted construction	3–10
Steel frame	5–6
Unconstrained viscoelastic layer on steel-concrete girder	4–5
Constrained viscoelastic layer on steel-concrete girder	5–8

9.10 Vibration Isolation

Vibration isolation is a procedure by which the undesirable effects of vibration are reduced [9.22–9.24]. Basically, it involves the insertion of a resilient member (or isolator) between the vibrating mass (or equipment or payload) and the source of vibration so that a reduction in the dynamic response of the system is achieved under specified conditions of vibration excitation. An isolation system is said to be active or passive depending on whether or not external power is required for the isolator to perform its function. A passive isolator consists of a resilient member (stiffness) and an energy dissipator (damping). Examples of passive isolators include metal springs, cork, felt, pneumatic springs, and elastomer (rubber) springs. Figure 9.17 shows typical spring and pneumatic mounts that can be used as passive isolators, and Fig. 9.18 illustrates the use of passive isolators in the mounting of a high-speed punch press [9.25]. The optimal synthesis of vibration isolators is presented in Refs. [9.26–9.30].

An active isolator is comprised of a servomechanism with a sensor, signal processor, and an actuator. The effectiveness of an isolator is stated in terms of its transmissibility. The transmissibility (T_r) is defined as the ratio of the amplitude of the force transmitted to that of the exciting force.

(a)

(b)

(c)

FIGURE 9.17   (a) Undamped spring mount; (b) damped spring mount; (c) pneumatic rubber mount. (Courtesy of Sound and Vibration.)

Vibration isolation can be used in two types of situations. In the first type, the foundation or base of a vibrating machine is protected against large unbalanced forces (as in the case of reciprocating and rotating machines) or impulsive forces (as in the case of forging and stamping presses). In these cases, if the system is modeled as a single degree of freedom system as shown in Fig. 9.19(a), the force is transmitted to the foundation through the spring and the damper. The force transmitted to the base (F_t) is given by

$$F_t(t) = kx(t) + c\dot{x}(t) \qquad (9.86)$$

If the force transmitted to the base $F_t(t)$ varies harmonically, as in the case of unbalanced reciprocating and rotating machines, the resulting stresses in the foundation bolts also vary harmonically, which might lead to fatigue failure. Even if the force transmitted is not harmonic, its magnitude is to be limited to safe permissible values.

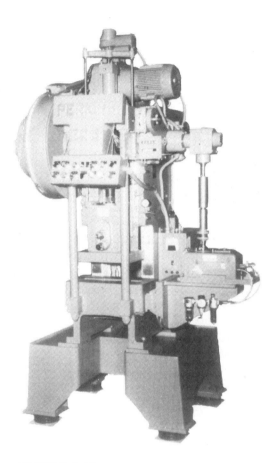

FIGURE 9.18   High-speed punch press mounted on pneumatic rubber mounts. (Courtesy of Sound and Vibration.)

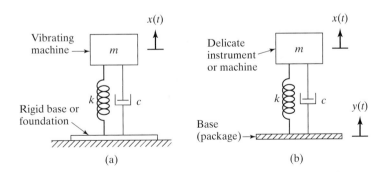

FIGURE 9.19   Vibration isolation.

In the second type, the system is protected against the motion of its foundation or base (as in the case of protection of a delicate instrument or equipment from the motion of its container). If the delicate instrument is modeled as a single degree of freedom system, as shown in Fig. 9.19(b), the force transmitted to the instrument (mass m in Fig. 9.19b) is given by

$$F_t(t) = m\ddot{x}(t) \equiv k[x(t) - y(t)] + c[\dot{x}(t) - \dot{y}(t)] \tag{9.87}$$

where $(x - y)$ and $(\dot{x} - \dot{y})$ denote the relative displacement and relative velocity of the spring and the damper, respectively. In many practical problems, the package is to be designed properly to avoid transmission of large forces to the delicate instrument to avoid damage.

9.10.1 Vibration Isolation System with Rigid Foundation

Reduction of the Force Transmitted to Foundation.   When a machine is bolted directly to a rigid foundation or floor, the foundation will be subjected to a harmonic load due to the unbalance in the machine in addition to the static load due to the weight of the machine. Hence an elastic or resilient member is placed between the machine and the rigid foundation to reduce the force transmitted to the foundation. The system can then be idealized as a single degree of freedom system, as shown in Fig. 9.20(a). The resilient member is assumed to have both elasticity and damping and is modeled as a spring k and a dashpot c, as shown in Fig. 9.20(b). It is assumed that the operation of the machine gives rise to a harmonically varying force $F(t) = F_0 \cos \omega t$. The equation of motion of the machine (of mass m) is given by

$$m\ddot{x} + c\dot{x} + kx = F_0\cos \omega t \tag{9.88}$$

Since the transient solution dies out after some time, only the steady-state solution will be left. The steady-state solution of Eq. (9.88) is given by (see Eq. 3.25)

$$x(t) = X \cos (\omega t - \phi) \tag{9.89}$$

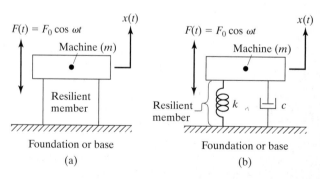

FIGURE 9.20   Machine and resilient member on rigid foundation.

where

$$X = \frac{F_0}{[(k - m\omega^2)^2 + \omega^2 c^2]^{1/2}} \qquad (9.90)$$

and

$$\phi = \tan^{-1}\left(\frac{\omega c}{k - m\omega^2}\right) \qquad (9.91)$$

The force transmitted to the foundation through the spring and the dashpot, $F_t(t)$, is given by

$$F_t(t) = kx(t) + c\dot{x}(t) = kX \cos{(\omega t - \phi)} - c\omega X \sin{(\omega t - \phi)} \qquad (9.92)$$

The magnitude of the total transmitted force (F_T) is given by

$$F_T = [(kx)^2 + (c\dot{x})^2]^{1/2} = X\sqrt{k^2 + \omega^2 c^2}$$

$$= \frac{F_0(k^2 + \omega^2 c^2)^{1/2}}{[(k - m\omega^2)^2 + \omega^2 c^2]^{1/2}} \qquad (9.93)$$

The transmissibility or transmission ratio of the isolator (T_r) is defined as the ratio of the magnitude of the force transmitted to that of the exciting force:

$$T_r = \frac{F_T}{F_0} = \left\{\frac{k^2 + \omega^2 c^2}{(k - m\omega^2)^2 + \omega^2 c^2}\right\}^{1/2}$$

$$= \left\{\frac{1 + (2\zeta r)^2}{[1 - r^2]^2 + (2\zeta r)^2}\right\}^{1/2} \qquad (9.94)$$

where $r = \frac{\omega}{\omega_n}$ is the frequency ratio. The variation of T_r with the frequency ratio $r = \frac{\omega}{\omega_n}$ is shown in Fig. 9.21. In order to achieve isolation, the force transmitted to the foundation needs to be less than the excitation force. It can be seen, from Fig. 9.21, that the forcing frequency has to be greater than $\sqrt{2}$ times the natural frequency of the system in order to achieve isolation of vibration.

Notes

1. The magnitude of the force transmitted to the foundation can be reduced by decreasing the natural frequency of the system (ω_n).
2. The force transmitted to the foundation can also be reduced by decreasing the damping ratio. However, since vibration isolation requires $r > \sqrt{2}$, the machine should pass through resonance during start-up and stopping. Hence, some damping is essential to avoid infinitely large amplitudes at resonance.

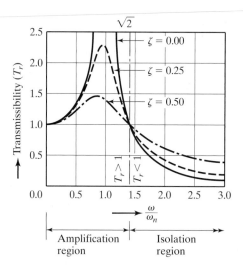

FIGURE 9.21   Variation of transmission ratio (T_r) with ω.

3. Although damping reduces the amplitude of the mass (X) for all frequencies, it reduces the maximum force transmitted to the foundation (F_t) only if $r < \sqrt{2}$. Above that value, the addition of damping increases the force transmitted.

4. If the speed of the machine (forcing frequency) varies, we must compromise in choosing the amount of damping to minimize the force transmitted. The amount of damping should be sufficient to limit the amplitude X and the force transmitted F_t while passing through the resonance, but not so much to increase unnecessarily the force transmitted at the operating speed.

Reduction of the Force Transmitted to the Mass.   If a sensitive instrument or machine of mass m is to be isolated from the unwanted harmonic motion of its base, the governing equation is given by Eq. (3.75):

$$m\ddot{z} + c\dot{z} + kz = -m\ddot{y} \tag{9.95}$$

where $z = x - y$ denotes the displacement of the mass relative to the base. If the base motion is harmonic, then the motion of the mass will also be harmonic. Hence the displacement transmissibility, $T_d = \frac{X}{Y}$, is given by Eq. (3.68)

$$T_d = \frac{X}{Y} = \left\{ \frac{1 + (2\zeta r)^2}{(1 - r^2)^2 + (2\zeta r)^2} \right\}^{1/2} \tag{9.96}$$

where the right-hand side expression in Eq. (9.96) can be identified to be the same as that in Eq. (9.94). Note that Eq. (9.96) is also equal to the ratio of the maximum steady-state accelerations of the mass and the base.

**9.10.2
Isolation of
Source of
Vibration from
Surroundings.**

The force transmitted to the base or ground by a source of vibration (vibrating mass) is given by Eq. (9.94) and is shown in Fig. 9.21 as a graph between $T_r = F_T/F_0$ and $r = \omega/\omega_n$. As noted earlier, vibration isolation—reduction of the force transmitted to the ground—can be achieved for $r > \sqrt{2}$. In the region $r > \sqrt{2}$, low values of damping are desired for more effective isolation. For large values of r and low values of ζ, the term $(2\zeta r)^2$ becomes very small and can be neglected in Eq. (9.94) for simplicity. Thus Eq. (9.94) can be approximated as

$$T_r = \frac{1}{r^2 - 1} \left(r > \sqrt{2}, \zeta = \text{small} \right) \tag{9.97}$$

By defining the natural frequency of vibration of the undamped system as

$$\omega_n = \sqrt{\frac{k}{m}} = \sqrt{\frac{g}{\delta_{st}}} \tag{9.98}$$

and the exciting frequency ω as

$$\omega = \frac{2\pi N}{60} \tag{9.99}$$

where δ_{st} is the static deflection of the spring and N is the frequency in cycles per minute or revolutions per minute (rpm) of rotating machines such as electric motors and turbines. Equations (9.97) to (9.99) can be combined to obtain

$$r = \frac{\omega}{\omega_n} = \frac{2\pi N}{60}\sqrt{\frac{\delta_{st}}{g}} = \sqrt{\frac{2 - R}{1 - R}} \tag{9.100}$$

where $R = 1 - T_r$ is used to indicate the quality of the isolator and denotes the percent reduction achieved in the transmitted force. Equation (9.100) can be rewritten as

$$N = \frac{30}{\pi}\sqrt{\frac{g}{\delta_{st}}\left(\frac{2 - R}{1 - R}\right)} = 29.9092\sqrt{\frac{2 - R}{\delta_{st}\,(1 - R)}} \tag{9.101}$$

Equation (9.101) can be used to generate a graph between $\log N$ and $\log \delta_{st}$ as a series of straight lines for different values of R, as shown in Fig. 9.22. This graph serves as a design chart for selecting a suitable spring for the isolation.

Reduction of the Force Transmitted to the Foundation Due to Rotating Unbalance.
The excitation force caused by a rotating unbalance is given by

$$F(t) = F_0 \sin \omega t \equiv me\omega^2 \sin \omega t \tag{9.102}$$

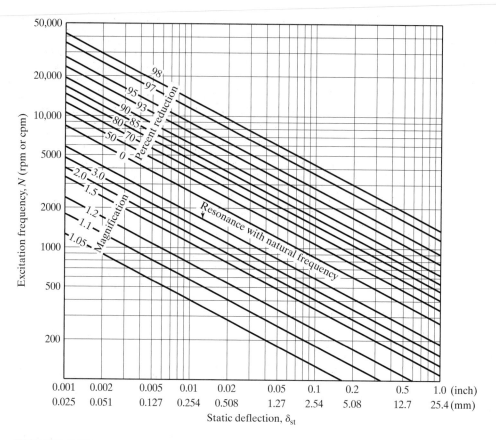

FIGURE 9.22   Isolation efficiency.

In this case, the force transmissibility, T_r, can be expressed as

$$T_r = \frac{F_T}{F_0} = \frac{F_T}{me\omega^2} = \frac{F_T}{mer^2\,\omega_n^2} \qquad (9.103)$$

Equations (9.103) and (9.94) yield

$$\frac{F_T}{me\omega_n^2} = r^2\left\{\frac{1 + (2\zeta r)^2}{(1 - r^2)^2 + (2\zeta r)^2}\right\}^{1/2} \qquad (9.104)$$

The plot of Eq. (9.104) will be identical to the one shown in Fig. 3.16.

**9.10.3
Vibration
Isolation System
with Flexible
Foundation**

In many practical situations, the structure or foundation to which the isolator is connected moves when the machine mounted on the isolator operates. For example, in the case of a turbine supported on the hull of a ship or an aircraft engine mounted on the wing of an airplane, the area surrounding the point of support also moves with the isolator. In such cases, the system can be represented as having two degrees of freedom. In Fig. 9.23, m_1 and m_2 denote the masses of the machine and the supporting structure that moves with the isolator, respectively. The isolator is represented by a spring k, and the damping is disregarded for the sake of simplicity. The equations of motion of the masses m_1 and m_2 are

$$m_1\ddot{x}_1 + k\,(x_1 - x_2) = F_0\cos\omega t \tag{9.105}$$

$$m_2\ddot{x}_2 + k\,(x_2 - x_1) = 0 \tag{9.106}$$

By assuming a harmonic solution of the form

$$x_j = X_j\cos\omega t, \qquad j = 1, 2 \tag{9.107}$$

Eqs. (9.105) and (9.106) give

$$\left.\begin{aligned}
X_1(k - m_1\omega^2) - X_2 k &= F_0\\
- X_1 k + X_2(k - m_2\omega^2) &= 0
\end{aligned}\right\} \tag{9.108}$$

The natural frequencies of the system are given by the roots of the equation

$$\begin{vmatrix} (k - m_1\omega^2) & -k \\ -k & (k - m_2\omega^2) \end{vmatrix} = 0 \tag{9.109}$$

The roots of Eq. (9.109) are given by

$$\omega_1^2 = 0, \qquad \omega_2^2 = \frac{(m_1 + m_2)\,k}{m_1 m_2} \tag{9.110}$$

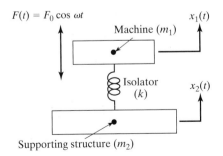

$F(t) = F_0\cos\omega t$

Machine (m_1) $x_1(t)$

Isolator
(k) $x_2(t)$

Supporting structure (m_2)

FIGURE 9.23   Machine with isolator on a flexible foundation.

The value $\omega_1 = 0$ corresponds to rigid-body motion since the system is unconstrained. In the steady state, the amplitudes of m_1 and m_2 are governed by Eq. (9.108), whose solution yields

$$X_1 = \frac{(k - m_2\omega^2) F_0}{[(k - m_1\omega^2)(k - m_2\omega^2) - k^2]} \qquad (9.111)$$

$$X_2 = \frac{kF_0}{[(k - m_1\omega^2)(k - m_2\omega^2) - k^2]} \qquad (9.112)$$

The force transmitted to the supporting structure (F_t) is given by the amplitude of $m_2\ddot{x}_2$:

$$F_t = -m_2\omega^2 X_2 = \frac{-m_2 k \omega^2 F_0}{[(k - m_1\omega^2)(k - m_2\omega^2) - k^2]} \qquad (9.113)$$

The transmissibility of the isolator (T_r) is given by

$$T_r = \frac{F_t}{F_0} = \frac{-m_2 k \omega^2}{[(k - m_1\omega^2)(k - m_2\omega^2) - k^2]}$$

$$= \frac{1}{\left(\dfrac{m_1 + m_2}{m_2} - \dfrac{m_1\omega^2}{k} \right)} = \frac{m_2}{(m_1 + m_2)} \left(\frac{1}{1 - \dfrac{\omega^2}{\omega_2^2}} \right) \qquad (9.114)$$

where ω_2 is the natural frequency of the system given by Eq. (9.110). Equation (9.114) shows, as in the case of an isolator on a rigid base, that the force transmitted to the foundation becomes less as the natural frequency of the system ω_2 is reduced.

Spring Support for Exhaust Fan

EXAMPLE 9.4

An exhaust fan, rotating at 1000 rpm, is to be supported by four springs, each having a stiffness of K. If only 10 percent of the unbalanced force of the fan is to be transmitted to the base, what should be the value of K? Assume the mass of the exhaust fan to be 40 kg.

Solution: Since the transmissibility has to be 0.1, we have, from Eq. (9.94),

$$0.1 = \left[\frac{1 + \left(2\zeta \dfrac{\omega}{\omega_n} \right)^2}{\left\{ 1 - \left(\dfrac{\omega}{\omega_n} \right)^2 \right\}^2 + \left(2\zeta \dfrac{\omega}{\omega_n} \right)^2} \right]^{1/2} \qquad (E.1)$$

where the forcing frequency is given by

$$\omega = \frac{1000 \times 2\pi}{60} = 104.72 \text{ rad/s} \tag{E.2}$$

and the natural frequency of the system by

$$\omega_n = \left(\frac{k}{m}\right)^{1/2} = \left(\frac{4K}{40}\right)^{1/2} = \frac{\sqrt{K}}{3.1623} \tag{E.3}$$

By assuming the damping ratio to be $\zeta = 0$, we obtain from Eq. (E.1),

$$0.1 = \frac{\pm 1}{\left\{1 - \left(\dfrac{104.72 \times 3.1623}{\sqrt{K}}\right)^2\right\}} \tag{E.4}$$

To avoid imaginary values, we need to consider the negative sign on the right-hand side of Eq. (E.4). This leads to

$$\frac{331.1561}{\sqrt{K}} = 3.3166$$

or

$$K = 9969.6365 \text{ N/m}$$

∎

Isolation of Vibrating System

EXAMPLE 9.5

A vibrating system is to be isolated from its supporting base. Find the required damping ratio that must be achieved by the isolator to limit the transmissibility at resonance to $T_r = 4$. Assume the system to have a single degree of freedom.

Solution: By setting $\omega = \omega_n$, Eq. (9.94) gives

$$T_r = \frac{\sqrt{1 + (2\zeta)^2}}{2\zeta}$$

or

$$\zeta = \frac{1}{2\sqrt{T_r^2 - 1}} = \frac{1}{2\sqrt{15}} = 0.1291$$

∎

EXAMPLE 9.6

Isolator for Stereo Turntable

A stereo turntable, of mass 1 kg, generates an excitation force at a frequency of 3 Hz. If it is supported on a base through a rubber mount, determine the stiffness of the rubber mount to reduce the vibration transmitted to the base by 80 percent.

Solution: Using $N = 3 \times 60 = 180$ cpm and $R = 0.80$, Eq. (9.101) gives

$$180 = 29.9092 \sqrt{\frac{2 - 0.80}{\delta_{st}(1 - 0.80)}}$$

or

$$\delta_{st} = 0.1657 \text{ m}$$

The static deflection of the rubber mount can be expressed, in terms of its stiffness (k), as

$$\delta_{st} = \frac{mg}{k}$$

which gives the stiffness of the rubber mount as

$$0.1657 = \frac{1(9.81)}{k} \qquad \text{or} \qquad k = 59.2179 \text{ N/m}$$

∎

9.10.4 Vibration Isolation System with Partially Flexible Foundation

Figure 9.24 shows a more realistic situation in which the base of the isolator, instead of being completely rigid or completely flexible, is partially flexible [9.34]. We can define the mechanical impedance of the base structure, $Z(\omega)$, as the force at frequency ω required to produce a unit displacement of the base (as in Section 3.5):

$$Z(\omega) = \frac{\text{Applied force of frequency } \omega}{\text{Displacement}}$$

The equations of motion are given by[6]

$$m_1 \ddot{x}_1 + k(x_1 - x_2) = F_0 \cos \omega t \tag{9.115}$$

$$k(x_2 - x_1) = -x_2 Z(\omega) \tag{9.116}$$

---

[6]If the base is completely flexible with an unconstrained mass of m_2, $Z(\omega) = -\omega^2 m_2$, and Eqs. (9.115) to (9.117) lead to Eq. (9.108).

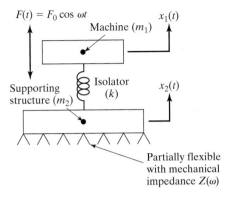

FIGURE 9.24   Machine with isolator
on a partially flexible foundation.

By substituting the harmonic solution

$$x_j(t) = X_j \cos \omega t, \qquad j = 1, 2 \tag{9.117}$$

into Eqs. (9.115) and (9.116), X_1 and X_2 can be obtained as in the previous case:

$$X_1 = \frac{[k + Z(\omega)] X_2}{k} = \frac{[k + Z(\omega)] F_0}{[Z(\omega)(k - m_1\omega^2) - km_1\omega^2]}$$

$$X_2 = \frac{kF_0}{[Z(\omega)(k - m_1\omega^2) - km_1\omega^2]} \tag{9.118}$$

The amplitude of the force transmitted is given by

$$F_t = X_2 Z(\omega) = \frac{kZ(\omega) F_0}{[Z(\omega)(k - m_1\omega^2) - km_1\omega^2]} \tag{9.119}$$

and the transmissibility of the isolator by

$$T_r = \frac{F_t}{F_0} = \frac{kZ(\omega)}{[Z(\omega)(k - m_1\omega^2) - km_1\omega^2]} \tag{9.120}$$

In practice, the mechanical impedance $Z(\omega)$ depends on the nature of the base structure. It can be found experimentally by measuring the displacement produced by a vibrator that applies a harmonic force on the base structure. In some cases—for example, if an isolator is resting on a concrete raft on soil—the mechanical impedance at any frequency ω can be found in terms of the spring-mass-dashpot model of the soil.

9.10.5
Shock Isolation

As stated earlier, a shock load involves the application of a force for a short duration, usually for a period of less than one natural time period of the system. The forces involved in forge hammers, punch presses, blasts, and explosions are examples of shock loads. Shock isolation can be defined as a procedure by which the undesirable effects of shock are reduced. We noted that vibration isolation under a harmonic disturbance (input) occurs for the frequency ratio $r > \sqrt{2}$, with a smaller value of the damping ratio (ζ) leading to better isolation. On the other hand, shock isolation must occur over a wide range of frequencies, usually with large values of ζ. Thus a good vibration isolation design proves to be a poor shock isolation design and vice versa. In spite of the differences, the basic principles involved in shock isolation are similar to those of vibration isolation; however the equations are different due to the transient nature of the shock.

A short-duration shock load $F(t)$, applied over a time period T, can be treated as an impulse $\underset{\sim}{F}$:

$$\underset{\sim}{F} = \int_0^T F(t)\, dt \tag{9.121}$$

Since this impulse acts on the mass m, the principle of impulse-momentum can be applied to find the velocity imparted to the mass (v) as

$$v = \frac{\underset{\sim}{F}}{m} \tag{9.122}$$

This indicates that the application of a short duration shock load can be considered as equivalent to giving an initial velocity to the system. Thus the response of the system under the shock load can be determined as the free vibration solution with a specified initial velocity. By assuming the initial conditions as $x(0) = x_0 = 0$ and $\dot{x}(0) = \dot{x}_0 = v$, the free vibration solution of a viscously damped single degree of freedom system (displacement of the mass m) can be found from Eq. (2.72) as

$$x(t) = \frac{v e^{-\zeta \omega_n t}}{\omega_d} \sin \omega_d t \tag{9.123}$$

where $\omega_d = \sqrt{1 - \zeta^2}\, \omega_n$ is the frequency of damped vibrations. The force transmitted to the foundation, $F_t(t)$, due to the spring and the damper is given by

$$F_t(t) = k x(t) + c \dot{x}(t) \tag{9.124}$$

Using Eq. (9.123), $F_t(t)$ can be expressed as

$$F_t(t) = \frac{v}{\omega_d} \sqrt{(k - c\zeta \omega_n)^2 + (c\omega_d)^2}\, e^{-\zeta \omega_n t} \sin(\omega_d t + \phi) \tag{9.125}$$

where

$$\phi = \tan^{-1} \left(\frac{c\omega_d}{k - c\zeta \omega_n} \right) \tag{9.126}$$

Equations (9.125) and (9.126) can be used to find the maximum value of the force transmitted to the foundation.

For longer-duration shock loads, the maximum transmitted force can occur while the shock is being applied. In such cases, the shock spectrum, discussed in Section 4.6, can be used to find the maximum force transmitted to the foundation.

The following examples illustrate different approaches that can be used for the design of shock isolators.

Isolation Under Shock

EXAMPLE 9.7

An electronic instrument of mass 20 kg is subject to a shock in the form of a step velocity of 2 m/s. If the maximum allowable values of deflection (due to clearance limit) and acceleration are specified as 20 mm and 25 g, respectively, determine the spring constant of an undamped shock isolator.

Solution: The electronic instrument supported on the spring can be considered as an undamped system subject to base motion (in the form of step velocity). The mass vibrates at the natural frequency of the system with the magnitudes of velocity and acceleration given by

$$\dot{x}_{max} = X \omega_n \tag{E.1}$$

$$\ddot{x}_{max} = -X \omega_n^2 \tag{E.2}$$

where X is the amplitude of displacement of the mass. Since the maximum value of (step) velocity is specified as 2 m/s and the maximum allowable value of X is given to be 0.02 m, Eq. (E.1) yields

$$X = \frac{\dot{x}_{max}}{\omega_n} < 0.02 \quad \text{or} \quad \omega_n > \frac{\dot{x}_{max}}{X} = \frac{2}{0.02} = 100 \text{ rad/s} \tag{E.3}$$

Similarly, using the maximum specified value of $\ddot{x}_{max}$ as 25 g, Eq. (E.2) gives

$$X \omega_n^2 \le 25 \,(9.81) = 245.25 \text{ m/s}^2 \quad \text{or} \quad \omega_n \le \sqrt{\frac{\ddot{x}_{max}}{X}} = \sqrt{\frac{245.25}{0.02}} = 110.7362 \text{ rad/s} \tag{E.4}$$

Equations (E.3) and (E.4) give 100 rad/s $\le \omega_n \le$ 110.7362 rad/s. By selecting the value of ω_n in the middle of the permissible range as 105.3681 rad/s, the stiffness of the spring (isolator) can be found as

$$k = m \omega_n^2 = 20 \,(105.3681)^2 = 2.2205 \times 10^5 \text{ N/m} \tag{E.5}$$

■

Isolation Under Step Load

EXAMPLE 9.8

A sensitive electronic instrument of mass 100 kg is supported on springs and packaged for shipment. During shipping, the package is dropped from a height that effectively applied a shock load of intensity F_0 to the instrument, as shown in Fig. 9.25(a). Determine the stiffness of the springs used in the package if the maximum deflection of the instrument is required to be less than 2 mm. The response spectrum of the shock load is shown in Fig. 9.25(b) with $F_0 = 1000$ N and $t_0 = 0.1$ s.

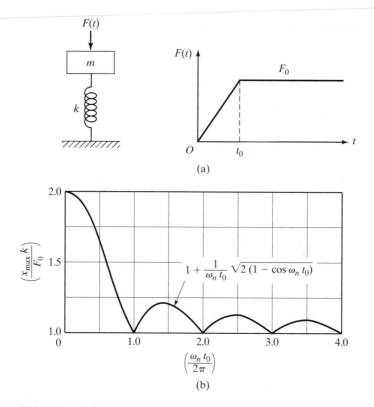

FIGURE 9.25   Shock load on electronic instrument.

Solution: The response spectrum, indicating the maximum response of an undamped single degree of freedom system subject to the given shock, is given by

$$\frac{x_{max}\, k}{F_0} = 1 + \frac{1}{\omega_n t_0}\sqrt{2\,(1 - \cos 2\,\omega_n t_0)} \qquad (E.1)$$

where ω_n is the natural frequency of the system:

$$\omega_n = \sqrt{\frac{k}{m}} = \sqrt{\frac{k}{100}} = 0.1\sqrt{k} \qquad (E.2)$$

$F_0 = 1000$ N, $t_0 = 0.1$ s, and k is the stiffness of the springs used in the package. Using the known data, Eq. (E.1) can be expressed as

$$\frac{x_{max}\, k}{1000} = 1 + \frac{1}{0.1\sqrt{k}\,(0.1)}\sqrt{2\,(1 - \cos 2\,(0.1\sqrt{k})\,(0.1))} \le \frac{2}{1000}\left(\frac{k}{1000}\right) \qquad (E.3)$$

By using the equality sign, Eq. (E.3) can be rearranged as

$$\frac{100}{\sqrt{k}}\sqrt{2(1-\cos 0.02\sqrt{k})} - 2\times 10^{-6}\,k + 1 = 0 \qquad\qquad \text{(E.4)}$$

The root of Eq. (E.4) gives the desired stiffness value as $k = 6.2615 \times 10^5$ N/m. The following MATLAB program can be used to find the root of Eq. (E.4):

```
>> x=1000:1:10000000;
>> f='(100/sqrt(x))*sqrt(2*(1-cos(0.02*sqrt(x))))-0.000002*x+1';
>> root=fzero(f,100000)

root =

  6.2615e+005

>>
```

■

9.10.6
Active Vibration
Control

A vibration isolation system is called active if it uses external power to perform its function. It consists of a servomechanism with a sensor, signal processor, and an actuator, as shown schematically in Fig. 9.26 [9.31–9.33]. This system maintains a constant distance (l) between the vibrating mass and the reference plane. As the force $F(t)$ applied to the system (mass) varies, the distance l tends to vary. This change in l is sensed by the sensor and a signal, proportional to the magnitude of the excitation (or response) of the vibrating body, is produced. The signal processor produces a command signal to the actuator based on the sensor signal it receives. The actuator develops a motion or force proportional to the command signal. The actuator motion or force will control the base displacement such that the distance l is maintained at the desired constant value.

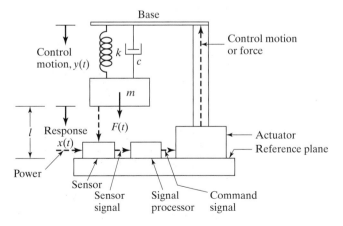

FIGURE 9.26   Active vibration isolation system.

Different types of sensors are available to create feedback signals based on the displacement, velocity, acceleration, jerk, or force. The signal processor may consist of a passive mechanism such as a mechanical linkage or an active electronic or fluidic network that can perform functions such as addition, integration, differentiation, attenuation, or amplification. The actuator may be a mechanical system such as a rack-and-pinion or ball screw mechanism, fluidic system or piezoelectric and electromagnetic force generating systems. Depending on the types of sensor, signal processor, and actuator used, an active vibration control system can be called electromechanical, electrofluidic, electromagnetic, piezoelectric, or fluidic.

9.11   Vibration Absorbers

A machine or system may experience excessive vibration if it is acted upon by a force whose excitation frequency nearly coincides with a natural frequency of the machine or system. In such cases, the vibration of the machine or system can be reduced by using a *vibration neutralizer* or *dynamic vibration absorber*, which is simply another spring mass system. The dynamic vibration absorber is designed such that the natural frequencies of the resulting system are away from the excitation frequency. We shall consider the analysis of a dynamic vibration absorber by idealizing the machine as a single degree of freedom system.

9.11.1 Undamped Dynamic Vibration Absorber

When we attach an auxiliary mass m_2 to a machine of mass m_1 through a spring of stiffness k_2, the resulting two degree of freedom system will look as shown in Fig. 9.27. The equations of motion of the masses m_1 and m_2 are

$$m_1\ddot{x}_1 + k_1 x_1 + k_2(x_1 - x_2) = F_0\sin \omega t$$
$$m_2\ddot{x}_2 + k_2(x_2 - x_1) = 0 \qquad (9.127)$$

By assuming harmonic solution,

$$x_j(t) = X_j\sin \omega t, \qquad j = 1, 2 \qquad (9.128)$$

we can obtain the steady-state amplitudes of the masses m_1 and m_2 as

$$X_1 = \frac{(k_2 - m_2\omega^2)F_0}{(k_1 + k_2 - m_1\omega^2)(k_2 - m_2\omega^2) - k_2^2} \qquad (9.129)$$

$$X_2 = \frac{k_2 F_0}{(k_1 + k_2 - m_1\omega^2)(k_2 - m_2\omega^2) - k_2^2} \qquad (9.130)$$

We are primarily interested in reducing the amplitude of the machine (X_1). In order to make the amplitude of m_1 zero, the numerator of Eq. (9.129) should be set equal to zero. This gives

$$\omega^2 = \frac{k_2}{m_2} \qquad (9.131)$$

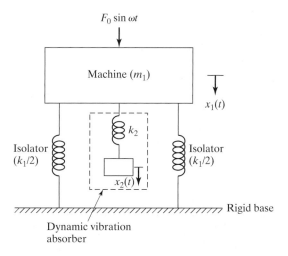

$F_0 \sin \omega t$

Machine (m_1)

$x_1(t)$

k_2

Isolator
($k_1/2$)

Isolator
($k_1/2$)

$x_2(t)$

Rigid base

Dynamic vibration
absorber

FIGURE 9.27   Undamped dynamic vibration
absorber.

If the machine, before the addition of the dynamic vibration absorber, operates near its resonance, $\omega^2 \simeq \omega_1^2 = k_1/m_1$. Thus if the absorber is designed such that

$$\omega^2 = \frac{k_2}{m_2} = \frac{k_1}{m_1} \tag{9.132}$$

the amplitude of vibration of the machine, while operating at its original resonant frequency, will be zero. By defining

$$\delta_{st} = \frac{F_0}{k_1}, \qquad \omega_1 = \left(\frac{k_1}{m_1}\right)^{1/2}$$

as the natural frequency of the machine or main system, and

$$\omega_2 = \left(\frac{k_2}{m_2}\right)^{1/2} \tag{9.133}$$

as the natural frequency of the absorber or auxiliary system, Eqs. (9.129) and (9.130) can be rewritten as

$$\frac{X_1}{\delta_{st}} = \frac{1 - \left(\dfrac{\omega}{\omega_2}\right)^2}{\left[1 + \dfrac{k_2}{k_1} - \left(\dfrac{\omega}{\omega_1}\right)^2\right]\left[1 - \left(\dfrac{\omega}{\omega_2}\right)^2\right] - \dfrac{k_2}{k_1}} \tag{9.134}$$

$$\frac{X_2}{\delta_{st}} = \frac{1}{\left[1 + \dfrac{k_2}{k_1} - \left(\dfrac{\omega}{\omega_1}\right)^2\right]\left[1 - \left(\dfrac{\omega}{\omega_2}\right)^2\right] - \dfrac{k_2}{k_1}} \tag{9.135}$$

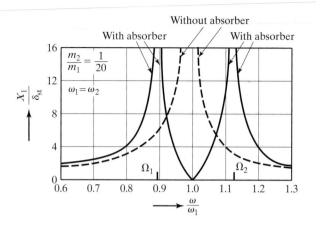

FIGURE 9.28   Effect of undamped vibration absorber on the response of machine.

Figure 9.28 shows the variation of the amplitude of vibration of the machine (X_1/δ_{st}) with the machine speed (ω/ω_1). The two peaks correspond to the two natural frequencies of the composite system. As seen before, $X_1 = 0$ at $\omega = \omega_1$. At this frequency, Eq. (9.135) gives

$$X_2 = -\frac{k_1}{k_2}\delta_{st} = -\frac{F_0}{k_2} \qquad (9.136)$$

This shows that the force exerted by the auxiliary spring is opposite to the impressed force ($k_2 X_2 = -F_0$) and neutralizes it, thus reducing X_1 to zero. The size of the dynamic vibration absorber can be found from Eqs. (9.136) and (9.132):

$$k_2 X_2 = m_2\omega^2 X_2 = -F_0 \qquad (9.137)$$

Thus the values of k_2 and m_2 depend on the allowable value of X_2.

It can be seen from Fig. 9.28 that the dynamic vibration absorber, while eliminating vibration at the known impressed frequency ω, introduces two resonant frequencies Ω_1 and Ω_2, at which the amplitude of the machine is infinite. In practice, the operating frequency ω must therefore be kept away from the frequencies Ω_1 and Ω_2. The values of Ω_1 and Ω_2 can be found by equating the denominator of Eq. (9.134) to zero. Noting that

$$\frac{k_2}{k_1} = \frac{k_2}{m_2}\frac{m_2}{m_1}\frac{m_1}{k_1} = \frac{m_2}{m_1}\left(\frac{\omega_2}{\omega_1}\right)^2 \qquad (9.138)$$

and setting the denominator of Eq. (9.134) to zero leads to

$$\left(\frac{\omega}{\omega_2}\right)^4\left(\frac{\omega_2}{\omega_1}\right)^2 - \left(\frac{\omega}{\omega_2}\right)^2\left[1 + \left(1 + \frac{m_2}{m_1}\right)\left(\frac{\omega_2}{\omega_1}\right)^2\right] + 1 = 0 \qquad (9.139)$$

The two roots of this equation are given by

$$\left.\begin{array}{r}\left(\dfrac{\Omega_1}{\omega_2}\right)^2 \\[2mm] \left(\dfrac{\Omega_2}{\omega_2}\right)^2\end{array}\right\} = \dfrac{\left\{\left[1 + \left(1 + \dfrac{m_2}{m_1}\right)\left(\dfrac{\omega_2}{\omega_1}\right)^2\right] \mp \left\{\left[1 + \left(1 + \dfrac{m_2}{m_1}\right)\left(\dfrac{\omega_2}{\omega_1}\right)^2\right]^2 - 4\left(\dfrac{\omega_2}{\omega_1}\right)^2\right\}^{1/2}\right\}}{2\left(\dfrac{\omega_2}{\omega_1}\right)^2} \tag{9.140}$$

which can be seen to be functions of (m_2/m_1) and (ω_2/ω_1).

Notes

1. It can be seen, from Eq. (9.140), that Ω_1 is less than and Ω_2 is greater than the operating speed (which is equal to the natural frequency, ω_1) of the machine. Thus the machine must pass through Ω_1 during start-up and stopping. This results in large amplitudes.

2. Since the dynamic absorber is tuned to one excitation frequency (ω), the steady-state amplitude of the machine is zero only at that frequency. If the machine operates at other frequencies or if the force acting on the machine has several frequencies, then the amplitude of vibration of the machine may become large.

3. The variations of Ω_1/ω_2 and Ω_2/ω_2 as functions of the mass ratio m_2/m_1 are shown in Fig. 9.29 for three different values of the frequency ratio ω_2/ω_1. It can be seen that the difference between Ω_1 and Ω_2 increases with increasing values of m_2/m_1.

▬▬▬▬▬▬▬▬▬▬ **Vibration Absorber for Diesel Engine**

EXAMPLE 9.9
_____

A diesel engine, weighing 3000 N, is supported on a pedestal mount. It has been observed that the engine induces vibration into the surrounding area through its pedestal mount at an operating speed of 6000 rpm. Determine the parameters of the vibration absorber that will reduce the vibration when mounted on the pedestal. The magnitude of the exciting force is 250 N, and the amplitude of motion of the auxiliary mass is to be limited to 2 mm.

Solution: The frequency of vibration of the machine is

$$f = \frac{6000}{60} = 100 \text{ Hz} \qquad \text{or} \qquad \omega = 628.32 \text{ rad/s}$$

Since the motion of the pedestal is to be made equal to zero, the amplitude of motion of the auxiliary mass should be equal and opposite to that of the exciting force. Thus from Eq. (9.137), we obtain

$$|F_0| = m_2 \omega^2 X_2 \tag{E.1}$$

Substitution of the given data yields

$$250 = m_2 \, (628.32)^2 (0.002)$$

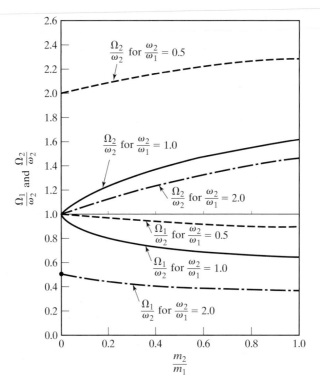

FIGURE 9.29   Variations of Ω_1 and Ω_2 given by Eq. (9.140)

Therefore $m_2 = 0.31665$ kg. The spring stiffness k_2 can be determined from Eq. (9.132):

$$\omega^2 = \frac{k_2}{m_2}$$

Therefore, $k_2 = (628.32)^2(0.31665) = 125009$ N/m.

■

EXAMPLE 9.10

■■■■■■■ **Absorber for Motor-Generator Set**

A motor-generator set, shown in Fig. 9.30 is designed to operate in the speed range of 2000 to 4000 rpm. However, the set is found to vibrate violently at a speed of 3000 rpm due to a slight unbalance in the rotor. It is proposed to attach a cantilever mounted lumped mass absorber system to eliminate the problem. When a cantilever carrying a trial mass of 2 kg tuned to 3000 rpm is attached to the set, the resulting natural frequencies of the system are found to be 2500 rpm and 3500 rpm. Design the absorber to be attached (by specifying its mass and stiffness) so that the natural frequencies of the total system fall outside the operating speed range of the motor-generator set.

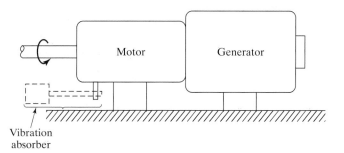

FIGURE 9.30   Motor-generator set.

Solution: The natural frequencies ω_1 of the motor-generator set and ω_2 of the absorber are given by

$$\omega_1 = \sqrt{\frac{k_1}{m_1}}, \qquad \omega_2 = \sqrt{\frac{k_2}{m_2}} \tag{E.1}$$

The resonant frequencies Ω_1 and Ω_2 of the combined system are given by Eq. (9.140). Since the absorber ($m = 2$ kg) is tuned, $\omega_1 = \omega_2 = 314.16$ rad/s (corresponding to 3000 rpm). Using the notation

$$\mu = \frac{m_2}{m_1}, \qquad r_1 = \frac{\Omega_1}{\omega_2}, \qquad \text{and} \qquad r_2 = \frac{\Omega_2}{\omega_2}$$

Eq. (9.140) becomes

$$r_1^2, r_2^2 = \left(1 + \frac{\mu}{2}\right) \mp \sqrt{\left(1 + \frac{\mu}{2}\right)^2 - 1} \tag{E.2}$$

Since Ω_1 and Ω_2 are known to be 261.80 rad/s (or 2500 rpm) and 366.52 rad/s (or 3500 rpm), respectively, we find that

$$r_1 = \frac{\Omega_1}{\omega_2} = \frac{261.80}{314.16} = 0.8333$$

$$r_2 = \frac{\Omega_2}{\omega_2} = \frac{366.52}{314.16} = 1.1667$$

Hence

$$r_1^2 = \left(1 + \frac{\mu}{2}\right) - \sqrt{\left(1 + \frac{\mu}{2}\right)^2 - 1}$$

or

$$\mu = \left(\frac{r_1^4 + 1}{r_1^2}\right) - 2 \tag{E.3}$$

Since $r_1 = 0.8333$, Eq. (E.3) gives $\mu = m_2/m_1 = 0.1345$ and $m_1 = m_2/0.1345 = 14.8699$ kg. The specified lower limit of Ω_1 is 2000 rpm or 209.44 rad/s, and so

$$r_1 = \frac{\Omega_1}{\omega_2} = \frac{209.44}{314.16} = 0.6667$$

With this value of r_1, Eq. (E.3) gives $\mu = m_2/m_1 = 0.6942$ and $m_2 = m_1(0.6942) = 10.3227$ kg. With these values, the second resonant frequency can be found from

$$r_2^2 = \left(1 + \frac{\mu}{2}\right) + \sqrt{\left(1 + \frac{\mu}{2}\right)^2 - 1} = 2.2497$$

which gives $\Omega_2 \simeq 4499.4$ rpm, larger than the specified upper limit of 4000 rpm. The spring stiffness of the absorber is given by

$$k_2 = \omega_2^2 m_2 = (314.16)^2 (10.3227) = 1.0188 \times 10^6 \text{ N/m}$$

∎

9.11.2
Damped
Dynamic
Vibration
Absorber

The dynamic vibration absorber described in the previous section removes the original resonance peak in the response curve of the machine but introduces two new peaks. Thus the machine experiences large amplitudes as it passes through the first peak during start-up and stopping. The amplitude of the machine can be reduced by adding a damped vibration absorber, as shown in Fig. 9.31. The equations of motion of the two masses are given by

$$m_1\ddot{x}_1 + k_1 x_1 + k_2 (x_1 - x_2) + c_2 (\dot{x}_1 - \dot{x}_2) = F_0 \sin \omega t \qquad (9.141)$$

$$m_2\ddot{x}_2 + k_2 (x_2 - x_1) + c_2 (\dot{x}_2 - \dot{x}_1) = 0 \qquad (9.142)$$

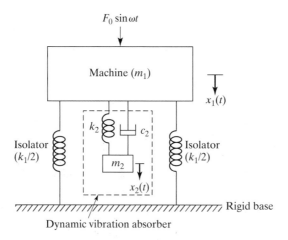

FIGURE 9.31   Damped dynamic vibration absorber.

By assuming the solution to be

$$x_j(t) = X_j e^{i\omega t}, \qquad j = 1, 2 \tag{9.143}$$

the steady-state solution of Eqs. (9.141) and (9.142) can be obtained:

$$X_1 = \frac{F_0(k_2 - m_2\omega^2 + ic_2\omega)}{[(k_1 - m_1\omega^2)(k_2 - m_2\omega^2) - m_2 k_2\omega^2] + i\omega c_2\,(k_1 - m_1\omega^2 - m_2\omega^2)} \tag{9.144}$$

$$X_2 = \frac{X_1(k_2 + i\omega c_2)}{(k_2 - m_2\omega^2 + i\omega c_2)} \tag{9.145}$$

By defining

$$\mu = m_2/m_1 = \text{Mass ratio} = \text{Absorber mass/main mass}$$

$$\delta_{st} = F_0/k_1 = \text{Static deflection of the system}$$

$$\omega_a^2 = k_2/m_2 = \text{Square of natural frequency of the absorber}$$

$$\omega_n^2 = k_1/m_1 = \text{Square of natural frequency of main mass}$$

$$f = \omega_a/\omega_n = \text{Ratio of natural frequencies}$$

$$g = \omega/\omega_n = \text{Forced frequency ratio}$$

$$c_c = 2m_2\omega_n = \text{Critical damping constant}$$

$$\zeta = c_2/c_c = \text{Damping ratio}$$

the magnitudes, X_1 and X_2, can be expressed as

$$\frac{X_1}{\delta_{st}} = \left[\frac{(2\zeta g)^2 + (g^2 - f^2)^2}{(2\zeta g)^2\,(g^2 - 1 + \mu g^2)^2 + \{\mu f^2 g^2 - (g^2 - 1)\,(g^2 - f^2)\}^2} \right]^{1/2} \tag{9.146}$$

and

$$\frac{X_2}{\delta_{st}} = \left[\frac{(2\zeta g)^2 + f^4}{(2\zeta g)^2(g^2 - 1 + \mu g^2)^2 + \{\mu f^2 g^2 - (g^2 - 1)(g^2 - f^2)\}^2} \right]^{1/2} \tag{9.147}$$

Equation (9.146) shows that the amplitude of vibration of the main mass is a function of μ, f, g, and ζ. The graph of

$$\left| \frac{X_1}{\delta_{st}} \right|$$

against the forced frequency ratio $g = \omega/\omega_n$ is shown in Fig. 9.32 for $f = 1$ and $\mu = 1/20$ for a few different values of ζ.

If damping is zero ($c_2 = \zeta = 0$), then resonance occurs at the two undamped resonant frequencies of the system, a result that is already indicated in Fig. 9.28. When the damping becomes infinite ($\zeta = \infty$), the two masses m_1 and m_2 are virtually clamped together,

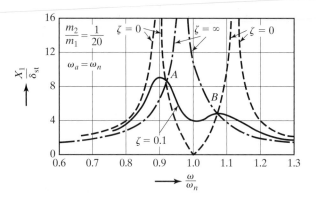

FIGURE 9.32   Effect of damped vibration absorber on the response of the machine.

and the system behaves essentially as a single degree of freedom system with a mass of $(m_1 + m_2) = (21/20)m$ and stiffness of k_1. In this case also, resonance occurs with $X_1 \rightarrow \infty$ at

$$g = \frac{\omega}{\omega_n} = \frac{1}{\sqrt{1 + \mu}} = 0.9759$$

Thus the peak of X_1 is infinite for $c_2 = 0$ as well as for $c_2 = \infty$. Somewhere in between these limits, the peak of X_1 will be a minimum.

Optimally Tuned Vibration Absorber.   It can be seen from Fig. 9.32 that all the curves intersect at points A and B regardless of the value of damping. These points can be located by substituting the extreme cases of $\zeta = 0$ and $\zeta = \infty$ into Eq. (9.146) and equating the two. This yields

$$g^4 - 2g^2 \left(\frac{1 + f^2 + \mu f^2}{2 + \mu} \right) + \frac{2f^2}{2 + \mu} = 0 \qquad (9.148)$$

The two roots of Eq. (9.148) indicate the values of the frequency ratio, $g_A = \omega_A/\omega$ and $g_B = \omega_B/\omega$, corresponding to the points A and B. The ordinates of A and B can be found by substituting the values of g_A and g_B, respectively, into Eq. (9.146). It has been observed [9.35] that the most efficient vibration absorber is one for which the ordinates of the points A and B are equal. This condition requires that [9.35]

$$f = \frac{1}{1 + \mu} \qquad (9.149)$$

An absorber satisfying Eq. (9.149) can be correctly called the *tuned vibration absorber.* Although Eq. (9.149) indicates how to tune an absorber, it does not indicate the optimal

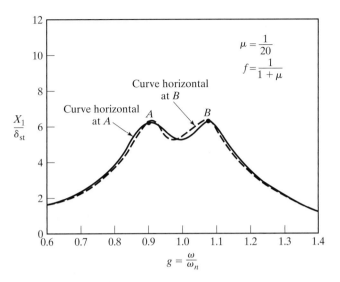

FIGURE 9.33   Tuned vibration absorber.

value of the damping ratio ζ and the corresponding value of X_1/δ_{st}. The optimal value of ζ can be found by making the response curve X_1/δ_{st} as flat as possible at peaks A and B. This can be achieved by making the curve horizontal at either A or B, as shown in Fig. 9.33. For this, first Eq. (9.149) is substituted into Eq. (9.146) to make the resulting equation applicable to the case of optimum tuning. Then the modified Eq. (9.146) is differentiated with respect to g to find the slope of the curve of X_1/δ_{st}. By setting the slope equal to zero at points A and B, we obtain

$$\zeta^2 = \frac{\mu\left\{3 - \sqrt{\dfrac{\mu}{\mu + 2}}\right\}}{8(1 + \mu)^3} \qquad \text{for point } A \qquad (9.150)$$

and

$$\zeta^2 = \frac{\mu\left\{3 + \sqrt{\dfrac{\mu}{\mu + 2}}\right\}}{8(1 + \mu)^3} \qquad \text{for point } B \qquad (9.151)$$

A convenient average value of ζ^2 given by Eqs. (9.150) and (9.151) is used in design so that

$$\zeta^2_{\text{optimal}} = \frac{3\mu}{8(1 + \mu)^3} \qquad (9.152)$$

The corresponding optimal value of $\left(\dfrac{X_1}{\delta_{st}}\right)$ becomes

$$\left(\frac{X_1}{\delta_{st}}\right)_{optimal} = \left(\frac{X_1}{\delta_{st}}\right)_{max} = \sqrt{1 + \frac{2}{\mu}} \qquad (9.153)$$

Notes

1. It can be seen from Eq. (9.147) that the amplitude of the absorber mass (X_2) is always much greater than that of the main mass (X_1). Thus the design should be able to accommodate the large amplitudes of the absorber mass.
2. Since the amplitudes of m_2 are expected to be large, the absorber spring (k_2) needs to be designed from a fatigue point of view.
3. Additional work relating to the optimum design of vibration absorbers can be found in Refs. [9.36–9.39].

9.12   Examples Using MATLAB

Plotting of Transmissibility

EXAMPLE 9.11

Using MATLAB, plot the variation of transmissibility of a single degree of freedom system with the frequency ratio, given by Eq. (9.94), corresponding to $\zeta = 0.0, 0.1, 0.2, 0.3, 0.4$ and 0.5.

Solution: The following MATLAB program plots the variation of transmissibility as a function of the frequency ratio using Eq. (9.94):

```
%Exam 9-2
for j = 1 : 5
    kesi = j * 0.1;
    for i = 1 : 1001
        w_wn(i) = 3 * (i - 1)/1000;
        T(i) = sqrt((1 + (2 * kesi * w_wn(i)) ^ 2)/((1 - w_wn(i) ^
        2) ^ 2 +
2 * kesi * w_wn(i) ^ 2));
    end;
    plot(w_wn, T);
    hold on;
end;

xlabel ('w/w_n');
ylabel('Tr');
gtext('zeta = 0.1');
gtext('zeta = 0.2');
gtext('zeta = 0.3');
gtext('zeta = 0.4');
gtext('zeta = 0.5');
title('Ex9.2');
grid on;
```

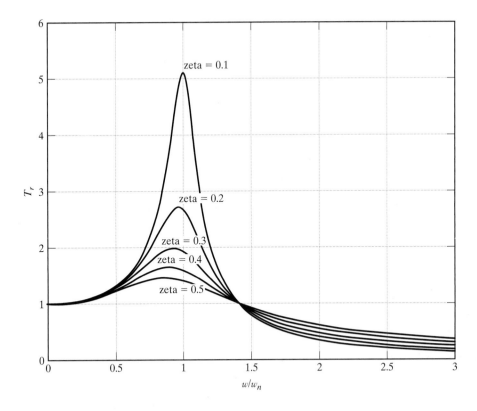

Vibration Amplitudes of Masses of Vibration Absorber

EXAMPLE 9.12

Using MATLAB, plot the variations of vibration amplitudes of the main and auxiliary masses of a vibration absorber, Eqs. (9.134) and (9.135), as functions of the frequency ratio.

Solution: Equations (9.134) and (9.135) are plotted for the following data: $f = \frac{\omega_a}{\omega_n} = 1, \zeta = 0.1$ and 0.5, $\mu = \frac{m_2}{m_1} = 0.05$ and 0.1.

```
f = 1;
%------    zeta = 0.1, mu=0.05 --------------------------------------- ⤶
-------
zeta = 0.1;
mu = 0.05;
g = 0.6 : 0.001 : 1.3;

tzg2 = (2.*zeta.*g).^2 ;%---    tzg2 = (2*zeta*g)^2
g2_f2_2 = (g.^2-f.^2).^2 ;% g2_f2_2 = (g^2-f^2)^2
g2_1mug2_2 = (g.^2-1+mu.*g.^2).^2;
muf2g2 = mu.*f.^2*g.^2 ;
g2_1 = g.^2-1 ;
g2_f2 = g.^2-f.^2 ;
```

```
x1r =sqrt((tzg2+g2_f2_2)./(tzg2.*g2_1mug2_2+(muf2g2-g2_1.*g2_f2).^2)));
x2r =sqrt((tzg2+f.^4)./(tzg2.*g2_1mug2_2+(muf2g2-g2_1.*g2_f2).^2)));
plot(g,x1r)
hold on
plot(g,x2r);
hold on
%------    zeta = 0.1, mu=0.1 -------------------------------------/
------
zeta = 0.1;
mu = 0.1; 0.001:1.3;
g = 0.6:


tzg2 = (2.*zeta.*g).^2 ;%---   tzg2 = (2*zeta*g)^2
g2_f2_2 = (g.^2-f.^2).^2 ;% g2_f2_2 = (g^2-f^2)^2
g2_1mug2_2 = (g.^2-1+mu.*g.^2).^2;
muf2g2 = mu.*f.^2*g.^2 ;
g2_1 = g.^2-1 ;
g2_f2 = g.^2-f.^2 ;


x1r =sqrt((tzg2+g2_f2_2)./(tzg2.*g2_1mug2_2+(muf2g2-g2_1.*g2_f2).^2)));
x2r =sqrt((tzg2+f.^4)./(tzg2.*g2_1mug2_2+(muf2g2-g2_1.*g2_f2).^2)));
plot(g,x1r,'-.');
hold on
plot(g,x2r,'-.');
hold on
%------    zeta = 0.5, mu=0.05 ------------------------------------/
------
zeta = 0.5;
mu = 0.05;
g = 0.6 : 0.001 : 1.3;


tzg2 = (2.*zeta.*g).^2 ;%---   tzg2 = (2*zeta*g)^2
g2_f2_2 = (g.^2-f.^2).^2) ;% g2_f2_2 = (g^2-f^2)^2
g2_1mug2_2 = (g.^2-1+mu.*g.^2).^2;
muf2g2 = mu.*f.^2*g.^2;
g2_1 = g.^2-1 ;
g2_f2 = g.^2-f.^2;


x1r =sqrt((tzg2+g2_f2_2)./(tzg2.*g2_1mug2_2+(muf2g2-g2_1.*g2_f2).^2));
x2r =sqrt((tzg2+f.^4)./(tzg2.*g2_1mug2_2+(muf2g2-g2_1.*g2_f2).^2));
plot(g,x1r,'--');
hold on
plot(g,x2r,'--');
hold on
%------    zeta = 0.5, mu=0.1 -------------------------------------/
------
zeta = 0.5;
mu = 0.1;
g = 0.6 : 0.001 : 1.3;


tzg2 = (2.*zeta.*g).^2 ;%---   tzg2 = (2*zeta*g)^2
g2_f2_2 = (g.^2-f.^2).^2 ;% g2_f2_2 = (g^2-f^2)^2
g2_1mug2_2 = (g.^2-1+mu.*g.^2).^2 ;
muf2g2 = mu.*f.^2*g.^2 ;
g2_1 = g.^2-1 ;
g2_f2 = g.^2-f.^2 ;


x1r =sqrt((tzg2+g2_f2_2)./(tzg2.*g2_1mug2_2+(muf2g2-g2_1.*g2_f2).^2));
x2r =sqrt((tzg2+f.^4)./(tzg2.*g2 1mug2 2+(muf2g2-g2 1.*g2 f2).^2));
plot(g,x1r,':');
hold on
plot(g,x2r,':');
```

```
xlabel('g')
ylabel('X1r and X2r')
axis ([0.6 1.3 0 16])
```

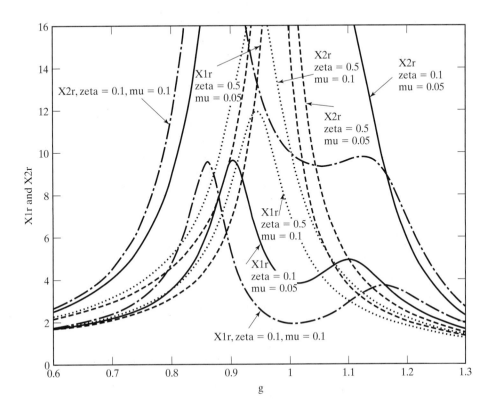

■

Resonant Frequencies of Vibration Absorber

EXAMPLE 9.13

Using MATLAB, plot the variations of the resonant frequency ratios given by Eq. (9.140) with the mass raio, $\frac{m_2}{m_1}$.

Solution: The ratios $\frac{\Omega_1}{\omega_2}$ and $\frac{\Omega_2}{\omega_2}$, given by Eq. (9.140), are plotted for $\frac{\omega_2}{\omega_1} = 0.5$, 1.0 and 2.0 over the range of $\frac{m_2}{m_1} = 0$ to 1.

```
%------ omega2/omega1=0.5 --------------------------------------------✓
---
omega21=0.5
```

```
m21 = 0:0.001:1.0
X11 = sqrt(((1 + (1+m21)*omega21.^2) +((1+(1+m21).*omega21.^2).^2-↙
4.*omega21.^2).^0.5)...
       /(2.*omega21.^2))
plot(m21,X11,':')
axis([0 1.0 0.0 2.6])
hold on


X12 = sqrt(((1+(1+m21)*omega21.^2) - ((1 + (1+m21).*omega21.^2).^2-↙
4.*omega21.^2).^0.5)...
       /(2.*omega21.^2))
plot(m21,X12,':')
hold on


%------ omega2/omega1=1.0 ------------------------------------↙
---


omega21=1.0
m21 = 0:0.001:1.0
X21 = sqrt(((1+(1+m21)*omega21.^2) +((1+(1+m21).*omega21.^2).^2-↙
4.*omega21.^2).^0.5)...
       /(2.*omega21.^2))
plot(m21,X21,'-')
axis([0 1.0 0.0 2.6])
hold on


X22 = sqrt(((1+(1+m21)*omega21.^2) -((1+(1+m21).*omega21.^2).^2-↙
4.*omega21.^2).^0.5)...
       /(2.*omega21.^2))
plot(m21,X22,'-')
hold on


%------ omega2/omega1=2.0-------------------------------------↙
---


omega21=2.0
m21 = 0 : 0.001 : 1.0
X31 = sqrt(((1+(1+m21)*omega21.^2) +((1+(1+m21).*omega21.^2).^2).↙
^2-4.*omega21.^2).^0.5)...
       / (2.*omega21.^2))
plot(m21,X31,'-.')
axis([0 1.0 0.0 2.6])
hold on


X32 = sqrt(((1+(1+m21)*omega21.^2) -((1+(1+m21).*omega21.^2).^2-4.↙
*omega21.^2) .^0.5)...
      /(2.*omega21.^2))
plot(m21,X32,'-.')
hold on
xlabel ('mr')
ylabel ('OM1 and OM2')
```

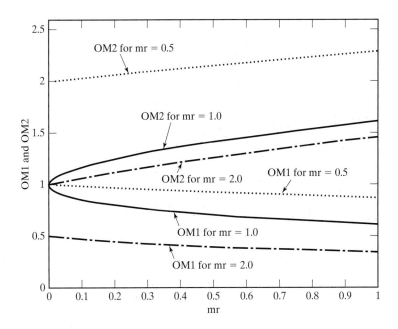

Two-Plane Balancing

EXAMPLE 9.14

Develop a general MATLAB program called **Program 13.m** for the two-plane balancing of rotating machines. Use the program to solve Example 9.2.

Solution: Program 13.m is developed to accept the vectors $\vec{V}_A$, $\vec{V}_B$, $\vec{V}'_A$, $\vec{V}'_B$, $\vec{V}''_A$, $\vec{V}''_B$, $\vec{W}_L$, and $\vec{W}_R$ as input in the form of two-dimensional arrays VA, VB, VAP, VBP, VAPP, VBPP, WL, and WR, respectively. The program gives the vectors B_L and B_R as output in the form of two-dimensional arrays BL and BR indicating the magnitude and position of the balancing weights in the left and right planes, respectively. The listing of the program and the output are given below.

```
%===================================================================
%
% Program13.m
% Two-plane balancing
%
%===================================================================
% Run "Program13" in MATLAB command window. Progrm13.m, balan.m,
vsub.m,
% vdiv.m and vmult.m should be in the same folder,and set the
MATLAB path
% to this folder.
% following 8 lines contain problem-dependent data
va=[8.5 60];
vap=[6 125];
```

```
wl=[10 270];
vb=[6.5 205];
vbp=[4.5 230];
vapp=[6 35];
vbpp=[10.5 160];
wr=[12 180];
% end of problem-dependent data
[bl,br]=balan(va,vb,vap,vbp,vapp,vbpp,wl,wr);
fprintf('                  Results of two-plane balancing \n\n');
fprintf('Left-plane balancing weight   Right-plane balancing weight');
fprintf('\n\n');
fprintf('Magnitude=%8.6f            Magnitude=%8.6f \n\n',bl(1),br(1));
fprintf('Angel=%8.6f               Angel=%8.6f   \n\n',bl(2),br(2));
%===================================================================
%
%Function Balan.m
%
%===================================================================
function [bl,br]=balan(va,vb,vap,vbp,vapp,vbpp,wl,wr);
pi=180/3.1415926;
va(2)=va(2)/pi;
p(1)=va(1);
p(2)=va(2);
va(1)=p(1)*cos(p(2));
va(2)=p(1)*sin(p(2));
vb(2)=vb(2)/pi;
p(1)=vb(1);
p(2)=vb(2);
vb(1)=p(1)*cos(p(2));
vb(2)=p(1)*sin(p(2));
vap(2)=vap(2)/pi;
p(1)=vap(1);
p(2)=vap(2);
vap(1)=p(1)*cos(p(2));
vap(2)=p(1)*sin(p(2));
vbp(2)=vbp(2)/pi;
p(1)=vbp(1);
p(2)=vbp(2);
vbp(1)=p(1)*cos(p(2));
vbp(2)=p(1)*sin(p(2));
vapp(2)=vapp(2)/pi;
p(1)=vapp(1);
p(2)=vapp(2);
vapp(1)=p(1)*cos(p(2));
vapp(2)=p(1)*sin(p(2));
vbpp(2)=vbpp(2)/pi;
p(1)=vbpp(1);
p(2)=vbpp(2);
vbpp(1)=p(1)*cos(p(2));
vbpp(2)=p(1)*sin(p(2));
wl(2)=wl(2)/pi;
p(1)=wl(1);
p(2)=wl(2);
wl(1)=p(1)*cos(p(2));
wl(2)=p(1)*sin(p(2));
wr(2)=wr(2)/pi;
p(1)=wr(1);
p(2)=wr(2);
wr(1)=p(1)*cos(p(2));
wr(2)=p(1)*sin(p(2));
[r]=vsub(vap,va);
[aal]=vdiv(r,wl);
[s]=vsub(vbp,vb);
[abl]=vdiv(s,wl);
[p]=vsub(vapp,va);
[aar]=vdiv(p,wr);
[q]=vsub(vbpp,vb);
```

```
[abr]=vdiv(q,wr);
[ar1]=sqrt(aar(1)^2+aar(2)^2);
[ar2]=atan(aar(2)/aar(1))*pi;
[al1]=sqrt(aal(1)^2+aal(2)^2);
[al2]=atan(aal(2)/aal(1))*pi;
[r]=vmult(abl,va);
[s]=vmult(aal,vb);
[vap]=vsub(r,s);
[r]=vmult(aar,abl);
[s]=vmult(aal,abr);
[vbp]=vsub(r,s);
[ur]=vdiv(vap,vbp);
[r]=vmult(abr,va);
[s]=vmult(aar,vb);
[vap]=vsub(r,s);
[r]=vmult(abr,aal);
[s]=vmult(aar,abl);
[vbp]=vsub(r,s);
[ul]=vdiv(vap,vbp);
bl(1)=sqrt(ul(1)^2+ul(2)^2);
a1=ul(2)/ul(1);
bl(2)=atan(ul(2)/ul(1));
br(1)=sqrt(ur(1)^2+ur(2)^2);
a2=ur(2)/ur(1);
br(2)=atan(ur (2)/ur (1));
bl(2)=bl(2)*pi;
br(2)=br(2)*pi;
bl(2)=bl(2)+180;
br(2)=br(2)+180;


%===================================================================
%
%Function vdiv.m
%
%===================================================================
function [c]=vdiv(a,b);
c(1)=(a(1)*b(1)+a(2)*b(2))/(b(1)^2+b(2)^2);
c(2)=(a(2)*b(1)-a(1)*b(2))/(b(1)^2+b(2)^2);
%===================================================================
%
%Function vmult.m
%
%===================================================================
function [c]=vmult(a,b);
c(1)=a(1)*b(1)-a(2)*b(2);
c(2)=a(2)*b(1)+a(1)*b(2);


%===================================================================
%
%Function vsub.m
%
%===================================================================
function [c]=vsub(a,b);
c(1)=a(1)-b(1);
c(2)=a(2)-b(2);
```

Results of two-plane balancing

Left-plane balancing weight	Right-plane balancing weight
Magnitude=10.056139	Magnitude=5.877362
Angel=145.554799	Angel=248.255931

■

9.13   C++ Program

A C++ program, called **Program13.cpp** is given for the two-plane balancing of rotating machines. The input and output of the program are similar to those of the program **Program13.m**.

<table>
<tr><td>EXAMPLE 9.15</td><td>Two-Plane Balancing Using Program13.cpp</td></tr>
</table>

Using **Program13.cpp**, solve Example 9.2.

Solution: The output of the program is given below.

RESULTS OF TWO-PLANE BALANCING

LEFT-PLANE BALANCING WEIGHT   RIGHT-PLANE BALANCING WEIGHT

MAGNITUDE = 10.05613891   MAGNITUDE = 5.87736171
ANGLE = 145.55479890   ANGLE = 248.25593054

■

9.14   Fortran Program

A Fortran program called **PROGRAM13.F** is given for the two-plane balancing of rotating machines. The input and output of **PROGRAM13.F** are similar to those of **Program13.m**.

<table>
<tr><td>EXAMPLE 9.16</td><td>Two-Plane Balancing Using PROGRAM13.F</td></tr>
</table>

Using **PROGRAM13.F**, solve Example 9.2.

Solution: The output of the program is given below.

RESULTS OF TWO-PLANE BALANCING

LEFT-PLANE BALANCING WEIGHT   RIGHT-PLANE BALANCING WEIGHT

MAGNITUDE= 0.10056140E+02   MAGNITUDE= 0.58773613E+01
ANGLE=   0.14555478E+03   ANGLE=   0.24825594E+03

■

REFERENCES

9.1 J. E. Ruzicka, "Fundamental concepts of vibration control," *Sound and Vibration*, Vol. 5, July 1971, pp. 16–22.

9.2 J. A. Macinante, *Seismic Mountings for Vibration Isolation*, Wiley, New York, 1984.

9.3 International Organization for Standardization, *Mechanical Vibration of Machines with Operating Speeds from 10 to 200 rev/s—Basis for Specifying Evaluation Standards*, ISO 2372, 1974.

9.4 International Organization for Standardization, *Evaluation and Measurement of Vibration in Buildings*, Draft Proposal, ISO DP 4866, 1975.

9.5 R. J. Steffens, "Some aspects of structural vibration," in *Proceedings of the Symposium on Vibrations in Civil Engineering*, B. O. Skipp (ed.), Butterworths, London, 1966, pp. 1–30.

9.6 International Organization for Standardization, *Guide for the Evaluation of Human Exposure to Whole-Body Vibration*, ISO 2631, 1974.

9.7 C. Zenz, *Occupational Medicine: Principles and Practical Application* (2nd ed.) Year Book Medical Publishers, Chicago, 1988.

9.8 R. L. Fox, "Machinery vibration monitoring and analysis techniques," *Sound and Vibration*, Vol. 5, November 1971, pp. 35–40.

9.9 D. G. Stadelbauer, "Dynamic balancing with microprocessors," *Shock and Vibration Digest*, Vol. 14, December 1982, pp. 3–7.

9.10 J. Vaughan, *Static and Dynamic Balancing* (2nd ed.), Bruel and Kjaer Application Notes, Naerum, Denmark.

9.11 R. L. Baxter, "Dynamic balancing," *Sound and Vibration*, Vol. 6, April 1972, pp. 30–33.

9.12 J. H. Harter and W. D. Beitzel, *Mathematics Applied to Electronics*, Reston Publishing, Reston, Virginia, 1980.

9.13 R. G. Loewy and V. J. Piarulli, "Dynamics of rotating shafts," *Shock and Vibration Monograph SVM-4*, Shock and Vibration Information Center, Naval Research Laboratory, Washington, D. C., 1969.

9.14 J. D. Irwin and E. R. Graf, *Industrial Noise and Vibration Control*, Prentice Hall, Englewood Cliffs, N.J., 1979.

9.15 T. Iwatsuba, "Vibration of rotors through critical speeds," *Shock and Vibration Digest*, Vol. 8, No. 2, February 1976, pp. 89–98.

9.16 R. J. Trivisonno, "Fortran IV computer program for calculating critical speeds of rotating shafts," NASA TN D-7385, 1973.

9.17 R. E. D. Bishop and G. M. L. Gladwell, "The vibration and balancing of an unbalanced flexible rotor," *Journal of Mechanical Engineering Science*, Vol. 1, 1959, pp. 66–77.

9.18 A. G. Parkinson, "The vibration and balancing of shafts rotating in asymmetric bearings," *Journal of Sound and Vibration*, Vol. 2, 1965, pp. 477–501.

9.19 C. E. Crede, *Vibration and Shock Isolation*, Wiley, New York, 1951.

9.20 W. E. Purcell, "Materials for noise and vibration control," *Sound and Vibration*, Vol. 16, July 1982, pp. 6–31.

9.21 B. C. Nakra, "Vibration control with viscoelastic materials," *Shock and Vibration Digest*, Vol. 8, No. 6, June 1976, pp. 3–12.

9.22 G. R. Tomlinson, "The use of constrained layer damping in vibration control," in *Modern Practice in Stress and Vibration Analysis*, J. E. Mottershead (ed.), Pergamon Press, Oxford, 1989, pp. 99–107.

9.23 D. E. Baxa and R. A. Dykstra, "Pneumatic isolation systems control forging hammer vibration," *Sound and Vibration*, Vol. 14, May 1980, pp. 22–25.

9.24 E. I. Rivin, "Vibration isolation of industrial machinery—Basic considerations," *Sound and Vibration*, Vol. 12, November 1978, pp. 14–19.

9.25 C. M. Salerno and R. M. Hochheiser, "How to select vibration isolators for use as machinery mounts," *Sound and Vibration*, Vol. 7, August 1973, pp. 22–28.

9.26 C. A. Mercer and P. L. Rees, "An optimum shock isolator," *Journal of Sound and Vibration*, Vol. 18, 1971, pp. 511–520.

9.27 M. L. Munjal, "A rational synthesis of vibration isolators," *Journal of Sound and Vibration*, Vol. 39, 1975, pp. 247–263.

9.28 C. Ng and P. F. Cunniff, "Optimization of mechanical vibration isolation systems with multi-degrees of freedom," *Journal of Sound and Vibration*, Vol. 36, 1974, pp. 105–117.

9.29 S. K. Hati and S. S. Rao, "Cooperative solution in the synthesis of multidegree of freedom shock isolation systems," *Journal of Vibration, Acoustics, Stress, and Reliability in Design*, Vol. 105, 1983, pp. 101–103.

9.30 S. S. Rao and S. K. Hati, "Optimum design of shock and vibration isolation systems using game theory," *Journal of Engineering Optimization*, Vol. 4, 1980, pp. 1–8.

9.31 J. E. Ruzicka, "Active vibration and shock isolation," Paper no. 680747, *SAE Transactions*, Vol. 77, 1969, pp. 2872–2886.

9.32 R. W. Horning and D. W. Schubert, "Air suspension and active vibration-isolation systems," in *Shock and Vibration Handbook* (3rd ed.), C. M. Harris (ed.) McGraw-Hill, New York, 1988.

9.33 O. Vilnay, "Active control of machinery foundation," *Journal of Engineering Mechanics, ASCE*, Vol. 110, 1984, pp. 273–281.

9.34 J. I. Soliman and M. G. Hallam, "Vibration isolation between non-rigid machines and non-rigid foundations," *Journal of Sound and Vibration*, Vol. 8, 1968, pp. 329–351.

9.35 J. Ormondroyd and J. P. Den Hartog, "The theory of the dynamic vibration absorber," *Transactions of ASME*, Vol. 50, 1928, p. APM-241.

9.36 H. Puksand, "Optimum conditions for dynamic vibration absorbers for variable speed systems with rotating and reciprocating unbalance," *International Journal of Mechanical Engineering Education*, Vol. 3, April 1975, pp. 145–152.

9.37 A. Soom and M.-S. Lee, "Optimal design of linear and nonlinear absorbers for damped systems," *Journal of Vibration, Acoustics, Stress, and Reliability in Design*, Vol. 105, 1983, pp. 112–119.

9.38 J. B. Hunt, *Dynamic Vibration Absorbers*, Mechanical Engineering Publications, London, 1979.

REVIEW QUESTIONS

9.1 Give brief answers to the following:

1. Name some sources of industrial vibration.
2. What are the various methods available for vibration control?
3. What is single-plane balancing?
4. Describe the two-plane balancing procedure.
5. What is whirling?
6. What is the difference between stationary damping and rotary damping?
7. How is the critical speed of a shaft determined?
8. What causes instability in a rotor system?

9. What considerations are to be taken into account for the balancing of a reciprocating engine?
10. What is the function of a vibration isolator?
11. What is a vibration absorber?
12. What is the difference between a vibration isolator and a vibration absorber?
13. Does spring mounting always reduce the vibration of the foundation of a machine?
14. Is it better to use a soft spring in the flexible mounting of a machine? Why?
15. Is the shaking force proportional to the square of the speed of a machine? Does the vibratory force transmitted to the foundation increase with the speed of the machine?
16. Why does dynamic balancing imply static balancing?
17. Explain why dynamic balancing can never be achieved by a static test alone.
18. Why does a rotating shaft always vibrate? What is the source of the shaking force?
19. Is it always advantageous to include a damper in the secondary system of a dynamic vibration absorber?
20. What is active vibration isolation?
21. Explain the difference between passive and active isolations.

9.2 Indicate whether each of the following statements is true or false:

1. Vibration can cause structural and mechanical failures.
2. The response of a system can be reduced by the use of isolators and absorbers.
3. Vibration control means the elimination or reduction of vibration.
4. The vibration caused by a rotating unbalanced disc can be eliminated by adding a suitable mass to the disc.
5. Any unbalanced mass can be replaced by two equivalent unbalanced masses in the end planes of the rotor.
6. The oil whip in the bearings can cause instability in a rotor system.
7. The natural frequency of a system can be changed by varying its damping.
8. The stiffness of a rotating shaft can be altered by changing the location of its bearings.
9. All practical systems have damping.
10. High loss factor of a material implies less damping.
11. Passive isolation systems require external power to function.
12. The transmissibility is also called the transmission ratio.
13. The force transmitted to the foundation of an isolator with rigid foundation can never be infinity.
14. Internal and external friction can cause instability in a rotating shaft at speeds above the first critical speed.

9.3 Fill in each of the following blanks with the appropriate word:

1. Even a small excitation force can cause an undesirably large response near _____.
2. The use of close tolerances and better surface finish for machine parts tends a machine _____ susceptible to vibration.
3. The presence of unbalanced mass in a rotating disc causes _____.
4. When the speed of rotation of a shaft equals one of the natural frequencies of the shaft, it is called _____ speed.
5. The moving elements of a reciprocating engine are the crank, the connecting rod, and the _____.

6. The vertical component of the inertia force of a reciprocating engine has primary and _____ parts.
7. Laminated structures have _____ damping.
8. Materials with a large value of the loss factor are subject to _____ stress.
9. Vibration isolation involves insersion of a resilient member between the vibrating mass and the _____ of vibration.
10. Cork is a _____ isolator.
11. An active isolator consists of a sensor, signal processor, and an _____.
12. Vibration neutralizer is also known as dynamic vibration _____.
13. Although an undamped vibration absorber removes the original resonance peak of the response, it introduces _____ new peaks.
14. The single-plane balancing is also known as _____ balancing.
15. Phase marks are used in _____ plane balancing using a vibration analyzer.
16. Machine errors can cause _____ in rotating machines.
17. The combustion instabilities are a source of _____ in engines.
18. The deflection of a rotating shaft becomes very large at the _____ speed.
19. Oil whip in bearings can cause _____ in a flexible rotor system.

9.4   Select the most appropriate answer out of the multiple choices given:

1. An example of a source of vibration that can not be altered is:
   (a) atmospheric turbulance
   (b) hammer blow
   (c) tire stiffness of an automobile.
2. The two-plane balancing is also known as:
   (a) static balancing
   (b) dynamic balancing
   (c) proper balancing
3. The unbalanced force caused by an eccentric mass m rotating at an angular speed ω and located at a distance r from the axis of rotation is
   (a) $m\, r^2 \omega^2$ (b) $m\, g\, \omega^2$ (c) $m\, r\, \omega^2$
4. The following material has high internal damping:
   (a) cast iron   (b) copper   (c) brass
5. Transmissibility is the ratio of
   (a) force transmitted and exciting force
   (b) force applied and the resulting displacement
   (c) input displacement and output displacement.
6. Mechanical impedance is the ratio of
   (a) force transmitted and exciting force
   (b) force applied and force transmitted
   (c) applied force and displacement
7. Vibration can be eliminated on the basis of theoretical analysis
   (a) sometimes   (b) always   (c) never
8. A long rotor can be balanced by adding weights in
   (a) a single plane   (b) any two planes   (c) two specific planes

9. The damping caused by the internal friction of a shaft material is called
   (a) stationary damping
   (b) external damping
   (c) rotary damping

10. The damping caused by the bearing support structure of a rotating shaft is called
    (a) stationary damping
    (b) internal damping
    (c) rotary damping

11. An undamped vibration absorber removes the original resonance peak but introduces
    (a) one new peak   (b) two new peaks   (c) several new peaks

9.5 Match the items in the two columns below.

1. Control natural frequency	(a) Introduce damping
2. Avoid excessive response at resonance	(b) Use vibration isolator
3. Reduce transmission of excitation force from one part to another	(c) Add vibration absorber
4. Reduce response of the system	(d) Avoid resonance

PROBLEMS

The problem assignments are organized as follows:

Problems	Section Covered	Topic Covered
9.1–9.2	9.2	Vibration criteria
9.3–9.14	9.4	Balancing of rotating machines
9.15–9.21	9.5	Critical speeds of rotating shafts
9.22–9.26	9.6	Balancing of reciprocating engines
9.27–9.50	9.10	Vibration isolation
9.51–9.64	9.11	Vibration absorbers
9.65–9.68	9.12	MATLAB programs
9.69	9.13	C++ program
9.70–9.71	9.14	Fortran program
9.72	—	Design project

9.1 An automobile moving on a rough road, in the form of a sinusoidal surface, is modeled as a spring-mass system, as shown in Fig. 9.34. The sinusoidal surface has a wave length of 5 m and an amplitude of $Y = 1$ mm. If the mass of the automobile, including the passengers, is 1500 kg and the stiffness of the suspension system (k) is 400 kN/m, determine the range of speed (v) of the automobile in which the passengers perceive the vibration. Suggest possible methods of improving the design for a more comfortable ride of the passengers.

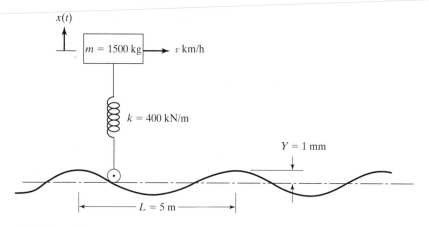

FIGURE 9.34

9.2 The root mean square value of a signal $x(t)$, x_{rms}, is defined as

$$x_{rms} = \left\{ \lim_{T \to \infty} \frac{1}{T} \int_0^T x^2(t) \, dt \right\}^{1/2}$$

Using this definition, find the root mean square values of the displacement (x_{rms}), velocity ($\dot{x}_{rms}$) and acceleration ($\ddot{x}_{rms}$) corresponding to $x(t) = X \cos \omega t$.

9.3 Two identical discs are connected by four bolts of different sizes and mounted on a shaft, as shown in Fig. 9.35. The masses and locations of three bolts are as follows:

$m_1 = 35$ grams, $r_1 = 110$ mm, and $\theta_1 = 40°$; $m_2 = 15$ grams, $r_2 = 90$ mm, and $\theta_2 = 220°$; and $m_3 = 25$ grams, $r_3 = 130$ mm, $\theta_3 = 290°$. Find the mass and location of the fourth bolt (m_c, r_c, and θ_c), which results in the static balance of the discs.

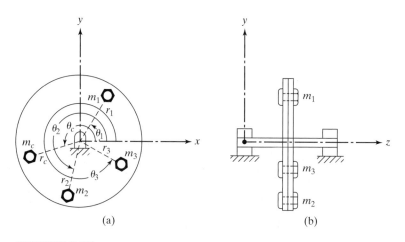

(a)   (b)

FIGURE 9.35

9.4 Four holes are drilled in a uniform circular disc at a radius of 4 in. and angles of 0°, 60°, 120°, and 180°. The weight removed at holes 1 and 2 is 4 oz each and the weight removed at holes 3 and 4 is 5 oz each. If the disc is to be balanced statically by drilling a fifth hole at a radius of 5 in., find the weight to be removed and the angular location of the fifth hole.

9.5 Three masses, weighing 0.5 lb, 0.7 lb, and 1.2 lb, are attached around the rim, of diameter 30 in., of a flywheel at the angular locations $\theta = 10°$, 100°, and 190°, respectively. Find the weight and the angular location of the fourth mass to be attached on the rim that leads to the dynamic balance of the flywheel.

9.6 The amplitude and phase angle due to original unbalance in a grinding wheel operating at 1200 rpm are found to be 10 mils and 40° counterclockwise from the phase mark. When a trial weight $W = 6$ oz is added at 65° clockwise from the phase mark and at a radial distance 2.5 in. from the center of rotation, the amplitude and phase angle are observed to be 19 mils and 150° counterclockwise. Find the magnitude and angular position of the balancing weight if it is to be located 2.5 in. radially from the center of rotation.

9.7 An unbalanced flywheel shows an amplitude of 6.5 mils and a phase angle of 15° clockwise from the phase mark. When a trial weight of magnitude 2 oz is added at an angular position 45° counterclockwise from the phase mark, the amplitude and the phase angle become 8.8 mils and 35° counterclockwise, respectively. Find the magnitude and angular position of the balancing weight required. Assume that the weights are added at the same radius.

9.8 In order to determine the unbalance in a grinding wheel, rotating clockwise at 2400 rpm, a vibration analyzer is used and an amplitude of 4 mils and a phase angle of 45° are observed with the original unbalance. When a trial weight $W = 4$ oz is added at 20° clockwise from the phase mark, the amplitude becomes 8 mils and the phase angle 145°. If the phase angles are measured counterclockwise from the right-hand horizontal, calculate the magnitude and location of the necessary balancing weight.

9.9 A turbine rotor is run at the natural frequency of the system. A stroboscope indicates that the maximum displacement of the rotor occurs at an angle 229° in the direction of rotation. At what angular position must mass be removed from the rotor in order to improve its balancing?

9.10 A rotor, having three eccentric masses in different planes, is shown in Fig. 9.36. The axial, radial, and angular locations of mass m_i are given by l_i, r_i, and θ_i, respectively, for $i = 1, 2, 3$. If the rotor is to be dynamically balanced by locating two masses m_{b1} and m_{b2} at radii r_{b1} and r_{b2} at the angular locations θ_{b1} and θ_{b2}, as shown in Fig. 9.36, derive expressions for $m_{b1}r_{b1}$, $m_{b2}r_{b2}$, θ_{b1}, and θ_{b2}.

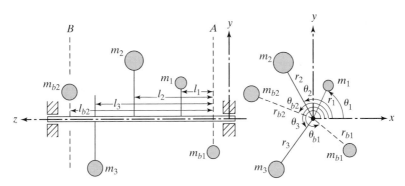

FIGURE 9.36

9.11 The rotor shown in Fig. 9.37(a) is balanced temporarily in a balancing machine by adding the weights $W_1 = W_2 = 0.2$ lb in the plane A and $W_3 = W_4 = 0.2$ lb in the plane D at a radius of 3 in., as shown in Fig. 9.37(b). If the rotor is permanently balanced by drilling holes at a radius of 4 in. in planes B and C, determine the position and amount of material to be removed from the rotor. Assume that the adjustable weights W_1 to W_4 will be removed from the planes A and D.

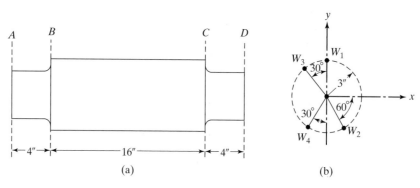

FIGURE 9.37

9.12 Weights of 2 lb, 4 lb, and 3 lb are located at radii 2 in., 3 in., and 1 in. in the planes C, D, and E, respectively, on a shaft supported at the bearings B and F, as shown in Fig. 9.38. Find the weights and angular locations of the two balancing weights to be placed in the end planes A and G so that the dynamic load on the bearings will be zero.

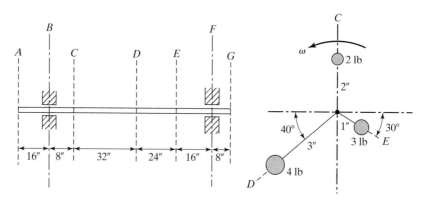

FIGURE 9.38

9.13 The data obtained in a two-plane balancing procedure are given in the table below. Determine the magnitude and angular position of the balancing weights, assuming that all angles are measured from an arbitrary phase mark and all weights are added at the same radius.

Condition	Amplitude (mils)		Phase Angle	
	Bearing A	**Bearing B**	**Bearing A**	**Bearing B**
Original unbalance	5	4	100°	180°
W_L = 2 oz added at 30° in the left plane	6.5	4.5	120°	140°
W_R = 2 oz added at 0° in the right plane	6	7	90°	60°

9.14 Figure 9.39 shows a rotating system in which the shaft is supported in bearings at A and B. The three masses m_1, m_2, and m_3 are connected to the shaft as indicated in the figure. (a) Find the bearing reactions at A and B if the speed of the shaft is 1000 rpm. (b) Determine the locations and magnitudes of the balancing masses to be placed at a radius of 0.25 m in the planes L and R, which can be assumed to pass through the bearings A and B.

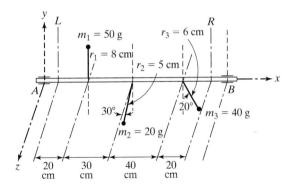

FIGURE 9.39

9.15 A flywheel, with a weight of 100 lb and an eccentricity of 0.5 in., is mounted at the center of a steel shaft of diameter 1 in. If the length of the shaft between the bearings is 30 in. and the rotational speed of the flywheel is 1200 rpm, find (a) the critical speed, (b) the vibration amplitude of the rotor, and (c) the force transmitted to the bearing supports.

9.16 Derive the expression for the stress induced in a shaft with an unbalanced concentrated mass located midway between two bearings.

9.17 A steel shaft of diameter 2.5 cm and length 1 m is supported at the two ends in bearings. It carries a turbine disc, of mass 20 kg and eccentricity 0.005 m, at the middle and operates at 6000 rpm. The damping in the system is equivalent to viscous damping with $\zeta = 0.01$. Determine the whirl amplitude of the disc at (a) operating speed, (b) critical speed, and (c) 1.5 times the critical speed.

9.18 Find the bearing reactions and the maximum bending stress induced in the shaft at (a) operating speed, (b) critical speed, and (c) 1.5 times the critical speed for the shaft-rotor system described in Problem 9.17.

9.19 Solve Problem 9.17 by assuming that the material of the shaft is aluminum rather than steel.

9.20 Solve Problem 9.18 by assuming that the material of the shaft is aluminum rather than steel.

9.21 A shaft, having a stiffness of 3.75 MN/m, rotates at 3600 rpm. A rotor, having a mass of 60 kg and an eccentricity of 2000 micron, is mounted on the shaft. Determine (a) the steady state whirl amplitude of the rotor and (b) the maximum whirl amplitude of the rotor during start-up and stopping conditions. Assume the damping ratio of the system as 0.05.

9.22 The cylinders of a four-cylinder in-line engine are placed at intervals of 12 in. in the axial direction. The cranks have the same length, 4 in., and their angular positions are given by 0°, 180°, 180°, and 0°. If the length of the connecting rod is 10 in. and the reciprocating weight is 2 lb for each cylinder, find the unbalanced forces and moments at a speed of 3000 rpm, using the center line through cylinder 1 as the reference plane.

9.23 The reciprocating mass, crank radius, and connecting rod length of each of the cylinders in a two-cylinder in-line engine is given by m, r, and l, respectively. The crank angles of the two cylinders are separated by 180°. Find the unbalanced forces and moments in the engine.

9.24 A 4-cylinder in-line engine has a reciprocating weight of 3 lb, a stroke of 6 in., and a connecting rod of length 10 in. in each cylinder. The cranks are separated by 4 in. axially and 90° radially, as shown in Fig. 9.40. Find the unbalanced primary and secondary forces and moments with respect to the reference plane shown in Fig. 9.40 at an engine speed of 1500 rpm.

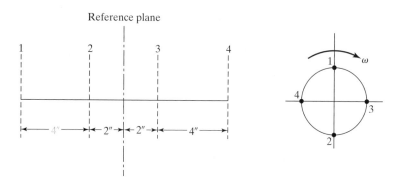

FIGURE 9.40

9.25 The arrangement of cranks in a 6-cylinder in-line engine is shown in Fig. 9.41. The cylinders are separated by a distance a in the axial direction, and the angular positions of the cranks are given by $\alpha_1 = \alpha_6 = 0°$, $\alpha_2 = \alpha_5 = 120°$, and $\alpha_3 = \alpha_4 = 240°$. If the crank length, connecting rod length, and the reciprocating mass of each cylinder is r, l, and m, respectively, find the primary and secondary unbalanced forces and moments with respect to the reference plane indicated in Fig. 9.41.

9.26 A single-cylinder engine has a total mass of 150 kg. Its reciprocating mass is 5 kg, and the rotating mass is 2.5 kg. The stroke ($2r$) is 15 cm, and the speed is 600 rpm. (a) If the engine

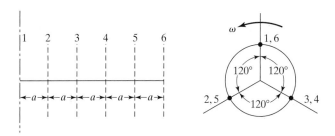

FIGURE 9.41

is mounted floating on very weak springs, what is the amplitude of vertical vibration of the engine? (b) If the engine is mounted solidly on a rigid foundation, what is the alternating force amplitude transmitted? Assume the connecting rod to be of infinite length.

9.27 An electronic instrument is to be isolated from a panel that vibrates at frequencies ranging from 25 Hz to 35 Hz. It is estimated that at least 80 percent vibration isolation must be achieved to prevent damage to the instrument. If the instrument weights 85 N, find the necessary static deflection of the isolator.

9.28* An exhaust fan, having a small unbalance, weights 800 N and operates at a speed of 600 rpm. It is desired to limit the response to a transmissibility of 2.5 as the fan passes through resonance during start-up. In addition, an isolation of 90 percent is to be achieved at the operating speed of the fan. Design a suitable isolator for the fan.

9.29* An air compressor of mass 500 kg has an eccentricity of 50 kg-cm, and operates at a speed of 300 rpm. The compressor is to be mounted on one of the following mountings: (a) An isolator consisting of a spring with negligible damping, and (b) a shock absorber having a damping ratio of 0.1 and negligible stiffness. Select a suitable mounting and specify the design details by considering the static deflection of the compressor, the transmission ratio, and the amplitude of vibration of the compressor.

9.30 The armature of a variable speed electric motor, of mass 200 kg, has an unbalance due to manufacturing errors. The motor is mounted on an isolator having a stiffness of 10 kN/m, and a dashpot having a damping ratio of 0.15. (a) Find the speed range over which the amplitude of the fluctuating force transmitted to the foundation will be larger than the exciting force. (b) Find the speed range over which the transmitted force amplitude will be less than 10 percent of the exciting force amplitude.

9.31 A dishwashing machine weighing 150 lb operates at 300 rpm. Find the minimum static deflection of an isolator that provides 60 percent isolation. Assume that the damping in the isolator is negligible.

9.32 A washing machine of mass 50 kg operates at 1200 rpm. Find the maximum stiffness of an isolator that provides 75 percent isolation. Assume that the damping ratio of the isolator is 7 percent.

9.33 It is found that an exhaust fan, of mass 80 kg and operating speed 1000 rpm, produces a repeating force of 10,000 N on its rigid base. If the maximum force transmitted to the base is to be limited to 2000 N using an undamped isolator, determine (a) the maximum permissible stiffness of the isolator that serves the purpose; (b) the steady-state amplitude of the exhaust

---

*The asterisk denotes a problem with no unique answer.

fan with the isolator that has the maximum permissible stiffness; and (c) the maximum amplitude of the exhaust fan with isolation during start-up.

9.34 It has been found that a printing press, of mass 300 kg and operating speed 3000 rpm, produces a repeating force of 30,000 N when attached to a rigid foundation. Find a suitable viscously damped isolator to satisfy the following requirements: (a) the static deflection should be as small as possible; (b) the steady-state amplitude should be less than 2.5 mm; (c) the amplitude during start-up conditions should not exceed 20 mm; and (d) the force transmitted to the foundation should be less than 10,000 N.

9.35 A compressor of mass 120 kg has a rotating unbalance of 0.2 kg-m. If an isolator of stiffness 0.5 MN/m and damping ratio 0.06 is used, find the range of operating speeds of the compressor over which the force transmitted to the foundation will be less than 2500 N.

9.36 An internal combustion engine has a rotating unbalance of 1.0 kg-m and operates between 800 and 2000 rpm. When attached directly to the floor, it transmitted a force of 7,018 N at 800 rpm and 43,865 N at 2000 rpm. Find the stiffness of the isolator that is necessary to reduce the force transmitted to the floor to 6,000 N over the operating speed range of the engine. Assume that the damping ratio of the isolator is 0.08, and the mass of the engine is 200 kg.

9.37 A small machine tool of mass 100 kg operates at 600 rpm. Find the static deflection of an undamped isolator that provides 90 percent isolation.

9.38 A diesel engine of mass 300 kg and operating speed 1800 rpm is found to have a rotating unbalance of 1 kg-m. It is to be installed on the floor of an industrial plant for purposes of emergency power generation. The maximum permissible force that can be transmitted to the floor is 8000 N and the only type of isolator available has a stiffness of 1 MN/m and a damping ratio of 5 percent. Investigate possible solutions to the problem.

9.39 The force transmitted by an i.c. engine of mass 500 kg, when placed directly on a rigid floor, is given by

$$F_t(t) = (18000 \cos 300\, t + 3600 \cos 600\, t)\ \text{N}$$

Design an undamped isolator so that the maximum magnitude of the force transmitted to the floor does not exceed 12,000 N.

9.40 Design the suspension of an automobile such that the maximum vertical acceleration felt by the driver is less than 2g at all speeds between 40 and 80 mph while traveling on a road whose surface varies sinusoidally as $y(u) = 0.5 \sin 2\, u$ ft where u is the horizontal distance in feet. The weight of the automobile, with the driver, is 1500 lb and the damping ratio of the suspension is to be 0.05. Use a single degree of freedom model for the automobile.

9.41 Consider a single degree of freedom system with Coulomb damping (which offers a constant friction force, F_c). Derive an expression for the force transmissibility when the mass is subjected to a harmonic force, $F(t) = F_0 \sin \omega t$.

9.42 Consider a single degree of freedom system with Coulomb damping (which offers a constant friction force, F_c). Derive expressions for the absolute and relative displacement transmissibilities when the base is subjected to a harmonic displacement, $y(t) = Y \sin \omega t$.

9.43 When a washing machine, of mass 200 kg and an unbalance 0.02 kg-m, is mounted on an isolator, the isolator deflects by 5 mm under the static load. Find (a) the amplitude of the washing machine and (b) the force transmitted to the foundation at the operating speed of 1200 rpm.

9.44 An electric motor, of mass 60 kg, rated speed 3000 rpm, and an unbalance 0.002 kg-m, is to be mounted on an isolator to achieve a force transmissibility of less than 0.25. Determine (a) the stiffness of the isolator, (b) the dynamic amplitude of the motor, and (c) the force transmitted to the foundation.

9.45 An engine is mounted on a rigid foundation through four springs. During operation, the engine produces an excitation force at a frequency of 3000 rpm. If the weight of the engine causes the springs to deflect by 10 mm, determine the reduction in the force transmitted to the foundation.

9.46 A sensitive electronic system, of mass 30 kg, is supported by a spring-damper system on the floor of a building that is subject to a harmonic motion in the frequency range 10–75 Hz. If the damping ratio of the suspension is 0.25, determine the stiffness of the suspension if the amplitude of vibration transmitted to the system is to be less than 15 percent of the floor vibration over the given frequency range.

9.47 A machine weighing 2600 lb is mounted on springs. A piston of weight $w = 60$ lb moves up and down in the machine at a speed of 600 rpm with a stroke of 15 in. Considering the motion to be harmonic, determine the maximum force transmitted to the foundation if (a) $k = 10000$ lb/in., and (b) $k = 25000$ lb/in.

9.48 A printed circuit board of mass 1 kg is supported to the base through an undamped isolator. During shipping, the base is subjected to a harmonic disturbance (motion) of amplitude 2 mm and frequency 2 Hz. Design the isolator so that the displacement transmitted to the printed circuit board is to be no more than 5 percent of the base motion.

9.49 An electronic instrument of mass 10 kg is mounted on an isolation pad. If the base of the isolation pad is subjected to a shock in the form of a step velocity of 10 mm/s, find the stiffness of the isolation pad if the maximum permissible values of deflection and acceleration of the instrument are specified as 10 mm and 20 g, respectively.

9.50 A water tank of mass 10^5 kg is supported on a reinforced cement concrete column, as shown in Fig. 9.42(a). When a projectile hits the tank, it causes a shock, in the form of a step force, as shown in Fig. 9.42(b). Determine the stiffness of the column if the maximum deflection of the tank is to be limited to 0.5 m. The response spectrum of the shock load is shown in Fig. 9.42(c).

9.51 An air compressor of mass 200 kg, with an unbalance of 0.01 kg-m, is found to have a large amplitude of vibration while running at 1200 rpm. Determine the mass and spring constant of the absorber to be added if the natural frequencies of the system are to be at least 20 percent from the impressed frequency.

9.52 An electric motor, having an unbalance of 2 kg-cm, is mounted at the end of a steel cantilever beam, as shown in Fig. 9.43. The beam is observed to vibrate with large amplitudes at the operating speed of 1500 rpm of the motor. It is proposed to add a vibration absorber to reduce the vibration of the beam. Determine the ratio of the absorber mass to the mass of the motor needed in order to have the lower frequency of the resulting system equal to 75 percent of the operating speed of the motor. If the mass of the motor is 300 kg, determine the stiffness and mass of the absorber. Also find the amplitude of vibration of the absorber mass.

9.53* The pipe carrying feedwater to a boiler in a thermal power plant has been found to vibrate violently at a pump speed of 800 rpm. In order to reduce the vibrations, an absorber consisting of a spring of stiffness k_2 and a trial mass m'_2 of 1 kg is attached to the pipe. This arrangement is found to give the natural frequencies of the system as 750 rpm and 1000 rpm. It is desired to keep the natural frequencies of the system outside the operating speed range

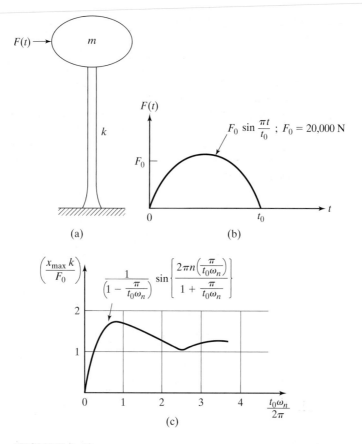

FIGURE 9.42

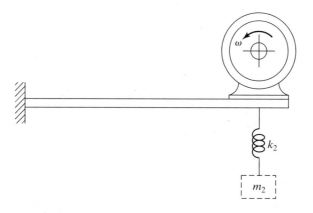

FIGURE 9.43

of the pump, which is 700 rpm to 1040 rpm. Determine the values of k_2 and m_2 that satisfy this requirement.

9.54 A reciprocating engine is installed on the first floor of a building, which can be modeled as a rigid rectangular plate resting on four elastic columns. The equivalent weight of the engine and the floor is 2000 lb. At the rated speed of the engine, which is 600 rpm, the operators experience large vibration of the floor. It has been decided to reduce these vibrations by suspending a spring-mass system from the bottom surface of the floor. Assume that the spring stiffness is $k_2 = 5000$ lb/in. (a) Find the weight of the mass to be attached to absorb the vibrations. (b) What will be the natural frequencies of the system after the absorber is added?

9.55* Find the values of k_2 and m_2 in Problem 9.54 in order to have the natural frequencies of the system at least 30 percent away from the forcing frequency.

9.56* A hollow steel shaft of outer diameter 2 in., inner diameter 1.5 in., and length 30 in. carries a solid disc of diameter 15 in. and weight 100 lb. Another hollow steel shaft of length 20 in., carrying a solid disc of diameter 6 in. and weight 20 lb, is attached to the first disc, as shown in Fig. 9.44. Find the inner and outer diameters of the shaft such that the attached shaft-disc system acts as an absorber.

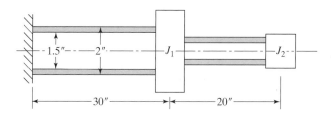

FIGURE 9.44

9.57* A rotor, having a mass moment of inertia $J_1 = 15$ kg-m^2, is mounted at the end of a steel shaft having a torsional stiffness of 0.6 MN-m/rad. The rotor is found to vibrate violently when subjected to a harmonic torque of $300 \cos 200t$ N-m. A tuned absorber, consisting of a torsional spring and a mass moment of inertia (k_{t2} and J_2), is to be attached to the first rotor to absorb the vibrations. Find the values of k_{t2} and J_2 such that the natural frequencies of the system are away from the forcing frequency by at least 20 percent.

9.58 Plot the graphs of (Ω_1/ω_2) against (m_2/m_1) and (Ω_2/ω_2) against (m_2/m_1) as (m_2/m_1) varies from 0 to 1.0 when $\omega_2/\omega_1 = 0.1$ and 10.0.

9.59 Determine the operating range of the frequency ratio ω/ω_2 for an undamped vibration absorber to limit the value of $|X_1/\delta_{st}|$ to 0.5. Assume that $\omega_1 = \omega_2$ and $m_2 = 0.1 m_1$.

9.60 When an undamped vibration absorber, having a mass 30 kg and a stiffness k, is added to a spring-mass system, of mass 40 kg and stiffness 0.1 MN/m, the main mass (40 kg mass) is found to have zero amplitude during its steady-state operation under a harmonic force of amplitude 300 N. Determine the steady-state amplitude of the absorber mass.

9.61 An electric motor, of mass 20 kg and operating speed 1350 rpm, is placed on a fixed-fixed steel beam of width 15 cm and depth 12 cm, as shown in Fig. 9.45. The motor has a rotating unbalance of 0.1 kg-m. The amplitude of vibration of the beam under steady-state operation of the motor is suppressed by attaching an undamped vibration absorber underneath the

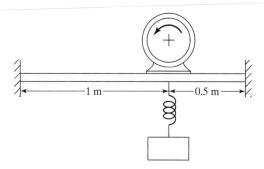

FIGURE 9.45

motor, as shown in Fig. 9.45. Determine the mass and stiffness of the absorber such that the amplitude of the absorber mass is less than 2 cm.

9.62 A bridge is found to vibrate violently when a vehicle, producing a harmonic load of magnitude 600 N, crosses it. By modeling the bridge as an undamped spring mass system with a mass 15,000 kg and a stiffness 2 MN/m, design a suitable tuned damped vibration absorber. Determine the improvement achieved in the amplitude of the bridge with the absorber.

9.63 A small motor, weighing 100 lb, is found to have a natural frequency of 100 rad/s. It is proposed that an undamped vibration absorber weighing 10 lb be used to suppress the vibrations when the motor operates at 80 rad/s. Determine the necessary stiffness of the absorber.

9.64 Consider the system shown in Fig. 9.46 in which a harmonic force acts on the mass m. Derive the condition under which the steady-state displacement of mass m will be zero.

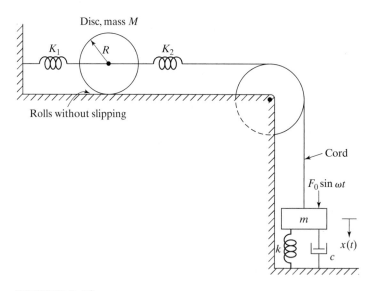

FIGURE 9.46

9.65 Using MATLAB, plot Eq. (9.94) for $\zeta = 0, 0.25, 0.5, 0.75$, and 1 over the range $0 \le r \le 3$.

9.66 Using MATLAB, plot Eqs. (9.134) and (9.135) for $f = 1$, $\zeta = 0.2, 0.3$, and 0.4, and $\mu = 0.2$ and 0.5 over the range $0.6 \le \frac{\omega}{\omega_1}$.

9.67 Using MATLAB, plot the ratios $\frac{\Omega_1}{\omega_2}$ and $\frac{\Omega_2}{\omega_2}$ given by Eq. (9.140) for $\frac{\omega_2}{\omega_1} = 1.5, 3.0$, and 4.5 and $\frac{m_2}{m_1} = 0$ to 1.

9.68 Using Program 13.m, solve Problem 9.13.

9.69 Using Program 13.cpp, solve Problem 9.13.

9.70 Using PROGRAM 13.F, solve Problem 9.13.

9.71 Write a computer program to find the displacement of the main mass and the auxiliary mass of a damped dynamic vibration absorber. Use this program to generate the results of Fig. 9.32.

DESIGN PROJECT

9.72 Ground vibrations from a crane operation, a forging press, and an air compressor are transmitted to a nearby milling machine and are found to be detrimental to achieving specified accuracies during precision milling operations. The ground vibrations at the locations of the

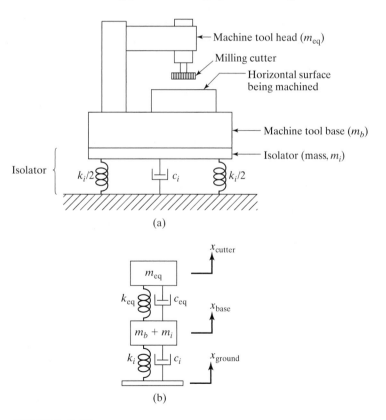

FIGURE 9.47

crane, forging press, and air compressor are given by $x_c(t) = A_c e^{-\omega_c \zeta_c t} \sin \omega_c t$, $x_f(t) = A_f \sin \omega_f t$, and $x_a(t) = A_a \sin \omega_a t$, respectively, where $A_c = 20 \, \mu\text{m}$, $A_f = 30 \, \mu\text{m}$, $A_a = 25 \, \mu\text{m}$, $\omega_c = 10 \, \text{Hz}$, $\omega_f = 15 \, \text{Hz}$, $\omega_a = 20 \, \text{Hz}$, and $\zeta_c = 0.1$. The ground vibrations travel at the shear wave velocity of the soil, which is equal to 980 ft/sec, and the amplitudes attenuate according to the relation $A_r = A_0 e^{-0.005r}$, where A_0 is the amplitude at the source and A_r is the amplitude at a distance of r ft from the source. The crane, forging press, and air compressor are located at a distance of 60 ft, 80 ft, and 40 ft, respectively, from the milling machine. The equivalent mass, stiffness, and damping ratio of the machine tool head in vertical vibration (at the location of the cutter) are experimentally determined to be 500 kg, 480 kN/m, and 0.15, respectively. The equivalent mass of the machine tool base is 1000 kg. It is proposed that an isolator for the machine tool be used, as shown in Fig. 9.47 to improve the cutting accuracies [9.2]. Design a suitable vibration isolator, consisting of a mass, spring, and damper, as shown in Fig. 9.47(b), for the milling machine such that the maximum vertical displacement of the milling cutter, relative to the horizontal surface being machined, due to ground vibration from all the three sources does not exceed 5 μm peak-to-peak.

C H A P T E R   1 0

Vibration Measurement and Applications

10.1 Introduction

In practice the measurement of vibration becomes necessary due to the following reasons:

1. The increasing demands of higher productivity and economical design lead to higher operating speeds of machinery[1] and efficient use of materials through lightweight structures. These trends make the occurrence of resonant conditions more frequent during the operation of machinery and reduce the reliability of the system. Hence the periodic measurement of vibration characteristics of machinery and structures becomes essential to ensure adequate safety margins. Any observed shift in the natural frequencies or other vibration characteristics will indicate either a failure or a need for maintenance of the machine.

2. The measurement of the natural frequencies of a structure or machine is useful in selecting the operational speeds of nearby machinery to avoid resonant conditions.

3. The theoretically computed vibration characteristics of a machine or structure may be different from the actual values due to the assumptions made in the analysis.

---

[1]According to Eshleman, in Ref. [10.12], the average speed of rotating machines doubled—from 1800 rpm to 3600 rpm—during the period between 1940 and 1980.

4. The measurement of frequencies of vibration and the forces developed is necessary in the design and operation of active vibration isolation systems.

5. In many applications, the survivability of a structure or machine in a specified vibration environment is to be determined. If the structure or machine can perform the expected task even after completion of testing under the specified vibration environment, it is expected to survive the specified conditions.

6. Continuous systems are often approximated as multidegree of freedom systems for simplicity. If the measured natural frequencies and mode shapes of a continuous system are comparable to the computed natural frequencies and mode shapes of the multidegree of freedom model, then the approximation will be proved to be a valid one.

7. The measurement of input and the resulting output vibration characteristics of a system helps in identifying the system in terms of its mass, stiffness, and damping.

8. The information about ground vibrations due to earthquakes, fluctuating wind velocities on structures, random variation of ocean waves, and road surface roughness are important in the design of structures, machines, oil platforms, and vehicle suspension systems.

Vibration Measurement Scheme.   Figure 10.1 illustrates the basic features of a vibration measurement scheme. In this figure, the motion (or dynamic force) of the vibrating body is converted into an electrical signal by the vibration transducer or pickup. In general, a transducer is a device that transforms changes in mechanical quantities (such as displacement, velocity, acceleration, or force) into changes in electrical quantities (such as voltage or current). Since the output signal (voltage or current) of a transducer is too small to be recorded directly, a signal conversion instrument is used to amplify the signal to the required value. The output from the signal conversion instrument can be presented on a display unit for visual inspection, or recorded by a recording unit or stored in a computer for later use. The data can then be analyzed to determine the desired vibration characteristics of the machine or structure.

Depending on the quantity measured, a vibration measuring instrument is called a vibrometer, a velocity meter, an accelerometer, a phase meter, or a frequency meter. If the instrument is designed to record the measured quantity, then the suffix "meter" is to be replaced by "graph" [10.1]. In some application, we need to vibrate a machine or structure to find its resonance characteristics. For this, electrodynamic vibrators, electrohydraulic vibrators, and signal generators (oscillators) are used.

The following considerations often dictate the type of vibration measuring instruments to be used in a vibration test: (1) expected ranges of the frequencies and amplitudes, (2) sizes of the machine/structure involved, (3) conditions of operation of the

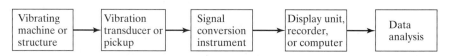

FIGURE 10.1   Basic vibration measurement scheme.

machine/equipment/structure, and (4) type of data processing used (such as graphical display or graphical recording or storing the record in digital form for computer processing).

10.2 Transducers

A transducer is a device that transforms values of physical variables into equivalent electrical signals. Several types of transducers are available; some of them are less useful than others due to their nonlinearity or slow response. Some of the transducers commonly used for vibration measurement are given below.

10.2.1 Variable Resistance Transducers

In these transducers, a mechanical motion produces a change in electrical resistance (of a rheostat, a strain gage, or a semiconductor), which in turn causes a change in the output voltage or current. The schematic diagram of an electrical resistance strain gage is shown in Fig. 10.2. An electrical resistance strain gage consists of a fine wire whose resistance changes when it is subjected to mechanical deformation. When the strain gage is bonded to a structure, it experiences the same motion (strain) as the structure and hence its resistance change gives the strain applied to the structure. The wire is sandwiched between two sheets of thin paper. The strain gage is bonded to the surface where the strain is to be measured. The most common gage material is a copper-nickel alloy known as Advance. When the surface undergoes a normal strain (ϵ), the strain gage also undergoes the same strain and the resulting change in its resistance is given by [10.6]

$$K = \frac{\Delta R/R}{\Delta L/L} = 1 + 2v + \frac{\Delta r}{r}\frac{L}{\Delta L} \approx 1 + 2v \tag{10.1}$$

where

$$K = \text{Gage factor for the wire}$$

$$R = \text{Initial resistance}$$

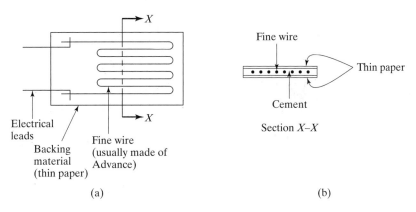

(a)   (b)

FIGURE 10.2   Electric resistance strain gage.

ΔR = Change in resistance

L = Initial length of the wire

ΔL = Change in length of the wire

v = Poisson's ratio of the wire

r = Resistivity of the wire

Δr = Change in resistivity of the wire $\approx$ 0 for Advance

The value of the gage factor K is given by the manufacturer of the strain gage and hence the value of ϵ can be determined, once ΔR and R are measured, as

$$\epsilon = \frac{\Delta L}{L} = \frac{\Delta R}{RK} \tag{10.2}$$

In a vibration pickup[2] the strain gage is mounted on an elastic element of a spring mass system, as shown in Fig. 10.3. The strain at any point on the cantilever (elastic member) is proportional to the deflection of the mass, $x(t)$, to be measured. Hence the strain indicated by the strain gage can be used to find $x(t)$. The change in resistance of the wire ΔR can be measured using a Wheatstone bridge, potentiometer circuit, or voltage divider. A typical Wheatstone bridge, representing a circuit which is sensitive to small changes in the resistance, is shown in Fig. 10.4. A d.c. voltage V is applied across the points a and c. The resulting voltage across the points b and d is given by [10.6]:

$$E = \left[\frac{R_1 R_3 - R_2 R_4}{(R_1 + R_2)(R_3 + R_4)} \right] V \tag{10.3}$$

Initially the resistances are balanced (adjusted) so that the output voltage E is zero. Thus, for initial balance, Eq. (10.3) gives

$$R_1 R_3 = R_2 R_4 \tag{10.4}$$

When the resistances (R_i) change by small amounts (ΔR_i), the change in the output voltage ΔE can be expressed as

$$\Delta E \approx V r_0 \left(\frac{\Delta R_1}{R_1} - \frac{\Delta R_2}{R_2} + \frac{\Delta R_3}{R_3} - \frac{\Delta R_4}{R_4} \right) \tag{10.5}$$

where

$$r_0 = \frac{R_1 R_2}{(R_1 + R_2)^2} = \frac{R_3 R_4}{(R_3 + R_4)^2} \tag{10.6}$$

---

[2]When a transducer is used in conjunction with other components that permit the processing and transmission of the signal, the device is called a *pickup*.

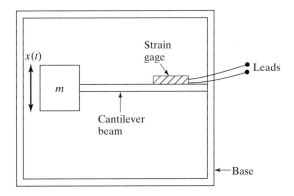

FIGURE 10.3 Strain gage as vibration pickup.

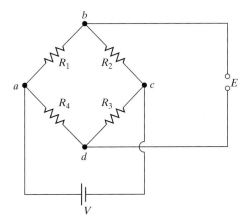

FIGURE 10.4 Wheatstone bridge.

If the strain gage leads are connected between the points a and b, $R_1 = R_g$, $\Delta R_1 = \Delta R_g$, and $\Delta R_2 = \Delta R_3 = \Delta R_4 = 0$, and Eq. (10.5) gives

$$\frac{\Delta R_g}{R_g} = \frac{\Delta E}{V r_0} \qquad (10.7)$$

where R_g is the initial resistance of the gage. Equations (10.2) and (10.7) yield

$$\frac{\Delta R_g}{R_g} = \epsilon K = \frac{\Delta E}{V r_0}$$

or

$$\Delta E = K V r_0 \epsilon \tag{10.8}$$

Since the output voltage is proportional to the strain, it can be calibrated to read the strain directly.

**10.2.2
Piezoelectric
Transducers**

Certain natural and manufactured materials like quartz, tourmaline, lithium sulfate, and Rochelle salt generate electrical charge when subjected to a deformation or mechanical stress (see Fig. 10.5a). The electrical charge disappears when the mechanical loading is removed. Such materials are called piezoelectric materials and the transducers, which take advantage of the piezoelectric effect, are known as piezoelectric transducers. The charge generated in the crystal due to a force F_x is given by

$$Q_x = k F_x = k A p_x \tag{10.9}$$

where k is called the piezoelectric constant, A is the area on which the force F_x acts, and p_x is the pressure due to F_x. The output voltage of the crystal is given by

$$E = v t p_x \tag{10.10}$$

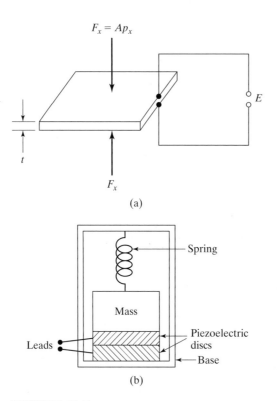

(a)

(b)

FIGURE 10.5   Piezoelectric accelerometer.

where v is called the voltage sensitivity and t is the thickness of the crystal. The values of the piezoelectric constant and voltage sensitivity for quartz are 2.25×10^{-12} C/N and 0.055 volt-meter/N, respectively [10.6]. These values are valid only when the perpendicular to the largest face is along the x-axis of the crystal. The electric charge developed and the voltage output will be different if the crystal slab is cut in a different direction.

A typical piezoelectric transducer (accelerometer) is shown in Fig. 10.5(b). In this figure, a small mass is spring loaded against a piezoelectric crystal. When the base vibrates, the load exerted by the mass on the crystal changes with acceleration and hence the output voltage generated by the crystal will be proportional to the acceleration. The main advantages of the piezoelectric accelerometer include compactness, ruggedness, high sensitivity, and high frequency range [10.5, 10.8].

Output Voltage of a Piezoelectric Transducer

EXAMPLE 10.1

A quartz crystal having a thickness of 0.1 inch is subjected to a pressure of 50 psi. Find the output voltage if the voltage sensitivity is 0.055 V-m/N.

Solution: With $t = 0.1$ in. $= 0.00254$ m, $p_x = 50$ psi $= 344.738$ N/m^2, and $v = 0.055$ V-m/N, Eq. (10.10) gives

$$E = (0.055)(0.00254)(344.738) = 48.1599 \text{ volts}$$

$\blacksquare$

10.2.3 Electrodynamic Transducers

When an electrical conductor, in the form of a coil, moves in a magnetic field as shown in Fig. 10.6, a voltage E is generated in the conductor. The value of E in volts is given by

$$E = Dlv \qquad (10.11)$$

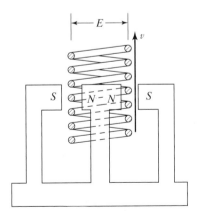

FIGURE 10.6 Basic idea behind electrodynamic transducer.

where D is the magnetic flux density (Telsas), l is the length of the conductor (meters), and v is the velocity of the conductor relative to the magnetic field (meters/second). The magnetic field may be produced by either a permanent magnet or an electromagnet. Sometimes, the coil is kept stationary and the magnet is made to move. Since the voltage output of an electromagnetic transducer is proportional to the relative velocity of the coil, they are frequently used in "velocity pickups." Equation (10.11) can be rewritten as

$$Dl = \frac{E}{v} = \frac{F}{I} \tag{10.12}$$

where F denotes the force (Newtons) acting on the coil while carrying a current I (amperes). Equation (10.12) shows that the performance of an electrodynamic transducer can be reversed. In fact, Eq. (10.12) forms the basis for using an electrodynamic transducer as a "vibration exciter" (see Section 10.5.2).

10.2.4 Linear Variable Differential Transformer Transducer

The schematic diagram of a linear variable differential transformer (LVDT) transducer is shown in Fig. 10.7. It consists of a primary coil at the center, two secondary coils at the ends, and a magnetic core that can move freely inside the coils in the axial direction. When an a.c. input voltage is applied to the primary coil, the output voltage will be equal to the difference of the voltages induced in the secondary coils. This output voltage depends on the magnetic coupling between the coils and the core, which in turn depends on the axial displacement of the core. The secondary coils are connected in phase opposition so that, when the magnetic core is in the exact middle position, the voltages in the two coils will be equal and 180° out of phase. This makes the output voltage of the LVDT as zero. When the core is moved to either side of the middle (zero) position, the magnetic coupling will be increased in one secondary coil and decreased in the other coil. The output polarity depends on the direction of the movement of the magnetic core.

The range of displacement for many LVDTs on the market is from 0.0002 cm to 40 cm. The advantages of an LVDT over other displacement transducers include insensitivity to

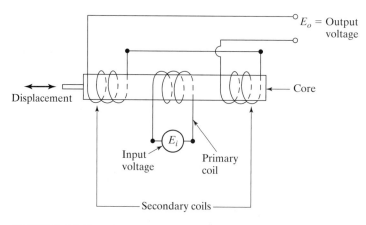

FIGURE 10.7   Schematic diagram of a LVDT transducer.

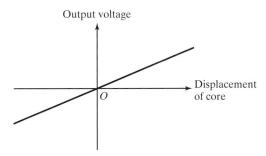

FIGURE 10.8 Linearity of voltage with displacement of core.

temperature and high output. The mass of the magnetic core restricts the use of the LVDT for high frequency applications [10.4].

As long as the core is not moved very far from the center of the coil, the output voltage varies linearly with the displacement of the core, as shown in Fig. 10.8; hence the name linear variable differential transformer.

10.3 Vibration Pickups

When a transducer is used in conjunction with another device to measure vibrations, it is called a *vibration pickup*. The commonly used vibration pickups are known as seismic instruments. A seismic instrument consists of a mass-spring-damper system mounted on the vibrating body, as shown in Fig. 10.9. Then the vibratory motion is measured by finding the displacement of the mass relative to the base on which it is mounted.

The instrument consists of a mass m, a spring k, and a damper c inside a cage, which is fastened to the vibrating body. With this arrangement, the bottom ends of the spring and the dashpot will have the same motion as the cage (which is to be measured, y) and their vibration excites the suspended mass into motion. Then the displacement of the mass relative to the cage, $z = x - y$, where x denotes the vertical displacement of the suspended mass, can be measured if we attach a pointer to the mass and a scale to the cage, as shown in Fig. 10.9.[3]

The vibrating body is assumed to have a harmonic motion:

$$y(t) = Y \sin \omega t \tag{10.13}$$

The equation of motion of the mass m can be written as

$$m\ddot{x} + c(\dot{x} - \dot{y}) + k(x - y) = 0 \tag{10.14}$$

---

[3]The output of the instrument shown in Fig. 10.9 is the relative mechanical motion of the mass, as shown by the pointer and the graduated scale on the cage. For high-speed operation and convenience, the motion is often converted into an electrical signal by a transducer.

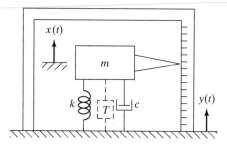

FIGURE 10.9   Seismic instrument.

By defining the relative displacement z as

$$z = x - y \qquad (10.15)$$

Eq. (10.14) can be written as

$$m\ddot{z} + c\dot{z} + kz = -m\ddot{y} \qquad (10.16)$$

Equations (10.13) and (10.16) lead to

$$m\ddot{z} + c\dot{z} + kz = m\omega^2 Y \sin \omega t \qquad (10.17)$$

This equation is identical to Eq. (3.75); hence the steady-state solution is given by

$$z(t) = Z \sin (\omega t - \phi) \qquad (10.18)$$

where Z and ϕ are given by (see Eqs. 3.76 and 3.77):

$$Z = \frac{Y\omega^2}{[(k - m\omega^2)^2 + c^2\omega^2]^{1/2}} = \frac{r^2 Y}{[(1 - r^2)^2 + (2\zeta r)^2]^{1/2}} \qquad (10.19)$$

$$\phi = \tan^{-1}\left(\frac{c\omega}{k - m\omega^2}\right) = \tan^{-1}\left(\frac{2\zeta r}{1 - r^2}\right) \qquad (10.20)$$

$$r = \frac{\omega}{\omega_n} \qquad (10.21)$$

and

$$\zeta = \frac{c}{2m\omega_n} \qquad (10.22)$$

The variations of Z and ϕ with respect to r are shown in Figs. 10.10 and 10.11. As will be seen later, the type of instrument is determined by the useful range of the frequencies, indicated in Fig. 10.10.

10.3.1
Vibrometer

A vibrometer or a seismometer is an instrument that measures the displacement of a vibrating body. It can be observed from Fig. 10.10 that $Z/Y \approx 1$ when $\omega/\omega_n \geq 3$ (range II). Thus the relative displacement between the mass and the base (sensed by the transducer)

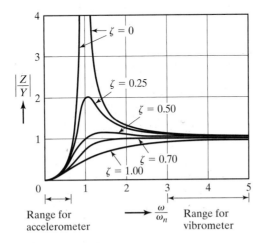

FIGURE 10.10   Response of a vibration-measuring instrument.

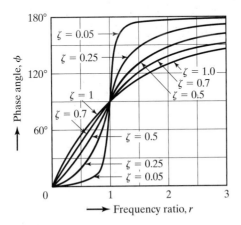

FIGURE 10.11   Variation of ϕ with r.

is essentially the same as the displacement of the base. For an exact analysis, we consider Eq. (10.19). We note that

$$z(t) \simeq Y \sin(\omega t - \phi) \tag{10.23}$$

if

$$\frac{r^2}{[(1 - r^2)^2 + (2\zeta r)^2]^{1/2}} \approx 1 \tag{10.24}$$

A comparison of Eq. (10.23) with $y(t) = Y \sin \omega t$ shows that $z(t)$ gives directly the motion $y(t)$ except for the phase lag ϕ. This phase lag can be seen to be equal to 180° for $\zeta = 0$. Thus the recorded displacement $z(t)$ lags behind the displacement being measured $y(t)$ by time $t' = \phi/\omega$. This time lag is not important if the base displacement $y(t)$ consists of a single harmonic component.

Since $r = \omega/\omega_n$ has to be large and the value of ω is fixed, the natural frequency $\omega_n = \sqrt{k/m}$ of the mass-spring-damper must be low. This means that the mass must be large and the spring must have a low stiffness. This results in a bulky instrument, which is not desirable in many applications. In practice, the vibrometer may not have a large value of r and hence the value of Z may not be equal to Y exactly. In such a case, the true value of Y can be computed by using Eq. (10.19), as indicated in the following example.

Amplitude by Vibrometer

EXAMPLE 10.2

A vibrometer having a natural frequency of 4 rad/s and $\zeta = 0.2$ is attached to a structure that performs a harmonic motion. If the difference between the maximum and the minimum recorded values is 8 mm, find the amplitude of motion of the vibrating structure when its frequency is 40 rad/s.

Solution: The amplitude of the recorded motion Z is 4 mm. For $\zeta = 0.2$, $\omega = 40.0$ rad/s, and $\omega_n = 4$ rad/s, $r = 10.0$, and Eq. (10.19) gives

$$Z = \frac{Y(10)^2}{[(1 - 10^2)^2 + \{2(0.2)(10)\}^2]^{1/2}} = 1.0093Y$$

Thus the amplitude of vibration of the structure is $Y = Z/1.0093 = 3.9631$ mm.

■

**10.3.2
Accelerometer**

An accelerometer is an instrument that measures the acceleration of a vibrating body (see Fig. 10.12). Accelerometers are widely used for vibration measurements [10.7] and also to record earthquakes. From the accelerometer record, the velocity and displacements are obtained by integration. Equations (10.18) and (10.19) yield

$$-z(t)\omega_n^2 = \frac{1}{[(1 - r^2)^2 + (2\zeta r)^2]^{1/2}}\{-Y\omega^2 \sin(\omega t - \phi)\} \tag{10.25}$$

(a)

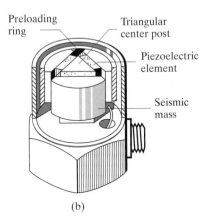

(b)

FIGURE 10.12   Accelerometers.
(Courtesy of Bruel and Kjaer Instruments,
Inc., Marlborough, Mass.)

This shows that if

$$\frac{1}{[(1 - r^2)^2 + (2\zeta r)^2]^{1/2}} \simeq 1 \qquad (10.26)$$

Eq. (10.25) becomes

$$-z(t)\omega_n^2 \simeq -Y\omega^2 \sin(\omega t - \phi) \qquad (10.27)$$

By comparing Eq. (10.27) with $\ddot{y}(t) = -Y\omega^2 \sin \omega t$, we find that the term $z(t)\omega_n^2$ gives the acceleration of the base $\ddot{y}$, except for the phase lag ϕ. Thus the instrument can be made to record (give) directly the value of $\ddot{y} = -z(t)\omega_n^2$. The time by which the record lags the acceleration is given by $t' = \phi/\omega$. If $\ddot{y}$ consists of a single harmonic component, the time lag will not be of importance.

The value of the expression on the left-hand side of Eq. (10.26) is shown plotted in Fig. 10.13. It can be seen that the left-hand side of Eq. (10.26) lies between 0.96 and 1.04 for $0 \leq r \leq 0.6$ if the value of ζ lies between 0.65 and 0.7. Since r is small, the natural frequency of the instrument has to be large compared to the frequency of vibration to be measured. From the relation $\omega_n = \sqrt{k/m}$, we find that the mass needs to be small and the spring needs to have a large value of k (i.e., short spring), so the instrument will be small in size. Due to their small size and high sensitivity, accelerometers are preferred in vibration

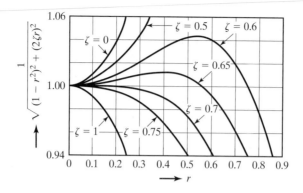

FIGURE 10.13   Variation of lefthand side of Eq. (10.26) with r.

measurements. In practice, Eq. (10.26) may not be satisfied exactly; in such cases the quantity

$$\frac{1}{[(1 - r^2)^2 + (2\zeta r)^2]^{1/2}}$$

can be used to find the correct value of the acceleration measured, as illustrated in the following example.

Design of an Accelerometer

EXAMPLE 10.3

An accelerometer has a suspended mass of 0.01 kg with a damped natural frequency of vibration of 150 Hz. When mounted on an engine undergoing an acceleration of 1 g at an operating speed of 6000 rpm, the acceleration is recorded as 9.5 m/s² by the instrument. Find the damping constant and the spring stiffness of the accelerometer.

Solution: The ratio of measured to true accelerations is given by

$$\frac{1}{[(1 - r^2)^2 + (2\zeta r)^2]^{1/2}} = \frac{\text{Measured value}}{\text{True value}} = \frac{9.5}{9.81} = 0.9684 \qquad (E.1)$$

which can be written as

$$[(1 - r^2)^2 + (2\zeta r)^2] = (1/0.9684)^2 = 1.0663 \qquad (E.2)$$

The operating speed of the engine gives

$$\omega = \frac{6000(2\pi)}{60} = 628.32 \text{ rad/s}$$

The damped natural frequency of vibration of the accelerometer is

$$\omega_d = \sqrt{1 - \zeta^2}\,\omega_n = 150(2\pi) = 942.48 \text{ rad/s}$$

Thus

$$\frac{\omega}{\omega_d} = \frac{\omega}{\sqrt{1 - \zeta^2}\,\omega_n} = \frac{r}{\sqrt{1 - \zeta^2}} = \frac{628.32}{942.48} = 0.6667 \tag{E.3}$$

Equation (E.3) gives

$$r = 0.6667\sqrt{1 - \zeta^2} \quad \text{or} \quad r^2 = 0.4444(1 - \zeta^2) \tag{E.4}$$

Substitution of Eq. (E.4) into (E.2) leads to a quadratic equation in ζ^2 as

$$1.5801\zeta^4 - 2.2714\zeta^2 + 0.7576 = 0 \tag{E.5}$$

The solution of Eq. (E.5) gives

$$\zeta^2 = 0.5260, 0.9115$$

or

$$\zeta = 0.7253, 0.9547$$

By choosing $\zeta = 0.7253$ arbitrarily, the undamped natural frequency of the accelerometer can be found as

$$\omega_n = \frac{\omega_d}{\sqrt{1 - \zeta^2}} = \frac{942.48}{\sqrt{1 - 0.7253^2}} = 1368.8889 \text{ rad/s}$$

Since $\omega_n = \sqrt{k/m}$, we have

$$k = m\omega_n^2 = (0.01)(1368.8889)^2 = 18738.5628 \text{ N/m}$$

The damping constant can be determined from

$$c = 2m\omega_n\zeta = 2(0.01)(1368.8889)(0.7253) = 19.8571 \text{ N-s/m}$$

∎

10.3.3
Velometer

A velometer measures the velocity of a vibrating body. Equation (10.13) gives the velocity of the vibrating body

$$\dot{y}(t) = \omega Y \cos \omega t \tag{10.28}$$

and Eq. (10.18) gives

$$\dot{z}(t) = \frac{r^2 \omega Y}{[(1 - r^2)^2 + (2\zeta r)^2]^{1/2}} \cos(\omega t - \phi) \tag{10.29}$$

If

$$\frac{r^2}{[(1 - r^2)^2 + (2\zeta r)^2]^{1/2}} \simeq 1 \tag{10.30}$$

then

$$\dot{z}(t) \simeq \omega Y \cos(\omega t - \phi) \tag{10.31}$$

A comparison of Eqs. (10.28) and (10.31) shows that, except for the phase difference ϕ, $\dot{z}(t)$ gives directly $\dot{y}(t)$, provided that Eq. (10.30) holds true. In order to satisfy Eq. (10.30), r must be very large. In case Eq. (10.30) is not satisfied, then the velocity of the vibrating body can be computed using Eq. (10.29).

■■■■■■■■■ Design of a Velometer

EXAMPLE 10.4 ——

Design a velometer if the maximum error is to be limited to 1 percent of the true velocity. The natural frequency of the velometer is to be 80 Hz and the suspended mass is to be 0.05 kg.

Solution: The ratio (R) of the recorded and the true velocities is given by Eq. (10.29):

$$R = \frac{r^2}{[(1 - r^2)^2 + (2\zeta r)^2]^{1/2}} = \frac{\text{Recorded velocity}}{\text{True velocity}} \tag{E.1}$$

The maximum of (E.1) occurs when (see Eq. (3.84))

$$r = r^* = \frac{1}{\sqrt{1 - 2\zeta^2}} \tag{E.2}$$

Substitution of Eq. (E.2) into (E.1) gives

$$\frac{\left(\dfrac{1}{1 - 2\zeta^2}\right)}{\sqrt{\left[1 - \left(\dfrac{1}{1 - 2\zeta^2}\right)\right]^2 + 4\zeta^2\left(\dfrac{1}{1 - 2\zeta^2}\right)}} = R$$

which can be simplified as

$$\frac{1}{\sqrt{4\zeta^2 - 4\zeta^4}} = R \tag{E.3}$$

For an error of 1 percent, $R = 1.01$ or 0.99, and Eq. (E.3) leads to

$$\zeta^4 - \zeta^2 + 0.245075 = 0 \tag{E.4}$$

and

$$\zeta^4 - \zeta^2 + 0.255075 = 0 \tag{E.5}$$

Equation (E.5) gives imaginary roots and Eq. (E.4) gives

$$\zeta^2 = 0.570178, 0.429821$$

or

$$\zeta = 0.755101, 0.655607$$

We choose the value $\zeta = 0.755101$ arbitrarily. The spring stiffness can be found as

$$k = m\omega_n^2 = 0.05(502.656)^2 = 12633.1527 \text{ N/m}$$

since

$$\omega_n = 80(2\pi) = 502.656 \text{ rad/s}$$

The damping constant can be determined from

$$c = 2\zeta\omega_n m = 2(0.755101)(502.656)(0.05) = 37.9556 \text{ N-s/m}$$

■

10.3.4
Phase Distortion

As shown by Eq. (10.18), all vibration-measuring instruments exhibit phase lag. Thus the response or output of the instrument lags behind the motion or input it measures. The time lag is given by the phase angle divided by the frequency ω. The time lag is not important if we measure a single harmonic component. But, occasionally, the vibration to be recorded is not harmonic but consists of the sum of two or more harmonic components. In such a case, the recorded graph may not give an accurate picture of the vibration because different harmonics may be amplified by different amounts and their phase shifts may also be different. The distortion in the wave form of the recorded signal is called the phase distortion or phase-shift error. To illustrate the nature of the phase-shift error, we consider a vibration signal of the form shown in Fig. 10.14(a) [10.10]:

$$y(t) = a_1 \sin \omega t + a_3 \sin 3\omega t \tag{10.32}$$

Let the phase shift be 90° for the first harmonic and 180° for the third harmonic of Eq. (10.32). The corresponding time lags are given by $t_1 = \theta_1/\omega = 90°/\omega$ and $t_2 = \theta_2/(3\omega) = 180°/(3\omega)$. The output signal is shown in Fig. 10.14(b). It can be seen that the output signal is quite different from the input signal due to phase distortion.

As a general case, let the complex wave being measured be given by the sum of several harmonics as

$$y(t) = a_1 \sin \omega t + a_2 \sin 2\omega t + \cdots \tag{10.33}$$

If the displacement is measured using a vibrometer, its response to each component of the series is given by an equation similar to Eq. (10.18) so that the output of the vibrometer becomes

$$z(t) = a_1 \sin(\omega t - \phi_1) + a_2 \sin(2\omega t - \phi_2) + \cdots \tag{10.34}$$

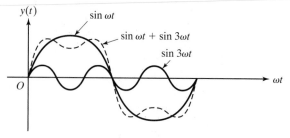

(a) Input signal

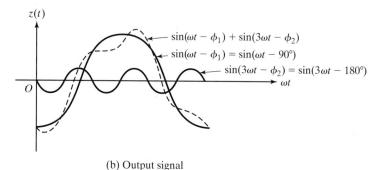

(b) Output signal

FIGURE 10.14 Phase-shift error.

where

$$\tan \phi_j = \frac{2\zeta\left(j\dfrac{\omega}{\omega_n}\right)}{1 - \left(j\dfrac{\omega}{\omega_n}\right)^2}, j = 1, 2, \ldots \tag{10.35}$$

Since ω/ω_n is large for this instrument, we can find from Fig. 10.11 that $\phi_j \simeq \pi$, $j = 1, 2, \ldots$ and Eq. (10.34) becomes

$$z(t) \simeq -[a_1 \sin \omega t + a_2 \sin 2\omega t + \cdots] \simeq -y(t) \tag{10.36}$$

Thus the output record will be simply opposite to the motion being measured. This is unimportant and can easily be corrected.

By using a similar reasoning, we can show, in the case of a velometer, that

$$\dot{z}(t) \simeq -\dot{y}(t) \tag{10.37}$$

for an input signal consisting of several harmonics. Next we consider the phase distortion for an accelerometer. Let the acceleration curve to be measured be expressed, using Eq. (10.33), as

$$\ddot{y}(t) = -a_1 \omega^2 \sin \omega t - a_2(2\omega)^2 \sin 2\omega t - \cdots \tag{10.38}$$

The response or output of the instrument to each component can be found as in Eq. (10.34), and so

$$\ddot{z}(t) = -a_1\omega^2\sin(\omega t - \phi_1) - a_2(2\omega)^2\sin(2\omega t - \phi_2) - \cdots \qquad (10.39)$$

where the phase lags ϕ_j are different for different components of the series in Eq. (10.39). Since the phase lag ϕ varies almost linearly from $0°$ at $r = 0$ to $90°$ at $r = 1$ for $\zeta = 0.7$ (see Fig. 10.11), we can express ϕ as

$$\phi \simeq \alpha r = \alpha\frac{\omega}{\omega_n} = \beta\omega \qquad (10.40)$$

where α and $\beta = \alpha/\omega_n$ are constants. The time lag is given by

$$t' = \frac{\phi}{\omega} = \frac{\beta\omega}{\omega} = \beta \qquad (10.41)$$

This shows that the time lag of the accelerometer is independent of the frequency for any component, provided that the frequency lies in the range $0 \le r \le 1$. Since each component of the signal has the same time delay or phase lag, we have, from Eq. (10.39),

$$-\omega^2\ddot{z}(t) = -a_1\omega^2\sin(\omega t - \omega\beta) - a_2(2\omega)^2\sin(2\omega t - 2\omega\beta) - \cdots$$
$$= -a_1\omega^2\sin\omega\tau - a_2(2\omega)^2\sin 2\omega\tau - \cdots \qquad (10.42)$$

where $\tau = t - \beta$. Note that Eq. (10.42) assumes that $0 \le r \le 1$—that is, even the highest frequency involved, $n\omega$, is less than ω_n. This may not be true in practice. Fortunately, no significant phase distortion occurs in the output signal even when some of the higher order frequencies are larger than ω_n. The reason is that, generally, only the first few components are important to approximate even a complex wave form; the amplitudes of the higher harmonics are small and contribute very little to the total wave form. Thus the output record of the accelerometer represents a reasonably true acceleration being measured [10.7, 10.11].

10.4   Frequency-Measuring Instruments

Most frequency-measuring instruments are of the mechanical type and are based on the principle of resonance. Two kinds of instruments are discussed in the following paragraphs: the Fullarton tachometer and the Frahm tachometer.

Single-Reed Instrument or Fullarton Tachometer.   This instrument consists of a variable-length cantilever strip with a mass attached at one of its ends. The other end of the strip is clamped, and its free length can be changed by means of a screw mechanism (see Fig. 10.15a). Since each length of the strip corresponds to a different natural frequency, the reed is marked along its length in terms of its natural frequency. In practice, the clamped end of the strip is pressed against the vibrating body, and the screw mechanism is manipulated to alter its free length until the free end shows the largest amplitude of vibration. At that instant, the excitation frequency is equal to the natural frequency of the cantilever; it can be read directly from the strip.

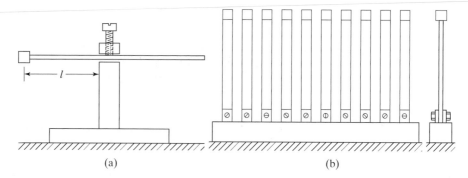

(a)   (b)

FIGURE 10.15   Frequency-measuring instruments.

Multireed-Instrument or Frahm Tachometer.   This instrument consists of a number of cantilevered reeds carrying small masses at their free ends (see Fig. 10.15b). Each reed has a different natural frequency and is marked accordingly. Using a number of reeds makes it possible to cover a wide frequency range. When the instrument is mounted on a vibrating body, the reed whose natural frequency is nearest the unknown frequency of the body vibrates with the largest amplitude. The frequency of the vibrating body can be found from the known frequency of the vibrating reed.

Stroboscope.   A stroboscope is an instrument that produces light pulses intermittently. The frequency at which the light pulses are produced can be altered and read from the instrument. When a specific point on a rotating (vibrating) object is viewed with the stroboscope, it will appear to be stationary only when the frequency of the pulsating light is equal to the speed of the rotating (vibrating) object. The main advantage of the stroboscope is that it does not make contact with the rotating (vibrating) body. Due to the persistence of vision, the lowest frequency that can be measured with a stroboscope is approximately 15 Hz. A typical stroboscope is shown in Fig. 10.16.

FIGURE 10.16   A stroboscope. (Courtesy of Bruel and Kjaer Instruments, Inc., Marlborough, Mass.)

10.5 Vibration Exciters

The vibration exciters or shakers can be used in several applications such as determination of the dynamic characteristics of machines and structures and fatigue testing of materials. The vibration exciters can be mechanical, electromagnetic, electrodynamic, or hydraulic type. The working principles of mechanical and electromagnetic exciters are described in this section.

10.5.1 Mechanical Exciters

As indicated in Section 1.10 (Fig. 1.38), a Scotch yoke mechanism can be used to produce harmonic vibrations. The crank of the mechanism can be driven either by a constant or a variable-speed motor. When a structure is to be vibrated, the harmonic force can be applied either as an inertia force, as shown in Fig. 10.17(a), or as an elastic spring force, as shown

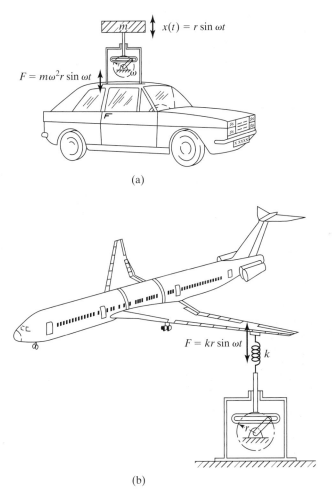

(a)

(b)

FIGURE 10.17 Vibration of a structure through (a) an inertia force and (b) an elastic spring force.

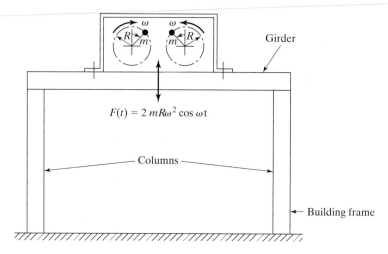

FIGURE 10.18 Vibration excitation due to unbalanced force.

in Fig. 10.17(b). These vibrators are generally used for frequencies less than 30 Hz and loads less than 700 N [10.1].

The unbalance created by two masses rotating at the same speed in opposite directions (see Fig. 10.18) can be used as a mechanical exciter. This type of shaker can be used to generate relatively large loads between 250 N and 25,000 N. If the two masses, of magnitude m each, rotate at an angular velocity ω at a radius R, the vertical force $F(t)$ generated is given by

$$F(t) = 2mR\omega^2 \cos \omega t \qquad (10.43)$$

The horizontal components of the two masses cancel and hence the resultant horizontal force will be zero. The force $F(t)$ will be applied to the structure to which the exciter is attached.

**10.5.2
Electrodynamic
Shaker**

The schematic diagram of an electrodynamic shaker, also known as the electromagnetic exciter, is shown in Fig. 10.19(a). As stated in Section 10.2.3, the electrodynamic shaker can be considered as the reverse of an electrodynamic transducer. When current passes through a coil placed in a magnetic field, a force F (in Newtons) proportional to the current I (in amperes) and the magnetic flux intensity D (in Telsas), is produced which accelerates the component placed on the shaker table:

$$F = DIl \qquad (10.44)$$

where l is the length of the coil (in meters). The magnetic field is produced by a permanent magnet in small shakers while an electromagnet is used in large shakers. The magnitude of acceleration of the table or component depends on the maximum current and the

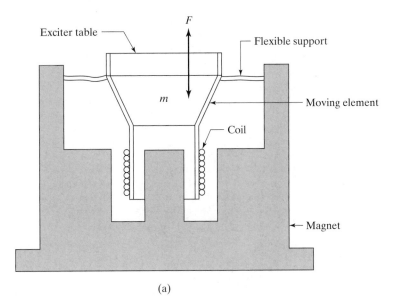

(a)

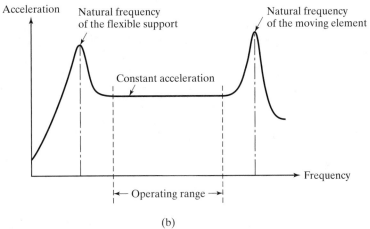

(b)

FIGURE 10.19   (a) Electrodynamic shaker. (b) Typical resonance characteristics of an electrodynamic exciter.

masses of the component and the moving element of the shaker. If the current flowing through the coil varies harmonically with time (a.c. current), the force produced also varies harmonically. On the other hand, if direct current is used to energize the coil, a constant force is generated at the exciter table. The electrodynamic exciters can be used in conjunction with an inertia or a spring as in the case of Figs. 10.17(a) and (b) to vibrate a structure.

FIGURE 10.20   An exciter with a general purpose head. (Courtesy of Bruel and Kjaer Instruments, Inc., Marlborough, Mass.)

Since the coil and the moving element should have a linear motion, they are suspended from a flexible support (having a very small stiffness) as shown in Fig. 10.19(a). Thus the electromagnetic exciter has two natural frequencies; one corresponding to the natural frequency of the flexible support and the other corresponding to the natural frequency of the moving element, which can be made very large. These two resonant frequencies are shown in Fig. 10.19(b). The operating frequency range of the exciter lies between these two resonant frequencies, as indicated in Fig. 10.19(b) [10.7].

The electrodynamic exciters are used to generate forces up to 30,000 N, displacements up to 25 mm, and frequencies in the range of 5 Hz to 20 kHz [10.1]. A practical electrodynamic exciter is shown in Fig. 10.20.

10.6   Signal Analysis

In signal analysis, we determine the response of a system under a known excitation and present it in a convenient form. Often, the time response of a system will not give much useful information. However, the frequency response will show one or more discrete frequencies around which the energy is concentrated. Since the dynamic characteristics of individual components of the system are usually known, we can relate the distinct frequency components (of the frequency response) to specific components [10.3].

For example, the acceleration-time history of a machine frame that is subjected to excessive vibration might appear as shown in Fig. 10.21(a). This figure cannot be used to

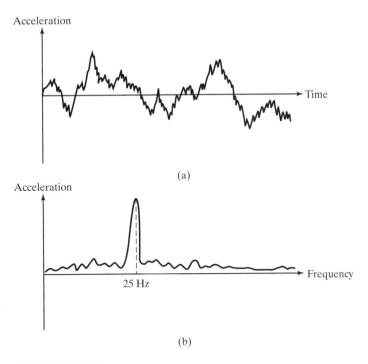

FIGURE 10.21   Acceleration history.

identify the cause of vibration. If the acceleration-time history is transformed to the frequency domain, the resulting frequency spectrum might appear as shown in Fig. 10.21(b), where, for specificness, the energy is shown concentrated around 25 Hz. This frequency can easily be related, for example, to the rotational speed of a particular motor. Thus the acceleration spectrum shows a strong evidence that the motor might be the cause of vibration. If the motor is causing the excessive vibrations, changing either the motor or its speed of operation might avoid resonance and hence the problem of excessive vibrations.

**10.6.1
Spectrum
Analyzers**

Spectrum or frequency analyzers can be used for signal analysis. A spectrum or frequency analyzer is a device that analyzes a signal in the frequency domain by separating the energy of the signal into various frequency bands. The separation of signal energy into frequency bands is accomplished through a set of filters. The analyzers are usually classified according to the type of filter employed. For example, if an octave band filter is used, the spectrum analyzer is called an *octave band analyzer*.

In recent years, digital analyzers have become quite popular for real-time signal analysis. In a real-time frequency analysis, the signal is continuously analyzed over all the frequency bands. Thus the calculation process must not take more time than the time taken to collect the signal data. Real-time analyzers are especially useful for machinery health monitoring since a change in the noise or vibration spectrum can be observed at the same time that change in the machine occurs. There are two types of real-time analysis procedures: the digital filtering method and the Fast Fourier Transform (FFT) method [10.13]. The digital filtering method is best suited for constant percent bandwidth analysis while the FFT method is most suitable for constant bandwidth analysis. Before we consider the difference between the constant percent bandwidth and constant bandwidth analyses, we first discuss the basic component of a spectrum analyzer—namely, the bandpass filter.

10.6.2 Bandpass Filter

A bandpass filter is a circuit that permits the passage of frequency components of a signal over a frequency band and rejects all other frequency components of the signal. A filter can be built by using, for example, resistors, inductors, and capacitors. Figure 10.22 illustrates the response characteristics of a filter whose lower and upper cutoff frequencies are f_l and f_u, respectively. A practical filter will have a response characteristic deviating from the ideal rectangle, as shown by the full line in Fig. 10.22. For a good bandpass filter, the ripples within the band will be minimum and the slopes of the filter skirts will be steep to maintain the actual bandwidth close to the ideal value, $B = f_u - f_l$. For a practical filter, the frequencies f_l and f_u at which the response is 3 dB[4] below its mean bandpass response are called the cutoff frequencies.

There are two types of bandpass filters used in signal analysis: the constant percent bandwidth filters and constant bandwidth filters. For a constant percent bandwidth filter, the ratio of the bandwidth to the center (tuned) frequency, $(f_u - f_l)/f_c$, is a constant. The octave,[5] one-half-octave, and one-third-octave band filters are examples of constant percent bandwidth filters. Some of the cutoff limits and center frequencies of octave bands used in signal analysis are shown in Table 10.1. For a constant bandwidth filter, the bandwidth, $f_u - f_l$, is independent of the center (tuned) frequency, f_c.

TABLE 10.1

Lower cutoff limit (Hz)	5.63	11.2	22.4	44.7	89.2	178	355	709	1410
Center frequency (Hz)	8.0	16.0	31.5	63.0	125	250	500	1000	2000
Upper cutoff limit (Hz)	11.2	22.4	44.7	89.2	178	355	709	1410	2820

---

[4]A decibel (dB) of a quantity (such as power, P) is defined as

$$\text{Quantity in dB} = 10 \log_{10}\left(\frac{P}{P_{\text{ref}}}\right)$$

where P is the power and P_{ref} is a reference value of the power.

[5]An octave is the interval between any two frequencies $(f_2 - f_1)$, whose frequency ratio (f_2/f_1), is 2. Two frequencies f_1 and f_2 are said to be separated by a number of octaves N when

$$\frac{f_2}{f_1} = 2^N \quad \text{or} \quad N \text{ (in octaves)} = \log_2\left(\frac{f_2}{f_1}\right)$$

where N can be an integer or a fraction. If $N = 1$, we have an octave; if $N = 1/3$, we get a one-third octave, and so on.

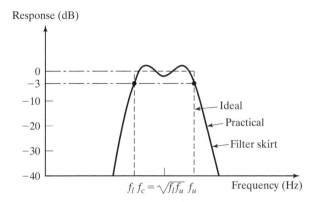

FIGURE 10.22 Response of a filter.

**10.6.3
Constant Percent
Bandwidth and
Constant
Bandwidth
Analyzers**

The primary difference between the constant percent bandwidth and constant bandwidth analyzers lies in the detail provided by the various bandwidths. The octave band filters, whose upper cutoff frequency is twice the lower cutoff frequency, give a less detailed (too coarse) analysis for practical vibration and noise encountered in machines. The one-half-octave band filter gives twice the information but requires twice the amount of time to obtain the data. A spectrum analyzer with a set of octave and one-third-octave filters can be used for noise (signal) analysis. Each filter is tuned to a different center frequency to cover the entire frequency range of interest. Since the lower cutoff frequency of a filter is equal to the upper cutoff frequency of the previous filter, the composite filter characteristic will appear as shown in Fig. 10.23. Figure 10.24 shows a real-time octave and fractional-octave digital frequency analyzer. A constant bandwidth analyzer is used to obtain a more detailed analysis than in the case of a constant percent bandwidth analyzer, especially in the high-frequency range of the signal. The constant

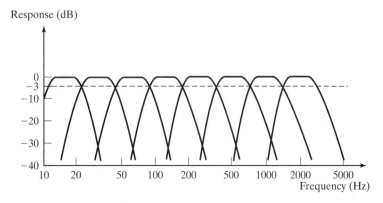

FIGURE 10.23 Response characteristic of a typical octave-band filter set.

bandwidth filter, when used with a continuously varying center frequency, is called a wave or heterodyne analyzer. Heterodyne analyzers are available with constant filter bandwidths ranging from one to several hundred Hertz. A practical heterodyne analyzer is shown in Fig. 10.25.

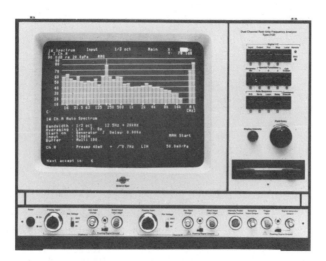

FIGURE 10.24   Octave and fractional-octave digital frequency analyzer. (Courtesy of Bruel and Kjaer Instruments, Inc., Marlborough, Mass.)

FIGURE 10.25   Heterodyne analyzer. (Courtesy of Bruel and Kjaer Instruments, Inc., Marlborough, Mass.)

10.7 Dynamic Testing of Machines and Structures

The dynamic testing of machines (structures) involves finding the deformation of the machines (structures) at a critical frequency. This can be done using the following two approaches [10.3].

10.7.1 Using Operational Deflection Shape Measurements

In this method, the forced dynamic deflection shape is measured under the steady-state (operating) frequency of the system. For the measurement, an accelerometer is mounted at some point on the machine (structure) as a reference, and another moving accelerometer is placed at several other points, and in different directions, if necessary. Then the magnitudes and the phase differences between the moving and reference accelerometers at all the points under steady-state operation of the system are measured. By plotting these measurements, we can find how the various parts of the machine (structure) move relative to one another and also absolutely.

The deflection shape measured is valid only for the forces/frequency associated with the operating conditions; as such, we cannot get information about deflections under other forces and/or frequencies. However, the measured deflection shape can be quite useful. For example, if a particular part or location is found to have excessive deflection, we can stiffen that part or location. This, in effect, increases the natural frequency beyond the operational frequency range of the system.

10.7.2 Using Modal Testing

Since any dynamic response of a machine (structure) can be obtained as a combination of its modes, a knowledge of the mode shapes, modal frequencies, and modal damping ratios constitute a complete dynamic description of the machine (structure). The experimental modal analysis procedure is described in the following section.

10.8 Experimental Modal Analysis

10.8.1 The Basic Idea

Experimental modal analysis, also known as modal analysis or modal testing, deals with the determination of natural frequencies, damping ratios, and mode shapes through vibration testing. Two basic ideas are involved:

1. When a structure, machine, or any system is excited, its response exhibits a sharp peak at resonance when the forcing frequency is equal to its natural frequency when damping is not large.
2. The phase of the response changes by $180°$ as the forcing frequency crosses the natural frequency of the structure or machine and the phase will be $90°$ at resonance.

10.8.2 The Necessary Equipment

The measurement of vibration requires the following hardware:

1. An exciter or source of vibration to apply a known input force to the structure or machine.
2. A transducer to convert the physical motion of the structure or machine into an electrical signal.

3. A signal conditioning amplifier to make the transducer characteristics compatible with the input electronics of the digital data acquisition system.
4. An analyzer to perform the tasks of signal processing and modal analysis using suitable software.

Exciter.   The exciter may be an electromagnetic shaker or an impact hammer. As explained in Section 10.5.2, the electromagnetic shaker can provide large input forces so that the response can be measured easily. Also the output of the shaker can be controlled easily if it is of electromagnetic type. The excitation signal is usually of a swept sinusoidal or a random type signal. In the swept sinusoidal input, a harmonic force of magnitude F is applied at a number of discrete frequencies over a specific frequency range of interest. At each discrete frequency, the structure or machine is made to reach a steady state before the magnitude and phase of the response are measured. If the shaker is attached to the structure or machine being tested, the mass of the shaker will influence the measured response (known as the mass loading effect). As such, care is to be taken to minimize the effect of the mass of the shaker. Usually the shaker is attached to the structure or machine through a short thin rod, called a *stringer*, to isolate the shaker, reduce the added mass and apply the force to the structure or machine along the axial direction of the stringer. This permits the control of the direction of the force applied to the structure or machine.

The impact hammer is a hammer with a built-in force transducer in its head, as indicated in Examples 4.4 and 4.5. The impact hammer can be used to hit or impact the structure or machine being tested to excite a wide range of frequencies without causing the problem of mass loading. The impact force caused by the impact hammer, which is nearly proportional to the mass of the hammer head and the impact velocity, can be found from the force transducer embedded in the head of the hammer. As shown in Section 6.15, the response of the structure or machine to an impulse is composed of excitations at each of the natural frequencies of the structure or machine.

Although the impact hammer is simple, portable, inexpensive and much faster to use than a shaker, it is often not capable of imparting sufficient energy to obtain adequate response signals in the frequency range of interest. It is also difficult to control the direction of the applied force with an impact hammer. A typical frequency response of a structure or machine obtained using an impact hammer is shown in Fig. 10.26. The shape of the frequency response is dependent on the mass and stiffness of both the hammer and the structure or machine. Usually, the useful range of frequency excitation is limited by a cutoff frequency, ω_c, which implies that the structure or machine did not receive sufficient energy to excite modes beyond ω_c. The value of ω_c is often taken as the frequency where the amplitude of the frequency response reduces by 10 to 20 dB from its maximum value.

Transducer.   Among the transducers, the piezoelectric transducers are most popular (see Section 10.2.2). A piezoelectric transducer can be designed to produce signals proportional to either force or acceleration. In an accelerometer, the piezoelectric material acts as a stiff spring that causes the transducer to have a resonant or natural frequency. Usually, the maximum measurable frequency of an accelerometer is a fraction of its natural frequency.

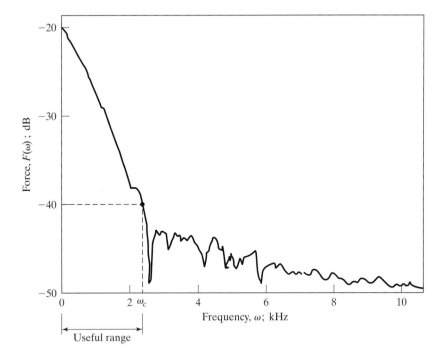

FIGURE 10.26 Frequency response of an impulse created by an impact hammer.

Strain gages can also be used to measure the vibration response of a structure or machine, as discussed in Section 10.2.1.

Signal Conditioner. Since the output impedance of transducers is not suitable for direct input into the signal analysis equipment, signal conditioners, in the form of charge or voltage amplifiers, are used to match and amplify the signals before signal analysis.

Analyzer. The response signal, after conditioning, is sent to an analyzer for signal processing. A commonly used analyzer is called a *fast Fourier transform (FFT) analyzer*. Such an analyzer receives analog voltage signals (representing displacement, velocity, acceleration, strain, or force) from a signal conditioning amplifier, filter, and digitizer for computations. It computes the discrete frequency spectra of individual signals as well as cross-spectra between the input and the different output signals. The analyzed signals can be used to find the natural frequencies, damping ratios, and mode shapes either in numerical or graphical form.

The general arrangement for the experimental modal analysis of a structural or mechanical system is shown in Fig. 10.27. Note that all the equipment is to be calibrated before it is used. For example, the built-in force transducer in an impact hammer is to be calibrated dynamically for each tip used. Similarly, the transducers, along with the signal conditioners, are to be calibrated with respect to magnitude and phase over the frequency range of interest.

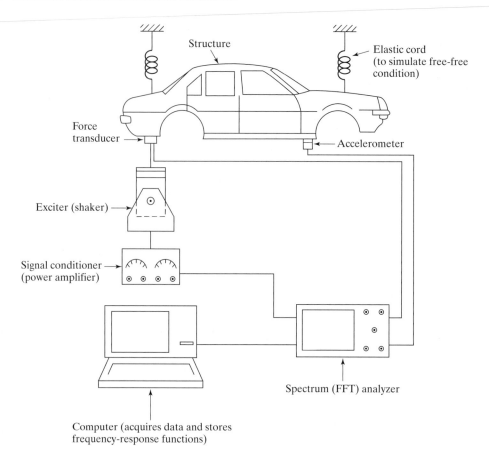

FIGURE 10.27   Experimental modal analysis.

10.8.3
Digital Signal
Processing

The analyzer converts the analog time-domain signals, $x(t)$, into digital frequency-domain data using Fourier series relations, given by Eqs. (1.97) to (1.99), to facilitate digital computation. Thus the analyzer accepts the analog output signals of accelerometers or force transducers, $x(t)$, and computes the spectral coefficients of these signals a_0, a_n and b_n using Eqs. (1.97) to (1.99) in the frequency domain. The process of converting analog signals into digital data is indicated in Fig. 10.28 for two representative signals. In Fig. 10.28, $x(t)$ denotes the analog signal and $x_i = x(t_i)$ represents the corresponding digital record, with t_i indicating the ith discrete value of time. This process is performed by an analog-to-digital (A/D) converter, which is part of a digital analyzer. If N samples of $x(t)$ are collected at discrete values of time, t_i, the data $[x_1(t_i), x_2(t_i), \ldots, x_N(t_i)]$ can be used to obtain the discrete form of Fourier transform as

$$x_j = x(t_j) = \frac{a_0}{2} + \sum_{i=1}^{N/2} \left(a_i \cos \frac{2\pi i t_j}{T} + b_i \sin \frac{2\pi i t_j}{T} \right); \qquad j = 1, 2, \ldots, N \quad (10.45)$$

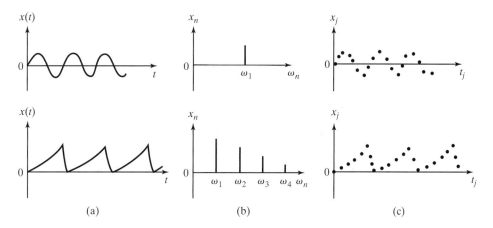

FIGURE 10.28   Representation of signals in different forms: (a) Signals in time domain. (b) Signals in frequency domain. (c) Digital records of $x(t)$.

where the digital spectral coefficients a_0, a_i, and b_i are given by (see Eqs. 1.97 to 1.99)

$$a_0 = \frac{1}{N} \sum_{j=1}^{N} x_j \tag{10.46}$$

$$a_i = \frac{1}{N} \sum_{j=1}^{N} x_j \cos \frac{2\pi i t_j}{N} \tag{10.47}$$

$$b_i = \frac{1}{N} \sum_{j=1}^{N} x_j \sin \frac{2\pi i t_j}{N} \tag{10.48}$$

with the number of samples N equal to some power of 2 (such as 256, 512, or 1024) which is fixed for a given analyzer. Equations (10.46) to (10.48) denote N algebraic equations for each of the N samples. The equations can be expressed in matrix form as

$$\vec{X} = [A]\vec{d} \tag{10.49}$$

where $\vec{X} = \{x_1 \, x_2 \cdots x_N\}^T$ is the vector of samples, $\vec{d} = \{a_0 \, a_1 \, a_2 \cdots a_{N/2} \, b_1 \, b_2 \cdots b_{N/2}\}^T$ is the vector of spectral coefficients, and $[A]$ is the matrix composed of the coefficients $\cos \frac{2\pi i t_j}{T}$ and $\sin \frac{2\pi i t_j}{T}$ of Eq. (10.46). The frequency content of the signal or response of the system can be determined from the solution

$$\vec{d} = [A]^{-1} \vec{X} \tag{10.50}$$

where $[A]^{-1}$ is computed efficiently using fast Fourier transform (FFT) by the analyzer.

10.8.4
Analysis of
Random Signals

The input and output data measured by the transducers usually contain some random component or noise that makes it difficult to analyze in a deterministic manner. Also, in some cases random excitation is used in vibration testing. Thus random signal analysis becomes necessary in vibration testing. If $x(t)$ is a random signal, as shown in Fig. 10.29, its average or mean, denoted as $\bar{x}$, is defined as[6]

$$\bar{x}(t) = \lim_{T \to \infty} \frac{1}{N} \int_0^T x(t) \, dt \tag{10.51}$$

which, for a digital signal, can be expressed as

$$\bar{x} = \lim_{N \to \infty} \frac{1}{N} \sum_{j=1}^{N} x(t_j) \tag{10.52}$$

Corresponding to any random signal $y(t)$, we can always define a new variable $x(t)$ as $x(t) = y(t) - \bar{y}(t)$ so that the mean value of $x(t)$ is zero. Hence, without loss of generality, we can assume the signal $x(t)$ to have a zero mean and define the mean square value or variance of $x(t)$, denoted as $\bar{x}^2(t)$, as

$$\bar{x}^2(t) = \lim_{N \to \infty} \frac{1}{T} \int_0^T x^2(t) \, dt \tag{10.53}$$

which, for a digital signal, takes the form

$$\bar{x}^2 = \lim_{n \to \infty} \frac{1}{N} \sum_{j=1}^{N} x^2(t_j) \tag{10.54}$$

The root mean square (RMS) value of $x(t)$ is given by

$$x_{\text{RMS}} = \sqrt{\bar{x}^2} \tag{10.55}$$

$x(t)$

FIGURE 10.29   A random signal, $x(t)$.

---

[6]A detailed discussion of random signals (processes) and random vibration is given in Chapter 14.

The autocorrelation function of a random signal $x(t)$, denoted as $R(t)$, gives a measure of the speed with which the signal changes in the time domain and is defined as

$$R(t) = \overline{x^2} = \lim_{T \to \infty} \frac{1}{T} \int_0^T x(\tau) \, x(\tau + t) \, d\tau \tag{10.56}$$

which, for a digital signal, can be written as

$$R(n, \Delta t) = \frac{1}{N - n} \sum_{j=0}^{N-n} x_j \, x_{j+n} \tag{10.57}$$

where N is the number of samples, Δt is the sampling interval, and n is an adjustable parameter that can be used to control the number of points used in the computation. It can be seen that $R(0)$ denotes the mean square value, $\overline{x^2}$, of $x(t)$. The autocorrelation function can be used to identify the presence of periodic components present (buried) in a random signal. If $x(t)$ is purely random, then $R(t) \to 0$ as $T \to \infty$. However, if $x(t)$ is periodic or has a periodic component, then $R(t)$ will also be periodic.

The power spectral density (PSD) of a random signal $x(t)$, denoted as $S(\omega)$, gives a measure of the speed with which the signal changes in the frequency domain and is defined as the Fourier transform of $R(t)$

$$S(\omega) = \frac{1}{2\pi} \int_{-\infty}^{\infty} R(\tau) e^{-i\omega\tau} \, d\tau \tag{10.58}$$

which, in digital form, can be expressed as

$$S(\Delta\omega) = \frac{|x(\omega)|^2}{N \, \Delta t} \tag{10.59}$$

where $|x(\omega)|^2$ represents the magnitude of the Fourier transform of the sampled data of $x(t)$. The definitions of autocorrelation and PSD functions can be extended for two different signals, such as a displacement signal $x(t)$ and an applied force signal $f(t)$. This leads to the cross-correlation function, $R_{xf}(t)$ and the cross-PSD $S_{xf}(\omega)$:

$$R_{xf}(t) = \lim_{T \to \infty} \frac{1}{T} \int_0^T x(\tau) \, f(\tau + t) \, d\tau \tag{10.60}$$

$$S_{xf}(\omega) = \frac{1}{2\pi} \int_{-\infty}^{\infty} R_{xf}(\tau) \, e^{-i\omega\tau} \, d\tau \tag{10.61}$$

Equations (10.60) and (10.61) permit the determination of the transfer functions of the structure or machine being tested. In Eq. (10.60), if $f(\tau + t)$ is replaced by $x(\tau + t)$, we obtain $R_{xx}(t)$, which when used in Eq. (10.61), leads to $S_{xx}(\omega)$. The freqency response function, $H(i\omega)$, is related to the PSD functions as

$$S_{xx}(\omega) = |H(i\omega)|^2 S_{ff}(\omega) \tag{10.62}$$

$$S_{fx}(\omega) = H(i\omega)\, S_{ff}(\omega) \tag{10.63}$$

$$S_{xx}(\omega) = H(i\omega)\, S_{xf}(\omega) \tag{10.64}$$

with $f(t)$ and $x(t)$ denoting the random force input and the resulting output response, respectively. $S_{xx}(\omega)$, given by Eq. (10.62), contains information about the magnitude of the transfer function of the system (structure or machine) while $S_{xf}(\omega)$ and $S_{xx}(\omega)$, given by Eqs. (10.63) and (10.64), contain information about both magnitude and phase. In vibration testing, the spectrum analyzer first computes different spectral density functions from the transducer output, and then computes the frequency response functuion $H(i\omega)$ of the system using Eqs. (10.63) and (10.64).

Coherence Function.   A function, known as coherence function (β), is defined as a measure of the noise present in the signals as

$$\beta(\omega) = \left(\frac{S_{fx}(\omega)}{S_{ff}(\omega)}\right)\left(\frac{S_{xf}(\omega)}{S_{xx}(\omega)}\right) = \frac{|S_{xf}(\omega)|^2}{S_{xx}(\omega)\, S_{ff}(\omega)} \tag{10.65}$$

Note that if the measurements of x and f are pure noises, then $\beta = 0$ and if the measurements of x and f are not contaminated at all with noise, then $\beta = 1$. The plot of a typical coherence function is shown in Fig. 10.30. Usually, $\beta \approx 1$ near the natural frequency of the system because the signals are large and are less influenced by the noise.

10.8.5 Determination of Modal Data from Observed Peaks

The frequency response function, $H(i\omega)$, computed from Eq. (10.63) or (10.64), can be used to find the natural frequencies, damping ratios, and mode shapes corresponding to all resonant peaks observed in the plot of $H(i\omega)$. Let the graph of the frequency-response function be as shown in Fig. 10.31, with its four peaks or resonances suggesting that the system being tested can be modeled as a four degree of freedom system. Sometimes it becomes difficult to assign the number of degrees of freedom to the system, especially when the resonant peaks are closely spaced in the graph of $H(i\omega)$, which can be plotted by applying a harmonic force of adjustable frequency at a specific point of the structure or machine, measuring the response (for example, displacement) at another point, and finding the value of the frequency response function using Eq. (10.63) or (10.64). The graph

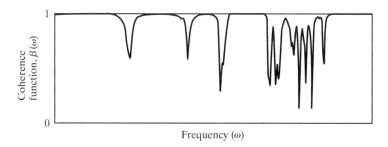

FIGURE 10.30   A typical coherence function.

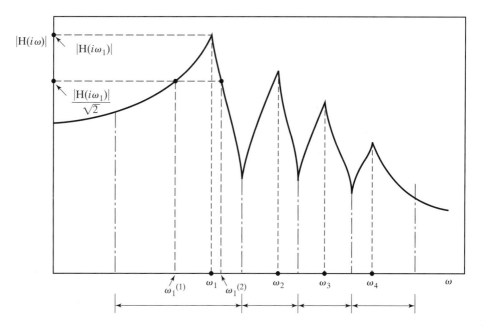

FIGURE 10.31   A typical graph of the frequency-response function of a structure or machine, obtained using Eq. (10.63) or (10.64).

of $H(i\omega)$, similar to Fig. 10.31, can be plotted by finding the values of $H(i\omega)$ at a number of frequencies of the applied harmonic force.

A simple method of finding the modal data involves the use of a single degree of freedom approach. In this method, the graph of $H(i\omega)$ is partitioned into several frequency ranges with each range bracketing one peak, as shown in Fig. 10.31. Each partitioned frequency range is then considered as the frequency-response function of a single degree of freedom system. This implies that the frequency-response function in each frequency range is dominated by that specific single mode. As observed in Section 3.4, a peak denotes a resonance point corresponding to a phase angle of 90°. Thus the resonant frequencies can be identified as the peaks in the graph of $H(i\omega)$, which can be confirmed from an observation of the values of the phase angle to be 90° at each of the peaks. The damping ratio corresponding to peak j, with resonant frequency ω_j, in Fig. 10.31 denotes the modal damping ratio, ζ_j. This ratio can be found, using Eq. (3.45), as

$$\zeta_j = \frac{\omega_j^{(2)} - \omega_j^{(1)}}{2\,\omega_j} \qquad (10.66)$$

where $\omega_j^{(1)}$ and $\omega_j^{(2)}$, known as half-power points, lie on either side of the resonant frequency ω_j and satisfy the relation

$$|H(i\omega_j^{(1)})| = |H(i\omega_j^{(2)})| = \frac{|H(i\omega_j)|}{\sqrt{2}} \qquad (10.67)$$

Note that ω_j actually represents the damped natural frequency of the system being tested. However, when damping is small, ω_j can be considered approximately equal to the undamped natural frequency of the system. When the system being tested is approximated as a k degree of freedom system ($k = 4$ for the system corresponding to Fig. 10.31), each peak observed in the graph of $H(i\omega)$ is assumed to be a single degree of freedom system, and the k resonant frequencies (peaks) and the corresponding damping ratios are determined by repeating the above procedure (and using an application of Eq. 10.66) k times.

EXAMPLE 10.5

Determination of Damping Ratio from Bode Diagram

The graphs showing the variations of the magnitude of the response and its phase angle with the frequency of a single degree of freedom system, as indicated in Fig. 3.11, provides the frequency response of the system. Instead of dealing with the magnitude curves directly, if the logarithms of the magnitude ratios (in decibels) are used, the resulting plots are called Bode diagrams. Find the natural frequency and the damping ratio of a system whose Bode diagram is shown in Fig. 10.32.

Solution: The natural frequency, which corresponds approximately to the peak response of the system, can be seen to be 10 Hz and the peak response to be -35 dB. The half-power points correspond to frequencies ω_1 and ω_2 where the response amplitude is equal to 0.707 times the peak response. From Fig. 10.32, the half-power points can be identified as $\omega_1 = 9.6$ Hz and $\omega_2 = 10.5$ Hz; thus the damping ratio can be determined by using Eq. (10.66) as

$$\zeta = \frac{\omega_2 - \omega_1}{2\omega_n} = \frac{10.5 - 9.6}{2(10.0)} = 0.045$$

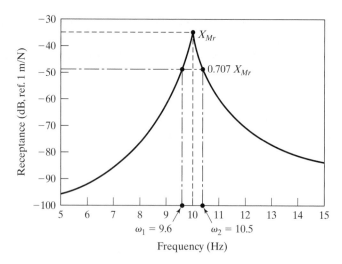

FIGURE 10.32   Bode diagram.

The procedure described in this section for finding the modal parameters is basically a visual approach. A more systematic, computer-based approach that can be implemented by the analyzer in conjunction with suitable programming is presented in the next section.

**10.8.6
Determination
of Modal Data
from Nyquist
Plot**

According to in this method, a single mode is also assumed to dominate in the neighborhood of its natural frequency in the frequency-response function. When the real and imaginary parts of the frequency-response function of a single degree of freedom system (given by Eq. 3.54) are plotted along the horizontal and vertical axes of a graph for a range of frequencies, the resulting graph will be in the form of a circle and that is known as the *Nyquist circle* or *Nyquist plot*. The frequency-response function, given by Eq. (3.54), can be written as

$$\alpha(i\omega) = \frac{1}{1 - r^2 + i2\zeta r} = u + iv \tag{10.68}$$

where

$$r = \frac{\omega}{\omega_n} \tag{10.69}$$

$$u = \text{Real part of } \alpha(i\omega) = \frac{1 - r^2}{(1 - r^2)^2 + 4\zeta^2 r^2} \tag{10.70}$$

$$v = \text{Imaginary part of } \alpha(i\omega) = \frac{-2\zeta r}{(1 - r^2)^2 + 4\zeta^2 r^2} \tag{10.71}$$

During vibration testing, the analyzer has the driving frequency values ω and the corresponding computed values of $u = \text{Re}(\alpha)$ and $v = \text{Im}(\alpha)$ from the measured data. The graph between u and v resembles a circle for large values of damping (ζ) while it increasingly assumes the shape of a circle as the damping becomes smaller and smaller, as shown in Fig. 10.33.

Properties of Nyquist Circle.   To identify the properties of the Nyquist circle, we first observe that large values of u and v are attained in the vicinity of resonance, $r = 1$. In that region, we can replace $1 - r^2$ in Eqs. (10.70) and (10.71) as

$$1 - r^2 = (1 + r)(1 - r) \approx 2(1 - r) \text{ and } 2\zeta r \approx 2\zeta$$

so that

$$u = \text{Re}(\alpha) \approx \frac{1 - r}{2[(1 - r)^2 + \zeta^2]} \tag{10.72}$$

$$v = \text{Im}(\alpha) \approx \frac{-\zeta}{2[(1 - r)^2 + \zeta^2]} \tag{10.73}$$

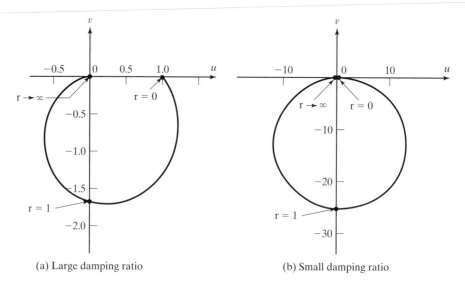

(a) Large damping ratio   (b) Small damping ratio

FIGURE 10.33   Nyquist circle.

It can be easily verified that u and v, given by Eqs. (10.72) and (10.73), satisfy the relation

$$u^2 + \left(v + \frac{1}{4\zeta}\right)^2 = \left(\frac{1}{4\zeta}\right)^2 \tag{10.74}$$

which denotes the equation of a circle with its center at $\left(u = 0, v = -\frac{1}{4\zeta}\right)$ and radius $\frac{1}{4\zeta}$. The half-power points occur at $r = 1 \pm \zeta$, which correspond to $u = \pm\frac{1}{4\zeta}$ and $v = \frac{1}{4\zeta}$. These points are located at the two ends of the horizontal diameter of the circle at which point u has its maximum magnitude.

These observations can be used to find ω_n ($r = 1$) and ζ. Once the measured values of the frequency-response function $H(i\omega)$ are available (with the applied force magnitude fixed) for a range of driving frequencies ω, instead of searching for the peak in the plot of $H(i\omega)$ versus ω, we can construct the Nyquist plot of Re $(H(i\omega))$ against Im $(H(i\omega))$ by using a least squares approach to fit a circle. This process also averages out the experimental errors. The intersection of the fitted circle with the negative imaginary axis will then correspond to $H(i\omega_n)$. The bandwidth, $(\omega^{(2)} - \omega^{(1)})$, is given by the difference of the frequencies at the two horizontal diametral points from which ζ can be found as $\zeta = \left(\dfrac{\omega^{(2)} - \omega^{(1)}}{2\omega_n}\right)$.

**10.8.7
Measurement of
Mode Shapes**

To determine the mode shapes from vibration testing, we need to express the equations of motion of the multidegree of freedom system in modal coordinates [10.18]. For this, we first consider an undamped system.

Undamped Multidegree of Freedom System.   The equations of motion of an undamped multidegree of freedom system in physical coordinates are given by

$$[m]\ddot{\vec{x}} + [k]\vec{x} = \vec{f} \tag{10.75}$$

For free harmonic vibration, Eq. (10.75) becomes

$$\left[[k] - \omega_i^2[m] \right]\vec{y}_i = \vec{0} \tag{10.76}$$

where ω_i is the ith natural frequency and $\vec{y}_i$ is the corresponding mode shape. The orthogonality relations for the mode shapes can be expressed as

$$[Y]^T[m][Y] = \text{diag}[M] \equiv \left[\diagup M_i \diagdown \right] \tag{10.77}$$

$$[Y]^T[k][Y] = \text{diag}[K] \equiv \left[\diagup K_i \diagdown \right] \tag{10.78}$$

where $[Y]$ is the modal matrix containing the modes $\vec{y}_1, \vec{y}_2, \ldots, \vec{y}_N$ as columns (N denotes the number of degrees of freedom of the system, also equal to the number of measured natural frequencies or peaks), M_i and K_i are the elements of diag $[M]$ and diag $[K]$, also called the *modal mass* and *modal stiffness*, respectively, corresponding to mode i, and

$$\omega_i^2 = \frac{K_i}{M_i} \tag{10.79}$$

When the forcing functions are harmonic, $\vec{f}(t) = \vec{F}e^{\tilde{i}\omega t}$, with $\tilde{i} = \sqrt{-1}$, Eq. (10.75) yields

$$\vec{x}(t) = \vec{X}e^{\tilde{i}\omega t} = \left[[k] - \omega^2[m] \right]^{-1}\vec{F}e^{\tilde{i}\omega t} \equiv [\alpha(\omega)]\vec{F}e^{\tilde{i}\omega t} \tag{10.80}$$

where $[\alpha(\omega)]$ is called the frequency-response function or receptance matrix of the system. Using the orthogonality relations of Eqs. (10.77) and (10.78), $[\alpha(\omega)]$ can be expressed as

$$[\alpha(\omega)] = [Y]\left[[K] - \omega^2[M] \right]^{-1}[Y]^T \tag{10.81}$$

An individual element of the matrix $[\alpha(\omega)]$ lying in row p and column q denotes the harmonic response of one coordinate, X_p, caused by a harmonic force applied at another coordinate, F_q (with no other forces), and can be written as

$$\alpha_{pq}(\omega) = [\alpha(\omega)]_{pq} = \left.\frac{X_p}{F_q}\right|_{\text{with }\ F_j=0;\ \ j=1,2,..,N;\ j\neq q}$$

$$= \sum_{i=1}^{N} \frac{(\vec{y}_i)_p(\vec{y}_i)_q}{K_i - \omega^2 M_i} \tag{10.82}$$

where $(\vec{y}_i)_j$ denotes the jth component of mode $\vec{y}_i$. If the modal matrix $[Y]$ is further normalized (rescaled or mass-normalized) as

$$[\Phi] \equiv \left[\vec{\phi}_1\, \vec{\phi}_2 \cdots \vec{\phi}_N \right] = [Y][M]^{-1/2} \tag{10.83}$$

the shape of the modes $\vec{\phi}_1, \vec{\phi}_2, \ldots, \vec{\phi}_N$ will not change, but Eq. (10.82) becomes

$$\alpha_{pq}(\omega) = \sum_{i=1}^{N} \frac{\left(\vec{\phi}_i\right)_p \left(\vec{\phi}_i\right)_q}{\omega_i^2 - \omega^2} \tag{10.84}$$

Damped Multidegree of Freedom System.   The equations of motion of a damped multidegree of freedom system in physical coordinates are given by

$$[m]\ddot{\vec{x}} + [c]\dot{\vec{x}} + [k]\vec{x} = \vec{f} \tag{10.85}$$

For simplicity, we assume proportional damping so that the damping matrix $[c]$ can be expressed as

$$[c] = a[k] + b[m] \tag{10.86}$$

where a and b are constants. Then the undamped mode shapes of the system, $\vec{y}_i$ and $\vec{\phi}_i$ not only diagonalize the mass and stiffness matrices, as indicated in Eqs. (10.77) and (10.78), but also the damping matrix:

$$[Y]^T[c][Y] = \text{diag } [C] = \left[\diagentry{C_i} \right] \tag{10.87}$$

Thus the mode shapes of the damped system will remain the same as those of the undamped system, but the natural frequencies will change and in general become complex. When the forcing vector $\vec{f}$ is assumed to be harmonic in Eq. (10.85), the frequency-response function or receptance can be derived as

$$\alpha_{pq}(\omega) = [\alpha(\omega)]_{pq} = \sum_{i=1}^{N} \frac{\left(\vec{y}_i\right)_p \left(\vec{y}_i\right)_q}{K_i - \omega^2 M_i + \tilde{i}\,\omega C_i} \tag{10.88}$$

When mass-normalized mode shapes are used (see Eq. 10.83), $\alpha_{pq}(\omega)$ becomes

$$\alpha_{pq}(\omega) = \sum_{i=1}^{N} \frac{\left(\vec{\phi}_i\right)_p \left(\vec{\phi}_i\right)_q}{\omega_i^2 - \omega^2 + 2\tilde{i}\zeta_i \omega_i \omega} \tag{10.89}$$

where ζ_i is the damping ratio in mode i.

As indicated earlier, the element of the matrix $[\alpha(\omega)]$ in row p and column q, $\alpha_{pq}(\omega) = [\alpha(\omega)]_{pq}$, denotes the transfer function between the displacement or response at point $p\,(X_p)$ and the input force at point $q\,(F_q)$ of the system being tested (with all other forces equal to zero). Since this transfer function denotes the ratio $\frac{X_p}{F_q}$, it is given by $H_{pq}(\omega)$. Thus

$$\alpha_{pq}(\omega) = H_{pq}(\omega) \tag{10.90}$$

If the peaks or resonant (natural) frequencies of the system are well separated, then the term corresponding to the particular peak (ith peak) dominates all other terms in the summation of Eq. (10.88) or (10.89). By substituting $\omega = \omega_i$ in Eq. (10.89), we obtain

$$\alpha_{pq}(\omega_i) = H_{pq}(\omega_i) = \frac{\left(\vec{\phi}_i\right)_p \left(\vec{\phi}_i\right)_q}{\omega_i^2 - \omega_i^2 + \widetilde{i}\,2\zeta_i\omega_i^2}$$

or

$$\left|\alpha_{pq}(\omega_i)\right| = \left|H_{pq}(\omega_i)\right| = \frac{\left|\left(\vec{\phi}_i\right)_p \left(\vec{\phi}_i\right)_q\right|}{2\,\zeta_i\omega_i^2}$$

or

$$\left|\left(\vec{\phi}_i\right)_p \left(\vec{\phi}_i\right)_q\right| = 2\zeta_i\omega_i^2 \left|H_{pq}(\omega_i)\right| \tag{10.91}$$

It can be seen that Eq. (10.91) permits the computation of the absolute value of $\left(\vec{\phi}_i\right)_p \left(\vec{\phi}_i\right)_q$ using the measured values of the natural frequency (ω_i), damping ratio (ζ_i) and the transfer function $\left|H_{pq}(\omega_i)\right|$ at peak i. To determine the sign of the element $\left(\vec{\phi}_i\right)_p \left(\vec{\phi}_i\right)_q$, the phase plot of $H_{pq}(\omega_i)$ can be used. Since there are only N independent unknown components of $\vec{\phi}_i$ in the N^2 elements of the matrix $\left[\left(\vec{\phi}_i\right)_p \left(\vec{\phi}_i\right)_q\right] = \left[\vec{\phi}_i\,\vec{\phi}_i^{\mathrm{T}}\right]_{pq}$, N measurements of $\left|H_{pq}(\omega_i)\right|$ are required to determine the mode shape $\vec{\phi}_i$ corresponding to the modal frequency ω_i. This can be achieved by measuring the displacement or response of the system at point q with input at point 1 first, at point 2 next, $\dots$, and at point N last.

10.9   Machine Condition Monitoring and Diagnosis

Most machines produce low levels of vibration when designed properly. During operation, all machines are subjected to fatigue, wear, deformation, and foundation settlement. These effects cause an increase in the clearances between mating parts, misalignments in shafts, initiation of cracks in parts and unbalances in rotors—all leading to an increase in the level of vibration, which causes additional dynamic loads on bearings. As time progresses, the vibration levels continue to increase, leading ultimately to the failure or breakdown of the machine. The common types of faults or operating conditions that lead to increased levels of vibration in machines include bent shafts, eccentric shafts, misaligned components, unbalanced components, faulty bearings, faulty gears, impellers with faulty blades, and loose mechanical parts.

10.9.1 Vibration Severity Criteria

The vibration severity charts, given by standards such as ISO 2372, can be used as a guide to determine the condition of a machine. In most cases, the root mean square (RMS) value of the vibratory velocity of the machine is compared against the criteria set by the standards. Although it is very simple to implement this procedure, the overall velocity signal used for comparison may not give sufficient warning of the imminent damage of the machine.

10.9.2 Machine Maintenance Techniques

The life of a machine follows the classic *bathtub curve* shown in Fig. 10.34. Since the failure of a machine is usually characterized by an increase in vibration and/or noise level, the vibration level also follows the shape of the same bathtub curve. The vibration level decreases during the initial running-in period, then increases very slowly during the normal operating period due to the normal wear, and finally increases rapidly due to excessive wear until failure or breakdown in the wearout period.

Three types of maintenance schemes can be used in practice:

1. *Breakdown maintenance.* The machine is allowed to fail, at which time the failed machine is replaced by a new one. This strategy can be used if the machine is inexpensive to replace and the breakdown does not cause any other damage. Otherwise, the cost of lost production, safety risks, and additional damage to other machines make this scheme unacceptable.

2. *Preventive maintenance.* Maintenance is performed at fixed intervals such as every 3000 operating hours or once a year. The maintenance intervals are usually determined statistically from past experience. Although this method reduces the chance of unexpected breakdowns, it has been found to be uneconomical. The stoppage for maintenance involves not only lost production time but also a high risk of introducing imperfections due to human error. In addition, the probability of failure of a machine component cannot be reduced by replacing it with a new one during the normal wearout period.

3. *Condition-based maintenance.* The fixed-interval overhauls are replaced by fixed-interval measurements that permit the observation of changes in the running condition of the machine regularly. Thus the onset of fault conditions can be detected and their developments closely followed. The measured vibration levels can be extrapolated in order to predict when the vibration levels reach unacceptable values and when the machine must be serviced. Hence this scheme is also known as predictive maintenance. In this method, the maintenance costs are greatly reduced due to fewer catastrophic failures, better utilization of spare parts, and elimination of the unnecessary preventive maintenance. The vibration level (and hence the failure probability) of the machine due to condition-based maintenance follows the shape indicated in Fig. 10.35.

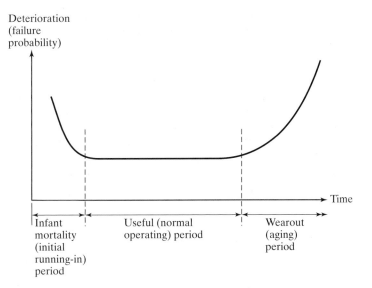

FIGURE 10.34   The bathtub curve for the life of a machine.

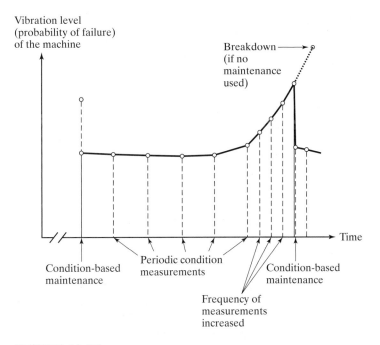

FIGURE 10.35   Condition-based maintenance.

**10.9.3
Machine
Condition
Monitoring
Techniques**

Several methods can be used to monitor the condition of a machine, as indicated in Fig. 10.36. Aural and visual methods are the basic forms of monitoring techniques in which a skilled technician, having an intimate knowledge of machines, can identify a failure simply by listening to the sounds and/or visually observing the large amplitudes of vibration produced by a damaged machine. Sometimes a microphone or a stroboscope is used to hear the machine noise. Similarly, devices ranging from magnifying glasses to stroboscopes are used to visually monitor the condition of a machine. Current and voltage monitoring can be used for the condition monitoring of electrical drives such as large generators and motors.

In the operational variables method of monitoring, also known as performance or duty cycle monitoring, the performance of a machine is observed with regard to its intended duty. Any deviation from the intended performance denotes a malfunction of the machine. Temperature monitoring involves measuring the operational or surface temperature of a machine. This method can be considered as a kind of operational variables method. A rapid increase in the temperature of a component, occurring mostly due to wear, is an indication of a malfunction such as inadequate lubricant in journal bearings. Temperature monitoring uses such devices as optical pyrometers, thermocouples, thermography, and resistance thermometers. In some cases, dye penetrants are used to identify cracks occurring on the surface of a machine. This procedure requires the use of heat-sensitive paints, known as thermographic paints, to detect surface cracks on hot surfaces. In such cases, the most suitable paint matching the expected surface temperature is selected.

Wear debris is generated at relative moving surfaces of load-bearing machine elements. The wear particles that can be found in the lubricating oils or grease can be used to assess the extent of damage. As wear increases, the particles of the material used to construct machine components such as bearings and gears can be found in increasing concentration. Thus the severity of the wear can be assessed by observing the concentration (quantity), size, shape, and color of the particles. Note that the color of the particles indicates how hot they have been.

Vibration analysis is most commonly used for machine condition monitoring. Vibration in machines is caused by cyclic excitation forces arising from imbalances, wear, or failure of parts. What type of changes occur in the vibration level, how these changes can be detected, and how the condition of the machine is interpreted has been the topic of several research studies in the past. The available vibration monitoring techniques can be classified as shown in Fig. 10.37. These techniques are described in the following section.

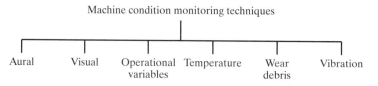

FIGURE 10.36   Machine condition monitoring techniques.

Machine vibration monitoring techniques

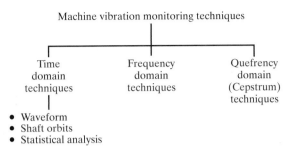

FIGURE 10.37   Machine vibration monitoring techniques.

10.9.4 Vibration Monitoring Techniques

Time Domain Analysis

Time Waveforms.   Time-domain analysis uses the time history of the signal (waveform). The signal is stored in an oscilloscope or a real-time analyzer and any nonsteady or transient impulses are noted. Discrete damages such as broken teeth in gears and cracks in inner or outer races of bearings can be identified easily from the waveform of the casing of a gearbox. As an example, Fig. 10.38 shows the acceleration signal of a single-stage gearbox. The pinion of the gear pair is coupled to a 5.6 kW, 2865 rpm, AC electric motor. Since the pinion (shaft) speed is 2865 rpm or 47.75 Hz, the period can be noted as 20.9 ms. The acceleration waveform indicates that pulses occur periodically with a period of 20 ms approximately. Noting that this period is the same as the period of the pinion, the origin of the pulses in the acceleration signal can be attributed to a broken gear tooth on the pinion.

Indices.   In some cases, indices such as the peak level, root mean square (RMS) level and the crest factor are used to identify damage in machine condition monitoring. Since the peak level occurs only once, it is not a statistical quantity and hence is not a reliable index

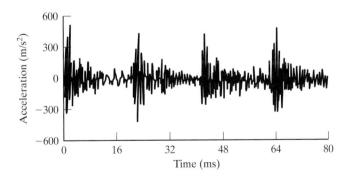

FIGURE 10.38   Time domain waveform of a faulty gearbox [10.23].

to detect damage in continuously operating systems. Although the RMS value is a better index to detect damage in steady-state applications, it may not be useful if the signal contains information from more than one component, as in the case of vibration of a complete gearbox that consists of several gears, shafts, and bearings. The crest factor, defined as the ratio of the peak to RMS level, includes information from both the peak and the RMS levels. However, it may also not be able to identify failure in certain cases. For example, if the failure occurs progressively, the RMS level of the signal might be increasing gradually although the crest factor might be showing a decreasing trend.

Orbits.   Sometimes, certain patterns known as Lissajous figures can be obtained by displaying time waveforms obtained from two transducers whose outputs are shifted by 90° in phase. Any change in the pattern of these figures or orbits can be used to identify faults such as misalignment in shafts, unbalance in shafts, shaft rub, wear in journal bearings, and hydrodynamic instability in lubricated bearings. Figure 10.39 illustrates a change in orbit caused by a worn bearing. The enlarged orbit diameter in the vertical direction indicates that the bearing has become stiffer in the horizontal direction—that is, it has more bearing clearance in the vertical direction.

Statistical Methods

Probability Density Curve.   All vibration signals will have a characteristic shape for its probability density curve. The probability density of a signal can be defined as the probability of finding its instantaneous amplitude within a certain range, divided by the range. Usually, the waveform corresponding to good components will have a bell-shaped probability density curve similar to normal distribution. Thus any significant deviation from the bell shape can be associated with the failure of a component. Since the use of the probability density curve involves the comparison of variations in shape rather than variations in amplitudes, it is very useful in the diagnosis of faults in machines.

Moments.   In some cases, the moments of the probability density curve can be used for the machine condition monitoring. The moments of the curve are similar to mechanical

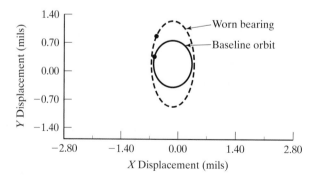

FIGURE 10.39   Change in orbit due to a bearing failure [10.23].

moments about the centroidal axis of the area. The first four moments of a probability density curve (with proper normalization) are known as the mean, standard deviation, skewness, and kurtosis, respectively. For practical signals, the odd moments are usually close to zero and the even moments denote the impulsiveness of the signal. The fourth-order moment, kurtosis, is commonly used in machine condition monitoring. The kurtosis is defined as

$$k = \frac{1}{\sigma^4} \int_{-\infty}^{\infty} (x - \bar{x})^4 f(x)\, dx \qquad (10.92)$$

where $f(x)$ is the probability density function of the instantaneous amplitude, $x(t)$, at time t, $\bar{x}$ is the mean value, and σ is the standard deviation of $x(t)$. Faults such as cracked races and spalling of rollers and balls in bearings cause relatively large pulses in the time-domain waveform of the signal, which in turn lead to large values of kurtosis. Thus an increase in the value of kurtosis can be attributed to the failure of a machine component.

Frequency Domain Analysis

Frequency Spectrum.   The frequency domain signal or frequency spectrum is a plot of the amplitude of vibration response versus the frequency and can be derived by using the digital fast Fourier analysis of the time waveform. The frequency spectrum provides valuable information about the condition of a machine. The vibration response of a machine is governed not only by its components but also by its assembly, mounting, and installation. Thus the vibration characteristics of any machine are somewhat unique to that particular machine; hence the vibration spectrum can be considered as the vibration signature of that machine. As long as the excitation forces are constant or vary by small amounts, the measured vibration level of the machine also remains constant or varies by small amounts. However, as the machine starts developing faults, its vibration level and hence the shape of the frequency spectrum changes. By comparing the frequency spectrum of the machine in damaged condition with the reference frequency spectrum corresponding to the machine in good condition, the nature and location of the fault can be detected. Another important characteristic of a spectrum is that each rotating element in a machine generates identifiable frequency, as illustrated in Fig. 10.40; thus the changes in the spectrum at a given frequency can be attributed directly to the corresponding machine component. Since such changes can be detected more easily compared to changes in the overall vibration levels, this characteristic will be very valuable in practice.

Since the peaks in the spectrum relate to various machine components, it is necessary to be able to compute the fault frequencies. A number of formulas can be derived to find the fault frequencies of standard components like bearings, gearboxes, pumps, fans, and pulleys. Similarly, certain standard fault conditions can be described for standard faults such as unbalance, misalignment, looseness, oil whirl, and resonance.

Quefrency Domain Analysis.   Quefrency serves as the abscissa (*x*-axis) for a parameter known as cepstrum similar to frequency that serves as the abscissa for the parameter spectrum. Several definitions are available for the term *cepstrum* in the literature. Originally,

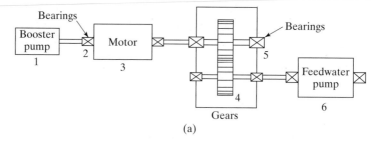

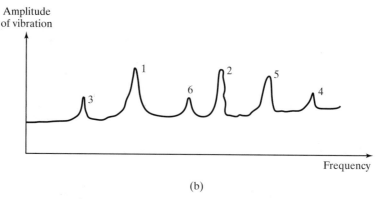

FIGURE 10.40   Relationship between machine components and the vibration spectrum.

cepstrum was defined as the power spectrum of the logarithm of the power spectrum. If
$x(t)$ denotes a time signal, its power spectrum, $S_X(\omega)$, is given by

$$S_X(\omega) = |F\{x(t)\}|^2 \tag{10.93}$$

where $F\{\}$ denotes the Fourier transform of $\{\}$:

$$F\{x(t)\} = \frac{1}{T}\int_{-\frac{T}{2}}^{\frac{T}{2}} x(t)\, e^{i\omega t}\, dt \tag{10.94}$$

Thus the cepstrum, $c(\tau)$, is given by

$$c(\tau) = |F\{\log S_X(\omega)\}|^2 \tag{10.95}$$

Later, the cepstrum was defined as the inverse Fourier transform of the logarithm of the
power spectrum so that $c(\tau)$ becomes

$$c(\tau) = F^{-1}\{\log S_X(\omega)\} \tag{10.96}$$

The word *cepstrum* is derived by rearranging the letters in the word *spectrum*. The reason
for this link is that the cepstrum is basically the spectrum of a spectrum. In fact, many of

the terms used in spectrum analysis have been modified for use in cepstrum analysis. A few examples are given below:

Quefrency—Frequency

Rahmonics—Harmonics

Gamnitude—Magnitude

Saphe—Phase

From this, it is logical to see why quefrency serves as the abscissa of the cepstrum.

In practice, the choice of the definition of cepstrum is not critical since both definitions—Eqs. (10.95) and (10.96)—show distinct peaks in the same location if there is strong periodicity in the (logarithmic) spectrum. The cepstrum is useful in machine condition monitoring and diagnosis since it can detect any periodicity in the spectrum caused by the failure of components, such as a blade in a turbine and a gear tooth in a gearbox. As an example, the spectra and cepstra of two truck gearboxes, one in good condition and the other in bad condition, running on a test stand with first gear in engagement, are shown in Figs. 10.41(a) to (d). Note that in Fig. 10.41(a), the good gearbox shows no marked periodicity in its spectrum while the bad gearbox indicates a large number of sidebands with an approximate spacing of 10 Hz in its spectrum (Fig. 10.41b). This spacing cannot be determined more accurately from Fig. 10.41(b). Similarly, the cepstrum of the good gearbox does not indicate any quefrencies prominently (Fig. 10.41d). However, the cepstrum of the bad gearbox (Fig. 10.41c) indicates three prominant quefrencies at 28.1 ms (35.6 Hz), 95.9 ms (10.4 Hz), and 191.0 ms (5.2 Hz). The first series of rahmonics corresponding to 35.6 Hz has been identified to correspond to the input speed of the gearbox. The theoretical output speed is 5.4 Hz. Thus the rahmonics corresponding to 10.4 Hz is not expected to be same as the second harmonic of the output speed, which would be 10.8 Hz. A careful examination revealed that the rahmonics corresponding to the frequency 10.4 Hz are same as the speed of the second gear. This indicates that the second gear was at fault although the first gear was in engagement.

10.9.5
Instrumentation
Systems

Based on their degree of sophistication, three types of instrumentation systems can be used for condition monitoring of machines—the basic system, the portable system, and the computer-based system. The first type, which can be labeled as the basic system, consists of a simple pocket-sized vibration meter, a stroboscope, and a headset. The vibration meter measures the overall vibration levels (RMS or peak values of acceleration or velocity) over suitable frequency ranges, the stroboscope indicates the speed of the machine, and the headset aids in hearing the machine vibration. The overall RMS velocity readings can be compared with published severity charts and any need for condition-based maintenance can be established. The overall vibration levels can also be plotted against time to find how rapidly the condition of the machine is changing. The vibration meter can also be used in conjunction with a pocket computer to collect and store the measurements. Sometimes, an experienced operator can hear the vibration (sound) of a machine over a period of time and find its condition. In some cases, faults such as misalignment, unbalance, or looseness of parts can be observed visually.

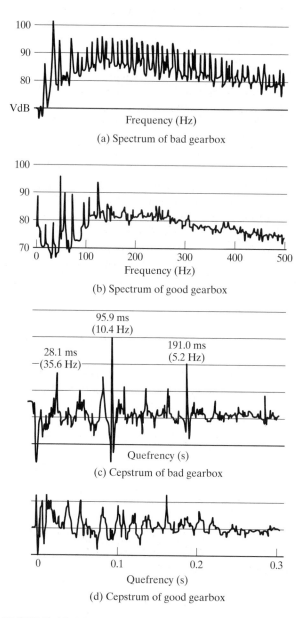

(a) Spectrum of bad gearbox

(b) Spectrum of good gearbox

(c) Cepstrum of bad gearbox

(d) Cepstrum of good gearbox

FIGURE 10.41   Spectrum and Cepstrum of a gearbox [10.24].

The portable condition monitoring system consists of a portable fast Fourier transform (FFT) vibration analyzer based on battery power. This vibration analyzer can be used for fault detection by recording and storing vibration spectra from each of the measurement points. Each newly recorded spectrum can be compared with a reference spectrum that was recorded at that particular measurement point when the machine was known to be in good condition. Any significant increase in the amplitudes in the new spectrum indicates a fault that needs further investigation. The vibration analyzer also has certain diagnostic capability to identify problems such as faulty belt drives and gearboxes and loose bearings. When the fault diagnosed requires a replacement of parts, it can be done by the operator. If a rotor requires balancing, the vibration analyzer can be used to compute the locations and magnitudes of the correction masses necessary to rebalance the rotor.

The computer-based condition monitoring system is useful and economical when the number of machines, the number of monitoring points, and the complexity of fault detection increases. It consists of an FFT vibration analyzer coupled with a computer for maintaining a centralized database that can also provide diagnostic capabilities. The data are stored on a disk, allowing them to be used for spectrum comparison or for three-dimensional plots (see Fig. 10.42). Certain computer-based systems use tape recorders to record vibration signals from each machine at all the measurement points. These measurements can be played back into the computer for storage and post processing.

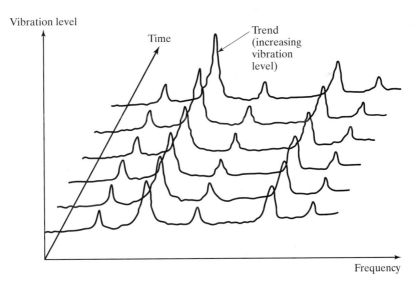

FIGURE 10.42   Three-dimensional plot of data.

**10.9.6
Choice of
Monitoring
Parameter**

Piezoelectric accelerometers are commonly used for measuring the vibration of machines. They are preferred because of their smaller size, superior frequency and dynamic range, reliability over long periods, and robustness. When an accelerometer is used as the vibration pickup, the velocity and displacements can be obtained from the integrators built in the analyzer. Thus the user can choose between acceleration, velocity, and displacement as the monitoring parameter. Although any of these three spectra can be used for the condition monitoring of a machine, usually the velocity spectrum will be the flattest one (indicating that the range of velocity amplitudes is the smallest). Since a change in the amplitude of velocity can be observed easily in a flatter spectrum, velocity is commonly used as the parameter for monitoring the condition of machines.

10.10   Examples Using MATLAB

▮▮▮▮▮▮▮▮▮▮ Plotting of Nyquist Circle

EXAMPLE   10.6
_____

Using MATLAB, plot the Nyquist circle for the following data:

a. $\zeta = 0.75$
b. $\zeta = 0.05$

Solution: Equations (10.70) and (10.71) are plotted along the horizontal and vertical axes. The MATLAB program to plot the Nyquist circle is given below.

```
%Ex10_6.m
zeta = 0.05;
for i = 1: 10001
    r(i) = 50 * (i-1) / 10000;
    Re1(i) = ( 1-r(i)^2 )/( (1-r(i)^2)^2 + 4*zeta^2*r(i)^2 );
    Im1(i) = -2*zeta*r(i)/( (1-r(i)^2)^2 + 4*zeta^2*r(i)^2 );
end
zeta = 0.75;
for i = 1: 10001
    r(i) = 50 *(i-1) / 10000;
    Re2(i) = ( 1-r(i)^2 )/( (1-r(i)^2)^2 + 4*zeta^2*r(i)^2 );
    Im2(i) = -2*zeta*r(i)/( (1-r(i)^2)^2 + 4*zeta^2*r(i)^2 );
end
plot(Re1, Im1);
title('Nyquist plot: zeta = 0.05');
ylabel('Imaginary axis');
xlabel('Real axis');
pause;
plot(Re2, Im2);
title('Nyquist plot: zeta = 0.75');
ylabel('Imaginary axis');
xlabel('Real axis');
```

■

Nyquist plot: zeta = 0.05

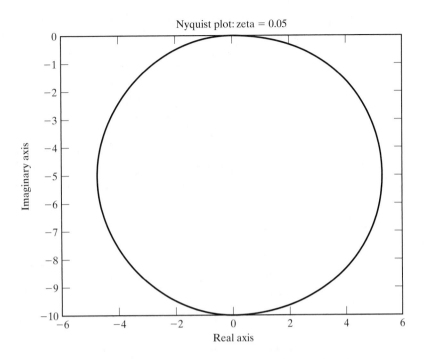

Nyquist plot: zeta = 0.75

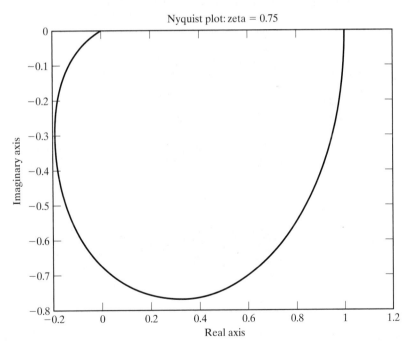

████████████ Plotting of Accelerometer Equation

EXAMPLE 10.7 ─────────────────────────────────

───────────── Using MATLAB, plot the ratio of measured to true accelerations, given by

$$f(r) = \frac{1}{\{(1 - r^2)^2 + (2\zeta r)^2\}^{1/2}} \tag{E.1}$$

for $\zeta = 0.0, 0.25, 0.5, 0.75,$ and 1.0.

Solution: The MATLAB program to plot Eq. (E.1) in the range $0 \le r \le 1$ is given below.

```
%Ex10_7.m
zeta = 0.0;
for i = 1: 101
    r(i) = (i-1)/100;
    f1(i) = 1/sqrt((1-r(i)^2)^2 + (2*zeta*r(i))^2);
end
zeta = 0.25;
for i = 1: 101
    r(i) = (i-1)/100;
    f2(i) = 1/sqrt( (1-r(i)^2)^2 + (2*zeta*r(i))^2 );
end
zeta = 0.5;
for i = 1: 101
    r(i) = (i-1)/100;
    f3(i) = 1/sqrt( (1-r(i)^2)^2 + (2*zeta*r(i))^2 );
end
zeta = 0.75;
for i = 1: 101
    r(i) = (i-1)/100;
    f4(i) = 1/sqrt( (1-r(i)^2)^2 + (2*zeta*r(i))^2 );
end
zeta = 1.0;
for i = 1: 101
    r(i) = (i-1)/100;
    f5(i) = 1/sqrt( (1-r(i)^2)^2 + (2*zeta*r(i))^2 );
end
plot(r,f1);
axis([0 1 0 5]);
gtext('zeta = 0.00');
hold on;
plot(r,f2);
gtext('zeta = 0.25');
hold on;
plot(r,f3);
gtext('zeta = 0.50');
hold on;
plot(r,f4);
gtext('zeta = 0.75');
hold on;
plot(r,f5);
gtext('zeta = 1.00');
xlabel('r');
ylabel('f(r)');
```

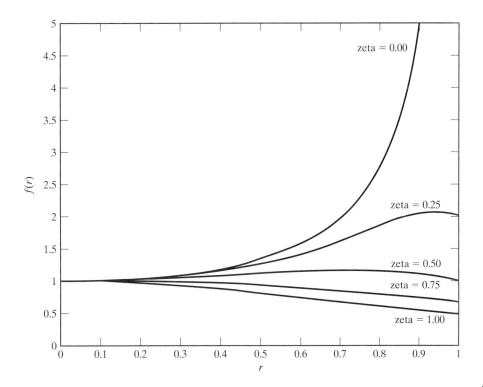

REFERENCES

10.1 G. Buzdugan, E. Mihailescu, and M. Rades, *Vibration Measurement*, Martinus Nijhoff, Dordrecht, The Netherlands, 1986.

10.2 *Vibration Testing*, Bruel & Kjaer, Naerum, Denmark, 1983.

10.3 O. Dossing, *Structural Testing. Part I. Mechanical Mobility Measurements*, Bruel & Kjaer, Naerum, Denmark, 1987.

10.4 D. N. Keast, *Measurements in Mechanical Dynamics*, McGraw-Hill, New York, 1967.

10.5 B. W. Mitchell (ed.), *Instrumentation and Measurement for Environmental Sciences* (2nd ed.), American Society of Agricultural Engineers, Saint Joseph, Mich., 1983.

10.6 J. P. Holman, *Experimental Methods for Engineers* (4th ed.), McGraw-Hill, New York, 1984.

10.7 J. T. Broch, *Mechanical Vibration and Shock Measurements*, Bruel & Kjaer, Naerum, Denmark, 1976.

10.8 R. R. Bouche, *Calibration of Shock and Vibration Measuring Transducers*, Shock and Vibration Information Center, Washington, D.C., SVM-11, 1979.

10.9 M. Rades, "Methods for the analysis of structural frequency-response measurement data," *Shock and Vibration Digest*, Vol. 8, No. 2, February 1976, pp. 73–88.

10.10 J. D. Irwin and E. R. Graf, *Industrial Noise and Vibration Control*, Prentice Hall, Englewood Cliffs, N.J., 1979.

10.11 R. K. Vierck, *Vibration Analysis*, Harper & Row, New York, 1979.

10.12 J. A. Macinante, *Seismic Mountings for Vibration Isolation*, Wiley, New York, 1984.

10.13 R. B. Randall and R. Upton, "Digital filters and FFT technique in real-time analysis," pp. 45–67, in *Digital Signal Analysis Using Digital Filters and FFT Techniques*, Bruel & Kjaer, Naerum, Denmark, 1985.

10.14 G. Dovel, "A modal analysis—a dynamic tool for design and troubleshooting," *Mechanical Engineering*, Vol. 111, No. 3, March 1989, pp. 82–86.

10.15 C. W. deSilva and S. S. Palusamy, "Experimental modal analysis—a modeling and design tool," *Mechanical Engineering*, Vol. 106, No. 6, June 1984, pp. 56–65.

10.16 K. Zaveri, *Modal Analysis of Large Structures—Multiple Exciter Systems*, Bruel & Kjaer, Denmark, 1984.

10.17 O. Dossing, *Structural Testing—Part 2: Modal Analysis and Simulation*, Bruel & Kjaer, Naerum, Denmark, 1988.

10.18 D. J. Ewins, "Modal analysis as a tool for studying structural vibration," in *Mechanical Signature Analysis: Theory and Applications*, S. Braun (ed.), Academic Press, London, pp. 217–261, 1986.

10.19 B. A. Brinkman and D. J. Macioce, "Understanding modal parameters and mode shape scaling," *Sound and Vibration*, Vol. 19, No. 6, pp. 28–30, June 1985.

10.20 N. Tandon and B. C. Nakra, "Vibration and acoustic monitoring techniques for the detection of defects in rolling element bearings—a review," *Shock and Vibration Digest*, Vol. 24, No. 3, March 1992, pp. 3–11.

10.21 S. Braun, "Vibration monitoring," in *Mechanical Signature Analysis: Theory and Applications*, S. Braun (ed.), Academic Press, London, 1986, pp. 173–216.

10.22 A. El-Shafei, "Measuring vibration for machinery monitoring and diagnostics," *Shock and Vibration Digest*, Vol. 25, No. 1, January 1993, pp. 3–14.

10.23 J. Mathew, "Monitoring the vibrations of rotating machine elements—an overview," in *Diagnostics, Vehicle Dynamics and Special Topics*, T. S. Sankar (ed.), American Society of Mechanical Engineers, New York, 1989, pp. 15–22.

10.24 R. B. Randall, "Advances in the application of cepstrum analysis to gearbox diagnosis," in *Second International Conference Vibrations in Rotating Machinery (1980)*. Institution of Mechanical Engineers, London, 1980, pp. 169–174.

REVIEW QUESTIONS

10.1 Give brief answers to the following:

1. What is the importance of vibration measurement?
2. What is the difference between a vibrometer and a vibrograph?
3. What is a transducer?
4. Discuss the basic principle on which a strain gage works.
5. Define the gage factor of a strain gage.
6. What is the difference between a transducer and a pickup?
7. What is a piezoelectric material? Give two examples of such material.
8. What is the working principle of an electrodynamic transducer?

9. What is an LVDT? How does it work?
10. What is a seismic instrument?
11. What is the frequency range of a seismometer?
12. What is an accelerometer?
13. What is phase-shift error? When does it become important?
14. Give two examples of a mechanical vibration exciter.
15. What is an electromagnetic shaker?
16. Discuss the advantage of using operational deflection shape measurement.
17. What is the purpose of experimental modal analysis?
18. Describe the use of frequency response function in modal analysis.
19. Name two frequency measuring instruments.
20. State three methods of representing the frequency response data.
21. How are Bode plots used?
22. How is a Nyquist diagram constructed?
23. What is the principle of mode superposition? What is its use in modal analysis?
24. State the three types of maintenance schemes used for machinery.
25. How are orbits used in machine diagnosis?
26. Define the terms *kurtosis* and *cepstrum*.

10.2 Indicate whether each of the following statements is true or false:

1. A strain gage is a variable resistance transducer.
2. The value of the gage factor of a strain gage is given by the manufacturer.
3. The voltage output of an electromagnetic transducer is proportional to the relative velocity of the coil.
4. The principle of electrodynamic transducer can be used in vibration exciters.
5. A seismometer is also known as a vibrometer.
6. All vibration-measuring instruments exhibit phase lag.
7. The time lag is important when measuring harmonic motion of frequency ω.
8. The Scotch yoke mechanism can be used as a mechanical shaker.
9. The time response of a system gives better information on energy distribution than does the frequency-response.
10. A spectrum analyzer is a device that analyzes a signal in the frequency domain.
11. The complete dynamic response of a machine can be determined through modal testing.
12. The damping ratio of a vibrating system can be found from the Bode diagram.
13. The spectrum analyzers are also known as fast Fourier transform (FFT) analyzers.
14. In breakdown maintenance, the machine is run until failure.
15. Time-domain waveforms can be used to detect discrete damages of machinery.

10.3 Fill in each of the following blanks with the appropriate word:

1. A device that transforms values of physical variables into equivalent electrical signals is called a _____ .
2. Piezoelectric transducers generate electrical _____ when subjected to mechanical stress.
3. A seismic instrument consists of a _____ system mounted on the vibrating body.
4. The instrument that measures the acceleration of a vibrating body is called _____ .
5. _____ can be used to record earthquakes.
6. The instrument that measures the velocity of a vibrating body is called a _____ .

7. Most mechanical frequency measuring instruments are based on the principle of _____.

8. The Frahm tachometer is a device consisting of several _____ carrying masses at free ends.

9. The main advantage of a stroboscope is that it can measure the speed without making _____ with the rotating body.

10. In real-time frequency analysis, the signal is continuously analyzed over all the _____ bands.

11. Real-time analyzers are useful for machinery _____ monitoring since a change in the noise or vibration spectrum can be observed immediately.

12. An _____ is the interval between any two frequencies $(f_2 - f_1)$ whose frequency ratio $\left(\dfrac{f_2}{f_1}\right)$ is 2.

13. The dynamic testing of a machine involves finding the _____ of the machine at a critical frequency.

14. For vibration testing, the machine is supported to simulate a _____ condition of the system so that rigid body modes can also be observed.

15. The excitation force is measured by a _____ cell.

16. The response of a system is usually measured by _____.

17. The frequency response of a system can be measured using _____ analyzers.

18. The condition of a machine can be determined using _____ severity charts.

19. The life of a machine follows the classic _____ curve.

20. The _____ observed in Lissajous figures can be used to identify machinery faults.

21. Cepstrum can be defined as the power spectrum of the logarithm of the _____.

10.4 Select the most appropriate answer out of the choices given:

1. When a transducer is used in conjunction with another device to measure vibration, it is called a
(a) vibration sensor   (b) vibration pickup   (c) vibration actuator

2. The instrument that measures the displacement of a vibrating body is called a
(a) seismometer   (b) transducer   (c) accelerometer

3. The circuit that permits the passage of frequency components of a signal over a frequency band and rejects all other frequency components is called a
(a) band pass filter   (b) frequency filter   (c) spectral filter

4. A decibel (dB) is a quantity, such as power (P), defined in terms of a reference value (P_{ref}), as

$$\text{(a)} \ 10 \log_{10}\left(\frac{P}{P_{ref}}\right) \qquad \text{(b)} \ \log_{10}\left(\frac{P}{P_{ref}}\right) \qquad \text{(c)} \ \frac{1}{P_{ref}} \log_{10}(P)$$

5. The following function plays an important role in the experimental modal analysis:
(a) time response function
(b) modal response function
(c) frequency response function

6. The method of subjecting a system to a known force as an initial condition and then releasing is known as
(a) step relaxation
(b) excitation by electromagnetic shaker
(c) impactor

7. The process of using an electrical signal, generalized by a spectrum analyzer, for applying a mechanical force on a system is known as
   (a) step relaxation
   (b) excitation by electromagnetic shaker
   (c) impactor

8. The procedure of using a hammer with a built-in load cell to apply load at different points of a system is known as
   (a) step relaxation
   (b) excitation by electromagnetic shaker
   (c) impactor

9. During the initial running-in period, usually the deterioration of a machine
   (a) decreases   (b) increases   (c) remains constant

10. During the normal operating period, the deterioration of a machine usually
    (a) decreases   (b) increases   (c) remains constant

11. During the aging or wearout period, the deterioration of a machine usually
    (a) decreases   (b) increases   (c) remains constant

10.5 Match the items in the two columns below:

1.	Piezoelectric accelerometer	(a)	produces light pulses intermittently
2.	Electrodynamic transducer	(b)	has high output and is insensitive to temperature
3.	LVDT transducer	(c)	frequently used in velocity pickups
4.	Fullarton tachometer	(d)	has high sensitivity and frequency range
5.	Stroboscope	(e)	variable length cantilever with a mass at its free end

PROBLEMS

The problem assignments are organized as follows:

Problems	Section Covered	Topic Covered
10.1	10.2	Transducers
10.2–10.18	10.3	Vibration Pickups
10.19	10.4	Frequency-measuring instruments
10.20–10.25	10.8	Experimental modal analysis
10.26–10.30	10.9	Machine condition monitoring and diagnosis
10.31–10.32	10.10	MATLAB programs
10.33–10.35	—	Design projects

10.1 A Rochelle salt crystal, having a voltage sensitivity of 0.098 V-m/N and thickness 2 mm, produced an output voltage of 200 volts under pressure. Find the pressure applied to the crystal.

10.2 A spring-mass system with $m = 0.5$ kg and $k = 10,000$ N/m, with negligible damping, is used as a vibration pickup. When mounted on a structure vibrating with an amplitude of 4 mm, the total displacement of the mass of the pickup is observed to be 12 mm. Find the frequency of the vibrating structure.

10.3 The vertical motion of a machine is measured by using the arrangement shown in Fig. 10.43. The motion of the mass m relative to the machine body is recorded on a drum. If the damping constant c is equal to $c_{cri}/\sqrt{2}$, and the vertical vibration of the machine body is given by $y(t) = Y \sin \omega t$, find the amplitude of motion recorded on the drum.

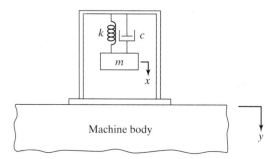

FIGURE 10.43

10.4 It is proposed that the vibration of the foundation of an internal combustion engine be measured over the speed range 500 rpm to 1500 rpm using a vibrometer. The vibration is composed of two harmonics, the first one caused by the primary inertia forces and the second one by the secondary inertia forces in the engine. Determine the maximum natural frequency of the vibrometer in order to have an amplitude distortion less than 2 percent.

10.5 Determine the maximum percent error of a vibrometer in the frequency ratio range $4 \le r < \infty$ with a damping ratio of $\zeta = 0$.

10.6 Solve Problem 10.5 with a damping ratio of $\zeta = 0.67$.

10.7 A vibrometer is used to measure the vibration of an engine whose operating speed range is from 500 to 2000 rpm. The vibration consists of two harmonics. The amplitude distortion must be less than 3 percent. Find the natural frequency of the vibrometer if (a) the damping is negligible and (b) the damping ratio is 0.6.

10.8 A spring-mass system, having a static deflection of 10 mm and negligible damping, is used as a vibrometer. When mounted on a machine operating at 4000 rpm, the relative amplitude is recorded as 1 mm. Find the maximum values of displacement, velocity, and acceleration of the machine.

10.9 A vibration pickup has a natural frequency of 5 Hz and a damping ratio of $\zeta = 0.5$. Find the lowest frequency that can be measured with a 1 percent error.

10.10 A vibration pickup has been designed for operation above a frequency level of 100 Hz without exceeding an error of 2 percent. When mounted on a structure vibrating at a frequency of 100 Hz, the relative amplitude of the mass is found to be 1 mm. Find the suspended mass of the pickup if the stiffness of the spring is 4000 N/m and damping is negligible.

10.11 A vibrometer has an undamped natural frequency of 10 Hz and a damped natural frequency of 8 Hz. Find the lowest frequency in the range to infinity at which the amplitude can be directly read from the vibrometer with less than 2 percent error.

10.12 Determine the maximum percent error of an accelerometer in the frequency ratio range $0 < r \le 0.65$ with a damping ratio of $\zeta = 0$.

10.13 Solve Problem 10.12 with a damping ratio of 0.75.

10.14 Determine the necessary stiffness and the damping constant of an accelerometer if the maximum error is to be limited to 3 percent for measurements in the frequency range of 0 to 100 Hz. Assume that the suspended mass is 0.05 kg.

10.15 An accelerometer is constructed by suspending a mass of 0.1 kg from a spring of stiffness 10,000 N/m with negligible damping. When mounted on the foundation of an engine, the peak-to-peak travel of the mass of the accelerometer has been found to be 10 mm at an engine speed of 1000 rpm. Determine the maximum displacement, maximum velocity, and maximum acceleration of the foundation.

10.16 A spring-mass-damper system, having an undamped natural frequency of 100 Hz and a damping constant of 20 N-s/m, is used as an accelerometer to measure the vibration of a machine operating at a speed of 3000 rpm. If the actual acceleration is 10 m/s^2 and the recorded acceleration is 9 m/s^2, find the mass and the spring constant of the accelerometer.

10.17 A machine shop floor is subjected to the following vibration due to electric motors running at different speeds:

$$x(t) = 20 \sin 4\pi t + 10 \sin 8\pi t + 5 \sin 12\pi t \text{ mm}$$

If a vibrometer having an undamped natural frequency of 0.5 Hz and a damped natural frequency of 0.48 Hz is used to record the vibration of the machine shop floor, what will be the accuracy of the recorded vibration?

10.18 A machine is subjected to the vibration

$$x(t) = 20 \sin 50t + 5 \sin 150t \text{ mm} \qquad (t \text{ in sec})$$

An accelerometer having a damped natural frequency of 80 rad/s and an undamped natural frequency of 100 rad/s is mounted on the machine to read the acceleration directly in mm/s^2. Discuss the accuracy of the recorded acceleration.

10.19 A variable-length cantilever beam of rectangular cross-section $\frac{1}{16}$ in. $\times$ 1 in., made of spring steel, is used to measure the frequency of vibration. The length of the cantilever can be varied between 2 in. and 10 in. Find the range of frequencies that can be measured with this device.

10.20 Show that the real component of the harmonic response of a viscously damped single degree of freedom system (from X in Eq. 3.54) attains a maximum at

$$R_1 = \frac{\omega_1}{\omega_n} = \sqrt{1 - 2\zeta}$$

and a minimum at

$$R_2 = \frac{\omega_2}{\omega_n} = \sqrt{1 + 2\zeta}$$

10.21 Find the value of the frequency at which the imaginary component of the harmonic response of a viscously damped single degree of freedom system (from X in Eq. 3.54) attains a minimum.

10.22 Construct the Nyquist diagram for a single degree of freedom system with hysteretic damping.

10.23 The Bode plot of shaft vibration of a turbine obtained during coast-down is shown in Fig. 10.44. Determine the damping ratio of the system when the static deflection of the shaft is equal to 0.05 mil.

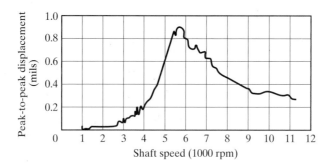

FIGURE 10.44

10.24 The vibratory response at the bearing of an internal combustion engine is shown in Fig. 10.45. Determine the equivalent viscous damping ratio of the system.

10.25 Suggest a method of using the Bode plot of phase angle versus frequency (Fig. 3.11b) to identify the natural frequency and the damping ratio of the system.

10.26 Two ball bearings, each with 16 balls, are used to support the shaft of a fan that rotates at 750 rpm. Determine the frequencies, in Hz, corresponding to the following defects:* cage, inner race, outer race, and ball. Assume that $d = 15$ mm, $D = 100$ mm and $\alpha = 30°$.

10.27 Determine the defect frequencies in Hertz* corresponding to roller, inner race, outer race and cage defects for a roller bearing with 18 rollers when installed in a machine that runs at a speed of 1000 rpm. Assume $d = 2$ cm, $D = 15$ cm, and $\alpha = 20°$.

10.28 An angular contact thrust bearing consists of 18 balls, each of diameter 10 mm, and is mounted on a shaft that rotates at 1500 rpm. If the contact angle of the bearing is 40° with a pitch diameter 80 mm, find the frequencies corresponding to cage, ball, inner race, and outer race faults.*

10.29 Find the value of kurtosis for a vibration signal that is uniformly distributed in the range 1–5 mm;

$$f(x) = \frac{1}{4}; 1 \le x \le 5 \text{ mm}$$

---

*Each type of failure in ball and roller bearings generates frequency of vibration f (impact rate per minute) as follows. Inner race defect: $f = \frac{1}{2}nN(1 + c)$; outer race defect: $f = \frac{1}{2}nN(1 - c)$; ball or roller defect: $f = \frac{DN}{d}c(2 - c)$; cage defect: $f = \frac{1}{2}N(1 - c)$ where d = ball or roller diameter, D = pitch diameter, α = contact angle, n = number of balls or rollers, N = speed (rpm), and $c = \frac{d}{D}\cos\alpha$.

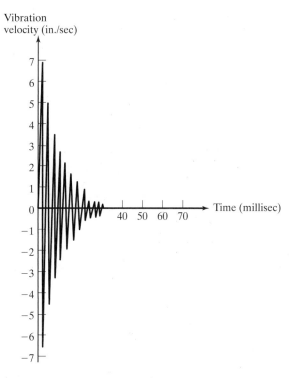

FIGURE 10.45   Response in time domain.

10.30 Find the value of kurtosis for a vibration amplitude that can be approximated as a discrete random variable with the following probability mass function:

x(mm)	1	2	3	4	5	6	7
$f(x)$	$\dfrac{1}{32}$	$\dfrac{3}{32}$	$\dfrac{3}{16}$	$\dfrac{6}{16}$	$\dfrac{3}{16}$	$\dfrac{3}{32}$	$\dfrac{1}{32}$

10.31 Figure 10.46 shows the experimental transfer function of a structure. Determine the approximate values of ω_i and ζ_i.

10.32 The experimental Nyquist circle of a structure is shown in Fig. 10.47. Estimate the modal damping ratio corresponding to this circle.

DESIGN PROJECTS

10.33 Design a vibration exciter to satisfy the following requirements:

    a. Maximum weight of the test specimen = 10 N

    b. Range of operating frequency = 10 to 50 Hz

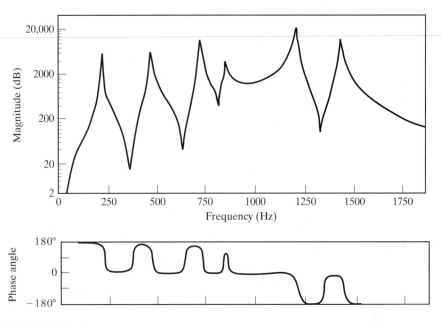

FIGURE 10.46

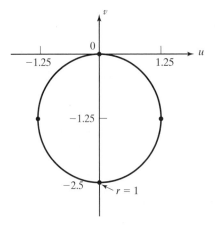

FIGURE 10.47

    c. Maximum acceleration level $= 20$ g

    d. Maximum vibration amplitude $= 0.5$ cm peak to peak.

10.34 Frahm tachometers are particularly useful to measure the speeds of engines whose rotating shafts are not easily accessible. When the tachometer is placed on the frame of a running engine, the vibration generated by the engine will cause one of the reeds to vibrate noticeably when the engine speed corresponds to the resonant frequency of a reed. Design a compact and lightweight Frahm tachometer with 12 reeds to measure engine speeds in the range 300–600 rpm.

10.35 A cantilever beam with an end mass m is fixed at the top of a multistory building to measure the acceleration induced at the top of the building during wind and earthquake loads (see Fig. 10.48). Design the beam (that is, determine the material, cross-sectional dimensions, and the length of the beam) such that the stress induced in the beam should not exceed the yield stress of the material under an acceleration of 0.2 g at the top of the building. Assume that the end mass m is equal to one-half of the mass of the beam.

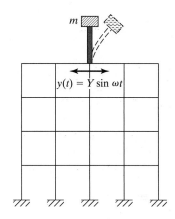

FIGURE 10.48

Nathan Newmark (1910–1981) was an American engineer and a professor of civil engineering at the University of Illinois at Champaign-Urbana. His research in earthquake resistant structures and structural dynamics is widely known. The numerical method he presented in 1959 for the dynamic response computation of linear and nonlinear systems is known as the Newmark β-method.

C H A P T E R   1 1

Numerical Integration Methods in Vibration Analysis

11.1 Introduction

When the differential equation of motion of a vibrating system cannot be integrated in closed form, a numerical approach must be used. Several numerical methods are available for the solution of vibration problems [11.1–11.3].[1] Numerical integration methods have two fundamental characteristics. First, they are not intended to satisfy the governing differential equation(s) at all time t but only at discrete time intervals Δt apart. Second, a suitable type of variation of the displacement x, velocity $\dot{x}$, and acceleration $\ddot{x}$ is assumed within each time interval Δt. Different numerical integration methods can be obtained, depending on the type of variation assumed for the displacement, velocity, and acceleration, within each time interval Δt. We shall assume that the values of x and $\dot{x}$ are known to be x_0 and $\dot{x}_0$, respectively, at time $t = 0$ and that the solution of the problem is required from $t = 0$ to $t = T$. In the following, we subdivide the time duration T into n equal steps Δt so that $\Delta t = T/n$ and seek the solution at $t_0 = 0$, $t_1 = \Delta t$, $t_2 = 2\,\Delta t,\ldots,$ $t_n = n\Delta t = T$. We shall derive formulas for finding the solution at $t_i = i\Delta t$ from the known solution at $t_{i-1} = (i - 1)\Delta t$ according to five different numerical integration schemes: (1) the finite difference method, (2) the Runge-Kutta method, (3) the Houbolt

---

[1]A numerical procedure using different types of interpolation functions for approximating the forcing function $F(t)$ was presented in Section 4.8.

method, (4) the Wilson method, and (5) the Newmark method. In the finite difference and Runge-Kutta methods, the current displacement (solution) is expressed in terms of the previously determined values of displacement, velocity, and acceleration, and the resulting equations are solved to find the current displacement. These methods fall under the category of explicit integration methods. In the Houbolt, Wilson, and Newmark methods, the temporal difference equations are combined with the current equations of motion and the resulting equations are solved to find the current displacement. These methods belong to the category of implicit integration methods.

11.2 Finite Difference Method

The main idea in the finite difference method is to use approximations to derivatives. Thus the governing differential equation of motion and the associated boundary conditions, if applicable, are replaced by the corresponding finite difference equations. Three types of formulas—forward, backward, and central difference formulas—can be used to derive the finite difference equations [11.4–11.6]. We shall consider only the central difference formulas in this chapter, since they are most accurate.

In the finite difference method, we replace the solution domain (over which the solution of the given differential equation is required) with a finite number of points, referred to as *mesh* or *grid points*, and seek to determine the values of the desired solution at these points. The grid points are usually considered to be equally spaced along each of the independent coordinates (see Fig. 11.1). By using Taylor's series expansion, x_{i+1} and x_{i-1} can be expressed about the grid point i as

$$x_{i+1} = x_i + h\dot{x}_i + \frac{h^2}{2}\ddot{x}_i + \frac{h^3}{6}\dddot{x}_i + \cdots \tag{11.1}$$

$$x_{i-1} = x_i - h\dot{x}_i + \frac{h^2}{2}\ddot{x}_i - \frac{h^3}{6}\dddot{x}_i + \cdots \tag{11.2}$$

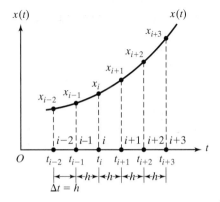

FIGURE 11.1   Grid points.

where $x_i = x(t = t_i)$ and $h = t_{i+1} - t_i = \Delta t$. By taking two terms only and subtracting Eq. (11.2) from Eq. (11.1), we obtain the central difference approximation to the first derivative of x at $t = t_i$:

$$\dot{x}_i = \frac{dx}{dt}\bigg|_{t_i} = \frac{1}{2h}(x_{i+1} - x_{i-1}) \tag{11.3}$$

By taking terms up to the second derivative and adding Eqs. (11.1) and (11.2), we obtain the central difference formula for the second derivative:

$$\ddot{x}_i = \frac{d^2x}{dt^2}\bigg|_{t_i} = \frac{1}{h^2}(x_{i+1} - 2x_i + x_{i-1}) \tag{11.4}$$

11.3   Central Difference Method for Single Degree of Freedom Systems

The governing equation of a viscously damped single degree of freedom system is

$$m\frac{d^2x}{dt^2} + c\frac{dx}{dt} + kx = F(t) \tag{11.5}$$

Let the duration over which the solution of Eq. (11.5) is required be divided into n equal parts of interval $h = \Delta t$ each. To obtain a satisfactory solution, we must select a time step Δt that is smaller than a critical time step Δt_{cri}.[2] Let the initial conditions be given by $x(t = 0) = x_0$ and $\dot{x}(t = 0) = \dot{x}_0$.

Replacing the derivatives by the central differences and writing Eq. (11.5) at grid point i gives

$$m\left\{\frac{x_{i+1} - 2x_i + x_{i-1}}{(\Delta t)^2}\right\} + c\left\{\frac{x_{i+1} - x_{i-1}}{2\,\Delta t}\right\} + kx_i = F_i \tag{11.6}$$

where $x_i = x(t_i)$ and $F_i = F(t_i)$. Solution of Eq. (11.6) for x_{i+1} yields

$$x_{i+1} = \left\{\frac{1}{\dfrac{m}{(\Delta t)^2} + \dfrac{c}{2\,\Delta t}}\right\}\left[\left\{\frac{2m}{(\Delta t)^2} - k\right\}x_i \right.$$

$$\left. + \left\{\frac{c}{2\,\Delta t} - \frac{m}{(\Delta t)^2}\right\}x_{i-1} + F_i\right] \tag{11.7}$$

---

[2]Numerical methods that require the use of a time step (Δt) smaller than a critical time step (Δt_{cri}) are said to be *conditionally stable* [11.7]. If Δt is taken to be larger than Δt_{cri}, the method becomes unstable. This means that the truncation of higher-order terms in the derivation of Eqs. (11.3) and (11.4) (or rounding-off in the computer) causes errors that grow and make the response computations worthless in most cases. The critical time step is given by $\Delta t_{\text{cri}} = \tau_n/\pi$, where τ_n is the natural period of the system or the smallest such period in the case of a multidegree of freedom system [11.8]. Naturally, the accuracy of the solution always depends on the size of the time step. By using an unconditionally stable method, we can choose the time step with regard to accuracy only, not with regard to stability. This usually allows a much larger time step to be used for any given accuracy.

This is called the *recurrence formula*. It permits us to calculate the displacement of the mass (x_{i+1}) if we know the previous history of displacements at t_i and t_{i-1}, as well as the present external force F_i. Repeated application of Eq. (11.7) yields the complete time history of the behavior of the system. Note that the solution of x_{i+1} is based on the use of the equilibrium equation at time t_i—that is, Eq. (11.6). For this reason, this integration procedure is called an *explicit integration method*. Certain care has to be exercised in applying Eq. (11.7) for $i = 0$. Since both x_0 and x_{-1} are needed in finding x_1, and the initial conditions provide only the values of x_0 and $\dot{x}_0$, we need to find the value of x_{-1}. Thus the method is not self-starting. However, we can generate the value of x_{-1} by using Eqs. (11.3) and (11.4) as follows. By substituting the known values of x_0 and $\dot{x}_0$ into Eq. (11.5), $\ddot{x}_0$ can be found:

$$\ddot{x}_0 = \frac{1}{m}[F(t = 0) - c\dot{x}_0 - kx_0] \tag{11.8}$$

Application of Eqs. (11.3) and (11.4) at $i = 0$ yields the value of x_{-1}:

$$x_{-1} = x_0 - \Delta t \dot{x}_0 + \frac{(\Delta t)^2}{2}\ddot{x}_0 \tag{11.9}$$

EXAMPLE   11.1

Response of Single Degree of Freedom System

Find the response of a viscously damped single degree of freedom system subjected to a force

$$F(t) = F_0\left(1 - \sin\frac{\pi t}{2t_0}\right)$$

with the following data: $F_0 = 1$, $t_0 = \pi$, $m = 1$, $c = 0.2$, and $k = 1$. Assume the values of the displacement and velocity of the mass at $t = 0$ to be zero.

Solution: The governing differential equation is

$$m\ddot{x} + c\dot{x} + kx = F(t) = F_0\left(1 - \sin\frac{\pi t}{2t_0}\right) \tag{E.1}$$

The finite difference solution of Eq. (E.1) is given by Eq. (11.7). Since the initial conditions are $x_0 = \dot{x}_0 = 0$, Eq. (11.8) yields $\ddot{x}_0 = 1$; hence Eq. (11.9) gives $x_{-1} = (\Delta t)^2/2$. Thus the solution of Eq. (E.1) can be found from the recurrence relation

$$x_{i+1} = \frac{1}{\left[\dfrac{m}{(\Delta t)^2} + \dfrac{c}{2\Delta t}\right]}\left[\left\{\frac{2m}{(\Delta t)^2} - k\right\}x_i \right.$$
$$\left. + \left\{\frac{c}{2\Delta t} - \frac{m}{(\Delta t)^2}\right\}x_{i-1} + F_i\right], \quad i = 0, 1, 2, \ldots \tag{E.2}$$

with $x_0 = 0$, $x_{-1} = (\Delta t)^2/2$, $x_i = x(t_i) = x(i\Delta t)$, and

$$F_i = F(t_i) = F_0\left(1 - \sin\frac{i\pi\Delta t}{2t_0}\right)$$

The undamped natural frequency and the natural period of the system are given by

$$\omega_n = \left(\frac{k}{m}\right)^{1/2} = 1 \tag{E.3}$$

and

$$\tau_n = \frac{2\pi}{\omega_n} = 2\pi \tag{E.4}$$

Thus the time step Δt must be less than $\tau_n/\pi = 2.0$. We shall find the solution of Eq. (E.1) by using the time steps $\Delta t = \tau_n/40$, $\tau_n/20$, and $\tau_n/2$. The time step $\Delta\tau = \tau_n/2 > \Delta t_{cri}$ is used to illustrate the unstable (diverging) behavior of the solution. The values of the response x_i obtained at different instants of time t_i are shown in Table 11.1.

This example can be seen to be identical to Example 4.17. The results obtained by idealization 4 (piecewise linear type interpolation) of Example 4.17 are shown in Table 11.1 up to time $t_i = \pi$

TABLE 11.1   Comparison of Solutions of Example 11.1

Time (t_i)	Values of $x_i = x(t_i)$ obtained with			Value of x_i given by idealization 4 of Example 4.17
	$\Delta t = \dfrac{\tau_n}{40}$	$\Delta t = \dfrac{\tau_n}{20}$	$\Delta t = \dfrac{\tau_n}{2}$	
0	0.00000	0.00000	0.00000	0.00000
$\pi/10$	0.04638	0.04935	—	0.04541
$2\pi/10$	0.16569	0.17169	—	0.16377
$3\pi/10$	0.32767	0.33627	—	0.32499
$4\pi/10$	0.50056	0.51089	—	0.49746
$5\pi/10$	0.65456	0.66543	—	0.65151
$6\pi/10$	0.76485	0.77491	—	0.76238
$7\pi/10$	0.81395	0.82185	—	0.81255
$8\pi/10$	0.79314	0.79771	—	0.79323
$9\pi/10$	0.70297	0.70340	—	0.70482
π	0.55275	0.54869	4.9348	0.55647
2π	0.19208	0.19898	−29.551	—
3π	2.7750	2.7679	181.90	—
4π	0.83299	0.83852	−1058.8	—
5π	−0.05926	−0.06431	6253.1	—

in the last column of the table. It can be observed that the finite difference method gives reasonably accurate results with time steps $\Delta t = \tau_n/40$ and $\tau_n/20$ (which are smaller than Δt_{cri}) but gives diverging results with $\Delta \tau = \tau_n/2$ (which is larger than Δt_{cri}).

∎

11.4   Runge-Kutta Method for Single Degree of Freedom Systems

In the Runge-Kutta method, the approximate formula used for obtaining x_{i+1} from x_i is made to coincide with the Taylor's series expansion of x at x_{i+1} up to terms of order $(\Delta t)^n$. The Taylor's series expansion of $x(t)$ at $t + \Delta t$ is given by

$$x(t + \Delta t) = x(t) + \dot{x} \, \Delta t + \ddot{x} \frac{(\Delta t)^2}{2!} + \dddot{x} \frac{(\Delta t)^3}{3!}$$

$$+ \ddddot{x} \frac{(\Delta t)^4}{4!} + \cdots \tag{11.10}$$

In contrast to Eq. (11.10), which requires higher order derivatives, the Runge-Kutta method does not require explicitly derivatives beyond the first [11.9–11.11]. For the solution of a second-order differential equation, we first reduce it to two first-order equations. For example, Eq. (11.5) can be rewritten as

$$\ddot{x} = \frac{1}{m}[F(t) - c\dot{x} - kx] = f(x, \dot{x}, t) \tag{11.11}$$

By defining $x_1 = x$ and $x_2 = \dot{x}$, Eq. (11.11) can be written as two first-order equations:

$$\dot{x}_1 = x_2$$
$$\dot{x}_2 = f(x_1, x_2, t) \tag{11.12}$$

By defining

$$\vec{X}(t) = \begin{Bmatrix} x_1(t) \\ x_2(t) \end{Bmatrix} \quad \text{and} \quad \vec{F}(t) = \begin{Bmatrix} x_2 \\ f(x_1, x_2, t) \end{Bmatrix}$$

the following recurrence formula is used to find the values of $\vec{X}(t)$ at different grid points t_i according to the fourth-order Runge-Kutta method

$$\vec{X}_{i+1} = \vec{X}_i + \frac{1}{6} [\vec{K}_1 + 2\vec{K}_2 + 2\vec{K}_3 + \vec{K}_4] \tag{11.13}$$

where

$$\vec{K}_1 = h\vec{F}(\vec{X}_i, t_i) \tag{11.14}$$

$$\vec{K}_2 = h\vec{F}(\vec{X}_i + \tfrac{1}{2}\vec{K}_1, t_i + \tfrac{1}{2}h) \tag{11.15}$$

$$\vec{K}_3 = h\vec{F}(\vec{X}_i + \tfrac{1}{2}\vec{K}_2, t_i + \tfrac{1}{2}h) \tag{11.16}$$

$$\vec{K}_4 = h\vec{F}(\vec{X}_i + \vec{K}_3, t_{i+1}) \tag{11.17}$$

The method is stable and self-starting—that is, only the function values at a single previous point are required to find the function value at the current point.

■■■■■■■■■■■ Response of Single Degree of Freedom System

EXAMPLE   11.2

Find the solution of Example 11.1 using the Runge-Kutta method.

Solution: We use a step size of $\Delta t = 0.3142$ and define

$$\vec{X}(t) = \begin{Bmatrix} x_1(t) \\ x_2(t) \end{Bmatrix} = \begin{Bmatrix} x(t) \\ \dot{x}(t) \end{Bmatrix}$$

and

$$\vec{F}(t) = \begin{Bmatrix} x_2 \\ f(x_1, x_2, t) \end{Bmatrix} = \begin{Bmatrix} \dot{x}(t) \\ \dfrac{1}{m}\left[F_0\left(1 - \sin\dfrac{\pi t}{2t_0} \right) - c\dot{x}(t) - kx(t) \right] \end{Bmatrix}$$

From the known initial conditions, we have

$$\vec{X}_0 = \begin{Bmatrix} 0 \\ 0 \end{Bmatrix}$$

The values of $\vec{X}_{i+1}, i = 0, 1, 2, \ldots$ obtained according to Eq. (11.13) are shown in Table 11.2.

TABLE 11.2

Step i	Time t_i	$x_1 = x$	$x_2 = \dot{x}$
1	0.3142	0.045406	0.275591
2	0.6283	0.163726	0.461502
3	0.9425	0.324850	0.547296
⋮			
19	5.9690	−0.086558	0.765737
20	6.2832	0.189886	0.985565

■

11.5   Central Difference Method for Multidegree of Freedom Systems

The equation of motion of a viscously damped multidegree of freedom system (see Eq. 6.119) can be expressed as

$$[m]\ddot{\vec{x}} + [c]\dot{\vec{x}} + [k]\vec{x} = \vec{F} \qquad\qquad (11.18)$$

where $[m]$, $[c]$, and $[k]$ are the mass, damping, and stiffness matrices, $\vec{x}$ is the displacement vector, and $\vec{F}$ is the force vector. The procedure indicated for the case of a single degree of freedom system can be directly extended to this case [11.12, 11.13]. The central difference formulas for the velocity and acceleration vectors at time $t_i = i\Delta t (\dot{\vec{x}}_i$ and $\ddot{\vec{x}}_i)$ are given by

$$\dot{\vec{x}}_i = \frac{1}{2\,\Delta t}(\vec{x}_{i+1} - \vec{x}_{i-1}) \tag{11.19}$$

$$\ddot{\vec{x}}_i = \frac{1}{(\Delta t)^2}(\vec{x}_{i+1} - 2\vec{x}_i + \vec{x}_{i-1}) \tag{11.20}$$

which are similar to Eqs. (11.3) and (11.4). Thus the equations of motion, Eq. (11.18), at time t_i can be written as

$$[m]\frac{1}{(\Delta t)^2}(\vec{x}_{i+1} - 2\vec{x}_i + \vec{x}_{i-1}) + [c]\frac{1}{2\,\Delta t}(\vec{x}_{i+1} - \vec{x}_{i-1}) + [k]\vec{x}_i = \vec{F}_i \tag{11.21}$$

where $\vec{x}_{i+1} = \vec{x}(t = t_{i+1})$, $\vec{x}_i = \vec{x}(t = t_i)$, $\vec{x}_{i-1} = \vec{x}(t = t_{i-1})$, $\vec{F}_i = \vec{F}(t = t_i)$, and $t_i = i\,\Delta t$. Equation (11.21) can be rearranged to obtain

$$\left(\frac{1}{(\Delta t)^2}[m] + \frac{1}{2\,\Delta t}[c]\right)\vec{x}_{i+1} + \left(-\frac{2}{(\Delta t)^2}[m] + [k]\right)\vec{x}_i$$

$$+ \left(\frac{1}{(\Delta t)^2}[m] - \frac{1}{2\,\Delta t}[c]\right)\vec{x}_{i-1} = \vec{F}_i$$

or

$$\left(\frac{1}{(\Delta t)^2}[m] + \frac{1}{2\,\Delta t}[c]\right)\vec{x}_{i+1} = \vec{F}_i - \left([k] - \frac{2}{(\Delta t)^2}[m]\right)\vec{x}_i$$

$$- \left(\frac{1}{(\Delta t)^2}[m] - \frac{1}{2\,\Delta t}[c]\right)\vec{x}_{i-1} \tag{11.22}$$

Thus Eq. (11.22) gives the solution vector $\vec{x}_{i+1}$ once $\vec{x}_i$ and $\vec{x}_{i-1}$ are known. Since Eq. (11.22) is to be used for $i = 1, 2, \ldots, n$, the evaluation of $\vec{x}_1$ requires $\vec{x}_0$ and $\vec{x}_{-1}$. Thus a special starting procedure is needed to find $\vec{x}_{-1} = \vec{x}(t = -\Delta t)$. For this, Eqs. (11.18) to (11.20) are evaluated at $i = 0$ to obtain

$$[m]\ddot{\vec{x}}_0 + [c]\dot{\vec{x}}_0 + [k]\vec{x}_0 = \vec{F}_0 = \vec{F}(t = 0) \tag{11.23}$$

$$\dot{\vec{x}}_0 = \frac{1}{2\,\Delta t}(\vec{x}_1 - \vec{x}_{-1}) \tag{11.24}$$

$$\ddot{\vec{x}}_0 = \frac{1}{(\Delta t)^2}(\vec{x}_1 - 2\vec{x}_0 + \vec{x}_{-1}) \tag{11.25}$$

Equation (11.23) gives the initial acceleration vector as

$$\ddot{\vec{x}}_0 = [m]^{-1}(\vec{F}_0 - [c]\dot{\vec{x}}_0 - [k]\vec{x}_0) \tag{11.26}$$

and Eq. (11.24) gives the displacement vector at t_1 as

$$\vec{x}_1 = \vec{x}_{-1} + 2\,\Delta t\dot{\vec{x}}_0 \tag{11.27}$$

Substituting Eq. (11.27) for $\vec{x}_1$, Eq. (11.25) yields

$$\ddot{\vec{x}}_0 = \frac{2}{(\Delta t)^2}[\Delta t\dot{\vec{x}}_0 - \vec{x}_0 + \vec{x}_{-1}]$$

or

$$\vec{x}_{-1} = \vec{x}_0 - \Delta t\dot{\vec{x}}_0 + \frac{(\Delta t)^2}{2}\ddot{\vec{x}}_0 \tag{11.28}$$

where $\ddot{\vec{x}}_0$ is given by Eq. (11.26). Thus $\vec{x}_{-1}$ needed for applying Eq. (11.22) at $i = 1$ is given by Eq. (11.28). The computational procedure can be described by the following steps.

1. From the known initial conditions $\vec{x}(t = 0) = \vec{x}_0$ and $\dot{\vec{x}}(t = 0) = \dot{\vec{x}}_0$, compute $\ddot{\vec{x}}(t = 0) = \ddot{\vec{x}}_0$ using Eq. (11.26).
2. Select a time step Δt such that $\Delta t < \Delta t_{\text{cri}}$.
3. Compute $\vec{x}_{-1}$ using Eq. (11.28).
4. Find $\vec{x}_{i+1} = \vec{x}(t = t_{i+1})$, starting with $i = 0$; from Eq. (11.22), as

$$\vec{x}_{i+1} = \left[\frac{1}{(\Delta t)^2}[m] + \frac{1}{2\,\Delta t}[c]\right]^{-1}\left[\vec{F}_i - \left([k] - \frac{2}{(\Delta t)^2}[m]\right)\vec{x}_i\right.$$
$$\left. - \left(\frac{1}{(\Delta t)^2}[m] - \frac{1}{2\,\Delta t}[c]\right)\vec{x}_{i-1}\right] \tag{11.29}$$

where

$$\vec{F}_i = (t = t_i) \tag{11.30}$$

If required, evaluate accelerations and velocities at t_i:

$$\ddot{\vec{x}}_i = \frac{1}{(\Delta t)^2}[\vec{x}_{i+1} - 2\vec{x}_i + \vec{x}_{i-1}] \tag{11.31}$$

and

$$\dot{\vec{x}}_i = \frac{1}{2\,\Delta t}[\vec{x}_{i+1} - \vec{x}_{i-1}] \tag{11.32}$$

Repeat Step 4 until $\vec{x}_{n+1}$ (with $i = n$) is determined. The stability of the finite dif-ference scheme for solving matrix equations is discussed in Ref. [11.14].

■■■■■ Central Difference Method for a Two Degree of Freedom System

EXAMPLE 11.3 ————————————————————————————————

Find the response of the two degree of freedom system shown in Fig. 11.2 when the forcing functions are given by $F_1(t) = 0$ and $F_2(t) = 10$. Assume the value of c as zero and the initial conditions as $\vec{x}(t = 0) = \dot{\vec{x}}(t = 0) = \vec{0}$.

Solution

Approach: Use $\Delta t = \tau/10$, where τ is the smallest time period in the central difference method.
The equations of motion are given by

$$[m]\ddot{\vec{x}}(t) + [c]\dot{\vec{x}}(t) + [k]\,\vec{x}(t) = \vec{F}(t) \tag{E.1}$$

where

$$[m] = \begin{bmatrix} m_1 & 0 \\ 0 & m_2 \end{bmatrix} = \begin{bmatrix} 1 & 0 \\ 0 & 2 \end{bmatrix} \tag{E.2}$$

$$[c] = \begin{bmatrix} c & -c \\ -c & c \end{bmatrix} = \begin{bmatrix} 0 & 0 \\ 0 & 0 \end{bmatrix} \tag{E.3}$$

$$[k] = \begin{bmatrix} k_1 + k & -k \\ -k & k + k_2 \end{bmatrix} = \begin{bmatrix} 6 & -2 \\ -2 & 8 \end{bmatrix} \tag{E.4}$$

$$\vec{F}(t) = \begin{Bmatrix} F_1(t) \\ F_2(t) \end{Bmatrix} = \begin{Bmatrix} 0 \\ 10 \end{Bmatrix} \tag{E.5}$$

and

$$\vec{x}(t) = \begin{Bmatrix} x_1(t) \\ x_2(t) \end{Bmatrix} \tag{E.6}$$

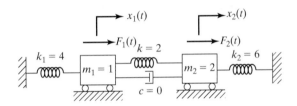

FIGURE 11.2   Two degree of freedom system.

The undamped natural frequencies and the mode shapes of the system can be found by solving the eigenvalue problem

$$\left[-\omega^2 \begin{bmatrix} 1 & 0 \\ 0 & 2 \end{bmatrix} + \begin{bmatrix} 6 & -2 \\ -2 & 8 \end{bmatrix}\right] \begin{Bmatrix} X_1 \\ X_2 \end{Bmatrix} = \begin{Bmatrix} 0 \\ 0 \end{Bmatrix} \tag{E.7}$$

The solution of Eq. (E.7) is given by

$$\omega_1 = 1.807747, \qquad \vec{X}^{(1)} = \begin{Bmatrix} 1.0000 \\ 1.3661 \end{Bmatrix} \tag{E.8}$$

$$\omega_2 = 2.594620, \qquad \vec{X}^{(2)} = \begin{Bmatrix} 1.0000 \\ -0.3661 \end{Bmatrix} \tag{E.9}$$

Thus the natural periods of the system are

$$\tau_1 = \frac{2\pi}{\omega_1} = 3.4757 \qquad \text{and} \qquad \tau_2 = \frac{2\pi}{\omega_2} = 2.4216$$

We shall select the time step (Δt) as $\tau_2/10 = 0.24216$. The initial value of $\ddot{\vec{x}}$ can be found as follows

$$\ddot{\vec{x}}_0 = [m]^{-1} \{\vec{F} - [k]\, \vec{x}_0\} = \begin{bmatrix} 1 & 0 \\ 0 & 2 \end{bmatrix}^{-1} \begin{Bmatrix} 0 \\ 10 \end{Bmatrix}$$

$$= \frac{1}{2} \begin{bmatrix} 2 & 0 \\ 0 & 1 \end{bmatrix} \begin{Bmatrix} 0 \\ 10 \end{Bmatrix} = \begin{Bmatrix} 0 \\ 5 \end{Bmatrix} \tag{E.10}$$

and the value of $\vec{x}_{-1}$ as follows

$$\vec{x}_{-1} = \vec{x}_0 - \Delta t\, \dot{\vec{x}}_0 + \frac{(\Delta t)^2}{2} \ddot{\vec{x}}_0 = \begin{Bmatrix} 0 \\ 0.1466 \end{Bmatrix} \tag{E.11}$$

Now Eq. (11.29) can be applied recursively to obtain $\vec{x}_1, \vec{x}_2, \ldots$. The results are shown in Table 11.3.

∎

11.6   Finite Difference Method for Continuous Systems

**11.6.1
Longitudinal
Vibration of
Bars**

Equation of Motion.   The equation of motion governing the free longitudinal vibration of a uniform bar (see Eqs. 8.49 and 8.20) can be expressed as

$$\frac{d^2U}{dx^2} + \alpha^2 U = 0 \tag{11.33}$$

where

$$\alpha^2 = \frac{\omega^2}{c^2} = \frac{\rho\omega^2}{E} \tag{11.34}$$

TABLE 11.3

Time ($t_i = i\Delta t$)	$\vec{x}_i = \vec{x}(t = t_i)$
t_1	$\begin{Bmatrix} 0 \\ 0.1466 \end{Bmatrix}$
t_2	$\begin{Bmatrix} 0.0172 \\ 0.5520 \end{Bmatrix}$
t_3	$\begin{Bmatrix} 0.0931 \\ 1.1222 \end{Bmatrix}$
t_4	$\begin{Bmatrix} 0.2678 \\ 1.7278 \end{Bmatrix}$
t_5	$\begin{Bmatrix} 0.5510 \\ 2.2370 \end{Bmatrix}$
t_6	$\begin{Bmatrix} 0.9027 \\ 2.5470 \end{Bmatrix}$
t_7	$\begin{Bmatrix} 1.2354 \\ 2.6057 \end{Bmatrix}$
t_8	$\begin{Bmatrix} 1.4391 \\ 2.4189 \end{Bmatrix}$
t_9	$\begin{Bmatrix} 1.4202 \\ 2.0422 \end{Bmatrix}$
t_{10}	$\begin{Bmatrix} 1.1410 \\ 1.5630 \end{Bmatrix}$
t_{11}	$\begin{Bmatrix} 0.6437 \\ 1.0773 \end{Bmatrix}$
t_{12}	$\begin{Bmatrix} 0.0463 \\ 0.6698 \end{Bmatrix}$

To obtain the finite difference approximation of Eq. (11.33), we first divide the bar of length l into $n - 1$ equal parts each of length $h = l/(n - 1)$ and denote the mesh points as $1, 2, 3, \ldots, i, \ldots, n$, as shown in Fig. 11.3. Then, by denoting the value of U at mesh point i as U_i and using a formula for the second derivative similar to Eq. (11.4), Eq. (11.33) for mesh point i can be written as

$$\frac{1}{h^2}(U_{i+1} - 2U_i + U_{i-1}) + \alpha^2 U_i = 0$$

or

$$U_{i+1} - (2 - \lambda)U_i + U_{i-1} = 0 \qquad (11.35)$$

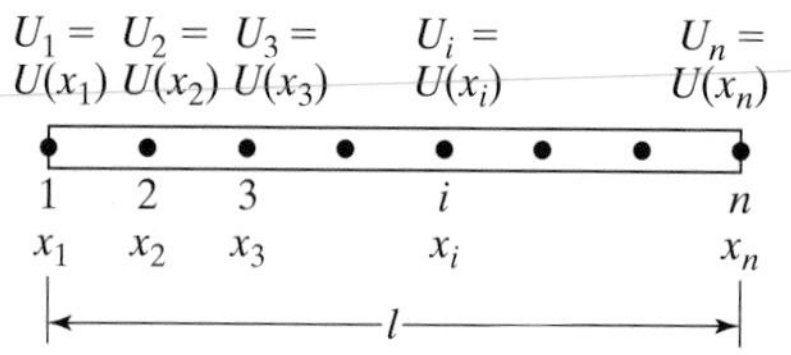

$$U_1 = U_2 = U_3 = \quad U_i = \quad U_n =$$
$$U(x_1)\ U(x_2)\ U(x_3) \quad U(x_i) \quad U(x_n)$$

FIGURE 11.3   Division of a bar for finite difference approximation.

where $\lambda = h^2\alpha^2$. The application of Eq. (11.35) at mesh points $i = 2, 3, \ldots, n - 1$ leads to the equations

$$U_3 - (2 - \lambda)U_2 + U_1 = 0$$
$$U_4 - (2 - \lambda)U_3 + U_2 = 0$$
$$\vdots$$
$$U_n - (2 - \lambda)U_{n-1} + U_{n-2} = 0 \tag{11.36}$$

which can be stated in matrix form as

$$
\begin{bmatrix}
-1 & (2-\lambda) & -1 & 0 & 0 & \cdots & 0 & 0 & 0 \\
0 & -1 & (2-\lambda) & -1 & 0 & \cdots & 0 & 0 & 0 \\
0 & 0 & -1 & (2-\lambda) & -1 & \cdots & 0 & 0 & 0 \\
\cdot & \cdot & \cdot & \cdot & \cdot & & \cdot & \cdot & \cdot \\
\cdot & \cdot & \cdot & \cdot & \cdot & \cdots & \cdot & \cdot & \cdot \\
\cdot & \cdot & \cdot & \cdot & \cdot & & \cdot & \cdot & \cdot \\
0 & 0 & 0 & 0 & 0 & \cdots & -1 & (2-\lambda) & -1 \\
\end{bmatrix}
\begin{Bmatrix}
U_1 \\ U_2 \\ U_3 \\ \cdot \\ \cdot \\ \cdot \\ U_n
\end{Bmatrix}
=
\begin{Bmatrix}
0 \\ 0 \\ 0 \\ \cdot \\ \cdot \\ \cdot \\ 0
\end{Bmatrix}
\tag{11.37}
$$

Boundary Conditions

Fixed End.   The deflection is zero at a fixed end. Assuming that the bar is fixed at $x = 0$ and $x = l$, we set $U_1 = U_n = 0$ in Eq. (11.37) and obtain the equation

$$[[A] - \lambda[I]]\vec{U} = \vec{0} \tag{11.38}$$

where

$$[A] = \begin{bmatrix} 2 & -1 & 0 & 0 & \cdots & 0 & 0 & 0 \\ -1 & 2 & -1 & 0 & \cdots & 0 & 0 & 0 \\ 0 & -1 & 2 & -1 & \cdots & 0 & 0 & 0 \\ \cdot & \cdot & \cdot & \cdot & \cdots & \cdot & \cdot & \cdot \\ \cdot & \cdot & \cdot & \cdot & \cdots & \cdot & \cdot & \cdot \\ \cdot & \cdot & \cdot & \cdot & \cdots & \cdot & \cdot & \cdot \\ 0 & 0 & 0 & 0 & \cdots & 0 & -1 & 2 \end{bmatrix} \tag{11.39}$$

$$\vec{U} = \begin{Bmatrix} U_2 \\ U_3 \\ \cdot \\ \cdot \\ \cdot \\ U_{n-1} \end{Bmatrix} \tag{11.40}$$

and $[I]$ = identity matrix of order $n - 2$.

Note that the eigenvalue problem of Eq. (11.38) can be solved easily, since the matrix $[A]$ is a tridiagonal matrix [11.15–11.17].

Free End.   The stress is zero at a free end, so $(dU)/(dx) = 0$. We can use a formula for the first derivative similar to Eq. (11.3). To illustrate the procedure, let the bar be free at $x = 0$ and fixed at $x = l$. The boundary conditions can then be stated as

$$\left. \frac{dU}{dx} \right|_1 \simeq \frac{U_2 - U_{-1}}{2h} = 0 \quad \text{or} \quad U_{-1} = U_2 \tag{11.41}$$

$$U_n = 0 \tag{11.42}$$

In order to apply Eq. (11.41), we need to imagine the function $U(x)$ to be continuous beyond the length of the bar and create a fictitious mesh point -1 so that U_{-1} becomes the fictitious displacement of the point x_{-1}. The application of Eq. (11.35) at mesh point $i = 1$ yields

$$U_2 - (2 - \lambda)U_1 + U_{-1} = 0 \tag{11.43}$$

By incorporating the condition $U_{-1} = U_2$ (Eq. 11.41), Eq. (11.43) can be written as

$$(2 - \lambda)U_1 - 2U_2 = 0 \tag{11.44}$$

By adding Eqs. (11.44) and (11.37), we obtain the final equations:

$$[[A] - \lambda[I]]\vec{U} = \vec{0} \tag{11.45}$$

where

$$[A] = \begin{bmatrix} 2 & -2 & 0 & 0 & \cdots & 0 & 0 & 0 \\ -1 & 2 & -1 & 0 & \cdots & 0 & 0 & 0 \\ 0 & -1 & 2 & -1 & \cdots & 0 & 0 & 0 \\ & \cdot & & & & & & \\ & \cdot & & & & & & \\ & \cdot & & & & & & \\ 0 & 0 & 0 & 0 & \cdots & -1 & 2 & -1 \\ 0 & 0 & 0 & 0 & \cdots & 0 & -1 & 2 \end{bmatrix} \tag{11.46}$$

and

$$\vec{U} = \begin{Bmatrix} U_1 \\ U_2 \\ \cdot \\ \cdot \\ \cdot \\ U_{n-1} \end{Bmatrix} \tag{11.47}$$

11.6.2 Transverse Vibration of Beams

Equation of Motion.   The governing differential equation for the transverse vibration of a uniform beam is given by Eq. (8.83)

$$\frac{d^4 W}{dx^4} - \beta^4 W = 0 \tag{11.48}$$

where

$$\beta^4 = \frac{\rho A \omega^2}{EI} \tag{11.49}$$

By using the central difference formula for the fourth derivative,[3] Eq. (11.48) can be written at any mesh point i as

$$W_{i+2} - 4W_{i+1} + (6 - \lambda)W_i - 4W_{i-1} + W_{i-2} = 0 \tag{11.50}$$

where

$$\lambda = h^4 \beta^4 \tag{11.51}$$

Let the beam be divided into $n - 1$ equal parts with n mesh points and $h = l/(n - 1)$. The application of Eq. (11.50) at the mesh points $i = 3, 4, \ldots, n - 2$ leads to the equations

---

[3]The central difference formula for the fourth derivative (see Problem 11.3) is given by

$$\left. \frac{d^4 f}{dx^4} \right|_i \simeq \frac{1}{h^4} (f_{i+2} - 4f_{i+1} + 6f_i - 4f_{i-1} + f_{i-2})$$

$$
\begin{bmatrix}
1 & -4 & (6-\lambda) & -4 & 1 & 0 & 0 & \cdots & 0 & 0 & 0 & & 0 & 0 \\
0 & 1 & -4 & (6-\lambda) & -4 & 1 & 0 & \cdots & 0 & 0 & 0 & & 0 & 0 \\
0 & 0 & 1 & -4 & (6-\lambda) & -4 & 1 & \cdots & 0 & 0 & 0 & & 0 & 0 \\
\cdot & & & & & & & & & & & & & \\
\cdot & & & & & & & & & & & & & \\
\cdot & & & & & & & & & & & & & \\
0 & 0 & 0 & & 0 & & 0 & 0 & 0 & \cdots & 1 & -4 & (6-\lambda) & -4 & 1
\end{bmatrix}
$$

$$
\begin{Bmatrix}
W_1 \\
W_2 \\
W_3 \\
\cdot \\
\cdot \\
\cdot \\
W_n
\end{Bmatrix}
=
\begin{Bmatrix}
0 \\
0 \\
0 \\
\cdot \\
\cdot \\
\cdot \\
0
\end{Bmatrix}
\tag{11.52}
$$

Boundary Conditions

Fixed End.   The deflection W and the slope $(dW)/(dx)$ are zero at a fixed end. If the end $x = 0$ is fixed, we introduce a fictitious node -1 on the left-hand side of the beam, as shown in Fig. 11.4, and state the boundary conditions, using the central difference formula for $(dW)/(dx)$, as

$$
W_1 = 0
$$

$$
\left. \frac{dW}{dx} \right|_1 = \frac{1}{2h}(W_2 - W_{-1}) = 0 \qquad \text{or} \qquad W_{-1} = W_2 \tag{11.53}
$$

where W_i denotes the value of W at node i. If the end $x = l$ is fixed, we introduce the fictitious node $n + 1$ on the right side of the beam, as shown in Fig. 11.4, and state the boundary conditions as

$$
W_n = 0
$$

$$
\left. \frac{dW}{dx} \right|_n = \frac{1}{2h}(W_{n+1} - W_{n-1}) = 0, \qquad \text{or} \qquad W_{n+1} = W_{n-1} \tag{11.54}
$$

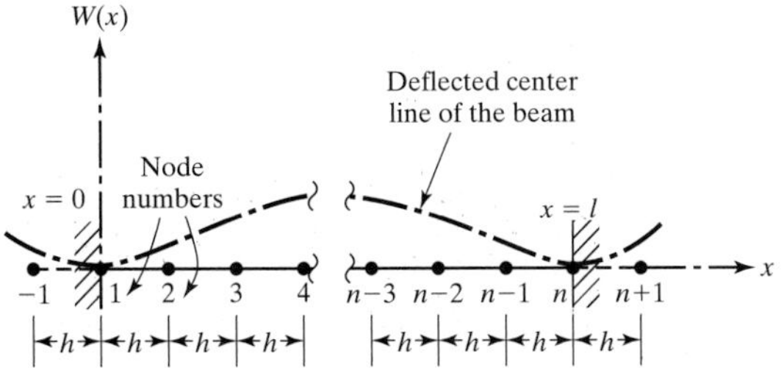

FIGURE 11.4   Beam with fixed ends.

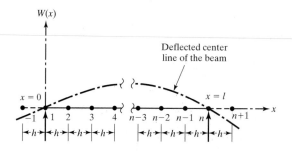

FIGURE 11.5   Beam with simply supported ends.

Simply Supported End.   If the end $x = 0$ is simply supported (see Fig. 11.5), we have

$$W_1 = 0$$

$$\left.\frac{d^2W}{dx^2}\right|_1 = \frac{1}{h^2}(W_2 - 2W_1 + W_{-1}) = 0, \qquad \text{or} \qquad W_{-1} = -W_2 \qquad (11.55)$$

Similar equations can be written if the end $x = l$ is simply supported.

Free End.   Since bending moment and shear force are zero at a free end, we introduce two fictitious nodes outside the beam, as shown in Fig. 11.6, and use central difference formulas for approximating the second and the third derivatives of the deflection W. For example, if the end $x = 0$ is free, we have

$$\left.\frac{d^2W}{dx^2}\right|_1 = \frac{1}{h^2}(W_2 - 2W_1 + W_{-1}) = 0$$

$$\left.\frac{d^3W}{dx^3}\right|_1 = \frac{1}{2h^3}(W_3 - 2W_2 + 2W_{-1} - W_{-2}) = 0 \qquad (11.56)$$

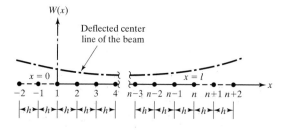

FIGURE 11.6   Beam with free ends.

■■■■■■ Pinned-Fixed Beam

EXAMPLE 11.4

Find the natural frequencies of the simply supported-fixed beam shown in Fig. 11.7. Assume that the cross section of the beam is constant along its length.

Solution: We shall divide the beam into four segments and express the governing equation

$$\frac{d^4W}{dx^4} - \beta^4 W = 0 \tag{E.1}$$

in finite difference form at each of the interior mesh points. This yields the equations

$$W_0 - 4W_1 + (6 - \lambda)W_2 - 4W_3 + W_4 = 0 \tag{E.2}$$

$$W_1 - 4W_2 + (6 - \lambda)W_3 - 4W_4 + W_5 = 0 \tag{E.3}$$

$$W_2 - 4W_3 + (6 - \lambda)W_4 - 4W_5 + W_6 = 0 \tag{E.4}$$

where W_0 and W_6 denote the values of W at the fictitious nodes 0 and 6, respectively, and

$$\lambda = h^4 \beta^4 = \frac{h^4 \rho A \omega^2}{EI} \tag{E.5}$$

The boundary conditions at the simply supported end (mesh point 1) are

$$W_1 = 0$$
$$W_0 = -W_2 \tag{E.6}$$

At the fixed end (mesh point 5) the boundary conditions are

$$W_5 = 0$$
$$W_6 = W_4 \tag{E.7}$$

With the help of Eqs. (E.6) and (E.7), Eqs. (E.2) to (E.4) can be reduced to

$$(5 - \lambda)W_2 - 4W_3 + W_4 = 0 \tag{E.8}$$

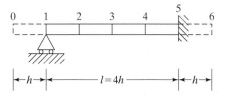

FIGURE 11.7   Simply supported-fixed beam.

$$-4W_2 + (6 - \lambda)W_3 - 4W_4 = 0 \tag{E.9}$$

$$W_2 - 4W_3 + (7 - \lambda)W_4 = 0 \tag{E.10}$$

Equations (E.8) to (E.10) can be written in matrix form as

$$\begin{bmatrix} (5 - \lambda) & -4 & 1 \\ -4 & (6 - \lambda) & -4 \\ 1 & -4 & (7 - \lambda) \end{bmatrix} \begin{Bmatrix} W_2 \\ W_3 \\ W_4 \end{Bmatrix} = \begin{Bmatrix} 0 \\ 0 \\ 0 \end{Bmatrix} \tag{E.11}$$

The solution of the eigenvalue problem (Eq. (E.11)) gives the following results:

$$\lambda_1 = 0.7135, \quad \omega_1 = \frac{0.8447}{h^2}\sqrt{\frac{EI}{\rho A}}, \quad \begin{Bmatrix} W_2 \\ W_3 \\ W_4 \end{Bmatrix}^{(1)} = \begin{Bmatrix} 0.5880 \\ 0.7215 \\ 0.3656 \end{Bmatrix} \tag{E.12}$$

$$\lambda_2 = 5.0322, \quad \omega_2 = \frac{2.2433}{h^2}\sqrt{\frac{EI}{\rho A}}, \quad \begin{Bmatrix} W_2 \\ W_3 \\ W_4 \end{Bmatrix}^{(2)} = \begin{Bmatrix} 0.6723 \\ -0.1846 \\ -0.7169 \end{Bmatrix} \tag{E.13}$$

$$\lambda_3 = 12.2543, \quad \omega_3 = \frac{3.5006}{h^2}\sqrt{\frac{EI}{\rho A}}, \quad \begin{Bmatrix} W_2 \\ W_3 \\ W_4 \end{Bmatrix}^{(3)} = \begin{Bmatrix} 0.4498 \\ -0.6673 \\ 0.5936 \end{Bmatrix} \tag{E.14}$$

∎

11.7   Runge-Kutta Method for Multidegree of Freedom Systems

In the Runge-Kutta method, the matrix equations of motion, Eq. (11.18), are used to express the acceleration vector as

$$\ddot{\vec{x}}(t) = [m]^{-1}(\vec{F}(t) - [c]\dot{\vec{x}}(t) - [k]\vec{x}(t)) \tag{11.57}$$

By treating the displacements as well as velocities as unknowns, a new vector, $\vec{X}(t)$, is defined as $\vec{X}(t) = \begin{Bmatrix} \vec{x}(t) \\ \dot{\vec{x}}(t) \end{Bmatrix}$ so that

$$\dot{\vec{X}} = \begin{Bmatrix} \dot{\vec{x}} \\ \ddot{\vec{x}} \end{Bmatrix} = \begin{Bmatrix} \dot{\vec{x}} \\ [m]^{-1}(\vec{F} - [c]\dot{\vec{x}} - [k]\vec{x}) \end{Bmatrix} \tag{11.58}$$

Equation (11.58) can be rearranged to obtain

$$\dot{\vec{X}}(t) = \begin{bmatrix} [0] & [I] \\ -[m]^{-1}[k] & -[m]^{-1}[c] \end{bmatrix} \begin{Bmatrix} \vec{x}(t) \\ \dot{\vec{x}}(t) \end{Bmatrix} + \begin{Bmatrix} 0 \\ [m]^{-1}\vec{F}(t) \end{Bmatrix}$$

that is,

$$\dot{\vec{X}}(t) = \vec{f}(\vec{X}, t) \tag{11.59}$$

where

$$\vec{f}(\vec{X}, t) = [A]\vec{X}(t) + \underset{\sim}{\vec{F}}(t) \tag{11.60}$$

$$[A] = \begin{bmatrix} [0] & [I] \\ -[m]^{-1}[k] & -[m]^{-1}[c] \end{bmatrix} \tag{11.61}$$

and

$$\underset{\sim}{\vec{F}}(t) = \left\{ \begin{array}{c} \vec{0} \\ [m]^{-1}\vec{F}(t) \end{array} \right\} \tag{11.62}$$

TABLE 11.4

Time $t_i = i\Delta t$	$\vec{x}_i = \vec{x}(t = t_i)$
t_1	$\left\{ \begin{array}{c} 0.0014 \\ 0.1437 \end{array} \right\}$
t_2	$\left\{ \begin{array}{c} 0.0215 \\ 0.5418 \end{array} \right\}$
t_3	$\left\{ \begin{array}{c} 0.0978 \\ 1.1041 \end{array} \right\}$
t_4	$\left\{ \begin{array}{c} 0.2668 \\ 1.7059 \end{array} \right\}$
t_5	$\left\{ \begin{array}{c} 0.5379 \\ 2.2187 \end{array} \right\}$
t_6	$\left\{ \begin{array}{c} 0.8756 \\ 2.5401 \end{array} \right\}$
t_7	$\left\{ \begin{array}{c} 1.2008 \\ 2.6153 \end{array} \right\}$
t_8	$\left\{ \begin{array}{c} 1.4109 \\ 2.4452 \end{array} \right\}$
t_9	$\left\{ \begin{array}{c} 1.4156 \\ 2.0805 \end{array} \right\}$
t_{10}	$\left\{ \begin{array}{c} 1.1727 \\ 1.6050 \end{array} \right\}$
t_{11}	$\left\{ \begin{array}{c} 0.7123 \\ 1.1141 \end{array} \right\}$
t_{12}	$\left\{ \begin{array}{c} 0.1365 \\ 0.6948 \end{array} \right\}$

With this, the recurrence formula to evaluate $\vec{X}(t)$ at different grid points t_i according to the fourth order Runge-Kutta method becomes [11.10]

$$\vec{X}_{i+1} = \vec{X}_i + \frac{1}{6} [\vec{K}_1 + 2\vec{K}_2 + 2\vec{K}_3 + \vec{K}_4] \tag{11.63}$$

where

$$\vec{K}_1 = h \vec{f}(\vec{X}_i, t_i) \tag{11.64}$$

$$\vec{K}_2 = h \vec{f}(\vec{X}_i + \frac{1}{2}\vec{K}_1, t_i + \frac{1}{2}h) \tag{11.65}$$

$$\vec{K}_3 = h \vec{f}(\vec{X}_i + \frac{1}{2}\vec{K}_2, t_i + \frac{1}{2}h) \tag{11.66}$$

$$\vec{K}_4 = h \vec{f}(\vec{X}_i + \vec{K}_3, t_{i+1}) \tag{11.67}$$

■■■■■■■■■■ **Runge-Kutta Method for a Two Degree of Freedom System**

EXAMPLE   11.5 —————————————————————————

———————————— Find the response of the two degree of freedom system considered in Example 11.3 using the fourth-order Runge-Kutta method.

Solution

Approach: Use the Runge-Kutta method with $\Delta t = 0.24216$.

Using the initial conditions $\vec{x}(t = 0) = \dot{\vec{x}}(t = 0) = \vec{0}$, Eq. (11.63) is sequentially applied with $\Delta t = 0.24216$ to obtain the results shown in Table 11.4.

■

11.8   Houbolt Method

We shall consider the Houbolt method with reference to a multidegree of freedom system. In this method, the following finite difference expansions are employed:

$$\dot{\vec{x}}_{i+1} = \frac{1}{6\Delta t}(11\vec{x}_{i+1} - 18\vec{x}_i + 9\vec{x}_{i-1} - 2\vec{x}_{i-2}) \tag{11.68}$$

$$\ddot{\vec{x}}_{i+1} = \frac{1}{(\Delta t)^2}(2\vec{x}_{i+1} - 5\vec{x}_i + 4\vec{x}_{i-1} - \vec{x}_{i-2}) \tag{11.69}$$

To derive Eqs. (11.68) and (11.69), consider the function $x(t)$. Let the values of x at the equally spaced grid points $t_{i-2} = t_i - 2\,\Delta t$, $t_{i-1} = t_i - \Delta t$, t_i, and $t_{i+1} = t_i + \Delta t$ be given by x_{i-2}, x_{i-1}, x_i, and x_{i+1}, respectively, as shown in Fig. 11.8 [11.18]. The Taylor's series expansion, with backward step, gives several possibilities.

- With Step Size = Δt:

$$x(t) = x(t + \Delta t) - \Delta t \, \dot{x}(t + \Delta t) + \frac{(\Delta t)^2}{2!}\ddot{x}(t + \Delta t) - \frac{(\Delta t)^3}{3!}\dddot{x}(t + \Delta t)$$

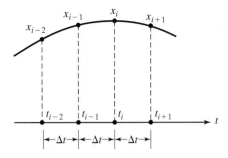

FIGURE 11.8   Equally spaced grid points.

or

$$x_i = x_{i+1} - \Delta t\, \dot{x}_{i+1} + \frac{(\Delta t)^2}{2}\ddot{x}_{i+1} - \frac{(\Delta t)^3}{6}\dddot{x}_{i+1} + \cdots \tag{11.70}$$

- With Step Size $= 2\,\Delta t$:

$$x(t - \Delta t) = x(t + \Delta t) - (2\,\Delta t)\dot{x}(t + \Delta t)$$
$$+ \frac{(2\,\Delta t)^2}{2!}\ddot{x}(t + \Delta t) - \frac{(2\,\Delta t)^3}{3!}\dddot{x}(t + \Delta t) + \cdots$$

or

$$x_{i-1} = x_{i+1} - 2\,\Delta t\,\dot{x}_{i+1} + 2(\Delta t)^2\ddot{x}_{i+1} - \tfrac{4}{3}(\Delta t)^3\,\dddot{x}_{i+1} + \cdots \tag{11.71}$$

- With Step Size $= 3\,\Delta t$:

$$x(t - 2\,\Delta t) = x(t + \Delta t) - (3\,\Delta t)\dot{x}(t + \Delta t)$$
$$+ \frac{(3\,\Delta t)^2}{2!}\ddot{x}(t + \Delta t) - \frac{(3\,\Delta t)^3}{3!}\dddot{x}(t + \Delta t) + \cdots$$

or

$$x_{i-2} = x_{i+1} - 3\,\Delta t\,\dot{x}_{i+1} + \tfrac{9}{2}(\Delta t)^2\ddot{x}_{i+1} - \tfrac{9}{2}(\Delta t)^3\,\dddot{x}_{i+1} + \cdots \tag{11.72}$$

By considering terms up to $(\Delta t)^3$ only, Eqs. (11.70) to (11.72) can be solved to express $\dot{x}_{i+1}$, $\ddot{x}_{i+1}$, and $\dddot{x}_{i+1}$ in terms of x_{i-2}, x_{i-1}, x_i, and x_{i+1}. This gives $\dot{x}_{i+1}$ and $\ddot{x}_{i+1}$ as [11.18]:

$$\dot{x}_{i+1} = \frac{1}{6(\Delta t)}(11x_{i+1} - 18x_i + 9x_{i-1} - 2x_{i-2}) \tag{11.73}$$

$$\ddot{x}_{i+1} = \frac{1}{(\Delta t)^2}(2x_{i+1} - 5x_i + 4x_{i-1} - x_{i-2}) \qquad (11.74)$$

Equations (11.68) and (11.69) represent the vector form of these equations.

To find the solution at step $i + 1(\vec{x}_{i+1})$, we consider Eq. (11.18) at t_{i+1}, so that

$$[m]\ddot{\vec{x}}_{i+1} + [c]\dot{\vec{x}}_{i+1} + [k]\vec{x}_{i+1} = \vec{F}_{i+1} \equiv \vec{F}(t = t_{i+1}) \qquad (11.75)$$

By substituting Eqs. (11.68) and (11.69) into Eq. (11.75), we obtain

$$\left(\frac{2}{(\Delta t)^2}[m] + \frac{11}{6\,\Delta t}[c] + [k]\right)\vec{x}_{i+1}$$

$$= \vec{F}_{i+1} + \left(\frac{5}{(\Delta t)^2}[m] + \frac{3}{\Delta t}[c]\right)\vec{x}_i$$

$$- \left(\frac{4}{(\Delta t)^2}[m] + \frac{3[c]}{2\,\Delta t}\right)\vec{x}_{i-1} + \left(\frac{1}{(\Delta t)^2}[m] + \frac{[c]}{3\,\Delta t}\right)\vec{x}_{i-2} \qquad (11.76)$$

Note that the equilibrium equation at time t_{i+1}, Eq. (11.75), is used in finding the solution $\vec{X}_{i+1}$ through Eq. (11.76). This is also true of the Wilson and Newmark methods. For this reason, these methods are called *implicit integration methods*.

It can be seen from Eq. (11.76) that a knowledge of $\vec{x}_i$, $\vec{x}_{i-1}$, and $\vec{x}_{i-2}$ is required to find the solution $\vec{x}_{i+1}$. Thus the values of $\vec{x}_{-1}$ and $\vec{x}_{-2}$ are to be found before attempting to find the vector $\vec{x}_1$ using Eq. (11.76). Since there is no direct method to find $\vec{x}_{-1}$ and $\vec{x}_{-2}$, we can not use Eq. (11.76) to find $\vec{x}_1$ and $\vec{x}_2$. This makes the method non–self-starting. To start the method, we can use the central difference method described in Section 11.5 to find $\vec{x}_1$ and $\vec{x}_2$. Once $\vec{x}_0$ is known from the given initial conditions of the problem and $\vec{x}_1$ and $\vec{x}_2$ are known from the central difference method, the subsequent solutions $\vec{x}_3, \vec{x}_4, \ldots$ can be found by using Eq. (11.76).

The step-by-step procedure to be used in the Houbolt method is as follows:

1. From the known initial conditions $\vec{x}(t = 0) = \vec{x}_0$ and $\dot{\vec{x}}(t = 0) = \dot{\vec{x}}_0$, find $\ddot{\vec{x}}_0 = \ddot{\vec{x}}(t = 0)$ using Eq. (11.26).
2. Select a suitable time step Δt.
3. Determine $\vec{x}_{-1}$ using Eq. (11.28).
4. Find $\vec{x}_1$ and $\vec{x}_2$ using the central difference equation (11.29).
5. Compute $\vec{x}_{i+1}$, starting with $i = 2$ and using Eq. (11.76):

$$\vec{x}_{i+1} = \left[\frac{2}{(\Delta t)^2}[m] + \frac{11}{6\,\Delta t}[c] + [k]\right]^{-1}$$

$$\times \left\{\vec{F}_{i+1} + \left(\frac{5}{(\Delta t)^2}[m] + \frac{3}{\Delta t}[c]\right)\vec{x}_i\right.$$

$$- \left(\frac{4}{(\Delta t)^2} [m] + \frac{3}{2 \, \Delta t} [c] \right) \vec{x}_{i-1}$$

$$+ \left(\frac{1}{(\Delta t)^2} [m] + \frac{1}{3 \, \Delta t} [c] \right) \vec{x}_{i-2} \bigg\}$$ (11.77)

If required, evaluate the velocity and acceleration vectors $\dot{\vec{x}}_{i+1}$ and $\ddot{\vec{x}}_{i+1}$ using Eqs. (11.68) and (11.69).

Houbolt Method for a Two Degree of Freedom System

EXAMPLE   11.6

Find the response of the two degree of freedom system considered in Example 11.3 using the Houbolt method.

TABLE 11.5

Time $t_i = i \Delta t$	$\vec{x}_i = \vec{x}(t = t_i)$
t_1	$\left\{ \begin{matrix} 0.0000 \\ 0.1466 \end{matrix} \right\}$
t_2	$\left\{ \begin{matrix} 0.0172 \\ 0.5520 \end{matrix} \right\}$
t_3	$\left\{ \begin{matrix} 0.0917 \\ 1.1064 \end{matrix} \right\}$
t_4	$\left\{ \begin{matrix} 0.2501 \\ 1.6909 \end{matrix} \right\}$
t_5	$\left\{ \begin{matrix} 0.4924 \\ 2.1941 \end{matrix} \right\}$
t_6	$\left\{ \begin{matrix} 0.7867 \\ 2.5297 \end{matrix} \right\}$
t_7	$\left\{ \begin{matrix} 1.0734 \\ 2.6489 \end{matrix} \right\}$
t_8	$\left\{ \begin{matrix} 1.2803 \\ 2.5454 \end{matrix} \right\}$
t_9	$\left\{ \begin{matrix} 1.3432 \\ 2.2525 \end{matrix} \right\}$
t_{10}	$\left\{ \begin{matrix} 1.2258 \\ 1.8325 \end{matrix} \right\}$
t_{11}	$\left\{ \begin{matrix} 0.9340 \\ 1.3630 \end{matrix} \right\}$
t_{12}	$\left\{ \begin{matrix} 0.5178 \\ 0.9224 \end{matrix} \right\}$

Solution

Approach: Use the Houbolt method with $\Delta t = 0.24216$.

The value of $\ddot{\vec{x}}_0$ can be found using Eq. (11.26):

$$\ddot{\vec{x}}_0 = \begin{Bmatrix} 0 \\ 5 \end{Bmatrix}$$

By using a value of $\Delta t = 0.24216$, Eq. (11.29) can be used to find $\vec{x}_1$ and $\vec{x}_2$, and then Eq. (11.77) can be used recursively to obtain $\vec{x}_3, \vec{x}_4, \ldots$, as shown in Table 11.5.

∎

11.9   Wilson Method

The Wilson method assumes that the acceleration of the system varies linearly between two instants of time. In particular, the two instants of time are taken as indicated in Fig. 11.9. Thus the acceleration is assumed to be linear from time $t_i = i\Delta t$ to time $t_{i+\theta} = t_i + \theta\,\Delta t$, where $\theta \geq 1.0$ [11.19]. For this reason, this method is also called the *Wilson θ method*. If $\theta = 1.0$, this method reduces to the linear acceleration scheme [11.20].

A stability analysis of the Wilson method shows that it is unconditionally stable provided that $\theta \geq 1.37$. In this section, we shall consider the Wilson method for a multi-degree of freedom system.

Since $\ddot{\vec{x}}(t)$ is assumed to vary linearly between t_i and $t_{i+\theta}$, we can predict the value of $\ddot{\vec{x}}$ at any time $t_i + \tau, 0 \leq \tau \leq \theta\,\Delta t$:

$$\ddot{\vec{x}}(t_i + \tau) = \ddot{\vec{x}}_i + \frac{\tau}{\theta\,\Delta t}(\ddot{\vec{x}}_{i+\theta} - \ddot{\vec{x}}_i) \tag{11.78}$$

By integrating Eq. (11.78), we obtain[4]

$$\dot{\vec{x}}(t_i + \tau) = \dot{\vec{x}}_i + \ddot{\vec{x}}_i\tau + \frac{\tau^2}{2\theta\,\Delta t}(\ddot{\vec{x}}_{i+\theta} - \ddot{\vec{x}}_i) \tag{11.79}$$

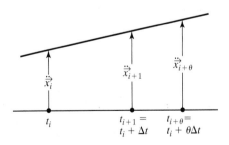

FIGURE 11.9   Linear acceleration assumption of the Wilson method.

---

[4] $\dot{\vec{x}}_i$ and $\vec{x}_i$ have been substituted in place of the integration constants in Eqs. (11.79) and (11.80), respectively.

and

$$\vec{x}(t_i + \tau) = \vec{x}_i + \dot{\vec{x}}_i\tau + \frac{1}{2}\ddot{\vec{x}}_i\tau^2 + \frac{\tau^3}{6\theta\,\Delta t}(\ddot{\vec{x}}_{i+\theta} - \ddot{\vec{x}}_i) \tag{11.80}$$

By substituting $\tau = \theta\,\Delta t$ into Eqs. (11.79) and (11.80), we obtain

$$\dot{\vec{x}}_{i+\theta} = \dot{\vec{x}}(t_i + \theta\,\Delta t) = \dot{\vec{x}}_i + \frac{\theta\,\Delta t}{2}(\ddot{\vec{x}}_{i+\theta} + \ddot{\vec{x}}_i) \tag{11.81}$$

$$\vec{x}_{i+\theta} = \vec{x}(t_i + \theta\,\Delta t) = \vec{x}_i + \theta\,\Delta t\,\dot{\vec{x}}_i + \frac{\theta^2(\Delta t)^2}{6}(\ddot{\vec{x}}_{i+\theta} + 2\ddot{\vec{x}}_i) \tag{11.82}$$

Equation (11.82) can be solved to obtain

$$\ddot{\vec{x}}_{i+\theta} = \frac{6}{\theta^2(\Delta t)^2}(\vec{x}_{i+\theta} - \vec{x}_i) - \frac{6}{\theta\,\Delta t}\dot{\vec{x}}_i - 2\ddot{\vec{x}}_i \tag{11.83}$$

By substituting Eq. (11.83) into Eq. (11.81), we obtain

$$\dot{\vec{x}}_{i+\theta} = \frac{3}{\theta\,\Delta t}(\vec{x}_{i+\theta} - \vec{x}_i) - 2\dot{\vec{x}}_i - \frac{\theta\,\Delta t}{2}\ddot{\vec{x}}_i \tag{11.84}$$

To obtain the value of $x_{i+\theta}$, we consider the equilibrium equation (11.18) at time $t_{i+\theta} = t_i + \theta\,\Delta t$ and write

$$[m]\ddot{\vec{x}}_{i+\theta} + [c]\dot{\vec{x}}_{i+\theta} + [k]\vec{x}_{i+\theta} = \underset{\sim}{F}_{i+\theta} \tag{11.85}$$

where the force vector $\underset{\sim}{F}_{i+\theta}$ is also obtained by using the linear assumption:

$$\underset{\sim}{F}_{i+\theta} = \vec{F}_i + \theta(\vec{F}_{i+1} - \vec{F}_i) \tag{11.86}$$

Substituting Eqs. (11.83), (11.84), and (11.86) for $\ddot{\vec{x}}_{i+\theta}$, $\dot{\vec{x}}_{i+\theta}$, and $\underset{\sim}{F}_{i+\theta}$, Eq. (11.85) gives

$$\left\{\frac{6}{\theta^2(\Delta t)^2}[m] + \frac{3}{\theta\,\Delta t}[c] + [k]\right\}\vec{x}_{i+1}$$

$$= \vec{F}_i + \theta(\vec{F}_{i+1} - \vec{F}_i) + \left\{\frac{6}{\theta^2(\Delta t)^2}[m] + \frac{3}{\theta\,\Delta t}[c]\right\}\vec{x}_i$$

$$+ \left\{\frac{6}{\theta\,\Delta t}[m] + 2[c]\right\}\dot{\vec{x}}_i + \left\{2[m] + \frac{\theta\,\Delta t}{2}[c]\right\}\ddot{\vec{x}}_i \tag{11.87}$$

which can be solved for $\vec{x}_{i+1}$.

The Wilson method can be described by the following steps:

1. From the known initial conditions $\vec{x}_0$ and $\dot{\vec{x}}_0$, find $\ddot{\vec{x}}_0$ using Eq. (11.26).
2. Select a suitable time step Δt and a suitable value of θ (θ is usually taken as 1.4).

3. Compute the effective load vector $\underset{\approx}{\vec{F}}_{i+\theta}$, starting with $i = 0$:

$$\underset{\approx}{\vec{F}}_{i+\theta} = \vec{F}_i + \theta(\vec{F}_{i+1} - \vec{F}_i) + [m]\left(\frac{6}{\theta^2(\Delta t)^2}\vec{x}_i + \frac{6}{\theta \Delta t}\dot{\vec{x}}_i + 2\ddot{\vec{x}}_i\right)$$

$$+ [c]\left(\frac{3}{\theta \Delta t}\vec{x}_i + 2\dot{\vec{x}}_i + \frac{\theta \Delta t}{2}\ddot{\vec{x}}_i\right) \tag{11.88}$$

4. Find the displacement vector at time $t_{i+\theta}$:

$$\vec{x}_{i+\theta} = \left[\frac{6}{\theta^2(\Delta t)^2}[m] + \frac{3}{\theta \Delta t}[c] + [k]\right]^{-1}\underset{\approx}{F}_{i+\theta} \tag{11.89}$$

5. Calculate the acceleration, velocity, and displacement vectors at time t_{i+1}:

$$\ddot{\vec{x}}_{i+1} = \frac{6}{\theta^3(\Delta t)^2}(\vec{x}_{i+\theta} - \vec{x}_i) - \frac{6}{\theta^2 \Delta t}\dot{\vec{x}}_i + \left(1 - \frac{3}{\theta}\right)\ddot{\vec{x}}_i \tag{11.90}$$

$$\dot{\vec{x}}_{i+1} = \dot{\vec{x}}_i + \frac{\Delta t}{2}(\ddot{\vec{x}}_{i+1} + \ddot{\vec{x}}_i) \tag{11.91}$$

$$\vec{x}_{i+1} = \vec{x}_i + \Delta t \dot{\vec{x}}_i + \frac{(\Delta t)^2}{6}(\ddot{\vec{x}}_{i+1} + 2\ddot{\vec{x}}_i) \tag{11.92}$$

▬▬▬▬▬▬ **Wilson Method for a Two Degree of Freedom System**

EXAMPLE   11.7
Find the response of the system considered in Example 11.3, using the Wilson θ method with $\theta = 1.4$.

Solution
Approach: Use Wilson method with $\Delta t = 0.24216$.
   The value of $\ddot{\vec{x}}_0$ can be obtained as in the case of Example 11.3:

$$\ddot{\vec{x}}_0 = \begin{Bmatrix} 0 \\ 5 \end{Bmatrix}$$

Then, by using Eqs. (11.90) to (11.92) with a time step of $\Delta t = 0.24216$, we obtain the results indicated in Table 11.6.

∎

11.10   Newmark Method

The Newmark integration method is also based on the assumption that the acceleration varies linearly between two instants of time. The resulting expressions for the velocity and displacement vectors $\dot{\vec{x}}_{i+1}$ and $\vec{x}_{i+1}$, for a multidegree of freedom system [11.21], are written as in Eqs. (11.79) and (11.80)

$$\dot{\vec{x}}_{i+1} = \dot{\vec{x}}_i + [(1 - \beta)\ddot{\vec{x}}_i + \beta\ddot{\vec{x}}_{i+1}]\Delta t \tag{11.93}$$

TABLE 11.6

Time $t_i = i\Delta t$	$\vec{x}_i = \vec{x}(t = t_i)$
t_1	$\left\{ \begin{matrix} 0.0033 \\ 0.1392 \end{matrix} \right\}$
t_2	$\left\{ \begin{matrix} 0.0289 \\ 0.5201 \end{matrix} \right\}$
t_3	$\left\{ \begin{matrix} 0.1072 \\ 1.0579 \end{matrix} \right\}$
t_4	$\left\{ \begin{matrix} 0.2649 \\ 1.6408 \end{matrix} \right\}$
t_5	$\left\{ \begin{matrix} 0.5076 \\ 2.1529 \end{matrix} \right\}$
t_6	$\left\{ \begin{matrix} 0.8074 \\ 2.4981 \end{matrix} \right\}$
t_7	$\left\{ \begin{matrix} 1.1035 \\ 2.6191 \end{matrix} \right\}$
t_8	$\left\{ \begin{matrix} 1.3158 \\ 2.5056 \end{matrix} \right\}$
t_9	$\left\{ \begin{matrix} 1.3688 \\ 2.1929 \end{matrix} \right\}$
t_{10}	$\left\{ \begin{matrix} 1.2183 \\ 1.7503 \end{matrix} \right\}$
t_{11}	$\left\{ \begin{matrix} 0.8710 \\ 1.2542 \end{matrix} \right\}$
t_{12}	$\left\{ \begin{matrix} 0.3897 \\ 0.8208 \end{matrix} \right\}$

$$\vec{x}_{i+1} = \vec{x}_i + \Delta t\, \dot{\vec{x}}_i + [(\tfrac{1}{2} - \alpha)\ddot{\vec{x}}_i + \alpha\ddot{\vec{x}}_{i+1}](\Delta t)^2 \tag{11.94}$$

where the parameters α and β indicate how much the acceleration at the end of the interval enters into the velocity and displacement equations at the end of the interval Δt. In fact, α and β can be chosen to obtain the desired accuracy and stability characteristics [11.22]. When $\beta = \frac{1}{2}$ and $\alpha = \frac{1}{6}$, Eqs. (11.93) and (11.94) correspond to the linear acceleration method (which can also be obtained using $\theta = 1$ in the Wilson method). When $\beta = \frac{1}{2}$ and $\alpha = \frac{1}{4}$, Eqs. (11.93) and (11.94) correspond to the assumption of constant acceleration between t_i and t_{i+1}. To find the value of $\ddot{\vec{x}}_{i+1}$, the equilibrium equation (11.18) is considered at $t = t_{i+1}$ so that

$$[m]\ddot{\vec{x}}_{i+1} + [c]\dot{\vec{x}}_{i+1} + [k]\vec{x}_{i+1} = \vec{F}_{i+1} \tag{11.95}$$

Equation (11.94) can be used to express $\ddot{\vec{x}}_{i+1}$ in terms of $\vec{x}_{i+1}$, and the resulting expression can be substituted into Eq. (11.93) to express $\dot{\vec{x}}_{i+1}$ in terms of $\vec{x}_{i+1}$. By substituting these expressions for $\dot{\vec{x}}_{i+1}$ and $\ddot{\vec{x}}_{i+1}$ into Eq. (11.95), we can obtain a relation for finding $\vec{x}_{i+1}$:

$$
\vec{x}_{i+1} = \left[\frac{1}{\alpha(\Delta t)^2}[m] + \frac{\beta}{\alpha\,\Delta t}[c] + [k] \right]^{-1}
$$

$$
\times \left\{ \vec{F}_{i+1} + [m]\left(\frac{1}{\alpha(\Delta t)^2}\vec{x}_i + \frac{1}{\alpha\,\Delta t}\dot{\vec{x}}_i + \left(\frac{1}{2\alpha} - 1 \right)\ddot{\vec{x}}_i \right) \right.
$$

$$
+ [c]\left(\frac{\beta}{\alpha\,\Delta t}\vec{x}_i + \left(\frac{\beta}{\alpha} - 1 \right)\dot{\vec{x}}_i \right.
$$

$$
\left. \left. + \left(\frac{\beta}{\alpha} - 2 \right)\frac{\Delta t}{2}\ddot{\vec{x}}_i \right) \right\} \tag{11.96}
$$

The Newmark method can be summarized in the following steps:

1. From the known values of $\vec{x}_0$ and $\dot{\vec{x}}_0$, find $\ddot{\vec{x}}_0$ using Eq. (11.26).
2. Select suitable values of Δt, α, and β.
3. Calculate the displacement vector $\vec{x}_{i+1}$, starting with $i = 0$ and using Eq. (11.96).
4. Find the acceleration and velocity vectors at time t_{i+1}:

$$
\ddot{\vec{x}}_{i+1} = \frac{1}{\alpha(\Delta t)^2}(\vec{x}_{i+1} - \vec{x}_i) - \frac{1}{\alpha\,\Delta t}\dot{\vec{x}}_i - \left(\frac{1}{2\alpha} - 1 \right)\ddot{\vec{x}}_i \tag{11.97}
$$

$$
\dot{\vec{x}}_{i+1} = \dot{\vec{x}}_i + (1 - \beta)\,\Delta t\,\ddot{\vec{x}}_i + \beta\,\Delta t\,\ddot{\vec{x}}_{i+1} \tag{11.98}
$$

It is important to note that unless β is taken as $\frac{1}{2}$, there is a spurious damping introduced, proportional to $(\beta - \frac{1}{2})$. If β is taken as zero, a negative damping results; this involves a self-excited vibration arising solely from the numerical procedure. Similarly, if β is greater than $\frac{1}{2}$, a positive damping is introduced. This reduces the magnitude of response even without real damping in the problem [11.21]. The method is unconditionally stable for $\alpha \geq \frac{1}{4}(\beta + \frac{1}{2})^2$ and $\beta \geq \frac{1}{2}$.

████████████████ **Newmark Method for a Two Degree of Freedom System**

EXAMPLE   11.8

Find the response of the system considered in Example 11.3, using the Newmark method with $\alpha = \frac{1}{6}$ and $\beta = \frac{1}{2}$.

Solution

Approach: Use the Newmark method with $\Delta t = 0.24216$.

The value of $\ddot{\vec{x}}_0$ can be found using Eq. (11.26):

$$
\ddot{\vec{x}}_0 = \begin{Bmatrix} 0 \\ 5 \end{Bmatrix}
$$

TABLE 11.7

Time $t_i = i\Delta t$	$\vec{x}_i = \vec{x}(t = t_i)$
t_1	$\begin{Bmatrix} 0.0026 \\ 0.1411 \end{Bmatrix}$
t_2	$\begin{Bmatrix} 0.0246 \\ 0.5329 \end{Bmatrix}$
t_3	$\begin{Bmatrix} 0.1005 \\ 1.0884 \end{Bmatrix}$
t_4	$\begin{Bmatrix} 0.2644 \\ 1.6870 \end{Bmatrix}$
t_5	$\begin{Bmatrix} 0.5257 \\ 2.2027 \end{Bmatrix}$
t_6	$\begin{Bmatrix} 0.8530 \\ 2.5336 \end{Bmatrix}$
t_7	$\begin{Bmatrix} 1.1730 \\ 2.6229 \end{Bmatrix}$
t_8	$\begin{Bmatrix} 1.3892 \\ 2.4674 \end{Bmatrix}$
t_9	$\begin{Bmatrix} 1.4134 \\ 2.1137 \end{Bmatrix}$
t_{10}	$\begin{Bmatrix} 1.1998 \\ 1.6426 \end{Bmatrix}$
t_{11}	$\begin{Bmatrix} 0.7690 \\ 1.1485 \end{Bmatrix}$
t_{12}	$\begin{Bmatrix} 0.2111 \\ 0.7195 \end{Bmatrix}$

With the values of $\alpha = \frac{1}{6}$, $\beta = 0.5$, and $\Delta t = 0.24216$, Eq. (11.96) gives the values of $\vec{x}_i = \vec{x}(t = t_i)$, as shown in Table 11.7.

■

11.11   Examples Using MATLAB

MATLAB Solution of a Single Degree of Freedom System

EXAMPLE 11.9

Using the MATLAB function **ode23**, solve Example 11.1.

Solution: Defining $x_1 = x$ and $x_2 = \dot{x}$, Eq. (E.1) of Example 11.1 can be expressed as a set of two first-order differential equations

$$\dot{x}_1 = x_2 \tag{E.1}$$

$$\dot{x}_2 = \frac{1}{m}\left[F_0\left(1 - \sin\frac{\pi t}{2 t_0}\right) - c\,x_2 - k\,x_1\right] \tag{E.2}$$

with initial conditions $x_1(0) = x_2(0) = 0$. The MATLAB program to solve Eqs. (E.1) and (E.2) is given below.

```
% Ex11_9.m
tspan = [0: 0.1: 5*pi];
x0 = [0; 0];
[t,x] = ode23 ('dfunc11_9', tspan, x0);
plot (t,x(:,1));
xlabel ('t');
ylabel ('x(t) and xd(t)');
gtext ('x(t)');
hold on;
plot (t,x(:,2), '--');
gtext ('xd(t)')

%dfunc11_9.m
function f = dfunc11_9(t,x)
m = 1;
k = 1;
c = 0.2;
t0 = pi;
F0 = 1;
f = zeros (2,1);
f(1) = x(2);
f(2) = (F0* (1 - sin(pi*t/(2*t0))) - c*x(2) - k*x(1) )/m;
```

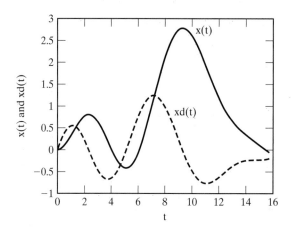

■

MATLAB Solution of Multidegree of Freedom System

EXAMPLE 11.10

Using the MATLAB function **ode23**, solve Example 11.3.

Solution: The equations of motion of the two degree of freedom system in Eq. (E.1) of Example 11.3 can be expressed as a system of four first-order differential equations in terms of

$$y_1 = x_1, \qquad y_2 = \dot{x}_1, \qquad y_3 = x_2, \qquad y_4 = \dot{x}_2$$

as

$$\dot{y}_1 = y_2 \tag{E.1}$$

$$\dot{y}_2 = \frac{1}{m_1} \{F_1(t) - c\, y_2 + c\, y_4 - (k_1 + k)\, y_1 + k\, y_3\} = -6\, y_1 + y_3 \tag{E.2}$$

$$\dot{y}_3 = y_4 \tag{E.3}$$

$$\dot{y}_4 = \frac{1}{m_2} \{F_2(t) + c\, y_2 - c\, y_4 + k\, y_1 - (k + k_2)\, y_3\}$$

$$= \frac{1}{2} \{10 + 2\, y_1 - 8\, y_3\} = 5 + y_1 - 4\, y_3 \tag{E.4}$$

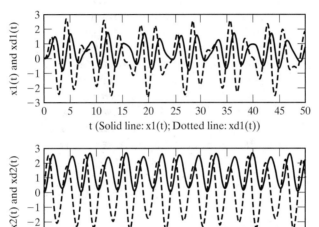

with initial conditions $y_i(0) = 0$, $i = 1, 2, 3, 4$. The MATLAB program to solve Eqs. (E.1) to (E.4) is given below.

```
% Ex11_10.m
tspan = [0: 0.05: 50];
y0 = [0; 0; 0; 0];
[t,y] = ode23 ('dfunc11_10', tspan, y0);
subplot (211);
plot (t,y(:,1));
xlabel ('t ( Solid line: x1 (t) Dotted line: xd1 (t) ) ');
ylabel ('x1 (t) amd xd1 (t)');
hold on;
plot (t,y(:, 2), '--');
subplot (212);
plot (t,y(:, 3));
xlabel ('t ( Solid line: x2 (t) Dotted line: xd2 (t) )');
ylabel ('x2 (t) amd xd2 (t) ');
hold on;
plot (t,y (:,4), '--');
```

```
%dfunc11_10.m
function f = dfunc11_10 (t,y)
m1 = 1;
m2 = 2;
k1 = 4;
k2 = 6;
k = 2;
c = 0;
F1 = 0;
F2 = 10;
f = zeros (4,1);
f(1) = y(2);
f(2) = ( F1 - c*y(2) + c*y(4) - (k1+k) *y(1) + k*y(3) )/m1;
f(3) = y(4);
f(4) = ( F2 + c*y(2) - c*y(4) + k*y(1) - (k + k2) *y(3) )/m2;
```

∎

Program to Implement Fourth-Order Runge-Kutta Method

EXAMPLE 11.11

Develop a general MATLAB program called **Program14.m** for solving a set of first-order differential equations using the fourth-order Runge-Kutta method. Use the program to solve Example 11.2.

Solution: **Program14.m** is developed to accept the following input data:

n = number of first-order differential equations

xx = initial values $x_i(0)$, a vector of size n

dt = time increment

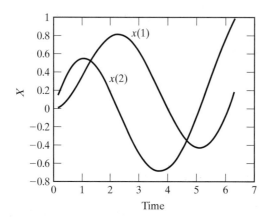

The program requires a subprogram to define the functions $f_i(\vec{x}, t), i = 1, 2, \ldots, n$. The program gives the values of $x_i(t)$, $i = 1, 2, \ldots, n$ at different values of time t.

I	Time(I)	x(1)	x(2)
1	1.570800e−001	1.186315e−002	1.479138e−001
2	3.141600e−001	4.540642e−002	2.755911e−001

3	4.712400e−001	9.725706e−002	3.806748e−001
4	6.283200e−001	1.637262e−001	4.615022e−001
5	7.854000e−001	2.409198e−001	5.171225e−001
.			
.			
.			
36	5.654880e+000	−2.868460e−001	5.040887e−001
37	5.811960e+000	−1.969950e−001	6.388500e−001
38	5.969040e+000	−8.655813e−002	7.657373e−001
39	6.126120e+000	4.301693e−002	8.821039e−001
40	6.283200e+000	1.898865e−001	9.855658e−001

■

Program for Central Difference Method

EXAMPLE 11.12

Using the central difference method, develop a general MATLAB program called **Program15.m** to find the dynamic response of a multidegree of freedom system. Use the program to find the solution of Example 11.3.

Solution: **Program15.m** is developed to accept the following input data:
   n = degree of freedom of the system
   m = mass matrix, of size n × n

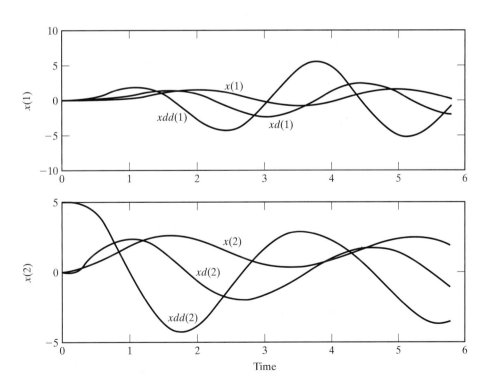

c = damping matrix, of size n × n

k = stiffness matrix, of size n × n

xi = initial values of x_i, a vector of size n

xdi = initial values of $\dot{x}_i$, a vector of size n

$nstep(nstp)$ = number of time steps at which solution is to be found

$delt$ = increment between time steps

The program requires a subprogram to define the forcing functions $f_i(t), i = 1, 2, \ldots, n$ at any time t. It gives the values of the response at different time steps i as $x_j(i)$, $\dot{x}_j(i)$, and $\ddot{x}_j(i)$, $j = 1, 2, \ldots, n$.

```
Solution by central difference method

Given data:

n= 2 nstp= 24 delt=2.421627e-001

Solution:
```

step	time	x(i,1)	xd(i,1)	xdd(i,1)	x(i,2)	xd(i,2)	xdd(i,2)
1	0.0000	0.0000e+000	0.0000e+000	0.0000e+000	0.0000e+000	0.0000e+000	5.0000e+000
2	0.2422	0.0000e+000	0.0000e+000	0.0000e+000	1.4661e−001	0.0000e+000	5.0000e+000
3	0.4843	1.7195e−002	3.5503e−002	2.9321e−001	5.5204e−001	1.1398e+000	4.4136e+000
4	0.7265	9.3086e−002	1.9220e−001	1.0009e+000	1.1222e+000	2.0143e+000	2.8090e+000
5	0.9687	2.6784e−001	5.1752e−001	1.6859e+000	1.7278e+000	2.4276e+000	6.0429e−001
.							
.							
.							
21	4.8433	1.6034e+000	1.7764e+000	−4.0959e+000	2.2077e+000	1.6763e+000	−1.0350e+000
22	5.0854	1.6083e+000	6.5025e−001	−5.2053e+000	2.4526e+000	1.2813e+000	−2.2272e+000
23	5.3276	1.3349e+000	−5.5447e−001	−4.7444e+000	2.5098e+000	6.2384e−001	−3.2023e+000
24	5.5697	8.8618e−001	−1.4909e+000	−2.9897e+000	2.3498e+000	−2.1242e−001	−3.7043e+000
25	5.8119	4.0126e−001	−1.9277e+000	−6.1759e−001	1.9837e+000	−1.0863e+000	−3.5128e+000

■

Program for Houbolt Method

EXAMPLE 11.13

Using the Houbolt method, develop a general MATLAB program called **Program16.m** to find the dynamic response of a multidegree of freedom system. Use the program to find the solution of Example 11.6.

Solution: Program16.m is developed to accept the following input data:

n = degree of freedom of the system

m = mass matrix, of size n × n

c = damping matrix, of size n × n

k = stiffness matrix, of size n × n

xi = initial values of x_i, a vector of size n

xdi = initial values of $\dot{x}_i$, a vector of size n

$nstep(nstp)$ = number of time steps at which solution is to be found

$delt$ = increment between time steps

The program requires a subprogram to define the forcing functions $f_i(t)$, $i = 1, 2, \ldots, n$ at any time t. It gives the values of the response at different time stations i as $x_j(i)$, $\dot{x}_j(i)$, and $\ddot{x}_j(i)$, $j = 1, 2, \ldots, n$.

```
Solution by Houbolt method

Given data:

n= 2 nstp= 24 delt=2.421627e-001

Solution:
```

step	time	x(i,1)	xd(i,1)	xdd(i,1)	x(i,2)	xd(i,2)	xdd(i,2)
1	0.0000	0.0000e+000	0.0000e+000	0.0000e+000	0.0000e+000	0.0000e+000	5.0000e+000
2	0.2422	0.0000e+000	0.0000e+000	0.0000e+000	1.4661e−001	0.0000e+000	5.0000e+000
3	0.4843	1.7195e−002	3.5503e−002	2.9321e−001	5.5204e−001	1.1398e+000	4.4136e+000
4	0.7265	9.1732e−002	4.8146e−001	1.6624e+000	1.1064e+000	2.4455e+000	6.6609e−001
5	0.9687	2.5010e−001	8.6351e−001	1.8812e+000	1.6909e+000	2.3121e+000	−1.5134e+000
.							
.							
.							
21	4.8433	8.7373e−001	1.7900e+000	−1.7158e+000	1.7633e+000	1.3850e+000	−1.1795e+000
22	5.0854	1.2428e+000	1.1873e+000	−3.3403e+000	2.0584e+000	1.0125e+000	−1.9907e+000
23	5.3276	1.4412e+000	3.6619e−001	−4.1553e+000	2.2460e+000	4.9549e−001	−2.5428e+000
24	5.5697	1.4363e+000	−4.8458e−001	−4.0200e+000	2.2990e+000	−9.6748e−002	−2.7595e+000
25	5.8119	1.2410e+000	−1.1822e+000	−3.0289e+000	2.2085e+000	−6.8133e−001	−2.5932e+000

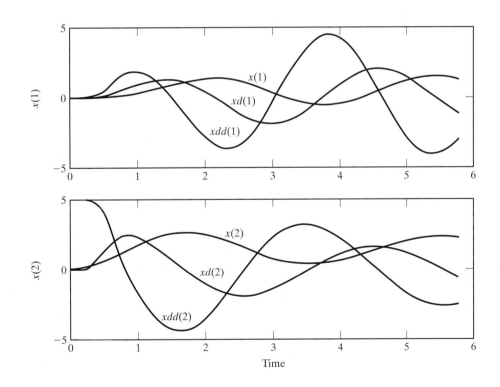

11.12   C++ Programs

Three interactive C++ programs, called **Program14.cpp**, **Program15.cpp**, and **Program16.cpp**, are given to implement different methods of numerical integration for vibration analysis. **Program14.cpp** is developed to implement the fourth-order Runge-Kutta method to solve a set of simultaneous first-order differential equations. **Program15.cpp** implements the central difference method for finding the dynamic response of multidegree of freedom systems. **Program16.cpp** implements the Houbolt method for finding the dynamic response of a multidegree of freedom system. The inputs and outputs of these programs are similar to those of **Program14.m**, **Program15.m**, and **Program16.m**.

Fourth-Order Runge-Kutta Method

EXAMPLE 11.14

Using **Program 14.cpp**, find the solution of Example 11.2.

Solution: The input data are to be given interactively. The output of the program is given below.

I	TIME(I)	X(1)	X(2)
1	0.15708000	0.01186315	0.14791379
2	0.31416000	0.04540642	0.27559111
3	0.47124000	0.09725706	0.38067484
4	0.62832000	0.16372624	0.46150222
5	0.78540000	0.24091978	0.51712252
.			
.			
.			
36	5.65488000	−0.28684602	0.50408868
37	5.81196000	−0.19699505	0.63884996
38	5.96904000	−0.08655813	0.76573735
39	6.12612000	0.04301693	0.88210391
40	6.28320000	0.18988652	0.98556585

∎

Central Difference Method

EXAMPLE 11.15

Using **Program15.cpp**, find the solution of Example 11.3.

Solution: The input data are to be entered interactively. The input and output of the program are given below.

```
Please input N and NSTEP:
2    24

Please input M matrix row by row
1.0    0.0
0.0    2.0

Please input K matrix row by row
6.0   −2.0
−2.0   8.0
```

```
Please input C matrix row by row
0.0     0.0
0.0     0.0

SOLUTION BY CENTRAL DIFFERENCE METHOD

GIVEN DATA:
N = 2 NSTEP = 24 DELT = 0.24216267

SOLUTION:
```

STEP	TIME	X(I,1)	XD(I,1)	XDD(I,1)	X(I,2)	XD(I,2)	XDD(I,2)
1	0.0000	0.0000	0.0000	0.0000	0.0000	0.0000	5.0000
2	0.2422	0.0000	0.0000	0.0000	0.1466	0.0000	5.0000
3	0.4843	0.0172	0.0355	0.2932	0.5520	1.1398	4.4136
4	0.7265	0.0931	0.1922	1.0009	1.1222	2.0143	2.8090
5	0.9687	0.2678	0.5175	1.6859	1.7278	2.4276	0.6043
.							
.							
.							
20	4.6011	1.2933	2.4742	−1.6667	1.8321	1.7822	0.1598
21	4.8433	1.6034	1.7764	−4.0959	2.2077	1.6763	−1.0350
22	5.0854	1.6083	0.6502	−5.2053	2.4526	1.2813	−2.2272
23	5.3276	1.3349	−0.5545	−4.7444	2.5098	0.6238	−3.2023
24	5.5697	0.8862	−1.4909	−2.9897	2.3498	−0.2124	−3.7043
25	5.8119	0.4013	−1.9277	−0.6176	1.9837	−1.0863	−3.5128

■

Houbolt Method

EXAMPLE 11.16

Using **Program16.cpp**, find the solution of Example 11.6.

Solution: The input data are to be given interactively. The input and output of the program are given below.

```
Please input N and NSTEP:
2  24

Please input M matrix row by row
1.0 0.0
0.0 2.0

Please input K matrix row by row
6.0  −2.0
−2.0   8.0

Please input C matrix row by row
0.0 0.0
0.0 0.0

SOLUTION BY HOUBOLT METHOD

GIVEN DATA:
N = 2 NSTEP = 24 DELT = 0.24216267

SOLUTION:
```

STEP	TIME	X(I,1)	XD(I,1)	XDD(I,1)	X(I,2)	XD(I,2)	XDD(I,2)
1	0.0000	0.0000	0.0000	0.0000	0.0000	0.0000	5.0000
2	0.2422	0.0000	0.0000	0.0000	0.1466	0.0000	5.0000

3	0.4843	0.0172	0.0355	0.2932	0.5520	1.1398	4.4136
4	0.7265	0.0917	0.4815	1.6624	1.1064	2.4455	0.6661
5	0.9687	0.2501	0.8635	1.8812	1.6909	2.3121	−1.5134
.						
.						
.						
21	4.8433	0.8737	1.7900	−1.7158	1.7633	1.3850	−1.1795
22	5.0854	1.2428	1.1873	−3.3403	2.0584	1.0125	−1.9907
23	5.3276	1.4412	0.3662	−4.1553	2.2460	0.4955	−2.5428
24	5.5697	1.4363	−0.4846	−4.0200	2.2990	−0.0967	−2.7595
25	5.8119	1.2410	−1.1822	−3.0289	2.2085	−0.6813	−2.5932

∎

11.13 Fortran Programs

Three Fortran programs—**PROGRAM14F**, **PROGRAM15.F** and **PROGRAM16.F**—are given for the numerical solution of vibration problems involving dynamic response computations. **PROGRAM14.F** implements the fourth-order Runge-Kutta method for solving a system of first-order differential equations and calls a subroutine called **RK4**. It requires a user-supplied subroutine **FUN** to evaluate the forcing functions F(1), F(2), ..., F(N) at any specified displacement vector $\vec{X}$ and time t. **PROGRAM15.F** implements the central difference method for solving a multidegree of freedom system and calls a subroutine called **CDIFF** and requires a user-supplied subroutine called **EXTFUN** to evaluate the forcing functions F(1), F(2), ..., F(N) at any time t. **PROGRAM16.F** implements the Houbolt method and calls a subroutine called **HOBOLT**. It requires a user-supplied subroutine called **EXTFUN** to evaluate the forcing functions F(1), F(2), ..., F(N) at any time t. The inputs and outputs of these programs are similar to those of **Program14.m**, **Program15.m**, and **Program16.m**.

■■■■■■■■■■ Fourth-Order Runge-Kutta Method

EXAMPLE 11.17 ──

Using **PROGRAM14.F**, find the solution of Example 11.2.

Solution: The output of the program is given below.

I	TIME(I)	X(1)	X(2)
1	0.1571	0.11863151E−01	0.14791380E+00
2	0.3142	0.45406420E−01	0.27559111E+00
3	0.4712	0.97257055E−01	0.38067484E+00
4	0.6283	0.16372624E+00	0.46150222E+00
5	0.7854	0.24091978E+00	0.51712251E+00
.			
.			
.			
36	5.6549	−0.28684574E+00	0.50408858E+00
37	5.8120	−0.19699481E+00	0.63884968E+00
38	5.9690	−0.86557955E−01	0.76573688E+00
39	6.1261	0.43017015E−01	0.88210326E+00
40	6.2832	0.18988648E+00	0.98556501E+00

∎

■■■■■■■■ Central Difference Method

EXAMPLE 11.18 _____

_____ Using **PROGRAM15.F**, find the solution of Example 11.3.

Solution: The output of the program is shown below.

SOLUTION BY CENTRAL DIFFERENCE METHOD

GIVEN DATA:
N= 2 NSTEP= 24 DELT= 0.24216267E+00

SOLUTION:

STEP	TIME	X(I,1)	XD(I,1)	XDD(I,1)	X(I,2)	XD(I,2)	XDD(I,2)
1	0.0000	0.0000E+00	0.0000E+00	0.0000E+00	0.0000E+00	0.0000E+00	0.5000E+01
2	0.2422	0.0000E+00	0.0000E+00	0.0000E+00	0.1466E+00	0.0000E+00	0.5000E+01
3	0.4843	0.1719E-01	0.3550E-01	0.2932E+00	0.5520E+00	0.1140E+01	0.4414E+01
4	0.7265	0.9309E-01	0.1922E+00	0.1001E+01	0.1122E+01	0.2014E+01	0.2809E+01
5	0.9687	0.2678E+00	0.5175E+00	0.1686E+01	0.1728E+01	0.2428E+01	0.6043E+00
.							
.							
.							
21	4.8433	0.1603E+01	0.1776E+01	−.4096E+01	0.2208E+01	0.1676E+01	−.1035E+01
22	5.0854	0.1608E+01	0.6502E+00	−.5205E+01	0.2453E+01	0.1281E+01	−.2227E+01
23	5.3276	0.1335E+01	−.5545E+00	−.4744E+01	0.2510E+01	0.6238E+00	−.3202E+01
24	5.5697	0.8862E+00	−.1491E+01	−.2990E+01	0.2350E+01	−.2124E+00	−.3704E+01
25	5.8119	0.4013E+00	−.1928E+01	−.6176E+00	0.1984E+01	−.1086E+01	−.3513E+01

■

■■■■■■■■ Houbolt Method

EXAMPLE 11.19 _____

_____ Using **PROGRAM16.F**, find the solution of Example 11.6.

Solution: The output of the program is given below.

SOLUTION BY HOUBOLT METHOD

GIVEN DATA:
N= 2 NSTEP= 24 DELT= 0.24216267E+00

SOLUTION:

STEP	TIME	X(I,1)	XD(I,1)	XDD(I,1)	X(I,2)	XD(I,2)	XDD(I,2)
1	0.0000	0.0000E+00	0.0000E+00	0.0000E+00	0.0000E+00	0.0000E+00	0.5000E+01
2	0.2422	0.0000E+00	0.0000E+00	0.0000E+00	0.1466E+00	0.0000E+00	0.5000E+01
3	0.4843	0.1719E-01	0.3550E-01	0.2932E+00	0.5520E+00	0.1140E+01	0.4414E+01
4	0.7265	0.9173E-01	0.4815E+00	0.1662E+01	0.1106E+01	0.2446E+01	0.6661E+00
5	0.9687	0.2501E+00	0.8635E+00	0.1881E+01	0.1691E+01	0.2312E+01	−.1513E+01
.							
.							
.							
21	4.8433	0.8737E+00	0.1790E+01	−.1716E+01	0.1763E+01	0.1385E+01	−.1179E+01
22	5.0854	0.1243E+01	0.1187E+01	−.3340E+01	0.2058E+01	0.1013E+01	−.1991E+01
23	5.3276	0.1441E+01	0.3662E+00	−.4155E+01	0.2246E+01	0.4955E+00	−.2543E+01
24	5.5697	0.1436E+01	−.4846E+00	−.4020E+01	0.2299E+01	−.9674E-01	−.2760E+01
25	5.8119	0.1241E+01	−.1182E+01	−.3029E+01	0.2209E+01	−.6813E+00	−.2593E+01

■

REFERENCES

11.1 G. L. Goudreau and R. L. Taylor, "Evaluation of numerical integration methods in elasto-dynamics," *Computational Methods in Applied Mechanics and Engineering*, Vol. 2, 1973, pp. 69–97.

11.2 S. W. Key, "Transient response by time integration: Review of implicit and explicit operations," in J. Donéa (ed.), *Advanced Structural Dynamics*, Applied Science Publishers, London, 1980.

11.3 R. E. Cornwell, R. R. Craig, Jr., and C. P. Johnson, "On the application of the mode-acceleration method to structural engineering problems," *Earthquake Engineering and Structural Dynamics*, Vol. 11, 1983, pp. 679–688.

11.4 T. Wah and L. R. Calcote, *Structural Analysis by Finite Difference Calculus*, Van Nostrand Reinhold, New York, 1970.

11.5 R. Ali, "Finite difference methods in vibration analysis," *Shock and Vibration Digest*, Vol. 15, March 1983, pp. 3–7.

11.6 P. C. M. Lau, "Finite difference approximation for ordinary derivatives," *International Journal for Numerical Methods in Engineering*, Vol. 17, 1981, pp. 663–678.

11.7 R. D. Krieg, "Unconditional stability in numerical time integration methods," *Journal of Applied Mechanics*, Vol. 40, 1973, pp. 417–421.

11.8 S. Levy and W. D. Kroll, "Errors introduced by finite space and time increments in dynamic response computation," *Proceedings, First U.S. National Congress of Applied Mechanics*, 1951, pp. 1–8.

11.9 A. F. D'Souza and V. K. Garg, *Advanced Dynamics. Modeling and Analysis*, Prentice Hall, Englewood Cliffs, N.J., 1984.

11.10 A Ralston and H. S. Wilf (eds.), *Mathematical Methods for Digital Computers*, Wiley, New York, 1960.

11.11 S. Nakamura, *Computational Methods in Engineering and Science*, Wiley, New York, 1977.

11.12 T. Belytschko, "Explicit time integration of structure-mechanical systems," in J. Donéa (ed.), *Advanced Structural Dynamics*, Applied Science Publishers, London, 1980, pp. 97–122.

11.13 S. Levy and J. P. D. Wilkinson, *The Component Element Method in Dynamics with Application to Earthquake and Vehicle Engineering*, McGraw-Hill, New York, 1976.

11.14 J. W. Leech, P. T. Hsu, and E. W. Mack, "Stability of a finite-difference method for solving matrix equations," *AIAA Journal*, Vol. 3, 1965, pp. 2172–2173.

11.15 S. D. Conte and C. W. DeBoor, *Elementary Numerical Analysis: An Algorithmic Approach* (2nd ed.), McGraw-Hill, New York, 1972.

11.16 C. F. Gerald and P. O. Wheatley, *Applied Numerical Analysis* (3rd ed.), Addison-Wesley, Reading, Mass., 1984.

11.17 L. V. Atkinson and P. J. Harley, *Introduction to Numerical Methods with PASCAL*, Addison-Wesley, Reading, Mass., 1984.

11.18 J. C. Houbolt, "A recurrence matrix solution for the dynamic response of elastic aircraft," *Journal of Aeronautical Sciences*, Vol. 17, 1950, pp. 540–550, 594.

11.19 E. L. Wilson, I. Farhoomand, and K. J. Bathe, "Nonlinear dynamic analysis of complex structures," *International Journal of Earthquake Engineering and Structural Dynamics*, Vol. 1, 1973, pp. 241–252.

11.20 S. P. Timoshenko, D. H. Young, and W. Weaver, Jr., *Vibration Problems in Engineering* (4th ed.), Wiley, New York, 1974.

11.21 N. M. Newmark, "A method of computation for structural dynamics," *ASCE Journal of Engineering Mechanics Division*, Vol. 85, 1959, pp. 67–94.

11.22 T. J. R. Hughes, "A note on the stability of Newmark's algorithm in nonlinear structural dynamics," *International Journal for Numerical Methods in Engineering*, Vol. 11, 1976, pp. 383–386.

REVIEW QUESTIONS

11.1 Give brief answers to the following:

  1. Describe the procedure of the finite difference method.
  2. Using Taylor's series expansion, derive the central difference formulas for the first and the second derivatives of a function,
  3. What is a conditionally stable method?
  4. What is the main difference between the central difference method and the Runge-Kutta method?
  5. Why is it necessary to introduce fictitious mesh points in the finite difference method of solution?
  6. Define a tridiagonal matrix.
  7. What is the basic assumption of the Wilson method?
  8. What is a linear acceleration method?
  9. What is the difference between explicit and implicit integration methods?
  10. Can we use the numerical integration methods discussed in this chapter to solve nonlinear vibration problems?

11.2 Indicate whether each of the following statements is true or false:

  1. The grid points in the finite difference methods are required to be uniformly spaced.
  2. The Runge-Kutta method is stable.
  3. The Runge-Kutta method is self-starting.
  4. The finite difference method is an implicit integration method.
  5. The Newmark method is an implicit integration method.
  6. For a beam with grid points $-1, 1, 2, 3, \ldots$, the central difference equivalence of the condition $\dfrac{dW}{dx}\bigg|_1 = 0$ is $W_{-1} = W_2$.
  7. For a beam with grid points $-1, 1, 2, 3, \ldots$, the central difference approximation of a simply supported end condition at grid point 1 is given by $W_{-1} = W_2$.
  8. For a beam with grid points $-1, 1, 2, 3, \ldots$, the central difference approximation of $\dfrac{d^2W}{dx^2}\bigg|_1 = 0$ yields $W_2 - 2W_1 + W_{-1} = 0$.

11.3 Fill in each of the following blanks with the appropriate word:

  1. Numerical methods are to be used when the equations of motion cannot be solved in _____ form.
  2. In finite difference methods, approximations are used for _____.
  3. Finite difference equations can be derived using _____ different approaches.

    4. In finite difference methods, the solution domain is to be replaced by _____ points.

    5. The finite difference approximations are based on _____ series expansion.

    6. Numerical methods that require the use of a time step (Δt) smaller than a critical value (Δt_{cri}) are said to be _____ stable.

    7. In a conditionally stable method, the use of Δt larger than Δt_{cri} makes the method _____.

    8. A _____ formula permits the computation of x_i from known values of x_{i-1}.

11.4 Select the most appropriate answer out of the choices given:

    1. The central difference approximation of dx/dt at t_i is given by

        (a) $\dfrac{1}{2h}(x_{i+1} - x_i)$

        (b) $\dfrac{1}{2h}(x_i - x_{i-1})$

        (c) $\dfrac{1}{2h}(x_{i+1} - x_{i-1})$

    2. The central difference approximation of d^2x/dt^2 at t_i is given by

        (a) $\dfrac{1}{h^2}(x_{i+1} - 2x_i + x_{i-1})$

        (b) $\dfrac{1}{h^2}(x_{i+1} - x_{i-1})$

        (c) $\dfrac{1}{h^2}(x_i - x_{i-1})$

    3. An integration method in which the computation of x_{i+1} is based on the equilibrium equation at t_i is known as

        (a) explicit method

        (b) implicit method

        (c) regular method

    4. In a non–self-starting method, we need to generate the value of the following quantity using the finite difference approximations of $\dot{x}_i$ and $\ddot{x}_i$:

        (a) $\dot{x}_{-1}$ (b) $\ddot{x}_{-1}$ (c) x_{-1}

    5. Runge-Kutta methods find the approximate solutions of

        (a) algebraic equations

        (b) differential equations

        (c) matrix equations

    6. The finite difference approximation of $d^2U/dx^2 + \alpha^2 U = 0$ at x_i is given by

        (a) $U_{i+1} - (2 - h^2\alpha^2)U_i + U_{i-1} = 0$

        (b) $U_{i+1} - 2U_i + U_{i+1} = 0$

        (c) $U_{i+1} - (2 - \alpha^2)U_i + U_{i-1} = 0$

    7. The finite difference method requires the use of finite difference approximations in

        (a) governing differential equation only

        (b) boundary conditions only

        (c) governing differential equation as well as boundary conditions

    8. If a bar under longitudinal vibration is fixed at node 1, the forward difference formula gives

        (a) $U_1 = 0$ (b) $U_1 = U_2$ (c) $U_1 = U_{-1}$

9. If a bar under longitudinal vibration is free at node 1, the forward difference formula gives
   (a) $U_1 = 0$ (b) $U_1 = U_2$ (c) $U_1 = U_{-1}$

10. The central difference approximation of $d^4W/dx^4 - \beta^4 W = 0$ at grid point i with step size h is
    (a) $W_{i+2} - 4W_{i+1} + (6 - h^4\beta^4)W_i - 4W_{i-1} + W_{i-2} = 0$
    (b) $W_{i+2} - 6W_{i+1} + (6 - h^4\beta^4)W_i - 6W_{i-1} + W_{i-2} = 0$
    (c) $W_{i+3} - 4W_{i+1} + (6 - h^4\beta^4)W_i - 4W_{i-1} + W_{i-3} = 0$

11.5 Match the items in the two columns below:

1. Houbolt method	(a) Assumes that acceleration varies linearly between t_i and $t_i + \theta \, \Delta t$; $\theta \geq 1$
2. Wilson method	(b) Assumes that acceleration varies linearly between t_i and t_{i+1}; can lead to negative damping
3. Newmark method	(c) Based on the solution of equivalent system of first-order equations
4. Runge-Kutta method	(d) Same as Wilson method with $\theta = 1$
5. Finite difference method	(e) Uses finite difference expressions for $\dot{x}_{i+1}$ and $\ddot{x}_{i+1}$ in terms of x_{i-2}, x_{i-1}, x_i and x_{i+1}
6. Linear acceleration method	(f) Conditionally stable

PROBLEMS

The problem assignments are organized as follows:

Problems	Section Covered	Topic Covered
11.1–11.3	11.2	Finite difference approach
11.4–11.10	11.3	Central difference method for single degree of freedom systems
11.11–11.17, 11.22	11.4	Runge-Kutta method for single degree of freedom systems
11.18–11.20	11.5	Central difference method for multidegree of freedom systems
11.21, 11.23–11.28	11.6	Central difference method for continuous systems
11.29–11.31	11.7, 11.11	Runge-Kutta method for multidegree of freedom systems and MATLAB
11.32–11.34	11.8, 11.11	Houbolt method and MATLAB
11.35–11.37	11.9	Wilson method
11.38–11.40	11.10	Newmark method
11.41–11.42	11.11	MATLAB programs
11.43–11.45	11.12	C++ programs
11.46–11.50	11.13	Fortran programs

11.1 The forward difference formulas make use of the values of the function to the right of the base grid point. Thus the first derivative at point $i(t = t_i)$ is defined as

$$\frac{dx}{dt} = \frac{x(t + \Delta t) - x(t)}{\Delta t} = \frac{x_{i+1} - x_i}{\Delta t}$$

Derive the forward difference formulas for $(d^2x)/(dt^2)$, $(d^3x)/(dt^3)$, and $(d^4x)/(dt^4)$ at t_i.

11.2 The backward difference formulas make use of the values of the function to the left of the base grid point. Accordingly, the first derivative at point $i(t = t_i)$ is defined as

$$\frac{dx}{dt} = \frac{x(t) - x(t - \Delta t)}{\Delta t} = \frac{x_i - x_{i-1}}{\Delta t}$$

Derive the backward difference formulas for $(d^2x)/(dt^2)$, $(d^3x)/(dt^3)$, and $(d^4x)/(dt^4)$ at t_i.

11.3 Derive the formula for the fourth derivative, $(d^4x)/(dt^4)$, according to the central difference method.

11.4 Find the free vibratory response of an undamped single degree of freedom system with $m = 1$ and $k = 1$, using the central difference method. Assume $x_0 = 0$ and $\dot{x}_0 = 1$. Compare the results obtained with $\Delta t = 1$ and $\Delta t = 0.5$ with the exact solution $x(t) = \sin t$.

11.5 Integrate the differential equation

$$-\frac{d^2x}{dt^2} + 0.1x = 0 \qquad \text{for} \qquad 0 \le t \le 10$$

using the backward difference formula with $\Delta t = 1$. Assume the initial conditions as $x_0 = 1$ and $\dot{x}_0 = 0$.

11.6 Find the free vibration response of a viscously damped single degree of freedom system with $m = k = c = 1$, using the central difference method. Assume that $x_0 = 0$, $\dot{x}_0 = 1$, and $\Delta t = 0.5$.

11.7 Solve Problem 11.6 by changing c to 2.

11.8 Solve Problem 11.6 by taking the value of c as 4.

11.9 Find the solution of the equation $4\ddot{x} + 2\dot{x} + 3000x = F(t)$, where $F(t)$ is as shown in Fig. 11.10 for the duration $0 \le t \le 1$. Assume that $x_0 = \dot{x}_0 = 0$ and $\Delta t = 0.05$.

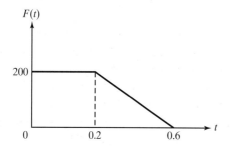

FIGURE 11.10

11.10 Find the solution of a spring-mass-damper system governed by the equation $m\ddot{x} + c\dot{x} + kx = F(t) = \delta F \cdot t$ with $m = c = k = 1$ and $\delta F = 1$. Assume the initial values of x and $\dot{x}$ to be zero and $\Delta t = 0.5$. Compare the central difference solution with the exact solution given in Example 4.9.

11.11 Express the following nth order differential equation as a system of n first-order differential equations:

$$a_n \frac{d^n x}{dt^n} + a_{n-1} \frac{d^{n-1} x}{dt^{n-1}} + \cdots + a_1 \frac{dx}{dt} = g(x, t)$$

11.12 Find the solution of the following equations by using the fourth-order Runge-Kutta method with $\Delta t = 0.1$:

(a) $\dot{x} = x - 1.5 e^{-0.5t}; x_0 = 1$

(b) $\dot{x} = -tx^2; x_0 = 1.$

11.13 The second-order Runge-Kutta formula is given by

$$\vec{X}_{i+1} = \vec{X}_i + \tfrac{1}{2}(\vec{K}_1 + \vec{K}_2)$$

where

$$\vec{K}_1 = h\vec{F}(\vec{X}_i, t_i) \qquad \text{and} \qquad \vec{K}_2 = h\vec{F}(\vec{X}_i + \vec{K}_1, t_i + h)$$

Using this formula, solve the problem considered in Example 11.2.

11.14 The third-order Runge-Kutta formula is given by

$$\vec{X}_{i+1} = \vec{X}_i + \tfrac{1}{6}(\vec{K}_1 + 4\vec{K}_2 + \vec{K}_3)$$

where

$$\vec{K}_1 = h\vec{F}(\vec{X}_i, t_i)$$

$$\vec{K}_2 = h\vec{F}(\vec{X}_i + \tfrac{1}{2}\vec{K}_1, t_i + \tfrac{1}{2}h)$$

and

$$\vec{K}_3 = h\vec{F}(\vec{X}_i - \vec{K}_1 + 2\vec{K}_2, t_i + h)$$

Using this formula, solve the problem considered in Example 11.2.

11.15 Using the second-order Runge-Kutta method, solve the differential equation $\ddot{x} + 1000x = 0$ with the initial conditions $x_0 = 5$ and $\dot{x}_0 = 0$. Use $\Delta t = 0.01$.

11.16 Using the third-order Runge-Kutta method, solve Problem 11.15.

11.17 Using the fourth-order Runge-Kutta method, solve Problem 11.15.

11.18 Using the central difference method, find the response of the two degree of freedom system shown in Fig. 11.2 when $c = 2$, $F_1(t) = 0$, $F_2(t) = 10$.

11.19 Using the central difference method, find the response of the system shown in Fig. 11.2 when $F_1(t) = 10 \sin 5t$ and $F_2(t) = 0$.

11.20 The equations of motion of a two degree of freedom system are given by $2\ddot{x}_1 + 6x_1 - 2x_2 = 5$ and $\ddot{x}_2 - 2x_1 + 4x_2 = 20\sin 5t$. Assuming the initial conditions as $x_1(0) = \dot{x}_1(0) = x_2(0) = \dot{x}_2(0) = 0$, find the response of the system, using the central difference method with $\Delta t = 0.25$.

11.21 The ends of a beam are elastically restrained by linear and torsional springs, as shown in Fig. 11.11. Using the finite difference method, express the boundary conditions.

11.22 Using the fourth-order Runge-Kutta method, solve Problem 11.20.

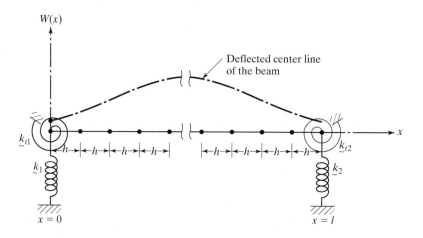

$W(x)$

Deflected center line of the beam

k_{t1} k_1 $x = 0$

k_{t2} k_2 $x = l$

FIGURE 11.11

11.23 Find the natural frequencies of a fixed-fixed bar undergoing longitudinal vibration, using three mesh points in the range $0 < x < l$.

11.24 Derive the finite difference equations governing the forced longitudinal vibration of a fixed-free uniform bar, using a total of n mesh points. Find the natural frequencies of the bar, using $n = 4$.

11.25 Derive the finite difference equations for the forced vibration of a fixed-fixed uniform shaft under torsion, using a total of n mesh points.

11.26 Find the first three natural frequencies of a uniform fixed-fixed beam.

11.27 Derive the finite difference equations for the forced vibration of a cantilever beam subjected to a transverse force $f(x, t) = f_0 \cos \omega t$ at the free end.

11.28 Derive the finite difference equations for the forced vibration analysis of a rectangular membrane, using m and n mesh points in the x and y directions, respectively. Assume the membrane to be fixed along all the edges. Use the central difference formula.

11.29 Using **Program14.m** (fourth-order Runge-Kutta method), solve Problem 11.18 with $c = 1$.

11.30 Using **Program14.m** (fourth-order Runge-Kutta method), solve Problem 11.19.

11.31 Using **Program15.m** (central difference method), solve Problem 11.20.

11.32 Using **Program15.m** (central difference method), solve Problem 11.18 with $c = 1$.

11.33 Using `Program16.m` (Houbolt method), solve Problem 11.19.

11.34 Using `Program16.m` (Houbolt method), solve Problem 11.20.

11.35 Using the Wilson method with $\theta = 1.4$, solve Problem 11.18.

11.36 Using the Wilson method with $\theta = 1.4$, solve Problem 11.19.

11.37 Using the Wilson method with $\theta = 1.4$, solve Problem 11.20.

11.38 Using the Newmark method with $\alpha = \frac{1}{6}$ and $\beta = \frac{1}{2}$, solve Problem 11.18.

11.39 Using the Newmark method with $\alpha = \frac{1}{6}$ and $\beta = \frac{1}{2}$, solve Problem 11.19.

11.40 Using the Newmark method with $\alpha = \frac{1}{6}$ and $\beta = \frac{1}{2}$, solve Problem 11.20.

11.41 Using MATLAB function `ode23`, solve the differential equation $5\ddot{x} + 4\dot{x} + 3x = 6 \sin t$ with $x(0) = \dot{x}(0) = 0$

11.42 The equations of motion of a two degree of freedom system are given by

$$\begin{bmatrix} 2 & 0 \\ 0 & 4 \end{bmatrix} \begin{Bmatrix} \ddot{x}_1 \\ \ddot{x}_2 \end{Bmatrix} + 5 \begin{bmatrix} 2 & -1 \\ -1 & 3 \end{bmatrix} \begin{Bmatrix} x_1 \\ x_2 \end{Bmatrix} = \begin{Bmatrix} F_1(t) \\ 0 \end{Bmatrix}$$

where $F_1(t)$ denotes a rectangular pulse of magnitude 5 acting over $0 \leq t \leq 2$. Find the solution of the equations using MATLAB.

11.43 Using `Program14.cpp`, solve Problem 11.18 with $c = 1$.

11.44 Using `Program15.cpp`, solve Problem 11.19.

11.45 Using `Program16.cpp`, solve Problem 11.20.

11.46 Using `PROGRAM14.F`, solve Problem 11.18 with $c = 1$.

11.47 Using `PROGRAM15.F`, solve Problem 11.19.

11.48 Using `PROGRAM16.F`, solve Problem 11.20.

11.49 Write a subroutine **WILSON** for implementing the Wilson method. Use this program to find the solution of Example 11.7.

11.50 Write a subroutine **NUMARK** for implementing the Newmark method. Use this subroutine to find the solution of Example 11.8.

Aurel Boreslav Stodola (1859–1942) was a Swiss engineer who joined the Swiss Federal Institute of Technology in Zurich in 1892 to occupy the chair of thermal machinery. He worked in several areas including machine design, automatic controls, thermodynamics, rotor dynamic, and steam turbines. He published one of the most outstanding books, namely, *Die Dampfturbin*, at the turn of the century. This book discussed not only the thermodynamic issues involved in turbine design but also the aspects of fluid flow, vibration, stress analysis of plates, shells and rotating discs, thermal stresses and stress concentrations at holes and fillets, and was translated into many languages. The approximate method he presented for the computation of natural frequencies of beams has become known as the Stodola method. (Photo courtesy of *Applied Mechanics Reviews*.)

CHAPTER 12

Finite Element Method

12.1 Introduction

The finite element method is a numerical method that can be used for the accurate solution of complex mechanical and structural vibration problems [12.1, 12.2]. In this method, the actual structure is replaced by several pieces or elements, each of which is assumed to behave as a continuous structural member called a *finite element*. The elements are assumed to be interconnected at certain points known as *joints* or *nodes*. Since it is very difficult to find the exact solution (such as the displacements) of the original structure under the specified loads, a convenient approximate solution is assumed in each finite element. The idea is that if the solutions of the various elements are selected properly, they can be made to converge to the exact solution of the total structure as the element size is reduced. During the solution process, the equilibrium of forces at the joints and the compatibility of displacements between the elements are satisfied so that the entire structure (assemblage of elements) is made to behave as a single entity.

The basic procedure of the finite element method, with application to simple vibration problems, is presented in this chapter. The element stiffness and mass matrices, and force vectors are derived for a bar element, a torsion element, and a beam element. The transformation of element matrices and vectors from the local to the global coordinate system is presented. The equations of motion of the complete system of finite elements and the incorporation of the boundary conditions are discussed. The concepts of consistent and lumped mass matrices are presented along with a numerical example. Finally, a computer program for the eigenvalue analysis of stepped beams is presented. Although the techniques presented in this chapter can be applied to more complex problems involving two- and three-dimensional finite elements, only the use of one-dimensional elements is considered in the numerical treatment.

12.2 Equations of Motion of an Element

For illustration, the finite element model of a plano-milling machine structure (Fig. 12.1a) is shown in Fig. 12.1(b). In this model, the columns and the overarm are represented by triangular plate elements and the cross slide and the tool holder are represented by beam elements [12.3]. The elements are assumed to be connected to each other only at the joints. The displacement within an element is expressed in terms of the displacements at the corners or joints of the element. In Fig. 12.1(b), the transverse displacement within a typical element e is assumed to be $w(x, y, t)$. The values of w, $(\partial w)/(\partial x)$, and $(\partial w)/(\partial y)$ at joints 1, 2, and 3—namely $w(x_1\ y_1, t)$, $(\partial w)/(\partial x)(x_1,\ y_1, t)$, $(\partial w)/(\partial y)(x_1, y_1, t)$, $\ldots$, $(\partial w)/(\partial y)(x_3, y_3, t)$—are treated as unknowns and are denoted as $w_1(t)$, $w_2(t)$, $w_3(t)$, $\ldots$, $w_9(t)$. The displacement $w(x, y, t)$ can be expressed in terms of the unknown joint displacements $w_i(t)$ in the form

$$w(x, y, t) = \sum_{i=1}^{n} N_i(x, y) w_i(t) \tag{12.1}$$

where $N_i(x, y)$ is called the *shape function* corresponding to the joint displacement $w_i(t)$ and n is the number of unknown joint displacements ($n = 9$ in Fig. 12.1b). If a distributed

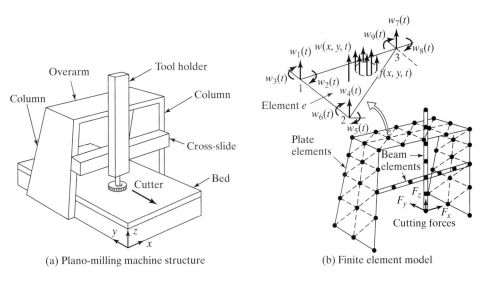

(a) Plano-milling machine structure   (b) Finite element model

FIGURE 12.1   Finite element modeling.

load $f(x, y, t)$ acts on the element, it can be converted into equivalent joint forces $f_i(t) (i = 1, 2, \ldots, 9)$. If concentrated forces act at the joints, they can also be added to the appropriate joint force $f_i(t)$. We shall now derive the equations of motion for determining the joint displacements $w_i(t)$ under the prescribed joint forces $f_i(t)$. By using Eq. (12.1), the kinetic energy T and the strain energy V of the element can be expressed as

$$T = \frac{1}{2} \vec{\dot{W}}^T [m] \vec{\dot{W}} \tag{12.2}$$

$$V = \frac{1}{2} \vec{W}^T [k] \vec{W} \tag{12.3}$$

where

$$\vec{W} = \begin{Bmatrix} w_1(t) \\ w_2(t) \\ \cdot \\ \cdot \\ \cdot \\ w_n(t) \end{Bmatrix}, \qquad \vec{\dot{W}} = \begin{Bmatrix} \dot{w}_1(t) \\ \dot{w}_2(t) \\ \cdot \\ \cdot \\ \cdot \\ \dot{w}_n(t) \end{Bmatrix} = \begin{Bmatrix} dw_1/dt \\ dw_2/dt \\ \cdot \\ \cdot \\ \cdot \\ dw_n/dt \end{Bmatrix}$$

and $[m]$ and $[k]$ are the mass and stiffness matrices of the element. By substituting Eqs. (12.2) and (12.3) into Lagrange's equations, Eq. (6.44), the equations of motion of the finite element can be obtained as

$$[m]\vec{\ddot{W}} + [k]\vec{W} = \vec{f} \tag{12.4}$$

where $\vec{f}$ is the vector of joint forces and $\vec{\ddot{W}}$ is the vector of joint accelerations given by

$$\vec{\ddot{W}} = \begin{Bmatrix} \ddot{w}_1 \\ \ddot{w}_2 \\ \cdot \\ \cdot \\ \cdot \\ \ddot{w}_n \end{Bmatrix} = \begin{Bmatrix} d^2w_1/dt^2 \\ d^2w_2/dt^2 \\ \cdot \\ \cdot \\ \cdot \\ d^2w_n/dt^2 \end{Bmatrix}$$

Note that the shape of the finite elements and the number of unknown joint displacements may differ for different applications. Although the equations of motion of a single element, Eq. (12.4), are not useful directly (as our interest lies in the dynamic response of the assemblage of elements), the mass matrix $[m]$, the stiffness matrix $[k]$, and the joint force vector $\vec{f}$ of individual elements are necessary for the final solution. We shall derive the element mass and stiffness matrices and the joint force vectors for some simple one-dimensional elements in the next section.

12.3   Mass Matrix, Stiffness Matrix, and Force Vector

12.3.1
Bar Element

Consider the uniform bar element shown in Fig. 12.2. For this one-dimensional element, the two end points form the joints (nodes). When the element is subjected to axial loads $f_1(t)$ and $f_2(t)$, the axial displacement within the element is assumed to be linear in x as

$$u(x, t) = a(t) + b(t)x \tag{12.5}$$

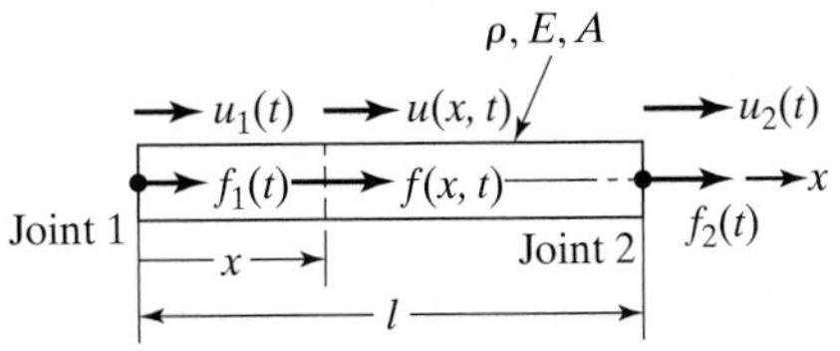

FIGURE 12.2   Uniform bar element.

When the joint displacements $u_1(t)$ and $u_2(t)$ are treated as unknowns, Eq. (12.5) should satisfy the conditions

$$u(0, t) = u_1(t), \qquad u(l, t) = u_2(t) \tag{12.6}$$

Equations (12.5) and (12.6) lead to

$$a(t) = u_1(t)$$

and

$$a(t) + b(t)l = u_2(t) \qquad \text{or} \qquad b(t) = \frac{u_2(t) - u_1(t)}{l} \tag{12.7}$$

Substitution for $a(t)$ and $b(t)$ from Eq. (12.7) into Eq. (12.5) gives

$$u(x, t) = \left(1 - \frac{x}{l}\right) u_1(t) + \frac{x}{l} u_2(t) \tag{12.8}$$

or

$$u(x, t) = N_1(x) u_1(t) + N_2(x) u_2(t) \tag{12.9}$$

where

$$N_1(x) = \left(1 - \frac{x}{l}\right), \qquad N_2(x) = \frac{x}{l} \tag{12.10}$$

are the shape functions.

The kinetic energy of the bar element can be expressed as

$$
\begin{aligned}
T(t) &= \frac{1}{2} \int_0^l \rho A \left\{ \frac{\partial u(x, t)}{\partial t} \right\}^2 dx \\
&= \frac{1}{2} \int_0^l \rho A \left\{ \left(1 - \frac{x}{l}\right) \frac{du_1(t)}{dt} + \left(\frac{x}{l}\right) \frac{du_2(t)}{dt} \right\}^2 dx \\
&= \frac{1}{2} \frac{\rho A l}{3} (\dot{u}_1^2 + \dot{u}_1 \dot{u}_2 + \dot{u}_2^2) \tag{12.11}
\end{aligned}
$$

where

$$\dot{u}_1 = \frac{du_1(t)}{dt}, \qquad \dot{u}_2 = \frac{du_2(t)}{dt},$$

ρ is the density of the material and A is the cross-sectional area of the element.
By expressing Eq. (12.11) in matrix form,

$$T(t) = \frac{1}{2}\dot{\vec{u}}(t)^T [m] \dot{\vec{u}}(t) \tag{12.12}$$

where

$$\dot{\vec{u}}(t) = \begin{Bmatrix} \dot{u}_1(t) \\ \dot{u}_2(t) \end{Bmatrix}$$

and the superscript T indicates the transpose, the mass matrix $[m]$ can be identified as

$$[m] = \frac{\rho A l}{6}\begin{bmatrix} 2 & 1 \\ 1 & 2 \end{bmatrix} \tag{12.13}$$

The strain energy of the element can be written as

$$V(t) = \frac{1}{2}\int_0^l EA\left\{\frac{\partial u(x,t)}{\partial x}\right\}^2 dx$$

$$= \frac{1}{2}\int_0^l EA\left\{-\frac{1}{l}u_1(t) + \frac{1}{l}u_2(t)\right\}^2 dx$$

$$= \frac{1}{2}\frac{EA}{l}(u_1^2 - 2u_1 u_2 + u_2^2) \tag{12.14}$$

where $u_1 = u_1(t)$, $u_2 = u_2(t)$, and E is Young's modulus. By expressing Eq. (12.14) in matrix form as

$$V(t) = \frac{1}{2}\vec{u}(t)^T [k] \vec{u}(t) \tag{12.15}$$

where

$$\vec{u}(t) = \begin{Bmatrix} u_1(t) \\ u_2(t) \end{Bmatrix} \qquad \text{and} \qquad \vec{u}(t)^T = \{u_1(t)\ u_2(t)\}$$

the stiffness matrix $[k]$ can be identified as

$$[k] = \frac{EA}{l}\begin{bmatrix} 1 & -1 \\ -1 & 1 \end{bmatrix} \tag{12.16}$$

The force vector

$$\vec{f} = \begin{Bmatrix} f_1(t) \\ f_2(t) \end{Bmatrix}$$

can be derived from the virtual work expression. If the bar is subjected to the distributed force $f(x, t)$, the virtual work δW can be expressed as

$$
\begin{aligned}
\delta W(t) &= \int_0^l f(x, t)\, \delta u(x, t)\, dx \\
&= \int_0^l f(x, t) \left\{ \left(1 - \frac{x}{l}\right) \delta u_1(t) + \left(\frac{x}{l}\right) \delta u_2(t) \right\} dx \\
&= \left(\int_0^l f(x, t) \left(1 - \frac{x}{l}\right) dx \right) \delta u_1(t) \\
&\quad + \left(\int_0^l f(x, t) \left(\frac{x}{l}\right) dx \right) \delta u_2(t)
\end{aligned}
\tag{12.17}
$$

By expressing Eq. (12.17) in matrix form as

$$\delta W(t) = \delta \vec{u}(t)^T \vec{f}(t) \equiv f_1(t)\, \delta u_1(t) + f_2(t)\, \delta u_2(t) \tag{12.18}$$

the equivalent joint forces can be identified as

$$
\left.
\begin{aligned}
f_1(t) &= \int_0^l f(x, t) \left(1 - \frac{x}{l}\right) dx \\
f_2(t) &= \int_0^l f(x, t) \left(\frac{x}{l}\right) dx
\end{aligned}
\right\}
\tag{12.19}
$$

**12.3.2
Torsion Element**

Consider a uniform torsion element with the x axis taken along the centroidal axis, as shown in Fig. 12.3. Let I_p denote the polar moment of inertia about the centroidal axis and

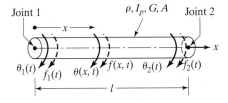

FIGURE 12.3   Uniform torsion element.

GJ represent the torsional stiffness ($J = I_p$ for a circular cross section). When the torsional displacement (rotation) within the element is assumed to be linear in x as

$$\theta(x, t) = a(t) + b(t)x \tag{12.20}$$

and the joint rotations $\theta_1(t)$ and $\theta_2(t)$ are treated as unknowns, Eq. (12.20) can be expressed, by proceeding as in the case of a bar element, as

$$\theta(x, t) = N_1(x)\theta_1(t) + N_2(x)\theta_2(t) \tag{12.21}$$

where $N_1(x)$ and $N_2(x)$ are the same as in Eq. (12.10). The kinetic energy, the strain energy, and the virtual work for pure torsion are given by

$$T(t) = \frac{1}{2} \int_0^l \rho I_p \left\{ \frac{\partial\theta(x, t)}{\partial t} \right\}^2 dx \tag{12.22}$$

$$V(t) = \frac{1}{2} \int_0^l GJ \left\{ \frac{\partial\theta(x, t)}{\partial x} \right\}^2 dx \tag{12.23}$$

$$\delta W(t) = \int_0^l f(x, t)\, \delta\theta(x, t)\, dx \tag{12.24}$$

where ρ is the mass density and $f(x, t)$ is the distributed torque per unit length. Using the procedures employed in Section 12.3.1, we can derive the element mass and stiffness matrices and the force vector:

$$[m] = \frac{\rho I_p l}{6} \begin{bmatrix} 2 & 1 \\ 1 & 2 \end{bmatrix} \tag{12.25}$$

$$[k] = \frac{GJ}{l} \begin{bmatrix} 1 & -1 \\ -1 & 1 \end{bmatrix} \tag{12.26}$$

$$\vec{f} = \begin{Bmatrix} f_1(t) \\ f_2(t) \end{Bmatrix} = \begin{Bmatrix} \int_0^l f(x, t)\left(1 - \frac{x}{l}\right) dx \\ \int_0^l f(x, t)\left(\frac{x}{l}\right) dx \end{Bmatrix} \tag{12.27}$$

12.3.3
Beam Element

We now consider a beam element according to the Euler-Bernoulli theory.[1] Figure 12.4 shows a uniform beam element subjected to the transverse force distribution $f(x, t)$. In this case, the joints undergo both translational and rotational displacements, so the unknown joint displacements are labeled as $w_1(t)$, $w_2(t)$, $w_3(t)$, and $w_4(t)$. Thus there will be linear joint forces $f_1(t)$ and $f_3(t)$ corresponding to the linear joint displacements $w_1(t)$ and $w_3(t)$ and rotational joint forces (bending moments) $f_2(t)$ and $f_4(t)$ corresponding to the rotational

---

[1]The beam element, according to the Timoshenko theory, was considered in Refs. [12.4–12.7].

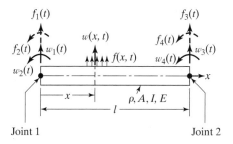

FIGURE 12.4   Uniform beam element.

joint displacements $w_2(t)$ and $w_4(t)$, respectively. The transverse displacement within the element is assumed to be a cubic equation in x (as in the case of static deflection of a beam):

$$w(x, t) = a(t) + b(t)x + c(t)x^2 + d(t)x^3 \tag{12.28}$$

The unknown joint displacements must satisfy the conditions

$$\left.\begin{array}{ll} w(0, t) = w_1(t), & \dfrac{\partial w}{\partial x}(0, t) = w_2(t) \\[2mm] w(l, t) = w_3(t), & \dfrac{\partial w}{\partial x}(l, t) = w_4(t) \end{array}\right\} \tag{12.29}$$

Equations (12.28) and (12.29) yield

$$a(t) = w_1(t)$$
$$b(t) = w_2(t)$$
$$c(t) = \frac{1}{l^2}[-3w_1(t) - 2w_2(t)l + 3w_3(t) - w_4(t)l]$$
$$d(t) = \frac{1}{l^3}[2w_1(t) + w_2(t)l - 2w_3(t) + w_4(t)l] \tag{12.30}$$

By substituting Eqs. (12.30) into Eq. (12.28), we can express $w(x, t)$ as

$$w(x, t) = \left(1 - 3\frac{x^2}{l^2} + 2\frac{x^3}{l^3}\right)w_1(t) + \left(\frac{x}{l} - 2\frac{x^2}{l^2} + \frac{x^3}{l^3}\right)lw_2(t)$$

$$+ \left(3\frac{x^2}{l^2} - 2\frac{x^3}{l^3}\right)w_3(t) + \left(-\frac{x^2}{l^2} + \frac{x^3}{l^3}\right)lw_4(t) \tag{12.31}$$

This equation can be rewritten as

$$w(x, t) = \sum_{i=1}^{4} N_i(x)w_i(t) \tag{12.32}$$

where $N_i(x)$ are the shape functions given by

$$N_1(x) = 1 - 3\left(\frac{x}{l}\right)^2 + 2\left(\frac{x}{l}\right)^3 \tag{12.33}$$

$$N_2(x) = x - 2l\left(\frac{x}{l}\right)^2 + l\left(\frac{x}{l}\right)^3 \tag{12.34}$$

$$N_3(x) = 3\left(\frac{x}{l}\right)^2 - 2\left(\frac{x}{l}\right)^3 \tag{12.35}$$

$$N_4(x) = -l\left(\frac{x}{l}\right)^2 + l\left(\frac{x}{l}\right)^3 \tag{12.36}$$

The kinetic energy, bending strain energy, and virtual work of the element can be expressed as

$$T(t) = \frac{1}{2}\int_0^l \rho A\left\{\frac{\partial w(x, t)}{\partial t}\right\}^2 dx \equiv \frac{1}{2}\dot{\vec{w}}(t)^T [m]\dot{\vec{w}}(t) \tag{12.37}$$

$$V(t) = \frac{1}{2}\int_0^l EI\left\{\frac{\partial^2 w(x, t)}{\partial x^2}\right\}^2 dx \equiv \frac{1}{2}\vec{w}(t)^T [k]\vec{w}(t) \tag{12.38}$$

$$\delta W(t) = \int_0^l f(x, t)\,\delta w(x, t)\,dx \equiv \delta\vec{w}(t)^T \vec{f}(t) \tag{12.39}$$

where ρ is the density of the beam, E is Young's modulus, I is the moment of inertia of the cross section, A is the area of cross section, and

$$\vec{w}(t) = \begin{Bmatrix} w_1(t) \\ w_2(t) \\ w_3(t) \\ w_4(t) \end{Bmatrix}, \qquad \dot{\vec{w}}(t) = \begin{Bmatrix} dw_1/dt \\ dw_2/dt \\ dw_3/dt \\ dw_4/dt \end{Bmatrix}$$

$$\delta\vec{w}(t) = \begin{Bmatrix} \delta w_1(t) \\ \delta w_2(t) \\ \delta w_3(t) \\ \delta w_4(t) \end{Bmatrix}, \qquad \vec{f}(t) = \begin{Bmatrix} f_1(t) \\ f_2(t) \\ f_3(t) \\ f_4(t) \end{Bmatrix}$$

By substituting Eq. (12.31) into Eqs. (12.37) to (12.39) and carrying out the necessary integrations, we obtain

$$[m] = \frac{\rho A l}{420}\begin{bmatrix} 156 & 22l & 54 & -13l \\ 22l & 4l^2 & 13l & -3l^2 \\ 54 & 13l & 156 & -22l \\ -13l & -3l^2 & -22l & 4l^2 \end{bmatrix} \tag{12.40}$$

$$[k] = \frac{EI}{l^3} \begin{bmatrix} 12 & 6l & -12 & 6l \\ 6l & 4l^2 & -6l & 2l^2 \\ -12 & -6l & 12 & -6l \\ 6l & 2l^2 & -6l & 4l^2 \end{bmatrix}$$

(12.41)

$$f_i(t) = \int_0^l f(x, t) \, N_i(x) \, dx, \qquad i = 1, 2, 3, 4$$

(12.42)

12.4   Transformation of Element Matrices and Vectors

As stated earlier, the finite element method considers the given dynamical system as an assemblage of elements. The joint displacements of an individual element are selected in a convenient direction, depending on the nature of the element. For example, for the bar element shown in Fig. 12.2, the joint displacements $u_1(t)$ and $u_2(t)$ are chosen along the axial direction of the element. However, other bar elements can have different orientations in an assemblage, as shown in Fig. 12.5. Here x denotes the axial direction of an individual element and is called a *local coordinate axis*. If we use $u_1(t)$ and $u_2(t)$ to denote the joint displacements of different bar elements, there will be one joint displacement at joint 1, three at joint 2, two at joint 3, and 2 at joint 4. However, the displacements of joints can be specified more conveniently using reference or global coordinate axes X and Y. Then the displacement components of joints parallel to the X and Y axes can be used as the joint displacements in the global coordinate system. These are shown as $U_i(t), i = 1, 2, \ldots, 8$ in Fig. 12.5. The joint displacements in the local and the global coordinate system for a

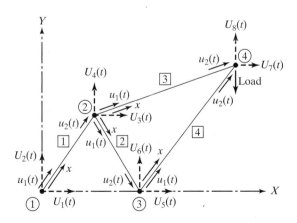

FIGURE 12.5   A dynamical system (truss) idealized as an assemblage of four bar elements.

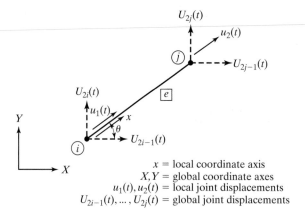

$$x = \text{local coordinate axis}$$
$$X, Y = \text{global coordinate axes}$$
$$u_1(t), u_2(t) = \text{local joint displacements}$$
$$U_{2i-1}(t), \dots, U_{2j}(t) = \text{global joint displacements}$$

FIGURE 12.6   Local and global joint displacements of element e.

typical bar element e are shown in Fig. 12.6. The two sets of joint displacements are related as follows:

$$u_1(t) = U_{2i-1}(t) \cos \theta + U_{2i}(t) \sin \theta$$
$$u_2(t) = U_{2j-1}(t) \cos \theta + U_{2j}(t) \sin \theta \qquad (12.43)$$

These can be rewritten as

$$\vec{u}(t) = [\lambda]\vec{U}(t) \qquad (12.44)$$

where $[\lambda]$ is the coordinate transformation matrix given by

$$[\lambda] = \begin{bmatrix} \cos \theta & \sin \theta & 0 & 0 \\ 0 & 0 & \cos \theta & \sin \theta \end{bmatrix} \qquad (12.45)$$

and $\vec{u}(t)$ and $\vec{U}(t)$ are the vectors of joint displacements in the local and the global coordinate system, respectively, and are given by

$$\vec{u}(t) = \begin{Bmatrix} u_1(t) \\ u_2(t) \end{Bmatrix}, \qquad \vec{U}(t) = \begin{Bmatrix} U_{2i-1}(t) \\ U_{2i}(t) \\ U_{2j-1}(t) \\ U_{2j}(t) \end{Bmatrix}$$

It is useful to express the mass matrix, stiffness matrix, and joint force vector of an element in terms of the global coordinate system while finding the dynamical response of the complete system. Since the kinetic and strain energies of the element must be independent of the coordinate system, we have

$$T(t) = \frac{1}{2}\dot{\vec{u}}(t)^T [m] \dot{\vec{u}}(t) = \frac{1}{2}\dot{\vec{U}}(t)^T [\overline{m}]\dot{\vec{U}}(t) \tag{12.46}$$

$$V(t) = \frac{1}{2}\vec{u}(t)^T [k]\vec{u}(t) = \frac{1}{2}\vec{U}(t)^T [\overline{k}]\vec{U}(t) \tag{12.47}$$

where $[\overline{m}]$ and $[\overline{k}]$ denote the element mass and stiffness matrices, respectively, in the global coordinate system and $\vec{U}(t)$ is the vector of joint velocities in the global coordinate system, related to $\vec{u}(t)$ as in Eq. (12.44):

$$\vec{u}(t) = [\lambda]\vec{U}(t) \tag{12.48}$$

By inserting Eqs. (12.44) and (12.48) into (12.46) and (12.47), we obtain

$$T(t) = \frac{1}{2}\dot{\vec{U}}(t)^T[\lambda]^T[m][\lambda]\dot{\vec{U}}(t) \equiv \frac{1}{2}\dot{\vec{U}}(t)^T[\overline{m}]\dot{\vec{U}}(t) \tag{12.49}$$

$$V(t) = \frac{1}{2}\vec{U}(t)^T[\lambda]^T[k][\lambda]\vec{U}(t) \equiv \frac{1}{2}\vec{U}(t)^T[\overline{k}]\vec{U}(t) \tag{12.50}$$

Equations (12.49) and (12.50) yield

$$[\overline{m}] = [\lambda]^T[m][\lambda] \tag{12.51}$$

$$[\overline{k}] = [\lambda]^T[k][\lambda] \tag{12.52}$$

Similarly, by equating the virtual work in the two coordinate systems,

$$\delta W(t) = \delta\vec{u}(t)^T \vec{f}(t) = \delta\vec{U}(t)^T \vec{\overline{f}}(t) \tag{12.53}$$

we find the vector of element joint forces in the global coordinate system $\vec{\overline{f}}(t)$:

$$\vec{\overline{f}}(t) = [\lambda]^T \vec{f}(t) \tag{12.54}$$

Equations (12.51), (12.52), and (12.54) can be used to obtain the equations of motion of a single finite element in the global coordinate system:

$$[\overline{m}]\ddot{\vec{U}}(t) + [\overline{k}]\vec{U}(t) = \vec{\overline{f}}(t) \tag{12.55}$$

Although this equation is not of much use, since our interest lies in the equations of motion of an assemblage of elements, the matrices $[\overline{m}]$ and $[\overline{k}]$ and the vector $\vec{\overline{f}}$ are useful in deriving the equations of motion of the complete system, as indicated in the following section.

12.5   Equations of Motion of the Complete System of Finite Elements

Since the complete structure is considered an assemblage of several finite elements, we shall now extend the equations of motion obtained for single finite elements in the global system to the complete structure. We shall denote the joint displacements of the complete structure in the global coordinate system as $U_1(t), U_2(t), \ldots, U_M(t)$ or, equivalently, as a column vector:

$$\underset{\sim}{\vec{U}}(t) = \begin{Bmatrix} U_1(t) \\ U_2(t) \\ \cdot \\ \cdot \\ \cdot \\ U_M(t) \end{Bmatrix}$$

For convenience, we shall denote the quantities pertaining to an element e in the assemblage by the superscript e. Since the joint displacements of any element e can be identified in the vector of joint displacements of the complete structure, the vectors $\vec{U}^{(e)}(t)$ and $\underset{\sim}{\vec{U}}(t)$ are related

$$\vec{U}^{(e)}(t) = [A^{(e)}]\underset{\sim}{\vec{U}}(t) \tag{12.56}$$

where $[A^{(e)}]$ is a rectangular matrix composed of zeros and ones. For example, for element 1 in Fig. 12.5, Eq. (12.56) becomes

$$\vec{U}^{(1)}(t) \equiv \begin{Bmatrix} U_1(t) \\ U_2(t) \\ U_3(t) \\ U_4(t) \end{Bmatrix} = \begin{bmatrix} 1 & 0 & 0 & 0 & 0 & 0 & 0 & 0 \\ 0 & 1 & 0 & 0 & 0 & 0 & 0 & 0 \\ 0 & 0 & 1 & 0 & 0 & 0 & 0 & 0 \\ 0 & 0 & 0 & 1 & 0 & 0 & 0 & 0 \end{bmatrix} \begin{Bmatrix} U_1(t) \\ U_2(t) \\ \cdot \\ \cdot \\ \cdot \\ U_8(t) \end{Bmatrix} \tag{12.57}$$

The kinetic energy of the complete structure can be obtained by adding the kinetic energies of individual elements

$$T = \sum_{e=1}^{E} \frac{1}{2} \vec{\dot{U}}^{(e)T} [\overline{m}] \vec{\dot{U}}^{(e)} \tag{12.58}$$

where E denotes the number of finite elements in the assemblage. By differentiating Eq. (12.56), the relation between the velocity vectors can be derived:

$$\vec{\dot{U}}^{(e)}(t) = [A^{(e)}]\underset{\sim}{\vec{\dot{U}}}(t) \tag{12.59}$$

Substitution of Eq. (12.59) into (12.58) leads to

$$T = \frac{1}{2} \sum_{e=1}^{E} \underset{\sim}{\vec{\dot{U}}}^{T} [A^{(e)}]^{T} [\overline{m}^{(e)}] [A^{(e)}] \underset{\sim}{\vec{\dot{U}}} \tag{12.60}$$

The kinetic energy of the complete structure can also be expressed in terms of joint velocities of the complete structure $\vec{U}$

$$T = \frac{1}{2} \, \dot{\vec{U}}^{(e)T} [M] \dot{\vec{U}} \tag{12.61}$$

where $[M]$ is called the mass matrix of the complete structure. A comparison of Eqs. (12.60) and (12.61) gives the relation[2]

$$[M] = \sum_{e=1}^{E} [A^{(e)}]^T [\overline{m}^{(e)}][A^{(e)}] \tag{12.62}$$

Similarly, by considering strain energy, the stiffness matrix of the complete structure, $[K]$, can be expressed as

$$[K] = \sum_{e=1}^{E} [A^{(e)}]^T [\overline{k}^{(e)}] [A^{(e)}] \tag{12.63}$$

Finally the consideration of virtual work yields the vector of joint forces of the complete structure, $\vec{F}$:

$$\vec{F} = \sum_{e=1}^{E} [A^{(e)}]^T \vec{f}^{(e)} \tag{12.64}$$

Once the mass and stiffness matrices and the force vector are known, Lagrange's equations of motion for the complete structure can be expressed as

$$[M]\ddot{\vec{U}} + [K]\vec{U} = \vec{F} \tag{12.65}$$

Note that the joint force vector $\vec{F}$ in Eq. (12.65) was generated by considering only the distributed loads acting on the various elements. If there is any concentrated load acting along the joint displacement $U_i(t)$, it must be added to the ith component of $\vec{F}$.

12.6   Incorporation of Boundary Conditions

In the preceding derivation, no joint was assumed to be fixed. Thus the complete structure is capable of undergoing rigid body motion under the joint forces. This means that $[K]$ is a singular matrix (see Section 6.12). Usually the structure is supported such that the

---

[2]An alternative procedure can be used for the assembly of element matrices. In this procedure, each of the rows and columns of the element (mass or stiffness) matrix is identified by the corresponding degree of freedom in the assembled structure. Then the various entries of the element matrix can be placed at their proper locations in the overall (mass or stiffness) matrix of the assembled system. For example, the entry belonging to the ith row (identified by the degree of freedom p) and the jth column (identified by the degree of freedom q) of the element matrix is to be placed in the pth row and qth column of the overall matrix. This procedure is illustrated in Example 12.3.

displacements are zero at a number of joints, to avoid rigid body motion of the structure. A simple method of incorporating the zero displacement conditions is to eliminate the corresponding rows and columns from the matrices $[M]$ and $[K]$ and the vector $\vec{F}$. The final equations of motion of the restrained structure can be expressed as

$$\underset{N \times N}{[M]} \underset{N \times 1}{\ddot{\vec{U}}} + \underset{N \times N}{[K]} \underset{N \times 1}{\vec{U}} = \underset{N \times 1}{\vec{F}} \tag{12.66}$$

where N denotes the number of free joint displacements of the structure.

Note the following points concerning finite element analysis:

1. The approach used in the above presentation is called the *displacement method* of finite element analysis because it is the displacements of elements that are directly approximated. Other methods, such as the force method, the mixed method, and hybrid methods, are also available [12.8, 12.9].
2. The stiffness matrix, mass matrix, and force vector for other finite elements, including two-dimensional and three-dimensional elements, can be derived in a similar manner, provided the shape functions are known [12.1, 12.2].
3. In the Rayleigh-Ritz method discussed in Section 8.8, the displacement of the continuous system is approximated by a sum of assumed functions, where each function denotes a deflection shape of the entire structure. In the finite element method, an approximation using shape functions (similar to the assumed functions) is also used for a finite element instead of the entire structure. Thus the finite element procedure can also be considered a Rayleigh-Ritz method.
4. Error analysis of the finite element method can also be conducted [12.10].

Analysis of a Bar

EXAMPLE 12.1

Consider a uniform bar, of length 0.5 m, area of cross section 5×10^{-4} m^2, Young's modulus 200 GPa, and density 7850 kg/m^3, which is fixed at the left end, as shown in Fig. 12.7.

a. Find the stress induced in the bar under an axial static load of 1000 N applied at joint 2 along u_2.
b. Find the natural frequency of vibration of the bar.

Use a one-element idealization.

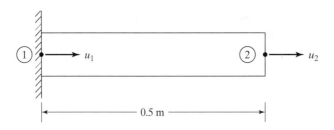

FIGURE 12.7   Uniform bar with two degrees of freedom.

Solution

a. Using the stiffness matrix of a bar element, Eq. (12.16), the equilibrium equations can be written as

$$\frac{AE}{l}\begin{bmatrix} 1 & -1 \\ -1 & 1 \end{bmatrix}\begin{Bmatrix} u_1 \\ u_2 \end{Bmatrix} = \begin{Bmatrix} f_1 \\ f_2 \end{Bmatrix} \tag{E.1}$$

With $A = 5 \times 10^{-4}$, $E = 2 \times 10^{11}$, $l = 0.5$, $f_2 = 1000$, Eq. (E.1) becomes

$$2 \times 10^8 \begin{bmatrix} 1 & -1 \\ -1 & 1 \end{bmatrix}\begin{Bmatrix} u_1 \\ u_2 \end{Bmatrix} = \begin{Bmatrix} f_1 \\ 1000 \end{Bmatrix} \tag{E.2}$$

where u_1 is the displacement and f_1 is the unknown reaction at joint 1. To incorporate the boundary condition $u_1 = 0$, we delete the first scalar equation (first row) and substitute $u_1 = 0$ in the resulting Eq. (E.2). This gives

$$2 \times 10^8 \, u_2 = 1000 \quad \text{or} \quad u_2 = 500 \times 10^{-8} \text{ m} \tag{E.3}$$

The stress (σ) - strain (ε) relation gives

$$\sigma = E\,\varepsilon = E\frac{\Delta l}{l} = E\left(\frac{u_2 - u_1}{l}\right) \tag{E.4}$$

where $\Delta l = u_2 - u_1$ denotes the change in length of the element and $\dfrac{\Delta l}{l}$ indicates the strain.

Equation (E.4) yields

$$\sigma = 2 \times 10^{11}\left(\frac{500 \times 10^{-8} - 0}{0.5}\right) = 2 \times 10^6 \text{ Pa} \tag{E.5}$$

b. Using the stiffness and mass matrices of the bar element, Eqs. (12.16) and (12.13), the eigenvalue problem can be expressed as

$$\frac{AE}{l}\begin{bmatrix} 1 & -1 \\ -1 & 1 \end{bmatrix}\begin{Bmatrix} U_1 \\ U_2 \end{Bmatrix} = \omega^2 \frac{\rho \, Al}{6}\begin{bmatrix} 2 & 1 \\ 1 & 2 \end{bmatrix}\begin{Bmatrix} U_1 \\ U_2 \end{Bmatrix} \tag{E.6}$$

where ω is the natural frequency and U_1 and U_2 are the amplitudes of vibration of the bar at joints 1 and 2, respectively. To incorporate the boundary condition $U_1 = 0$, we delete the first row and first column in each of the matrices and vectors and write the resulting equation as

$$\frac{AE}{l} U_2 = \omega^2 \frac{\rho \, Al}{6} (2)U_2$$

or

$$\omega = \sqrt{\frac{3E}{\rho l^2}} = \sqrt{\frac{3(2 \times 10^{11})}{7850 \, (0.5)^2}} = 17{,}485.2076 \text{ rad/s} \tag{E.7}$$

■

■■■■■■■■ Natural Frequencies of a Simply Supported Beam

EXAMPLE 12.2

Find the natural frequencies of the simply supported beam shown in Fig. 12.8(a) using one finite element.

Solution: Since the beam is idealized using only one element, the element joint displacements are the same in both local and global systems, as indicated in Fig. 12.8(b). The stiffness and mass matrices of the beam are given by

$$[K] = [K^{(1)}] = \frac{EI}{l^3} \begin{bmatrix} 12 & 6l & -12 & 6l \\ 6l & 4l^2 & -6l & 2l^2 \\ -12 & -6l & 12 & -6l \\ 6l & 2l^2 & -6l & 4l^2 \end{bmatrix} \tag{E.1}$$

$$[M] = [M^{(1)}] = \frac{\rho A l}{420} \begin{bmatrix} 156 & 22l & 54 & -13l \\ 22l & 4l^2 & 13l & -3l^2 \\ 54 & 13l & 156 & -22l \\ -13l & -3l^2 & -22l & 4l^2 \end{bmatrix} \tag{E.2}$$

and the vector of joint displacements by

$$\vec{W} = \begin{Bmatrix} W_1 \\ W_2 \\ W_3 \\ W_4 \end{Bmatrix} \equiv \begin{Bmatrix} w_1^{(1)} \\ w_2^{(1)} \\ w_3^{(1)} \\ w_4^{(1)} \end{Bmatrix} \tag{E.3}$$

The boundary conditions corresponding to the simply supported ends ($W_1 = 0$ and $W_3 = 0$) can be incorporated[3] by deleting the rows and columns corresponding to W_1 and W_3 in Eqs. (E.1) and (E.2).

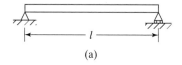

(a)

(b)

FIGURE 12.8   Simply supported beam.

---

[3]The bending moment cannot be set equal to zero at the simply supported ends explicitly, since there is no degree of freedom (joint displacement) involving the second derivative of the displacement w.

This leads to the overall matrices

$$[K] = \frac{2EI}{l}\begin{bmatrix} 2 & 1 \\ 1 & 2 \end{bmatrix} \tag{E.4}$$

$$[M] = \frac{\rho Al^3}{420}\begin{bmatrix} 4 & -3 \\ -3 & 4 \end{bmatrix} \tag{E.5}$$

and the eigenvalue problem can be written as

$$\left[\frac{2EI}{l}\begin{bmatrix} 2 & 1 \\ 1 & 2 \end{bmatrix} - \frac{\rho Al^3\omega^2}{420}\begin{bmatrix} 4 & -3 \\ -3 & 4 \end{bmatrix}\right]\begin{Bmatrix} W_2 \\ W_4 \end{Bmatrix} = \begin{Bmatrix} 0 \\ 0 \end{Bmatrix} \tag{E.6}$$

By multiplying throughout by $l/(2EI)$, Eq. (E.6) can be expressed as

$$\begin{bmatrix} 2 - 4\lambda & 1 + 3\lambda \\ 1 + 3\lambda & 2 - 4\lambda \end{bmatrix}\begin{Bmatrix} W_2 \\ W_4 \end{Bmatrix} = \begin{Bmatrix} 0 \\ 0 \end{Bmatrix} \tag{E.7}$$

where

$$\lambda = \frac{\rho Al^4\omega^2}{840EI} \tag{E.8}$$

By setting the determinant of the coefficient matrix in Eq. (E.7) equal to zero, we obtain the frequency equation

$$\begin{vmatrix} 2 - 4\lambda & 1 + 3\lambda \\ 1 + 3\lambda & 2 - 4\lambda \end{vmatrix} = (2 - 4\lambda)^2 - (1 + 3\lambda)^2 = 0 \tag{E.9}$$

The roots of Eq. (E.9) give the natural frequencies of the beam as

$$\lambda_1 = \frac{1}{7} \quad \text{or} \quad \omega_1 = \left(\frac{120EI}{\rho Al^4}\right)^{1/2} \tag{E.10}$$

$$\lambda_2 = 3 \quad \text{or} \quad \omega_2 = \left(\frac{2520EI}{\rho Al^4}\right)^{1/2} \tag{E.11}$$

These results can be compared with the exact values (see Fig. 8.15):

$$\omega_1 = \left(\frac{97.41EI}{\rho Al^4}\right)^{1/2}, \quad \omega_2 = \left(\frac{1558.56EI}{\rho Al^4}\right)^{1/2} \tag{E.12}$$

■

EXAMPLE 12.3

Stresses in a Two-Bar Truss

Find the stresses developed in the two members of the truss shown in Fig. 12.9(a), under a vertical load of 200 lb at joint 3. The areas of cross section are 1 in.^{2} for member 1 and 2 in.^{2} for member 2, and the Young's modulus is 30×10^6 psi.

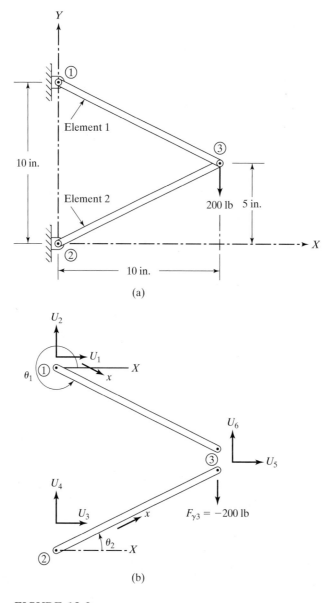

FIGURE 12.9   Two bar truss.

Solution

Approach: Derive the static equilibrium equations and solve them to find the joint displacements. Use the elasticity relations to find the element stresses. Each member is to be treated as a bar element. From Fig. 12.9(a), the coordinates of the joints can be found as

$$(X_1, Y_1) = (0, 10) \text{ in.}; \quad (X_2, Y_2) = (0, 0) \text{ in.}; \quad (X_3, Y_3) = (10, 5) \text{ in.}$$

The modeling of the truss as an assemblage of two bar elements and the displacement degrees of freedom of the joints are shown in Fig. 12.9(b). The lengths of the elements can be computed from the coordinates of the ends (joints) as

$$l^{(1)} = \{(X_3 - X_1)^2 + (Y_3 - Y_1)^2\}^{1/2} = \{(10 - 0)^2 + (5 - 10)^2\}^{1/2}$$

$$= 11.1803 \text{ in.}$$

$$l^{(2)} = \{(X_3 - X_2)^2 + (Y_3 - Y_2)^2\}^{1/2} = \{(10 - 0)^2 + (5 - 0)^2\}^{1/2}$$

$$= 11.1803 \text{ in.} \tag{E.1}$$

The element stiffness matrices in the local coordinate system can be obtained as

$$[k^{(1)}] = \frac{A^{(1)}E^{(1)}}{l^{(1)}} \begin{bmatrix} 1 & -1 \\ -1 & 1 \end{bmatrix} = \frac{(1)(30 \times 10^6)}{11.1803} \begin{bmatrix} 1 & -1 \\ -1 & 1 \end{bmatrix}$$

$$= 2.6833 \times 10^6 \begin{bmatrix} 1 & -1 \\ -1 & 1 \end{bmatrix}$$

$$[k^{(2)}] = \frac{A^{(2)}E^{(2)}}{l^{(2)}} \begin{bmatrix} 1 & -1 \\ -1 & 1 \end{bmatrix} = \frac{(2)(30 \times 10^6)}{11.1803} \begin{bmatrix} 1 & -1 \\ -1 & 1 \end{bmatrix}$$

$$= 5.3666 \times 10^6 \begin{bmatrix} 1 & -1 \\ -1 & 1 \end{bmatrix} \tag{E.2}$$

The angle between the local x-coordinate and the global X-coordinate is given by

$$\left. \begin{aligned} \cos \theta_1 &= \frac{X_3 - X_1}{l^{(1)}} = \frac{10 - 0}{11.1803} = 0.8944 \\ \sin \theta_1 &= \frac{Y_3 - Y_1}{l^{(1)}} = \frac{5 - 10}{11.1803} = -0.4472 \end{aligned} \right\} \text{ for element 1} \tag{E.3}$$

$$\left. \begin{aligned} \cos \theta_2 &= \frac{X_3 - X_2}{l^{(2)}} = \frac{10 - 0}{11.1803} = 0.8944 \\ \sin \theta_2 &= \frac{Y_3 - Y_2}{l^{(2)}} = \frac{5 - 0}{11.1803} = 0.4472 \end{aligned} \right\} \text{ for element 2} \tag{E.4}$$

The stiffness matrices of the elements in the global (X, Y) coordinate system can be derived as

$$[\bar{k}^{(1)}] = [\lambda^{(1)}]^T [k^{(1)}][\lambda^{(1)}]$$

$$= 2.6833 \times 10^6 \begin{array}{c} \\ \begin{bmatrix} 0.8 & -0.4 & -0.8 & 0.4 \\ -0.4 & 0.2 & 0.4 & -0.2 \\ -0.8 & 0.4 & 0.8 & -0.4 \\ 0.4 & -0.2 & -0.4 & 0.2 \end{bmatrix} \begin{array}{l} 1 \\ 2 \\ 5 \\ 6 \end{array} \end{array} \quad \text{(E.5)}$$

with column headings $1 \quad 2 \quad 5 \quad 6$

$$[\bar{k}^{(2)}] = [\lambda^{(2)}]^T [k^{(2)}][\lambda^{(2)}]$$

$$= 5.3666 \times 10^6 \begin{array}{c} \\ \begin{bmatrix} 0.8 & 0.4 & -0.8 & -0.4 \\ 0.4 & 0.2 & -0.4 & -0.2 \\ -0.8 & -0.4 & 0.8 & 0.4 \\ -0.4 & -0.2 & 0.4 & 0.2 \end{bmatrix} \begin{array}{l} 3 \\ 4 \\ 5 \\ 6 \end{array} \end{array} \quad \text{(E.6)}$$

with column headings $3 \quad 4 \quad 5 \quad 6$

where

$$[\lambda^{(1)}] = \begin{bmatrix} \cos\theta_1 & \sin\theta_1 & 0 & 0 \\ 0 & 0 & \cos\theta_1 & \sin\theta_1 \end{bmatrix}$$

$$= \begin{bmatrix} 0.8944 & -0.4472 & 0 & 0 \\ 0 & 0 & 0.8944 & -0.4472 \end{bmatrix} \quad \text{(E.7)}$$

$$[\lambda^{(2)}] = \begin{bmatrix} \cos\theta_2 & \sin\theta_2 & 0 & 0 \\ 0 & 0 & \cos\theta_2 & \sin\theta_2 \end{bmatrix}$$

$$= \begin{bmatrix} 0.8944 & 0.4472 & 0 & 0 \\ 0 & 0 & 0.8944 & 0.4472 \end{bmatrix} \quad \text{(E.8)}$$

Note that the top and right-hand sides of Eqs. (E.5) and (E.6) denote the global degrees of freedom corresponding to the rows and columns of the respective stiffness matrices. The assembled stiffness matrix of the system, $[\underset{\sim}{K}]$ can be obtained, by placing the elements of $[\bar{k}^{(1)}]$ and $[\bar{k}^{(2)}]$ at their proper places in $[\underset{\sim}{K}]$, as

$$[\underset{\sim}{K}] = 2.6833 \times 10^6 \begin{array}{c} \\ \begin{bmatrix} 0.8 & -0.4 & & & -0.8 & 0.4 \\ -0.4 & 0.2 & & & 0.4 & -0.2 \\ & & 1.6 & 0.8 & -1.6 & -0.8 \\ & & 0.8 & 0.4 & -0.8 & -0.4 \\ -0.8 & 0.4 & -1.6 & -0.8 & (0.8 & (-0.4 \\ & & & & +1.6) & +0.8) \\ 0.4 & -0.2 & -0.8 & -0.4 & (-0.4 & (0.2 \\ & & & & +0.8) & +0.4) \end{bmatrix} \begin{array}{l} 1 \\ 2 \\ 3 \\ 4 \\ 5 \\ \\ 6 \\ \end{array} \end{array} \quad \text{(E.9)}$$

with column headings $1 \quad 2 \quad 3 \quad 4 \quad 5 \quad 6$

The assembled force vector can be written as

$$\vec{F} = \begin{Bmatrix} F_{X1} \\ F_{Y1} \\ F_{X2} \\ F_{Y2} \\ F_{X3} \\ F_{Y3} \end{Bmatrix} \tag{E.10}$$

where, in general, (F_{Xi}, F_{Yi}) denote the forces applied at joint i along (X, Y) directions. Specifically, (F_{X1}, F_{Y1}) and (F_{X2}, F_{Y2}) represent the reactions at joints 1 and 2, while $(F_{X3}, F_{Y3}) = (0, -200)$ lb shows the external forces applied at joint 3. By applying the boundary conditions $U_1 = U_2 = U_3 = U_4 = 0$ (i.e., by deleting the rows and columns 1, 2, 3, and 4 in Eqs. E.9 and E.10), we get the final assembled stiffness matrix and the force vector as

$$\begin{matrix} & 5 & 6 \end{matrix} \tag{E.11}$$
$$[K] = 2.6833 \times 10^6 \begin{bmatrix} 2.4 & 0.4 \\ 0.4 & 0.6 \end{bmatrix} \begin{matrix} 5 \\ 6 \end{matrix}$$

$$\vec{F} = \begin{Bmatrix} 0 \\ -200 \end{Bmatrix} \begin{matrix} 5 \\ 6 \end{matrix} \tag{E.12}$$

The equilibrium equations of the system can be written as

$$[K]\vec{U} = \vec{F} \tag{E.13}$$

where $\vec{U} = \begin{Bmatrix} U_5 \\ U_6 \end{Bmatrix}$. The solution of Eq. (E.13) can be found as

$$U_5 = 23.2922 \times 10^{-6} \text{ in.}, \qquad U_6 = -139.7532 \times 10^{-6} \text{ in.} \tag{E.14}$$

The axial displacements of elements 1 and 2 can be found as

$$\begin{Bmatrix} u_1 \\ u_2 \end{Bmatrix}^{(1)} = [\lambda^{(1)}] \begin{Bmatrix} U_1 \\ U_2 \\ U_5 \\ U_6 \end{Bmatrix}$$

$$= \begin{bmatrix} 0.8944 & -0.4472 & 0 & 0 \\ 0 & 0 & 0.8944 & -0.4472 \end{bmatrix} \begin{Bmatrix} 0 \\ 0 \\ 23.2922 \times 10^{-6} \\ -139.7532 \times 10^{-6} \end{Bmatrix}$$

$$= \begin{Bmatrix} 0 \\ 83.3301 \times 10^{-6} \end{Bmatrix} \text{ in.} \tag{E.15}$$

$$\left\{ \begin{matrix} u_1 \\ u_2 \end{matrix} \right\}^{(2)} = [\lambda^{(2)}] \left\{ \begin{matrix} U_3 \\ U_4 \\ U_5 \\ U_6 \end{matrix} \right\}$$

$$= \begin{bmatrix} 0.8944 & 0.4472 & 0 & 0 \\ 0 & 0 & 0.8944 & 0.4472 \end{bmatrix} \left\{ \begin{matrix} 0 \\ 0 \\ 23.2922 \times 10^{-6} \\ -139.7532 \times 10^{-6} \end{matrix} \right\}$$

$$= \left\{ \begin{matrix} 0 \\ -41.6651 \times 10^{-6} \end{matrix} \right\} \text{ in.} \tag{E.16}$$

The stresses in elements 1 and 2 can be determined as

$$\sigma^{(1)} = E^{(1)}\epsilon^{(1)} = E^{(1)}\frac{\Delta l^{(1)}}{l^{(1)}} = \frac{E^{(1)}(u_2 - u_1)^{(1)}}{l^{(1)}}$$

$$= \frac{(30 \times 10^6)(83.3301 \times 10^{-6})}{11.1803} = 223.5989 \text{ psi} \tag{E.17}$$

$$\sigma^{(2)} = E^{(2)}\epsilon^{(2)} = \frac{E^{(2)}\Delta l^{(2)}}{l^{(2)}} = \frac{E^{(2)}(u_2 - u_1)^{(2)}}{l^{(2)}}$$

$$= \frac{(30 \times 10^6)(-41.6651 \times 10^{-6})}{11.1803} = -111.7996 \text{ psi} \tag{E.18}$$

where $\sigma^{(i)}$ denotes the stress, $\varepsilon^{(i)}$ represents the strain, and $\Delta l^{(i)}$ indicates the change in length of element i ($i = 1, 2$).

∎

12.7   Consistent and Lumped Mass Matrices

The mass matrices derived in Section 12.3 are called *consistent mass matrices*. They are consistent because the same displacement model that is used for deriving the element stiffness matrix is used for the derivation of mass matrix. It is of interest to note that several dynamic problems have been solved with simpler forms of mass matrices. The simplest form of the mass matrix, known as the lumped mass matrix, can be obtained by placing point (concentrated) masses m_i at node points i in the directions of the assumed displacement degrees of freedom. The concentrated masses refer to translational and rotational inertia of the element and are calculated by assuming that the material within the mean locations on either side of the particular displacement behaves like a rigid body while the remainder of the element does not participate in the motion. Thus this assumption excludes the dynamic coupling that exists between the element displacements and hence the resulting element mass matrix is purely diagonal [12.11].

12.7.1 Lumped Mass Matrix for a Bar Element

By dividing the total mass of the element equally between the two nodes, the lumped mass matrix of a uniform bar element can be obtained as

$$[m] = \frac{\rho Al}{2} \begin{bmatrix} 1 & 0 \\ 0 & 1 \end{bmatrix} \tag{12.67}$$

12.7.2 Lumped Mass Matrix for a Beam Element

In Fig. 12.4, by lumping one half of the total beam mass at each of the two nodes, along the translational degrees of freedom, we obtain the lumped mass matrix of the beam element as

$$[m] = \frac{\rho Al}{2} \begin{bmatrix} 1 & 0 & 0 & 0 \\ 0 & 0 & 0 & 0 \\ 0 & 0 & 1 & 0 \\ 0 & 0 & 0 & 0 \end{bmatrix} \tag{12.68}$$

Note that the inertia effect associated with the rotational degrees of freedom has been assumed to be zero in Eq. (12.68). If the inertia effect is to be included, we compute the mass moment of inertia of half of the beam segment about each end and include it at the diagonal locations corresponding to the rotational degrees of freedom. Thus, for a uniform beam, we have

$$I = \frac{1}{3}\left(\frac{\rho Al}{2}\right)\left(\frac{l}{2}\right)^2 = \frac{\rho Al^3}{24} \tag{12.69}$$

and hence the lumped mass matrix of the beam element becomes

$$[m] = \frac{\rho Al}{2} \begin{bmatrix} 1 & 0 & 0 & 0 \\ 0 & \left(\dfrac{l^2}{12}\right) & 0 & 0 \\ 0 & 0 & 1 & 0 \\ 0 & 0 & 0 & \left(\dfrac{l^2}{12}\right) \end{bmatrix} \tag{12.70}$$

12.7.3 Lumped Mass Versus Consistent Mass Matrices

It is not obvious whether the lumped mass matrices or consistent mass matrices yield more accurate results for a general dynamic response problem. The lumped mass matrices are approximate in the sense that they do not consider the dynamic coupling present between the various displacement degrees of freedom of the element. However, since the lumped mass matrices are diagonal, they require less storage space during computation. On the other hand, the consistent mass matrices are not diagonal and hence require more storage space. They too are approximate in the sense that the shape functions, which are derived using static displacement patterns, are used even for the solution of dynamics problems.

The following example illustrates the application of lumped and consistent mass matrices in a simple vibration problem.

Consistent and Lumped Mass Matrices of a Bar

EXAMPLE 12.4

Find the natural frequencies of the fixed-fixed uniform bar shown in Fig. 12.10 using consistent and lumped mass matrices. Use two bar elements for modeling.

Solution: The stiffness and mass matrices of a bar element are

$$[k] = \frac{AE}{l}\begin{bmatrix} 1 & -1 \\ -1 & 1 \end{bmatrix} \tag{E.1}$$

$$[m]_c = \frac{\rho Al}{6}\begin{bmatrix} 2 & 1 \\ 1 & 2 \end{bmatrix} \tag{E.2}$$

$$[m]_l = \frac{\rho Al}{2}\begin{bmatrix} 1 & 0 \\ 0 & 1 \end{bmatrix} \tag{E.3}$$

where the subscripts c and l to the mass matrices denote the consistent and lumped matrices, respectively. Since the bar is modeled by two elements, the assembled stiffness and mass matrices are given by

$$[\underset{\sim}{K}] = \frac{AE}{l}\begin{bmatrix} \overset{1}{1} & \overset{2}{-1} & \overset{3}{0} \\ -1 & 1 & +1 & -1 \\ 0 & & -1 & 1 \end{bmatrix}\begin{matrix} 1 \\ 2 \\ 3 \end{matrix} = \frac{AE}{l}\begin{bmatrix} 1 & -1 & 0 \\ -1 & 2 & -1 \\ 0 & -1 & 1 \end{bmatrix} \tag{E.4}$$

$$[\underset{\sim}{M}]_c = \frac{\rho Al}{6}\begin{bmatrix} \overset{1}{2} & \overset{2}{1} & \overset{3}{0} \\ 1 & 2 & +2 & 1 \\ 0 & & 1 & 2 \end{bmatrix}\begin{matrix} 1 \\ 2 \\ 3 \end{matrix} = \frac{\rho Al}{6}\begin{bmatrix} 2 & 1 & 0 \\ 1 & 4 & 1 \\ 0 & 1 & 2 \end{bmatrix} \tag{E.5}$$

$$[\underset{\sim}{M}]_l = \frac{\rho Al}{2}\begin{bmatrix} \overset{1}{1} & \overset{2}{0} & \overset{3}{0} \\ 0 & 1 & +1 & 0 \\ 0 & & 0 & 1 \end{bmatrix}\begin{matrix} 1 \\ 2 \\ 3 \end{matrix} = \frac{\rho Al}{2}\begin{bmatrix} 1 & 0 & 0 \\ 0 & 2 & 0 \\ 0 & 0 & 1 \end{bmatrix} \tag{E.6}$$

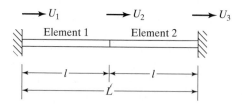

FIGURE 12.10   Fixed-fixed uniform bar.

The dashed boxes in Eqs. (E.4) through (E.6) enclose the contributions of elements 1 and 2. The degrees of freedom corresponding to the columns and rows of the matrices are indicated at the top and the right-hand side of the matrices. The eigenvalue problem, after applying the boundary conditions $U_1 = U_3 = 0$, becomes

$$[[K] - \omega^2[M]]\,\{U_2\} = \{0\} \tag{E.7}$$

The eigenvalue ω^2 can be determined by solving the equation

$$|[K] - \omega^2[M]| = 0 \tag{E.8}$$

which, for the present case, becomes

$$\left| \frac{AE}{l}[2] - \omega^2 \frac{\rho Al}{6}[4] \right| = 0 \qquad \text{with consistent mass matrices} \tag{E.9}$$

and

$$\left| \frac{AE}{l}[2] - \omega^2 \frac{\rho Al}{2}[2] \right| = 0 \qquad \text{with lumped mass matrices} \tag{E.10}$$

Equations (E.9) and (E.10) can be solved to obtain

$$\omega_c = \sqrt{\frac{3E}{\rho l^2}} = 3.4641 \sqrt{\frac{E}{\rho L^2}} \tag{E.11}$$

$$\omega_l = \sqrt{\frac{2E}{\rho l^2}} = 2.8284 \sqrt{\frac{E}{\rho L^2}} \tag{E.12}$$

These values can be compared with the exact value (see Fig. 8.7)

$$\omega_1 = \pi \sqrt{\frac{E}{\rho L^2}} \tag{E.13}$$

∎

12.8   Examples Using MATLAB

Finite Element Analysis of a Stepped Bar

EXAMPLE 12.5

Consider the stepped bar shown in Fig. 12.11 with the following data: $A_1 = 16 \times 10^{-4}$ m^2, $A_2 = 9 \times 10^{-4}$ m^2, $A_3 = 4 \times 10^{-4}$ m^2, $E_i = 20 \times 10^{10}$ Pa, $i = 1, 2, 3$, $\rho_i = 7.8 \times 10^3$ kg/m^3, $i = 1, 2, 3$, $l_1 = 1$ m, $l_2 = 0.5$ m, $l_3 = 0.25$ m.
Write a MATLAB program to determine the following:

a.   Displacements $u_1, u_2,$ and u_3 under load $p_3 = 1000$ N
b.   Natural frequencies and mode shapes of bar

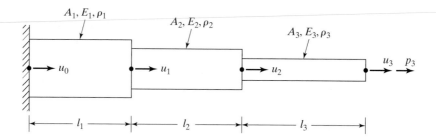

FIGURE 12.11   Stepped bar.

Solution: The assembled stiffness and mass matrices of the stepped bar are given by

$$
[\underset{\sim}{K}] =
\begin{bmatrix}
\dfrac{A_1 E_1}{l_1} & -\dfrac{A_1 E_1}{l_1} & 0 & 0 \\[3mm]
-\dfrac{A_1 E_1}{l_1} & \dfrac{A_1 E_1}{l_1} + \dfrac{A_2 E_2}{l_2} & -\dfrac{A_2 E_2}{l_2} & 0 \\[3mm]
0 & -\dfrac{A_2 E_2}{l_2} & \dfrac{A_2 E_2}{l_2} + \dfrac{A_3 E_3}{l_3} & -\dfrac{A_3 E_3}{l_3} \\[3mm]
0 & 0 & -\dfrac{A_3 E_3}{l_3} & \dfrac{A_3 E_3}{l_3}
\end{bmatrix}
\tag{E.1}
$$

$$
[\underset{\sim}{M}] = \frac{1}{6}
\begin{bmatrix}
2\rho_1 A_1 l_1 & \rho_1 A_1 l_1 & 0 & 0 \\[2mm]
\rho_1 A_1 l_1 & 2\rho_1 A_1 l_1 + 2\rho_2 A_2 l_2 & \rho_2 A_2 l_2 & 0 \\[2mm]
0 & \rho_2 A_2 l_2 & 2\rho_2 A_2 l_2 + 2\rho_3 A_3 l_3 & \rho_3 A_3 l_3 \\[2mm]
0 & 0 & \rho_3 A_3 l_3 & 2\rho_3 A_3 l_3
\end{bmatrix}
\tag{E.2}
$$

The system matrices $[K]$ and $[M]$ can be obtained by incorporating the boundary condition $u_0 = 0$—that is, by deleting the first row and first column in Eqs. (E.1) and (E.2).

a.   The equilibrium equations under the load $p_3 = 1000$ N are given by

$$
[K]\vec{U} = \vec{P}
\tag{E.3}
$$

where

$$
[K] =
\begin{bmatrix}
\dfrac{A_1 E_1}{l_1} + \dfrac{A_2 E_2}{l_2} & -\dfrac{A_2 E_2}{l_2} & 0 \\[3mm]
-\dfrac{A_2 E_2}{l_2} & \dfrac{A_2 E_2}{l_2} + \dfrac{A_3 E_3}{l_3} & -\dfrac{A_3 E_3}{l_3} \\[3mm]
0 & -\dfrac{A_3 E_3}{l_3} & \dfrac{A_3 E_3}{l_3}
\end{bmatrix}
\tag{E.4}
$$

$$\vec{U} = \begin{Bmatrix} u_1 \\ u_2 \\ u_3 \end{Bmatrix}, \qquad \vec{P} = \begin{Bmatrix} 0 \\ 0 \\ 1000 \end{Bmatrix}$$

b.   The eigenvalue problem can be expressed as

$$\left[[K] - \omega^2 [M] \right] \vec{U} = \vec{0} \qquad \text{(E.5)}$$

where $[K]$ is given by Eq. (E.4) and $[M]$ by

$$[M] = \frac{1}{6} \begin{bmatrix} 2\rho_1 A_1 l_1 + 2\rho_2 A_2 l_2 & \rho_2 A_2 l_2 & 0 \\ \rho_2 A_2 l_2 & 2\rho_2 A_2 l_2 + 2\rho_3 A_3 l_3 & \rho_3 A_3 l_3 \\ 0 & \rho_3 A_3 l_3 & 2\rho_3 A_3 l_3 \end{bmatrix} \qquad \text{(E.6)}$$

The MATLAB solution of Eqs. (E.3) and (E.5) is given below.

```
%------ Program Ex12_5.m
%------Initialization of values-------------------------
A1 = 16e-4 ;
A2 = 9e-4 ;
A3 = 4e-4 ;

E1 = 20e10 ;
E2 = E1 ;
E3 = E1 ;

R1 = 7.8e3 ;
R2 = R1 ;
R3 = R1 ;

L1 = 1 ;
L2 = 0.5 ;
L3 = 0.25 ;

%------Definition of [K]--------------------------------

K11 = A1*E1/L1+A2*E2/L2 ;
K12 = -A2*E2/L2 ;
K13 = 0 ;

K21 = K12 ;
K22 = A2*E2/L2+A3*E3/L3 ;
K23 = -A3*E3/L3 ;

K31 = K13 ;
K32 = K23 ;
K33 = A3*E3/L3 ;

K = [ K11 K12 K13; K21 K22 K23; K31 K32 K33 ]

%-------- Calculation of matrix

P = [ 0 0 1000]'

U = inv(K)*P
```

```
%------- Definition of [M] -------------------------

M11 = (2*R1*A1*L1+2*R2*A2*L2) / 6;
M12 = (R2*A2*L2) / 6;
M13 = 0;

M21 = M12;
M22 = (2*R2*A2*L2+2*R3*A3*L3) / 6;
M23 = R3*A3*L3;

M31 = M13;
M32 = M23;
M33 = 2*M23;

M= [M11 M12 M13; M21 M22 M23; M31 M32 M33 ]

MI   = inv (M)

KM = MI*K

%-------------Calculation of eigenvector and eigenvalue--------------

[L, V] = eig (KM)


>> Ex12_5
K =

    680000000   -360000000           0
   -360000000    680000000  -320000000
           0    -320000000   320000000

P =

           0
           0
        1000

U =

   1.0e-005 *

   0.3125
   0.5903
   0.9028

M =

    5.3300     0.5850          0
    0.5850     1.4300     0.7800
         0     0.7800     1.5600

MI =
    0.2000    -0.1125     0.0562
   -0.1125     1.0248    -0.5124
    0.0562    -0.5124     0.8972

KM =

   1.0e+008 *

    1.7647    -1.6647     0.5399
   -4.4542     9.0133    -4.9191
    2.2271    -6.5579     4.5108

L =

   -0.1384     0.6016     0.3946
    0.7858    -0.1561     0.5929
   -0.6028    -0.7834     0.7020
```

```
V =

   1.0e+009 *

      1.3571        0        0
           0   0.1494        0
           0        0   0.0224
>>
```

■

Program for Eigenvalue Analysis of a Stepped Beam

EXAMPLE 12.6

Develop a MATLAB program called **Program17.m** for the eigenvalue analysis of a fixed-fixed stepped beam of the type shown in Fig. 12.12.

Solution: Program17.m is developed to accept the following input data:

xl(i) = length of element (step) i

xi(i) = moment of inertia of element i

a(i) = area of cross section of element i

bj(i, j) = global degree of freedom number corresponding to the local jth degree of freedom of element i

e = Young's modulus

rho = mass density

The program gives the natural frequencies and mode shapes of the beam as output.

```
     Natural frequencies of the stepped beams
     1.6008e+002   6.1746e+002   2.2520e+003   7.1266e+003
     Mode shapes
     1    1.0333e-002     1.8915e-004    1.4163e-002   4.4518e-005
     2   -3.7660e-003     2.0297e-004    4.7109e-003   2.5950e-004
     3    1.6816e-004    -1.8168e-004    1.3570e-003   2.0758e-004
     4    1.8324e-004     6.0740e-005    3.7453e-004   1.6386e-004
```

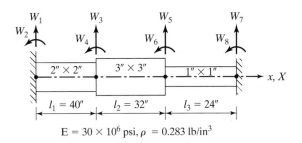

$E = 30 \times 10^6$ psi, $\rho = 0.283$ lb/in^3

FIGURE 12.12 Stepped beam.

■

12.9   C++ Program

An interactive C++ program, called **Program17.cpp**, is given for the eigenvalue solution of a stepped beam. The input and output of the program are similar to those of **Program17.m**.

▄▄▄▄▄▄▄▄▄▄▄▄▄▄▄▄▄ Eigenvalue Solution of a Stepped Beam

EXAMPLE 12.7 ───

Find the natural frequencies and mode shapes of the stepped beam shown in Fig. 12.12 using **Program17.cpp**.

Solution: The input data are to be entered interactively. The output of the program is shown below.

```
NATURAL FREQUENCIES OF THE STEPPED BEAM

  160.083080   617.459700   2251.975785   7126.595358

MODE SHAPES

   1     0.0103332469    0.0001891485    0.0141625494    0.0000445137
   2    -0.0037659939    0.0002029733    0.0047108916    0.0002594971
   3     0.0001681571   -0.0001816828    0.0013569926    0.0002075837
   4     0.0001832390    0.0000607403    0.0003745272    0.0001638648
```

■

12.10   Fortran Program

A Fortran program called **PROGRAM17.F** is given for the eigenvalue analysis of a stepped beam. The input and output of the program are similar to those of **Program17.m**.

▄▄▄▄▄▄▄▄▄▄▄▄▄▄▄▄▄ Eigenvalue Analysis of a Stepped Beam

EXAMPLE 12.8 ───

Find the natural frequencies and mode shapes of the stepped beam shown in Fig. 12.12 using **PROGRAM17.F**.

Solution: The output of the program is given below.

```
NATURAL FREQUENCIES OF THE STEPPED BEAM

  0.160083E+03    0.617460E+03    0.225198E+04    0.712653E+04

MODE SHAPES
   1     0.103333E-01    0.189147E-03    0.141626E-01    0.445258E-04
   2    -0.376593E-02    0.202976E-03    0.471097E-02    0.259495E-03
   3     0.167902E-03   -0.181687E-03    0.135672E-02    0.207580E-03
   4     0.182648E-03    0.607247E-04    0.373775E-03    0.163869E-03
```

■

REFERENCES

12.1 O. C. Zienkiewicz, *The Finite Element Method* (4th ed.), McGraw-Hill, London, 1987.

12.2 S. S. Rao, *The Finite Element Method in Engineering* (3rd ed.), Butterworth-Heinemann, Boston, 1999.

12.3 G. V. Ramana and S. S. Rao, "Optimum design of plano-milling machine structure using finite element analysis," *Computers and Structures*, Vol. 18, 1984, pp. 247–253.

12.4 R. Davis, R. D. Henshell, and G. B. Warburton, "A Timoshenko beam element," *Journal of Sound and Vibration*, Vol. 22, 1972, pp. 475–487.

12.5 D. L. Thomas, J. M. Wilson, and R. R. Wilson, "Timoshenko beam finite elements," *Journal of Sound and Vibration*, Vol. 31, 1973, pp. 315–330.

12.6 J. Thomas and B. A. H. Abbas, "Finite element model for dynamic analysis of Timoshenko beams," *Journal of Sound and Vibration*, Vol. 41, 1975, pp. 291–299.

12.7 R. S. Gupta and S. S. Rao, "Finite element eigenvalue analysis of tapered and twisted Timoshenko beams," *Journal of Sound and Vibration*, Vol. 56, 1978, pp. 187–200.

12.8 T. H. H. Pian, "Derivation of element stiffness matrices by assumed stress distribution," *AIAA Journal*, Vol. 2, 1964, pp. 1333–1336.

12.9 H. Alaylioglu and R. Ali, "Analysis of an automotive structure using hybrid stress finite elements," *Computers and Structures*, Vol. 8, 1978, pp. 237–242.

12.10 I. Fried, "Accuracy of finite element eigenproblems," *Journal of Sound and Vibration*, Vol. 18, 1971, pp. 289–295.

12.11 P. Tong, T. H. H. Pian, and L. L. Bucciarelli, "Mode shapes and frequencies by the finite element method using consistent and lumped matrices," *Computers and Structures*, Vol. 1, 1971, pp. 623–638.

REVIEW QUESTIONS

12.1 Give brief answers to the following:

    1. What is the basic idea behind the finite element method?

    2. What is a shape function?

    3. What is the role of transformation matrices in the finite element method?

    4. What is the basis for the derivation of transformation matrices?

    5. How are fixed boundary conditions incorporated in the finite element equations?

    6. How do you solve a finite element problem having symmetry in geometry and loading by modeling only half of the problem?

    7. Why is the finite element approach presented in this chapter called the displacement method?

    8. What is a consistent mass matrix?

    9. What is a lumped mass matrix?

    10. What is the difference between the finite element method and the Rayleigh-Ritz method?

    11. How is the distributed load converted into equivalent joint force vector in the finite element method?

12.2 Indicate whether each of the following statements is true or false:

    1. For a bar element of length l with two nodes, the shape function corresponding to node 2 is given by x/l.

2. The element stiffness matrices are always singular.
3. The element mass matrices are always singular.
4. The system stiffness matrix is always singular unless the boundary conditions are incorporated.
5. The system mass matrix is always singular unless the boundary conditions are incorporated.
6. The lumped mass matrices are always diagonal.
7. The coordinate transformation of element matrices is required for all systems.
8. The element stiffness matrix in the global coordinate system, $[\overline{k}]$, can be expressed in terms of the local matrix $[k]$ and the coordinate transformation matrix $[\lambda]$ as $[\lambda]^T[k][\lambda]$.
9. The derivation of system matrices involves the assembly of element matrices.
10. Boundary conditions are to be imposed to avoid rigid body motion of the system.

12.3 Fill in each of the following blanks with appropriate word:

1. In the finite element method, the solution domain is replaced by several _____.
2. In the finite element method, the elements are assumed to be interconnected at certain points known as _____.
3. In the finite element method, an _____ solution is assumed within each element.
4. The displacement within a finite element is expressed in terms of _____ functions.
5. For a thin beam element, _____ degrees of freedom are considered at each node.
6. For a thin beam element, the shape functions are assumed to be polynomials of degree _____.
7. In the displacement method, the _____ of elements is directly approximated.
8. If the displacement model used in the derivation of the element stiffness matrices is also used to derive the element mass matrices, the resulting mass matrix is called _____ mass matrix.
9. If the mass matrix is derived by assuming point masses at node points, the resulting mass matrix is called _____ mass.
10. The lumped mass matrices do not consider the _____ coupling between the various displacement degrees of freedom of the element.
11. Different orientations of finite elements require _____ of element matrices.

12.4 Select the most appropriate answer out of the choices given:

1. For a bar element of length l with two nodes, the shape function corresponding to node 1 is given by

   (a) $\left(1 - \dfrac{x}{l}\right)$ (b) $\dfrac{x}{l}$ (c) $\left(1 + \dfrac{x}{l}\right)$

2. The simplest form of mass matrix is known as
   (a) lumped mass matrix
   (b) consistent mass matrix
   (c) global mass matrix
3. The finite element method is
   (a) an approximate analytical method
   (b) a numerical method
   (c) an exact analytical method

4. The stiffness matrix of a bar element is given by

(a) $\dfrac{EA}{l}\begin{bmatrix} 1 & 1 \\ 1 & 1 \end{bmatrix}$ (b) $\dfrac{EA}{l}\begin{bmatrix} 1 & -1 \\ -1 & 1 \end{bmatrix}$ (c) $\dfrac{EA}{l}\begin{bmatrix} 1 & 0 \\ 0 & 1 \end{bmatrix}$

5. The consistent mass matrix of a bar element is given by

(a) $\dfrac{\rho Al}{6}\begin{bmatrix} 2 & 1 \\ 1 & 2 \end{bmatrix}$ (b) $\dfrac{\rho Al}{6}\begin{bmatrix} 2 & -1 \\ -1 & 2 \end{bmatrix}$ (c) $\dfrac{\rho Al}{6}\begin{bmatrix} 1 & 0 \\ 0 & 1 \end{bmatrix}$

6. The finite element method is similar to
   (a) Rayleigh's method
   (b) the Rayleigh-Ritz method
   (c) the Lagrange method

7. The lumped mass matrix of a bar element is given by

(a) $\rho Al \begin{bmatrix} 1 & 0 \\ 0 & 1 \end{bmatrix}$ (b) $\dfrac{\rho Al}{6}\begin{bmatrix} 2 & 1 \\ 1 & 2 \end{bmatrix}$ (c) $\dfrac{\rho Al}{2}\begin{bmatrix} 1 & 0 \\ 0 & 1 \end{bmatrix}$

8. The element mass matrix in the global coordinate system, $[\overline{m}]$, can be expressed in terms of the element mass matrix in local coordinate system $[m]$ and the coordinate transformation matrix $[\lambda]$, as

(a) $[\overline{m}] = [\lambda]^T[m]$ (b) $[\overline{m}] = [m][\lambda]$ (c) $[\overline{m}] = [\lambda]^T[m][\lambda]$

12.5 Match the items in the two columns below. Assume a fixed-fixed bar with one middle node:

Element matrices: $[k] = \dfrac{AE}{l}\begin{bmatrix} 1 & -1 \\ -1 & 1 \end{bmatrix}$, $[m]_c = \dfrac{\rho Al}{6}\begin{bmatrix} 2 & 1 \\ 1 & 2 \end{bmatrix}$, $[m]_l = \dfrac{\rho Al}{2}\begin{bmatrix} 1 & 0 \\ 0 & 1 \end{bmatrix}$

Steel bar: $E = 30 \times 10^6$ lb/in.^{2}, $\rho = 0.0007298$ lb $-$ sec^2/in.^{4}, $L = 12$ in.

Aluminum bar: $E = 10.3 \times 10^6$ lb/in.^{2}, $\rho = 0.0002536$ lb $-$ sec^2/in.^{4}, $L = 12$ in.

1. Natural frequency of steel bar given by lumped mass matrices   (a) 58,528.5606 rad/sec
2. Natural frequency of aluminum bar given by consistent mass matrices   (b) 47,501.0898 rad/sec
3. Natural frequency of steel bar given by consistent mass matrices   (c) 58,177.2469 rad/sec
4. Natural frequency of aluminum bar given by lumped mass matrices   (d) 47,787.9336 rad/sec

PROBLEMS

The problem assignments are organized as follows:

Problems	Section Covered	Topic Covered
12.1, 12.2, 12.4	12.3	Derivation of element matrices and vectors
12.5, 12.7	12.4	Transformation matrix
12.6, 12.9	12.5	Assembly of matrices and vectors
12.4, 12.8, 12.10–12.30	12.6	Application of boundary conditions and solution of problem

Problems	Section Covered	Topic Covered
12.31, 12.32	12.7	Consistent and lumped mass matrices
12.33–12.35	12.8	MATLAB programs
12.36	12.9	C++ program
12.3, 12.37–12.40	12.10	Fortran programs
12.41–12.42	—	Design projects

12.1   Derive the stiffness matrix and the consistent and lumped mass matrices of the tapered bar element shown in Fig. 12.13. The diameter of the bar decreases from D to d over its length.

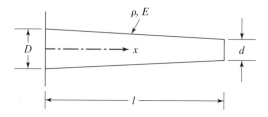

FIGURE 12.13

12.2   Derive the stiffness matrix of the bar element in longitudinal vibration whose cross-sectional area varies as $A(x) = A_0 e^{-(x/l)}$, where A_0 is the area at the root (see Fig. 12.14).

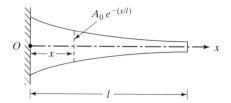

FIGURE 12.14

12.3   Write a computer program for finding the stresses in a planar truss.

12.4   The tapered cantilever beam shown in Fig. 12.15 is used as a spring to carry a load P. (a) Derive the stiffness matrix of the beam element. (b) Use the result of (a) to find the stress induced in the beam when $B = 25$ cm, $b = 10$ cm, $t = 2.5$ cm, $l = 2$ m, $E = 2.07 \times 10^{11}$ N/m², and $P = 1000$ N. Use one beam element for idealization.

12.5   Find the global stiffness matrix of each of the four bar elements of the truss shown in Fig. 12.5 using the following data:

Nodal coordinates: $(X_1, Y_1) = (0, 0)$, $(X_2, Y_2) = (50, 100)$ in., $(X_3, Y_3) = (100, 0)$ in., $(X_4, Y_4) = (200, 150)$ in.

Cross-sectional areas: $A_1 = A_2 = A_3 = A_4 = 2$ in.².

Young's modulus of all members: 30×10^6 lb/in.².

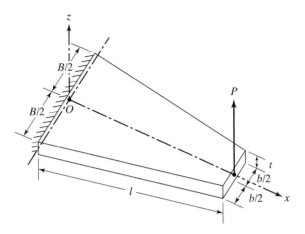

FIGURE 12.15

12.6 Using the result of Problem 12.5, find the assembled stiffness matrix of the truss and formulate the equilibrium equations if the vertical downward load applied at node 4 is 1000 lb.

12.7 Derive the stiffness and mass matrices of the planar frame element (general beam element) shown in Fig. 12.16 in the global XY-coordinate system.

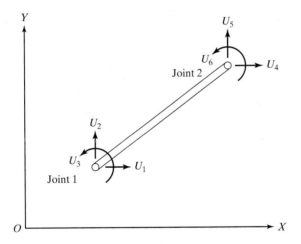

FIGURE 12.16   A frame element in global system.

12.8 A multiple-leaf spring used in automobiles is shown in Fig. 12.17. It consists of five leaves, each of thickness $t = 0.25$ in. and width $w = 1.5$ in. Find the deflection of the leaves under a load of $P = 2000$ lb. Model only a half of the spring for the finite element analysis. The Young's modulus of the material is 30×10^6 psi.

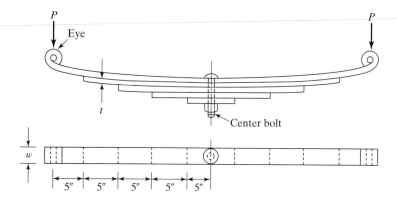

FIGURE 12.17   A multiple-leaf spring.

12.9 Derive the assembled stiffness and mass matrices of the multiple-leaf spring of Problem 12.8 assuming a specific weight of 0.283 lb/in.³ for the material.

12.10 Find the nodal displacements of the crane shown in Fig. 12.18 when a vertically downward load of 1000 lb is applied at node 4. The Young's modulus is 30×10^6 psi and the cross-sectional area is 2 in.² for elements 1 and 2 and 1 in.² for elements 3 and 4.

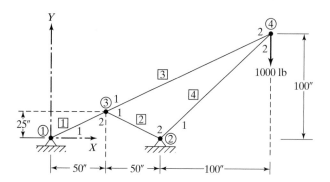

FIGURE 12.18

12.11 Find the tip deflection of the cantilever beam shown in Fig. 12.19 when a vertical load of $P = 500$ N is applied at point Q using (a) a one element approximation and (b) a two-element approximation. Assume $l = 0.25$ m, $h = 25$ mm, $b = 50$ mm, $E = 2.07 \times 10^{11}$ Pa, and $k = 10^5$ N/m.

12.12 Find the stresses in the stepped beam shown in Fig. 12.20 when a moment of 1000 N-m is applied at node 2 using a two-element idealization. The beam has a square cross section 50×50 mm between nodes 1 and 2 and 25×25 mm between nodes 2 and 3. Assume the Young's modulus as 2.1×10^{11} Pa.

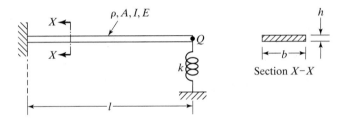

FIGURE 12.19

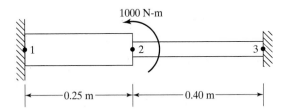

FIGURE 12.20

12.13 Find the transverse deflection and slope of node 2 of the beam shown in Fig. 12.21 using a two element idealization. Compare the solution with that of simple beam theory.

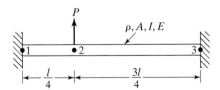

FIGURE 12.21

12.14 Find the natural frequencies of a cantilever beam of length l, cross-sectional area A, moment of inertia I, Young's modulus E, and density ρ, using one finite element.

12.15 Using one beam element, find the natural frequencies of the uniform pinned-free beam shown in Fig. 12.22.

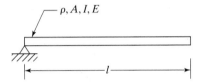

FIGURE 12.22

12.16 Using one beam element and one spring element, find the natural frequencies of the uniform, spring-supported cantilever beam shown in Fig. 12.19.

12.17 Using one beam element and one spring element, find the natural frequencies of the system shown in Fig. 12.23.

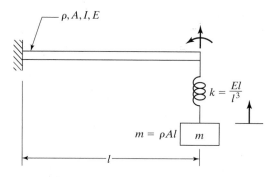

FIGURE 12.23

12.18 Using two beam elements, find the natural frequencies and mode shapes of the uniform fixed-fixed beam shown in Fig. 12.24.

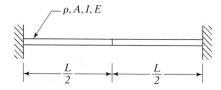

FIGURE 12.24

12.19* An electric motor, of mass $m = 100$ kg and operating speed $= 1800$ rpm, is fixed at the middle of a clamped-clamped steel beam of rectangular cross section, as shown in Fig. 12.25. Design the beam such that the natural frequency of the system exceeds the operating speed of the motor.

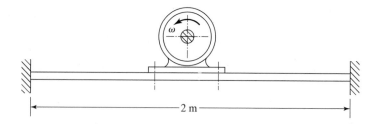

FIGURE 12.25

---

*An asterisk denotes a problem with no unique answer.

12.20 Find the natural frequencies of the beam shown in Fig. 12.26, using three finite elements of length l each.

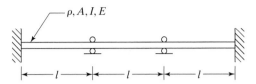

FIGURE 12.26

12.21 Find the natural frequencies of the cantilever beam carrying an end mass M shown in Fig. 12.27, using a one beam element idealization.

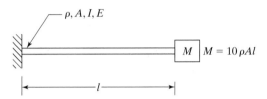

FIGURE 12.27

12.22 Find the natural frequencies of vibration of the beam shown in Fig. 12.28, using two beam elements. Also find the load vector if a uniformly distributed transverse load p is applied to element 1.

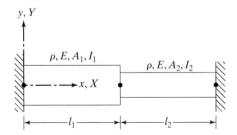

FIGURE 12.28

12.23 Find the natural frequencies of a beam of length l, which is pin connected at $x = 0$ and fixed at $x = l$, using one beam element.

12.24 Find the natural frequencies of torsional vibration of the stepped shaft shown in Fig. 12.29. Assume that $\rho_1 = \rho_2 = \rho, G_1 = G_2 = G, I_{p1} = 2I_{p2} = 2I_p, J_1 = 2J_2 = 2J,$ and $l_1 = l_2 = l$.

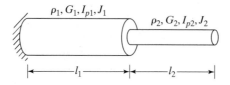

ρ_1, G_1, I_{p1}, J_1

ρ_2, G_2, I_{p2}, J_2

l_1 l_2

FIGURE 12.29

12.25 Find the dynamic response of the stepped bar shown in Fig. 12.30(a) when its free end is subjected to the load given in Fig. 12.30(b).

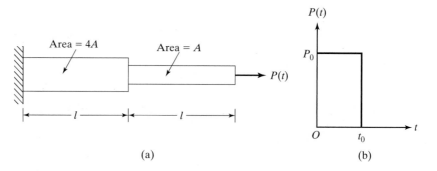

Area = $4A$

Area = A

$P(t)$

l l

$P(t)$

P_0

O t_0 t

(a)   (b)

FIGURE 12.30

12.26 Find the displacement of node 3 and the stresses in the two members of the truss shown in Fig. 12.31. Assume that the Young's modulus and the cross-sectional areas of the two members are the same with $E = 30 \times 10^6$ psi and $A = 1$ in.^{2}.

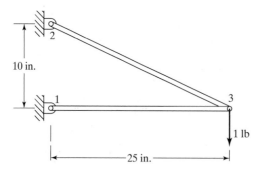

10 in.

25 in.

1 lb

FIGURE 12.31

12.27 The simplified model of a radial drilling machine structure is shown in Fig. 12.32. Using two beam elements for the column and one beam element for the arm, find the natural frequencies and mode shapes of the machine. Assume the material of the structure as steel.

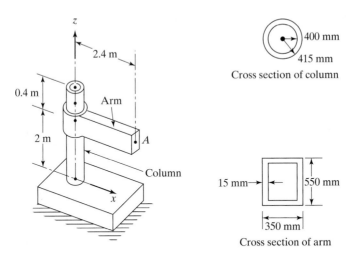

FIGURE 12.32   A radial drilling machine structure.

12.28 If a vertical force of 5000 N along the z-direction and a bending moment of 500 N-m in the xz-plane are developed at point A during a metal cutting operation, find the stresses developed in the machine tool structure shown in Fig. 12.32.

12.29 The crank in the slider-crank mechanism shown in Fig. 12.33 rotates at a constant clockwise angular speed of 1000 rpm. Find the stresses in the connecting rod and the crank when the

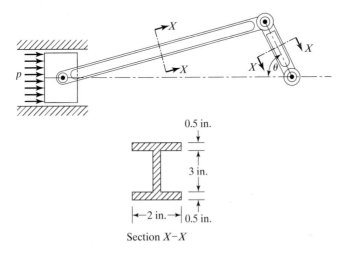

FIGURE 12.33   A slider-crank mechanism.

pressure acting on the piston is 200 psi and $\theta = 30°$. The diameter of the piston is 12 in. and the material of the mechanism is steel. Model the connecting rod and the crank by one beam element each. The lengths of the crank and connecting rod are 12 in. and 48 in., respectively.

12.30 A water tank of weight W is supported by a hollow circular steel column of inner diameter d, wall thickness t, and height l. The wind pressure acting on the column can be assumed to vary linearly from 0 to p_{max}, as shown in Fig. 12.34. Find (a) the bending stress induced in the column under the loads, and (b) the natural frequencies of the water tank using a one beam element idealization. Data: $W = 10{,}000$ lb, $l = 40$ ft, $d = 2$ ft, $t = 1$ in., and $p_{max} = 100$ psi.

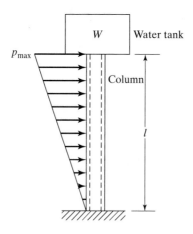

FIGURE 12.34

12.31 Find the natural frequencies of the stepped bar shown in Fig. 12.35 with the following data using consistent and lumped mass matrices: $A_1 = 2$ in.^{2}, $A_2 = 1$ in.^{2}, $E = 30 \times 10^6$ psi, $\rho = 0.283$ lb/in.^{3}, and $l_1 = l_2 = 50$ in.

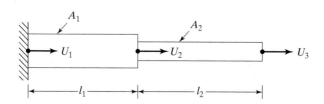

FIGURE 12.35

12.32 Find the undamped natural frequencies of longitudinal vibration of the stepped bar shown in Fig. 12.36 with the following data using consistent and lumped mass matrices: $l_1 = l_2 = l_3 = 0.2$ m, $A_1 = 2A_2 = 4A_3 = 0.4 \times 10^{-3}$ m^2, $E = 2.1 \times 10^{11}$ N/m^2, and $\rho = 7.8 \times 10^3$ kg/m^3.

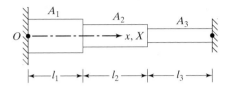

FIGURE 12.36

12.33 Consider the stepped bar shown in Fig. 12.11 with the following data: $A_1 = 25 \times 10^{-4}\,\mathrm{m}^2$, $A_2 = 16 \times 10^{-4}\,\mathrm{m}^2$, $A_3 = 9 \times 10^{-4}\,\mathrm{m}^2$, $E_i = 2 \times 10^{11}$ Pa, $i = 1, 2, 3$, $\rho_i = 7.8 \times 10^3\,\mathrm{kg/m}^3$, $i = 1, 2, 3$, $l_1 = 3$ m, $l_2 = 2$ m, $l_3 = 1$ m. Using MATLAB, find the axial displacements u_1, u_2, and u_3 under the load $p_3 = 500$ N.

12.34 Using MATLAB, find the natural frequencies and mode shapes of the stepped bar described in Problem 12.33.

12.35 Use **Program17.m** to find the natural frequencies of a fixed-fixed stepped beam, similar to the one shown in Fig. 12.12, with the following data:

Cross sections of elements: 1, 2, 3: $4'' \times 4''$, $3'' \times 3''$, $2'' \times 2''$

Lengths of elements: 1, 2, 3: $30''$, $20''$, $10''$

Young's modulus of all elements: 10^7 lb/in.^{2}

Weight density of all elements: 0.1 lb/in.^{3}

12.36 Use **Program17.cpp** to solve Problem 12.35.

12.37 Use **PROGRAM17.F** to solve Problem 12.35.

12.38 Write a computer program for finding the assembled stiffness matrix of a general planar truss.

12.39 Generalize the computer program of Section 12.10 to make it applicable to the solution of any stepped beam having a specified number of steps.

12.40 Find the natural frequencies and mode shapes of the beam shown in Fig. 12.12 with $l_1 = l_2 = l_3 = 10$ in. and a uniform cross section of 1 in. $\times$ 1 in. throughout the length, using the computer program of Section 12.10. Compare your results with those given in Chapter 8. (*Hint:* Only the data XL, XI, and A need to be changed.)

DESIGN PROJECTS

12.41 Derive the stiffness and mass matrices of a uniform beam element in transverse vibration rotating at an angular velocity of Ω rad/sec about a vertical axis as shown in Fig. 12.37(a). Using these matrices, find the natural frequencies of transverse vibration of the rotor blade of a helicopter (see Fig. 12.37b) rotating at a speed of 300 rpm. Assume a uniform rectangular cross section $1'' \times 12''$ and a length $48''$ for the blade. The material of the blade is aluminum.

12.42 An electric motor weighing 1000 lb operates on the first floor of a building frame that can be modeled by a steel girder supported by two reinforced concrete columns, as shown in Fig. 12.38. If the operating speed of the motor is 1500 rpm, design the girder and the columns such that the fundamental frequency of vibration of the building frame is greater than the operating speed

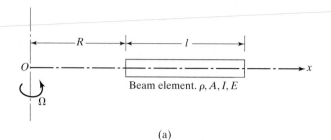

(a)

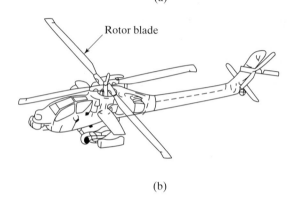

Rotor blade

(b)

FIGURE 12.37

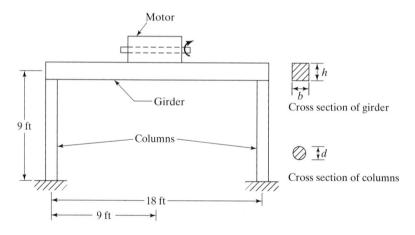

FIGURE 12.38

of the motor. Use two beam and two bar elements for the idealization. Assume the following data:

$$\text{Girder: } E = 30 \times 10^6 \text{ psi}, \qquad \rho = 8.8 \times 10^{-3} \text{ lbm/in.}^3, \qquad h/b = 2$$

$$\text{Columns: } E = 4 \times 10^6 \text{ psi}, \qquad \rho = 2.7 \times 10^{-3} \text{ lbm/in.}^3$$

Jules Henri Poincaré (1854–1912) was a French mathematician and professor of celestial mechanics at the University of Paris and of mechanics at the Ecole Polytechnique. His contributions to pure and applied mathematics, particularly to celestial mechanics and electrodynamics, are outstanding. His classification of singular points of nonlinear autonomous systems is important in the study of nonlinear vibrations. (Photo courtesy of Dirk J. Struik, *A Concise History of Mathematics*, 2nd ed. Dover Publications, New York, 1948)

CHAPTER 13

Nonlinear Vibration

13.1 Introduction

In the preceding chapters, the equation of motion contained displacement or its derivatives only to the first degree, and no square or higher powers of displacement or velocity were involved. For this reason, the governing differential equations of motion and the corresponding systems were called *linear*. For convenience of analysis, most systems are modeled as linear systems, but real systems are actually more often nonlinear than linear [13.1–13.6]. Whenever finite amplitudes of motion are encountered, nonlinear analysis becomes necessary. The superposition principle, which is very useful in linear analysis, does not hold true in the case of nonlinear analysis. Since mass, damper, and spring are the basic components of a vibratory system, nonlinearity into the governing differential equation may be introduced through any of these components. In many cases, linear analysis is insufficient to describe the behavior of the physical system adequately. One of the main reasons for modeling a physical system as a nonlinear one is that totally unexpected phenomena sometimes occur in nonlinear systems—phenomena that are not predicted or even hinted at by linear theory. Several methods are available for the solution of nonlinear vibration problems. Some of the exact methods, approximate analytical techniques, graphical procedures, and numerical methods are presented in this chapter.

13.2   Examples of Nonlinear Vibration Problems

The following examples are given to illustrate the nature of nonlinearity in some physical systems.

**13.2.1
Simple
Pendulum**

Consider a simple pendulum of length l, having a bob of mass m, as shown in Fig. 13.1(a). The differential equation governing the free vibration of the pendulum can be derived from Fig. 13.1(b):

$$ml^2 \ddot{\theta} + mgl \sin \theta = 0 \tag{13.1}$$

For small angles, $\sin \theta$ may be approximated by θ and Eq. (13.1) reduces to a linear equation:

$$\ddot{\theta} + \omega_0^2 \theta = 0 \tag{13.2}$$

where

$$\omega_0 = (g/l)^{1/2} \tag{13.3}$$

The solution of Eq. (13.2) can be expressed as

$$\theta(t) = A_0 \sin(\omega_0 t + \phi) \tag{13.4}$$

where A_0 is the amplitude of oscillation, ϕ is the phase angle, and ω_0 is the angular frequency. The values of A_0 and ϕ are determined by the initial conditions and the angular frequency ω_0 is independent of the amplitude A_0. Equation (13.4) denotes an approximate solution of the simple pendulum. A better approximate solution can be obtained by using a two-term approximation for $\sin \theta$ near $\theta = 0$ as $\theta - \theta^3/6$ in Eq. (13.1)

$$ml^2 \ddot{\theta} + mgl\left(\theta - \frac{\theta^3}{6}\right) = 0$$

or

$$\ddot{\theta} + \omega_0^2(\theta - \tfrac{1}{6}\theta^3) = 0 \tag{13.5}$$

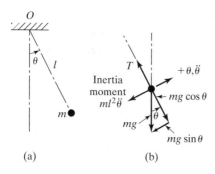

(a)   (b)

FIGURE 13.1   Simple pendulum.

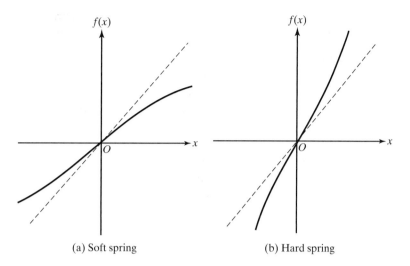

(a) Soft spring   (b) Hard spring

FIGURE 13.2   Nonlinear spring characteristics.

It can be seen that Eq. (13.5) is nonlinear because of the term involving θ^3 (due to geometric nonlinearity). Equation (13.5) is similar to the equation of motion of a spring-mass system with a nonlinear spring. If the spring is nonlinear (due to material nonlinearity), the restoring force can be expressed as $f(x)$, where x is the deformation of the spring and the equation of motion of the spring-mass system becomes

$$m\ddot{x} + f(x) = 0 \qquad (13.6)$$

If $df/dx(x) = k = $ constant, the spring is linear. If df/dx is a strictly increasing function of x, the spring is called a hard spring, and if df/dx is a strictly decreasing function of x, the spring is called a soft spring as shown in Fig. 13.2. Due to the similarity of Eqs. (13.5) and (13.6), a pendulum with large amplitudes is considered, in a loose sense, as a system with a nonlinear elastic (spring) component.

**13.2.2
Mechanical
Chatter, Belt
Friction System**

Nonlinearity may be reflected in the damping term as in the case of Fig. 13.3(a). The system behaves nonlinearly because of the dry friction between the mass m and the moving belt. For this system, there are two friction coefficients: the static coefficient of friction (μ_s), corresponding to the force required to initiate the motion of the body held by dry friction; and the kinetic coefficient of friction (μ_k), corresponding to the force required to maintain the body in motion. In either case, the component of the applied force tangent to the friction surface (F) is the product of the appropriate friction coefficient and the force normal to the surface.

The sequence of motion of the system shown in Fig. 13.3(a) is as follows [13.7]. The mass is initially at rest on the belt. Due to the displacement of the mass m along with the belt, the spring elongates. As the spring extends, the spring force on the mass increases until the static friction force is overcome and the mass begins to slide. It slides rapidly towards the right, thereby relieving the spring force until the kinetic friction force halts it. The spring then begins to build up the spring force again. The variation of the damping

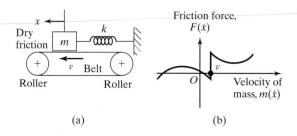

FIGURE 13.3   Dry friction damping.

force with the velocity of the mass is shown in Fig. 13.3(b). The equation of motion can be expressed as

$$m\ddot{x} + F(\dot{x}) + kx = 0 \tag{13.7}$$

where the friction force F is a nonlinear function of $\dot{x}$, as shown in Fig. 13.3(b).

For large values of $\dot{x}$, the damping force is positive (the curve has a positive slope) and energy is removed from the system. On the other hand, for small values of $\dot{x}$, the damping force is negative (the curve has a negative slope) and energy is put into the system. Although there is no external stimulus, the system can have an oscillatory motion; it corresponds to a nonlinear self-excited system. This phenomenon of self-excited vibration is called *mechanical chatter*.

**13.2.3
Variable Mass
System**

Nonlinearity may appear in the mass term as in the case of Fig. 13.4 [13.8]. For large deflections, the mass of the system depends on the displacement x, and so the equation of motion becomes

$$\frac{d}{dt}(m\dot{x}) + kx = 0 \tag{13.8}$$

Note that this is a nonlinear differential equation due to the nonlinearity of the first term.

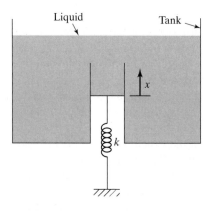

FIGURE 13.4   Variable mass system.

13.3   Exact Methods

An exact solution is possible only for a relatively few nonlinear systems whose motion is governed by specific types of second-order nonlinear differential equations. The solutions are exact in the sense that they are given either in closed form or in the form of an expression that can be numerically evaluated to any degree of accuracy. In this section, we shall consider a simple nonlinear system for which the exact solution is available. For a single degree of freedom system with a general restoring (spring) force $F(x)$, the free vibration equation can be expressed as

$$\ddot{x} + a^2 F(x) = 0 \tag{13.9}$$

where a^2 is a constant. Equation (13.9) can be rewritten as

$$\frac{d}{dx}(\dot{x}^2) + 2a^2 F(x) = 0 \tag{13.10}$$

Assuming the initial displacement as x_0 and the velocity as zero at $t = t_0$, Eq. (13.10) can be integrated to obtain

$$\dot{x}^2 = 2a^2 \int_x^{x_0} F(\eta) d\eta \quad \text{or} \quad |\dot{x}| = \sqrt{2}a \left\{ \int_x^{x_0} F(\eta) d\eta \right\}^{1/2} \tag{13.11}$$

where η is the integration variable. Equation (13.11), when integrated again, gives

$$t - t_0 = \frac{1}{\sqrt{2}a} \int_0^x \frac{d\xi}{\left\{ \int_\xi^{x_0} F(\eta) d\eta \right\}^{1/2}} \tag{13.12}$$

where ξ is the new integration variable and t_0 corresponds to the time when $x = 0$. Equation (13.12) thus gives the exact solution of Eq. (13.9) in all those situations where the integrals of Eq. (13.12) can be evaluated in closed form. After evaluating the integrals of Eq. (13.12), one can invert the result and obtain the displacement-time relation. If $F(x)$ is an odd function,

$$F(-x) = -F(x) \tag{13.13}$$

By considering Eq. (13.12) from zero displacement to maximum displacement, the period of vibration τ can be obtained:

$$\tau = \frac{4}{\sqrt{2}a} \int_0^{x_0} \frac{d\xi}{\left\{ \int_\xi^{x_0} F(\eta) d\eta \right\}^{1/2}} \tag{13.14}$$

For illustration, let $F(x) = x^n$. In this case Eqs. (13.12) and (13.14) become

$$t - t_0 = \frac{1}{a}\sqrt{\frac{n+1}{2}} \int_0^{x_0} \frac{d\xi}{(x_0^{n+1} - \xi^{n+1})^{1/2}} \tag{13.15}$$

and

$$\tau = \frac{4}{a}\sqrt{\frac{n+1}{2}}\int_0^{x_0}\frac{d\xi}{\left(x_0^{n+1}-\xi^{n+1}\right)^{1/2}} \tag{13.16}$$

By setting $y = \xi/x_0$, Eq. (13.16) can be written as

$$\tau = \frac{4}{a}\frac{1}{\left(x_0^{n-1}\right)^{1/2}}\sqrt{\frac{n+1}{2}}\int_0^1\frac{dy}{\left(1-y^{n+1}\right)^{1/2}} \tag{13.17}$$

This expression can be evaluated numerically to any desired level of accuracy.

13.4   Approximate Analytical Methods

In the absence of an exact analytical solution to a nonlinear vibration problem, we wish to find at least an approximate solution. Although both analytical and numerical methods are available for approximate solution of nonlinear vibration problems, the analytical methods are more desirable [13.6, 13.9]. The reason is that once the analytical solution is obtained, any desired numerical values can be substituted and the entire possible range of solutions can be investigated. We shall now consider four analytical techniques in the following subsections.

13.4.1
Basic
Philosophy

Let the equations governing the vibration of a nonlinear system be represented by a system of n first order differential equations[1]

$$\vec{\dot{x}}(t) = \vec{f}(\vec{x},t) + \alpha\vec{g}(\vec{x},t) \tag{13.18}$$

where the nonlinear terms are assumed to appear only in $\vec{g}(\vec{x},t)$ and α is a small parameter. In Eq. (13.18)

$$\vec{x} = \begin{Bmatrix} x_1 \\ x_2 \\ \cdot \\ \cdot \\ \cdot \\ x_n \end{Bmatrix}, \quad \vec{\dot{x}} = \begin{Bmatrix} dx_1/dt \\ dx_2/dt \\ \cdot \\ \cdot \\ \cdot \\ dx_n/dt \end{Bmatrix}, \quad \vec{f}(\vec{x},t) = \begin{Bmatrix} f_1(x_1, x_2, \ldots, x_n, t) \\ f_2(x_1, x_2, \ldots, x_n, t) \\ \cdot \\ \cdot \\ \cdot \\ f_n(x_1, x_2, \ldots, x_n, t) \end{Bmatrix}$$

and

$$\vec{g}(\vec{x},t) = \begin{Bmatrix} g_1(x_1, x_2, \ldots, x_n, t) \\ g_2(x_1, x_2, \ldots, x_n, t) \\ \cdot \\ \cdot \\ \cdot \\ g_n(x_1, x_2, \ldots, x_n, t) \end{Bmatrix}$$

---

[1]Systems governed by Eq. (13.18), in which the time appears explicitly, are known as *nonautonomous* systems. On the other hand, systems for which the governing equations are of the type

$$\vec{\dot{x}}(t) = \vec{f}(x) + \alpha\vec{g}(x)$$

where time does not appear explicitly are called *autonomous* systems.

The solution of differential equations having nonlinear terms associated with a small parameter was studied by Poincaré [13.6]. Basically, he assumed the solution of Eq. (13.18) in series form as

$$\vec{x}(t) = \vec{x}_0(t) + \alpha \vec{x}_1(t) + \alpha^2 \vec{x}_2(t) + \alpha^3 \vec{x}_3(t) + \cdots \tag{13.19}$$

The series solution of Eq. (13.19) has two basic characteristics:

1. As $\alpha \rightarrow 0$, Eq. (13.19) reduces to the exact solution of the linear equations $\dot{\vec{x}} = \vec{f}(\vec{x}, t)$.
2. For small values of α, the series converges fast so that even the first two or three terms in the series of Eq. (13.19) yields a reasonably accurate solution.

The various approximate analytical methods presented in this section can be considered to be modifications of the basic idea contained in Eq. (13.19). Although Poincaré's solution, Eq. (13.19), is valid for only small values of α, the method can still be applied to systems with large values of α. The solution of the pendulum equation, Eq. (13.5), is presented to illustrate the Poincaré's method.

Solution of Pendulum Equations.   Equation (13.5) can be rewritten as

$$\ddot{x} + \omega_0^2 x + \alpha x^3 = 0 \tag{13.20}$$

where $x = \theta$, $\omega_0 = (g/l)^{1/2}$, and $\alpha = -\omega_0^2/6$. Equation (13.20) is known as the free Duffing's equation. Assuming weak nonlinearity (i.e., α is small), the solution of Eq. (13.20) is expressed as

$$x(t) = x_0(t) + \alpha x_1(t) + \alpha^2 x_2(t) + \cdots + \alpha^n x_n(t) + \cdots \tag{13.21}$$

where $x_i(t), i = 0, 1, 2, \ldots, n$, are functions to be determined. By using a two-term approximation in Eq. (13.21), Eq. (13.20) can be written as

$$(\ddot{x}_0 + \alpha \ddot{x}_1) + \omega_0^2(x_0 + \alpha x_1) + \alpha(x_0 + \alpha x_1)^3 = 0$$

that is,

$$(\ddot{x}_0 + \omega_0^2 x_0) + \alpha(\ddot{x}_1 + \omega_0^2 x_1 + x_0^3) + \alpha^2(3x_0^2 x_1)$$
$$+ \alpha^3(3x_0 x_1^2) + \alpha^4 x_1^3 = 0 \tag{13.22}$$

If terms involving α^2, α^3, and α^4 are neglected (since α is assumed to be small), Eq. (13.22) will be satisfied if the following equations are satisfied:

$$\ddot{x}_0 + \omega_0^2 x_0 = 0 \tag{13.23}$$

$$\ddot{x}_1 + \omega_0^2 x_1 = -x_0^3 \tag{13.24}$$

The solution of Eq. (13.23) can be expressed as

$$x_0(t) = A_0 \sin(\omega_0 t + \phi) \tag{13.25}$$

In view of Eq. (13.25), Eq. (13.24) becomes

$$\ddot{x}_1 + \omega_0^2 x_1 = -A_0^3 \sin^3(\omega_0 t + \phi)$$

$$= -A_0^3 \left[\frac{3}{4} \sin(\omega_0 t + \phi) - \frac{1}{4} \sin 3(\omega_0 t + \phi) \right] \qquad (13.26)$$

The particular solution of Eq. (13.26) is (and can be verified by substitution)

$$x_1(t) = \frac{3}{8\omega_0} t A_0^3 \cos(\omega_0 t + \phi) - \frac{A_0^3}{32\,\omega_0^2} \sin 3(\omega_0 t + \phi) \qquad (13.27)$$

Thus the approximate solution of Eq. (13.20) becomes

$$x(t) = x_0(t) + \alpha x_1(t)$$

$$= A_0 \sin(\omega_0 t + \phi) + \frac{3\alpha t}{8\omega_0} A_0^3 \cos(\omega_0 t + \phi) - \frac{A_0^3 \alpha}{32\,\omega_0^2} \sin 3(\omega_0 t + \phi) \qquad (13.28)$$

The initial conditions on $x(t)$ can be used to evaluate the constants A_0 and ϕ.

Notes

1. It can be seen that even a weak nonlinearity (i.e., small value of α) leads to a nonperiodic solution since Eq. (13.28) is not periodic due to the second term on the right-hand side of Eq. (13.28). In general, the solution given by Eq. (13.21) will not be periodic if we retain only a finite number of terms.

2. In Eq. (13.28), the second term, and hence the total solution, can be seen to approach infinity as t tends to infinity. However, the exact solution of Eq. (13.20) is known to be bounded for all values of t. The reason for the unboundedness of the solution, Eq. (13.28), is that only two terms are considered in Eq. (13.21). The second term in Eq. (13.28) is called a *secular term*. The infinite series in Eq. (13.21) leads to a bounded solution of Eq. (13.20) because the process is a convergent one. To illustrate this point, consider the Taylor's series expansion of the function $\sin(\omega t + \alpha t)$:

$$\sin(\omega + \alpha)t = \sin \omega t + \alpha t \cos \omega t$$

$$- \frac{\alpha^2 t^2}{2!} \sin \omega t - \frac{\alpha^3 t^3}{3!} \cos \omega t + \cdots \qquad (13.29)$$

If only two terms are considered on the right-hand side of Eq. (13.29), the solution approaches infinity as $t \rightarrow \infty$. However, the function itself and hence its infinite series expansion can be seen to be a bounded one.

13.4.2 Lindstedt's Perturbation Method

This method assumes that the angular frequency along with the solution varies as a function of the amplitude A_0. This method eliminates the secular terms in each step of the approximation [13.5] by requiring the solution to be periodic in each step. The solution and the angular frequency are assumed as

$$x(t) = x_0(t) + \alpha x_1(t) + \alpha^2 x_2(t) + \cdots \qquad (13.30)$$

$$\omega^2 = \omega_0^2 + \alpha\omega_1(A_0) + \alpha^2\omega_2(A_0) + \cdots \tag{13.31}$$

We consider the solution of the pendulum equation, Eq. (13.20), to illustrate the perturbation method. We use only linear terms in α in Eqs. (13.30) and (13.31):

$$x(t) = x_0(t) + \alpha x_1(t) \tag{13.32}$$

$$\omega^2 = \omega_0^2 + \alpha\omega_1(A_0) \qquad \text{or} \qquad \omega_0^2 = \omega^2 - \alpha\omega_1(A_0) \tag{13.33}$$

Substituting Eqs. (13.32) and (13.33) into Eq. (13.20), we get

$$\ddot{x}_0 + \alpha\ddot{x}_1 + [\omega^2 - \alpha\omega_1(A_0)][x_0 + \alpha x_1] + \alpha[x_0 + \alpha x_1]^3 = 0$$

that is,

$$\ddot{x}_0 + \omega_0^2 x_0 + \alpha(\omega^2 x_1 + x_0^3 - \omega_1 x_0 + \ddot{x}_1)$$
$$+ \alpha^2(3x_1 x_0^2 - \omega_1 x_1) + \alpha^3(3x_1^2 x_0) + \alpha^4(x_1^3) = 0 \tag{13.34}$$

Setting the coefficients of various powers of α to zero and neglecting the terms involving α^2, α^3, and α^4 in Eq. (13.34), we obtain

$$\ddot{x}_0 + \omega^2 x_0 = 0 \tag{13.35}$$

$$\ddot{x}_1 + \omega^2 x_1 = -x_0^3 + \omega_1 x_0 \tag{13.36}$$

Using the solution of Eq. (13.35),

$$x_0(t) = A_0 \sin(\omega t + \phi) \tag{13.37}$$

into Eq. (13.36), we obtain

$$\ddot{x}_1 + \omega^2 x_1 = -[A_0 \sin(\omega t + \phi)]^3 + \omega_1[A_0 \sin(\omega t + \phi)]$$
$$= -\tfrac{3}{4}A_0^3 \sin(\omega t + \phi) + \tfrac{1}{4}A_0^3 \sin 3(\omega t + \phi)$$
$$+ \omega_1 A_0 \sin(\omega t + \phi) \tag{13.38}$$

It can be seen that the first and the last terms on the right-hand side of Eq. (13.38) lead to secular terms. They can be eliminated by taking ω_1 as

$$\omega_1 = \tfrac{3}{4}A_0^2, \qquad A_0 \neq 0 \tag{13.39}$$

With this, Eq. (13.38) becomes

$$\ddot{x}_1 + \omega^2 x_1 = \tfrac{1}{4}A_0^3 \sin 3(\omega t + \phi) \tag{13.40}$$

The solution of Eq. (13.40) is

$$x_1(t) = A_1 \sin(\omega t + \phi_1) - \frac{A_0^3}{32\,\omega^2} \sin 3(\omega t + \phi) \qquad (13.41)$$

Let the initial conditions be $x(t = 0) = A$ and $\dot{x}(t = 0) = 0$. Using Lindstedt's method, we force the solution $x_0(t)$ given by Eq. (13.37) to satisfy the initial conditions so that

$$x(0) = A = A_0 \sin \phi, \qquad \dot{x}(0) = 0 = A_0\omega \cos \phi$$

or

$$A_0 = A \qquad \text{and} \qquad \phi = \frac{\pi}{2}$$

Since the initial conditions are satisfied by $x_0(t)$ itself, the solution $x_1(t)$ given by Eq. (13.41) must satisfy zero initial conditions.[2] Thus

$$x_1(0) = 0 = A_1 \sin \phi_1 - \frac{A_0^3}{32\,\omega^2} \sin 3\phi$$

$$\ddot{x}_1(0) = 0 = A_1\omega \cos \phi_1 - \frac{A_0^3}{32\,\omega^2}(3\omega) \cos 3\phi$$

In view of the known relations $A_0 = A$ and $\phi = \pi/2$, the above equations yield

$$A_1 = -\left(\frac{A^3}{32\,\omega^2} \right) \qquad \text{and} \qquad \phi_1 = \frac{\pi}{2}.$$

Thus the total solution of Eq. (13.20) becomes

$$x(t) = A_0 \sin(\omega t + \phi) - \frac{\alpha A_0^3}{32\,\omega^2} \sin 3(\omega t + \phi) \qquad (13.42)$$

with

$$\omega^2 = \omega_0^2 + \alpha \frac{3}{4} A_0^2 \qquad (13.43)$$

For the solution obtained by considering three terms in the expansion of Eq. (13.30), see Problem 13.13. It is to be noted that the Lindstedt's method gives only the periodic solutions of Eq. (13.20); it cannot give any nonperiodic solutions, even if they exist.

13.4.3
Iterative Method

In the basic iterative method, first the equation is solved by neglecting certain terms. The resulting solution is then inserted in the terms that were neglected at first to obtain a second, improved, solution. We shall illustrate the iterative method to find the solution of Duffing's equation, which represents the equation of motion of a damped, harmonically excited, single degree of freedom system with a nonlinear spring. We begin with the solution of the undamped equation.

---

[2]If $x_0(t)$ satisfies the initial conditions, each of the solutions $x_1(t)$, $x_2(t)$, ... appearing in Eq. (13.30) must satisfy zero initial conditions.

Solution of the Undamped Equation.   If damping is disregarded, Duffing's equation becomes

$$\ddot{x} + \omega_0^2 \pm \alpha x^3 = F \cos \omega t$$

or

$$\ddot{x} = -\omega_0^2 x \mp \alpha x^3 + F \cos \omega t \tag{13.44}$$

As a first approximation, we assume the solution to be

$$x_1(t) = A \cos \omega t \tag{13.45}$$

where A is an unknown. By substituting Eq. (13.45) into Eq. (13.44), we obtain the differential equation for the second approximation:

$$\ddot{x}_2 = -A\omega_0^2 \cos \omega t \mp A^3 \alpha \cos^3 \omega t + F \cos \omega t \tag{13.46}$$

By using the identity

$$\cos^3 \omega t = \tfrac{3}{4} \cos \omega t + \tfrac{1}{4} \cos 3\omega t \tag{13.47}$$

Eq. (13.46) can be expressed as

$$\ddot{x}_2 = -(A\omega_0^2 \pm \tfrac{3}{4} A^3 \alpha - F) \cos \omega t \mp \tfrac{1}{4} A^3 \alpha \cos 3\omega t \tag{13.48}$$

By integrating this equation and setting the constants of integration to zero (so as to make the solution harmonic with period $\tau = 2\pi/\omega$), we obtain the second approximation:

$$x_2(t) = \frac{1}{\omega^2} (A\omega_0^2 \pm \tfrac{3}{4} A^3 \alpha - F) \cos \omega t \pm \frac{A^3 \alpha}{36\omega^2} \cos 3\omega t \tag{13.49}$$

Duffing [13.7] reasoned at this point that if $x_1(t)$ and $x_2(t)$ are good approximations to the solution $x(t)$, the coefficients of $\cos \omega t$ in the two equations (13.45) and (13.49) should not be very different. Thus by equating these coefficients, we obtain

$$A = \frac{1}{\omega^2} \left(A\omega_0^2 \pm \frac{3}{4} A^3 \alpha - F \right)$$

or

$$\omega^2 = \omega_0^2 \pm \frac{3}{4} A^2 \alpha - \frac{F}{A} \tag{13.50}$$

For present purposes, we will stop the procedure with the second approximation. It can be verified that this procedure yields the exact solution for the case of a linear spring (with $\alpha = 0$)

$$A = \frac{F}{\omega_0^2 - \omega^2} \tag{13.51}$$

where A denotes the amplitude of the harmonic response of the linear system.

For a nonlinear system (with $\alpha \neq 0$), Eq. (13.50) shows that the frequency ω is a function of α, A, and F. Note that the quantity A, in the case of a nonlinear system, is not the amplitude of the harmonic response but only the coefficient of the first term of its solution. However, it is commonly taken as the amplitude of the harmonic response of the system.[3] For the free vibration of the nonlinear system, $F = 0$ and Eq. (13.50) reduces to

$$\omega^2 = \omega_0^2 \pm \tfrac{3}{4} A^2 \alpha \tag{13.52}$$

This equation shows that the frequency of the response increases with the amplitude A for the hardening spring and decreases for the softening spring. The solution, Eq. (13.52), can also be seen to be same as the one given by Lindstedt's method, Eq. (13.43).

For both linear and nonlinear systems, when $F \neq 0$ (forced vibration), there are two values of the frequency ω for any given amplitude $|A|$. One of these values of ω is smaller and the other larger than the corresponding frequency of free vibration at that amplitude. For the smaller value of ω, $A > 0$ and the harmonic response of the system is in phase with the external force. For the larger value of ω, $A < 0$ and the response is 180° out of phase with the external force. Note that only the harmonic solutions of Duffing's equation—that is, solutions for which the frequency is the same as that of the external force $F \cos \omega t$—have been considered in the present analysis. It has been observed [13.2] that oscillations whose frequency is a fraction, such as $\tfrac{1}{2}, \tfrac{1}{3}, \ldots, \tfrac{1}{n}$, of that of the applied force, are also possible for Duffing's equation. Such oscillations, known as subharmonic oscillations, are considered in Section 13.5.

Solution of the Damped Equation. If we consider viscous damping, we obtain Duffing's equation:

$$\ddot{x} + c\dot{x} + \omega_0^2 x \pm \alpha x^3 = F \cos \omega t \tag{13.53}$$

For a damped system, it was observed in earlier chapters that there is a phase difference between the applied force and the response or solution. The usual procedure is to prescribe the applied force first and then determine the phase of the solution. In the present case, however, it is more convenient to fix the phase of the solution and keep the phase of the applied force as a quantity to be determined. We take the differential equation, Eq. (13.53), in the form

$$\ddot{x} + c\dot{x} + \omega_0^2 x \pm \alpha x^3 = F \cos(\omega t + \phi)$$
$$= A_1 \cos \omega t - A_2 \sin \omega t \tag{13.54}$$

in which the amplitude $F = (A_1^2 + A_2^2)^{1/2}$ of the applied force is considered fixed, but the ratio $A_1/A_2 = \tan^{-1} \phi$ is left to be determined. We assume that c, A_1, and A_2 are all small, of order α. As with Eq. (13.44), we assume the first approximation to the solution to be

$$x_1 = A \cos \omega t \tag{13.55}$$

---

[3]The first approximate solution, Eq. (13.45), can be seen to satisfy the initial conditions $x(0) = A$ and $\dot{x}(0) = 0$.

where A is assumed fixed and ω to be determined. By substituting Eq. (13.55) into Eq. (13.54) and making use of the relation (13.47), we obtain

$$\left[(\omega_0^2 - \omega^2)A \pm \frac{3}{4}\alpha A^3\right] \cos \omega t - c\omega A \sin \omega t \pm \frac{\alpha A^3}{4} \cos 3\omega t$$

$$= A_1 \cos \omega t - A_2 \sin \omega t \qquad (13.56)$$

By disregarding the term involving $\cos 3\omega t$ and equating the coefficients of $\cos \omega t$ and $\sin \omega t$ on both sides of Eq. (13.56), we obtain the following relations:

$$(\omega_0^2 - \omega^2)A \pm \tfrac{3}{4}\alpha A^3 = A_1$$
$$c\omega A = A_2 \qquad (13.57)$$

The relation between the amplitude of the applied force and the quantities A and ω can be obtained by squaring and adding the equations (13.57):

$$\left[(\omega_0^2 - \omega^2)A \pm \tfrac{3}{4}\alpha A^3\right]^2 + (c\omega A)^2 = A_1^2 + A_2^2 = F^2 \qquad (13.58)$$

Equation (13.58) can be rewritten as

$$S^2(\omega, A) + c^2\omega^2 A^2 = F^2 \qquad (13.59)$$

where

$$S(\omega, A) = (\omega_0^2 - \omega^2)A \pm \tfrac{3}{4}\alpha A^3 \qquad (13.60)$$

It can be seen that for $c = 0$, Eq. (13.59) reduces to $S(\omega, A) = F$, which is the same as Eq. (13.50). The response curves given by Eq. (13.59) are shown in Fig. 13.5.

Jump Phenomenon. As mentioned earlier, nonlinear systems exhibit phenomena that cannot occur in linear systems. For example, the amplitude of vibration of the system

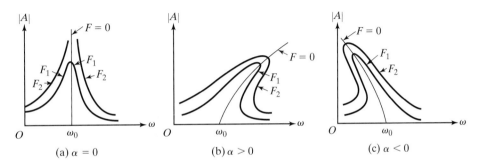

FIGURE 13.5 Response curves of Duffing's equation.

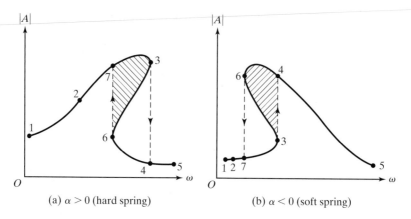

(a) $\alpha > 0$ (hard spring)   (b) $\alpha < 0$ (soft spring)

FIGURE 13.6   Jump phenomenon.

described by Eq. (13.54) has been found to increase or decrease suddenly as the excitation frequency ω is increased or decreased, as shown in Fig. 13.6. For a constant magnitude of F, the amplitude of vibration will increase along the points 1, 2, 3, 4, 5 on the curve when the excitation frequency ω is slowly increased. The amplitude of vibration jumps from point 3 to 4 on the curve. Similarly, when the forcing frequency ω is slowly decreased, the amplitude of vibration follows the curve along the points 5, 4, 6, 7, 2, 1 and makes a jump from point 6 to 7. This behavior is known as the *jump phenomenon*. It is evident that two amplitudes of vibration exist for a given forcing frequency, as shown in the shaded regions of the curves of Fig. 13.6. The shaded region can be thought of as unstable in some sense. Thus an understanding of the jump phenomena requires a knowledge of the mathematically involved stability analysis of periodic solutions [13.24, 13.25]. The jump phenomena were also observed experimentally by several investigators [13.26, 13.27].

13.4.4 Ritz-Galerkin Method

In the Ritz-Galerkin method, an approximate solution of the problem is found by satisfying the governing nonlinear equation in the average. To see how the method works, let the nonlinear differential equation be represented as

$$E[x] = 0 \tag{13.61}$$

An approximate solution of Eq. (13.61) is assumed as

$$x(t) = a_1\phi_1(t) + a_2\phi_2(t) + \cdots + a_n\phi_n(t) \tag{13.62}$$

where $\phi_1(t), \phi_2(t), \ldots, \phi_n(t)$ are prescribed functions of time and $a_1, a_2, \ldots, a_n$ are weighting factors to be determined. If Eq. (13.62) is substituted in Eq. (13.61), we get a function $E[x(t)]$. Since $x(t)$ is not, in general, the exact solution of Eq. (13.61), $E(t) = E[x(t)]$ will not be zero. However, the value of $E[t]$ will serve as a measure of the accuracy of the approximation; in fact, $E[t] \rightarrow 0$ as $x \rightarrow x$.

The weighting factors a_i are determined by minimizing the integral

$$\int_0^\tau \tilde{E}^2[t]\, dt \tag{13.63}$$

where τ denotes the period of the motion. The minimization of the function of Eq. (13.63) requires

$$\frac{\partial}{\partial a_i}\left(\int_0^\tau \tilde{E}^2[t]\, dt\right) = 2\int_0^\tau \tilde{E}[t]\frac{\partial \tilde{E}[t]}{\partial a_i}\, dt = 0,$$

$$i = 1, 2, \ldots, n \tag{13.64}$$

Equation (13.64) represents a system of n algebraic equations that can be solved simultaneously to find the values of $a_1, a_2, \ldots, a_n$. The procedure is illustrated with the following example.

EXAMPLE 13.1

Solution of Pendulum Equation Using the Ritz-Galerkin Method

Using a one-term approximation, find the solution of the pendulum equation

$$E[x] = \ddot{x} + \omega_0^2 x - \frac{\omega_0^2}{6}x^3 = 0 \tag{E.1}$$

by the Ritz-Galerkin method.

Solution: By using a one-term approximation for $x(t)$ as

$$\tilde{x}(t) = A_0 \sin \omega t \tag{E.2}$$

Eqs. (E.1) and (E.2) lead to

$$E[\tilde{x}(t)] = -\omega^2 A_0 \sin \omega t + \omega_0^2\left[A_0 \sin \omega t - \frac{1}{6}\sin^3 \omega t\right]$$

$$= \left(\omega_0^2 - \omega^2 - \frac{1}{8}\omega_0^2 A_0^2\right)A_0 \sin \omega t + \frac{\omega_0^2}{24}A_0^3 \sin 3\omega t \tag{E.3}$$

The Ritz-Galerkin method requires the minimization of

$$\int_0^\tau E^2[\tilde{x}(t)]\, dt \tag{E.4}$$

for finding A_0. The application of Eq. (13.64) gives

$$\int_0^\tau \tilde{E}\frac{\partial \tilde{E}}{\partial A_0}\, dt = \int_0^\tau \left[\left(\omega_0^2 - \omega^2 - \frac{1}{8}\omega_0^2 A_0^2\right)A_0 \sin \omega t + \frac{\omega_0^2}{24}A_0^3 \sin 3\omega t\right]$$

$$\times \left[\left(\omega_0^2 - \omega^2 - \frac{3}{8}\omega_0^2 A_0^2\right)\sin \omega t + \frac{1}{8}\omega_0^2 A_0^2 \sin 3\omega t\right]dt = 0$$

that is,

$$A_0\left(\omega_0^2 - \omega^2 - \frac{1}{8}\omega_0^2 A_0^2\right)\left(\omega_0^2 - \omega^2 - \frac{3}{8}\omega_0^2 A_0^2\right)\int_0^\tau \sin^2 \omega t \, dt$$

$$+ \frac{\omega_0^2 A_0^3}{24}\left(\omega_0^2 - \omega^2 - \frac{3}{8}\omega_0^2 A_0^2\right)\int_0^\tau \sin \omega t \sin 3\omega t \, dt$$

$$+ \frac{1}{8}\omega_0^2 A_0^3\left(\omega_0^2 - \omega^2 - \frac{1}{8}\omega_0^2 A_0^2\right)\int_0^\tau \sin \omega t \sin 3\omega t \, dt$$

$$+ \frac{\omega_0^4 A_0^5}{192}\int_0^\tau \sin^2 3\omega t \, dt = 0$$

that is,

$$A_0\left[\left(\omega_0^2 - \omega^2 - \frac{1}{8}\omega_0^2 A_0^2\right)\left(\omega_0^2 - \omega^2 - \frac{3}{8}\omega_0^2 A_0^2\right) + \frac{\omega_0^4 A_0^4}{192}\right] = 0 \qquad \text{(E.5)}$$

For a nontrivial solution, $A_0 \neq 0$, and Eq. (E.5) leads to

$$\omega^4 + \omega^2\omega_0^2\left(\frac{1}{2}A_0^2 - 2\right) + \omega_0^4\left(1 - \frac{1}{2}A_0^2 + \frac{5}{96}A_0^4\right) = 0 \qquad \text{(E.6)}$$

The roots of the quadratic equation in ω^2, Eq. (E.5), can be found as

$$\omega^2 = \omega_0^2(1 - 0.147938 \, A_0^2) \qquad \text{(E.7)}$$

$$\omega^2 = \omega_0^2(1 - 0.352062 \, A_0^2) \qquad \text{(E.8)}$$

It can be verified that ω^2 given by Eq. (E.7) minimizes the quantity of (E.4), while the one given by Eq. (E.8) maximizes it. Thus the solution of Eq. (E.1) is given by Eq. (E.2) with

$$\omega^2 = \omega_0^2(1 - 0.147938 \, A_0^2) \qquad \text{(E.9)}$$

This expression can be compared with Lindstedt's solution and the iteration methods (Eqs. 13.43 and 13.52):

$$\omega^2 = \omega_0^2(1 - 0.125 \, A_0^2) \qquad \text{(E.10)}$$

The solution can be improved by using a two-term approximation for $x(t)$ as

$$\underset{\sim}{x}(t) = A_0 \sin \omega t + A_3 \sin 3\omega t \qquad \text{(E.11)}$$

The application of Eq. (13.64) with the solution of Eq. (E.11) leads to two simultaneous algebraic equations that must be numerically solved for A_0 and A_3.

■

Other approximate methods, such as the equivalent linearization scheme and the harmonic balance procedure, are also available for solving nonlinear vibration problems [13.10–13.12]. Specific solutions found using these techniques include the free vibration response of single degree of freedom oscillators [13.13, 13.14], two degree of freedom systems [13.15], and elastic beams [13.16, 13.17], and the transient response of forced systems [13.18, 13.19]. Several nonlinear problems of structural dynamics have been discussed by Crandall [13.30].

13.5   Subharmonic and Superharmonic Oscillations

We noted in Chapter 3 that for a linear system, when the applied force has a certain frequency of oscillation, the steady-state response will have the same frequency of oscillation. However, a nonlinear system will exhibit subharmonic and superharmonic oscillations. Subharmonic response involves oscillations whose frequencies (ω_n) are related to the forcing frequency (ω) as

$$\omega_n = \frac{\omega}{n} \tag{13.65}$$

where n is an integer ($n = 2, 3, 4, \ldots$). Similarly, superharmonic response involves oscillations whose frequencies (ω_n) are related to the forcing frequency (ω) as

$$\omega_n = n\omega \tag{13.66}$$

where $n = 2, 3, 4, \ldots$.

13.5.1 Subharmonic Oscillations

In this section, we consider the subharmonic oscillations of order $\frac{1}{3}$ of an undamped pendulum whose equation of motion is given by (undamped Duffing's equation):

$$\ddot{x} + \omega_0^2 x + \alpha x^3 = F \cos 3\omega t \tag{13.67}$$

where α is assumed to be small. We find the response using the perturbation method [13.4, 13.6]. Accordingly, we seek a solution of the form

$$x(t) = x_0(t) + \alpha x_1(t) \tag{13.68}$$

$$\omega^2 = \omega_0^2 + \alpha\omega_1 \quad \text{or} \quad \omega_0^2 = \omega^2 - \alpha\omega_1 \tag{13.69}$$

where ω denotes the fundamental frequency of the solution (equal to the third subharmonic frequency of the forcing frequency). Substituting Eqs. (13.68) and (13.69) into Eq. (13.67) gives

$$\ddot{x}_0 + \alpha\ddot{x}_1 + \omega^2 x_0 + \omega^2 \alpha x_1 - \alpha\omega_1 x_0 - \alpha^2 x_1 \omega_1$$
$$+ \alpha(x_0 + \alpha x_1)^3 = F \cos 3\omega t \tag{13.70}$$

If terms involving α^2, α^3, and α^4 are neglected, Eq. (13.70) reduces to

$$\ddot{x}_0 + \omega^2 x_0 + \alpha\ddot{x}_1 + \alpha\omega^2 x_1 - \alpha\omega_1 x_0 + \alpha x_0^3 = F \cos 3\omega t \tag{13.71}$$

We first consider the linear equation (by setting $\alpha = 0$):

$$\ddot{x}_0 + \omega^2 x_0 = F \cos 3\omega t \tag{13.72}$$

The solution of Eq. (13.72) can be expressed as

$$x_0(t) = A_1 \cos \omega t + B_1 \sin \omega t + C \cos 3\omega t \tag{13.73}$$

If the initial conditions are assumed as $x(t = 0) = A$ and $\dot{x}(t = 0) = 0$, we obtain $A_1 = A$ and $B_1 = 0$ so that Eq. (13.73) reduces to

$$x_0(t) = A \cos \omega t + C \cos 3\omega t \tag{13.74}$$

where C denotes the amplitude of the forced vibration. The value of C can be determined by substituting Eq. (13.74) into Eq. (13.72) and equating the coefficients of $\cos 3\omega t$ on both sides of the resulting equation, which yields

$$C = -\frac{F}{8\omega^2} \tag{13.75}$$

Now we consider the terms involving α in Eq. (13.71) and set them equal to zero:

$$\alpha(\ddot{x}_1 + \omega^2 x_1 - \omega_1 x_0 + x_0^3) = 0$$

or

$$\ddot{x}_1 + \omega^2 x_1 = \omega_1 x_0 - x_0^3 \tag{13.76}$$

The substitution of Eq. (13.74) into Eq. (13.76) results in

$$\ddot{x}_1 + \omega^2 x_1 = \omega_1 A \cos \omega t + \omega_1 C \cos 3\omega t - A^3 \cos^3 \omega t$$
$$- C^3 \cos^3 3\omega t - 3A^2 C \cos^2 \omega t \cos 3\omega t$$
$$- 3AC^2 \cos \omega t \cos^2 3\omega t \tag{13.77}$$

By using the trigonometric relations

$$\left.\begin{array}{l} \cos^2 \theta = \frac{1}{2} + \frac{1}{2}\cos 2\theta \\[4pt] \cos^3 \theta = \frac{3}{4}\cos \theta + \frac{1}{4}\cos 3\theta \\[4pt] \cos \theta \cos \phi = \frac{1}{2}\cos(\theta - \phi) + \frac{1}{2}\cos(\theta + \phi) \end{array}\right\} \tag{13.78}$$

Eq. (13.77) can be expressed as

$$\ddot{x}_1 + \omega^2 x_1 = A\left(\omega_1 - \frac{3}{4}A^2 - \frac{3}{2}C^2 - \frac{3}{4}AC\right)\cos \omega t$$
$$+ \left(\omega_1 C - \frac{A^3}{4} - \frac{3}{4}C^3 - \frac{3}{2}A^2 C\right)\cos 3\omega t$$
$$- \frac{3}{4}AC(A + C)\cos 5\omega t - \frac{3AC^2}{4}\cos 7\omega t - \frac{C^3}{4}\cos 9\omega t \tag{13.79}$$

The condition to avoid a secular term in the solution is that the coefficient of $\cos \omega t$ in Eq. (13.79) must be zero. Since $A \neq 0$ in order to have a subharmonic response,

$$\omega_1 = \tfrac{3}{4}(A^2 + AC + 2C^2) \tag{13.80}$$

Equations (13.80) and (13.75) give

$$\omega_1 = \frac{3}{4}\left(A^2 - \frac{AF}{8\omega^2} + \frac{2F^2}{64\omega^4}\right) \tag{13.81}$$

Substituting Eq. (13.81) into Eq. (13.69) and rearranging the terms, we obtain the equation to be satisfied by A and ω in order to have subharmonic oscillation as

$$\omega^6 - \omega_0^2\omega^4 - \frac{3\alpha}{256}(64A^2\omega^4 - 8AF\omega^2 + 2F^2) = 0 \tag{13.82}$$

Equation (13.82) can be seen to be a cubic equation in ω^2 and a quadratic in A. The relationship between the amplitude (A) and the subharmonic frequency (ω), given by Eq. (13.82), is shown graphically in Fig. 13.7. It has been observed that the curve PQ, where the slope is positive, represents stable solutions while the curve QR, where the slope is negative, denotes unstable solutions [13.4, 13.6]. The minimum value of amplitude for the

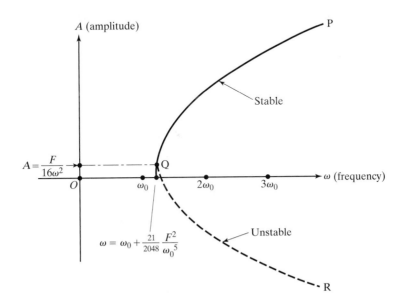

FIGURE 13.7   Subharmonic oscillations.

existence of stable subharmonic oscillations can be found by setting $d\omega^2/dA = 0$ as $A = (F/16\omega^2)$.[4]

13.5.2 Superharmonic Oscillations

Consider the undamped Duffing's equation

$$\ddot{x} + \omega_0^2 x + \alpha x^3 = F \cos \omega t \tag{13.83}$$

The solution of this equation is assumed as

$$x(t) = A \cos \omega t + C \cos 3\omega t \tag{13.84}$$

where the amplitudes of the harmonic and superharmonic components, A and C, are to be determined. The substitution of Eq. (13.84) into Eq. (13.83) gives, with the use of the trigonometric relations of Eq. (13.78),

$$
\begin{aligned}
\cos \omega t & \left[-\omega^2 A + \omega_0^2 A + \tfrac{3}{4}\alpha A^3 + \tfrac{3}{4}\alpha A^2 C + \tfrac{3}{2}\alpha AC^2 \right] \\
+ \cos 3\omega t & \left[-9\omega^2 C + \omega_0^2 C + \tfrac{1}{4}\alpha A^3 + \tfrac{3}{4}\alpha C^3 + \tfrac{3}{2}\alpha A^2 C \right] \\
+ \cos 5\omega t & \left[\tfrac{3}{4}\alpha A^2 C + \tfrac{3}{4}\alpha AC^2 \right] + \cos 7\omega t \left[\tfrac{3}{4}\alpha AC^2 \right] \\
+ \cos 9\omega t & \left[\tfrac{1}{4}\alpha C^3 \right] = F \cos \omega t
\end{aligned}
\tag{13.85}
$$

Neglecting the terms involving $\cos 5\omega t$, $\cos 7\omega t$, and $\cos 9\omega t$, and equating the coefficients of $\cos \omega t$ and $\cos 3\omega t$ on both sides of Eq. (13.85), we obtain

$$\omega_0^2 A - \omega^2 A + \tfrac{3}{4}\alpha A^3 + \tfrac{3}{4}\alpha A^2 C + \tfrac{3}{2}\alpha AC^2 = F \tag{13.86}$$

$$\omega_0^2 C - 9\omega^2 C + \tfrac{1}{4}\alpha A^3 + \tfrac{3}{4}\alpha C^3 + \tfrac{3}{2}\alpha A^2 C = 0 \tag{13.87}$$

Equations (13.86) and (13.87) represent a set of simultaneous nonlinear equations that can be solved numerically for A and C.

As a particular case, if C is assumed to be small compared to A, the terms involving C^2 and C^3 can be neglected and Eq. (13.87) gives

$$C \approx \frac{-\tfrac{1}{4}\alpha A^3}{\left(\tfrac{3}{2}\alpha A^2 + \omega_0^2 - 9\omega^2\right)} \tag{13.88}$$

and Eq. (13.86) gives

$$C \approx \frac{F - \omega_0^2 A + \omega^2 A - \tfrac{3}{4}\alpha A^3}{\tfrac{3}{4}\alpha A^2} \tag{13.89}$$

---

[4]Equation (13.82) can be rewritten as

$$(\omega^2)^3 - \omega_0^2(\omega^2)^2 - \frac{3\alpha}{4}A^2(\omega^2)^2 + \frac{3\alpha F}{32}A(\omega^2) - \frac{3\alpha F^2}{128} = 0$$

which, upon differentiation, gives

$$3(\omega^2)^2\, d\omega^2 - 2\omega_0^2\omega^2\, d\omega^2 - \frac{3\alpha}{4}(2A\, dA)(\omega^2)^2 - \frac{3\alpha}{2}A^2\omega^2\, d\omega^2 + \frac{3\alpha F}{32}\omega^2\, dA + \frac{3\alpha F}{32}A\, d\omega^2 = 0$$

By setting $d\omega^2/dA = 0$, we obtain $A = (F/16\omega^2)$.

Equating C from Eqs. (13.88) and (13.89) leads to

$$(-\tfrac{1}{4}\alpha A^3)(\tfrac{3}{4}\alpha A^2) = (\tfrac{3}{2}\alpha A^2 + \omega_0^2 - 9\omega^2)$$
$$\times (F - \omega_0^2 A + \omega^2 A - \tfrac{3}{4}\alpha A^3) \tag{13.90}$$

which can be rewritten as

$$-A^5(\tfrac{15}{16}\alpha^2) + A^3(\tfrac{33}{4}\alpha\omega^2 - \tfrac{9}{4}\alpha\omega_0^2) + A^2(\tfrac{3}{2}\alpha F)$$
$$+ A(10\,\omega^2\omega_0^2 - 9\omega^4 - \omega_0^4) + (\omega_0^2 F - 9\omega^2 F) = 0 \tag{13.91}$$

Equation (13.88), in conjunction with Eq. (13.91), gives the relationship between the amplitude of superharmonic oscillations (C) and the corresponding frequency (3ω).

13.6 Systems with Time-Dependent Coefficients (Mathieu Equation)

Consider the simple pendulum shown in Fig. 13.8(a). The pivot point of the pendulum is made to vibrate in the vertical direction as

$$y(t) = Y \cos \omega t \tag{13.92}$$

where Y is the amplitude and ω is the frequency of oscillation. Since the entire pendulum accelerates in the vertical direction, the net acceleration is given by $g - \ddot{y}(t) = g - \omega^2 Y \cos \omega t$. The equation of motion of the pendulum can be derived as

$$ml^2\ddot{\theta} + m(g - \ddot{y})\,l \sin \theta = 0 \tag{13.93}$$

For small deflections near $\theta = 0$, $\sin \theta \approx \theta$ and Eq. (13.93) reduces to

$$\ddot{\theta} + \left(\frac{g}{l} - \frac{\omega^2 Y}{l} \cos \omega t\right)\theta = 0 \tag{13.94}$$

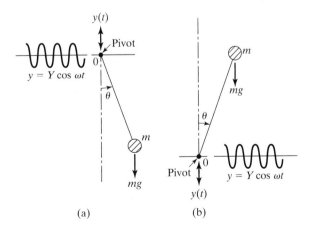

(a)   (b)

FIGURE 13.8 Simple pendulum with oscillations of pivot.

If the pendulum is inverted as shown in Fig. 13.8(b), the equation of motion becomes

$$ml^2\ddot{\theta} - mgl \sin \theta = 0$$

or

$$\ddot{\theta} - \frac{g}{l} \sin \theta = 0 \tag{13.95}$$

where θ is the angle measured from the vertical (unstable equilibrium) point. If the pivot point 0 vibrates as $y(t) = Y \cos \omega t$, the equation of motion becomes

$$\ddot{\theta} + \left(-\frac{g}{l} + \frac{\omega^2 Y}{l} \cos \omega t \right) \sin \theta = 0 \tag{13.96}$$

For small angular displacements around $\theta = 0$, Eq. (13.96) reduces to

$$\ddot{\theta} + \left(-\frac{g}{l} + \frac{\omega^2 Y}{l} \cos \omega t \right) \theta = 0 \tag{13.97}$$

Equations (13.94) and (13.97) are particular forms of an equation called the *Mathieu equation* for which the coefficient of θ in the differential equation varies with time to form a nonautonomous equation. We shall study the periodic solutions and their stability characteristics of the system for small values of Y in this section.

Periodic Solutions Using Lindstedt's Perturbation Method [13.7].   Consider the Mathieu equation in the form

$$\frac{d^2 y}{dt^2} + (a + \epsilon \cos t)y = 0 \tag{13.98}$$

where ϵ is assumed to be small. We approximate the solution of Eq. (13.98) as

$$y(t) = y_0(t) + \epsilon y_1(t) + \epsilon^2 y_2(t) + \cdots \tag{13.99}$$

$$a = a_0 + \epsilon a_1 + \epsilon^2 a_2 + \cdots \tag{13.100}$$

where $a_0, a_1, \ldots$ are constants. Since the periodic coefficient $\cos t$ in Eq. (13.98) varies with a period of 2π, it was found that only two types of solutions exist—one with period 2π and the other with period 4π [13.7, 13.28]. Thus we seek the functions $y_0(t), y_1(t), \ldots$ in Eq. (13.99) in such a way that $y(t)$ is a solution of Eq. (13.98) with period 2π or 4π. Substituting Eqs. (13.99) and (13.100) into Eq. (13.98) results in

$$(\ddot{y}_0 + a_0 y_0) + \epsilon(\ddot{y}_1 + a_1 y_0 + y_0 \cos t + a_0 y_1)$$
$$+ \epsilon^2(\ddot{y}_2 + a_2 y_0 + a_1 y_1 + y_1 \cos t$$
$$+ a_0 y_2) + \cdots = 0 \tag{13.101}$$

where $\ddot{y}_i = d^2 y_i/dt^2$, $i = 0, 1, 2, \ldots$ Setting the coefficients of various powers of ϵ in Eq. (13.101) equal to zero, we obtain

$$\epsilon^0: \quad \ddot{y}_0 + a_0 y_0 = 0 \tag{13.102}$$

$$\epsilon^1: \quad \ddot{y}_1 + a_0 y_1 + a_1 y_0 + y_0 \cos t = 0 \tag{13.103}$$

$$\epsilon^2: \quad \ddot{y}_2 + a_0 y_2 + a_2 y_0 + a_1 y_1 + y_1 \cos t = 0 \tag{13.104}$$

$$\cdot$$
$$\cdot$$
$$\cdot$$

where each of the functions y_i is required to have a period of 2π or 4π. The solution of Eq. (13.102) can be expressed as

$$y_0(t) = \begin{cases} \cos \sqrt{a_0}\, t \\ \sin \sqrt{a_0}\, t \end{cases} \equiv \begin{cases} \cos \dfrac{n}{2} t \\ \sin \dfrac{n}{2} t \end{cases}, \quad n = 0, 1, 2, \ldots \tag{13.105}$$

and

$$a_0 = \frac{n^2}{4}, \quad n = 0, 1, 2, \ldots$$

Now, we consider the following specific values of n.

When $n = 0$: Equation (13.105) gives $a_0 = 0$ and $y_0 = 1$ and Eq. (13.103) yields

$$\ddot{y}_1 + a_1 + \cos t = 0 \quad \text{or} \quad \ddot{y}_1 = -a_1 - \cos t \tag{13.106}$$

In order to have y_1 as a periodic function, a_1 must be zero. When Eq. (13.106) is integrated twice, the resulting periodic solution can be expressed as

$$y_1(t) = \cos t + \alpha \tag{13.107}$$

where α is a constant. With the known values of $a_0 = 0$, $a_1 = 0$, $y_0 = 1$ and $y_1 = \cos t + \alpha$, Eq. (13.104) can be rewritten as

$$\ddot{y}_2 + a_2 + (\cos t + \alpha)\cos t = 0$$

or

$$\ddot{y}_2 = -\frac{1}{2} - a_2 - \alpha \cos t - \frac{1}{2} \cos 2t \tag{13.108}$$

In order to have y_2 as a periodic function, $(-\frac{1}{2} - a_2)$ must be zero (i.e., $a_2 = -\frac{1}{2}$). Thus, for $n = 0$, Eq. (13.100) gives

$$a = -\frac{1}{2}\epsilon^2 + \cdots \tag{13.109}$$

When n = 1:   For this case, Eq. (13.105) gives $a_0 = \frac{1}{4}$ and $y_0 = \cos(t/2)$ or $\sin(t/2)$. With $y_0 = \cos(t/2)$, Eq. (13.103) gives

$$\ddot{y}_1 + \frac{1}{4}y_1 = \left(-a_1 - \frac{1}{2}\right)\cos\frac{t}{2} - \frac{1}{2}\cos\frac{3t}{2} \qquad (13.110)$$

The homogeneous solution of Eq. (13.110) is given by

$$y_1(t) = A_1 \cos\frac{t}{2} + A_2 \sin\frac{t}{2}$$

where A_1 and A_2 are constants of integration. Since the term involving $\cos(t/2)$ appears in the homogeneous solution as well as the forcing function, the particular solution will contain a term of the form $t\cos(t/2)$, which is not periodic. Thus the coefficient of $\cos(t/2)$—namely $(-a_1 - 1/2)$, must be zero in the forcing function to ensure periodicity of $y_1(t)$. This gives $a_1 = -1/2$, and Eq. (13.110) becomes

$$\ddot{y}_1 + \frac{1}{4}y_1 = -\frac{1}{2}\cos\frac{3t}{2} \qquad (13.111)$$

By substituting the particular solution $y_1(t) = A_2\cos(3t/2)$ into Eq. (13.111), we obtain $A_2 = \frac{1}{4}$, and hence $y_1(t) = \frac{1}{4}\cos(3t/2)$. Using $a_0 = \frac{1}{4}, a_1 = -\frac{1}{2}$ and $y_1 = \frac{1}{4}\cos(3t/2)$, Eq. (13.104) can be expressed as

$$\ddot{y}_2 + \frac{1}{4}y_2 = -a_2\cos\frac{t}{2} + \frac{1}{2}\left(\frac{1}{4}\cos\frac{3t}{2}\right) - \left(\frac{1}{4}\cos\frac{3t}{2}\right)\cos t$$

$$= \left(-a_2 - \frac{1}{8}\right)\cos\frac{t}{2} + \frac{1}{8}\cos\frac{3t}{2} - \frac{1}{8}\cos\frac{5t}{2} \qquad (13.112)$$

Again, since the homogeneous solution of Eq. (13.112) contains the term $\cos(t/2)$, the coefficient of the term $\cos(t/2)$ on the right-hand side of Eq. (13.112) must be zero. This leads to $a_2 = -\frac{1}{8}$ and hence Eq. (13.100) becomes

$$a = \frac{1}{4} - \frac{\epsilon}{2} - \frac{\epsilon^2}{8} + \cdots \qquad (13.113a)$$

Similarly, by starting with the solution $y_0 = \sin(t/2)$, we obtain the relation (see Problem 13.17)

$$a = \frac{1}{4} + \frac{\epsilon}{2} - \frac{\epsilon^2}{8} + \cdots \qquad (13.113b)$$

When n = 2:   Equation (13.105) gives $a_0 = 1$ and $y_0 = \cos t$ or $\sin t$. With $a_0 = 1$ and $y_0 = \cos t$, Eq. (13.103) can be written as

$$\ddot{y}_1 + y_1 = -a_1\cos t - \frac{1}{2} - \frac{1}{2}\cos 2t \qquad (13.114)$$

Since $\cos t$ is a solution of the homogeneous equation, the term involving $\cos t$ in Eq. (13.114) gives rise to $t\cos t$ in the solution of y_1. Thus, to impose periodicity of y_1, we set $a_1 = 0$. With this, the particular solution of $y_1(t)$ can be assumed as $y_1(t) = A_3 + B_3 \cos 2t$. When this solution is substituted into Eq. (13.114), we obtain $A_3 = -\frac{1}{2}$ and $B_3 = \frac{1}{6}$. Thus Eq. (13.104) becomes

$$\ddot{y}_2 + y_2 + a_2 \cos t + y_1 \cos t = 0$$

or

$$\ddot{y}_2 + y_2 = -a_2 \cos t - (-\tfrac{1}{2} + \tfrac{1}{6} \cos 2t)\cos t$$

$$= \cos t(-a_1 + \tfrac{1}{2} - \tfrac{1}{12}) + \tfrac{1}{2} \cos 3t \qquad (13.115)$$

For periodicity of $y_2(t)$, we set the coefficient of $\cos t$ equal to zero in the forcing function of Eq. (13.115). This gives $a_2 = \frac{5}{12}$, and hence

$$a = 1 + \tfrac{5}{12}\epsilon^2 + \cdots \qquad (13.116a)$$

Similarly, by proceeding with $y_0 = \sin t$, we obtain (see Problem 13.17)

$$a = 1 - \tfrac{1}{12}\epsilon^2 + \cdots \qquad (13.116b)$$

To observe the stability characteristics of the system, Eqs. (13.109), (13.113), and (13.116) are plotted in the (a, ϵ) plane as indicated in Fig. 13.9. These equations represent curves

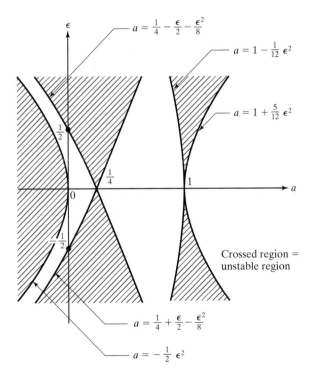

FIGURE 13.9   Stability of periodic solutions.

that are known as the *boundary* or *transition curves* that divide the (a, ϵ) plane into regions of stability and instability. These boundary curves are such that a point belonging to any one curve represents a periodic solution of Eq. (13.98). The stability of these periodic solutions can be investigated [13.7, 13.25, 13.28]. In Fig. 13.9, the points inside the shaded region denote unstable motion. It can be noticed from this figure that stability is also possible for negative values of a, which correspond to the equilibrium position $\theta = 180°$. Thus with the right choice of the parameters, the pendulum can be made to be stable in the upright position by moving its support harmonically.

13.7   Graphical Methods

13.7.1
Phase Plane
Representation

Graphical methods can be used to obtain qualitative information about the behavior of the nonlinear system and also to integrate the equations of motion. We shall first consider a basic concept known as the *phase plane*. For a single degree of freedom system, two parameters are needed to describe the state of motion completely. These parameters are usually taken as the displacement and velocity of the system. When the parameters are used as coordinate axes, the resulting graphical representation of the motion is called the *phase plane representation*. Thus each point in the phase plane represents a possible state of the system. As time changes, the state of the system changes. A typical or representative point in the phase plane (such as the point representing the state of the system at time $t = 0$) moves and traces a curve known as the *trajectory*. The trajectory shows how the solution of the system varies with time.

▬▬▬▬▬▬ Trajectories of a Simple Harmonic Oscillator

EXAMPLE   13.2

Find the trajectories of a simple harmonic oscillator.

Solution: The equation of motion of an undamped linear system is given by

$$\ddot{x} + \omega_n^2 x = 0 \tag{E.1}$$

By setting $y = \dot{x}$, Eq. (E.1) can be written as

$$\frac{dy}{dt} = -\omega_n^2 x$$

$$\frac{dx}{dt} = y \tag{E.2}$$

from which we can obtain

$$\frac{dy}{dx} = -\frac{\omega_n^2 x}{y} \tag{E.3}$$

Integration of Eq. (E.3) leads to

$$y^2 + \omega_n^2 x^2 = c^2 \tag{E.4}$$

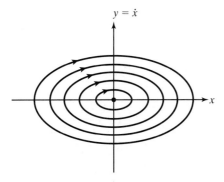

$y = \dot{x}$

x

FIGURE 13.10   Trajectories of a simple harmonic oscillator.

where c is a constant. The value of c is determined by the initial conditions of the system. Equation (E.4) shows that the trajectories of the system in the phase plane (x-y plane) are a family of ellipses, as shown in Fig. 13.10. It can be observed that the point $(x = 0, y = 0)$ is surrounded by closed trajectories. Such a point is called a *center*. The direction of motion of the trajectories can be determined from Eq. (E.2). For instance, if $x > 0$ and $y > 0$, Eq. (E.2) implies that $dx/dt > 0$ and $dy/dt < 0$; therefore, the motion is clockwise.

■

Phase-plane of an Undamped Pendulum

EXAMPLE 13.3

Find the trajectories of an undamped pendulum.

Solution: The equation of motion is given by Eq. (13.1):

$$\ddot{\theta} = -\omega_0^2 \sin\theta \qquad (E.1)$$

where $\omega_0^2 = g/l$. Introducing $x = \theta$ and $y = \dot{x} = \dot{\theta}$, Eq. (E.1) can be rewritten as

$$\frac{dx}{dt} = y, \qquad \frac{dy}{dt} = -\omega_0^2 \sin x$$

or

$$\frac{dy}{dx} = -\frac{\omega_0^2 \sin x}{y}$$

or

$$y\, dy = -\omega_0^2 \sin x \, dx \qquad (E.2)$$

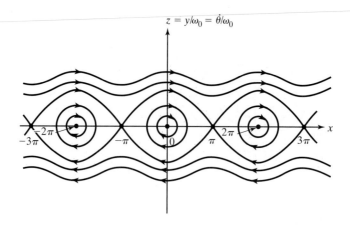

$z = y/\omega_0 = \dot{\theta}/\omega_0$

FIGURE 13.11   Trajectories of an undamped pendulum.

Integrating Eq. (E.2) and using the condition that $\dot{x} = 0$ when $x = x_0$ (at the end of the swing), we obtain

$$y^2 = 2\omega_0^2(\cos x - \cos x_0) \tag{E.3}$$

Introducing $z = y/\omega_0$ Eq. (E.3) can be expressed as

$$z^2 = 2(\cos x - \cos x_0) \tag{E.4}$$

The trajectories given by Eq. (E.4) are shown in Fig. 13.11.

■

EXAMPLE   13.4   ━━━━━ **Phase-plane of an Undamped Nonlinear System**

Find the trajectories of a nonlinear spring-mass system governed by the equation

$$\ddot{x} + \omega_0^2(x - 2\alpha x^3) = 0 \tag{E.1}$$

Solution: The nonlinear pendulum equation can be considered as a special case of Eq. (E.1). To see this, we use the approximation $\sin\theta \simeq \theta - \theta^3/6$ in the neighborhood of $\theta = 0$ in Eq. (E.1) of Example 13.3 to obtain

$$\ddot{\theta} + \omega_0^2\left(\theta - \frac{\theta^3}{6}\right) = 0$$

which can be seen to be a special case of Eq. (E.1). Equation (E.1) can be rewritten as

$$\frac{dx}{dt} = y, \qquad \frac{dy}{dt} = -\omega_0^2(x - 2\alpha x^3)$$

or

$$\frac{dy}{dx} = -\frac{\omega_0^2(x - 2\alpha x^3)}{y}$$

or

$$y \, dy = -\omega_0^2(x - 2\alpha x^3) \, dx \tag{E.2}$$

Integration of Eq. (E.2), with the condition $\dot{x} = 0$ when $x = x_0$ (at the end of the swing in the case of a pendulum), gives

$$z^2 + x^2 - \alpha x^4 = A^2 \tag{E.3}$$

where $z = y/\omega_0$ and $A^2 = x_0^2(1 - \alpha x_0^2)$ is a constant. The trajectories, in the phase-plane, given by Eq. (E.3), are shown in Fig. 13.12 for several values of α.

It can be observed that for $\alpha = 0$, Eq. (E.3) denotes a circle of radius A and corresponds to a simple harmonic motion. When $\alpha < 0$, Eq. (E.3) represents ovals within the circle given by $\alpha = 0$ and the ovals touch the circle at the points $(0, \pm A)$. When $\alpha = (1/4) A^2$, Eq. (E.3) becomes

$$y^2 + x^2 - \frac{x^4}{4A^2} - A^2 = \left[y - \left(A - \frac{x^2}{2A} \right) \right]\left[y + \left(A - \frac{x^2}{2A} \right) \right] = 0 \tag{E.4}$$

Equation (E.4) indicates that the trajectories are given by the parabolas

$$y = \pm \left(A - \frac{x^2}{2A} \right) \tag{E.5}$$

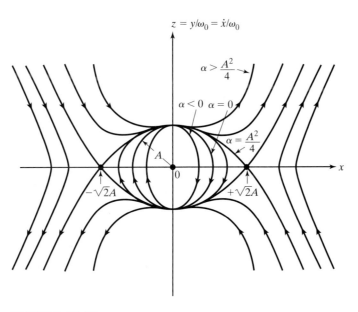

FIGURE 13.12   Trajectories of a nonlinear system.

These two parabolas intersect at points $(x = \pm\sqrt{2}A, y = 0)$, which correspond to the points of unstable equilibrium.

When $(1/4)\,A^2 \geq \alpha \geq 0$, the trajectories given by Eq. (E.3) will be closed ovals lying between the circle given by $\alpha = 0$ and the two parabolas given by $\alpha = (1/4)A^2$. Since these trajectories are closed curves, they represent periodic vibrations. When $\alpha > (1/4)A^2$, the trajectories given by Eq. (E.3) lie outside the region between the parabolas and extend to infinity. These trajectories correspond to the conditions that permit the body to escape from the center of force.

■

To see some of the characteristics of trajectories, consider a single degree of freedom nonlinear oscillatory system whose governing equation is of the form

$$\ddot{x} + f(x, \dot{x}) = 0 \tag{13.117}$$

By defining

$$\frac{dx}{dt} = \dot{x} = y \tag{13.118}$$

and

$$\frac{dy}{dt} = \dot{y} = -f(x, y) \tag{13.119}$$

we obtain

$$\frac{dy}{dx} = \frac{dy/dt}{dx/dt} = -\frac{f(x, y)}{y} = \phi(x, y), \text{ say.} \tag{13.120}$$

Thus there is a unique slope of the trajectory at every point (x, y) in the phase plane, provided that $\phi(x, y)$ is not indeterminate. If $y = 0$ and $f \neq 0$ (that is, if the point lies on the x axis), the slope of the trajectory is infinite. This means that all trajectories must cross the x axis at right angles. If $y = 0$ and $f = 0$, the point is called a *singular point*, and the slope is indeterminate at such points. A singular point corresponds to a state of equilibrium of the system—the velocity $y = \dot{x}$ and the force $\ddot{x} = -f$ are zero at a singular point. Further investigation is necessary to establish whether the equilibrium represented by a singular point is stable or unstable.

13.7.2
Phase Velocity

The velocity $\vec{v}$ with which a representative point moves along a trajectory is called the *phase velocity*. The components of phase velocity parallel to the x and y axes are

$$v_x = \dot{x}, \qquad v_y = \dot{y} \tag{13.121}$$

and the magnitude of $\vec{v}$ is given by

$$|\vec{v}| = \sqrt{v_x^2 + v_y^2} = \sqrt{\left(\frac{dx}{dt}\right)^2 + \left(\frac{dy}{dt}\right)^2} \tag{13.122}$$

We can note that if the system has a periodic motion, its trajectory in the phase plane is a closed curve. This follows from the fact that the representative point, having started its motion along a closed trajectory at an arbitrary point (x, y), will return to the same point after one period. The time required to go around the closed trajectory (the period of oscillation of the system) is finite because the phase velocity is bounded away from zero at all points of the trajectory.

**13.7.3
Method of
Constructing
Trajectories**

We shall now consider a method known as the *method of isoclines* for constructing the trajectories of a dynamical system with one degree of freedom. By writing the equations of motion of the system as

$$\frac{dx}{dt} = f_1(x, y)$$

$$\frac{dy}{dt} = f_2(x, y) \tag{13.123}$$

where f_1 and f_2 are nonlinear functions of x and $y = \dot{x}$, the equation for the integral curves can be obtained as

$$\frac{dy}{dx} = \frac{f_2(x, y)}{f_1(x, y)} = \phi(x, y), \text{ say.} \tag{13.124}$$

The curve

$$\phi(x, y) = c \tag{13.125}$$

for a fixed value of c is called an *isocline*. An isocline can be defined as the locus of points at which the trajectories passing through them have the constant slope c. In the method of isoclines we fix the slope $(dy)/(dx)$ by giving it a definite number c_1 and solve Eq. (13.125) for the trajectory. The curve $\phi(x, y) - c_1 = 0$ thus represents an isocline in the phase plane. We plot several isoclines by giving different values $c_1, c_2, \ldots$ to the slope $\phi = (dy)/(dx)$. Let $h_1, h_2, \ldots$ denote these isoclines in Fig. 13.13(a). Suppose that we

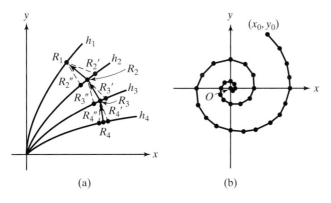

(a)   (b)

FIGURE 13.13   Method of isoclines.

are interested in constructing the trajectory passing through the point R_1 on the isocline h_1. We draw two straight line segments from R_1: one with a slope c_1, meeting h_2 at R_2', and the other with a slope c_2 meeting h_2 at R_2''. The middle point between R_2' and R_2'' lying on h_2 is denoted as R_2. Starting at R_2, this construction is repeated, and the point R_3 is determined on h_3. This procedure is continued until the polygonal trajectory with sides R_1R_2, R_2R_3, R_3R_4, ... is taken as an approximation to the actual trajectory passing through the point R_1. Obviously, the larger the number of isoclines, the better is the approximation obtained by this graphical method. A typical final trajectory looks like the one shown in Fig. 13.13(b).

Trajectories Using the Method of Isoclines

EXAMPLE   13.5

Construct the trajectories of a simple harmonic oscillator by the method of isoclines.

Solution: The differential equation defining the trajectories of a simple harmonic oscillator is given by Eq. (E.3) of Example 13.2. Hence the family of isoclines is given by

$$c = -\frac{\omega_n^2 x}{y} \qquad \text{or} \qquad y = \frac{-\omega_n^2}{c}x \tag{E.1}$$

This equation represents a family of straight lines passing through the origin, with c representing the slope of the trajectories on each isocline. The isoclines given by Eq. (E.1) are shown in Fig. 13.14. Once the isoclines are known, the trajectory can be plotted as indicated above.

∎

**13.7.4
Obtaining Time
Solution from
Phase Plane
Trajectories**

The trajectory plotted in the phase plane is a plot of $\dot{x}$ as a function of x, and time (t) does not appear explicitly in the plot. For the qualitative analysis of the system, the trajectories are enough, but in some cases we may need the variation of x with time t. In such cases, it is possible to obtain the time solution $x(t)$ from the phase plane diagram, although the original

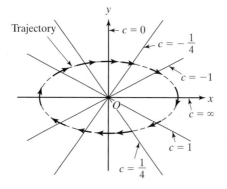

FIGURE 13.14   Isoclines of a simple harmonic oscillator.

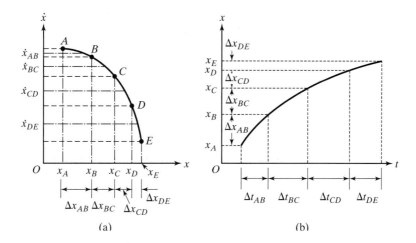

FIGURE 13.15   Obtaining time solution from a phase plane plot.

differential equation cannot be solved for x and $\dot{x}$ as functions of time. The method of obtaining a time solution is essentially a step-by-step procedure; several different schemes may be used for this purpose. In this section, we shall present a method based on the relation $\dot{x} = (\Delta x)/(\Delta t)$.

For small increments of displacement and time (Δx and Δt), the average velocity can be taken as $\dot{x}_{av} = (\Delta x)/(\Delta t)$, so that

$$\Delta t = \frac{\Delta x}{\dot{x}_{av}} \tag{13.126}$$

In the phase plane trajectory shown in Fig. 13.15, the incremental time needed for the representative point to traverse the incremental displacement Δx_{AB} is shown as Δt_{AB}. If $\dot{x}_{AB}$ denotes the average velocity during Δt_{AB}, we have $\Delta t_{AB} = \Delta x_{AB}/\dot{x}_{AB}$. Similarly, $\Delta t_{BC} = \Delta x_{BC}/\dot{x}_{BC}$, etc. Once $\Delta t_{AB}, \Delta t_{BC}, \ldots$ are known, the time solution $x(t)$ can be plotted easily, as shown in Fig. 13.15(b). It is evident that for good accuracy, the incremental displacements $\Delta x_{AB}, \Delta x_{BC}, \ldots$ must be chosen small enough that the corresponding incremental changes in $\dot{x}$ and t are reasonably small. Note that Δx need not be constant; it can be changed depending on the nature of the trajectories.

13.8   Stability of Equilibrium States

13.8.1
Stability
Analysis

Consider a single degree of freedom nonlinear vibratory system described by two first-order differential equations

$$\frac{dx}{dt} = f_1(x, y)$$

$$\frac{dy}{dt} = f_2(x, y) \tag{13.127}$$

where f_1 and f_2 are nonlinear functions of x and $y = \dot{x} = dx/dt$. For this system, the slope of the trajectories in the phase plane is given by

$$\frac{dy}{dx} = \frac{\dot{y}}{\dot{x}} = \frac{f_2(x, y)}{f_1(x, y)} \tag{13.128}$$

Let (x_0, y_0) be a singular point or an equilibrium point so that $(dy)/(dx)$ has the form 0/0:

$$f_1(x_0, y_0) = f_2(x_0, y_0) = 0 \tag{13.129}$$

A study of Eqs. (13.127) in the neighborhood of the singular point provides us with answers as to the stability of equilibrium. We first note that there is no loss of generality if we assume that the singular point is located at the origin (0, 0). This is because the slope $(dy)/(dx)$ of the trajectories does not vary with a translation of the coordinate axes x and y to x' and y':

$$x' = x - x_0$$

$$y' = y - y_0$$

$$\frac{dy}{dx} = \frac{dy'}{dx'} \tag{13.130}$$

Thus we assume that $(x = 0, y = 0)$ is a singular point, so that

$$f_1(0, 0) = f_2(0, 0) = 0$$

If f_1 and f_2 are expanded in terms of Taylor's series about the singular point (0, 0), we obtain

$$\dot{x} = f_1(x, y) = a_{11}x + a_{12}y + \text{Higher order terms}$$

$$\dot{y} = f_2(x, y) = a_{21}x + a_{22}y + \text{Higher order terms} \tag{13.131}$$

where

$$a_{11} = \left.\frac{\partial f_1}{\partial x}\right|_{(0, 0)}, \qquad a_{12} = \left.\frac{\partial f_1}{\partial y}\right|_{(0, 0)}, \qquad a_{21} = \left.\frac{\partial f_2}{\partial x}\right|_{(0, 0)}, \qquad a_{22} = \left.\frac{\partial f_2}{\partial y}\right|_{(0, 0)}$$

In the neighborhood of (0, 0), x and y are small; f_1 and f_2 can be approximated by linear terms only, so that Eqs. (13.131) can be written as

$$\begin{Bmatrix} \dot{x} \\ \dot{y} \end{Bmatrix} = \begin{bmatrix} a_{11} & a_{12} \\ a_{21} & a_{22} \end{bmatrix} \begin{Bmatrix} x \\ y \end{Bmatrix} \tag{13.132}$$

The solutions of Eq. (13.132) are expected to be geometrically similar to those of Eq. (13.127). We assume the solution of Eq. (13.132) in the form

$$\begin{Bmatrix} x \\ y \end{Bmatrix} = \begin{Bmatrix} X \\ Y \end{Bmatrix} e^{\lambda t} \tag{13.133}$$

where X, Y, and λ are constants. Substitution of Eq. (13.133) into Eq. (13.132) leads to the eigenvalue problem

$$\begin{bmatrix} a_{11} - \lambda & a_{12} \\ a_{21} & a_{22} - \lambda \end{bmatrix} \begin{Bmatrix} X \\ Y \end{Bmatrix} = \begin{Bmatrix} 0 \\ 0 \end{Bmatrix} \tag{13.134}$$

The eigenvalues λ_1 and λ_2 can be found by solving the characteristic equation

$$\begin{vmatrix} a_{11} - \lambda & a_{12} \\ a_{21} & a_{22} - \lambda \end{vmatrix} = 0$$

as

$$\lambda_1, \lambda_2 = \tfrac{1}{2}(p \pm \sqrt{p^2 - 4q}) \tag{13.135}$$

where $p = a_{11} + a_{22}$ and $q = a_{11}a_{22} - a_{12}a_{21}$. If

$$\begin{Bmatrix} X_1 \\ Y_1 \end{Bmatrix} \qquad \text{and} \qquad \begin{Bmatrix} X_2 \\ Y_2 \end{Bmatrix}$$

denote the eigenvectors corresponding to λ_1 and λ_2, respectively, the general solution of Eqs. (13.127) can be expressed as (assuming $\lambda_1 \neq 0$, $\lambda_2 \neq 0$, and $\lambda_1 \neq \lambda_2$):

$$\begin{Bmatrix} x \\ y \end{Bmatrix} = C_1 \begin{Bmatrix} X_1 \\ Y_1 \end{Bmatrix} e^{\lambda_1 t} + C_2 \begin{Bmatrix} X_2 \\ Y_2 \end{Bmatrix} e^{\lambda_2 t} \tag{13.136}$$

where C_1 and C_2 are arbitrary constants. We can note the following:

   If $(p^2 - 4q) < 0$, the motion is oscillatory.

   If $(p^2 - 4q) > 0$, the motion is aperiodic.

   If $p > 0$, the system is unstable.

   If $p < 0$, the system is stable.

If we use the transformation

$$\begin{Bmatrix} x \\ y \end{Bmatrix} = \begin{bmatrix} X_1 & X_2 \\ Y_1 & Y_2 \end{bmatrix} \begin{Bmatrix} \alpha \\ \beta \end{Bmatrix} = [T] \begin{Bmatrix} \alpha \\ \beta \end{Bmatrix}$$

where $[T]$ is the modal matrix and α and β are the generalized coordinates, Eqs. (13.132) will be uncoupled:

$$\begin{Bmatrix} \dot{\alpha} \\ \dot{\beta} \end{Bmatrix} = \begin{bmatrix} \lambda_1 & 0 \\ 0 & \lambda_2 \end{bmatrix} \begin{Bmatrix} \alpha \\ \beta \end{Bmatrix} \qquad \text{or} \qquad \begin{matrix} \dot{\alpha} = \lambda_1 \alpha \\ \dot{\beta} = \lambda_2 \beta \end{matrix} \tag{13.137}$$

The solution of Eqs. (13.137) can be expressed as

$$\alpha(t) = e^{\lambda_1 t}$$
$$\beta(t) = e^{\lambda_2 t} \tag{13.138}$$

13.8.2
Classification of
Singular Points

Depending on the values of λ_1 and λ_2 in Eq. (13.135), the singular or equilibrium points can be classified as follows [13.20, 13.23].

Case (i) — λ_1 and λ_2 Are Real and Distinct ($p^2 > 4q$).   In this case, Eq. (13.138) gives

$$\alpha(t) = \alpha_0 e^{\lambda_1 t} \quad \text{and} \quad \beta(t) = \beta_0 e^{\lambda_2 t} \tag{13.139}$$

where α_0 and β_0 are the initial values of α and β, respectively. The type of motion depends on whether λ_1 and λ_2 are of the same sign or of opposite signs. If λ_1 and λ_2 have the same sign ($q > 0$), the equilibrium point is called a *node*. The phase plane diagram for the case $\lambda_2 < \lambda_1 < 0$ (when λ_1 and λ_2 are real and negative or $p < 0$) is shown in Fig. 13.16(a). In this case, Eq. (13.139) shows that all the trajectories tend to the origin as $t \to \infty$ and hence the origin is called a *stable node*. On the other hand, if $\lambda_2 > \lambda_1 > 0$ ($p > 0$), the arrow heads change in direction and the origin is called an *unstable node* (see Fig. 13.16b). If λ_1 and λ_2 are real but of opposite signs ($q < 0$ irrespective of the sign of p), one solution tends to the origin while the other tends to infinity. In this case the origin is called a *saddle point* and it corresponds to unstable equilibrium (see Fig. 13.16d).

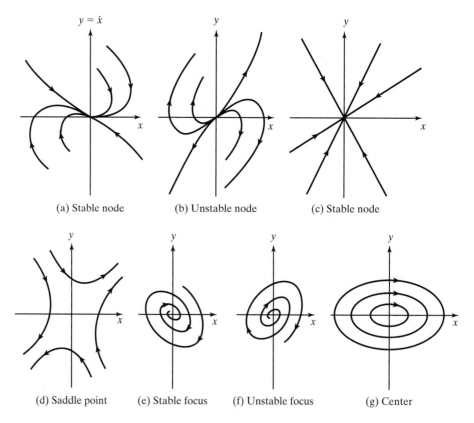

(a) Stable node   (b) Unstable node   (c) Stable node

(d) Saddle point   (e) Stable focus   (f) Unstable focus   (g) Center

FIGURE 13.16   Types of equilibrium points.

Case (ii)—λ_1 and λ_2 Are Real and Equal ($p^2 = 4q$).   In this case Eq. (13.138) gives

$$\alpha(t) = \alpha_0 e^{\lambda_1 t} \qquad \text{and} \qquad \beta(t) = \beta_0 e^{\lambda_1 t} \tag{13.140}$$

The trajectories will be straight lines passing through the origin and the equilibrium point (origin) will be a *stable node* if $\lambda_1 < 0$ (see Fig. 13.16c) and an *unstable node* if $\lambda_1 > 0$.

Case (iii)—λ_1 and λ_2 Are Complex Conjugates ($p^2 < 4q$).   Let $\lambda_1 = \theta_1 + i\theta_2$ and $\lambda_2 = \theta_1 - i\theta_2$, where θ_1 and θ_2 are real. Then Eq. (13.137) gives

$$\dot{\alpha} = (\theta_1 + i\theta_2)\alpha \qquad \text{and} \qquad \dot{\beta} = (\theta_1 - i\theta_2)\beta \tag{13.141}$$

This shows that α and β must also be complex conjugates. We can rewrite Eq. (13.138) as

$$\alpha(t) = (\alpha_0 e^{\theta_1 t})e^{i\theta_2 t}, \qquad \beta(t) = (\beta_0 e^{\theta_1 t})e^{-i\theta_2 t} \tag{13.142}$$

These equations represent logarithmic spirals. In this case the equilibrium point (origin) is called a *focus* or a *spiral point*. Since the factor $e^{i\theta_2 t}$ in $\alpha(t)$ represents a vector of unit magnitude rotating with angular velocity θ_2 in the complex plane, the magnitude of the complex vector $\alpha(t)$, and hence the stability of motion, is determined by $e^{\theta_1 t}$. If $\theta_1 < 0$, the motion will be asymptotically stable and the focal point will be stable ($p < 0$ and $q > 0$). If $\theta_1 < 0$, the focal point will be unstable ($p > 0$ and $q > 0$). The sign of θ_2 merely gives the direction of rotation of the complex vector, counterclockwise for $\theta_2 > 0$ and clockwise for $\theta_2 < 0$.

If $\theta_1 = 0$ ($p = 0$), the magnitude of the complex radius vector $\alpha(t)$ will be constant and the trajectories reduce to circles with the center as the equilibrium point (origin). The motion will be periodic and hence will be stable. The equilibrium point in this case is called a *center* or *vertex point* and the motion is simply stable and not asymptotically stable. The stable focus, unstable focus, and center are shown in Figs. 13.16(e) to (g).

13.9 Limit Cycles

In certain vibration problems involving nonlinear damping, the trajectories, starting either very close to the origin or far away from the origin, tend to a single closed curve, which corresponds to a periodic solution of the system. This means that every solution of the system tends to a periodic solution as $t \to \infty$. The closed curve to which all the solutions approach is called a *limit cycle*.

For illustration, we consider the following equation, known as the *van der Pol equation*:

$$\ddot{x} - \alpha(1 - x^2)\dot{x} + x = 0, \qquad \alpha > 0 \tag{13.143}$$

This equation exhibits, mathematically, the essential features of some vibratory systems like certain electrical feedback circuits controlled by valves where there is a source of power that increases with the amplitude of vibration. Van der Pol invented Eq. (13.143) by introducing a type of damping that is negative for small amplitudes but becomes positive for large amplitudes. In this equation, he assumed the damping term to be a multiple of $-(1 - x^2)\dot{x}$ in order to make the magnitude of the damping term independent of the sign of x.

The qualitative nature of the solution depends upon the value of the parameter α. Although the analytical solution of the equation is not known, it can be represented using the method of isoclines, in the phase plane. Equation (13.143) can be rewritten as

$$y = \dot{x} = \frac{dx}{dt} \tag{13.144}$$

$$\dot{y} = \frac{dy}{dt} = \alpha(1 - x^2)y - x \tag{13.145}$$

Thus the isocline corresponding to a specified value of the slope $dy/dx = c$ is given by

$$\frac{dy}{dx} = \frac{dy/dt}{dx/dt} = \frac{\alpha(1 - x^2)y - x}{y} = c$$

or

$$y + \left[\frac{x}{-\alpha(1 - x^2) + c} \right] = 0 \tag{13.146}$$

By drawing the curves represented by Eq. (13.146) for a set of values of c, as shown in Fig. 13.17, the trajectories can be sketched in with fair accuracy, as shown in the same figure. The isoclines will be curves since the equation, Eq. (13.146), is nonlinear. An infinity of isoclines pass through the origin, which is a singularity.

An interesting property of the solution can be observed from Fig. 13.17. All the trajectories, irrespective of the initial conditions, approach asymptotically a particular closed

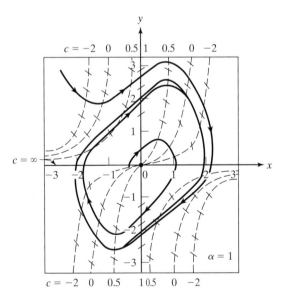

FIGURE 13.17   Trajectories and limit cycle for van der Pol's equation.

curve, known as the limit cycle, which represents a steady-state periodic (but not harmonic) oscillation. This is a phenomenon that can be observed only with certain nonlinear vibration problems and not in any linear problem. If the initial point is inside the limit cycle, the ensuing solution curve spirals outward. On the other hand, if the initial point falls outside the limit cycle, the solution curve spirals inward. As stated above, the limit cycle in both the cases is attained finally. An important characteristic of the limit cycle is that the maximum value of x is always close to 2 irrespective of the value of α. This result can be seen by solving Eq. (13.143) using the perturbation method (see Problem 13.34).

13.10   Chaos

Chaos represents the behavior of a system that is inherently unpredictable. In other words, chaos refers to the dynamic behavior of a system whose response, although described by a deterministic equation, becomes unpredictable because the nonlinearities in the equation enormously amplify the errors in the initial conditions of the system.

Attractor.   Consider a pendulum whose amplitude of oscillation decreases gradually due to friction, which means that the system loses part of its energy in each cycle and eventually comes to a rest position. This is indicated in the phase plane shown in Fig. 13.18(a). The rest position is called an attractor. Thus the pendulum has just one attractor. If the pendulum is given a push at the end of each swing to replenish the energy lost due to friction, the resulting motion can be indicated as a closed loop in the phase plane (see Fig. 13.18b). In general, for a dynamic system, an attractor is a point (or object) toward which all nearby solutions move as time progresses.

Poincaré Section.   A pendulum is said to have two degrees of freedom—namely, x and $\dot{x}$. In general, a phase space of a system can be defined with as many axes as there are degrees of freedom. Thus, for a system with three degrees of freedom, the phase space might appear (as a spiral converging in the z-direction) as shown in Fig. 13.19(a). Since the points are displaced from one another and never coincide in Fig. 13.19(a), the system does not have a periodic motion. The intersection of the phase space with the yz plane

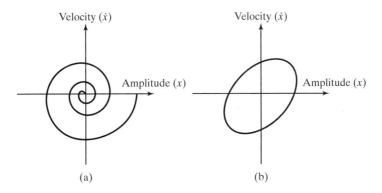

FIGURE 13.18   Attractor.

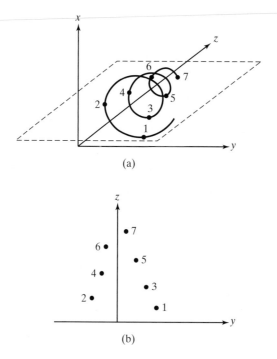

(a)

(b)

FIGURE 13.19   Phase space for a three degree of freedom system.

appears as shown in Fig. 13.19(b). This diagram is known as the Poincaré section or Poincaré map, and it represents points that occur at equal intervals of time, $nT (n = 1, 2, \dots)$, where T denotes the fundamental period of the forcing function. Note that, if the system is periodic, all the dots would be at the same location in the Poincaré section.

13.10.1 Functions with Stable Orbits

Consider the sequence of numbers generated by the following equation:

$$x_{n+1} = \sqrt{x_n};\, n = 1, 2, \dots \tag{13.147}$$

For any two initial values of x_1, which differ by a small amount, the values of x_{n+1} converge to 1. For example, with $x_1 = 10.0$ and $x_1 = 10.2$, the sequences of numbers given by Eq. (13.147) are shown below:

$$10.0 \rightarrow 3.1623 \rightarrow 1.7783 \rightarrow 1.3335 \rightarrow 1.1548 \rightarrow 1.0746 \rightarrow 1.0366 \rightarrow 1.0182$$

$$\rightarrow 1.0090 \rightarrow 1.0045 \rightarrow 1.0023 \rightarrow 1.0011 \rightarrow 1.0006 \rightarrow 1.0003 \rightarrow 1.0001 \rightarrow$$

$$1.0001 \rightarrow 1.0000$$

and

$$10.2 \to 3.1937 \to 1.7871 \to 1.3368 \to 1.1562 \to 1.0753 \to 1.0370 \to 1.0183$$
$$\to 1.0091 \to 1.0045 \to 1.0023 \to 1.0011 \to 1.0006 \to 1.0003 \to 1.0001 \to$$
$$1.0001 \to 1.0000$$

Note that the influence of a change in the initial value of x_1 (by 0.2) is lost very soon and the pattern converges to 1. Any starting value between 0 and 1 also iterates to 1. Thus the functional relation, Eq. (13.147), is said to have a stable orbit at $x = 1$.

**13.10.2
Functions with
Unstable Orbits**

Consider the sequence of numbers generated by the following equation:

$$x_{n+1} = a x_n (1 - x_n); n = 1, 2, \ldots \tag{13.148}$$

where a is a constant. Equation (13.148) has been used as a simple model for population growth with no predators such as fish and fowl. In such cases, the constant a denotes the rate of growth of the population, x_n represents the population in generation n, and the factor $(1 - x_n)$ acts as a stabilizing factor. It can be seen that Eq. (13.148) has the following limitations [13.31]:

1. x_1 has to lie between 0 and 1. If x_1 exceeds 1, the iterative process diverges to $-\infty$, implying that the population becomes extinct.
2. x_{n+1} attains a maximum value of $\frac{a}{4}$ at $x_n = \frac{1}{2}$. This implies that $a < 4$.
3. If $a < 1$, x_{n+1} converges to zero.
4. Thus, for a nontrivial dynamic behavior (to avoid population from becoming extinct), a has to satisfy the relation $1 \leq a \leq 4$.

The system will have an equilibrium condition if the birth rate replenishes the loss due to death or migration. The population can be seen to stabilize or reach a definite limiting value (predictable) for some values of a—such as $a = 3.0$. For some other values of a—such as $a = 4.0$ with $x_1 = 0.5$—the species can be seen to disappear after only two generations, as indicated below:

Equation (13.148) with $a = 4.0$ and $x_1 = 0.5$:

$$0.5 \to 1.0 \to 0.0 \to 0.0 \to 0.0 \to \cdots$$

However, for $a = 4.0$ with $x_1 = 0.4$, the population count can be seen to be completely random, as indicated below:

Equation (13.148) with $a = 4.0$ and $x_1 = 0.4$:

$$0.4 \to 0.96 \to 0.154 \to 0.520 \to 0.998 \to 0.006 \to 0.025 \to 0.099 \to$$
$$0.358 \to 0.919 \to 0.298 \to 0.837 \to 0.547 \to 0.991 \to 0.035 \to 0.135 \to$$
$$0.466 \to 0.996 \to 0.018 \to \cdots$$

This indicates that the system is a chaotic one; even small changes in the deterministic equation, Eq. (13.148), can lead to unpredictable results. Physically, this implies that the system has become chaotic with population varying wildly from year to year. In fact, as will be shown in the following paragraph, the system, Eq. (13.148), has unstable orbits.

Bifurcations.   Equation (13.148) exhibits a phenomenon known as *bifurcation*. To see this, we start with $a = 2$ and a few different values of x_1. With this, x_{n+1} can be seen to converge to 0.5. When we start with $a = 2.5$ and a few different values of x_1, the process converges to 0.6. If we use $a = 3.0$ and $x_1 = 0.1$, the process converges to a single value, but while getting there, oscillates between two separate values (namely, 0.68... and 0.65...). If we use $a = 3.25$ and $x_1 = 0.5$, then the value of x_{n+1} oscillates between the two values $x^{(1)} = 0.4952$ and $x^{(2)} = 0.8124$. At that point the system is said to have a periodicity of 2. The solution, in this case, moves into a two-pronged fork-type of state with two equilibrium points. If we use $a = 3.5$ and $x_1 = 0.5$, the system will have a period 4—that is, the equilibrium state will oscillate between the four values $x^{(3)} = 0.3828$, $x^{(4)} = 0.5008$, $x^{(5)} = 0.8269$, and $x^{(6)} = 0.8749$. This implies that the stable behavior of each of the previous two solutions has been broken into further bifurcation paths. In fact, the system continues to bifurcate with the range of a needed for each successive birth of bifurcations becoming smaller as a increases, as shown in Fig. 13.20. Figure 13.20 is known as a bifurcation plot or Feigenbaum diagram [13.31, 13.35]. It can be observed that the system has reached a chaotic state through a series of bifurcations, with the number of values assumed by Eq. (13.148) doubling at each stage.

Strange Attractors.   For several years, it was believed that the attractors toward which physical systems tend are equilibrium or rest points (as in the case of the rest position of a pendulum), limit cycles, or repeating configurations. However, in recent years it has been found that the attractors associated with chaotic systems are more complex than the rest points and limit cycles. The geometric points in state space to which chaotic trajectories are attracted are called *strange attractors*.

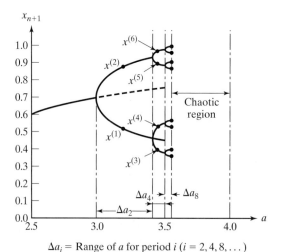

Δa_i = Range of a for period i ($i = 2, 4, 8, \dots$)

FIGURE 13.20   Bifurcation plot.

**13.10.3
Chaotic
Behavior of
Duffing's
Equation
Without the
Forcing Term**

Consider a single degree of freedom system with a nonlinear spring and a harmonic forcing function. The equation of motion (Duffing's equation) can be expressed as

$$\ddot{x} + \mu\dot{x} - \alpha x + \beta x^3 = F_0 \cos \omega t \qquad (13.149)$$

First, we consider the free, undamped vibration of the system with $\alpha = \beta = 0.5$:

$$\ddot{x} - 0.5\,x + 0.5\,x^3 = 0 \qquad (13.150)$$

The static equilibrium positions of this system can be found by setting the spring force equal to zero, as $x = 0, +1, -1$. It can be easily verified that the equilibrium solution $x = 0$ is unstable (saddle point) with respect to infinitesimal disturbances, while the equilibrium solutions -1 and $+1$ are stable (centers) with respect to infinitesimal disturbances. The stability of the system about the three equilibrium positions can be seen more clearly from the graph of its potential energy. To find the potential energy of the system, we multiply Eq. (13.150) by $\dot{x}$ and integrate the resulting equation to obtain

$$\frac{1}{2}(\dot{x})^2 + \frac{1}{8}x^4 - \frac{1}{4}x^2 = C \qquad (13.151)$$

where C is a constant. The first term on the left-hand side of Eq. (13.151) represents the kinetic energy and the second and third terms denote the potential energy (P) of the system. Equation (13.151) indicates that the sum of the kinetic and potential energies is a constant (conservative system). A plot of the potential energy, $P = \frac{1}{8}x^4 - \frac{1}{4}x^2$, is shown in Fig. 13.21.

Next, we consider the free, damped vibration of the system. The governing equation is

$$\ddot{x} + \mu\dot{x} - 0.5\,x + 0.5\,x^3 = 0 \qquad (13.152)$$

Let the boundary conditions be given by

$$x(t = 0) = x_0,\ \dot{x}(t = 0) = \dot{x}_0 \qquad (13.153)$$

It can be expected from Fig. 13.21 that the static equilibria, $x = +1$ and $x = -1$, are unstable with respect to finite disturbances. Physically, when a finite disturbance is given

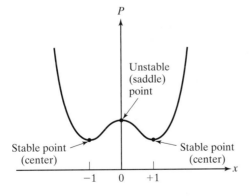

FIGURE 13.21   Plot of potential energy.

about one static equilibrium point, the system could be driven to the other static equilib-rium point. In fact, the steady-state solution can be shown to be extremely sensitive to the initial conditions, exhibiting a form of chaos. It can be verified easily [13.32] that for $x_0 = 1$ and $0 < \dot{x}_0 < 0.521799$, the steady-state solution is $x(t \rightarrow \infty) = +1$. The phase plane trajectory for $\dot{x}_0 = 0.52$ is shown in Fig. 13.22(a). Note that $x > 0$ for all values of t. For $x_0 = 1$ and $0.521799 < \dot{x}_0 < 0.5572$, the steady-state solution is $x(t \rightarrow \infty) = -1$. Figure 13.22(b) shows the phase plane trajectory for $\dot{x}_0 = 0.54$, indi-cating a single crossing of the $x = 0$ axis. For $0.5572 < \dot{x}_0 < 0.5952$, the steady-state solution is $x(t \rightarrow \infty) = +1$. The phase plane trajectory for $\dot{x}_0 = 0.56$ is shown in Fig. 13.22(c), which indicates two crossings of the $x = 0$ axis.

In fact, by giving a series of values to $\dot{x}_0$, we can construct several phase plane tra-jectories, from which a composite plot, known as the shell plot, can be constructed as shown in Fig. 13.23 [13.33, 13.34]. Here also, the steady-state solution can be seen to be $x = +1$ or -1 depending on the initial conditions, x_0 and $\dot{x}_0$. It can be seen that the var-ious regions are identified by the numbers 0, 1, 2, 3, ... First, consider the region labeled "0" with $x_0 \geq 0$. If the initial conditions fall in this region, the solution spirals into $x = +1$ as $t \rightarrow \infty$ and the solution crosses the axis $x = 0$ zero times (similar to Fig. 13.22a). Next, consider the region labeled "1." If the initial conditions fall in this region, the solution moves clockwise, crosses the axis $x = 0$ once, and settles into $x = -1$ as $t \rightarrow \infty$ (similar to Fig. 13.22b). Next, consider the region labeled "2." If the system starts with the initial conditions falling in this region, the phase plane moves clock-wise, crosses the axis $x = 0$, continues to move clockwise, crosses the axis $x = 0$ again, moves into the region "0" for $x > 0$, and spirals into $x = +1$ as $t \rightarrow \infty$ (similar to Fig. 13.22c). This pattern continues with other regions as well, with the labeled region number indicating the number of crossings of the axis $x = 0$ by the phase plane trajectory.

Figure 13.23 indicates that if there is sufficient uncertainty in the initial conditions x_0 and $\dot{x}_0$, the final state of the system, $x = +1$ or -1, is unpredictable or uncertain. If damp-ing is reduced further, the width of each region in Fig. 13.23 (except the regions labeled "0") becomes even smaller and vanishes as $\mu \rightarrow 0$. Thus the final steady state of the sys-tem is unpredictable as $\mu \rightarrow 0$ for any finite uncertainty in x_0, $\dot{x}_0$, or both. This means that the system exhibits chaos.

13.10.4 Chaotic Behavior of Duffing's Equation with the Forcing Term

Consider Duffing's equation with $\mu = 0.2$, and $\alpha = \beta = 1$ including the forcing term. Within the forcing term, small variations in the frequency (ω) or the amplitude (F_0) can lead to chaos, as indicated below.

When ω Is Changed.   For a fixed value of F_0, the phase plane response of Eq. (13.149) can be periodic or chaotic, depending on the value of ω. For example, Figs. 13.24 and 13.25 indicate the two situations that are described by the equations

$$\ddot{x} + 0.2\,\dot{x} - x + x^3 = 0.3 \cos 1.4\,t \text{ (periodic, Fig. 13.24)} \qquad (13.154)$$

$$\ddot{x} + 0.2\,\dot{x} - x + x^3 = 0.3 \cos 1.29\,t \text{ (chaotic, Fig. 13.25)} \qquad (13.155)$$

where $F_0 = 0.3$ has been assumed. Figure 13.24 has been plotted using an approximate analysis, known as the harmonic balance method. On the other hand, Fig. 13.25 represents

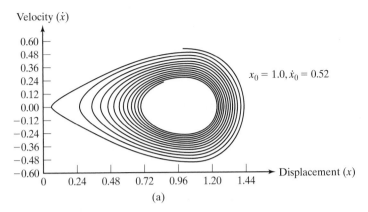

(a)

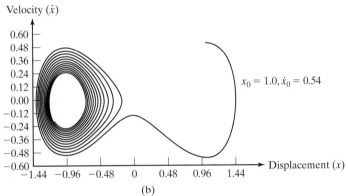

(b)

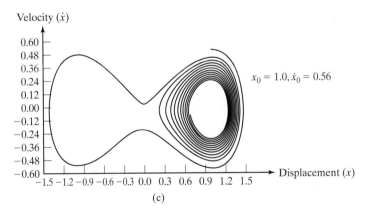

(c)

FIGURE 13.22   Phase-plane trajectories for different initial velocities. (From [13.32]; reprinted with permission of Society of Industrial and Applied Mathematics and E. H. Dowell and C. Pierre.)

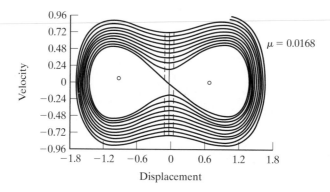

FIGURE 13.23   Shell plot. (From [13.33]; reprinted with permission of American Society of Mechanical Engineers.)

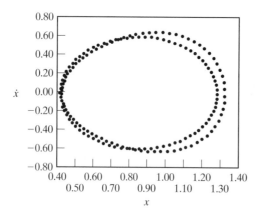

FIGURE 13.24   Phase plane of Eq. (13.154). (From [13.34]; reprinted with permission of Academic Press.)

a Poincaré map, which indicates points that occur at equal intervals of time $T_0, 2T_0, 3T_0, \ldots$ where T_0 is the fundamental period of excitation, $T_0 = \frac{2\pi}{\omega} = \frac{2\pi}{1.29}$.

When F_0 Is Changed.   Chaos can also be observed when the amplitude of the force changes. To illustrate, we consider the following equation [13.33]:

$$\ddot{x} + 0.168\,\dot{x} - 0.5\,x + 0.5\,x^3 = F_0 \sin \omega t \equiv F_0 \sin t \qquad (13.156)$$

For definiteness, we assume the initial conditions as $x_0 = 1$ and $\dot{x}_0 = 0$. When F_0 is sufficiently small, the response of the system will be a simple harmonic motion (that is, the phase plane will be an ellipse) about its static equilibrium position, $x = +1$. When F_0 is increased, additional harmonics beyond the fundamental are detected and the phase plane will be distorted from a simple ellipse, as shown in Fig. 13.26(a) for $F_0 = 0.177$. Note that

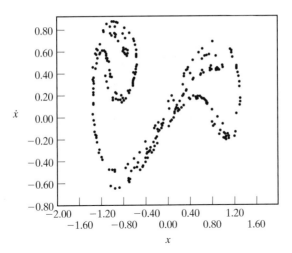

FIGURE 13.25   Poincaré map of Eq. (13.155). (From [13.34]; reprinted with permission of Academic Press.)

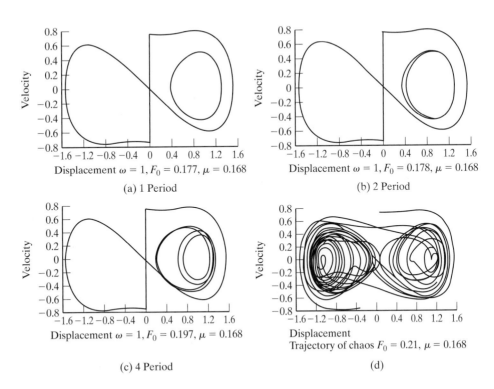

FIGURE 13.26   Distortion of phase plane. (From [13.33]; reprinted with permission of American Society of Mechanical Engineers.)

the boundary of the region labeled "0" in Fig. 13.23 is also shown in Fig. 13.26(a) for a comparison. The response is called a 1 period motion for $0 \leq F_0 \leq 0.177$, implying that the response oscillates through one period while the force oscillates through one period. For $F_0 = 0.178$, the phase plane is shown in Fig. 13.26(b), which indicates that the response is a 2 period motion. Thus, for the response to oscillate through one period, the force must oscillate through two periods. This change from a 1 to a 2 period response as F_0 changes from 0.177 to 0.178 is called a *bifurcation*. When $F_0 = 0.197$, the response will be a 4 period motion (see Fig. 13.26c). As F_0 is increased, 8, 16, ... period motions occur and finally, for $F_0 \geq 0.205$, chaos can be observed with no apparent periodicity, as indicated in Fig. 13.26(d).

13.11   Numerical Methods

Most of the numerical methods described in the earlier chapters can be used for finding the response of nonlinear systems. The Runge-Kutta method described in Section 11.4 is directly applicable for nonlinear systems and is illustrated in Section 13.12. The central difference, Houbolt, Wilson, and Newmark methods considered in Chapter 11 can also be used for solving nonlinear multidegree of freedom vibration problems with slight modification. Let a multidegree of freedom system be governed by the equation

$$[m]\ddot{\vec{x}}(t) + [c]\dot{\vec{x}}(t) + \vec{P}(\vec{x}(t)) = \vec{F}(t) \qquad (13.157)$$

where the internal set of forces opposing the displacements $\vec{P}$ are assumed to be nonlinear functions of $\vec{x}$. For the linear case, $\vec{P} = [k]\vec{x}$. In order to find the displacement vector $\vec{x}$ that satisfies the nonlinear equilibrium in Eq. (13.157), it is necessary to perform an equilibrium iteration sequence in each time step. In implicit methods (Houbolt, Wilson, and Newmark methods), the equilibrium conditions are considered at the same time for which solution is sought. If the solution is known for time t_i and we wish to find the solution for time t_{i+1}, then the following equilibrium equations are considered

$$[m]\ddot{\vec{x}}_{i+1} + [c]\dot{\vec{x}}_{i+1} + \vec{P}_{i+1} = \vec{F}_{i+1} \qquad (13.158)$$

where $\vec{F}_{i+1} = \vec{F}(t = t_{i+1})$ and $\vec{P}_{i+1}$ is computed as

$$\vec{P}_{i+1} = \vec{P}_i + [k_i]\Delta\vec{x} = \vec{P}_i + [k_i](\vec{x}_{i+1} - \vec{x}_i) \qquad (13.159)$$

where $[k_i]$ is the linearized or tangent stiffness matrix computed at time t_i. Substitution of Eq. (13.159) in Eq. (13.158) gives

$$[m]\ddot{\vec{x}}_{i+1} + [c]\dot{\vec{x}}_{i+1} + [k_i]\vec{x}_{i+1} = \hat{\vec{F}}_{i+1} = \vec{F}_{i+1} - \vec{P}_i + [k_i]\vec{x}_i \quad (13.160)$$

Since the right-hand side of Eq. (13.160) is completely known, it can be solved for $\vec{x}_{i+1}$ using any of the implicit methods directly. The $\vec{x}_{i+1}$ found is only an approximate vector due to the linearization process used in Eq. (13.159). To improve the accuracy of the solution and to avoid the development of numerical instabilities, an iterative process has to be used within the current time step [13.21].

13.12   Examples Using MATLAB

Solution of the Pendulum Equation

EXAMPLE   13.6

Using MATLAB, find the solution of the following forms of the pendulum equation with

$$\omega_0 = \sqrt{\frac{g}{l}} = 0.09.$$

(a) $\quad \ddot{\theta} + \omega_0^2 \theta = 0$ $\qquad$ (E.1)

(b) $\ddot{\theta} + \omega_0^2 \theta - \frac{1}{6}\omega_0^2 \theta^3 = 0$ $\qquad$ (E.2)

(c) $\quad \ddot{\theta} + \omega_0^2 \sin \theta = 0$ $\qquad$ (E.3)

Use the following initial conditions:

(i) $\theta(0) = 0.1, \qquad \dot{\theta}(0) = 0$ $\qquad$ (E.4)

(ii) $\theta(0) = \frac{\pi}{4}, \qquad \dot{\theta}(0) = 0$ $\qquad$ (E.5)

(iii) $\theta(0) = \frac{\pi}{2}, \qquad \dot{\theta}(0) = 0$ $\qquad$ (E.6)

Solution: Using $x_1 = \theta$ and $x_2 = \dot{\theta}$, each of the Eqs. (E.1) to (E.3) can be rewritten as a system of two first-order differential equations as follows:

(a) $\dot{x}_1 = x_2$

$\dot{x}_2 = -\omega_0^2 x_1 \quad$ (Linear equation) $\qquad$ (E.7)

(b) $\dot{x}_1 = x_2$

$\dot{x}_2 = -\omega_0^2 x_1 + \frac{1}{6}\omega_0^2 x_1^3 \quad$ (Nonlinear equation) $\qquad$ (E.8)

(c) $\dot{x}_1 = x_2$

$\dot{x}_2 = -\omega_0^2 \sin x_1 \quad$ (Nonlinear equation) $\qquad$ (E.9)

Equations (E.7) to (E.9) are solved using the MATLAB program **ode23** for each of the initial conditions given by Eqs. (E.4) to (E.6). The solutions $\theta(t)$ given by Eqs. (E.7), (E.8), and (E.9) for a specific initial conditions, are plotted in the same graph.

```
% Ex13_6.m
% This program will use the function dfunc1_a.m, dfunc1_b.m and
dfun1_c.m
% they should be in the same folder
tspan = [0: 1: 250];
```

```
x0 = [0.1; 0.0];
x0_1 = [0.7854; 0.0];
x0_2 = [1.5708; 0.0];
[t, xa] = ode23 ('dfunc1_a', tspan, x0);
[t, xb] = ode23 ('dfunc1_b', tspan, x0);
[t, xc] = ode23 ('dfunc1_c', tspan, x0);
[t, xa1] = ode23 ('dfunc1_a', tspan, x0_1);
[t, xb1] = ode23 ('dfunc1_b', tspan, x0_1);
[t, xc1] = ode23 ('dfunc1_c', tspan, x0_1);
[t, xa2] = ode23 ('dfunc1_a', tspan, x0_2);
[t, xb2] = ode23 ('dfunc1_b', tspan, x0_2);
[t, xc2] = ode23 ('dfunc1_c', tspan, x0_2);
plot (t, xa(:, 1));
ylabel ('Theta(t)');
xlabel ('t');
ylabel ('i.c. = [0.1; 0.0]');
title. . .
('Function a: solid line, Function b: dashed line, Function c: dotted
line');
hold on;
plot (t, xb(:, 1), '--');
hold on;
plot (t, xc(:, 1), ':');
pause;
hold off;
plot (t, xa1(:, 1));
ylabel ('Theta(t)');
xlabel ('t');
ylabel ('i.c. = [0.7854; 0.0]');
title. . .
('Function a: solid line, Function b: dashed line, Function c: dotted
line');
hold on;
plot (t, xb1(:, 1), '--');
hold on;
plot (t, xc1(:, 1), ':');
pause;
hold off;
plot(t,xa2(:,1));
hold on;
ylabel('Theta(t)');
xlabel('t');
ylabel('i.c. = [1.5708; 0.0]')
title. . .
('Function a: solid line, Function b: dashed line, Function c: dotted
line');
plot(t, xb2(:,1),'--');
hold on;
plot(t,xc2(:,1),':');

% dfunc1_a.m
function f = dfunc1_a(t,x);
f = zeros(2,1);
f(1) = x(2);
f(2) = -0.0081 * x(1);

% dfunc1_b.m
function f = dfunc1_b(t,x);
f = zeros(2,1);
f(1) = x(2);
f(2) = 0.0081 * ((x(1))^3) / 6.0 - 0.0081 * x(1);

% dfunc1_c.m
function f = dfunc1_c(t,x);
f = zeros(2,1);
f(1) = x(2);
f(2) = -0.0081 * sin(x(1));
```

Function a: solid line, Function b: dashed line, Function c: dotted line

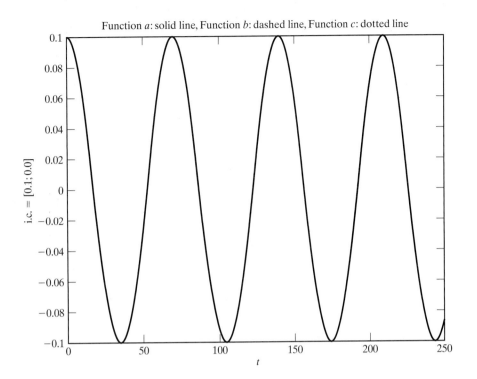

Function a: solid line, Function b: dashed line, Function c: dotted line

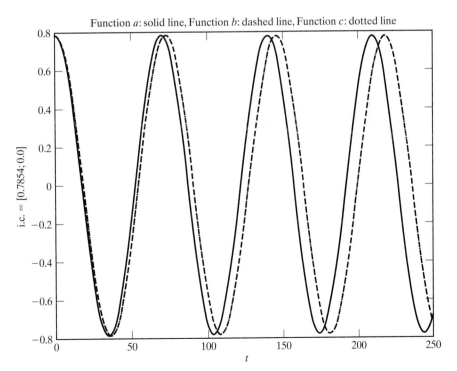

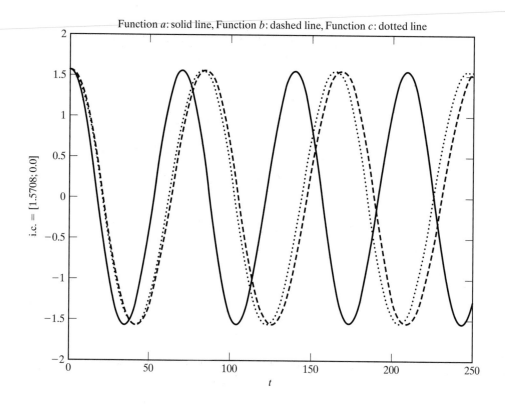

Function a: solid line, Function b: dashed line, Function c: dotted line

i.c. = [1.5708; 0.0]

t

■

Solution of Nonlinearly Damped System

EXAMPLE   13.7

Using MATLAB, find the solution of a single degree of freedom system with velocity squared damping.

Governing equation:

$$m\,\ddot{x} + c(\dot{x})^2 \operatorname{sign}(\dot{x}) + k\,x = F_0 \sin \omega t \tag{E.1}$$

Data: $m = 10$, $c = 0.01$, $k = 4000$, $F_0 = 200$, $\omega = 10$ and 20, $x(0) = 0.5$, $\dot{x}(0) = 1.0$

Also find the solution of the system using the equivalent viscous damping constant (c_{eq})

$$m\,\ddot{x} + c_{\text{eq}}\,\dot{x} + k\,x = F_0 \sin \omega t \tag{E.2}$$

where c_{eq} is given by Eq. (E.4) of Example 3.7 as

$$c_{\text{eq}} = \frac{8\,c\,\omega\,X}{3\,\pi} \tag{E.3}$$

Solution: By introducing $x_1 = x$ and $x_2 = \dot{x}$, Eqs. (E.1) and (E.2) are written as systems of two first-order differential equations as

$$(a) \ \dot{x}_1 = x_2$$

$$\dot{x}_2 = \frac{F_0}{m} \sin \omega t - \frac{c}{m} x_2^2 \ \text{sign} \ (x_2) - \frac{k}{m} x_1 \qquad \text{(Nonlinear equation)} \qquad \text{(E.4)}$$

$$(b) \ \dot{x}_1 = x_2$$

$$\dot{x}_2 = \frac{F_0}{m} \sin \omega t - \frac{c_{eq}}{m} x_2 - \frac{k}{m} x_1 \qquad \text{(Linear equation)} \qquad \text{(E.5)}$$

and X, in Eq. (E.3), is taken as the steady-state or static deflection of the system as $X = \frac{F_0}{k}$. The MATLAB solutions given by Eqs. (E.4) and (E.5) are plotted in the same graph for a specific value of ω.

```
% Ex13_7.m
% This program will use the function dfunc3_a.m, dfunc3_b.m
% dfunc3_a1.m, dfunc3_b1.m, they should be in the same folder
tspan = [0: 0.005: 10];
x0 = [0.5; 1.0];
[t,xa] = ode23 ('dfunc3_a', tspan, x0);
[t,xb] = ode23 ('dfunc3_b', tspan, x0);
[t,xa1] = ode23 ('dfunc3_a1', tspan, x0);
[t,xb1] = ode23 ('dfunc3_b1', tspan, x0);
subplot (211);
plot (t,xa (:,1));
title ('Theta(t): function a (Solid line), function b (Dashed
line)');
ylabel ('w = 10 ');
hold on;
plot (t,xb(:,1), '--');
subplot (212);
plot (t,xa1(:,1));
ylabel ('w = 20 ');
hold on;
plot (t,xb1 (:,1), '--');
xlabel ('t');

% dfunc3_a.m
function f = dfunc3_a (t,x);
f0 = 200;
m = 10;
a = 0.01;
k = 4000;
w = 10;
f = zeros (2,1);
f(1) = x(2);
f(2) = f0* sin (w*t) /m - a* x(2)^2 * sign(x(2)) /m - k*x(1) /m;

% dfunc3_a1.m
function f = dfunc3_a1 (t,x);
f0 = 200;
m = 10;
a = 0.01;
k = 4000;
w = 20;
f = zeros (2,1);
```

```
f(1) = x(2);
f(2) = f0* sin (w*t) /m - a* x(2)^2 * sign (x(2)) /m - k*x(1) /m;

% dfunc3_b.m
function f = dfunc3_b (t,x);
f0 = 200;
m = 10;
a = 0.01;
k = 4000;
ceq = sqrt (8*a*f0 / (3*pi));
w = 10;
f = zeros (2,1);
f(1) = x(2);
f(2) = f0* sin (w*t) /m - ceq * x(2) /m - k*x(1) /m;

% dfunc3_b1.m
function f = dfunc3_b1 (t,x);
f0 = 200;
m = 10;
a = 0.01;
k = 4000;
ceq = sqrt (8*a*f0 / (3*pi));
w = 20;
f = zeros (2,1);
f(1) = x(2);
f(2) = f0* sin (w*t) /m - ceq * x(2) /m - k*x(1) /m;
```

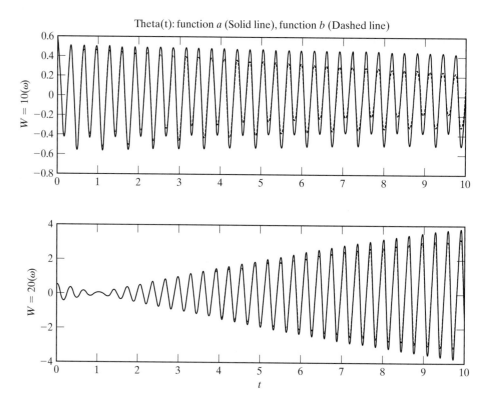

Theta(t): function a (Solid line), function b (Dashed line)

Solution of Nonlinear System Under Pulse Loading

EXAMPLE   13.8

Using MATLAB, find the solution of a nonlinear single degree of freedom system governed by the equation

$$m\,\ddot{x} + k_1\,x + k_2\,x^3 = f(t) \tag{E.1}$$

where $F(t)$ is a rectangular pulse load of magnitude F_0 applied over $0 \le t \le t_0$. Assume the following data: $m = 10$, $k_1 = 4000$, $F_0 = 1000$, $t_0 = 1.0$, $x(0) = 0.05$, $\dot{x}(0) = 5$. Solve Eq. (E.1) for two cases: one with $k_2 = 0$ and the other with $k_2 = 500$.

Solution: Using $x_1 = x$ and $x_2 = \dot{x}$, Eq. (E.1) is rewritten, as a set of two first-order differential equations as

$$\dot{x}_1 = x_2$$

$$\dot{x}_2 = \frac{F(t)}{m} - \frac{k_1}{m}x_1 - \frac{k_2}{m}x_1^3 \tag{E.2}$$

The responses, $x(t)$, found with $k_2 = 0$ (linear system) and $k_2 = 500$ (nonlinear system) are plotted in the same graph.

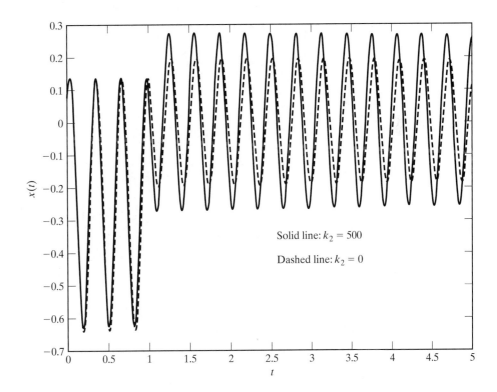

Solid line: $k_2 = 500$

Dashed line: $k_2 = 0$

```
% Ex13_8.m
% This program will use the function dfunc13_8_1.m and dfunc13_8_2.m
% they should be in the same folder
tspan = [0: 0.01: 5];
x0 = [0.05; 5];
[t,x] = ode23('dfunc13_8_1', tspan, x0);
plot(t,x(:,1));
xlabel('t');
ylabel('x(t)');
hold on;
[t,x] = ode23('dfunc13_8_2', tspan, x0);
plot(t,x(:,1),'--');
gtext('Solid line: k_2 = 500');
gtext('Dashed line: k_2 = 0')

% dfunc13_8_1.m
function f = dfunc13_8_1(t,x)
f = zeros(2,1);
m = 10;
k1 = 4000;
k2 = 500;
F0 = 1000;
F = F0 * (stepfun(t, 0) - stepfun(t, 1));
f(1) = x(2);
f(2) = -F/m - k1 * x(1)/m - k2 * (x(1))^3/m;

% dfunc13_8_2.m
function f = dfunc13_8_2(t,x)
f = zeros(2,1);
m = 10;
k1 = 4000;
k2 = 0;
F0 = 1000;
F = F0 * (stepfun(t, 0) - stepfun(t, 1));
f(1) = x(2);
f(2) = -F/m - k1 * x(1)/m - k2 * (x(1))^3/m;
```

■

Solution of Nonlinear Differential Equation

EXAMPLE   13.9

Using the fourth-order Runge-Kutta method, develop a general MATLAB program called **Program18.m** to find the solution of a single degree of freedom equation of the form

$$m\,\ddot{x} + c\,\dot{x} + k\,x + k^*\,x^3 = 0 \qquad (E.1)$$

Use the program to solve Eq. (E.1) for the following data: $m = 0.01$, $c = 0.1$, $k = 2.0$, $k^* = 0.5$, $x(0) = 7.5$, $\dot{x}(0) = 0$.

Solution: Equation (E.1) is rewritten as

$$\dot{x}_1 = f_1(x_1, x_2) = x_2$$

$$\dot{x}_2 = f_2(x_1, x_2) = -\frac{c}{m}x_2 - \frac{k}{m}x_1 - \frac{k^*}{m}x_1^3 \qquad (E.2)$$

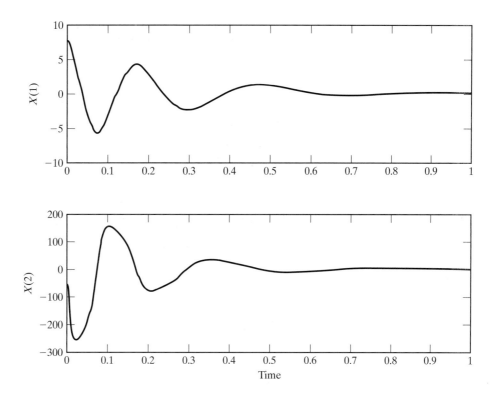

Program18.m is developed to accept the values of *m, c, k*, and *k** as *YM, YC, YK*, and *YKS*, respectively. The time step (Δt) and the number of time steps (NSTEP) are specified as 0.0025 and 400, respectively. A subprogram, called **fun;** is to be given to define $f_1(x_1, x_2)$ and $f_2(x_1, x_2)$. The program gives the values of t_i, $x(t_i)$ and $\dot{x}(t_i)$, $i = 1, 2, \ldots$, NSTEP as output. The program also plots $x(t) = x_1(t)$ and $\dot{x}(t) = x_2(t)$.

```
Solution of nonlinear vibration problem
by fourth order Runge-Kutta method

Data:
 ym = 1.000000e-002
 yc = 1.000000e-001
 yk = 2.00000000e+000
yks = 5.00000000e-001

Results

      i          time(i)            x(i,1)            x(i,2)

      1       2.500000e-003      7.430295e+000     -5.528573e+001
      6       1.500000e-002      5.405670e+000     -2.363166e+002
     11       2.750000e-002      2.226943e+000     -2.554475e+002
     16       4.000000e-002     -8.046611e-001     -2.280328e+002
     21       5.250000e-002     -3.430513e+000     -1.877713e+002
     26       6.500000e-002     -5.296623e+000     -1.002752e+002
      .
      .
      .
```

371	9.275000e−001	1.219287e−001	7.673075e−002
376	9.400000e−001	1.209954e−001	−2.194914e−001
381	9.525000e−001	1.166138e−001	−4.744062e−001
386	9.650000e−001	1.093188e−001	−6.853283e−001
391	9.775000e−001	9.966846e−002	−8.512093e−001
396	9.900000e−001	8.822462e−002	−9.724752e−001

■

13.13   C++ Program

An interactive C++ program called **Program18.cpp** is given for finding the solution of a single degree of freedom system with a nonlinear spring using the fourth-order Runge-Kutta method. The input and output of the program are similar to those of the MATLAB program, **Program18.m**, given in Example 13.9.

▬▬▬▬▬▬▬▬▬ Solution of Nonlinear Spring-Mass-Damper System

EXAMPLE 13.10 ──

Use **Program18.cpp** to solve the problem described in Example 13.9.

Solution: The input data are to be entered interactively. The input and output of the program are given below.

```
SOLUTION OF NONLINEAR VIBRATION PROBLEM
BY FOURTH ORDER RUNGE-KUTTA METHOD

DATA:
YM = 0.010000
YC = 0.100000
YK = 2.000000
YKS = 0.500000

RESULTS:
```

I	TIME(I)	X(I,1)	X(I,2)
1	0.00250000	7.43029541	−55.28573134
2	0.00500000	7.22711455	−106.35020086
3	0.00750000	6.90397627	−150.94361827
4	0.01000000	6.47900312	−187.66152131
5	0.01250000	5.97267903	−216.01444338
.			
.			
.			
395	0.98750000	0.09063023	−0.95172604
396	0.99000000	0.08822462	−0.97247522
397	0.99250000	0.08576929	−0.99150470
398	0.99500000	0.08326851	−1.00883370
399	0.99750000	0.08072652	−1.02448309
400	1.00000000	0.07814748	−1.03847532

■

13.14 Fortran Program

A Fortran program, called **PROGRAM18.F**, is given for the numerical solution of the non-linear free vibration problem of a single degree of freedom system using the fourth-order Runge-Kutta method. The input and output of the program are similar to those of the MATLAB program, **Program18.m**, given in Example 13.9.

Solution of Nonlinear Vibration Problem

EXAMPLE 13.11

Find the solution of the problem described in Example 13.9 using **PROGRAM18.F**.

Solution: The output of the program is given below.

```
SOLUTION OF NONLINEAR VIBRATION PROBLEM
BY FOURTH ORDER RUNGE-KUTTA METHOD

DATA:
YM  = 0.100000E-01
YC  = 0.100000E+00
YK  = 0.200000E+01
YKS = 0.500000E+00

RESULTS:

    I      TIME(I)       X(I,1)          X(I,2)

    1     0.002500     0.743030E+01    -0.552857E+02
    2     0.005000     0.722711E+01    -0.106350E+03
    3     0.007500     0.690398E+01    -0.150944E+03
    4     0.010000     0.647900E+01    -0.187662E+03
    5     0.012500     0.597268E+01    -0.216014E+03
    .
    .
    .
  395     0.987511     0.906302E-01    -0.951725E+00
  396     0.990011     0.882246E-01    -0.972474E+00
  397     0.992511     0.857693E-01    -0.991504E+00
  398     0.995011     0.832685E-01    -0.100883E+01
  399     0.997511     0.807265E-01    -0.102448E+01
  400     1.000011     0.781475E-01    -0.103847E+01
```

REFERENCES

13.1 C. Hayashi, *Nonlinear Oscillations in Physical Systems*, McGraw-Hill, New York, 1964.

13.2 A. A. Andronow and C. E. Chaikin, *Theory of Oscillations* (English language edition), Princeton University Press, Princeton, N.J., 1949.

13.3 N. V. Butenin, *Elements of the Theory of Nonlinear Oscillations*, Blaisdell Publishing, New York, 1965.

13.4 A. Blaquiere, *Nonlinear System Analysis*, Academic Press, New York, 1966.

13.5 Y. H. Ku, *Analysis and Control of Nonlinear Systems*, Ronald Press, New York, 1958.

13.6 J. N. J. Cunningham, *Introduction to Nonlinear Analysis*, McGraw-Hill, New York, 1958.

13.7 J. J. Stoker, *Nonlinear Vibrations in Mechanical and Electrical Systems*, Interscience Publishers, New York, 1950.

13.8   J. P. Den Hartog, *Mechanical Vibrations* (4th ed.), McGraw-Hill, New York, 1956.

13.9   N. Minorsky, *Nonlinear Oscillations*, D. Van Nostrand, Princeton, N.J., 1962.

13.10 R. E. Mickens, "Perturbation solution of a highly nonlinear oscillation equation," *Journal of Sound and Vibration*, Vol. 68, 1980, pp. 153–155.

13.11 B. V. Dasarathy and P. Srinivasan, "Study of a class of nonlinear systems reducible to equivalent linear systems," *Journal of Sound and Vibration*, Vol. 7, 1968, pp. 27–30.

13.12 G. L. Anderson, "A modified perturbation method for treating nonlinear oscillation problems," *Journal of Sound and Vibration*, Vol. 38, 1975, pp. 451–464.

13.13 B. L. Ly, "A note on the free vibration of a nonlinear oscillator," *Journal of Sound and Vibration*, Vol. 68, 1980, pp. 307–309.

13.14 V. A. Bapat and P. Srinivasan, "Free vibrations of nonlinear cubic spring mass systems in the presence of Coulomb damping," *Journal of Sound and Vibration*, Vol. 11, 1970, pp. 121–137.

13.15 H. R. Srirangarajan, P. Srinivasan, and B. V. Dasarathy, "Analysis of two degrees of freedom systems through weighted mean square linearization approach," *Journal of Sound and Vibration*, Vol. 36, 1974, pp. 119–131.

13.16 S. R. Woodall, "On the large amplitude oscillations of a thin elastic beam," *International Journal of Nonlinear Mechanics*, Vol. 1, 1966, pp. 217–238.

13.17 D. A. Evenson, "Nonlinear vibrations of beams with various boundary conditions," *AIAA Journal*, Vol. 6, 1968, pp. 370–372.

13.18 M. E. Beshai and M. A. Dokainish, "The transient response of a forced nonlinear system," *Journal of Sound and Vibration*, Vol. 41, 1975, pp. 53–62.

13.19 V. A. Bapat and P. Srinivasan, "Response of undamped nonlinear spring mass systems subjected to constant force excitation," *Journal of Sound and Vibration*, Vol. 9, 1969, Part I: pp. 53–58 and Part II: pp. 438–446.

13.20 W. E. Boyce and R. C. DiPrima, *Elementary Differential Equations and Boundary Value Problems* (4th ed.), Wiley, New York, 1986.

13.21 D. R. J. Owen, "Implicit finite element methods for the dynamic transient analysis of solids with particular reference to nonlinear situations," in *Advanced Structural Dynamics*, J. Donéa (ed.), Applied Science Publishers, London, 1980, pp. 123–152.

13.22 B. van der Pol, "Relaxation oscillations," *Philosophical Magazine*, Vol. 2, pp. 978–992, 1926.

13.23 L. A. Pipes and L. R. Harvill, *Applied Mathematics for Engineers and Physicists* (3rd ed.), McGraw-Hill, New York, 1970.

13.24 N. N. Bogoliubov and Y. A. Mitropolsky, *Asymptotic Methods in the Theory of Nonlinear Oscillations*, Hindustan Publishing, Delhi, 1961.

13.25 A. H. Nayfeh and D. T. Mook, *Nonlinear Oscillations*, Wiley, New York, 1979.

13.26 G. Duffing, "Erzwungene Schwingungen bei veranderlicher Eigenfrequenz und ihre technische Bedeutung," Ph.D. thesis (Sammlung Vieweg, Braunschweig, 1918).

13.27 C. A. Ludeke, "An experimental investigation of forced vibrations in a mechanical system having a nonlinear restoring force," *Journal of Applied Physics*, Vol. 17, pp. 603–609, 1946.

13.28 D. W. Jordan and P. Smith, *Nonlinear Ordinary Differential Equations* (2nd ed.), Clarendon Press, Oxford, 1987.

13.29 R. E. Mickens, *An Introduction to Nonlinear Oscillations*, Cambridge University Press, Cambridge, 1981.

13.30 S. H. Crandall, "Nonlinearities in structural dynamics," *The Shock and Vibration Digest*, Vol. 6, No. 8, August 1974, pp. 2–14.

13.31 R. M. May, "Simple mathematical models with very complicated dynamics," *Nature*, Vol. 261, June 1976, pp. 459–467.

13.32 E. H. Dowell and C. Pierre, "Chaotic oscillations in mechanical systems," in *Chaos in Nonlinear Dynamical Systems*, J. Chandra (ed.), SIAM, Philadelphia, 1984, pp. 176–191.

13.33 E. H. Dowell and C. Pezeshki, "On the understanding of chaos in Duffing's equation including a comparison with experiment," *ASME Journal of Applied Mechanics*, Vol. 53, March 1986, pp. 5–9.

13.34 B. H. Tongue, "Existence of chaos on a one-degree-of-freedom system," *Journal of Sound and Vibration*, Vol. 110, No. 1, October, 1986, pp. 69–78.

13.35 M. Cartmell, *Introduction to Linear, Parametric, and Nonlinear Vibrations*, Chapman and Hall, London, 1990.

REVIEW QUESTIONS

13.1 Give brief answers to the following:

1. How do you recognize a nonlinear vibration problem?
2. What are the various sources of nonlinearity in a vibration problem?
3. What is the source of nonlinearity in Duffing's equation?
4. How is the frequency of the solution of Duffing's equation affected by the nature of the spring?
5. What are subharmonic oscillations?
6. Explain the jump phenomenon.
7. What principle is used in the Ritz-Galerkin method?
8. Define these terms: *phase plane, trajectory, singular point, phase velocity.*
9. What is the method of isoclines?
10. What is the difference between a hard spring and a soft spring?
11. Explain the difference between subharmonic and superharmonic oscillations.
12. What is a secular term?
13. Give an example of a system that leads to an equation of motion with time-dependent coefficients.
14. Explain the significance of the following: stable node, unstable node, saddle point, focus, and center.
15. What is a limit cycle?
16. Give two examples of physical phenomena that can be represented by van der Pol's equation.

13.2 Indicate whether each of the following statements is true or false:

1. Nonlinearity can be introduced into the governing differential equation through mass, spring and/or dampers.
2. Nonlinear analysis of a system can reveal several unexpected phenomena.
3. The Mathieu equation is an autonomous equation.
4. A singular point corresponds to a state of equilibrium of the system.
5. Jump phenomenon is exhibited by both linear and nonlinear systems.

   6. The Ritz-Galerkin method finds the approximate solution by satisfying the nonlinear equation in the average.

   7. Dry friction can introduce nonlinearity in the system.

   8. Poincaré's solution of nonlinear equations is in the form of a series.

   9. The secular term appears in the solution of free Duffing's equation.

   10. According to Lindstedt's perturbation method, the angular frequency is assumed to be a function of the amplitude.

   11. An isocline is the locus of points at which the trajectories passing through them have a constant slope.

   12. Time does not appear explicitly in a trajectory plotted in the phase plane.

   13. The time variation of the solution can be found from the phase plane trajectories.

   14. A limit cycle denotes a steady-state periodic oscillation.

   15. Approximate solutions of nonlinear vibration problems can be found using numerical methods such as Houbolt, Wilson, and Neumark method.

13.3   Fill in each of the following blanks with the appropriate words:

   1. When finite amplitudes of motion are involved, _____ analysis becomes necessary.

   2. _____ principle is not applicable in nonlinear analysis.

   3. _____ equation involves time-dependent coefficients.

   4. The governing equation of a simple pendulum whose pivot is subjected to vertical vibration is called _____ equation.

   5. The representation of the motion of a system in the displacement-velocity plane is known as _____ plane representation.

   6. The curve traced by a typical point in the phase plane is called a _____.

   7. The velocity with which a representative point moves along a trajectory is called the _____ velocity.

   8. The phenomenon of realizing two amplitudes for the same frequency is known as _____ phenomenon.

   9. The forced vibration solution of Duffing's equation has _____ values of the frequency ω for any given amplitude $|A|$.

   10. The Ritz-Galerkin method involves the solution of _____ equations.

   11. Mechanical chatter is a _____ vibration.

   12. If time does not appear explicitly in the governing equation, the corresponding system is said to be _____.

   13. The method of _____ can be used to construct the trajectories of a one degree of freedom dynamical system.

   14. Van der Pol's equation exhibits _____ cycles.

13.4   Select the most appropriate answer out of the choices given:

   1. Each term in the equation of motion of a linear system involves displacement, velocity, and acceleration of the
   (a) first degree   (b) second degree   (c) zero degree

   2. A nonlinear stress-strain curve can lead to nonlinearity of the
   (a) mass   (b) spring   (c) damper

   3. If the rate of change of force with respect to displacement, df/dx, is an increasing function of x, the spring is called a

   (a) soft spring   (b) hard spring   (c) linear spring

4. If the rate of change of force with respect to displacement, df/dx, is a decreasing function of k, the spring is called a
   (a) soft spring   (b) hard spring   (c) linear spring

5. The point surrounded by closed trajectories is called a
   (a) center   (b) mid-point   (c) focal point

6. For a system with periodic motion, the trajectory in the phase plane is a
   (a) closed curve   (b) open curve   (c) point

7. In subharmonic oscillations, the natural frequency (ω_n) and the forcing frequency (ω) are related as
   (a) $\omega_n = \omega$
   (b) $\omega_n = n\omega; n = 2, 3, 4, \cdots$
   (c) $\omega_n = \dfrac{\omega}{n}; n = 2, 3, 4, \cdots$

8. In superharmonic oscillations, the natural frequency (ω_n) and the forcing frequency (ω) are related as
   (a) $\omega_n = \omega$
   (b) $\omega_n = n\omega; n = 2, 3, 4, \cdots$
   (c) $\omega_n = \dfrac{\omega}{n}; n = 2, 3, 4, \cdots$

9. If time appears explicitly in the governing equation, the corresponding system is called
   (a) an autonomous system
   (b) a nonautonomous system
   (c) a linear system.

10. Duffing's equation is given by
    (a) $\ddot{x} + \omega_0^2 x + \alpha x^3 = 0$
    (b) $\ddot{x} + \omega_0^2 x = 0$
    (c) $\ddot{x} + \alpha x^3 = 0$

11. Lindstedt's perturbation method gives
    (a) periodic and nonperiodic solutions
    (b) periodic solutions only
    (c) nonperiodic solutions only

13.5 Match the items in the two columns below for the nature of equilibrium points in the context of the stability analysis of equilibrium states with λ_1 and λ_2 as eigenvalues:

(1) λ_1 and λ_2 with same sign (λ_1, λ_2: real and distinct)	(a) Unstable node
(2) λ_1 and $\lambda_2 < 0$ (λ_1, λ_2: real and distinct)	(b) Saddle point
(3) λ_1 and $\lambda_2 > 0$ (λ_1, λ_2: real and distinct)	(c) Node
(4) λ_1 and λ_2: real with opposite signs	(d) Focus or spiral point
(5) λ_1 and λ_2: complex conjugates	(e) Stable node

13.6   Match the items in the two columns below:

(1) $\ddot{x} + f \dfrac{\dot{x}}{\lvert \dot{x} \rvert} + \omega_n^2\, x = 0$	(a) Nonlinearity in mass
(2) $\ddot{x} + \omega_0^2 \left(x - \dfrac{x^3}{6} \right) = 0$	(b) Nonlinearity in damping
(3) $ax\ddot{x} + kx = 0$	(c) Linear equation
(4) $\ddot{x} + c\dot{x} + kx = ax^3$	(d) Nonlinearity in spring

---

PROBLEMS

The problem assignments are organized as follows:

Problems	Section Covered	Topic Covered
13.1	13.1	Introduction
13.2–13.8	13.2	Examples of nonlinear vibration
13.9–13.11	13.3	Exact methods of solution
13.12, 13.13	13.4	Approximate analytical methods
13.14–13.16	13.5	Subharmonic and superharmonic oscillations
13.17	13.6	Mathieu equation
13.18–13.22	13.7	Graphical methods
13.23–13.33	13.8	Stability of equilibrium states
13.34	13.9	Limit cycles
13.35, 13.36	13.10	Chaos
13.37–13.42	13.12	MATLAB programs
13.43–13.45	13.13	C++ program
13.46–13.49	13.14	Fortran program
13.50–13.51	—	Design projects

13.1   The equation of motion of a simple pendulum, subjected to a constant torque, $M_t = ml^2 f$, is given by

$$\ddot{\theta} + \omega_0^2 \sin \theta = f \tag{E.1}$$

If $\sin \theta$ is replaced by its two-term approximation, $\theta - (\theta^3/6)$, the equation of motion becomes

$$\ddot{\theta} + \omega_0^2 \theta = f + \frac{\omega_0^2}{6}\theta^3 \tag{E.2}$$

Let the solution of the linearized equation

$$\ddot{\theta} + \omega_0^2 \theta = f \qquad\qquad (E.3)$$

be denoted as $\theta_1(t)$, and the solution of the equation

$$\ddot{\theta} + \omega_0^2 \theta = \frac{\omega_0^2}{6}\theta^3 \qquad\qquad (E.4)$$

be denoted as $\theta_2(t)$. Discuss the validity of the total solution. $\theta(t)$, given by $\theta(t) = \theta_1(t) + \theta_2(t)$, for Eq. (E.2).

13.2 Two springs, having different stiffnesses k_1 and k_2 with $k_2 > k_1$, are placed on either side of a mass m, as shown in Fig. 13.27. When the mass is in its equilibrium position, no spring is in contact with the mass. However, when the mass is displaced from its equilibrium position, only one spring will be compressed. If the mass is given an initial velocity $\dot{x}_0$ at $t = 0$, determine (a) the maximum deflection and (b) the period of vibration of the mass.

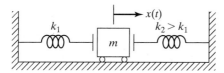

FIGURE 13.27

13.3 Find the equation of motion of the mass shown in Fig. 13.28. Draw the spring force versus x diagram.

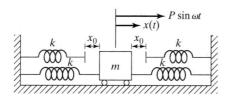

FIGURE 13.28

13.4 Two masses m_1 and m_2 are attached to a stretched wire, as shown in Fig. 13.29. If the initial tension in the wire is P, derive the equations of motion for large transverse displacements of the masses.

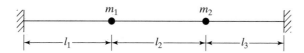

FIGURE 13.29

13.5 A mass m, connected to an elastic rubber band of unstretched length l and stiffness k, is permitted to swing as a pendulum bob, as shown in Fig. 13.30. Derive the nonlinear equations of motion of the system using x and θ as coordinates. Linearize the equations of motion and determine the natural frequencies of vibration of the system.

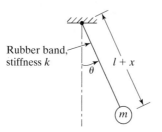

Rubber band, stiffness k

$l + x$

θ

m

FIGURE 13.30

13.6 A uniform bar of length l and mass m is hinged at one end $(x = 0)$, supported by a spring at $x = \frac{2l}{3}$, and acted by a force at $x = l$, as shown in Fig. 13.31. Derive the nonlinear equation of motion of the system.

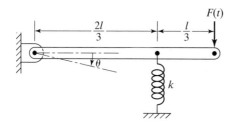

$F(t)$

$\frac{2l}{3}$

$\frac{l}{3}$

θ

k

FIGURE 13.31

13.7 Derive the nonlinear equation of motion of the spring-mass system shown in Fig. 13.32

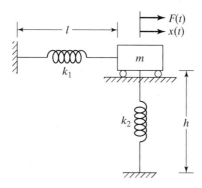

$F(t)$

$x(t)$

l

m

k_1

k_2

h

FIGURE 13.32

13.8 Derive the nonlinear equations of motion of the system shown in Fig. 13.33. Also, find the linearized equations of motion for small displacements, $x(t)$ and $\theta(t)$.

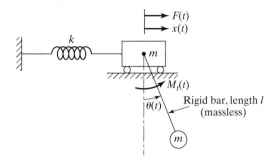

FIGURE 13.33

13.9 Find the natural time period of oscillation of the pendulum shown in Fig. 13.1(a) when it oscillates between the limits $\theta = -\pi/2$ and $\theta = \pi/2$, using Eqs. (13.1) and (13.12).

13.10 A simple pendulum of length 30 in. is released from the initial position of 80° from the vertical. How long does it take to reach the position $\theta = 0°$?

13.11 Find the exact solution of the nonlinear pendulum equation

$$\ddot{\theta} + \omega_0^2 \left(\theta - \frac{\theta^3}{6} \right) = 0$$

with $\dot{\theta} = 0$ when $\theta = \theta_0$, where θ_0 denotes the maximum angular displacement.

13.12 Find the solution of Example 13.1 using the following two-term approximation for $x(t)$:

$$x(t) = A_0 \sin \omega t + A_3 \sin 3 \omega t$$

13.13 Using a three-term expansion in Lindstedt's perturbation method [Eq. (13.30)], find the solution of the pendulum equation, Eq. (13.20).

13.14 The equation of motion for the forced vibration of a single degree of freedom nonlinear system can be expressed as

$$\ddot{x} + c\dot{x} + k_1 x + k_2 x^3 = a_1 \cos 3\omega t - a_2 \sin 3\omega t$$

Derive the conditions for the existence of subharmonics of order 3 for this system.

13.15 The equation of motion of a nonlinear system is given by

$$\ddot{x} + c\dot{x} + k_1 x + k_2 x^2 = a \cos 2\omega t$$

Investigate the subharmonic solution of order 2 for this system.

13.16 Prove that, for the system considered in Section 13.5.1, the minimum value of ω^2 for which the amplitude of subharmonic oscillations A will have a real value is given by

$$\omega_{\min} = \omega_0 + \frac{21}{2048} \frac{F^2}{\omega_0^5}$$

Also, show that the minimum value of the amplitude, for stable subharmonic oscillations, is given by

$$A_{min} = \frac{F}{16\omega^2}$$

13.17 Derive Eqs. (13.113b) and (13.116b) for the Mathieu equation.

13.18 The equation of motion of a single degree of freedom system is given by

$$2\ddot{x} + 0.8\dot{x} + 1.6x = 0$$

with initial conditions $x(0) = -1$ and $\dot{x}(0) = 2$. (a) Plot the graph $x(t)$ versus t for $0 \le t \le 10$. (b) Plot a trajectory in the phase plane.

13.19 Find the equilibrium position and plot the trajectories in the neighborhood of the equilibrium position corresponding to the following equation:

$$\ddot{x} + 0.1\left(x^2 - 1\right)\dot{x} + x = 0$$

13.20 Obtain the phase trajectories for a system governed by the equation

$$\ddot{x} + 0.4\dot{x} + 0.8x = 0$$

with the initial conditions $x(0) = 2$ and $\dot{x}(0) = 1$ using the method of isoclines.

13.21 Plot the phase-plane trajectories for the following system:

$$\ddot{x} + 0.1\dot{x} + x = 5$$

The initial conditions are $x(0) = \dot{x}(0) = 0$.

13.22 A single degree of freedom system is subjected to Coulomb friction so that the equation of motion is given by

$$\ddot{x} + f\frac{\dot{x}}{|\dot{x}|} + \omega_n^2 x = 0$$

Construct the phase plane trajectories of the system using the initial conditions $x(0) = 10(f/\omega_n^2)$ and $\dot{x}(0) = 0$.

13.23 The equation of motion of a simple pendulum subject to viscous damping can be expressed as

$$\ddot{\theta} + c\dot{\theta} + \sin\theta = 0$$

If the initial conditions are $\theta(0) = \theta_0$ and $\dot{\theta}(0) = 0$, show that the origin in the phase plane diagram represents (a) a stable focus for $c > 0$ and (b) an unstable focus for $c < 0$.

13.24 The equation of motion of a simple pendulum, subjected to external force, is given by

$$\ddot{\theta} + 0.5\dot{\theta} + \sin\theta = 0.8$$

Find the nature of singularity at $\theta = \sin^{-1}(0.8)$.

13.25 The phase plane equation of a single degree of freedom system is given by

$$\frac{dy}{dx} = \frac{-cy - (x - 0.1x^3)}{y}$$

Investigate the nature of singularity at $(x, y) = (0, 0)$ for $c > 0$.

13.26 Identify the singularity and find the nature of solution near the singularity for van der Pol's equation:

$$\ddot{x} - \alpha\left(1 - x^2\right)\dot{x} + x = 0$$

13.27 Identify the singularity and investigate the nature of solution near the singularity for an undamped system with a hard spring:

$$\ddot{x} + \omega_n^2\left(1 + k^2x^2\right)x = 0$$

13.28 Solve Problem 13.27 for an undamped system with a soft spring:

$$\ddot{x} + \omega_n^2\left(1 - k^2x^2\right)x = 0$$

13.29 Solve Problem 13.27 for a simple pendulum:

$$\ddot{\theta} + \omega_n^2\sin\theta = 0$$

13.30 Determine the eigenvalues and eigenvectors of the following equations:

a. $\dot{x} = x - y,$ $\dot{y} = x + 3y$
b. $\dot{x} = x + y,$ $\dot{y} = 4x + y$

13.31 Find the trajectories of the system governed by the equations

$$\dot{x} = x - 2y, \qquad \dot{y} = 4x - 5y$$

13.32 Find the trajectories of the system governed by the equations

$$\dot{x} = x - y, \qquad \dot{y} = x + 3y$$

13.33 Find the trajectories of the system governed by the equations

$$\dot{x} = 2x + y, \qquad \dot{y} = -3x - 2y$$

13.34 Using Lindstedt's perturbation method, find the solution of the van der Pol's equation, Eq. (13.143).

13.35 Verify that the following equation exhibits chaotic behavior:

$$x_{n+1} = k\,x_n\left(1 - x_n\right)$$

Hint: Give values of 3.25, 3.5 and 3.75 to k and observe the sequence of values generated with $x_1 = 0.5$.

13.36 Verify that the following equation exhibits chaotic behavior:

$$x_{n+1} = 2.0\, x_n (x_n - 1)$$

Hint: Observe the sequence of values generated using $x_1 = 1.001$, 1.002 and 1.003.

13.37 Using MATLAB, solve the simple pendulum equations, Eqs. (E.1) to (E.3) given in Example 13.6, for the following data:

a. $\omega_0 = 0.1$, $\theta(0) = 0.01$, $\dot{\theta}(0) = 0$

b. $\omega_0 = 0.1$, $\theta(0) = 0.01$, $\dot{\theta}(0) = 10$

13.38 Using MATLAB, find the solution of the nonlinearly damped system, Eq. (E.1) of Example 13.7, for the following data: $m = 10$, $c = 0.1$, $k = 4000$, $F_0 = 200$, $\omega = 20$, $x(0) = 0.5$, $\dot{x}(0) = 1.0$.

13.39 Using MATLAB, find the solution of a nonlinear single degree of freedom system governed by Eq. (E.1) of Example 13.8 under a pulse load for the following data: $m = 10$, $k_1 = 4000$, $k_2 = 1000$, $F_0 = 1000$, $t_0 = 5$, $x(0) = 0.05$, $\dot{x}(0) = 5$.

13.40 Solve the equation of motion $\ddot{x} + 0.5\dot{x} + x + 1.2x^3 = 1.8 \cos 0.4t$, using the Runge-Kutta method with $\Delta t = 0.05$, $t_{max} = 5.0$, and $x_0 = \dot{x}_0 = 0$. Plot the variation of x with t. Use **Program18.m** for the solution.

13.41 Find the time variation of the angular displacement of a simple pendulum (i.e., the solution of Eq. 13.5) for $g/l = 0.5$, using the initial conditions $\theta_0 = 45°$ and $\dot{\theta}_0 = 0$. Use the Runge-Kutta method given in **Program18.m**.

13.42 In the static firing test of a rocket, the rocket is anchored to a rigid wall by a nonlinear spring-damper system and fuel is burnt to develop a thrust, as shown in Fig. 13.34. The thrust acting on the rocket during the time period $0 \le t \le t_0$ is given by $F = m_0 v$, where m_0 is the constant rate at which fuel is burned and v is the velocity of the jet stream. The initial mass of the rocket is M, so that its mass at any time t is given by $m = M - m_0 t$, $0 \le t \le t_0$. The data is: spring force $= 8 \times 10^5 x + 6 \times 10^3 x^3$ N, damping force $= 10\dot{x} + 20\dot{x}^2$ N, $m_0 = 10$ kg/s, $v = 2000$ m/s, $M = 2000$ kg, and $t_0 = 100$ s. (a) Using the Runge-Kutta method of numerical integration, derive the equation of motion of the rocket and (b) Find the variation of the displacement of the rocket. Use **Program18.m**.

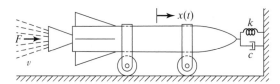

FIGURE 13.34

13.43 Using **Program18.cpp**, solve Problem 13.40.

13.44 Using **Program18.cpp**, solve Problem 13.41.

13.45 Using **Program18.cpp**, solve Problem 13.42.

13.46 Using **PROGRAM18.F**, solve Problem 13.40.

13.47 Using **PROGRAM18.F**, solve Problem 13.41.

13.48 Using **PROGRAM18.F**, solve Problem 13.42.

13.49 Write a computer program for finding the period of vibration corresponding to Eq. (13.14). Use a suitable numerical integration procedure. Using this program, find the solution of Problem 13.41.

DESIGN PROJECTS

13.50 In some periodic vibratory systems, external energy is supplied to the system over part of a period and dissipated within the system in another part of the period. Such periodic oscillations are known as *relaxation oscillations*. Van der Pol [13.22] indicated several instances of occurrence of relaxation oscillations such as a pneumatic hammer, the scratching noise of a knife on a plate, the squeaking of a door, and the fluctuation of populations of animal species. Many relaxation oscillations are governed by van der Pol's equation:

$$\ddot{x} - \alpha \left(1 - x^2 \right) \dot{x} + x = 0 \tag{E.1}$$

a. Plot the phase plane trajectories for three values of α: $\alpha = 0.1$, $\alpha = 1$, and $\alpha = 10$. Use the initial conditions (i) $x(0) = 0.5$, $\dot{x}(0) = 0$ and (ii) $x(0) = 0$ and $\dot{x}(0) = 5$.

b. Solve Eq. (E.1) using the fourth-order Runge-Kutta method using the initial conditions stated in (a) for $\alpha = 0.1$, $\alpha = 1$, and $\alpha = 10$.

13.51 A machine tool is mounted on two nonlinear elastic mounts, as shown in Fig. 13.35. The equations of motion, in terms of the coordinates $x(t)$ and $\theta(t)$, are given by

$$m\ddot{x} + k_{11} \left(x - l_1\theta \right) + k_{12} \left(x - l_1\theta \right)^3 + k_{21} \left(x + l_2\theta \right)$$
$$+ k_{22} \left(x + l_2\theta \right)^3 = 0 \tag{E.1}$$

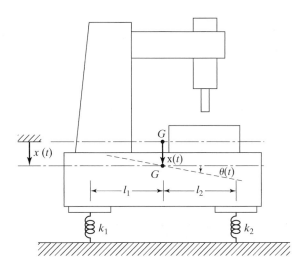

FIGURE 13.35

$$J_0 \ddot{\theta} - k_{11} (x - l_1 \theta) l_1 - k_{12} (x - l_1 \theta)^3 l_1$$
$$+ k_{21} (x + l_2 \theta) l_2 + k_{22} (x + l_2 \theta)^3 l_2 = 0 \qquad \text{(E.2)}$$

where m is the mass and J_0 is the mass moment of inertia about G of the machine tool. Using the Runge-Kutta method, find $x(t)$ and $\theta(t)$ for the following data: $m = 1000$ kg, $J_0 = 2500$ kg-m^2, $l_1 = 1$ m, $l_2 = 1.5$ m, $k_1 = 40x_1 + 10x_1^3$ kN/m, and $k_2 = 50x_2 + 5x_2^3$ kN/m.

CHAPTER 14

Random Vibration

14.1 Introduction

If vibrational response characteristics such as displacement, acceleration, and stress are known precisely as functions of time, the vibration is known as *deterministic vibration*. This implies a deterministic system (or structure) and a deterministic loading (or excitation); deterministic vibration exists only if there is perfect control over all the variables that influence the structural properties and the loading. In practice, however, there are many processes and phenomena whose parameters cannot be precisely predicted. Such processes are therefore called *random processes* [14.1–14.4]. An example of a random process is pressure fluctuation at a particular point on the surface of an aircraft flying in air. If several records of these pressure fluctuations are taken under the same flight speed, altitude, and load factor, they might look as indicated in Fig. 14.1. The records are not identical even though the measurements are taken under seemingly identical conditions. Similarly, a building subjected to ground acceleration due to an earthquake, a water tank under wind loading, and a car running on a rough road represent random processes. An elementary treatment of random vibration is presented in this chapter.

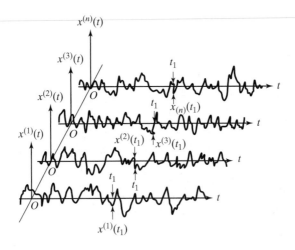

FIGURE 14.1   Ensemble of a random process ($x^{(i)}(t)$ is the ith sample function of the ensemble).

14.2   Random Variables and Random Processes

Most phenomena in real life are nondeterministic. For example, the tensile strength of steel and the dimensions of a machined part are nondeterministic. If many samples of steel are tested, their tensile strengths will not be the same—they will fluctuate about a mean or average value. Any quantity, like the tensile strength of steel, whose magnitude cannot be precisely predicted, is known as a *random variable* or a *probabilistic quantity*. If experiments are conducted to find the value of a random variable x, each experiment will give a number that is not a function of any parameter. For example, if 20 samples of steel are tested, their tensile strengths might be $x^{(1)} = 284, x^{(2)} = 302, x^{(3)} = 269, \ldots, x^{(20)} = 298$ N/mm^2. Each of these outcomes is called a *sample point*. If n experiments are conducted, all the n possible outcomes of the random variable constitute what is known as the *sample space* of the random variable.

There are other types of probabilistic phenomena for which the outcome of an experiment is a function of some parameter such as time or a spatial coordinate. Quantities such as the pressure fluctuations shown in Fig. 14.1 are called *random processes*. Each outcome of an experiment, in the case of a random process, is called a *sample function*. If n experiments are conducted, all the n possible outcomes of a random process constitute what is known as the *ensemble* of the process [14.5]. Notice that if the parameter t is fixed at a particular value t_1, $x(t_1)$ is a random variable whose sample points are given by $x^{(1)}(t_1), x^{(2)}(t_1), \ldots, x^{(n)}(t_1)$.

14.3   Probability Distribution

Consider a random variable x such as the tensile strength of steel. If n experimental values of x are available as $x_1, x_2, \ldots, x_n$, the probability of realizing the value of x smaller than some specified value $\underset{\sim}{x}$ can be found as

$$\text{Prob}(x \le \underset{\sim}{x}) = \frac{\underset{\sim}{n}}{n} \tag{14.1}$$

where $\underset{\sim}{n}$ denotes the number of x_i values which are less than or equal to $\underset{\sim}{x}$. As the number of experiments $n \to \infty$, Eq. (14.1) defines the probability distribution function of x, $P(x)$:

$$P(x) = \lim_{n \to \infty} \frac{\underset{\sim}{n}}{n} \tag{14.2}$$

The probability distribution function can also be defined for a random time function. For this, we consider the random time function shown in Fig. 14.2. During a fixed time span t, the time intervals for which the value of $x(t)$ is less than $\underset{\sim}{x}$ are denoted as Δt_1, Δt_2, Δt_3, and Δ_4. Thus the probability of realizing $x(t)$ less than or equal to $\underset{\sim}{x}$ is given by

$$\text{Prob}\,[x(t) \le \underset{\sim}{x}] = \frac{1}{t} \sum_i \Delta t_i \tag{14.3}$$

As $t \to \infty$, Eq. (14.3) gives the probability distribution function of $x(t)$:

$$P(x) = \lim_{t \to \infty} \frac{1}{t} \sum_i \Delta t_i \tag{14.4}$$

If $x(t)$ denotes a physical quantity, the magnitude of $x(t)$ will always be a finite number, so $\text{Prob}[x(t) < -\infty] = P(-\infty) = 0$ (impossible event), and $\text{Prob}[x(t) < \infty] = P(\infty) = 1$ (certain event). The typical variation of $P(x)$ with x is shown in Fig. 14.3(a). The function $P(x)$ is called the *probability distribution function* of x. The derivative of $P(x)$ with respect to x is known as the *probability density function* and is denoted as $p(x)$. Thus

$$p(x) = \frac{dP(x)}{dx} = \lim_{\Delta x \to 0} \frac{P(x + \Delta x) - P(x)}{\Delta x} \tag{14.5}$$

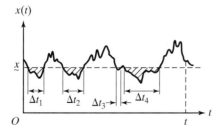

FIGURE 14.2   Random time function.

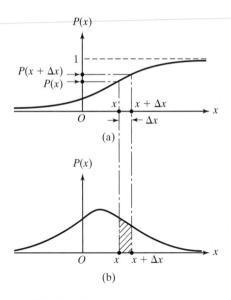

FIGURE 14.3   Probability distribution and density functions.

where the quantity $P(x + \Delta x) - P(x)$ denotes the probability of realizing $x(t)$ between the values x and $x + \Delta x$. Since $p(x)$ is the derivative of $P(x)$, we have

$$P(x) = \int_{-\infty}^{x} p(x') \, dx' \tag{14.6}$$

As $P(\infty) = 1$, Eq (14.6) gives

$$P(\infty) = \int_{-\infty}^{\infty} p(x') \, dx' = 1 \tag{14.7}$$

which means that the total area under the curve of $p(x)$ is unity.

14.4   Mean Value and Standard Deviation

If $f(x)$ denotes a function of the random variable x, the expected value of $f(x)$, denoted as μ_f or $E[f(x)]$ or $\overline{f(x)}$, is defined as

$$\mu_f = E[f(x)] = \overline{f(x)} = \int_{-\infty}^{\infty} f(x) p(x) \, dx \tag{14.8}$$

If $f(x) = x$, Eq. (14.8) gives the expected value, also known as the *mean value* of x:

$$\mu_x = E[x] = \overline{x} = \int_{-\infty}^{\infty} x p(x) \, dx \tag{14.9}$$

Similarly, if $f(x) = x^2$, we get the mean square value of x:

$$\mu_{x^2} = E[x^2] = \overline{x^2} = \int_{-\infty}^{\infty} x^2 p(x)\,dx \tag{14.10}$$

The variance of x, denoted as σ_x^2, is defined as the mean square value of x about the mean,

$$\sigma_x^2 = E[(x - \overline{x})^2] = \int_{-\infty}^{\infty} (x - \overline{x})^2\, p(x)\,dx = (\overline{x^2}) - (\overline{x})^2 \tag{14.11}$$

The positive square root of the variance, $\sigma(x)$, is called the *standard deviation* of x.

■■■■■■■■ Probabilistic Characteristics of Eccentricity of a Rotor

EXAMPLE 14.1

The eccentricity of a rotor (x), due to manufacturing errors, is found to have the following distribution:

$$p(x) = \begin{cases} kx^2, & 0 \le x \le 5 \text{ mm} \\ 0, & \text{elsewhere} \end{cases} \tag{E.1}$$

where k is a constant. Find (a) the mean, standard deviation, and the mean square value of the eccentricity and (b) the probability of realizing x less than or equal to 2 mm.

Solution: The value of k in Eq. (E.1) can be found by normalizing the probability density function:

$$\int_{-\infty}^{\infty} p(x)\,dx = \int_0^5 kx^2\,dx = 1$$

That is,

$$k\left(\frac{x^3}{3}\right)_0^5 = 1$$

That is,

$$k = \frac{3}{125} \tag{E.2}$$

(a) The mean value of x is given by Eq. (14.9):

$$\overline{x} = \int_0^5 p(x)x\,dx = k\left(\frac{x^4}{4}\right)_0^5 = 3.75 \text{ mm} \tag{E.3}$$

The standard deviation of x is given by Eq. (14.11):

$$\sigma_x^2 = \int_0^5 (x - \overline{x})^2 p(x)\,dx$$

$$= \int_0^5 (x^2 + \overline{x}^2 - 2\overline{x}x)\, p(x)\,dx$$

$$= \int_0^5 kx^4 \, dx - (\bar{x})^2$$

$$= k \left(\frac{x^5}{5} \right)_0^5 - (\bar{x})^2$$

$$= k \left(\frac{3125}{5} \right) - (3.75)^2 = 0.9375$$

$$\therefore \quad \sigma_x = 0.9682 \text{ mm} \tag{E.4}$$

The mean square value of x is

$$\overline{x^2} = k \left(\frac{3125}{5} \right) = 15 \text{ mm}^2 \tag{E.5}$$

(b)

$$\text{Prob } [x \le 2] = \int_0^2 p(x) \, dx = k \int_0^2 x^2 \, dx$$

$$= k \left(\frac{x^3}{3} \right)_0^2 = \frac{8}{125} = 0.064 \tag{E.6}$$

∎

14.5   Joint Probability Distribution of Several Random Variables

When two or more random variables are being considered simultaneously, their joint behavior is determined by a *joint probability distribution function*. For example, while testing the tensile strength of steel specimens, we can obtain the values of yield strength and ultimate strength in each experiment. If we are interested in knowing the relation between these two random variables, we must know the joint probability density function of yield strength and ultimate strength. The probability distributions of single random variables are called *univariate distributions*; the distributions that involve two random variables are called *bivariate distributions*. In general, if a distribution involves more than one random variable, it is called a *multivariate distribution*.

The bivariate density function of the random variables x_1 and x_2 is defined by

$$p(x_1, x_2) \, dx_1 \, dx_2 = \text{Prob } [x_1 < x_1' < x_1 + dx_1, x_2 < x_2' < x_2 + dx_2] \tag{14.12}$$

that is, the probability of realizing the value of the first random variable between x_1 and $x_1 + dx_1$ and the value of the second random variable between x_2 and $x_2 + dx_2$. The joint probability density function has the property that

$$\int_{-\infty}^{\infty}\int_{-\infty}^{\infty} p(x_1, x_2) \, dx_1 \, dx_2 = 1 \tag{14.13}$$

The joint distribution function of x_1 and x_2 is

$$P(x_1, x_2) = \text{Prob } [x_1' < x_1, x_2' < x_2]$$

$$= \int_{-\infty}^{x_1}\int_{-\infty}^{x_2} p(x_1', x_2') \, dx_1' \, dx_2' \tag{14.14}$$

The marginal or individual density functions can be obtained from the joint probability density function as

$$p(x) = \int_{-\infty}^{\infty} p(x, y) \, dy \tag{14.15}$$

$$p(y) = \int_{-\infty}^{\infty} p(x, y) \, dx \tag{14.16}$$

The variances of x and y can be determined as

$$\sigma_x^2 = E[(x - \mu_x)^2] = \int_{-\infty}^{\infty} (x - \mu_x)^2 p(x) \, dx \tag{14.17}$$

$$\sigma_y^2 = E[(y - \mu_y)^2] = \int_{-\infty}^{\infty} (y - \mu_y)^2 p(y) \, dy \tag{14.18}$$

The covariance of x and y, σ_{xy}, is defined as the expected value or average of the product of the deviations from the respective mean values of x and y. It is given by

$$\sigma_{xy} = E[(x - \mu_x)(y - \mu_y)] = \int_{-\infty}^{\infty}\int_{-\infty}^{\infty} (x - \mu_x)(y - \mu_y)p(x, y) \, dx \, dy$$

$$= \int_{-\infty}^{\infty}\int_{-\infty}^{\infty} (xy - x\mu_y - y\mu_x + \mu_x\mu_y)p(x, y) \, dx \, dy$$

$$= \int_{-\infty}^{\infty}\int_{-\infty}^{\infty} xyp(x, y) \, dx \, dy - \mu_y \int_{-\infty}^{\infty}\int_{-\infty}^{\infty} xp(x, y) \, dx \, dy$$

$$- \mu_x \int_{-\infty}^{\infty}\int_{-\infty}^{\infty} yp(x, y) \, dx \, dy + \mu_x\mu_y \int_{-\infty}^{\infty}\int_{-\infty}^{\infty} p(x, y) \, dx \, dy$$

$$= E[xy] - \mu_x\mu_y \tag{14.19}$$

The correlation coefficient between x and y, ρ_{xy}, is defined as the normalized covariance:

$$\rho_{xy} = \frac{\sigma_{xy}}{\sigma_x \sigma_y} \tag{14.20}$$

It can be seen that the correlation coefficient satisfies the relation $-1 \leq \rho_{xy} \leq 1$.

14.6   Correlation Functions of a Random Process

If $t_1, t_2, \ldots$ are fixed values of t, we use the abbreviations $x_1, x_2, \ldots$ to denote the values of $x(t)$ at $t_1, t_2, \ldots$, respectively. Since there are several random variables $x_1, x_2, \ldots$, we form the products of the random variables $x_1, x_2, \ldots$ (values of $x(t)$ at different times) and average the products over the set of all possibilities to obtain a sequence of functions:

$$K(t_1, t_2) = E[x(t_1)x(t_2)] = E[x_1 x_2]$$

$$K(t_1, t_2, t_3) = E[x(t_1)x(t_2)x(t_3)] = E[x_1 x_2 x_3] \tag{14.21}$$

and so on. These functions describe the statistical connection between the values of $x(t)$ at different times $t_1, t_2, \ldots$ and are called *correlation functions* [14.6, 14.7].

Autocorrelation Function.   The mathematical expectation of $x_1 x_2$—the correlation function $K(t_1, t_2)$—is also known as the *autocorrelation function*, designated as $R(t_1, t_2)$. Thus

$$R(t_1, t_2) = E[x_1 x_2] \tag{14.22}$$

If the joint probability density function of x_1 and x_2 is known to be $p(x_1, x_2)$, the auto-correlation function can be expressed as

$$R(t_1, t_2) = \int_{-\infty}^{\infty} \int_{-\infty}^{\infty} x_1 x_2 p(x_1, x_2) \, dx_1 \, dx_2 \tag{14.23}$$

Experimentally, we can find $R(t_1, t_2)$ by taking the product of $x^{(i)}(t_1)$ and $x^{(i)}(t_2)$ in the ith sample function and averaging over the ensemble:

$$R(t_1, t_2) = \frac{1}{n} \sum_{i=1}^{n} x^{(i)}(t_1) x^{(i)}(t_2) \tag{14.24}$$

where n denotes the number of sample functions in the ensemble (see Fig. 14.4). If t_1 and t_2 are separated by τ (with $t_1 = t$ and $t_2 = t + \tau$), we have $R(t + \tau) = E[x(t)x(t + \tau)]$.

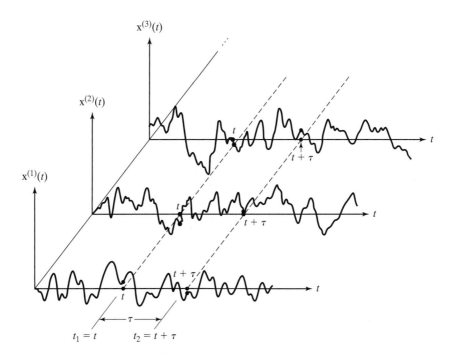

FIGURE 14.4   Ensemble of a random process.

14.7   Stationary Random Process

A stationary random process is one for which the probability distributions remain invariant under a shift of the time scale; the family of probability density functions applicable now also applies five hours from now or 500 hours from now. Thus the probability density function $p(x_1)$ becomes a universal density function $p(x)$, independent of time. Similarly, the joint density function $p(x_1, x_2)$, to be invariant under a shift of the time scale, becomes a function of $\tau = t_2 - t_1$, but not a function of t_1 or t_2 individually. Thus $p(x_1, x_2)$ can be written as $p(t, t + \tau)$. The expected value of stationary random process $x(t)$ can be written as

$$E[x(t_1)] = E[x(t_1 + t)] \qquad \text{for any } t \qquad (14.25)$$

and the autocorrelation function becomes independent of the absolute time t and will depend only on the separation time τ:

$$R(t_1, t_2) = E[x_1 x_2] = E[x(t)x(t + \tau)] = R(\tau) \qquad \text{for any } t \qquad (14.26)$$

where $\tau = t_2 - t_1$. We shall use subscripts to R to identify the random process when more than one random process is involved. For example, we shall use $R_x(\tau)$ and $R_y(\tau)$ to denote the autocorrelation functions of the random processes $x(t)$ and $y(t)$, respectively. The autocorrelation function has the following characteristics [14.2, 14.4]:

1. If $\tau = 0$, $R(\tau)$ gives the mean square value of $x(t)$:

$$R(0) = E[x^2] \tag{14.27}$$

2. If the process $x(t)$ has a zero mean and is extremely irregular, as shown in Fig. 14.5(a), its autocorrelation function $R(\tau)$ will have small values except at $\tau = 0$, as indicated in Fig. 14.5(b).

3. If $x(t) \simeq x(t + \tau)$, the autocorrelation function $R(\tau)$ will have a constant value as shown in Fig. 14.6.

4. If $x(t)$ is stationary, its mean and standard deviations will be independent of t:

$$E[x(t)] = E[x(t + \tau)] = \mu \tag{14.28}$$

and

$$\sigma_{x(t)} = \sigma_{x(t+\tau)} = \sigma \tag{14.29}$$

The correlation coefficient, ρ, of $x(t)$ and $x(t + \tau)$ can be found as

$$\begin{aligned}
\rho &= \frac{E[\{x(t) - \mu\}\{x(t + \tau) - \mu\}]}{\sigma^2} \\
&= \frac{E[x(t)x(t + \tau)] - \mu E[x(t + \tau)] - \mu E[x(t)] + \mu^2}{\sigma^2} \\
&= \frac{R(\tau) - \mu^2}{\sigma^2}
\end{aligned} \tag{14.30}$$

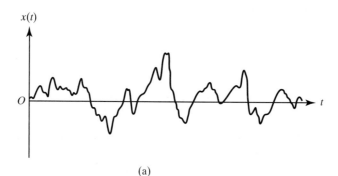

(a)

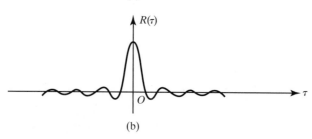

(b)

FIGURE 14.5   Autocorrelation function.

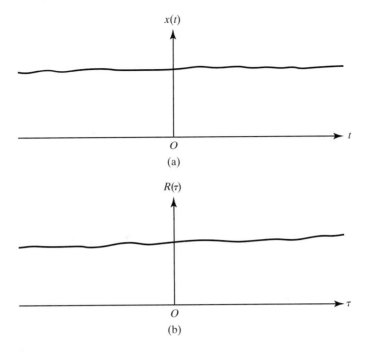

$x(t)$

O

(a)

$R(\tau)$

O

(b)

FIGURE 14.6   Constant values of function.

that is,

$$R(\tau) = \rho\sigma^2 + \mu^2 \qquad (14.31)$$

Since $|\rho| \leq 1$, Eq. (14.31) shows that

$$-\sigma^2 + \mu^2 \leq R(\tau) \leq \sigma^2 + \mu^2 \qquad (14.32)$$

This shows that the autocorrelation function will not be greater than the mean square value, $E[x^2] = \sigma^2 + \mu^2$.

5. Since $R(\tau)$ depends only on the separation time τ and not on the absolute time t for a stationary process,

$$R(\tau) = E[x(t)x(t + \tau)] = E[x(t)x(t - \tau)] = R(-\tau) \qquad (14.33)$$

Thus $R(\tau)$ is an even function of τ.

6. When τ is large ($\tau \to \infty$), there will not be a coherent relationship between the two values $x(t)$ and $x(t + \tau)$. Hence the correlation coefficient $\rho \to 0$ and Eq. (14.31) gives

$$R(\tau \to \infty) \to \mu^2 \qquad (14.34)$$

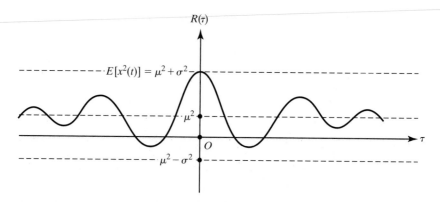

FIGURE 14.7   Autocorrelation function.

A typical autocorrelation function is shown in Fig. 14.7.

Ergodic Process.   An ergodic process is a stationary random process for which we can obtain all the probability information from a single sample function and assume that it is applicable to the entire ensemble. If $x^{(i)}(t)$ represents a typical sample function of duration T, the averages can be computed by averaging with respect to time along $x^{(i)}(t)$. Such averages are called *temporal averages*. By using the notation $\langle x(t) \rangle$ to represent the temporal average of $x(t)$ (the time average of x), we can write

$$E[x] = \langle x(t) \rangle = \lim_{T \to \infty} \frac{1}{T} \int_{-T/2}^{T/2} x^{(i)}(t)\, dt \qquad (14.35)$$

where $x^{(i)}(t)$ has been assumed to be defined from $t = -T/2$ to $t = T/2$ with $T \to \infty$ (T very large). Similarly,

$$E[x^2] = \langle x^2(t) \rangle = \lim_{T \to \infty} \frac{1}{T} \int_{-T/2}^{T/2} [x^{(i)}(t)]^2\, dt \qquad (14.36)$$

and

$$R(\tau) = \langle x(t)x(t + \tau) \rangle = \lim_{T \to \infty} \frac{1}{T} \int_{-T/2}^{T/2} x^{(i)}(t)x^{(i)}(t + \tau)\, dt \qquad (14.37)$$

14.8   Gaussian Random Process

The most commonly used distribution for modeling physical random processes is called the *Gaussian* or *normal random process*. The Gaussian process has a number of remarkable properties that permit the computation of the random vibration characteristics in a simple manner. The probability density function of a Gaussian process $x(t)$ is given by

$$p(x) = \frac{1}{\sqrt{2\pi}\sigma_x}e^{-\frac{1}{2}\left(\frac{x-\bar{x}}{\sigma_x}\right)^2}$$ (14.38)

where $\bar{x}$ and σ_x denote the mean value and standard deviation of x. The mean $(\bar{x})$ and standard deviation (σ_x) of $x(t)$ vary with t for a nonstationary process but are constants (independent of t) for a stationary process. A very important property of the Gaussian process is that the forms of its probability distributions are invariant with respect to linear operations. This means that if the excitation of a linear system is a Gaussian process, the response is generally a different random process, but still a normal one. The only changes are that the magnitude of the mean and standard deviations of the response are different from those of the excitation.

The graph of a Gaussian probability density function is a bell-shaped curve, symmetric about the mean value; its spread is governed by the value of the standard deviation, as shown in Fig. 14.8. By defining a standard normal variable z as

$$z = \frac{x - \bar{x}}{\sigma_x}$$ (14.39)

Eq. (14.38) can be expressed as

$$p(x) = \frac{1}{\sqrt{2\pi}}e^{-\frac{1}{2}z^2}$$ (14.40)

The probability of $x(t)$ lying in the interval $-c\sigma$ and $+c\sigma$ where c is any positive number can be found, assuming $\bar{x} = 0$:

$$\text{Prob}\,[-c\sigma \le x(t) \le c\sigma] = \int_{-c\sigma}^{c\sigma} \frac{1}{\sqrt{2\pi}\sigma}e^{-\frac{1}{2}\frac{x^2}{\sigma^2}}\,dx$$ (14.41)

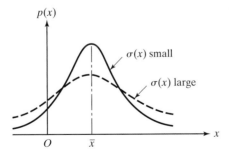

FIGURE 14.8   Gaussian probability density function.

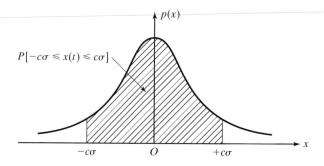

FIGURE 14.9   Graphical representation of Prob[$-c\sigma \leq x(t) \leq c\sigma$].

The probability of $x(t)$ lying outside the range $\mp c\sigma$ is one minus the value given by Eq. (14.41). This can also be expressed as

$$\text{Prob}\,[|x(t)| > c\sigma] = \frac{2}{\sqrt{2\pi}\sigma}\int_{c\sigma}^{\infty} e^{-\frac{1}{2}\frac{x^2}{\sigma^2}}\,dx \tag{14.42}$$

The integrals in Eqs. (14.41) and (14.42) have been evaluated numerically and tabulated [14.5]; some typical values are indicated in the following table (see Fig. 14.9 also).

Value of c	1	2	3	4
Prob[$-c\sigma \leq x(t) \leq c\sigma$]	0.6827	0.9545	0.9973	0.999937
Prob[$|x(t)| > c\sigma$]	0.3173	0.0455	0.0027	0.000063

14.9   Fourier Analysis

**14.9.1
Fourier Series**

We saw in Chapter 1 that any periodic function $x(t)$, of period τ, can be expressed in the form of a complex Fourier series

$$x(t) = \sum_{n=-\infty}^{\infty} c_n e^{in\omega_0 t} \tag{14.43}$$

where ω_0 is the fundamental frequency given by

$$\omega_0 = \frac{2\pi}{\tau} \tag{14.44}$$

and the complex Fourier coefficients c_n can be determined by multiplying both sides of Eq. (14.43) with $e^{-im\omega_0 t}$ and integrating over one time period:

$$\int_{-\tau/2}^{\tau/2} x(t)e^{-im\omega_0 t}\, dt = \sum_{n=-\infty}^{\infty} \int_{-\tau/2}^{\tau/2} c_n e^{i(n-m)\omega_0 t}\, dt$$

$$= \sum_{n=-\infty}^{\infty} c_n \int_{-\tau/2}^{\tau/2} [\cos(n-m)\omega_0 t + i\sin(n-m)\omega_0 t]\, dt \quad (14.45)$$

Equation (14.45) can be simplified to obtain (see Problem 14.27)

$$c_n = \frac{1}{\tau}\int_{-\tau/2}^{\tau/2} x(t)e^{-in\omega_0 t}\, dt \quad (14.46)$$

Equation (14.43) shows that the function $x(t)$ of period τ can be expressed as a sum of an infinite number of harmonics. The harmonics have amplitudes given by Eq. (14.46) and frequencies which are multiples of the fundamental frequency ω_0. The difference between any two consecutive frequencies is given by

$$\omega_{n+1} - \omega_n = (n+1)\omega_0 - n\omega_0 = \Delta\omega = \frac{2\pi}{\tau} = \omega_0 \quad (14.47)$$

Thus the larger the period τ, the denser the frequency spectrum becomes. Equation (14.46) shows that the Fourier coefficients c_n are, in general, complex numbers. However, if $x(t)$ is a real and even function, then c_n will be real. If $x(t)$ is real, the integrand of c_n in Eq. (14.46) can also be identified as the complex conjugate of that of c_{-n}. Thus

$$c_n = c^*_{-n} \quad (14.48)$$

The mean square value of $x(t)$—that is, the time average of the square of the function $x(t)$—can be determined as

$$\overline{x^2(t)} = \frac{1}{\tau}\int_{-\tau/2}^{\tau/2} x^2(t)\, dt = \frac{1}{\tau}\int_{-\tau/2}^{\tau/2}\left(\sum_{n=-\infty}^{\infty} c_n e^{in\omega_0 t}\right)^2 dt$$

$$= \frac{1}{\tau}\int_{-\tau/2}^{\tau/2}\left(\sum_{n=-\infty}^{-1} c_n e^{in\omega_0 t} + c_0 + \sum_{n=1}^{\infty} c_n e^{in\omega_0 t}\right)^2 dt$$

$$= \frac{1}{\tau}\int_{-\tau/2}^{\tau/2}\left\{\sum_{n=1}^{\infty}(c_n e^{in\omega_0 t} + c^*_n e^{-in\omega_0 t}) + c_0\right\}^2 dt$$

$$= \frac{1}{\tau}\int_{-\tau/2}^{\tau/2}\left\{\sum_{n=1}^{\infty} 2c_n c^*_n + c_0^2\right\} dt$$

$$= c_0^2 + \sum_{n=1}^{\infty} 2|c_n|^2 = \sum_{n=-\infty}^{\infty} |c_n|^2 \quad (14.49)$$

Thus the mean square value of $x(t)$ is given by the sum of the squares of the absolute values of the Fourier coefficients. Equation (14.49) is known as *Parseval's formula* for periodic functions [14.1].

Complex Fourier Series Expansion

EXAMPLE 14.2

Find the complex Fourier series expansion of the function shown in Fig. 14.10(a).

Solution: The given function can be expressed as

$$x(t) = \begin{cases} A\left(1 + \dfrac{t}{a}\right), & -\dfrac{\tau}{2} \le t \le 0 \\[2mm] A\left(1 - \dfrac{t}{a}\right), & 0 \le t \le \dfrac{\tau}{2} \end{cases} \tag{E.1}$$

where the period (τ) and the fundamental frequency (ω_0) are given by

$$\tau = 2a \qquad \text{and} \qquad \omega_0 = \frac{2\pi}{\tau} = \frac{\pi}{a} \tag{E.2}$$

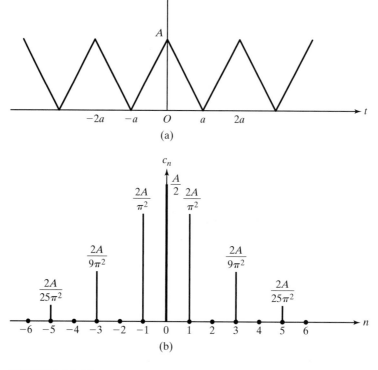

FIGURE 14.10   Complex Fourier series representation.

The Fourier coefficients can be determined as

$$c_n = \frac{1}{\tau} \int_{-\tau/2}^{\tau/2} x(t) e^{-in\omega_0 t}\, dt$$

$$= \frac{1}{\tau}\left[\int_{-\tau/2}^{0} A\left(1 + \frac{t}{a}\right) e^{-in\omega_0 t}\, dt + \int_{0}^{\tau/2} A\left(1 - \frac{t}{a}\right) e^{-in\omega_0 t}\, dt \right] \tag{E.3}$$

Using the relation

$$\int te^{kt}\, dt = \frac{e^{kt}}{k^2}(kt - 1) \tag{E.4}$$

c_n can be evaluated as

$$c_n = \frac{1}{\tau}\left[\frac{A}{-in\omega_0} e^{-in\omega_0 t}\Big|_{-\tau/2}^{0} + \frac{A}{a}\left\{ \frac{e^{-in\omega_0 t}}{(-in\omega_0)^2}[-in\omega_0 t - 1] \right\}\Big|_{-\tau/2}^{0} \right.$$

$$\left. + \frac{A}{-in\omega_0} e^{-in\omega_0 t}\Big|_{0}^{\tau/2} - \frac{A}{a}\left\{ \frac{e^{-in\omega_0 t}}{(-in\omega_0)^2}[-in\omega_0 t - 1] \right\}\Big|_{0}^{\tau/2} \right] \tag{E.5}$$

This equation can be reduced to

$$c_n = \frac{1}{\tau}\left[\frac{A}{in\omega_0} e^{in\pi} + \frac{2A}{a}\frac{1}{n^2\omega_0^2} - \frac{A}{in\omega_0} e^{-in\pi} - \frac{A}{a}\frac{1}{n^2\omega_0^2} e^{in\pi} - \frac{A}{a}\frac{1}{n^2\omega_0^2} e^{-in\pi} \right.$$

$$\left. + \frac{A}{a}\frac{1}{n^2\omega_0^2}(in\pi)e^{in\pi} - \frac{A}{a}\frac{1}{n^2\omega_0^2}(in\pi)e^{-in\pi} \right] \tag{E.6}$$

Noting that

$$e^{in\pi} \quad \text{or} \quad e^{-in\pi} = \begin{cases} 1, & n = 0 \\ -1, & n = 1, 3, 5, \ldots \\ 1, & n = 2, 4, 6, \ldots \end{cases} \tag{E.7}$$

Eq. (E.6) can be simplified to obtain

$$c_n = \begin{cases} \dfrac{A}{2}, & n = 0 \\[2mm] \left(\dfrac{4A}{a\tau n^2\omega_0^2}\right) = \dfrac{2A}{n^2\pi^2}, & n = 1, 3, 5, \ldots \\[2mm] 0, & n = 2, 4, 6, \ldots \end{cases} \tag{E.8}$$

The frequency spectrum is shown in Fig. 14.10(b).

■

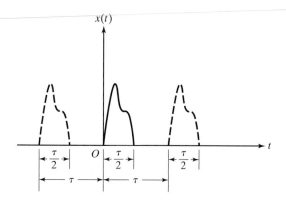

FIGURE 14.11   Nonperiodic function.

**14.9.2
Fourier Integral**

A nonperiodic function, such as the one shown by the solid curve in Fig. 14.11, can be treated as a periodic function having an infinite period ($\tau \to \infty$). The Fourier series expansion of a periodic function is given by Eqs. (14.43), (14.44) and (14.46):

$$x(t) = \sum_{n=-\infty}^{\infty} c_n e^{in\omega_0 t} \tag{14.50}$$

with

$$\omega_0 = \frac{2\pi}{\tau} \tag{14.51}$$

and

$$c_n = \frac{1}{\tau} \int_{-\tau/2}^{\tau/2} x(t) e^{-in\omega_0 t} \, dt \tag{14.52}$$

As $\tau \to \infty$, the frequency spectrum becomes continuous and the fundamental frequency becomes infinitesimal. Since the fundamental frequency ω_0 is very small, we can denote it as $\Delta\omega$, $n\omega_0$ as ω, and rewrite Eq. (14.52) as

$$\lim_{\tau \to \infty} \tau c_n = \lim_{\tau \to \infty} \int_{-\tau/2}^{\tau/2} x(t) e^{-i\omega t} \, dt = \int_{-\infty}^{\infty} x(t) e^{-i\omega t} \, dt \tag{14.53}$$

By defining $X(\omega)$ as

$$X(\omega) = \lim_{\tau \to \infty} (\tau c_n) = \int_{-\infty}^{\infty} x(t) e^{-i\omega t} \, dt \tag{14.54}$$

we can express $x(t)$ from Eq. (14.50) as

$$x(t) = \lim_{\tau \to \infty} \sum_{n=-\infty}^{\infty} c_n e^{i\omega t} \frac{2\pi\tau}{2\pi\tau}$$

$$= \lim_{\tau \to \infty} \sum_{n=-\infty}^{\infty} (c_n \tau) e^{i\omega t} \left(\frac{2\pi}{\tau}\right) \frac{1}{2\pi}$$

$$= \frac{1}{2\pi} \int_{-\infty}^{\infty} X(\omega) e^{i\omega t}\, d\omega \tag{14.55}$$

This equation indicates the frequency decomposition of the nonperiodic function $x(t)$ in a continuous frequency domain, similar to Eq. (14.50) for a periodic function in a discrete frequency domain. The equations

$$x(t) = \frac{1}{2\pi} \int_{-\infty}^{\infty} X(\omega) e^{i\omega t}\, d\omega \tag{14.56}$$

and

$$X(\omega) = \int_{-\infty}^{\infty} x(t) e^{-i\omega t}\, dt \tag{14.57}$$

are known as the (integral) Fourier transform pair for a nonperiodic function $x(t)$, similar to Eqs. (14.50) and (14.52) for a periodic function $x(t)$ [14.9, 14.10].

The mean square value of a nonperiodic function $x(t)$ can be determined from Eq. (14.49):

$$\frac{1}{\tau} \int_{-\tau/2}^{\tau/2} x^2(t)\,dt = \sum_{n=-\infty}^{\infty} |c_n|^2$$

$$= \sum_{n=-\infty}^{\infty} c_n c_n^* \frac{\tau\omega_0}{\tau\omega_0} = \sum_{n=-\infty}^{\infty} c_n c_n^* \frac{\tau\omega_0}{\tau\left(\dfrac{2\pi}{\tau}\right)}$$

$$= \frac{1}{\tau} \sum_{n=-\infty}^{\infty} (\tau c_n)(c_n^* \tau) \frac{\omega_0}{2\pi} \tag{14.58}$$

Since $\tau c_n \to X(\omega)$, $\tau c_n^* \to X^*(\omega)$, and $\omega_0 \to d\omega$ as $\tau \to \infty$, Eq. (14.58) gives the mean square value of $x(t)$ as

$$\overline{x^2(t)} = \lim_{\tau \to \infty} \frac{1}{\tau} \int_{-\tau/2}^{\tau/2} x^2(t)\, dt = \int_{-\infty}^{\infty} \frac{|X(\omega)|^2}{2\pi\tau}\, d\omega \tag{14.59}$$

Equation (14.59) is known as Parseval's formula for nonperiodic functions [14.1].

▬▬▬▬▬▬▬ **Fourier Transform of a Triangular Pulse**

EXAMPLE   14.3

Find the Fourier transform of the triangular pulse shown in Fig. 14.12(a).

Solution: The triangular pulse can be expressed as

$$x(t) = \begin{cases} A\left(1 - \dfrac{|t|}{a}\right), & |t| \le a \\ 0, & \text{otherwise} \end{cases} \tag{E.1}$$

The Fourier transform of $x(t)$ can be found, using Eq. (14.57), as

$$X(\omega) = \int_{-\infty}^{\infty} A\left(1 - \frac{|t|}{a}\right)e^{-i\omega t}\, dt$$

$$= \int_{-\infty}^{0} A\left(1 + \frac{t}{a}\right)e^{-i\omega t}\, dt + \int_{0}^{\infty} A\left(1 - \frac{t}{a}\right)e^{-i\omega t}\, dt \tag{E.2}$$

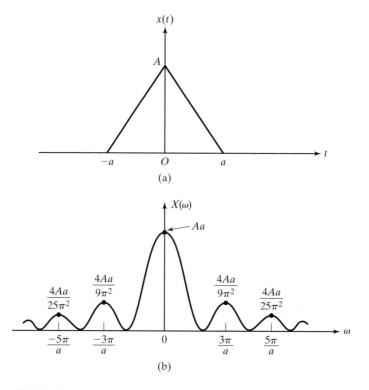

(a)

(b)

FIGURE 14.12   Fourier transform of a triangular pulse.

Since $x(t) = 0$ for $|t| > 0$, Eq. (E.2) can be expressed as

$$X(\omega) = \int_{-a}^{0} A\left(1 + \frac{t}{a}\right)e^{-i\omega t}\, dt + \int_{0}^{a} A\left(1 - \frac{t}{a}\right)e^{-i\omega t}\, dt$$

$$= \left(\frac{A}{-i\omega}\right)e^{-i\omega t}\bigg|_{-a}^{0} + \frac{A}{a}\left\{\frac{e^{-i\omega t}}{(-i\omega)^2}[-i\omega t - 1]\right\}\bigg|_{-a}^{0}$$

$$+ \left(\frac{A}{-i\omega}\right)e^{-i\omega t}\bigg|_{0}^{a} - \frac{A}{a}\left\{\frac{e^{-i\omega t}}{(-i\omega)^2}[-i\omega t - 1]\right\}\bigg|_{0}^{a} \qquad \text{(E.3)}$$

Equation (E.3) can be simplified to obtain

$$X(\omega) = \frac{2A}{a\omega^2} + e^{i\omega a}\left(-\frac{A}{a\omega^2}\right) + e^{-i\omega a}\left(-\frac{A}{a\omega^2}\right)$$

$$= \frac{2A}{a\omega^2} - \frac{A}{a\omega^2}(\cos\omega a + i\sin\omega a) - \frac{A}{a\omega^2}(\cos\omega a - i\sin\omega a)$$

$$= \frac{2A}{a\omega^2}(1 - \cos\omega a) = \frac{4A}{a\omega^2}\sin^2\left(\frac{\omega a}{2}\right) \qquad \text{(E.4)}$$

Equation (E.4) is plotted in Fig. 14.12(b). Notice the similarity of this figure with the discrete Fourier spectrum shown in Fig. 14.10(b).

■

14.10   Power Spectral Density

The power spectral density $S(\omega)$ of a stationary random process is defined as the Fourier transform of $R(\tau)/2\pi$

$$S(\omega) = \frac{1}{2\pi}\int_{-\infty}^{\infty} R(\tau)e^{-i\omega\tau}\, d\tau \qquad \text{(14.60)}$$

so that

$$R(\tau) = \int_{-\infty}^{\infty} S(\omega)e^{i\omega t}\, d\omega \qquad \text{(14.61)}$$

Equations (14.60) and (14.61) are known as the Wiener-Khintchine formulas [14.1]. The power spectral density is more often used in random vibration analysis than the autocorrelation function. The following properties of power spectral density can be observed:

1. From Eqs. (14.27) and (14.61), we obtain

$$R(0) = E[x^2] = \int_{-\infty}^{\infty} S(\omega)\, d\omega \qquad \text{(14.62)}$$

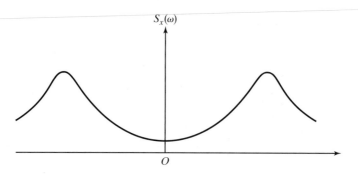

FIGURE 14.13 Typical power spectral density function.

If the mean is zero, the variance of $x(t)$ is given by

$$\sigma_x^2 = R(0) = \int_{-\infty}^{\infty} S(\omega)\, d\omega \qquad (14.63)$$

If $x(t)$ denotes the displacement, $R(0)$ represents the average energy. From Eq. (14.62), it is clear that $S(\omega)$ represents the energy density associated with the frequency ω. Thus $S(\omega)$ indicates the spectral distribution of energy in a system. Also, in electrical circuits, if $x(t)$ denotes random current, then the mean square value indicates the power of the system (when the resistance is unity). This is the origin of the term *power spectral density*.

2. Since $R(\tau)$ is an even function of τ and real, $S(\omega)$ is also an even and real function of ω. Thus $S(-\omega) = S(\omega)$. A typical power spectral density function is shown in Fig. 14.13.

3. From Eq. (14.62), the units of $S(\omega)$ can be identified as those of x^2/unit of angular frequency. It can be noted that both negative and positive frequencies are counted in Eq. (14.62). In experimental work, for convenience, an equivalent one-sided spectrum $W_x(f)$ is widely used [14.1, 14.2].[1]

   The spectrum $W_x(f)$ is defined in terms of linear frequency (i.e., cycles per unit time) and only the positive frequencies are counted. The relationship between $S_x(\omega)$ and $W_x(f)$ can be seen with reference to Fig. 14.14. The differential frequency $d\omega$ in Fig. 14.14(a) corresponds to the differential frequency $df = d\omega/2\pi$ in Fig. 14.14(b). Since $W_x(f)$ is the equivalent spectrum defined over positive values of f only, we have

$$E[x^2] = \int_{-\infty}^{\infty} S_x(\omega)\, d\omega \equiv \int_{0}^{\infty} W_x(f)\, df \qquad (14.64)$$

---

[1]When several random processes are involved, a subscript is used to identify the power spectral density function (or simply the spectrum) of a particular random process. Thus $S_x(\omega)$ denotes the spectrum of $x(t)$.

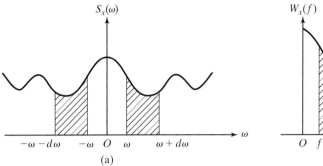
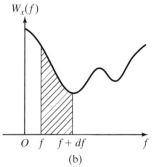

FIGURE 14.14   Two- and one-sided spectrum.

In order to have the contributions of the frequency bands $d\omega$ and df to the mean square value to be same, the shaded areas in both Figs. 14.14(a) and (b) must be the same. Thus

$$2S_x(\omega)\, d\omega = W_x(f)\, df \tag{14.65}$$

which gives

$$W_x(f) = 2S_x(\omega)\frac{d\omega}{df} = 2S_x(\omega)\frac{d\omega}{d\omega/2\pi} = 4\pi S_x(\omega) \tag{14.66}$$

14.11 Wide-Band and Narrow-Band Processes

A wide-band process is a stationary random process whose spectral density function $S(\omega)$ has significant values over a range or band of frequencies that is approximately the same order of magnitude as the center frequency of the band. An example of a wide-band random process is shown in Fig. 14.15. The pressure fluctuations on the surface of a rocket due to acoustically transmitted jet noise or due to supersonic boundary layer turbulence are examples of physical processes that are typically wide-band. A narrow-band random process is a stationary process whose spectral density function $S(\omega)$ has significant values only in a range or band of frequencies whose width is small compared to the magnitude of the center frequency of the process. Figure 14.16 shows the sample function and the corresponding spectral density and autocorrelation functions of a narrow-band process.

A random process whose power spectral density is constant over a frequency range is called *white noise*, an analogy with the white light that spans the visible spectrum more or less uniformly. It is called *ideal white noise* if the band of frequencies $\omega_2 - \omega_1$ is infinitely wide. Ideal white noise is a physically unrealizable concept, since the mean square value of such a random process would be infinite because the area under the spectrum would be infinite. It is called *band-limited white noise* if the band of frequencies has finite cut-off

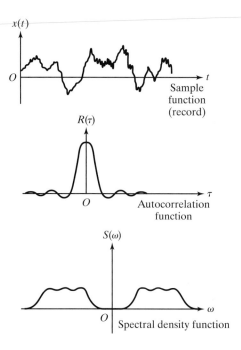

FIGURE 14.15   Wide-band stationary random process.

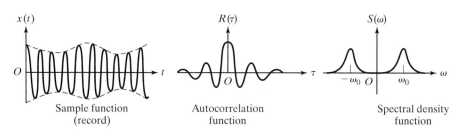

FIGURE 14.16   Narrow-band stationary random process.

frequencies ω_1 and ω_2 [14.8]. The mean square value of a band-limited white noise is given by the total area under the spectrum—namely, $2S_0(\omega_2 - \omega_1)$, where S_0 denotes the constant value of the spectral density function.

Autocorrelation and Mean Square Value of a Stationary Process

EXAMPLE   14.4

The power spectral density of a stationary random process $x(t)$ is shown in Fig. 14.17(a). Find its autocorrelation function and the mean square value.

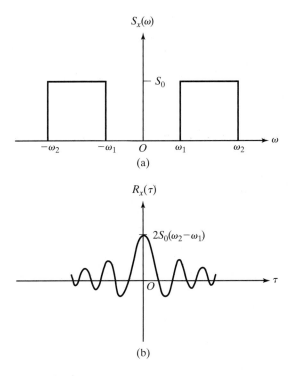

FIGURE 14.17   Autocorrelation function of a
stationary process.

Solution:
(a)   Since $S_x(\omega)$ is real and even in ω, Eq. (14.61) can be rewritten as

$$R_x(\tau) = 2 \int_0^\infty S_x(\omega)\cos \omega\tau \, d\omega = 2S_0 \int_{\omega_1}^{\omega_2} \cos \omega\tau \, d\omega$$

$$= 2S_0 \left(\frac{1}{\tau}\sin \omega\tau\right)\bigg|_{\omega_1}^{\omega_2} = \frac{2S_0}{\tau}(\sin \omega_2\tau - \sin \omega_1\tau)$$

$$= \frac{4S_0}{\tau}\cos \frac{\omega_1 + \omega_2}{2}\tau \sin \frac{\omega_2 - \omega_1}{2}\tau$$

This function is shown graphically in Fig. 14.17(b).
(b)   The mean square value of the random process is given by

$$E[x^2] = \int_{-\infty}^{\infty} S_x(\omega) \, d\omega = 2S_0 \int_{\omega_1}^{\omega_2} d\omega = 2S_0(\omega_2 - \omega_1)$$

■

14.12   Response of a Single Degree of Freedom System

The equation of motion for the system shown in Fig. 14.18 is

$$\ddot{y} + 2\zeta\omega_n\dot{y} + \omega_n^2 y = x(t) \tag{14.67}$$

where

$$x(t) = \frac{F(t)}{m}, \qquad \omega_n = \sqrt{\frac{k}{m}}, \qquad \zeta = \frac{c}{c_c}, \qquad \text{and} \qquad c_c = 2km$$

The solution of Eq. (14.67) can be obtained by using either the impulse response approach or the frequency response approach.

**14.12.1
Impulse
Response
Approach**

Here we consider the forcing function $x(t)$ to be made up of a series of impulses of varying magnitude, as shown in Fig. 14.19(a) (see Section 4.5.2). Let the impulse applied at time τ be denoted as $x(\tau)\,d\tau$. If $y(t) = h(t - \tau)$ denotes the response to the unit impulse[2] excitation $\delta(t - \tau)$, it is called the impulse response function. The total response of the system at time t can be found by superposing the responses to impulses of magnitude $x(\tau)\,d\tau$ applied at different values of $t = \tau$. The response to the excitation $x(\tau)\,d\tau$ will

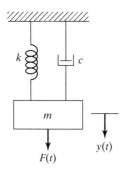

FIGURE 14.18
Single degree of
freedom system.

---

[2]The unit impulse applied at $t = \tau$ is denoted as

$$x(t) = \delta(t - \tau)$$

where $\delta(t - \tau)$ is the Dirac delta function with (see Fig. 14.19b)

$$\delta(t - \tau) \rightarrow \infty \qquad \text{as } t \rightarrow \tau$$

$$\delta(t - \tau) = 0 \qquad \text{for all } t \text{ except at } t = \tau$$

$$\int_{-\infty}^{\infty} \delta(t - \tau)\,dt = 1 \text{ (area under the curve is unity)}$$

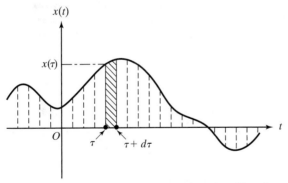

(a) Forcing function in the form of a series of impulses

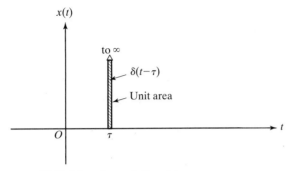

(b) Unit impulse excitation at $t = \tau$

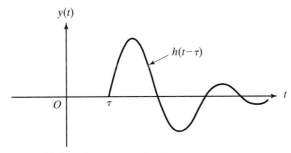

(c) Impulse response function

FIGURE 14.19   Impulse response approach.

be $[x(\tau)\, d\tau]h(t - \tau)$, and the response to the total excitation will be given by the super-position or convolution integral:

$$y(t) = \int_{-\infty}^{t} x(\tau)h(t - \tau)\, d\tau \tag{14.68}$$

14.12.2
Frequency
Response
Approach

The transient function $x(t)$ can be expressed in terms of its Fourier transform $X(\omega)$ as

$$x(t) = \frac{1}{2\pi} \int_{\omega=-\infty}^{\infty} X(\omega) e^{i\omega t} \, d\omega \tag{14.69}$$

Thus $x(t)$ can be considered as the superposition of components of different frequencies ω. If we consider the forcing function of unit modulus as

$$\underset{\sim}{x}(t) = e^{i\omega t} \tag{14.70}$$

its response can be denoted as

$$\underset{\sim}{y}(t) = H(\omega) e^{i\omega t} \tag{14.71}$$

where $H(\omega)$ is called the *complex frequency response function* (see Section 3.5). Since the actual excitation is given by the superposition of components of different frequencies (Eq. 14.69), the total response of the system can also be obtained by superposition as

$$y(t) = H(\omega)x(t) = \int_{-\infty}^{\infty} H(\omega) \frac{1}{2\pi} X(\omega) e^{i\omega t} \, d\omega$$

$$= \frac{1}{2\pi} \int_{-\infty}^{\infty} H(\omega) X(\omega) e^{i\omega t} d\omega \tag{14.72}$$

If $Y(\omega)$ denotes the Fourier transform of the response function $y(t)$, we can express $y(t)$ in terms of $Y(\omega)$ as

$$y(t) = \frac{1}{2\pi} \int_{-\infty}^{\infty} Y(\omega) e^{i\omega t} \, d\omega \tag{14.73}$$

Comparison of Eqs. (14.72) and (14.73) yields

$$Y(\omega) = H(\omega) X(\omega) \tag{14.74}$$

14.12.3
Characteristics
of the Response
Function

The following characteristics of the response function can be noted:

1. Since $h(t - \tau) = 0$ when $t < \tau$ or $\tau > t$ (i.e., the response before the application of the impulse is zero), the upper limit of integration in Eq. (14.68) can be replaced by ∞ so that

$$y(t) = \int_{-\infty}^{\infty} x(\tau) h(t - \tau) \, d\tau \tag{14.75}$$

2. By changing the variable from τ to $\theta = t - \tau$, Eq. (14.75) can be rewritten as

$$y(t) = \int_{-\infty}^{\infty} x(t - \theta)h(\theta)\,d\theta \tag{14.76}$$

3. The superposition integral, Eq. (14.68) or (14.75) or (14.76), can be used to find the response of the system $y(t)$ for any arbitrary excitation $x(t)$ once the impulse-response function of the system $h(t)$ is known. The Fourier integral, Eq. (14.72), can also be used to find the response of the system once the complex frequency response of the system, $H(\omega)$, is known. Although the two approaches appear to be different, they are intimately related to one another. To see their inter-relationship, consider the excitation of the system to be a unit impulse $\delta(\tau)$ in Eq. (14.72). By definition, the response is $h(t)$ and Eq. (14.72) gives

$$y(t) = h(t) = \frac{1}{2\pi} \int_{-\infty}^{\infty} X(\omega)H(\omega)e^{i\omega t}\,d\omega \tag{14.77}$$

where $X(\omega)$ is the Fourier transform of $x(t) = \delta(t)$:

$$X(\omega) = \int_{-\infty}^{\infty} x(t)e^{-i\omega t}\,dt = \int_{-\infty}^{\infty} \delta(t)e^{-i\omega t}\,dt = 1 \tag{14.78}$$

since $\delta(t) = 0$ everywhere except at $t = 0$ where it has a unit area and $e^{-i\omega t} = 1$ at $t = 0$. Equations (14.77) and (14.78) give

$$h(t) = \frac{1}{2\pi} \int_{-\infty}^{\infty} H(\omega)e^{i\omega t}\,d\omega \tag{14.79}$$

which can be recognized as the Fourier integral representation of $h(t)$ in which $H(\omega)$ is the Fourier transformation of $h(t)$:

$$H(\omega) = \int_{-\infty}^{\infty} h(t)e^{-i\omega t}\,dt \tag{14.80}$$

14.13   Response Due to Stationary Random Excitations

In the previous section, the relationships between excitation and response were derived for arbitrary known excitations $x(t)$. In this section, we consider similar relationships when the excitation is a stationary random process. When the excitation is a stationary random process, the response will also be a stationary random process [14.15, 14.16]. We consider the relation between the excitation and the response using the impulse response (time domain) as well as the frequency response (frequency domain) approaches.

**14.13.1
Impulse
Response
Approach**

Mean Values.   The response for any particular sample excitation is given by Eq. (14.76):

$$y(t) = \int_{-\infty}^{\infty} x(t - \theta)h(\theta) \, d\theta \qquad (14.81)$$

For ensemble average, we write Eq. (14.81) for every (x, y) pair in the ensemble and then take the average to obtain[3]

$$E[y(t)] = E\left[\int_{-\infty}^{\infty} x(t - \theta)h(\theta) \, d\theta \right]$$

$$= \int_{-\infty}^{\infty} E[x(t - \theta)]h(\theta) \, d\theta \qquad (14.82)$$

Since the excitation is assumed to be stationary, $E[x(\tau)]$ is a constant independent of τ, Eq. (14.82) becomes

$$E[y(t)] = E[x(t)] \int_{-\infty}^{\infty} h(\theta) \, d\theta \qquad (14.83)$$

The integral in Eq. (14.83) can be obtained by setting $\omega = 0$ in Eq. (14.80) so that

$$H(0) = \int_{-\infty}^{\infty} h(t) \, dt \qquad (14.84)$$

Thus a knowledge of either the impulse response function $h(t)$ or the frequency response function $H(\omega)$ can be used to find the relationship between the mean values of the excitation and the response. It is to be noted that both $E[x(t)]$ and $E[y(t)]$ are independent of t.

Autocorrelation.   We can use a similar procedure to find the relationship between the autocorrelation functions of the excitation and the response. For this, we first write

$$y(t)y(t + \tau) = \int_{-\infty}^{\infty} x(t - \theta_1)h(\theta_1) \, d\theta_1$$

$$\cdot \int_{-\infty}^{\infty} x(t + \tau - \theta_2)h(\theta_2)d\,\theta_2$$

$$= \int_{-\infty}^{\infty}\int_{-\infty}^{\infty} x(t - \theta_1)x(t + \tau - \theta_2)$$

$$\cdot h(\theta_1)h(\theta_2) \, d\theta_1 \, d\theta_2 \qquad (14.85)$$

---

[3]In deriving Eq. (14.82), the integral is considered as a limiting case of a summation and hence the average of a sum is treated to be same as the sum of the averages, that is,

$$E[x_1 + x_2 + \cdots] = E[x_1] + E[x_2] + \cdots$$

where θ_1 and θ_2 are used instead of θ to avoid confusion. The autocorrelation function of $y(t)$ can be found as

$$
\begin{aligned}
R_y(\tau) &= E[y(t)y(t + \tau)] \\
&= \int_{-\infty}^{\infty} \int_{-\infty}^{\infty} E[x(t - \theta_1)x(t + \tau - \theta_2)]h(\theta_1)h(\theta_2)d\theta_1\, d\theta_2 \\
&= \int_{-\infty}^{\infty} \int_{-\infty}^{\infty} R_x(\tau + \theta_1 - \theta_2)h(\theta_1)h(\theta_2)d\theta_1\, d\theta_2
\end{aligned}
\tag{14.86}
$$

14.13.2 Frequency Response Approach

Power Spectral Density.   The response of the system can also be described by its power spectral density, which by definition, is (see Eq. 14.60)

$$
S_y(\omega) = \frac{1}{2\pi} \int_{-\infty}^{\infty} R_y(\tau)e^{-i\omega\tau}\, d\tau
\tag{14.87}
$$

Substitution of Eq. (14.86) into Eq. (14.87) gives

$$
\begin{aligned}
S_y(\omega) &= \frac{1}{2\pi} \int_{-\infty}^{\infty} e^{-i\omega\tau}\, d\tau \\
&\times \int_{-\infty}^{\infty} \int_{-\infty}^{\infty} R_x(\tau + \theta_1 - \theta_2)h(\theta_1)h(\theta_2)d\theta_1\, d\theta_2
\end{aligned}
\tag{14.88}
$$

Introduction of

$$
e^{i\omega\theta_1}e^{-i\omega\theta_2}e^{-i\omega(\theta_1-\theta_2)} = 1
\tag{14.89}
$$

into Eq. (14.88) results in

$$
\begin{aligned}
S_y(\omega) &= \int_{-\infty}^{\infty} h(\theta_1)e^{i\omega\theta_1}\, d\theta_1 \int_{-\infty}^{\infty} h(\theta_2)e^{-i\omega\theta_2}\, d\theta_2 \\
&\times \frac{1}{2\pi} \int_{-\infty}^{\infty} R_x(\tau + \theta_1 - \theta_2)e^{-i\omega(\theta_1-\theta_2)}\, d\tau
\end{aligned}
\tag{14.90}
$$

In the third integral on the right-hand side of Eq. (14.90), θ_1 and θ_2 are constants and the introduction of a new variable of integration η as

$$
\eta = \tau + \theta_1 - \theta_2
\tag{14.91}
$$

leads to

$$\frac{1}{2\pi} \int_{-\infty}^{\infty} R_x(\tau + \theta_1 - \theta_2) e^{-i\omega(\tau + \theta_1 - \theta_2)} d\tau$$

$$= \frac{1}{2\pi} \int_{-\infty}^{\infty} R_x(\eta) e^{-i\omega\eta} d\eta \equiv S_x(\omega) \tag{14.92}$$

The first and the second integrals on the right-hand side of Eq. (14.90) can be recognized as the complex frequency response functions $H(\omega)$ and $H(-\omega)$, respectively. Since $H(-\omega)$ is the complex conjugate of $H(\omega)$, Eq. (14.90) gives

$$S_y(\omega) = |H(\omega)|^2 S_x(\omega) \tag{14.93}$$

This equation gives the relationship between the power spectral densities of the excitation and the response.

Mean Square Response.   The mean square response of the stationary random process $y(t)$ can be determined either from the autocorrelation function $R_y(\tau)$ or from the power spectral density $S_y(\omega)$:

$$E[y^2] = R_y(0) = \int_{-\infty}^{\infty}\int_{-\infty}^{\infty} R_x(\theta_1 - \theta_2) h(\theta_1) h(\theta_2) \, d\theta_1 \, d\theta_2 \tag{14.94}$$

and

$$E[y^2] = \int_{-\infty}^{\infty} S_y(\omega) \, d\omega = \int_{-\infty}^{\infty} |H(\omega)|^2 S_x(\omega) \, d\omega \tag{14.95}$$

Note: Equations (14.93) and (14.95) form the basis for the random vibration analysis of single- and multidegree of freedom systems [14.11, 14.12]. The random vibration analysis of road vehicles is given in Refs. [14.13, 14.14].

■■■■■■■■■■ **Mean Square Value of Response**

EXAMPLE   14.5 ————————————————

A single degree of freedom system (Fig. 14.20a) is subjected to a force whose spectral density is a white noise $S_x(\omega) = S_0$. Find the following:

(a) Complex frequency response function of the system
(b) Power spectral density of the response
(c) Mean square value of the response

Solution

(a) To find the complex frequency response function $H(\omega)$, we substitute the input as $e^{i\omega t}$ and the corresponding response as $y(t) = H(\omega)e^{i\omega t}$ in the equation of motion

$$m\ddot{y} + c\dot{y} + ky = x(t)$$

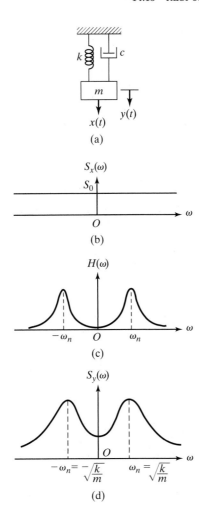

FIGURE 14.20   Single degree of freedom system.

to obtain

$$(-m\omega^2 + ic\omega + k)H(\omega)e^{i\omega t} = e^{i\omega t}$$

and

$$H(\omega) = \frac{1}{-m\omega^2 + ic\omega + k} \tag{E.1}$$

(b) The power spectral density of the output can be found as

$$S_y(\omega) = |H(\omega)|^2 S_x(\omega) = S_0 \left| \frac{1}{-m\omega^2 + ic\omega + k} \right|^2 \tag{E.2}$$

(c) The mean square value of the output is given by[4]

$$E[y^2] = \int_{-\infty}^{\infty} S_y(\omega)\, d\omega$$

$$= S_0 \int_{-\infty}^{\infty} \left| \frac{1}{-m\omega^2 + k + ic\omega} \right|^2 d\omega = \frac{\pi S_0}{kc} \qquad (E.3)$$

which can be seen to be independent of the magnitude of the mass m. The functions $H(\omega)$ and $S_y(\omega)$ are shown graphically in Fig. 14.20(b).

■

Design of the Columns of a Building

EXAMPLE 14.6

A single-story building is modeled by four identical columns of Young's modulus E and height h and a rigid floor of weight W, as shown in Fig. 14.21(a). The columns act as cantilevers fixed at the ground. The damping in the structure can be approximated by an equivalent viscous damping constant c. The ground acceleration due to an earthquake is approximated by a constant spectrum S_0. If each column has a tubular cross section with mean diameter d and wall thickness $t = d/10$, find the mean diameter of the columns such that the standard deviation of the displacement of the floor relative to the ground does not exceed a specified value δ.

Solution

Approach: Model the building as a single degree of freedom system. Use the relation between the power spectral densities of excitation and output.

The building can be modeled as a single degree of freedom system as shown in Fig. 14.21(b) with

$$m = W/g \qquad (E.1)$$

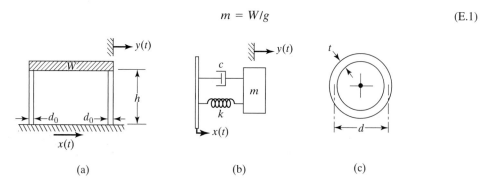

(a)   (b)   (c)

FIGURE 14.21   Single-story building.

---

[4]The values of this and other similar integrals have been found in the literature [14.1]. For example, if

$$H(\omega) = \frac{i\omega B_1 + B_0}{-\omega^2 A_2 + i\omega A_1 + A_0}, \quad \int_{-\infty}^{\infty} |H(\omega)|^2\, d\omega = \pi \left\{ \frac{(B_0^2/A_0) A_2 + B_1^2}{A_1 A_2} \right\}$$

and

$$k = 4\left(\frac{3EI}{h^3}\right) \tag{E.2}$$

since the stiffness of one cantilever beam (column) is equal to $(3EI/h^3)$, where E is the Young's modulus, h is the height, and I is the moment of inertia of the cross section of the columns given by (see Fig. 14.21(c)):

$$I = \frac{\pi}{64}(d_0^4 - d_i^4) \tag{E.3}$$

Equation (E.3) can be simplified, using $d_0 = d + t$ and $d_i = d - t$, as

$$I = \frac{\pi}{64}(d_0^2 + d_i^2)(d_0 + d_i)(d_0 - d_i)$$

$$= \frac{\pi}{64}[(d + t)^2 + (d - t)^2][(d + t) + (d - t)][(d + t) - (d - t)]$$

$$= \frac{\pi}{8} dt(d^2 + t^2) \tag{E.4}$$

With $t = d/10$, Eq. (E.4) becomes

$$I = \frac{101\pi}{8000}d^4 = 0.03966d^4 \tag{E.5}$$

and hence Eq. (E.2) gives

$$k = \frac{12\,E(0.03966d^4)}{h^3} = \frac{0.47592\,Ed^4}{h^3} \tag{E.6}$$

When the base of the system moves, the equation of motion is given by (see Section 3.6)

$$m\ddot{z} + c\dot{z} + kz = -m\ddot{x} \tag{E.7}$$

where $z = y - x$ is the displacement of the mass (floor) relative to the ground. Equation (E.7) can be rewritten as

$$\ddot{z} + \frac{c}{m}\dot{z} + \frac{k}{m}z = -\ddot{x} \tag{E.8}$$

The complex frequency response function $H(\omega)$ can be obtained by making the substitution

$$\ddot{x} = e^{i\omega t} \quad \text{and} \quad z(t) = H(\omega)e^{i\omega t} \tag{E.9}$$

so that

$$\left[-\omega^2 + i\omega\frac{c}{m} + \frac{k}{m}\right]H(\omega)e^{i\omega t} = -e^{i\omega t}$$

which gives

$$H(\omega) = \frac{-1}{\left(-\omega^2 + i\omega\dfrac{c}{m} + \dfrac{k}{m}\right)} \tag{E.10}$$

The power spectral density of the response $z(t)$ is given by

$$S_z(\omega) = |H(\omega)|^2 S_{\dot{x}}(\omega) = S_0 \left| \frac{-1}{\left(-\omega^2 + i\omega\dfrac{c}{m} + \dfrac{k}{m}\right)} \right|^2 \tag{E.11}$$

The mean square value of the response $z(t)$ can be determined, using Eq. (E.4) of Example 14.5, as

$$E[z^2] = \int_{-\infty}^{\infty} S_z(\omega)\,d\omega$$

$$= S_0 \int_{-\infty}^{\infty} \left| \frac{-1}{\left(-\omega^2 + i\omega\dfrac{c}{m} + \dfrac{k}{m}\right)} \right|^2 d\omega$$

$$= S_0 \left(\frac{\pi m^2}{kc}\right) \tag{E.12}$$

Substitution of the relations (E.1) and (E.6) into Eq. (E.12) gives

$$E[z^2] = \pi S_0 \frac{W^2 h^3}{g^2 c(0.47592\ Ed^4)} \tag{E.13}$$

Assuming the mean value of $z(t)$ to be zero, the standard deviation of z can be found as

$$\sigma_z = \sqrt{E[z^2]} = \sqrt{\frac{\pi S_0 W^2 h^3}{0.47592 g^2 cEd^4}} \tag{E.14}$$

Since $\sigma_z \leq \delta$, we find that

$$\frac{\pi S_0 W^2 h^3}{0.47592 g^2 cEd^4} \leq \delta^2$$

or

$$d^4 \geq \frac{\pi S_0 W^2 h^3}{0.47592 g^2 cE\delta^2} \tag{E.15}$$

Thus the required mean diameter of the columns is given by

$$d \geq \left\{ \frac{\pi S_0 W^2 h^3}{0.47592 g^2 c E \delta^2} \right\}^{1/4} \tag{E.16}$$

■

14.14   Response of a Multidegree of Freedom System

The equations of motion of a multidegree of freedom system with proportional damping can be expressed, using the normal mode approach, as (see Eq. 6.128)

$$\ddot{q}_i(t) + 2 \zeta_i \omega_i \dot{q}_i(t) + \omega_i^2 q_i(t) = Q_i(t); \quad i = 1, 2, \ldots, n \tag{14.96}$$

where n is the number of degrees of freedom, ω_i is the ith natural frequency, $q_i(t)$ is the ith generalized coordinate, and $Q_i(t)$ is the ith generalized force. The physical and generalized coordinates are related as

$$\vec{x}(t) = [X]\vec{q}(t)$$

or

$$x_i(t) = \sum_{j=1}^{n} X_i^{(j)} q_j(t) \tag{14.97}$$

where $[X]$ is the modal matrix and $X_i^{(j)}$ is the ith component of jth modal vector. The physical and generalized forces are related as

$$\vec{Q}(t) = [X]^T \vec{F}(t)$$

or

$$Q_i(t) = \sum_{j=1}^{n} X_j^{(i)} F_j(t) \tag{14.98}$$

where $F_j(t)$ is the force acting along the coordinate $x_j(t)$. Let the applied forces be expressed as

$$F_j(t) = f_j \tau(t) \tag{14.99}$$

so that Eq. (14.98) becomes

$$Q_i(t) = \left(\sum_{j=1}^{n} X_j^{(i)} f_j \right) \tau(t) = N_i \tau(t) \tag{14.100}$$

where

$$N_i = \sum_{j=1}^{n} X_j^{(i)} f_j \qquad (14.101)$$

By assuming a harmonic force variation

$$\tau(t) = e^{i\omega t} \qquad (14.102)$$

the solution of Eq. (14.96) can be expressed as

$$q_i(t) = \frac{N_i}{\omega_i^2} H_i(\omega) \tau(t) \qquad (14.103)$$

where $H_i(\omega)$ denotes the frequency response function

$$H_i(\omega) = \frac{1}{1 - \left(\dfrac{\omega}{\omega_i}\right)^2 + i\, 2\, \zeta_i \dfrac{\omega}{\omega_i}} \qquad (14.104)$$

The mean square value of the physical displacement, $x_i(t)$, can be obtained from Eqs. (14.97) and (14.103) as

$$\overline{x_i^2(t)} = \lim_{t \to \infty} \frac{1}{2T} \int_{-T}^{T} x_i^2(t)\, dt$$

$$= \sum_{r=1}^{n} \sum_{s=1}^{n} X_i^{(r)} X_i^{(s)} \frac{N_r\, N_s}{\omega_r^2\, \omega_s^2} \lim_{T \to \infty} \int_{-T}^{T} H_r(\omega)\, H_s(\omega)\, \tau^2(t)\, dt \qquad (14.105)$$

From Eq. (3.56), $H_r(\omega)$ can be expressed as

$$H_r(\omega) = |H_r(\omega)| e^{-i\phi_r} \qquad (14.106)$$

where the magnitude of $H_r(\omega)$, known as the magnification factor, is given by

$$|H_r(\omega)| = \left\{ \left[1 - \left(\frac{\omega}{\omega_r}\right)^2 \right]^2 + \left(2\zeta_r \frac{\omega}{\omega_r}\right)^2 \right\}^{-1/2} \qquad (14.107)$$

and the phase angle, ϕ_r, by

$$\phi_r = \tan^{-1}\left\{ \frac{2\zeta_r \dfrac{\omega}{\omega_r}}{1 - \left(\dfrac{\omega}{\omega_r}\right)^2} \right\} \qquad (14.108)$$

By neglecting the phase angles, the integral on the right-hand side of Eq. (14.105) can be expressed as

$$\lim_{T \to \infty} \frac{1}{2T} \int_{-T}^{T} H_r(\omega) \, H_s(\omega) \, \tau^2(t) \, dt$$

$$\approx \lim_{T \to \infty} \frac{1}{2T} \int_{-T}^{T} |H_r(\omega)| |H_s(\omega)| \tau^2(t) \, dt \qquad (14.109)$$

For a stationary random process, the mean square value of $\tau^2(t)$ can be expressed in terms of its power spectral density function, $S_\tau(\omega)$, as

$$\overline{\tau^2(t)} = \lim_{T \to \infty} \frac{1}{2T} \int_{-T}^{T} \tau^2(t) \, dt = \int_{-\infty}^{\infty} S_\tau(\omega) \, d\omega \qquad (14.110)$$

Combining Eqs. (14.109) and (14.110) gives

$$\lim_{T \to \infty} \frac{1}{2T} \int_{-T}^{T} H_r(\omega) \, H_s(\omega) \, \tau^2(t) \, dt$$

$$\approx \int_{-\infty}^{\infty} |H_r(\omega)| |H_s(\omega)| S_\tau(\omega) \, d\omega \qquad (14.111)$$

Substitution of Eq. (14.111) into Eq. (14.105) yields the mean square value of $x_i(t)$ as

$$\overline{x_i^2(t)} \approx \sum_{r=1}^{n} \sum_{s=1}^{n} X_i^{(r)} X_i^{(s)} \frac{N_r \, N_s}{\omega_r^2 \, \omega_s^2} \int_{-\infty}^{\infty} |H_r(\omega)| |H_s(\omega)| S_\tau(\omega) \, d\omega \qquad (14.112)$$

The magnification factors, $|H_r(\omega)|$ and $|H_s(\omega)|$, are shown in Fig. 14.22. It can be seen that the product, $|H_r(\omega)| |H_s(\omega)|$ for $r \neq s$, is often negligible compared to $|H_r(\omega)|^2$ and $|H_s(\omega)|^2$. Hence Eq. (14.112) can be rewritten as

$$\overline{x_i^2(t)} \approx \sum_{r=1}^{n} \left(X_i^{(r)} \right)^2 \frac{N_r^2}{\omega_r^4} \int_{-\infty}^{\infty} |H_r(\omega)|^2 S_\tau(\omega) \, d\omega \qquad (14.113)$$

For lightly damped systems, the integral in Eq. (14.113) can be evaluated by approximating the graph of $S_\tau(\omega)$ to be flat with $S_\tau(\omega) = S_\tau(\omega_r)$ as

$$\int_{-\infty}^{\infty} |H_r(\omega)|^2 S_\tau(\omega) \, d\omega \approx S_\tau(\omega_r) \int_{-\infty}^{\infty} |H_r(\omega)|^2 \, d\omega = \frac{\pi \omega_r \, S_\tau(\omega_r)}{2 \zeta_r} \qquad (14.114)$$

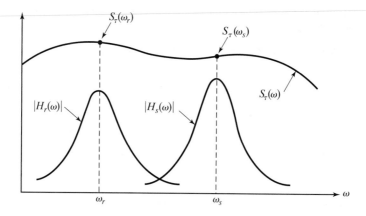

FIGURE 14.22   Magnification factors.

Equations (14.113) and (14.114) yield

$$\overline{x_i^2(t)} = \sum_{r=1}^{n} (X_i^{(r)})^2 \frac{N_r^2}{\omega_r^4} \left(\frac{\pi \omega_r S_\tau(\omega_r)}{2\zeta_r} \right)$$ (14.115)

The following example illustrates the computational procedure.

Response of a Building Frame Under an Earthquake

EXAMPLE   14.7

The three-story building frame shown in Fig. 14.23 is subjected to an earthquake. The ground acceleration during the earthquake can be assumed to be a stationary random process with a power spectral density $S(\omega) = 0.05 \, (\text{m}^2/\text{s}^4)/(\text{rad/s})$. Assuming a modal damping ratio of 0.02 in each mode,

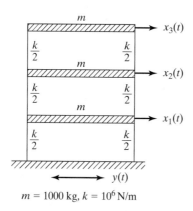

$m = 1000$ kg, $k = 10^6$ N/m

FIGURE 14.23   Three-story
building frame.

determine the mean square values of the responses of the various floors of the building frame under the earthquake.

Solution: The stiffness and mass matrices of the building frame can be found as

$$[k] = k \begin{bmatrix} 2 & -1 & 0 \\ -1 & 2 & -1 \\ 0 & -1 & 1 \end{bmatrix} \tag{E.1}$$

$$[m] = m \begin{bmatrix} 1 & 0 & 0 \\ 0 & 1 & 0 \\ 0 & 0 & 1 \end{bmatrix} \tag{E.2}$$

From Examples 6.10 and 6.11, the eigenvalues and the eigenvectors (normalized with respect to the mass matrix $[m]$) can be computed using the values $k = 10^6$ N/m and $m = 1000$ kg, as

$$\omega_1 = 0.44504 \sqrt{\frac{k}{m}} = 14.0734 \text{ rad/s} \tag{E.3}$$

$$\omega_2 = 1.2471 \sqrt{\frac{k}{m}} = 39.4368 \text{ rad/s} \tag{E.4}$$

$$\omega_3 = 1.8025 \sqrt{\frac{k}{m}} = 57.0001 \text{ rad/s} \tag{E.5}$$

$$\vec{Z}^{(1)} = \frac{0.3280}{\sqrt{m}} \begin{Bmatrix} 1.0000 \\ 1.8019 \\ 2.2470 \end{Bmatrix} = \begin{Bmatrix} 0.01037 \\ 0.01869 \\ 0.02330 \end{Bmatrix} \tag{E.6}$$

$$\vec{Z}^{(2)} = \frac{0.7370}{\sqrt{m}} \begin{Bmatrix} 1.0000 \\ 0.4450 \\ -0.8020 \end{Bmatrix} = \begin{Bmatrix} 0.02331 \\ 0.01037 \\ -0.01869 \end{Bmatrix} \tag{E.7}$$

$$\vec{Z}^{(3)} = \frac{0.5911}{\sqrt{m}} \begin{Bmatrix} 1.0000 \\ -1.2468 \\ 0.5544 \end{Bmatrix} = \begin{Bmatrix} 0.01869 \\ -0.02330 \\ 0.01036 \end{Bmatrix} \tag{E.8}$$

Note that the notation $\vec{Z}^{(i)}$ is used to denote the ith mode shape instead of $\vec{X}^{(i)}$ since the relative displacements, $z_i(t)$, will be used instead of the absolute displacements, $x_i(t)$, in the solution.

By denoting the ground motion as $y(t)$, the relative displacements of the floors, $z_i(t)$, can be expressed as $z_i(t) = x_i(t) - y(t)$, $i = 1, 2, 3$. The equations of motion can be expressed as

$$[m]\ddot{\vec{x}} + [c]\dot{\vec{z}} + [k]\vec{z} = \vec{0} \tag{E.9}$$

which can be rewritten as

$$[m]\ddot{\vec{z}} + [c]\dot{\vec{z}} + [k]\vec{z} = -[m]\ddot{\vec{y}} \tag{E.10}$$

where $\ddot{\vec{y}} = \begin{Bmatrix} \ddot{y} \\ \ddot{y} \\ \ddot{y} \end{Bmatrix}$. By expressing the vector $\vec{z}$ in terms of normal modes, we have

$$\vec{z} = [Z]\vec{q} \tag{E.11}$$

where $[Z]$ denotes the modal matrix. By substituting Eq. (E.11) into Eq. (E.10) and premultiplying the resulting equation by $[Z]^T$, we can derive the uncoupled equations of motion. Assuming a damping ratio $\zeta_i(\zeta_i = 0.02)$ in mode i, the uncoupled equations of motion are given by

$$\ddot{q}_i + 2\zeta_i\omega_i + \omega_i^2 q_i = Q_i; i = 1, 2, 3 \tag{E.12}$$

where

$$Q_i = \sum_{j=1}^{n} Z_j^{(i)} F_j(t) \tag{E.13}$$

and

$$F_j(t) = -m_j \ddot{y}(t) = -m \ddot{y}(t) \tag{E.14}$$

with $m_j = m$ denoting the mass of the jth floor. By representing $F_j(t)$ as

$$F_j(t) = f_j \tau(t) \tag{E.15}$$

we note that

$$f_j = -m_j = -m \tag{E.16}$$

and

$$\tau(t) = \ddot{y}(t) \tag{E.17}$$

The mean square values, $\overline{z_i^2(t)}$, can be determined from Eq. (14.115):

$$\overline{z_i^2(t)} = \sum_{r=1}^{3} (Z_i^{(r)})^2 \frac{N_r^2}{\omega_r^3}\left(\frac{\pi}{2\zeta_r}\right)S_\tau(\omega_r) \tag{E.18}$$

Nothing that $\vec{Z}^{(1)}, \vec{Z}^{(2)}$, and $\vec{Z}^{(3)}$ are given by Eqs. (E.6), (E.7), and (E.8), respectively, and

$$N_1 = \sum_{j=1}^{3} Z_j^{(1)} f_j = -m \sum_{j=1}^{3} Z_j^{(1)} = -1000 \, (0.05236) = -52.36 \tag{E.19}$$

$$N_2 = \sum_{j=1}^{3} Z_j^{(2)} f_j = -m \sum_{j=1}^{3} Z_j^{(2)} = -1000 \, (0.05237) = -52.37 \tag{E.20}$$

$$N_3 = \sum_{j=1}^{3} Z_j^{(3)} f_j = -m \sum_{j=1}^{3} Z_j^{(3)} = -1000 \,(0.05235) = -52.35 \qquad \text{(E.21)}$$

Eq. (E.18) yields the mean square values of the relative displacements of the various floors of the building frame as

$$\overline{z_1^2(t)} = 0.00053132 \text{ m}^2 \qquad \text{(E.22)}$$

$$\overline{z_2^2(t)} = 0.00139957 \text{ m}^2 \qquad \text{(E.23)}$$

$$\overline{z_3^2(t)} = 0.00216455 \text{ m}^2 \qquad \text{(E.24)}$$

∎

Probability of Relative Displacement Exceeding a Specified Value

EXAMPLE 14.8

Find the probability of the magnitude of the relative displacement of the various floors exceeding 1, 2, 3, and 4 standard deviations of the corresponding relative displacement for the building frame of Example 14.7.

Solution

Approach: Assume the ground acceleration to be a normally distributed random process with zero mean and use standard normal tables.

Since the ground acceleration, $\ddot{y}(t)$, is assumed to be normally distributed with zero mean value, the relative displacements of the various floors can also be assumed to be normally distributed with zero mean values. Thus the standard deviations of the relative displacements of the floors are given by

$$\sigma_{zi} = \sqrt{\overline{z_i^2(t)}}; \, i = 1, 2, 3$$

The probability of the absolute value of the relative displacement, $z_i(t)$, exceeding a specified number of standard deviations can be found from standard normal tables as (see Section 14.8):

$$P[|z_i(t)| > p\sigma_{zi}] = \begin{cases} 0.31732 \text{ for } p = 1 \\ 0.04550 \text{ for } p = 2 \\ 0.00270 \text{ for } p = 3 \\ 0.00006 \text{ for } p = 4 \end{cases}$$

∎

14.15   Examples Using MATLAB

Plotting of Autocorrelation Function

EXAMPLE 14.9

Using MATLAB, plot the autocorrelation function corresponding to white noise with spectral density S_0 for the following cases:

(a)   band limited white noise with $\omega_1 = 0$ and $\omega_2 = 4$ rad/s, 6 rad/s, 8 rad/s
(b)   band limited white noise with $\omega_1 = 2$ rad/s and $\omega_2 = 4$ rad/s, 6 rad/s, 8 rad/s
(c)   ideal white noise

Solution

For (a) and (b), the autocorrelation function, $R(\tau)$, can be expressed as (from Example 14.4)

$$\frac{R(\tau)}{S_0} = \frac{2}{\tau}(\sin \omega_2\tau - \sin \omega_1\tau) \tag{E.1}$$

For (c) it can be expressed as $\tau \rightarrow 0$,

$$R(0) = \lim_{\tau\to 0}\left\{2\,S_0\left(\frac{\omega_2\sin\omega_2\tau}{\omega_2\tau}\right) - 2\,S_0\left(\frac{\omega_1\sin\omega_1\tau}{\omega_1\tau}\right)\right\} = 2\,S_0(\omega_2 - \omega_1)$$

For an ideal white noise, we let $\omega_1 = 0$ and $\omega_2 \rightarrow \infty$, which yields $R = 2S_0\,\delta(\tau)$ where $\delta(\tau)$ is the Dirac delta function. The MATLAB program to plot Eq. (E.1) is given below.

```
% Ex14_9.m
w1 = 0;
w2 = 4;
for i = 1:101
    t(i) =    -5   + 10* (i-1)/100;
    R1(i) = 2   * ( sin(w2 *t(i)) - sin(w1*t(i)) )/t(i);
end
w1 = 0;
w2 = 6;
for i = 1:101
    t(i) =    -5 + 10*(i-1)/100;
    R2(i) = 2 * ( sin(w2 *t(i)) - sin(w1*t(i)) )/t(i);
end
w1 = 0;
w2 = 8;
for i = 1:101
    t(i) =   - 5 + 10*(i-1)/100;
    R3(i) =   2 * ( sin(w2 *t(i)) - sin(w1*t(i)) )/t(i);
end
for i = 1:101
    t1(i) = 0.0001 + 4.9999*(i-1)/100;
    R3_1(i) = 2 * ( sin(w2 *t(i)) - sin(w1*t(i)) )/t(i);
end
xlabel ('t');
ylabel('R/S_0');
plot(t,R1);
hold on;
gtext('Solid line: w1 = 0, w2 = 4')
gtext('Dashed line: w1 = 0, w2 = 6');
plot(t,R2,'--');
gtext('Dotted line: w1 = 0, w2 = 8');
plot(t,R3,':');
w1 = 2;
w2 = 4;
for i = 1:101
    t(i) = -5 + 10*(i-1)/100;
    R4(i) = 2 * ( sin(w2 *t(i)) - sin(w1*t(i)) )/t(i);
end
w1 = 2;
w2 = 6;
for i = 1:101
    t(i) = -5 + 10*(i-1)/100;
    R5(i) = 2 * ( sin(w2 *t(i)) - sin(w1*t(i)) )/t(i);
end
w1 = 2;
w2 = 8;
for i = 1 : 101
    t (i) = -5 + 10* (i-1) / 100;
    R6 (i) = 2 * ( sin (w2 *t (i) ) - sin (w1 *t (i) ) ) / t (i);
end
pause
```

```
hold off;
xlabel ('t');
ylabel ('R/S_0');
plot (t, R4);
hold on;
gtext ('Solid line: w1 = 2, w2 = 4')
gtext ('Dashed line: w1 = 2, w2 = 6');
plot (t, R5, '--');
gtext ('Dotted line: w1 = 2, w2 = 8');
plot (t, R6, ':');
```

Solid line: w1 = 0, w2 = 4

Dashed line: w1 = 0, w2 = 6

Dotted line: w1 = 0, w2 = 8

Solid line: w1 = 2, w2 = 4

Dashed line: w1 = 2, w2 = 6

Dotted line: w1 = 2, w2 = 8

███████████ Evaluation of a Gaussian Probability Distribution Function

EXAMPLE 14.10 ───

──────────── Using MATLAB, evaluate the following probability for $c = 1, 2$ and 3:

$$\text{Prob}\left[|x(t)| \geq c\,\sigma\right] = \frac{2}{\sqrt{2\pi}\sigma} \int_{c\,\sigma}^{\infty} e^{-\frac{1}{2}\frac{x^2}{\sigma^2}}\, dx \qquad \text{(E.1)}$$

Assume the mean value of $x(t)$ to be zero and standard deviation of $x(t)$ to be one.

Solution

Equation (E.1) can be rewritten, for $\sigma = 1$, as

$$\text{Prob}\left[|x(t)| \geq c\right] = 2\left\{\frac{1}{\sqrt{2\pi}} \int_{-\infty}^{c} e^{-0.5x^2}\, dx\right\} \qquad \text{(E.2)}$$

The MATLAB program to evaluate Eq. (E.2) is given below.

```
Ex14_10.m

>> q = quad ('normp', -7, 1);
>> prob1 = 2 * q

prob1 =

    1.6827

>> q = quad ('normp', -7, 2);
>> prob2 = 2 * q

prob2 =

    1.9545

>> q = quad ('normp', -7, 3);
>> prob3 = 2 * q

prob3 =

    1.9973

%normp.m
function pdf = normp(x)
pdf = exp(-0.5*x.^2)/sqrt(2.0 * pi);
```

■

REFERENCES

14.1 S. H. Crandall and W. D. Mark, *Random Vibration in Mechanical Systems*, Academic Press, New York, 1963.

14.2 D. E. Newland, *An Introduction to Random Vibrations and Spectral Analysis*, Longman, London, 1975.

14.3 J. D. Robson, *An Introduction to Random Vibration*, Edinburgh University Press, Edinburgh, 1963.

14.4 C. Y. Yang, *Random Vibration of Structures*, Wiley, New York, 1986.

14.5 A. Papoulis, *Probability, Random Variables and Stochastic Processes*, McGraw-Hill, New York, 1965.

14.6 J. S. Bendat and A. G. Piersol, *Engineering Applications of Correlation and Spectral Analysis*, Wiley, New York, 1980.

14.7 P. Z. Peebles, Jr., *Probability, Random Variables, and Random Signal Principles*, McGraw-Hill, New York, 1980.

14.8 J. B. Roberts, "The response of a simple oscillator to band-limited white noise," *Journal of Sound and Vibration*, Vol. 3, 1966, pp. 115–126.

14.9 M. H. Richardson, "Fundamentals of the discrete Fourier transform," *Sound and Vibration*, Vol. 12, March 1978, pp. 40–46.

14.10 E. O. Brigham, *The Fast Fourier Transform*, Prentice Hall, Englewood Cliffs, N.J., 1974.

14.11 J. K. Hammond, "On the response of single and multidegree of freedom systems to non-stationary random excitations," *Journal of Sound and Vibration*, Vol. 7, 1968, pp. 393–416.

14.12 S. H. Crandall, G. R. Khabbaz, and J. E. Manning, "Random vibration of an oscillator with nonlinear damping," *Journal of the Acoustical Society of America*, Vol. 36, 1964, pp. 1330–1334.

14.13 S. Kaufman, W. Lapinski, and R. C. McCaa, "Response of a single-degree-of-freedom isolator to a random disturbance," *Journal of the Acoustical Society of America*, Vol. 33, 1961, pp. 1108–1112.

14.14 C. J. Chisholm, "Random vibration techniques applied to motor vehicle structures," *Journal of Sound and Vibration*, Vol. 4, 1966, pp. 129–135.

14.15 Y. K. Lin, *Probabilistic Theory of Structural Dynamics*, McGraw-Hill, New York, 1967.

14.16 I. Elishakoff, *Probabilistic Methods in the Theory of Structures*, Wiley, New York, 1983.

14.17 H. W. Liepmann, "On the application of statistical concepts to the buffeting problem," *Journal of the Aeronautical Sciences*, Vol. 19, No. 12, 1952, pp. 793–800, 822.

14.18 W. C. Hurty and M. F. Rubinstein, *Dynamics of Structures*, Prentice Hall, Englewood Cliffs, N.J., 1964.

REVIEW QUESTIONS

14.1   Give brief answers to the following:

   1. What is the difference between a sample space and an ensemble?
   2. Define probability density function and probability distribution function.
   3. How are the mean value and variance of a random variable defined?
   4. What is a bivariate distribution function?
   5. What is the covariance between two random variables X and Y?
   6. Define the correlation coefficient, ρ_{XY}.

7. What are the bounds on the correlation coefficient?
8. What is a marginal density function?
9. What is autocorrelation function?
10. Explain the difference between a stationary process and a nonstationary process?
11. What are the bounds on the autocorrelation function of a stationary random process?
12. Define an ergodic process.
13. What are temporal averages?
14. What is a Gaussian random process? Why is it frequently used in vibration analysis?
15. What is Parseval's formula?
16. Define the following terms: *power spectral density function, white noise, band-limited white noise, wide-band process,* and *narrow-band process.*
17. How are the mean square value, autocorrelation function, and the power spectral density function of a stationary random process related?
18. What is an impulse response function?
19. Express the response of a single degree of freedom system using the Duhamel integral.
20. What is complex frequency response function?
21. How are the power spectral density functions of input and output of a single degree of freedom system related?
22. What are Wiener-Khintchine formulas?

14.2   Indicate whether each of the following statements is true or false:

1. A deterministic system requires deterministic system properties and loading.
2. Most phenomena in real life are deterministic.
3. A random variable is a quantity whose magnitude cannot be predicated precisely.
4. The expected value of x, in terms of its probability density function, $p(x)$, is given by $\int_{-\infty}^{\infty} x p(x)\, dx$.
5. The correlation coefficient ρ_{XY} satisfies the relation $|\rho_{XY}| \le 1$.
6. The autocorrelation function $R(t_1, t_2)$ is the same as $E[x(t_1)x(t_2)]$.
7. The mean square value of $x(t)$ can be determined as $E[x^2] = R(0)$.
8. If $x(t)$ is stationary, its mean will be independent of t.
9. The autocorrelation function $R(\tau)$ is an even function of τ.
10. The Wiener-Khintchine formulas relate the power spectral density to the autocorrelation function.
11. The ideal white noise is a physically realizable concept.

14.3   Fill in each of the following blanks with the appropriate word:

1. When the vibrational response of a system is known precisely, the vibration is called _____ vibration.
2. If any parameter of a vibrating system is not known precisely, the resulting vibration is called _____ vibration.
3. The pressure fluctuation at a point on the surface of an aircraft flying in the air is a _____ process.
4. In a random process, the outcome of an experiment will be a function of some _____ such as time.
5. The standard deviation is the positive square root of _____.

6. The joint behavior of several random variables is described by the _____ probability distribution function.

7. Univariate distributions describe the probability distributions of _____ random variables.

8. The distribution of two random variables is known as _____ distribution.

9. The distribution of several random variables is called _____ distribution.

10. The standard deviation of a stationary random process $x(t)$ will be independent of _____.

11. If all the probability information of a stationary random process can be obtained from a single sample function, the process is said to be _____.

12. The Gaussian density function is a symmetric _____-shaped curve about the mean value.

13. The standard normal variable has mean of _____ and standard deviation of _____.

14. A nonperiodic function can be treated as a periodic function having an _____ period.

15. The _____ spectral density function is an even function of ω.

16. If $S(\omega)$ has significant values over a wide range of frequencies, the process is called a _____ process.

17. If $S(\omega)$ has significant values only over a small range of frequencies, the process is called a _____ process.

18. The power spectral density $S(\omega)$ of a stationary random process is defined as the _____ transform of $R(\tau)/2\pi$.

19. If the band of frequencies has finite cut-off frequencies for a white noise, it is called _____ white noise.

14.4 Select the most appropriate answer out of the choices given:

1. Each outcome of an experiment for a random variable is called
   (a) a sample point   (b) a random point   (c) an observed value

2. Each outcome of an experiment, in the case of a random process, is called a
   (a) sample point   (b) sample space   (c) sample function

3. The probability distribution function, $P(\tilde{x})$, denotes
   (a) $P(x \le \tilde{x})$
   (b) $P(x > \tilde{x})$
   (c) $P(\tilde{x} \le x \le \tilde{x} + \Delta x)$

4. The probability density function, $p(\tilde{x})$, denotes
   (a) $P(x \le \tilde{x})$.
   (b) $P(x > \tilde{x})$
   (c) $P(\tilde{x} \le x \le \tilde{x} + \Delta x)$

5. Normalization of probability distribution implies
   (a) $P(\infty) = 1$ (b) $\int_{-\infty}^{\infty} p(x) = 1$ (c) $\int_{-\infty}^{\infty} p(x) = 0$

6. The variance of x is given by
   (a) $\overline{x^2}$ (b) $(\overline{x^2}) - (\overline{x})^2$ (c) $(\overline{x})^2$

7. The marginal density function of x can be determined form the bivariate density function $p(x,y)$ as
   (a) $p(x) = \int_{-\infty}^{\infty} p(x, y)\, dy$

$$\text{(b) } p(x) = \int_{-\infty}^{\infty} p(x, y)\, dx$$

$$\text{(c) } p(x) = \int_{-\infty}^{\infty} \int_{-\infty}^{\infty} p(x, y)\, dx\, dy$$

8. The correlation coefficient of x and y is given by

   (a) σ_{xy} (b) $\sigma_{xy}/(\sigma_x \sigma_y)$ (c) $\sigma_x \sigma_y$

9. The standard normal variable, z, corresponding to the normal variable x, is defined as

   (a) $z = \dfrac{\overline{x}}{\sigma_x}$ (b) $z = \dfrac{x - \overline{x}}{\sigma_x}$ (c) $z = \dfrac{x}{\sigma_x}$

10. If the excitation of a linear system is a Gaussian process, the response will be
    (a) a different random process
    (b) a Gaussian process
    (c) an ergodic process

11. For a normal probability density function, Prob$[-3\sigma \le x(t) \le 3\sigma]$ is
    (a) 0.6827   (b) 0.999937   (c) 0.9973

12. The mean square response of a stationary random process can be determined from the:
    (a) autocorrelation function only
    (b) power spectral density only
    (c) autocorrelation function or power spectral density

14.5 Match the items in the two columns below:

1. All possible outcomes of a random variable	(a)	Correlation functions in an experiment
2. All possible outcomes of a random process	(b)	Nonstationary process
3. Statistical connections between the values of $x(t)$ at times $t_1, t_2, \ldots$	(c)	Sample space
4. A random process invariant under a shift of the time scale	(d)	White noise
5. Mean and standard deviations of $x(t)$ vary with t	(e)	Stationary process
6. Power spectral density is constant over a frequency range	(f)	Ensemble

PROBLEMS

The problem assignments are organized as follows:

Problems	Section Covered	Topic Covered
14.1, 14.10	14.3	Probability distribution
14.3, 14.11	14.4	Mean value and standard deviation

Problems	Section Covered	Topic Covered
14.2, 14.4	14.5	Joint probability distribution
14.7, 14.9	14.6	Correlation functions
14.5, 14.26	14.7	Stationary random process
14.12	14.8	Gaussian random process
14.8, 14.13–14.16, 14.27	14.9	Fourier analysis
14.6, 14.17–14.22	14.10	Power spectral density
14.23–14.25, 14.28–14.30	14.12	Response of a single degree of freedom system
14.31, 14.32	14.14	Response of a multidegree of freedom system
14.33–14.35	14.15	MATLAB programs
14.36	—	Design project

14.1 The strength of the foundation of a reciprocating machine (x) has been found to vary between 20 and 30 kips/ft^2 according to the probability density function:

$$p(x) = \begin{cases} k\left(1 - \dfrac{x}{30}\right), & 20 \le x \le 30 \\ 0, & \text{elsewhere} \end{cases}$$

What is the probability of the foundation carrying a load greater than 28 kips/ft^2?

14.2 The joint density function of two random variables X and Y is given by

$$p_{X,Y}(x, y) = \begin{cases} \dfrac{xy}{9}, & 0 \le x \le 2, 0 \le y \le 3 \\ 0, & \text{elsewhere} \end{cases}$$

(a) Find the marginal density functions of X and Y. (b) Find the means and standard deviations of X and Y. (c) Find the correlation coefficient $\rho_{X,Y}$.

14.3 The probability density function of a random variable x is given by

$$p(x) = \begin{cases} 0 & \text{for } x < 0 \\ 0.5 & \text{for } 0 \le x \le 2 \\ 0 & \text{for } x > 2 \end{cases}$$

Determine $E[x]$, $E[x^2]$, and σ_x.

14.4 If x and y are statistically independent, then $E[xy] = E[x]E[y]$. That is, the expected value of the product xy is equal to the product of the separate mean values. If $z = x + y$, where x and y are statistically independent, show that $E[z^2] = E[x^2] + E[y^2] + 2E[x]E[y]$.

14.5 The autocorrelation function of a random process $x(t)$ is given by

$$R_x(\tau) = 20 + \frac{5}{1 + 3\tau^2}$$

Find the mean square value of $x(t)$.

14.6   The autocorrelation function of a random process is given by

$$R_x(\tau) = A \cos \omega\tau; \quad -\frac{\tau}{2\omega} \leq \tau \leq \frac{\pi}{2\omega}$$

where A and ω are constants. Find the power spectral density of the random process.

14.7   Find the autocorrelation functions of the periodic functions shown in Fig. 14.24.

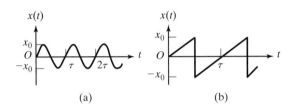

(a)   (b)

FIGURE 14.24

14.8   Find the complex form of the Fourier series for the wave shown in Fig. 14.24(b).

14.9   Compute the autocorrelation function of a periodic square wave with zero mean value and compare this result with that of a sinusoidal wave of the same period. Assume the amplitudes to be the same for both waves.

14.10 The life T in hours of a vibration transducer is found to follow exponential distribution

$$p_T(t) = \begin{cases} \lambda e^{-\lambda t}, & t \geq 0 \\ 0, & t < 0 \end{cases}$$

where λ is a constant. Find (a) the probability distribution function of T, (b) mean value of T, and (c) standard deviation of T.

14.11 Find the temporal mean value and the mean square value of the function $x(t) = x_0 \sin(\pi t/2)$.

14.12 An air compressor of mass 100 kg is mounted on an undamped isolator and operates at an angular speed of 1800 rpm. The stiffness of the isolator is found to be a random variable with mean value $\overline{k} = 2.25 \times 10^6$ N/m and standard deviation $\sigma_k = 0.225 \times 10^6$ N/m following normal distribution. Find the probability of the natural frequency of the system exceeding the forcing frequency.

14.13–
14.16 Find the Fourier transform of the functions shown in Figs. 14.25–14.28 and plot the corresponding spectrum.

14.17 A periodic function $F(t)$ is shown in Fig. 14.29. Use the values of the function $F(t)$ at ten equally spaced time stations t_i to find (a) the spectrum of $F(t)$ and (b) the mean square value of $F(t)$.

14.18 The autocorrelation function of a stationary random process $x(t)$ is given by

$$R_x(\tau) = ae^{-b|\tau|}$$

where a and b are constants. Find the power spectral density of $x(t)$.

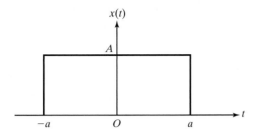

FIGURE 14.25

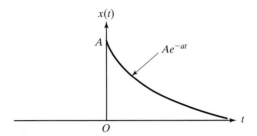

FIGURE 14.26

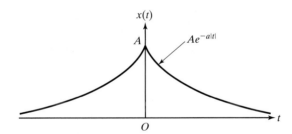

FIGURE 14.27

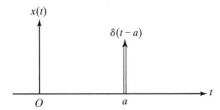

FIGURE 14.28

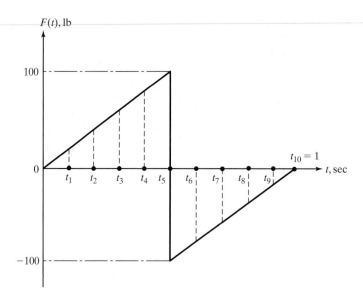

FIGURE 14.29

14.19 Find the autocorrelation function of a random process whose power spectral density is given by $S(\omega) = S_0 = $ constant between the frequencies ω_1 and ω_2.

14.20 The autocorrelation function of a Gaussian random process representing the unevenness of a road surface is given by

$$R_x(\tau) = \sigma_x^2 e^{-\alpha|v\tau|} \cos \beta v\tau$$

where σ_x^2 is the variance of the random process and v is the velocity of the vehicle. The values of σ_x, α, and β for different types of road are as follows:

Type of road	σ_x	α	β
Asphalt	1.1	0.2	0.4
Paved	1.6	0.3	0.6
Gravel	1.8	0.5	0.9

Compute the spectral density of the road surface for the different types of road.

14.21 Compute the autocorrelation function corresponding to the ideal white noise spectral density.

14.22 Starting from Eqs. (14.60) and (14.61), derive the relations

$$R(\tau) = \int_0^\infty S(f) \cos 2\pi f\tau \cdot df$$

$$S(f) = 4 \int_0^\infty R(\tau) \cos 2\pi f\tau \cdot d\tau$$

14.23 Write a computer program to find the mean square value of the response of a single degree of freedom system subjected to a random excitation whose power spectral density function is given as $S_x(\omega)$.

14.24 A machine, modeled as a single degree of freedom system, has the following parameters: $mg = 2000$ lb, $k = 4 \times 10^4$ lb/in., and $c = 1200$ lb-in./sec. It is subjected to the force shown in Fig. 14.29. Find the mean square value of the response of the machine (mass).

14.25 A mass, connected to a damper as shown in Fig. 14.30, is subjected to a force $F(t)$. Find the frequency response function $H(\omega)$ for the velocity of the mass.

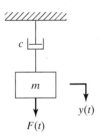

FIGURE 14.30

14.26 The spectral density of a random signal is given by

$$S(f) = \begin{cases} 0.0001 \text{ m}^2/\text{cycle/sec}, & 10 \text{ Hz} \le f \le 1000 \text{ Hz} \\ 0, & \text{elsewhere} \end{cases}$$

Find the standard deviation and the root mean square value of the signal by assuming its mean value to be 0.05 m.

14.27 Derive Eq. (14.46) from Eq. (14.45).

14.28 A simplified model of a motor cycle traveling over a rough road is shown in Fig. 14.31. It is assumed that the wheel is rigid, the wheel does not leave the road surface, and the cycle moves at a constant speed v. The cycle has a mass m and the suspension system has a spring constant k and a damping constant c. If the power spectral density of the rough road surface is taken as S_0, find the mean square value of the vertical displacement of the motor cycle (mass, m).

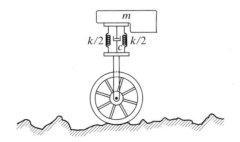

FIGURE 14.31

14.29 The motion of a lifting surface about the steady flight path due to atmospheric turbulence can be represented by the equation

$$\ddot{x}(t) + 2\zeta\omega_n\dot{x}(t) + \omega_n^2 x(t) = \frac{1}{m}F(t)$$

where ω_n is the natural frequency, m is the mass, and ζ is the damping coefficient of the system. The forcing function $F(t)$ denotes the random lift due to the air turbulence and its spectral density is given by [14.17]

$$S_F(\omega) = \frac{S_T(\omega)}{\left(1 + \dfrac{\pi\omega c}{v}\right)}$$

where c is the chord length and v is the forward velocity of the lifting surface and $S_T(\omega)$ is the spectral density of the upward velocity of air due to turbulence, given by

$$S_T(\omega) = A^2 \frac{1 + \left(\dfrac{L\omega}{v}\right)^2}{\left\{1 + \left(\dfrac{L\omega}{v}\right)^2\right\}^2}$$

where A is a constant and L is the scale of turbulence (constant). Find the mean square value of the response $x(t)$ of the lifting surface.

14.30 The wing of an airplane flying in gusty wind has been modeled as a spring-mass-damper system, as shown in Fig. 14.32. The undamped and damped natural frequencies of the wing are found to be ω_1 and ω_2, respectively. The mean square value of the displacement of m_{eq} (i.e., the wing) is observed to be δ under the action of the random wind force whose power spectral density is given by $S(\omega) = S_0$. Derive expressions for the system parameters m_{eq}, k_{eq}, and c_{eq} in terms of ω_1, ω_2, δ, and S_0.

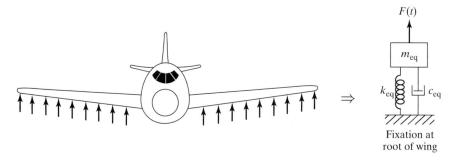

FIGURE 14.32

14.31 If the building frame shown in Fig. 14.23 has a structural damping coefficient of 0.01 (instead of the modal damping ratio 0.02), determine the mean square values of the relative displacements of the various floors.

14.32 The building frame shown in Fig. 14.23 is subjected to a ground acceleration whose power spectral density is given by

$$S(\omega) = \frac{1}{4 + \omega^2}$$

Find the mean square values of the relative displacements of the various floors of the building frame. Assume a modal damping ratio of 0.02 in each mode.

14.33 Using MATLAB, plot the Gaussian probability density function

$$f(x) = \frac{1}{\sqrt{2\pi}} e^{-0.5x^2}$$

over $-7 \le x \le 7$.

14.34 Plot the Fourier transform of a triangular pulse:

$$X(\omega) = \frac{4A}{a\,\omega^2} \sin^2 \frac{\omega a}{2}, \quad -7 \le \frac{\omega a}{\pi} \le 7$$

(See Fig. 14.12.)

14.35 The mean square value of the response of a machine, $E[y^2]$, subject to the force shown in Fig. 14.29, is given by (see Problem 14.24):

$$E[y^2] = \sum_{n=0}^{N-1} \frac{|c_n|^2}{\left(k - m\omega_n^2\right)^2 + c^2\omega_n^2}$$

where

$$c_n = \frac{1}{N} \sum_{j=1}^{N} F_j \left\{ \cos \frac{2\pi n j}{N} - i \sin \frac{2\pi n j}{N} \right\}$$

with F_j = 0, 20, 40, 60, 80, 100, −80, −60, −40, −20, 0 for j = 0, 1, 2, 3, 4, 5, 6, 7, 8, 9, 10; $k = 4 \times 10^4$, $c = 1200$, $m = 5.1760$ and $\omega_n = 2\pi n$.
Using MATLAB, find the value of $E[y^2]$ with $N = 10$.

DESIGN PROJECT

14.36 The water tank shown in Fig. 14.33 is supported by a hollow circular steel column. The tank, made of steel, is in the form of a thin-walled pressure vessel and has a capacity of 10,000 gallons. Design the column to satisfy the following specifications: (a) The undamped natural frequency of vibration of the tank, either empty or full, must exceed a value of 1 Hz. (b) The mean square value of the displacement of the tank, either empty or full, must not exceed a value of 16 in² when subjected to an earthquake ground acceleration whose power spectral density is given by

$$S(\omega) = 0.0002 \frac{m^2/s^4}{rad/s}$$

Assume damping to be 10 percent of the critical value.

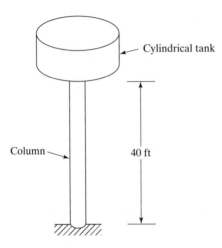

Cylindrical tank

Column

40 ft

FIGURE 14.33

Jean Le Rond D'Alembert (1717–1783), a French mathematician and physicist, was abandoned by his mother as a newborn infant near the church of Saint Jean Le Rond in Paris. In 1741, he published his famous *Traite de Dynamique*, which contained the method that has become known as d'Alembert's principle. D'Alembert was the first to use partial differential equations for the solution of vibrating string problems. His early brilliance led to his appointment as the *secretaire perpetuel* (permanent secretary) of the French Academy, a position that secured his place as the most influential man of science in France. (Photo courtesy of Dirk J. Struik, *A Concise History of Mathematics*, 2nd ed., Dover Publications, New York, 1948.)

A P P E N D I X A

Mathematical Relationships

Some of the relationships from trigonometry, algebra, and calculus that are frequently used in vibration analysis are given below.

$$\sin (\alpha \pm \beta) = \sin \alpha \cos \beta \pm \cos \alpha \sin \beta$$

$$\cos (\alpha \pm \beta) = \cos \alpha \cos \beta \mp \sin \alpha \sin \beta$$

$$\sin (\alpha + \beta) \sin (\alpha - \beta) = \sin^2\alpha - \sin^2\beta = \cos^2\beta - \cos^2\alpha$$

$$\cos (\alpha + \beta) \cos (\alpha - \beta) = \cos^2\alpha - \sin^2\beta = \cos^2\beta - \sin^2\alpha$$

$$\sin \alpha \sin \beta = \frac{1}{2}[\cos (\alpha - \beta) - \cos (\alpha + \beta)]$$

$$\cos \alpha \cos \beta = \frac{1}{2}[\cos (\alpha - \beta) + \cos (\alpha + \beta)]$$

$$\sin \alpha \cos \beta = \frac{1}{2}[\sin (\alpha + \beta) + \sin (\alpha - \beta)]$$

$$\sin \alpha + \sin \beta = 2 \sin \left(\frac{\alpha + \beta}{2}\right) \cos \left(\frac{\alpha - \beta}{2}\right)$$

$$\sin \alpha - \sin \beta = 2 \cos \left(\frac{\alpha + \beta}{2} \right) \sin \left(\frac{\alpha - \beta}{2} \right)$$

$$\cos \alpha + \cos \beta = 2 \cos \left(\frac{\alpha + \beta}{2} \right) \cos \left(\frac{\alpha - \beta}{2} \right)$$

$$\cos \alpha - \cos \beta = -2 \sin \left(\frac{\alpha + \beta}{2} \right) \sin \left(\frac{\alpha - \beta}{2} \right)$$

$$A \sin \alpha + B \cos \alpha = \sqrt{A^2 + B^2} \cos (\alpha - \phi_1)$$

$$= \sqrt{A^2 + B^2} \sin (\alpha + \phi_2)$$

$$\text{where } \phi_1 = \tan^{-1} \frac{A}{B}, \ \phi_2 = \tan^{-1} \frac{B}{A}$$

$$\sin^2\alpha + \cos^2\alpha = 1$$

$$\cos 2\alpha = 1 - 2 \sin^2 \alpha = 2 \cos^2 \alpha - 1 = \cos^2 \alpha - \sin^2 \alpha$$

Law of cosines for triangles: $c^2 = a^2 + b^2 - 2 a b \cos C$

$$\pi = 3.14159265 \text{ rad}, 1 \text{ rad} = 57.29577951°, 1° = 0.017453292 \text{ rad}$$

$$e = 2.71828183$$

$$\log a^b = b \log a, \ \log_{10} x = 0.4343 \log_e x, \ \log_e x = 2.3026 \log_{10} x$$

$$e^{ix} = \cos x + i \sin x$$

$$\sin x = \frac{e^{ix} - e^{-ix}}{2i}, \ \cos ix = \frac{e^{ix} + e^{-ix}}{2}$$

$$\sinh x = \frac{1}{2}(e^x - e^{-x}), \ \cosh x = \frac{1}{2}(e^x + e^{-x})$$

$$\cosh^2 x - \sinh^2 x = 1$$

$$\frac{d}{dx}(uv) = u \frac{dv}{dx} + v \frac{du}{dx}$$

$$\frac{d}{dx}\left(\frac{u}{v} \right) = \frac{1}{v} \frac{du}{dx} - \frac{u}{v^2} \frac{dv}{dx} = \frac{v \frac{du}{dx} - u \frac{dv}{dx}}{v^2}$$

$$\int e^{ax} \, dx = \frac{1}{a} e^{ax}$$

$$\int u \, v \, dx = u \int v \, dx - \int \left(\frac{du}{dx} \int v \, dx \right) dx$$

Complex algebra:

$$z = x + i\,y \equiv A\,e^{i\theta} \text{ with } A = \sqrt{x^2 + y^2} \text{ and } \theta = \tan^{-1}\left(\frac{y}{x}\right)$$

$$\text{If } z_1 = x_1 + i\,y_1 \text{ and } z_2 = x_2 + i\,y_2,$$

$$z_1 \pm z_2 = (x_1 \pm x_2) + i(y_1 \pm y_2)$$

$$z_1\,z_2 = (x_1\,x_2 - y_1\,y_2) + i(x_1\,y_1 + x_2\,y_2)$$

$$\frac{z_1}{z_2} = \frac{(x_1\,x_2 + y_1\,y_2) + i(x_2\,y_1 - x_1\,y_2)}{\sqrt{x_2^2 + y_2^2}}$$

$$\text{If } z_1 = A_1\,e^{i\theta_1} \qquad \text{and} \qquad z_2 = A_2\,e^{i\theta_2},$$

$$z_1 + z_2 = A\,e^{i\theta}$$

$$\text{with } A = [A_1^2 + A_2^2 - 2\,A_1\,A_2\cos(\theta_1 - \theta_2)]^{\frac{1}{2}}$$

$$\text{and } \theta = \tan^{-1}\left[\frac{A_1\sin\theta_1 + A_2\sin\theta_2}{A_1\cos\theta_1 + A_2\cos\theta_2}\right]$$

$$z_1\,z_2 = A_1\,A_2\,e^{i(\theta_1 + \theta_2)}$$

$$\frac{z_1}{z_2} = \frac{A_1}{A_2}e^{i(\theta_1 - \theta_2)}$$

Carl Gustov Jacob Jacobi (1804–1851), a German mathematician, was educated at the University of Berlin and became a full professor at the University of Konigsberg in 1832. The method he developed for finding the eigen solution of real symmetric matrices has become known as the Jacobi method. He made significant contributions to the fields of elliptic functions, number theory, differential equations, and mechanics and introduced the definition of Jacobian in the theory of determinants. (Photo courtesy of Dirk J. Struik, *A Concise History of Mathematics*, 2nd ed., Dover Publications, New York, 1948.)

A P P E N D I X   B

Deflection of Beams and Plates

Cantilever Beam

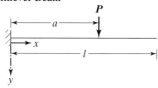

$$y(x) = \begin{cases} \dfrac{Px^2}{6\,EI}(3a - x);\ 0 \le x \le a \\[2ex] \dfrac{Pa^2}{6\,EI}(3x - a);\ a \le x \le l \end{cases}$$

Simply Supported Beam

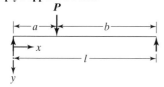

$$y(x) = \begin{cases} \dfrac{Pbx}{6\,EIl}(l^2 - x^2 - b^2);\ 0 \le x \le a \\[2ex] \dfrac{Pa(l - x)}{6\,EIl}(2lx - x^2 - a^2);\ a \le x \le l \end{cases}$$

Fixed-fixed Beam

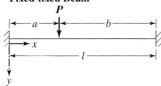

$$y(x) = \begin{cases} \dfrac{Pb^2x^2}{6\,EIl^3}[3\,al - x(3a + b)]; 0 \le x \le a \\ \dfrac{Pa^2(l - x)^2}{6\,EIl^3}[3\,bl - (l - x)(3b + a)]; a \le x \le l \end{cases}$$

Simply Supported Beam with Overhang

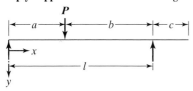

$$y(x) = \begin{cases} \text{Same as in case of Simply Supported beam} \\ \text{for } 0 \le x \le a \text{ and } a \le x \le l \\ \dfrac{Pa}{6\,EIl}(l^2 - a^2)(x - l); l \le x \le l + c \end{cases}$$

Simply Supported Beam with Overhanging load

$$y(x) = \begin{cases} \dfrac{Pax}{6\,EIl}(x^2 - l^2); 0 \le x \le l \\ \dfrac{P(x - l)}{6\,EIl}[a(3x - l) - (x - l)^2]; l \le x \le l + a \end{cases}$$

Fixed-fixed Beam with End Displacement

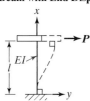

$$y(x) = \frac{P}{12\,EI}(3\,lx^2 - 2\,x^3)$$

Simply Supported Circular Plate

$$y_{\text{center}} = \frac{Pr^2\,(3 + v)}{16\,\pi D\,(1 + v)}$$

$$\text{where } D = \frac{Et^3}{12\,(1 - v^2)}, t = \text{plate thickness,}$$

$$\text{and } v = \text{Poisson's ratio}$$

Fixed Circular Plate

$$y_{center} = \frac{Pr^2}{16\,\pi D}$$

Square Plate Simply Supported on All Sides

$$y_{center} = \frac{\alpha\,P\,a^2}{E\,t^3} \quad \text{with} \quad \alpha = 0.1267 \quad \text{for} \quad \nu = 0.3$$

Square Plate Fixed on All Sides

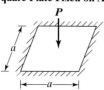

$$y_{center} = \frac{\alpha\,P\,a^2}{E\,t^3} \quad \text{with} \quad \alpha = 0.0611 \quad \text{for} \quad \nu = 0.3$$

Arthur Cayley (1821–1895) was a British mathematician and professor of mathematics at Cambridge University. His greatest work, produced with James Joseph Sylvester, was the development of the theory of invariants, which played a crucial role in the theory of relativity. He made many important contributions to n-dimensional geometry and invented and developed the theory of matrices. (Photo courtesy of Dirk J. Struik, *A Concise History of Mathematics*, 2nd ed., Dover Publications, New York, 1948.)

A P P E N D I X   C

Matrices

C.1 Definitions

Matrix. A matrix is a rectangular array of numbers. An array having m rows and n columns enclosed in brackets is called an m-by-n matrix. If $[A]$ is an $m \times n$ matrix, it is denoted as

$$[A] = [a_{ij}] = \begin{bmatrix} a_{11} & a_{12} & \cdot & \cdot & \cdot & a_{1n} \\ a_{21} & a_{22} & \cdot & \cdot & \cdot & a_{2n} \\ \cdot & & & & & \\ \cdot & & & & & \\ \cdot & & & & & \\ a_{m1} & a_{m2} & \cdot & \cdot & \cdot & a_{mn} \end{bmatrix} \qquad (C.1)$$

where the numbers a_{ij} are called the *elements* of the matrix. The first subscript i denotes the row and the second subscript j specifies the column in which the element a_{ij} appears.

Square Matrix. When the number of rows (m) is equal to the number of columns (n), the matrix is called a *square matrix of order n*.

Column Matrix. A matrix consisting of only one column—that is, an $m \times 1$ matrix—is called a *column matrix* or more commonly a *column vector*. Thus if $\vec{a}$ is a column vector having m elements, it can be represented as

$$\vec{a} = \begin{Bmatrix} a_1 \\ a_2 \\ \cdot \\ \cdot \\ \cdot \\ a_m \end{Bmatrix} \tag{C.2}$$

Row Matrix. A matrix consisting of only one row—that is a $1 \times n$ matrix—is called a *row matrix* or a *row vector*. If $\lfloor b \rfloor$ is a row vector, it can be denoted as

$$\lfloor b \rfloor = [b_1 \; b_2 \; \cdots \; b_n] \tag{C.3}$$

Diagonal Matrix. A square matrix in which all the elements are zero except those on the principal diagonal is called a *diagonal matrix*. For example, if $[A]$ is a diagonal matrix of order n, it is given by

$$[A] = \begin{bmatrix} a_{11} & 0 & 0 & \cdot & \cdot & \cdot & 0 \\ 0 & a_{22} & 0 & \cdot & \cdot & \cdot & 0 \\ 0 & 0 & a_{33} & \cdot & \cdot & \cdot & 0 \\ \cdot & & & & & & \\ \cdot & & & & & & \\ \cdot & & & & & & \\ 0 & 0 & 0 & \cdot & \cdot & \cdot & a_{nn} \end{bmatrix} \tag{C.4}$$

Identity Matrix. If all the elements of a diagonal matrix have a value 1, then the matrix is called an *identity matrix* or *unit matrix* and is usually denoted as $[I]$.

Zero Matrix. If all the elements of a matrix are zero, it is called a *zero* or *null matrix* and is denoted as $[0]$. If $[0]$ is of order 2×4, it is given by

$$[0] = \begin{bmatrix} 0 & 0 & 0 & 0 \\ 0 & 0 & 0 & 0 \end{bmatrix} \tag{C.5}$$

Symmetric Matrix. If the element in ith row and jth column is the same as the one in jth row and ith column in a square matrix, it is called a *symmetric matrix*. This means that if $[A]$ is a symmetric matrix, we have $a_{ji} = a_{ij}$. For example,

$$[A] = \begin{bmatrix} 4 & -1 & -3 \\ -1 & 0 & 7 \\ -3 & 7 & 5 \end{bmatrix} \tag{C.6}$$

is a symmetric matrix of order 3.

Transpose of a Matrix. The transpose of an $m \times n$ matrix $[A]$ is the $n \times m$ matrix obtained by interchanging the rows and columns of $[A]$ and is denoted as $[A]^T$. Thus if

$$[A] = \begin{bmatrix} 2 & 4 & 5 \\ 3 & 1 & 8 \end{bmatrix} \tag{C.7}$$

then $[A]^T$ is given by

$$[A]^T = \begin{bmatrix} 2 & 3 \\ 4 & 1 \\ 5 & 8 \end{bmatrix} \tag{C.8}$$

Note that the transpose of a column matrix (vector) is a row matrix (vector), and vice versa.

Trace. The sum of the main diagonal elements of a square matrix $[A] = [a_{ij}]$ is called the *trace* of $[A]$ and is given by

$$\text{Trace}[A] = a_{11} + a_{22} + \cdots + a_{nn} \tag{C.9}$$

Determinant. If $[A]$ denotes a square matrix of order n, then the determinant of $[A]$ is denoted as $|[A]|$. Thus

$$|[A]| = \begin{vmatrix} a_{11} & a_{12} & \cdot & \cdot & \cdot & a_{1n} \\ a_{21} & a_{22} & \cdot & \cdot & \cdot & a_{2n} \\ \cdot & & & & & \\ \cdot & & & & & \\ \cdot & & & & & \\ a_{n1} & a_{n2} & \cdot & \cdot & \cdot & a_{nn} \end{vmatrix} \tag{C.10}$$

The value of a determinant can be found by obtaining the minors and cofactors of the determinant.

The *minor* of the element a_{ij} of the determinant $|[A]|$ of order n is a determinant of order $(n - 1)$ obtained by deleting the row i and the column j of the original determinant. The minor of a_{ij} is denoted as M_{ij}.

The *cofactor* of the element a_{ij} of the determinant $|[A]|$ of order n is the minor of the element a_{ij}, with either a plus or a minus sign attached; it is defined as

$$\text{Cofactor of } a_{ij} = \beta_{ij} = (-1)^{i+j} M_{ij} \tag{C.11}$$

where M_{ij} is the minor of a_{ij}. For example, the cofactor of the element a_{32} of

$$\det[A] = \begin{vmatrix} a_{11} & a_{12} & a_{13} \\ a_{21} & a_{22} & a_{23} \\ a_{31} & a_{32} & a_{33} \end{vmatrix} \tag{C.12}$$

is given by

$$\beta_{32} = (-1)^5 M_{32} = -\begin{vmatrix} a_{11} & a_{13} \\ a_{21} & a_{23} \end{vmatrix} \tag{C.13}$$

The value of a second-order determinant $\|[A]\|$ is defined as

$$\det[A] = \begin{vmatrix} a_{11} & a_{12} \\ a_{21} & a_{22} \end{vmatrix} = a_{11}a_{22} - a_{12}a_{21} \tag{C.14}$$

The value of an nth order determinant $\|[A]\|$ is defined as

$$\det[A] = \sum_{j=1}^{n} a_{ij}\beta_{ij} \text{ for any specific row } i$$

or

$$\det[A] = \sum_{i=1}^{n} a_{ij}\beta_{ij} \text{ for any specific column } j \tag{C.15}$$

For example, if

$$\det[A] = \|[A]\| = \begin{vmatrix} 2 & 2 & 3 \\ 4 & 5 & 6 \\ 7 & 8 & 9 \end{vmatrix} \tag{C.16}$$

then, by selecting the first column for expansion, we obtain

$$\det[A] = 2\begin{vmatrix} 5 & 6 \\ 8 & 9 \end{vmatrix} - 4\begin{vmatrix} 2 & 3 \\ 8 & 9 \end{vmatrix} + 7\begin{vmatrix} 2 & 3 \\ 5 & 6 \end{vmatrix}$$

$$= 2(45 - 48) - 4(18 - 24) + 7(12 - 15) = -3 \tag{C.17}$$

Properties of Determinants

1. The value of a determinant is not affected if rows (or columns) are written as columns (or rows) in the same order.
2. If all the elements of a row (or a column) are zero, the value of the determinant is zero.
3. If any two rows (or two columns) are interchanged, the value of the determinant is multiplied by -1.
4. If all the elements of one row (or one column) are multiplied by the same constant a, the value of the new determinant is a times the value of the original determinant.

5. If the corresponding elements of two rows (or two columns) of a determinant are proportional, the value of the determinant is zero. For example,

$$\det[A] = \begin{vmatrix} 4 & 7 & -8 \\ 2 & 5 & -4 \\ -1 & 3 & 2 \end{vmatrix} = 0 \tag{C.18}$$

Adjoint Matrix. The adjoint matrix of a square matrix $[A] = [a_{ij}]$ is defined as the matrix obtained by replacing each element a_{ij} by its cofactor β_{ij} and then transposing. Thus

$$\text{Adjoint}[A] = \begin{bmatrix} \beta_{11} & \beta_{12} & \cdot & \cdot & \cdot & \beta_{1n} \\ \beta_{21} & \beta_{22} & \cdot & \cdot & \cdot & \beta_{2n} \\ \cdot & & & & & \\ \cdot & & & & & \\ \cdot & & & & & \\ \beta_{n1} & \beta_{n2} & \cdot & \cdot & \cdot & \beta_{nn} \end{bmatrix}^T = \begin{bmatrix} \beta_{11} & \beta_{21} & \cdot & \cdot & \cdot & \beta_{n1} \\ \beta_{12} & \beta_{22} & \cdot & \cdot & \cdot & \beta_{n2} \\ \cdot & & & & & \\ \cdot & & & & & \\ \cdot & & & & & \\ \beta_{1n} & \beta_{2n} & \cdot & \cdot & \cdot & \beta_{nn} \end{bmatrix} \tag{C.19}$$

Inverse Matrix. The inverse of a square matrix $[A]$ is written as $[A]^{-1}$ and is defined by the following relationship:

$$[A]^{-1}[A] = [A][A]^{-1} = [I] \tag{C.20}$$

where $[A]^{-1}[A]$, for example, denotes the product of the matrix $[A]^{-1}$ and $[A]$. The inverse matrix of $[A]$ can be determined (see Ref. [A.1]):

$$[A]^{-1} = \frac{\text{adjoint}[A]}{\det[A]} \tag{C.21}$$

when $\det[A]$ is not equal to zero. For example, if

$$[A] = \begin{bmatrix} 2 & 2 & 3 \\ 4 & 5 & 6 \\ 7 & 8 & 9 \end{bmatrix} \tag{C.22}$$

its determinant has a value $\det[A] = -3$. The cofactor of a_{11} is

$$\beta_{11} = (-1)^2 \begin{vmatrix} 5 & 6 \\ 8 & 9 \end{vmatrix} = -3 \tag{C.23}$$

In a similar manner, we can find the other cofactors and determine

$$[A]^{-1} = \frac{\text{adjoint}[A]}{\det[A]} = \frac{1}{-3} \begin{bmatrix} -3 & 6 & -3 \\ 6 & -3 & 0 \\ -3 & -2 & 2 \end{bmatrix} = \begin{bmatrix} 1 & -2 & 1 \\ -2 & 1 & 0 \\ 1 & 2/3 & -2/3 \end{bmatrix} \tag{C.24}$$

Singular Matrix. A square matrix is said to be singular if its determinant is zero.

C.2 Basic Matrix Operations

Equality of Matrices.   Two matrices [A] and [B], having the same order, are equal if and only if $a_{ij} = b_{ij}$ for every i and j.

Addition and Subtraction of Matrices.   The sum of the two matrices [A] and [B], having the same order, is given by the sum of the corresponding elements. Thus if $[C] = [A] + [B] = [B] + [A]$, we have $c_{ij} = a_{ij} + b_{ij}$ for every i and j. Similarly, the difference between two matrices [A] and [B] of the same order, [D], is given by $[D] = [A] - [B]$ with $d_{ij} = a_{ij} - b_{ij}$ for every i and j.

Multiplication of Matrices.   The product of two matrices [A] and [B] is defined only if they are conformable—that is, if the number of columns of [A] is equal to the number of rows of [B]. If [A] is of order $m \times n$ and [B] is of order $n \times p$, then the product $[C] = [A][B]$ is of order $m \times p$ and is defined by $[C] = [c_{ij}]$, with

$$c_{ij} = \sum_{k=1}^{n} a_{ik}b_{kj} \tag{C.25}$$

This means that c_{ij} is the quantity obtained by multiplying the ith row of [A] and the jth column of [B] and summing these products. For example, if

$$[A] = \begin{bmatrix} 2 & 3 & 4 \\ 1 & -5 & 6 \end{bmatrix} \quad \text{and} \quad [B] = \begin{bmatrix} 8 & 0 \\ 2 & 7 \\ -1 & 4 \end{bmatrix} \tag{C.26}$$

then

$$\begin{aligned} [C] = [A][B] &= \begin{bmatrix} 2 & 3 & 4 \\ 1 & -5 & 6 \end{bmatrix} \begin{bmatrix} 8 & 0 \\ 2 & 7 \\ -1 & 4 \end{bmatrix} \\ &= \begin{bmatrix} 2 \times 8 + 3 \times 2 + 4 \times (-1) & 2 \times 0 + 3 \times 7 + 4 \times 4 \\ 1 \times 8 + (-5) \times 2 + 6 \times (-1) & 1 \times 0 + (-5) \times 7 + 6 \times 4 \end{bmatrix} \\ &= \begin{bmatrix} 18 & 37 \\ -8 & -11 \end{bmatrix} \end{aligned} \tag{C.27}$$

If the matrices are conformable, the matrix multiplication process is associative

$$([A][B])[C] = [A]([B][C]) \tag{C.28}$$

and is distributive

$$([A] + [B])[C] = [A][C] + [B][C] \tag{C.29}$$

The product $[A][B]$ denotes the premultiplication of $[B]$ by $[A]$ or the postmultiplication of $[A]$ by $[B]$. It is to be noted that the product $[A][B]$ is not necessarily equal to $[B][A]$.

The transpose of a matrix product can be found to be the product of the transposes of the separate matrices in reverse order. Thus, if $[C] = [A][B]$,

$$[C]^T = ([A][B])^T = [B]^T[A]^T \tag{C.30}$$

The inverse of a matrix product can be determined from the product of the inverse of the separate matrices in reverse order. Thus if $[C] = [A][B]$,

$$[C]^{-1} = ([A][B])^{-1} = [B]^{-1}[A]^{-1} \tag{C.31}$$

REFERENCE

C.1 Barnett, *Matrix Methods for Engineers and Scientists*, McGraw-Hill, New York, 1982.

Pierre Simon Laplace (1749–1827) was a French mathematician remembered for his fundamental contributions to probability theory, mathematical physics, and celestial mechanics; the name Laplace occurs in both mechanical and electrical engineering. Much use is made of Laplace transforms in vibrations and applied mechanics, and the Laplace equation is applied extensively in the study of electric and magnetic fields. (Photo courtesy of Dirk J. Struik, *A Concise History of Mathematics*, 2nd ed., Dover Publications, New York, 1948.)

A P P E N D I X   D

Laplace Transform Pairs

Laplace Domain $$\bar{f}(s) = \int_0^\infty f(t)e^{-st}\,dt$$	**Time Domain** $f(t)$
1. $\quad c_1\bar{f}(s) + c_2\bar{g}(s)$	$c_1 f(t) + c_2 g(t)$
2. $\quad \bar{f}\left(\dfrac{s}{a}\right)$	$f(a \cdot t)a$
3. $\quad \bar{f}(s)\bar{g}(s)$	$\displaystyle\int_0^t f(t-\tau)g(\tau)\,d\tau = \int_0^t f(\tau)g(t-\tau)\,d\tau$
4. $\quad s^n\bar{f}(s) - \displaystyle\sum_{j=1}^n s^{n-j}\dfrac{d^{j-1}f}{dt^{j-1}}(0)$	$\dfrac{d^n f}{dt^n}(t)$
5. $\quad \dfrac{1}{s^n}\bar{f}(s)$	$\displaystyle\underbrace{\int_0^t \cdots \int_0^t f(\tau)\,d\tau \cdots d\tau}_{n}$
6. $\quad \bar{f}(s+a)$	$e^{-at}f(t)$

1044

Laplace Domain	**Time Domain**
$\bar{f}(s) = \int_0^\infty f(t)e^{-st}\,dt$	$f(t)$

7. $\dfrac{a}{s(s+a)}$ $1 - e^{-at}$

8. $\dfrac{s+a}{s^2}$ $1 + at$

9. $\dfrac{a^2}{s^2(s+a)}$ $at - (1 - e^{-at})$

10. $\dfrac{s+b}{s(s+a)}$ $\dfrac{b}{a}\left\{1 - \left(1 - \dfrac{a}{b}\right)e^{-at}\right\}$

11. $\dfrac{a}{s^2+a^2}$ $\sin at$

12. $\dfrac{s}{s^2+a^2}$ $\cos at$

13. $\dfrac{a^2}{s(s^2+a^2)}$ $1 - \cos at$

14.* $\dfrac{1}{s^2+2\zeta\omega_n s+\omega_n^2}$ $\dfrac{1}{\omega_d}e^{-\zeta\omega_n t}\sin\omega_d t$

15.* $\dfrac{s}{s^2+2\zeta\omega_n s+\omega_n^2}$ $-\dfrac{\omega_n}{\omega_d}e^{-\zeta\omega_n t}\sin(\omega_d t-\phi_1)$

16.* $\dfrac{s+2\zeta\omega_n s}{s^2+2\zeta\omega_n s+\omega_n^2}$ $\dfrac{\omega_n}{\omega_d}e^{-\zeta\omega_n t}\sin(\omega_d t+\phi_1)$

17.* $\dfrac{\omega_n^2}{s(s^2+2\zeta\omega_n s+\omega_n^2)}$ $1 - \dfrac{\omega_n}{\omega_d}e^{-\zeta\omega_n t}\sin(\omega_d t+\phi_1)$

18.* $\dfrac{s+\zeta\omega_n}{s(s^2+2\zeta\omega_n s+\omega_n^2)}$ $e^{-\zeta\omega_n t}\sin(\omega_d t+\phi_1)$

19. 1 Unit impulse at $t = 0$

20. $\dfrac{e^{-as}}{s}$ Unit step function at $t = a$

*$\omega_d = \omega_n\sqrt{1-\zeta^2}$; $\zeta < 1$

$\phi_1 = \cos^{-1}\zeta$; $\zeta < 1$

Heinrich Rudolf Hertz (1857–1894), a German physicist and a professor of physics at the Polytechnic Institute in Karlsruhe and later at the University of Bonn, gained fame through his experiments on radio waves. His investigations in the field of elasticity formed a relatively small part of his achievements but are of vital importance to engineers. His work on the analysis of elastic bodies in contact is referred to as "Hertzian stresses," and is very important in the design of ball and roller bearings. The unit of frequency of periodic phenomena, measured in cycles per second, is named Hertz in SI units. (Photo courtesy of *Applied Mechanics Reviews*.)

A P P E N D I X   E

Units

The English system of units is now being replaced by the International System of units (SI). The SI system is the modernized version of the metric system of units. Its name in French is Système International; hence the abbreviation *SI*. The SI system has seven basic units. All other units can be derived from these seven [E.1–E.2]. The three basic units of concern in the study of vibrations are meter for length, kilogram for mass, and second for time.

The common prefixes for multiples and submultiples of SI units are given in Table E.1. In the SI system, the combined units must be abbreviated with care. For example, a torque of 4 N $\times$ 2 m must be stated as 8 N m or 8 N $\cdot$ m with either a space or a dot between N and m. It should not be written as Nm. Another example is 8 m $\times$ 5 s = 40 m s or 40 m $\cdot$ s or 40 meter-seconds. If it is written as 40 ms, it means 40 milliseconds.

Conversion of Units

To convert the units of any given quantity from one system to another, we use the equivalence of units given in Table E.2. The following examples illustrate the procedure.

TABLE E.1   Prefixes for Multiples and Submultiples of SI Units

Multiple	Prefix	Symbol	Submultiple	Prefix	Symbol
10	deka	da	10^{-1}	deci	d
10^2	hecto	h	10^{-2}	centi	c
10^3	kilo	k	10^{-3}	milli	m
10^6	mega	M	10^{-6}	micro	μ
10^9	giga	G	10^{-9}	nano	n
10^{12}	tera	T	10^{-12}	pico	p

TABLE E.2   Conversion of units

Quantity	SI Equivalence	English Equivalence
Mass	$1\ lb_f - sec^2/ft\ (slug) = 14.5939\ kg$ $= 32.174\ lb_m$ $1\ lb_m = 0.45359237\ kg$	$1\ kg = 2.204623\ lb_m$ $= 0.06852178\ slug$ $(lb_f - sec^2/ft)$
Length	$1\ in = 0.0254\ m$ $1\ ft = 0.3048\ m$ $1\ mile = 5280\ ft = 1.609344\ km$	$1\ m = 39.37008\ in.$ $= 3.28084\ ft$ $1\ km = 3280.84\ ft = 0.621371\ mile$
Area	$1\ in^2 = 0.00064516\ m^2$ $1\ ft^2 = 0.0929030\ m^2$	$1\ m^2 = 1550.0031\ in^2$ $= 10.76391\ ft^2$
Volume	$1\ in^3 = 16.3871 \times 10^{-6}\ m^3$ $1\ ft^3 = 28.3168 \times 10^{-3}\ m^3$ $1\ US\ gallon = 3.7853\ litres$ $= 3.7853 \times 10^{-3}\ m^3$	$1\ m^3 = 61.0237 \times 10^3\ in^3$ $= 35.3147\ ft^3$ $= 10^3\ litres = 0.26418\ US\ gallon$
Force or weight	$1\ lb_f = 4.448222\ N$	$1\ N = 0.2248089\ lb_f$
Torque or moment	$1\ lb_f - in. = 0.1129848\ N \cdot m$ $1\ lb_f - ft = 1.355818\ N \cdot m$	$1\ N \cdot m = 8.850744\ lb_f - in.$ $= 0.737562\ lb_f - ft$
Stress, pressure, or elastic modulus	$1\ lb_f/in^2\ (psi) = 6894.757\ Pa$ $1\ lb_f/ft^2 = 47.88026\ Pa$	$1\ Pa = 1.450377 \times 10^{-4}\ lb_f/in^2\ (psi)$ $= 208.8543 \times 10^{-4}\ lb_f/ft^2$
Mass density	$1\ lb_m/in^3 = 27.6799 \times 10^3\ kg/m^3$ $1\ lb_m/ft^3 = 16.0185\ kg/m^3$	$1\ kg/m^3 = 36.127 \times 10^{-6}\ lb_m/in^3$ $= 62.428 \times 10^{-3}\ lb_m/ft^3$

EXAMPLE E.1

Mass moment of inertia:

$$\begin{pmatrix} \text{Mass moment of} \\ \text{inertia in SI units} \end{pmatrix} = \begin{pmatrix} \text{Mass moment of inertia} \\ \text{in English units} \end{pmatrix} \times \begin{pmatrix} \text{Multiplication} \\ \text{factor} \end{pmatrix}$$

$$(\text{kg} \cdot \text{m}^2) \equiv (\text{N} \cdot \text{m} \cdot \text{s}^2) = \left(\frac{\text{N}}{\text{lb}_f} \cdot \text{lb}_f \right) \left(\frac{\text{m}}{\text{in.}} \cdot \text{in.} \right) (\text{sec}^2)$$

$$= (\text{N per 1 lb}_f)(\text{m per 1 in.})(\text{lb}_f\text{-in.-sec}^2)$$

$$= (4.448222)(0.0254)(\text{lb}_f\text{-in.-sec}^2)$$

$$= 0.1129848 \ (\text{lb}_f\text{-in.-sec}^2)$$

∎

EXAMPLE E.2

Stress:

$$(\text{Stress in SI units}) = (\text{Stress in English units}) \times \begin{pmatrix} \text{Multiplication} \\ \text{factor} \end{pmatrix}$$

$$(\text{Pa}) \equiv (\text{N/m}^2) = \left(\frac{\text{N}}{\text{lb}_f} \cdot \text{lb}_f \right) \frac{1}{\left(\frac{\text{m}}{\text{in.}} \cdot \text{in.} \right)^2} = \frac{\text{N}}{\text{lb}_f} \frac{1}{\left(\frac{\text{m}}{\text{in.}} \right)^2} (\text{lb}_f/\text{in}^2)$$

$$= \frac{(\text{N per 1 lb}_f)}{(\text{m per 1 in.})^2} (\text{lb}_f/\text{in}^2)$$

$$= \frac{(4.448222)}{(0.0254)^2} (\text{lb}_f/\text{in}^2)$$

$$= 6894.757 \ (\text{lb}_f/\text{in}^2)$$

∎

REFERENCES

E.1 E. A. Mechtly, "The International System of Units" (2nd rev.), NASA SP-7012, 1973.

E.2 C. Wandmacher, *Metric Units in Engineering—Going SI*, Industrial Press, New York, 1978.

APPENDIX   F

Introduction to MATLAB

MATLAB, derived from MATrix LABoratory, is a software package that can be used for the solution of a variety of scientific and engineering problems including linear algebraic equations, nonlinear equations, numerical differentiation and integration, curve fitting, ordinary and partial differential equations, optimization, and graphics. It uses matrix notation extensively; in fact, the only data type in MATLAB is a complex-valued matrix. Thus it handles scalars, vectors, real- and integer-valued matrices as special cases of complex matrices. The software can be used to execute a single statement or a list of statements, called a script file. MATLAB provides excellent graphing and programming capabilities. It can also be used to solve many types of problems symbolically. Simple computations can be done by entering an instruction, similar to what we do on a calculator, at the prompt. The symbols to be used for the basic arithmetic operations of addition, subtraction, multiplication, division, and exponentiation are $+$, $-$, $*$, $/$ and $\wedge$, respectively. In any expression, the computations are performed from left to right with exponentiation having the highest priority, followed by multiplication and division (with equal priority) and then the addition and subtraction (with equal priority). It uses the symbol *log* to denote the natural logarithm (ln). MATLAB uses double precision during computations but prints results on the screen using a shorter format. This default can be changed by using the *format* command.

F.1   Variables

Variable names in MATLAB should start with a letter and can have a length of up to 31 characters in any combination of letters, digits, and underscores. Upper- and lowercase letters are treated separately. As stated earlier, MATLAB treats all variables as matrices, although scalar quantities need not be given as arrays.

F.2   Arrays and Matrices

The name of a matrix must start with a letter and may be followed by any combination of letters or digits. The letters may be upper- or lowercase. Before performing arithmetic operations such as addition, subtraction, multiplication, and division on matrices, the matrices must be created using statements such as the following:

Row vector

$$\gg A = [1 \quad 2 \quad 3];$$

A row vector is treated as a 1-by-n matrix; its elements are enclosed in brackets and are separated by either spaces or commas. Note that the command-line prompt in the professional version of MATLAB is $\gg$ while it is EDU$\gg$ in the student edition of MATLAB. If a semicolon is not put at the end of the line, MATLAB displays the results of the line on the screen.

Column vector

$$\gg A = \begin{matrix} [1 \\ 2 \\ 3] \end{matrix} \quad \text{or} \quad A = [1; \; 2; \; 3] \quad \text{or} \quad A = [1 \quad 2 \quad 3]';$$

A column vector is treated as an n-by-1 matrix. Its elements can be entered in different lines or in a single line using a semicolon to separate them or in a single line using a row vector with a prime on the right bracket (to denote the transpose).

Matrix

To define the matrix

$$[A] = \begin{bmatrix} 1 & 2 & 3 \\ 4 & 5 & 6 \\ 7 & 8 & 6 \end{bmatrix}$$

the following specification can be used:

$$\gg A = \begin{matrix} [1 & 2 & 3 \\ 4 & 5 & 6 \\ 7 & 6 & 9] \end{matrix} \quad \text{or} \quad A = [1 \quad 2 \quad 3; \; 4 \quad 5 \quad 6; \; 7 \quad 8 \quad 9];$$

F.3   Arrays with Special Structure

In some cases, the special structure of an array is used to specify the array in a simpler manner. For example, $A = 1{:}10$ denotes a row vector

$$A = [1 \quad 2 \quad 3 \quad 4 \quad 5 \quad 6 \quad 7 \quad 8 \quad 9 \quad 10]$$

and $A = 2 : 0.5 : 4$ represents the row vector

$$A = [2.5 \quad 3.0 \quad 3.5 \quad 4.0]$$

F.4   Special Matrices

Some of the special matrices are identified as follows:

$\gg A =$ eye (3); implies an identity matrix of order 3,

$$A = \begin{bmatrix} 1 & 0 & 0 \\ 0 & 1 & 0 \\ 0 & 0 & 1 \end{bmatrix}$$

$\gg A =$ ones (3); implies a square matrix of order 3 with all elements equal to one,

$$A = \begin{bmatrix} 1 & 1 & 1 \\ 1 & 1 & 1 \\ 1 & 1 & 1 \end{bmatrix}$$

$\gg A =$ zeros $(2, 3)$; implies a 2×3 matrix with all elements equal to zero,

$$A = \begin{bmatrix} 0 & 0 & 0 \\ 0 & 0 & 0 \end{bmatrix}$$

F.5   Matrix Operations

To add the matrices $[A]$ and $[B]$ to get $[C]$, we use the statement

$$\gg C = A + B;$$

To solve a system of linear equations $[A]\vec{X} = \vec{B}$, we define the matrix A and the vector B and use the following statement:

$$\gg X = A \backslash B$$

F.6   Functions in MATLAB

MATLAB has a large number of built-in functions such as the following:

Square root of x: sqrt(x)

Sine of x: sin(x)

Logarithm of x to base 10: log10(x)

Gamma function of x: gamma(x)

To generate a new vector y having 11 values given by the function $y = e^{-2x} \cos x$ with $x = 0.0, 0.1, \ldots, 1.0$, we type the following:

$$\gg x = [0: 0.1: 1];$$

$$\gg y = \exp{(-2{}^*x)}.{}^*\cos(x).;$$

F.7   Complex Numbers

MATLAB considers complex number algebra automatically. The symbol i or j can be used to represent the imaginary part with no need of an asterisk between i or j and a number. For example, $a = 1 - 3i$ is a complex number with real and imaginary parts equal to 1 and -3, respectively. The magnitude and angle of a complex number can be found using the statements

```
>> a = 1 - 3i;
>> abs (a)
ans =

    . . .

>> angle (a)
ans =

    . . . (in radians)
```

F.8   *M*-files

MATLAB can be used in an interactive mode by typing each command from the keyboard. In this mode, MATLAB performs the operations much like an extended calculator. However, there are situations in which this mode of operation is inefficient. For example, if the same set of commands is to be repeated a number of times with different values of the input parameters, developing a MATLAB program will be quicker and more efficient.

A MATLAB program consists of a sequence of MATLAB instructions written outside MATLAB and then executed in MATLAB as a single block of commands. Such a program is called a *script file* or *M-file*. It is necessary to give a name to the script file. The name should end with **.m** (a dot (.) followed by the letter **m**). A typical **M**-file (called **fibo.m**) is given below:

```
file "fibo.m"

% m-file to compute Fibonacci numbers
```

```
f=[1 1];
i=1;
while f(i)+f(i+1)<1000
    f(i+2)=f(i)+f(i+1);
    i=i+1;
end
```

An *M*-file can also be used to write function subroutines. For example, the solution of a quadratic equation

$$Ax^2 + Bx + C = 0$$

can be determined using the following program:

```
% roots_quadra.m (Note: Line starting with % denotes a
comment line)
function [x1, x2] = roots_quadra(A, B, C)
% det = determinant
det = B^2 - 4 * A * C;
if (det < 0.0);
    x1 = (-B + j * sqrt(-det))/(2 * A);
    x2 = (-B - j * sqrt(-det))/(2 * A);
    disp('Roots are complex conjugates');
elseif (abs(det) < 1e-8); % det = 0.0
    x1 = -B / (2 * A);
    x2 = -B / (2 * A);
    disp('Roots are identical');
else (det > 0);
    x1 = (-B + sqrt(det))/(2 * A);
    x2 = (-B - sqrt(det))/(2 * A);
    disp('Roots are real and distinct');
end
```

The program **roots_quadra.m** can be used to find the roots of a quadratic with $A = 2$, $B = 2$, and $C = 1$, for example, as follows:

```
>> [x1,x2]=roots_quadra(2, 2, 1)
Roots are complex conjugates

x1 =

 -0.5000 + 0.5000i

x2 =

 -0.5000 - 0.5000i
```

F.9 Plotting of Graphs

To plot a graph in MATLAB, we define a vector of values of the independent variable x (array x) and a vector of values of the dependent variable y corresponding to the values of x (array y). Then the x-y graph can be plotted using the command:

```
plot (x,y)
```

As an example, the following commands can be used to plot the function $y = x^2 + 1$ in the range $0 \le x \le 3$:

```
x = 0 : 0.2 : 3;
y = x^2 + 1;
plot (x,y);
hold on
x1 = [0 3];
y1 = [0 0];
plot (x1,y1);
grid on
hold off
```

Note that the first two lines are used to generate the arrays x and y (using increments of 0.2 for x); the third line plots the graph (using straight lines between the indicated points); the next six lines permit the plotting of x and y axes along with the setting up of the grid (using the **grid on** command).

F.10   Roots of Nonlinear Equations

To find the roots of a nonlinear equation, the MATLAB function **fzero(y,x1)** can be used. Here **y** defines the nonlinear function and **x1** denotes the initial estimate (starting value) of the root. The roots of polynomials can be determined using the function **roots(p)** where **p** is a row vector of coefficients of the polynomial in descending order of the power of the variable.

```
>> f='tan (x)-tanh (x)'

f =

tan (x)-tanh (x)

>> root=fzero(f,1.0)

root =

      1.5708

>> roots([1 0 0 0 0 0 -2])

ans =

    -1.1225
    -0.5612 + 0.9721i
    -0.5612 - 0.9721i
     0.5612 + 0.9721i
     0.5612 - 0.9721i
     1.1225

>>
```

F.11   Solution of Linear Algebraic Equations

A set of simultaneous linear algebraic equations $[A]\vec{x} = \vec{b}$ can be solved using MATLAB in two different ways: Find $\vec{x}$ as $[A]^{-1}\vec{b}$ or find $\vec{x}$ directly as indicated by the following examples:

```
>> A=[4 -3 2; 2 3 1; 5 4 7]

A =

     4    -3     2
     2     3     1
     5     4     7

>> b=[16; -1; 18]

b =

    16
    -1
    18

>> C=inv(A)

C =

    0.2099    0.3580   -0.1111
   -0.1111    0.2222    0.0000
   -0.0864   -0.3827    0.2222

>> x=C*b

x =

    1.0000
   -2.0000
    3.0000

>> x=A\b

x =

    1.0000
   -2.0000
    3.0000

>>
```

F.12   Solution of Eigenvalue Problem

An algebraic eigenvalue problem is defined by $[A]\vec{X} = \lambda \vec{X}$, where $[A]$ is a square matrix of size $n \times n$, $\vec{X}$ is a column vector of size n, and λ is a scalar. For any given matrix $[A]$, the solution can be found using two types of commands. The use of the command **b = eig(A)** gives the eigenvalues of the matrix $[A]$ as elements of the vector d. The use of the command **[V,D] = eig(A)** gives the eigenvalues as diagonal elements of the matrix $[D]$ and the eigenvectors as corresponding columns of the matrix $[V]$. The following example illustrates the procedure:

```
>> A=[2 1 3 4; 1 -3 1 5; 3 1 6 -2; 4 5 -2 -1]

A =

     2     1     3     4
     1    -3     1     5
     3     1     6    -2
     4     5    -2    -1
```

```
>> b=eig(A)

b=

       7.9329
       5.6689
      -1.5732
      -8.0286

>> [V, d] = eig(A)

V =

    0.5601    0.3787    0.6880    0.2635
    0.2116    0.3624   -0.6241    0.6590
    0.7767   -0.5379   -0.2598   -0.1996
    0.1954    0.6602   -0.2638   -0.6756

d =

    7.9329         0         0         0
         0    5.6689         0         0
         0         0   -1.5732         0
         0         0         0   -8.0286

>>
```

F.13   Solution of Differential Equations

MATLAB has several functions or solvers, based on the use of Runge-Kutta methods, that can be used for the solution of a system of first-order ordinary differential equations. Note that an nth order ordinary differential equation is to be converted into a system of n first-order ordinary differential equations before using MATLAB functions. The MATLAB function **ode23** implements a combination of second- and third-order Runge-Kutta methods while the function **ode45** is based on a combination of fourth- and fifth-order Runge-Kutta methods. To solve a system of first-order differential equations $\dot{y} = f(t, y)$ using the MATLAB function **ode23**, the following command can be used:

```
>>[t,y] = ode('dfunc',tspan,y0)
```

where **'dfunc'** is the name of the function **m-file** whose input must be **t** and **y** and whose output must be a column vector denoting dy/dt, that is, **f(t,y)**. The number of rows in the column vector must be equal to the number of first-order equations. The vector **tspan** should contain the initial and final values of the independent variable **t,** and optionally, any intermediate values of **t** at which the solution is desired. The vector **y0** should contain the initial values of **y(t)**. Note that the function **m-file** should have two input arguments **t** and **y** even if the function **f(t,y)** does not involve **t**. A similar procedure can be used with the MATLAB function **ode45**.

As an example, consider the solution of the differential equation with $c = 0.1$ and $k = 10.0$

$$\frac{d^2y}{dt^2} + c\,\frac{dy}{dt} + k\,y = 0; \qquad y(0) = 0, \frac{dy}{dt}(0) = 0$$

This equation can be written as a set of two first-order differential equations by introducing

$$y_1 = y$$

and

$$y_2 = \frac{dy}{dt} = \frac{dy_1}{dt}$$

as

$$\frac{d\vec{y}}{dt} = \vec{f} = \left\{ \begin{matrix} f_1(t, \vec{y}) \\ f_2(t, \vec{y}) \end{matrix} \right\} = \left\{ \begin{matrix} y_2 \\ -c\, y_2 - k\, y_1 \end{matrix} \right\}$$

with

$$\vec{y}(0) = \left\{ \begin{matrix} 1 \\ 0 \end{matrix} \right\}$$

The following MATLAB program finds the solution of the above differential equations:

```
% ProbappendixF.m
tspan = [0: 0.05: 3];
y0 = [1; 0];
[t,y] = ode23 ('dfunc', tspan, y0);
[t y]
plot (t, y(:,1));
xlabel ('t');
ylabel ('y(1) and y(2)')
gtext ('y(1)');
hold on
plot (t,y (:,2));
gtext ('y(2)');

%dfunc.m
function f = dfunc(t,y)
f = zeros (2,1);

f(1) = y(2);
f(2) = -0.1 * y(2) - 10.0 * y(1);

>> ProbappendixF

ans =

          0    1.0000         0
     0.0500    0.9875   -0.4967
     0.1000    0.9505   -0.9785
     0.1500    0.8901   -1.4335
     0.2000    0.8077   -1.8505
     0.2500    0.7056   -2.2191
     0.3000    0.5866   -2.5308
     0.3500    0.4534   -2.7775
     0.4000    0.3098   -2.9540
     0.4500    0.1592   -3.0561
     0.5000    0.0054   -3.0818
```

```
0.5500   -0.1477   -3.0308
  .
  .
  .
2.7500   -0.6380   -1.8279
2.8000   -0.7207   -1.4788
2.8500   -0.7851   -1.0949
2.9000   -0.8296   -0.6858
2.9500   -0.8533   -0.2617
3.0000   -0.8556    0.1667
```

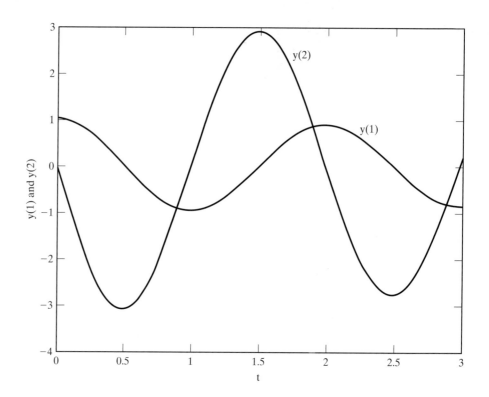

Answers to Selected Problems

Chapter 1

1.7 $k_{eq} = \dfrac{k_2\,k_3\,k_4\,k_5 + 2\,k_1\,k_3\,k_4\,k_5 + k_1\,k_2\,k_4\,k_5 + 2\,k_1\,k_2\,k_3\,k_5}{(k_2\,k_3\,k_4 + k_2\,k_3\,k_5 + 2\,k_1\,k_3\,k_4 + 2\,k_1\,k_3\,k_5 + k_1\,k_2\,k_4 + k_1\,k_2\,k_5 + 2\,k_1\,k_2\,k_3)}$

1.10 **(a)** $k = 37.08 \times 10^7$ N/m, **(b)** $k = 12.36 \times 10^7$ N/m, **(c)** $k = 4.12 \times 10^7$ N/m

1.12 $k_{eq} = 3\,k\cos^2\alpha$ **1.14** $l_{eq} = \dfrac{4t\,(d+t)}{D\,d}$ **1.16** $k = \dfrac{p\gamma A^2}{v}$

1.19 $F(x) = (32000\,x - 80)$ N   **1.22** $k_{eq} = \dfrac{1}{l}\left(E_s\,A_s + E_a\,A_a\right)$

1.24 **(a)** $k_{t_{eq}} = 5.54811 \times 10^6$ N-m/rad, **(b)** $k_{t_{eq}} = 5.59597 \times 10^6$ N-m/rad

1.26 **(a)** $k_{eq} = 89.931$ lb/in, **(b)** $k_{eq} = 3.0124$ lb/in

1.28 $k_{axial} = 16{,}681.896$ lb/in; $k_{torsion} = 139.1652$ lb-in/rad

1.30 $m_{eq} = m_1\left(\dfrac{a}{b}\right)^2 + m_2 + J_0\left(\dfrac{1}{b^2}\right)$

1.33 $m_{eq} = m_h + \dfrac{J_b}{l_3^2} + J_c\left[\dfrac{l_2}{l_3\,r_c}\right]^2$

1.35 (a) $c_{eq} = c_1 + c_2 + c_3$,   (c) $c_{eq} = c_1 + c_2\left(\dfrac{l_2}{l_1}\right)^2 + c_3\left(\dfrac{l_3}{l_1}\right)^2$,

(b) $\dfrac{1}{c_{eq}} = \dfrac{1}{c_1} + \dfrac{1}{c_2} + \dfrac{1}{c_3}$,   (d) $c_{teq} = c_{t1} + c_{t2}\left(\dfrac{n_1}{n_2}\right)^2 + c_{t_3}\left(\dfrac{n_1}{n_3}\right)^2$

1.38 $c_t = \dfrac{\pi \mu D^2(l - h)}{2d} + \dfrac{\pi \mu D^3}{32h}$ **1.41** $c = 4205.64$ N-s/m   **1.44** $A = 4.4721, \theta = -26.5651°$

1.46 $z = 11.1803\, e^{0.1798\,i}$ **1.49** $X = 9.8082$ mm, $Y = 9.4918$ mm, $\phi = 39.2072°$

1.53 $x_2(t) = 6.1966 \sin(\omega t + 83.7938°)$ **1.55** Not harmonic

1.58 $X = 2.5$ mm, $\omega = 5.9092$ rad/sec, $\omega + \delta\omega = 6.6572$ rad/sec

1.60 $A = 0.5522$ mm, $\dot{x}_{max} = 52.04$ mm/sec   **1.62** $x_{rms} = X/\sqrt{2}$

1.66 $x(t) = \dfrac{A}{\pi} + \dfrac{A}{2}\sin \omega t - \dfrac{2A}{\pi}\displaystyle\sum_{n=2,4,6,\ldots}^{\infty}\dfrac{\cos n\omega t}{(n^2 - 1)}$ **1.68** $x(t) = \dfrac{8A}{\pi^2}\displaystyle\sum_{n=1,3,5,\ldots}^{\infty}(-1)^{\frac{n-1}{2}}\dfrac{\sin n\omega t}{n^2}$

1.72 $p(t) = \dfrac{a_0}{2} + \displaystyle\sum_{m=1}^{\infty}[a_m \cos m\omega t + b_m \sin m\omega t]$ lb/in^2 where $a_0 = 50$, $a_1 = 31.8309$, $a_2 = 0$,

$a_3 = -10.6103$, $b_1 = 31.8309$, $b_2 = 31.8309$, $b_3 = 10.6103$

1.75 $a_0 = 19.92$, $a_1 = -20.16$, $a_2 = 3.31$, $a_3 = 3.77$; $b_1 = 23.52$, $b_2 = 12.26$, $b_3 = -0.41$

1.88 $a_0 = -0.38$, $a_1 = -0.62$, $a_2 = 0.46$, $a_3 = 0.41$; $b_1 = -0.35$, $b_2 = 0.92$, $b_3 = -0.17$

Chapter 2

2.2 (a) 0.1715 sec,   (b) 0.2970 sec   **2.4** 0.0993 sec

2.6 (a) $A = 0.03183$ m,   (c) $\ddot{x}_{max} = 0.31415$ m/s^2,   **2.8** $\omega_n = 22.1472$ rad/sec

(b) $\dot{x}_0 = 0.07779$ m/s,   (d) $\phi_0 = 51.0724°$ **2.10** $\omega_n = 4.8148$ rad/sec

2.13 $\omega_n = [k/(4m)]^{1/2}$ **2.15** (a) $\omega_n = \sqrt{\dfrac{4k}{M}}$,   (b) $\omega_n = \sqrt{\dfrac{4k}{m + M}}$

2.17 $\omega_n = \sqrt{\dfrac{g}{W}\left(\dfrac{3E_1 I_1}{l_1^3} + \dfrac{48 E_2 I_2}{l_2^3}\right)}$ **2.19** $k = 52.6381$ N/m, $m = 1/3$ kg

2.21 (a) $\omega_n = \sqrt{\dfrac{k\,g\,\cosec^2\theta}{W}}$,   (b) $\omega_n = \sqrt{\dfrac{k\,g}{W}}$ **2.23** (a) $\omega_n = \sqrt{\dfrac{k}{2m}}$,   (b) $\omega_n = \sqrt{\dfrac{8m}{b^2}\left[l^2 - \dfrac{b^2}{4}\right]}$

2.26 (a) $m\ddot{x} + \left(\dfrac{1}{a} + \dfrac{1}{b}\right)Tx = 0$,   (b) $\omega_n = \sqrt{\dfrac{T(a + b)}{mab}}$ **2.28** $T = 1656.3147$ lb

2.30 (a) $N = 81.914$ rpm,   (b) $\omega_n = 37.5851$ rad/sec   **2.32** $\omega_n = \sqrt{\dfrac{2g}{L}}$

2.34 $A = 0.9536 \times 10^{-4}$ m^2 **2.37** Torsion about z-axis   **2.39** $\omega_n = 2578.9157$ rad/sec

2.42 $\mu = \sqrt{\left(\dfrac{\omega^2 Wc - 2kgc}{Wg + Wa\omega^2 - 2kga}\right)}$ **2.44** $m\ddot{x} + (k_1 + k_2)x = 0$

2.47 $\left(m + \dfrac{J_0}{r^2}\right)\ddot{x} + 16kx = 0$ **2.49** $\omega_n = 359.6872$ rad/sec

2.51 $x(t) = 0.1 \cos 15.8114\, t + 0.3162 \sin 15.8114\, t$ m   **2.53** $x_0 = 0.007864$ m; $\dot{x}_0 = -0.013933$ m/s

2.55 $\dot{x}_0 = 4$ m/s   **2.57** $d = 0.1291$ in, $N = 29.58$ **2.59** $\omega_n = 2$ rad/s, $l = 2.4525$ m

2.61 $\tau_n = 1.4185$ sec   **2.63** $\omega_n = 13.4841$ rad/sec   **2.65** $\tau_n = 0.04693$ sec

2.67 $\omega_n = 17.7902$ rad/sec   **2.69** $\omega_n = \left\{ \dfrac{(k_1 + k_2)(R + a)^2}{1.5 m R^2} \right\}^{1/2}$

2.71 $\frac{1}{3} m l^2 \ddot{\theta} + (k_t + k_1 a^2 + k_2 l^2)\theta = 0$ **2.74** $m_{eff} = \dfrac{17}{35} m$ **2.76** $\omega_n = \sqrt{\dfrac{k}{4m}}$

2.79 45.1547 rad/sec   **2.81** $\omega_n = \sqrt{\dfrac{\rho_0 g}{\rho_w h}}$ **2.83** $\omega_n = \sqrt{\dfrac{16 k r^2}{m r^2 + J_0}}$

2.85 (a) 14265.362,   (b) 3.8296 **2.87** $x_{max} = \left(x_0 + \dfrac{\dot{x}_0}{\omega_n} \right) e^{-(\dot{x}_0/(\dot{x}_0 + \omega_n x_0))}$

2.90 (a) $c_c = 1000$ N-s/m,   (b) $\omega_d = 8.6603$ rad/sec,   (c) $\delta = 3.6276$ **2.92** $\theta = 0.09541°$

2.94 $\zeta = 0.013847$ **2.96** $m = 500$ kg, $k = 27066.403$ N/m   **2.99** $\omega_n = \sqrt{\dfrac{2k}{3m}}$

2.103 $\dfrac{3}{2} m \ddot{x} + c \dot{x} + 2 k x = 0$ **2.105** $\rho_0 = 2682.8816$ kg/m^3

2.107 (a) $J_0 = 1.9436 \times 10^{-4}$ N-m-s^2,   (b) $\tau_n = 1.8297$ sec,

   (c) $c_t = 5.3887 \times 10^{-4}$ N-m-s/rad, (d) $k_t = 2.2917 \times 10^{-3}$ N-m/rad

2.108 (a) $\zeta = 0.75, \omega_d = 6.6144$ rad/s,   (b) $\zeta = 1.0, \omega_d = 0$,   (c) $\zeta = 1.25$

2.110 (a) 60.8368 J,   (b) 124.6784 J   **2.111** Coulomb, 5N, 14.1421 rad/sec   **2.113** 5.8 mm

2.115 (a) 5   (b) 0.7025 sec   (c) 1.9620 cm   **2.117** $c_{eq} = \dfrac{4\,\mu N}{\pi \omega X}$ **2.120** 1.40497 s

2.122 1.7022 sec, 0.004 m   **2.124** $\beta = 0.03032, c_{eq} = 0.04288$ N-s/m, $\Delta W = 19.05 \times 10^{-6}$ N-m

2.126 $h = 0.583327$ N/m

Chapter 3

3.2 5 sec

3.4 (a) $x(t) = 0.1 \cos 20\, t + t \sin 20\, t$ **3.6** (a) $x(t) = 0.18 \cos 20\, t - 0.08 \cos 30\, t$,

   (b) $x(t) = (0.5 + t)\sin 20\, t$ (b) $x(t) = 0.08 \cos 20\, t + 0.5 \sin 20\, t - 0.08 \cos 30\, t$,

   (c) $x(t) = 0.1 \cos 20\, t + (0.5 + t)\sin 20\, t$ (c) $x(t) = 0.18 \cos 20\, t + 0.5 \sin 20\, t - 0.08 \cos 30\, t$

3.8 9.1189 kg   **3.11** $X = \left| \dfrac{m r l^3 N^2}{22.7973\, E b a^3 - 0.2357\, \rho a b l^4 N^2} \right|$ **3.13** $\omega = 743.7442$ Hz

3.17 0.676 sec   **3.19** $\theta_p(t) = \Theta \sin \omega t$ with $\Theta = -8.5718 \times 10^{-4}$ rad and $\omega = 104.72$ rad/sec

3.21 $x_p(t) = 0.06610 \cos(10\, t - 0.1325)$ m

   $x_{total}(t) = 0.0345\, e^{-2t} \cos(19.8997\, t + 0.0267) + 0.0661 \cos(10\, t - 0.1325)$ m

3.23 $x_p(t) = 0.25 \cos\left(20\, t - \dfrac{\pi}{2}\right)$ m

   $x_{total}(t) = 0.2611\, e^{-2t} \cos(19.8997\, t + 1.1778) + 0.25 \cos\left(20\, t - \dfrac{\pi}{2}\right)$ m

3.25 $k = 6.6673 \times 10^4$ lb/in, c $= 2.3983$ lb-sec/in

3.27 $r = \sqrt{1 - 2\zeta^2}$, $X_{max} = \dfrac{\delta_{st}}{2\zeta\sqrt{1 - \zeta^2}}$

3.29 $\zeta = 0.1180$ **3.32**   **(a)** 64.16 rad/sec,   **(b)** 967.2 N-m

3.34   **(a)** $\zeta = 0.25$,   **(b)** $\omega_1 = 22.2145$ rad/sec, $\omega_2 = 38.4766$ rad/sec   **3.36** 169.5294×10^{-6} m

3.38 $k = 1.0070 \times 10^5$ N/m, c $= 633.4038$ N-s/m   **3.42** $0.3339 \sin 25\, t$ mm

3.44 $X = 0.106$ m, s $= 246.73$ km/hr   **3.46** $c = (k - m\omega^2)/\omega$

3.48 $\theta(t) = 0.01311 \sin(10\, t - 0.5779)$ rad   **3.51** $x_p(t) = 110.9960 \times 10^{-6} \sin(314.16\, t + 0.07072)$ m

3.54 0.4145×10^{-3} m, 1.0400×10^{-3} m   **3.56**   1.4195 N-m   **3.58** $\zeta = 0.1364$

3.61   Maximum force $= 26.68$ lb   **3.64** $\mu = 0.1$ **3.67**   **(a)** 10.2027 lb/in,   **(b)** 40.8108 lb-in

3.70   **(c)** $\dfrac{1}{\left\{\dfrac{4\,\mu N}{\pi X k} + \dfrac{3}{4k}c\omega^3\, X^2\right\}}$ **3.73**   **(a)** 1.0623 Hz,   **(b)** 1.2646 m/s,   **(c)** 5.557×10^{-4} m

Chapter 4

4.2 $x(t) = \dfrac{F_0}{2k} - \dfrac{4F_0}{\pi^2 k} \displaystyle\sum_{n=1,3,\ldots}^{\infty} \dfrac{1}{n^2}\dfrac{1}{\sqrt{(1 - r^2 n^2)^2 + (2\zeta n r)^2}} \cos(n\,\omega t - \phi_n)$

with $r = \omega/\omega_n$ and $\phi_n = \tan^{-1}\left(\dfrac{2\zeta n r}{1 - n^2 r^2}\right)$

4.6 $\theta(t) = 0.0023873 + \displaystyle\sum_{n=1}^{\infty}\left\{\dfrac{318.3091 \sin 5.8905\, n \cos n\,\omega t + 318.3091\, (1 - \cos 5.8905\, n) \sin n\,\omega t}{n(392700.0 - 1096.6278\, n^2)}\right\}$ rad

4.12 $x_p(t) = 6.6389 \times 10^{-4} - 13.7821 \times 10^{-4}\cos(10.472t - 0.0172)$

$+\ 15.7965 \times 10^{-4}\sin(10.472t - 0.0172) +\ \ldots$ m

4.15 $x(t) = \dfrac{F_0}{k}\left\{1 + \dfrac{\sin \omega_n(t - t_0) - \sin \omega_n t}{\omega_n t_0}\right\}$;   for $t \geq t_0$

4.19 $x(t) = \dfrac{F_0}{2k\left(1 - \dfrac{\omega^2}{\omega_n^2}\right)}\left[2 - \dfrac{\omega^2}{\omega_n^2}\left(1 - \cos\dfrac{\omega_n \pi}{\omega}\right)\right] + \dfrac{F_0}{k}\left[1 - \cos \omega_n\left(t - \dfrac{\pi}{\omega}\right)\right]$ for $t > \pi/\omega$.

4.25 $x(t) = 1.7689 \sin 6.2832\,(t - 0.018) - 0.8845 \sin 6.2832t$

$-\ 0.8845 \sin 6.2832\,(t - 0.036)$m; for $t > 0.036$ sec

4.29 $x_p(t) = 0.002667$ m

4.32 $\theta(t) = 0.3094\, e^{-t} + 0.05717 \sin 5.4127\, t - 0.3094 \cos 5.4127\, t$ rad

4.35 $x(t) = 0.04048 e^{-t} + 0.01266 \sin 3.198\, t - 0.04048 \cos 3.198\, t$ m

4.37 $x(t) = 0.5164\, e^{-t} \sin 3.8729\, t$ m

4.42 $x_m = \dfrac{F_0}{k\omega_n t_0}[(1 - \cos \omega_n t_0)^2 + (\omega_n t_0 - \sin \omega_n t_0)^2]^{1/2}$; for $t > t_0$

4.45 $d = 0.6$ in   **4.48** $k = 12771.2870$ lb/in

4.51 $x(t) = \begin{cases} \dfrac{F_0}{m\omega_n^2}(1 - \cos\omega_n t); & 0 \le t \le t_0 \\[3mm] \dfrac{F_0}{m\omega_n^2}[\cos\omega_n(t - t_0) - \cos\omega_n t]; & t \ge t_0 \end{cases}$

4.54 $\dot{x}_i(t_i = \pi) = \begin{cases} -0.549289, & \text{Eq. (4.68)} \\ -0.551730, & \text{Eq. (4.71)} \end{cases}$

Chapter 5

5.1 $\omega_1 = 3.6603$ rad/sec, $\omega_2 = 13.6603$ rad/sec   **5.3** $\omega_1 = \sqrt{\dfrac{k}{m}}, \omega_2 = \sqrt{\dfrac{2k}{m}}$

5.5 1.1 in^2 **5.6** $\omega_1 = 7.3892$ rad/s, $\omega_2 = 58.2701$ rad/s

5.7 $\omega_{1,2}^2 = \dfrac{48}{7}\dfrac{EI}{m_1 m_2}\left[(m_1 + 8 m_2) \mp \sqrt{(m_1 - 8 m_2)^2 + 25\, m_1 m_2}\right]$

5.9 $\omega_1 = 0.7654\sqrt{\dfrac{g}{l}}, \omega_2 = 1.8478\sqrt{\dfrac{g}{l}}$ **5.12** $\omega_1 = 12.8817$ rad/sec, $\omega_2 = 30.5624$ rad/sec

5.15 $x_1(t) = 0.1046 \sin 40.4225t + 0.2719 \sin 58.0175t$,
$x_2(t) = 0.1429 \sin 40.4225t - 0.09952 \sin 58.0175t$

5.17 $\omega_1 = 3.7495\sqrt{\dfrac{EI}{mh^3}}, \omega_2 = 9.0524\sqrt{\dfrac{EI}{mh^3}}$ **5.19** $\vec{X}^{(1)} = \begin{Bmatrix} 1.0 \\ 2.3029 \end{Bmatrix}, \vec{X}^{(2)} = \begin{Bmatrix} 1.0 \\ -1.3028 \end{Bmatrix}$

5.21 $x_2(0) = r_1 x_1(0) = \dfrac{x_1(0)}{\sqrt{3} - 1}, \dot{x}_2(0) = r_1 \dot{x}_1(0) = \dfrac{\dot{x}_1(0)}{\sqrt{3} - 1}$

5.25 $x_1(t) = 0.5 \cos 2t + 0.5 \cos\sqrt{12}\,t; x_2(t) = 0.5 \cos 2t - 0.5 \cos\sqrt{12}\,t$

5.28 $\omega_1 = 0.5176\sqrt{k_t/J_0}, \omega_2 = 1.9319\sqrt{k_t/J_0}$ **5.31** $\omega_1 = 0.38197\sqrt{k_t/J_0}, \omega_2 = 2.61803\sqrt{k_t/J_0}$

5.33 Frequency equation:

$$\omega^4 (m_1 m_2 l_1^2 l_2^2) - \omega^2 \left\{ m_2 l_2^2 (W_1 l_1 + k l_1^2) + m_1 l_1^2 (W_2 l_2 + k l_2^2) \right\}$$

$$+ (W_1 l_1 W_2 l_2 + W_2 l_2 k l_1^2 + W_1 l_1 k l_2^2) = 0$$

5.35 $\omega_{1,2}^2 = \left\{ \dfrac{(J_0 k + mk_t) \pm \sqrt{(J_0 k + mk_t)^2 - 4(J_0 - me^2)mkk_t}}{2m(J_0 - me^2)} \right\}$

5.38 $1000\,\ddot{x} + 40000\,x + 15000\,\theta = 900 \sin 8.7267t$
$\qquad + 1100 \sin(8.7267\,t - 1.5708)$
$\qquad 810\,\ddot{\theta} + 15000\,x + 67500\,\theta = 1650 \sin(8.7267\,t - 1.5708) - 900 \sin 8.7267\,t$

5.41 (a) $\begin{bmatrix} m & 0 \\ 0 & J_0 \end{bmatrix}\begin{Bmatrix} \ddot{x} \\ \ddot{\theta} \end{Bmatrix} + \begin{bmatrix} 3k & kl/6 \\ kl/6 & 17kl^2/36 \end{bmatrix}\begin{Bmatrix} x \\ \theta \end{Bmatrix} = \begin{Bmatrix} F(t) \\ l\,F(t)/3 \end{Bmatrix}$ where $J_0 = ml^2/12$ and $F(t) = F_0 \sin \omega t$,

(b) Static coupling

5.45 (a) $\omega_1 = 12.2474$ rad/sec, $\omega_2 = 38.7298$ rad/sec

5.48 $x_j(t) = X_j e^{i\omega t}$

with $X_1 = (-40.0042 - 0.01919\,i) \times 10^{-4}$ in,

$X_2 = (0.9221 + 0.2948\,i) \times 10^{-4}$ in

5.49 $k_2 = m_2 \omega^2$ **5.50** $x_2(t) = \left\{ \dfrac{k_2 F_0}{(-m_1 \omega^2 + k_1 + k_2)(-m_2 \omega^2 + k_2) - k_2^2} \right\} \sin \omega t$

5.52 $x_1(t) = (17.2915\,F_0 \cos \omega t + 6.9444\,F_0 \sin \omega t)10^{-4}$

$x_2(t) = (17.3165\,F_0 \cos \omega t + 6.9684\,F_0 \sin \omega t)10^{-4}$

5.54 $x_1(t) = 0.009773 \sin 4\pi t$ m, $x_2(t) = 0.016148 \sin 4\pi t$ m

5.56 $x_2(t) = (\frac{1}{60} - \frac{1}{40} \cos 10t + \frac{1}{120} \cos 10\sqrt{3}t)u(t)$

5.58 $\omega_1 = 0, \omega_2 = \sqrt{\dfrac{4k}{3m}}$ **5.59** $b_1 c_2 - c_1 b_2 = 0$

5.61 $\ddot{\alpha} + \left(\dfrac{k_t}{J_1} + \dfrac{k_t}{J_2} \right)\alpha = 0$ where $\alpha = \theta_1 - \theta_2$ **5.63** $\omega_1 = 0, \omega_2 = \sqrt{\dfrac{6k(m+M)}{mM}}$

5.66 $k \geq \dfrac{mg}{2l}$

Chapter 6

6.1 $\begin{bmatrix} m_1 & 0 & 0 \\ 0 & m_2 & 0 \\ 0 & 0 & m_3 \end{bmatrix} \begin{Bmatrix} \ddot{x}_1 \\ \ddot{x}_2 \\ \ddot{x}_3 \end{Bmatrix} + k \begin{bmatrix} 7 & -1 & -5 \\ -1 & 2 & -1 \\ -5 & -1 & 7 \end{bmatrix} \begin{Bmatrix} x_1 \\ x_2 \\ x_3 \end{Bmatrix} = \begin{Bmatrix} F_1(t) \\ F_2(t) \\ F_3(t) \end{Bmatrix}$

6.3 $\dfrac{m}{3} \begin{bmatrix} 1 & 0 & 2 \\ 2 & 0 & 1 \\ 0 & 15 & 0 \end{bmatrix} \begin{Bmatrix} \ddot{x}_1 \\ \ddot{x}_2 \\ \ddot{x}_3 \end{Bmatrix} + \dfrac{c}{25} \begin{bmatrix} 6 & -10 & 4 \\ 9 & -15 & 6 \\ -15 & 25 & -10 \end{bmatrix} \begin{Bmatrix} \dot{x}_1 \\ \dot{x}_2 \\ \dot{x}_3 \end{Bmatrix}$

$\qquad + \dfrac{k}{25} \begin{bmatrix} 6 & -10 & 29 \\ 34 & -15 & 6 \\ -15 & 25 & -10 \end{bmatrix} \begin{Bmatrix} x_1 \\ x_2 \\ x_3 \end{Bmatrix} = \begin{Bmatrix} F_3(t) \\ F_1(t) \\ F_2(t) \end{Bmatrix}$

6.5 $I_1 \ddot{\theta}_1 + k_{t_1}(\theta_1 - \theta_2) = M_1 \cos \omega t$

$\left(I_2 + I_3 \dfrac{n_2^2}{n_3^2} \right)\ddot{\theta}_2 + k_{t_1}(\theta_2 - \theta_1) + k_{t_2}\dfrac{n_2}{n_3}\left(\theta_2 \dfrac{n_2}{n_3} - \theta_3 \right) = 0$

$\left(I_4 + I_5 \dfrac{n_4^2}{n_5^2} \right)\ddot{\theta}_3 + k_{t_2}(\theta_3 - \theta_2 \dfrac{n_2}{n_3}) + k_{t_3}\dfrac{n_4}{n_5}\left(\theta_3 \dfrac{n_4}{n_5} - \theta_4 \right) = 0$

$I_6 \ddot{\theta}_4 + k_{t_3}(\theta_4 - \theta_3 \dfrac{n_4}{n_5}) = 0$

6.7 $k \begin{bmatrix} 7 & -1 & -5 \\ -1 & 2 & -1 \\ -5 & -1 & 7 \end{bmatrix}$ **6.9** $\dfrac{k}{25} \begin{bmatrix} 34 & -15 & 6 \\ -15 & 25 & -10 \\ 6 & -10 & 29 \end{bmatrix}$

6.11
$$\begin{bmatrix} k_{t_1} & -k_{t_1} & 0 & 0 \\ -k_{t_1} & k_{t_1} + k_{t_2}\left(\dfrac{n_2}{n_3}\right)^2 & -k_{t_2}\left(\dfrac{n_2}{n_3}\right) & 0 \\ 0 & -k_{t_2}\left(\dfrac{n_2}{n_3}\right) & k_{t_2} + k_{t_3}\left(\dfrac{n_4}{n_5}\right)^2 & -k_{t_3}\left(\dfrac{n_4}{n_5}\right) \\ 0 & 0 & -k_{t_3}\left(\dfrac{n_4}{n_5}\right) & k_{t_3} \end{bmatrix}$$

6.13 $\begin{bmatrix} \dfrac{k_1 + k_2}{k_1 k_2} & \dfrac{1}{k_1 r} \\ \dfrac{1}{k_1 r} & \dfrac{1}{k_1 r^2} \end{bmatrix}$ **6.15** $\begin{bmatrix} \dfrac{2}{3k} & -\dfrac{1}{3 kl} \\ -\dfrac{1}{3 kl} & \dfrac{2}{3 kl^2} \end{bmatrix}$ **6.17** $\begin{bmatrix} m & 0 \\ 0 & 4ml^2 \end{bmatrix}$

6.19 $[k] = \begin{bmatrix} (k_1 + k_2) & -k_2 & 0 \\ -k_2 & (k_2 + k_3) & -k_3 \\ 0 & -k_3 & (k_3 + k_4) \end{bmatrix}$ **6.21** $[a] = \dfrac{l^3}{EI}\begin{bmatrix} 9/64 & 1/6 & 13/192 \\ 1/6 & 1/3 & 1/6 \\ 13/192 & 1/6 & 9/64 \end{bmatrix}$

6.25 $2\,k$ **6.27** $\begin{bmatrix} m_1 & 0 & 0 \\ 0 & m_2 & 0 \\ 0 & 0 & m_3 \end{bmatrix}$ **6.29** $\dfrac{m}{3}\begin{bmatrix} 2 & 0 & 1 \\ 0 & 15 & 0 \\ 1 & 0 & 2 \end{bmatrix}$

6.31 $\begin{bmatrix} I_1 & 0 & 0 & 0 \\ 0 & I_2 + I_3\left(\dfrac{n_2}{n_3}\right)^2 & 0 & 0 \\ 0 & 0 & I_4 + I_5\left(\dfrac{n_4}{n_5}\right)^2 & 0 \\ 0 & 0 & 0 & I_6 \end{bmatrix}$

6.34 $2\,m\ddot{x} + kx = 0, \; l\,\ddot{\theta} + g\,\theta = 0$

6.36 $m_1\,\ddot{x}_1 + (k_1 + k_2)x_1 - k_2\,x_2 = 0$
$m_2\,\ddot{x}_2 - k_2\,x_1 + (k_2 + k_3)\,x_2 - k_3\,x_3 = 0$
$m_3\,\ddot{x}_3 - k_3\,x_2 + (k_3 + k_4)\,x_3 = 0$

6.39 $m_1\,\ddot{x}_1 + 7\,k\,x_1 - k\,x_2 - 5\,k\,x_3 = F_1(t)$
$m_2\,\ddot{x}_2 - k\,x_1 + 2\,k\,x_2 - k\,x_3 = F_2(t)$
$m_3\,\ddot{x}_3 - 5\,k\,x_1 - k\,x_2 + 7\,k\,x_3 = F_3(t)$

6.42 $\left(M + \dfrac{J_0}{9\,r^2}\right)\ddot{x}_1 - \dfrac{J_0}{9\,r^2}\ddot{x}_2 + \dfrac{41}{9}\,k\,x_1 - \dfrac{8}{9}\,k\,x_2 - \dfrac{8}{3}\,k\,x_3 = F_1(t)$

$-\dfrac{J_0}{9\,r^2}\ddot{x}_1 + \left(3\,m + \dfrac{J_0}{9\,r^2}\right)\ddot{x}_2 - \dfrac{8}{9}\,k\,x_1 + \dfrac{2}{9}\,k\,x_2 + \dfrac{2}{3}\,k\,x_3 = F_2(t)$

$m\,\ddot{x}_3 - \dfrac{8}{3}\,k\,x_1 + \dfrac{2}{3}\,k\,x_2 + 5\,k\,x_3 = F_3(t)$

6.44 $\omega_1 = 0.44504\sqrt{k/m}, \; \omega_2 = 1.2471\sqrt{k/m}, \; \omega_3 = 1.8025\sqrt{k/m}$

6.47 $\omega_1 = 0.533399\sqrt{k/m}, \; \omega_2 = 1.122733\sqrt{k/m}, \; \omega_3 = 1.669817\sqrt{k/m}$

6.50 $\lambda_1 = 2.21398, \; \lambda_2 = 4.16929, \; \lambda_3 = 10.6168$

6.53 $\omega_1 = 0.644798\sqrt{g/l}, \; \omega_2 = 1.514698\sqrt{g/l}, \; \omega_3 = 2.507977\sqrt{g/l}$

6.56 $\omega_1 = 0.562587\sqrt{\dfrac{P}{ml}}, \; \omega_2 = 0.915797\sqrt{\dfrac{P}{ml}}, \; \omega_3 = 1.584767\sqrt{\dfrac{P}{ml}}$

6.59 $[X] = \dfrac{1}{2}\begin{Bmatrix} 1 & 1 & 0 \\ -1 & 1 & \sqrt{2/3} \\ 1 & 1 & \sqrt{8/3} \end{Bmatrix}$ **6.62** $\omega_1 = 0.7653\sqrt{\dfrac{k}{m}},\ \omega_2 = 1.8478\sqrt{\dfrac{k}{m}},\ \omega_3 = 3.4641\sqrt{\dfrac{k}{m}}$

6.64 $\omega_1 = 0,\ \omega_2 = 0.752158\sqrt{k/m},\ \omega_3 = 1.329508\sqrt{k/m}$

6.66 $x_3(t) = x_{10}\left\{0.5\cos 0.4821\sqrt{\dfrac{k}{m}}\,t - 0.3838\cos\sqrt{\dfrac{k}{m}}\,t \right.$

$\left. + 0.8838\cos 1.1976\sqrt{\dfrac{k}{m}}\,t\right\}$

6.68 $x_3(t) = x_{20}\left\{0.1987\cos 0.5626\sqrt{\dfrac{P}{lm}}\,t - 0.06157\cos 0.9158\sqrt{\dfrac{P}{lm}}\,t \right.$

$\left. - 0.1372\cos 1.5848\sqrt{\dfrac{P}{lm}}\,t\right\}$

6.71 $x_1(t) = \dot{x}_0\left\{\dfrac{t}{3} + \sqrt{\dfrac{m}{4k}}\sin\sqrt{\dfrac{k}{m}}\,t + \sqrt{\dfrac{m}{108k}}\sin\sqrt{\dfrac{3k}{m}}\,t\right\}$

6.73 $x_1(t) = \dfrac{1}{2}\left[\cos 2t + \dfrac{1}{2}\sin 2t + \cos\sqrt{12}\,t - \dfrac{1}{\sqrt{12}}\sin\sqrt{12}\,t\right]$

$x_2(t) = \dfrac{1}{2}\left[\cos 2t + \dfrac{1}{2}\sin 2t - \cos\sqrt{12}\,t + \dfrac{1}{\sqrt{12}}\sin\sqrt{12}\,t\right]$

6.75 **(a)** $\omega_1 = 0.44497\sqrt{k_t/J_0},\ \omega_2 = 1.24700\sqrt{k_t/J_0},\ \omega_3 = 1.80194\sqrt{k_t/J_0}$

(b) $\vec{\theta}(t) = \begin{Bmatrix} -0.0000025 \\ 0.0005190 \\ -0.0505115 \end{Bmatrix}\cos 100t$ radians   **6.77** $\vec{x}(t) = \begin{Bmatrix} 5.93225 \\ 10.28431 \\ 12.58863 \end{Bmatrix}\dfrac{F_0}{k}\cos\omega t$

6.80 $\vec{x}(t) = \begin{Bmatrix} 0.03944\,(1 - \cos 18.3013\,t) + 0.01057\,(1 - \cos 68.3015\,t) \\ 0.05387\,(1 - \cos 18.3013\,t) - 0.00387\,(1 - \cos 68.3015\,t) \end{Bmatrix}$

6.83 $x_3(t) = 0.0256357\cos(\omega t + 0.5874^{\circ})$ m

Chapter 7

7.1 **(a)** $\omega_1 \simeq 2.6917\sqrt{\dfrac{EI}{ml^3}},$ **(b)** $\omega_1 \simeq 2.7994\sqrt{\dfrac{EI}{ml^3}}$ **7.3** $3.5987\sqrt{\dfrac{EI}{ml^3}}$

7.5 $0.3015\sqrt{k/m}$ **7.7** $0.4082\sqrt{k/m}$ **7.9** $1.0954\sqrt{\dfrac{T}{lm}}$

7.13 $\omega_1 = 0,\ \omega_2 \simeq 6.2220$ rad/s, $\omega_3 \simeq 25.7156$ rad/s   **7.16** $\omega_1 = \sqrt{k/m}$

7.18 $\omega_1 = 0.3104,\ \omega_2 = 0.4472,\ \omega_3 = 0.6869$ where $\omega_i = 1/\sqrt{\lambda_i}$

7.21 $\tilde{\omega}_1 = 0.765366,\ \tilde{\omega}_2 = 1.414213,\ \tilde{\omega}_3 = 1.847759$ with $\omega_i = \tilde{\omega}_i\sqrt{\dfrac{GJ}{lJ_0}}$

7.24 $\omega_1 = 0.2583,\ \omega_2 = 3.0,\ \omega_3 = 7.7417$

7.27 $[U]^{-1} = \begin{bmatrix} 0.44721359 & 0.083045475 & -0.12379687 \\ 0 & 0.41522738 & 1.1760702 \\ 0 & 0 & 1.7950547 \end{bmatrix}$

7.45 $\omega_1 = 5.8694,\ \omega_2 = 85.5832,\ \omega_3 = 293.5470$ **7.47** $\omega_1 = 0.2430,\ \omega_2 = 0.5728,\ \omega_3 = 7.1842$

Chapter 8

8.1 28.2843 m/sec   **8.3** $\omega_3 = 9000$ Hz, both increased by 9.54%

8.6 **(a)** 0.1248×10^6 N,   **8.8** $w(x, t) = \dfrac{8\,al}{\pi^3 c} \displaystyle\sum_{n=1,3,5,\ldots} (-1)^{\frac{n-1}{2}} \dfrac{1}{n^3} \sin \dfrac{n\pi x}{l} \sin \dfrac{n\pi ct}{l}$
(b) 3.12×10^6 N

8.11 $w\left(x, \dfrac{l}{c}\right) = -\dfrac{\sqrt{3}\,9h}{2\pi^2} \sin\dfrac{\pi x}{l} + \dfrac{\sqrt{3}\,9h}{8\pi^2} \sin\dfrac{2\pi x}{l} - \dfrac{\sqrt{3}\,9h}{32\,\pi^2} \sin\dfrac{4\pi x}{l} + \dfrac{\sqrt{3}\,9h}{50\,\pi^2} \sin\dfrac{5\pi x}{l}$

8.15 $\tan\dfrac{\omega l}{c} = \dfrac{A\,E\,\omega\,c(k - M\,\omega^2)}{A^2\,E^2\,\omega^2 - M\,\omega^2\,k\,c^2}$ **8.18** $\tan\dfrac{\omega l_1}{c_1} \tan\dfrac{\omega l_2}{c_2} = \dfrac{A_1 E_1 c_2}{A_2 E_2 c_1}$

8.21 $\omega_n = \dfrac{n\pi}{l} \sqrt{\dfrac{G}{\rho}};\ n = 1, 2, 3, \ldots$ **8.23** $\omega_n = \dfrac{(2n+1)\pi}{2} \sqrt{\dfrac{G}{\rho l^2}};\ n = 0, 1, 2, \ldots$

8.26 5030.59 rad/sec   **8.29** $\cos\beta l \cosh\beta l = -1$ **8.32** $\tan\beta l - \tanh\beta l = 0$ **8.34** 20.2328 N-m

8.37 $\cos\beta l \cosh\beta l = 1$, and $\tan\beta l - \tanh\beta l = 0$ **8.39** $\omega \approx \sqrt{120} \left(\dfrac{EI_0}{\rho A_0\,l^4}\right)^{1/2}$

8.44 $w(x,t) = \dfrac{F_0}{2\,\rho\,Ac^2} \left\{ \cos\beta x + \cosh\beta x + \tan\dfrac{\beta l}{2} \sin\beta x - \tanh\dfrac{\beta l}{2} \sinh\beta x - 2 \right\} \sin\omega t$

8.47 $w(x, t) = \displaystyle\sum_{n=1}^{\infty} W_n(x)\,q_n(t)$ where $q_n(t) = \dfrac{M_0}{\rho\,A\,l\,\omega_n^2} \dfrac{dW_n}{dx}\bigg|_{x=l} (1 - \cos\omega_n t)$

8.50 $\omega_{mn}^2 = \dfrac{\gamma_n P}{\rho}$, where $J_m(\gamma_n R) = 0;\ m = 0, 1, 2, \ldots;\ n = 1, 2, \ldots$

8.54 $w(x, y, t) = \dfrac{\dot{w}_0}{\omega_{12}} \sin\dfrac{\pi x}{a} \sin\dfrac{2\pi y}{b} \sin\omega_{12}t$ **8.57** $22.4499\sqrt{\dfrac{EI}{\rho\,A\,l^4}}$

8.59 $\omega = 15.4510\sqrt{\dfrac{EI}{\rho\,A\,l^4}}$ **8.61** $7.7460\sqrt{\dfrac{EI_0}{\rho\,A_0\,l^4}}$ **8.64** $2.4146\sqrt{\dfrac{EA_0}{m_0\,l^2}}$

8.66 $\omega \approx 13867.3328$ rad/sec   **8.68** **(a)** $1.73205\sqrt{\dfrac{E}{\rho l^2}}$,   **(b)** $1.57669\sqrt{\dfrac{E}{\rho l^2}},\ 5.67280\sqrt{\dfrac{E}{\rho l^2}}$

8.71 $\omega_1 = 3.142\sqrt{\dfrac{P}{\rho l^2}},\ \omega_2 = 10.12\sqrt{\dfrac{P}{\rho l^2}}$

Chapter 9

9.1 Around 46.78 km/hour   **9.3** $m_c r_c = 3354.6361$ g-mm, $\theta_c = -25.5525°$
9.5 $m_4 = 0.99$ oz, $\theta_4 = -35°$ **9.8** 1.6762 oz, $\alpha = 75.6261°$ CW

9.11 Remove 0.1336 lb at 10.8377° CCW in plane B and 0.2063 lb at 1.3957° CCW in plane C at radii 4 in.

9.14 (a) $\vec{R}_A = -28.4021\vec{j} - 3.5436\,\vec{k}$, $\vec{R}_B = 13.7552\,\vec{j} + 4.7749\,\vec{k}$,   (b) $m_L = 10.44$ g, $\theta_L = 7.1141^0$

9.17 (a) 0.005124 m, (b) 0.06074 m, (c) 0.008457 m

9.20 (a) 0.5497×10^8 N/m^2 (b) 6.4698×10^8 N/m^2 (c) 0.9012×10^8 N/m^2

9.22 $F_{xp} = 0$, $F_{xs} = 3269.4495$ lb, $M_{zp} = M_{zs} = 0$

9.25 The engine is completely force and moment balanced.

9.27 0.2385 mm   **9.30** (a) $\omega < 95.4927$ rpm   (b) $\omega > 276.7803$ rpm

9.32 $k = 152243.1865$ N/m   **9.35** 79.7808 rad/sec $-$ 1419.8481 rad/sec   **9.37** $\delta_{st} = 0.02733$ m

9.40 $k = 1332.6646$ lb/ft   **9.43** (a) $X = 11.4188 \times 10^{-3}$ m   (b) $F_T = 44.8069$ N   **9.45** 98.996%

9.47 (a) 2,775.66 lb,   (b) 40,145.81 lb   **9.49** 49,752.86 N/m

9.52 $\mu = 0.3403$; $m_2 = 102.09$ kg, $k_2 = 2.519$ MN/m; $X_2 = -0.1959$ mm

9.54 (a) 487.379 lb   (b) $\Omega_1 = 469.65$ rpm, $\Omega_2 = 766.47$ rpm

9.56 For $D/d = 4/3$, $d = 0.5732$ in, $D = 0.7643$ in

9.59 $0.9764 \leq \dfrac{\omega}{\omega_2} \leq 1.05125$ **9.61** $m_2 = 10$ kg, $k_2 = 0.19986$ MN/m   **9.63** 165.6315 lb/in

Chapter 10

10.2 18.3777 Hz   **10.4** 3.6935 Hz   **10.6** 0.53%   **10.9** 35.2635 Hz

10.12 73.16%   **10.14** $k = 33623.85$ N/m, $c = 50.55$ N-sec/m   **10.16** $m = 19.41$ g, $k = 7622.8$ N/m

10.19 111.20 rad/sec $-$ 2780.02 rad/sec   **10.21** $r \approx 1$ **10.23** $\zeta = 0.1111$

10.26 Cage (51.93 Hz), Inner race (1078.97 Hz), Outer race (830.88 Hz), Ball (193.31 Hz)

10.29 1.8   **10.30** 2.9630   **10.32** $\zeta = 0.2$

Chapter 11

11.2 $\left.\dfrac{d^4x}{dt^4}\right|_i = \dfrac{x_i - 4x_{i-1} + 6x_{i-2} - 4x_{i-3} + x_{i-4}}{(\Delta t)^4}$

11.4 $x(t = 5) = -1$ with $\Delta t = 1$ and -0.9733 with $\Delta t = 0.5$

11.6 $x_{10} = -0.0843078$, $x_{15} = 0.00849639$

11.9 $x(t = 0.1) = 0.131173$, $x(t = 0.4) = -0.0215287$, $x(t = 0.8)$
$= -0.0676142$

11.14 With $\Delta t = 0.07854$, $x_1 = x$ and $x_2 = \dot{x}$, $x_1(t = 0.2356) = 0.100111$, $x_2(t = 0.2356) = 0.401132$,
$x_1(t = 1.5708) = 1.040726$, $x_2 = (t = 1.5708) = -0.378066$

11.20

t	x_1	x_2
0.25	0.07813	1.1860
1.25	2.3360	-0.2832
3.25	-0.6363	2.3370

11.23 $\omega_1 = 3.06147 \sqrt{\dfrac{E}{\rho l^2}}, \omega_2 = 5.65685 \sqrt{\dfrac{E}{\rho l^2}}, \omega_3 = 7.39103 \sqrt{\dfrac{E}{\rho l^2}}$

11.26 $\omega_1 = 17.9274 \sqrt{\dfrac{EI}{\rho Al^4}}, \omega_2 = 39.1918 \sqrt{\dfrac{EI}{\rho Al^4}}, \omega_3 = 57.1193 \sqrt{\dfrac{EI}{\rho Al^4}}$

11.38 With $\Delta t = 0.24216267$,

t	x_1	x_2
0.2422	0.01776	0.1335
2.4216	0.7330	1.8020
4.1168	0.1059	0.8573

Chapter 12

12.2 $[k] = \dfrac{EA_0}{l}(0.6321)\begin{bmatrix} 1 & -1 \\ -1 & 1 \end{bmatrix}$ **12.4** $3.3392 \times 10^7 \text{ N/m}^2$

12.8 5.184 in. under load   **12.10** 0.05165 in. under load

12.13 Deflection $= 0.002197 \dfrac{Pl^3}{EI}$, slope $= 0.008789 \dfrac{Pl^3}{EI}$

12.17 $\omega_1 = 0.8587 \sqrt{\dfrac{EI}{\rho Al^4}}, \omega_2 = 4.0965 \sqrt{\dfrac{EI}{\rho Al^4}}, \omega_3 = 34.9210 \sqrt{\dfrac{EI}{\rho Al^4}}$

12.20 $\omega_1 = 15.1357 \sqrt{\dfrac{EI}{\rho Al^4}}, \omega_2 = 28.9828 \sqrt{\dfrac{EI}{\rho Al^4}}$ **12.23** $\omega_1 = 20.4939 \sqrt{\dfrac{EI}{\rho Al^4}}$

12.26 $\sigma^{(1)} = -2.5056 \text{ psi}, \sigma^{(2)} = 2.6936 \text{ psi}$

12.29 Maximum bending stresses: -37218 psi (in both connecting rod and crank), maximum axial stresses: -6411 psi (in connecting rod), -5649 psi (in crank)

12.31 $\omega_1 = 6445 \text{ rad/sec}, \omega_2 = 12451 \text{ rad/sec}$

Chapter 13

13.2 (a) $\sqrt{\dfrac{m}{k_1}} \dot{x}_0$ (b) $\tau_n = \pi \left(\sqrt{\dfrac{m}{k_1}} + \sqrt{\dfrac{m}{k_2}} \right)$ **13.5** $\sqrt{\dfrac{k}{m}}, \sqrt{\dfrac{g}{l}}$

13.7 $m\ddot{x} + k_1 x + k_2 x^3/(2h^2) = F(t)$ **13.9** $\tau = 4\sqrt{\dfrac{l}{g}} \displaystyle\int_0^{\pi/2} \dfrac{d\phi}{\sqrt{1 - k^2 \sin^2 \phi}}$, where $k = \sin(\theta_0/2)$

13.11 $\dfrac{4}{\omega_0 \left(1 - \dfrac{\theta_0^2}{12}\right)} F\left(a, \dfrac{\pi}{2}\right)$, where $F(a, \beta)$ is an incomplete elliptic integral of the first kind

13.13 $x(t) = A_0 \cos \omega t - \dfrac{A_0^3 \alpha}{32\omega^2}(\cos \omega t - \cos 3\omega t) - \dfrac{A_0^5 \alpha^2}{1024\omega^4}(\cos \omega t - \cos 5\omega t);$

$\omega^2 = \omega_0^2 + \dfrac{3}{4} A_0^2 \alpha - \dfrac{3}{128} \dfrac{A_0^4}{\omega^2} \alpha^2$

13.18 (a) $x(t) = e^{-0.2t}(-\cos 0.87178\, t + 1.7708 \sin 0.87178\, t)$

13.21 $x(t) = 5[1 - 1.0013e^{-0.05t}\{\cos(0.9987t - 2.8681°)\}]$

13.25 For $0 < c < 2$: stable focus, for $c \geq 2$: stable nodal point   **13.27** The equilibrium point is a center.

13.30 (a) $\lambda_1 = \lambda_2 = 2$ (b) $\lambda_1 = -1, \lambda_2 = 3$ **13.32** $x(t) = c_1\begin{Bmatrix} 1 \\ -1 \end{Bmatrix}e^{2t} + c_2\begin{Bmatrix} 1 \\ -1 \end{Bmatrix}te^{2t} + c_2\begin{Bmatrix} 0 \\ -1 \end{Bmatrix}e^{2t}$

13.34 $x(t) = 2\cos \omega t + \dfrac{\alpha}{4\omega}\sin 3\,\omega t + \dfrac{3\alpha^2}{32\omega^2}\cos 3\,\omega t + \dfrac{5\alpha^2}{96\omega^2}\cos 5\,\omega t; \quad \omega^2 = 1 + \dfrac{\alpha^2}{8}$

Chapter 14

14.1 0.04   **14.3** 1.0, 1.3333, 0.5773   **14.5** 25

14.10 (a) $1 - e^{-\lambda t}$,   (b) $\dfrac{1}{\lambda}$,   (c) $\dfrac{1}{\lambda}$ **14.12** 0.3316×10^{-8}

14.14 $X(\omega) = \left(\dfrac{Aa}{a^2 + \omega^2}\right) - i\left(\dfrac{A\omega}{a^2 + \omega^2}\right)$ **14.17** (b) 3400.0

14.19 $\dfrac{2S_0}{\tau}(\sin \omega_2\tau - \sin \omega_1\tau)$ **14.26** $\sigma = 0.3106$ m   **14.28** $E[z^2] = \dfrac{\pi S_0\omega^4}{2\zeta\omega_n^3}$

14.30 $m_{eq} = \left\{\dfrac{\pi S_0}{2\,\delta\omega_1^2(\omega_1^2 - \omega_2^2)^{1/2}}\right\}^{1/2}$,   **14.32** $\overline{z_1^2(t)} = 42.4744 \times 10^{-6}$ m^2,

$k_{eq} = \left\{\dfrac{\pi S_0\omega_1^2}{2\,\delta(\omega_1^2 - \omega_2^2)^{1/2}}\right\}^{1/2}$, $\overline{z_2^2(t)} = 133.9971 \times 10^{-6}$ m^2,

$C_{eq} = \left\{\dfrac{2\,\pi S_0(\omega_1^2 - \omega_2^2)^{1/2}}{\delta\omega_1^2}\right\}^{1/2}$ $\overline{z_3^2(t)} = 208.3902 \times 10^{-6}$ m^2

Index